Mechanism Machine Theory

Ashok G. Ambekar

Director and Former Principal
Swami Vivekanand College of Engineering, Indore

and

Former Professor and Head
Shri Govindram Seksaria Institute of Technology and Science (SGSITS)
Indore

PHI Learning Private Limited

New Delhi-110001
2012

Rs. 525.00

MECHANISM AND MACHINE THEORY

Ashok G. Ambekar

© 2007 by PHI Learning Private Limited, New Delhi. All rights reserved. No part of this book may be reproduced in any form, by mimeograph or any other means, without permission in writing from the publisher.

ISBN-978-81-203-3134-1

The export rights of this book are vested solely with the publisher.

Fourth Printing   ...   ...   **May, 2012**

Published by Asoke K. Ghosh, PHI Learning Private Limited, M-97, Connaught Circus, New Delhi-110001 and Printed by Rajkamal Electric Press, Plot No. 2, Phase IV, HSIDC, Kundli-131028, Sonepat, Haryana.

CONTENTS

8. Gears

9.   Gear Trains

10.   Gyroscopic Effects

11. Friction Gears 443–512

12. Belt, Rope and Chain Drives 513–558

15. Governors

PREFACE

This book covers that field of engineering studies which conventionally goes under the name 'Theory of Machines'. Over the years, knowledge in this area has expanded manifold both in width as well as in depth. This is invariably reflected in the theme and scope of some of the prominent text/reference books published over the last 40 years or so. Departing from traditionally common titles, some of the relatively recent book titles are *Applied Kinematics, Mechanics of Machinery, Kinematics and Dynamics of Plane Mechanisms* and *Synthesis and Analysis of Mechanisms*. Details of these titles appear in the Bibliography provided at the end of this book. Needless to say that in some way, may be in different proportions, these trends should get reflected in our course curriculum.

On the analysis side, there is a visible shift of emphasis from graphical to analytical approach. Modern high-speed applications have greatly enhanced the importance of acceleration analysis for accessing the effect of inertia forces in the overall design of machine components. This is further accentuated by the availability of high speed computers. Besides accuracy and speed of computations, analytical methods help in establishing motion characteristics (e.g. velocity and acceleration) for large number of configurations of a mechanism (e.g. for different crank positions in an I.C. engine) with minimal extra time. This feature enables to establish maximum value of inertia force and corresponding crank angle of rotation, thereby rendering design calculations more rationale. In view of the importance of this area of learning, a separate chapter has been devoted to the analytical methods of velocity and acceleration analysis.

The above comments, however, should not be taken to mean that traditional graphical methods need to be replaced by analytical methods. It is generally believed that graphical methods provide a better insight into the subject matter. This is in fact true. A separate chapter on graphical analysis of velocity and acceleration analysis has, therefore, been included in this book.

A practising mechanical engineer is required to design a mechanism that satisfies the prescribed motion characteristics. The synthesis method calls for exposure to type synthesis, number synthesis or dimensional synthesis. An obvious outcome of number synthesis can be seen in the 'Enumeration of distinct chains of given number of links and degrees of freedom'. This aspect has been discussed in the chapter on planar mechanisms. Function and path generation problems involve dimensional synthesis. Coupler curves provide a very useful tool in mechanism synthesis. A separate chapter has been dedicated to the studies on elements of mechanism synthesis in kinematics and it includes discussions on the coupler curves and cognate linkages.

In all conventional topics, attempt has been made to present the subject matter in a way commensurate with state-of-the-art presentation in the published literature. While teaching dynamic force analysis, a common observation is that the students do not have enough exposure to the application part of D'Alembert's principle. Therefore, in chapters involving dynamic force analysis, special emphasis has been laid to explain as to how the problem is converted into an equivalent problem of static equilibrium. Special efforts have been made to explain how a positive drive is really achieved through gearing action. Similarly, extra efforts have been made to convey conceptual meaning of 'interference' of involute gear through the principle of conjugate teeth.

Wherever deemed necessary, applications of 'principle of inversion' and 'instantaneous centre of rotation' have been highlighted. Throughout the text, special emphasis has been laid on developing conceptual understanding. This will be especially visible in the chapter on governors, cams and gyroscopes.

The book is designed to serve as a textbook for undergraduate engineering students studying in mechanical, production and aeronautical streams. The course content is intended to be covered as a two-semester course, requiring about 90 lecture hours. While preparing the manuscript of this book, I had in my mind needs of B.E./B.Tech. students of various universities and institutes of higher learning. This is also true in respect of students preparing for various technical boards, professional examining bodies (e.g. Institution of Engineers and Institution of Mechanical Engineers) and national level examinations like GATE and IES. A good number of solved problems in S.I. units appear in each chapter, through which theory is amplified and extended. I believe that providing good number of representative solved problems is an effective way of clarifying doubts, if any, and it also helps in presenting the subject matter in a more comprehensive way. Ample number of unsolved problems, with answers, have been provided at the end of each chapter as Review Questions.

The course material is largely based on my lectures delivered as a faculty at Shri G.S. Institute of Technology and Science, Indore for a period of over 35 years at undergraduate and postgraduate levels and also at Swami Vivekanand College of Engineering, Indore. I feel highly indebted to my teachers Dr. S.S. Rao and Dr. J.N. Chakraborty (during my studies in M.Tech. at IIT Kanpur) and to Dr. V.P. Agrawal and Dr. Raghavacharlu (during my Ph.D. studies at IIT Delhi) for enrichment of my knowledge. Without their classroom teaching and general guidance, a book writing project of this type would have been impossible. I acknowledge with thanks all other teachers and fellow students at both the IITs who have directly or indirectly contributed in refining my level of conceptual understanding in the subject.

I feel highly indebted to (Late) Dr. J.P. Shrivastava, the then Director and my teacher at S.G.S. Institute of Technology and Science, Indore, without whose inspiration this project would not have started. Words are inadequate to put on record the motivating role of my father, late G.W. Ambekar, in this endeavour. This project turned out to be his last desire. I also desire to place on record my heartfelt thanks to the publisher Prentice-Hall of India and their expert editorial team for the quality conscious editing work for this book. I am especially thankful to

Ms. Swaty Prakash, the copy editor, and to the acquisition editor Mr. Malay Ranjan Parida for his general guidance and encouraging replies regularly.

The preface will remain incomplete if I do not acknowledge the patience with which my wife Mrs. Sushama Ambekar, spared her valuable time and energy to share my day-to-day responsibilities during the period of book writing. Thanks are also due to my sons Chi. Aniruddha and Chi. Ashwin for their encouragement and an ever helping hand in this stupendous task.

Special efforts have been made to satisfy quest of knowledge of readers through this book. Also, great care has been taken to make the book 'error-free'. I will be grateful for constructive criticism and suggestions for further improvements in the book.

Ashok G. Ambekar

Chapter 1

INTRODUCTION TO KINEMATICS AND MECHANISMS

1.1 INTRODUCTION

A more traditional name of this subject is *Theory of Machines*. The subject comprises study of *Kinematics* and *Dynamics of Machines*. The former deals with relationship between the geometry of machine parts and their relative motion. Dynamics of machines, on the other hand, deals with the forces which produce these motions. The study of dynamics comprises *Statics* and *Kinetics*. In statics, one deals with the forces that act on the machine parts which are assumed to be mass-less. Kinetics, on the other hand, deals with the inertia forces arising out of the combined effect of mass and the motion of the parts.

An elementary definition of mechanism describes it as a mechanical device which has the purpose of transferring motion and/or force from the source to the output. Mechanisms pervades all walks of life. A few illustrative applications of mechanisms can be found in domestic appliances (e.g., sewing machines, washing machines, refrigerator, mixers, air-coolers, door-locks/latches, etc.), toys (e.g., peddling cars for kids, walking toys, etc.), agricultural implements (e.g., sprinkler mechanism, tractors, corn drills, ploughs, cultivators, mowers, forage harvesters, potato diggers, etc.), transport vehicles (on air, land and sea, incorporating steering, braking and transmission mechanism, I.C. engine mechanism, etc.), construction equipment (e.g., concrete mixers, cranes, excavators, rock-crushers, bulldozers, etc.), military equipment (e.g., armoured tank, recoil-less guns, self-loading guns, etc.), manufacturing machines and machines used in production processes (e.g., lathe, shaping, drilling, milling and grinding machines, lifting tackles, overhead cranes, conveyors, press, rivet and punching machines, robots and manipulators, etc). Mechanisms also find applications in bio-medical engineering in the form of artificial limbs (e.g., prosthetic knee mechanism, pace maker mechanism, etc.). In short, there is hardly any place where mechanisms are not present. Some of the representative mechanisms are illustrated in Figs. 1.17(a) through 1.17(w). In big cities readers can have a look at different types of construction equipment in operation at any construction cite of tall buildings or clusters of such buildings.

The studies under kinematics and mechanisms are fundamental to the studies in a much broader discipline called *Machine Design*. Developments in engineering applications over the last few decades indicate a consistent trend towards increasing the machine speed. This trend can be readily seen in automobiles. As a direct sequel to this trend, consideration of inertia forces in design calculations have become much more essential. In a number of cases, inertia force consideration compels the designer to cut down the weight of the components. This necessitates a greater emphasis on acceleration part of motion analysis.

1.2 THE FOUR-BAR MECHANISM

Mechanisms play a key role in transferring motion from one part of a machine to some other part. Individual parts/components of a mechanism are called *bars* or, more precisely *links*, which are connected to one another through pin joints ensuring a relative motion of rotation. Based on similarity of relative motion between door panels and the frame in houses, these pin joints are also called hinges. A four-bar mechanism constitutes the simplest type of closed loop linkage. It consists of three mobile links and a fixed link called frame. One of the links is connected to power source/prime mover and is called the input link. An another link, called output link, delivers required type of motion to some other part of the machine. The input and output links are connected to the frame through pin joints called pivots. The input and output links are interconnected at their other ends through a link called coupler/floating link through pin joints. The coupler, input and output links are called the mobile links in this mechanism. A simple application of a four-bar mechanism is shown in the form of a treadle-operated grinding machine in Fig. 1.17(a). The foot pressure is indicated by F in the figure.

1.3 MOTION OF A PARTICLE

A particle, of the size of a point, is supposed to have no dimensions. In kinematics, the word 'particle' is reserved for those objects whose motion is not affected in any way by the dimensions of the object. The concept of 'motion of a particle' is of far-reaching consequence and therefore needs special attention.

When a particle moves in space, it occupies successive positions in the coordinate system with respect to time. When the successive positions of the particle are connected by a smooth curve, it produces a curved line. This is obvious because the particle does not have any dimension. This line, representing successive positions of the point, is called the path or locus of the point.

Spatial and Planar Motion

When all the three coordinates are necessary to describe the path or locus, the particle is said to have *spatial motion*. When only two coordinates are necessary (the third coordinate always being zero or constant) to describe the path, the motion of particle is said to be a *planar motion*.

In the above definitions of planar and spatial motions of a particle, it is implied that frame of coordinate axis is so chosen in orientation that minimum possible number of coordinates is needed to describe the motion.

Rectilinear or Straight Line Motion

In the same spirit, a particle is said to execute straight line or rectilinear motion when a single coordinate is required to describe the motion. For instance, let us assume that a particle moves along a straight path *AB* as shown in Fig. 1.1. In such a case, re-orientation of axes of reference, with *x*-axis lying along *AB*, necessitates only one coordinate x' along *AB*. Therefore, it is a rectilinear motion.

In general, a plane curved path requires a minimum of two coordinates while a space (skew) curve requires three coordinates.

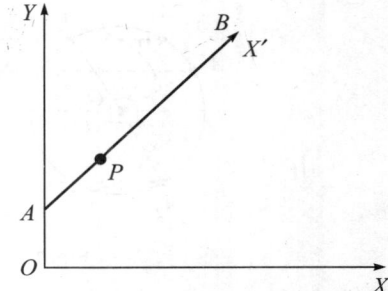

Fig. 1.1 Particle moving along straight path.

1.4 MOTION OF A RIGID BODY

When dimensions of a body become important in studying intricacies of motion, the motion can be idealised as a 'rigid-body motion'. The body in such a case is considered to consist of a large number of small particles.

To understand the concept of *motion of a particle*, consider the motion executed by a piston of an I.C. engine. The piston reciprocates along a straight line, but does not have any other motion (*e.g.,* a motion of rotation about its own axis) associated with it. Thus, each and every particle of the piston, lying on the cylindrical surface or on the axis, has same motion characteristics. In such a motion, it is immaterial whether a piston is represented by a cylinder of specific dimensions or by a particle located at the piston pin on the axis. The resulting simplicity in analysis on account of idealisation of 'motion of a particle' is quite significant. As against this, motion of a connecting rod cannot be idealised as 'motion of a particle.' This is mainly because every particle in the body can have its own motion characteristics.

A very common assumption in studying motion of a rigid body is that any two particles in the body of the object always remain apart at the same distance, howsoever large the deforming forces can be. This is the concept of rigid body motion.

1.5 MOTION OF TRANSLATION

When each particle of a body has exactly the same motion as every other particle of which it is composed, the body is said to execute *motion of translation*. This implies that a straight line, drawn joining two particles, remains parallel to itself throughout the motion.

Motion of translation can be of two types.

(A) *Rectilinear Motion of Translation.* When all the particles of a rigid body have exactly the same motion, and that motion is along a straight line, the motion is called rectilinear motion of translation. Piston of I.C. engine/steam engine and cross-head of a steam engine execute rectilinear motion of translation.

(B) *Curvilinear Motion of Translation.* When motion of all the particles of a rigid body are exactly the same but take place along curved path, the motion of body is called curvilinear motion of translation. Figure 1.2 describes a double crank and slider mechanism in which the slider executes rectilinear motion of translation but the coupler *AB* executes curvilinear motion of translation. It can be verified that as cranks *OA* and *CB* have equal lengths, the coupler *AB* always moves parallel to itself, and any point *K* on this coupler actually moves along the arc of a circle of radius *CB* or *OA,* but with centre at O_k.

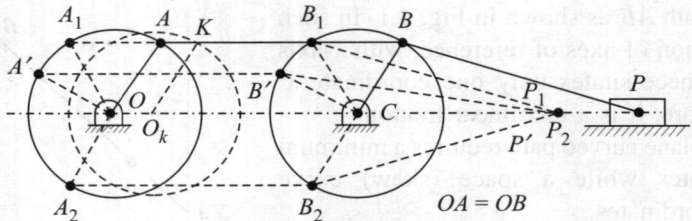

Fig. 1.2 Double crank-slider mechanism.

It follows therefore that if motion of slider P or coupler AB is required to be analysed, the entire shape and size can be ignored and the body be replaced by a point for the purpose of analysis.

1.6 MOTION OF ROTATION

A rigid body is said to have motion of rotation when any point in the body remains at the same distance from a point, called centre of rotation. Motion of links OA and CB in Fig. 1.2 are called motion of rotation. It can be verified that when a body rotates, a straight line joining any two points in the rotating body, does not remain parallel to itself, except perhaps for those lines lying along axis of rotation.

1.7 PLANAR MOTION AND EULER'S THEOREM

Motion of connecting rod PB in Fig. 1.2 is truely representative of a general complex motion of a rigid body moving always in a plane. The motion of connecting rod is a combination of motion of translation and motion of rotation. This is easy to understand because one end B of the connecting rod always moves along a circle, while the other end P reciprocates along the line of stroke. This is in accordance with **Euler's theorem** which states that–

Any displacement of a rigid body is equivalent to the sum of a net translation of any point P and a net rotation of the body about that point.

It follows that displacement difference between any two points on the same rigid body is attributed to motion of the rotation of the body, and not to the motion of translation.

1.8 DEGREES OF FREEDOM

Minimum number of independent coordinates or parameters needed to describe configuration of a body in motion is called its *degrees of freedom*. While specifying degrees of freedom, it is common to assume that lengths of links are known. A crank-slider mechanism constitutes a simple example for this purpose.

In case of slider-crank mechanism, shown in Fig. 1.3, if angle θ made by the crank with line of stroke is known, the entire mechanism can be described in position. For instance, one may choose crank shaft centre O and line of stroke arbitrarily and, at an angle θ to the line of stroke, draw the crank OA of known length. From A as centre, with compass set for a radius equal to length of connecting rod *AP*, an arc can be drawn to cut line of stroke at *P*. This gives position of slider *P* as also the position of the connecting rod *AP*. Thus a single coordinate θ is required to describe configuration and hence, degrees of freedom (abbreviated as d.o.f.) equals 1.

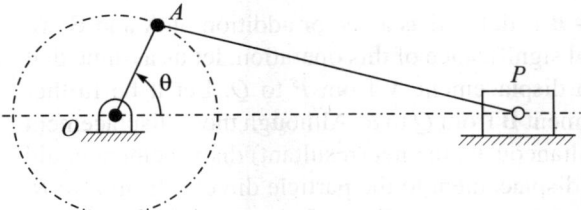

Fig. 1.3 A slider-crank mechanism.

Degrees of freedom can also be defined on the basis of number of independent motions which a body can have. In general, a body in space can have six independent motions; three of translation along *x*, *y* and *z* axes and three of rotation about *x*, *y* and *z* axes, as shown in Fig. 1.4. The body can have motion of translation *u*, *v* and *w* parallel to *x*, *y* and *z* axes respectively. The body can also rotate through α, β and γ about lines parallel to *x*, *y* and *z* axes respectively. In forming chains and mechanism, some of these degrees of freedom are supressed.

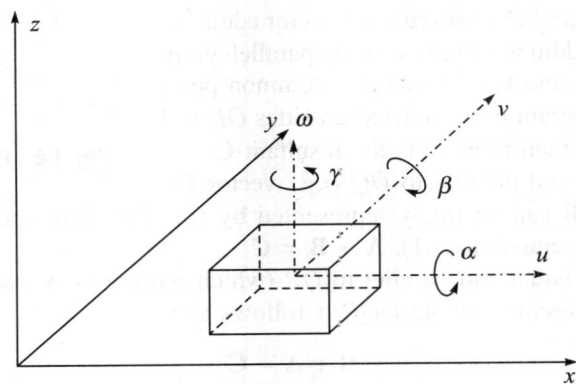

Fig. 1.4 Independent motions of a body in space.

1.9 VECTORS AND THEIR TREATMENT

The readers must be already conversant with the definitions of scalars and vectors. As distinct from scalars, a vector quantity has both a magnitude and direction. Thus, mass of a body is a scalar quantity but its weight is a vector quantity. A vector is also defined as a directed line segment which has specified length (magnitude) and a specified direction. The line segment for a vector must be drawn to some scale, with its length representing the magnitude of the vector quantity. An arrow, with its head pointing in the direction of vector quantity, is placed at the appropriate end of the above line segment.

(A) *Vector addition.* (Refer Fig. 1.5) While using letters of the alphabet to designate vectors, boldface capital characters will be reserved for vectors and italic characters for scalars. Thus, italic letters *a, b, c* will be used for scalar component of vectors **A**, **B**, **C** respectively.

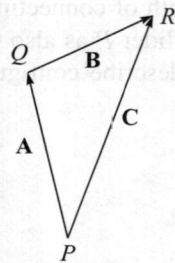

The operation **A** + **B** is defined as a vector addition of **A** and **B**. To understand the physical significance of this operation, let us assume that a particle undergoes a displacement **A** from P to Q. Let it be further subjected to a displacement **B** from Q to R. Although these displacement do not take place simultaneously, the net (resultant) displacement could be obtained by giving displacement to the particle directly from P to R.

Fig. 1.5 Vector addition.

The displacements PQ and QR in Fig. 1.5 are represented by vectors **A** and **B** respectively, and the net displacement PR is represented by vector **C**, so that vector equation, indicating vector addition, is

$$\mathbf{A} + \mathbf{B} = \mathbf{C} \tag{1.1}$$

Equation (1.1), when examined in the light of Fig. 1.5, leads to the conclusion that the resultant vector **C** originates at the tail of the vector **A** and terminates at the tip of the arrowhead of vector **B**.

It follows from vector algebra that *commutative law* holds good in vectorial addition and subtraction. This can be further verified using parallelogram rule for vector addition. Fig. 1.6 demonstrates addition of vectors using parallelogram rule. Let the vectors **A** and **B** originate at a common point O, and let the parallelogram be completed on sides OP and OQ. The diagonal OR then represents the resultant **C**.

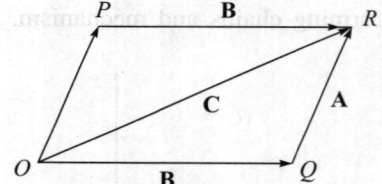

Fig. 1.6 Parallelogram of vectors.

Since PR is equal and parallel to OQ (*i.e.,* vector **B**), it follows that vector **B** can be truely represented by side PR. Thus the vector triangle OPR, confirms earlier vector equation (1.1), **A** + **B** = **C**

Again, side QR is equal and parallel to OP (which represents **A**), and as such, represents vector **A**. Thus, from vector triangle OQR it follows that

$$\mathbf{B} + \mathbf{A} = \mathbf{C} \tag{1.2}$$

Comparing equations (1.1) and (1.2), it follows that *in vector addition, order of vectors is unimportant*. Again, it is concluded from vector triangle OPR of Fig. 1.6 that to find the vector sum **A** + **B**, tail of vector **B** be placed at the tip of the arrow-head of vector **A**. Similarly, for finding the vector sum **B** + **A** from vector triangle OQR, one needs to keep tail of the vector **A** at the tip of the arrow-head of vector **B**, and the resulting vector **C** will have tip of its arrow-head touching tip of the arrowhead of **A**.

(B) *Vector sum of more than two vectors.* When a vector sum of say three vectors **A**, **B** and **C** is to be obtained, one may consider the sum (**A** + **B**) + **C**. An expression of this type implies that the resultant of vector sum **A** + **B** and vector **C** are to be added vectorially. Let the vectors **A**, **B** and **C** be as shown in Fig. 1.7(a). Then, the resultant of vector sum **A** + **B** is represented by vector **D** in Fig. 1.7(b), and Fig. 1.7(c) shows the addition of vectors (**A** + **B**) and **C**.

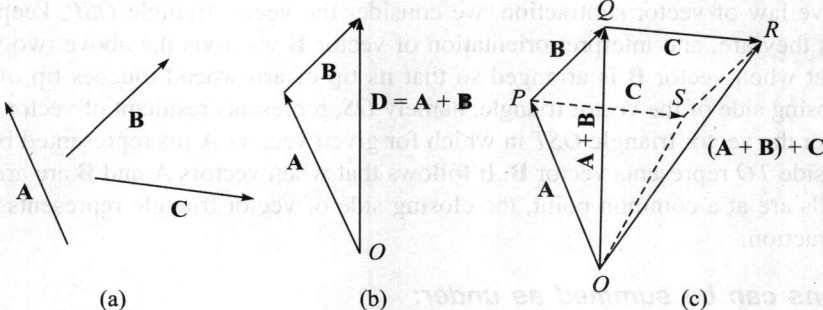

Fig. 1.7 Vector sum of three vectors.

It can be seen in Fig. 1.7(c) that the resultant vector $(A + B) + C$ originates at the tail of the first vector A and terminates at the tip of the arrowhead of the last vector C.

By completing parallelogram $PQRS$, where PQ and QR represent vectors B and C respectively, it is possible to prove that commutative law is valid for this vector operation. For instance, in the parallelogram $PQRS$, the sides PS and SR can be taken to represent vectors C and B. Then, from vector polygon $OPSR$, line joining OS represents vector sum $A + C$ and therefore vector $OR = (A + C) + B$.

It follows that

$$(A + B) + C = (A + C) + B \tag{1.3}$$

In other words, in vector addition, order of vectors is unimportant.

(C) *Vector subtraction.* It is more convenient to represent vector subtraction $A - B$ as a problem of addition of a negative of a vector B to vector A. Thus, mathematically

$$A - B \equiv A + (-B) \tag{1.4}$$

Physically, negative of a vector B implies a vector pointing out in diametrically opposite side. Fig. 1.8 may be referred for complete understanding.

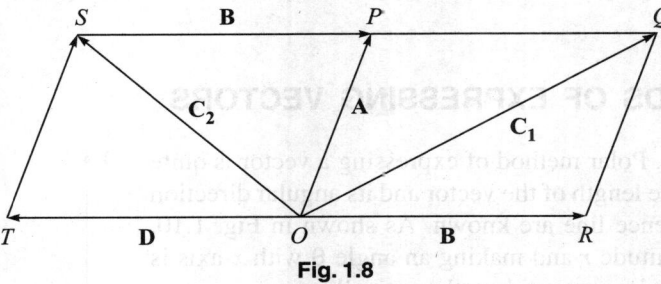

Fig. 1.8

The vector C_1 in Fig. 1.8 is the resultant of vector summation $A + B$ while the vector C_2 represents resultant of vector addition $A + D$, where $D \equiv -B$. Thus, in effect, vector C_1 and C_2 represent resultant of vector addition and subtraction respectively of vectors A and B. Since side OT represents $-B$, the side SP in vector triangle OSP represents vector B.

To evolve law of vector subtraction, we consider the vector triangle *OSP*, keeping vectors **A** and **C₂** as they are, and interpret orientation of vector **B** vis-á-vis the above two vectors. We conclude that when vector **B** is arranged so that its tip of arrowhead touches tip of arrowhead of **A**, the closing side of the vector triangle, namely *OS*, represents resultant of vector subtraction. Next consider the vector triangle *OST* in which for given vectors **A** (as represented by sense *TS*) and **C₂**, the side *TO* represents vector **B**. It follows that when vectors **A** and **B** are arranged such that their tails are at a common point, the closing side of vector triangle represents resultant of vector subtraction.

Conclusions can be summed as under:

(1) For vector addition, vectors **A** and **B** should be so arranged that tail of one vector touches tip of the arrowhead of the other. The closing side of the vector triangle then represents resultant of vector addition. The arrowhead for the resultant vector **C₁** should be so placed that it opposes sense of the two vectors Thus, in Fig 1.9 (b) vectors **A** and **B** are in a clockwise sense and the resultant **C₁** should therefore be in counterclockwise sense.

(2) For vector subtraction, vectors **A** and **B** must be so arranged that either (a) their tips of arrowhead touch at a common point or (b) the tails of the two vectors touch at a common point. This is shown in Figs. 1.9(d) and (e).

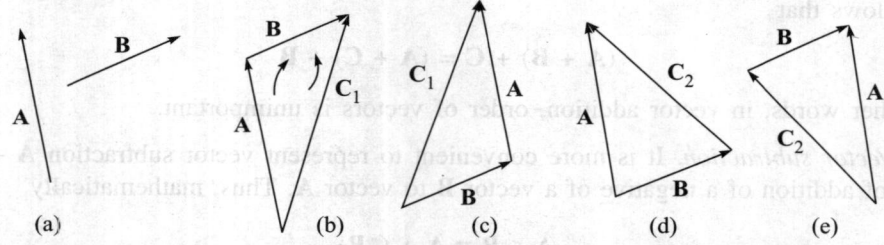

Fig. 1.9 Vector Addition (b) and (c) Vector Subtractions (d) and (e)

Students are advised to study above vector operations and their graphical depiction carefully. This is basic for the proper understanding of velocity and acceleration polygons to be dealt with later.

1.10 METHODS OF EXPRESSING VECTORS

(A) *Polar Notation*. Polar method of expressing a vector is quite convenient when the length of the vector and its angular direction from a given reference line are known. As shown in Fig. 1.10, a vector **R** of magnitude r and making an angle θ with x-axis is shown. This vector is expressed mathematically as:

$$\mathbf{R} = r \angle \theta \qquad (1.5)$$

(B) *Rectangular Notation*. This method consists in defining a vector in terms of its components along x and y axes or, if necessary, along any other set of rectangular axes. This can be expressed mathematically as:

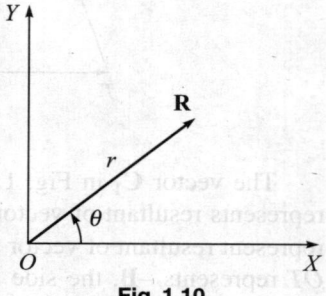

Fig. 1.10

$$\mathbf{R} = R^x + R^y \tag{1.6}$$

A vector $\mathbf{R} = r \angle \theta$ in polar form can be expressed in terms of rectangular coordinates. Thus, the vector $r \angle \theta$ has components along x-axes and y-axes as

$$r^x = r\cos\theta \text{ and } r^y = r\sin\theta$$

Thus,
$$r = \sqrt{(r^x)^2 + (r^y)^2} \tag{1.7}$$

and
$$\theta = \tan^{-1}\left(\frac{r^y}{r^x}\right) \tag{1.8}$$

where r^x and r^y are the magnitudes of R^x and R^y.

(C) *Complex Rectangular Notation.* The convention adopted in this case is that a vector always originates from the origin of co-ordinate system. It follows, therefore, that a vector will always act away from the origin. Any system, which specifies length and direction, defines a vector. Hence a complex rectangular notation, indicated on next page, will also define a vector. The vector $\mathbf{A}$ is thus expressed as,

$$\mathbf{A} = a + jb \tag{1.9}$$

where a represents real part and b represents imaginary part.

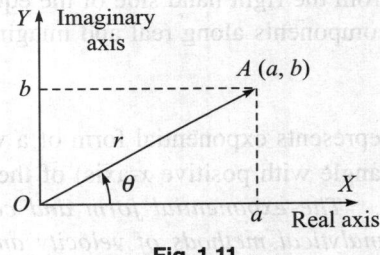

Fig. 1.11

It is usual to define x-axis as real axis and y-axis as imaginary axis. When j is prefixed to b, it indicates 90° c.c.w. rotation with respect to real axis along which a is measured. So, here j appears as an operator which indicates that b is to be measured along y-axis. Physical significance of equation (1.9) is illustrated in Fig. (1.11), where the magnitude of vector is,

$$r = \sqrt{a^2 + b^2}, \text{ and}$$
$$\theta = \tan^{-1}(b/a)$$

When the operator j has a positive sign, it indicates positive y-axis direction, but when j is negative, it indicates negative side of y-axis. *As a rule, whenever the imaginary constant* ($j = \sqrt{-1}$) *is prefixed to a number, it implies rotation through right angle in c.c.w. sense.* The operator j may be used repeatedly, but each time its use implies a rotation through 90° in c.c.w. sense. For instance, consider versions R_2, R_3, R_4, as obtained by multiplying vector R_1 successively by j. The result is depicted in Fig. 1.12.

(D) *Complex Polar Notation.* From De Moivre's theorem, we know that

$$e^{j\theta} = \cos\theta + j\sin\theta \tag{1.10}$$

where,
$$j = \sqrt{-1}$$

Multiplying equation (1.10) by r on the either sides, we find that

$$re^{j\theta} = r(\cos\theta + j\sin\theta) \tag{1.11}$$

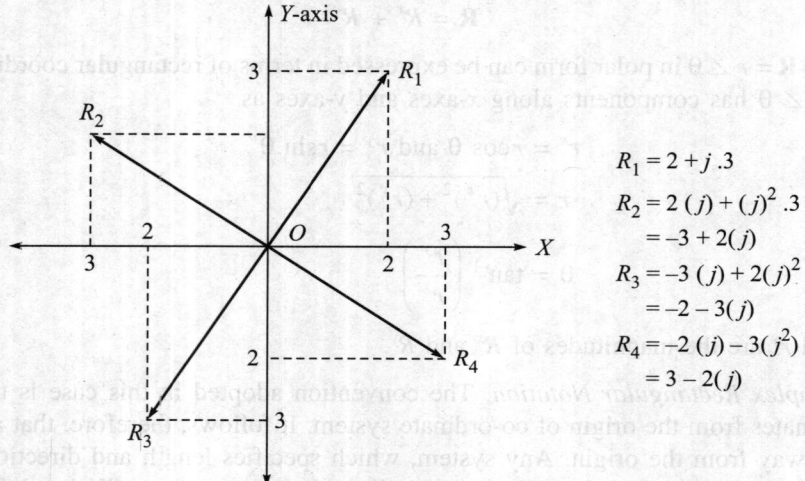

$$R_1 = 2 + j\,.3$$
$$R_2 = 2\,(j) + (j)^2\,.3$$
$$\quad = -3 + 2(j)$$
$$R_3 = -3\,(j) + 2(j)^2$$
$$\quad = -2 - 3(j)$$
$$R_4 = -2\,(j) - 3(j^2)$$
$$\quad = 3 - 2(j)$$

Fig. 1.12 Effect of successive multiplication by j on a vector.

In view of exposure to the concept of real and imaginary axes, it is rather simple to conclude from the right hand side of the equal to sign, in equation (1.11), that $r \cos \theta$ and $r \sin \theta$ are the components along real and imaginary axes respectively. Hence,

$$\mathbf{R} = re^{j\theta} \tag{1.12}$$

represents exponential form of a vector, in which r is the magnitude and $\angle\theta$ is the orientation (angle with positive x-axis) of the above vector.

The exponential form and complex rectangular notation of vectors are very useful in analytical methods of velocity and acceleration analysis and, therefore, students should get themselves conversant with these notations.

1.11 POSITION VECTORS

Location of a point is given by its co-ordinates. These co-ordinates can be either rectangular or polar. A position vector is defined as a vector which originates from origin of co-ordinate frame and defines the location or position of a point in reference frame. Position of a point is pivotal in defining displacement and other motion characteristics. And this speaks of importance of the concept of position vector. In Fig. 1.13 the vector $\mathbf{R}_Q$ defines the position of point Q in the co-ordinate frame. If $OQ = q$ and θ is the angle made by OQ with x-axis and if x' and y' are the co-ordinates of point Q, the position vector can be expressed with all the four notations as,

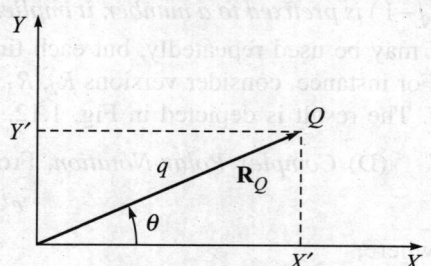

Fig. 1.13 Illustration for a position vector.

$$\mathbf{R}_Q = q \angle \theta$$
$$\mathbf{R}_Q = R_Q^x + R_Q^y$$
$$\mathbf{R}_Q = x' + jy'$$
$$\mathbf{R}_Q = q \cdot e^{j\theta} \tag{1.13}$$

1.12 DISPLACEMENT OF A PARTICLE

A particle may travel along a straight line or a curved path. Distance measured along the given path between two locations A and B is the distance moved through by the particle in the given time interval. Displacement, on the other hand, is treated as a vector, and is defined to be the net change of position of the particle over the given period of time interval. For instance in Fig. 1.14, although distance moved is as along R_1PQR_2, the displacement vector is $\mathbf{R}_1\mathbf{R}_2 = \Delta\mathbf{R}$.

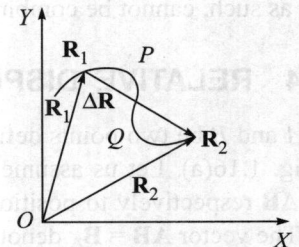

Fig. 1.14 Illustration for displacement vector.

Displacement as a vector can be conveniently expressed in terms of position vector of initial position $\mathbf{R}_1$ and final position $\mathbf{R}_2$. Thus vector equation involving position vectors $\mathbf{R}_1$ and $\mathbf{R}_2$ and displacement vector $\Delta\mathbf{R}$ is given by

$$\mathbf{R}_2 = \mathbf{R}_1 + \Delta\mathbf{R} \tag{1.14}$$

or

$$\Delta\mathbf{R} = \mathbf{R}_2 - \mathbf{R}_1 \tag{1.15}$$

This defines displacement mathematically.

1.13 RIGID BODY DISPLACEMENT

A general planar rigid body motion may be conceived to consist of two component motions:

1. Motion of translation, and
2. Motion of rotation.

In motion of translation, every particle has identical motion and as such, displacement of a particle in body (say, its C.G.) also describes the translatory motion of the rigid body.

As a direct sequel to the definition of motion of rotation in article 1.4, a straight line AB joining two particles in the body, does not remain parallel to itself. A rigid body displacement involving motion of translation and rotation, is shown in Fig. 1.15. It can be seen that due to translation alone, there is no relative motion between different particles of the body AB. However, the angular-rotation component from $A_1 B_1$ to $A_2 B_1$ does produce relative displace-

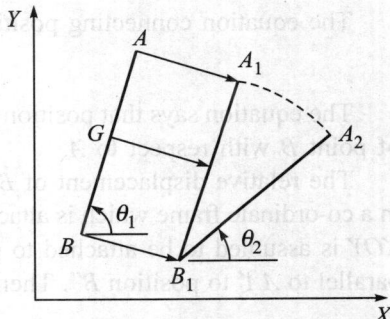

Fig. 1.15 Rigid body displacement.

ment between different points of the body. We, therefore, introduce the concept of angular displacement $\Delta\theta$ and define it as the net angular change in the position of a line (such as *AB*) in the rigid body. It must be noted, however, that we are not concerned here with the actual path points *A* and *B* have traversed.

As seen in figure, the net angular displacement is,

$$\Delta\theta = \theta_1 - \theta_2 \qquad \text{in clockwise sense in radians.} \tag{1.16}$$

It is important to note that angular displacements do not follow commutative rule to addition and, as such, cannot be combined vectorially. It can be treated as scalar with c.w or c.c.w. sense.

1.14   RELATIVE DISPLACEMENT

Let *A* and *B* be two points defined with respect to origin by position vectors **A** and **B**, as shown in Fig. 1.16(a). Let us assume further that points *A* and *B* undergo small displacements of $\Delta\mathbf{A}$ and $\Delta\mathbf{B}$ respectively to positions *A′* and *B′* in small interval of time Δt.

The vector $\mathbf{AB} = \mathbf{B_A}$ denotes relative position *B* with respect to *A*. Complete parallelograms, first with sides *AB* and *BB′* and second one with sides *AB* and *AA′*. Join *A′A″* and *B′B″* as shown in Fig. 1.16(b).

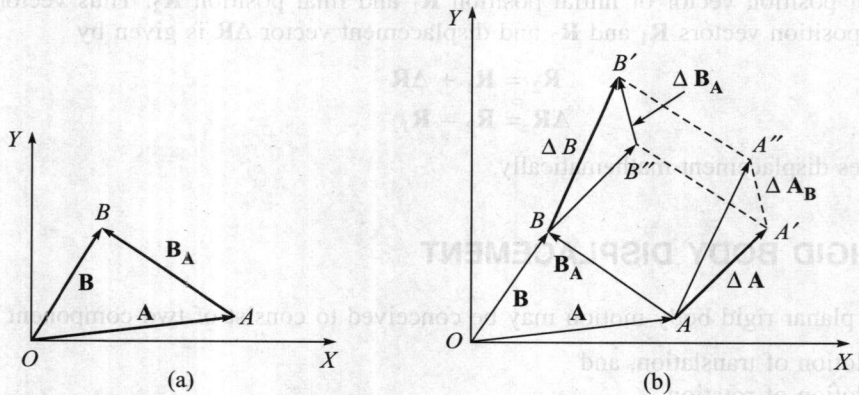

Fig. 1.16   Relative displacement.

The equation connecting position vectors of *A* and B and their relative position vector is

$$\mathbf{B} = \mathbf{A} + \mathbf{B_A} \tag{1.17}$$

The equation says that position of point *B* is equal to the position of point *A* plus the position of point *B* with respect to *A*.

The relative displacement of *B* with respect to *A* is the displacement which *B* would have in a co-ordinate frame which is attached to *A* and moves with it. Thus, in Fig. 1.16(b), if the frame *XOY* is assumed to be attached to point *A* and moving with it, point *B* will move through $\Delta\mathbf{A}$ parallel to *AA′* to position *B″*. Then *B″B′* gives displacement of *B* relative to *A*. In other words,

$$\Delta\mathbf{B}_A = B''B'$$

Since, BB'' is equal and parallel to AA', it represents $\Delta \mathbf{A}$ truely. Hence from vector triangle $B'BB''$ we have

$$\Delta \mathbf{B} = \Delta \mathbf{A} + \Delta \mathbf{B}_A \qquad (1.18)$$

or

$$\Delta \mathbf{B}_A = \Delta \mathbf{B} - \Delta \mathbf{A} \qquad (1.19)$$

Similarly, relative displacement of A with respect to B is the displacement of A in a co-ordinate frame, which is attached to B and moves with it. Thus, when co-ordinate frame XOY is attached to B and moves it, the point A in the co-ordinate frame moves through a distance ΔB to the position A''. As AA'' is equal and parallel to BB' it follows that AA'' truely represents vector ΔB. Then $A''A'$ represents displacement of A relative to B. Thus from vector triangle $A A'A''$ we have.

$$\Delta \mathbf{A} = \Delta \mathbf{B} + \Delta \mathbf{A}_B \qquad (1.20)$$

and

$$\Delta \mathbf{A}_B = \Delta \mathbf{A} - \Delta \mathbf{B} \qquad (1.21)$$

Students are advised to get themselves familiarised with the above concept before proceeding to the chapter on velocity and acceleration analysis.

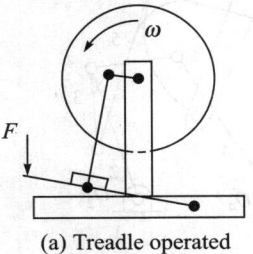

(a) Treadle operated
grinding machine

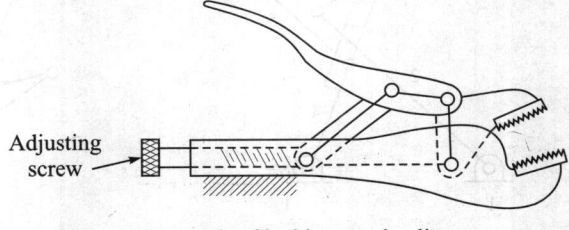

(b) A pair of locking toggle pliers

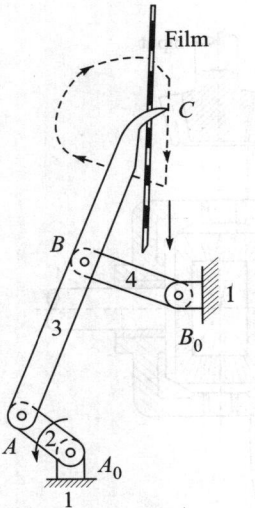

(c) A film advance mechanism of a
movie camera/projector

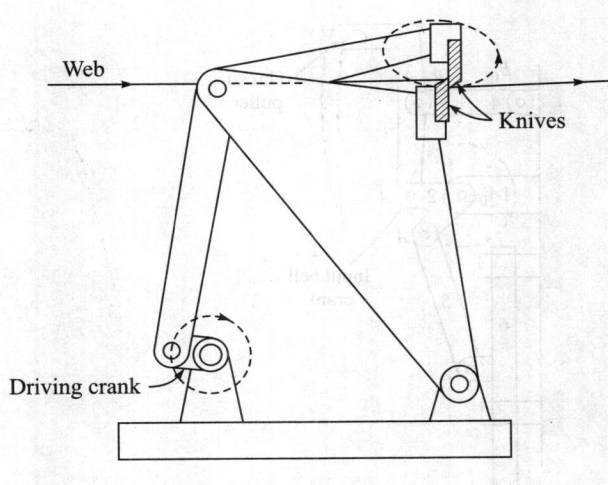

(d) A web-cutter mechanism

(Contd.)

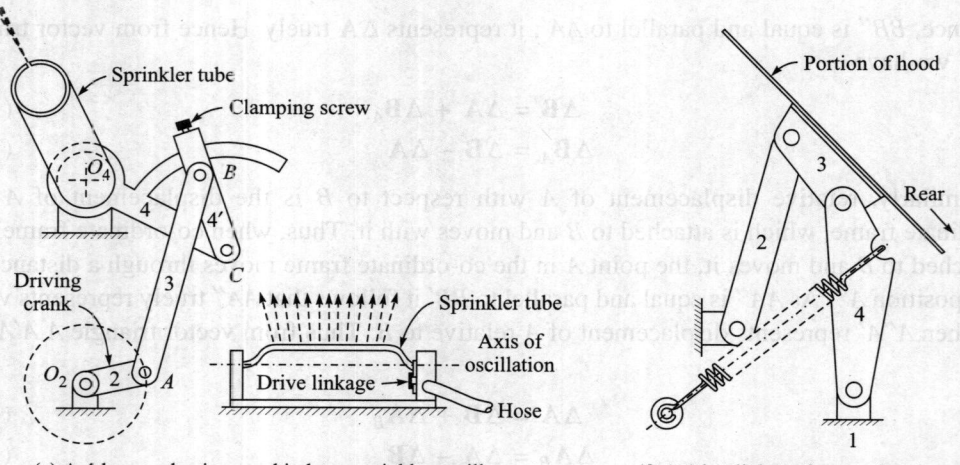

(e) A 4-bar mechanism used in lawn-sprinkler oscillator

(f) A 4-bar linkage for auto-hood

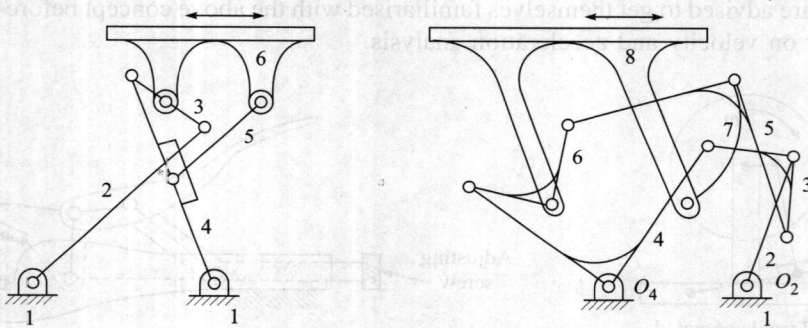

(g) Straight line motion mechanisms on 6 links & 8 links

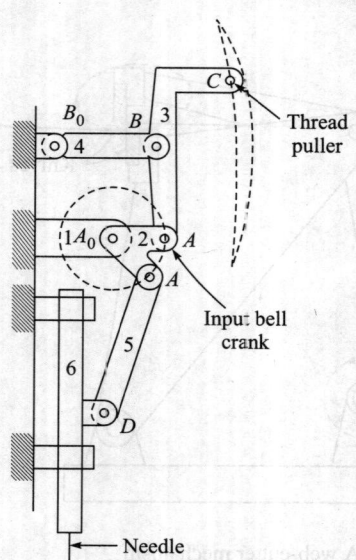

(h) A sewing machine mechanism

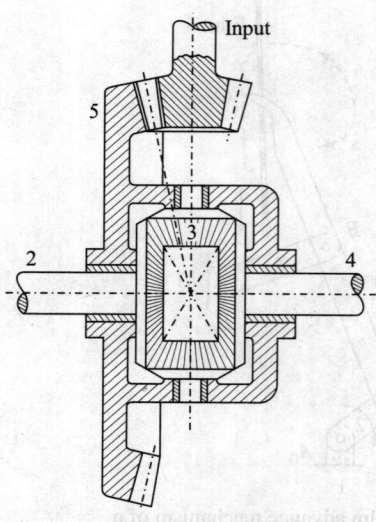

(i) A bevel gear differential mechanism for a 4-wheeler

(Contd.)

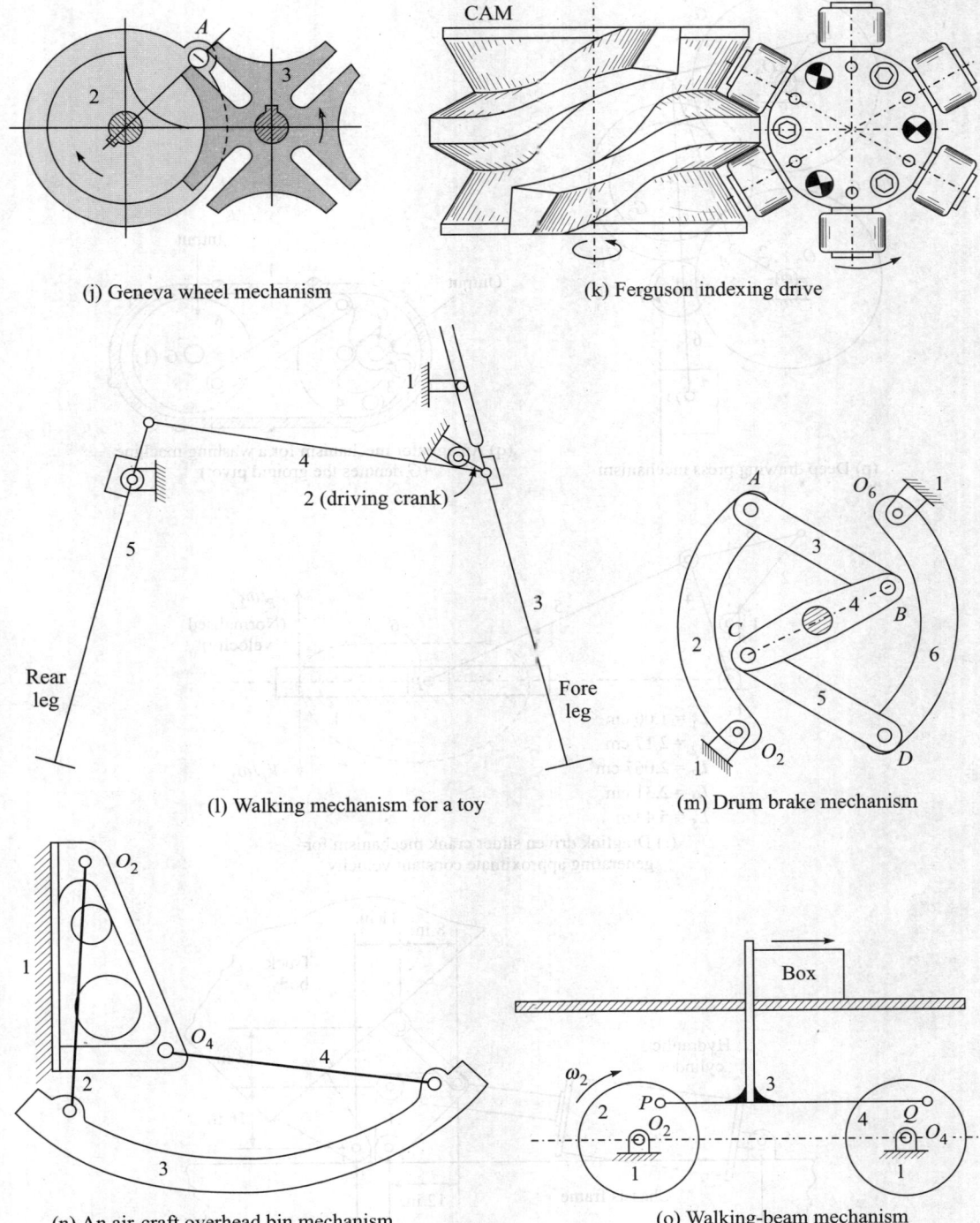

(j) Geneva wheel mechanism

(k) Ferguson indexing drive

(l) Walking mechanism for a toy

(m) Drum brake mechanism

(n) An air-craft overhead bin mechanism

(o) Walking-beam mechanism

(Contd.)

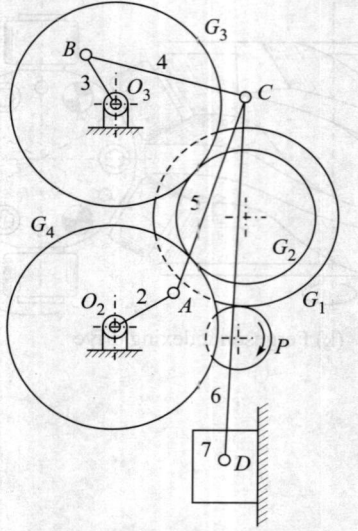

(p) Deep drawing press mechanism

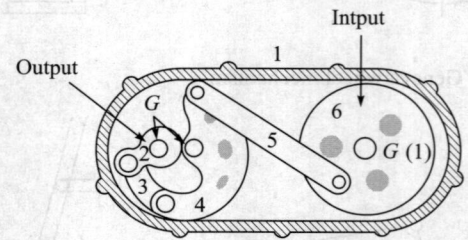

(q) An agitator mechanism for a washing machine
(*G* denotes the ground pivot)

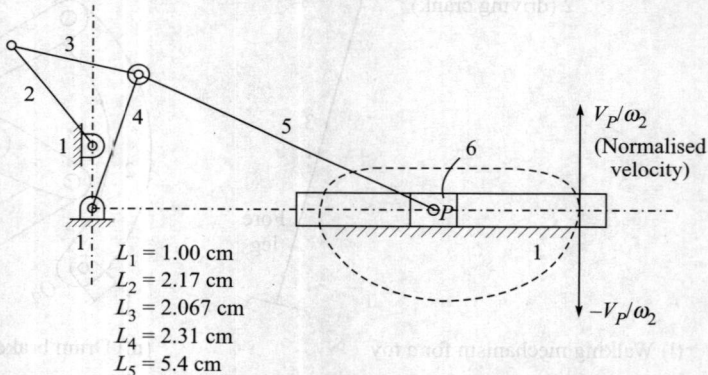

$L_1 = 1.00$ cm
$L_2 = 2.17$ cm
$L_3 = 2.067$ cm
$L_4 = 2.31$ cm
$L_5 = 5.4$ cm

(r) Draglink driven slider crank mechanism for
generating approximate constant velocity

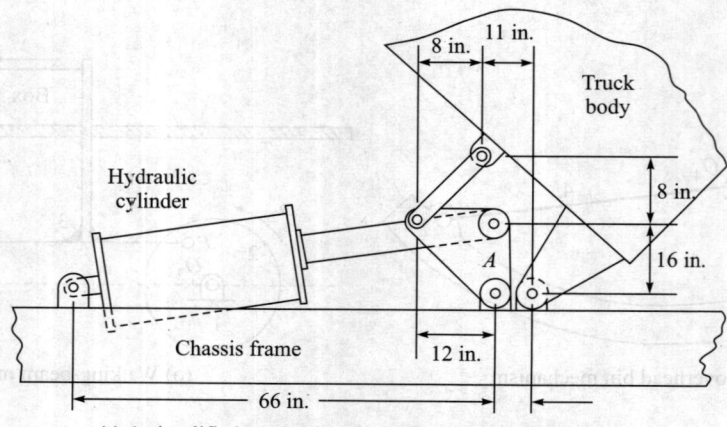

(s) A simplified mechanism for hydraulic dump truck

(Contd.)

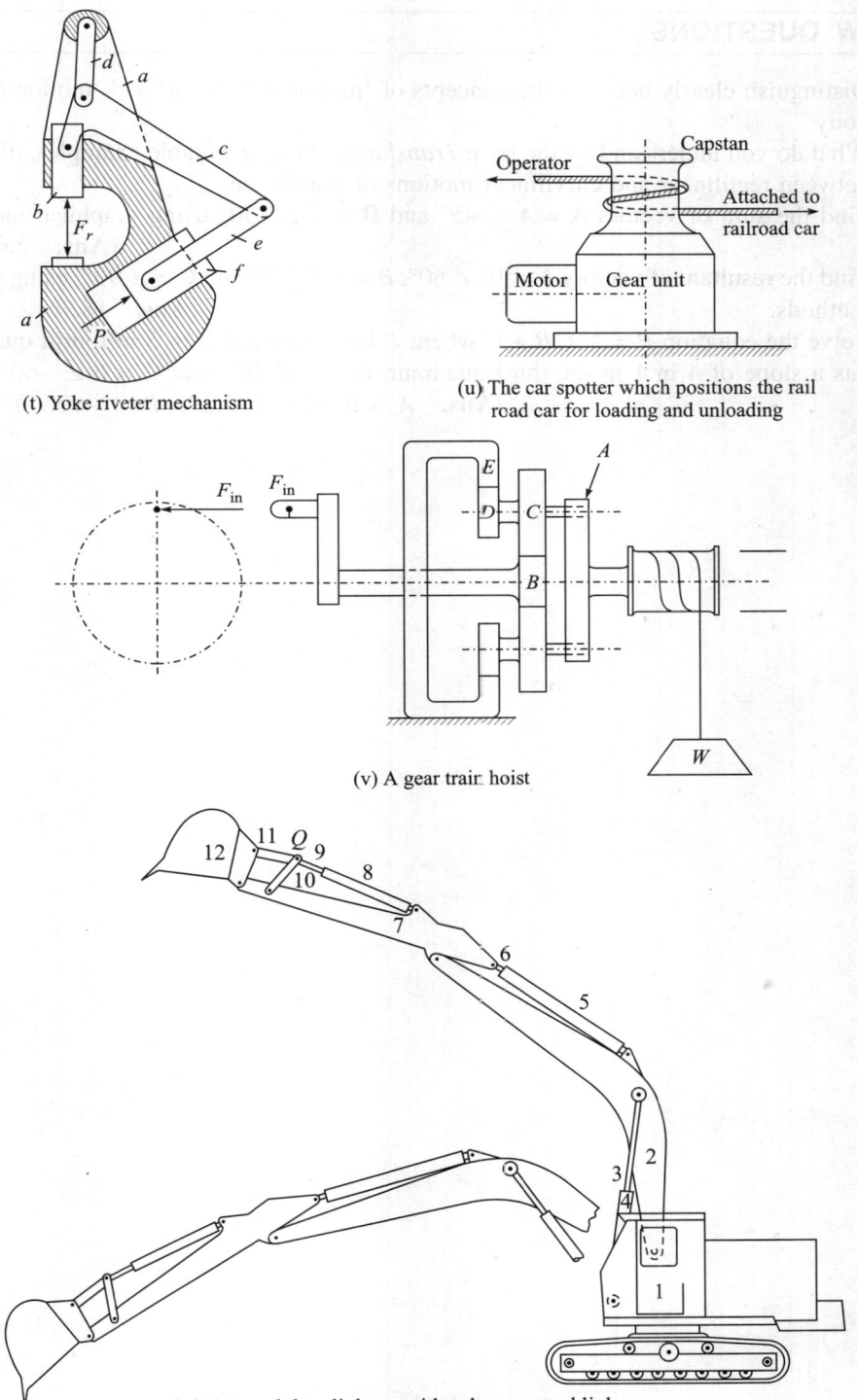

(t) Yoke riveter mechanism

(u) The car spotter which positions the rail road car for loading and unloading

(v) A gear train hoist

(w) A trench hoe linkage with cab as ground link

Fig. 1.17 A few applications of mechanisms.

REVIEW QUESTIONS

1.1 Distinguish clearly between the concepts of 'motion of a point' and 'motion of a rigid body.'

1.2 What do you understand by the term *Translation*. Giving suitable examples, distinguish between rectilinear and curvilinear motions of translation.

1.3 Find the sum of vectors $\mathbf{A} = 4 \angle 45°$ and $\mathbf{B} = 3 \angle -30°$ using graphical methods.

(**Ans.** $5.6 \angle 14°$)

1.4 Find the resultant of vectors $\mathbf{A} = 10 \angle 60°$, $B = 8 \angle 150°$ and $C = \angle 90°$, using graphical methods. (**Ans.** $18.76 \angle -84.12°$)

1.5 Solve the equation $P = A + B + C$ when, P has a slope of 2 in 5 in fourth quadrant, $\mathbf{A}$ has a slope of 4 in 1 in the third quadrant, $B = 9 \angle 30°$, and $C = 6 \angle -60°$.

(**Ans.** $A = (0.82) - j (3.29)$; $P = -(9.97) - j (3.99)$

Chapter 2

PLANAR MECHANISMS AND
GEOMETRY OF MOTION

2.1 INTRODUCTION

As the book is devoted to the study of mechanisms and machines, it is necessary to understand what these terms are. To a beginner, the term *machine* may be defined as a device/contrivance which receives energy in some available form and uses it to do certain particular kind of work. In a similar way, *mechanism* may be defined as a contrivance which transforms motion from one form to another.

2.2 DEFINITIONS AND BASIC CONCEPTS

A Link. A link may be defined as a single resistant part (or an assembly of rigidly connected parts) of a machine, having a motion relative to other parts.

Even if two or more connected parts are manufactured separately, they cannot be treated as different links unless there is a relative motion between them. For instance, piston, piston-rod and cross-head of a steam engine constitute a single link as there is no relative motion between them. From the point of view of a kinematician therefore, slider-crank mechanisms of a steam engine and I.C. engine are just the same.

A link need not necessarily be rigid but it must be resistant. It may be recalled that a *resistant body is one which transmits force with negligible deformation in the direction of force transmission*. Based on above considerations, a spring which has no effect on the kinematics of a device, and has significant deformation in the direction of applied force, is not treated as a link but only as a device to apply force. They are usually ignored during kinematic analysis, and their 'force-effects' are introduced during dynamic analysis.

There are machine members which possess one-way rigidity. For instance, because of their resistance to deformation under tensile load, belts, ropes and chains are treated as links only when they are in tension. Similarly, liquids on account of their incompressibility can be treated as links only when transmitting compressive force. Figure 2.1 shows a few representative mechanisms in which links have been numbered to bring home the above concept.

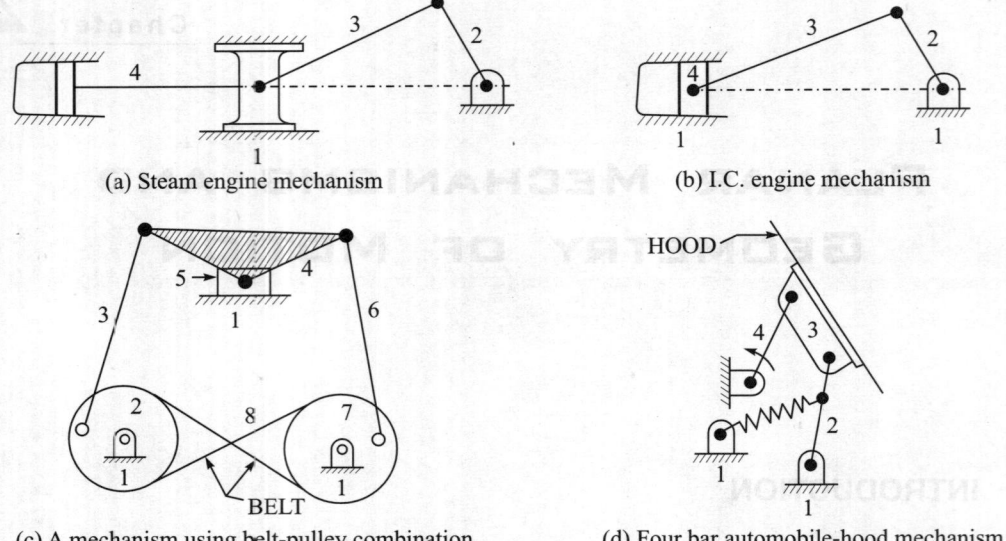

(a) Steam engine mechanism (b) I.C. engine mechanism

(c) A mechanism using belt-pulley combination (d) Four bar automobile-hood mechanism

Fig. 2.1

An Element. The kind of relative motion between links of a mechanism is controlled by the form of the contacting surfaces of the adjacent (connected) links. These contacting surfaces may be thought of as 'working surfaces' of the connection between adjacent links. For instance, the connection between a lathe carriage and its bed is through 'working surfaces' (ways) which are so shaped that only motion of translation is possible. Similarly, the working surfaces of I.C. engine piston and connecting rod at piston pin are so shaped that relative motion of rotation alone is possible. Each of these working surface is also called an *element.*

An element may therefore be defined as a geometrical form provided on a link so as to ensure a working surface that permits desired relative motion between connected links.

A Pair. The two contacting elements of a connection constitute a kinematic pair. *A pair may also be defined as a connection between two adjacent links that permits a definite relative motion* between them. It may be noted that the above statement is generally true. In the case of multiple joint, however, more than two links can be connected at a kinematic pair (also known as joint). Cylindrical contacting surfaces between I.C. engine cylinder and piston constitute a pair. Similarly, cylindrical contacting surfaces of a rotating shaft and a journal bearing also constitute a pair.

When all the points in different links in a chain move in planes which are mutually parallel, the chain is said to have a planar motion. A motion other than planar motion is a spatial motion.

Kinematic Diagram (Skeleton Diagram). When the links are assumed to be rigid in kinematics, there can be no change in relative positions of any two arbitrarily chosen points on the same link. In particular, relative position(s) of pairing elements on the same link does not change. As a consequence of assumption of rigidity, many of the intricate details, shape and size of the actual part (link) become unimportant in kinematic analysis. For this reason, it is customary to draw highly simplified schematic diagrams which contain only the important features in respect of the shape of each link (e.g., relative locations of pairing elements). This necessarily requires to completely suppress the information about real geometry of manufactured parts. Schematic diagrams of various links, showing relative locations of pairing elements, are shown in Fig. 2.5. Conventions followed in drawing kinematic diagram are also shown there.

In drawing a kinematic diagram, it is customary to draw the parts (links) in the most simplified form so that only those dimensions are considered which affect the relative motion. One such simplified kinematic diagram of slider-crank mechanism of an I.C. engine is shown in Fig. 2.2 in which connecting rod 3 and crank 2 are represented by lines joining their respective pairing

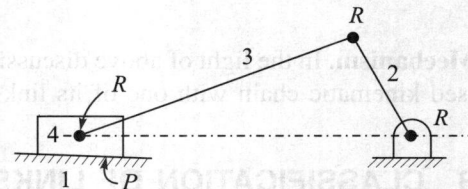

Fig. 2.2   Kinematic Diagram of an I.C. engine Mechanism.

elements. The piston has been represented by the slider 4 while cylinder (being a stationary member) has been represented by frame link 1.

It may be noted, however, that these schematics have a limitation in that they have little resemblance to the physical hardware. And, one should remember that kinematic diagrams are particularly useful in kinematic analysis and synthesis but they have very little significance in designing the machine components of such a mechanism.

Kinematic Chain. A kinematic chain can be defined as an assemblage of links which are interconnected through pairs, permitting relative motion between links. A chain is called a closed chain when links are so connected in sequence that first link is connected to the last, ensuring that all pairs are complete because of mated elements forming working surfaces at joints. As against this, when links are connected in a sequence, with first link not connected to the last (leaving incomplete pairs), the chain is called an open chain. Examples of planar open loop chain are not many but they have many applications in the area of robotics and manipulators as space mechanisms. An example of a planar open-loop chain, which permits the use of a singular link (a link with only one element on it), is the common weighing scale shown in Fig. 2.3. Various links are numbered in the figure. Links 3, 1 and 4 are singular links.

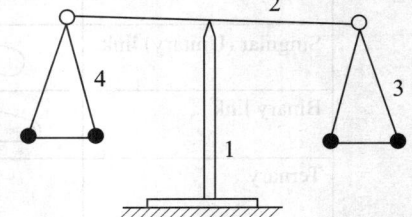

Fig. 2.3   Common weighing scale (An example of an open chain).

Structure. Geometric considerations sometimes prevent relative motion between links of a chain after closure. In such a situation the chain is called a structure. The structure can be statically determinate or indeterminate (*see* Fig. 2.4). It may be observed that a minimum of three links with three kinematic pairs are necessary to form a simple closed chain. However, if all the three pairs are such that no space-motion is possible, and one of the links is fixed, there cannot be any relative movement of links, such an arrangement is called 'structure'.

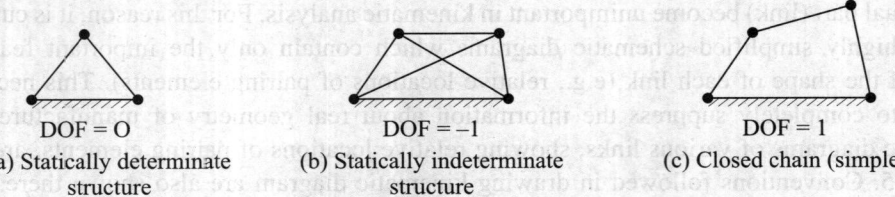

DOF = 0	DOF = –1	DOF = 1
(a) Statically determinate structure	(b) Statically indeterminate structure	(c) Closed chain (simple)

Fig. 2.4

A Mechanism. In the light of above discussions, a mechanism may now be defined as a movable closed kinematic chain with one of its links stationary (fixed).

2.3 CLASSIFICATION OF LINKS

A link can be called singular (unitary), binary, ternary, quaternary (etc.) link depending on the number of elements it has for pairing with other links. Thus a link carrying a single element is called a singular (unitary) link and a link with two elements is called a binary link. Similarly, a link having three elements is called a ternary link while a link having four elements is called a quaternary link. These links, along with their conventional representation, are shown in Fig. 2.5.

Type of Link	Typical Form	Schematic Representation
Single link (Typical shapes)	 1 1	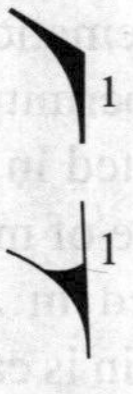 1 1
Singular (Unitary) link		
Binary link	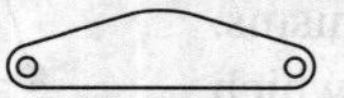	
Ternary	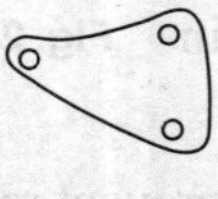	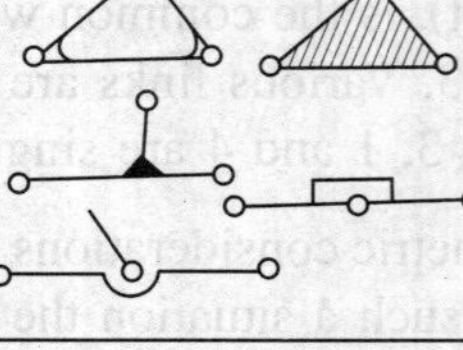
Quaternary link	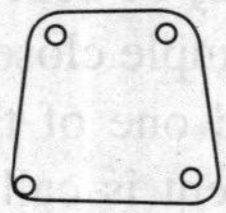	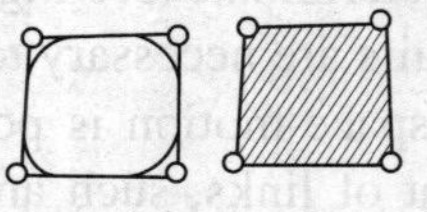

Fig. 2.5 Conventional Representation of different types/shapes of Links in Kinematic Diagram.

2.4 CLASSIFICATION OF PAIRS

Kinematic pairs are classified on the basis of any of the following characteristics:

(a) Type of relative motion between contacting elements.

(b) Type of contact between contacting elements.

(c) Number of degrees of freedom.

(d) Type of closure (*i.e.* whether self-closed or force-closed).

2.4.1 Classification of Pairs Based on Type of Relative Motion

The relative motion of a point on one element relative to the other on mating element can be that of turning, sliding, screw (helical direction), planar, cylindrical or spherical. The controlling factor that determines the relative motions allowed by a given joint is the shapes of the mating surfaces or elements. Each type of joint has its own characteristic shapes for the elements, and each permits a particular type of motion, which is determined by the possible ways in which these elemental surfaces can move with respect to each other. The shapes of mating elemental surfaces restrict the totally arbitrary motion of two unconnected links to some prescribed type of relative motion.

(i) *Turning Pair*. (Also called a hinge, a pin joint or a revolute pair).   This is the most common type of kinematic pair and is designated by the letter R.

A pin joint has cylindrical element surfaces and assuming that the links cannot slide axially, these surfaces permit relative motion of rotation only. A pin joint allows the two connected links to experience relative rotation about the pin centre. Thus, the pair permits only one degree of freedom. Kinematic pairs, marked R in Fig. 2.2, represent turning or revolute pairs. Thus, the pair at piston pin, the pair at crank pin and the pair formed by rotating crank-shaft in bearing are all examples of turning pairs.

(ii) *Sliding or Prismatic Pair*.   This is also a common type of pair and is designated as P. This type of pair permits relative motion of sliding only in one direction (along a line) and as such has only one degree of freedom. Pairs between piston and cylinder, cross-head and guides, die-block and slot of slotted lever are all examples of sliding pairs.

(iii) *Screw Pair*.   This pair permits a relative motion between concident points, on mating elements, along a helix curve. Both axial sliding and rotational motions are involved. But as the sliding and rotational motions are related through helix angle α, the pair has only one degree of freedom. The pair is commonly designated by the letter S. Examples of such pairs are to be found in translatory screws operating against rotating nuts to transmit large forces at comparatively low speed, *e.g.* in screw-jacks, screw-presses, valves and pressing screw of rolling mills. Other examples are rotating lead screws operating in nuts to transmit motion accurately as in lathes, machine tools, measuring instruments, etc.

(iv) *Cylindrical Pair*.   A cylindrical pair permits a relative motion which is a combination of rotation θ and translation s parallel to the axis of rotation between the contacting elements. The pair has thus a degree of freedom of two and is designated by a letter C. A shaft free to rotate in a bearing and also free to slide axially inside the bearing provides example of a cylindrical pair.

(v) *Globular or Spherical Pair.* Designated by the letter *G*, the pair permits relative motion such that coincident points on working surfaces of elements move along spherical surface. In other words, for a given position of spherical pair, the joint permits relative rotation about three mutually perpendicular axes. It has thus three degrees of freedom. A ball and socket joint (*e.g.,* the shoulder joint at arm-pit of a human being) is the best example of spherical pair.

(vi) *Flat pair (Planar Pair).* A flat or planar pair is seldom, if ever, found in mechanisms. The pair permits a planar relative motion between contacting elements. This relative motion can be described in terms of two translatory motions in *x* and *y* directions and a rotation θ about third direction *z*, – *x, y, z* being mutually perpendicular directions. The pair is designated as *F* and has a degree of freedom of 3.

All the above six types of pairs, illustrated in Fig. 2.6, are representative of a particular class.

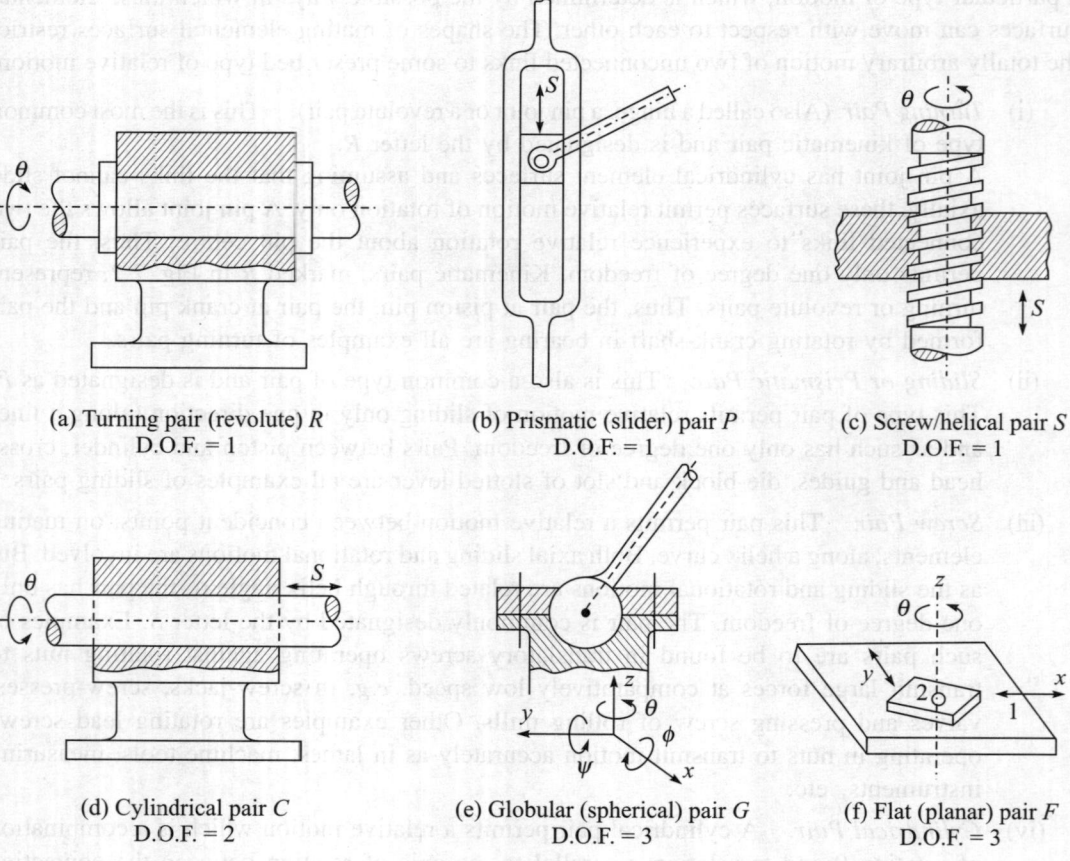

(a) Turning pair (revolute) *R*
D.O.F. = 1

(b) Prismatic (slider) pair *P*
D.O.F. = 1

(c) Screw/helical pair *S*
D.O.F. = 1

(d) Cylindrical pair *C*
D.O.F. = 2

(e) Globular (spherical) pair *G*
D.O.F. = 3

(f) Flat (planar) pair *F*
D.O.F. = 3

Fig. 2.6

(vii) *Rolling Pair*. When surfaces of mating elements have a relative motion of rolling, the pair is called a rolling pair. Castor wheel of trolleys, ball and roller bearings, wheels of locomotive/wagon and rail are a few examples of this type.

2.4.2 Classification of Pairs Based on Type of Contact

This is the best known classification of kinematic pairs on the basis of nature of contact and was suggested by the great kinematician Reuleaux. He classified kinematic pairs in two categories:

(i) *Lower Pair*. Kinematic pairs in which there is a surface (area) contact between the contacting elements are called lower pairs. All revolute pairs, sliding pairs, screw pairs, globular pairs, cylindrical pairs and flat pairs fall in this category.

(ii) *Higher Pair*. Kinematic pairs in which there is a point or line contact between the contacting elements are called higher pairs. Meshing gear-teeth, cam follower pair, wheel rolling on a surface, ball and roller bearings and pawl and ratchet are a few examples of higher pairs.

Since lower pairs involve surface contact rather than line or point contact, it follows that lower pairs can be more heavily loaded for the same unit pressure. They are considerably more wear-resistant. For this reason, development in kinematics have involved more and more number of lower pairs. As against this, use of higher pairs implies lesser friction.

According to Hartenberg R.S. and Denavit J., the real concept of lower pairs lies in the particular kind of relative motion permitted by the connected links. For instance, let us assume that two mating elements P and Q form a kinematic pair. If the path traced by any point on the element P, releative to element Q, is identical to the path traced by a corresponding (coincident) point in the element Q relative to element P, then the two elements P and Q are said to form a lower pair. Elements not satisfying the above condition obviously form the higher pairs.

Since a turning pair involves relative motion of rotation about pin-axis, coincident points on the two contacting elements will have circular areas of same radius as their path. Similarly elements of sliding pair will have straight lines as the path for coincident points. In the case of screw pair, the coincident points on mating elements will have relative motion along helices. As against this a point on periphery of a disk rolling along a straight line generates cycloidal path, but the coincident point on straight line generates involute path when the straight line rolls over the disk. The two paths are thus different and the pair is a higher pair. *As a direct sequel to the above consideration, unlike a lower pair, a higher pair cannot be inverted. That is, the two elements of the pair cannot be interchanged with each other without affecting the overall motion of the mechanism.*

Lower pairs are further subdivided into linear motion and surface motion pairs. The distinction between these two sub-categories is based on the number of degrees of freedom of the pair. Linear motion lower pairs are those having one degree of freedom, *i.e.* each point on one element of the pair can move only along a single line or curve relative to the other element. This category includes turning pairs, prismatic pairs and screw pairs.

Surface-motion lower pairs have two or more degrees of freedom. This category includes cylindrical pair, spherical pair and the planar (Flat) pair.

2.4.3 Classification of Pairs Based on Degrees of Freedom

A free body in space has six degrees of freedom. In forming a kinematic pair, one or more degrees of freedom are lost. The remaining degrees of freedom of the pair can then be used to classify pairs. Thus,

$$\text{d.o.f. of a pair} = 6 - \text{(Number of restraints)}.$$

A kinematic pair can therefore be classified on the basis of number of restraints imposed on the relative motion of connected links. This is done in Table 2.1 for different forms of pairing element shown in Fig. 2.7.

Table 2.1

S.No. in Fig. 2.7	Geometrical shapes of elements in contact	Number of Restraints on		Total Number of Restraints	Class of pair
		Translatory motion	Rotary motion		
(a)	Sphere and plane	1	0	1	I
(b)	Sphere inside a cylinder	2	0	2	II
(c)	Cylinder on plane	1	1	2	II
(d)	Sphere in spherical socket	3	0	3	III
(e)	Sphere in slotted cylinder	2	1	3	III
(f)	Prism on a plane	1	2	3	III
(g)	Spherical ball in slotted socket	3	1	4	IV
(h)	Cylinder in cylindrical hollow	2	2	4	IV
(i)	Collared cylinder in hollow cylinder	3	2	5	V
(j)	Prism in prismatic hollow	2	3	5	V

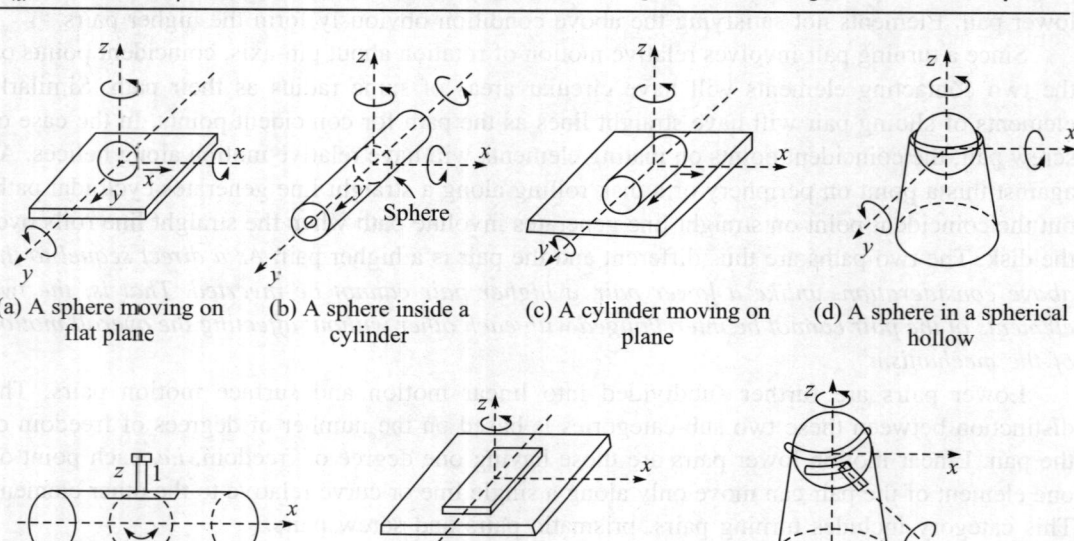

(a) A sphere moving on flat plane

(b) A sphere inside a cylinder

(c) A cylinder moving on plane

(d) A sphere in a spherical hollow

(e) A spherical element of link 1 inside a cylinder, with an axial slot

(f) A prism on another flat surface

(g) A spherical element in a spherical hollow with a slot to clear link

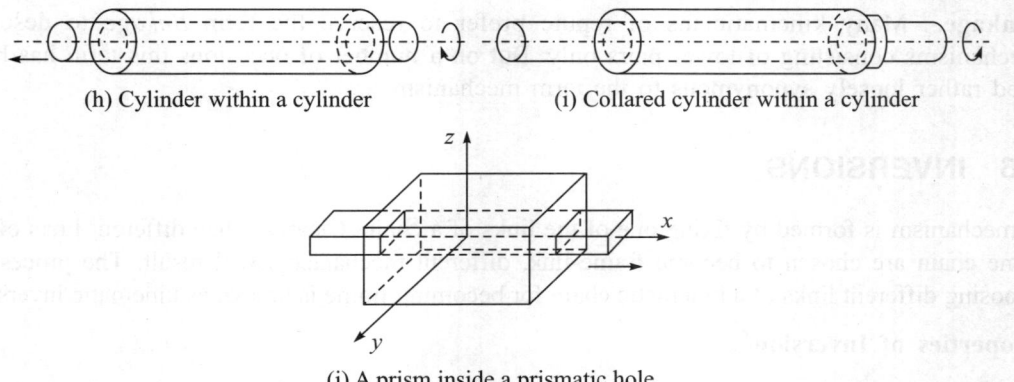

(h) Cylinder within a cylinder　　　(i) Collared cylinder within a cylinder

(j) A prism inside a prismatic hole

Fig. 2.7

2.4.4 Classification of Pairs Based on Type of Closure

Another important way of classifying pairs is to group them as (a) closed kinematic pairs and (b) open kinematic pairs.

In closed pairs, one element completely surrounds the other so that it is held in place in all possible positions. Restraint is achieved only by the form of pair and, therefore, the pair is called closed or self-closed pair. The other possible definition of closed pair is: "closed pairs are those pairs in which elements are held together mechanically." All the lower pairs and a few higher pairs fall in the category of closed pairs.

As against this, open kinematic pairs maintain relative positions only when there is some external means to prevent separation of contacting elements. Open pairs are also sometimes called as 'unclosed pairs'. A cam and roller-follower mechanism, held in contact due to spring force, is an example of this type.

2.5 MECHANISM AND MACHINE

A mechanism can be defined as a combination of resistant bodies, so shaped and connected that as they move upon each other, a definite relative motion results. A beginner usually finds it difficult to conceive the term *mechanism* as distinct from a machine. Basically, a *mechanism* is a device to transform one motion into another. If the device transmits substantial forces as well, it can be called a machine. It follows, therefore, that all machines are mechanisms in spirit.

	A Machine	*A Mechanism*
(i)	It is an assemblage of links having relative motion and is capable of modifying available energy in a suitable way.	It is an assemblage of links used to transmit and modify motion.
(ii)	A machine may consist of one or more mechanisms for accomplishing the above task.	It represents skeleton outline of a machine. Only those link-dimensions are considered which are important from the point of view of motion transmission.
(iii)	All machines are mechanisms in spirit.	If the device transmits substantial forces as well, it can be called a machine.

Linkage. Many kinematicians of repute prefer to reserve the term *linkage* to describe mechanisms consisting of lower pairs only. But on a number of occasions this term has been used rather loosely synonymous to the term mechanism.

2.6 INVERSIONS

A mechanism is formed by fixing one of the links of a chain. Clearly, when different links of the same chain are chosen to become frame-link, different mechanisms will result. The process of choosing different links of a kinematic chain for becoming frame is known as kinematic inversion.

Properties of Inversion

1. Number of inversions possible for a kinematic chain equals the number of links in the parent kinematic chain.

2. Relative motion (displacement, velocity and acceleration) between any two links does not change with inversion. This is simply because relative motion between different links is a property of parent kinematic chain.

3. Absolute motion of points on various links (measured with respect to the frame-link) may, however, change drastically from one inversion to the other, even in direct inversion.

As explained in Section 2.4 *unlike lower pairs a higher pairs cannot be inverted. This is simply because the two elements of a higher pair cannot be interchanged with each other without affecting overall motion of the mechanism.*

Importance of Inversion

Important aspects of the concept of inversion can be summarised as under:

1. The concept of inversion enables us to categorise a group of mechanisms arising out of inversions of a parent kinematic chain as a family of mechanisms. Members of this family have a common characteristic in respect of relative motion.

2. In case of *direct inversions*, as relative velocity and relative acceleration between two links remain the same, it follows that complex problems of velocity/acceleration analysis may often be simplified by considering a direct inversion of the original mechanism. Such procedure is the basis of Goodman's ingenious method of indirect acceleration analysis. The concept is also useful in converting motion analysis problem of an epicyclic gear-train to that of a simple reverted gear train by fixing arm and freeing the fixed member.

   A Direct Inversion is one in which relative link-dimensions remain same in every inversion.

3. In many cases of inversions, by changing proportions of lengths of links, desirable features of the inversion may be accentuated and many useful mechanisms may be developed.

2.7 QUADRIC CYCLE CHAIN AND ITS INVERSIONS

A kinematic chain consisting of turning pairs only (*i.e.*, a linkage) must have a minimum of four links and four pairs. It is customary to call a fully rotating link a '**crank**', an oscillating link a

'**Rocker/lever**' and the connecting link a '**coupler**' or '**connecting rod**'. The coupler is the link which is not connected to the frame directly.

An important consideration when designing a mechanism is to see whether an electric motor is a driving member. In such a case, naturally there must be a crank member in the mechanism to receive power from motor. For a four-bar linkage, Grashoff's law provides a very simple test to check whether any of the links in the chain can be a crank.

Grashoff's Law.   *Grashoff's law states that for a planar four-bar linkage, sum of the shortest and longest link-lengths must be less than or equal to the sum of the remaining two link-lengths, if there is to be a continuous relative rotation between two members.* Thus, if s and l be the lengths of shortest and longest links respectively and p and q be the remaining two link-lengths, then one of the links, in particular the shortest link, will rotate continuously relative to the other three links, if and only if,

$$s + l \le p - q \tag{2.1}$$

If this inequality is not satisfied, the chain is called non-Grashoff chain in which none of the links can have complete revolution relative to other links.

It is important to note that the Grashoff's law does not specify the order in which the links are to be connected. Thus any of the links having length l, p and q can be the link opposite to the link of length s. A chain satisfying Grashoff's law generates three distinct inversions only. *A non-Grashoff chain, on the other hand, generates only one distinct inversion, namely the "Rocker-Rocker mechanism."*

Inversions of a Grashoff's Chain

1. *Drag link or Double-crank Mechanism.* When the shortest link of a Grashoff-chain is fixed, it gives rise to a drag-link or double-crank mechanism in which both the links connected to the frame rotate continuously. [Fig. 2.8(a)]

2. *Double Rocker/Lever Mechanism.* When the link opposite to the shortest link is fixed, a double rocker/lever mechanism results. None of the two links (driver and driven) connected to the frame can have complete revolution but the coupler link can have full revolution. [Fig. 2.8(b)]

3. *Crank Rocker/Lever Mechanism.* When any of the two remaining links (adjacent to the shortest link) is fixed, a crank-lever/rocker mechanism results and one of the two links (driver or driven) directly connected to the frame, is capable of having full revolution. [Fig. 2.8(c)]

In Fig. 2.8(b) OP_1Q_1C and OP_2Q_2C represent extreme positions of the double-rocker mechanism. Similarly in Fig. 2.8(c) it is seen that as crank OP continues to rotate c.c.w. from position OP_1 onwards, the lever CQ_1 reverses its direction of motion and at crank position OP_2 takes back the position CQ_2.

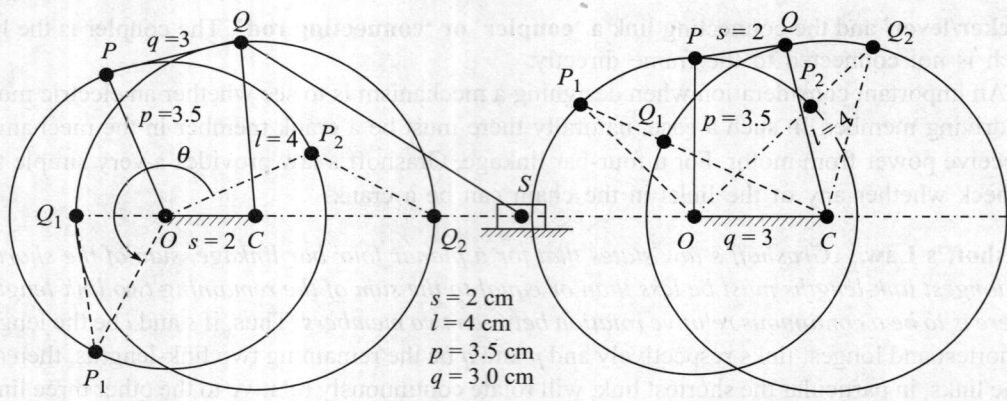

(a) Double crank mechanism inversion-I

(b) Rocker-rocker mechanism inversion-II

s = 2 cm
l = 4 cm
p = 3.5 cm
q = 3.0 cm

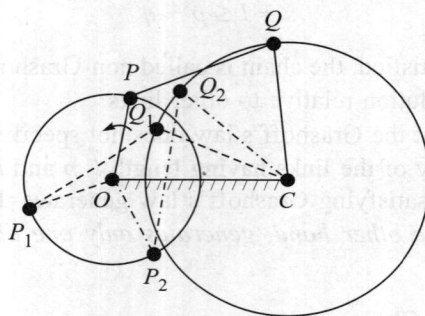

(c) Crank-rocker mechanism inversion-III

Fig. 2.8 Inversions of Quadric cycle chain.

EXAMPLE 2.1 Figure 2.9 shows a planar mechanism with link-lengths given in some unit. If slider A is the driver, will link CG revolve or oscillate? Justify your answer. (A.M.I.E. Summer 1993)

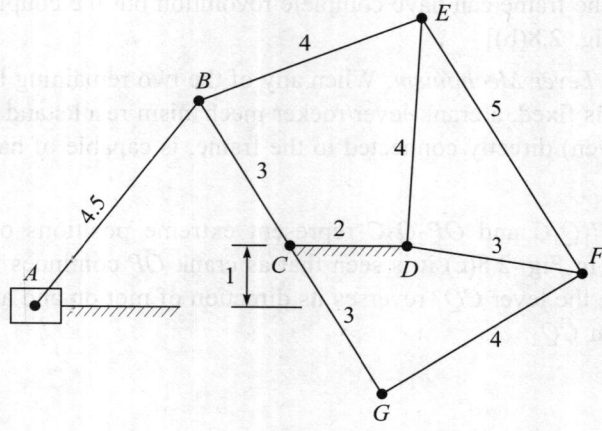

Fig. 2.9

Solution: The loop formed by three links *DE*, *EF* and *FD* represents a structure. Thus the loop can be taken to represent a ternary link.

Now in the 4-link loop *CDEB*, $s = 2$; $l = 4$; and $p + q = 7$. Thus the 4-link loop portion *CDEB* satisfies Grashoff's criterion. And as the shortest link *CD* is fixed, link *CB* is capable of complete revolution. Also, 4-link loop *CDFG* satisfies Grashoff's criterion $(l + s = p + q)$ and the shortest link *CD* is fixed. Thus whether considered a part of 4-link loop *CDEB* or that of *CDFG*, link *BCG* is capable of full revolution. **Ans.**

EXAMPLE 2.2 In a 4-bar mechanism, the lengths of driver crank, coupler and follower link are 150 mm, 250 mm and 300 mm respectively. The fixed link-length is L_o. Find the range of values for L_o, so as to make it a –

(i) Crank-rocker mechanism (ii) Crank-crank mechanism. (SGSITS: 1999)

Solution: (i) For a crank-rocker mechanism the conditions to be satisfied are:

(a) Link adjacent to fixed link must be the smallest link and, (b) $s + l \le p + q$

We have to consider both the possibilities, namely, (i) when L_o is the longest link and (ii) when L_o is not the longest link.

When L_o is the longest link, from Grashoff's criterion,

$$L_o + 150 \le 250 + 300$$

or

$$L_o \le 400 \text{ mm}$$

When L_o is not the longest link, from Grashoff's criterion,

$$300 + 150 \le L_o + 250$$

or

$$L_o \ge 200$$

Thus, for crank-rocker mechanism, range of values for L_o is

$$200 \le L_o \le 400 \text{ mm}$$ **Ans.**

(ii) For crank-crank mechanism, the conditions to be satisfied are
(a) shortest link must be the frame link, and

(b) $$s + l \le p + q.$$

Thus, $$L_o + 300 \le 150 + 250$$

or $$L_o \le 100 \text{ mm}$$ **Ans.**

Applications (Inversions of Quadric Cycle Chain):

First Inversion (Drag Link Quick Return Mechanism). For understanding the application of drag link mechanism, consider Fig. 2.8(a) which shows its extension in the form of a Drag Link Quick Return Mechanism. For this purpose, let the reciprocating ram *S* (representing cutting tool-post) be connected through a coupler *QC* to the quadric cycle chain through turning pairs at *Q* and *S*.

For extreme left hand position of ram *S*, the driven link will be located in position CQ_1 and the driving link *OP* in position OP_1. As the crank *OP* rotates c.c.w., the ram moves to the right and as the driven link takes up position CQ_2, the ram completes its forward stroke taking up extreme right hand position. For this position, the driver link is located in position OP_2. Thus for completing forward stroke of ram, the crank OP has to rotate through say 136° (which is less than 180°). For completing return stroke, clearly the driver crank will have to rotate from OP_2 to OP_1 c.c.w. through an angle say 224° (which is larger than 180°). It must be noted that crank rotation for completing forward and return stroke of ram together, cannot take more than 360°.

Thus it is seen that as crank OP (which is driven by a constant speed drive) requires smaller angle of rotation for moving slider from left to right, and the stroke will require smaller time, and hence may be used as a return (idle) stroke. The other stroke, requiring larger time, may be reserved for cutting stroke in a shaper machine tool. This arrangement results in a better utilization of available time for cutting operation.

Angle θ_C, the larger of the two angles, corresponds to cutting/working stroke while the smaller one θ_R corresponds to the return/idle stroke. Since the crank rotates uniformly, the time of cutting stroke bears a ratio to time of idle stroke, given by

$$\frac{\theta_C}{\theta_R} = \frac{\text{Time of cutting stroke}}{\text{Time of Return stroke}} \qquad (2.2)$$

Second Inversion. There are a number of applications of double lever/rocker mechanisms. Some of them are Pantograph, Ackerman steering gear mechanism and Watt's approximate straight line motion mechanism. These are shown in Figs. 2.10(a), (b) and (c).

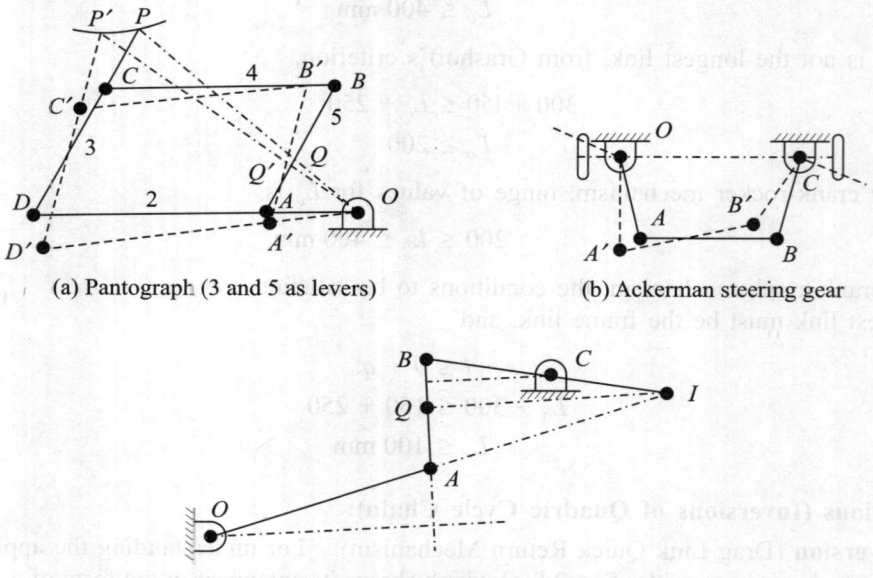

(a) Pantograph (3 and 5 as levers) (b) Ackerman steering gear

(c) Watt's approximate straight line motion mechanism

Fig. 2.10

Third Inversion (Crank-Rocker Mechanism). Beam engine mechanism is used for converting rotary motion of crank into reciprocating motion of piston. The line diagram explains the principle of a beam engine (Fig. 2.11). The link 1 is the crank while link 3 is the rocker.

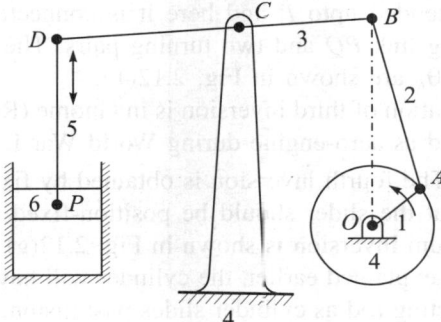

Fig. 2.11 Beam engine mechanism.

Limiting condition of Grashof's chain-comments:

Experience reveals that mechanisms with proportions approaching the limits of Grashof's rule are not satisfactory in practice. During a motion-cycle, links assume relative positions approaching to "toggling" (dead centre positions), producing chatter and possibly excessive stresses. Further, such linkages require tighter machining tolerances on links, if undesirable effects on motion-characteristics are to be avoided, and as such are more expensive to manufacture.

2.8 INVERSION OF SLIDER CRANK CHAIN

A slider crank chain is as shown in Fig. 2.12(a).

1. **First Inversion.** It is obtained by fixing link 1 of the chain and the result is the crank-slider mechanism as shown in Fig. 2.12(b). This mechanism is very commonly used in I.C. engines, steam engines and reciprocating compressor mechanism.

2. **Second Inversion.** It is obtained by fixing link 3, the connecting rod. The mechanism obtained by 'verbatum inversion', as shown in Fig. 2.12(c), has some practical difficulties. For instance, the oscillating cylinder will have to be slotted for clearing the pin through which slider is pivoted to frame. The problem is resolved, if one remembers that any suitable alteration in shapes of members, ensuring same type of pairs between links 3 and 4 and also between links 1 and 4, is permissible. This gives rise to an inversion at Fig. 2.12(d). The resulting mechanism is oscillating cylinder engine mechanism. It is used in hoisting engine mechanism and also in toys. In hoisting purposes its chief advantage lies in its compactness of construction as it permits simple scheme of supplying steam to the cylinder.

   Second application of the above inversion lies in 'Slotted Lever Quick Return Mechanism', shown in Fig. 2.12(e). The extremum positions of lever 4 is decided by the tangents drawn from lever-pivot to the crank-circle on either side. Corresponding positions of crank 1 include angles θ_C and θ_R, which correspond to cutting stroke angle and return stroke angle.

3. **Third Inversion.** The third inversion is obtained by fixing crank 2. It is the slider-crank equivalent of Drag-link mechanism and forms the basis of Whit-Worth Quick Return Mechanism. Basic inversion is given by portion OAS. To derive advantage however, the

slotted link 1 is extended upto P and here it is connected to reciprocating tool-post through a connecting link PQ and two turning pairs. The cutting stroke angle θ_C and return stroke angle θ_R are shown in Fig. 2.12(f).

A yet another application of third inversion is in Gnome (Rotary cylinders) engine. This was extensively used as aero-engine during World War I.

Fourth Inversion. The fourth inversion is obtained by fixing slider, the link 4. Fixing of slider implies that the slider should be position-fixed and also fixed in respect of rotation. The verbatum inversion is shown in Fig. 2.12(g). This form has certain practical difficulties. As explained earlier, the cylinder will have to be slotted so as to clear piston pin of connecting rod as cylinder slides past piston. To overcome this difficulty, the shapes of piston and cylinder are exchanged as shown in Fig. 2.12(h). This gives a hand pump mechanism. Lever 2 is extended

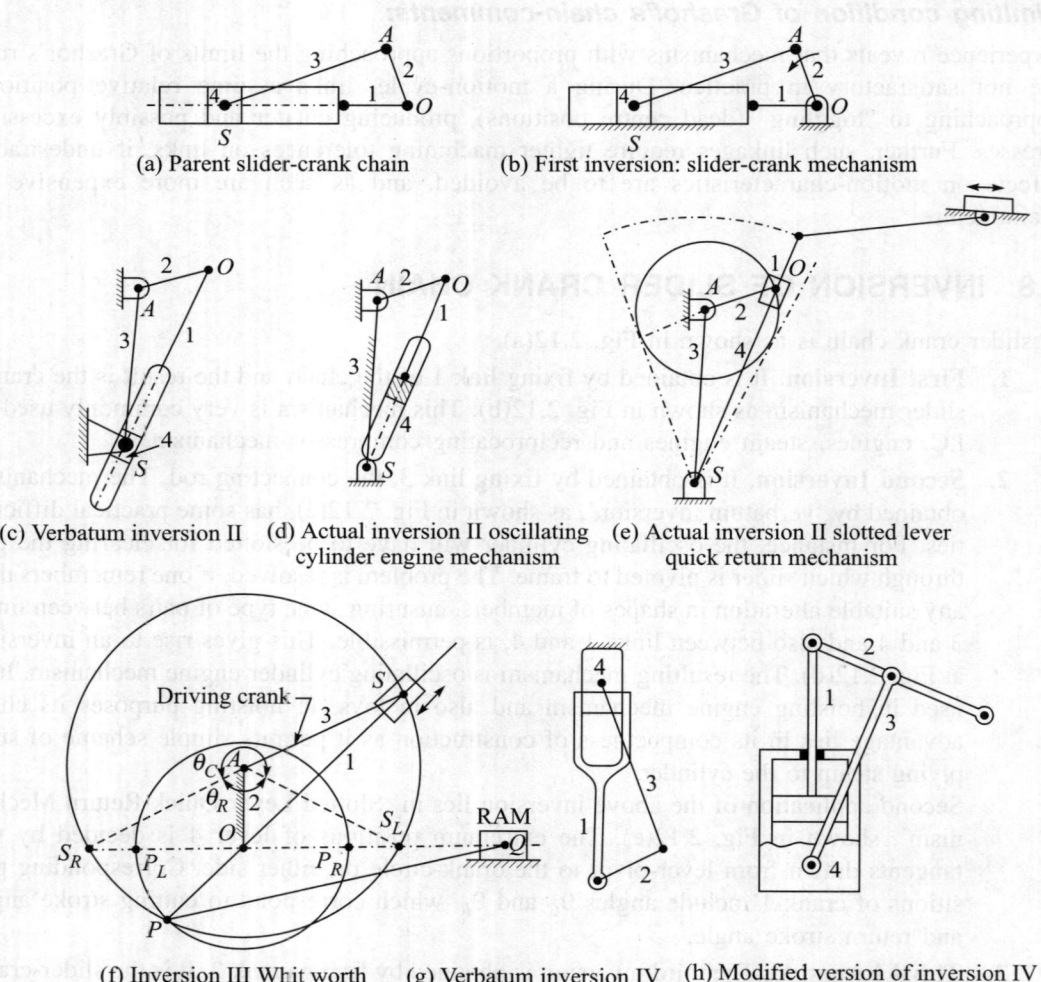

(a) Parent slider-crank chain

(b) First inversion: slider-crank mechanism

(c) Verbatum inversion II

(d) Actual inversion II oscillating cylinder engine mechanism

(e) Actual inversion II slotted lever quick return mechanism

(f) Inversion III Whit worth quick return mechanism

(g) Verbatum inversion IV

(h) Modified version of inversion IV hand pump mechanism

Fig. 2.12 Inversions of a slider crank chain.

2.9 THE DOUBLE SLIDER CRANK CHAIN AND ITS INVERSION

This consists of four binary links with two revolute and two sliding pairs placed in the order stated. There are three distinct inversions.

First Inversion. The first inversion occurs when the link, consisting of two perpendicular slots, becomes the frame-link. The result is an elliptical trammel. As the two sliders which are connected to the binary link 3 continue to slide in the slots, any point S on the extension of this link describes an ellipse. Clearly QS and PS are the semi-major and minor axes of the ellipse. It can be proved as under:

$$x_s = (QS)\sin\theta$$

or,

$$\sin\theta = (x_s/QS)$$

Also,

$$y_s = (PS)\cos\theta$$

or,

$$\cos\theta = (y_s/PS)$$

Squaring and adding on either side,

$$\sin^2\theta + \cos^2\theta = \left(\frac{x_s}{QS}\right)^2 + \left(\frac{y_s}{PS}\right)^2 \qquad (2.3)$$

Thus $\left(\dfrac{x_s}{QS}\right)^2 + \left(\dfrac{y_s}{PS}\right)^2 = 1$, which is an equation to ellipse with semi-major axis $= QS$ and semi-minor axis $= PS$.

This inversion is shown in Fig. 2.13(a).

Second Inversion. The second inversion occurs when one of the two sliding blocks P and Q is fixed. This is shown in Fig. 2.13(b). The block P is shown fixed so that PQ can be treated as a crank, pivoted to frame at P and carrying a sliding block Q at the other end of this crank. This causes the frame to reciprocate. The fixed block P guides the frame. This results in scotch yoke mechanism. Due to guidance of sliding block P, the mechanism converts rotary motion of PQ into reciprocating motion of cross.

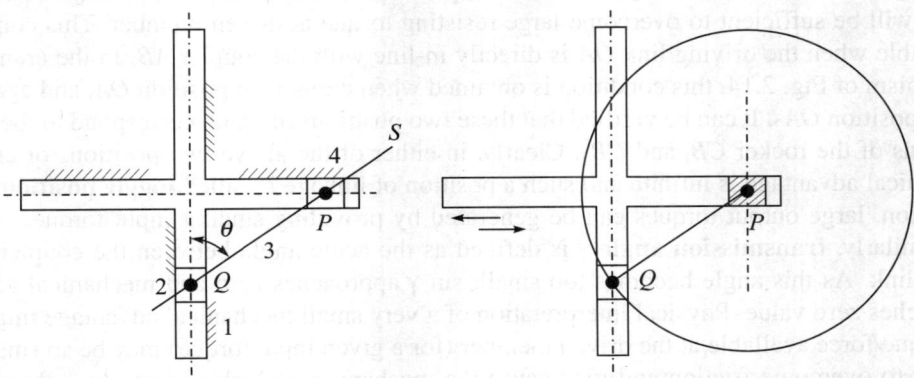

(a) First inversion: elliptical trammel   (b) Second inversion: scotch yoke mechanism

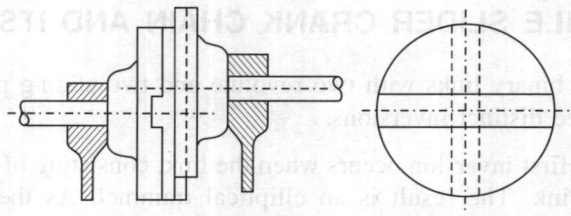

(c) Third inversion oldham's coupling

Fig. 2.13 Inversions of Double slider-crank chains

Third Inversion. The third inversion is obtained when connecting link 3 is fixed. This inversion leads to oldham's coupling, which is used to transmit uniform angular velocity when shaft axes are parallel but eccentric. This is shown in Fig. 2.13(c).

2.10 TRANSMISSION OF TORQUE AND FORCE IN MECHANISMS

Ability to transmit torque or force effectively is one of the major criteria which a designer must consider. Four-bar linkages are being widely used in practice and, as such, we need to develop a criteria to judge the quality of such a linkage for its intended application. Some mechanisms, such as a gear train, transmit a constant torque ratio from input to output shaft. Apparently, this is possible because there is a constant speed ratio between input and output shaft. In the case of a normal four-bar linkage however, this is not possible because torque ratio is a function of geometric parameters which generally change during the course of mechanism motion.

Mechanical advantage of a linkage may be defined as the ratio of the output torque, supplied by the driven link, to the input torque, required to be supplied to the driver link. It will be shown later that the mechanical advantage of the four-bar linkage varies directly as the sine of the angle γ between the coupler and the follower and varies inversely as the sine of the angle β between coupler and driving link. Needless to say that as the mechanism goes on changing positions, both these angles change continuously.

When angle β becomes quite small its sine also approaches zero value, and the mechanical advantage becomes infinite. Physically, this implies that at this position of linkage a small input torque will be sufficient to overcome large resisting torque at driven member. This condition is obtainable when the driving link OA is directly in-line with the coupler AB. In the crank-rocker mechanism of Fig. 2.14, this condition is obtained when crank is in position OA_1 and again when it is in position OA_2. It can be verified that these two positions of crank correspond to the extreme positions of the rocker CB_1 and CB_2. Clearly, in either of the above two positions of crank, the mechanical advantage is infinite and such a position of linkage is called **toggle position**. In such a position, large output torques can be generated by providing smaller input torque.

Similarly, **transmission angle** γ is defined as the acute angle between the coupler and the driven link. As this angle becomes too small, $\sin\gamma$ approaches zero and mechanical advantage approaches zero value. Physical interpretation of a very small mechanical advantage implies that the torque/force available at the driven member (for a given input torque) may be so small as not enough to overcome friction and may cause the mechanism to lock or jam. As a thumb rule a four-bar mechanism should neither be designed for intended use, nor be used in regions where

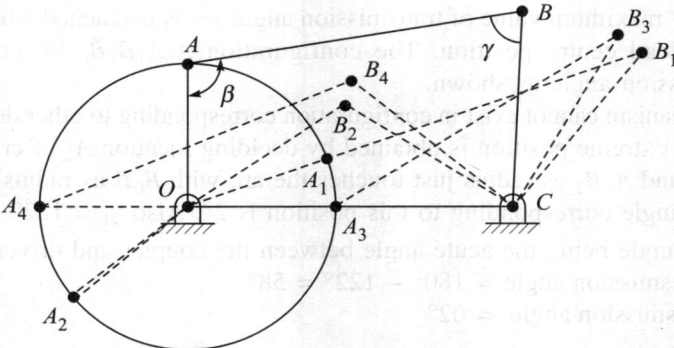

Fig. 2.14 Index of mechanical advantage: Transmission angle γ.

transmission angle γ is less than 45–50°. The smallest transmission angle occurs when the crank OA lies along line of centres OC in position OA_3, maximum value of transmission angle occurs when crank takes up position OA_4 along line of centres.

Transmission angle criteria as an index of assessing mechanical advantage has become more popular on account of ease with which it can be inspected visually.

Besides transmission angle, in some of the mechanisms like meshing gear teeth or a cam-follower system, the **pressure angle** is used as an index of merit. In a four-bar mechanism, pressure angle is taken to be the angle which is compliment of transmission angle.

EXAMPLE 2.3 Mechanism shown in Fig. 2.15 is driven by turning link $A_6 A$. Find out geometrically the maximum and minimum transmission angles. (A.M.I.E. Summer 1993)

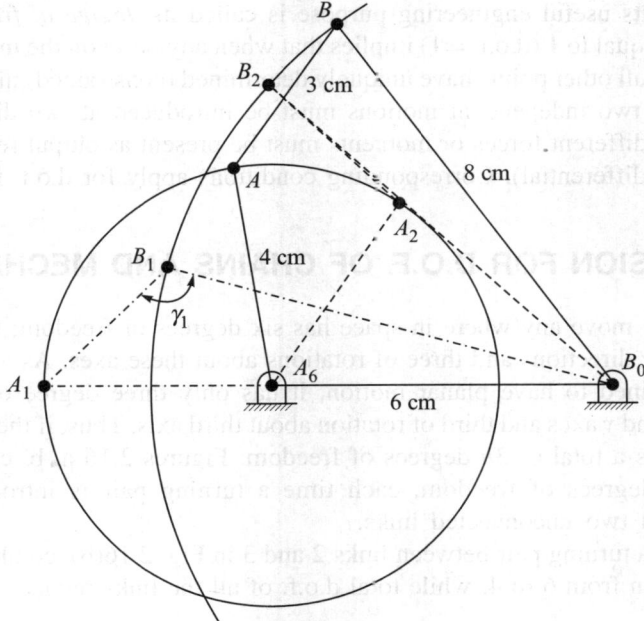

Fig. 2.15

Solution: Clearly maximum value of transmission angle $\gamma = \gamma_1$ is reached when the input crank A_6A_1 lies at the dead centre position. The configuration $A_6A_1B_1B_0$ (a) corresponds to the maximum transmission angle as shown.

Since the mechanism cannot exist in configuration corresponding to other dead centre position of crank, the other extreme position is obtained by deciding location A_2 of crank for which arc with A_2 as centre and A_2B_2 as radius just touches the arc with B_0B as radius and B_0 as centre. The transmission angle corresponding to this position is 2°. Also $\gamma_1 = 122°$.

Transmission angle being the acute angle between the coupler and driven link,

Maximum transmission angle = $180° - 122° = 58°$ **Ans.**

Minimum transmission angle = 02°   **Ans.**

2.11 CONSTRAINED MOTION AND DEGREES OF FREEDOM OF A MECHANISM

Constrained motion is defined as that motion in which all points move in predetermined paths, irrespective of the directions and magnitudes of the applied forces. Mechanisms may be categorised in a number of ways to emphasise their similarities and differences. One such grouping can be to divide mechanisms into planar, spherical and spatial categories. As seen earlier, a planar mechanism is one in which all particles, on any link of a mechanism, describe plane curves in space and all these curves lie in parallel planes.

Mobility/Degree of Freedom of Mechanisms. In the design or analysis of a mechanism, one of the prime concerns is the number of degrees of freedom or mobility of the mechanism. The number of independent input parameters which must be controlled independently so that a mechanism fulfills its useful engineering purpose is called its *degree of freedom* or *mobility*. Degree of freedom equal to 1 (d.o.f. = 1) implies that when any point on the mechanism is moved in a prescribed way, all other points have uniquely determined (constrained) motions. When d.o.f. = 2, it follows that two independent motions must be introduced at two different points in a mechanism, or two different forces or moments must be present as output resistances (as is the case in automotive differential). Corresponding conditions apply for d.o.f. more than two.

2.12 EXPRESSION FOR D.O.F. OF CHAINS AND MECHANISMS

A rigid body free to move any where in space has six degrees of freedom; three of translation parallel to x, y and z directions and three of rotations about these axes. As against this, when a rigid body is restrained to have planar motion, it has only three degree of freedom; two of translation along x and y axes and third of rotation about third axis. Thus, if there are unconnected n links, they possess a total of $3n$ degrees of freedom. Figures 2.16 a, b, c, d, e illustrate the process of losing degrees of freedom, each time a turning pair is introduced, *i.e.* adding constraints, between two unconnected links.

By introducing a turning pair between links 2 and 3 in Fig. 2.16(b), combined d.o.f. of links 2 and 3 are cut down from 6 to 4, while total d.o.f. of all the links reduces to 10. This is easy to understand as two coordinates (x_1, y_1) and angle ψ, measured with respect to say, x axis is required to fix link 2 in position. Additionally, only angle θ between links 2 and 3 will be required to fix link 3 in position relative to link 2.

By introducing turning pairs between 2 and 3 and also between 3 and 4, the total d.o.f. are further cut down from 12 to 8. This is easy to verify in Fig. 2.16(c). Besides coordinates (x_1, y_1) and angle ψ for fixing link 2 in position and angle θ for fixing position of link 3 w.r. to 2 we now require angle ϕ for fixing position of link 4 in relation to 2 and 3.

In Fig. 2.16(d), after inserting 3rd turning pair we require x_1, y_1, ψ, θ, ϕ and α for locating connected links 2, 3, 4 in two-dimensional space and hence the combined d.o.f. is 6, which tallies with the figure worked out in Fig. 2.16(d).

With the insertion of the 4th turning pair between links 1 and 4 loop closure is complete. The completed chain requires four parameters for uniquely describing its members in x-y plane; these are x_1, y_1, ϕ and θ only. This gives four d.o.f. to the chain which tallies with the figures worked out in Fig. 2.16(e).

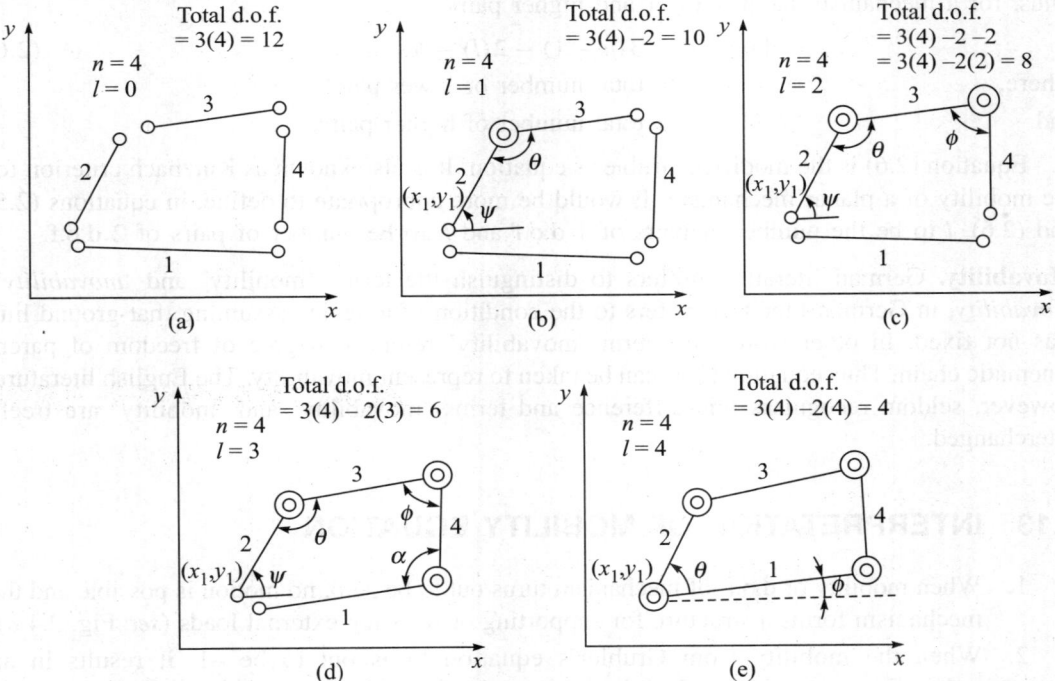

Fig. 2.16 Figures illustrating process of losing degrees of freedom each time a constraint is added by way of a turning pair between two unconnected links.

We thus conclude that introduction of each lower pair (more precisely, turning pairs) between a pair of links cut down degree of freedom by 2 from an assembly of n unconnected mobile links. The process of adding kinematic pairs between links thus amounts to constraining the motion so as to produce a resultant motion which is useful for some purpose. Based on above discussions, expression for degree of freedom of a planar kinematic chain, consisting of lower pairs (of d.o.f. = 1) only, is given by –

$$F = 3n - 2(l) \tag{2.4}$$

where n = number of mobile links

and l = total number of lower pairs.

In case of a mechanism which is obtained from a chain by fixing one link, number of mobile links reduces to $(n - 1)$ and therefore, expression for degrees of freedom of a mechanism, consisting of lower pairs only, is given by–

$$\text{d.o.f. of a mechanism, } F = 3(n - 1) - 2(l) \tag{2.5}$$

Equation (2.5) is known as Grubler's equation, and is one of the most popular mobility equations.

Effect of Higher Pair. Just as a lower pair (linear motion lower pair) cuts down 2 d.o.f., a higher pair cuts only 1 d.o.f. (this is because invariably rolling is associated with slipping, permitting 2 d.o.f.). Hence equation (2.5) can be further modified to include the effect of higher pairs also. Thus, for a mechanism having lower and higher pairs,

$$\text{d.o.f., } F = 3(n - 1) - 2(l) - h \tag{2.6}$$

where, l = total number of lower pair

and h = total number of higher pairs.

Equation (2.6) is the modified Grubler's equation. It is also known as Kutzbach criterion for the mobility of a planar mechanism. It would be more appropriate to define, in equations (2.5) and (2.6), l to be the number of pairs of 1 d.o.f and h to be number of pairs of 2 d.o.f.

Movability. German literature prefers to distinguish the terms 'mobility' and '*movability*'. *Movability*, in German literature, refers to the condition of a device assuming that ground link was not fixed. In other words, the term 'movability' refers to degree of freedom of parent kinematic chain. Thus equation (2.4) can be taken to represent movability. The English literature, however, seldom recognises this difference and terms 'movability' and 'mobility' are freely interchanged.

2.13 INTERPRETATION OF MOBILITY EQUATION

1. When mobility or d.o.f. of mechanism turns out to be zero, no motion is possible and the mechanism forms a structure for supporting or resisting external loads (*see* Fig. 2.4 *a*).

2. When the mobility from Grubler's equation turns out to be −1, it results in an indeterminate structure. It follows that the linkage has an additional (redundant) constraint, beyond what is necessary to establish uniquely the position of the assemblage. Therefore, with the technique of replacing internal forces of a device with binary links (*see* Fig. 2.17) in the associated linkage, $F = -1$ is characteristic of both pre-loaded structure and internal force exerting devices (*e.g.* clamps).

3. If the mobility F turns out to be +1, the mechanism can be driven by a single input motion. If $F = 2$, two separate input motions are necessary to produce constrained motion of the mechanism.

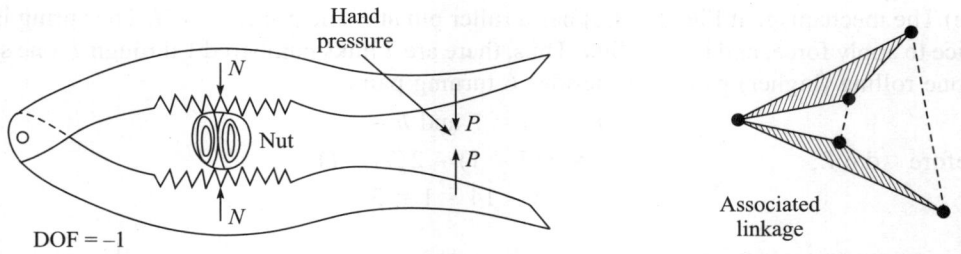

Fig. 2.17   A Nut Cracker. In associated linkage, the dashed line indicate imaginary binary links which represent the effect of forces N and P.

Effect of Spring Connection.   Purpose of a spring is to produce a mutual force between the two links which it connects but, in no way, does it kinematically constrain the relative motion between the two links. The action of spring, to elongate or compress, can also be simulated by a pair of binary links and three pin joints which form the instantaneous-velocity-equivalent lower-pair model to the spring, allowing same degrees of freedom of relative motion between the connected links.

Effect of Multiple Joint.   A multiple joint is called a p-tuple joint when a given link is connected to p other links at that joint. (See Fig. 16.49a, where C is a triple joint). Thus, a double joint connects a given link to two other links, and a triple joint connects a given link to three other links and so on. Many a time a multiple joint is preferred in practice on account of smaller construction cost and, quite often, smaller space requirement. While finding degree of freedom of a mechanism with multiple joints, a double joint should be considered equivalent to two simple joints and a triple joint be considered equivalent to three simple joints, etc.

EXAMPLE 2.4   Find out degrees of freedom of mechanism shown in Figs. 2.18(a), (b), (c), (d) and (e).

Solution: (a) Here $n = 9$; $l = 11$
Therefore   d.o.f. $= 3(9 - 1) - 2(11) = 2$
    (b) Here $n = 8$
                $l = 9 + 2$ (on account of a double joint)
                    $= 11$
Therefore $=$ d.o.f. $= 3(8 - 1) - 2(11)$
                    $= 21 - 22 = -1$ **Ans.**

    i.e. the mechanism at Fig. 2.18(b) is a statically indeterminate structure.
    (c) As in case (b), here too there are double joints at A & B. Hence
                $n = 10$; $l = 9 + 2(2) = 13$
Therefore   d.o.f. $= 3(10 - 1) - 2(13) = 1$ **Ans.**

    (d) The mechanism at Fig. 2.18(d) has three ternary links (links 2, 3 and 4) and 5 binary links (links 1, 5, 6, 7 and 8) and one slider. It has 9 simple turning pairs marked R, one sliding pair marked P and one double joint at J. Since the double joint J joins 3 links, it may be taken equivalent to two simple turning pairs. Thus,
                $n = 9$; $l = 11$
Therefore   d.o.f. $= 3(9 - 1) - 2(11) = 2$ **Ans.**

(e) The mechanism at Fig. 2 .18(e) has a roller pin at E and a spring at H. The spring is only a device to apply force, and is not a link. Thus, there are 7 links numbered 1 through 7, one sliding pair, one rolling (higher) pairs at E besides 6 turning pairs

Thus $\qquad$ $n = 7;\ l = 7$ and $h = 1$

Therefore d.o.f., $\qquad F = 3(7 - 1) - 2(7) - (1)$
$$= 18 - 14 - 1 = 3 \qquad \textbf{Ans.}$$

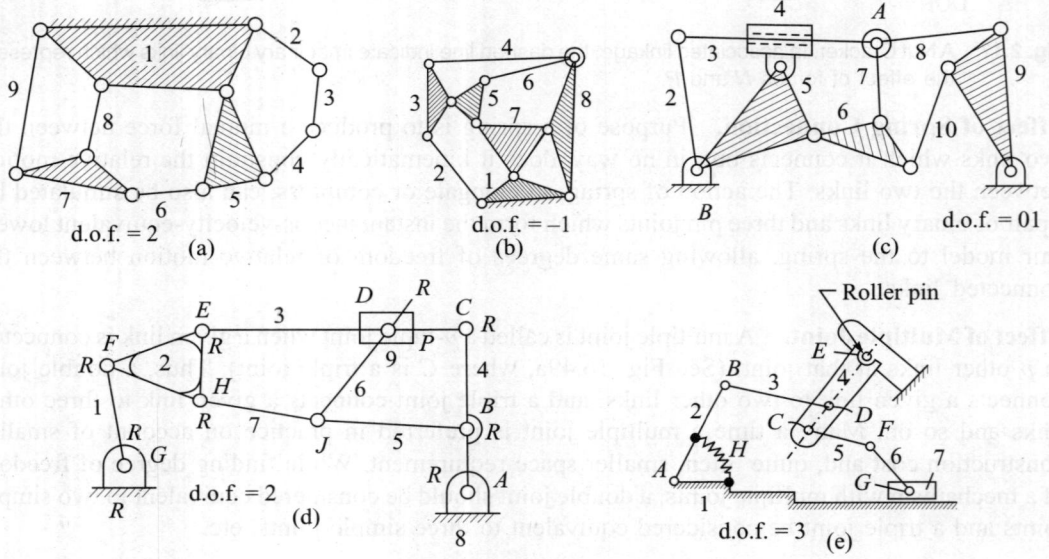

d.o.f. = 2
(a)

d.o.f. = –1
(b)

d.o.f. = 01
(c)

d.o.f. = 2
(d)

Roller pin

d.o.f. = 3
(e)

Fig. 2.18

EXAMPLE 2.5 Find out degrees of freedom of the mechanism shown in Figs. 2.19(a), (b) and (c).

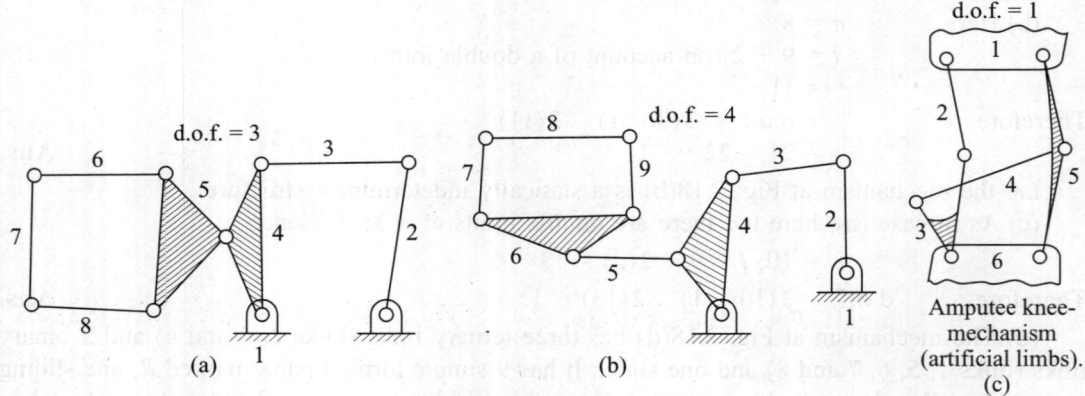

d.o.f. = 3
(a)

d.o.f. = 4
(b)

d.o.f. = 1

Amputee knee-mechanism (artificial limbs).
(c)

Fig. 2.19

Solution: (a) $n = 8;\ l = 9$

Therefore d.o.f. $= 3(8 - 1) - 2(9) = 3$ $\qquad$ **Ans.**

(b) $n = 9$; $l = 10$

d.o.f. $= 3 (9 - 1) - 2 (10) = 24 - 20 = 4$ **Ans.**

(c) $n = 6$; $l = 7$

Therefore   d.o.f. $= 3 (6 - 1) - 2 (7) = 1$.   **Ans.**

EXAMPLE 2.6   Show that the automobile window glass guiding mechanism in Fig. 2.20 has a single degree of freedom.

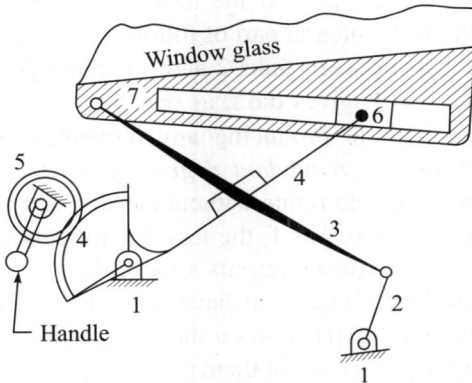

Fig. 2.20   Automobile window guidance linkage.

Solution: As numbered, there are total 7 links. There are seven revolute pairs between link pairs (1, 2), (2, 3), (3, 4), (3, 7), (4, 6), (4, 1) and (1, 5). Besides, there is one sliding pair between links 6 and 7 and a geared pair between links 4 and 5.

Thus, $l = 8$ and $h = 1$

Hence,   d.o.f. $= 3 (7 - 1) - 2 (8) - 1$

$= 18 - 16 - 1 = 1$ **Ans.**

2.14 INCONSISTENCIES OF GRUBLER'S EQUATION

In a good number of cases, Grubler's equation appears to give incorrect results, particularly when–

(1) the mechanism has a lower pair which could be replaced by a higher pair, without influencing output motion.

(2) the mechanism has a kinematically redundant pair, and

(3) there is a link with redundant degree of freedom.

Inconsistency at (1) may be illustrated with the help of Figs. 2.21(a) and (b). Figure 2.21(a) depicts a mechanism with three links and three sliding pairs. According to Grubler's theory, this combination of links has a degree of freedom of zero. But by inspection, it is clear that the links have a constrained motion, because as the link 2 is pushed to the left, link 3 is lifted due to wedge action. A little consideration shows that the sliding pair between links 2 and 3 can be replaced by a slip rolling pair [Fig. 2.21(b)], ensuring constrained motion. In the latter case, $n = 3$, $l = 2$ and $h = 1$ which, according to Grubler's equation, gives d.o.f. = 1.

Figure 2.21(c) demonstrates inconsistency at (2). The cam follower mechanism has 4 links, 3 turning pairs and a rolling pair, giving d.o.f. as 2. However, a close scrutiny reveals that as a function generator, oscillatory motion of follower is a unique function of cam rotation, *i.e.* $\phi = f(\theta)$. In other words, d.o.f. of the above mechanism is only 1. It may be noted, however, that the function of roller in this case is to minimise friction, it does not in any way influence the motion of follower. For instance, even if the turning pair between follower and roller is eliminated (rendering roller to be an integral part of follower), the motion of follower will not be affected. Thus the kinematic pair between links 2 and 3 is redundant. Therefore, with this pair eliminated, $n = 3$, $: l = 2$ and $h = 1$, gives d.o.f. as one.

If a link can be moved without producing any movement in the remaining links of mechanism, the link is said to have *redundant degree of freedom*. Link 3 in mechanism of Fig. 2.21(d), for instance, can slide and rotate without causing any movement in links 2 and 4. Since the Grubler's equation gives d.o.f. as 1, the loss due to redundant d.o.f. of link 3 implies effective d.o.f. as zero, and Fig. 2.21(d) represents a locked system. However, if link 3 is bent, as shown in Fig. 2.21(e), the link 3 ceases to have redundant d.o.f. and constrained motion results for the mechanism. Figure 2.21(f) shows a mechanism in which one of the two parallel links *AB* and *PQ* is a redundant link, as none of them produces additional constraint. By removing any of the two links, motion remains the same. It is logical therefore to consider only one of the two links in calculating degrees of freedom. Another example where Grubler's equation gives zero mobility is the mechanism shown in Fig. 2.21(g), which has a constrained motion.

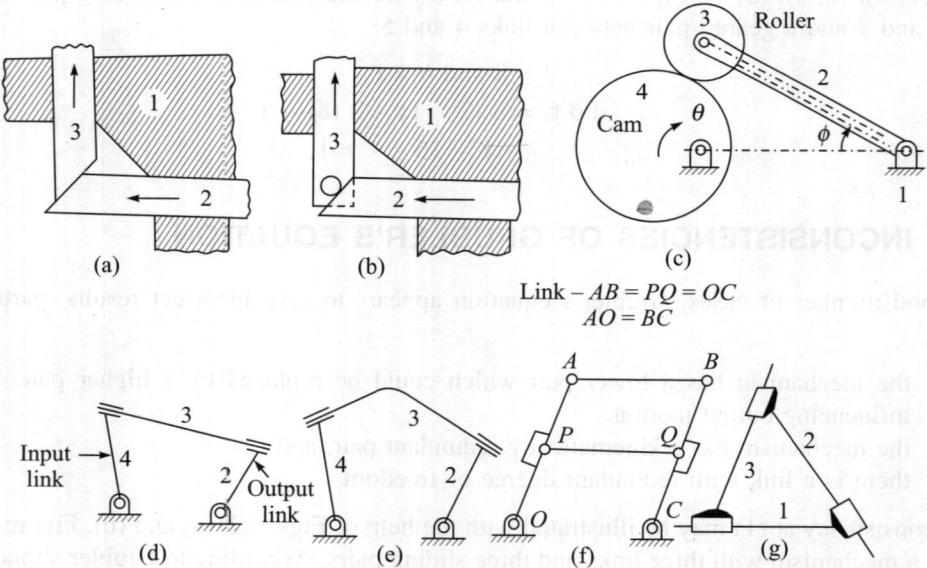

Link $- AB = PQ = OC$
$AO = BC$

Fig. 2.21 Examples of inconsistencies of Grubler's equation.

2.15 DEGREES OF FREEDOM PERMITTED BY JOINTS OTHER THAN TURNING AND SLIDING

As against one degree of freedom of relative motion permitted by turning and sliding pairs, higher pairs may permit a higher number of degrees of freedom of relative motion. *Each such higher pair is equivalent to as many lower pairs as the number of degrees of freedom of relative motion permitted by the given higher pair.* This is elaborated for different types of higher pairs, as discussed below:

(a) *Rolling Contact Without Sliding.* This allows only one d.o.f. of relative motion as only relative motion of rotation exists. A pure rolling type of joint can therefore be taken equivalent to a lower pair with one d.o.f. The lower pair equivalent for instantaneous velocity is given by a simple hinge joint at the relative instant centre, which is the point of contact between rolling links. Note that instantaneous velocity implies that in case a higher pair is replaced by a lower pair equivalent, the instantaneous relative velocity between the connecting links remains the same, but the relative acceleration may, in general, change.

(b) *Roll-Slide Contact.* Due to sliding motion associated with rolling, only one out of three planar motions is constrained. Thus a lower pair equivalence for instantaneous velocity is given by a slider and pin joint combination between the connected links. This implies that such a pair can be replaced by a link and two lower pairs. This allows for two degrees of freedom of relative motion. Such a joint is also taken care of, in Grubler's equation, by making contribution to the term h.

(c) *Gear-Tooth Contact* (*Roll-Slide*). Gear tooth contact is a roll-slide pair and therefore makes a contribution to the term h in Grubler's equation. Thus, on account of two turning pairs at gear centres together with a higher pair at contacting teeth,

$$\text{d.o.f.} = 3\,(3 - 1) - 2\,(2) - 1 = 1$$

Lower pair equivalent for instantaneous velocity of such a pair is a 4-bar mechanism with fixed pivots at gear centres and moving pivots at the centres of curvature of contacting tooth profiles. *In case of involute teeth, these centres of curvature will coincide with points of tangency of common tangent drawn to base circles of the two gears. Such a 4-bar mechanism retains that d.o.f. equal to 1.*

(d) *A Spring Connection.* Purpose of a spring is to exert force on the connected links, but it does not participate in relative motion between connected links actively. Since the spring permits elongation and contraction in length, a pair of binary links, with a turning pair connecting them, can be considered to constitute instantaneous velocity equivalent lower pair mechanism. A pair of binary links with a turning pair permits variation in distance between their other ends (unconnected), and allows same degree of freedom of relative motion between links connected by the spring (for $n = 4$, $l = 3$, $F = +\,3$). It may be noted that in the presence of spring, ($n = 2$, $l = 0$, $h = 0$) the d.o.f. would be 3.

(e) *The Belt and Pulley or Chain and Sprockets Connection.* When the belt or chain is maintained tight, it provides planar connections. Instantaneous velocity, lower pair equivalent can be found in a ternary link with three pin joints (sliding is not allowed). It can be verified from Table 2.2 that d.o.f. of equivalent six bar linkage is

$$F = 3 (6 - 1) - 2 (7) = 1$$

Table 2.2 Lower pair equivalents of higher pairs.

Name of Joint	Diagram original representation	Lower pair equivalent for instantaneous velocity
Rolling contact (without sliding)	$N = 2$ $F = 1$ 	$N = 2$ $l = 1$ $F = 1$
Rolling-sliding contact	$N = 2$ $l = 0$ $h = 1$ $F = 2$ 	$N = 3$ $l = 2$ $h = 0$ $F = 2$
Gear contact (involving roll-slide contact)	$N = 3$ $l = 2$ $h = 1$ $F = 1$ 	$N = 4$ $l = 4$ $F = 1$
Spring connection	$N = 2$ $l = 0$ $h = 0$ $F = 3$ 	$N = 4$ $l = 3$ $h = 0$ $F = 3$
Belt-pulley (no sliding) contact on chain-sprocket connection	$F = 1$ 	$N = 6$ $l = 7$ $F = 1$

EXAMPLE 2.7 Find out degrees of freedom of mechanisms shown in Figs. 2.22(a), (b), and (c).

Solution: (a) In the case of undercarriage mechanism of aircraft in Fig. 2.22(a), we note that

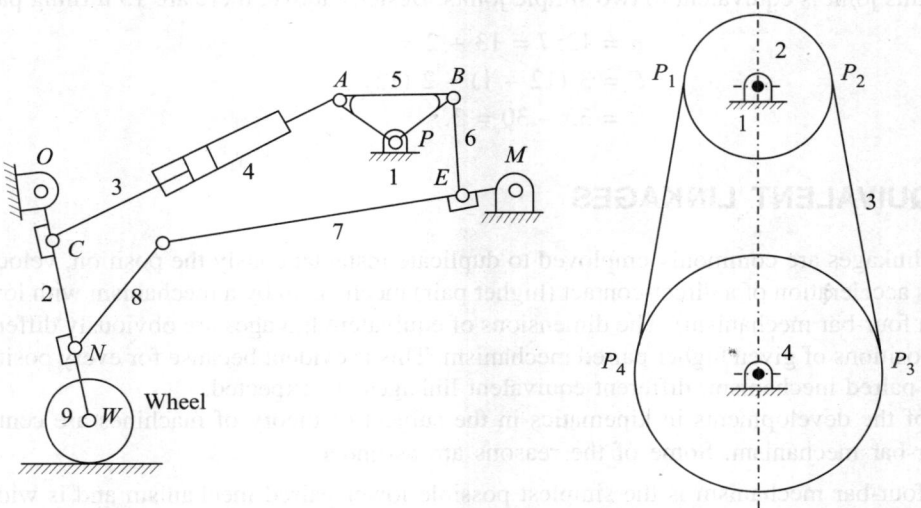

(a) Undercarriage mechanism of an aircraft   (b) Belt-pulley mechanism (assume belt to be tight)

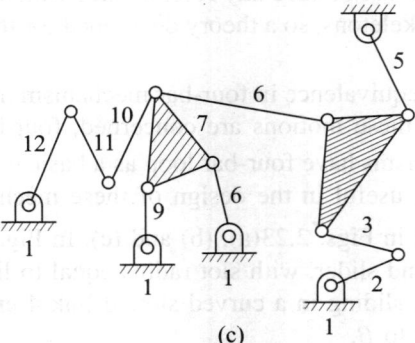

(c)

Fig. 2.22

Total number of links $n = 9$
Total number of pairs of single d.o.f. = 11
Higher pair of 2 d.o.f. (between wheel and runway) = 01

∴
$$\text{d.o.f. } F = 3 (9 - 1) - 2 (11) - 1 (1)$$
$$= 24 - 22 - 1 = 01$$
$$F = 1$$

(b) In the case of belt-pulley drive, assuming the belt to be tight, the four links are marked as 1, 2, 3 and 4. The two distinct lower (turning) pairs are pivots of pully 2 and 4. The points P_1, P_2, P_3 and P_4, at which belt enters/leaves pulley, constitute 4 higher pairs. Thus

$$n = 4 \; ; \; l = 2 \; ; \; h = 4$$

Therefore
$$F = 3 \, (4 - 1) - 2 \, (2) - 4 = 1.$$

(c) In the case of mechanism at Fig. 2.22, there is a double joint between links 6, 7 and 10. Therefore, this joint is equivalent to two simple joints. Besides above, there are 13 turning pairs.

Hence,
$$n = 12; \; l = 13 + 2 = 15$$

Therefore
$$F = 3 \, (12 - 1) - 2 \, (15)$$
$$= 33 - 30 = 3.$$

2.16   EQUIVALENT LINKAGES

Equivalent linkages are commonly employed to duplicate instantaneously the position, velocity, and perhaps acceleration of a direct-contact (higher pair) mechanism by a mechanism with lower pairs (say, a four-bar mechanism). The dimensions of equivalent linkages are obviously different at various positions of given higher paired mechanism. This is evident because for every position of a higher paired mechanism, different equivalent linkages are expected.

Much of the developments in kinematics in the subject of theory of machines are centred around four-bar mechanism. Some of the reasons are as under:

(1)   A four-bar mechanism is the simplest possible lower paired mechanism and is widely used.

(2)   Many mechanisms which do not have any resemblance with a four-bar mechanism have four bars for their basic skeletons, so a theory developed for the four-bar applies to them also.

(3)   Many mechanisms have equivalence in four-bar mechanism in respect of certain motion aspects, Thus, as far as these motions are concerned, four-bar theory is applicable.

(4)   Several complex mechanisms have four-bar loop as a basic element. Theory of four-bar mechanism is, therefore, useful in the design of these mechanisms.

Point (2) above, is illustrated in Figs. 2.23(a), (b) and (c). In Fig. 2.23(b), the link 4 in Fig (a) is replaced by a curved slot and slider, with slot radius equal to link length. In Fig. 2.23(c) the link 3 is replaced by a slider, sliding in a curved slotted link 4 ensuring relative motion of rotation of pinned end A relative to B.

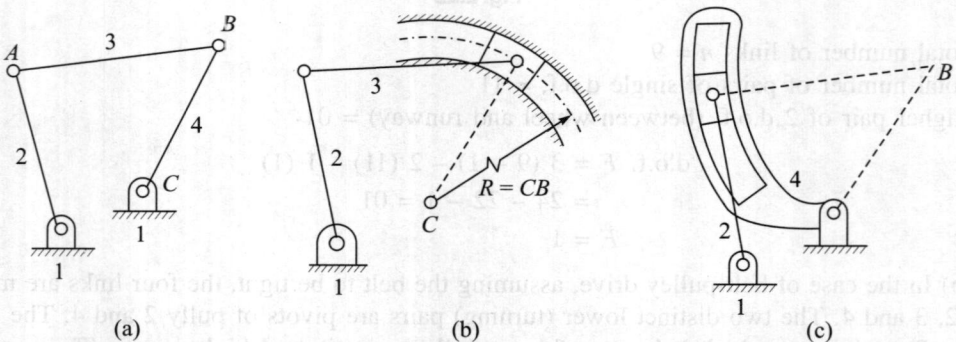

Fig. 2.23   Equivalent mechanisms (kinematically identical mechanisms having the 4-bar as the basic skeleton).

Point (3) is illustrated in Figs. 2.24(a), (b) and (c). Mechanisms in which relative motion between driver and driven links 2 and 4 is identical are illustrated in Fig. 2.24.

In Fig. 2.24(b) the centres of curvature of circular cam and roller constitute the end point of link *AB*; link 3 becomes roller and link 2 become circular cam. For d.o.f. = 1 however, the rolling pair in (b) should be without slip.

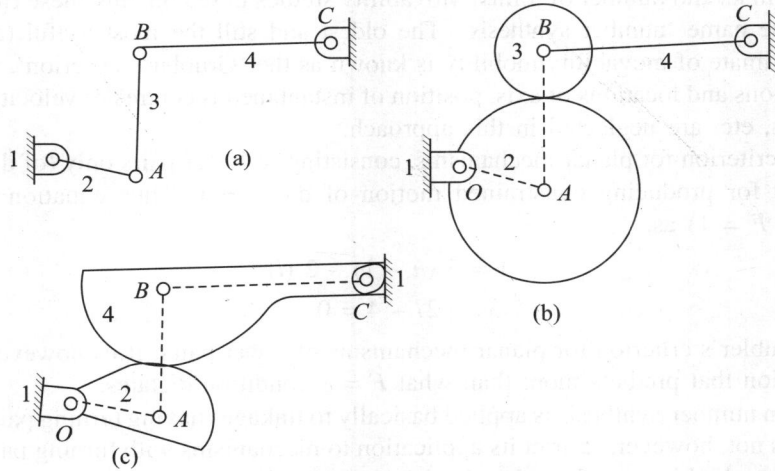

Fig. 2.24   Mechanisms having identical relative motions between links 2 and 4.

Use of Spring to Replace Turning Pairs. Extension and compression in a spring is comparable to variation in length between the turning pairs accomplished by a pair of binary links connected through an another turning pair. For instance pair of binary links 4 and 5 of a Stephenson's chain can be replaced by a spring to obtain an equivalent mechanism. This is shown in Figs. 2.25(a) and (b).

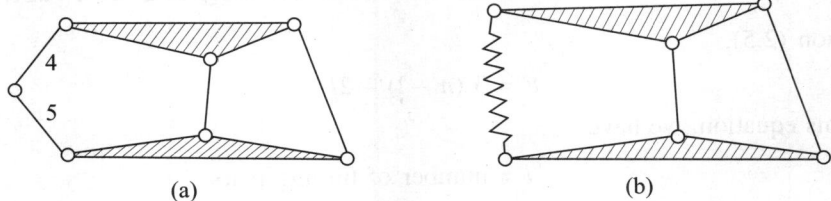

Fig. 2.25   Spring to replace a pair of binary links and ternary pairs.

A Ternary with Turning Pairs to Replace Belt/Pulley or Chain/Sprocket Pair. When the belt or chain is maintained tight, a ternary link with three turning pairs is the instantaneous-velocity equivalent lower pair connection to the belt and pulley (sliding/slipping is disallowed).

2.17   NUMBER SYNTHESIS

Whereas kinematic analysis aims at analysing the motion inherent in a given machine or mechanism, kinematic synthesis aims at determining mechanisms that are required to fulfill certain motion specifications. Kinematic synthesis can, therefore, be thought of as a reverse problem to kinematic analysis of mechanisms. Synthesis is very fundamental of a design as it represents creation of a new hardware to meet particular needs of motion, namely displacement, velocity or acceleration—singly or in combination.

Probably, the most obvious external characteristics of a kinematic chain or mechanism are: the number of links and number of joints. Movability studies based on only these two parameters come under the name 'number synthesis.' The oldest and still the most useful (although with limitations) estimate of movability/mobility is known as the 'Grubler's criterion'. Effect of link lengths, directions and locations of axis, position of instantaneous centres of velocity, complexity of connections, etc. are neglected in this approach.

Grubler's criterion for planar mechanisms, consisting of lower pairs only (of d.o.f. = 1), can be remodelled for producing constrained motion of d.o.f. = 1. Thus equation (2.5) can be rewritten (For $F = 1$) as,

$$1 = 3 (n - 1) - 2 (l)$$

or,
$$3n - 2l - 4 = 0 \qquad (2.7)$$

This is Grubler's criterion for planar mechanisms of lower pairs. It is, however, possible to derive a criterion that predicts more than what $F = 1$ condition dictates.

Mechanism number-synthesis is applied basically to linkages having turning pairs (pin joints) only. This does not, however, restrict its application to mechanisms with turning pairs alone. For it has been shown by kinematicians that, having once developed complete variety of pin jointed mechanisms, the method can most readily be converted to accomodate cams, gears, belt drives, hydraulic cylinder mechanisms and clamping devices.

Following deductions will be useful in deriving possible link combinations of a given number of links for a given degree of freedom. It is assumed that all joints are simple and there is no singular link.

2.17.1 Effect of Even/Odd Number of Links on Degrees of Freedom

From equation (2.5),

$$F = 3 (n - 1) - 2l$$

Rewriting this equation, we have

$$l = \text{number of turning pairs}$$

$$= \left[\frac{3 (n - 1)}{2} - \frac{F}{2} \right] \qquad (2.8)$$

Since total number of turning pairs must be an integer number, it follows that either $(n - 1)$ and F should be both even or both odd. Thus, for l to be an integer number:

(1) If d.o.f. F is odd (say, 1,3,5 ...), $(n - 1)$ should also be odd. In other words, n must be even.

(2) If d.o.f. F is even (say 2,4 ...), $(n - 1)$ should also be even. In other words for F to be even, n must be odd.

Summing up, for d.o.f. F to be even, n must be odd and for F to be odd, n must be even.

2.17.2 Minimum Number of Binary Links in a Mechanism

Let

n_2 = number of binary links

n_3 = number of ternary links

n_4 = number of quaternary links

$\vdots$

n_k = number of k-nary links.

The above number of links must add up to the total number of links in the mechanism. Thus,

$$n = n_2 + n_3 + n_4 + \ldots + n_k \qquad (2.9)$$

or

$$n = \sum_{i=2}^{k} n_i$$

Since discussions are limited to simple jointed chains, each joint/pair consists of two elements. Thus, if e is total number of elements in the mechanism, then

$$e = 2(l) \qquad (2.10)$$

By definition binary, ternary, quaternary, etc. links consist of 2, 3, 4 elements respectively. Hence, total number of elements are also given by

$$e = 2n_2 + 3n_3 + 4n_4 + \ldots + k(n_k) \qquad (2.11)$$

Comparing right hand side of equations (2.10) and (2.11), we have

$$2l = 2n_2 + 3n_3 + 4n_4 + \ldots + k(n_k) \qquad (2.12)$$

Substituting for n and $2l$ from (2.9) and (2.12) in (2.5), we have

$$F = 3\ [(n_2 + n_3 + n_4 + \ldots + n_k) - 1] - [2n_2 + 3n_3 + 4n_4 + \ldots + kn_k]$$

Simplifying further,

$$F = [n_2 - n_4 - 2n_5 - 3n_6 - \ldots - (k - 3)\ n_k] - 3 \qquad (2.13)$$

or rearranging,

$$n_2 = (F + 3) + [n_4 + 2n_5 + 3n_6 + \ldots + (k - 3)\ n_k] \qquad (2.14)$$

Thus, number of binary links required in a mechanism depend on d.o.f. and also on the number of links having elements > 3. Sacrificing exactness for fear of a complex relation, minimum number of binary links can be deduced from eq. (2.14) as:

$$n_2 \geq 4, \text{ for d.o.f.} = 1$$

$$n_2 \geq 5, \text{ for d.o.f.} = 2$$

$$n_2 \geq 6, \text{ for d.o.f.} = 3, \text{ etc.} \qquad (2.15)$$

This proves that minimum number of binary links for d.o.f. = 1 is 4, while the minimum number of binary links required for d.o.f. = 2 is 5.

2.17.3 Maximum Possible Number of Turning Pairs on any of the *n* Links in a Mechanism

The problem is approached in an indirect manner. We pose the problem to be that of finding minimum number of links *n* required for closure when one of the links has largest number of elements = *k*. An attempt is now made to close the chain in Fig. 2.26 having a link *A* of *k* elements.

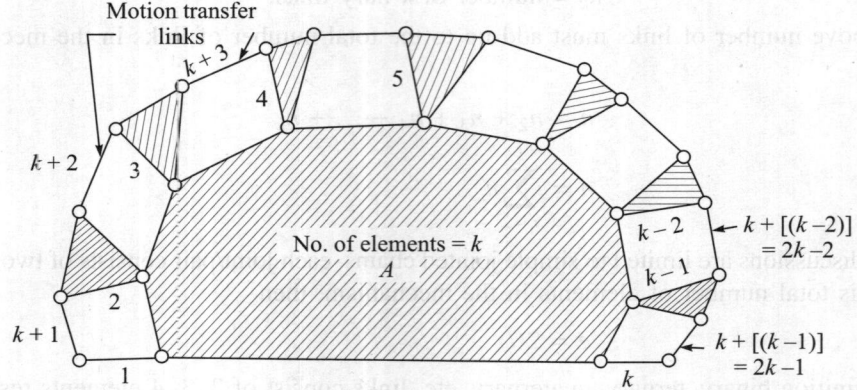

Fig. 2.26 Minimum number of link's required for closure.

For completing the chain with a minimum number of links involving no multiple joint, it is necessary to interconnect ternary links at all the elements of link *A* except the first and last element. Connecting ternaries at intermediate elements ensures a continuity of motion from link 1 to link *k*. Links directly connected to link *A* are labelled 1 through *k*, while the motion transfer links shown in Fig. 2.26 are numbered as $(k + 1)$, $(k + 2)$, $(k + 3)$, ... $[k + (k - 2)]$, $[k + (k - 1)]$. The last motion transfer link is thus numbered as $(2k - 1)$. Clearly, minimum number of links required to complete the chain is $(2k - 1)$, besides the link of highest elements.

In other words, for a given number of links $n = 2k$, a link can have a maximum of *k* elements. Hence,

$$(k) = \frac{n}{2} \tag{2.16}$$

Thus when *n* is even, maximum possible number of elements which a link can have is *n*/2.

An important conclusion emerging out of eq. (2.13) is that the number of ternary links does not have any influence on degrees of freedom of a mechanism. According to B. Paul, for a mechanism with d.o.f. = 1,

$$C = \frac{1}{2}(n) - 1 \tag{2.17}$$

where C = number of independent circuit or loops
and, n = number of links.

2.18 ENUMERATION OF KINEMATIC CHAINS

Let N be the number of links and F degree of freedom, for which all possible planar chains are needed to be established.

Step 1. For given N and F establish the number of joints (hinges) using

$$l = \frac{3N - (F + 3)}{2} \qquad (2.18)$$

For d.o.f. $= 1$, this reduces to

$$l = \frac{3N - 4}{2}$$

Step 2. For given N, establish maximum number of elements permissible on any link, using

$$k = \frac{N}{2}, \text{ for } F = 1,3,5$$

and

$$k = \frac{(N + 1)}{2}, \text{ for } F = 2,4,\ldots$$

Step 3. Substituting expression for N and $2l$, namely,

$$N = n_2 + n_3 + n_4 + \ldots + n_k$$

and

$$2l = 2n_2 + 3n_3 + 4n_4 + \ldots + k(n_k)$$

in Grubler's eq. (2.5),

$$F = 3\,[(n_2 + n_3 + n_4 + \ldots + n_k) - 1] - [2n_2 + 3n_3 + 4n_4 + \ldots kn_k] \qquad (2.19)$$

we get,

$$F = [n_2 - (n_4 + 2n_5 + \ldots + (k - 3)\, n_k) - 3] \qquad (2.20)$$

Thus for $F = 1$,

$$n_2 - n_4 - 2n_5 + \ldots + (k - 3)\, n_k = 4 \qquad (2.21)$$

Above equations may be used to list all possible combinations of $n_2, n_4, n_5, \ldots$ which satisfy above conditions.

EXAMPLE 2.8 Enumerate all chains possible with $N = 6$ and $F = 1$.

Solution: Total number of hinges $l = \dfrac{3\,(6) - 4}{2} = 7$

Also, for even number of links ($N = 6$), maximum number of hinges on any link $= 6/2 = 3$. Thus the chains will consist of binary and terary links only. Hence, we have from eqs. (2.9) and (2.12),

$$n_2 + n_3 = N = 6 \qquad \text{and} \qquad 2n_2 + 3n_3 = 2l = 14 \qquad (2.22)$$

Substituting in Grubler's criterion, $3N - 2l - 4 = 0$,

we have, $3(n_2 + n_3) - (2n_2 + 3n_3) - 4 = 0$ or $n_2 - 4 = 0$

Thus, $n_2 = 4$ and from eq. (2.22), $n_3 = 2$.

We begin by considering the ways in which links of highest degree (i.e., links having largest number of elements) can be interconnected. The two ternaries can be either connected directly through a common pair or can be connected only through one or two binary links.

In Fig. 2.27 we consider the first possibility. The two ternaries of Figs. 2.27(a) and (b) can not be connected through a single link as it amounts to forming a structural loop (3-link loop). The only way to connect them through 4 binaries (avoiding formation of a 3-link loop) is therefore, as shown at Figs. 2.27(c), which gives Watt's chain.

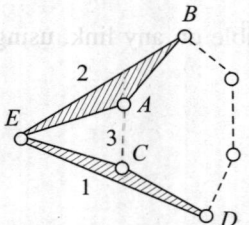 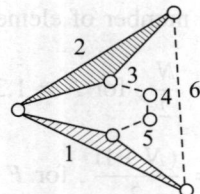 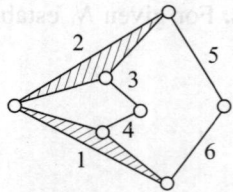

(a) Undesirable due to existance of 3-link loop 1-2-3   (b) Undesirable due to formation of 3-link loop 1-2-6   (c) Permissible combination as no 3-link loop exists

Fig. 2.27

Considering the second alternative, Fig. 2.28(a) shows the two ternary links 1 and 2 being connected through a single binary link 3. Then, between one of remaining two pairs of elements of links 1 and 2, we may introduce a single binary and between the other pair of elements, two remaining binaries. The resulting arrangements are as at Fig. 2.28(b) and (c). It is easy to verify that arrangements at Figs. 2.28(b) and (c) are structurally the same.

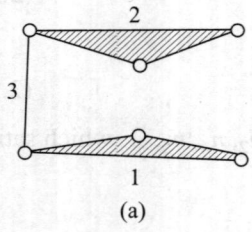 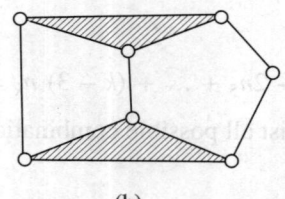 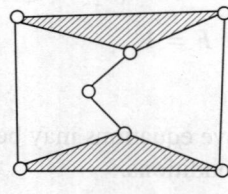

(a)   (b) Valid chain Stephenson's chain   (c) Valid chain (same as that at b)

Fig. 2.28

EXAMPLE 2.9 Enumerate all possible chains of $N = 7$ and d.o.f. = 2.

Solution: Total number of hinges $l = \dfrac{3(N) - (F + 3)}{2} = \dfrac{3(7) - (2 + 3)}{2} = 8$

Maximum number of elements on any link : $\leq (N + 1)/2 = 4$.

Hence, only binary, ternary and quaternary links are possible. Thus, from eqs. (2.9) and (2.12),

$$n_2 + n_3 + n_4 = N(= 7)$$
$$2n_2 + 3n_3 + 4n_4 = 2l(= 16)$$

Substituting in Grubler's equation,

$$F = 3(N - 1) - 2l$$

we have

$$2 = 3[(n_2 + n_3 + n_4) - 1] - (2n_2 + 3n_3 + 4n_4)$$

or
$$n_2 - n_4 = 5$$

Thus the possible combinations are (Note that for any mechanism with $F = 2$, $n_2 \geq 5$):

$$n_4 = 1; \quad n_2 = 6$$
$$n_4 = 0; \quad n_2 = 5, \quad \text{(Also check that } n_2 + n_3 + n_4 = 7\text{)}.$$

Obviously, remaining links in above combinations will be the ternaries. Thus the two combinations possible are:

n_4	n_3	n_2	Total N
1	0	6	7
0	2	5	7

Different chains that can be formed are as shown in Figs. (a), (b), (c) and (d) of 2.29. Chain at Fig. 2.29(a) involves a quaternary link with remaining 6 binary links forming two independent loops of d.o.f. = 1. A mechanism of 2 d.o.f. is possible only when any link other than quaternary is fixed. Chain at Fig. 2.29(b) involves two ternary links that are directly connected. A binary cannot be used singly to connect these ternaries at any of the remaining pairs of elements as that leads to a 3-link loop. Therefore, the only option is to connect these ternaries through two binaries and through three binaries at remaining pairs of elements. This is shown in Fig. 2.29(b).

When the two ternaries are connected through a single binary, the two possible ways of incorporating remaining 4 binaries are shown at Figs. 2.29(c) and (d).

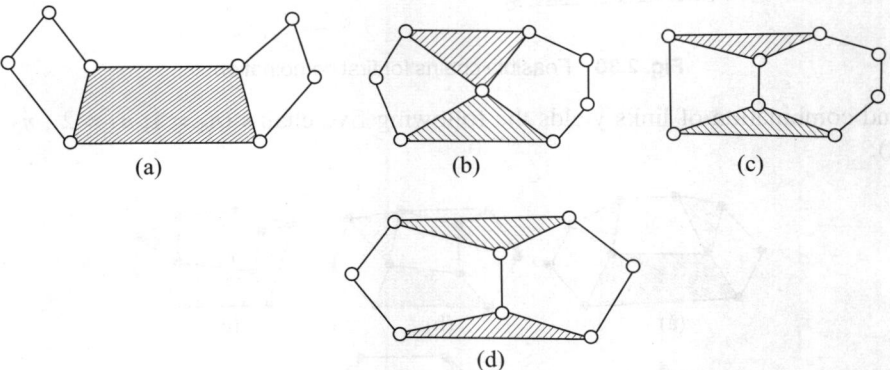

Fig. 2.29 Feasible chains on sevan links.

EXAMPLE 2.10 Enumerate combination of links possible in case of a 8-link chain with $F = 1$.

Solution: No. of hinges $l = \dfrac{3(8) - 4}{2} = 10$

Maximum number of elements on one link $= \dfrac{8}{2} = 4$.

Hence the chains can have binary, ternary and quaternary links only. From equations (2.9) and (2.12).

$$n_2 + n_3 + n_4 = 8 \quad \text{and} \quad 2n_2 + 3n_3 + 4n_4 = 20$$

Substituting in Grubler's criterion,

$$3\,(n_2 + n_3 + n_4) - (2n_2 + 3n_3 + 4n_4) - 4 = 0 \text{ or } (n_2 - n_4) = 4$$

We study various combinations indicated by above expression in respect of their viability:

n_4 (assumed)	$n_2 = (n_4 + 4)$	$n_3 = 8 - (n_2 + n_4)$	Remark
4	8	–	Not acceptable as $n_2 + n_4 > N$
3	7	–	Not acceptable as $n_2 + n_4 > N$
2	6	–	acceptable
1	5	2	acceptable
0	4	4	acceptable

Thus the three valid combination of links are:

   (i) $n_4 = 2$; $n_3 = 0$; $n_2 = 6$
  (ii) $n_4 = 1$; $n_3 = 2$; $n_2 = 5$
 (iii) $n_4 = 0$; $n_3 = 4$; $n_2 = 4$

The first combination yields the following two chains ($n_4 = 2$; $n_3 = 0$; $n_2 = 6$). (Fig. 2.30)

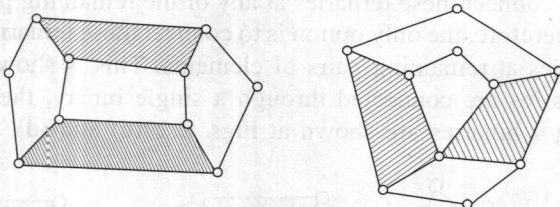

Fig. 2.30 Feasible chains for first combination.

The second combination of links yields the following five chains ($n_4 = 1$; $n_3 = 2$; $n_2 = 5$) (*see* Fig. 2.31).

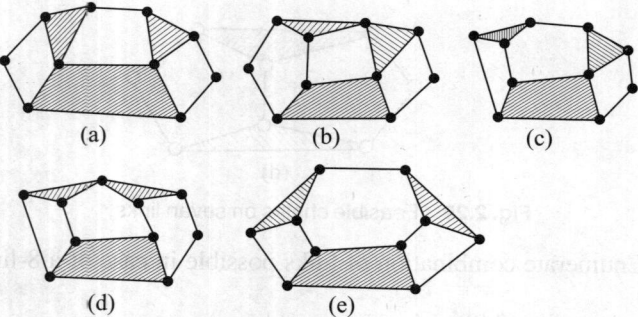

(a)   (b)   (c)

(d)   (e)

Fig. 2.31 Feasible chains for second combination.

The third combination ($n_4 = 0$; $n_3 = 4$; $n_2 = 4$) of links yields following chains (*see* Fig. 2.32).

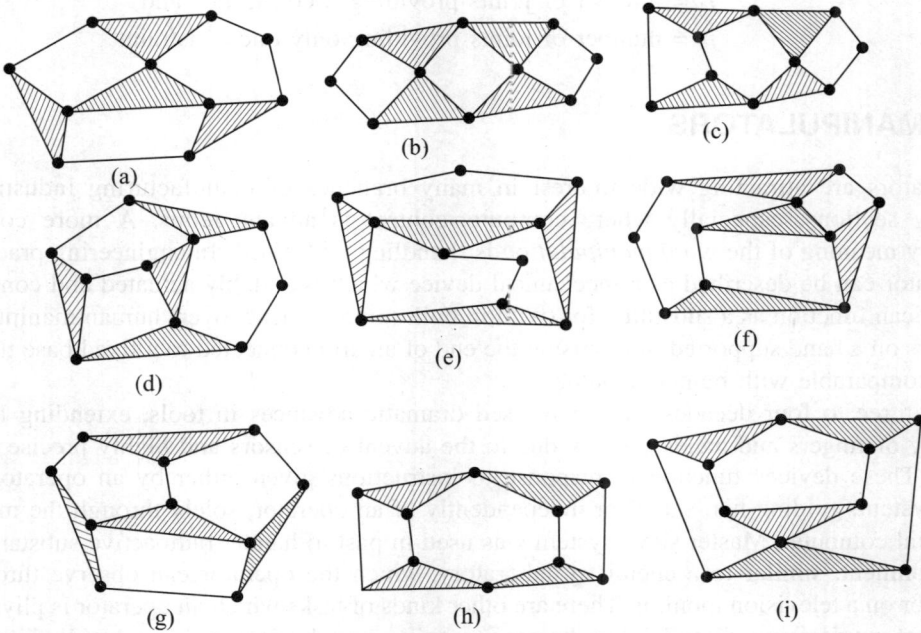

Fig. 2.32 Feasible chains for third combination.

2.19 SPATIAL MECHANISMS

As distinct from a planar mechanism, in a spherical mechanism, each link has some point which remains stationary as the links move and in which the stationary points of all the links lie at a common location (the centre of an imaginary sphere). The locus of each point on a link is a curve lying on an imaginary spherical surface. These spherical mechanisms are constructed out of revolute pairs, and the axes of all revolute pairs must intersect at a point.

Spatial mechanisms do not incorporate any restriction on the relative motions of the particles. A spatial mechanism may have particles describing paths of double curvature.

Grubler's criterion was originally developed for planar mechanisms. If similar criterion is to be developed for spatial mechanisms, we must remember that an unconnected link has six d.o.f. in place of 3. As such, by fixing one link of a chain the total d.o.f. of ($n - 1$) links separately will be $6(n - 1)$. Again a revolute and prismatic pair would provide 5 constraints (permitting 1 d.o.f.), rolling pairs will provide 4 constraints, and so on. Hence, an expression for d.o.f. of a closed spatial mechanism can be written as:

$$F = 6\,(N - 1) - 5j_1 - 4j_2 - 3j_3 - 2j_4 - j_5 \qquad (2.23)$$

where
N = total number of links

j_1 = number of joints providing 5 constraints

j_2 = number of joints providing 4 constraints

j_3 = number of joints providing 3 constraints

j_4 = number of joints providing 2 constraints and

j_5 = number of joints providing only one constraint.

2.20 MANIPULATORS

Manipulators are attracting wide interest in many branches of manufacturing industry and assembly sections, especially where computer-control is advantageous. A more common dictionary meaning of the word *manipulation* is 'handling with skill'. In engineering practice, a manipulator can be described as a mechanical device which is suitably actuated and controlled so that it can function as a substitute for (with in built improvements over) human manipulation by fingers on a hand supported or a wrist at the end of an arm connected to a fixed base through a joint, comparable with human shoulder.

Last three to four decades have witnessed dramatic advances in tools, extending limited capability of fingers and hands, mainly due to the advent of sensors and highly precise servo-devices. These devices function in response to instructions given either by an operator on a master system, guiding his 'slave' or independently of an operator, solely through the medium of a digital computer. Master slave system was used in past to handle radioactive substances in an environment, similar to a chemistry laboratory, which the operator can observe through a window or on a television monitor. There are other kinds of tasks which an operator is physically capable of performing directly, but being repetitive and boring, can be handled through manipulators.

Subject of kinematics relates closely to manipulators because, their mechanical functioning must be based on a system of relatively moving bodies. Components of manipulators are assumed to be rigid and the joints between them are kinematic pair – lower pairs in particular. Kinematic behaviour, especially the freedom and constraints of joints, divides the capability and potential of a manipulator-arm.

REVIEW QUESTIONS

2.1 Define the term *Link*. Can spring, belt, liquids be treated as Links? Justify your answer.

2.2 Explain how I.C. engine mechanism and steam engine mechanism are kinematically identical.

2.3 Define the terms *Element* and *Pair*.

2.4 Explain the term *kinematic* with a schematic diagram. What is its role in kinematic analysis and synthesis?

2.5 Define the terms *machine* and *mechanism*. How do they differ?

2.6 Explain various methods of classifying pairs, giving three examples of each category.

2.7 Distinguish between higher and lower pairs. Is there some more convincing definition than the conventional one, based on point/line and area contact basis?

2.8 'Slider crank mechanism is a special case of 4-bar mechanism'. Justify the statement.

2.9 Define the term *Inversion*. What are the properties of Inversion? Explain advantages arising out of the concept of Inversion.

2.10 Explain with suitable sketches 'Inversions of slider crank chain.'

2.11 Explain Grashof's criterion and describe inversions of 4-bar chain giving suitable sketches.

2.12 Define terms *degrees of freedom* and *constrained motion* of a mechanism.

2.13 Distinguish between a chain structure and a mechanism.

2.14 Proceeding systematically, derive expression for degrees of freedom of a chain and mechanism.

2.15 Distinguish between the terms *mobility* and *moveability*. State Grubler's equation for mobility of a mechanism. What is the interpretation when mobility turns out to be –1, 0, + 1 and + 2?

2.16 In a slotted lever quick-return mechanism, distance between the fixed centres is 7.5 cm and radius of driving crank is 4 cm. Find the ratio of the time taken during cutting stroke and return stroke. **(Ans.** 2.105)

2.17 Find out degrees of freedom of mechanisms shown in Fig. 2.33.

[**Ans.** (*a*) = 3 ; (*b*) 3; (*c*) = 2]

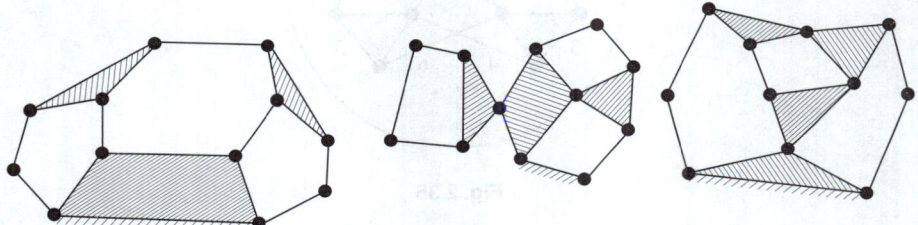

Fig. 2.33

2.18 Determine mobility of the mechanisms shown in Fig. 2.34.

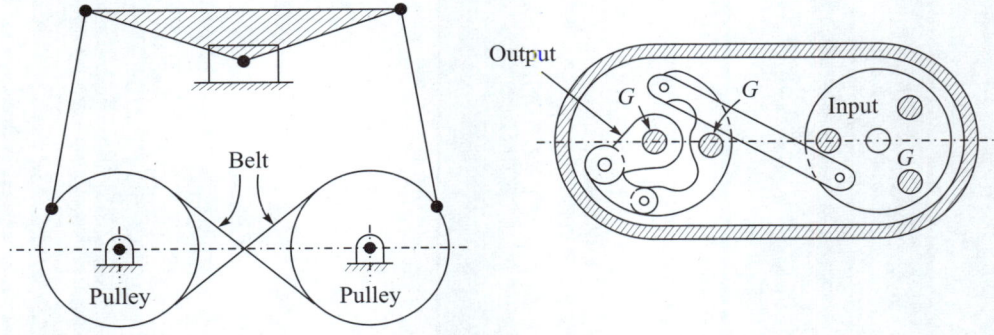

Output

G   G   Input

G

Belt

Pulley   Pulley

Fig. 2.34

2.19 What do you understand by mechanical advantage in a mechanism? How is it measured and what are the desirable limits on range of operation of mechanisms?

2.20 For *N* = 8 and d.o.f. = 3, list all possible combination of links.

2.21 Enumerate all possible chains for–

(i) $N = 6$ and d.o.f. = 1

(ii) $N = 7$ and d.o.f. = 2

2.22 Show that number of ternary links has no effect on d.o.f. of a mechanism.

2.23 Show that maximum possible number of elements on any link of a chain on N links is $N/2$.

2.24 A grab-bucket, shown diagramatically in Fig. 2.35, consists of eight links. Find its d.o.f.

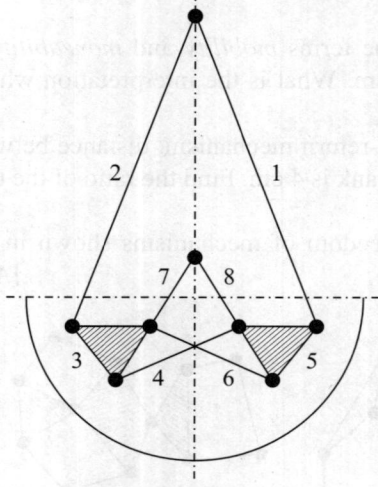

Fig. 2.35

Chapter 3

VELOCITY AND ACCELERATION

ANALYSIS (GRAPHICAL APPROACHES)

3.1 INTRODUCTION

Design and analysis of machinery and mechanisms relies heavily on the skill and ability of a designer to visualise relative motion between machine components. With the ever increasing use of high speed machines, velocity and acceleration analysis have become indispensable in machine design process. Kinematic analysis usually aims at determining motion characteristics (like displacement, velocity and acceleration) of various links for a given input motion. But in yet another category of problems, a kinematician may also be required to establish the type of input motion needed for providing a desired output motion.

Present chapter is devoted to velocity and acceleration analysis of lower paired mechanisms only. The methods can, however, be applied to equivalent lower paired mechanisms in case kinematic analysis is needed for higher paired mechanisms. With the advent of high speed digital computers, coupled with ease in availability of PCs in technical institutions, analytical methods of velocity and acceleration analysis are becoming more and more popular. But this does not reduce the importance of graphical methods. Graphical methods are essential in developing a conceptual understanding about the subject matter. Besides, they provide the fastest method of checking results of machine computation.

3.2 LINEAR AND ANGULAR VELOCITY

(A) Linear Velocity. Consider a particle R to move along a curved path and let it shift its position from R_1 to R_2 along the curved path during a time interval Δt. The small displacemet ΔR occurring during this interval can be expressed in terms of position vectors as

$$\Delta R = R_2 - R_1$$

If displacement ΔR occurs over a small interval of time Δt, (Fig. 3.1), the rate of change of displacement or the instantaneous velocity is given by

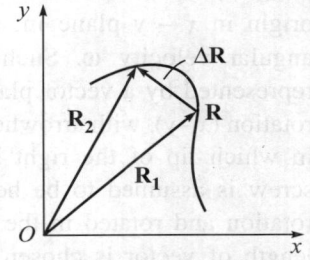

Fig. 3.1 Displacement of a particle.

$$V = \lim_{\Delta t \to 0} \left(\frac{\Delta R}{\Delta t} \right) = \left(\frac{dR}{dt} \right) \tag{3.1}$$

The instantaneous value of velocity (which will be called simply as 'velocity' henceforth) is thus given by time derivative of position vector.

Many a time it is quite convenient to use Newtonian notations for time derivatives. Thus,

$$V = \frac{dR}{dt} = \dot{R} \tag{3.2}$$

when direction of motion is implied in the problem, it may be sufficient to deal with magnitudes only. In such cases, the velocity is given by

$$v = \lim_{\Delta t \to 0} \left(\frac{\Delta r}{\Delta t} \right) = \frac{dr}{dt} \tag{3.3}$$

$$\therefore \qquad\qquad v = \dot{r}$$

(B) Angular Velocity. Discussions for linear velocity can be extended to cover angular velocity aspect also. Let $\Delta\theta$ be the angular displacement in time interval Δt of a rotating line, rotating position vector or of a rotating rigid body. The average angular velocity during this time Δt is,

$$(\omega)_{\text{average}} = \left(\frac{\Delta\theta}{\Delta t} \right)$$

The instantaneous value of angular velocity may be obtained by taking a limit. Thus,

$$\omega = \lim_{\Delta t \to 0} \left(\frac{\Delta\theta}{\Delta t} \right)$$

or

$$\omega = \frac{d\theta}{dt} = \dot{\theta} \tag{3.4}$$

The angular velocity ω can act either in clockwise or counterclockwise sense. When the c.w. and c.c.w. sense of rotation is taken care of by assigning positive and negative signs, it is not necessary to treat them as vectors. The usual unit of angular velocity is radians per second.

At times, it becomes appropriate to treat angular velocity also as a vector. In such a case, the angular velocity vector is represented by right hand screw rule. Let us assume that the line or body is rotating about origin in $x - y$ plane in, say, clockwise sense with angular velocity ω. Such angular velocity is then represented by a vector placed normal to the plane of rotation ($x-y$), with arrowhead pointing in the direction in which tip of the right handed screw moves. The screw is assumed to be held normal to the plane of rotation and rotated in the sense of rotation $\Delta\theta$. The length of vector is chosen to represent magnitude to some scale. (Fig. 3.2)

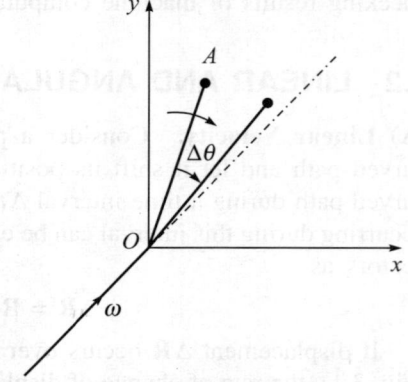

Fig. 3.2 Vector representation of angular velocity.

3.3 VELOCITY OF A POINT ON ROTATING RIGID BODY

In Section 1.4 motion of rotation of a body has been defined as a motion of a rigid body such that a line within the body does not remain parallel to itself. Fig. 3.3 shows a rigid body rotating in c.c.w. direction about an axis through 0 perpendicular to the plane of the paper. Let A be a point with coordinates $x = r_a \cos \theta$ and $y = r_a \sin \theta$ and the position vector of this point be

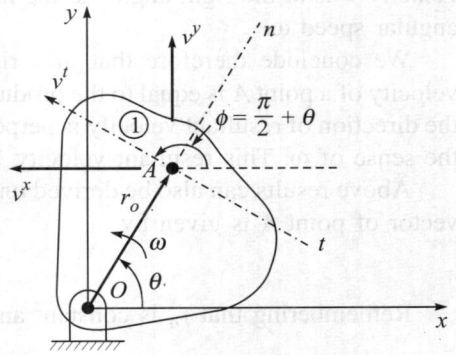

Fig. 3.3   Rotating rigid body.

$$R_a = r_a e^{j\theta}$$

Differentiating x and y coordinates, the rectangular velocity components are

$$v^x = \frac{d}{dt}(r_a \cos \theta) \quad \text{and} \quad v^y = \frac{d}{dt}(r_a \sin \theta)$$

or

$$v^x = -r_a\left(\frac{d\theta}{dt}\right)\sin \theta \quad \text{and} \quad v^y = r_a\left(\frac{d\theta}{dt}\right)\cos \theta$$

Noting that $\dfrac{d\theta}{dt} = \omega$, the angular speed of rotation of body 1 w.r. to 2, we have

$$v^x = -r_a \omega \sin \theta \tag{3.5}$$

and

$$v^y = r_a \omega \cos \theta \tag{3.6}$$

The resultant of the above two vectors give resultant velocity as

$$v = \sqrt{(v^x)^2 + (v^y)^2}$$

or

$$v = \sqrt{(-r_a \omega \sin \theta)^2 + (r_a \omega \cos \theta)^2}$$

$$= (r_a \omega)\sqrt{\sin^2 \theta + \cos^2 \theta}$$

or

$$v = (r_a)\,\omega \tag{3.7}$$

The resultant velocity v makes an angle ϕ with x-axis given by

$$\tan \phi = \left(\frac{v^y}{v^x}\right) = \frac{(r_a\,\omega)\cos \theta}{-(r_a\,\omega)\sin \theta}$$

or

$$\tan \phi = -\cot \theta \tag{3.8}$$

Thus

$$\tan \phi = \tan\left(\frac{\pi}{2} + \theta\right) \tag{3.9}$$

Therefore

$$\phi = \left(\frac{\pi}{2} + \theta\right)$$

Since this is the angle measured with respect to positive direction of x-axis, the resultant velocity V is at the right angles to the line OA and its sense will be commensurate with that of angular speed ω.

We conclude therefore that in a rigid body, rotating about an axis through O, resultant velocity of a point A is equal to the product of radius r_a at point A and angular velocity ω. Further, the direction of resultant velocity is perpendicular to the radial line in a sense commensurate with the sense of ω. This resultant velocity V can therefore be treated as tangential velocity.

Above results can also be derived on the basis of complex polar notations. Thus, the position vector of point A is given by

$$\boldsymbol{R} = (r_a)e^{j\theta}$$

Remembering that r_a is constant and $j = \sqrt{-1}$, differentiating w.r. to time,

$$V = \frac{d}{dt}(R) = \frac{d}{dt}(r_a e^{j\theta})$$

$$= (0)e^{j\theta} + r_a\left(j\frac{d\theta}{dt}\right)e^{j\theta}$$

or
$$V = j(r_e\,\omega)\,e^{j\theta} \tag{3.10}$$

Magnitude of resultant velocity is given by $(r_a\omega)$ while $j(e^{j\theta})$ represents direction of this vector. Clearly, the direction indicated is the one obtained by rotating position vector through right angle in c.c.w. sense.

3.4 GRAPHICAL DIFFERENTIATION

When displacement as a function of time is available in the form of analytical expression, analytical method of differentiation can be conveniently used to obtain velocity. Sometimes, however, displacement v/s time data are available in a graphical or numerical form. Numerical methods of differentiation are best suited if data are available in numerical form.

There are two types of situations in which graphical differentiation can be used:

(a) When displacement-time relationship $x = f(t)$ is known and,
(b) When displacement-crank angle of rotation, *i.e.*, $x = f(\theta)$, relationship is known.

In the latter category of the problem, graphical differentiation is of use only if crank speed ω is assumed to be constant. For instance, consider the relation

$$x = f(\theta)$$

Differentiating w.r. to time,

$$\frac{dx}{dt} = \frac{df}{d\theta}\frac{d\theta}{dt}$$

or
$$v = f'(\theta) \times \omega \tag{3.11}$$

As crank speed ω is constant, it becomes a scaling factor and equation (3.11) can also be written as,

$$v \propto f'(\theta) \tag{3.12}$$

Differentiating (3.11) again w.r. to time,

$$\frac{dv}{dt} = \frac{d}{d\theta}[f'(\theta)] \times \frac{d\theta}{dt} \times \omega + f'(\theta) \times \frac{d\omega}{dt}$$

or
$$a = \omega^2 f''(\theta) + (0)f'(\theta)$$

or
$$a = \omega^2 f''(\theta) \qquad\qquad (3.13)$$

This can also be rewritten as:

$$a \propto f''(\theta) \qquad\qquad (3.14)$$

where
$$f'(\theta) = \frac{df}{d\theta} \quad \text{and} \quad f''(\theta) = \frac{d^2 f}{d\theta^2}$$

In such problems, slope values of θ v/s $f(\theta)$ diagram can only give values which are proportional to velocity values on a common scale, but do not give actual values of velocities. This is apparent from eq. (3.11) in which ω is not known. In a similar way, slopes of θ v/s $f'(\theta)$ diagram do not give actual acceleration values.

First category of problems are, however, comparatively straight forward. This is because,

$$x = f(t)$$

and
$$\frac{dx}{dt} = \text{velocity} = \frac{df}{dt} \qquad\qquad (3.15)$$

$$= \text{slope of displacement curve}$$

Thus, slope of the time v/s displacement curve itself gives actual velocity at the given time.

As seen in Fig. 3.4, if two ordinates KE and LF enclose a strip on the displacement–time curve, the slope of the chord EF of the curve will be approximately equal to the slope of the tangent to the curve on a point mid-way between K and L. And therefore, the slope of the chord EF also represents velocity approximately. Smaller is the error involved in this approximation, shorter the interval KL is. On the other hand, smaller the interval KL is, greater is the percentage error in determining difference in length of the ordinates KE and LF. A better method of drawing velocity/time diagram and acceleration-time diagram is therefore as under:

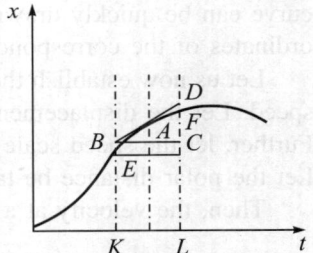

Fig. 3.4 Displacement v/s time diagram.

For the displacement-time curve, as shown in Fig. 3.5(b), choose a pole distance Pa in Fig. 3.5(a) as per convenience along time axis, and draw a line ab parallel to displacement axis. Now, if from the pole P lines are drawn parallel to tangents at various points on the displacement time curve, we get triangles on same base Pa. For any point Q on the curve, Pq is the line drawn parallel to tangent to the displacement curve. In that case aq/Pa gives slope of the displacement curve at Q and therefore, the speed at point Q. Similarly, for point R on the displacement curve Pr is the line drawn parallel to the slope of curve at R and so ar/Pa gives the slope of curve and the velocity at point R on displacement curve.

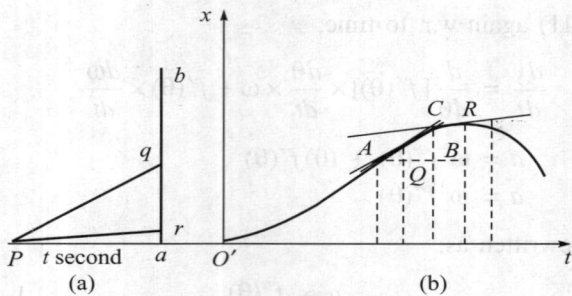

Fig. 3.5 Graphical Differentiation.

In Fig. 3.5(b) the strip width AB and the line of tangency at Q encloses a triangle ABC. In Fig. 3.5(a), line Pq is drawn parallel to tangent AC in Fig. 3.5(b). Hence, triangles Pqa and ACB are similar. Therefore,

$$v_Q = \frac{\delta x}{\delta t} = \frac{BC}{AB} = \left(\frac{aq}{Pa}\right) \qquad (3.16)$$

And so,

$$aq = (Pa)V_Q$$

The method of graphical differentiation thus consists in dividing. Dividing the speed diagram into a number of vertical strips, not necessarily of same width, lines radiating from P may be drawn parallel to tangent at mid-point of strip on the curve. Since the base (Pa) remains same in Fig. 3.5(a), the intercepts made on x-axis by all such radiating lines are clearly proportional to the mean speeds for respective strips. Therefore, using information in Fig. 3.5(a) a speed-time curve can be quickly drawn by projecting these intercepts on ab in Fig. 3.5(a) on to the mid-ordinates of the corresponding strip.

Let us now establish the scale to which the ordinates of the speed-time curve represents the speed. Let the displacement scale be c_x m per cm and the time scale be C_t seconds per cm. Further, let the speed scale to which speeds are plotted in speed-time curve be C_v m/s per cm. Let the polar distance be taken as equal to t seconds.

Then, the velocity at a mid-point of a strip like Q,

$$V_Q = \left(\frac{\delta x}{\delta t}\right) = \left(\frac{C_x\, BC}{C_t\, AB}\right) \qquad (3.17)$$

But

$$\frac{BC}{AB} = \left(\frac{aq}{Pa}\right) \qquad (3.18)$$

As C_t is the scale along time axis,

$$(C_t)(Pa) = t$$

Therefore

$$(Pa) = (t/C_t) \text{ cm}$$

Substituting for (BC/AB) and (Pa) in equation (3.17),

$$V_Q = \left(\frac{C_x}{C_t}\right)\left(\frac{aq}{t/C_t}\right)$$

Therefore $$V_Q = \left(\frac{C_x}{t}\right) aq \qquad\qquad (3.19)$$

But, $$V_Q = (C_v) aq$$

Substituting for V_Q in eq. (3.19),

$$(C_v) aq = \left(\frac{C_x}{t}\right) aq \quad \text{or} \quad C_v = \left(\frac{C_x}{t}\right) \qquad (3.20)$$

Advantage of the above method lies in its flexibility. The method permits variation in the strip width (*i.e.* time interval) depending on the curvature of displacement-time curve. Thus, a shorter strip width may be employed where radius of curvature of displacement curve is small. This increases accuracy. Similar discussions are possible for acceleration-time curve also.

3.5 RELATIVE VELOCITY

One of the dilemmas faced while preparing write-up of this chapter was in respect of the need to draw distinction between motion of two different points on the same body (*e.g.*, between two pin axes of a connecting rod) and motion between two different points on different bodies (*e.g.*, motion between two coincident points on sliding elements of slider and slotted lever of quick return mechanism). It is probably more appropriate to call the former as 'motion-difference' and the latter as 'apparent motion', rather than including both of them in a single common term 'relative motion'. However, it is rather too early to expose students to this subtle difference at U.G. level. Hence, we propose to continue with the use of conventional terms *Relative motion*, *Relative velocity*, etc., in chapters on velocity and acceleration analysis.

Figure 3.6 shows two independent particles A and B which are capable of moving with velocities $\mathbf{V_A}$ and $\mathbf{V_B}$ in the directions shown. Then, velocity of B relative to A is the velocity with which point B appears to move in a coordinate frame which is attached to A and moves with it. Relative velocity of B w.r. to A can also be defined as the velocity with which B appears to move to an observer situated at A and moving with it. Apart from the method illustrated in section (1.10), there is another way to illustrate the concept of relative velocity. In this method, particle A is brought to standstill by applying a velocity equal and opposite to $\mathbf{V_A}$ to both points A and B. In doing so, the relative motion between A and B is not changed but absolute motion of A is brought to zero. With A brought to standstill condition, resultant of velocities $\mathbf{V_B}$ and $(-\mathbf{V_A})$ applied at B gives relative velocity of B w.r. to A. This is shown by vector $\mathbf{V_{BA}}$ in Fig. 3.6. It follows that vector sum of $\mathbf{V_B}$ and $(-\mathbf{V_A})$ gives $\mathbf{V_{BA}}$, the relative velocity of B w.r. to A, *i.e.*

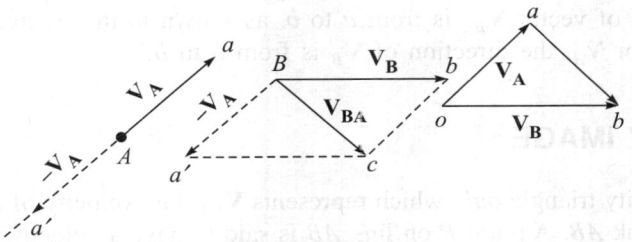

Fig. 3.6 Relative velocity of particle B w.r. to particle A.

$$\mathbf{V}_{BA} = \mathbf{V}_B + (-\mathbf{V}_A)$$

or $$\mathbf{V}_{BA} = \mathbf{V}_B - \mathbf{V}_A \qquad (3.21)$$

Figure 3.6 also shows a triangle *oab* in which *ob* and *ab* represent velocities $\mathbf{V}_B$ and $\mathbf{V}_{BA}$ and the side *oa* represents $\mathbf{V}_A$ in positive direction. It can be verified that this vector triangle represents equation (3.21).

3.6 RELATIVE VELOCITY BETWEEN TWO POINTS ON THE SAME LINK

When *A* and *B* are two points on a given link, the velocities $\mathbf{V}_A$ and $\mathbf{V}_B$ cannot be chosen arbitrarily for both points *A* and *B*. This is simply because points *A* and *B* remain apart at a fixed distance throughout the motion transmission. Thus relative motion between points *A* and *B* is not possible along the line *AB*. In other words, relative motion between points *A* and *B* can occur only in a direction perpendicular to *AB*, which is possible only when the body *AB* has a motion of rotation w.r. to point *A*.

This can also be explained in yet another way. In Section 1.9 it was shown that any displacement of a rigid body *AB* consists partly of translation and partly of rotation. By definition, motion of translation must be the same for points *A* and *B*. Hence, the relative motion between points *A* and *B* can be attributed only to the motion of rotation of link *AB*.

In view of the above explanation, problem of finding velocity of point *B* for a given velocity of *A* resolves itself in drawing a vector triangle *oab* in which side *oa* represents velocity $\mathbf{V}_A$ in magnitude and direction. A line from point '*a*' parallel to the direction of velocity $\mathbf{V}_{BA}$ and a line from parallel to direction of $\mathbf{V}_B$ completes the triangle. The side *ob* represents $\mathbf{V}_B$ while side *ab* represents velocity $\mathbf{V}_{BA}$.

Figure 3.7 shows a link *AB*. Velocity of point *A* is completely known but velocity of point *B* is known only in direction, which coincides with the tangent to the path followed by point *B*. So equation (3.21),

$$\mathbf{V}_B = \mathbf{V}_A + \mathbf{V}_{BA} \qquad (3.22)$$

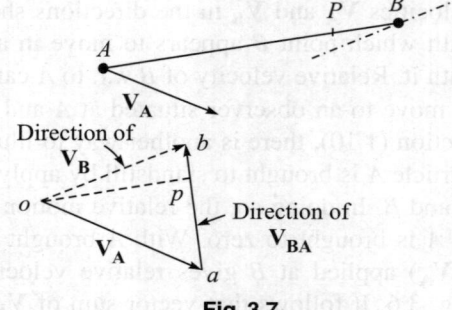

Fig. 3.7

In Fig. 3.7, vector *oa* is drawn to scale to represent velocity $\mathbf{V}_A$. The direction of absolute velocity vector $\mathbf{V}_B$ is drawn from *o* while line parallel to the direction of $\mathbf{V}_{BA}$ is drawn from the tip of vector $\mathbf{V}_A$, to complete the triangle *oab*. Since velocity $\mathbf{V}_{BA}$ adds to the velocity vector $\mathbf{V}_A$ in equation (3.22), the direction of vector $\mathbf{V}_{BA}$ is from *a* to *b*, as shown in the figure. Also, as vector $\mathbf{V}_B$ subtracts from vector $\mathbf{V}_A$, the direction of $\mathbf{V}_B$ is from *o* to *b*.

3.7 VELOCITY IMAGE

Side *ab* of the velocity triangle *oab*, which represents $\mathbf{V}_{BA}$, *i.e.*, velocity of *B* w.r. to *A*, is called *velocity image* of link *AB*. A point *P* on link *AB* is said to have a velocity image at point *p* on *ab,* if *p* divides *ab* in proportion same as *P* does in the link *AB*. Thus a necessary condition for point *p* on *ab* to become velocity image of point *P* is,

$$ap : ab = AP : AB$$

The above property of velocity image is justified on the basis of the discussion, below. *P*, being a point on link AB, has a motion of rotation similar to *B* about point *A*. Clearly, $\mathbf{V}_{PA}$ can act only in a direction perpendicular to *AB* and therefore, like *b*, point *p* must lie on line *ab*. Similarly, for a given angular speed of rotation ω of link *AB* about *A*,

$$(\mathbf{V}_{BA}/\mathbf{V}_{PA}) = (ab/ap) = (AB/AP)$$

It follows therefore that,

$$ap : ab = AP : AB \tag{3.23}$$

An interesting case of velocity image occurs when a point *X* on link *AB* does not lie on the line joining *AB*. This is illustrated in Fig. 3.8. Point *X* on link *AB* subtends, angles α and β at *A* and *B* respectively. To complete velocity polygon *oab*, draw *ob* and *oa* parallel to $\mathbf{V}_B$ and $\mathbf{V}_A$ respectively, with their lengths representing magnitudes of the respective vectors. The relative velocity vector $\mathbf{V}_{BA}$ represented by vector *ab* will be perpendicular to the link centre line *AB*. Remembering that velocity of *X* w.r. to *A* can exist in a direction perpendicular to *AX*, *ax* has been drawn perpendicular to *AX*. Similarly, velocity of *X* w.r. to *B* can exist only in a direction perpendicular to *BX*. Hence line *bx* is drawn perpendicular to *BX*. The intersection point defines velocity image *x* of point *X*.

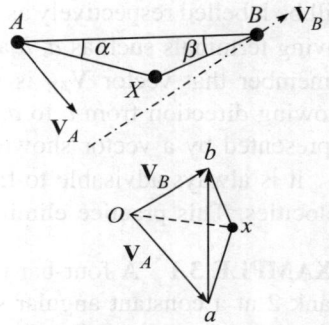

Fig. 3.8 Velocity Image of Point X.

Since, lines *ab*, *ax* and *bx* have been drawn in directions perpendicular to sides *AB*, *AX* and *BX* respectively, triangles *abx* and *ABX* are similar so that ∠*bax* = α and ∠*abx* = β. Again, if link *AB* is assumed to rotate with angular speed ω about *A*, then

$$\left(\frac{\mathbf{V}_{BA}}{AB}\right) = \left(\frac{\mathbf{V}_{XA}}{AX}\right)$$

Therefore $\dfrac{AX}{AB} = \left(\dfrac{\mathbf{V}_{XA}}{\mathbf{V}_{BA}}\right)$; V_{BA} and V_{XA} being the magnitudes of corresponding velocities.

Also, from similar triangles *abx* and *ABX*,

$$\frac{AX}{AB} = \frac{ax}{ab}$$

or

$$ax = ab\left(\frac{AX}{AB}\right)$$

or

$$\mathbf{V}_{XA} = \left(\frac{AX}{AB}\right)\mathbf{V}_{BA} \tag{3.24}$$

Join *o* to *x*. Now as *oa* = $\mathbf{V}_A$ and *ax* = $\mathbf{V}_{XA}$,

$$\mathbf{V}_X = \mathbf{V}_A + \mathbf{V}_{XA} \tag{3.25}$$

It follows that *ox* represents absolute velocity $\mathbf{V}_X$ of point *X*.

3.8 VELOCITY POLYGON

A *vector polygon* for velocities is a graphical depiction of vector equation, relating velocities of two or more points. When constructed to a scale, a velocity polygon contains information about magnitudes and directions of velocities of respective points.

Constructing velocity polygon in thick black lines makes them easy to read. However, when velocity polygon is intended to give graphical solution to vector equation, use of thick lines may result in inaccuracies. So, it is advisable to draw the polygon using thin sharp lines, made with hard drawing pencils. The solution in the form of vector polygon is started invariably by choosing a suitable velocity scale and a convenient location for a point O_v called *velocity pole*. This point represents zero velocity. In general, absolute velocity vectors like $\mathbf{V}_A$, $\mathbf{V}_B$, $\mathbf{V}_C$ are constructed with their origin at this velocity pole O_V. Terminals of absolute velocity vectors $\mathbf{V}_A$, $\mathbf{V}_B$, $\mathbf{V}_C$ etc. will be labelled respectively as a, b, c and so on. The lines not originating from velocity pole and having terminals such as a, b and c then represent relative velocity vectors. It may be useful to remember that vector $\mathbf{V}_{AB}$ is read as velocity of A w.r. to B and is represented by a vector showing direction from b to a. Conversely, vector $\mathbf{V}_{BA}$ is read as velocity of B w.r. to A and is represented by a vector showing direction from a to b in the velocity polygon.

It is always advisable to first write down vector equation connecting unknown and known velocities. This practice eliminates chances of error due to any confusion.

EXAMPLE 3.1 A four-bar mechanism, with dimensions as shown in Fig. 3.9, is driven by a crank 2 at a constant angular speed of $\omega_2 = 600$ r.p.m. in c.c.w. sense. Find the instantaneous velocities of coupler point P and angular velocities of links 3 and 4 in the position shown.

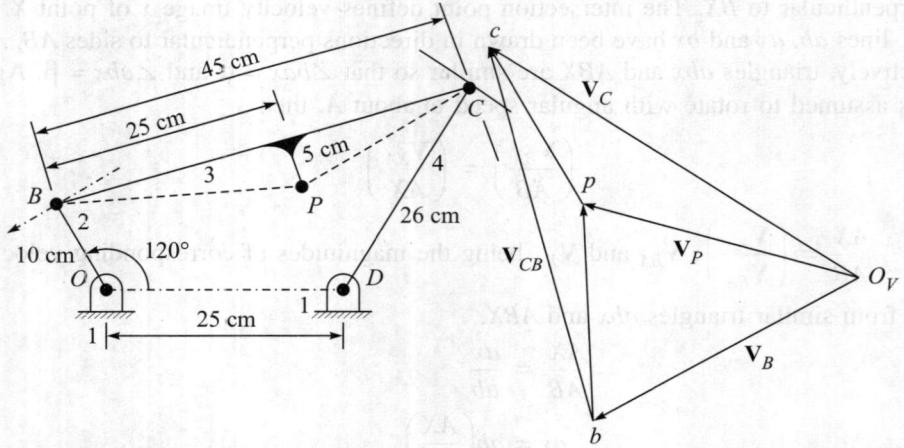

(a) Mechanism configuration, Scale: 1 cm = 7.5 cm   (b) Velocity polygon, Scale: 1 cm = 148 cm/s

Fig. 3.9

Solution: $\omega_2 = \dfrac{2\pi N}{60} = \dfrac{2\pi \times 600}{60} = 62.83$ rad/s

The vector equation connecting known velocity $\mathbf{V}_B$ to unknown velocity $\mathbf{V}_C$ on same link BC is

$$\mathbf{V}_C = \mathbf{V}_B + \mathbf{V}_{CB} \qquad (3.26)$$

(a) Velocity of crank pin B is known both in magnitude and direction.

$$\mathbf{V}_B = (10)\,(62.83) = 628.3 \text{ cm/s}$$

(b) Velocity of point C is not known in magnitude but it acts in a direction tangential to path of C, *i.e.* perpendicular to DC.

(c) Velocity of point C w.r. to B is not known in magnitude, but its direction must be along a perpendicular to link BC.

Construction. For velocity polygon in Fig. 3.9(b), select a suitable velocity pole o_v and draw first velocity vector which is completely known. Thus, draw $o_v\, b$ in a direction perpendicular to OB, to represent velocity $\mathbf{V}_B = 628.3$ cm/s to suitable scale. From o_v next draw the direction of absolute velocity vector $\mathbf{V}_C$ which is known only in direction. Thus draw $o_v c$ from o_v which is in a direction perpendicular to crank DC. Finally from the tip b of vector $o_v b$, draw a line in a direction perpendicular to BC to cut above line at c. Note that bc represents $\mathbf{V}_{CB}$ which is now known in magnitude & direction both. This completes velocity polygon $o_v bc$. The directions of the velocity vectors $\mathbf{V}_C$ and $\mathbf{V}_{CB}$ are obtained by referring back to velocity equation (3.26).

According to the velocity equation, relative velocity vector $\mathbf{V}_{CB}$ is to be additive to $\mathbf{V}_B$. Hence the arrowhead on line bc should be from b to c. Also, from the equation, $\mathbf{V}_C$ is subtractive from $\mathbf{V}_B$. Hence arrowhead on line $o_v c$ should be from o_v to c.

By measurement $\qquad\qquad\qquad o_v c = V_C = 915$ cm/s

And so, $\qquad\qquad\qquad \omega_4 = \dfrac{V_C}{DC} = \dfrac{915}{26} = 35.19$ rad/s $\qquad$ **Ans.**

Also from measurement, $\quad bc = V_{CB} = 835$ cm/s

Therefore $\qquad\qquad\qquad \omega_3 = \dfrac{V_{CB}}{BC} = \dfrac{835}{45} = 18.55$ rad/s $\qquad$ **Ans.**

For obtaining linear velocity of coupler point P, we use the principle of velocity image. Thus, from b draw a line bp in a direction perpendicular to the line joining BP. Similarly, from point c draw a line cp in a direction perpendicular to the line joining CP. The intersection of the two lines define velocity image p of coupler point P. It can be verified that triangles BPC and bpc are similar.

Line $o_v p$ on velocity polygon represents absolute velocity $\mathbf{V}_P$ of coupler point P. The sense of $\mathbf{V}_P$ is decided by referring to the velocity equation:

$$\mathbf{V}_P = \mathbf{V}_B + \mathbf{V}_{PB}$$

As $\mathbf{V}_P$ and $\mathbf{V}_B$ are to be subtractive in nature, direction of $\mathbf{V}_P$ is, as shown, from o_v to P. Velocity of coupler point P measures:

$$o_v p = \mathbf{V}_P = 600 \text{ cm/s} \qquad\qquad\qquad \textbf{Ans.}$$

EXAMPLE 3.2 In the slider-crank mechanism as shown in Fig. 3.10, the crank 2 makes 80 r.p.m. in c.w. sense. Determine the linear velocity of slider and angular velocity of connecting rod 3. Also find out the linear velocity of point Q on connecting rod. Lengths of crank and connecting rod are 8 cm and 32 cm respectively.

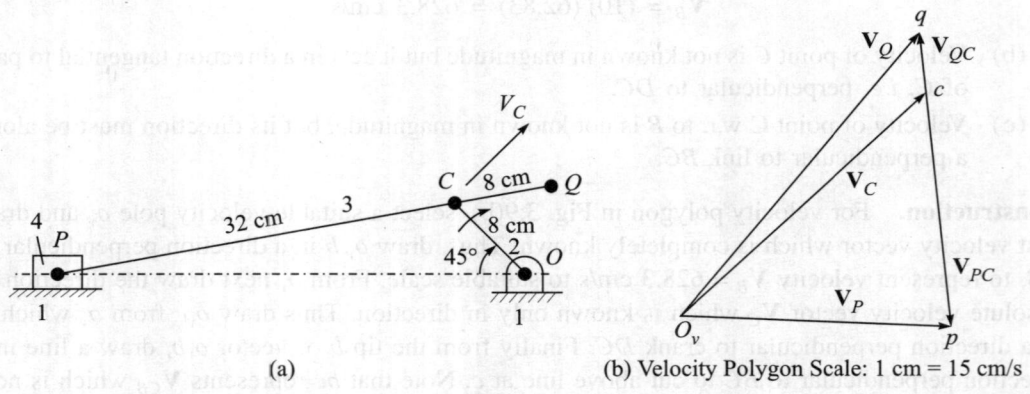

(a)

(b) Velocity Polygon Scale: 1 cm = 15 cm/s

Fig. 3.10

Solution: Crank speed $= \dfrac{2\pi \times 80}{60} = 8.377$ rad/s

Tangential (linear) velocity of crank pin $C = 8 \times 8.377 = 67.0$ cm/s

The velocity equation connecting known velocity of point C to unknown velocity of piston pin P on the connecting rod is

$$\mathbf{V}_P = \mathbf{V}_C + \mathbf{V}_{PC}$$

The velocity $\mathbf{V}_C$ has a magnitude of 67.0 cm/sec and direction of $\mathbf{V}_C$ is at right angles to OC as shown. Velocity of P w.r. to C (*i.e.* $\mathbf{V}_{PC}$) is not known in magnitude but its direction is known to be perpendicular to PC. Similarly, velocity of piston pin P is not known in magnitude, but its direction must be along the line of stroke.

Construction. For velocity polygon in Fig. 3.10(b), select a suitable point o_v as velocity pole. From o_v draw $o_v c$ to represent $\mathbf{V}_C = 67.0$ cm/s to scale in a direction perpendicular to crank OC. After representing the velocity vector, which is completely known, from o_v draw line $o_v p$ parallel to line of stroke. Note that $\mathbf{V}_P$ is known only in direction. Similarly from c draw a line cp in a direction perpendicular to CP. Note that $\mathbf{V}_{PC}$ was known only in direction. Intersection of the two lines locate point p.

Directions of velocities $\mathbf{V}_P$ and $\mathbf{V}_{PC}$ will be obtained by referring back to the velocity equation. Since $\mathbf{V}_{PC}$ is additive to $\mathbf{V}_C$, the arrowhead should be as shown in figure. Again, $\mathbf{V}_P$ being on the other side of equal to sign, should be subtractive from $\mathbf{V}_C$. Hence the arrowhead on $\mathbf{V}_P$ and $\mathbf{V}_{PC}$ should be as shown.

By measurement $\qquad\qquad\qquad \mathbf{V}_P = 54.0$ cm/s $\qquad\qquad\qquad$ **Ans.**

$$\mathbf{V}_{PC} = 47 \text{ cm/s}$$

The angular velocity of link CP is obtained thus: velocity $\mathbf{V}_{PC}$ represents velocity of P w.r. to C. Hence connecting rod can be assumed to rotate about C as centre. So,

$$\omega_{PC} = \left(\frac{V_{PC}}{PC}\right) = \frac{48}{32} = 1.5 \text{ rad/s} \qquad \textbf{Ans.}$$

Velocity of point Q is obtained by using the principle of velocity image. Extend pc to q such that $pq: pc = PQ : PC$.

Velocity of Q is then given by line joining $O_v q$.

By measurement $\qquad\qquad\qquad \mathbf{V}_Q = 74 \text{ cm/s} \qquad\qquad \textbf{Ans.}$

Direction of $\mathbf{V}_Q$, which is as shown in figure, is decided based on equation $\mathbf{V}_Q = \mathbf{V}_C + \mathbf{V}_{QC}$.

Readers are advised to verify that the direction of vector $\mathbf{V}_Q$, shown in velocity polygon, confirms with the above equation.

EXAMPLE 3.3 For the mechanism shown in Fig. 3.11, determine velocities of points C, E and P and the angular velocities of links 3, 4 and 5. Crank OB rotates at 120 r.p.m. c.c.w.

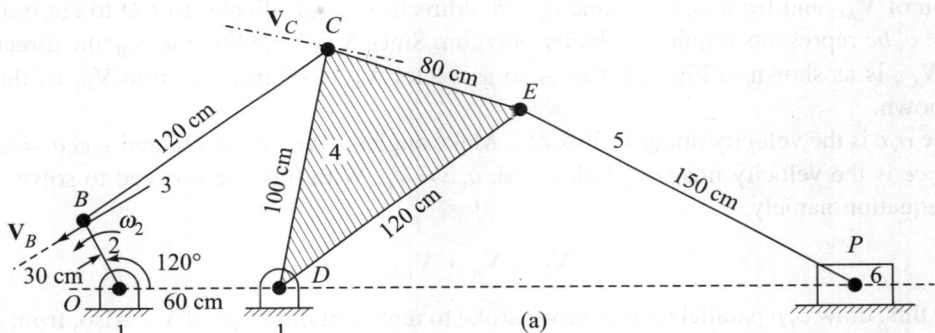

Scale 1 cm = 29.5 cm

(a)

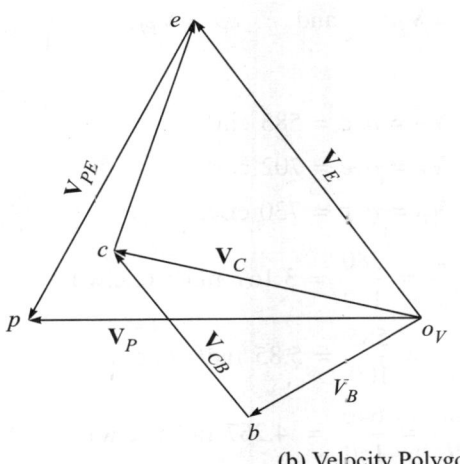

Scale 1 cm = 122 cm/s

(b) Velocity Polygon.

Fig. 3.11

Solution:
$$\omega_2 = \frac{2\pi(120)}{60} = 12.570 \text{ rad/s}$$

$$V_B = 376.99 \text{ cm/s} = 377.0 \text{ cm/s, say}$$

There are two four-link loops 1–2–3–4 and 1–4–5–6. Therefore, we may solve velocity polygon problem in two stages. In the first stage, we solve velocity polygon using vector equation,

$$V_C = V_B + V_{CB}$$

This connects unknown velocity V_C on link 3 to known velocity V_B.

In the second stage, after establishing velocity of C, and hence of E (using velocity image concept), we solve vector equation,

$$V_P = V_E + V_{PE}.$$

For solving first vector equation graphically, draw line $o_v b$ from velocity pole to represent V_B in magnitude and direction. From b draw a line in a direction perpendicular to BC to represent direction of V_{BC} and from o_v draw line $o_v c$ in a direction perpendicular to CD to cut it at point c. Then, $o_v bc$ represents required velocity polygon. Since V_{CB} is additive to V_B, the direction of vector V_{CB} is as shown in Fig. 3.11(b). Also as vector V_C is subtractive from V_B, its direction is as shown.

Line $o_v c$ is the velocity image of link DC. Make angles $\angle co_v e = \angle CDE$ and $\angle ceo_v = \angle DCE$. Then $o_v ce$ is the velocity image of link 4 and, $o_v e = V_E$. With this, we proceed to solve second vector equation namely,

$$V_P = V_B + V_{PE}$$

For this, draw $o_v p$ parallel to the line of stroke to represent direction of V_P. Also, from e draw ep in a direction perpendicular to EP. The two lines intersect at p. Thus

$$o_v p = V_P \qquad \text{and} \qquad ep = V_{PE}.$$

By measurement:

$$V_C = o_v c = 585 \text{ cm/s} \qquad \textbf{Ans.}$$

$$V_E = o_v e = 702 \text{ cm/s} \qquad \textbf{Ans.}$$

$$V_P = o_v p = 730 \text{ cm/s} \qquad \textbf{Ans.}$$

Also,
$$\omega_3 = \frac{V_{CB}}{BC} = \frac{380}{120} = 3.167 \text{ rad/s (c.c.w.)} \qquad \textbf{Ans.}$$

$$\omega_4 = \frac{V_C}{CD} = \frac{585}{100} = 5.85 \text{ rad/s (c.c.w.)} \qquad \textbf{Ans.}$$

$$\omega_5 = \frac{V_{PE}}{PE} = \frac{640}{150} = 4.267 \text{ rad/s (c.w)} \qquad \textbf{Ans.}$$

3.9 VELOCITY OF RUBBING

Elements on two links forming a turning pair are shown in Fig. 3.12. As can be seen, a pin forms an integral part of one element and this pin fits into the mating hole in the second link, permitting a relative motion of rotation.

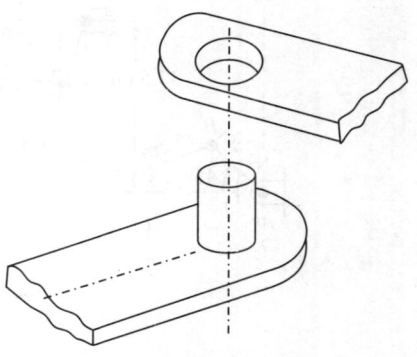

   When the two links are connected at the elements, surface of hole of one link rubs against the surface of pin of the other link. The rubbing velocity between the two depends on the angular velocity of one link relative to the other. This is illustrated below for pin joints O, B, C and D of the four-bar mechanism in Fig. 3.9.

Fig. 3.12

Pin at O. One of the connecting links OD is fixed. Hence, angular velocity at O between connected link is only due to link OB which rotates in c.c.w. sense with angular velocity ω_2. Therefore, rubbing velocity at O is $(r_0\omega_2)$, where r_0 = pin radius at O.

Pin at B. Link 2 rotates in c.c.w. sense. Therefore, relative to link BC, link BO rotates in c.c.w., *i.e.* ω_{OB} is c.c.w.

   Also, from velocity polygon of Fig. 3.9(b), $\mathbf{V}_{cb}$ acts upwards, *i.e.* angular velocity of C w.r. to B, ω_{CB} is c.c.w.

   Since ω_{OB} and ω_{CB} both act c.c.w. about B, their relative angular speed is $(\omega_{OB} \sim \omega_{CB})$ and so, for a pin radius of r_B, the velocity of rubbing is

$$= r_B\,(\omega_{OB} \sim \omega_{CB}) \tag{3.27}$$

Pin at C. From velocity polygon of Fig. 3.9(b), it is clear from the direction of $\mathbf{V}_C$ that angular velocity of link 4, ω_{CD} is c.c.w. Hence angular velocity of D w.r. to C, *i.e.* ω_{DC} is also c.c.w.

   Again from velocity polygon $\mathbf{V}_{BC}$ must be opposite to $\mathbf{V}_{CB}$. Hence $\mathbf{V}_{BC}$ acts downwards and therefore, ω_{BC} is c.c.w. Since ω_{DC} and ω_{BC} are both c.c.w. at C, the rubbing velocity,

$$v = (\omega_{DC} \sim \omega_{BC}) \times r_C$$

where, $\qquad\qquad\qquad\qquad r_C$ = pin radius at C.

Pin at D. Link OD is stationary and velocity $\mathbf{V}_{CD}$ from velocity polygon of Fig. 3.9(b) has a direction given by $\mathbf{V}_C$. Hence ω_{CD} acts c.c.w., and the velocity of rubbing at D,

$$= (r_D)\,\omega_{CD}$$

   In all the above cases, either only one of the two links was rotating or both the links were rotating in the same sense. In case both the links rotate in opposite sense, rubbing velocity $= r \times (\omega_1 + \omega_2)$.

EXAMPLE 3.4 In the mechanism shown in Fig. 3.13, the crank AB rotates about A at a uniform speed of 120 r.p.m. c.w. The lever DC oscillates about a fixed point D, which is connected to AB by the coupler BC. The sliding block P moves in the horizontal guides being driven by the link EP. Determine:

   (a)   Velocity of block P; (b) Angular velocity of lever DC, and (c) Rubbing speed of the pin C which is 5 cm in diameter.

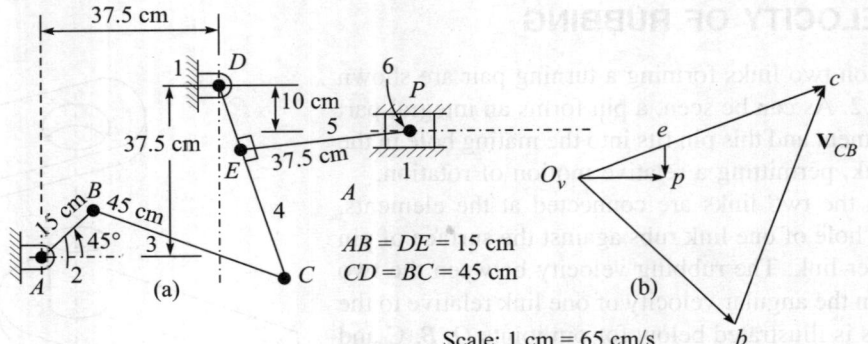

AB = DE = 15 cm
CD = BC = 45 cm

Scale: 1 cm = 65 cm/s

Fig. 3.13

Solution: Angular speed of crank $\omega_2 = \dfrac{2\pi \times 120}{60} = 12.567$ rad/s

Therefore   Tangential speed of B, $\mathbf{V}_B = 15 \times 12.567 = 188.5$ cm/s

The velocity polygon will be solved in two stages.

In the first stage, velocity polygon for links 1, 2, 3 and 4 will be constructed using equation,

$$\mathbf{V}_C = \mathbf{V}_B + \mathbf{V}_{CB} \qquad \text{(a)}$$

In the second stage with $\mathbf{V}_C$ known, velocity polygon will be solved for links 1, 4, 5 and 6 using velocity equation,

$$\mathbf{V}_P = \mathbf{V}_E + \mathbf{V}_{PE} \qquad \text{(b)}$$

Construction. From arbitrarily selected velocity pole o_v draw line $o_v b$ perpendicular to direction AB and measure velocity $\mathbf{V}_B = 188.5$ cm/s along it, to a suitable scale. Since $\mathbf{V}_{CB}$ is additive to $\mathbf{V}_B$, from b draw a line in a direction perpendicular to link BC. Similarly from o_v draw a line in direction perpendicular to DC (to represent $\mathbf{V}_C$) to cut above line at c. Then, in accordance with equation (a) line 'be' represents $\mathbf{V}_{CB}$ and $o_v c$ represents $\mathbf{V}_C$. Locate point e on $o_v C$ such that $o_v e : o_v c = DE : DC$. Then $o_v e$ represents $\mathbf{V}_E$. Now, in accordance with equation (b), draw ep in a direction perpendicular to EP and from o_v draw $o_v p$ parallel to $\mathbf{V}_P$, to meet above line at p. Then $o_v P$ represents piston velocity $\mathbf{V}_P$.

By measurements $\mathbf{V}_P = 74$ cm/s   **Ans.**

and $\mathbf{V}_{CB} = 220$ cm/s

Also $\mathbf{V}_C = 224$ cm/s

$\therefore$ $\omega_4 = \dfrac{224}{45} = 4.98$ rad/s   **Ans.**

Also $\omega_3 = \dfrac{\mathbf{V}_{CB}}{BC} = \dfrac{220}{45} = 4.89$

To find robbing speed at pin C, we proceed as under:

Since $\omega_4 = \dfrac{\mathbf{V}_C}{45}$ is c.c.w., angular speed of CD about c is also c.c.w.

Also, $\omega_{CB} = \dfrac{\mathbf{V}_{CB}}{BC}$ and is c.c.w. and so angular speed of B about C is also c.c.w.

Thus about point C, both ω_3 and ω_4 are c.c.w. Hence rubbing velocity at C,

$$\mathbf{V}_{rub} = r_C \,(\omega_3 \sim \omega_4)$$

$$\mathbf{V}_{rub} = \frac{5}{2}(4.89 \sim 4.98) = 0.225 \text{ cm/s} \qquad \textbf{Ans.}$$

EXAMPLE 3.5   Find out the rubbing velocity at pin point E in the mechanism shown in Fig. 3.11(a). Assume pin diameter at E to be 2 cm.

Solution: From the velocity polygon of Fig. 3.11(b),

$$\omega_{ED} = \frac{702}{120} = 5.85 \text{ rad/s (c.c.w.)}$$

Therefore $\qquad\qquad \omega_{DE} = 5.85 \text{ rad/s (c.c.w.)}$

Also $\qquad\qquad\qquad \omega_{PE} = \dfrac{\mathbf{V}_{PE}}{150} = \dfrac{640}{150} = 4.267 \text{ rad/s (c.w.)}$

Nett (relative) angular velocity between links 4 and 5 at $E = \omega_{DE} + \omega_{PE}$

$$= 5.85 + 4.267 = 10.117 \text{ rad/s}$$

Therefore   rubbing velocity at $E = \left(\dfrac{2}{2}\right) \times 10.117 = 10.117 \text{ cm/s}$ $\qquad$ **Ans.**

3.10   MECHANICAL ADVANTAGE AND POWER TRANSMISSION

Mechanical advantage is defined as the ratio of the magnitude of output force to input force. Sometimes, mechanisms are needed for applying large forces for doing desired work. Examples of this type are toggle mechanisms used in stone crushing mechanisms, presses, pump mechanism, etc. In all such cases velocity ratio may be sacrificed for attaining higher values of mechanical advantage.

EXAMPLE 3.6   A toggle mechanism has dimensions in mm as shown in Fig. 3.14. Find the velocities at pins B and P and the angular velocities of links AB, BC and BP. Assume the crank to rotate at 60 r.p.m. in counter clockwise direction. Also find the mechanical advantage.

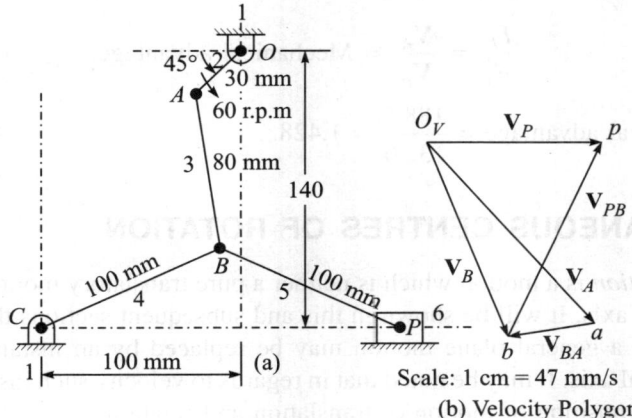

Fig. 3.14

Solution: Velocity of pin A, $\mathbf{V}_A = 30\left(\dfrac{2\pi \times 60}{60}\right) = 188.5$ mm/s. We first solve velocity polygon involving links 1, 2, 3, 4. Velocity equation for this portion of linkage is

$$\mathbf{V}_B = \mathbf{V}_A + \mathbf{V}_{BA}$$

From velocity pole o_V, therefore draw $o_v a$ perpendicular to OA and measure along it a velocity of 188.5 mm/s to some scale (say, 1 cm = 47 mm/s) From a draw a line ab in a direction perpendicular to AB and from o_v draw another line parallel to direction of $\mathbf{V}_B$ (which is perpendicular to CB), to cut earlier line at b. Then $ab = \mathbf{V}_{BA} = 69$ mm/s.

Therefore $\qquad\qquad\qquad \omega_{BA} = \dfrac{69}{80} = 0.863$ rad/s $\qquad\qquad\qquad$ **Ans.**

Also, by measurement $\mathbf{V}_B = o_v b = 162$ mm/s $\qquad\qquad\qquad$ **Ans.**

Therefore $\qquad\qquad\qquad \omega_4 = \dfrac{162}{100} = 1.62$ rad/s (c.w.) $\qquad\qquad$ **Ans.**

With $\mathbf{V}_b$ known, we can proceed to solve velocity polygon for the remaining part of linkage. Thus, velocity equation is

$$\mathbf{V}_P = \mathbf{V}_B + \mathbf{V}_{PB}$$

Hence, from point b on velocity polygon, draw bp perpendicular to BP to represent direction of $\mathbf{V}_{PB}$ and from o_v, draw o_{vp} to intersect it in a direction parallel to the line of stroke. Directions of vectors are to be assigned so as to satisfy the above velocity equation.

By measurement,

$$\mathbf{V}_P = o_v p = 132.0 \text{ mm/s} \qquad\qquad\qquad \textbf{Ans.}$$
and $\qquad\qquad\qquad\qquad \mathbf{V}_{PB} = 165 \text{ mm/s} \qquad\qquad\qquad\qquad\quad \textbf{Ans.}$

Therefore $\qquad\qquad\qquad \omega_{PB} = \dfrac{165}{100} = 1.65$ rad/s (c.c.w.) $\qquad\qquad$ **Ans.**

Now assuming 100% efficiency, work input must equal work output. Assuming F_A to be the tangential force at crank pin A and F_P the linear force at P, we have, for 100% efficiency,

$$F_A \mathbf{V}_A = F_P \mathbf{V}_P$$

or $\qquad\qquad\qquad\qquad \dfrac{F_P}{F_A} = \dfrac{\mathbf{V}_A}{\mathbf{V}_P} = $ Mechanical advantage.

Hence, mechanical advantage $= \dfrac{188.5}{132.0} = 1.428$. $\qquad\qquad\qquad$ **Ans.**

3.11 INSTANTANEOUS CENTRES OF ROTATION

A *general planar motion* is a motion which is neither a pure translatory motion nor a pure rotary motion about a fixed axis. It will be shown in this and subsequent sections that when velocities alone are concerned, a general plane motion may be replaced by an instantaneous motion of rotation about a virtual axis. It may be noted that in regards to velocity such instantaneous rotation is equivalent to superposition of motion of translation and rotation.

It is quite interesting to conceive of relative motion between two rigid bodies to be that of rotation about an instantaneous (but imaginary) axis which is common to both the bodies. Since we intend to mainly deal with planar motions, and these axes are normal to plane of motion, they will be referred to as *instantaneous centres*, centro or *velocity poles*.

As seen in Fig. 3.15, finite displacement of a link from an initial position A_1B_1 to the position A_2B_2, accomplished by moving A_1 and B_1 along prescribed paths, could also be brought about by rotation of the entire body through an angle $\Delta\theta$ about an axis through O' which is the point of intersection of normal bisectors of A_1A_2 and B_1B_2. It must be noted, however, that in view of comparatively large displacements involved, the discrepancy between actual paths of motion of A_1 and B_1 and corresponding circular arcs can be appreciable. For closest approximation between actual paths and corresponding arcs of circle, we may have to go for the limiting case of an infinitesimal displacement, indicated by a condition $\Delta\theta \rightarrow 0$. The virtual axis of rotation, under such limiting condition, is called *velocity pole* or *instant centre* of rotation,

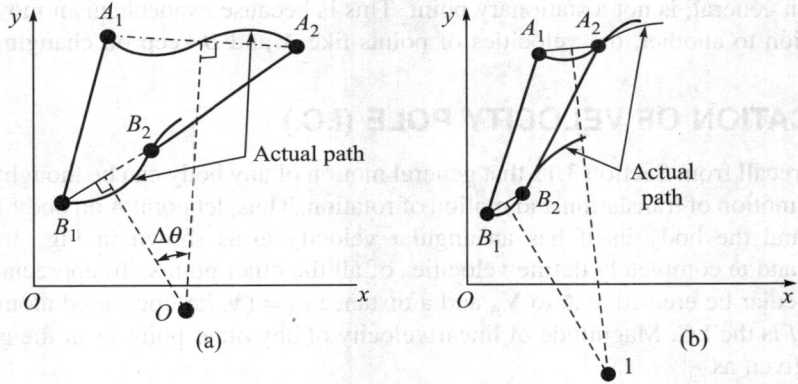

Fig. 3.15   Finite and Infinitesimal displacements.

Definition.   For two bodies having relative motion with respect to one another, instantaneous centre (abbreviated as I.C.) of rotation is an imaginary point, common to two bodies such that any of the two bodies can be assumed to have motion of rotation with respect to the other about the imaginary point.

3.12   PROPERTIES OF INSTANTANEOUS CENTRE

Let x-y coordinate frame in Fig. 3.16 represent plane (body) 1 and another plane (body) 2 have motion with respect to body 1. Let the relative motion of body 2 w.r. to 1 be represented by velocities $\mathbf{V}_A$ and $\mathbf{V}_B$ of two points A and B on the body 2. Let the perpendicular to the two velocity vectors $\mathbf{V}_A$ and $\mathbf{V}_B$ at A and B meet point I, the instantaneous centre of rotation of body 2 w.r. to 1.

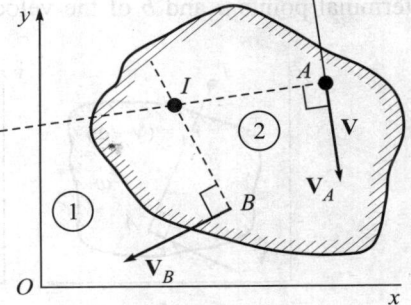

Fig. 3.16   Bodies in relative motion.

In order to bring out subtle meaning of I.C., let us consider I to define position of two coincident points, one on each body, having no relative motion instantaneously. Thus I may be considered to be a point on body 1 about which body 2 is instantaneously rotating; or it may be considered as a point on body 2 about which body 1 is instantaneously rotating. Again as the two coincident points at P (belonging to bodies 1 and 2) lie on the axis of relative rotation, they have zero relative velocity at P. In other words, when bodies 1 and 2 move with respect to a third body, keeping zero relative velocity at coincident points at I, they have the same velocity of I w.r. to body 3, whether it is considered to be a point on body 1 or 2.

Thus, the properties of I.C. can be summed up as under:

1. An I.C. of rotation is a point common to two links (bodies extended, if necessary) having a relative motion;
2. I.C. is a point about which one body can be assumed to rotate with respect to the other;
3. It is an imaginary point at which the two bodies have same absolute velocity. It follows that the two bodies have zero relative velocity at the coincident points I;
4. I.C. in general, is not a stationary point. This is because as mechanism moves from one position to another, the velocities of points like A and B keep on changing.

3.13 LOCATION OF VELOCITY POLE (I.C.)

Readers may recall from Section 3.11 that general motion of any body can be thought of as super-imposition of motion of translation and motion of rotation. Thus, let point A on body 1 has a linear velocity $\mathbf{V}_A$ and the body itself has an angular velocity ω as shown in Fig. 3.17(a). Then, velocities $\mathbf{V}_A$ and ω completely define velocities of all the other points. To appreciate this point, let a perpendicular be erected at A to $\mathbf{V}_A$ and a distance $r_A = (\mathbf{V}_A/\omega)$ measured along it to locate point I. Then I is the I.C. Magnitude of linear velocity of any other point D, at the given instant, will then be given as

$$\mathbf{V}_D = (ID) \times \omega$$

Again, when linear velocities $\mathbf{V}_A$ and $\mathbf{V}_C$ of any two points A and C on body 1 are known, the I.C. is located by point of intersection of perpendiculars drawn at A and C to $\mathbf{V}_A$ and $\mathbf{V}_C$ respectively [see Fig. 3.17(b)].

When velocities of two points A and B are parallel [as at Fig. 3.17(c)], the I.C. is located at a point of intersection of a line perpendicular to both $\mathbf{V}_A$ and $\mathbf{V}_B$ and an another line joining terminal points a and b of the velocity vectors $\mathbf{V}_A$ and $\mathbf{V}_B$.

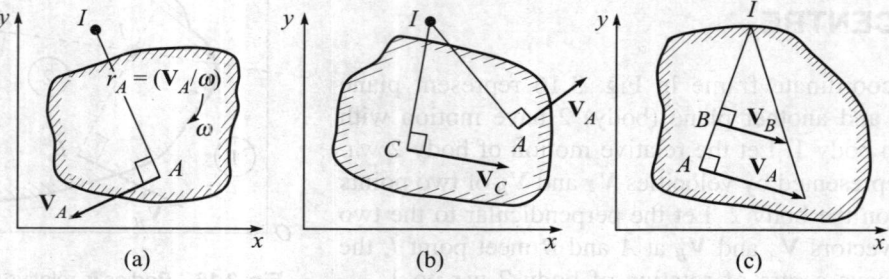

Fig. 3.17 Location of Instantaneous centres.

Figure 3.18 illustrates a few typical combinations of link motions and corresponding I.C.'s. A slider, sliding along a curved surface, is equivalent to motion of rotation about centre of curvature. Hence centre of curvature in Fig. 3.18(a) becomes an I.C. In Fig. 3.18(b) a slider moves along a straight path. Again, a straight line is an arc of a circle of infinite radius of curvature. Hence I.C. in such a case lies at infinity along a line perpendicular to the direction of motion. In Fig. 3.18(c) the slider slides along cylindrical surface. This motion is equivalent to the motion of rotation of slider about centre of circle. Therefore, I.C. lies at centre. In Fig. 3.18(d), the cylindrical disk rolls without slipping along straight line. For small angle of oscillation, any point P on disk moves in arc of circle with point of contact I as the centre. Hence the point I becomes an I.C.

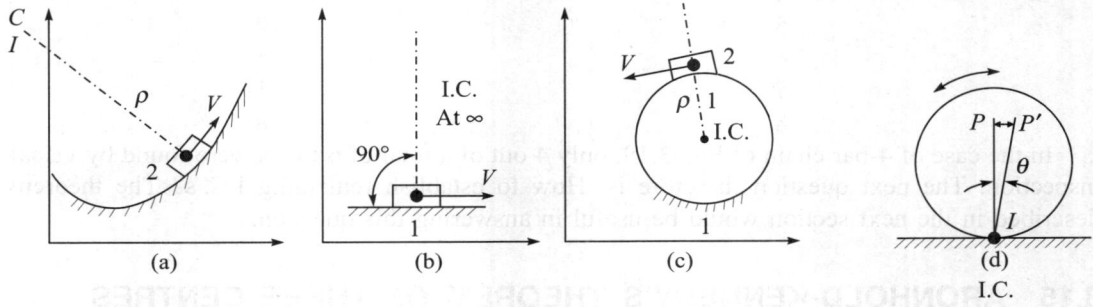

Fig. 3.18

3.14 INSTANT CENTRES OF 4-BAR MECHANISM

In the case of mechanisms and chains, the pin joints (hinges) become obvious choice for instant centres. This is because hinges are common to two bodies about which one body can be assumed to have motion of rotation with respect to the other. This is further explained as under.

For studying relative motion of any link A w.r. to B a convenient approach is to assume link B as fixed and link A as free to move. Thus in Fig. 3.19 relative motion of link 2 w.r. to 1 is obtained by assuming link 1 to be fixed. In that case, clearly, link 2 will revolve with pin A as centre. Similarly, assuming 2 to be stationary and link 1 to be free, link 1 can rotate w.r. to 2 about A as centre. Thus pin joint A is common to links 1 and 2. Further link 1 can revolve w.r. to 2 with A as centre. Similarly, link 2 can revolve w.r. to link 1 with A as centre. Hence pin joint A becomes I.C. for links 1 and 2. Arguing in a similar way, it can be shown that B is I.C. for links 2 and 3 and C is an I.C. for links 3 and 4. Finally pin D is an I.C. for links 1 and 4.

Notation. The most convenient way to represent an instantaneous centre is to denote it by a pair of digits that indicates corresponding links. Thus, in Fig. 3.19 the I.C.'s will be denoted as under:

Pin joint	Notation for I.C.
A	12 or 21
B	23 or 32
C	34 or 43
D	41 or 14

Fig. 3.19

Total Number of I.C.'s. Since we were able to identify four I.C.'s of a 4-bar chain, it is quite obvious to ask the question: How many I.C.'s can be reasonably expected in the above case? To answer this question, we realise that an instantaneous centre exists for a pair of links. Hence, there can be as many I.C.'s as are the possible number of pairs of links. Hence, total number of I.C.'s in chain of n links is–

$$\text{No. of I.C.'s} = {}^nC_2 = \frac{n(n-1)}{2} \qquad (3.28)$$

Thus with $n = 4, 5, 6, 7, 8$, total number of I.C.'s are:

No. of links	Total No. of I.C.'s
4	6
5	10
6	15
7	21
8	28

In the case of 4-bar chain of Fig. 3.19, only 4 out of a total of 6 I.C.'s were found by visual inspection. The next question therefore is: How to establish remaining I.C.'s? The theorem described in the next section would be useful in answering this question.

3.15 ARONHOLD-KENNEDY'S THEOREM OF THREE CENTRES

This theorem, which is sometimes called Kennedy's Rule, states that *three bodies, having relative motion with respect to one another, have three instantaneous centres, all of which lie on the same straight line.* When extended to chains and mechanisms, the above statement implies that with every combination of 3 links, there are 3 I.C.'s and, if two of them are known, the third one will lie on line joining them. In effect, the above theorem suggests us to identify combinations involving 3 links in such a way that unknown I.C. is common to both the combinations and the remaining 4 I.C.'s are known. For instance, in Fig. 3.19, if one selects combination of links (1,2,3) and (1,4,3) then I.C.'s 12, 23, 14, 34 are known but unknown I.C. 13 is common to both the groups. Hence, intersection of lines joining 12 – 23 and 14 – 34 will yield unknown I.C. 13.

Proof. Consider three links 1, 2 and 3 to have relative motion with respect to one another. Let link 1 be fixed and links 2 and 3 be pivoted at points O and C respectively. This is shown in Fig. 3.20. Pivots O and C obviously become I.C.'s 12 and 13. It remains to show that the third I.C. 23 lies on line joining I.C.'s 12 and 13, extended if necessary. This is done in an indirect way by considering situation that would exist when all the three I.C.'s do not lie on the same straight line.

Let us assume that the third I.C. lies at the actual point of contact P. Now if P has to be an I.C., the velocity of P as a point on link 2 and as a point on link 3 must be equal. Thus, velocity of P as a point on link 2,

$$\mathbf{V}p_2 = (OP)\omega_2$$

which is perpendicular to OP.

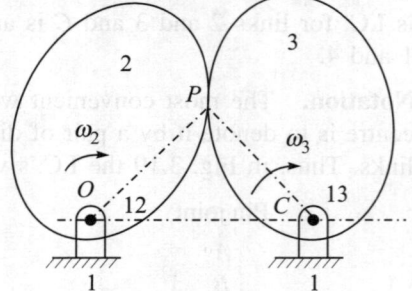

Fig. 3.20

Similarly, velocity of P as a point on body 3 is,

$$\mathbf{V}_{p_3} = (CP) \times \omega_3$$

which is perpendicular to CP.

As vectors $\mathbf{V}_{p_2}$ and $\mathbf{V}_{p_3}$ are not parallel, they cannot be equal. And this is true for any position of P which does not lie on line OC. In the given case, where ω_2 and ω_3 are acting in same sense, the I.C. will lie on OC extended. But when ω_2 and ω_3 are in opposite sense, any point P on line OC will produce same direction for $\mathbf{V}_{p_2}$ and $\mathbf{V}_{p_3}$. It proves, therefore, that the third I.C. 23 must lie on the line joining the other two I.C.'s.

3.16 LOCATING I.Cs. IN MECHANISMS

Figures 3.21 (a) and (b) show a 4-bar mechanism and a slider crank mechanism with links numbered as 1, 2, 3 and 4. As we know, pin/hinge axes give I.C.'s by inspection. Pin joints between links have been marked as 12, 23, 34 and 14. In the case of slider crank mechanism of Fig. 3.21(b), the instantaneous centre 14 is shown at infinity in a direction perpendicular to line of stroke.

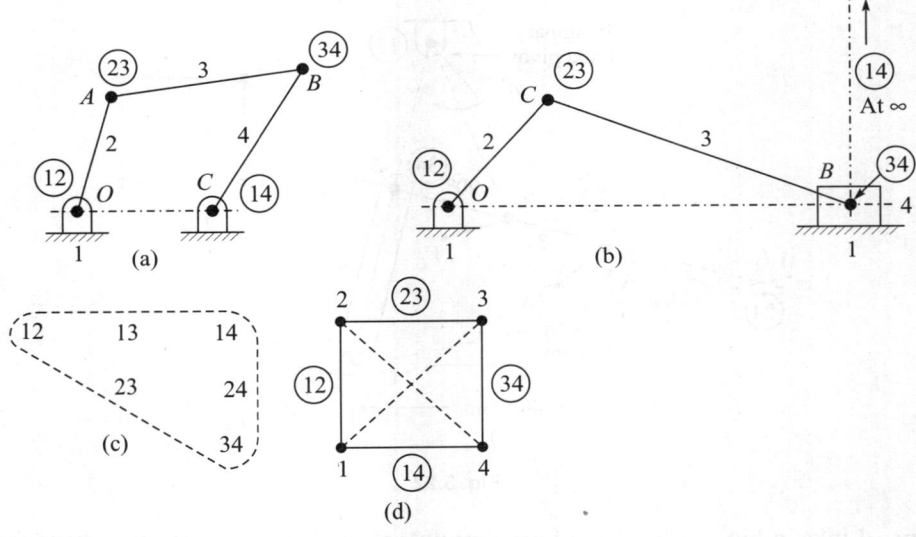

Fig. 3.21

For identifying the remaining I.C.'s one may opt for some sort of book keeping system. Such a system is particularly needed when number of links exceeds four. Thus the I.C.'s may be tabulated [Fig. 3.21 (c)] in the form of upper triangular square matrix, by considering possible combination of each link with every other link of higher index. For instance, after listing combinations of link 1 with links 2 to 4 (*e.g.* 12, 13, 14), combinations for links 2 will be 23, 24. This means that combinations (like 11, 22, 33, ...) of a link with itself are to be avoided. Further, as I.C. 12 was listed, the I.C. 21 (which is same as 12) is to be omitted from second row. This applies to row 3 and 4 also. The I.C.'s may be checked off as they are located. However, the polygon representation of I.Cs. shown in Fig. 3.21 (d) is much more useful. In this representation each vertex (dot) represents a link and line joining any pair of vertices represents corresponding I.C. The process begins by arranging dots in an ordered sequence (mainly for the sake of neatness) and the I.C.'s located by inspection are indicated by full lines connecting two dots, denoting corresponding links. The I.C.'s which have to be located in the process can be denoted by dotted lines.

In the light of Kennedy's theorem, polygon method of identifying unknown I.C.'s becomes more aggressive than tabular array method. Any dotted line that can become a common side of two triangles (whose other sides are full lines) can be used to locate unknown I.C. using theorem of three centres. Following examples illustrate the method.

EXAMPLE 3.7 Figure 3.22 shows a machine linkage in part where O and C are pivots and G is the centre of gravity of AB. The crank OA rotates uniformly at 30 rad/s clockwise. $OA =$ 10 cm; $AB = 28$ cm; $BC = 24$ cm; $CE = 12$ cm; $BG = 16$ cm. For the position shown, find velocity of G and E and indicate their direction. Also find the angular velocities of the link AB and the bell-crank lever BCE. Find the kinetic energy of bell-crank lever if its moment of inertia about an axis through C perpendicular to its plane of motion is 2.1 kg·cm².

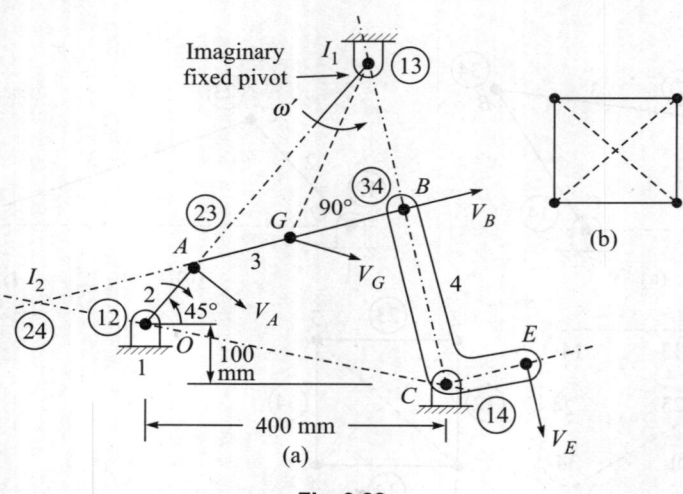

Fig. 3.22

Solution: Links in Fig. 3.22(a) have four pin joints at O, A, B and C and they have four I.C.'s 12, 23, 34 and 14 respectively, which can be located easily by visual inspection. These are shown by four full lines of polygon in Fig. 3.22(b). The unconnected diagonals 13 and 24 denote unknown I.C.'s. To locate the I.C.: 13, shown by dotted line, one needs to consider two triangles 123 and 143 in Fig. 3.22(b), to which line 13 can be a common side. Thus, point of intersection of lines joining 12 with 23 and 14 with 34 gives I.C. 13, which will be called as I_1.

Similarly, in locating I.C.: 24, shown by dotted line in Fig. 3.22(b), one needs to consider triangles 234 and 412 to which line 24 is a common side. Point of intersection of lines joining 12 with 14 and 23 with 34 gives I.C. 24, which will be referred to as I_2. Then,

$$V_A = 10 \times \omega_2 = 10 \times 30 = 300 \text{ cm/s}$$

As link 1 is frame, instant centre 13 can be considered to be a fixed pivot with link AB swinging about it. Hence, the angular velocity ω', with which link oscillates about I.C. 13, is also the angular speed of rotation of link 3. Thus,

$$\omega' = \frac{V_A}{I_1 A} = \frac{300}{I_1 A}$$

By measurement, from Fig. 3.22(a), $I_1 A = 4.38 \times 8 = 35.04$ cm

$$I_1 B = 2.3 \times 8 = 18.4 \text{ cm}; \quad I_1 G = 3.2 \times 8 = 25.6 \text{ cm}$$

And, as A, B and G lie on the same link 3,

$$\frac{V_A}{I_1 A} = \frac{V_B}{I_1 B} = \frac{V_G}{I_1 G} = \omega'$$

$\therefore$

$$\omega' = \frac{300}{35.04} = 8.56 \text{ rad/s (c.c.w.)}$$

$\therefore$

$$V_B = I_1 B (\omega') = 18.4 \times 8.56 = 157.5 \text{ cm/s}$$

$$V_G = I_1 G (\omega') = 25.6 (8.56) = 219.14 \text{ cm/s}$$

Also

$$\omega_4 = \frac{V_B}{CB} = \frac{157.5}{24} = 6.56 \text{ rad/s (c.w.)}$$

$\therefore$

$$V_E = (CE) \times \omega_4 = 12 \times 6.56 = 78.72 \text{ cm/s}$$

K.E. of angular motion of bell-crank lever

$$= \frac{1}{2} I \omega_4^2 = \frac{1}{2}(2.1)(6.56)^2 = 45.19 \text{ N·cm} \qquad \textbf{Ans.}$$

EXAMPLE 3.8 Crank of a slider-crank mechanism is 48 cm long and rotates at 190 r.p.m. in c.c.w. direction. The connecting rod is 144 cm long. When the crank has turned through 50 degrees from the inner dead centre position, determine, using instantaneous centre method, the velocity of point E located at distance of 45 cm on the connecting rod extended. Also find point F on the connecting rod having the least absolute velocity and angular velocity of the connecting rod.

Solution: It may be noted that 14 is located at infinity along a line perpendicular to line of stroke from P. However, the distance OP is too small in comparison with infinity and as such the distance OP can be neglected in comparison to infinity for all practical purposes. Thus a line drawn from O perpendicular to line of stroke can be assumed to locate I.C. 14 at infinity. Hence point of intersection of line joining 23 with 34 and the line joining 12 with 14 along OK locates I.C. 24 at I_2. Similarly line joining I.C.'s 12 and 23 intersects the line joining I.C.'s 34 and 14 at 13 (= I_1).

Any of the instant centres 13 or 24 can be used to find velocity of the slider, but only I.C. 13 can be used to find the velocity of any point on link AP. First we will see as to how I.C. 24 can help in finding V_P.

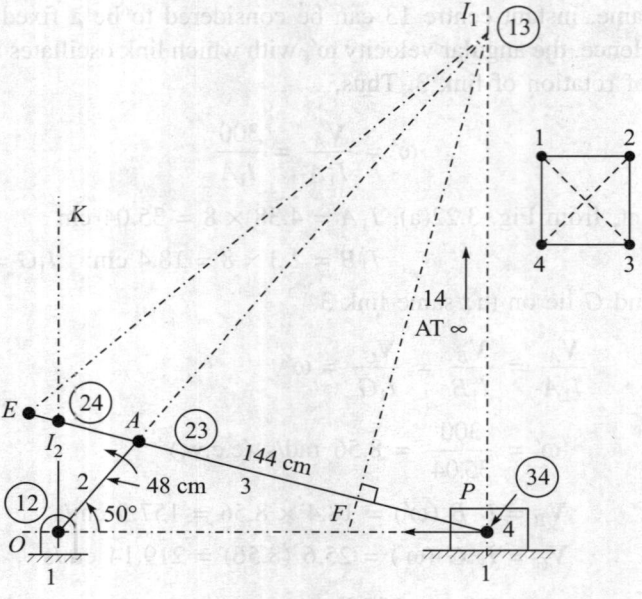

Fig. 3.23

I.C. 24 is a point common to links 2 and 4 (extended if necessary) by definition. Thus point 24 can be considered to be a point lying on and moving with either of the links 2 and 4. Thus, as a point on link 2, point 24 can be assumed to be rotating about O with ω_2. Thus, $V_{24} = \omega_2$ (OI_2). Again as a point on slider 4, point 24 can be assumed to be moving with velocity V_P parallel to the line of stroke. Naturally, 24 being a point on both links 2 and 4,

$$(OI_2)\omega_2 = V_P$$

Thus in the given problem as $OI_2 = 48$ cm

Therefore $$V_P = 48 \times \frac{2\pi \times 190}{60} = 955.0 \text{ cm/s}$$

As link 2 is revolving in c.c.w. direction, a point 24 on link 2 will have the direction of V_{24} to the left. Hence V_P also acts to the left.

The advantage of locating I.C. 24 is that it gives velocity of slider or any other point on link 4 (if it is a link other than a slider) in the fastest and most convenient way. However, if velocity of any point on coupler/connecting rod 3 is required, I.C. 24 is hardly of any use.

As against this, I.C. 13 in such problems can give velocity of any point on coupler/connecting-rod and this includes points like E, F and P. Hence, I.C. 13 becomes more useful in a majority of problems. *However, as a piece of general advice, one must ascertain beforehand as to which I.C. is actually required. Lot of time can be saved either in examination hall or even otherwise, if one decides to determine only those I.C.'s which are actually required.*

The instant centre 13, being a point common to frame link 1 and coupler 3, can be treated as a fixed pivot about which link 3 is instantaneously oscillating. If ω' be the angular velocity of link 3 about I_1, linear absolute velocity of any point on link 3 will be proportional to the distance of that point from I_1. Clearly, point of least velocity on link 3 will be the foot of perpendicular F dropped from I_1 on AP.

Now, from measurements from Fig. 3.23,

$$I_1 A = 220.8 \text{ cm}; \quad I_1 E = 266.4 \text{ cm}$$
$$I_1 F = 201.6 \text{ cm}; \quad I_1 P = 208.8 \text{ cm}$$

Then,
$$\omega' = \frac{V_A}{I_1 A} = \left(48 \times \frac{2\pi \times 190}{60} \right) \times \frac{1}{220.8}$$

$$= 4.32 \text{ rad/s (clock-wise)}$$

Hence,
$$V_E = (I_1 E) \times \omega' = 266.4 \times 4.32 = 1150.85 \text{ cm/s}$$
$$V_F = (I_1 F) \times \omega' = 201.6 \times 4.32 = 870.9 \text{ cm/s}$$
$$V_P = (I_1 P) \times \omega' = 208.8 \times 4.32 = 902.0 \text{ cm/s in the direction shown.}$$

3.17 ACCELERATION IN MECHANISMS

In Section 3.2, instantaneous velocity was defined as the time rate of change of position vector of a point, *i.e.*, $V = d\mathbf{R}/dt$. In a similar way we shall define *instantaneous acceleration* (or simply acceleration) as the time rate of change of velocity, and shall be represented by equation,

$$A = \left(\frac{dV}{dt} \right) = \frac{d}{dt} \left(\frac{dR}{dt} \right) \tag{3.29}$$

or
$$A = \ddot{R}$$

In a similar way, angular acceleration of a rotating body is defined by the equation

$$\alpha = \lim_{\Delta t \to 0} \left(\frac{\Delta \omega}{\Delta t} \right) = \frac{d\omega}{dt}$$

or
$$\alpha = \frac{d}{dt} \left(\frac{d\theta}{dt} \right) = \ddot{\theta} \tag{3.30}$$

When direction of linear acceleration is implied in the problem (this is particularly so in rectilinear motion), it may be sufficient to deal with only the scalar quantities. Thus, linear acceleration,

$$a = \lim_{\Delta t \to 0} \left(\frac{\Delta V}{\Delta t} \right) = \frac{dV}{dt}$$

or
$$a = \frac{d}{dt} \left(\frac{dr}{dt} \right) = \ddot{r}$$

3.18 MOTION OF A PARTICLE ALONG CURVED PATH

As a particle P moves along a curved path, its position with respect to origin changes. This change involves change in magnitude as well as in direction as shown in Fig. 3.24. However, if the curvature of path does not change abruptly, the position vector originating at the centre of curvature can be associated with velocity and acceleration, due to change in position of P. Thus

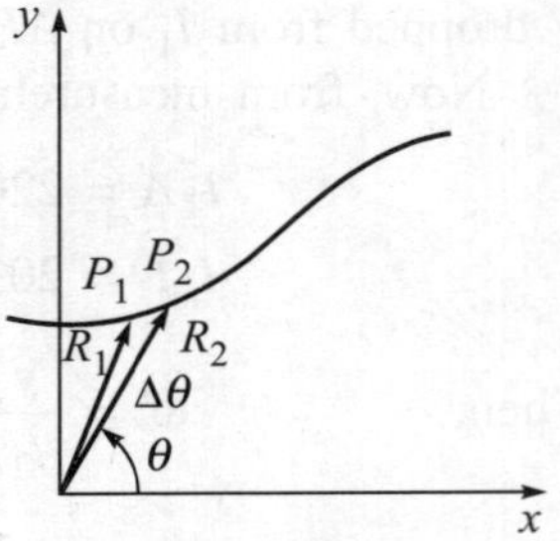

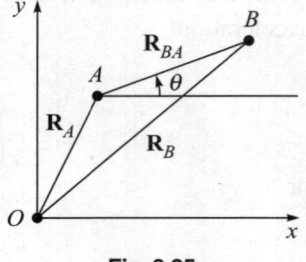

Fig. 3.24

$$\mathbf{R} = re^{j\theta}, \text{ where } r \text{ is constant.}$$

Therefore

$$\mathbf{V} = \frac{dR}{dt} = r(j\dot{\theta})\, e^{j\theta} = j(r\dot{\theta})\, e^{j\theta} \qquad (3.31)$$

and

$$\mathbf{A} = \frac{dV}{dt} = jr\,(\ddot{\theta})\, e^{j\theta} + j(r\dot{\theta})\,(j\dot{\theta})\, e^{j\theta} = [j(r\ddot{\theta}) - r(\dot{\theta})^2]e^{j\theta} \qquad (3.32)$$

It follows from equation (3.31) that magnitude of velocity is $(r\dot{\theta})$ *i.e.*, $(r\omega)$ and the direction is obtained from $(je^{j\theta})$ which denotes a direction obtained by rotating the position vector $\mathbf{R}$ through right angles. Here $\dot{\theta} \equiv \omega$ is considered a scalar and depending on its sign (positive or negative), the vector $\mathbf{R}$ will be rotated c.w. or c.c.w.

Equation (3.32) suggests that the total acceleration $\mathbf{A}$ consists of two components. The component $j(r\ddot{\theta})\, e^{j\theta}$ acts in a direction perpendicular to position vector $\mathbf{R}$ but the sense of rotation through right angle will depend on sign of angular acceleration $\ddot{\theta} = \alpha$. This component acts in a direction perpendicular (transverse or tangential) to position vector and is called transverse or tangential component of acceleration. The other component $(r\dot{\theta}^2)\, e^{j\theta}$ has a magnitude of $r(\dot{\theta})^2 = r\,\omega^2$ and its sense is 180° out of phase (as $j^2 = -1$) with position vector. In other words, this component is directed towards the centre of curvature and is therefore called radial (more specifically, centripetal) or normal component of acceleration. Thus,

$$\mathbf{A} = \mathbf{A}^t + \mathbf{A}^r \qquad (3.33)$$

where,

$$\mathbf{A}^r = r\ddot{\theta} = r\alpha$$

and

$$\mathbf{A}^r = (-r\omega^2)$$

3.19 ACCELERATION OF A RIGID LINK

Let the line AB in Fig. 3.25 represent a rigid body in motion. Let points A and B be located in co-ordinate frame by position vectors $\mathbf{R}_A$ and $\mathbf{R}_B$. Then, the relative position vector $\mathbf{R}_{BA}$ represents position of link AB. This is given by equation,

$$\mathbf{R}_{BA} = (r_{BA})\, e^{j\theta} \qquad (3.34)$$

AB being a rigid link r_{BA} is constant, differentiation of the above equation gives velocity of B w.r. to A. Thus

Fig. 3.25

$$\mathbf{V}_{BA} = \frac{d}{dt} \, (\mathbf{R}_{BA}) = r_{BA} \, (j\dot{\theta}) \, e^{j\theta}$$

$$\mathbf{V}_{BA} = j(r_{BA} \, \dot{\theta}) \, e^{j\theta} \qquad\qquad (3.35)$$

The above expression shows that magnitude of velocity of B w.r. to A, $\mathbf{V}_{BA} = (r_{BA}\dot{\theta})$ and, its direction is given by $j\,(e^{j\theta})$ which is at right angles to relative position vector $\mathbf{R}_{BA}$ in the sense of $\dot{\theta}$.

Differentiation of equation (3.35) again w.r. to time gives acceleration of B w.r. to A.

Thus,
$$\mathbf{A}_{BA} = \frac{d}{dt} (\mathbf{V}_{BA})$$

$$= r_{BA} \, [j\ddot{\theta} + (j\dot{\theta}) \, (j\dot{\theta})] \; e^{j\theta}$$

$$\therefore \qquad\qquad \mathbf{A}_{BA} = r_{BA} \, [j(\alpha_{BA}) \, + \, (j)^2 \, (\omega_{BA})^2] e^{j\theta} \qquad\qquad (3.36)$$

where, $\ddot{\theta} = \alpha_{BA}$ is the angular acceleration of B w.r. to A and

$\dot{\theta} = \omega_{BA}$ = angular velocity of B w.r. to A.

As in the case of acceleration of a particle, even in the case of rigid body we find that acceleration of a point B w.r. to A has two components in mutually perpendicular directions.

Thus,
$$\mathbf{A}_{BA} = \mathbf{A}^t_{BA} + \mathbf{A}^r_{BA} \qquad\qquad (3.37)$$

where
$$|\mathbf{A}^t_{BA}| = (r_{BA})\,\ddot{\theta}_{BA}$$

and,
$$|\mathbf{A}^r_{BA}| = -\, r_{BA}\,\omega^2_{BA} = \text{centripetal acceleration.}$$

Relative Acceleration : *Concept*

Key to acceleration analysis for a lower paired mechanism of single degree of freedom lies in the acceleration analysis of 'motion-transfer link' (*e.g.* connecting rod of a crank-slider mechanism). Since a motion transfer link (3) is pin connected to input link (2), the point A on pin-connection has same velocity and acceleration, when it is considered a point on link (2) and also when it is considered a point on link (3). Thus the problem of acceleration analysis reduces to finding out acceleration of a point B on pin-connection between link 3 and output link, when velocity and acceleration of point A are given.

By definition of a rigid link, distance between any two points like A and B on the motion transfer link can neither increase nor decrease. Hence relative motion between points A and B cannot have any component of motion along the line of centres $A - B$. In other words, relative motion of point B with respect to point A can be that of rotation with A as centre.

Consider a motion-transfer link AB and assume that position of point A along with its absolute velocity and acceleration is known. *Acceleration of point B can then be established by assuming link AB to be instantaneously rotating about A. Such rotational motion of link AB about A as the centre involves two components of acceleration of B relative to A. These components are:*

(a) Centripetal (radial) component of acceleration of B relative to A, given by

$$\mathbf{A}^r_{BA} = (\mathbf{V}^2_{BA}/AB)$$
$$= (AB) \, (\omega^2_{BA}), \text{ and}$$

(b) Transverse or tangential component of acceleration of B with respect to A, given by

$$\mathbf{A}^t_{BA} = (AB)\alpha_{BA}$$

where α_{BA} is the angular acceleration of AB and ω_{BA} is the angular velocity of link AB. Readers are advised to compare these results with those obtained earlier in this section.

3.20 ACCELERATION IMAGE OF A LINK

Figure 3.26 shows a link PQ having an angular velocity ω and angular acceleration α both acting c.c.w on the link. Replacing BA by QP in equation (3.37), the acceleration equation for link PQ can be expressed as,

$$\mathbf{A}_{QP} = \mathbf{A}^r_{QP} + \mathbf{A}^t_{QP} \qquad (3.38)$$

where $\quad a^r_{QP} = |\mathbf{A}^r_{QP}| = (PQ)\omega^2$,

is represented by vector pq'

and $\quad a^t_{QP} = |\mathbf{A}^t_{QP}| = (PQ)\alpha$,

is represented by vector $q'q$

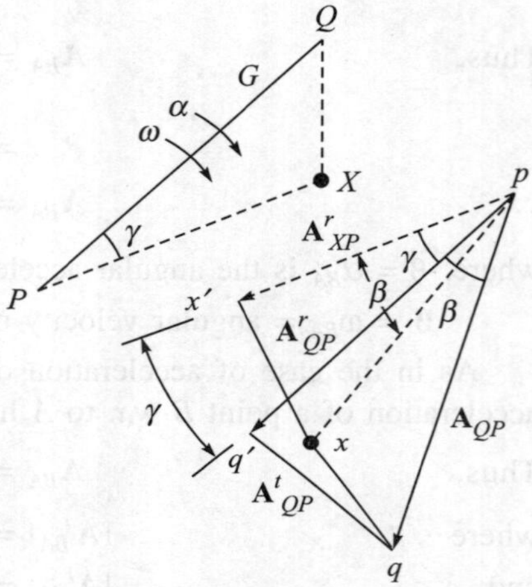

Since a^r_{QP} and a^t_{QP} are mutually at right angles, their resultant vector $a_{QP} \equiv pq$ (called total acceleration of Q w.r. to P) is

$$a_{QP} = \sqrt{(a^r_{QP})^2 + (a^t_{QP})^2}$$

Fig. 3.26

Therefore substituting for a^r_{QP} and a^t_{QP}, $a_{QP} = PQ\sqrt{\omega^4 + \alpha^2}$ $\qquad (3.39)$

Also, the vector pq, in acceleration polygon is inclined to the direction of link PQ (represented by vector $\mathbf{A}^r_{QP} \equiv pq'$) at angle β given by

$$\tan\beta = \frac{a^t_{QP}}{a^r_{QP}} = \frac{(PQ)\alpha}{(PQ)\omega^2} = \left(\frac{\alpha}{\omega^2}\right)$$

Hence, $\qquad \angle qpq' \equiv \beta = \tan^{-1}(\alpha/\omega^2) \qquad (3.40)$

The line pq in acceleration polygon is then called the acceleration image of link PQ. This is because, any point G on link PQ will have tangential and radial component of acceleration w.r. to P given by

$$a^r_{GP} = (PG)\omega^2 \quad \text{and} \quad a^t_{GP} = (PG)\alpha \qquad (3.41)$$

and, magnitude of total acceleration is given by

$$a_{GP} = \sqrt{(a^r_{GP})^2 + (a^t_{GP})^2} = PG\left(\sqrt{\omega^4 + \alpha^2}\right) \qquad (3.42)$$

It follows from equations (3.39) and (3.42) that,

$$a_{GP} : a_{QP} = (PG) : (PQ) \qquad (3.43)$$

Again, from equation (3.41),

$$\frac{(a^t_{GP})}{(a^r_{GP})} = \left\{ \frac{(PG)\alpha}{(PG)\omega^2} \right\} = \left(\frac{\alpha}{\omega^2} \right)$$

Thus, inclination of lines pg and pq with link PQ is same and is equal to $\tan^{-1}(\alpha/\omega)^2$. Hence we conclude from this and equation (3.43) that acceleration image g of point G divides line pq in proportion same as G divides line PQ. Mathematically,

$$pg : pq = PG : PQ$$

Any other point X (not lying on the line joining points P and Q) will have acceleration components w.r. to P as

$$a^r_{XP} = (PX)\omega^2 \quad \text{and} \quad a^t_{XP} = (PX)\alpha \qquad (3.44)$$

Vector px, representing total acceleration of X w.r. to P, will not be along line pq. Again from equation (3.44), angle of inclination of total acceleration A_{XP} with side PX

$$\beta = \tan^{-1}\left(\frac{\alpha}{\omega^2} \right) \qquad (3.45)$$

Comparing it with equation (3.40), it follows that total acceleration vector px is inclined to side PX (and hence to side px') at the same angle between pq' and px' is the same angle β. Further, as pq' is parallel to PQ and px' is parallel to PX, angle between pq' and px' is

$$\angle q'px' = \angle QPX = \gamma$$

Hence,
$$\angle q'px = \beta - \gamma$$

and, therefore,
$$\angle xpq = \beta - (\beta - \gamma) = \gamma$$

Next join xq. Using equation (3.44) from triangles pxq and PXQ of Fig. 3.26, we have

$$px = PX\sqrt{\omega^4 + \alpha^2}$$

and
$$pq = PQ\sqrt{\omega^4 + \alpha^2}$$

Therefore,
$$(px/pq) = (PX/PQ)$$

Further,
$$\angle QPX = \angle qpx = \gamma$$

Hence, triangles XPQ and xpq are similar. Therefore, x is the acceleration image of point X and can be obtained by constructing on line pq a triangle pqx similar to PQX. Then vector px represents total acceleration of X w.r. to P and qx represents total acceleration of X w.r. to Q.

3.21 GENERAL ACCELERATION EQUATION FOR A LINK

In acceleration analysis of simple mechanisms, three distinct cases can arise.

Case (A). A link may rotate about a fixed pivot and it may be required to find out acceleration of some fixed point, say a hinge pin, on it. This is a simple case and a student is very well exposed to such cases [see Fig. 3.27(a)].

We know that the hinge pin P will rotate in an arc of a circle with radius OP; O being the fixed pivot. The pin therefore has two acceleration components:

the radially inward acceleration component called centripetal acceleration

$$A^r_P = (OP)\omega^2 \tag{3.46}$$

and the tangential or transverse acceleration component, which is dependent on angular acceleration α of the link

$$A^t_P = (OP)\alpha \tag{3.47}$$

The two components are mutually perpendicular and can be compounded to give total acceleration A_P of the hinge pin.

Case (B). A link in question can be a floating link e.g., a coupler link PQ in 4-bar mechanism or a connecting rod in a slider-crank mechanism. It may be required to find out the acceleration of a point Q (which can be a hinge pin) when velocity and acceleration of point P (say, an another hinge pin) is known completely. This is a case commonly encountered in the analysis of simple mechanisms, and is illustrated in Fig. 3.27(b). When P and Q are hinge pins, they represent coincident points on connected links OP an CQ respectively. In such a case, velocity and acceleration of P, as a point on link PQ, is same as the velocity and acceleration of point P on link OP. In a similar way, velocity and acceleration of Q, as a point of link PQ, is same as the velocity and acceleration of point Q on output link CQ.

Since PQ is a rigid link, it follows from velocity equation 3.21 that,

$$\mathbf{V}_{QP} = \mathbf{V}_Q - \mathbf{V}_P \quad \text{or} \quad \mathbf{V}_Q = \mathbf{V}_P + \mathbf{V}_{QP}$$

Differentiating further w.r. to time, we have the acceleration equation as

$$\mathbf{A}_Q = \mathbf{A}_p + \mathbf{A}_{QP} \tag{3.48}$$

Unlike velocity, which primarily depends on displacement, relative acceleration between two fixed points P and Q on a rigid link can exist in radial (centripetal) as well as tangential (transverse) directions. This is because acceleration is a force dependent quantity.

In section 3.19, we have seen that the relative motion of Q w.r. to P is that of rotation about P. The centripetal (radial) and tangential components of acceleration are then given by

$$\mathbf{A}^r_{QP} = (PQ)\,\omega^2_{QP} \quad \text{and} \quad \mathbf{A}^t_{QP} = (PQ)\alpha_{QP}$$

Since points P and Q can also have radial and tangential acceleration components, equation 3.48 reduces to

$$(\mathbf{A}^r_Q + \mathbf{A}^t_Q) = (\mathbf{A}^r_P + \mathbf{A}^t_P) + (\mathbf{A}^r_{QP} + \mathbf{A}^t_{QP}) \tag{3.49}$$

Case (C). A link rotating about a fixed pivot may carry a sliding element through which the link may be connected to yet another link [Fig. 3.27(c)]. As the slider moves along the rotating link, we are required to deal with a case having a link OQ of variable length.

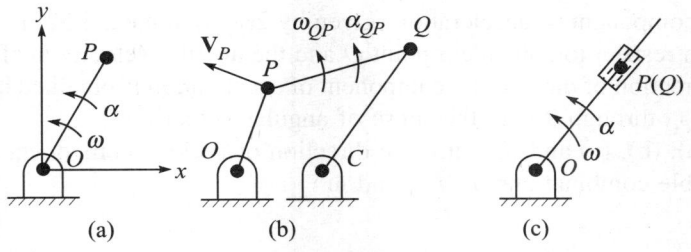

Fig. 3.27

To derive a more generalised expression, consider a rotating link OR pivoted at O and let a slider, carrying a pin joint P, slide along it (Fig. 3.28). Let Q be a point fixed to link OR, which is momentarily in coincidence with P. Thus, point P is sliding along OR w.r. to a fixed point Q on link OR. The point P is simultaneously rotating along with link OR. Thus as Q is fixed to link 2, its motion is completely decided by the motion of link 2. But this is not true for point P, which additionally slides along link 2. Thus the length $r = OQ$ of rotating position vector is variable and this vector is expressed in complex polar notation by

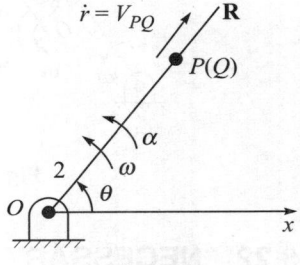

Fig. 3.28

$$\mathbf{R} = r\,e^{j\theta} \tag{3.50}$$

Remembering that r is variable with respect to time, first and second time derivatives give the velocity and total acceleration of sliding point P. Thus, using Newtonian notations,

$$\mathbf{V}_P = \frac{d}{dt}(\mathbf{R}) = [\dot{r} + r\,(j\dot{\theta})]\,e^{j\theta} \tag{3.50(a)}$$

As point Q is fixed to, and rotating with ling OR, $|\mathbf{V}_Q| = r\dot{\theta}$

Hence,
$$\mathbf{V}_P = \mathbf{V}_{PQ} + \mathbf{V}_Q$$

and, from equation [3.50(a)], $A_P = \dfrac{d^2}{dt^2}(\mathbf{R}) = [\ddot{r} + \dot{r}(j\dot{\theta}) + r(j\ddot{\theta})]\,e^{j\theta} + [\dot{r} + r(j\dot{\theta})]\,(j\dot{\theta}) \times e^{j\theta}$

or
$$A_P = [\ddot{r} + 2(\dot{r}\dot{\theta})\,j + j(r\ddot{\theta}) + r(j\dot{\theta})^2]\,e^{j\theta}$$

Separating real and imaginary parts,

$$\mathbf{A}_P = (\ddot{r} - r\dot{\theta}^2)\,e^{j\theta} + j(r\ddot{\theta} + 2\dot{r}\dot{\theta})\,e^{j\theta} \tag{3.51}$$

Noting that $\ddot{r} = a_{PQ}$; $\dot{r} = v_{PQ}$; $\dot{\theta} = \omega$ and $\ddot{\theta} = \alpha$, we have from equation (3.51),

$$A_p = (a_{PQ} - r\omega^2)\,e^{j\theta} + j(r\alpha + 2\,v_{PQ}\omega)\,e^{j\theta} \tag{3.52}$$

As is clear from equation (3.52), the component $(a_{PQ} - r\omega^2)$ acts along the line OR and is therefore called radial component; $r\omega^2$ being the centripetal acceleration of coincident point Q. The component $(r\alpha + 2\,v_{PQ}\omega)$ acts perpendicular to OR, i.e. it represents transverse or tangential component; $r\alpha$ being the tangential component of acceleration of conincident point Q. Equation (3.52) can also be rewritten in terms of acceleration components of Q and of P w.r. to Q. Thus,

$$\mathbf{A}_P = (\mathbf{A}_Q^r + \mathbf{A}_Q^t) + (\mathbf{A}_{PQ}^r + \mathbf{A}^{cor}) \tag{3.53}$$

The coriolis component of acceleration (given by $2(v_{PQ})\omega$ in eq. 3.52) is equal to twice the velocity of P with respect to coincident point Q and the angular velocity ω of the link to which Q is fixed. The direction of the coriolis component of acceleration is obtained by rotating relative velocity vector $\mathbf{V}_{PQ}$ through $90°$ in the sense of angular velocity ω.

Figure 3.29(a), (b), (c) and (d) illustrate direction of coriolis component of acceleration in all the four possible combinations of $\mathbf{V}_{PQ}$ and ω.

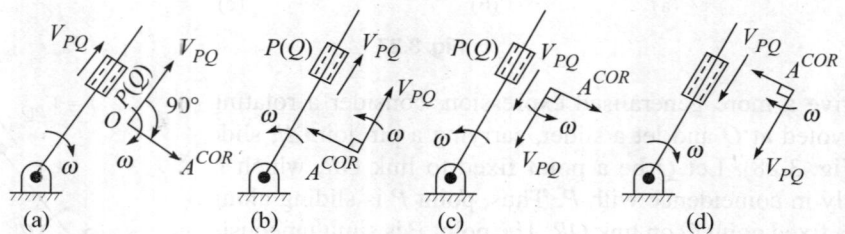

Fig. 3.29 Direction of coriolis component of acceleration.

3.22 NECESSARY CONDITIONS FOR A^{cor} TO EXIST

From the expression of coriolis component of acceleration, it is evident that for $\mathbf{A}^{cor}$ to exist both v_{PQ} and ω must exist. In other words, there must be a rotating link which carries a sliding pair through which it is connected to the other link.

Particular cases. It follows from equation (3.52) that

(a)   If $\omega = 0$, *i.e.* when link 2 is not free to rotate,

$$A_P = (a_{PQ})\, e^{j\theta} + 0$$

which gives acceleration of slider along a fixed link.

(b)   When $\mathbf{V}_{PQ} = a_{PQ} = 0$, *i.e.* when there is no slider on rotating link,

$$A_P = (-r\omega^2)\, e^{j\theta} + (r\alpha)\, e^{j\theta}$$

which gives acceleration of a fixed point P on a rotating link.

(c)   When the rotating link carrying sliding element does not have angular acceleration, i.e. when $\alpha = 0$, the term $(r\alpha)$ representing tangential component of acceleration of point Q will vanish from equation 3.52.

3.23 ACCELERATION POLYGON

Acceleration polygon is a graphical display of acceleration equation given in (3.49) or (3.53) as the case may be. Drawing acceleration polygon amounts to solving acceleration equation graphically. It thus provides a very useful means for solving problems on acceleration analysis of simple mechanisms. Of the two, equations (3.49) is simpler and is quite convenient for explaining construction procedure of acceleration polygon.

Contrary to the general apprehension among students, drawing acceleration polygon is not at all difficult provided they are systematic in their approach to solution of problems and remember following important rules.

Rules. 1. A common rule in drawing velocity and acceleration polygon is: as far as possible absolute velocity/acceleration vectors are to be drawn from velocity/acceleration pole, as the case may be.

2. Completely known vectors must be drawn first, irrespective of whether they are on left or right hand side of the equal to sign of velocity/acceleration equation.

3. Vector addition and subtraction follow commutative law. It is thus immaterial as to what sequence of vectors is adopted for vector addition and subtraction. However, in order to be able to make use of concept of acceleration image, it is desirable to arrange radial and tangential components of a given acceleration vector in sequence, one after the other.

4. It is very important to remember at this stage the rules of vector addition and vector subtraction. For vector addition, vectors A and B must be so placed that tail of one vector touches tip of arrowhead of the other. For vector subtraction however, vectors A and B must be so arranged that either the tips of arrowhead of both touch at a point or the tails of the two vectors touch at a common point.

Procedure to Draw Acceleration Polygon

The procedure to draw acceleration polygon can be summarised in the following steps:

1. Draw mechanism to some scale in the given configuration.

2. Write down velocity equation to connect known velocity of one point P with unknown velocity of second point Q on the same link through their relative velocity. Keep unknown velocity vector V_Q on the left hand side of the equal to sign.

3. Select suitable location for velocity pole O_v. Choose suitable scale. From O_v draw a line in a direction parallel to the direction of the known absolute velocity. From the tip of this vector, draw a line parallel to the direction of (partially known) relative velocity V_{QP} on R.H.S. Similarly from O_v draw a line parallel to the line of action of (partially known) absolute velocity vector V_Q on the L.H.S. The two lines intersect at a point completing the polygon.

4. Measure the magnitudes of velocity vectors which were known partially. Fix their directions of arrow heads so as to satisfy velocity equation.

5. Write down acceleration equation to connect unknown acceleration A_Q of one point with known acceleration A_P of the other point on the same link. Write down all acceleration quantities in terms of their radial and tangential components.

6. Using information of velocities from velocity polygon, identify acceleration components that are partially or completely known. Usually centripetal (radial) components of acceleration, which depend on velocity, are known. But tangential components of acceleration ($= R\alpha$) which depend on angular acceleration of a link, are not known in magnitude, but are known in direction.

7. Select location of acceleration pole O_a suitably, and choose a convenient scale for acceleration polygon. Starting from O_a plot all 'completely known' absolute acceleration, vectors. From the tip of these vectors, draw other 'completely known' vectors one after

the other, in such a way as to satisfy acceleration equation. When all the known (known in magnitude and direction) vectors are exhausted on R.H.S., draw lines parallel to the direction of the remaining (partially known) acceleration components. The two lines meet at a point defining solution. Assign arrowheads on each acceleration vector so as to ensure vector addition and subtraction as per acceleration equation. With this, construction of acceleration polygon is complete. Using the scale of acceleration diagram already chosen, unknown accelerations can be measured from the corresponding side of the polygon.

EXAMPLE 3.9 The crank of a steam engine mechanism is 25 cm long and the connecting rod is 100 cm long. At the given instant, the crank has turned through an angle of 210 degrees clockwise from inner dead centre. Assuming the crank to rotate at 180 r.p.m., find the velocity and acceleration of piston and angular velocity and angular acceleration of connecting rod when (a) crank rotates uniformly, and (b) when the crank is subject to an angular acceleration of 100 rad/s^2 in a direction opposite to that of rotation.

Solution: Angular speed ω of crank OC, is

$$\omega = \frac{2\pi \times 180}{60} = 18.85 \text{ rad/s}$$

The tangential velocity of crank pin C,

$$V_C = OC \times \omega = 25 \times 18.85 = 471.2 \text{ cm/s}$$

Velocity equation is:

$$V_P = V_C + V_{PC}$$

where, V_C is known completely, its magnitude is 471.2 cm/s and the direction is tangent to crank circle as shown.

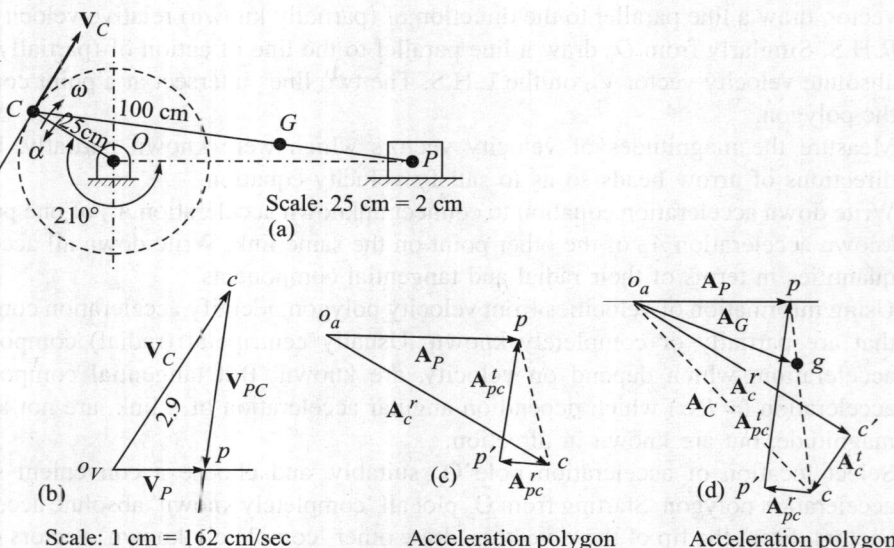

Scale: 25 cm = 2 cm
(a)

Scale: 1 cm = 162 cm/sec
(b)

(c) Acceleration polygon

(d) Acceleration polygon

Fig. 3.30

Velocity $\mathbf{V}_C$ and $\mathbf{V}_P$ are absolute velocity vectors and are therefore drawn from velocity pole o_V. Velocity vector $\mathbf{V}_C$ is known completely and is plotted to scale 162 cm/s = 1 cm in a direction normal to OC and is represented by $o_v C$. Since velocity of P w.r. to C is perpendicular to PC and, as per velocity equation, is to be added to vector $\mathbf{V}_C$, a line is drawn from point c perpendicular to direction PC. Since magnitudes of $\mathbf{V}_P$ and $\mathbf{V}_{pc}$ are not known, intersection of lines parallel to direction of $\mathbf{V}_{PC}$ and $\mathbf{V}_P$ define point p. Note that $\mathbf{V}_P$ can exist only along line of stroke. Since line cp was drawn parallel to $\mathbf{V}_{PC}$ and is additive to vector $\mathbf{V}_C$ (see velocity equation), the arrowhead on line cp is from c to p as shown. Similarly, as velocity $\mathbf{V}_P$ is subtractive from $\mathbf{V}_C$ (see again velocity equation), the arrowhead on line $o_v p$ should indicate direction o_v to p as shown. Thus $o_v p$ gives $\mathbf{V}_P$ in magnitude and direction, while cp gives velocity $\mathbf{V}_{PC}$ in magnitude and direction.

By measurement from velocity polygon,

$$\mathbf{V}_P = 195 \text{ cm/s} \quad \text{and} \quad \mathbf{V}_{PC} = 405 \text{ cm/s}$$

Angular speed ω_{PC} is obtained by assuming link PC to be rotating about C. Then ω_{PC} should act in a way consistent with direction of $\mathbf{V}_{PC}$. Thus

$$\omega_{PC} = \frac{|\mathbf{V}_{PC}|}{PC} = \frac{405}{100} = 4.05 \text{ rad/s (c.w.)}$$

Now, the acceleration equation is,

$$\mathbf{A}_P = \mathbf{A}_C + \mathbf{A}_{PC}$$

Each of the acceleration component can be considered to consist of radial and transverse component. Above equation can then be written as,

$$\mathbf{A}_P = (\mathbf{A}^r_C + \mathbf{A}^t_C) + (\mathbf{A}^r_{PC} + \mathbf{A}^t_{PC})$$

Note that as slider reciprocates, it has only a tangential acceleration component A_P along the line of stroke. The three completely known acceleration components are:

$\mathbf{A}^r_C$ = centripetal component = $(OC)\ \omega^2 = (25)\ (18.85)^2 = 8883 \text{ cm/s}^2$

$\mathbf{A}^t_C$ = tangential component of acceleration will exist only if crank OC has some angular acceleration = $(OC)\ \alpha$

Thus, $\mathbf{A}^t_C = 0$, in case (a) where crank rotates uniformly

and $\mathbf{A}^t_C = (OC)\alpha = 25 \times 100 = 2500 \text{ cm/s}^2$ in a direction opposite to $\mathbf{V}_C$, in case (b).

Also, $\mathbf{A}^r_{PC}$ = centripetal component of acceleration for assumed rotation of P about C, acting in a sense P to C

$$= \frac{|\mathbf{V}_{PC}|^2}{PC} = \frac{(405)^2}{100} = 1640.2 \text{ cm/s}^2$$

The other two accelerations $\mathbf{A}_P$ and $\mathbf{A}^t_{PC}$ are known only in directions. $\mathbf{A}^t_{PC}$ is not known in magnitude because angular acceleration α_{PC} is not known. $\mathbf{A}_P$ acts along line of stroke while $\mathbf{A}^t_{PC}$ acts in a direction perpendicular to PC.

Case. (a) Figure 3.30(b) shows velocity polygon, while Figs. 3.30(c) and (d) show acceleration polygons for cases (a) and (b). In Fig. 3.30(c), as $\alpha = 0$, $A_C^t = 0$, acceleration equation reduces to

$$\mathbf{A}_P = \mathbf{A}_C^r + \mathbf{A}_{PC}^r + \mathbf{A}_{PC}^t$$

Therefore, $\mathbf{A}_C^r$ is completely known and so is the case in $\mathbf{A}_{PC}^r$ which acts radially inwards from P to C and is equal to 1640.2 cm/s². Remaining two vectors $\mathbf{A}_P$ and $\mathbf{A}_{PC}^t$ are known only in direction. $\mathbf{A}_P$ acts along line of stroke and $\mathbf{A}_{PC}^t$ acts in a direction perpendicular to PC.

From acceleration pole o_a, a vector $o_a c$ is drawn parallel to OC and equal to 8883 cm/s² to represent $\mathbf{A}_C^r$. From acceleration equation, $\mathbf{A}_{PC}^r$ is additive to $\mathbf{A}_C^r$. Hence from c draw a line cp' parallel to PC so as to represent magnitude 1640.2 cm/s² to scale. The line cp' is drawn to the left of c so that for the direction of centripetal acceleration $\mathbf{A}_{PC}^r$, the vector is additive to $\mathbf{A}_C^r$. From o_a draw the line to represent direction of absolute acceleration $\mathbf{A}_P$ and from p' draw a line perpendicular to line PC to represent direction of $\mathbf{A}_{PC}^t$. The intersection of the two lines define point p.

Since $\mathbf{A}_{PC}^t$ is additive to $\mathbf{A}_{PC}^r$, the arrowhead on $p'p$ must be as shown in Fig. 3.30(c). Similarly, as $\mathbf{A}_P$ is subtractive from $\mathbf{A}_{PC}^t$, the vector $o_a p$ must point to the right. By measurement,

$$\mathbf{A}_P = 6560 \text{ cm/s}^2; \qquad \mathbf{A}_{PC}^t = 4240 \text{ cm/s}^2$$

$$\alpha_{PC} = \left(\frac{A_{PC}^t}{PC}\right) = 42.4 \text{ rad/s}^2 \qquad \qquad \textbf{Ans.}$$

Case. (b) When crank does not rotate uniformly, we have to consider component A_C^t also.

$$\mathbf{A}_C^t = (OC) \times \alpha = 25 \times 100 = 2500 \text{ cm/s}^2$$

In Fig. 3.30(d), therefore, from the tip of vector $o_a c'$ ($= \mathbf{A}_C^r$) = 8883 cm/s², draw a line $c'c$ perpendicular to OC and equal in length to represent $\mathbf{A}_C^t = 2500$ cm/s². Then as,

$$\mathbf{A}_C = \mathbf{A}_C^r + \mathbf{A}_C^t,$$

the line joining $o_a c$ represents total acceleration $\mathbf{A}_C$. From c then draw a line cp' parallel to PC to represent magnitude of 1640.2 cm/s². Note that direction of vector cp' is so chosen that for known direction $\mathbf{A}_{PC}^r$, the vector cp' is additive to $\mathbf{A}_C^t$. From p' draw a line perpendicular to PC to represent direction of $\mathbf{A}_{PC}^t$, and from o_a draw a line parallel to that of $\mathbf{A}_P$ to intersect above line at p. Then as $\mathbf{A}_{PC}^t$ is additive to $\mathbf{A}_{PC}^r$, direction of vector $\mathbf{A}_{PC}^t$ is from p' to p. Finally as $\mathbf{A}_P$ is subtractive to $\mathbf{A}_{PC}^t$, the direction is from o_a to p. By measurement,

$$\mathbf{A}_P = 5440 \text{ cm/s}^2; \qquad \mathbf{A}_{PC}^t = 6560 \text{ cm/s}^2$$

$$\alpha_{PC} = \frac{|\mathbf{A}_{PC}|}{PC} = \frac{6560}{100} = 65.6 \text{ rad/s}^2 \text{ (c.c.w.)} \qquad \qquad \textbf{Ans.}$$

Side pc in acceleration polygons of Figs. 3.30(c) and (d) represents acceleration image of link PC. Hence, if acceleration of any point G on PC is required the line pc is to be divided at g such that,

$$PG : PC = pg : pc$$

Therefore
$$pg = (PG/PC)pc$$

Then line joining $o_a g$ represents total acceleration of G in direction and magnitude.

EXAMPLE 3.10 The dimensions and configuration of a four-bar mechanism shown in Fig. 3.31 are as follows:

$P_1A = 300$ mm, $P_2B = 360$ mm, $AB = 360$ mm and $BG = 120$ mm

$P_1P_2 = 600$ mm. The angle $\angle AP_1P_2 = 60°$.

The crank P_1A has an angular velocity of 10 rad/s and angular acceleration of 30 rad/s^2, both clockwise. Determine the angular velocities and angular accelerations of P_2B and AB and the velocity and acceleration of the points B and G.

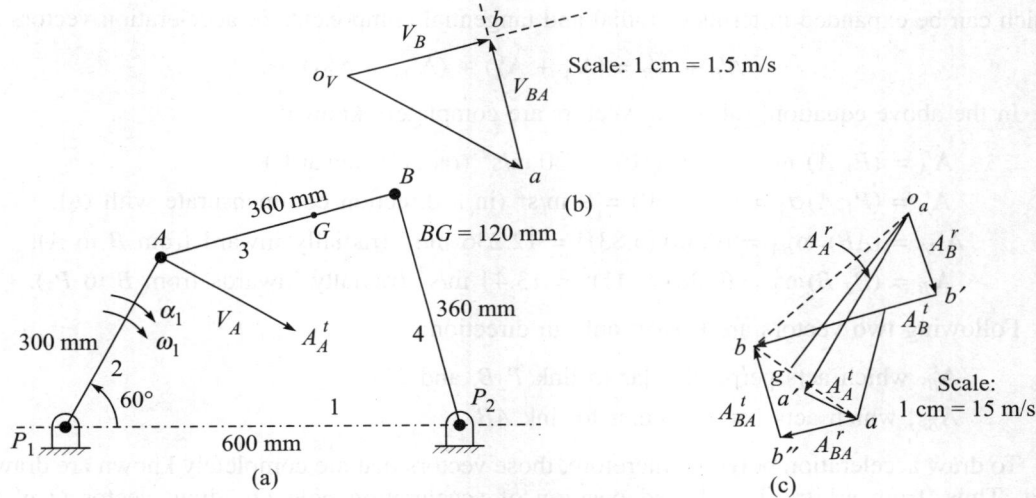

Fig. 3.31

Solution: Since $\omega_2 = 10$ rad/s, the tangential velocity $\mathbf{V}_A$ of A is,

$$\mathbf{V}_A = (P_1 A) \omega_1 = 300 \times 10 = 3000 \text{ mm/s}$$
$$= 3.0 \text{ m/s } (\perp^r \text{ to } P_1 A \text{ at } A)$$

The velocity equation is, $\quad \mathbf{V}_B = \mathbf{V}_A + \mathbf{V}_{BA}$

Here, $\mathbf{V}_A$ is completely known. Velocity $\mathbf{V}_B$ and relative velocity $\mathbf{V}_{BA}$ are not known in magnitude but act perpendicular to P_2B and AB respectively.

As a first step, configuration of mechanism is drawn to scale 15 mm: 1 mm in Fig. 3.31(a). To draw velocity polygon at Fig. 3.31(b), next we select velocity pole O_v and draw $O_v a$ parallel to the direction of $\mathbf{V}_A$ to a scale 1 cm = 1.5 m/s. From O_v draw a line parallel to direction of $\mathbf{V}_B$ (which is $\perp^r$ to $P_2 B$) and from point a draw another line parallel to $\mathbf{V}_{BA}$ (which is $\perp^r$ to AB) to intersect the above line at b. Now as $\mathbf{V}_{BA}$ is additive to $\mathbf{V}_A$ (vide, velocity equation), vector $\mathbf{V}_{BA}$

is to point from a to b in velocity polygon. Similarly, as $\mathbf{V}_B$ is to be subtractive from $\mathbf{V}_{BA}$, the velocity vector $\mathbf{V}_B$ should point from O_V to b. Measurements from velocity polygon show that

$$\mathbf{V}_B = 2.2 \text{ m/s} \quad \text{and} \quad \mathbf{V}_{BA} = 2.1 \text{ m/s}$$

Thus,
$$\omega_4 = \frac{|\mathbf{V}_B|}{P_2 B} = \frac{2.2}{0.36} = 6.11 \text{ rad/s (c.w.)}$$

and,
$$\omega_{BA} = \frac{|\mathbf{V}_{BA}|}{AB} = \frac{2.1}{0.36} = 5.83 \text{ rad/s (c.c.w.)}$$

It may be verified that the above sense of ω_4 and ω_{BA} is consistent with the direction of velocities $\mathbf{V}_B$ and $\mathbf{V}_{BA}$ respectively.

The acceleration polygon can now be drawn using acceleration equation,

$$\mathbf{A}_B = \mathbf{A}_A + \mathbf{A}_{BA}$$

which can be expanded in terms of radial and tangential components of acceleration vectors as

$$\mathbf{A}_B^r + \mathbf{A}_B^t = (\mathbf{A}_A^r + \mathbf{A}_A^t) + (\mathbf{A}_{BA}^r + \mathbf{A}_{BA}^t)$$

In the above equation, following vectors are completely known:

$\mathbf{A}_A^r = (P_1 A) \omega_1^2 = (0.3)(10)^2 = 30 \text{ m/s}^2$ (radially inwards).

$\mathbf{A}_A^t = (P_1 A)\alpha_1 = (0.3)(30) = 9 \text{ m/s}^2$ (in a direction commensurate with α_1).

$\mathbf{A}_{BA}^r = (AB) \omega_{BA}^2 = (0.36)(5.83)^2 = 12.236 \text{ m/s}^2$ (radially inward from B to A).

$\mathbf{A}_B^r = (P_2 B)\omega_4^2 = (0.36)(6.11)^2 = 13.44 \text{ m/s}^2$ (radially inwards from B to P_2).

Following two vectors are known only in direction:

$\mathbf{A}_B^t$, which acts perpendicular to link $P_2 B$, and

$\mathbf{A}_{BA}^t$, which acts perpendicular to link AB.

To draw acceleration polygon, therefore, those vectors that are completely known are drawn first. Thus, from arbitrarily selected position of acceleration pole O_a, draw vector $O_a a'$ to represent $\mathbf{A}_A^r$ in magnitude and direction and from a' draw $a'a$ to represent $\mathbf{A}_A^t$. Then $O_a a$ represents total acceleration $\mathbf{A}_A$. From the tip a of vector $\mathbf{A}_A^t$, draw a line ab'' parallel to link AB to represent acceleration $\mathbf{A}_{BA}^r$ to scale. From tip b'' now draw a line perpendicular to AB to represent direction of $\mathbf{A}_{BA}^t$. From O_a now draw $O_a b'$ parallel to $P_2 B$ to represent $\mathbf{A}_B^r$ to scale. From b' draw another line perpendicular to $P_2 B$ to represent direction or vector $\mathbf{A}_B^t$. Lines of action of $\mathbf{A}_B^t$ and $\mathbf{A}_{BA}^t$ intersect at point b. The total acceleration of B is then represented by $O_a b$ and measures 30 m/s^2. Also by measurement,

$$\mathbf{A}_{BA}^t = \text{represented by } b''b = 14 \text{ m/s}^2$$
$$\mathbf{A}_A^t = \text{represented by } b'b = 28 \text{ m/s}^2$$

The directions of vectors $\mathbf{A}_B^t$, $\mathbf{A}_{BA}^t$ are decided by using acceleration equation. $\mathbf{A}_{BA}^t$ is to be additive to $\mathbf{A}_{BA}^r$ and $\mathbf{A}_B^t$ is to be additive to $\mathbf{A}_B^r$. Hence directions of these vectors are as shown in Fig. 3.31(c).

Line ab in Fig. 3.31(c) is the acceleration image of link AB. Divide ab at g such that

$$\frac{bg}{ba} = \frac{BG}{BA} = \frac{120}{360} = \frac{1}{3}$$

Then g is the acceleration image of point G and $O_a g$ gives total acceleration of G,

$$\mathbf{A}_G = 29 \text{m/s}^2$$

3.24 COMBINED FOUR-BAR CHAIN AND SLIDER-CRANK MECHANISM

In the case of such problems, one is required to solve the problem in two stages; each stage requiring graphical solution of one velocity and one acceleration equation. Generally the connecting rod of slider-crank portion is connected to coupler and output link at respective joint. Velocity and acceleration of this joint can therefore be used conveniently to initiate graphical solution to velocity and acceleration equation of slider-crank portion. This is further illustrated in the following example problem:

EXAMPLE 3.11 In the mechanism shown in Fig. 3.32(a), the crank OA rotates uniformly at 5 rad/s. Determine, for the position shown,

    (a)   the linear acceleration of slider, and (b) angular acceleration of links AB, BCO_4 and CD.

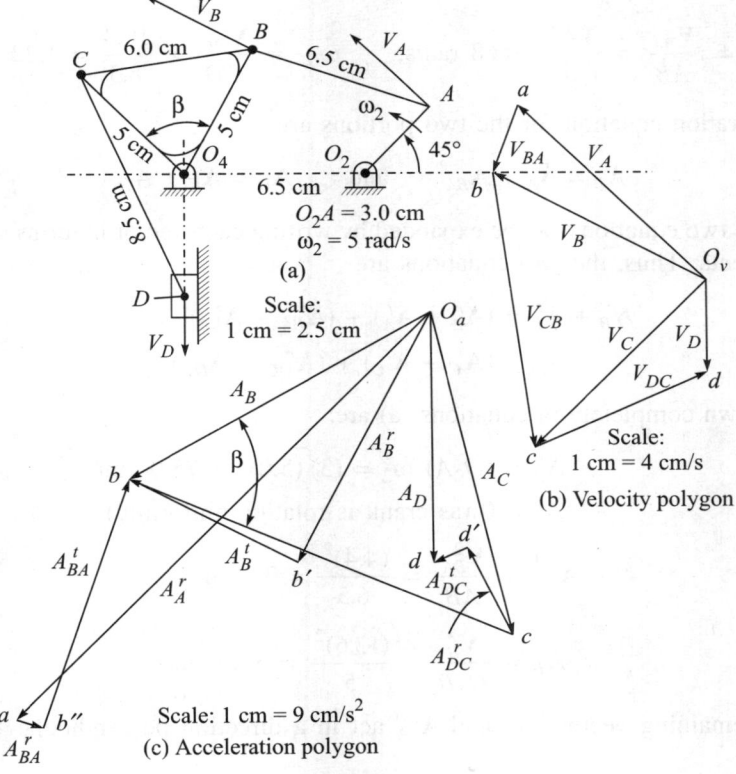

Fig. 3.32

Solution: The velocity equations are:

$$\mathbf{V}_B = \mathbf{V}_A + \mathbf{V}_{BA} \qquad \text{and} \qquad \mathbf{V}_D = \mathbf{V}_C + \mathbf{V}_{DC}$$

Since $\mathbf{V}_A = 3 \times 5.0 = 15$ cm/s, the velocity polygon is drawn first by drawing $O_v a$ to represent velocity $\mathbf{V}_A = 7.5$ cm/s, to scale in a direction perpendicular to O_2A. From point a then draw a line in a direction perpendicular to AB and from O_v draw another line perpendicular to line O_4B (to represents direction of $\mathbf{V}_B$) to cut it at point b. Then $O_v b$ represent to scale magnitude and direction of $\mathbf{V}_B$ while ab represents velocity $\mathbf{V}_{BA}$ to scale in magnitude and direction. Now on $O_v b$ as base describe triangle $O_v bc$ similar to triangle O_4BC on base O_4B. Then $O_v c$ represents velocity $\mathbf{V}_C$ to scale.

Second velocity equation can now be used to obtain $\mathbf{V}_D$. From O_v draw a line parallel to the line of stroke of D and from c draw another line perpendicular to CD (to represent direction of $\mathbf{V}_{DC}$) to cut it at d. Then $O_v d$ represents $\mathbf{V}_D$ and cd represents $\mathbf{V}_{DC}$ in magnitude and direction.

By measurement,

$$|\mathbf{V}_B| = 13.6 \text{ cm/s} \qquad |\mathbf{V}_D| = 5.4 \text{ cm/s}$$

and

$$|\mathbf{V}_{BA}| = 4.4 \text{ cm/s} \qquad |\mathbf{V}_{DC}| = 10.4 \text{ cm/s}$$

Hence,

$$\omega_4 = \frac{\mathbf{V}_B}{O_4B} = \frac{13.6}{5} = 2.72 \text{ rad/s}$$

$$\omega_{BA} = \frac{\mathbf{V}_{BA}}{AB} = \frac{4.4}{6.5} = 0.68 \text{ rad/s}; \qquad \omega_{DC} = \frac{\mathbf{V}_{DC}}{CD} = \frac{10.4}{8.5} = 1.22 \text{ rad/s}$$

The acceleration equation for the two portions are

$$\mathbf{A}_B = \mathbf{A}_A + \mathbf{A}_{BA} \qquad \text{and} \qquad \mathbf{A}_D = \mathbf{A}_C + \mathbf{A}_{DC}$$

Each of the two equations can be expanded by writing each vector in terms of tangential and radial components. Thus, the two equations are

$$\mathbf{A}_B^r + \mathbf{A}_B^t = (\mathbf{A}_A^r + \mathbf{A}_A^t) + (\mathbf{A}_{BA}^r + \mathbf{A}_{BA}^t) \qquad\qquad (a)$$

and

$$\mathbf{A}_D = (\mathbf{A}_C^r + \mathbf{A}_C^t) + (\mathbf{A}_{DC}^r + \mathbf{A}_{DC}^t) \qquad\qquad (b)$$

Vector known completely in equations (a) are:

$$\mathbf{A}_A^r = (O_2A)\,\omega_2^2 = (3)(5.0)^2 = 75.0 \text{ cm/s}^2$$

$$\mathbf{A}_A^t = 0 \text{ (as crank is rotating uniformly)}$$

$$\mathbf{A}_{BA}^r = \frac{\mathbf{V}_{BA}^2}{AB} = \frac{(4.4)^2}{6.5} = 3.0 \text{ cm/s}^2$$

$$\mathbf{A}_B^r = \frac{\mathbf{V}_B^2}{O_4B} = \frac{(13.6)^2}{5} = 37.0 \text{ cm/s}^2$$

The two remaining vectors $\mathbf{A}_B^t$ and $\mathbf{A}_{BA}^t$ act in a direction perpendicular to O_4B and AB respectively.

The acceleration polygon is drawn by drawing a line $O_a a$, parallel to $O_2 A$, to scale and the centripetal acceleration $\mathbf{A}_A^r = 75.0$ cm/s^2 in a direction $A O_2$. From point a now draw a line ab' in a direction parallel to AB and to represent $\mathbf{A}_{BA}^r = 3.0$ cm/s^2 to scale. Note that $\mathbf{A}_{BA}^r$ is a centripetal component and must act in a direction A to B. Further, as per acceleration equation, $\mathbf{A}_{BA}^r$ must be additive to $\mathbf{A}_A^r$. Hence the vector ab' represent $\mathbf{A}_{BA}^r$. From b' draw a line in a direction perpendicular to AB to represent transverse acceleration $\mathbf{A}_{BA}^t$. Similarly from O_a draw line $O_a b''$ to represent centripetal acceleration $\mathbf{A}_B^r = 37$ cm/s^2 (parallel to $O_4 B$.) From b'' draw a line perpendicular to $O_4 B$ to represent direction of $\mathbf{A}_B^t$ so as to cut line of action of $\mathbf{A}_{BA}^t$ at b. Then for $\mathbf{A}_B^t$ to the additive to $\mathbf{A}_B^r$, the vector must point in direction b'' to b. Similarly $b'b$ represents $\mathbf{A}_{BA}^t$.

By measurement,

$$\mathbf{A}_{BA}^t = 35.1 \text{ cm/s}^2 \quad \text{and} \quad \mathbf{A}_B^t = 23.4 \text{ cm/s}^2$$

Therefore   Angular accelerations,

$$\alpha_{BA} = \frac{35.1}{6.5} = 5.4 \text{ rad/s}^2 \quad \text{and} \quad \alpha_4 = \frac{23.4}{5} = 4.68 \text{ rad/s}^2$$

As is evident from the direction of vector $b''b \ (= \mathbf{A}_B^t)$, α_4 acts c.c.w. and from the direction of vector $b'b \ (= \mathbf{A}_{BA}^t)$, it is clear that α_{BA} is c.w. in sense.

Acceleration of c is obtained by using principle of acceleration image. On $O_a b$, which is the acceleration image of $O_4 B$, construct a triangle $O_a bc$ which is similar to $O_4 BC$. Then $o_a c$ represents total acceleration of C, which is

$$\mathbf{A}_C = \mathbf{A}_C^r + \mathbf{A}_C^t$$

The second part of acceleration polygon is completed using second accelertion equation. Thus, from c draw cd' parallel to CD and of magnitude

$$\mathbf{A}_{DC}^r = \frac{\mathbf{V}_{DC}^2}{CD} = \frac{(10.4)^2}{8.5} = 12.7 \text{ cm/s}^2$$

pointing in the centripetal direction D to C. From d' now, draw a line perpedicular to CD to represent direction of $\mathbf{A}_{DC}^t$, to cut another line from O_a, drawn parallel to line of stroke of D, at point d. Then $O_a d$ gives acceleration $\mathbf{A}_D$ and $d'd$ represent $\mathbf{A}'_{DC}$

By measurements, $\mathbf{A}_{DC}^t = 4.95$ cm/s^2 and $\mathbf{A}_D = 32.4$ cm/s^2

Therefore $\alpha_{DC} = \dfrac{\mathbf{A}_{DC}^t}{DC} = \dfrac{4.95}{8.5} = 0.58$ rad/s^2

and piston acceleration $\mathbf{A}_D = 32.4$ cm/s^2 **Ans.**

3.25 ACCELERATION POLYGON INVOLVING CORIOLIS COMPONENT OF ACCELERATION

Construction of acceleration polygon for quick-return mechanism necessiates consideration of coriolis components of acceleration. One major difference between the present case and accelertion analysis in Sections 3.23 and 3.24 is that unlike previous cases, effective lengths of rotating links need not remain constant. Consider for example, slotted lever quick-return mechanism of Fig. 3.33. As the slotted lever *ck* oscillates, a slider hinged to a rotating crank *OP*, slides past the slotted lever. Let *Q* be a point on link *CK* instantly in coincidence with *P*. Then as slider slides part slotted lever, the length *CQ* of the lever goes on changing.

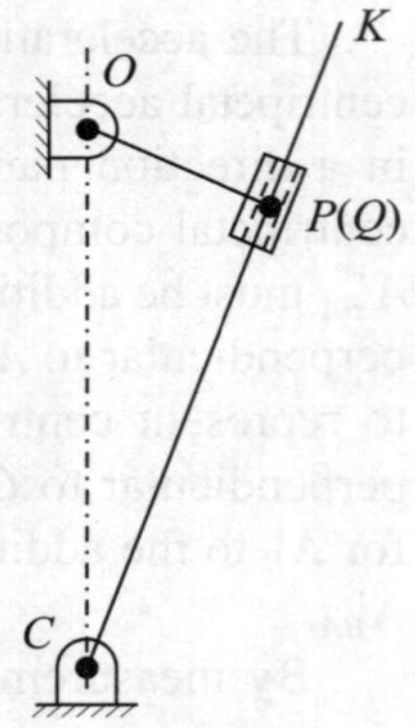

In acceleration equations of all the previous cases, it was common to arrange unknown acceleration quantity on L.H.S. of the equal to sign. But in case of acceleration equation involving coriolis component, acceleration of crank pin, which is usually known is arranged on L.H.S. of the equal to sign. The reason for this difference is to be seen in the way the equation (3.53) was derived. While writing down acceleration equation students are advised to keep this difference in mind.

Method of constructing acceleration polygon for problems involving coriolis component of acceleration will be clear through following example problems.

Fig. 3.33

EXAMPLE 3.12 The driving crank *AB* of the quick-return mechanism shown in Fig. 3.34 is 7.5 cm long and revolves at a uniform speed of 200 r.p.m. Find the velocity and acceleration of the tool box *R* in the position shown when the crank makes an angle of 60° with the vertical line of centres *PA*. What is the acceleration of sliding block at *B* along the slotted lever *PQ*?

(A.M.I.E. 1980)

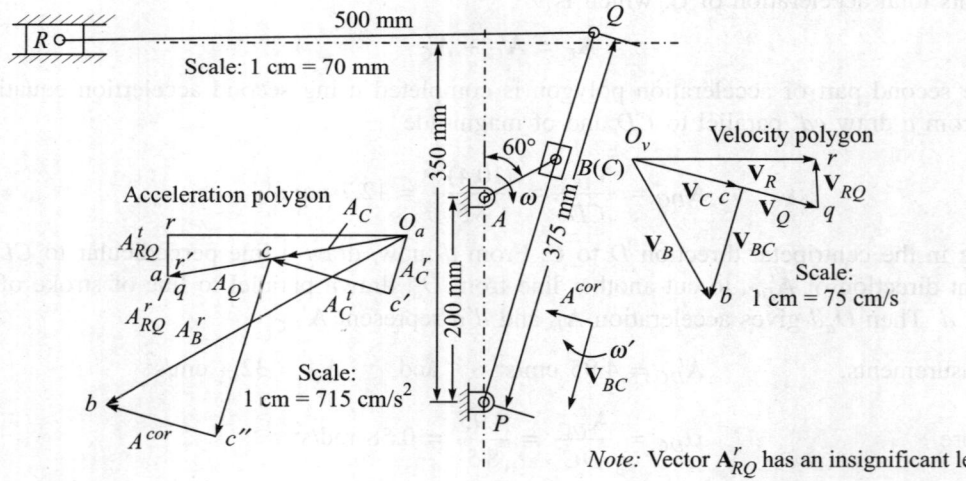

Note: Vector $\mathbf{A}_{RQ}^r$ has an insignificant length

Fig. 3.34

Solution: Assuming crank to rotate c.w., its angular speed

$$\omega = \frac{2\pi \times 200}{60} = 20.94 \text{ rad/s (c.w.)}$$

Therefore Point *B* rotates in arc of a circle with *AB* as radius. Tangential speed of crank pin

$$\mathbf{V}_B = (AB) \times \omega = 7.5 \times 20.94 = 157.05 \text{ cm/s}$$

From equation 3.50(a), the velocity of slider B in terms of coincident point C (on link PQ) is

$$\mathbf{V}_B = \mathbf{V}_C + \mathbf{V}_{BC} \tag{a}$$

Point C, as a point on link PQ, is rotating about P and hence, its velocity $\mathbf{V}_C$ is tangential to arc of circle with PC as radius. Thus, $\mathbf{V}_C$ is not known in magnitude (as angular speed of link PQ is not known). But $\mathbf{V}_B$ is known in magnitude and direction. Velocity $\mathbf{V}_{BC}$ acts along the link PQ, as slider can slide only along the slotted lever PQ.

Velocity triangle. Select O_v suitably as at Fig. 3.34(b). Draw a line $O_v b$ in a direction perpendicular to AB to represent direction of $\mathbf{V}_B$ and measure length $O_v b$ to represent velocity $\mathbf{V}_B = 157.05$ cm/s to scale. From the tip of vector $O_v b$ draw a line parallel to $\mathbf{V}_{BC}$, i.e. parallel to PQ and from O_v draw a line parallel to $\mathbf{V}_C$, i.e. perpendicular to PQ to meet the above line at c. Then triangle $O_v bc$ gives solution of above velocity equation at (a). To satisfy the above equation, $\mathbf{V}_{BC}$ must be subtractive from $\mathbf{V}_B$ and therefore, vector $\mathbf{V}_{BC}$ must point from c to b as shown. Similarly $\mathbf{V}_C$ must point O_v to c as shown. Extend $O_v c$ to q such that,

$$PC : PQ = O_v c : O_v q$$

Then $O_v q$ gives velocity of q. The velocity equation connecting $\mathbf{V}_R$ and $\mathbf{V}_Q$ is,

$$\mathbf{V}_R = \mathbf{V}_Q + \mathbf{V}_{RQ} \tag{b}$$

Hence, to solve this velocity equation, from q draw a line in a direction perpendicular to QR and from O_v draw another line parallel to line of action of R to meet it at r. $O_v r$ gives velocity of R and qr gives velocity $\mathbf{V}_{RQ}$.

Thus by measurement

$$|\mathbf{V}_{BC}| = cb = 110 \text{ cm/s}; \qquad |\mathbf{V}_C| = 110 \text{ cm/s}.$$

$$|\mathbf{V}_Q| = O_v q = 185 \text{ cm/s}; \qquad |\mathbf{V}_R| = 175 \text{ cm/s} \quad \text{and} \quad |\mathbf{V}_{RQ}| = qr = 45 \text{ cm/s}$$

Hence, angular speed ω' of $PQ = \dfrac{|\mathbf{V}_Q|}{PQ} = \dfrac{185}{37.5} = 4.93$ rad/s (c.w.)

and $\qquad$ angular speed of $QR = \omega_{RQ} = \dfrac{45}{50.0} = 0.9$ rad/s

To construct acceleration polygon, write down the equation connecting accelerations of B and C first.

Thus, $\qquad\qquad\qquad \mathbf{A}_B = (\mathbf{A}_C^r + \mathbf{A}_C^t) + \mathbf{A}_{BC} + \mathbf{A}^{cor} \tag{b}$

Since, crank AB revolves uniformly at 200 r.p.m.,

$$\mathbf{A}_B^t = 0 \quad \text{and,} \quad \mathbf{A}_B^r = (AB) \, \omega^2 = 7.5 \left(\frac{2\pi \times 200}{60} \right)^2$$

or $\quad \mathbf{A}_B^r = 3298.87$ cm/s^2 $= 3299$ cm/s^2 (from B to A)

$$\mathbf{A}_{RQ}^r = \frac{(\mathbf{V}_{RQ})^2}{RQ} = \frac{(45)^2}{50.0} = 40.5 \text{ cm/s}^2 \text{ (from } R \text{ to } Q)$$

$\mathbf{A}_C^r$ = centripetal acceleration of $C = (PC)\, \omega'^2 = (25.0)\,(4.93)^2 = 607.6$ cm/s^2 (from C to P)

$\mathbf{A}_C^t$ = known in direction (perpendicular to PQ) but not in magnitude.

$\mathbf{A}_{BC}$ = known in direction (along PQ) but not in magnitude.

$\mathbf{A}^{cor} = 2\,(\mathbf{V}_{BC})\,\omega' = 2\,(110) \times 4.93 = 1084.6$ cm/s^2 (in the direction shown in figure)

Thus, we shall prefer to draw acceleration components $\mathbf{A}_B^r$, $\mathbf{A}_C^r$, $\mathbf{A}^{cor}$ (which are completely known) first. This is done in Fig. 3.34(c). Thus, from acceleration pole O_a draw $O_a b$ to represent $\mathbf{A}_B^r = 3299$ cm/s^2 to scale. Again from O_a draw $O_a c'$ parallel to QP to represent $\mathbf{A}_C^r = 607.6$ cm/s^2 to scale. From c' it is preferable to draw $\mathbf{A}_C^t$ so that total acceleration vector $\mathbf{A}_C$ can be drawn. But unfortunately magnitude of $\mathbf{A}_C^t$ is not known. Hence from c' draw a line perpendicular to PQ. Next subtract $\mathbf{A}^{cor}$ and $\mathbf{A}_{BC}$ from $\mathbf{A}_B^r$ in accordance with equation (b).

Thus from b draw a line parallel to direction of $\mathbf{A}^{cor}$ and measure bc'' along it to represent $\mathbf{A}^{cor} = 1084.6$ cm/s^2 to scale. From c'' draw a line parallel to PQ to represent direction of $\mathbf{A}_{BC}^r$ and from c' draw another line perpendicular to PQ to represent direction of $\mathbf{A}_c^t$ to meet it at c. Then cc' represent $\mathbf{A}_C^t$ and $O_a c$ represents total acceleration $\mathbf{A}_C$. Again cc'' represents $\mathbf{A}_{BC}^r$. Extend $O_a c$ to q such that,

$$O_a c : O_a q = PC : PQ$$

Then $O_a q$ gives total acceleration of Q, i.e. $\mathbf{A}_Q$. From here onward, we use the following acceleration equation:

$$\mathbf{A}_R = \mathbf{A}_Q + \mathbf{A}_{RQ} \tag{c}$$

Thus from q draw line qr' parallel to RQ and equal to $(\mathbf{V}_{RQ}^2/RQ) = 40.5$ cm/s^2 to scale. Note, that this turns out to be very small compared to others, and difficult to show in figure. From r' draw another line perpendicular to QR to represent direction of $\mathbf{A}_{RQ}^t$. Similarly, from O_a draw another line parallel to the line of stroke of R to represent direction of $\mathbf{A}_R$. The intersection of these lines occur at r. The directions of vectors are marked so as to satisfy equations (b) and (c). By measurement

$$\mathbf{A}_R = O_a r = 2250 \text{ cm/s}^2 \qquad \text{and} \qquad \mathbf{A}_{BC}^r = cc'' = 1800 \text{ cm/s}^2$$

EXAMPLE 3.13 A whit-worth quick return motion mechanism is shown in Fig. 3.35. OA is a crank rotating at 30 r.p.m. in clockwise direction $OA = 15$ cm long; $OC = 10$ cm; $CD = 12.5$ cm and $DR = 50$ cm. Determine the acceleration of the sliding block R and the angular acceleration of the slotted lever CA.

Solution: Tangential velocity of crank pin A,

$$\mathbf{V}_A = \frac{2\pi \times 30 \times 15}{60} = 47.12 \text{ cm/s}$$

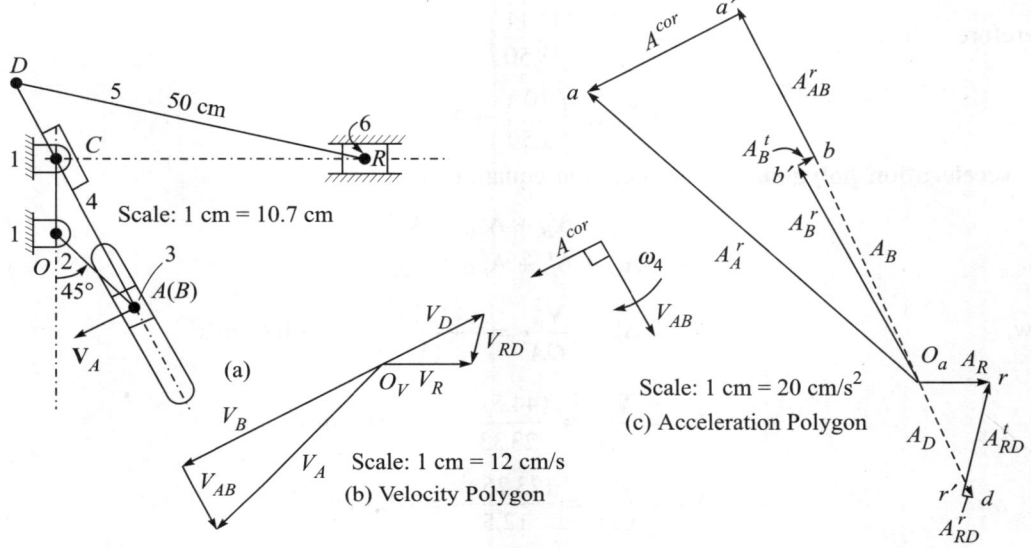

Fig. 3.35

The two velocity equations are:

$$\mathbf{V}_A = \mathbf{V}_B + \mathbf{V}_{AB} \tag{a}$$

and

$$\mathbf{V}_R = \mathbf{V}_D + \mathbf{V}_{RD} \tag{b}$$

Steps involved in drawing a velocity diagram are as under. Students are advised to reason out the steps for clear understanding.

(i) Draw line $O_v a$ from velocity pole in a direction perpendicular to OA to represent velocity $\mathbf{V}_A = 47.12$ cm/s to scale.

(ii) From velocity pole O_v draw another line parallel to the direction of $\mathbf{V}_B$ which is perpendicular to link CB. Also, from a draw a line parallel to $\mathbf{V}_{AB}$ which is parallel to link CB. The intersection gives point b. Using velocity equation (a), the direction of velocity vectors are as shown at Fig. 3.35(b).

(iii) Extend bO_v to d such that,

$$\mathbf{V}_D = O_v d = \left(\frac{CD}{CB}\right) \times \mathbf{V}_B = \left(\frac{1.85}{3.5}\right) \times 44 = 23.26 \text{ cm/s}$$

represents $\mathbf{V}_D$ to scale. From O_v draw a line parallel to the line of stroke of R and from d draw another line parallel to $\mathbf{V}_{RD}$ (which is perpendicular to line DR) to cut it at r. Then, using velocity equation (b), the vectors have directions as at Fig. 3.35(b). By measurement,

$$\mathbf{V}_{AB} = 13.5 \text{ cm/s}; \qquad \mathbf{V}_B = 44.5 \text{ cm/s}$$
$$\mathbf{V}_R = 18.5 \text{ cm/s}; \qquad \mathbf{V}_{RD} = 11.0 \text{ cm/s}$$

Therefore
$$\omega_{RD} = \left(\frac{V_{RD}}{RD}\right)\left(\frac{11}{50}\right) = 0.22 \text{ rad/s}$$

$$CB = 3.5 \times \left(\frac{10}{1.5}\right) = 23.33 \text{ cm}$$

Acceleration polygon. The acceleration equations are:

$$A_A = A_B^r + A_B^t + A_{AB}^r + A^{cor} \tag{c}$$

and,
$$A_R = A_D^r + A_D^t + A_{RD}^r + A_{RD}^t \tag{d}$$

Now,
$$A_A = A_A^r = \frac{V_A^2}{OA} = \frac{(47.12)^2}{15} = 148.02 \text{ cm/s}^2$$

$$A_B^r = \frac{V_B^2}{CB} = \frac{(44.5)^2}{23.33} = 84.88 \text{ cm/s}^2$$

$$A_D^r = \frac{V_D^2}{CD} = \frac{(23.26)^2}{12.5} = 43.3 \text{ cm/s}^2$$

$$A^{cor} = 2\,(V_{AB})\omega' = 2\,(13.5)\left(\frac{47.12}{23.33}\right) = 54.53 \text{ cm/s}^2$$

and acts in a direction as shown in figure. By measuremnt $A_B^t = 6 \text{ cm/s}^2$ and

$$A_{RD}^r = \frac{V_{RD}^2}{RD} = \frac{(11)^2}{50} = 2.42 \text{ cm/s}^2$$

Above vectors in equation (c) are fully known, while vectors A_{AB}^r and A_B^t are known only in direction.

Construction. The steps can be summarised as under for drawing acceleration polygon.

(1) Select O_a suitably and draw $O_a\,a$ parallel to OA to represent $A_A^r = 148.02 \text{ cm/s}^2$ to scale.

(2) From point a draw line aa' perpendicular to link CB to represent $A^{cor} = 54.53 \text{ cm/s}^2$ in the direction shown. Note that A^{cor} is subtractive to A_A^r, and acts in the direction shown.

(3) From the point a', draw a line parallel to link CB to represent direction of A_{AB}^r.

(4) From O_a draw $O_a\,b'$ parallel to OA to represent $A_B^r = 84.88 \text{ cm/s}^2$ to the scale and from b' draw another line perpendicular to BC to represent direction of A_B^t. The intersection of this line with direction of A_{AB}^r defines the point b.

(5) Then, $ba' = A_{AB}^r$ and, $b'b = A_B^t = 6 \text{ cm/s}^2$. Join $O_a\,b$, which then represents total acceleration of B, i.e. A_B. Extend $O_a\,b$ on the opposite side of O_a and locate point d' on it such that,

$$CD : CB = O_a d' : O_a b$$

(6) From point d' draw $d'd$ parallel to link DR to represent centripetal acceleration $A_{RD}^r = 2.42 \text{ cm/s}^2$ to scale.

(7) From point d draw a line perpendicular to link DR to represent $\mathbf{A}_{RD}^t$ in direction and from O_a draw another line parallel to line of stroke of slider R, to represent direction of $\mathbf{A}_R$. The intersection point is given by r.

Then, using equation (d), as $\mathbf{A}_{RD}^r$ and $\mathbf{A}_{RD}^t$ are additive to $\mathbf{A}_D$, their directions are given by dr' and $r'r$. The direction of acceleration $\mathbf{A}_R$ represented by $O_a r$ satisfy acceleration equation. By measurement,

$$\mathbf{A}_R = O_a r = 24 \text{ cm/s}^2 \qquad \textbf{Ans.}$$

and

$$\mathbf{A}_B^t = bb' = 6 \text{ cm/s}^2$$

Therefore

$$\alpha_{BC} = \frac{\mathbf{A}_B^t}{BC} = \frac{6}{24.6} = 0.24 \text{ rad/s}^2 \qquad \textbf{Ans.}$$

3.26 KLEIN'S CONSTRUCTION

Klein's construction provides a simple graphical method for determining velocity triangle and acceleration quadrilateral mainly for a slider-crank mechanism. The method is named after its proposer Prof. Klein. The method of construction and proof are described as follows.

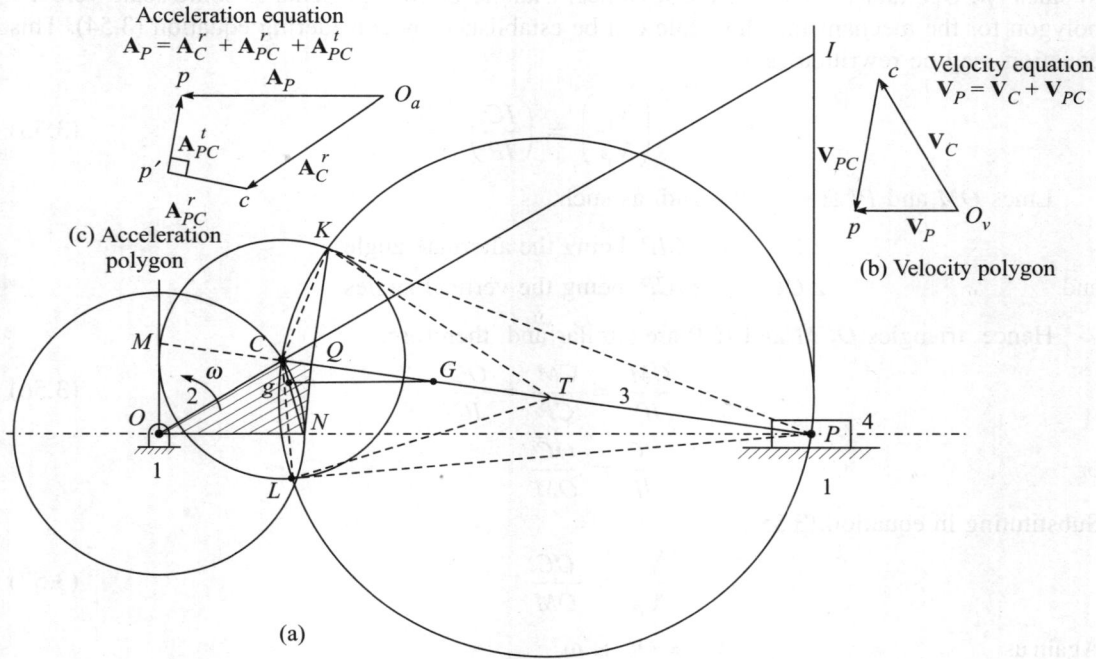

Fig. 3.36 Klein's construction.

Construction. Describe the slider-crank mechanism in the required configuration to some scale. Let OC be the crank, CP the connecting rod and P the slider. Let the crank OC rotate uniformly in c.c.w. sense with angular velocity ω rad/s.

Erect a perpendicular to the line of stroke of slider at O and produce (if necessary) line of centres of connecting rod PC to meet it at M. Construct a circle on CP as diameter and with C as centre, describe another circle of radius CM to cut it at points K and L. Join K and L and produce it, if necessary, to cut CP at Q and the line of stroke OP at N. Join O, C, Q, N. Then, triangle OCM represents velocity triangle to some scale and quadrilateral $OCQN$ represents acceleration polygon to some other scale.

Proof. Erect a perpendicular to the line of stroke at P and let the line OC, when extended, meet it at I. It is left as an exercise for the readers to show that the point of intersection I represents the instantaneous centre 13 of rotation of link 3 with respect of link 1. Link 1 being the frame, link 3 can be considered to have motion of rotation about an imaginary pivot at I.

Thus,

$$(\mathbf{V}_C/IC) = (\mathbf{V}_P/IP) \tag{3.54}$$

Comparing the velocity triangle $O_v cp$ with the triangle OCM, it is observed that the sides cp, $O_v c$ and $O_v p$ are perpendicular to sides CM, OC and OM respectively. When the triangle OCM is bodily rotated through $90°$ in c.c.w. direction, the sides CM, OC and OM will become parallel to sides cp, $O_v c$ and $O_v p$ respectively. Hence, triangle OCM represents to some scale velocity polygon for the mechanism. This scale can be established by considering equation (3.54). This equation may be rewritten as –

$$\left(\frac{\mathbf{V}_C}{\mathbf{V}_P}\right) = \left(\frac{IC}{IP}\right) \tag{3.55}$$

Lines OM and PI are parallel and, as such,

$$\angle COM = \angle CIP, \text{ being the alternate angles}$$

and $$\angle OCM = \angle ICP, \text{ being the vertical angles.}$$

Hence, triangles OCM and ICP are similar and, therefore,

$$\frac{OM}{IP} = \frac{CM}{CP} = \frac{OC}{IC} \tag{3.56}$$

or

$$\frac{IC}{IP} = \frac{OC}{OM}$$

Substituting in equation (3.55),

$$\frac{\mathbf{V}_C}{\mathbf{V}_P} = \frac{OC}{OM} \tag{3.57}$$

Again as $$\mathbf{V}_C = OC \times \omega,$$

it follows that, $$\mathbf{V}_P = OM \times \omega \text{ and } \mathbf{V}_{PC} = MC \times \omega$$

In other words, sides of triangle OCM represent velocity triangle to a scale ω.

To prove that the quadrilateral $OCQN$ represents acceleration polygon to some other scale, we proceed as follows:

It is easy to understand that the side $O_a c$ in acceleration polygon of Fig. 3.36(c) represents centripetal acceleration of c and is therefore parallel to crank OC. Again, side $O_a p$ of Fig. 3.36(c) represents piston acceleration and is, parallel to ON. Finally, $p'c$ and $p'p$ represent centripetal and transverse acceleration of P w.r. to C and, therefore, they are parallel to sides CQ and QN respectively. Thus, each side of quadrilateral $OCQN$ is parallel to only one side of the acceleration polygon. However, this in itself is not sufficient to say that quadrilateral $OCQN$ is similar to acceleration polygon $O_a cp'p$.

In order to say conclusively that quadrilateral $OCQN$ is similar to acceleration polygon, one must prove additionally that the ratio of two adjacent sides of quadrilateral $OCQN$ equals the ratio of corresponding adjacent sides of acceleration polygon $O_a cp'p$. In other words, we now intend to prove that,

$$(CQ/CO) = (cp'/co_a) \tag{3.58}$$

The right hand side of equation (3.58) will be shown to be equal to L.H.S. Thus,

$$\text{R.H.S.} = \frac{cp'}{co_a} = \frac{\text{Centripetal accen. of } P \text{ w.r. to } C}{\text{Centripetal accen. of } C}$$

$$= \frac{(\mathbf{V}_{PC})^2/CP}{(OC)\,\omega^2} \tag{3.59}$$

From the triangle OMC of Klein's construction,

$$\mathbf{V}_{PC} = (CM)\omega$$

Substituting in (3.59),

$$\text{R.H.S.} = \frac{(CM)^2\omega^2}{(CP)(OC)\omega^2}$$

Cancelling out ω^2 from numerator and denominator,

$$\text{R.H.S.} = \frac{(CM)^2}{(CP)(OC)}$$

But $CM = CK$, being radii of the same circle.
Hence,

$$\text{R.H.S.} = \frac{(CK)^2}{(CP)(OC)} \tag{3.60}$$

Join CK, KT, CL and LT.
Then triangles CKT and CLT are identical because,

$$CK = CL, \text{ being radii of the same circle,}$$

$$KT = LT, \text{ being radii of the same circle and } CT \text{ is common to both.}$$

Thus,
$$\angle KCP = \angle LCP$$

Again, triangles CKQ and CLQ are identical because,

$$CK = CL$$
$$\angle KCP = \angle LCP$$

and CQ is common.

Hence,
$$KQ = LQ$$
and
$$\angle KQC = \angle LQC = 90°$$

Further, triangles KCQ and KCP are similar because,

$$\angle KQC = 90° = \angle CKP;$$

$$\angle KCQ \text{ is common to both}$$

Hence, from proportion of sides,

$$\frac{CK}{CP} = \frac{CQ}{CK}$$

or
$$(CK)^2 = (CP)(CQ)$$

Substituting in (3.60), for $(CK)^2$,

$$\text{R.H.S.} = \frac{(CP)(CQ)}{(CP)(OC)}$$

Cancelling out CP in numerator and denominator,

$$\text{R.H.S.} = \left(\frac{CQ}{OC}\right) = \text{L.H.S.}$$

Hence proved.

It is thus concluded that quadrilateral $OCQN$ is similar to acceleration polygon $Q_a \, cp'p$ and, therefore, represents acceleration polygon to some scale.

Scale of Acceleration Polygon OCQN

Since, $p'c$ is parallel to CQ and $O_a c$ is parallel to OC, it follows that the side OC represents acceleration $\mathbf{A}_c^r \, (= OC \cdot \omega^2)$. Hence, the scale of the diagram is ω^2. Stated in other words,

$$\mathbf{A}_C^r = (OC) \, \omega^2 \; ; \; \mathbf{A}_{PC}^r = (CQ) \, \omega^2$$
$$\mathbf{A}_{PC}^t = (QN) \, \omega^2 \text{ and } \mathbf{A}_P = (ON) \, \omega^2$$

The acceleration of any point G on the rod may be obtained by drawing a line Gg parallel to the line of stroke to cut diagonal CN at point g. The total acceleration of point G is given by—

$$\mathbf{A}_G = (Og) \, \omega^2$$

For correct solution, a reader must not forget to take into account the scale to which the mechanism configuration is drawn, while measuring side ON, NQ etc. An another interesting point to note is that $\mathbf{A}_C^r$ being a centripetal acceleration is always directed towards O and, similarly, as the slider describes theoretically S.H.M., the acceleration of piston $\mathbf{A}_P$, represented by NO is always directed towards O.

Special Cases

(A) *Klein's Construction for Non-uniform Crank Rotation*
Let the crank OC rotate with angular velocity ω together with an angular acceleration α

acting in the same sense. The total acceleration of C now consists of two components in radial and tangential directions. Thus,

$$\mathbf{A}_C = \mathbf{A}_C^r + \mathbf{A}_C^t$$

where $\mathbf{A}_C^r$ and $\mathbf{A}_C^t$ are in mutually perpendicular direction given by

$$\mathbf{A}_C^r = (OC)\,\omega^2 \qquad \text{and} \qquad \mathbf{A}_c^t = (OC)\,\alpha$$

The angle θ made by total acceleration vector $\mathbf{A}_C$ with the radial component, is given by

$$\tan \theta = (\alpha/\omega^2)$$

or

$$\theta = \tan^{-1}(\alpha/\omega^2)$$

Construction procedure in this case consists in drawing a line CO_1 at an angle θ to CO and an another line OO_1 from O perpendicular to CO to meet it at O_1. (see Fig. 3.37)

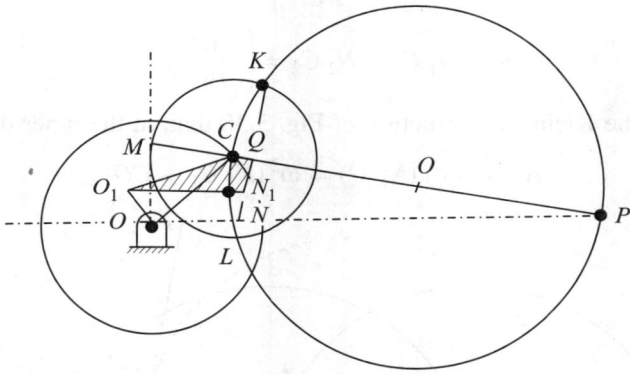

Fig. 3.37

Draw a line $O_1 N_1$ parallel to line of stroke to cut common chord KL at a point N_1. Then, the quadrilateral $Q_1 CQN_1$ represents acceleration polygon.

(B) *Outer Dead Centre and Inner Dead Centre Positions of the Crank*

Let C_1 and C_2 be the crank position at *i.d.c.* and *o.d.c.* respectively. As the crank OC and connecting rod CP lie along the line of centres, the point M coincides with crank centre O and the radius of circle $CM = CO$ and OC_1 and OC_2 represent the inner dead and outer dead centre positions of the crank respectively. Let P_1 and P_2 be the piston positions at inner and outer dead centre positions. Then, as per method of construction, points M_1 and M_2 will coincide with crank centre O. Further the common chord $K_1 L_1$ of circle on $P_1 C_1$ as diameter and that with C_1 as centre and $C_1 M_1$ as radius, will intersect line of stroke $C_1 P_1$ at point N_1 which coincides with Q_1.

Similarly, the common chord $K_2 L_2$ of circle on diameter $C_2 P_2$ and that with C_2 as centre and radius as $C_2 M_2$ will intersect the line of stroke. $C_2 P_2$ at point N_2 will coincides with Q_2. Now since,

$$C_1 M_1 = C_1 O = C_2 M_2$$

it follows that

$$K_1 L_1 = K_2 L_2$$

and

$$C_1 N_1 = C_2 N_2$$

Again as $\qquad$ $\mathbf{A}^r_{PC} = (CQ)\omega^2$, $\qquad$ and $\qquad$ $\mathbf{A}^r_{PC} = \dfrac{\mathbf{V}^2_{PC}}{PC} = \dfrac{(CM\omega^2)}{PC}$

comparing R.H.S.,

$$(CQ)\omega^2 = \frac{(CM)^2\omega^2}{PC}$$

or

$$CQ = \frac{CM^2}{PC}$$

Also as Q coincides with N and M coincides with O, we have

$$CN = \frac{CO^2}{PC}$$

Therefore, $\qquad\qquad\qquad N_1\, C_1 = N_2\, C_2 = \left(\dfrac{CO^2}{PC}\right)$ $\hfill$ (3.61)

It follows from the Klein's construction of Fig. 3.38 that, at the inner dead centre position,

$$\mathbf{A}_P = (\omega^2)\,(N_1\,O) = \omega^2\,(N_1\,C_1 + CO)$$

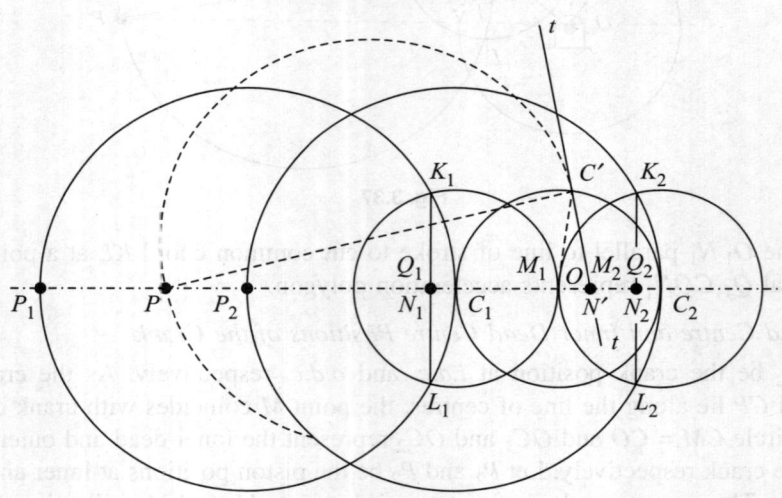

Fig. 3.38

Substituting from equation (3.61) for $N_1 C_1$

$$\mathbf{A}_P = \omega^2\left(\frac{CO^2}{PC} + CO\right) = \omega^2(CO)\left(\frac{CO}{PC} + 1\right)$$

or

$$\mathbf{A}_P = \omega^2(CO)\left(\frac{1}{n} + 1\right) \hfill (3.62)$$

where, $\qquad\qquad \dfrac{CO}{CP} = \dfrac{\text{radius of crank}}{\text{length of connecting rod}} = \dfrac{1}{n}$

Similarly for outer dead centre position,

$$\mathbf{A}_P = \omega^2 (N_2\,O) = \omega^2 (CO - N_2\,C_2)$$

or
$$\mathbf{A}_P = \omega^2 \left(CO - \frac{CO^2}{PC} \right) = (\omega^2)\,CO \left(1 - \frac{1}{n} \right) \qquad (3.63)$$

(c) When the crank OC is placed at right angles to the line to stroke, as in position OC', the point M coincides with C' and, as such, radius of circle $C'\,M = 0$. Hence, the common chord KL now becomes tangent to circle, with $C'\,P'$ as diameter at point C'. This common chord cuts line of stroke at point N'. Hence, acceleration of piston is this position is given by

$$\mathbf{A}_P = (ON')\,\omega^2 \qquad (3.64)$$

3.27 APPROXIMATE ANALYTICAL EXPRESSION FOR DISPLACEMENT, VELOCITY AND ACCELERATION OF PISTON OF RECIPROCATING ENGINE MECHANISM

Figure 3.39 shows crank OC of a slider-crank mechanism in a position making angle θ with the inner dead centre position OC_1. Let the crank be of length r and let it rotate uniformly in c.w. direction with angular speed ω rad/s. Let $l = (n)r$ be the length of the connecting rod.

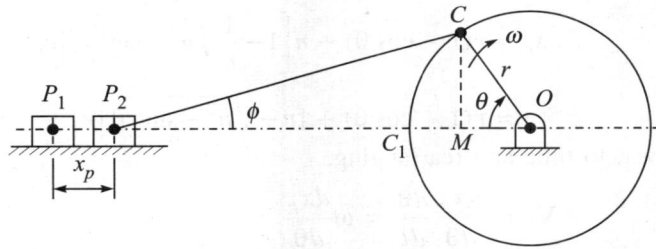

Fig. 3.39

When the crank is at the inner dead centre position OC_1, the connecting rod C_1P_1 lies along the line of stroke and for this position,

$$OP_1 = (l + r)$$

When the crank rotates through θ in c.w. direction, the distance of piston from crank shaft centre is given by

$$OP_2 = (l\cos\phi + r\cos\theta)$$

Thus, displacement x of the piston P from inner dead centre position is given by

$$\begin{aligned} x_p &= (OP_1 - OP_2) \\ &= r(1 - \cos\theta) + l(1 - \cos\phi) \\ &= r[(1 - \cos\theta) + l/r(1 - \cos\phi)] \qquad (3.65) \end{aligned}$$

Again from right angled triangles OCM and CMP_2,

$$CM = l\sin\phi = r\sin\theta$$

or

$$\sin\phi = (r/l)\sin\theta$$

Also,

$$\cos\phi = \sqrt{1 - \sin^2\phi} = \frac{r}{l}\sqrt{(l/r)^2 - \sin^2\theta}$$

or

$$\cos\phi = \frac{1}{n}\sqrt{n^2 - \sin^2\theta} = \sqrt{1 - \left(\frac{\sin\theta}{n}\right)^2} \qquad (3.66)$$

Remembering the binomial expansion,

$$(1 - x)^n = 1 - nx + \frac{n(n-1)}{2!}x^2 - \frac{n(n-1)(n-2)}{3!}x^3 + \ldots$$

We have the R.H.S. of equation (3.66) as

$$\left(1 - \frac{1}{n^2}\sin^2\theta\right)^{1/2} = 1 - \frac{1}{2n^2}\sin^2\theta - \frac{1}{8n^4}\sin^4\theta - \frac{1}{16n^6}\sin^6\theta \ldots$$

And so,

$$(1 - \cos\phi) = \frac{1}{2n^2}\sin^2\theta + \frac{1}{8n^4}\sin^4\theta + \frac{1}{16n^6}\sin^6\theta \ldots \qquad (3.67)$$

Again, substituting for $\cos\phi$ from (3.66) in (3.65),

$$x_p = r(1 - \cos\theta) + n\left[1 - \frac{1}{n}\sqrt{n^2 - \sin^2\theta}\right]r$$

$$= r(1 - \cos\theta) + [n - \sqrt{n^2 - \sin^2\theta}]r \qquad (3.68)$$

Differentiating w.r. to time and rearranging,

$$V_P = \frac{dx_p}{d\theta}\frac{d\theta}{dt} = \omega\frac{dx_p}{d\theta}$$

$$= r\omega\left[\sin\theta + \frac{\sin 2\theta}{2\sqrt{n^2 - \sin^2\theta}}\right] \qquad (3.69)$$

Again, differentiating further,

$$A_P = \frac{dV_P}{dt} = \frac{dV_P}{d\theta}\frac{d\theta}{dt}$$

$$= r\omega^2\left[\cos\theta + \frac{n^2\cos 2\theta + \sin^4\theta}{(n^2 - \sin^2\theta)^{3/2}}\right] \qquad (3.70)$$

Again, when $n\,(= l/r)$ is large, $\sin^2\theta$ becomes negligible in comparison to n^2 and from equation (3.68),

$$= x_p = r(1 - \cos\theta) + [n - \sqrt{n^2 - 0}\,]$$

$$= r(1 - \cos\theta), \text{ which is a simple harmonic motion.}$$

It follows that only when l/r ratio is sufficiently large, obliquity effects of connecting rod become negligible, and motion of the slider becomes simple harmonic.

Again, for large n, from (3.69) and (3.70), we have

$$\mathbf{V}_p \approx r\omega\left[\sin\theta + \frac{\sin 2\theta}{2n}\right] \qquad (3.71)$$

and

$$\mathbf{A}_P \approx r\omega^2\left[\cos\theta + \frac{\cos 2\theta}{n}\right] \qquad (3.72)$$

Also, from $\sin\phi = \dfrac{r}{l}\sin\theta$

differentiating on both sides w.r. to time, we have

$$(\cos\phi)\frac{d\phi}{dt} = \left(\frac{r}{l}\right)\cos\theta\,\frac{d\theta}{dt}$$

$$\therefore \qquad \frac{d\phi}{dt} = \omega_r = \left(\frac{r}{l}\right)\omega\,\frac{\cos\theta}{\cos\phi}$$

Substituting for $\cos\phi$ from (3.66)

$$\omega_r = \left(\frac{r}{l}\right)\omega\,\frac{n\cos\theta}{\sqrt{n^2 - \sin^2\theta}}$$

$$\omega_r = \frac{\omega\cos\theta}{\sqrt{n^2 - \sin^2\theta}}$$

and differentiating further, the angular acceleration of connecting rod is given by

$$\alpha_r = \frac{d\omega_r}{dt} = \frac{d\omega_r}{d\theta}\frac{d\theta}{dt}$$

$$= \omega\left(\frac{d\omega_r}{d\theta}\right)$$

$$= \frac{-\omega^2(n^2 - 1)\sin\theta}{(n^2 - \sin^2\theta)^{3/2}}$$

When n^2 is large in comparison to unity and also in comparison to $\sin^2\theta$, we have

$$\omega_r \approx \left(\frac{\omega\cos\theta}{n}\right)$$

$$\alpha_r \approx -\left(\frac{\omega^2\sin\theta}{n}\right)$$

Negative direction indicates that α always acts so as to reduce inclination ϕ of the connecting rod.

3.28 KINEMATIC ANALYSIS OF COMPLEX MECHANISMS

A mechanism is said to be kinematically simple if its acceleration analysis can be carried out completely by applying successively the acceleration equation $A_Q = A_P + A_{QP}$ combined, if necessary, with acceleration-image principle. In the above acceleration equation, P and Q are called motion transfer points of one and the same link. The equation is capable of being solved directly only if the radii of curvature of the paths of both the motion transfer points P and Q are known. If any of the two radii is not determinable by inspection, special methods of analysis are required and the mechanism is classed as kinematically complex.

Complex mechanisms are characterised by a multi-paired floating link, *i.e.* a link connected to at least three other moving links. A floating link is one which is not directly connected to the frame. Figures 3.40(a) and (b) represent two different classes of complex mechanisms. In Fig. 3.40(a), a triple-paired (ternary) floating link is shown where the movement of ternary link 3 is completely constrained by two of the three attached links *viz.* 2 and 4. Here the only unknown radius of curvature is that of point B. Hence the mechanism has a low degree of complexity. In such mechanisms if the input link is 6, the velocity or acceleration of point B cannot be determined from the velocity and acceleration of point A. However, if the link 2 or 4 is the input link, the velocity and acceleration of C or D can be determined. Using principle of velocity and acceleration image then, velocity and acceleration of point B can next be determined. Thus, the mechanism is rendered simple by change of input link. Such mechanisms are said to have low degree of complexity.

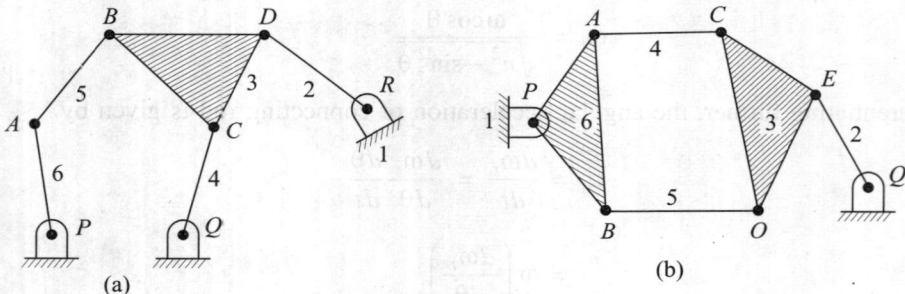

(a)   (b)

Fig. 3.40

Such approach, however, fails in the type of mechanisms shown in Fig. 3.40(b). In this mechanism, all the three attached links are essential for the constraint of the link 3 and also, the radii of curvature of paths of C and D are not known. This mechanism has therefore, high degree of complexity. It does not matter whether 2 is the input link or link 6 is the input link. Such mechanisms are analysed by considering a kinematically simple inversion. For instance, releasing link 1 and fixing link 5 instead, in Fig. 3.40(b), gives a simpler mechanism for establishing relative velocities.

EXAMPLE 3.14   Find out the velocity of slider D and angular velocity of CD in the mechanism shown in Fig. 3.41. Take $A_oA = B_oB = 2$ cm; $A_oB_o = 4$ cm; $CD = 6$ cm and $\omega = 1$ rad/sec.

(A.M.I.E., Summer 1993)

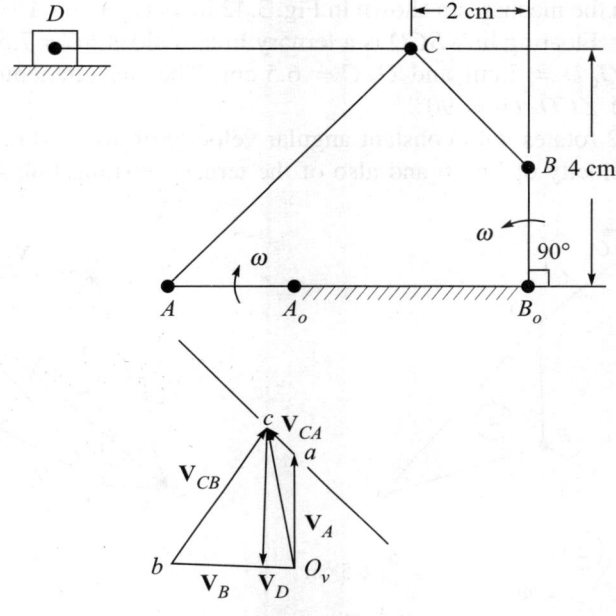

Fig. 3.41

Solution: $V_A = (AA_o) \times 1 = 2$ cm/s ($\perp^r$ to $A_o A$)

$\quad\quad\quad V_B = (B_o B) \times 1 = 2$ cm/s ($\perp^r$ to BB_o)

Now the two velocity equation are:

$$V_C = V_A + V_{CA} \tag{a}$$

and

$$V_C = V_B + V_{CB} \tag{b}$$

These vector equations are plotted in velocity polygon.

From the velocity pole O_v draw lines $O_v a$ and $O_v b$ to represent respectively the velocities V_A and V_B. Note that $V_A = V_B = 2 \times 1 = 2$ cm/s.

From the point a in velocity polygon draw a line perpendicular to AC to represent direction of V_{CA} and from point b draw another line perpendicular to BC to represent V_{CB}. The two lines meet at point c. The line $O_v c$ then represents V_C. Again,

$$V_D = V_C + V_{DC}.$$

Hence, from c draw a line perpendicular to CD to represent direction of dV_{DC} and from O_v draw another line parallel to the direction of V_D so as to meet it at d. Then $O_v d$ represents V_D in magnitude and direction. Similarly cd represents velocity V_{DC} in magnitude and direction.

By measurement $\quad\quad V_D = O_v d = 0.5$ cm/s

Also, $\quad\quad\quad\quad\quad\quad V_{DC} = cd = 2.4$ cm/s

And, therefore, $\quad\quad \omega_{DC} = \dfrac{V_{DC}}{DC} = \dfrac{2.4}{6} = 0.4$ rad/s (c.c.w).   **Ans.**

EXAMPLE 3.15 In the mechanism shown in Fig. 3.42 links $O_2 A$ and AB are of lengths 7.5 cm and 5 cm respectively. Floating link BCD is a ternary link of sides $BC = 7.5$ cm; $BD = 5$ cm and $CD = 10$ cm. Link $O_6 D = 5$ cm and $O_5 C = 6.5$ cm. The instantaneous configuration has $\angle AO_2 O_5 = 100°$ and $\angle CO_5 O_2 = 90°$.

The input crank 2 rotates with constant angular velocity of $\omega_2 = 10$ rad/s in c.w. direction. Determine angular velocity of link 6 and also of the ternary floating link 4, i.e. BCD.

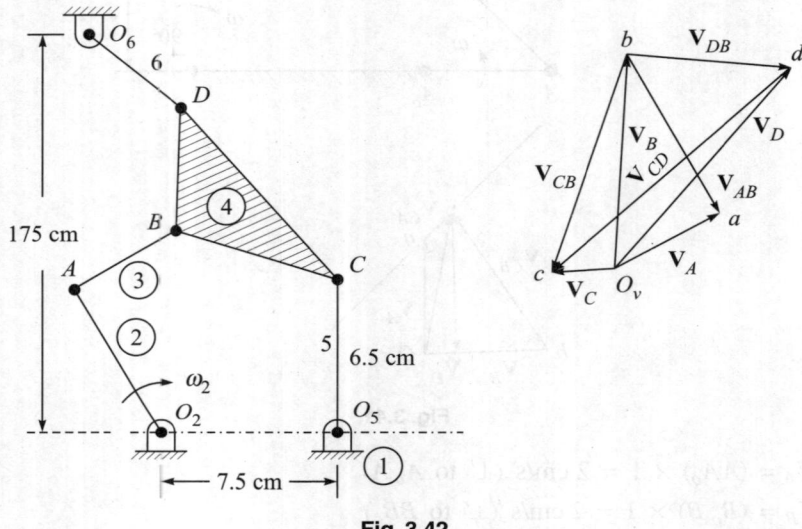

Fig. 3.42

Solution: Since $O_6 DCO_5$ forms a 4-bar mechanism, it could be much more easier to begin with velocity of link 5 or link 6. However, in the given problem velocity of D or C is not given. Hence, we attempt the problem in an indirect way. While drawing velocity polygon, to begin with, we do not assume scale. Assuming a length for vectors V_D and V_C then, does not amount to assuming velocity V_D and V_C arbitrarily. On establishing the length of velocity vector for V_A in velocity polygon, we try to establish the scale to which obtained length represents given velocity V_A. Once the velocity scale is established, other velocities are easily determined using the velocity polygon.

Construction. From velocity pole O_V draw $O_V d$ in a direction parallel to V_D and of arbitrary length equal to 6 cm. Since scale of velocity polygon is not decided, above step is permissible. From point d draw a line perpendicular to DC to represent direction of V_{CD} and from O_v draw another line parallel to V_C to meet it at c. Then O_{Vc} represents V_C and dc represents V_{CD} to some unknown scale. On line cd construct triangle cdb similar to CBD so as to get its velocity image.

From point b draw a line ba perpendicular to BA so as to represent direction of V_{AB}. From O_v draw another line perpendicular to $O_2 A$ to represent direction of V_A. The intersection defines point a. Then $O_v a$ represents velocity V_A to some scale and line ba represents velocity V_{AB} to the same scale.

Now, $$V_A = (O_2 A) \times \omega_2$$

and from velocity polygon $O_v a = 1.675$ cm and the sense of $\mathbf{V}_A = O_v a$, tallies with the sense ω_1 of link 2. Thus,

$$(O_v a) \times \text{scale of velocity polygon} = O_2 A \times \omega_2.$$

$\therefore$ Scale of velocity polygon $= \dfrac{7.5 \times 10}{1.675} = 44.8$

Hence, using this scale, lengths of corresponding sides in velocity polygon and velocities are:

$O_v b = 2.975$ cm;	$\mathbf{V}_B = 133.33$
$O_v c = 0.8$ cm;	$\mathbf{V}_C = 36.11$
$O_v d = 3.72$ cm;	$\mathbf{V}_D = 166.67$
$bc = 3.16$ cm;	$\mathbf{V}_{CB} = 141.668$
$bd = 2.23$ cm;	$\mathbf{V}_{DB} = 100.00$
$cd = 4.28$ cm;	$\mathbf{V}_{DC} = 191.668$
$ab = 2.54$ cm;	$\mathbf{V}_{AB} = 113.89$

$\therefore$ Angular velocity of link 6

$$\omega_6 = \frac{\mathbf{V}_D}{O_6 D} = \frac{166.67}{5} = 33.33 \text{ rad/s}$$

Angular velocity of ternary floating link 4,

$$\omega_4 = \frac{\mathbf{V}_{CB}}{BC} = \frac{141.668}{7.5} = 18.889 \text{ c.w.}$$

$$= \frac{\mathbf{V}_{DB}}{DB} = \frac{100.00}{5} = 20.00 \text{ c.w.}$$

$$= \frac{\mathbf{V}_{DC}}{DC} = \frac{191.668}{10} = 19.16 \text{ c.w.}$$

These are all approximately equal allowing for constructional errors.

REVIEW QUESTIONS

3.1 In the mechanism shown in Fig. 3.43, the crank OA rotates at 60 r.p.m. Determine (a) the linear acceleration of the slider at B, (b) the angular accelerations of the links AC, CQD and BD.

(**Ans.** 7 m/s^2; 23 rad/s^2; 14 rad/s^2)

3.2 In the mechanism shown in Fig. 3.44, QA is the driving crank rotating at 200 r.p.m. The lengths of various links are $OA = 2.5$ cm; $AB = 18$ cm; $AB = DB$, $DE = 10$ cm; $BC = 6$ cm; $EF = 10$ cm, horizontal distance between O and C is 15 cm and the vertical distance between F and O or

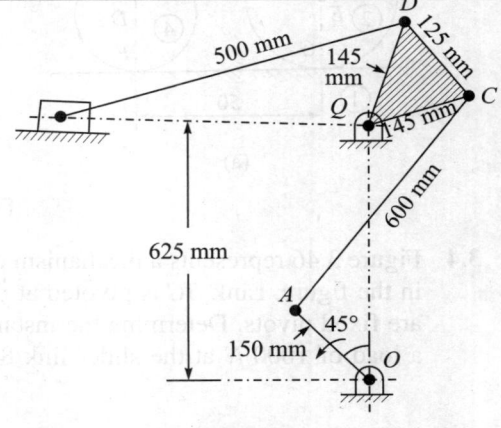

Fig. 3.43

C is 12 cm. Determine, for the configuration when *OA* makes 45° with the horizontal,

(a) Velocity of the slider block *F*;

(b) Angular velocity of link *DE*;

(c) Velocity of sliding of the link *DE* in the swivel block. *(A.M.I.E.)*

(**Ans.** 39 cm/sec; 8.25 rad/s c.c.w. 30.5 cm/s)

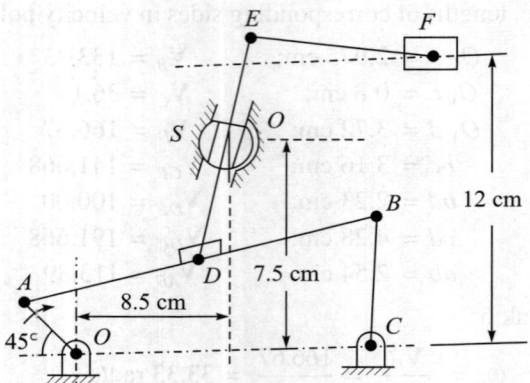

Fig. 3.44

3.3 Mechanism shown in Fig. 3.45 bears all dimensions in mm. The wheels 2 and 4 roll on 1. The uniform angular velocity of wheel 2 is 10 rad/s. Determine the angular velocity of links 3 and 4, and also the relative velocity of point *D* with respect to point *E*.

(**Hint.** The rollers 2 and 4 have their instantaneous centres at *G* and *F* and as such velocity polygon may be constructed for equivalent 4-bar mechanism as shown in Fig. 3.45(b). (**Ans.** $\omega_3 = 2.5$ rad/s; $\omega_4 = 13$ rad/s; $V_{DE} = 1.5$ cm/s)

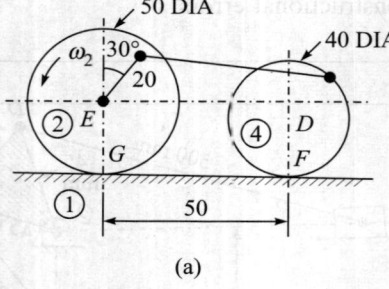

(a)

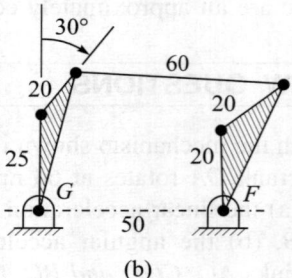

(b)

Fig. 3.45

3.4 Figure 3.46 represents a mechanism of a toggle press. The various dimensions are given in the figure. Link *EC* is pivoted at *O* and link *DPE* is pivoted at *P*. Pivots *O*, *Q* and *P* are fixed pivots. Determine the instantaneous torque required at crank *OA* to overcome a load of 1000 *N* at the slider link 8. (**Ans.** $TQ = 175$ N·m)

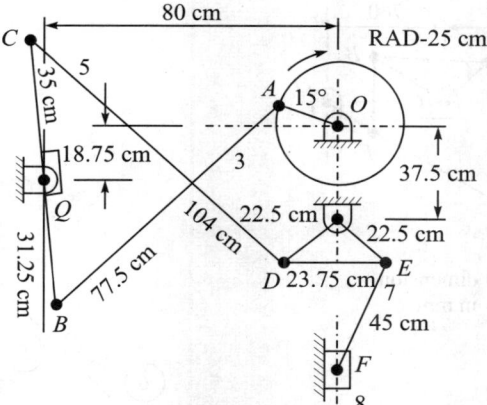

Fig. 3.46

3.5 Figure 3.47 represents mechanism used in the head of a sewing machine to operate the needle bar. The crank O_2A rotates uniformly in the clockwise direction, with 400 r.p.m. and the needle bar link 6 reciprocates vertically. From the lowest position, the needle bar moves up a short distance to create loop in the thread head in position by the needle, and then descends down to complete stitch. After doing the stitch, the needle again ascends up and finally descends to its lowest position. This completes the cycle. Find the velocity of needle for the given configuration when D lies vertically below O_2. Vertical distance between $O_2O_4 = 4.0$ cm and the horizontal distance between O_2 and $O_4 = 13$ mm.

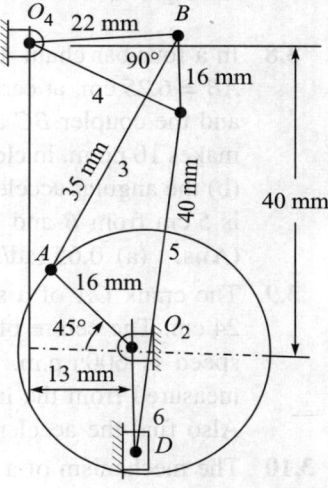

Fig. 3.47

(**Ans.** 94.2 cm/s)

3.6 Figure 3.48 shows an undercarriage mechanism for an aircraft. The various dimensions of the links are: QD = 365 mm; QG = 1065 mm; FG = 670 mm; OF = 1400 mm; OE = 300 mm; BE = 485 mm; BC = 365 mm; AB = 445 mm; AC = 145 mm.
Oil under pressure is supplied to the space under the piston for retracting the under carriage. If the velocity of the piston relative to the cylinder wall, *i.e.*, the sliding velocity of the piston in the cylinder is 10 cm/sec. Find the angular velocity of the arm QW.
(**Ans.** 950.0 rad/s)

3.7 In combined four-bar chain and slider-crank mechanism of Fig. 3.49, the angular velocity of the crank OA is 600 r.p.m. Determine the linear velocity and acceleration of slider 6 as also the angular velocity and acceleration of link 5, when the crank is inclined at an angle of 75° to the vertical. The dimensions of various links are: $OA = 28$ mm; AB = 44 mm; $BC = 49$ mm; $BD = 46$ mm and the centre distance between the centres of rotation O and C is 65 mm. Path of travel of slider is 11 mm vertically below fixed point C.   (**Ans.** 1.475 m/s; 31.75 m/s²: 34.8 rad/s c.c.w.; 27.8 rad/s²)

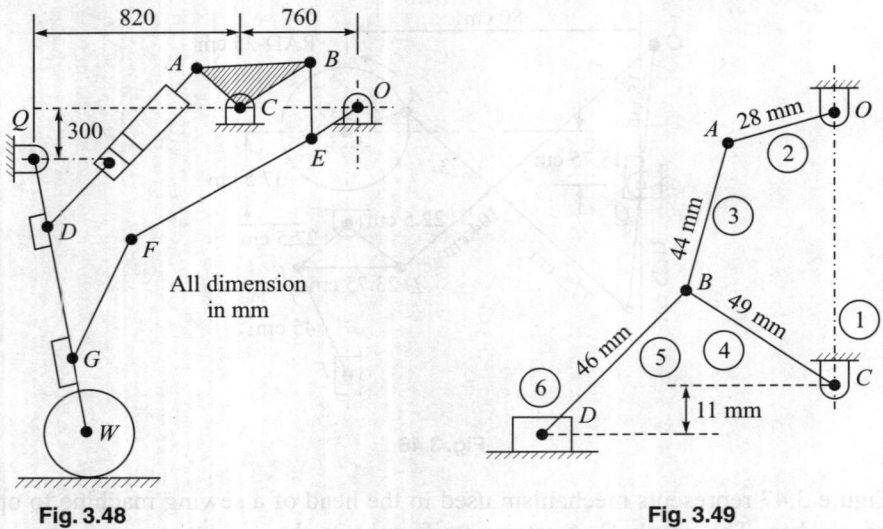

Fig. 3.48   Fig. 3.49

3.8 In a four-bar chain *ABCD*, *A* and *D* are fixed centres 12.5 cm apart. The driving crank *AB* = 6.25 cm, at certain instant makes an angle of 60° with *AD*. The driven crank *CD* and the coupler *BC* are 7.5 cm each. Determine the following when the driving crank makes 10 r.p.m. in clockwise direction: (a) The angular velocity of the links *CD* and *CB*. (b) the angular acceleration of the link *CD*, (c) the velocity and acceleration of *E* which is 5 cm from *B* and *C* and lying outside *ABCD*.

(**Ans.**   (a) 0.65 rad/s; 0.623 rad/s. (b) 1.03 rad/s^2; (c) 3.33 cm/s and 8.65 rad/s^2)

3.9 The crank *OA* of a steam engine is 8 cm and the length of the connecting rod *AB* is 24 cm. The centre of gravity of the rod is at *G*, 8 cm from the crank pin. The engine speed is 600 r.p.m. For the position when the crank makes 45° to the horizontal measured from the inner dead centre, find the velocity and acceleration of the piston. Also find the acceleration of the centre of gravity of the connecting rod.

3.10 The mechanism of a small air compressor is shown in Fig. 3.50. The centres *O* and *Q* are fixed and the yoke *AB* pivots on the crank pin *C*. The end *B* of the yoke is attached to the pin *Q* by the link *BQ* while the end *A* drives the piston *P* through the connecting rod *AP*. If the crank *OC* makes 480 r.p.m., find for the given position of the mechanism (1) the velocity of the piston and the angular velocity of the yoke. (2) the torque which must be exerted on crank to over come a resistance of 4903 Newton at the piston when the friction is neglected, (3) acceleration of the piston.

(**Ans.** 3.45 mm/s; 0.3 rad/s; 3.43 kg/m)

3.11 Figure 3.51 shows a quick-return motion mechanism in which the driving crank *OA* rotates at 120 r.p.m. in a clockwise direction. For the position shown, determine the magnitude and direction of (i) the acceleration of the block *D* (ii) the angular acceleration of the slotted bar *QB*.

(**Ans.** 705 cm/s^2; 17 rad/s^2)

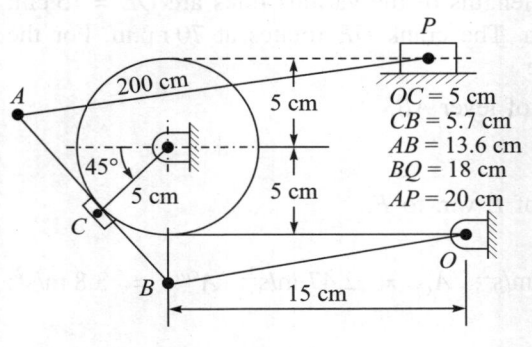

OC = 5 cm
CB = 5.7 cm
AB = 13.6 cm
BQ = 18 cm
AP = 20 cm

Fig. 3.50

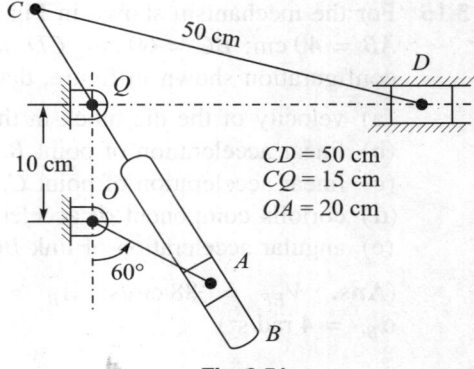

CD = 50 cm
CQ = 15 cm
OA = 20 cm

Fig. 3.51

3.12 In a quick-return mechanism shown in Fig. 3.52, the driving crank *OA* is 6 cm long and rotates at a uniform speed of 200 r.p.m. in a c.w. direction. For the position shown, find (i) Velocity of ram *R*; (ii) Acceleration of ram *R*; and (iii) Acceleration of sliding block *A* along with slotted arm *CD*. (**Ans.** 1.3 m/s; 9 m/s^2; 15 m/s^2)

3.13 The rotating cylinder mechanism is shown in Fig. 3.53 in which *OA* = 5 cm and *AB* = 12.5 cm. Determine, for the given configuration, the acceleration of piston inside the cylinder if the cylinder rotates at 300 r.p.m. in a clockwise sense.

(**Ans.** 9250 cm/s^2)

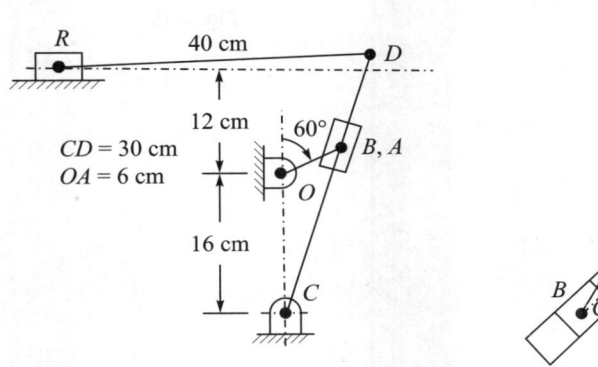

CD = 30 cm
OA = 6 cm

Fig. 3.52

Fig. 3.53

3.14 In the mechanism shown in Fig. 3.54, the sliders *B* and *D* move in straight paths as shown and are driven by the crank *OA* through connecting links *AB* and *CD*. *OA* rotates in a c.w. direction at, 60 r.p.m. and has an angular retardation of 8 rad/sec^2. Determine, for the position shown,

(a) velocity and acceleration of slider *B* and slider *D*, and

(b) angular velocity and acceleration of the link *CD*.

Take *OA* = 7.5 cm; *AB* = 15 cm; *AC* = 5 cm; *CD* = 22.5 cm.

3.15 For the mechanism shown in Fig. 3.55, lengths of the various links are $OE = 15$ cm; $AB = 40$ cm; $BC = 60$ cm; $CD = 20$ cm. The crank OE rotates at 70 r.p.m. For the configuration shown in figure, determine

(a) velocity of the die-block in the slot of lever AB.
(b) linear acceleration of point B.
(c) linear acceleration of point C.
(d) coriolis component of acceleration of E w.r. to F,
(e) angular acceleration of link BC.

(**Ans.** $V_{EF} = 88$ cm/s; $A_B = 4.25$ m/s^2; $A_C = 2.47$ m/s^2; $A^{cor} = 3.8$ m/s^2; $\alpha_{BC} = 4$ rad/s^2)

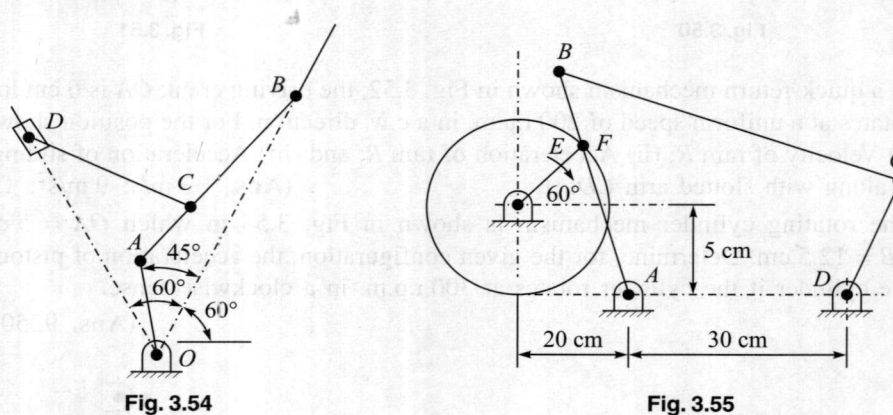

Fig. 3.54   Fig. 3.55

Chapter 4

VELOCITY AND ACCELERATION ANALYSIS (ANALYTICAL APPROACH)

4.1 INTRODUCTION

Kinematic analysis aims at establishing relationship between the motion of various links of a mechanism. Though in Chapter 3 and the current chapters only lower paired mechanisms are considered, the methods of analysis can be extended to higher paired mechanisms also. For this purpose, one is required to convert a given higher paired mechanism into an equivalent lower paired mechanism, and then apply the existing methods of kinematic analysis to it. Graphical method of kinematic analysis is preferred in many cases as it gives a better insight into the problem and can yield motion characteristics of all intermediate links. Graphical methods, however, become tedious when the analysis has to be carried out repeatedly for a number of positions of mechanisms. In view of amenability of computer programme, the analytical methods become very convenient in calculating motion characteristics for large number of mechanism configurations. Besides, the graphical methods have their limitation in respect of degree of accuracy. As against this, analytical methods can have a high degree of accuracy in computing motion characteristics.

4.2 VECTOR METHOD

Here a link is represented by a vector, using complex notation, and a displacement equation is obtained by considering the closed loop formed by the vectors representing the links. The number of such loops varies from one mechanism to another. We adopt a convention that the angle of orientation of any vector will be measured positive with respect to positive x-axis in counter clockwise direction.

Using the well-known Euler's equation from trigonometry,

$$e^{\pm j\theta} = \cos \theta \pm j \sin \theta \qquad (4.1)$$

Vector $\mathbf{R}$ can be written in complex polar form as

$$\mathbf{R} = r\, e^{j\theta} \qquad (4.2)$$

In order to get familiarised with useful manipulation techniques, les us consider vector equation representing vector triangle of Fig. 4.1.

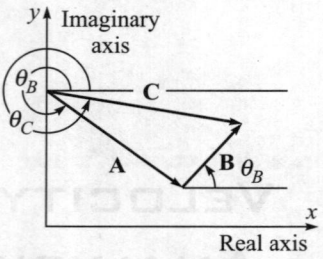

$$C = A + B \qquad (4.3)$$

This equation can be rewritten in complex polar form as,

$$Ce^{j\theta_C} = Ae^{j\theta_A} + Be^{j\theta_B} \qquad (4.4)$$

To evaluate the two unknowns C and θ_C, we may separate real and imaginary parts using Euler's equation. Thus,

Fig. 4.1 Convention of representation of vector equation.

$$C(\cos\,\theta_C + j\,\sin\,\theta_C) = A(\cos\,\theta_A + j\,\sin\,\theta_A) + B(\cos\,\theta_B + j\,\sin\,\theta_B) \qquad (4.5)$$

Equating real and imaginary parts separately,

$$C\cos\,\theta_C = A\cos\,\theta_A + B\cos\,\theta_B \qquad (4.6)$$

and

$$C\sin\,\theta_C = A\,\sin\,\theta_A + B\,\sin\,\theta_B \qquad (4.7)$$

squaring and adding on the two sides of equations (4.6) and (4.7), θ_C is eliminated and a solution for C is obtained as,

$$C = \sqrt{(A\cos\theta_A + B\cos\theta_B)^2 + (A\sin\theta_A + B\sin\theta_B)^2}$$

Therefore

$$C = \sqrt{A^2 + B^2 + 2AB\cos(\theta_B - \theta_A)} \qquad (4.8)$$

The choice for positive value of square root is arbitrary; a negative value of square root would yield a negative value for C which has values of θ_C differing by 180° from the previous value.

The angle θ_C is obtained by dividing equation (4.7) by (4.6) on respective sides. Thus,

$$\tan\,\theta_C = \frac{A\,\sin\,\theta_A + B\,\sin\,\theta_B}{A\cos\,\theta_A + B\cos\,\theta_B}$$

or

$$\theta_C = \tan^{-1}\left(\frac{A\,\sin\,\theta_A + B\,\sin\,\theta_B}{A\cos\,\theta_A + B\cos\,\theta_B}\right) \qquad (4.9)$$

In equation (4.9) signs of numerator and denominator must be considered separately in ascertaining proper quadrant of θ_C.

4.3 TYPES OF ANALYSIS PROBLEMS

Depending on the forms of two unknowns, Milton A, Chace, in his Ph. D. thesis (1964), has classified all problems faced in velocity and acceleration analysis, in four distinct cases:

Case (A): When magnitude and direction of same vector is required, *i.e.* C and θ_C,
Case (B): When magnitude alone of two different vectors (say C and B) are required,
Case (C): When magnitude of one vector and direction of another (say, C, θ_B) are required,
Case (D): When direction of two vectors (say θ_B and θ_C) are required.

Solution for case (A) is already discussed in section 4.2.

Case (B): When magnitudes B and C of two vectors are unknown, a convenient way of arriving at a solution is to divide (4.4) first by $e^{j\theta}{}_B$. Thus

$$C\,e^{j(\theta_C - \theta_B)} = A\,e^{j\theta(\theta_A - \theta_B)} + B \tag{4.10}$$

Remembering the convention that angles θ_A, θ_B, θ_C, etc. are measured in c.c.w. sense from real axis, it follows from Fig. 4.2 (a) that angle $(\theta_C - \theta_B)$ is obtained by adding angle $(-\theta_B)$ to the angle $(+\theta_C)$, which amounts to rotating real and imaginary axes through θ_B such that the real axis lies along the vector B.

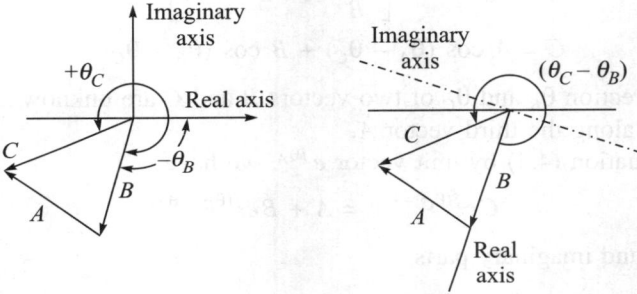

Fig. 4.2

Separating real and imaginary parts of equation (4.10) using Euler's equation,

$$C \cos (\theta_C - \theta_B) = A \cos (\theta_A - \theta_B) + B \tag{4.11}$$
$$C \sin (\theta_C - \theta_B) = A \sin (\theta_A - \theta_B) \tag{4.12}$$

It is seen that by dividing by unit vector $e^{j\theta_B}$ we have succeeded in eliminating vector **B** from one of the two equations (4.11) and (4.12).

Thus, from equation (4.12),

$$C = A\,\frac{\sin (\theta_A - \theta_B)}{\sin (\theta_C - \theta_B)} \tag{4.13}$$

Solution for the other unknown magnitude B is obtained in a similar way by dividing equation (4.4) by $e^{j\theta_c}$. Thus

$$\mathbf{C} = A\,e^{j(\theta_A - \theta_C)} + B\,e^{j(\theta_B - \theta_C)} \tag{4.14}$$

This implies rotation of real axis through θ_C such that it coincides with vector **C**.

Separating real and imaginary parts now,

$$C = A \cos (\theta_A - \theta_C) + B \cos (\theta_B - \theta_C) \tag{4.15}$$

and
$$0 = A \sin (\theta_A - \theta_C) + B \sin (\theta_B - \theta_C) \tag{4.16}$$

Thus, from equation (4.16),

$$B = -A\,\frac{\sin (\theta_A - \theta_C)}{\sin (\theta_B - \theta_C)} \tag{4.17}$$

Case (C): When magnitude of vector **C** and angle θ_B are unknown, we try to align real axis along vector **C**. Dividing by unit vector $e^{j\theta_C}$ the vector equation (4.4),

$$\mathbf{C} = A e^{j(\theta_A - \theta_C)} + B e^{j(\theta_B - \theta_C)} \qquad (4.18)$$

Separating out real and imaginary parts,

$$C = A \cos (\theta_A - \theta_C) + B \cos (\theta_B - \theta_C) \qquad (4.19)$$

and

$$0 = A \sin (\theta_A - \theta_C) + B \sin (\theta_B - \theta_C) \qquad (4.20)$$

From equation (4.20),

$$\theta_B = \theta_C + \sin^{-1}\left[\frac{-A}{B} \sin (\theta_A - \theta_C) \right] \qquad (4.21)$$

and

$$C = A \cos (\theta_A - \theta_C) + B \cos (\theta_B - \theta_C) \qquad (4.22)$$

Case (D): When direction θ_B and θ_C of two vectors B and C are unknown, we try to align the real axis along the third vector A.

Dividing the equation (4.4) by unit vector $e^{j\theta_A}$, we have

$$C e^{j(\theta_C - \theta_A)} = A + B e^{j(\theta_B - \theta_A)} \qquad (4.23)$$

Separating real and imaginary parts,

$$C \cos (\theta_C - \theta_A) = A + B \cos (\theta_B - \theta_A) \qquad (4.24)$$

and

$$C \sin (\theta_C - \theta_A) = B \sin (\theta_B - \theta_A) \qquad (4.25)$$

Squaring and adding on the two sides of equal to sign,

$$C^2 = A^2 + B^2 + 2\, AB \cos (\theta_B - \theta_A) \qquad (4.26)$$

Rearranging equation (4.26),

$$\cos (\theta_B - \theta_A) = \left(\frac{C^2 - A^2 - B^2}{2\, AB} \right)$$

Therefore

$$\theta_B = \theta_A \pm \cos^{-1}\left[\frac{C^2 - A^2 - B^2}{2\, AB} \right] \qquad (4.27)$$

Also, transferring A on the other side of equal to sign in equation (4.24) and then squaring and adding with equation (4.25),

$$B^2 = C^2 + A^2 - 2\, C A \cos (\theta_C - \theta_A)$$

or

$$\theta_C = \theta_A \pm \cos^{-1}\left[\frac{C^2 + A^2 - B^2}{2\, C A} \right] \qquad (4.28)$$

The presence of plus or minus sign serves as a reminder to us that arc cosines are each double valued and as such θ_B and θ_C have two solutions for each. Angles θ_B, θ_B' and θ_C, θ_C' must be so paired that together they satisfy equation (4.25). Thus case (D) has two distinct solutions.

4.4 THE LOOP CLOSURE EQUATION

In vector method of motion analysis, a link is represented by a vector using complex notation. The four pin joints A, B, C, D in Fig. 4.3 are shown as coincident points on connected links. Thus, A_1 and A_4 are the coincident points on links 1 and 4 at A. Similarly B_1, B_2; C_2, C_3; D_3, D_4 are the coincident points on connected links at pins B, C and D respectively.

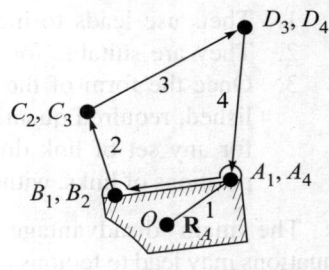

Fig. 4.3

Clearly, as (A_1, A_4), (B_1, B_2), (C_2, C_3) and (D_3, D_4) are pairs of coincident points, their relative position vectors are equal to zero.

Thus,
$$\mathbf{R}_{A_1 A_4} = \mathbf{R}_{B_2 B_1} = \mathbf{R}_{C_3 C_2} = \mathbf{R}_{D_4 D_3} = 0 \tag{4.29}$$

Again, with respect to point O on the frame-line 1

$$\mathbf{R}_{A_4} = \mathbf{R}_{A_1} + \mathbf{R}_{A_4 A_1}$$

or
$$\mathbf{R}_{A_4} = \mathbf{R}_{A_1} = \mathbf{R}_A \tag{4.30}$$

The other point B on link 1 can now be connected to pin A through the vector equation

$$\mathbf{R}_B = \mathbf{R}_A + \mathbf{R}_{BA} \tag{4.31}$$

In a similar way, vector equations connecting points C to B and D to C are

$$\mathbf{R}_C = \mathbf{R}_B + \mathbf{R}_{CB} \tag{4.32}$$

and,
$$\mathbf{R}_D = \mathbf{R}_C + \mathbf{R}_{DC} \tag{4.33}$$

And finally, vector equation connecting point D to A is,

$$\mathbf{R}_A = \mathbf{R}_D + \mathbf{R}_{AD} \tag{4.34}$$

Summing up equations (4.31), (4.32), (4.33) and (4.34) on respective sides, we have

$$\mathbf{R}_B + \mathbf{R}_C + \mathbf{R}_D + \mathbf{R}_A = (\mathbf{R}_A + \mathbf{R}_B + \mathbf{R}_C + \mathbf{R}_D) + \mathbf{R}_{BA} + \mathbf{R}_{CB} + \mathbf{R}_{DC} + \mathbf{R}_{AD} \tag{4.35}$$

Cancelling common terms on the two sides of equal to sign,

$$\mathbf{R}_{AD} + \mathbf{R}_{DC} + \mathbf{R}_{CB} + \mathbf{R}_{BA} = 0 \tag{4.36}$$

Equation (4.36) is an important vector equation called loop-closure equation for the mechanism shown. The equation emphasises the fact that the mechanism forms a close-loop and as such, the polygon formed by relative position vectors through successive links and joints must remain closed even when the mechanism goes on assuming different configurations. The constant lengths associated with the relative position vectors ensures that the joint centres remain at constant distance apart.

4.5 ALGEBRAIC POSITION ANALYSIS

There are three main advantages of algebraic methods of position analysis of planar mechanisms, over the graphical methods:

1. Their use leads to increased accuracy.
2. They are suitable for evaluations using computer or calculator.
3. Once the form of the solution has been established, required quantities can be evaluated for any set of link dimensions or for different positions of links, without starting all over again.

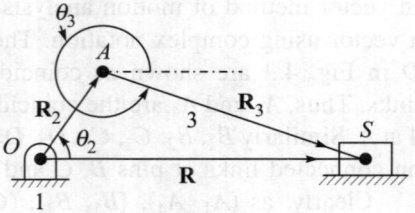

The main disadvantage is that the nature of equations may lead to tedious algebraic manipulations in finding the form of the solution.

Fig. 4.4

4.5.1 Case of Slider-Crank Mechanism

Consider the slider-crank mechanism of Fig. 4.4 for position analysis.
Vector loop-closure equation for Fig. (4.4) is

$$\mathbf{R} = \mathbf{R}_2 + \mathbf{R}_3 \tag{4.37}$$

This can be expressed in complex polar form as

$$re^{j(0)} = r_2 e^{j\theta_2} + r_3 e^{j\theta_3}$$

or,
$$r = r_2 e^{j\theta_2} + r_3 e^{j\theta_3} \tag{4.38}$$

Using Euler's formula, this yields

$$r = r_2 (\cos \theta_2 + j \sin \theta_2) + r_3 (\cos \theta_3 + j \sin \theta_3)$$

Separating real and imaginary parts,

$$r_2 \cos \theta_2 + r_3 \cos \theta_3 = r \tag{4.39}$$

and
$$r_2 \sin \theta_2 + r_3 \sin \theta_3 = 0 \tag{4.40}$$

From equation (4.40),

$$\sin \theta_3 = \left(-\frac{r_2}{r_3} \right) \sin \theta_2$$

This determines θ_3 as θ_2 is known. Also, as $\cos \theta_2$ and $\cos \theta_3$ are known, length of vector **R** can be obtained as

$$r = r_2 \cos \theta_2 + r_3 \cos \theta_3$$

Angle θ_3 and length r of vector **R** can also be obtained alternatively as under. By transposition of terms in equations (4.39) and (4.40)

$$r_3 \cos \theta_3 = r - r_2 \cos \theta_2 \tag{A}$$

and
$$r_3 \sin \theta_3 = (- r_2 \sin \theta_2) \tag{4.41}$$

Squaring and adding on respective sides,

$$r_3^2 = r^2 + r_2^2 - 2r\, r_2 \cos \theta_2 \tag{4.42}$$

Above equation can be solved either for θ_2, if slider displacement (and hence r) is given, or it can be solved for r if crank angle θ_2 is known. When r is known and θ_2 is required,

$$\pm\, \theta_2 = \cos^{-1}\left(\frac{r^2 + r_2^2 - r_3^2}{2 r r_2}\right) \tag{4.43}$$

Again, when θ_2 is known but r is unknown, rearranging equation (A),

$$r = r_3 \cos\theta_3 + r_2 \cos\theta_2 \tag{4.44}$$

Also, dividing equation (4.41) by equation (A) on respective sides,

$$\tan\theta_3 = \left(\frac{-r_2 \sin\theta_2}{r - r_2 \cos\theta_2}\right)$$

And so,

$$\theta_3 = \tan^{-1}\left(\frac{-r_2 \sin\theta_2}{r - r_2 \cos\theta_2}\right) \tag{4.45}$$

The two signs of θ_2 in equation (4.43) appear because $\cos\theta_2 \equiv \cos(-\theta_2)$, and correspond to two distinct configurations of slider-crank mechanism. While the positive sign for θ_2 ensures $\frac{3\pi}{2} < \theta_2 < 2\pi$, the negative sign ensures $\frac{\pi}{2} < \theta_2 < \pi$. Correct configuration may be identified by verifying sign of $\sin\theta_2$ or $\tan\theta_2$.

4.5.2 Case of Four-Bar Mechanism

This is a classic problem to demonstrate working of algebraic position analysis using diagonal distance r of AC. To understand the complexity of problem involved first, let us proceed as under. The loop-closure equation in complex polar form for the 4-bar mechanism may be written as

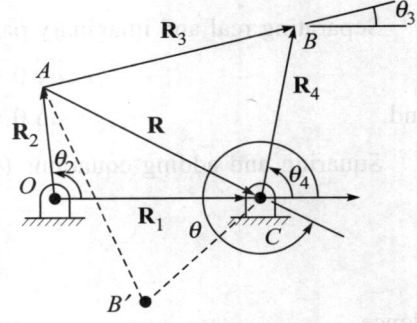

$$\mathbf{R_2} + \mathbf{R_3} = \mathbf{R_1} + \mathbf{R_4}$$

or

$$r_2\, e^{j\theta_2} + r_3\, e^{j\theta_3} = r_1\, e^{j\theta_1} + r_4\, e^{j\theta_4}$$

Since, $\theta_1 = 0$, the above equation reduces to

$$r_2\, e^{j\theta_2} + r_3\, e^{j\theta_3} = r_1 + r_4\, e^{j\theta_4} \tag{4.46}$$

Fig. 4.5

Using Euler's formula, equation (4.46) is rewritten as

$$r_2\,(\cos\theta_2 + j\sin\theta_2) + r_3\,(\cos\theta_3 + j\sin\theta_3) = r_1 + r_4\,(\cos\theta_4 + j\sin\theta_4)$$

Separating real and imaginary parts,

$$r_2 \cos\theta_2 + r_3 \cos\theta_3 = r_1 + r_4 \cos\theta_4 \tag{4.47}$$

and

$$r_2 \sin\theta_2 + r_3 \sin\theta_3 = r_4 \sin\theta_4 \tag{4.48}$$

Above equations involve two unknowns in θ_3 and θ_4. To solve the equations, let us rewrite them so as to isolate θ_3 term. Thus

$$r_3 \cos \theta_3 = (r_1 + r_4 \cos \theta_4) - r_2 \cos \theta_2 \qquad (4.49)$$

and
$$r_3 \sin \theta_3 = r_4 \sin \theta_4 - r_2 \sin \theta_2 \qquad (4.50)$$

Squaring and adding equations (4.49) and (4.50) on the two sides,

$$r_3^2 = [r_1^2 + r_4^2 \cos^2 \theta_4 + 2r_1 r_4 \cos \theta_4]$$
$$+ [r_2^2 \cos^2 \theta_2 - 2r_1 r_2 \cos \theta_2 - 2r_2 r_4 \cos \theta_2 \cos \theta_4]$$
$$+ [r_4^2 \sin^2 \theta_4 + r_2^2 \sin^2 \theta_2 - 2r_2 r_4 \sin \theta_2 \sin \theta_4]$$

Simplifying,

$$r_3^2 = (r_1^2 + r_2^2 + r_4^2) - 2r_1 (r_2 \cos \theta_2 - r_4 \cos \theta_4) - 2r_2 r_4 \cos (\theta_4 - \theta_2) \qquad (4.51)$$

Above equation involves unknown θ_4 as a non-linear variable and does not permit a direct closed form solution, as may be evident from the presence of two cosine terms involving θ_4. A numerical solution, using computer, is of course possible. Indirect solution in closed form for the above equations can, however, be obtained by considering a diagonal vector **R** which provides two distinct three-link loops namely *OACO* and *ABCA*. (vide Fig. 4.5.)

The loop closure equation for loop *OACO*, may be written as

$$\mathbf{R} = \mathbf{R}_1 - \mathbf{R}_2$$

Rewriting in complex polar form

$$r e^{j\theta} = r_1 e^{j\theta_1} - r_2 e^{j\theta_2}$$

Separating real and imaginary parts,

$$r \cos \theta = r_1 \cos \theta_1 - r_2 \cos \theta_2 \qquad (4.52)$$

and,
$$r \sin \theta = r_1 \sin \theta_1 - r_2 \sin \theta_2 \qquad (4.53)$$

Squaring and adding equations (4.52) and (4.53) on the two sides,

$$r^2 = r_1^2 + r_2^2 - 2r_1 r_2 (\cos \theta_1 \cos \theta_2 + \sin \theta_1 \sin \theta_2)$$
$$= r_1^2 + r_2^2 - 2r_1 r_2 \cos (\theta_1 - \theta_2)$$

Hence,
$$r = \sqrt{r_1^2 + r_2^2 - 2r_1 r_2 \cos (\theta_1 - \theta_2)} \qquad (4.54)$$

Also, dividing (4.53) by (4.54) on respective sides,

$$\tan \theta = \frac{r_1 \sin \theta_1 - r_2 \sin \theta_2}{r_1 \cos \theta_1 - r_2 \cos \theta_2}$$

Therefore
$$\theta = \tan^{-1} \left(\frac{r_1 \sin \theta_1 - r_2 \sin \theta_2}{r_1 \cos \theta_1 - r_2 \cos \theta_2} \right) \qquad (4.55)$$

Since $\theta_1 = 0$ and θ_2 the input angle, is known, the magnitude r and direction θ of the diagonal vector can be worked out.

Again from the three-link loop $ABCA$, the loop-closure equation is

$$\mathbf{R}_3 = \mathbf{R} + \mathbf{R}_4$$

which can be rewritten as,

$$r_3\, e^{j\theta_3} = r\, e^{j\theta} + r_4\, e^{j\theta_4} \qquad (4.56)$$

Comparing real and imaginary parts of equation (4.56),

$$r_3 \cos\theta_3 = r \cos\theta + r_4 \cos\theta_4 \qquad (4.57)$$

and, $\qquad\qquad r_3 \sin\theta_3 = r \sin\theta + r_4 \sin\theta_4 \qquad (4.58)$

Squaring and adding on respective sides,

$$r_3^2 = r^2 + r_4^2 + 2rr_4\,(\cos\theta\cos\theta_4 + \sin\theta\sin\theta_4)$$
$$= r^2 + r_4^2 + 2rr_4 \cos(\theta - \theta_4)$$

Hence, $\qquad\qquad (\theta - \theta_4) = \cos^{-1}\left(\dfrac{r_3^3 - r^2 - r_4^2}{2r\,r_4} \right)$

Since, $\qquad\qquad \cos A = \cos(-A),$

there are two values of $(\theta - \theta_4)$ which can satisfy the above equation.

Thus, $\qquad\qquad \pm\,(\theta - \theta_4) = \cos^{-1}\left(\dfrac{r_3^2 - r^2 - r_4^2}{2r\,r_4} \right) \qquad (4.59)$

Using equations (4.55) and (4.59), one can establish angle θ_4 for the given configuration. The configuration, other than the one shown in Fig. 4.5., which is possible from equation (4.59), is $OAB'C$, with $CB' = CB$ and $AB' = AB$ and point B' forming mirror image of B w.r. to AC. Note that in both the cases, the diagonal vector $\mathbf{R}$ remains the same.

Again, dividing equation (4.58) by (4.57) on respective sides,

$$\tan\theta_3 = \frac{r \sin\theta + r_4 \sin\theta_4}{r \cos\theta + r_4 \cos\theta_4}$$

Therefore $\qquad\qquad \theta_3 = \tan^{-1}\left(\dfrac{r \sin\theta + r_4 \sin\theta_4}{r \cos\theta + r_4 \cos\theta_4} \right) \qquad (4.60)$

With angles θ, θ_3 and θ_4 established, for a given crank angle θ_2, the configuration of mechanism is fully defined.

4.6 VELOCITY AND ACCELERATION ANALYSIS USING COMPLEX ALGEBRA (RAVEN'S APPROACH)

Velocity and acceleration analysis using complex algebra formula is more accurate. The method is amenable to solution by digital computers, providing solutions for a large number of input angles once the computer programme is ready. This method leads to a set of linear equations which can be solved conveniently.

4.7 APPLICATION TO SLIDER-CRANK MECHANISM

The loop closure equation for the slider-crank mechanism of Fig. 4.4 as given in (4.38), is

$$r = r_2\, e^{j\theta_2} + r_3\, e^{j\theta_3}$$

For establishing velocity relationship, the loop closure equation is differentiated w.r. to time. Thus,

$$\dot{r} = r_2\, (j\dot{\theta}_2)\, e^{j\theta_2} + r_3\, (j\dot{\theta}_3)\, e^{j\theta_3} \tag{4.61}$$

Note that link lengths r_2 and r_3 have been treated as constant but the length r of link 1 is variable one. Also, angle θ_1 is constant and equal to zero always. The term $\dot{r}$ represents velocity of slider, while $\dot{\theta}_2$ and $\dot{\theta}_3$ represent angular velocities of links 2 and 3 respectively. Hence, $r_2\, \dot{\theta}_2$ gives V_A while $r_3\, \dot{\theta}_3$ gives V_{SA} or V_{AS}. This, however, is not required in velocity or acceleration analysis.

Using Euler's formula, equation (4.61) is rewritten as

$$\dot{r} = r_2\, \dot{\theta}_2\, (j)\, (\cos\theta_2 + j\sin\theta_2) + r_3\, \dot{\theta}_3\, (j)\, (\cos\theta_3 + j\sin\theta_3)$$

or

$$\dot{r} = r_2\, \dot{\theta}_2\, (j\cos\theta_2 - \sin\theta_2) + r_3\, \dot{\theta}_3\, (j\cos\theta_3 - \sin\theta_3)$$

Comparing real and imaginary parts,

$$\dot{r} = -\, (r_2\, \dot{\theta}_2)\sin\theta_2 - (r_3\, \dot{\theta}_3)\sin\theta_3 \tag{4.62}$$

and

$$0 = (r_2\, \dot{\theta}_2)\cos\theta_2 + (r_3\, \dot{\theta}_3)\cos\theta_3 \tag{4.63}$$

Equations (4.62) and (4.63) are linear in variables $\dot{r}$, $\dot{\theta}_3$ and can be solved simultaneously. To eliminate $\dot{\theta}_3$ term, multiply (4.62) by $\cos\theta_3$ and equation (4.63) by $\sin\theta_3$ respectively and add them.

$$\dot{r}\,\cos\theta_3 = (r_2\, \dot{\theta}_2)\,[-\sin\theta_2\cos\theta_3 + \cos\theta_2\sin\theta_3]$$

or,

$$\dot{r}\,\cos\theta_3 = (r_2\, \dot{\theta}_2)\sin(\theta_3 - \theta_2)$$

Hence,

$$\dot{r} = V_S = \frac{(r_2\, \dot{\theta}_2)\sin(\theta_3 - \theta_2)}{\cos\theta_3} \tag{4.64}$$

Also, from equation (4.63),

$$\dot{\theta}_3 = \omega_3 = \frac{-\,(r_2\, \dot{\theta}_2)\cos\theta_2}{(r_3\cos\theta_3)} \tag{4.65}$$

For acceleration analysis, equation (4.61) is further differentiated w.r. to time. Thus,

$$\ddot{r} = r_2\, (j\ddot{\theta}_2)\, e^{j\theta_2} + r_2\, (j\dot{\theta}_2)^2\, e^{j\theta_2} + r_3\, (j\ddot{\theta}_3)\, e^{j\theta_3} + r_3\, (j\dot{\theta}_3)^2\, e^{j\theta_3}$$

Using Euler's formula, we have

$$\ddot{r} = r_2\, \ddot{\theta}_2\, (j\cos\theta_2 - \sin\theta_2) - r_2\, (\dot{\theta}_2)^2\, (\cos\theta_2 + j\sin\theta_2) + r_3\, \ddot{\theta}_3\, (j\cos\theta_3 - \sin\theta_3)$$
$$-\, r_3\, (\dot{\theta}_3)^2(\cos\theta_3 + j\sin\theta_3)$$

Comparing real and imaginary parts,

$$\ddot{r} = - (r_2\,\ddot{\theta}_2)\sin\theta_2 - r_2\,(\dot{\theta}_2)^2\cos\theta_2 - (r_3\,\ddot{\theta}_3)\sin\theta_3 - r_3(\dot{\theta}_3)^2\cos\theta_3$$

or $$\ddot{r} = - [r_2\,\ddot{\theta}_2\sin\theta_2 + \dot{\theta}_2^2\cos\theta_2) + r_3\,\ddot{\theta}_3\sin\theta_3 + \dot{\theta}_3^2\cos\theta_3)] \tag{4.66}$$

and $$0 = (r_2\,\ddot{\theta}_2)\cos\theta_2 - (r_2\,\dot{\theta}_2^2)\sin\theta_2 + (r_3\,\ddot{\theta}_3)\cos\theta_3 - (r_3\,\dot{\theta}_3^2)\sin\theta_3$$

or, $$0 = r_2\,(\ddot{\theta}_2\cos\theta_2 - \dot{\theta}_2^2\sin\theta_2) + r_3\,(\ddot{\theta}_3\cos\theta_3 - \dot{\theta}_3^2\sin\theta_3) \tag{4.67}$$

since $\dot{\theta}_2$ and $\dot{\theta}_3$ are established in velocity equation and $\ddot{\theta}_2$ may be either zero or may have some finite value in the problem, the angular acceleration $\ddot{\theta}_3$ of link 3 can be obtained from equation (4.67) as

$$\ddot{\theta}_3 = \left[\left(-\frac{r_2}{r_3}\right)(\ddot{\theta}_2\cos\theta_2 - \dot{\theta}_2^2\sin\theta_2) + (\dot{\theta}_3)^2\sin\theta_3 \right] / (\cos\theta_3) \tag{4.68}$$

Substituting for $(\ddot{\theta}_3)$ in equation (4.66), $\ddot{r}$ the acceleration of slider can be determined.

4.8 APPLICATION TO FOUR-BAR MECHANISM

The loop-closure equation for a 4-bar mechanism, as given in equation (4.46) for the mechanism in Fig. (4.5), is

$$r_2\,e^{j\theta_2} + r_3\,e^{j\theta_3} = r_1 + r_4\,e^{j\theta_4}$$

For velocity analysis, above loop-closure equation is differentiated. Remembering that r_1, r_2, r_3 and r_4 are invariants with time being the link lengths, differentiation gives

$$r_2\,(j\dot{\theta}_2)\,e^{j\theta_2} + r_3\,(j\dot{\theta}_3)\,e^{j\theta_3} = 0 + r_4\,(j\dot{\theta}_4)\,e^{j\theta_4} \tag{4.69}$$

Remembering that $\dot{\theta}_2$, $\dot{\theta}_3$ and $\dot{\theta}_4$ represent angular velocities ω_2, ω_3 and ω_4 of links 2, 3 and 4 respectively, the above equation is rewritten using Euler's formula as

$$r_2\,(\dot{\theta}_2)\,(j\cos\theta_2 - \sin\theta_2) + r_3\dot{\theta}_3\,(j\cos\theta_3 - \sin\theta_3) = r_4\dot{\theta}_4\,(j\cos\theta_4 - \sin\theta_4)$$

Comparing real and imaginary parts of the above equation,

$$r_2\,(\dot{\theta}_2)\sin\theta_2 + r_3\,(\dot{\theta}_3)\sin\theta_3 = r_4\,(\dot{\theta}_4)\sin\theta_4 \tag{4.70}$$

and, $$r_2\,(\dot{\theta}_2)\cos\theta_2 + r_3\,(\dot{\theta}_3)\cos\theta_3 = r_4\,(\dot{\theta}_4)\cos\theta_4 \tag{4.71}$$

Equations (4.70) and (4.71) are linear algebraic equations in variables $\dot{\theta}_3$ and $\dot{\theta}_4$. To establish $\dot{\theta}_4$ by eliminating unknown variable $\dot{\theta}_3$, multiply equation (4.70) by $\cos\theta_3$ and equation (4.71) by $\sin\theta_3$. Subtraction gives,

$$(r_2\dot{\theta}_2)(\sin\theta_2\cos\theta_3 - \cos\theta_2\sin\theta_3) + 0 = (r_4\,\dot{\theta}_4)(\sin\theta_4\cos\theta_3 - \cos\theta_4\sin\theta_3)$$

or $$(r_2\dot{\theta}_2)\sin(\theta_2 - \theta_3) = (r_4\,\dot{\theta}_4)\sin(\theta_4 - \theta_3)$$

Hence, $$\dot{\theta}_4 = \omega_4 = \left(\frac{r_2}{r_4}\right)\dot{\theta}_2\,\frac{\sin(\theta_2 - \theta_3)}{\sin(\theta_4 - \theta_3)} \tag{4.72}$$

Angular velocity $\dot{\theta}_2 = \omega_2$ of input link 2 being known, the angular velocity $\dot{\theta}_4$ can be established using expression (4.72).

Again, multiplying equation (4.70) by $\cos \theta_4$ and equation (4.71) by $\sin \theta_4$ and subtracting

$$(r_2 \dot{\theta}_2) [\sin \theta_2 \cos \theta_4 - \cos \theta_2 \sin \theta_4] + (r_3 \dot{\theta}_3) [\sin \theta_3 \cos \theta_4 - \cos \theta_3 \sin \theta_4] = 0$$

or
$$(r_2 \dot{\theta}_2) \sin (\theta_2 - \theta_4) + (r_3 \dot{\theta}_3) \sin (\theta_3 - \theta_4) = 0$$

or
$$\dot{\theta}_3 = \left(-\frac{r_2}{r_3}\right) \dot{\theta}_2 \frac{\sin (\theta_2 - \theta_4)}{\sin (\theta_3 - \theta_4)} \qquad (4.73)$$

Expressions (4.72) and (4.73) thus determine angular velocities $\dot{\theta}_3$ and $\dot{\theta}_4$.

For acceleration analysis equation (4.69) is differentiated once more, yielding

$$[(j\ddot{\theta}_2) + (j\dot{\theta}_2)^2] \, r_2 \, e^{j\theta_2} + [(j\ddot{\theta}_3) + (j\dot{\theta}_3)^2] \, r_3 \, e^{j\theta_3} = 0 + [(j\ddot{\theta}_4) + (j\dot{\theta}_4)^2] \, r_4 \, e^{j\theta_4}$$

Simplifying, using Euler's formula, we have

$$(r_2 \ddot{\theta}_2) (j \cos \theta_2 - \sin \theta_2) - r_2 \, \dot{\theta}_2^2 (\cos \theta_2 + j \sin \theta_2) + (r_3 \ddot{\theta}_3) (j \cos \theta_3 - \sin \theta_3)$$
$$- r_3 \, \dot{\theta}_3^2 (\cos \theta_3 + j \sin \theta_3) = (r_4 \ddot{\theta}_4) (j \cos \theta_4 - \sin \theta_4) - r_4 \dot{\theta}_4^2 (\cos \theta_4 + j \sin \theta_4)$$

Comparing real and imaginary parts and eliminating common negative sign,

$$(r_2 \ddot{\theta}_2) \sin \theta_2 + (r_2 \dot{\theta}_2^2) \cos \theta_2 + (r_3 \ddot{\theta}_3) \sin \theta_3 + (r_3 \dot{\theta}_3^2) \cos \theta_3$$
$$= (r_4 \ddot{\theta}_4) \sin \theta_4 + (r_4 \dot{\theta}_4^2) \cos \theta_4 \qquad (4.74)$$

and
$$(r_2 \ddot{\theta}_2) \cos \theta_2 - (r_2 \dot{\theta}_2^2) \sin \theta_2 + (r_3 \ddot{\theta}_3) \cos \theta_3 - (r_3 \dot{\theta}_3^2) \sin \theta_3$$
$$= (r_4 \ddot{\theta}_4) \cos \theta_4 - (r_4 \dot{\theta}_4^2) \sin \theta_4 \qquad (4.75)$$

Equations (4.74) and (4.75) are linear algebraic equations in variables $\ddot{\theta}_3$ ($= \alpha_3$) and $\ddot{\theta}_4$ ($= \alpha_4$), all other parameters like (θ_2, θ_3 and θ_4) are known. These equations can be solved simultaneously to obtain $\ddot{\theta}_3$ and $\ddot{\theta}_4$. Thus multiplying equation (4.74) by $\cos \theta_3$ and equation (4.75) by $\sin \theta_3$ and subtracting

$$(r_2 \ddot{\theta}_2) \sin (\theta_2 - \theta_3) + (r_2 \dot{\theta}_2^2) \cos (\theta_2 - \theta_3) + (r_3 \dot{\theta}_3^2) (\cos^2 \theta_3 + \sin^2 \theta_3)$$
$$= (r_4 \ddot{\theta}_4) \sin (\theta_4 - \theta_3) + (r_4 \dot{\theta}_4^2) \cos (\theta_4 - \theta_3)$$

Hence, by transposition of terms

$$\ddot{\theta}_4 = \frac{1}{r_4 \sin (\theta_4 - \theta_3)} [(r_2 \ddot{\theta}_2) \sin (\theta_2 - \theta_3) + (r_2 \dot{\theta}_2^2) \cos (\theta_2 - \theta_3)$$
$$+ (r_3 \dot{\theta}_3^2) - (r_4 \dot{\theta}_4^2) \cos (\theta_4 - \theta_3)] \qquad (4.76)$$

Also, multiplying equation (4.74) by $\cos \theta_4$ and equation (4.75) by $\sin \theta_4$ and then subtracting

$$(r_2 \ddot{\theta}_2) \sin (\theta_2 - \theta_4) + (r_2 \dot{\theta}_2^2) \cos (\theta_2 - \theta_4) + (r_3 \ddot{\theta}_3) \sin (\theta_3 - \theta_4)$$
$$+ (r_3 \dot{\theta}_3^2) \cos (\theta_3 - \theta_4) = 0 + (r_4 \dot{\theta}_4^2) (\cos^2 \theta_4 + \sin^2 \theta_4)$$

or
$$\ddot{\theta}_3 = \frac{1}{r_3 \sin(\theta_3 - \theta_4)} [r_4 \dot{\theta}_4^2 - (r_2 \ddot{\theta}_2) \sin(\theta_2 - \theta_4) - (r_2 \dot{\theta}_2^2) \cos(\theta_2 - \theta_4)$$

$$- (r_3 \dot{\theta}_3^2) \cos(\theta_3 - \theta_4)] \tag{4.77}$$

Angular accelerations α_3 and α_4 of links 3 and 4 can be evaluated using expressions (4.77) and (4.76) respectively.

4.9 APPLICATION TO QUICK-RETURN MECHANISM

The loop-closure equation for links 1, 2, 3 and 4 of quick-return mechanism shown in Fig. 4.6 (a) is

$$\mathbf{R}_4 = \mathbf{R}_1 + \mathbf{R}_2 \tag{4.78}$$

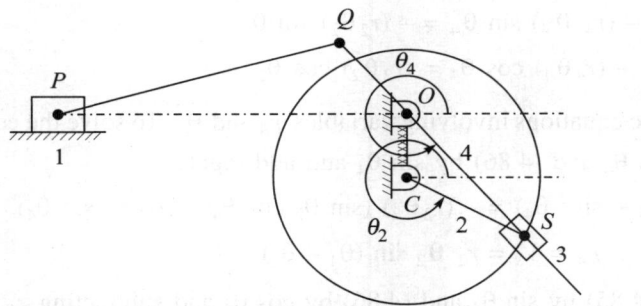
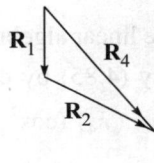

Fig. 4.6

In complex polar notation, the equation is

$$r_4 e^{j\theta_4} = r_1 e^{j\theta_1} + r_2 e^{j\theta_2} \tag{4.79}$$

Link 1 (OQ) is located at 270° w.r. to real axis. Also, r_1 and r_2 are constant but r_4 varies with respect to time. Rewriting equation (4.79) using Euler's formula

$$r_4 (\cos\theta_4 + j\sin\theta_4) = r_1 (\cos\theta_1 + j\sin\theta_1) + r_2 (\cos\theta_2 + j\sin\theta_2)$$

Comparing real and imaginary parts,

$$r_4 \cos\theta_4 = r_1 \cos\theta_1 + r_2 \cos\theta_2$$

and
$$r_4 \sin\theta_4 = r_1 \sin\theta_1 + r_2 \sin\theta_2$$

But as
$$\sin\theta_1 = -1 \quad \text{and} \quad \cos\theta_1 = 0,$$

above equations reduce to

$$r_4 \cos\theta_4 = r_2 \cos\theta_2 \tag{4.80}$$

and
$$r_4 \sin\theta_4 = -r_1 + r_2 \sin\theta_2 \tag{4.81}$$

Squaring and adding on respective sides of Eq. (4.80) and (4.81), and rearranging

$$r_4^2 = r_1^2 + r_2^2 - 2r_1 r_2 \sin\theta_2$$

Therefore
$$r_4 = \sqrt{r_1^2 + r_2^2 - 2r_1 r_2 \sin \theta_2} \qquad (4.82)$$

Also, dividing (4.81) by (4.80) and rearranging

$$\theta_4 = \tan^{-1}\left(\frac{r_2 \sin \theta_2 - r_1}{r_2 \cos \theta_2}\right) \qquad (4.83)$$

Remembering that r_1 and θ_1 are both time invariants and r_4, θ_4 and θ_2 are the time variants, equation (4.79) is differentiated w.r. to time for velocity analysis. Thus,

$$(\dot{r}_4)\, e^{j\theta_4} + r_4\, (j\dot{\theta}_4)\, e^{j\theta_4} = 0 + r_2\, (j\dot{\theta}_2)\, e^{j\theta_2} \qquad (4.84)$$

Rewriting using Euler's formula,

$$(\dot{r}_4)\, (\cos \theta_4 + j \sin \theta_4) + (r_4 \dot{\theta}_4)\, (j \cos \theta_4 - \sin \theta_4) = (r_2 \dot{\theta}_2)\, (j \cos \theta_2 - \sin \theta_2)$$

Comparing real and imaginary parts,

$$(\dot{r}_4)\, \cos \theta_4 - (r_4\, \dot{\theta}_4)\, \sin \theta_4 = - (r_2 \dot{\theta}_2)\, \sin \theta_2 \qquad (4.85)$$

and
$$(\dot{r}_4)\, \sin \theta_4 + (r_4 \dot{\theta}_4)\, \cos \theta_4 = (r_2 \dot{\theta}_2)\, \cos \theta_2 \qquad (4.86)$$

These are linear algebraic equations involving variables $\dot{r}_4$ and $\dot{\theta}_4$. To solve the equations for $(\dot{r}_4)$, multiply (4.85) by $\cos \theta_4$ and (4.86) by $\sin \theta_4$ and add together

$$(\dot{r}_4)\, (\cos^2 \theta_4 + \sin^2 \theta_4) = - (r_2 \dot{\theta}_2)\, (\sin \theta_2 \cos \theta_4 - \cos \theta_2 \sin \theta_4)$$

or
$$\dot{r}_4 = V_s = r_2\, \dot{\theta}_2 \sin (\theta_4 - \theta_2) \qquad (4.87)$$

Similarly, multiplying (4.85) by $\sin \theta_4$ and (4.86) by $\cos \theta_4$ and subtracting

$$(r_4 \dot{\theta}_4)\, (\cos^2 \theta_4 + \sin^2 \theta_4) = (r_2 \dot{\theta}_2)\, (\cos \theta_2 \cos \theta_4 + \sin \theta_2 \sin \theta_4)$$

or
$$\dot{\theta}_4 = \omega_4 = \left(\frac{r_2}{r_4}\right) \dot{\theta}_2\, \cos (\theta_2 - \theta_4) \qquad (4.88)$$

With $\dot{\theta}_4 \equiv \omega_4$ established, it is quite easy to obtain tangential velocity of point Q as

$$V_Q = (OQ)\omega_4$$

Velocity of point P, if desired, can be obtained by considering velocity analysis of slider-crank mechanism OQP.

For acceleration analysis, equation (4.84) is differentiated further w.r. to time. Thus,

$$[\ddot{r}_4 + \dot{r}_4\, (j\dot{\theta}_4)]\, e^{j\theta_4} + [\dot{r}_4\, (j\dot{\theta}_4) + r_4\, (j\ddot{\theta}_4) + r_4\, (j\dot{\theta}_4)^2] e^{j\theta_4} = [r_2\, (j\ddot{\theta}_2) + r_2\, (j\dot{\theta}_2)^2] e^{j\theta_2}$$

Simplifying, this becomes

$$[\ddot{r}_4 + 2j\, (\dot{r}_4\, \dot{\theta}_4) + j\, (r_4\, \ddot{\theta}_4) - r_4\, \dot{\theta}_4^2]\, e^{j\theta_4} = r_2\, (j\ddot{\theta}_2 - \dot{\theta}_2^2)\, e^{j\theta_2} \qquad (4.89)$$

Comparing real and imaginary parts of (4.89),

$$\ddot{r}_4 \cos \theta_4 - 2\, (\dot{r}_4 \dot{\theta}_4)\, \sin \theta_4 - (r_4 \ddot{\theta}_4)\, \sin \theta_4 - (r_4 \dot{\theta}_4^2)\, \cos \theta_4$$
$$= - (r_2 \ddot{\theta}_2 \sin \theta_2 + r_2\, \dot{\theta}_2^2 \cos \theta_2) \qquad (4.90)$$

and
$$\ddot{r}_4 \sin \theta_4 + 2\,(\dot{r}_4 \dot{\theta}_4) \cos \theta_4 + (r_4 \ddot{\theta}_4) \cos \theta_4 - (r_4 \dot{\theta}_4^2) \sin \theta_4$$
$$= (r_2 \ddot{\theta}_2 \cos \theta_2 - r_2\,\dot{\theta}_2^2 \sin \theta_2) \tag{4.91}$$

Equations (4.90) and (4.91) are linear algebraic in variables $\ddot{r}_4\ (\equiv A_S)$ and $\ddot{\theta}_4\ (\equiv \alpha_4)$. To eliminate $\ddot{r}_4$ terms, multiply (4.90) by $\sin \theta_4$ and (4.91) by $\cos \theta_4$. Subtraction gives

$$2\,(\dot{r}_4\,\dot{\theta}_4)\,(\cos^2 \theta_4 + \sin^2 \theta_4) + (r_4 \ddot{\theta}_4)\,(\cos^2 \theta_4 + \sin^2 \theta_4)$$
$$- (r_4 \dot{\theta}_4^4)\,(\sin \theta_4 \cos \theta_4 - \cos \theta_4 \sin \theta_4)$$
$$= (r_2 \ddot{\theta}_2)\,(\cos \theta_2 \cos \theta_4 + \sin \theta_2 \sin \theta_4) - r_2\,\dot{\theta}_2^2\,(\sin \theta_2 \cos \theta_4 + \cos \theta_2 \sin \theta_4)$$

or
$$\ddot{\theta}_4 = \frac{1}{r_4}\,[r_2 \ddot{\theta}_2 \cos (\theta_2 - \theta_4) - r_2\,\dot{\theta}_2^2 \sin (\theta_2 + \theta_4) - 2\,(\dot{r}_4 \dot{\theta}_4)] \tag{4.92}$$

Similarly, to eliminate terms involving $\ddot{\theta}_4$ multiply (4.90) by $\cos \theta_4$ and (4.91) by $\sin \theta_4$ and add together

$$\ddot{r}_4 + 0 + 0 - r_4 \dot{\theta}_4^2 = + r_2 \ddot{\theta}_2 \sin (\theta_4 - \theta_2) - r_2 \dot{\theta}_2^2 \cos (\theta_4 - \theta_2)$$

Therefore
$$\ddot{r}_4 = A_S = r_4 \dot{\theta}_4^2 + r_2 \ddot{\theta}_2 \sin (\theta_4 - \theta_2) - r_2 \dot{\theta}_2^2 \cos (\theta_4 - \theta_2)$$

Knowing $\ddot{\theta}_4$, $(\equiv \alpha_4)$, A_Q^t can be obtained as

$$(A_Q^t) = (OQ)\alpha_4 \text{ in the sense commensurate with that of } \alpha_4.$$

Also, knowing $\mathbf{A}_Q^r = (V_Q)^2/OQ$ and $\mathbf{A}_Q^t\ (OQ)\alpha_4$, the acceleration of slider P can be established by the following procedure described as "slider-crank mechanism."

EXAMPLE 4.1 In the slider-crank mechanism shown in Fig. 4.7, the line of stroke of slider P is off-set a perpendicular distance of 5 cm from the crank centre C. Length of connecting rods PB and crank BC are 75 cm, and 20 cm. respectively. Crank BC rotates clockwise at a uniform speed of 200 r.p.m. Find analytically linear velocity of slider and angular velocity of connecting rod. Also find out acceleration of slider and angular acceleration of PB, Take angle $DCB = 135°$, the datum line DC being parallel to the line of stroke.

Solution: Measured with respect to positive x-axis direction, the orientation of vectors $\mathbf{R}_2$, $\mathbf{R}_3$, $\mathbf{R}$ and $\mathbf{R}_1$ are θ_2, θ_3, θ and θ_1 respectively as shown in Figure. Note that $\theta = 90°$, $\theta_1 = 180°$ and, except for vector $\mathbf{R}_1$, all vectors, namely $\mathbf{R}_1$, $\mathbf{R}_2$ and $\mathbf{R}_3$ have constant magnitudes.

The loop-closure equations can be written as

$$\mathbf{R} + \mathbf{R}_1 = \mathbf{R}_2 + \mathbf{R}_3$$

This can be expressed in complex polar form as

$$re^{j\theta} + r_1 e^{j\theta_1} = r_2 e^{j\theta_2} + r_3 e^{j\theta_3} \tag{4.93}$$

Rewriting using Euler's formula and comparing real and imaginary terms,

$$r \cos \theta + r_1 \cos \theta_1 = r_2 \cos \theta_2 + r_3 \cos \theta_3$$
and
$$r \sin \theta + r_1 \sin \theta_1 = r_2 \sin \theta_2 + r_3 \sin \theta_3$$

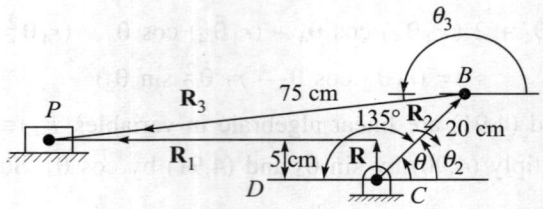

Fig. 4.7

Noting that $\theta = 90°$ and $\theta_1 = 180°$, above equations reduce to

$$- r_1 = r_2 \cos \theta_2 + r_3 \cos \theta_3 \tag{4.94}$$

and

$$r = r_2 \sin \theta_2 + r_3 \sin \theta_3 \tag{4.95}$$

For given configuration $\angle DCB = 135°$ and $\theta = 45°$. Rewriting above equations,

$$\sin \theta_3 = \frac{r - r_2 \sin \theta_2}{r_3} = \frac{5 - 20 \sin 45°}{75} = -0.122$$

Therefore

$$\theta_3 = 187°$$

Also, from Eq. (4.94),

$$r_1 = - (20 \cos 45° + 75 \cos 187°)$$
$$= 60.3 \text{ cm}$$

Noting that r, r_2, r_3 and θ, θ_1 are constant, differentiating equation (4.93) w.r. to time,

$$0 + (\dot{r}_1) \, e^{j\theta_1} = r_2 (j\dot{\theta}_2) \, e^{j\theta_2} + r_3 (j\dot{\theta}_3) \, e^{j\theta_3} \tag{4.96}$$

Using Euler's formula, this may be written as

$$\dot{r}_1 (\cos \theta_1 + j \sin \theta_1) = r_2 \dot{\theta}_2 (j \cos \theta_2 - \sin \theta_2) + r_3 \dot{\theta}_3 (j \cos \theta_3 - \sin \theta_3)$$

Comparing real and imaginary parts,

$$\dot{r}_1 \cos \theta_1 = - (r_2 \dot{\theta}_2) \sin \theta_2 - (r_3 \dot{\theta}_3) \sin \theta_3 \tag{4.97}$$

and

$$\dot{r}_1 \sin \theta_1 = + (r_2 \dot{\theta}_2) \cos \theta_2 + (r_3 \dot{\theta}_3) \cos \theta_3 \tag{4.98}$$

Since $\theta_1 = 180°$, from equation (4.98)

$$0 = (r_2 \dot{\theta}_2) \cos \theta_2 + (r_3 \dot{\theta}_3) \cos \theta_3$$

and so,

$$\dot{\theta}_3 = - \left(\frac{r_2}{r_3}\right) \dot{\theta}_2 \left(\frac{\cos \theta_2}{\cos \theta_3}\right)$$

$$= - \left(\frac{20}{75}\right) \left(\frac{2 \pi \times 200}{60}\right) \left(\frac{\cos 45°}{\cos 187°}\right)$$

$$= 3.98 \text{ rad/s}$$

Also, from equation (4.97),

$$- (\dot{r}_1) = - (r_2 \dot{\theta}_2) \sin \theta_2 - (r_3 \dot{\theta}_3) \sin \theta_3$$

or

$$\dot{r}_1 = -(20) \left(\frac{2 \pi \times 200}{60} \right) \sin 45° - (75 \times 3.98) \sin 187°$$

Therefore

$$V_P = \dot{r}_1 = 259.814 \text{ cm/s}$$

For acceleration, differentiating (4.96) further w.r. to time

$$(\ddot{r}_1) \, e^{j\theta_1} = r_2 [(j \ddot{\theta}_2) + (j \dot{\theta}_2)^2] \, e^{j\theta_2} + r_3 [(j \ddot{\theta}_3) + (j \dot{\theta}_3)^2] \, e^{j\theta_3}$$

Separating real and imaginary parts,

$$\ddot{r}_1 (\cos \theta_1 + j \sin \theta_1) = r_2 \ddot{\theta}_2 (j \cos \theta_2 - \sin \theta_2) - (r_2 \dot{\theta}_2^2) (\cos \theta_2 + j \sin \theta_2)$$

$$+ r_3 \ddot{\theta}_3 (j \cos \theta_3 - \sin \theta_3) - r_3 \dot{\theta}_3^2 (\cos \theta_3 + j \sin \theta_3)$$

Comparing real and imaginary parts,

$$(\ddot{r}_1) \cos \theta_1 = - (r_2 \ddot{\theta}_2) \sin \theta_2 - (r_2 \dot{\theta}_2^2) \cos \theta_2 - (r_3 \ddot{\theta}_3) \sin \theta_3 - (r_3 \dot{\theta}_3^2) \cos \theta_3 \quad (4.99)$$

and

$$(\ddot{r}_1) \sin \theta_1 = (r_2 \ddot{\theta}_2) \cos \theta_2 - (r_2 \dot{\theta}_2^2) \sin \theta_2 + (r_3 \ddot{\theta}_3) \cos \theta_3 - (r_3 \dot{\theta}_3^2) \sin \theta_3 \quad (4.100)$$

Remembering that for uniform crank rotation $\ddot{\theta}_2 = 0$, and substituting values of $\theta_1, \theta_2, \theta_3$ and r_2, r_3, the two equations reduce to

$$- (\ddot{r}_1) = - (r_2 \dot{\theta}_2^2 \cos \theta_2 + r_3 \ddot{\theta}_3 \sin \theta_3 + r_3 \dot{\theta}_3^2 \cos \theta_3) \quad (4.101)$$

and

$$0 = - (r_2 \dot{\theta}_2^2) \sin \theta_2 + (r_3 \ddot{\theta}_3) \cos \theta_3 - (r_3 \dot{\theta}_3^2) \sin \theta_3 \quad (4.102)$$

From (4.102), by rearranging,

$$\ddot{\theta}_3 = \frac{(r_2 \dot{\theta}_2^2) \sin \theta_2 + (r_3 \dot{\theta}_3^2) \sin \theta_3}{r_3 \cos \theta_3}$$

or

$$\ddot{\theta}_3 = \frac{(20) \left(\frac{2 \pi \times 200}{60} \right)^2 \sin 45° - (75) (3.98)^2 \sin 187°}{75 \cos 187°}$$

Thus,

$$\ddot{\theta}_3 = - 85.278 \text{ rad/s}^2 \ (i.e. \text{ c.c.w.}) \qquad \textbf{Ans.}$$

Also, from (4.101)

$$\ddot{r}_1 = (20) \left(\frac{2 \pi \times 200}{60} \right)^2 \cos 45° + (75) (- 85.278) \sin 187° + (75) (3.98)^2 \cos 187°$$

Therefore

$$V_P = 5803.72 \text{ cm/s}^2 \text{ (In the sense of vector } \mathbf{R}_1). \qquad \textbf{Ans.}$$

EXAMPLE 4.2 A pump is driven from an engine crank shaft by the mechanism shown in Fig. 4.8. The pump piston shown at F is 30 cm diameter and the uniform crank speed is 120 r.p.m. Determine for the position shown: (a) the velocity of cross-head E., (b) the torque required at the shaft to overcome a pressure of 29.4 N/sq cm, (c) acceleration of cross-head E and (d) angular acceleration of bell-crank lever BCD, take

$$OA = 30 \text{ cm} ; AB = 120 \text{ cm} ; BC = 90 \text{ cm} ; CD = 30 \text{ cm and } DE = 150 \text{ cm}.$$

Solution: Assume the crank to rotate in c.c.w. sense. We take this as positive sense. Representing links by vectors $\mathbf{R}_1$, $\mathbf{R}_2$, $\mathbf{R}_3$, $\mathbf{R}_4$, $\mathbf{R}_5$, $\mathbf{R}_6$, and $\mathbf{R}_7$, the loop closure equations, for parts $OABC$ and CDE of mechanism, are

$$\mathbf{R}_2 + \mathbf{R}_3 + \mathbf{R}_4 = \mathbf{R}_1 \tag{a}$$

and
$$\mathbf{R}_5 + \mathbf{R}_6 = \mathbf{R}_7 + \mathbf{R}_8 \tag{b}$$

For position analysis, we need to consider a diagonal position vector $\mathbf{R}$ as shown. The loop-closure equation for loop OAC is then,

$$\mathbf{R}_2 + \mathbf{R} = \mathbf{R}_1$$

which, in complex polar form, is

$$r_2 e^{j\theta_2} + r e^{j\theta} = r_1 e^{j\theta_1} \tag{c}$$

Similarly, loop closure equation for portion ACB is

$$\mathbf{R}_3 + \mathbf{R}_4 = \mathbf{R}$$

which in complex polar form is

$$r_3 e^{j\theta_3} + r_4 e^{j\theta_4} = r e^{j\theta} \tag{d}$$

where, $r_1 = OC = \sqrt{(120)^2 + (90)^2} = 150$ cm

$r_2 = OA = 30$ cm

$r_3 = AB = 120$ cm

$r_4 = BC = 90$ cm

$r_5 = CD = 30$ cm

$r_6 = DE = 150$ cm

$r_8 = 15$ cm

$r_7 = ?$

Also

$$(360 - \theta_1) = \tan^{-1}\left(+\frac{90}{120}\right) = 36.87°$$

$\therefore$
$$\theta_1 = 360 - 36.87$$
$$= 323.13°$$

And
$$\theta_2 = 180 + 45 = 225°$$
$$\theta_8 = 0 \text{ and } \theta_7 = 270°$$

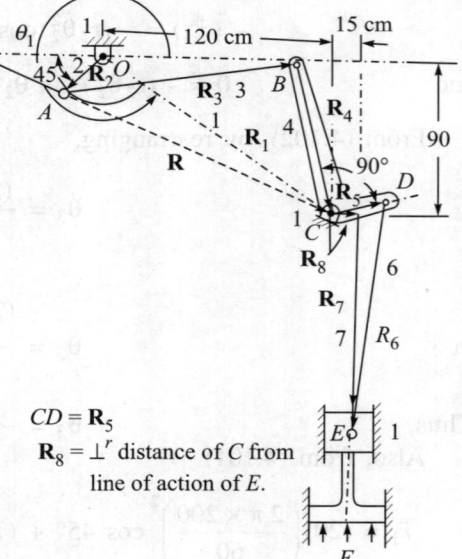

$CD = \mathbf{R}_5$

$\mathbf{R}_8 = \perp^r$ distance of C from line of action of E.

Fig. 4.8 Pump driving mechanism.

From equation (c), separating real & imaginary parts,

$$r \cos \theta = r_1 \cos \theta_1 - r_2 \cos \theta_2$$

and
$$r \sin \theta = r_1 \sin \theta_1 - r_2 \sin \theta_2$$

Squaring and adding under radical sign,

$$r = \sqrt{r_1^2 + r_2^2 - 2r_1 r_2 (\cos \theta_1 \cos \theta_2 + \sin \theta_1 \sin \theta_2)}$$

or

$$r = \sqrt{r_1^2 + r_2^2 - 2r_1 r_2 \cos (\theta_1 - \theta_2)}$$

or

$$r = \sqrt{(150)^2 + (30)^2 - 2(150)(30) \cos (323.13 - 225)}$$

Thus, $r = 157.07$ cm.

Also, by division,

$$\tan \theta = \frac{r_1 \sin \theta_1 - r_2 \sin \theta_2}{r_1 \cos \theta_1 - r_2 \cos \theta_2} = \frac{(150) \sin 323.13° - (30) \sin 225°}{(150) \cos 323.13° - (30) \cos 225°}$$

or

$$\theta = \tan^{-1} \left(\frac{-68.79}{141.213} \right) = - \tan^{-1} (- 0.4871) = - 27.47°$$

or

$$\theta = (360 - 27.47) = 332.53°$$

Again, from equation (d), separating real and imaginary parts,

$$\left. \begin{array}{l} r_3 \cos \theta_3 = r \cos \theta - r_4 \cos \theta_4 \\ r_3 \sin \theta_3 = r \sin \theta - r_4 \sin \theta_4 \end{array} \right] \qquad \text{(e)}$$

Squaring and adding

$$r_3^2 = r^2 + r_4^2 - 2rr_4 \cos (\theta - \theta_4)$$

Therefore $$\cos (\theta - \theta_4) = \frac{r^2 + r_4^2 - r_3^2}{2(r)(r_4)} = \frac{18370.98}{28272.6} = 0.64978$$

Therefore $\pm (\theta - \theta_4) = 49.475°$

As $\theta = 334.53°$,

either $\theta_4 = (334.53 - 49.475) = 285.055°$

or $\theta_4 = 334.53 + 49.475 = 384.00°$

From geometrical considerations of the mechanism configuration, $\theta_4 = 285°$ is more appropriate. Hence, $\theta_4 = 285°$.

Substituting in equation (e),

$$\sin \theta_3 = \left(\frac{r \sin \theta - r_4 \sin \theta_4}{r_3} \right)$$

or

$$\sin \theta_3 = \frac{(157.07) \sin 334.53 - (90) \sin 285}{120}$$

Thus

$$\sin \theta_3 = \frac{19.387}{120} = 0.1615$$

Therefore

$$\theta_3 = \sin^{-1} (0.1615) = 9.294°$$

Also, as $\angle BCD = 90°$, it follows that side CD of bell-crank lever (represented by vector $\mathbf{R}_5$) makes an angle θ_5 with positive x-axis, such that,

$$\theta_5 = 285 + 90 = 375° \quad i.e., \quad \theta_5 = 15°$$

So from the loop closure equation of portion CDE of mechanism,

$$\mathbf{R}_7 + \mathbf{R}_8 = \mathbf{R}_5 + \mathbf{R}_6$$

we have

$$r_7\, e^{j\theta_7} + r_8\, e^{j\theta_8} = r_5\, e^{j\theta_5} + r_6\, e^{j\theta_6}$$

Separating real and imaginary parts through Euler's formula

$$r_7 \cos\theta_7 + r_8 \cos\theta_8 = r_5 \cos\theta_5 + r_6 \cos\theta_6 \tag{f}$$

and $\qquad r_7 \sin\theta_7 + r_8 \sin\theta_8 = r_5 \sin\theta_5 + r_6 \sin\theta_6$

As $\qquad\qquad\qquad \theta_7 = 270°$ and $\theta_8 = 0$, above equations may be rewritten as

$$r_8 = r_5 \cos\theta_5 + r_6 \cos\theta_6 \tag{g}$$

and $\qquad\qquad -r_7 = r_5 \sin\theta_5 + r_6 \sin\theta_6 \tag{h}$

From equation (g)

$$\cos\theta_6 = \left(\frac{r_8 - r_5 \cos\theta_5}{r_6}\right) = \frac{15 - 30 \cos 15}{150}$$

Therefore $\qquad\qquad \pm (\theta_6) = \cos^{-1}(-0.093185) = 95.347°$

As should be clear from the configuration, $\theta_6 = -95.347$ (i.e. $\theta_6 = 360 - 95.347$) is acceptable.

Thus $\qquad\qquad\qquad\qquad \theta_6 = 264.653°$

Hence, from equation (h)

$$r_7 = -(30 \sin 15 + 150 \sin 264.653) = 141.58 \text{ cm,}$$

Thus, $\qquad r_1 = 150$ cm; and $\qquad \theta_1 = 323.13°$

$\qquad\qquad r_2 = 30$ cm; $\qquad\qquad\qquad \theta_2 = 225.0°$

$\qquad\qquad r_3 = 120$ cm; $\qquad\qquad\qquad \theta_3 = 9.2970°$

$\qquad\qquad r_4 = 90$ cm; $\qquad\qquad\qquad \theta_4 = 285°$

$\qquad\qquad r_5 = 30$ cm; $\qquad\qquad\qquad \theta_5 = 15°$

$\qquad\qquad r_6 = 150$ cm; $\qquad\qquad\qquad \theta_6 = 264.657°$

$\qquad\qquad r_7 = 141.58$ cm; $\qquad\qquad \theta_7 = 270°$

$\qquad\qquad r_8 = 15$ cm; $\qquad\qquad\qquad \theta = 334.53°$

$\qquad\qquad r = 157.87$ cm; $\qquad\qquad \theta_8 = 0$

Also, input velocity $\omega_2 = \theta_2 = \dfrac{2\pi \times 120}{60} = 4\,\pi$ rad/s

For velocity analysis, let us differentiate the loop closure equation

$$r_1 e^{j\theta_1} = r_2 e^{j\theta_2} + r_3 e^{j\theta_3} + r_4 e^{j\theta_4}$$

which gives

$$O = r_2 \, (j\dot{\theta}_2) \, e^{j\theta_2} + (r_3) \, (j\dot{\theta}_3)e^{j\theta_3} + r_4 \, (j\dot{\theta}_4) \, e^{j\theta_4} \tag{1}$$

(Note that r_1, r_2, r_3, r_4 and θ_1 are all time invariants.)

Separating real and imaginary parts of equation (1), using Euler's formula,

$$- r_2 \, \dot{\theta}_2 \sin \theta_2 - r_3 \dot{\theta}_3 \sin \theta_3 - r_4 \dot{\theta}_4 \sin \theta_4 = 0,$$

and

$$r_2 \dot{\theta}_2 \cos \theta_2 + r_3 \dot{\theta}_3 \cos \theta_3 + r_4 \dot{\theta}_4 \cos \theta_4 = 0$$

Rewriting above equations,

$$(r_4 \dot{\theta}_4) \sin \theta_4 = - (r_2 \dot{\theta}_2 \sin \theta_2 + r_3 \dot{\theta}_3 \sin \theta_3) \tag{2}$$

and

$$(r_4 \dot{\theta}_4) \cos \theta_4 = - (r_2 \dot{\theta}_2 \cos \theta_2 + r_3 \dot{\theta}_3 \cos \theta_3) \tag{3}$$

Multiplying equation (2) by $\cos \theta_3$, equation (3) by $\sin \theta_3$ and subtracting

$$(r_4 \dot{\theta}_4) \sin (\theta_4 - \theta_3) = r_2 \dot{\theta}_2 \sin (\theta_3 - \theta_2)$$

or

$$\dot{\theta}_4 = \omega_4 = \frac{(30) \, (4\pi) \sin (9.297 - 225)}{(90) \sin (285 - 9.297)}$$

or

$$\dot{\theta}_4 = -2.458 \text{ rad/s (c.w.)}$$

$\mathbf{R}_4$ and $\mathbf{R}_5$ being the arms of bell crank lever, $\dot{\theta}_5 = \dot{\theta}_4$

(Note that a sense c.c.w. of ω_2 was considered positive)

Rewriting equation (2), and substituting for $\dot{\theta}_4$

$$\dot{\theta}_3 = \frac{- (r_4 \dot{\theta}_4 \sin \theta_4 + r_2 \dot{\theta}_2 \sin \theta_2)}{r_3 \sin \theta_3}$$

$$\dot{\theta}_3 = \omega_3 = \frac{- [(90) \, (- 2.458) \sin 285 + (30) \, (4\pi) \sin 225]}{120 \sin (9.297)}$$

$$= 2.728 \text{ rad/s (c.c.w.)}$$

For velocity analysis of portion CDE of mechanism, loop closure equation (b) may be rewritten as

$$r_5 \, e^{j\theta_5} + r_6 \, e^{j\theta_6} = r_7 \, e^{j\theta_7} + r_8 \, e^{j\theta_8}$$

Remembering that r_5, r_6, r_8, θ_7 and θ_8 are time invariants, differentiating the above equation

$$r_5(j\dot{\theta}_5)e^{j\theta_5} + r_6(j\dot{\theta}_6)e^{j\theta_6} = (\dot{r}_7) \, e^{j\theta_7} + 0 \tag{4}$$

Separating real and imaginary parts and comparing them,

$$- (r_5 \dot{\theta}_5) \sin \theta_5 - (r_6 \dot{\theta}_6) \sin \theta_6 = (\dot{r}_7) \cos \theta_7 \tag{5}$$

and

$$(r_5 \dot{\theta}_5) \cos \theta_5 + (r_6 \dot{\theta}_6) \cos \theta_6 = (\dot{r}_7) \sin \theta_7 \tag{6}$$

Multiplying equation (5) by $\cos \theta_6$, equation (6) by $\sin \theta_6$ and adding,

$$(r_5 \dot{\theta}_5) \sin (\theta_6 - \theta_5) = \dot{r}_7 \cos (\theta_6 - \theta_7)$$

or
$$V_E \equiv V_F = \dot{r}_7 = \frac{r_5(\dot{\theta}_5) \sin(\theta_6 - \theta_5)}{\cos(\theta_6 - \theta_7)}$$

since
$$\dot{\theta}_5 = \dot{\theta}_4 = -2.458 \text{ rad/s}$$

$$V_F = \dot{r}_7 = \frac{30(-2.458) \sin(264.653 - 15)}{\cos(264.653 - 270)}$$

or
$$V_F = 69.44 \text{ cm/s} \qquad \textbf{Ans.}$$

Substituting for $\dot{r}_7$ in equation (6), and rearranging

$$\dot{\theta}_6 = \frac{(\dot{r}_7) \sin \theta_7 - (r_5 \dot{\theta}_5) \cos \theta_5}{(r_6) \cos \theta_6}$$

or
$$\dot{\theta}_6 = -\frac{(69.44) \sin 270 - (30)(-2.458) \cos 15}{(150) \cos 264.657} = -0.128 \text{ rad/s}$$

or
$$\dot{\theta}_6 = 0.128 \text{ rad/s (c.w.)}$$

Neglecting friction and other losses,

$$P \times V_F = T \times \omega_2$$

where P is the force acting on piston and T is the torque to be exerted on crank OA,

Now
$$P = \frac{\pi}{4}(30)^2 \times 29.4 = 20,781.635 \text{ Newton}$$

Hence,
$$T = \frac{(P V_F)}{\omega_2} = \frac{20,781.635 \times 69.44}{4\pi}$$

$$= 114,836.39 \text{ N-cm}$$
$$= 1148.364 \text{ N-m} \qquad \textbf{Ans.}$$

For acceleration analysis, differentiating equation (1) once more (note that $\dot{\theta}_2 = 0$, as crank rotates uniformly)

$$0 = r_2 [0 + (j\dot{\theta}_2)^2] e^{j\theta_2} + r_3 [(j\ddot{\theta}_3) + (j\dot{\theta}_3)^2] e^{j\theta_3} + r_4 [(j\ddot{\theta}_4) + (j\dot{\theta}_4)^2]e^{j\theta_4}$$

Separating real and imaginary parts and comparing them,

$$0 = -r_2(\dot{\theta}_2)^2 \cos \theta_2 - (r_3\ddot{\theta}_3) \sin \theta_3 - (r_3\dot{\theta}_3^2) \cos \theta_3 - (r_4\ddot{\theta}_4) \sin \theta_4 - (r_4\dot{\theta}_4^2) \cos \theta_4$$

and,
$$0 = -(r_2\dot{\theta}_2^2) \sin \theta_2 + (r_3\ddot{\theta}_3) \cos \theta_3 - (r_3\dot{\theta}_3^2) \sin \theta_3 + (r_4\ddot{\theta}_4) \cos \theta_4 - r_4(\dot{\theta}_4)^2 \sin \theta_4$$

Rearranging above equations,

$$-(r_4\ddot{\theta}_4) \sin \theta_4 = (r_2\dot{\theta}_2^2) \cos \theta_2 + (r_3\ddot{\theta}_3) \sin \theta_3 + (r_3\dot{\theta}_3^2) \cos \theta_3 + (r_4\dot{\theta}_4^2) \cos \theta_4$$

$$(r_4\ddot{\theta}_4) \cos \theta_4 = (r_2\dot{\theta}_2^2) \sin \theta_2 - (r_3\ddot{\theta}_3) \cos \theta_3 + (r_3\dot{\theta}_3^2) \sin \theta_3 + (r_4\dot{\theta}_4^2) \sin \theta_4$$

To eliminate terms involving $\ddot{\theta}_3$ multiply former equation by $\cos \theta_3$ and the latter one by $\sin \theta_3$ and add

$$(r_4 \ddot{\theta}_4) \sin (\theta_3 - \theta_4) = (r_2 \dot{\theta}_2^2) \cos (\theta_2 - \theta_3) + (r_3 \dot{\theta}_3^2) + (r_4 \dot{\theta}_4^2) \cos (\theta_3 - \theta_4)$$

or
$$\ddot{\theta}_4 = \frac{(r_2 \dot{\theta}_2^2) \cos (\theta_2 - \theta_3) + r_3 (\dot{\theta}_3)^2 + (r_4 \dot{\theta}_4^2) \cos (\theta_3 - \theta_4)}{r_4 \sin (\theta_3 - \theta_4)}$$

or $\ddot{\theta}_4 = \alpha_4 = \dfrac{(30)(4\pi)^2 \cos (225 - 9.297) + (120)(2.728)^2 + (90)(-2.458)^2 \cos (9.297 - 285)}{(90) \sin (9.297 - 285)}$

or
$$\ddot{\theta}_4 = \frac{-3847.028 + 893.038 + 54.0343}{89.554} = \frac{-2899.95}{89.554}$$

Thus $\qquad \ddot{\theta}_4 = -32.382 \text{ (c.w.) rad/s}^2$

Hence, $\qquad A_B^t = (BC)\, \alpha_4 = (90)(32.382)$

or $\qquad A_B^t = 2914.38 \text{ cm/s}^2$

Hence $\qquad \ddot{\theta}_5 = \ddot{\theta}_4 = 32.382 \text{ rad/s}^2$ **Ans.**

For acceleration analysis of portion *CDE* of mechanism, differentiate equation (4) w.r. to time

$$r_5 [(j\ddot{\theta}_5) + (j\dot{\theta}_5)^2]\, e^{j\theta_5} + r_6 [(j\ddot{\theta}_6) + (j\dot{\theta}_6)^2]\, e^{j\theta_6} = (\ddot{r}_7) e^{j\theta_7}$$

Separating real and imaginary parts and comparing

$$- (r_5 \ddot{\theta}_5) \sin \theta_5 - r_5 \dot{\theta}_5^2 \cos \theta_5 - r_6 \ddot{\theta}_6 \sin \theta_6 - r_6 \dot{\theta}_6^2 \cos \theta_6 = \ddot{r}_7 \cos \theta_7 \qquad (7)$$

and $\qquad (r_5 \ddot{\theta}_5) \cos \theta_5 - r_5 \dot{\theta}_5^2 \sin \theta_5 + r_6 \ddot{\theta}_6 \cos \theta_6 - r_6 \dot{\theta}_6^2 \sin \theta_6 = \ddot{r}_7 \sin \theta_7 \qquad (8)$

Since $\ddot{\theta}_5$ is known, to eliminate $\ddot{\theta}_6$ multiply (7) by $\cos \theta_6$ and (8) by $\sin \theta_6$ and add

$$(r_5 \ddot{\theta}_5) \sin (\theta_6 - \theta_5) - (r_5 \dot{\theta}_5^2) \cos (\theta_6 - \theta_5) - r_6 \dot{\theta}_6^2 = \ddot{r}_7 \cos (\theta_6 - \theta_7)$$

Hence $\qquad \ddot{r}_7 = \dfrac{(r_5 \ddot{\theta}_5) \sin (\theta_6 - \theta_5) - (r_5 \dot{\theta}_5^2) \cos (\theta_6 - \theta_5) - r_6 \dot{\theta}_6^2}{\cos (\theta_6 - \theta_7)}$

Therefore

$$\ddot{r}_7 = \frac{(30)(-32.382) \sin (264.657 - 15) - (30)(-2.458)^2 \cos (264.657 - 15) - (150)(-0.128)^2}{\cos (264.657 - 270)}$$

Therefore $\qquad \ddot{r}_7 = \dfrac{971.42}{0.9956} = 975.7 \text{ cm/s}^2$

Thus, acceleration of cross-head $A_E = \ddot{r}_7 = 975.7 \text{ cm/s}^2$ **Ans.**

EXAMPLE 4.3 The driving crank *OP* of quick-return mechanism, shown in Fig. 4.9, resolves at a uniform speed of 200 r.p.m. Find the velocity and acceleration of the tool post *R*, in the position shown, when the crank makes an angle of 60 degrees with the vertical line of centres *CO*. What is the acceleration of the sliding block at *P* along the slotted lever *CQ*?

Solution: $\omega_2 = \dot{\theta}_2 = 2\pi \times 200/60 = 20.94 \text{ rad/s}$ (assumed c.w.). Take *C* as origin.

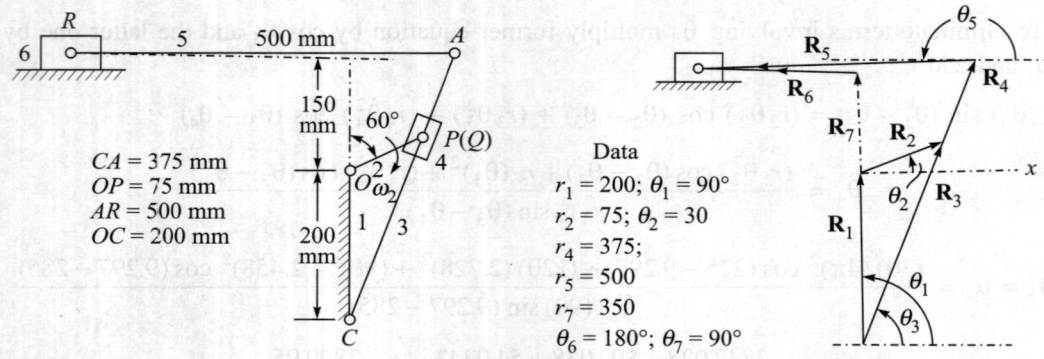

CA = 375 mm
OP = 75 mm
AR = 500 mm
OC = 200 mm

Data
$r_1 = 200$; $\theta_1 = 90°$
$r_2 = 75$; $\theta_2 = 30$
$r_4 = 375$;
$r_5 = 500$
$r_7 = 350$
$\theta_6 = 180°$; $\theta_7 = 90°$

Fig. 4.9   **Fig. 4.10**

Let OP be the crank and Q the point fixed on AC with which sliding point P is momentarily in coincidence. The loop closure equation for portion COP of mechanism is

$$\mathbf{R}_1 + \mathbf{R}_2 = \mathbf{R}_3$$

which can be rewritten in complex polar form as

$$r_1 e^{j\theta_1} + r_2 e^{j\theta_2} = r_3 e^{j\theta_3} \tag{1}$$

Separating real and imaginary parts and comparing

$$r_1 \cos\theta_1 + r_2 \cos\theta_2 = r_3 \cos\theta_3 \tag{2}$$

and

$$r_1 \sin\theta_1 + r_2 \sin\theta_2 = r_3 \sin\theta_3 \tag{3}$$

Squaring and adding

$$r_3 = \sqrt{r_1^2 + r_2^2 + 2r_1 r_2 \cos(\theta_1 - \theta_2)}$$

or

$$r_3 = \sqrt{(200)^2 + (75)^2 + 2(200)(75)\cos(90 - 30)}$$

Thus

$$r_3 = 246.2 \text{ mm}$$

Dividing (3) by (2)

$$\tan\theta_3 = \frac{r_1 \sin\theta_1 + r_2 \sin\theta_2}{r_1 \cos\theta_1 + r_2 \cos\theta_2}$$

Therefore

$$\theta_3 = \tan^{-1}\left(\frac{200\sin 90 + 75\sin 30}{200\cos 90 + 75\cos 30}\right)$$

or

$$\theta_3 = \tan^{-1}(3.6565) = 74.7°$$

Also, the loop closure equation for the portion CAR of mechanism (see Fig. 4.10) is

$$\mathbf{R}_4 + r_5 = \mathbf{R}_7 + \mathbf{R}_6$$

which may be rewritten as

$$r_4 \, e^{j\theta_4} + r_5 e^{j\theta_5} = r_6 e^{j\theta_6} + r_7 e^{j\theta_7} \tag{4}$$

Separating real and imaginary parts and comparing them

$$r_4 \cos \theta_4 + r_5 \cos \theta_5 = r_6 \cos \theta_6 + r_7 \cos \theta_7$$

and

$$r_4 \sin \theta_4 + r_5 \sin \theta_5 = r_6 \sin \theta_6 + r_7 \sin \theta_7$$

Remembering that r_6 and θ_5 are unknown and that $\theta_6 = 180°$ and $\theta_7 = 90°$, above equations can be reduced to

$$r_4 \cos \theta_4 + r_5 \cos \theta_5 = - r_6 \tag{5}$$

and

$$r_4 \sin \theta_4 + r_5 \sin \theta_5 = r_7$$

From equations (5), as $\theta_4 \equiv \theta_3 = 74.7°$

$$\theta_5 = \sin^{-1}\left[\frac{350 - (375) \sin 74.7}{500} \right] = - 1.342°$$

i.e.

$$\theta_5 = 180 + 1.342 = 181.342°$$

Substituting in equation (5)

$$r_6 = - [375 \cos 74.7 + 500 \cos 181.342°]$$
$$= 400.91 \text{ mm}$$

For velocity analysis, differentiate equation (1) w.r. to time,
(note that r_1, r_2 and θ_1 are constant but r_3, θ_2 and θ_3 are variables)

$$0 + r_2 \, (j \dot{\theta}_2) \, e^{j\theta_2} = [\dot{r}_3 + (j \, \dot{\theta}_3) \, r_3] \, e^{j\theta_3} \tag{7}$$

Separating real and imaginary parts and comparing

$$- r_2 \dot{\theta}_2 \sin \theta_2 = \dot{r}_3 \cos \theta_3 - r_3 \dot{\theta}_3 \sin \theta_3 \tag{8}$$

and

$$r_2 \dot{\theta}_2 \cos \theta_2 = \dot{r}_3 \sin \theta_3 + r_3 \dot{\theta}_3 \cos \theta_3 \tag{9}$$

To eliminate $\dot{r}_3$, multiplying equation (8) by $\sin \theta_3$ and (9) by $\cos \theta_3$ and subtracting

$$(r_2 \dot{\theta}_2) \cos (\theta_2 - \theta_3) = r_3 \dot{\theta}_3 \, (\cos^2 \theta_3 + \sin^2 \theta_3)$$

Therefore
$$\dot{\theta}_3 = \left(\frac{r_2}{r_3} \right) \dot{\theta}_2 \, \cos (\theta_2 - \theta_3) = 4.844 \text{ rad/s (c.w.)}$$

Substituting in equation (9),

$$\dot{r}_3 = V_Q = \frac{r_2 \dot{\theta}_2 \cos \theta_2 - r_3 \dot{\theta}_3 \cos \theta_3}{\sin \theta_3}$$

or
$$\dot{r}_3 = \frac{(75) \, (20.94) \cos 30 - (246.2) \, (4.844) \cos 74.7}{\sin (74.7)}$$

Hence
$$\dot{r}_3 = 1083.8 \text{ mm/s}$$

Differentiating equation (4) and remembering that r_4, r_5, r_7, θ_6 and θ_7 are constant,

$$r_4(j\dot{\theta}_4)\, e^{j\theta_4} + r_5(j\dot{\theta}_5)\, e^{j\theta_5} = \dot{r}_6\, e^{j\theta_6} + 0 \tag{10}$$

Separating real and imaginary parts and comparing

$$-r_4\dot{\theta}_4 \sin\theta_4 - r_5\dot{\theta}_5 \sin\theta_5 = \dot{r}_6 \cos\theta_6$$

and

$$r_4\dot{\theta}_4 \cos\theta_4 + r_5\dot{\theta}_5 \cos\theta_5 = \dot{r}_6 \sin\theta_6$$

Remembering that $\theta_6 = 180°$, above equations reduce to

$$r_4\dot{\theta}_4 \sin\theta_4 + r_5\dot{\theta}_5 \sin\theta_5 = \dot{r}_6 \tag{11}$$

and

$$r_4\dot{\theta}_4 \cos\theta_4 + r_5\dot{\theta}_5 \cos\theta_5 = 0 \tag{12}$$

From equation (12),

$$\dot{\theta}_5 = -\left(\frac{r_4}{r_5}\right)\frac{\dot{\theta}_4 \cos\theta_4}{\cos\theta_5}$$

Remembering that $\dot{\theta}_4 \equiv \dot{\theta}_3 = 4.844$ rad/s and $\theta_4 \equiv \theta_3$

$$\dot{\theta}_5 = -\left(\frac{375}{500}\right)\frac{4.844 \cos 74.7}{\cos 181.342}$$

or

$$\dot{\theta}_5 = 0.959 \text{ rad/s (c.w.)}$$

Substituting in equation (11),

$$\dot{r}_6 = 375\,(4.844) \sin 74.7 + (500)(0.959) \sin 181.342$$

Therefore

$$\mathbf{V}_R = \dot{r}_6 = 1740.9 \text{ mm/s} \qquad \textbf{Ans.}$$

For acceleration analysis, differentiate equation (7). Note that r_2 in constant.

$$r_2\,[(j\ddot{\theta}_2) + (j\dot{\theta}_2)^2]\, e^{j\theta_2} = [\ddot{r}_3 + \dot{r}_3\,(j\dot{\theta}_3) + (j\ddot{\theta}_3)\, r_3 + \dot{r}_3\,(j\dot{\theta}_3) + r_3\,(j\dot{\theta}_3)^2]\, e^{j\theta_3}$$

Now, as crank revolves uniformly $\ddot{\theta}_2 = 0$ and therefore,

$$-r_2\,(\dot{\theta}_2)^2\, e^{j\theta_2} = (\ddot{r}_3)\, e^{j\theta_3} + 2j\,(\dot{r}_3)(\dot{\theta}_3)\, e^{j\theta_3} + j\,(r_3\ddot{\theta}_3)\, e^{j\theta_3} - r_3\,(\dot{\theta}_3)^2\, e^{j\theta_3}$$

Separating out real and imaginary parts and comparing

$$-r_2\,(\dot{\theta}_2)^2 \cos\theta_2 = \ddot{r}_3 \cos\theta_3 - 2\,(\dot{r}_3)\,\dot{\theta}_3 \sin\theta_3 - r_3\ddot{\theta}_3 \sin\theta_3 - r_3\dot{\theta}_3^2 \cos\theta_3 \tag{13}$$

and $\quad -r_2\,(\dot{\theta}_2)^2 \sin\theta_2 = \ddot{r}_3 \sin\theta_3 - 2\,(\dot{r}_3)(\dot{\theta}_3) \cos\theta_3 + r_3\ddot{\theta}_3 \cos\theta_3 - r_3\,(\dot{\theta}_3)^2 \sin\theta_3 \tag{14}$

To eliminate terms involving $\ddot{r}_3$, multiply (13) by $\sin\theta_3$ and (14) by $\cos\theta_3$ and subtract

$$r_2\,(\dot{\theta}_2)^2 \sin(\theta_3 - \theta_2) = +2\,(\dot{r}_3)(\dot{\theta}_3)\,(\sin^2\theta_3 - \cos^2\theta_3) + r_3\,\ddot{\theta}_3$$

or,

$$\ddot{\theta}_3 = \frac{r_2\,(\dot{\theta}_2)^2 \sin(\theta_3 - \theta_2) + 2\,(\dot{r}_3)(\dot{\theta}_3) \cos 2\theta_3}{r_3}$$

or,

$$\ddot{\theta}_3 = \frac{(75)\,(20.94)^2 \sin(74.7 - 30) + 2\,(1083.8)(4.844) \cos 149.4}{246.2}$$

or

$$\ddot{\theta}_3 = 57.25 \text{ rad/s}^2 \qquad \textbf{Ans.}$$

Substituting in equation (13),

$$\ddot{r}_3 = \frac{[r_3(\dot{\theta}_3)^2 \cos\theta_3 + r_3(\ddot{\theta}_3)\sin\theta_3 + 2(\dot{r}_3)\dot{\theta}_3\sin\theta_3 - r_2(\dot{\theta}_2)^2\cos\theta_2]}{\cos\theta_3}$$

or $\quad \ddot{r}_3 = (246.2)(4.844)^2 + (246.2)(57.25)\tan 74.7 + 2(1083.8)(4.844)\tan 74.7$

$$- (75)(20.94)^2\frac{\cos 30}{\cos 74.7}$$

Hence $\qquad\qquad\qquad\qquad \ddot{r}_3 = -12{,}251.6$ mm/s

Therefore $\qquad\qquad\qquad A_{QP} = -12.2516$ mts./s $\qquad\qquad\qquad$ **Ans.**

Differentiating equation (10) further, for acceleration analysis, (Remember that $\ddot{\theta}_4 = \ddot{\theta}_3$; $\dot{\theta}_4$ = $\dot{\theta}_3$, and r_4, r_1, θ_7, r_5, θ_6 are constant) we have

$$r_4[j(\ddot{\theta}_4) + (j\dot{\theta}_4)^2]\,e^{j\theta_4} + r_5[j\ddot{\theta}_5 + (j\dot{\theta}_5)^2] = \ddot{r}_6\,e^{j\theta_6}$$

Comparing real and imaginary parts as before,

$$- r_4\ddot{\theta}_4\sin\theta_4 - r_4\dot{\theta}_4^2\cos\theta_4 - r_5\ddot{\theta}_5\sin\theta_5 - r_5\dot{\theta}_5^2\cos\theta_5 = \ddot{r}_6\cos\theta_6 \qquad (13)$$

$$r_4\ddot{\theta}_4\cos\theta_4 - r_4\dot{\theta}_4^2\sin\theta_4 + r_5\ddot{\theta}_5\cos\theta_5 - r_5\dot{\theta}_5^2\sin\theta_5 = \ddot{r}_6\sin\theta_6 \qquad (14)$$

To eliminate $\ddot{\theta}_5$ term, multiply (13) by $\cos\theta_5$ and (14) by $\sin\theta_5$ and add,

$$r_4\ddot{\theta}_4\sin(\theta_5 - \theta_4) - r_4\dot{\theta}_4^2\cos(\theta_5 - \theta_4) - r_5\dot{\theta}_5^2 = \ddot{r}_6\cos(\theta_5 - \theta_6)$$

As $\quad \theta_6 = 180$, we have

$$\ddot{r}_6 = \frac{375(57.25)\sin(181.342 - 74.7) - 375(4.844)^2\cos(181.342 - 74.7) - 500(0.959)^2}{\cos(181.342 - 180)}$$

or $\quad \ddot{r}_6 = \dfrac{17589.67}{\cos 1.342} = 17{,}594.5$ mm/s^2

Therefore $\qquad\qquad\qquad A_R = 17.595$ mts/s^2. $\qquad\qquad\qquad$ **Ans.**

REVIEW QUESTIONS

4.1 Figure 4.11 shows an antiparallel or crossed-bar linkage. If link 2 rotates at 1 rad/s c.c.w., find velocities of points P and Q.

(**Ans.** $V_P = 0.402$ m/s at 151°; $V_Q = 0.29$ m/s at 249°)

4.2 Figure 4.12 shows inversion of slider crank-mechanism. Driving crank 2 rotates uniformly at 60 rad/s c.c.w. Find the velocity of point B and angular velocities of links 3 and 4.

(**Ans.** $V_B = 4.77$ m/s at 96°; $\omega_3 = \omega_4 = 22$ rad/s c.c.w.)

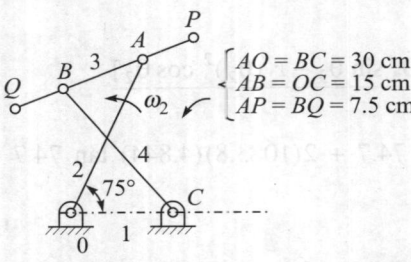

$$\begin{cases} AO = BC = 30 \text{ cm} \\ AB = OC = 15 \text{ cm} \\ AP = BQ = 7.5 \text{ cm} \end{cases}$$

Fig. 4.11

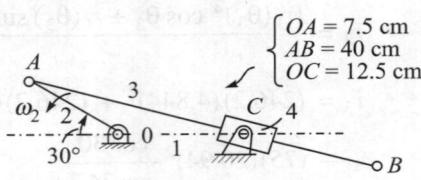

$$\begin{cases} OA = 7.5 \text{ cm} \\ AB = 40 \text{ cm} \\ OC = 12.5 \text{ cm} \end{cases}$$

Fig. 4.12

4.3 Driving link 2 of the linkage shown in Figure 4.13 has an angular velocity of 10 rad/s. c.c.w. Find the angular velocity of link 6 and the velocities of points P, C and D.

(**Ans.** $\omega_6 = 4$ rad/s (ccw); $V_P = 23.11$ cm/s at $180°$; $V_C = 48.48$ cm/s at $208°$; $V_D = 48.24$ cm/s at $205°$)

4.4 A two-cylinder $60°$ V-engine consisting in part of an articulated connecting rod is show in Fig. 4.14. If crank OA rotates at 2000 r.p.m. c.w., find the velocities of points P, C and S.

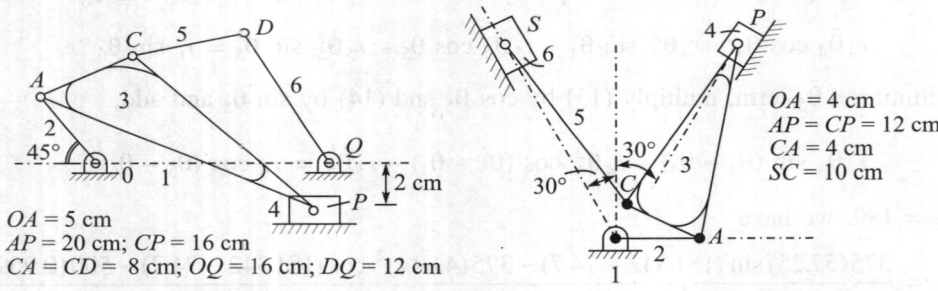

$OA = 5$ cm
$AP = 20$ cm; $CP = 16$ cm
$CA = CD = 8$ cm; $OQ = 16$ cm; $DQ = 12$ cm

Fig. 4.13

$OA = 4$ cm
$AP = CP = 12$ cm
$CA = 4$ cm
$SC = 10$ cm

Fig. 4.14

(**Ans.** $V_P = 852$ cm/s at $240°$; $V_C = 981.6$ cm/s at $267°$; $V_S = 758.4$ cm/s at $60°$)

4.5 The dimensions of the Atkinson-cycle engine mechanism, shown in Fig. 4.15 are—OA = 12 cm; QB = 16 cm; AB = 30 cm; AC = 32 cm; BC = 5 cm; CP = 36 cm. If the crank OA makes 150 r.p.m., find for given configuration, the velocity of the piston P and angular velocities of the links ABC and CP.

(**Ans.** 116.88 cm/s; 3.69 rad/s; and 3.52 rad/s)

4.6 In the 4-bar mechanism shown in Fig. 4.16, link AB rotates uniformly at 2 r.p.s. in c.w. sense. Find angular acceleration of links BC and CD and acceleration of point E in link BC. Take $AB = 7.5$ cm; $BC = 17.5$ cm; $EC = 5$ cm; $CD = 15$ cm; $DA = 10$ cm; and $\angle BAD = 90°$. (D.A.V. Indore: Nov. 1992).

(**Ans.** 11.42 rad/s^2; 15 rad/s^2; 11.25 m/s^2)

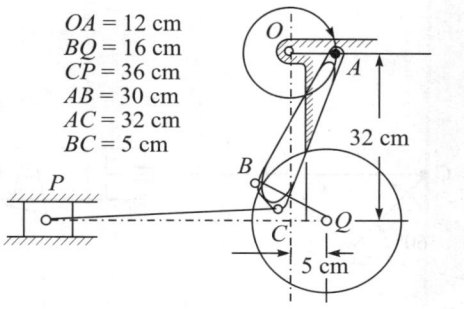

OA = 12 cm
BQ = 16 cm
CP = 36 cm
AB = 30 cm
AC = 32 cm
BC = 5 cm

32 cm

5 cm

Fig. 4.15

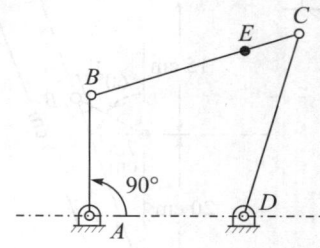

Fig. 4.16

4.7   The mechanism shown in Fig. 4.17 is a marine steering gear called Rapson's slide. O_2P is the tiller and AC is the actuating rod. If the velocity of AC is 25 mm/min to the left, find the angular acceleration of the tiller.

4.8   In the mechanism shown in Fig. 4.18, crank $OA = 10$ cm and rotates in c.w. direction at a speed of 100 r.p.m. The straight rod BCD rocks on a fixed pivot at C. BC and CD are each 20 cm long and the link

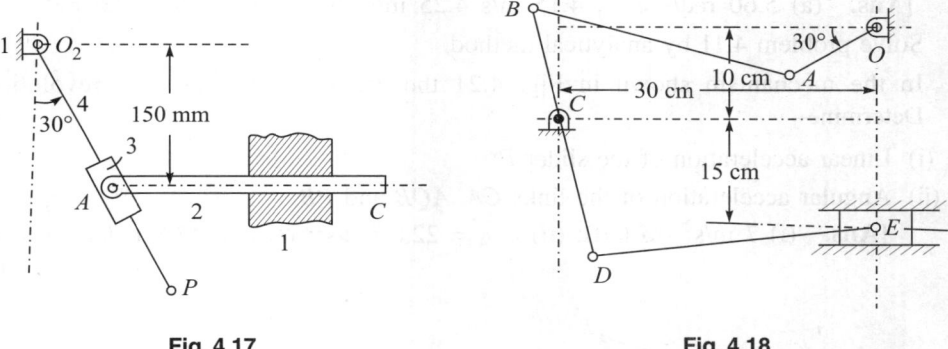

Fig. 4.17   **Fig. 4.18**

$AB = 30$ cm long. The slider E is driven by the rod DE which is 25 cm long. Find the velocity of point E. (D.A.V. Indore; Aug. 1991)

4.9   An engine mechanism has a crank $CB = 10$ cm, and a connecting rod $AB = 40$ cm. The crank rotates clockwise at 10 rad/s. For the position when crank makes an angle of 45° to i.d.c. in clockwise direction, find by exact analytical method: (a) Velocity of slider A and angular velocity of connecting rod AB, and (b) Acceleration of slider and angular acceleration of connecting rod. (D.A.V. Indore: March 1991; July 1993)

   (**Ans.** $V_A = 84$ cm/s; $A_A = 710$ cm/s²; $\omega_{AB} = 1.8$ rad/s; $\alpha_{AB} = 17.5$ rad/s²)

4.10   The driving crank AB of the quick-return mechanism shown in Fig. 4.19 revolves at a uniform speed of 200 r.p.m. Find the velocity and acceleration of the tool box R in the position shown, when the crank makes an angle of 60° with the vertical line of centres PA. What is the acceleration of sliding of the block at B along the slotted lever PQ?

   (**Ans.** $V_R = 161$ cm/s; $A_R = 2022$ cm/s²; $A_B = 1872$ cm/s²)

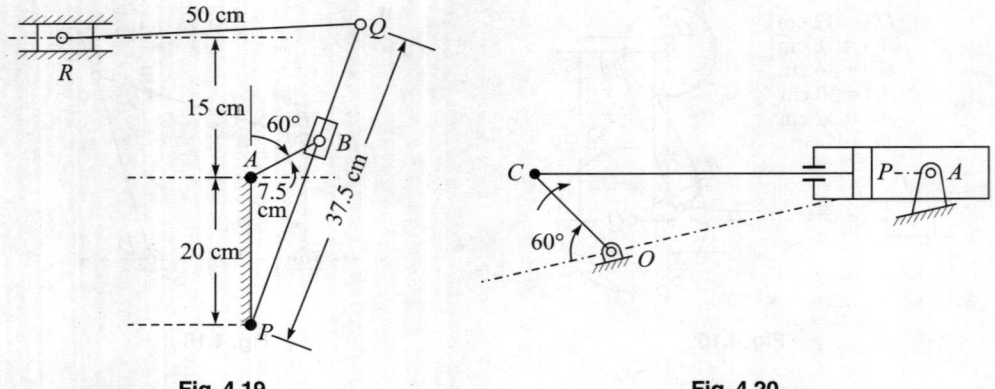

Fig. 4.19   **Fig. 4.20**

4.11 In the pump mechanism shown in Fig. 4.20, $OC = 32$ cm, $AP = 68$ cm and $OA = 65$ cm. For the given configuration, determine: (a) angular velocity of cylinder; the sliding velocity of the plunges and absolute velocity of the plunger, (b) linear (slider) acceleration of slider P and (c) angular acceleration of the piston rod. The crank OC rotates at 20 rad/s. clockwise. Solve by vector polygon method.

[**Ans.** (a) 5.60 rad/s c.w.; 4.15 m/s 4.25 m/s; (b) 72 m/s^2; (c) 20.6 rad/s^2 c.c.w.]

4.12 Solve problem 4.11 by analytical method.

4.13 In the mechanism shown in Fig. 4.21 the crank OC rotates at 1 revolution/sec. Determine–

(i) Linear acceleration of the slider P;

(ii) Angular acceleration of the links CA, AQB and PB.

[**Ans.** (i) 7 m/s^2 (to left); (ii) $\alpha_{CA} = 22.8$ rad/s^2; $\alpha_{AQB} = 13.8$ rad/s^2; $\alpha_{PB} = 14.4$ rad/s^2]

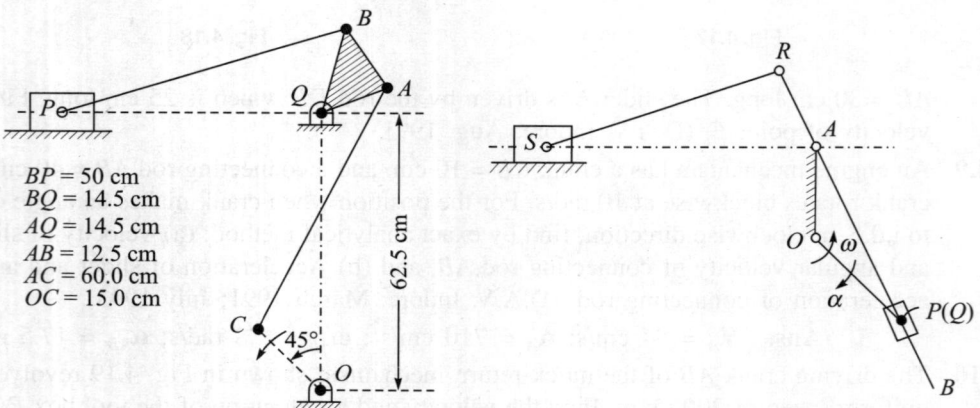

$BP = 50$ cm
$BQ = 14.5$ cm
$AQ = 14.5$ cm
$AB = 12.5$ cm
$AC = 60.0$ cm
$OC = 15.0$ cm

Fig. 4.21   **Fig. 4.22**

4.14 Figure 4.22 shows a Withworth Quick Return Mechanism. Link dimensions are crank $OP = 24$ cm, $OA = 15$ cm; $AR = 16.5$ cm and $RS = 43$ cm. Determine (a) the acceleration of the slider S, and (b) the angular acceleration of links AR and RS, if the crank OP rotates at an angular velocity of 2.5 rad/s and also has an angular acceleration of 20 rad/s^2. Assume the crank OP to make an angle of 45° with the vertical.

[**Ans.** (a) 0.392 m/s^2, (b) 1.56 rad/s^2 (c.w.) and 0.557 rad/s^2]

Chapter **5**

MECHANISMS WITH LOWER PAIRS

5.1 INTRODUCTION

As discussed in section 2.4, *a linear motion lower pair* (*e.g.*, turning pairs, prismatic pairs and screw pairs) *has a degree of freedom of one, as each point on one element of the pair can move only along a single line/curve relative to the other element*. Unless otherwise mentioned, the term *lower pair* will be used to describe linear motion lower pairs. Since lower pairs involve surface contact rather than line or point contact, it follows that lower pairs can be more heavily loaded for the same unit pressure. They are considerably more wear-resistant. For this reason, development in kinematics has centred more and more around lower pairs. It may be pertinent to make a passing reference to another common term *Linkage* which denotes a mechanism consisting only of lower pairs.

5.2 OFFSET SLIDER-CRANK MECHANISM AS A QUICK RETURN MECHANISM

In many situations, mechanisms are required to perform repetitive operations such as pushing parts along an assembly line and folding cardboard boxes in an automated packaging machine. Besides above there are applications like a shaper machine and punching/rivetting press in which working stroke is completed under load and must be executed slowly compared to the return stroke. This results in smaller work done per unit time. A quick return motion mechanism is useful in all such applications.

A quick return motion mechanism is essentially a slider-crank mechanism in which the slider has different average velocities in forward and return stroke. Thus even if crank rotates uniformly, the slider completes one stroke quickly compared to the other stroke. An offset slider-crank mechanism, shown in Fig. 5.1, can be used conveniently to achieve the above objective.

In Fig. 5.1, the crank centre O is offset by an amount e with respect to the line of stroke of slider. If r be the radius of crank OC and l the length of connecting rod CP, the extreme right-hand position P_R and extreme left-hand position P_L of slider P occur when the crank and connecting rod are positioned along a common line. The extreme right-hand position P_R of slider is obtained when the crank OC_R and connecting rod $C_R P_R$ of slider add up to give farthest possible position of slider at a distance of $(r + l)$ from crank centre O. Thus,

$$OP_R = (OC_R) + (C_R P_R) = (r + l) \tag{5.1}$$

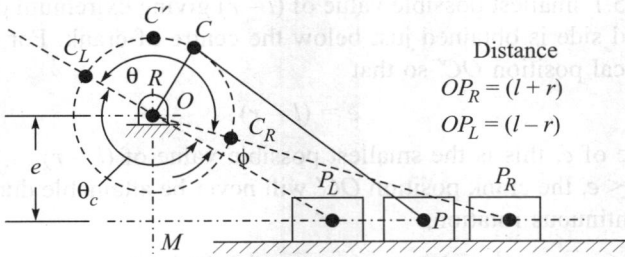

Fig. 5.1 Offset slider-crank mechanism as a quick return mechanism.

Similarly the closest possible position of slider P_L is obtained when crank is at C_L and the crank radius r subtracts from connecting rod length l so that

$$OP_L = (C_L P_L) - OC_L = (l - r) \tag{5.2}$$

From right angled triangles $P_L\,OM$ and $P_R\,OM$ remembering that $\angle P_L O P_R = \phi$, we have

$$\cos\angle MOP_L = \frac{OM}{OP_L} = \frac{e}{(l-r)}$$

and

$$\cos\angle MOP_R = \frac{OM}{OP_R} = \frac{e}{(r+l)}$$

Hence

$$\phi = \cos^{-1}\left(\frac{e}{r+l}\right) - \cos^{-1}\left(\frac{e}{l-r}\right) \tag{5.3}$$

Thus, return stroke angle,

$$\theta_R = (180 - \phi) \tag{5.4}$$

and cutting stroke angle,

$$\theta_C = (180 + \phi) \tag{5.5}$$

When it is desirable to design a mechanism, where slider moves much faster during the return stroke compared to the working/cutting stroke, a measure of suitability of the mechanism is given by a ratio of advance to return stroke time expressed mathematically as

$$Q = \frac{\text{Time of advance stroke}}{\text{Time of return stroke}}$$

When Q is greater than one, the mechanism is called *quick-return mechanism*.

Assuming that the driving motor rotates at constant r.p.m. N, the time of advance stroke and return stroke can be obtained as

$$\text{Time of advance stroke} = \left(\frac{\theta_c}{2\pi N}\right), \text{ and time of return stroke} = \left(\frac{\theta_R}{2\pi N}\right)$$

Hence
$$Q = \left(\frac{\theta_c}{\theta_R} \right) \qquad (5.6)$$

Condition of Rotatibility of Crank

As is clear from Fig. 5.1 smallest possible value of $(l - r)$ giving extremum possible position for slider on the left-hand side is obtained just below the centre of crank. For this position, crank OC will occupy vertical position OC' so that

$$e = (l - r)$$

For a given value of e, this is the smallest possible value of $(l - r)$.

Clearly, if $(l - r) < e$, the crank position OC' will never be attainable during rotation. Hence for crank to have continuous rotation,

$$(l - r) \geq e \qquad \text{or} \qquad l \geq (r + e) \qquad (5.7)$$

5.3 THE PANTOGRAPH

A pantograph is used to reproduce path described by a point either to an enlarged scale or a reduced scale. This mechanism has a peculiarity in that instead of one fixed link, only a point is fixed as a pivot while input motion is given by moving a point on some link along some given planar curve. Applications of pantograph include profile grinding (in which the part obtains its form from a greatly enlarged pattern), in engraving machine and also in guiding cutting torch to generate contour similar to that of a template. Sometimes the pantograph is also used as an indicator rig with a view to reproduce to a smaller scale the displacement of the cross-head and piston of a reciprocating engine. Yet another application of pantograph is in pencil mechanisms of engine indicators, where displacement of piston (against spring force), caused by variation in steam/gas pressure inside the cylinder, is reproduced to an enlarged scale.

Construction and Principle of Working

Figure 5.2(a) illustrates one form of pantograph while Figs. 5.2(b), (c) and (d) show three other variations of pantograph. The principle of working in all these cases, however, remains the same. As shown in Fig. 5.2(a), links OAB, BCP, CD and AD are pin-connected at A, B, C and D. Links are so proportioned that $AB = CD$ and $BC = AD$ with pins A, B, C, D constituting corners of a parallelogram. Link OAB is pivoted to frame at O. A point P on the extension of link BC and another point Q on link AD are so chosen to be the fixed points on respective links, such that points O, Q and P always lie on a common straight line. There are two important things to be proved. (i) Points O, P and Q lie always on a straight line and (ii) the radii of curvatures at P and Q, namely OP and OQ bear same ratio always.

Since lines AD and BC are parallel and lines OQP and OAB cut them, triangles OAQ and OBP are similar.
Hence,

$$\frac{OP}{OQ} = \frac{OB}{OA} = \frac{BP}{AQ} = k, \text{say} \qquad (5.8)$$

Let the point P be now displaced to position P' along the curved path PP'.

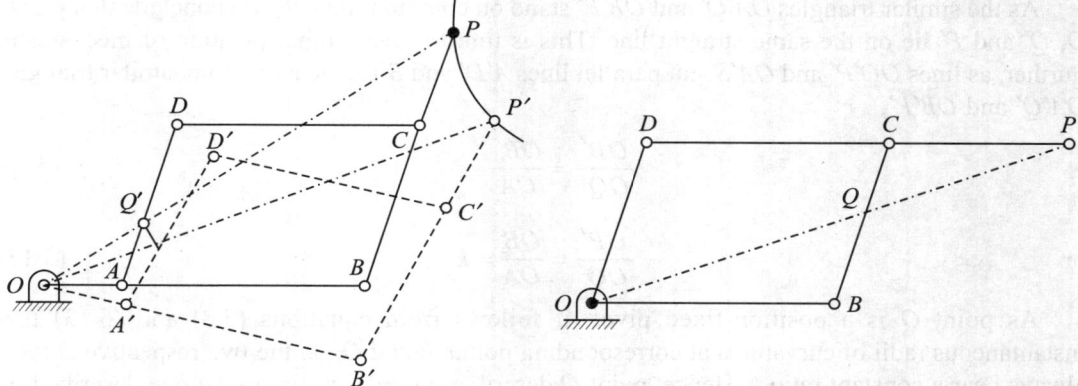

Fig. 5.2(a) One variation of pantograph.   **Fig. 5.2(b)** Second variation of pantograph.

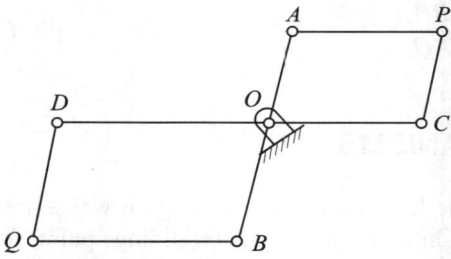   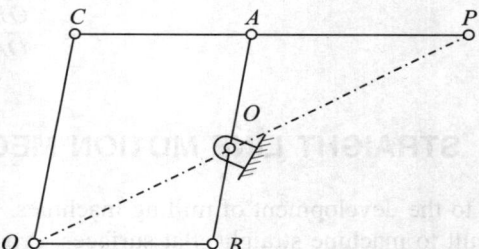

Fig. 5.2(c) Third variation of pantograph.   **Fig. 5.2(d)** Fourth variation of pantograph.

The mechanism configuration for this new position P' of point P is shown in broken lines as $OA'B'C'D'$. Point Q now shifts to position Q'. **In the displaced position first we must show that points O, P' and Q' still lie on the same straight line.**

Remembering that points Q and P were fixed points on links AD and BC (extended), and that $ABCD$ was a parallelogram mechanism, the link $A'B'$ is still parallel to $D'C'$ and $A'D'$ is parallel to $B'C'$. Further, as the link lengths do not change,

$$\frac{OB'}{OA'} = \frac{OB}{OA} \tag{5.9}$$

and

$$\frac{B'P'}{A'Q'} = \frac{BP}{AQ} \tag{5.10}$$

Comparing equations (5.9) and (5.10) with (5.8), we have

$$\frac{OB'}{OA'} = \frac{B'P'}{A'Q'} = \frac{OP}{OQ} \tag{5.11}$$

Since links $B'C'$ and $A'D'$ are parallel, triangles $OB'P'$ and $OA'Q'$ are similar.

As the similar triangles $OA'Q'$ and $OB'P'$ stand on common line OB' we conclude that points O, Q' and P' lie on the same straight line. This is true for every other position of mechanism. Further, as lines $OQ'P'$ and $OA'S'$ cut parallel lines $A'D'$ and $B'C'$, we have from similar triangles $OA'Q'$ and $OB'P'$,

$$\frac{OP'}{OQ'} = \frac{OB'}{OA} \tag{512}$$

or

$$\frac{OP'}{OQ'} = \frac{OB}{OA} = k \tag{5.13}$$

As point O is a position-fixed pivot, it follows from equations (5.8) and (5.13) that instantaneous radii of curvatures at corresponding points P and Q, on the two respective curves, always bear a constant ratio k. Hence, point Q describes a curve similar to the one described by point P, to a scale $1/k$. Similarly, if point Q is made to trace a given curve, point P will trace a similar curve to a scale k. It may be noted that the scaling factor k is given by

$$k = \frac{OB}{OA} = \frac{BP}{AQ} \tag{5.14}$$

5.4 STRAIGHT LINE MOTION MECHANISMS

Prior to the development of milling machines, in the late seventeenth century, it was extremely difficult to machine straight, flat surfaces. Production of good prismatic (sliding) pairs, free of backlash was therefore considered to be a difficult proposition. It was during this period that the problem of obtaining approximate straight line motion, using only turning pairs, attracted great attention of kinematicians. In a majority of cases, a reference was made to the flat segments of coupler curves, generated by mechanisms with turning pairs only. Such mechanisms are known as "approximate straight line motion mechanisms" and they are capable of tracing straight line during a finite range of traverse only.

An obvious way to have constrained motion of a point in a mechanism along a straight line is to use a sliding pair. Sliding pairs, however, have a limitation in that they are bulky and are always susceptible to comparatively rapid wear. In certain situations, therefore, it is necessary to obtain straight line motion using turning pairs only. In rest of the cases, the modern practice is to produce straight line motions using sliders.

Besides self-recording instruments in indicator mechanism, requiring a point to move along a straight line, a few other cases requiring a point to move along approximate straight path are illustrated in Fig. 5.3. The figures are self-explanatory.

Constraining a point in a plane to follow a straight line was one of the stipulations which led L. Burmster to lay the foundations of modern mechanism synthesis. The design of straight line trajectory crank mechanisms has always attracted great attention of kinematicians and Kemp dedicated a whole book to this topic by the year 1877. More recent work in this direction has led to the establishment of coupler curves and tables from which the best four-bar linkages for straight line guidance can be extracted.

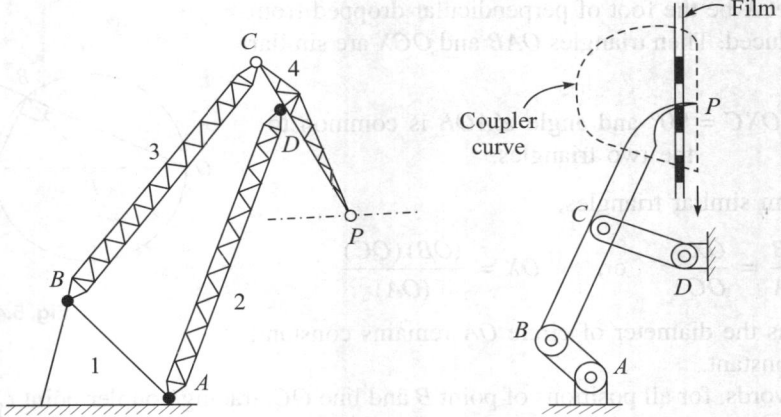

Fig. 5.3 **(a)** The level luffing crane. The path tracer point P traces approximately a horizontal straight line, **(b)** A four-bar mechanism used for advancing film of a movie camera. The coupler point P generates a straight line for a small portion.

Mechanisms for exact and approximate straight line motion may either use all turning pairs, or turning pairs with one sliding pair. The straight line motion mechanisms can be classified in two important groups of paths:

    (a) Mechanisms in which coincidence of the coupler curve arcs with a straight line is theoretically exact (called Exact straight line motion mechanisms), and

    (b) Mechanisms in which coincidence of the coupler curve arcs with a straight line is only approximate (called Approximate straight line motion mechanisms).

Compared to exact straight line motion mechanisms, approximate straight line motion mechanisms have fewer links and this turns out to be an advantage of major importance. A large number of 4-bar mechanisms exist in this category through which remarkable accuracy can be achieved. As against this, most 'exact straight-line trajectories' are obtainable using mechanisms with six or more links. It must be noted that increase in the number of links and joints increases the error due to production tolerances and clearances. This necessitates calculation of tolerances in cases where extreme accuracy is desired. Chebyshev proposed and carried through design of mechanisms, for straight line guidance, such that the segment of coupler curve, used for generating approximate straight line, lies between two selected tolerance limits on linearity.

5.5   EXACT STRAIGHT LINE MOTION MECHANISMS

5.5.1   Condition for Generating Exact Straight Line Motion

Statement: As illustrated in Fig. 5.4, let O, B and C represent three distinct points of a mechanism which are always constrained to lie along a common straight line as the line OC turns about O as centre. As the line OC turns, let the point B move along the circumference of a circle with OA as diameter and, at the same time, let the position of point C be such that the product (OB) (OC) is constant. Then the coupler point C will move along a straight line CX perpendicular to the diameter OA.

Proof: Let X be the foot of perpendicular dropped from C on OA produced. Then triangles OAB and OCX are similar because

$$\angle OBA = \angle OXC = 90°$$ and angle $\angle AOB$ is common to the two triangles.

Thus, from similar triangles,

$$\frac{OB}{OA} = \frac{OX}{OC} \quad \text{or,} \quad OX = \frac{(OB)(OC)}{(OA)}$$

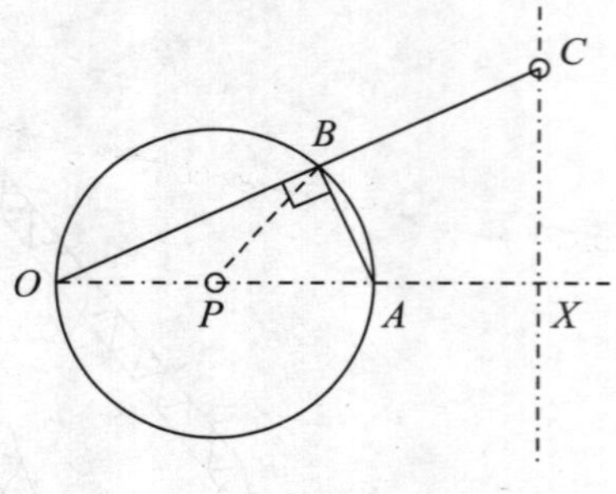

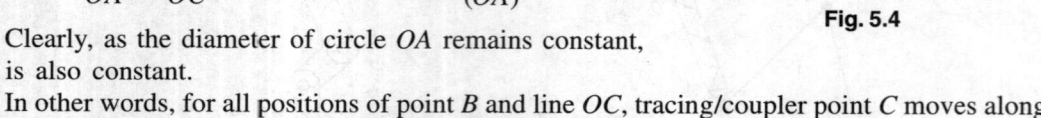

Fig. 5.4

Clearly, as the diameter of circle OA remains constant, OX is also constant.

In other words, for all positions of point B and line OC, tracing/coupler point C moves along straight line in a direction perpendicular to OA produced.

Several mechanisms have been proposed to connect points O, C and B in such a way that for all the configurations of mechanism, condition $(OB)(OC)/(OA)$ = constant, is satisfied. This is explained further through Peaucellier and Hart mechanisms.

5.5.2 Peaucellier Mechanism

In order to satisfy conditions stipulated in Section 5.5.1, an eight-link mechanism was proposed by Peaucellier. The frame link OP also constitutes the radius of circle, along the circumference of which point B is constrained to move through crank PB. The fixed link PO equals the radius of crank PB. Points B and C represent pins at opposite corners of a rhombus $BQCR$ whose pins Q and R are connected through equal links OQ and OR to fixed pivot O.

The tracing point C will describe a straight line CX perpendicular to the line of centres OP extended, if and only if,

$$(OB)(OC) = \text{constant.}$$

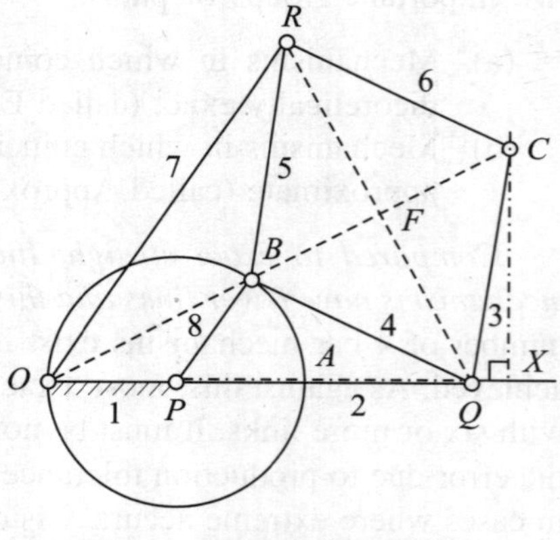

Fig. 5.5 Peaucellier Mechanism.

To prove this, join the diagonal RQ of the rhombus, which will bisect the other diagonal BC. From the right angled triangles ORF and CRF,

$$OR^2 = OF^2 + RF^2 \qquad (5.15)$$

and

$$CR^2 = CF^2 + RF^2 \qquad (5.16)$$

Subtracting

$$(OR^2 - CR^2) = OF^2 - CF^2$$
$$= (OF + CF)(OF - FC) \qquad (5.17)$$

Noting that $OF + CF = OC$

and $OF - FC = OB$

We have from Eq. (5.17),

$$(OR^2 - CR^2) = (OC)(OB) \tag{5.18}$$

Since OR and CR are both constant link lengths, the R.H.S. is also constant. Hence, $(OC)(OB)$ = constant for all configurations.

5.5.3   The Hart Mechanism

As against eight links in Peaucellier's mechanism, this mechanism consists of six links only. As seen in Fig. 5.6, link OP is the frame 1 while link PB ($= OP$) is the driving crank, through which it transmits motion to link DG. Links DE, EF, FG and DG constitutes a crossed parallelogram, so that, links $DE = FG$ and $DG = EF$. Under this situation, for any configuration, lines joining DF and EG are parallel. Lines joining D, E, G and F thus forms a trapezium.

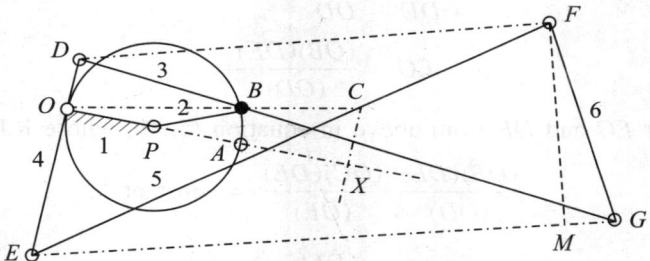

Fig. 5.6   Hart exact straight line motion mechanism.

Link DE is pivoted at O to the frame link. Clearly, if points B and C on link EF are selected on a straight line from O, parallel to DF, the line will remain parallel to DF and EG in all other configurations. Note that crank PB is connected at B to link DG through a turning pair, while C is a tracing point on link EF.

Being the end point of crank PB, point B is constrained to move along circle of radius PB. Draw FM perpendicular to EG. It remains to be shown next that for this mechanism too,

$$(OB)(OC) = \text{constant}.$$

From triangle EFG,

$$EF^2 = FG^2 + EG^2 - 2(FG)(EG)\cos\angle EGF \tag{5.19}$$

Now, $DEGF$ being a trapezium $MG = (EF - DF)/2$, and by symmetry

$$\cos\angle EGF = \frac{(EG - DF)}{2(FG)}$$

Substituting in equation (5.19),

$$EF^2 = FG^2 + EG^2 - 2(FG)(EG)\,\frac{(EG - DF)}{2(FG)}$$

or
$$EF^2 = FG^2 + EG^2 - [(EG)^2 - (EG)(DF)]$$

or
$$EF^2 = FG^2 + (EG)(DF)$$

or
$$(EG)(DF) = (EF^2 - FG^2) \tag{5.20}$$

EF and FG being link lengths, the R.H.S. of equation (5.20) is constant. It remains to be shown now that left hand side of equation (5.20) equals $(OB)(OC)$. Triangles DEF and OEC *are similar since all the three sides are parallel. Hence,*

$$\frac{DF}{DE} = \frac{OC}{OE}$$

and therefore
$$DF = \frac{(OC)(DE)}{(OE)}$$

Also, triangles DEG and ODB are similar, as all the three sides are parallel. Hence,

$$\frac{EG}{DE} = \frac{OB}{OD}$$

and hence
$$EG = \frac{(OB)(DE)}{(OD)}$$

Substituting for EG and DF from above in equation (5.20), whose R.H.S. is constant,

$$\frac{(OB)(DE)}{(OD)} \frac{(OC)(DE)}{(OE)} = \text{constant}$$

or
$$(OB)(OC) \frac{DE^2}{(OD)(OE)} = \text{constant} \tag{5.21}$$

As link lengths DE, OD and OE are constant it follows from Eq. (5.21) that $(OB)(OC) = $ constant. Thus, as all the conditions of Section 5.4.1 are fulfilled, point C traces a straight line perpendicular to diameter OA.

Though Hart's mechanism appears to be preferable to Peaucellier mechanism due to smaller number of links, it suffers on account of large space requirement. Even for a short path of C, the Hart's mechanism requires larger space and this constitutes a great practical disadvantage.

5.5.4 The Scott-Russel Mechanism

This exact straight line motion mechanism differs from earlier two cases in that it incorporates a slider also. As will be appreciated by readers during the course of discussions, this mechanism does not generate but merely copies straight line motion.

The mechanism consists of a slider-crank mechanism with crank OB equal in length to that of connecting rod BS (Fig. 5.7) and the connecting rod length SB extended to include coupler point C such that

$$BC = BS = OB.$$

The end S of the connecting rod is constrained to move along straight path OS using a slider element and guides. It can be shown that with this arrangement, C moves along a straight line through O perpendicular to OS. For this purpose it is enough to show that under all configurations, line joining C with O makes right angle with OS.

As $BC = BO = BS$, a circle drawn with B as centre and CS as diameter will necessarily pass through point O in all the configurations of the mechanism. Further, O being a point on the semicircle, $\angle COS = 90°$ always.

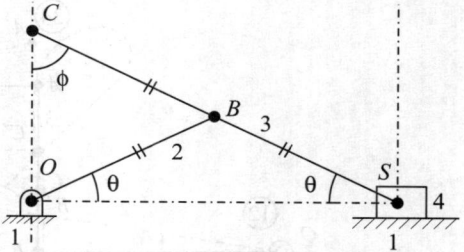

Fig. 5.7 Scott Russel Mechanism.

Hence point C always moves along a straight path perpendicular to OS through O.

This can also be proved by considering isosceles triangles OBS and OBC, in which,

$$\angle BOS = \angle BSO = \theta, \quad \text{and} \quad \angle BCO = \angle BOC = \phi$$

Since, the exterior angle

$$\angle OBC = \angle BOS + \angle BSO,$$
$$\angle OBC = 2\theta$$

Also, since the exterior angle,

$$\angle OBS = \angle OCB + \angle COB = 2\phi,$$

and $\angle OBC + \angle OBS = 180°$, it follows that $2\theta + 2\phi = 180°$ or $(\theta + \phi) = 90°$

This shows that angle

$$\angle COS = 90° \text{ in all configurations.}$$

The main problem with Scott-Russel mechanism is that exactness with which point C follows a straight line obviously depends upon exactness with which slider S is guided along the straight path. Needless to say that friction and wear of slider pair is much larger than the turning pair and as such the resulting straight line motion of point C is of little practical significance.

5.6 THE APPROXIMATE STRAIGHT LINE MOTION

5.6.1 The Watt Mechanism

This approximate straight line motion mechanism was used by J. Watt to guide piston rod along straight path in many of his early steam engines. In the absence of any means to machine straight flat surfaces, he had to rely on approximate straight line motion mechanism.

As shown in Fig. 5.8 it is a crossed four-bar mechanism $OABQ$ with cranks OA and QB of unequal lengths and pivoted to frame at O and Q respectively. The coupler (tracing) point is so chosen on link AB that it traces a figure of eight as shown in dotted line. The length of link AB is such that when links QB and OA are parallel, AB is perpendicular to them.

We select a position for which the links QB and OA are parallel and on opposite sides of the link AB. In this position the instantaneous centre of link AB is at infinity and accordingly if link QB be given a small rotation ϕ, the link OA will rotate through θ such that

$$\text{arc}BB' = \text{arc}AA'$$

or

$$(QB)\phi = (OA)\theta$$

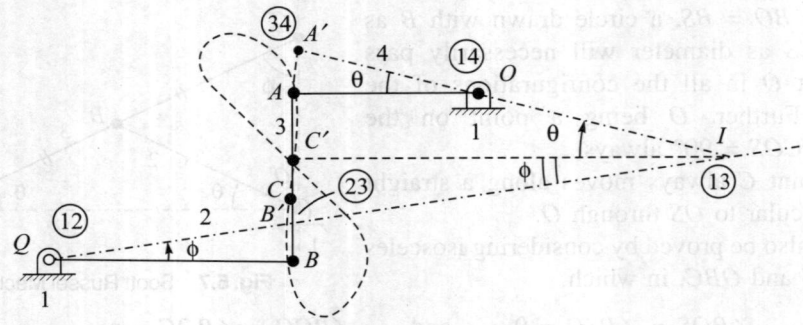

Fig. 5.8 Watt mechanism.

Such a rotation will shift the instantaneous centre to position I as shown. Best position for tracing point C is obtained by drawing a horizontal line IC' to meet corresponding position $A'B'$ of link at C'. I being the instantaneous centre (13), point C' will move in vertical direction along a straight line. For the position $QBAO$, all the points in link AB move up in vertical direction. For small angles θ and ϕ,

$$\frac{QB}{OA} \approx \frac{\theta}{\phi} \qquad (5.22)$$

Again $\qquad\qquad A'C' = (IC')\theta \qquad$ and $\qquad B'C' = (IC')\phi$

Therefore $\qquad\qquad\qquad \dfrac{A'C'}{B'C'} = \dfrac{\theta}{\phi}$

Substituting for θ/ϕ from equation (5.22)

$$(A'C'/B'C') = QB/OA$$

Thus, the tracing point C divides the link AB in two parts whose lengths vary inversely as the lengths of adjacent links.

5.6.2 The Grasshopper Mechanism

The Grasshopper mechanism is also known as *modified Scott Russel mechanism*. This name follows because the slider of Scott Russel mechanism in Fig. 5.7 is replaced by a lever QB which oscillates about centre Q with radius QB. In Grasshopper mechanism O and Q are the fixed pivots and points B and A are constrained to move along arcs of circle with radii QB and OA respectively. Tracing point C, which is located on extension of coupler link BA, then describes a straight path perpendicular to line OB. For this, necessary condition is that the instantaneous centre I ($\equiv 13$) of link 3 w.r. to 1 be such that IC is parallel to OB as shown in Fig. 5.9

There is one more deviation which this mechanism has compared to the Scott and Russel mechanism. Unlike Scott Russel mechanism, the path of coupler point C in Grasshopper mechanism does not pass through the fixed pivot O.

The point C will describe an approximate straight line if,

$$AC = \left(\frac{AB^2}{OA}\right) \qquad (5.23)$$

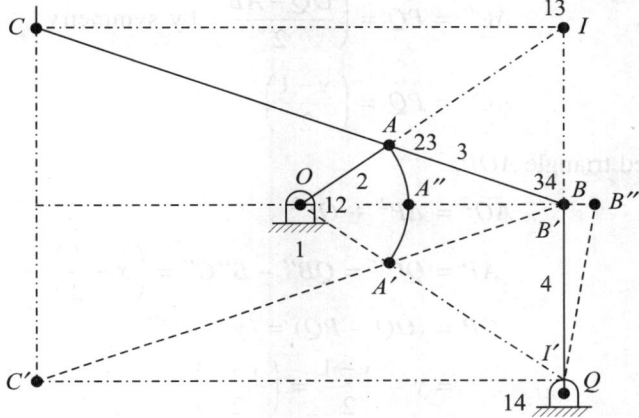

Fig. 5.9 The grasshopper mechanism (modified Scott-Russel mechanism).

5.6.3 The Tchebicheff Straight Line Motion Mechanism

This is again an approximate straight line motion mechanism consisting of a 4-bar rocker-rocker mechanism in which the crossed links OA and QB are of equal lengths (see Fig. 5.10). The tracing point C is located at the middle of the coupler link AB.

The link-lengths are so proportioned that in the extreme positions of mechanism (namely, configurations $OA'B'C'$ and $OA''B''C''$), the coupler point C'' is directly above A'' when AB lies along link QB and coupler point C' is directly above B' when link AB lies along link OA. The proportions of the links can be established as follows.

Let us consider coupler AB to be of unit length and find out proportionate lengths of other links. Thus, Let,

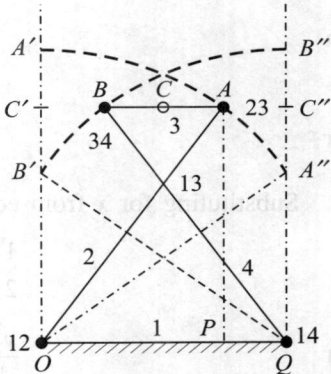

Fig. 5.10 Tchebicheff mechanism.

$$AB = 1$$
$$OA = QB = x$$

and
$$OQ = y$$

Now from right angled triangle $OA''Q$ of Fig. 5.10,

$$(OA'')^2 = (QA'')^2 + (OQ)^2$$

or
$$x^2 = (x - 1)^2 + y^2$$

or
$$2x = 1 + y^2$$

or $$x = \left(\frac{1+y^2}{2}\right) \qquad (5.24)$$

From A drop perpendicular AP on OQ. Then,

$$AC'' = PQ = \left(\frac{OQ - AB}{2}\right) \text{ by symmetry}$$

or $$AC'' = PQ = \left(\frac{y-1}{2}\right) \qquad (5.25)$$

From right angled triangle AOP,

$$AO^2 = AP^2 + OP^2 \qquad (5.26)$$

Now $$AP = QC'' = QB'' - B''C'' = \left(x - \frac{1}{2}\right)$$

and $$OP = (OQ - PQ) = (y - AC'')$$

$$= y - \frac{y-1}{2} = \left(\frac{y+1}{2}\right)$$

Substituting for OP and AP in (5.26),

$$x^2 = \left(x - \frac{1}{2}\right)^2 + \left(\frac{y+1}{2}\right)^2$$

or $$x^2 = \left(x^2 - x + \frac{1}{4}\right) + \left(\frac{y^2}{4} + \frac{y}{2} + \frac{1}{4}\right)$$

or $$x = \left(\frac{1}{2} + \frac{y}{2} + \frac{y^2}{4}\right)$$

Substituting for x from equation (5.24),

$$\frac{1}{2} + \frac{y^2}{2} = \frac{1}{2} + \frac{y}{2} + \frac{y^2}{4}$$

or $$\frac{y^2}{4} - \frac{y}{2} = 0$$

or $$\frac{y}{4}(y - 2) = 0$$

or $$y = 2, \text{ the only feasible solution.}$$

Substituting in (5.24), $$x = \frac{1 + (2)^2}{2} = 2.5$$

Therefore $$AP = \left(x - \frac{1}{2}\right) = 2.0$$

It follows therefore that the proportion of link lengths should be

$$AB : OQ : OA : AP = 1 : 2 : 2.5 : 2.0$$

For the configuration $OABQ$, the location of I.C. 13, which is just below C, justifies the motion of point C in horizontal direction.

5.6.4 The Roberts Straight Line Motion Mechanism

This is again a 4-bar rocker-rocker mechanism, with links connected through turning pairs only. As seen in Fig. 5.11, links OA and QB are of equal lengths while the coupler carries the tracing point C on an extension of coupler in the direction of perpendicular bisector of coupler AB. The exact location of tracing point on the perpendicular bisector PC can be decided by considering instantaneous centre I of rotation of coupler w.r. to frame link.

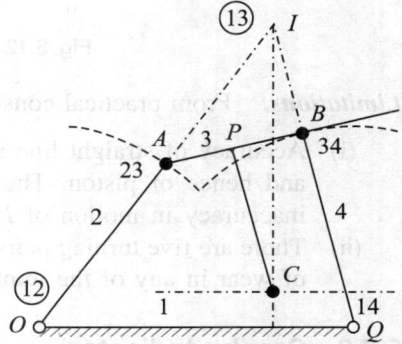

Fig. 5.11 Roberts mechanism.

In Fig. 5.11 the I.C. 13 of link 3 w.r. to 1 is located at I. Hence, if the tracing point C is to describe a horizontal straight line the point C must also lie on a vertical line from I. This is shown in Fig. 5.11. It can be verified that for this mechanism good dimensions can be $OQ : AB : OA : PC = 1 : 0.593 : 0.584 : 1.112$.

5.7 ENGINE INDICATORS

Pencile mechanisms used in engine indicators demonstrate one of the most interesting applications of straight line motion. A small piston is acted upon by steam or gas pressure in the cylinder. Variations in steam/gas pressure cause displacement of piston against a spring force. The amount of displacement of piston is directly proportional to the steam/gas pressure. The displacement of piston is communicated to the pencil which traces variation of pressure in cylinder on a graph paper (or sheet of paper) wrapped around indicator drum which rotates slowly about its axis so as to help plotting graph along x-direction.

5.7.1 Simplex Indicator

This is the simplest of all engine indicator mechanisms and consists of a pantograph with fixed pivot attached to the body of indicator and the point Q attached to the piston. The links AB, BC, CQ and AQ constitute a parallelogram with link BC extended upto the tracing point P. As required in pantograph, points O, Q and P lie on a straight line. This is shown in Fig. 5.12.

As seen in Section 5.3 point P will reproduce motion of point Q to an enlarged scale given by scale factor,

$$\frac{OP}{OQ} = \frac{OB}{OA} = \frac{BP}{BC} = k$$

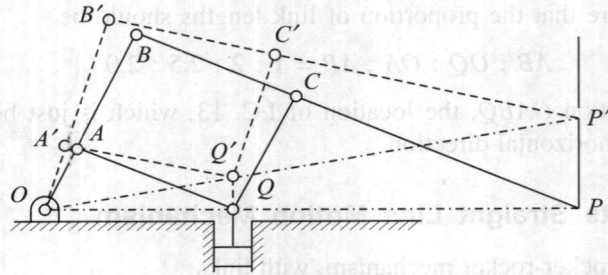

Fig. 5.12 Simplex engine indicator mechanism.

Limitations: From practical considerations, the simplex indicator has following limitations:

(i) Accuracy of straight line motion of point *P* depends on the accuracy of motion of *Q* and hence of piston. Thus any deviation of *Q* from straight line motion magnifies inaccuracy in motion of *P* by a factor *k*.

(ii) There are five turning pairs (viz. at *O*, *A*, *B*, *C* and *Q*), hence any inaccuracy on account of wear in any of the joints will adversely affect the accuracy of motion of point P.

5.7.2 Crosby Indicator

Crosby indicator is a modified form of Pantograph and is shown in Fig. 5.13. It uses six links in order to obtain motion of pencil point *C* similar to that of *Q* lying on piston of indicating mechanism.

In order that this mechanism serves the purpose of an indicator mechanism, it is necessary that

(a) Tracing point *C* lies on extension of line joining *O* to *Q*,

(b) The velocity ratio (v_c/v_Q) must be constant; v_c being the velocity of tracing point *C* and v_Q being the velocity of point *Q* on piston. Using instantaneous centres it will be shown that both the conditions are satisfied by crosby mechanism.

For the links numbered 1 through 6 as shown in Fig. 5.13, the instantaneous centres obtained by inspection, indicated against respective joints, are (0 : 15), (A : 45), (B : 34), (P : 23), (Q : 36), (D : 12) and the I.C. 16 is located at infinity in a direction perpendicular to the line of stroke of slider 6. Now point *O* being the intersection point of lines joining 12 with 23 and of 16 with 36, the point *O* is also the instantaneous centre 13.

Again, line joining I.C. 13 with 34 and the line joining I.C. 45 with 15 intersect at *O*. Hence *O* is also the I.C. 14. Thus, the point *O* represents three instantaneous centres: 13, 14 and 15.

Hence, both links 3 and 4 can be assumed to have motion of rotation instantaneously about *O*. Thus, when points *O*, *Q* and *C* lie on the same straight line, motion of *Q* and *C* is perpendicular to *OC*, satisfying condition (*a*) above.

Now as *O* and I.C. 13 overlap,

$$\frac{V_B}{OB} = \frac{V_P}{OP} = \frac{V_Q}{OQ} \qquad (5.27)$$

Again as *O* and I.C. 14 overlap,

$$\frac{V_B}{OB} = \frac{V_C}{OC} \qquad (5.28)$$

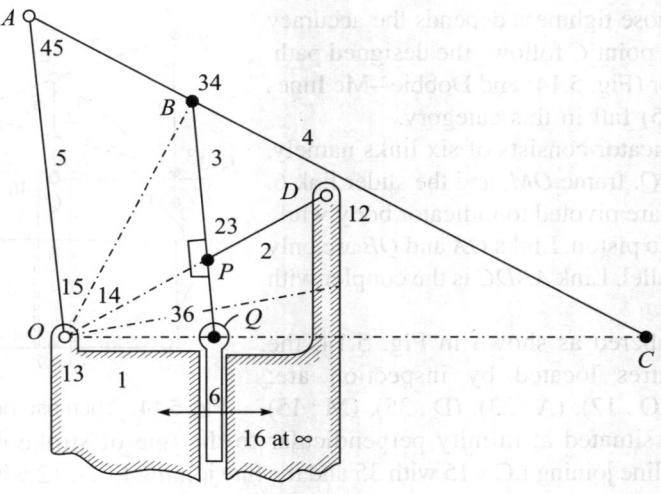

Fig. 5.13 Crosby indicator mechanism.

Comparing equations (5.27) and (5.28),

$$\frac{V_Q}{OQ} = \frac{V_C}{OC} \quad \text{or} \quad \frac{V_C}{V_Q} = \frac{OC}{OQ} \tag{5.29}$$

Since OA and BQ are parallel, from similar triangles, OAC and QBC,

$$\frac{OC}{QC} = \frac{OA}{QB} \tag{5.30}$$

But

$$QC = (OC - OQ)$$

Hence rewriting equation (5.30) and substituting for QC,

$$\frac{OC - OQ}{OC} = \frac{QB}{OA}$$

or

$$1 - \frac{OQ}{OC} = \frac{QB}{OA}$$

or

$$\frac{OQ}{OC} = 1 - \frac{QB}{OA} = \text{constant}$$

Hence, from equation (5.29),

$$\frac{V_c}{V_Q} = \frac{OC}{OQ} = \text{constant. This proves the condition } (b).$$

5.7.3 Thompson Indicator

It is desirable to have an indicator mechanism so arranged that the recording pencil (*i.e.*, the tracing point C) is made to move along a straight line or an approximate straight line quite independent of the path followed by slider point Q. This feature enables to cut down the number of pin joints on whose tightness depends the accuracy with which tracing point C follows the designed path. Thompson indicator (Fig. 5.14) and Dobbie –Mc Innes indicator (Fig. 5.15) fall in this category.

Thompson indicator consists of six links namely, OA, $ABDC$, DM, BQ, frame OM, and the slider link 6. Links OA and DM are pivoted to indicator body while link BQ is pivoted to piston. Links OA and QB are only approximately parallel. Link $ABDC$ is the coupler with tracing point at C.

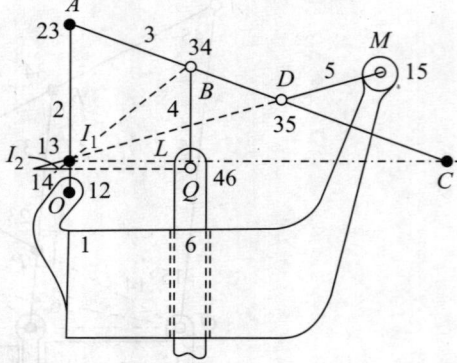

Fig. 5.14 Thompson indicator mechanism.

For links numbered as shown in Fig. 5.14, the instantaneous centres located by inspection are: (B : 34), (Q : 46), (O : 12), (A : 23), (D : 35), (M : 15) and the I.C. 16 is situated at infinity perpendicular to the line of stroke of piston 6.

Intersection of line joining I.C's 15 with 35 and the line joining I.C's. 12 with 23 locates I.C. 13 at I_1. In a similar way, intersection between line joining I.C. 46 with 16, and the line joining I.C's 13 and 34, locates I.C. 14 at I_2. The tracing point is to be located on line ABD such that I_1C is perpendicular to line of stroke of piston 6. This step ensures that the instantaneous motion of point C is parallel to that of piston 6.

In practice, links QB and OA cannot be exactly parallel. Also, the line I_1C cannot be exactly perpendicular to the line of stroke of the piston for all positions of the mechanism. It follows therefore that the ratio of the displacement of C to the displacement of Q cannot be exactly constant. This variation, for all practical purposes, is however negligible.

To show that the ratio of velocity of C to velocity of Q is very nearly constant, we proceed as under.

As points B and Q lie on link 4 for which the I.C. is I_2 (*i.e.*, 14), we have

$$\left(\frac{V_B}{V_Q}\right) = (I_2B)/(I_2Q) \tag{5.31}$$

Since I_1LC and I_2Q are drawn perpendicular to the line of stroke of slider 6, the triangles I_1BL and I_2BQ are similar. Therefore,

$$(I_2B)/(I_2Q) = (I_1B)/(I_LL) \tag{5.32}$$

Again points B and C are the points on link 3, for which the I.C. is I_1 (13). Hence,

$$\frac{V_c}{V_B} = \frac{I_1C}{I_1B} \tag{5.33}$$

Substituting for (I_2B/I_2Q) from (5.32) in (5.31) and multiplying with (5.33) on corresponding sides,

$$\left(\frac{V_c}{V_Q}\right) = \left(\frac{I_1C}{I_1L}\right) \tag{5.34}$$

Considering links QB and OA to be nearly parallel, the triangles CAI_1 and CBL are similar, and therefore,

$$\frac{CL}{I_1C} = \frac{BC}{AC}$$

or

$$\frac{I_1C - I_1L}{I_1C} = \frac{BC}{AC} = \text{constant}$$

or

$$1 - \left(\frac{I_1L}{I_1C}\right) = \text{constant}$$

Thus

$$(I_1L/I_1C) = \text{constant}$$

Hence from equation (5.34), $\dfrac{V_c}{V_Q} = \text{constant}$

5.7.4 Dobbie-McInnes Indicator

Except for minor changes this indicator is similar to Thomson indicator. For the difference, link BQ is connected not to the coupler link 3 but to link 5. This implies that the motion of slider is not transmitted to coupler directly.

For link lables shown in Fig. 5.15, the instantaneous centres located by inspection are:

(0 : 12), (A : 23), (D : 35), (B : 45), (M : 15), (Q : 46) and 16 which is located at infinity in a direction perpendicular to the line of stroke of piston 6. The instantaneous centre I_1 (*i.e.*, 13) is located at the point of intersection of lines joining I.C.'s (35, 15) and (12, 23). Similarly the instantaneous centre I_2 (*i.e.*, 14) is located at the point of intersection of lines joining I.C.'s (15, 45) and (46, 16). For tracing a path parallel to that of the slider 6, clearly, the tracing point C must be located on a line perpendicular to the line of stroke from I_1. Hence the point of intersection of horizontal line from I_1 and line joining AD defines the location of tracing point C. Now,

$$\frac{V_c}{V_Q} = \frac{V_c}{V_D} \frac{V_D}{V_B} \frac{V_B}{V_Q} \tag{5.35}$$

As the link DB rotates about point M (15),

$$\frac{V_D}{V_B} = \frac{MD}{MB} \tag{5.36}$$

and since link 4 instantaneously rotates about I.C. 14 (I_2),

$$\frac{V_B}{V_Q} = \frac{I_2B}{I_2Q} \tag{5.37}$$

Let BQ intersects I_1C at point F and let DL be the perpendicular dropped from D on I_1C. Now the triangles I_1BF and I_2BQ are similar and therefore,

$$\frac{I_2B}{I_2Q} = \frac{I_1B}{I_1F} \tag{5.38}$$

Again link 3 instantaneously rotates about I.C. I_1 (*i.e.*, 13) and therefore,

$$\frac{V_c}{V_D} = \frac{I_1C}{I_1D} \tag{5.39}$$

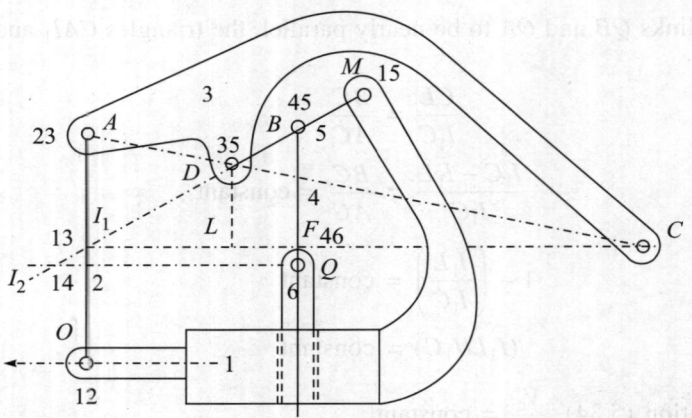

Fig. 5.15 Dobbie-McInnes indicator.

Substituting from equations (5.36), (5.37), (5.38), and (5.39), in equation (5.35),

$$\left(\frac{V_c}{V_Q}\right) = \left(\frac{I_1C}{I_1D}\right)\left(\frac{MD}{MB}\right)\left(\frac{I_1B}{I_1F}\right) \tag{5.40}$$

Again, triangles I_1DL and I_1BF are similar and as such,

$$\left(\frac{I_1B}{I_1F}\right) = \left(\frac{I_1D}{I_1L}\right) \tag{5.41}$$

Substituting for $\left(\dfrac{I_1B}{I_1F}\right)$ in equation (5.40),

$$\left(\frac{V_c}{V_Q}\right) = \left(\frac{I_1C}{I_1L}\right)\left(\frac{MD}{MB}\right) \tag{5.42}$$

But

$$\left(\frac{I_1L}{I_1C}\right) = \frac{(I_1C - CL)}{I_1C} = 1 - \left(\frac{CL}{I_1C}\right) \tag{5.43}$$

and from similar triangles CDL and CAI_1,

$$\left(\frac{CL}{I_1C}\right) = \left(\frac{CD}{CA}\right) = \text{constant}$$

and therefore, from equation (5.43)

$$\left(\frac{I_1L}{I_1C}\right) = \text{constant}$$

As MD and MB are constant it follows from Eq. (5.42) that

$$\left(\frac{V_c}{V_Q}\right) = \text{constant}$$

5.8 MOTOR CAR STEERING GEAR

Steering gear of an automobile is a mechanism which is used in automobile for turning axes of rotation of two front wheels, with reference to the chasis, so as to cause the automobile to follow some curved path. In the steering mechanism of a motor vehicle the front wheels are mounted on short axles, called stub axles. The stub axles are pivoted at separate points, permitting their rotation through different angles. The pivot points are fixed to the chasis. The rear wheels are placed over the back axle, at the two ends of the differential tube. As against this, in certain types of vehicles such as trailers or steam traction engines, the forward wheels are mounted on a turntable, free to swivel, so that the forward axle is directed towards the instantaneous centre I which lies on the line of the rear axle.

In general, when the vehicle takes a turn the front wheels, along with their respective axles, turn about the respective pivot points. The rear wheels remain straight and do not turn. Thus, the steering is done by means of front wheels only.

5.8.1 Condition of Correct Steering

Condition of correct steering requires that the relative motion between the wheels and the road surface should be that of pure rolling while taking a turn, avoiding lateral slip (skidding). This is possible when the wheels move along concentric arcs of circle about an instantaneous centre (I) of rotation. Under this condition the motion of the entire vehicle can be conceived to be a motion of body rotation about a vertical axis through I.

Let r be the wheel radius while R_i, ω_i, and R_o, ω_o be the radial distances and angular speeds of rotation of inner and outer wheels respectively. If ω' be the angular speed of body rotation of the vehicle about I, then

$$\omega' = \frac{r\omega_o}{R_o} = \frac{r\omega_i}{R_i}$$

The condition of correct steering therefore dictates that, while turning, the wheels should be in planes which are tangential to the respective concentric paths. As seen above, this is possible only when the axes of two front stub-axles, when produced, intersect the common axis of rear wheels at the same point which is the instantaneous centre I.

From the geometry of Fig. 5.16, $\tan \theta = \dfrac{CF}{FI}$; and $\tan \phi = \dfrac{AE}{EI}$

Therefore
$$\cot \phi - \cot \theta = \frac{EI}{AE} - \frac{FI}{CF} \qquad (5.44)$$

But $AE = CF = H$, the wheel base
Substituting in equation (5.44),

$$(\cot \phi - \cot \theta = \frac{(EI - FI)}{H} = \frac{EF}{H}$$

But $EF = W$, the distance between pivots

Hence, for correct steering, the inner and outer stub-axles should rotate through angles θ and ϕ respectively such that

$$(\cot \phi - \cot \theta) = \frac{W}{H} \qquad (5.45)$$

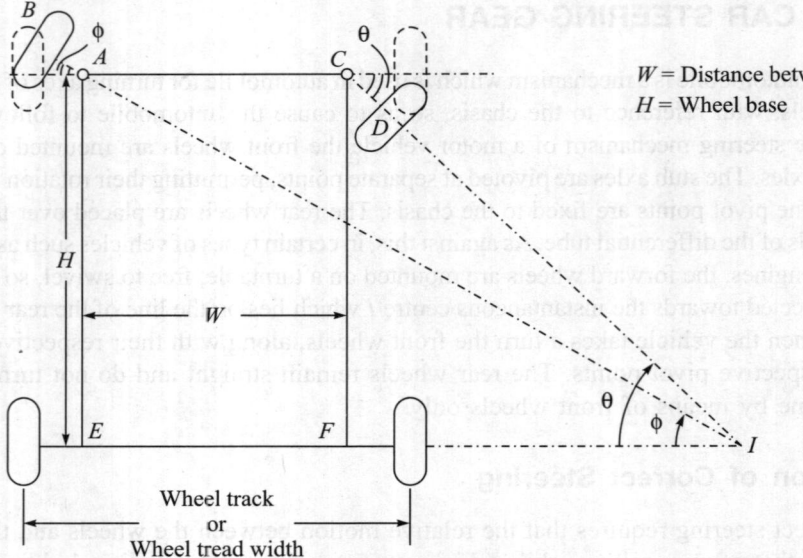

W = Distance between pivots
H = Wheel base

Fig. 5.16 Automobile steering gear.

5.8.2 Davis Steering Gear

Special features of this steering mechanism (Fig. 5.17) are:

1. The bell-crank levers *BAK* and *DCL*, which consists of slotted arms *AK* and *CL* respectively, while the other arms constitute stub-axles for front wheels.
2. The angles of the bell-crank levers namely, $\angle BAK$ and $\angle DCL$, are equal.
3. The die-blocks *M* and *N*, pin-connected at the ends of sliding link *MN*, provides a sliding pair between die-blocks and the slotted arms.
4. The link *MN* is constrained to move in a direction parallel to link *AC* by two guides G_1 and G_2 as shown.
5. The bell-crank levers *BAK* and *DCL* are pivoted to the front axle at *A* and *C* called king-pins.

The sliding is effected by sliding link *MN* either to the left or to the right. When the steering gear is in the mid-position and the car is moving straight, each of the slotted arms *AK* and *CL* is inclined at an angle α to the longitudinal axis of the car. If the link *MN* is now moved to the right through a distance *x*, relative to the chasis, the bell-crank levers *BAK* and *DCL* are moved to the position shown by dotted lines. The stub axle lines *BA* and *CD* in new position *B'A* and *CD'* when produced, intersect at *P*.

Let ϕ and θ be the angles through which the arms *AK* and *CL* are turned as a result of displacement *x* of link *MN* to the right. Let the new position of sliding link *MN* be *M'N'*, and corresponding position of bell-crank levers be *K'AB'* and *L'CD'* as shown in Fig. 5.17. Let *h* be the distance of *MN* from wheel axis *AC* and *2b* be the difference between lengths *MN* and *AC*. When the vehicle moves straight $\angle KAR = \angle LCR' = \alpha$ and

$$\tan\alpha = \left(\frac{b}{h}\right)$$

(5.46)

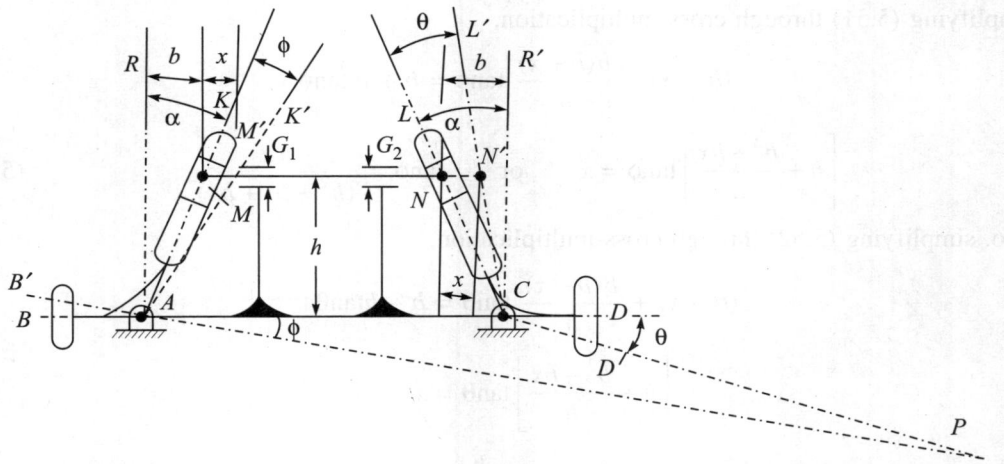

Fig. 5.17 Davis steering gear.

With link *MN* shifted to next position *M'N'*, angles of inclination of arms *KA* and *LC* with longitudinal direction change to $(\alpha + \phi)$ and $(\alpha - \theta)$ respectively. Thus,

$$\angle K'AR = (\alpha + \phi) \qquad \text{and} \qquad \angle L'CR' = (\alpha - \theta)$$

From the geometry of figure,

$$\tan(\alpha + \phi) = \left(\frac{b + x}{h}\right) \tag{5.47}$$

and

$$\tan(\alpha - \theta) = \left(\frac{b - x}{h}\right) \tag{5.48}$$

But

$$\tan(\alpha + \phi) = \frac{\tan\alpha + \tan\phi}{1 - \tan\alpha \tan\phi}$$

and

$$\tan(\alpha - \theta) = \frac{\tan\alpha - \tan\theta}{1 + \tan\alpha \tan\theta}$$

Substituting for $\tan\alpha$, we have

$$\tan(\alpha + \phi) = \frac{(b/h) + \tan\phi}{1 - (b/h)\tan\phi} \tag{5.49}$$

and

$$\tan(\alpha - \theta) = \frac{(b/h) - \tan\theta}{1 + (b/h)\tan\theta} \tag{5.50}$$

Substituting for L.H.S. from equations (5.47) and (5.48), in above equations,

$$\left(\frac{b + x}{h}\right) = \frac{(b/h) + \tan\phi}{1 - (b/h)\tan\phi} \tag{5.51}$$

$$\left(\frac{b - x}{h}\right) = \frac{(b/h) - \tan\theta}{1 + (b/h)\tan\theta} \tag{5.52}$$

Simplifying (5.51) through cross-multiplication,

$$(b + x) - \frac{b(b + x)}{h} \tan\phi = b + h \tan\phi$$

or $$\left[h + \frac{b^2 + bx}{h}\right] \tan\phi = x \qquad \text{or} \qquad \tan\phi = \frac{hx}{(h^2 + b^2 + bx)} \qquad (5.53)$$

Also, simplifying (5.52) through cross-multiplication,

$$(b - x) + \frac{b(b - x)}{h} \tan\theta = b - h\tan\theta$$

or $$\left[h + \frac{b^2 - bx}{h}\right] \tan\theta = x$$

or $$\tan\theta = \frac{hx}{(h^2 + b^2 - bx)} \qquad (5.54)$$

From equations (5.53) and (5.54), therefore

$$\cot\phi - \cot\theta = \frac{(h^2 + b^2 + bx)}{hx} - \frac{(h^2 + b^2 - bx)}{hx}$$

$$= \frac{2(bx)}{hx} = 2\left(\frac{b}{h}\right)$$

But $$\left(\frac{b}{h}\right) = \tan\alpha$$

Therefore, for Davis steering gear, $\cot\phi - \cot\theta = 2(\tan\alpha)$

However for correct steering condition, from (5.45), $\cot\phi - \cot\theta = \dfrac{W}{H}$

Hence, correct steering with Davis steering gear is obtained when

$$2(\tan\alpha) = \frac{W}{H} \qquad \text{or} \qquad \tan\alpha = \left(\frac{W}{2H}\right) \qquad (5.55)$$

In this case, the point of intersection P of front stub axles in new position will lie on the extension of rear axle line. With this condition, namely, $\tan\alpha = W/2H$, Davis steering gear will be equally effective at all the angles of steering in either direction.

Davis steering gear, however, has an inherent weakness. It can, at best, be described as a theoretically correct steering gear. The presence of sliding pair involves more of friction. Wear at the sliding surfaces produce slackness and, as a result, original accuracy is soon lost. For this reason, even though Davis steering gear gives correct steering at all angles of turn, it is rarely used in practice.

5.8.3 Ackerman Steering Gear

Let us now consider Ackerman steering gear in which sliding pairs are eliminated and are replaced by turning pairs. The gear is more commonly used and gives an approximation good enough for practical purposes.

This steering mechanism consists of a four-bar mechanism involving turning pairs only. The two opposite links AC and KL, of unequal lengths, parallel in normal position when the vehicle moves straight. The other two links AK and CL, which are the arms of bell-crank levers, are of equal lengths. These arms are inclined at an angle of α to the longitudinal axis when the vehicle is moving straight. Under this condition, the four-bar mechanism $AKCL$ constitutes a trapezium with parallel and unequal sides AC and KL.

In order to steer the car to, say right, the link CL is turned clockwise so as to increase angle α. This results in c.c.w. rotation of arm AK, reducing angle α at the other end. For a given value of ratio (AK/AC) and angle α, a unique value of angle θ may be obtained for a given value φ either graphically or analytically.

From the geometry of Fig. 5.18,

$$\cot\phi - \cot\theta = \frac{AQ - CQ}{PQ}$$

$$(\cot\phi - \cot\theta) = \left(\frac{AC}{PQ}\right) \qquad (5.56)$$

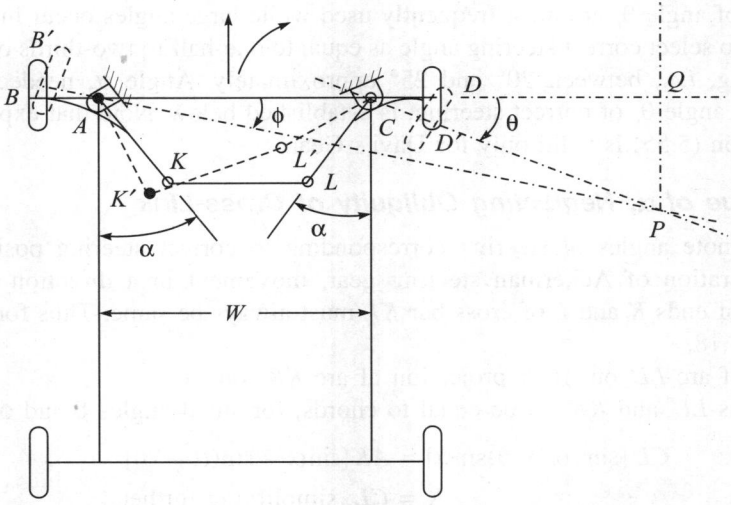

Fig. 5.18 Plan view of Ackerman steering gear.

where P is the point of intersection of stub-axles $B'A$ and CD' produced. Clearly, for correct steering, the point of intersection P of front stub axle lines should lie on rear axle line produced if necessary. In other words, equation (5.56) should conform with

$$(\cot\phi - \cot\theta) = \frac{W}{H}$$

For private cars with $AC : AK = 8.5 : 1$ and $\alpha = 18°$, if the distance AC is 0.4 times the wheel-base, Ackerman steering mechanism would give correct steering for $\theta \approx 24°$. Thus with above proportions, Ackerman steering gear provides correct steering in the following three positions only

(i)   When moving straight, *i.e.*, when $\theta = 0°$.
(ii) When turning to the left at $\theta \approx 24°$.
(iii) When turning to the right at $\theta \approx 24°$.

At all other values of angle θ, therefore, wheels tend to follow path along circular arcs that do not have a common centre. In all such cases, the instantaneous centre I does not lie on the axis of rear wheels but lies on a line parallel to the rear axle towards the front side. It follows that some amount of skidding is bound to occur at steering angles other than $\theta = 0$ and $24°$. Hence, angle α and the proportions of the mechanism should be so chosen as to reduce the inherent tendency to skid to a minimum, especially for turning radii that could be followed at high speeds, where the centrifugal action also contributes to skidding.

Despite the limitation stated above, Ackerman steering gear is still the most commonly used steering gear of the two. The advantage of the Ackerman steering gear lies in the use of revolute (pin) joint rather than the sliding pairs. Resulting simplicity facilitates in upkeeping. Further, as will be shown later, resulting deviation from condition of correct steering is small in the case of Ackerman steering mechanism.

The position of cross-arm (*i.e.*, the arm of bell-crank lever) corresponding to correct angle θ_c of steering, relative to the maximum angle of steering are governed by the design requirements. Smaller values of angle θ_c are most frequently used while large angles occur infrequently. It is usual therefore to select correct steering angle as equal to one-half to two-thirds of the maximum angle of steering, *i.e.*, between 20° and 25° approximately. Angle α, needed for providing specified values angle θ_c of correct steering, is established below. Note that expression $\tan \alpha = W/2H$ of equation (5.55) is valid only for Davi's Gear.

Required Value of α, Neglecting Obliquity of Cross-Link

Let θ and ϕ denote angles of steering corresponding to correct steering position. Now, for satisfactory operation of Ackerman steering gear, movement in a direction parallel to AC (*see* Fig. 5.18) of ends K and L of cross-bar KL must always be same. Thus for displacements shown in Fig. 5.18,

projection of arc LL' on AC = projection of arc KK' on AC.

Treating arcs LL' and KK' to be equal to chords, for small angles θ and ϕ, we have

$$CL \,[\sin (\alpha + \theta)\sin\alpha] = AK[\sin\alpha - \sin(\alpha - \phi)]$$

As $AK = CL$, simplifying further,

$$(\sin\alpha\cos \theta + \cos\alpha\sin\theta) - \sin\alpha = \sin\alpha - (\sin\alpha\cos\phi - \cos\alpha\sin\phi)$$

or $\sin\alpha(\cos\theta + \cos\phi - 2) = \cos\alpha(\sin\phi - \sin\theta)$

or $$\tan\alpha = \frac{\sin \phi - \sin\theta}{(\cos\theta + \cos\phi - 2)} \qquad (5.57)$$

where θ and ϕ are steering angles for the two wheels in correct steering position.

Unlike Davis' steering gear, thus, the expression for α is not quite simple (compare with equation 5.55). Thus to obtain desired value of α, one needs to know about the dimensions, like wheel base H and distance between pivots W of a 4-wheeler. With (W/H) ratio thus known, value of correct angle of steering ϕ for outer wheel, corresponding to each arbitrarily assumed values of angle θ for inner wheel, can then be calculated from,

$$\cot\phi - \cot\theta = \frac{W}{H}$$

With the set of values θ and ϕ for correct steering established, corresponding value of required angle α can then be calculated from equation (5.57). These calculated values of ϕ and α for a set of values, $\theta = 10°$, $15°$, $20°$, $25°$, and $30°$, are listed in Table 5.1.

Table 5.1 Table for optimum value of α for each value steering angle Q.; for W/H = 0.4 satisfying condition of correct steering

S.No.	Set of values for θ and ϕ to satisfy condition of correct steering: $\cot\phi - \cot\theta = W/H = 0.4$		Angle α required for providing correct steering at a given set of values θ and ϕ	
	θ	$\phi = \tan^{-1}\left(\dfrac{1}{0.4 + \cot\theta}\right)$	$\tan\alpha = \dfrac{\sin\phi - \sin\theta}{\cos\theta + \cos\phi - 2}$	$\alpha°$
1.	10°	9.35°	0.3927	21.44°
2.	15°	13.605°	0.3797	20.79°
3.	20°	17.626°	0.3658	20.09°
4.	25°	21.455°	0.3488	19.228°
5.	30°	25.128°	0.3296	18.244°

EXAMPLE 5.1 In a Davis steering gear, the distance between the pivots of front axle is 1 metre and the wheel base is 2.5 metre. Find the inclination of the track arm to the longitudinal axis of the car when it is moving along a straight path. (DAV Indore, Nov.–Dec. 1986)

Solution: Refer to Figs. 5.16 and 5.17.

$$W = 1 \text{ m} = 100 \text{ cm}$$
$$H = \text{wheel base} = 2.5 \text{ m} = 250 \text{ cm}$$

As shown in the discussions on Davis steering gear, for correct steering

$$\cot\phi - \cot\theta = \frac{W}{H}$$

which leads to the condition for Davis gear as

$$\tan\alpha = \frac{W}{2H}$$

$$= \frac{100}{(250)2} = \frac{1}{5} = 0.2$$

Therefore $\qquad\qquad \alpha = 11.309° = 11° - 18.54'$ **Ans.**

EXAMPLE 5.2 For a Davis steering gear, compute and plot steering angles of inner and outer wheels of a car with wheel base $H = 2 \times T$, where T is the wheel tread width (*i.e.,* wheel track). The distance between the king-pins at which wheels are pivoted is $(0.9) T$. Plot the angles of outer wheel versus inner wheel angles for a range 5° to 45°. Determine the radius of the path of travel in terms of T and W of the outer king pin, when the inner wheel is at 45°.

Solution: With reference to Fig. 5.16,

$$\cot\phi - \cot\theta = \frac{EI - FI}{H}$$

$$= \frac{W}{H} = \frac{(0.9)T}{2T} = 0.45$$

Hence, outer wheel angle ϕ is given by

$$\cot\phi = 0.45 + \cot\theta$$

Values of $\cot\phi$ and hence of ϕ for different values of θ are as under

θ	$\cot\theta$	$\cot\phi = 0.45 + \cot\theta$	$\phi°$
5	11.43	11.88	4.8115
10	5.67	6.1213	9.278
15	3.732	4.182	13.448
20	2.747	3.197	17.367
25	2.144	2.594	21.078
30	1.732	2.182	24.621
35	1.428	1.878	28.033
40	1.192	1.642	31.346
45	1.0	1.45	34.592

At
$$\theta = 45°, \ \phi = 34.592°$$

Hence, from Fig. 5.16,

$$IA = \frac{H}{\sin\phi} = \frac{2T}{\sin(34.592)} = (3.52)T$$

Also
$$W = (0.45)H$$

Substituting for H above,

$$IA = \frac{W}{(0.45)\sin 34.592} = (3.914)W$$

Thus the radius of path of outer wheel at $\theta = 45°$ is

$$R = (3.52)T \qquad\qquad \textbf{Ans.}$$

and
$$R = (3.914)W \qquad\qquad \textbf{Ans.}$$

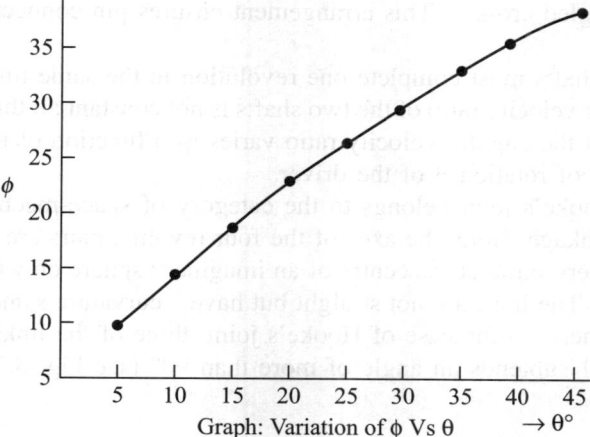

Graph: Variation of φ Vs θ

5.9 HOOKE'S (CARDAN) JOINT OR UNIVERSAL COUPLING

A Hooke's joint is used to connect two intersecting shafts. The axes of the shafts usually make small angle α with each other which may vary during operation. The Hooke's joint finds its use in automobiles for power transmission from the gear box to the back axle. *The joint permit some amount of angular misalignment in connected shafts. This becomes a very desirable feature on account of possible misalignment arising out of elastic frame and also due to spring in suspension.* Sometimes two such joints are used in the same drive system, as in automobiles, where one joint is used at the power transmission and the second joint at the rear axle. Hook joint also finds application in the transmission of the drive to the spindles of multi-spindle drilling machine.

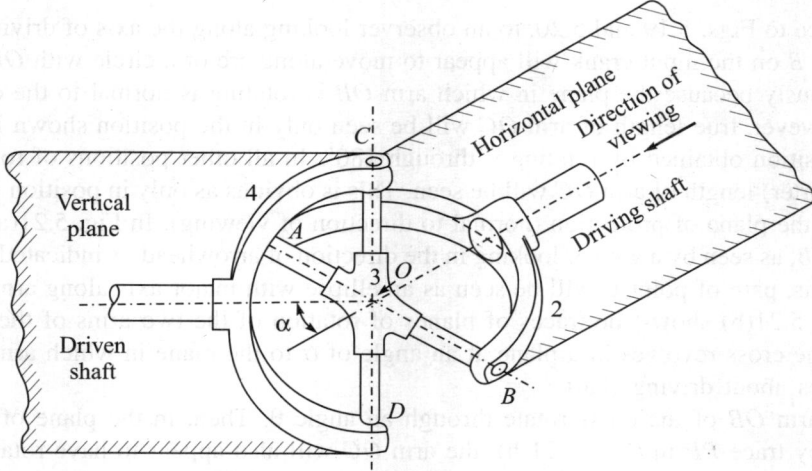

Fig. 5.19 Hooke's (cardan) joint.

Essentially a Hooke's joint consists of two semi-circular yokes (forks) 2 and 4 formed at the ends of driving and driven shafts (see Fig. 5.19). Bearings are provided in the yokes to receive the arms of right angled cross 3. This arrangement ensures pin connection between the cross-arms and yokes.

Although both shafts must complete one revolution in the same time-interval, it is easy to show that the angular velocity ratio of the two shafts is not constant all throughout the revolution. It will be shown that the angular velocity ratio varies as a function of the angle α between the shafts and the angle of rotation θ of the driver.

Essentially a Hooke's joint belongs to the category of space-mechanism and represents a spherical four-bar linkage. Here, the axes of the four revolute pairs are not parallel, but have a common point of intersection at the centre of an imaginary sphere on whose surface all the four revolute pairs move. The links are not straight but have a curvature same as that of great circles of the imaginary sphere. In the case of Hooke's joint, three of the links subtend angles of $90°$ and the fourth link 1 subtends an angle of more than $90°$ (see Fig. 5.20).

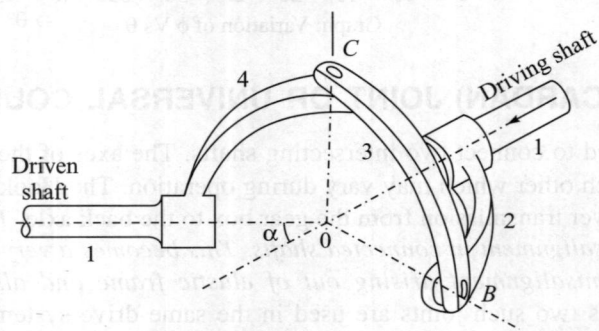

Fig 5.20   Hooke's joint as a spherical 4-revolute mechanism.

5.9.1   Transmission Characteristics

With reference to Figs. 5.19 and 5.20, to an observer looking along the axis of driving shaft (as shown) point B on the input crank will appear to move along arc of a circle with OB as radius. This is obviously because the plane in which arm OB is rotating is normal to the direction of viewing. However, true length of arm OC will be seen only in the position shown in Fig. 5.19 as also in position obtained by rotating it through $180°$. In all other positions of rotation, only apparent (shorter) length of arm OC will be seen. This is obvious as only in position OC the arm is parallel to the plane of projection (normal to direction of viewing). In Fig. 5.21(a) therefore, path of point B, as seen by a viewer looking in the direction of arrowhead, is indicated by a circle. As against this, path of point C will be seen as an ellipse with major axis along arm OC of the cross. Figure 5.21(b) shows the traces of planes of rotation of the two arms of the cross. The arm OC of the cross revolves in a plane at an angle of α to the plane in which arm OB of the cross revolves about driving shaft.

Let the arm OB of the cross rotate through an angle θ. Then, in the plane of projection, represented by trace PP in (Fig. 5.21 b), the arm OC will also appear to have rotated through angle θ (this is because OB and OC are the arms of the same rigid cross), and point C will move along the elliptical path to occupy new position OC_1. Actual angle of rotation (true angle) of arm OC, however, has to be obtained in its own plane of rotation represented by trace QQ. Since the

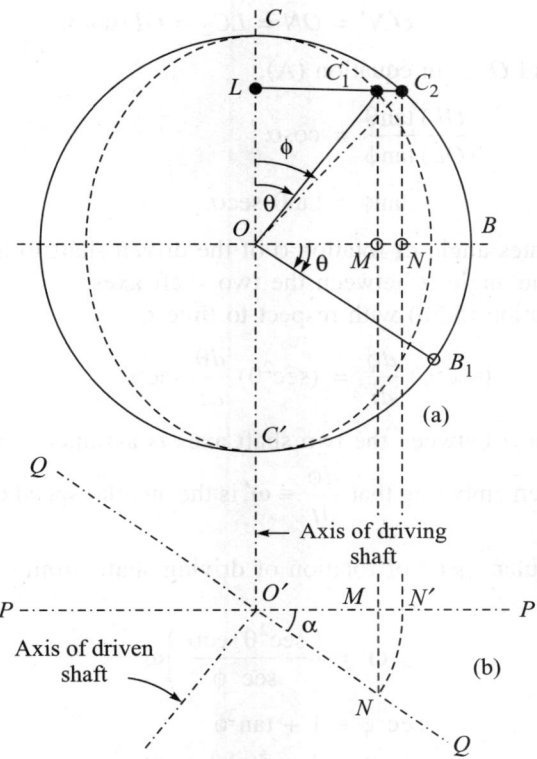

Fig. 5.21 Relation between θ and ϕ in Hooke's joint.

arms OB and OC are of the same length, actual path of point C is identical to the path of point B, which is shown as circle in Fig. 5.21(a). A line C_1C_2 drawn parallel to the line OB to meet the circle at C_2 then identifies OC_2 as the actual position of line OC. Angle $\angle COC_2 = \phi$ therefore, gives actual angle of rotation of arm OC in its own plane of rotation.

In order to establish relation between input angle of rotation θ and the output angle ϕ, let us project points C_1 and C_2 on the traces PP and QQ as also on vertical diameter CC'. Let M and N' be the foots of perpendicular on trace PP from points C_1 and C_2. Point N is obtained on trace QQ by drawing an arc with O' as centre and $O'N'$ as radius. Clearly NM is at right angles to the trace PP. Also drop perpendicular C_1L from point C_1 on vertical diameter CC'.

From right angled triangle $O'MN$ of Fig. 5.21(b),

$$\frac{O'M}{O'N} = \cos\alpha \tag{A}$$

But
$$O'N = O'N', \text{ and hence,}$$

$$\frac{O'M}{O'N'} = \cos\alpha$$

Again, from Fig. 5.21(a) and (b),

$$O'M = OM = LC_1 = OL\,(\tan\theta)$$

And
$$O'N' = ON = LC_2 = OL(\tan\phi)$$

Substituting for $O'M$ and $O'N'$ in equation (A),

$$\frac{(OL)\tan\theta}{(OL)\tan\phi} = \cos\alpha$$

or
$$\tan\phi = \tan\theta \sec\alpha \qquad (5.58)$$

Equation (5.58) relates angle of rotation ϕ of the driven shaft to the angle of rotation θ of driving shaft through the angle α between the two shaft axes.

Differentiating equation (5.58) with respect to time t,

$$(\sec^2\phi)\frac{d\phi}{dt} = (\sec^2\theta)\frac{d\theta}{dt}\sec\alpha \qquad (5.59)$$

Angle of inclination α between the two shaft axes is assumed constant for the purpose of differentiation above. Remembering that $\dfrac{d\phi}{dt} \equiv \omega'$ is the angular speed of rotation of driven shaft and $\dfrac{d\theta}{dt} \equiv \omega$ is the angular speed of rotation of driving shaft, from equation (5.59) we have

$$\omega' = \left(\frac{\sec^2\theta \sec\alpha}{\sec^2\phi}\right)\omega \qquad (5.60)$$

But
$$\sec^2\phi \equiv 1 + \tan^2\phi$$

or, using Eq. (5.58),
$$\sec^2\phi = 1 + (\tan\theta\sec\alpha)^2$$

Simplifying
$$\sec^2\phi = \frac{\cos^2\theta \cos^2\alpha + \sin^2\theta}{\cos^2\theta \cos^2\alpha}$$

or
$$\sec^2\phi = \frac{(1 - \sin^2\alpha)\cos^2\theta + \sin^2\theta}{\cos^2\theta \cos^2\alpha}$$

or
$$\sec^2\phi = \left(\frac{1 - \sin^2\alpha \cos^2\theta}{\cos^2\alpha \cos^2\theta}\right)$$

Substituting for $\sec^2\phi$ in equation (5.60),

$$\omega' = \left(\frac{\sec^2\theta \sec\alpha}{1}\right)\frac{\cos^2\alpha \cos^2\theta}{1 - \sin^2\alpha \cos^2\theta}\omega$$

or
$$\omega' = \omega\left(\frac{\cos\alpha}{1 - \sin^2\alpha \cos^2\theta}\right) \qquad (5.61)$$

It follows from equation (5.61) that for a constant input speed ω, output speed ω' changes with angle of rotation θ of driving crank.

Polar Angular Velocity Diagram

For a maximum value of ω', it follows from equation (5.61) that the denominator should be minimum, which requires that for given α, $\cos^2\theta$ be maximum. Hence maximum value of ω' occurs when $\theta = 0$, and π, for which

$$\omega' = \omega\left(\frac{\cos\alpha}{1-\sin^2\alpha}\right)$$

Thus

$$(\omega')_{max} = \left(\frac{\omega}{\cos\alpha}\right) \tag{5.62}$$

and occurs at

$$\theta = 0 \text{ and } \pi.$$

Again, for minimum value of ω', $\cos^2\theta$ should be minimum. This occurs when $\theta = \dfrac{\pi}{2}$ or $\dfrac{3\pi}{2}$,

Hence

$$(\omega')_{min} = \omega\left(\frac{\cos\alpha}{1}\right) = (\omega)\cos\alpha \tag{5.63}$$

Further $\omega' = \omega$, when the ratio $\dfrac{\omega'}{\omega}$ in equation (5.61) is equal to 1. This requires that,

$$\frac{\omega'}{\omega} = 1 = \frac{\cos\alpha}{1-\sin^2\cos^2\theta}$$

or

$$\cos\alpha = 1 - \sin^2\alpha\,\cos^2\theta$$

or

$$\cos^2\theta = \frac{1-\cos\alpha}{\sin^2\alpha} = \frac{(1-\cos\alpha)}{(1-\cos\alpha)\,(1+\cos\alpha)}$$

or

$$\cos^2\theta = \frac{1}{(1+\cos\alpha)}$$

Hence

$$\sin^2\theta = 1 - \left(\frac{1}{1+\cos\alpha}\right) = \left(\frac{\cos\alpha}{1+\cos\alpha}\right)$$

And, therefore,

$$\tan^2\theta = \cos\alpha \quad \text{or,} \quad \tan\theta = \pm\sqrt{\cos\alpha}$$

Taking

$$\alpha = 30°, \text{ and } \omega = 3 \text{ rad/s}$$

$$\omega = \omega' \text{ requires } \theta = \tan^{-1}(\pm\sqrt{\cos 30})$$

or, when

$$\theta = 42.94, (180 + 42.94) \text{ and } \theta = -42.94 \text{ and } (180 - 42.94)$$

i.e., at

$$\theta = 42.94;\ 222.94;\ -42.94 \text{ and } 137.058°$$

Further

$$(\omega')_{min} = \omega\cos 30$$

or

$$(\omega')_{min} = 3 \times 0.866 = 2.598 \text{ rad/s which occurs at } \theta = \frac{\pi}{2} \text{ and } \frac{3\pi}{2}$$

and

$$(\omega')_{max} = \frac{\omega}{\cos 30°} = 3.46 \text{ rad/s which occurs at } \theta = 0 \text{ and } \pi$$

The elliptical curve, shown in broken line in Fig. 5.22, is the polar diagram representing variable output velocity ω'. The circle in full line represents input velocity ω which is constant at all θ.

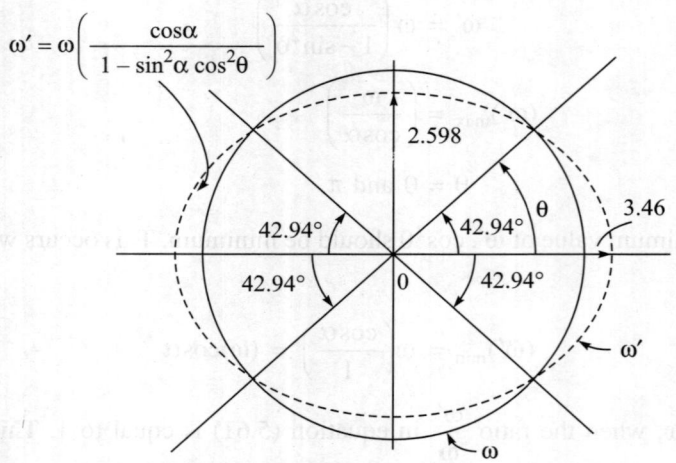

$$\omega' = \omega\left(\frac{\cos\alpha}{1 - \sin^2\alpha\,\cos^2\theta}\right)$$

Fig. 5.22 Polar velocity diagram for $\alpha = 30°$.

Angular Accelerations in Hooke's joint

From equation (5.61),

$$\omega' = \omega\left(\frac{\cos\alpha}{1 - \sin^2\alpha\,\cos^2\theta}\right)$$

The angular acceleration, for constant α, is given by

$$\frac{d\omega'}{dt} = \frac{d\omega'}{d\theta}\frac{d\theta}{dt} = \omega\,\frac{d\omega'}{d\theta}$$

Now

$$\frac{d\omega'}{d\theta} = \omega\left[\frac{-\cos\alpha\,\{0 + 2\cos\theta\,\sin^2\alpha\,\sin\theta\}}{(1 - \sin^2\alpha\,\cos^2\theta)^2}\right]$$

$$= \frac{-(\omega\cos\alpha)\,\sin^2\alpha\,\sin2\theta}{(1 - \sin^2\alpha\,\cos^2\theta)^2}$$

Hence

$$\frac{d\omega'}{dt} = \frac{\omega\,d\omega'}{d\theta} = \frac{-(\omega^2\cos\alpha)\,\sin^2\alpha\,\sin2\theta}{(1 - \sin^2\alpha\,\cos^2\theta)^2}$$

Thus

Angular acceleration of driven shaft,

$$\frac{d\omega'}{dt} = \frac{-(\omega^2\cos\alpha)\,\sin^2\alpha\,\sin2\theta}{(1 - \sin^2\alpha\,\cos^2\theta)^2} \tag{5.64}$$

The value of θ at which the acceleration is maximum, is obtained by differentiating equation (5.64) w.r. to θ and equating the same to zero. The resulting equation will not be simple. A simplified expression for θ at which acceleration is maximum is—

$$\cos 2\theta \approx \left(\frac{2\sin^2\alpha}{2-\sin^2\alpha}\right) \tag{5.65}$$

Even for values of α as high as 30°, the expression (5.65) is sufficiently accurate within a few minutes of angular measure. It may be of interest to have a look into the effect of angle α, included between shaft axes, on the percentage variation of angular speed of driven shaft. From equations (5.62) and (5.63), it follows that

Percentage of speed variation

$$= \frac{\omega'_{max} - \omega'_{min}}{\omega} \times 100$$

$$= \frac{(\omega/\cos\alpha) - (\omega\cos\alpha)}{\omega} \times 100 = \frac{(1-\cos^2\alpha)}{\cos\alpha} \times 100$$

$$= \left(\frac{\sin^2\alpha}{\cos\alpha}\right) \times 100 \tag{5.66}$$

Thus percentage speed variation purely depends on the angle α between shaft axes. Percentage speed variation of driven shaft with angle α is shown in Fig. 5.23. A good recommendation is to keep down α, the shaft angle of inclination, below 15°. Angles of 45° may be used for hand operated applications or very slow speed operations.

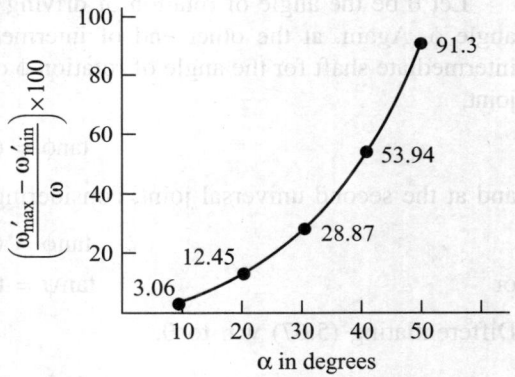

Fig. 5.23 Effect of angle α on % variation in output speed.

Effects of Large Angular Acceleration

Hooke's joint involves a rigid connection between the connected shafts and as such from the point of view of torque transmission, moment of inertia of connected parts should be small. In case this is not so, very high alternating torques, and therefore very high alternating stresses may be set up in the parts of the joint. For instance, the angular acceleration expression (5.64) for $\alpha = 20°$ modifies as under. For $\alpha = 20°$,

$$\frac{d\omega'}{dt} = \frac{-\omega^2(0.1099)\sin 2\theta}{(1-0.117\cos^2\theta)^2}$$

Thus, the angular accelerations at different values of θ are:

$\theta =$	0°	30°	60°	90°	120°	150°	180°	210°	240°
$\dfrac{d\omega'}{dt} =$	0	$-0.114\omega^2$	$-0.101\omega^2$	0	$0.101\omega^2$	$0.114\omega^2$	0	$-0.114\omega^2$	$-0.101\omega^2$
$\theta =$	270°	300°	330°	360°					
$\dfrac{d\omega'}{dt} =$	0	$0.101\omega^2$	$0.114\omega^2$	0					

It is concluded that the angular acceleration varies cyclically between positive and negative maximum value through zero. Thus frequency of fluctuation is twice during one revolution. Further, angular acceleration is directly proportional to the square of angular speed ω. Thus at high speeds, the problem becomes much more severe.

Again at α = 30°, the expression 5.64 reduces to

$$\frac{d\omega'}{dt} = -\frac{\omega^2 (0.2165) \sin 2\theta}{(1 - 0.117 \cos^2 \theta)^2}$$

A comparison with the corresponding expression for α = 20° reveals that larger the shaft angle α, higher is the acceleration and, therefore, this angle should be kept as small as possible.

5.9.2 Double Hooke's Joint

It is possible to connect two shafts by two Hooke's couplings through an intermediate shaft such that the uneven velocity ratio of the first coupling will be cancelled out by the other one.

Thus, two parallel or intersecting shafts may be connected by a double universal joint and have uniform output motions, provided that the intermediate shaft makes equal angles with the connected shafts and that the forks on the intermediate shaft are in the same plane.

Let θ be the angle of rotation of driving shaft and let the intermediate shaft rotate through angle ϕ. Again, at the other end of intermediate shaft, let ψ be the angle of rotation of the intermediate shaft for the angle of rotation ϕ of the intermediate shaft. Then at the first universal joint,

$$\tan\phi = (\tan\theta \ \sec\alpha) \qquad (5.67)$$

and at the second universal joint, considering ψ as the input angle and ϕ as output angles

$$\tan\phi = (\tan\psi \ \sec\alpha)$$

or $$\tan\psi = \tan\phi \ \cos\alpha \qquad (5.68)$$

Differentiating (5.67) w.r. to θ,

$$\left(\frac{d\phi}{d\theta}\right) = \frac{\sec^2\theta}{\sec^2\phi \cos\alpha} \qquad (5.69)$$

and differentiating (5.68) w.r. to ϕ,

$$\frac{d\psi}{d\phi} = \frac{\sec^2\phi \cos\alpha}{\sec^2\psi} \qquad (5.70)$$

Multiplying (5.69) by (5.70) on respective sides,

$$\frac{d\psi}{d\theta} = \frac{\sec^2\theta}{\sec^2\psi} = \left(\frac{1 + \tan^2\theta}{1 + \tan^2\psi}\right) \qquad (5.71)$$

Substituting for tanθ from (5.67) and for tanψ from (5.68),

$$\frac{d\psi}{d\theta} = \frac{1 + (\tan\phi\cos\alpha)^2}{1 + (\tan\phi\cos\alpha)^2} = 1$$

Thus the angular speed of input shaft is equal to the angular speed of output shaft and that the second Hooke's joint exactly neutralises the effect of the first one. Necessary conditions to be fulfilled by a double Hooke's joint of this type are as under.

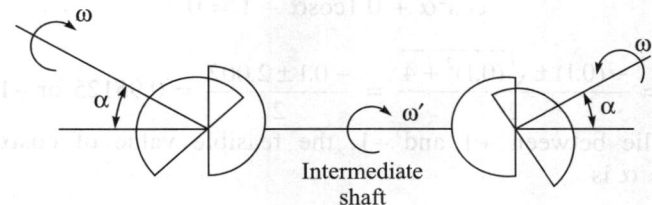

Fig. 5.24(a) Double Hooke's joint with non-parallel connected shafts.

Two shafts that are neither parallel nor intersecting may be connected by a double Hooke's joint and will have uniform angular speed ratio, provided that

1. intermediate shaft makes equal angles with the connected shafts (the input and output), and
2. forks on the intermediate shaft are so arranged that they simultaneously lie in the plane of shaft axes at the respective ends.

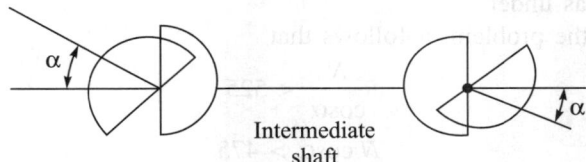

Fig. 5.24(b) Double Hooke's joint with parallel connected shaft.

Clearly, the intermediate shaft is the only shaft in a double Hooke's joint which is subjected to maximum accelerations and retardations and as such, in order to cut down high alternating inertia torques to and high values of alternating stresses, the intermediate shaft should be as light as possible.

EXAMPLE 5.3   Two shafts are to be connected by a Hooke's joint. The driving shaft rotate at a uniform speed of 500 r.p.m. and the speed of driven shaft must lie between 475 and 525 r.p.m. Determine the maximum permissible angle between the shafts (Univ. of Indore, April 1976, May 1982; DAVV, March 1990).

Solution: When the driving shaft rotates uniformly at N r.p.m., the maximum and minimum speed of driven shaft are given by,

$$\left(\frac{N}{\cos\alpha}\right) \quad \text{and} \quad (N\cos\alpha)$$

since, $N = 500$ r.p.m., it follows that maximum variation of speed is given by

$$N\left(\frac{1}{\cos\alpha} - \cos\alpha\right) = (525 - 475) = 50$$

Therefore $\dfrac{1}{\cos\alpha} - \cos\alpha = \dfrac{50}{500}$ or $\dfrac{1-\cos^2\alpha}{\cos\alpha} = 0.1$

or $\cos^2\alpha + 0.1\cos\alpha - 1 = 0$

Therefore $\cos\alpha = \dfrac{-(0.1)\pm\sqrt{(0.1)^2+4}}{2} = \dfrac{-0.1\pm2.002}{2} = 0.95125$ or -1.051

Since, $\cos\alpha$ must lie between $+1$ and -1, the feasible value of $\cos\alpha$ is 0.95125, and corresponding angle α is

$$\alpha = 17.964°$$

A little consideration, however, shows that this approach is not satisfactory. For instance,

$$\dfrac{N}{\cos\alpha} = \dfrac{500}{0.95125} = 525.624 \text{ r.p.m., which is outside the range}$$

and $$(N\cos\alpha) = 500 \times 0.95125 = 475.625 \text{ r.p.m.}$$

This clearly shows that the above procedure of establishing angle α is not satisfactory as the maximum speed exceeds maximum permissible speed of 525 r.p.m. A more satisfactory approach is therefore as under.

From the data of the problem, it follows that

$$\dfrac{N}{\cos\alpha} < 525 \qquad (1)$$

and $$N\cos\alpha > 475 \qquad (2)$$

Condition (1) simplifies to,

$$\dfrac{N}{525} < \cos\alpha \quad \text{or,} \quad \cos\alpha > 0.9523809$$

Condition (2) simplifies to,

$$\cos\alpha > \dfrac{475}{500} \quad \text{or,} \quad \cos\alpha > 0.95$$

Clearly, if both the conditions are to be satisfied,

$$\cos\alpha > 0.9523809 \quad \text{for which, } \alpha \le 17.7528°$$

It may be verified that for the limiting value of $\cos\alpha = 0.9523809$

$$N_{max} = \dfrac{500}{0.9523809} = 525.00002 \text{ r.p.m.}$$

and $$N_{min} = (500)(0.9523809) = 476.19 \text{ r.p.m.}$$

Thus, with this value of α, both the conditions are satisfied (As a cross-check students are advised to verify that if $\cos\alpha$ is slightly greater than above value, say 0.9524, the conditions are satisfied in a still better way as $\dfrac{N}{\cos\alpha} = 524.989$ r.p.m. and $N\cos\alpha = 476.2$ r.p.m.)

Hence, we adopt $\cos\alpha = 0.9523809$ for which $\alpha = 17° - 45' - 10''$ **Ans.**

EXAMPLE 5.4 Two shafts are connected by a Hooke's joint. The driving shaft revolves uniformly at 500 r.p.m. If the total permissible variation in speed of the driven shaft is not to exceed ± 6% of the mean speed, find the greatest permissible angle between the centre lines of the shafts. (Univ. of Indore, May 1974)

Solution: As per data, if N be the mean speed of 500 r.p.m.

$$N_{max} = 1.06N = 530 \text{ r.p.m.} \qquad \text{and} \qquad N_{min} = 0.94N = 470 \text{ r.p.m.}$$

We, therefore, adopt more fundamental approach. Thus,

$$\frac{N}{\cos\alpha} < 530 \text{ r.p.m.} \qquad (1)$$

and

$$N\cos\alpha > 470 \text{ r.p.m.} \qquad (2)$$

Condition (1) simplifies to $\cos\alpha > \dfrac{500}{530}$ *i.e.,* $\cos\alpha > 0.9433962$

and condition (2) simplifies to,

$$\cos\alpha > \frac{470}{500} \qquad \text{or} \qquad \cos\alpha > 0.94$$

Clearly, the smallest possible value of $\cos\alpha$ to satisfy both the conditions is

$$\cos\alpha \geq 0.9433962 \qquad \text{which gives} \qquad \alpha \leq 19.370038$$

With this, $\qquad N'_{max} = \dfrac{500}{0.9433962} = 529.9999 \text{ r.p.m.}$

and $\qquad\qquad N'_{min} = 500 \times 0.9433962 = 471.698 \text{ r.p.m.}$

Clearly $\qquad\qquad \alpha = 19.37$ will satisfy both the conditions.

Hence $\qquad\qquad \alpha = 19° - 22' - 6''$

This gives greatest permissible angle. **Ans.**

EXAMPLE 5.5 A Hooke's joint is used to connect two shafts whose axes are inclined at 20 degrees. The driving shaft rotates uniformly at 6000 r.p.m. What are the extreme angular velocities of the driven shaft? Find the maximum value of retardation or acceleration and state the angle where both will occur. (DAV Indore, May–June 1987; July 1988)

Solution: Speed of driving shaft $N = 6000$ r.p.m.

Therefore, maximum speed of driven shaft

$$= \frac{N}{\cos\alpha} = \frac{6000}{\cos 20} = 6385.0 \text{ r.p.m} \qquad \textbf{Ans.}$$

and occurs at $\qquad \theta = 0° \quad \text{and} \quad \pi$

Minimum speed of driven shaft

$$= N\cos\alpha = 6000 \cos 20 = 5638.156 \text{ r.p.m.} \qquad \textbf{Ans.}$$

and occurs at $\qquad \theta = 90° \quad \text{and} \quad 270°$

The angle θ of driving shaft at which acceleration/retardation is maximum is given by

$$\cos 2\theta \approx \left(\frac{2\sin^2\alpha}{2-\sin^2\alpha}\right) = \left(\frac{2\sin^2 20}{2-\sin^2 20}\right) = 0.1242447$$

Therefore $\qquad \pm 2\theta = 82.862859°$

or $\qquad 2\theta = (82.863°);\ (360 - 82.863);\ (360 + 82.863)\ \text{and}\ (720 - 82.863);$

Therefore $\qquad \theta = 41.43°;\ 138.568°;\ 221.43°\ \text{and}\ 318.568°$

Maximum acceleration is given by

$$\frac{d\omega'}{dt} = \frac{-\omega^2\cos\alpha\,(\sin^2\alpha)\sin^2 2\theta}{(1-\sin^2\alpha\cos^2\theta)^2}$$

For $\qquad \alpha = 20°\qquad$ and $\qquad \omega = \dfrac{2\pi \times 6000}{60} = (200\,\pi)\ \text{rad/s}$

$$\frac{d\omega'}{dt} = \frac{-43395.921\sin^2 2\theta}{(1-0.11697\cos^2\theta)^2}$$

Thus at $\qquad\qquad\qquad \theta = 41.43°,$

$$\frac{d\omega'}{dt} = \frac{43059.403}{0.8728149} = -49333.94\ \text{rad/s}^2$$

and, at $\qquad\qquad \theta = 138.568°,\ \dfrac{d\omega'}{dt} = 49333.94\ \text{rad/s}^2$

At $\qquad\qquad\qquad \theta = 221.43°,\ \dfrac{d\omega'}{dt} = -49333.943$

and at $\qquad\qquad \theta = 318.568°,\ \dfrac{d\omega'}{dt} = 49333.947$

Thus, maximum acceleration of 49333.947 rad/s^2 occurs at $\theta = 138.568°$ and $318.568°$ And, maximum retardation of 49333.947 rad/s^2 occurs at $\theta = 41.43°$ and $221.43°$ **Ans.**

EXAMPLE 5.6 Two shafts P and Q connected together by Hooke's coupling have their axes inclined at 20 degrees. The shaft P revolves at a uniform speed of 1500 r.p.m. and the shaft Q carries a flywheel of mass 15 kg and radius of gyration equal to 10 cm. Find the maximum torque in shaft Q if it is assumed that shafts are torsionally rigid.

Solution: Here, $\alpha = 20°$

Hence, maximum acceleration/retardation will occur at driving shaft angle θ, given by

$$\cos\theta = \left(\frac{2\sin^2 20}{2-\sin^2 20}\right) = 0.1232447 \qquad \text{or} \qquad \pm 2\,\theta = 82.8628$$

Therefore maximum value of acceleration or retardation

$$\frac{d\omega'}{dt} = -\frac{(\omega^2\cos\alpha\sin^2\alpha)\sin^2 2\theta}{(1-\sin^2\alpha\cos^2\theta)^2}$$

As
$$\omega = \frac{2\pi \times 1500}{60} = 50\,\pi$$

Therefore
$$\frac{d\omega'}{dt} = -(50\,\pi)^2 \frac{(\cos 20)\,(\sin^2 20)\,\sin^2(82.8628)}{(1 - \sin^2 20 \times \cos^2 41.4314)^2}$$

$$= -(50\,\pi)^2 \times (0.12564) = -3100.12 \text{ rad/s}^2$$

Therefore
$$\text{Inertia torque} = I\,\frac{d\omega'}{dt} \equiv mk^2 \times \frac{d\omega'}{dt}$$

$$= 15 \times \left(\frac{10}{100}\right)^2 \times (-3100.12) \text{ N·m} = 465.019 \text{ N·m} \qquad \textbf{Ans.}$$

EXAMPLE 5.7 A Hooke's joint is used to couple two shafts together. The driving shaft rotates at a uniform speed of 1000 r.p.m. Working from first principle determine the greatest permissible angle between the shaft axes so that the total fluctuation of speed of the driven shaft does not exceed 150 r.p.m. What will be the maximum speed of the driven shaft? (DAVV Indore, April 1977).

Solution: Note that here maximum and minimum speeds of driven shaft are not separately given. Hence α cannot be obtained on the basis of the better one of the two namely, $N/\cos\alpha$ and $N\cos\alpha$.

The maximum fluctuation of speed of driven shaft is

$$\left(\frac{N}{\cos\alpha} - N\cos\alpha\right) = 150 \text{ r.p.m.}$$

or
$$N\left(\frac{1}{\cos\alpha} - \cos\alpha\right) = 150$$

i.e.,
$$\frac{1 - \cos^2\alpha}{\cos\alpha} = \frac{150}{1000} = 0.15$$

Hence, the quadratic in $\cos^2\alpha$ is,

$$\cos^2\alpha + 0.15\cos\alpha - 1 = 0$$

or
$$\cos\alpha = \frac{-0.15 \pm \sqrt{(0.15)^2 + 4}}{2} = 1.0778 \text{ or } 0.9278$$

Neglecting negative value of $\cos\alpha$, as non-feasible (Note that $-1 < \cos\alpha < 1.0$),

$$\cos\alpha = 0.9278 \qquad \text{or} \qquad \alpha = 21° - 54' - 20'' \qquad \textbf{Ans.}$$

$$\text{Max. speed of driven shaft} = \frac{1000}{\cos\alpha} = 1077.82 \text{ r.p.m.} \qquad \textbf{Ans.}$$

5.10 TOGGLE MECHANISM

As described earlier, through Fig. 2.14 mechanical advantage of a linkage is the ratio of the output torque, exerted by the driven link, to the necessary input required at the driven-link. It can also be shown that the mechanical advantage of the 4-bar mechanism is directly proportional to the sine of the angle γ between the coupler and the follower link and inversely proportional to the angle β between the coupler and the driving link. As these angles go on changing, mechanical advantage also goes on changing with configuration of linkage. Clearly, mechanical advantage becomes infinite when the sine of the angle between input link and coupler becomes zero. For such a position of driver and coupler (making β = 0 or 180°), only a small input torque is sufficient to overcome a large output load. In either of these position, the mechanical advantage is infinite and a toggle position is said to be achieved for linkage.

Note that these positions also correspond to extreme positions of rocker, which is the driven link. Figure 5.25 shows one such toggle mechanism. As the slider *M* approaches its end of stroke to the right its velocity approaches zero. This is the position when β → 180° and mechanical advantage approaches to infinity and the slider is capable of exerting very large force *F* when the applied force *P* is small. Taking moments about *Q* of force *F'* in connecting rod and tangential force *P* on crank,

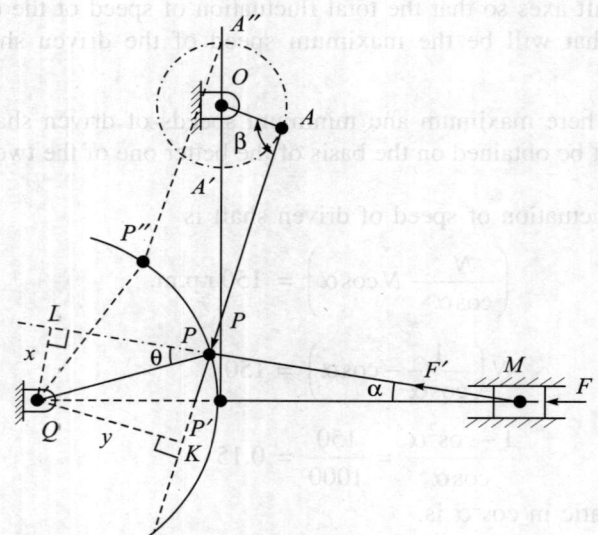

Fig. 5.25 A toggle mechanism.

$$F'(x) = P(y) \qquad (5.72)$$

But
$$F' = F/\cos\alpha \qquad (5.73)$$

Substituting in (5.72),

$$\frac{Fx}{\cos\alpha} = Py$$

or
$$\frac{F}{P} = \left(\frac{y}{x}\right)\cos\alpha \qquad (5.74)$$

If the links *PQ* and *PM* are taken equal, *QPM* is an isosceles slider-crank mechanism. Again, if mechanism dimensions are so chosen that force *P* is perpendicular to the line of action of force *F*, then *PQ* being equal to *PM*, line of action of force *P* bisects *QM*. For this position θ = 2α.

Then taking moments of F' and P about Q,

$$F' \times (PQ) \sin 2\alpha = P(PQ)\cos\alpha \qquad (5.75)$$

But $$F' = (F/\cos\alpha)$$

Hence, substituting in equation (5.75),

$$\frac{F}{\cos\alpha} \times (PQ)\, 2\sin\alpha\cos\alpha = P(PQ)\cos\alpha$$

This simplifies to, $\left(\dfrac{F}{P}\right) 2\tan\alpha = 1$ or $(F/P) = \dfrac{1}{2\tan\alpha}$

Many variations of this mechanism are used in stone crushing mechanism, presses, pneumatic riveters, clutches and other applications in which one needs to develop large force out of a small force.

5.11 SCOTCH YOKE MECHANISM

A scotch yoke mechanism consists of two revolute and two sliding pairs that can be used to transform uniform rotation of crank 2 into harmonic translatory motion of yoke 4. Figure 5.26 shows a scotch yoke mechanism. Measuring the angles of rotation θ of crank 2 w.r. to radius CO, the displacement of yoke w.r. to m.e.p. C,

$$x = (CP)\cos\theta$$

But $$\theta = \omega t, \text{ and hence}$$

$$x = r\cos\omega t$$

which is a simple harmonic motion.

Total displacement of yoke is $2r$; r being the radius of crank CP.

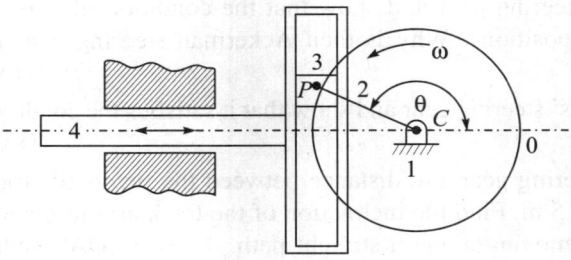

Fig. 5.26 Scotch yoke mechanism.

REVIEW QUESTIONS

5.1 Write a comprehensive note on 'Pantograph' with a neat sketch. (AMIE, Summer 1974)

5.2 For what purpose is Pantograph used? Sketch one form of Pantograph and show that it satisfies the required conditions. (D.A.V. Indore, 1980 Jan.) (AMIE, Summer 1982)

5.3 Sketch a Pantograph and explain how the mechanism would be used to enlarge a drawing. (DAV Indore, April 77) (AMIE, Winter 1999)

5.4 Describe an engine indicator mechanism and show that the pencile point traces a path which is approximately parallel to that of the displacement of pencile to the displacement of the indicator piston. (DAV Indore, March 1980)

5.5 What is the difference between copied and generated straight line motions? Give example for each of them. (Univ. of Indore, Dec. 75)

5.6 Show that in Watt's straight line mechanism the tracing point P on the coupler divides it in the ratio of the length of oscillating links which are connected by it.
(Univ. of Indore, Dec. 1975)

5.7 Derive the condition of exact straight line motion. Describe a mechanism consisting of turning pairs only giving a straight line motion to a point. (DAV Indore, 1982)

5.8 Draw a neat sketch of Peaucellier straight line motion mechanism. Explain with a proof how the tracing point describes straight path. (AMIE, Winter 1983, Winter 1997)

5.9 (a) Sketch the Dobbie-McInnes engine indicator mechanism and find the ratio of displacement of pencil to the displacement of the piston of the indicator mechanism.

(b) Sketch the Hart straight line motion mechanism and prove that the tracing point describes straight path. (AMIE, Summer 1983)

5.10 Describe one form of mechanism consisting of turning pairs only which will generate an exact straight line motion of a given point. Prove that the path followed by the point is a true straight line. (Engg. Services, 1976)

5.11 Describe an indicator mechanism and establish that in its optimum position, the magnification factor is constant. What is the purpose of such mechanism?
(DAV Indore, Feb. 1982)

5.12 Derive the condition of correct steering in the case of an automobile. Draw a neat sketch of Ackerman steering gear and show that the condition of correct steering is satisfied for only three positions. Why is then Ackerman steering gear preferred over Davis' steering gear? (DAV Indore, Feb. 1978)

5.13 Sketch the Davis' steering gear and show that it satisfies the condition for correct steering. (DAV lndore, Feb. 1978)

5.14 In a Davis' steering gear, the distance between the pivots of front axle is 1 m and the wheel-base is 2.5 m. Find the inclination of the track arm to the longitudinal axis of the car, when it is moving along a straight path. (DAV Indore, Nov.–Dec. 1986)

5.15 Two shafts whose axes are out of line by an angle α, are connected by a Hooke's joint. The driving shaft rotates at a uniform speed of ω rad/s. Show that the speed of the driven shaft for any given angular position θ of the driving shaft is

$$\left(\frac{\omega \cos\alpha}{1 - \cos^2\theta \sin^2\alpha} \right)$$

where θ is measured from the position in which the driving shaft fork lies in the plane of the shafts.

(Engg. Services: 1975, DAV Indore, Nov. 1975, Feb. 1981, Dec. 1982, Jan. 1987, July 1993)

5.16 What are the limitations of a single Hooke's joint? How are these limitations overcome in a double Hooke's joint? State the conditions which must be fulfilled by a double Hooke's joint for this purpose.

(DAV Indore, March 91, Nov.–Dec. 1986; April–May 1983)

5.17 (a) What is a Pantograph and what are its uses? Explain the working of a Pantograph Mechanism. (AMIE, Winter 1999)

(b) What are straight line mechanisms? How are they classified? Describe one type of exact straight line motion mechanism with the help of a sketch.

(AMIE, Summer 1981)

5.18 Write a short note on Scott-Russel Mechanism.

5.19 What is Pantograph and what are its uses? Give a neat sketch of Pantograph and explain its working principle. (AMIE, Winter 1979)

5.20 A circle has OR as its diameter and a point Q lies on its circumference. Another point P lies on the line OQ produced. If OQ twins about O as centre and the product $(OQ)(OP)$ remains constant show that the point P moves along a straight line perpendicular to the diameter OR.

5.21 Two shafts are connected through a Hooke's joint. The driving shaft rotates at a uniform speed of 500 r.p.m. and speed of driven shaft must lie between 480–520 r.p.m. Find maximum permissible angle between the shafts.

(DAV Indore, Jan. 1987) (**Ans.** $\alpha = 15° - 56' - 32''$)

5.22 Two shafts are coupled together by a Hooke's joint, the driving shaft rotating uniformly at 600 r.p.m. Working from first principles, determine the greatest permissible angle between the shafts if the maximum speed of the follower shafts is 630 r.p.m. What is then the minimum speed of the shaft? (DAV Indore, July 1977, 1986)

5.23 The angle between the axes of two horizontal shafts to be connected by Hooke's joint is 150°. The speed of driving shaft is 150 r.p.m. The driven shaft carries a flywheel weighing 98.0 N and having a radius of gyration of 10 cm. If the forked end of the driving shaft rotates 30° from the horizontal plane, find the torque required to drive the shaft to overcome the inertia of the flywheel.

(DAV Indore, 1975); (**Ans.** 5.68 N·m)

5.24 The driving shaft of a Hooke's joint runs at a uniform speed 240 r.p.m. and the angle α between shaft axes is 20°. The driven shaft with attached masses has a weight 540 N at a radius of gyration of 15 cm. (a) If a steady torque of 196 Nm resists rotation of the driven shaft, find and torque required at the driving shaft, when $\theta = 45°$, (b) At what value of α will the total fluctuation of speed of the driven shaft be limited to 24 r.p.m.?

$$\left[\text{Hint: Angular accen. } \frac{d\omega}{dt} = -78.4 \text{ rad/s}^2 \text{(compute)}\right]$$

Torque reqd. to accelerate driven shaft $I\,\dfrac{d\omega'}{dt} = -\dfrac{540}{9.81}\left(\dfrac{15}{100}\right)^2 \times 78.4 = -97.1$ N·m.

∴ Total torque reqd. = 196 − 97.1 = 98.9 N·m.

From $\quad T \cdot \omega = T' \omega'; \; T = T' \left(\dfrac{\omega'}{\omega} \right) = T' \left(\dfrac{\cos\alpha}{1 - \sin^2\alpha \cdot \cos^2\theta} \right)$

$$= -97.1 \times 0.988 = -96.91 \text{ N·m} \qquad \textbf{Ans.}$$

Again, $240 \left(\dfrac{1}{\cos\alpha} - \cos\alpha \right) = 24$, which gives $\alpha = 18.2°$

$\textbf{\textit{Ans.}}$ (AMIE, Summer 1981)

Chapter 6

ELEMENTS OF KINEMATIC
SYNTHESIS OF MECHANISMS
(GRAPHICAL AND ALGEBRAIC METHODS)

6.1 KINEMATIC SYNTHESIS

Ampere rightly defined kinematics as *the study of the motion of mechanisms and methods of creating them.* The first part of this definition refers to kinematic analysis which has been dealt with exhaustively in chapters 3 and 4. It was shown in these chapters that kinematic analysis aims at determining performance of a given mechanism. Kinematic synthesis may be treated as a reverse problem of kinematic analysis, and aims at designing a mechanism which can satisfy prescribed motion characteristics, such as displacement, velocity and acceleration—either singly or in combination. Obtaining cam profile for generating specified type of follower motion falls in the category of synthesis of mechanisms. Synthesis process may be accomplished, in general, in three interrelated phases:

1. The form or type synthesis of mechanisms,
2. The number synthesis, and
3. The dimensional synthesis.

Type Synthesis

Type synthesis refers to the kind of mechanism selected. Option can go in favour of a gear combination, a belt pulley combination, a linkage or a cam mechanism. During this early stage of synthesis, one may have to consider design aspects like space consideration, safety consideration, overall economics, manufacturing process, materials-selection, etc. Type synthesis is thus an involved process and makes it very difficult to lay down a systematic procedure aimed at unique determination of mechanism that assures specified motion characteristics. Type synthesis, in a majority of cases therefore, is beyond the scope of this book.

Number Synthesis

Number synthesis is based on the most obvious external characteristics of a kinematic chain, namely, the number of links together with the number and type of joints required for a specified motion. This aspect has already been covered under Sections 2.18 and 2.19.

Dimensional Synthesis

Dimensional synthesis aims at determining significant dimensions of links and the starting position of links in a mechanism, to accomplish specified task and prescribed motion characteristics. Hinge pin to hinge pin distances on binary, ternary, quaternary (and so on) links; angles between arms of bell-crank levers, cam contour dimension, diameter of roller of follower, gear ratios, eccentricities are just a few examples of what is meant by the term *significant dimensions*.

Although inventiveness and intuition play a major role in mechanism's design, there exist certain fundamental problems of synthesis which can be solved in a rational way. These problems include:

1. Guiding a point along a prescribed path, *i.e.* path generation,
2. Function generation involving coordination of the positions of input and output links, and
3. Motion generation which requires guiding a body through a number of prescribed positions.

In *path generation,* a point on the coupler link (also known as a floating link) is constrained to describe a path with reference to a fixed frame of reference. Generally, the portion of the path is an arc of a circle, an ellipse or a straight line. Sometimes the tracing point may be required to follow a path which crosses over itself in the form of the figure of eight. A linkage mechanism whose point C on coupler is required to follow a path $y = f(x)$, as the crank rotates, is an example of path generation. Other typical examples include 'the thread-guiding eye' on a sewing machine and mechanism to advance film on camera (Fig. 5.3b). Watt and Robert straight line motion mechanisms are the other examples of special kind of path generators in which certain geometric relationships assure the generation of straight line segments.

Function generators represent a major category of synthesis problems. It is frequently required in kinematic synthesis that an output link must rotate, oscillate or reciprocate according to a prescribed function of time or of input motion. This is called *function generation*. For instance, in a 4-bar mechanism, for generating a function $y = f(x)$, x would represent the motion of input crank and the proportions of link lengths of the mechanism are to be so chosen that the motion of output link represents the function $y = f(x)$ approximately.

A typical example of a function generator is shown in Fig. 6.1, which gives schematic representation of a 4-bar mechanism used as the impact printing machine in an electric

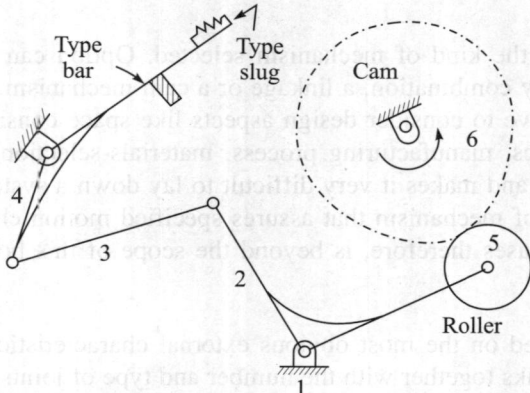

Fig. 6.1   Impact printing mechanism in an electric typewriter.

typewriter. Here the requirement is that the type slug should hit the platen roller with impact. To achieve this, the type bar must move first by smaller and later by larger angles per increment of input crank. Note that links 2 and 4 are bell-crank levers.

Another example of function generation can be seen in linkage mechanism to correlate steering angles of axes of front wheels of an all-terrain vehicle with the relative speed at which each driven wheel should rotate avoiding skidding or excessive wear.

Yet another example of application of function generator can be seen in a conveyor line for operations like bottle capping. The output members of a mechanism in such a case must move at a constant speed equal to that of conveyor while performing the operation (like capping), return, pick-up the next cap and repeat the operation. Similar other examples can be cited.

Motion generation or *Rigid body guidance* requires that an entire body be guided through a pre-selected sequence of motion. Such a body is guided usually as a part of coupler. For example, in construction industry, heavy parts such as buckets and blades of bulldozer must be moved through a series of prescribed positions. In case of buckets of 'bucket loader mechanism', the tip of bucket on coupler must have a prescribed path. Path to be followed by tip of the bucket is very important because the tip is supposed to have a scooping trajectory followed by lifting and dumping trajectory.

It must be noted that, whereas in path generation one is concerned with the path of a tracer point, in motion generation (rigid body guidance) the entire motion of the coupler-link is of concern.

6.2 APPROXIMATE AND EXACT SYNTHESIS

Function generation and path generation problems can be solved either by approximate or exact approach. The difference between the two approaches become clear at once if we compare Watt approximate straight line motion mechanism with Hart and Peaucellier straight line mechanisms. Being an approximate straight line motion mechanism, the tracing point of Watt's mechanism will have a wavy path, intersecting the desired path of straight line $y = mx$, at a finite number of points. Thus, except for the points of intersection, all other points traced by the coupler do not lie on exact straight path. The points of intersection of desired path with actual path are aptly defined as *accuracy* or *precision points*. Obviously, the theme of approximate approach of synthesis should be:

(a) there should be as many accuracy points as possible, and
(b) there should be a minimum deviation (or error) between the desired path and the actual path.

Exact method of synthesis has a limitation in the sense that it can handle only a few arbitrary functions. Further, *as in the case of exact straight line motion generation, a minimum of six links is required for path generation*. In view of the above, only the approximate method of synthesis will be discussed in this text book.

6.3 CHEBYSHEV'S SPACING OF ACCURACY POINTS

Discussions above lead us to the conclusion that when a mechanism is designed to generate a specified function or trace a given curve $y = F(x)$ over a given range $x_s \leq x \leq x_f$, *the function is exactly generated at a finite number of points only. These are known as accuracy (precision) points* and are represented in Fig. 6.2(a) by points of intersection of generated function $y = f(x)$ with desired function $y = F(x)$. These points are indicated as P_1, P_2, P_3,....., P_n in the same figure. *The number of these accuracy points is equal to the number of fixed parameters that may be used for synthesis and varies between three and six.* Let x_1, x_2, x_3, ..., x_n be the value of independent variables at precision points P_1, P_2, P_3, ..., P_n in the working range $x_s \leq x \leq x_f$.

The difference between the two functions is called structural error. The structural error $\varepsilon (x)$ is shown plotted against independent variable x in Fig. 6.2(b). It must be clearly understood that *structural error* is inherent in an approximate synthesis.

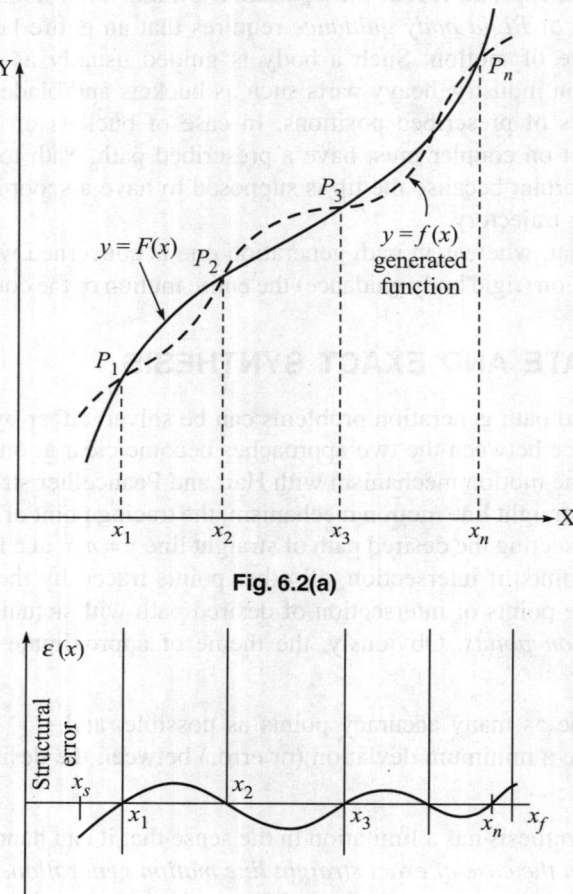

Fig. 6.2(a)

Fig. 6.2(b)   Illustration for structural error.

Best approximation between function to be generated and function actually generated can be obtained when the absolute value of maximum structural error between any two precision points equals corresponding value at the ends x_s and x_f. Quite often Chebyshev spacing of precision points is used as a first guess to minimise the structural error. It must be remembered, however, that this approach is applicable in special cases when the function is symmetric.

For the first guess, the best spacing of accuracy points, called *Chebyshev's spacing*, are given for the range $x_s \leq x \leq x_f$ by

$$x_j = \frac{1}{2}(x_s + x_f) - \frac{1}{2}(x_f - x_s)\cos\frac{\pi\,(2j-1)}{2n} \qquad (6.1)$$

where, $x_j(j = 1, 2, 3,...)$ are the accuracy points. Thus for first, second and third accuracy points, j is equal to 1, 2, and 3 respectively.

$$n = \text{total number of accuracy points.}$$

For instance, let us assume that using Chebyshev's spacing, three precision points are to be obtained for generating a function $y = x^{1.5}$ in the range $1 \leq x \leq 3$.

Here
$$x_s = 1; \quad x_f = 3$$
and
$$n = \text{number of accuracy points} = 3$$

Thus, first, second and third accuracy points are:

$$x_1 = \frac{1}{2}(1 + 3) - \frac{1}{2}(3 - 1)\cos\frac{\pi\,(2-1)}{6}$$

$$= 2 - 1\cos\frac{\pi}{6} = 2 - \frac{\sqrt{3}}{2} = 1.134$$

$$x_2 = \frac{1}{2}(1 + 3) - \frac{1}{2}(3 - 1)\cos\frac{\pi\,(4-1)}{6}$$

$$= 2 - 1\cos\frac{\pi}{2} = 2.0$$

and
$$x_3 = \frac{1}{2}(1 + 3) - \frac{1}{2}(3 - 1)\cos\frac{\pi\,(6-1)}{6}$$

$$= 2 - 1\left(\cos\frac{5\pi}{6}\right)$$

$$= 2.866$$

Corresponding values of dependent variable y, as obtained by substituting x_j values ($j = 1, 2, 3$) in the function $y = x^{1.5}$, are

$$y_1 = (1.134)^{1.5} = 1.207$$
$$y_2 = (2.0)^{1.5} = 2.828$$
$$y_3 = (2.863)^{1.5} = 4.844$$

The Chebychev accuracy points can also be obtained using a simple geometric construction. The method consists in drawing a circle on a diameter equal to the range of function of independent variable. Thus,

$$\Delta x = \text{Diameter of circle} = (x_f - x_s)$$

The next step is to inscribe in it a regular polygon of sides $2n$ where n is the number of accuracy points. The polygon must be so inscribed that at least two sides are perpendicular to x-axis. Feets of perpendiculars dropped from the corners of this polygon on x-axis then give accuracy points.

All throughout the text of this chapter Chebychev accuracy points will be used for solving problems on function generation. Readers must note that at best, the Chebychev spacing gives only the best first approximation and depending upon the accuracy requirement of the problem, this is good enough. For many functions the greatest error can be restricted to less than 4 per cent.

6.4 GRAPHICAL METHODS OF DIMENSIONAL SYNTHESIS

Graphical methods have advantage in that they are relatively fast in producing results and, at the same time, they maintain touch with physical reality to a much larger extent than do the algebraic methods. Also, geometric methods are easier to understand and the degree of accuracy is adequate for all purposes.

MOTION GENERATION

6.5 POLES AND RELATIVE POLES

Consider a rigid body, represented by a link A_1B_1 to undergo a finite displacement so as to occupy a position A_2B_2 as shown in Fig. 6.3. Mid-normals drawn to the lines joining point A_1 to A_2 and B_1 to B_2 meet at point P_{12} which is called pole for the finite displacement of body from position 1 to position 2. The rigid body A_1B_1 can be considered to have motion of body rotation about the pole P_{12} in describing finite displacement from position 1 to 2.

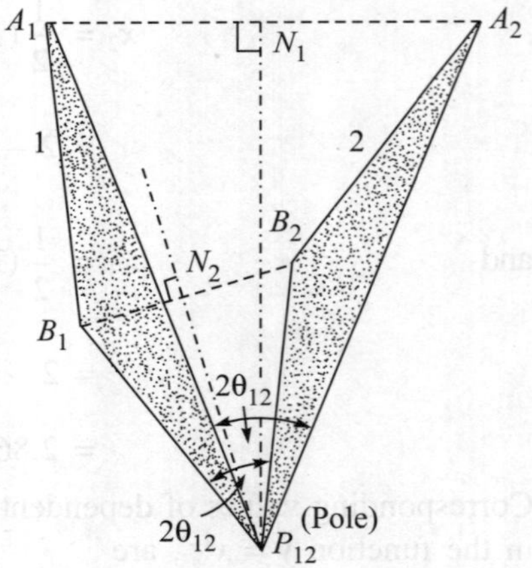

Join points A_1, B_1 and A_2, B_2 to pole P_{12}. Body rotation of link A_1B_1 about pole P_{12} can now be conceived as rotation of triangle $P_{12} B_1 A_1$ about imaginary pivot P_{12}. With this concept, clearly

$$\angle B_1 P_{12} B_2 = \angle A_1 P_{12} A_2 = 2\ \theta_{12} \qquad (6.2)$$

Again, as triangles $B_1 N_2 P_{12}$ and $B_2 N_2 P_{12}$ are identical,

$$\angle B_1 P_{12} N_2 = \angle B_2 P_{12} N_2 = \theta_{12}$$

Also triangles $A_1 P_{12} N_1$ and $A_2 P_{12} N_2$ are identical, and as such,

$$\angle A_1 P_{12} N_1 = \angle A_2 P_{12} N_1 = \theta_{12} \qquad (6.3)$$

Fig. 6.3 Pole and Rigid body rotation.

Readers should note that rigid body rotation concept described above has no connection with the actual path which is followed by points A and B in reaching position A_2B_2 from A_1B_1. The position of pole P_{12} therefore, depends only on the initial and final positions of the rigid body, represented by line AB. Further, as distinct from the concept of infinitesimally small displacement used in developing concept of instantaneous centre of rotation, discussions above refer to finite displacement. Clearly, *when angle $2\theta_{12}$ of rotation is infinitesimally small, the pole P_{12} becomes the instantaneous centre of rotation.*

An important rule follows from above discussions. Considering two finitely separated positions A_1B_1 and A_2B_2 of line AB, let O_A and O_B be arbitrarily selected positions of pivots along mid-normals of lines A_1A_2 and B_1B_2 respectively. Let O_BN_2 and O_AN_1 be the mid-normals of lines B_1B_2 and A_1A_2, intersecting at point P_{12} to give pole of rotation. From discussions above, $\angle B_1P_{12}N_2 = \angle A_1P_{12}N_1 = (\theta_{12}/2)$. Join A_1 to P_{12}.

Adding common angle $\angle A_1P_{12}N_2$ to equal angles $\angle B_1P_{12}N_2$ and $\angle A_1P_{12}N_1$ each equal to $\theta_{12/2}$), we conclude that $\angle A_1P_{12}B_1 = \angle O_AP_{12}O_B$.

This leads to the following generalised rule: *The coupler and the frame link subtends angles at the pole P_{12} which are either equal or differ from each other by an angle of 180°.*

In continuation to discussions in this section, it must be stated that if the pole P_{12} happens to fall too far away from the frame, as in Fig. 6.4, the link AB can be guided as a coupler of 4-bar mechanism with pivots located suitably at O_A and O_B on the mid normals of lines joining A_1, A_2 and B_1, B_2. *The point A is called a 'circle point' because an arc of circle can be passed through the corresponding positions A_1 and A_2.* Corresponding centre-point, in the form of fixed pivot O_A, can infact be located anywhere along the mid-normal of line A_1A_2 and will be known as a *centre point-conjugate to the circle point A.* A link joining centre point to the circle point can guide point A from A_1 to A_2. Similar discussion is valid for circle point B and its conjugate centre point (fixed pivot) O_B.

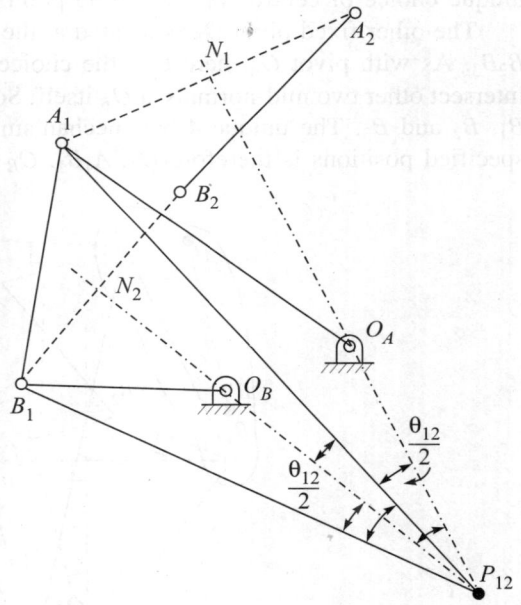

Fig. 6.4 Rigid body guidance as a coupler.

In rigid body guidance, *i.e.* motion generation, a designer has the choice of selecting position of line AB anywhere in the body or its extension. Thus, in guiding a rigid body through two positions, three free choices exist for selecting a pair of circle point and the corresponding centre point : two choices in respect of independent variables x and y for point A in the coordinate frame and one in respect of location of fixed pivot O_A anywhere along mid-normal of line joining two positions A_1 and A_2 of point A. Therefore, three infinites of solution are possible for fixing a pair of circle point and centre point in constructing a four-bar linkage for rigid body guidance. Even if it is not possible to locate fixed pivot O_B on the mid-normal of B_1B_2, it does not matter. For, another point C can be fixed on body, forming a triangle ABC. Pivot position O_C can then be obtained similarly on the mid-normal of C_1C_2.

6.6 MOTION GENERATION: THREE PRESCRIBED POSITIONS

Let A_1B_1, A_2B_2 and A_3B_3 be the three positions of line AB, chosen conveniently in a rigid body, which moves through three positions. As in the earlier case intersection of mid-normals of A_1A_2 with that of B_1B_2 locates the pole P_{12}. Similarly, intersection of mid-normals of A_2A_3 with that of B_2B_3 locates pole P_{23}. Finally intersection of mid-normals of A_1A_3 and that of B_1B_3 locates pole P_{13}. Unlike the previous case, however, the rigid body can not be rotated about any of the imaginary pivots P_{12}, P_{23} or P_{13} for obtaining positions A_1B_1, A_2B_2 and A_3B_3 even if the poles are accessible on the machine frame (see Fig. 6.5.). Instead of choosing fixed pivots arbitrarily anywhere on the mid-normals of A_1A_2 and B_1B_2, as in the previous case, unique choice O_A is obtained at the point of intersection of mid-normals of A_1A_2 and A_2A_3. It may be verified that the mid-normal of A_1A_3 will intersect above mid-normals at the same point O_A, which is the unique choice of centre-point of circle points A_1, A_2 and A_3.

The other fixed pivot O_B is located at the point of intersection of mid-normals of B_1B_2 and B_2B_3. As with pivot O_A, here too, the choice of O_B is unique since the mid-normals of B_1B_3 intersect other two mid-normals at O_B itself. So, O_B is the unique centre point for the circle points B_1, B_2 and B_3. The unique 4-bar mechanism that can guide the rigid body through the three specified positions is therefore O_A, A_1B_1, O_B as shown in Fig. 6.5.

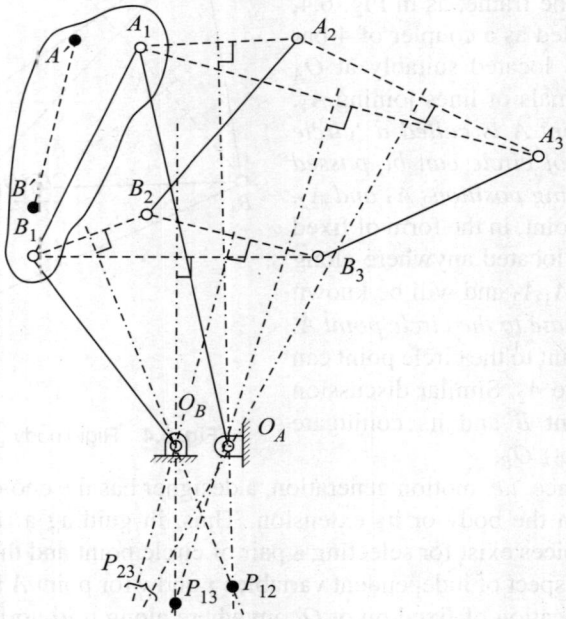

Fig. 6.5 A 4-bar mechanism for rigid body guidance through three positions.

Despite unique choice for pivots O_A and O_B, note that for guiding the rigid body through the three positions, the choice of AB in the rigid body in itself is arbitrary. Infact, point A can have any other set of x, y coordinates and similarly point B may also be chosen with any other set of x, y coordinates. Thus, for each of the circling points A and B, there are two infinites of possibilities. But, for every choice of line AB in the body, the choice, of fixed pivots O_A and O_B will be unique.

EXAMPLE 6.1 Synthesise a four-bar, mechanism to guide a rod AB through three consecutive positions A_1B_1, A_2B_2 and A_3B_3 as shown in Fig. 6.6. (AMIE, Summer 1993)

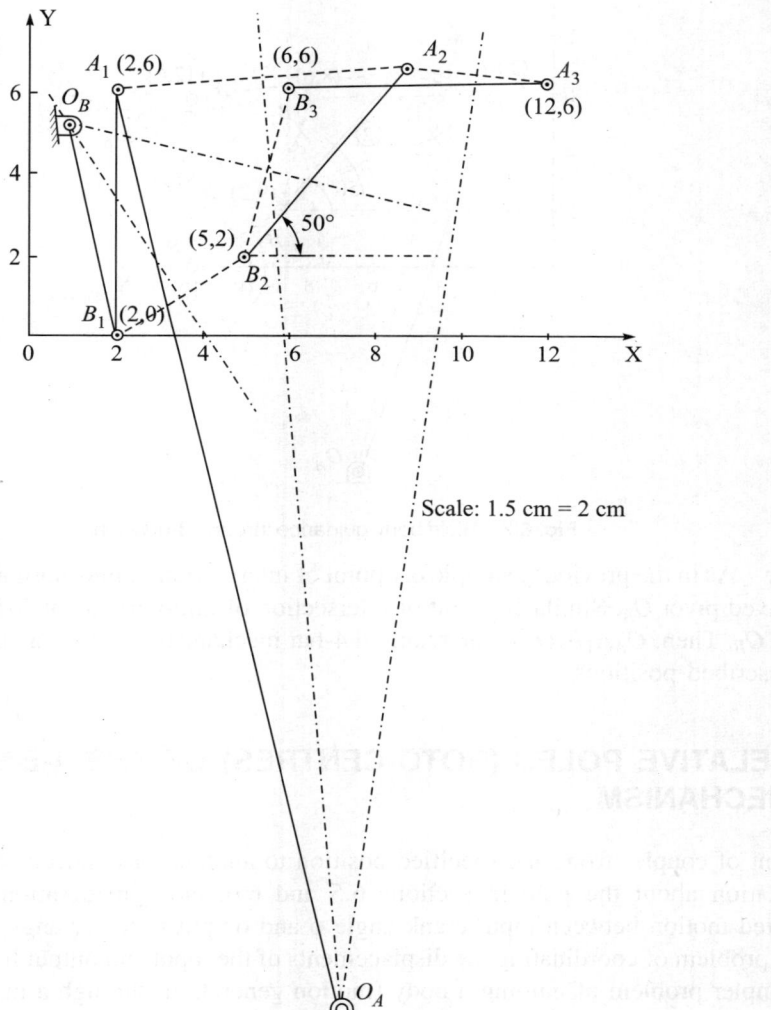

Fig. 6.6 Rigid body guidance through 3-positions.

Solution: *Construction:* Draw mid-normals of A_1A_2 and A_2A_3 to meet at O_A. Similarly, draw mid-normals of B_1B_2 and B_2B_3 to meet at O_B. Then O_A and O_B are the required fixed supports. The four-bar mechanism $O_AA_1B_1O_B$ represents the required 4-bar mechanism with coupler link A_1B_1 capable of taking positions A_2B_2 and A_3B_3.

EXAMPLE 6.2 Synthesise a mechanism to move AB successively through positions 1, 2 and 3 as shown in Fig. 6.7.

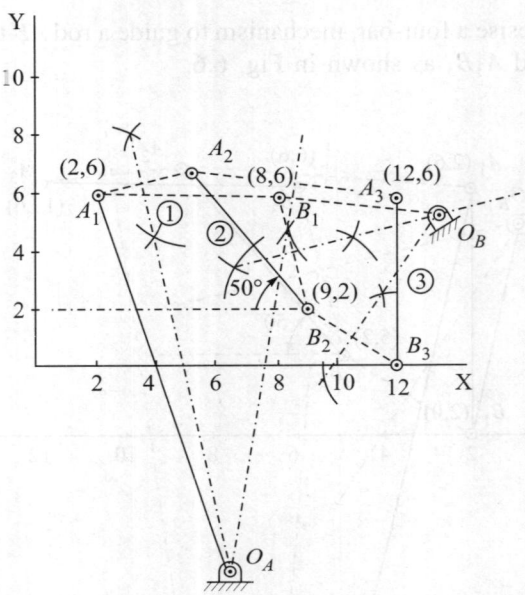

Fig. 6.7 Rigid body guidance through 3-positions.

Solution: As in the previous example 6.1 point of intersection of mid-normals of A_1A_2 and A_2A_3 locates fixed pivot O_A. Similarly, point of intersection of mid-normals of B_1B_2 and B_2B_3 locates the pivot O_B. Then, $O_AA_1B_1O_B$ is the required 4-bar mechanism which can guide AB through the three prescribed positions.

6.7 RELATIVE POLES (ROTO-CENTRES) OF THE 4-BAR MECHANISM

Movement of coupler from one specified position to another was conceived to be a motion of body rotation about the pole in sections 6.5 and 6.6. No consideration was given to the coordinated motion between input crank angle ϕ and output follower angle of rotation ψ. The complex problem of coordinating the displacements of the input and output links may be reduced to the simpler problem of guiding a body (motion generation) through a number of prescribed positions. Consequently, it will be possible for us to use methods discussed earlier. Principle of inversion simplifies such problems to a great extent. Thus, one of the two links (input or output link) is temporarily regarded as fixed and the relative motion of other links is next considered. The first step consists of finding the relative poles (roto centres,) which will be denoted as R_{ij} to distinguish them from poles P_{ij}.

Figure 6.8(a) shows a four-bar mechanism with link O_AA and O_BB as input and output links and O_AO_B as frame link. The same mechanism is shown in full lines in original position, and in dashed lines in new position. The angle turned through by the input link $O_A A$ is ϕ_{12} (c.w.) while the corresponding angle turned through by the output (follower) link $O_B B$ is ψ_{12} (c.w.). Thus $\angle AO_A A_1 = \phi_{12}$ and $\angle BO_B B_1 = \psi_{12}$. To obtain inversion, input link $O_A A$ is fixed and the fixed link $O_A O_B$ is released and rotated so as to have same inclination ϕ with the link $O_A A$. To ensure this, link $O_A O_B$ has to be rotated through $(-\phi_{12})$ about O_A keeping $\angle O_A O_B' B'$ same as $\angle O_A O_B B$. This is shown in Fig. 6.8(b). The mechanism in this position is represented by $O_A O_B' B_1' A$.

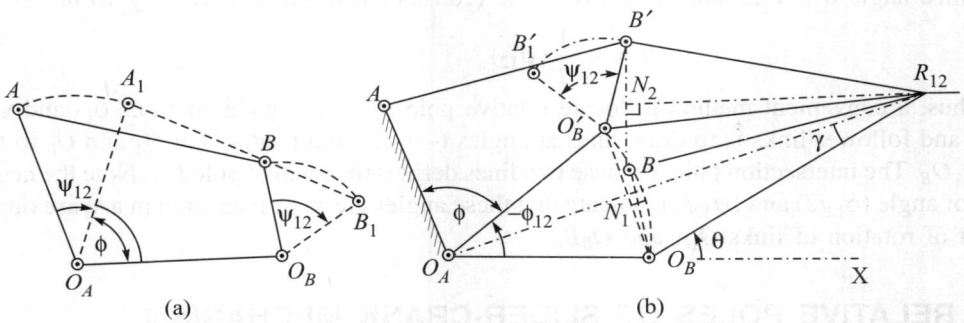

Fig. 6.8 Relative poles of 4-bar mechanism.

Finally link $O_B' B_1'$ is rotated through ψ_{12} in c.w. sense to position $O_B' B$, which gives final position of mechanism $O_A O_B' B' A'$, as shown in Fig. 6.8(b). Thus, net rotation of link $O_B B$ from position $O_B B$ to the position $O_B' B'$ is $(\psi_{12} - \phi_{12})$. It is important to note the convention followed in respect of angles. Angles measured in clockwise sense are being treated as positive, while the angles measured in counter clockwise sense are considered to be negative.

The relative pole in Fig. 6.8(b) is located at the point of intersection of mid-normals of lines joining B to B' and O_B to O_B'. In triangle $O_A O_B O_B'$, side $O_A O_B = O_A O_B'$ and $O_A R_{12}$ is the mid-normal to $O_B O_B'$ and, therefore,

$$\angle O_B O_A R_{12}, = -\left(\frac{\phi_{12}}{2}\right) \quad i.e., \text{ in } c.c.w. \text{ sense} \qquad (6.4)$$

Also from the triangle $O_A O_B R_{12}$, the exterior angle

$$\angle R_{12} O_B X = \theta = \angle O_B O_A R_{12} + \angle O_A R_{12} O_B$$

or

$$\theta = \left(\frac{\phi_{12}}{2}\right) + \gamma \qquad (6.5)$$

or

$$(\theta - \gamma) = (\phi_{12}/2)$$

To co-relate the motion ϕ_{12} of crank and ψ_{12} of the follower, the relative pole R_{12} can now be looked upon as a centre about which coupler $O_B B$ (with $O_A A$ as frame) has a motion of rotation. Thus, imagine triangular lamina $R_1 O_B B$ to revolve about R_{12} to take up final position $R_{12} O_B' B'$; the angle of rotation being $(\psi_{12} - \phi_{12})$ Thus, $R_{12} N_1 O_A$ being the mid-normal of $O_B O_B'$,

$$\angle O_B R_{12} O_A = \gamma = \frac{1}{2} \; (\psi_{12} - \phi_{12})$$

Substituting for γ in equation (6.5),

$$\theta = \left(\frac{\phi_{12}}{2}\right) + \frac{1}{2}(\psi_{12} - \phi_{12})$$

or

$$\theta = \frac{1}{2}(\psi_{12}) \tag{6.6}$$

Since angle θ is measured in c.c.w. sense (considered negative), we may write

$$\theta = -\frac{1}{2}(\psi_{12}) \tag{6.7}$$

Thus, a convenient method to locate relative pole R_{12} for angular motions ϕ_{12} and ψ_{12} of crank and follower links is to draw lines at angles $(-\phi_{12}/2)$ and $(-\psi_{12}/2)$ at O_A and O_B to frame link $O_A O_B$. The intersection point of these two lines defines the relative pole R_{12}. Note the negative signs of angle $(\phi_{12}/2)$ and $(\psi_{12}/2)$ indicate that these angles are to be measured in a sense opposite to that of rotation of links $O_A A$ and $O_B B$.

6.8   RELATIVE POLES OF SLIDER-CRANK MECHANISM

It may be recalled that a slider-crank mechanism is obtained as a limiting case of four-bar mechanism where one of the revolute pair of the follower is replaced by a sliding pair. In other words, the revolute pair O_B, connecting follower (slider) to the frame link, is perpendicular to the line of stroke of slider. Thus the line of centres joining O to O_B can also be assumed to be perpendicular to line of stroke through O.

The concept developed for 4-bar mechanism can now be extended to obtain relative pole for slider-crank mechanism too. For rotation of crank through ϕ_{12} let the displacement of slider be s_{12}. The relative pole R_{12} is then obtained as a point of intersection of line drawn parallel to OO_B at a distance of $-\left(\dfrac{s_{12}}{2}\right)$ and the line drawn at an angle of $\left(-\dfrac{\phi_{12}}{2}\right)$ to the line OO_B. This point is denoted in Fig. 6.9 by relative pole R_{12}.

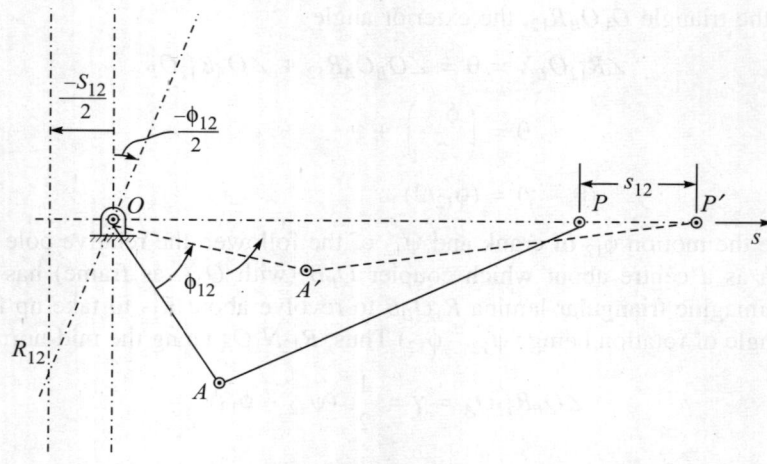

Fig. 6.9   Relative poles of slider crank mechanism.

EXAMPLE 6.3   Design a 4-bar linkage in which 30° of crank rotation produces 45° of follower rotation, both in the c.w. sense.

Solution: Select pivot points O_A and O_B arbitrarily, say at a distance 3.5 cm apart. Draw lines $O_B R_{12}$ and $O_A R_{12}$ (Fig. 6.10) making angles $\psi_{12/2} = 22\frac{1°}{2}$ and $\phi_{12/2} = 15°$ with $O_A O_B$ in c.c.w. sense (*i.e.* opposite to the sense of rotation) to meet at R_{12}, the relative pole as shown in Fig. 6.10.

Noting the rule that the coupler AB must subtend angle at R_{12} same as that subtended by $O_A O_B$, measure the angle $\angle O_A R_{12} O_B$ ($= \gamma$). In the present case, $\gamma = 8°$. Select point A arbitrarily and join $R_{12}A$. Draw another line $R_{12}B$ making an angle of $\gamma = 8°$ with $R_{12}A$. Select point B also arbitrarily on line $R_{12}A$. Then, $O_A ABO_B$ is the required 4-bar mechanism.

Since the choice of distance $O_A O_B$, and of points A and B was arbitrary, number of solutions possible are ∞^3. Note that the method will fail if any of the angles ψ_{12} or ϕ_{12} is equal to zero and also when $\phi_{12} = \psi_{12}$ for obvious reasons.

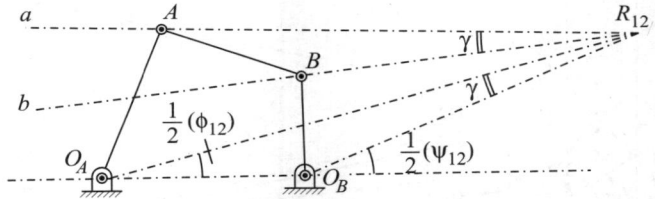

Fig. 6.10

EXAMPLE 6.4   Design a 4-bar mechanism for the following coordinated motion of crank and follower

$$\phi_{12} = 90°; \ \psi_{12} = 40°;$$
$$\phi_{13} = 150°; \ \psi_{13} = 75°; \text{ all in c.w. sense.}$$

Solution: *Construction:* Select suitable length $O_A O_B$ for frame link arbitrarily. In the present case, $O_A O_B$ has been selected as 4.5 cm. From O_A describe lines $O_A R_{12}$ and $O_A R_{13}$ at angles – $\phi_{12}/2$ ($= 45°$) and $-\phi_{13}/2$ ($= 75°$) in c.c.w. sense. Similarly, draw lines $O_B R_{12}$ and $O_B R_{13}$ at angles $-\psi_{12}/2$ ($= 20°$) and $-\Psi_{13}/2$ ($= 37.5°$) to $O_A O_B$ in c.c.w. sense to locate relative poles R_{12} and R_{13}. This is shown in Fig. 6.11.

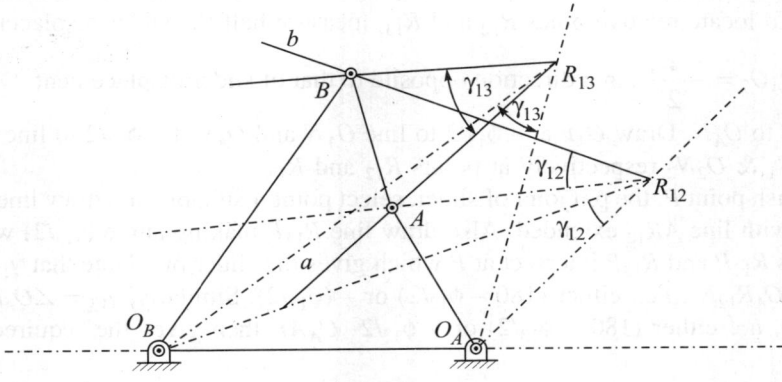

Fig. 6.11

Select a point A arbitrarily. Join $R_{12}A$ and $R_{13}A$. Describe angles γ_{12} and γ_{13} w.r. to lines $R_{12}A$ and $R_{13}A$ respectively to intersect at B. Then O_AO_BBA is the required 4-bar mechanism. Note that here choice of length O_AO_B and that of point A was arbitrary and therefore ∞^2 solutions are possible.

EXAMPLE 6.5 Design a slider-crank mechanism to have the following coordinated motion of slider and the crank:

$$\phi_{12} = 50°; \qquad s_{12} = 2 \text{ cm};$$
$$\phi_{13} = 110°; \qquad s_{13} = 4 \text{ cm}$$

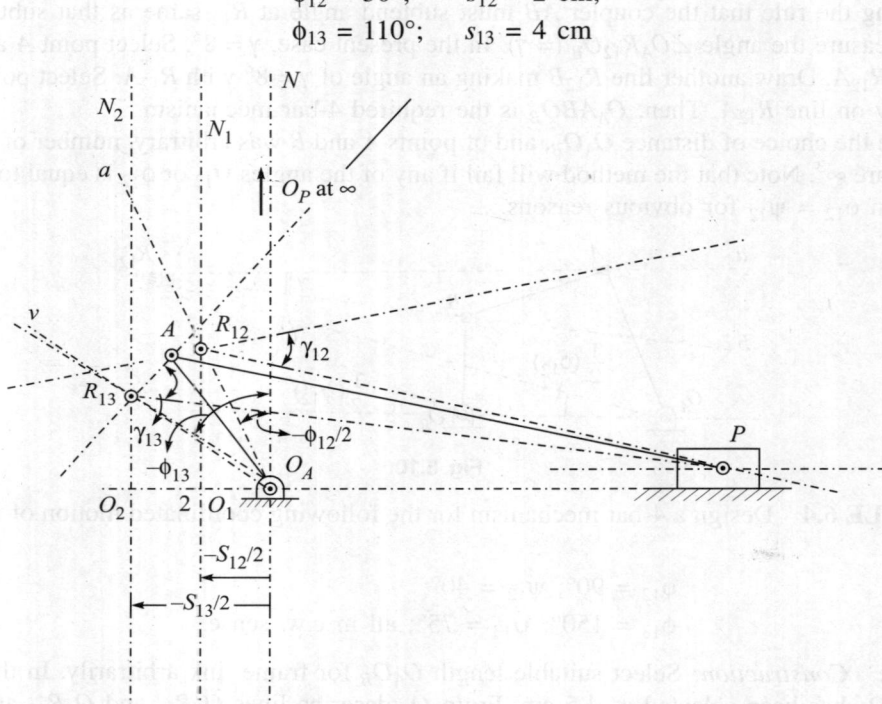

Fig. 6.12

Solution: Let O_A be the crank pivot as designed in Fig. 6.12. Then, for the shown line of action of slider P, the pivot centre O_P of equivalent follower can be assumed to be located at infinity along O_AN. To locate relative poles R_{12} and R_{13}, measure half the slider displacements $O_AO_1 = \dfrac{-s_{12}}{2}$; and $O_AO_2 = \dfrac{-s_{13}}{2}$, in a direction opposite to that of slider displacement. Draw O_1N_1 and O_2N_2 parallel to O_AN. Draw O_Au at $-\phi_{12}/2$ to line O_AN and O_Av at $-\phi_{13}/2$ to line O_AN, so as to cut lines O_1N_1 & O_2N_2 respectively at points R_{12} and R_{13}.

To establish point P, the pin joint of slider, select point A suitably and draw line $R_{12}P$ making angle $(\gamma_{12}/2)$ with line AR_{12} extended. Also draw line $R_{13}P$ making angle $(\gamma_{13}/2)$ with line AR_{13}. Both the lines $R_{12}P$ and $R_{13}P$ intersect at P which gives the slider pin. Note that $\gamma_{12} = \angle O_AR_{12}N_1$ or $(180° - \angle O_AR_{12}N)$, i.e., either $(180 - \phi_{12}/2)$ or $-(\phi_{12}/2)$. Similarly, $\gamma_{13} = \angle O_AR_{13}N_2$ or $(180 - \angle O_AR_{13}N_2)$, i.e. either $(180 - \phi_{13}/2)$ or $-\phi_{13}/2 \cdot O_AAP$ then gives the required slider-crank mechanism.

6.9 FUNCTION GENERATION (THREE PRECISION POINTS)

Graphical method for synthesising function generation problem, using three accuracy points, is similar to the one used for motion generation. Here too, kinematic inversion and the intersection of mid-normals are used to locate poles and relative poles.

Function generation synthesis attempts at establishing dimensions of link lengths, etc., for mechanisms whose output motion is a specified function of input motion. Very few functions can be generated exactly by a 4-bar linkage. Shaffer and Cochin {Trans. ASME, Oct. 1954) have proposed an equation, called 'compatibility equation' that can be used to determine whether or not a function can be generated exactly over a given range. Alternatively, Freudenstein's equation can be used to

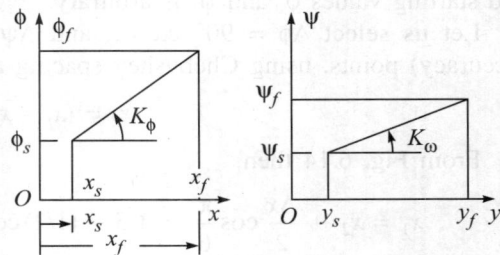

Fig. 6.13 Diagramatic illustration for relation between variables.

design a 4-bar linkage for generating a function accurately only at a finite number of points (called accuracy points). At every other point the generated and desired function will not be same. That is, the designed linkage will be compatible with the displacement equation (to be derived later) only at the accuracy points.

The input and output members can be either a crank or a slider. It is always possible to design a mechanism to correlate output and input motions through a smooth continuous function, with very marginal error over a limited range of variable. The input rotation ϕ of the input link is proportional to the independent variable x, while the output function ψ is proportional to the dependent variable y (see Fig. 6.13). A restriction on the range of independent variable x then automatically puts restriction on dependent variable $y = f(x)$. In other words, a range $x_s \leq x \leq x_f$, implies a range $f(x_s) \leq y \leq f(x_f)$ for the dependent variable. Also, as ϕ is proportional to x and ψ to y, this implies fixing a range for ϕ and ψ. Thus, $\phi_s \leq \phi \leq \phi_f$ and $\psi_s \leq \psi \leq \Psi_f$.

As explained in section 6.3, the deviation of designed (generated) function from desired function $y = f(x)$ is called structural error. As a first guess, Chebyshev accuracy points can be used to minimize structural error. The method of synthesis requires two scaling factors k_ϕ and k_ψ, defined as

$$k_\phi = (\Delta\phi/\Delta x) = \frac{(\phi_f - \phi_s)}{(x_f - x_s)} \tag{6.8}$$

and

$$k_\psi = (\Delta\psi/\Delta y) = \frac{(\psi_f - \psi_s)}{(y_f - y_s)} \tag{6.9}$$

It is thus possible to correlate any value of input angle ϕ_j to the corresponding output angle ψ_j using relations,

$$\phi_j = \phi_s + k_\phi(x_j - x_s) \tag{6.10}$$

and

$$\psi_j = \psi_s + k_\psi(y_j - y_s) \tag{6.11}$$

together with relation $y = f(x)$.

The procedure for designing a mechanism for generating a specified function $y = f(x)$, is explained below through two example problems.

EXAMPLE 6.6 Design a 4-bar linkage to generate the function $y = \log_{10}(x)$ in the interval $1 \leq x \leq 2$.

Solution: For the range $x_s = 1$ and $x_f = 2$, we have $y_s = \log_{10}(1) = 0$ and, $y_f = \log_{10}(2) = 0.3010$ and, therefore, $\Delta x = (2 - 1) = 1$ and $\Delta y = 0.3010 - 0 = 0.3010$. The choice for ranges $\Delta\phi$, $\Delta\psi$ and starting values ϕ_s and ψ_s is arbitrary.

Let us select $\Delta\phi = 90°$ c.c.w., and $\Delta\psi = 60°$ c.c.w. (Fig. 6.14). The three precision (accuracy) points, using Chebyshev spacing are:

$$x_2 = (x_f - x_s)/2 + x_s = 1.5$$

From Fig. 6.14 then,

$$x_1 = x_2 - \frac{\Delta x}{2}\cos\frac{\pi}{6} = 1.5 - (1/2)\cos\frac{\pi}{6} = 1.067$$

and

$$x_3 = x_2 + \frac{\Delta x}{2}\cos\frac{\pi}{6} = 1.5 + (1/2)\cos\frac{\pi}{6} = 1.933$$

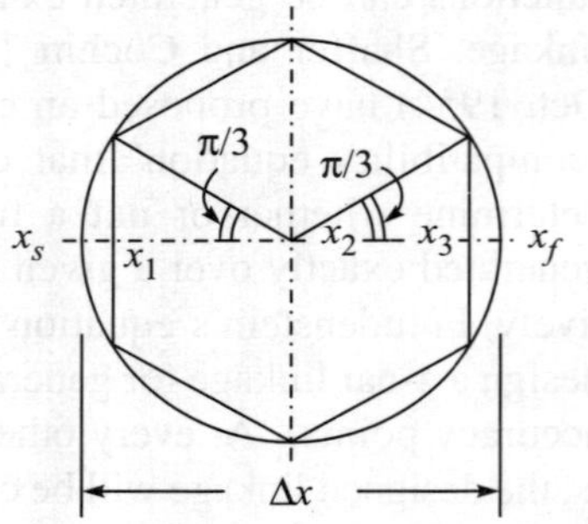

Hence
$$y_1 = \log_{10}(1.067) = 0.0282$$
$$y_2 = \log_{10}(1.5) = 0.1761 \text{ and}$$
$$y_3 = \log_{10}(1.933) = 0.2862$$

Here
$$\Delta\phi = 90°; \ \Delta\psi = 60°$$
and
$$\Delta x = 1 \text{ and } \Delta y = 0.301$$

Now
$$\phi_{12} = \phi_2 - \phi_1 = \frac{\Delta\phi}{\Delta x}(x_2 - x_1)$$

or
$$\phi_{12} = \frac{90}{1}(1.5 - 1.067) = 38.97° \simeq 39°$$

$$\phi_{23} = \phi_3 - \phi_2 = \frac{90}{1}(x_3 - x_2)$$

or
$$\phi_{23} = 90(1.933 - 1.5) = 38.97 \simeq 39°$$

Similarly
$$\psi_{12} = \psi_2 - \psi_1 = \frac{\Delta\psi}{\Delta y}(y_2 - y_1)$$

or
$$\psi_{12} = \frac{60.0}{0.301}(0.1761 - 0.0282) = 29.48° \approx 29.5°$$

$$\psi_{23} = \psi_3 - \psi_2 = \frac{\Delta\psi}{\Delta y}(y_3 - y_2)$$

or,
$$\psi_{23} = \frac{60.0}{0.301}(0.2862 - 0.1761) = 21.9468 \approx 22°$$

Let us select value of input angle $\phi_1 = 60°$, then

$$\phi_2 = 60 + 39 = 99° \text{ and } \phi_3 = 99 + 39 = 138°$$

Fig. 6.14 Distribution of accuracy points.

and
$$\phi_{s_1} = \frac{\Delta\phi}{\Delta x}(x_1 - x_s) = \frac{90}{1}(1.067 - 1.0) = 6.0°$$

Therefore
$$\phi_s = \phi_1 - \phi_{s_1} = 60° - 6°$$

Therefore
$$\phi_s = 54°$$

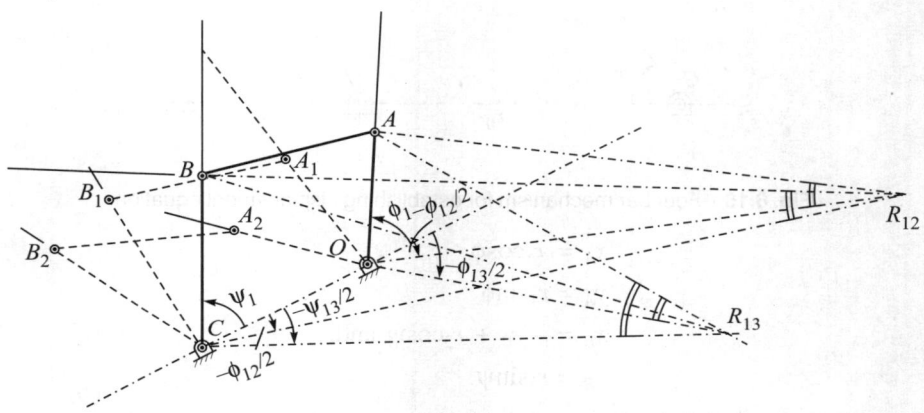

Fig. 6.15 Mechanism to generate function $y = \log_{10}(x)$.

Construction: Select length of frame link OC arbitrarily as 4.0 cm. Also select the link length $OA = 3$ cm and angle $\phi_1 = 60°$, both arbitrarily. Treating OA as the input crank, draw lines from O and C at angles $-\phi_{12}/2 = 19.5°$ and $-\psi_{12}/2 = 14.75°$ respectively in c.w. sense w.r. to line CO produced to meet at relative pole R_{12}. Similarly, draw lines from O and C making angles $-\phi_{13}/2 = 39°$ and $-\psi_{13}/2 = 25.75°$ respectively in c.w. sense w.r. to line CO produced so as to meet at relative pole R_{13}.

Join CR_{12} and CR_{13}. Draw lines from relative pole R_{12} and R_{13} making angles with lines $R_{12}C$ and $R_{13}C$ equal to the angles subtended by link OA at R_{12} and R_{13} respectively. These lines intersect at B giving required 4-bar mechanism $OABC$ as shown in Fig. 6.15.

6.10 ALGEBRAIC METHOD OF FUNCTION GENERATION

The algebraic method of synthesising a 4-bar mechanism for function generation aims at establishing a displacement equation which relates the output angle ψ to the input crank angle ϕ. To establish the displacement equation, consider a 4-bar mechanism $OABC$ shown in Fig. 6.16. The mechanism consists of four revolute pairs, axes of which are parallel and perpendicular to the plane of paper. Let the parameters r_0, r_1, r_2 and r_3 denote the distances CO, OA, AB and BC between corresponding revolute pairs. As the links are rigid, the parameters r_0, r_1, r_2 and r_3 are also constant.

Let the origin of coordinate axes be selected at O. Let ϕ and ψ be the angles made by links OA and CB with positive X-axis direction.

Then, the x and y coordinates of points may be expressed in terms of parameters, r_0, r_1, r_2, r_3, ϕ and ψ as under:

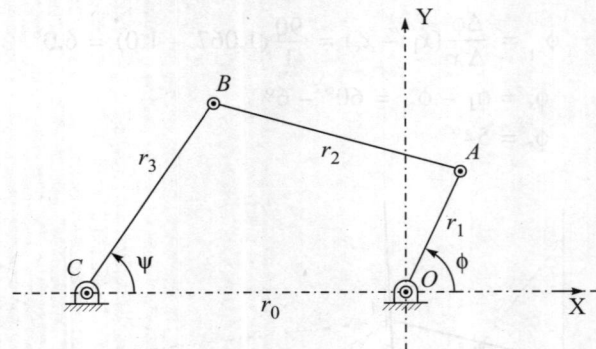

Fig. 6.16 Four bar mechanism for establishing displacement equation.

$$x_A = r_1\cos\phi$$
$$y_A = r_1\sin\phi$$
$$x_B = -r_0 + r_3\cos\psi \quad \text{and,}$$
$$y_B = r_3\sin\psi.$$

Now AB being a rigid link, distance between the revolute joints A and B is constant and is equal to r_2. Thus,

$$\begin{aligned}
r_2^2 &= (x_A - x_B)^2 + (y_A - y_B)^2 \\
&= (r_1\cos\phi + r_0 - r_3\cos\Psi)^2 + (r_1\sin\phi - r_3\sin\Psi)^2 \\
&= (r_1^2\cos^2\phi + r_0^2 + 2r_0r_1\cos\phi + r_3^2\cos^2\Psi - 2r_0r_3\cos\Psi \\
&\quad - 2r_1r_3\cos\phi\cos\Psi) + (r_1^2\sin^2\phi + r_3^2\sin^2\Psi - 2r_1r_3\sin\phi\sin\Psi)
\end{aligned}$$

Simplifying using the relation $\cos^2 A + \sin^2 A = 1$,

$$r_2^2 = r_0^2 + r_1^2 + r_3^2 - 2r_1r_3(\cos\phi\cos\Psi + \sin\phi\sin\Psi) + 2r_0(r_1\cos\phi - r_3\cos\Psi) \tag{6.12}$$

or $\quad r_2^2 = (r_0^2 + r_1^2 + r_3^2) - 2r_1r_3\cos(\phi - \Psi) + 2r_0(r_1\cos\phi - r_3\cos\Psi) \tag{6.13}$

Simplifying further, equation (6.13) may be rewritten as,

$$(r_0^2 + r_1^2 + r_3^2 - r_2^2) + 2r_0r_1\cos\phi - 2r_0r_3\cos\psi = 2r_1r_3\cos(\phi - \psi)$$

Dividing throughout by $2r_1r_3$,

$$\left(\frac{r_0}{r_3}\right)\cos\phi - \left(\frac{r_0}{r_1}\right)\cos\psi + \frac{(r_0^2 + r_1^2 + r_3^2 - r_2^2)}{2r_1r_3} = \cos(\phi - \psi)$$

Letting $\quad K_1 = (r_0/r_3); \ K_2 = \left(\dfrac{r_0}{r_1}\right) \quad$ and $\quad K_3 = \dfrac{(r_0^2 + r_1^2 + r_3^2 - r_2^2)}{2r_1r_3}$

Thus the first form of displacement equation (also known as Freudenstein equation) can be written as:

$$K_1\cos\phi - K_2\cos\psi + K_3 = \cos(\phi - \psi) \tag{6.14}$$

Second form of displacement equation is obtained by rewriting equation (6.12) as:

$$2r_1r_3(\cos\phi\,\cos\psi + \sin\phi\,\sin\psi) - 2r_0(r_1\cos\phi - r_3\cos\psi) = (r_0^2 + r_1^2 + r_3^2 - r_2^2)$$

Dividing out by $2r_1r_3$ and simplifying

$$(\sin\phi)\sin\psi + \left(\cos\phi + \frac{r_0}{r_1}\right)\cos\psi = \frac{(r_0^2 + r_1^2 + r_3^2 - r_2^2)}{2r_1r_3} + \frac{r_0}{r_3}\cos\phi$$

or

$$A\sin\psi + B\cos\psi = C \tag{6.15}$$

where

$$A = \sin\phi; \ B = \left(\frac{r_0}{r_1} + \cos\phi\right)$$

and

$$C = \left(\frac{r_0}{r_3}\right)\cos\phi + \frac{(r_0^2 + r_1^2 + r_3^2 - r_2^2)}{2r_1\,r_3}$$

Equation (6.15) gives the second form of displacement equation. This equation enables a user to establish the output angle ψ for every value of input angle ϕ when the link lengths r_0, r_1, r_2 and r_3 are known. A more convenient expression is obtained by using trignometric relations,

$$\sin\psi = \frac{2\tan(\psi/2)}{1 + \tan^2(\psi/2)}; \quad \text{and} \quad \cos\psi = \frac{1 - \tan^2(\psi/2)}{1 + \tan^2(\psi/2)}$$

Substituting in equation (6.15),

$$A\left(\frac{2\tan\psi/2}{1 + \tan^2\psi/2}\right) + B\left(\frac{1 - \tan^2\psi/2}{1 + \tan^2\psi/2}\right) = C$$

Simplifying in the form of a quadratic in $\tan\psi/2$,

$$(C + B)\tan^2\psi/2 - 2A\tan\psi/2 + (C - B) = 0$$

or

$$\tan\psi/2 = \frac{A \pm \sqrt{A^2 + B^2 - C^2}}{(B + C)} \tag{6.16}$$

Equation (6.16) represents two distinct solutions for $\psi/2$ (and therefore, two distinct configurations of the mechanism), depending on the sign used outside the radical sign. While using equation (6.16), care must be taken to avoid mixing of two branches represented by equation (6.16)

EXAMPLE 6.7 Design a 4-bar mechanism to generate a function $y = \log_{10}(x)$ in the range $1 \le x \le 2$.

Solution: Let us assume ranges for input and output angles be $\Delta\phi = 90°$; $\Delta\psi = 60°$, both c.c.w. and let $\phi_s = 60°$, $\psi_s = 60°$ be the starting values of input and output links.

The three accuracy points as per Chebychev spacing are:

$$x_2 = 1 + \frac{(2-1)}{2} = 1.5;$$

$$x_1 = 1.5 - 1\cos\pi/6 = 1.067$$

$$x_3 = 1.5 + 1\cos\pi/6 = 1.933$$

Values of dependent and independent variables at the three accuracy points are:

$$x_1 = 1.067; \ y_1 = \log_{10}(1.067) = 0.0282$$
$$x_2 = 1.5; \ y_2 = \log_{10}(1.5) = 0.1761$$
$$x_3 = 1.933; \ y_3 = \log_{10}(1.933) = 0.2862$$

Also

$$x_s = 1.0; \ y_s = 0.0$$
$$x_f = 2.0; \ y_f = 0.30103$$

Therefore
$$\Delta x = (2 - 1) = 1 \quad \text{and} \quad \Delta y = (y_f - y_s) = 0.301$$

and
$$\Delta\phi = 90° \ ; \ \Delta\psi = 60°$$
$$\phi_s = 60°; \ \psi_s = 60°$$

The displacement equations for the three accuracy points are:

$$K_1\cos\phi_1 - K_2\cos\psi_1 + K_3 = \cos(\phi_1 - \psi_1)$$
$$K_1\cos\phi_2 - K_2\cos\psi_2 + K_3 = \cos(\phi_2 - \psi_2)$$

and
$$K_1\cos\phi_3 - K_2\cos\psi_3 + K_3 = \cos(\phi_3 - \psi_3)$$

Eliminating K_3 by solving equations in pairs, we have

$$K_1(\cos\phi_1 - \cos\phi_2) - K_2(\cos\psi_1 - \cos\psi_2) = \cos(\phi_1 - \psi_1) - \cos(\phi_2 - \psi_2) \quad (6.17)$$

and
$$K_1(\cos\phi_1 - \cos\phi_3) - K_2(\cos\psi_1 - \cos\psi_3) = \cos(\phi_1 - \psi_1) - \cos(\phi_3 - \psi_3) \quad (6.18)$$

Values of input and output angles at accuracy points are computed as under:

$$(\phi_2 - \phi_1) = (x_2 - x_1)\frac{\Delta\phi}{\Delta x} = 38.97°$$

$$(\phi_3 - \phi_1) = (x_3 - x_1)\frac{\Delta\phi}{\Delta x} = 77.94°$$

$$(\psi_2 - \psi_1) = (y_2 - y_1)\frac{\Delta\psi}{\Delta y} = 29.48°$$

$$(\psi_3 - \psi_1) = (y_3 - y_1)\frac{\Delta\psi}{\Delta y} = 51.429°$$

Also
$$(\phi_1 - \phi_s) = (x_1 - x_s)\frac{\Delta\phi}{\Delta x} = 6.03°$$

and
$$(\psi_1 - \psi_s) = (y_1 - y_s)\frac{\Delta\psi}{\Delta y} = 5.621°$$

Hence
$$\phi_1 = 66.03°; \ \psi_1 = 65.621°$$
$$\phi_2 = 105.0°; \ \psi_2 = 95.101°$$
$$\phi_3 = 143.97°; \ \psi_3 = 117.05°$$

Substituting in equations (6.17) and (6.18),

$$(0.665)K_1 - (0.5017)K_2 = 0.01486$$
$$(1.215)K_1 - (0.8675)K_2 = 0.1083$$

Simplifying the above equation,

$$K_1 - (0.754)K_2 = 0.0223$$
$$K_1 - (0.714)K_2 = 0.0891$$

Solving simultaneously,

$$K_2 = 1.67; \qquad K_1 = 1.281$$

And, so, from displacement equation,

$$K_3 = 1.169$$

Hence, choosing frame-link length $r_0 = 4$ cm;

$$r_3 = \frac{4}{K_1} = 3.123 \text{ cm}; \qquad r_1 = \frac{4}{K_2} = 2.395 \text{ cm}$$

Also

$$r_2^2 = r_0^2 + r_1^2 + r_3^2 - 2(K_3)(r_1 r_3)$$
$$= (4)^2 + (2.395)^2 + (3.123)^2$$
$$- 2(1.169)(2.395)(3.123) = 14.00$$

Therefore $r_2 = 3.74$ cm

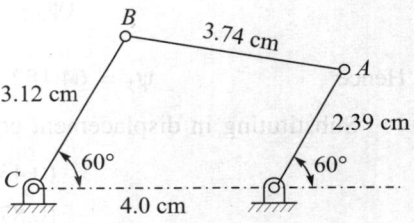

The dimensioned mechanism, as the function generator, is shown in Fig. 6.17.

Fig. 6.17 Required Function Generator for $y = \log_{10}(x)$.

EXAMPLE 6.8 A 4-bar linkage is required to generate a function $y = x^{1.6}$ for $1 \le x \le 3$. The crank rotates from an angle 60° to 120° whereas the follower rotates from an angle 60° to 150°. Use three accuracy points approximation based on Chebychev spacing.

Solution: Here, $\Delta\phi = 60°$; $\Delta\psi = 90°$

$$\phi_s = 60°; \quad \psi_s = 60°$$
$$x_s = 1.0; \quad \text{and} \quad x_f = 3.0 \quad i.e., \quad \Delta x = 3 - 1 = 2.0$$

And so $\quad y_s = (1)^{1.6} = 1.0 \quad$ and $\quad y_f = (3)^{1.6} = 5.799$

∴ $$\Delta y = y_f - y_s = 4.799$$

The three accuracy points based on Chebychev spacing and corresponding values of y are,

$$x_2 = 1 + \frac{(3-1)}{2} = 2.0; \qquad y_2 = (2.0)^{1.6} = 3.031$$

$$x_1 = 2 - 1\cos\frac{\pi}{6} = 1.134; \qquad y_1 = (1.134)^{1.6} = 1.223$$

$$x_3 = 2 + 1\cos\frac{\pi}{6} = 2.866; \qquad y_3 = (2.866)^{1.6} = 5.391$$

Thus

$$(\phi_1 - \phi_s) = (x_1 - x_s)\frac{\Delta\phi}{\Delta x} = 4.02°$$

$$(\phi_2 - \phi_1) = (x_2 - x_1)\frac{\Delta\phi}{\Delta x} = 25.98°$$

$$(\phi_3 - \phi_1) = (x_3 - x_1)\frac{\Delta\phi}{\Delta x} = 51.96°$$

Thus $\phi_1 = 64.02;$ $\phi_2 = 90°;$ $\phi_3 = 115.98°$

Also

$$(\psi_1 - \psi_s) = (y_1 - y_s)\frac{\Delta\psi}{\Delta y} = 4.182°$$

$$(\psi_2 - \psi_1) = (y_2 - y_1)\frac{\Delta\psi}{\Delta y} = 33.907$$

$$(\psi_3 - \psi_1) = (y_3 - y_1)\frac{\Delta\psi}{\Delta y} = 78.166°$$

Hence $\psi_1 = 64.182;$ $\psi_2 = 98.089;$ $\psi_3 = 142.348°$

Substituting in displacement equations (6.17) and (6.18),

$$(0.438)K_1 - (0.576)K_2 = 0.009945$$
$$(0.876)K_1 - (1.227)K_2 = 0.104036$$

or $K_1 - (1.315)K_2 = 0.0227$

and $K_1 - (1.4007)K_2 = 0.1188$

Hence by subtraction, $K_2 = 1.27$ and $K_1 = 1.897$

Therefore $K_3 = \cos(\phi_1 - \psi_1) - K_1\cos\phi_1 + K_2\cos\psi_1 = 0.7221$

Therefore $r_3 = (r_0/K_1)$ and $r_1 = (r_o/K_2)$

Selecting lengths of frame link $r_o = 5$ cm, the other link-lengths are:

$$r_3 = (5/1.897) = 2.636 \text{ cm}$$
$$r_1 = (5/1.27) = 3.937 \text{ cm}$$

And as such

$$r_2^2 = r_0^2 + r_1^2 + r_3^2 - 2(K_3)(r_1r_3)$$
$$= 32.46$$

Hence $r_2 = 5.697 \approx 5.7$ cm

The dimensioned function generating mechanism for function $y = x^{1.6}$ is as shown in Fig. 6.18.

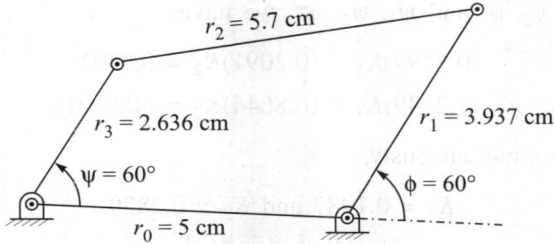

Fig. 6.18 Function generator for function $y = x^{1.6}$.

EXAMPLE 6.9 Synthesise a 4-bar linkage using Freudenstein's equation to generate the function $y = x^{1.5}$ for the interval $1 \le x \le 4$. The input crank is to start from $\phi_s = 30°$ and have a range of 90°. The output follower is to start at $\psi_s = 0°$ and have a range of 90°. Take three accuracy points.   (AMIE, Winter 1993)

Solution: The range of x, y, ϕ and ψ are:

$$\Delta x = 4 - 1 = 3$$
$$\Delta y = (4)^{1.5} - (1)^{1.5} = 8 - 1 = 7$$

The accuracy points are:

$$x_2 = 1 + \frac{(4-1)}{2} = 2.5; \qquad y_2 = (2.5)^{1.5} = 3.953$$

$$x_1 = 2.5 - \frac{3}{2}\cos\frac{\pi}{6} = 1.201; \qquad y_1 = (1.201)^{1.5} = 1.316$$

$$x_3 = 2.5 + \frac{3}{2}\cos\frac{\pi}{6} = 3.799; \qquad y_3 = (3.799)^{1.5} = 7.405$$

If ϕ_1, ϕ_2, ϕ_3 and ψ_1, ψ_2, ψ_3 are the input and output angles at the accuracy points, we have

$$\phi_2 - \phi_1 = \frac{90}{3}(2.5 - 1.201) = 38.97°; \; \psi_2 - \psi_1 = \frac{90}{7}(3.953 - 1.316) = 33.904°$$

$$\phi_3 - \phi_1 = \frac{90}{3}(3.799 - 1.201) = 77.94°; \; \psi_3 - \psi_1 = \frac{90}{7}(7.405 - 1.316) = 78.287°$$

$$\phi_1 - \phi_s = \frac{90}{3}(1.201 - 1.0) = 6.03°; \; \psi_1 - \psi_s = \frac{90}{7}(1.316 - 1.0) = 4.063°$$

Hence $\phi_1 = 36.03°; \qquad \phi_2 = 75.00°; \qquad \phi_3 = 113.97°$

and $\psi_1 = 4.063°; \qquad \psi_2 = 37.967°; \qquad \psi_3 = 82.350°$

Reduced form of displacement equations for the three accuracy points are:

$$K_1(\cos\phi_1 - \cos\phi_2) - K_2(\cos\psi_1 - \cos\psi_2) = \cos(\phi_1 - \psi_1) - \cos(\phi_2 - \psi_2)$$

and $$K_1(\cos\phi_1 - \cos\phi_3) - K_2(\cos\psi_1 - \cos\psi_3) = \cos(\phi_1 - \psi_1) - \cos(\phi_3 - \psi_3)$$

Substituting for ϕ_1, ϕ_2, ϕ_3 and ψ_1, ψ_2, ψ_3, we have

$$(0.5499)K_1 - (0.2091)K_2 = 0.0501$$
$$(1.2149)K_1 - (0.8644)K_2 = -0.00319$$

Solving above equations simultaneously,

$$K_1 = 0.1987 \text{ and } K_2 = 0.2829$$

Substituting in the displacement equation,

$$K_1\cos(36.03) - K_2\cos(4.063) + K_3$$
$$= \cos(36.03 - 4.063)$$

Therefore $\qquad K_3 = 0.9698$

Hence, assuming $r_0 = 5$ cm;

$$r_3 = (5/0.1987) = 25.16 \text{ cm}$$
$$r_1 = 5/(0.2829) = 17.67 \text{ cm}$$

Therefore $\qquad r_2^2 = r_0^2 + r_1^2 + r_3^2 - 2(K_3)r_1r_3 \quad \text{or} \quad r_2 = 10.39 \text{ cm}$

Dimensioned function generator is shown in Fig. 6.19.

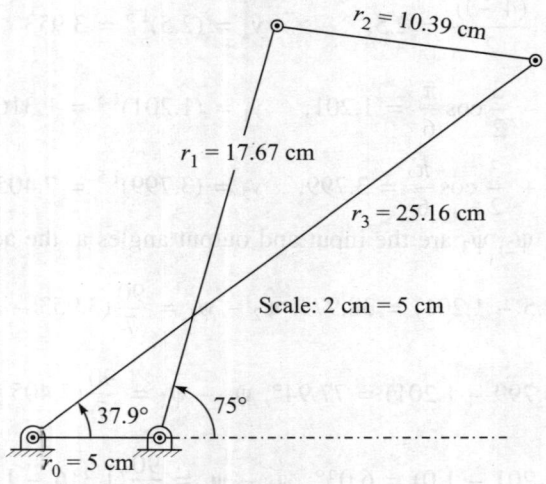

Fig. 6.19 Function Generator for function $y = x^{1.5}$.

6.11 COMMENTS ON DESIGN PARAMETERS AND SPECIAL NATURE OF RESULTS

It follows from Section 6.10 that, within the specified range, the number of independent design parameters involved in the mechanisation of a given function is seven. These are:

(i)   The three link-ratios (r_1/r_0) ; (r_2/r_0) and (r_3/r_0)

(ii) Two scaling factors $k_\phi = (\Delta\phi/\Delta x)$ and $k_\psi = \Delta\psi/\Delta y$ (or, the ranges of angle $\Delta\phi$ and $\Delta\psi$)

(iii) Two starting values of ϕ and ψ, namely ϕ_s and ψ_s.

Thus, the maximum possible accuracy points at which desired and actual (generated) functions match perfectly is seven. But the amount of work involved in synthesising a mechanism for seven-point approximation is prohibitive. In practice, problems of function generation, demanding higher degree of accuracy are therefore solved usually by 5-point approximation. For this purpose, either both the scaling factors k_ϕ and k_ψ are assumed or one scale factor and one initial value of angle ϕ_s or ψ_s is used. Besides, it is also possible to assume a combination of one scale factor and one link ratio. There is no rational approach to choose the parameters and, therefore, a designer must be guided either by past experience or should hope for a stroke of good luck.

For rough guidance, following rules of thumb may be suggested:

1. Ratio of ranges for output and input angles may be selected as $0.5 < (\Delta\psi/\Delta\phi) < 2$
2. Maximum values of range of angles should be $\Delta\phi < 120°$ and $\Delta\psi < 120°$.
3. Maximum value of ratio of link lengths should be less than 3.

If a function generator is required to be used as a device for indicating corresponding values of x and y, one could use a pointer and a double scale as shown in Fig. 6.20. Generally the x-scale is uniform, but y-scale is not. A more important requirement is to design a mechanism having uniform scales for both x and y. Advantages in having a uniform y-scale are as follows. Such a scale is easier to graduate. Further, in control instruments used in industrial processes, the effects produced by the instruments are usually proportional to their movements.

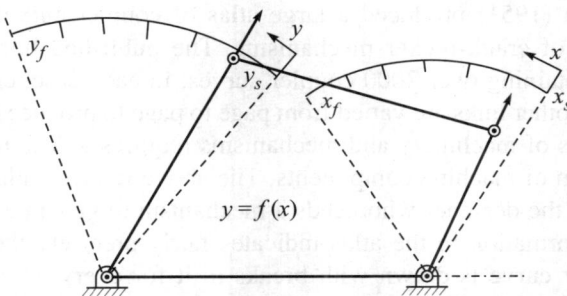

Fig. 6.20 Function Generator for indicating values of x and y.

Implication of Negative Dimension. While carrying out dimensional synthesis, in quite a few cases dimension of a link turns out to be a negative quantity. For instance, let us assume that the link dimension r_1 or r_3 turns out to be negative. Such a result indicates that the link extends opposite to that shown in Fig. 6.21. In other words, the pointer indicating functions x and y will be placed diametrically opposite to links r_1 and r_3.

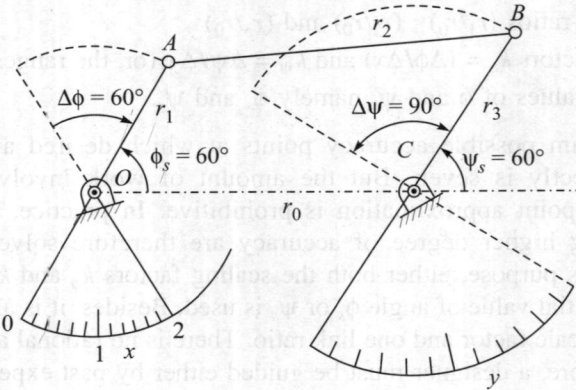

Fig. 6.21 Effect of negative dimension.

6.12 COUPLER CURVES

A four-link mechanism is capable of providing two different types of output motion. These output motions can be derived either from the follower link or the coupler link. This section is devoted to the study of the path traced out by a point located in the plane of the coupler link. The coupler of four-bar linkage may be thought of as a plane extending in all directions but connected to the input and output links through turning pairs. When the linkage is put into motion, any point attached to the plane of coupler generates some path/curve with respect to frame link, this path or curve is called *coupler curve*. It can be easily verified that coupler points, coincident with the pin connections, will generate simple circles with fixed pivots as centres. The other points generally trace much more complex curves.

Hrones and Nelson (1951) produced a large atlas of coupler curves for different coupler points C for hundreds of crank-rocker mechanisms. The published book consists of a set of 11×17 inch charts containing over 7000 coupler curves. In each case, crank-length is taken as unity and the lengths of other links are varied from page to page to produce different combinations.

Design and analysis of machinery and mechanisms requires a designer to use his ability to visualise relative motion of machine components. The above referred atlas of coupler curves is, therefore, invaluable to the designer who needs a mechanism to generate a curve with specified characteristics. The information in the atlas indicates fairly precisely the speed of the coupler point, since the coupler curve is drawn with breaks in it for every 5° of crank angle.

In general, a sixth order algebraic equation is expected for coupler curves. They have a wide variety of shapes and incorporate many interesting features. Some of the coupler curves have sections which are nearly straight line segments, while others are very nearly in the form of circular arcs. Some of the coupler curves exhibit one or more cusps or cross-over themselves representing a figure of eight. Because of very large spectrum of coupler curves available with 4-bar mechanisms, usually it not necessary to go for mechanisms with large number of links.

A very curious and interesting fact about the coupler curve equation is that the same coupler curve can always be generated by three different linkage called cognates. This is known as Roberts Chebychev's Theorem of cognate linkages.

Equation of Coupler Curves

Analytic geometry can be used to derive equation for coupler point curve for a 4-bar mechanism. We shall prefer to write the equation in Cartesian coordinates with x-axis selected along line of centres of fixed pivots and y-axis perpendicular to it through fixed pivot O (see Fig. 6.22).

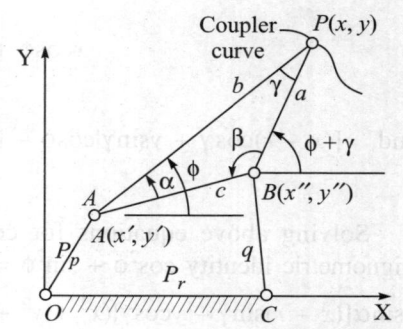

Fig. 6.22 Coordinate system and coupler-mechanism.

Let $OABC$ be the four-bar mechanism and P be the coupler point formed on coupler with sides of coupler $AP = b$, $BP = a$ and $AB = c$. Let the coordinates of points A, B and P be (x', y'), (x'', y'') and (x, y) respectively. Let angles included between the sides of coupler be α, β and γ as shown. Also, let inclination of side AP with x-axis be ϕ, so that angle made by PB with x-axis is $\phi + \gamma$.

Then
$$x' = x - b\cos\phi;$$
$$y' = y - b\sin\phi$$

and
$$x'' = x - a\cos(\phi + \gamma);$$
$$y'' = y - a\sin(\phi + \gamma) \tag{6.19}$$

Again, since A and B describe circles of radii p and q, we have,

$$x'^2 + y'^2 = p^2 \quad \text{and} \quad (x'' - r)^2 + y''^2 = q^2 \tag{6.20}$$

Substituting for (x', y') and (x'', y'') from equations (6.19) in (6.20), we have

$$(x - b\cos\phi)^2 + (y - b\sin\phi)^2 = p^2 \tag{6.21}$$

and
$$[x - a\cos(\phi + \gamma) - r]^2 + [y - a\sin(\phi + \gamma)]^2 = q^2 \tag{6.22}$$

Simplifying equation (6.21), we have

$$x^2 + y^2 + b^2 - 2b(x\cos\phi + y\sin\phi) = p^2$$

or
$$x\cos\phi + y\sin\phi = \frac{x^2 + y^2 + b^2 - p^2}{2b}$$

Simplifying equation (6.22),

$$x^2 + [a\cos(\phi + \gamma) + r]^2 - 2x\,[a\cos(\phi + \gamma) + r] + y^2 + a^2\sin^2(\phi + \gamma) - 2ay\sin(\phi + \gamma) = q^2$$

or
$$x^2 + a^2 + r^2 + 2ar\cos(\phi + \gamma) - 2ax\cos(\phi + \gamma) - 2xr + y^2 - 2ay\sin(\phi + \gamma) = q^2$$

or
$$(x - r)^2 + a^2 - q^2 + y^2 + 2ar[\cos\phi\cos\gamma - \sin\phi\sin\gamma]$$
$$- 2ax[\cos\phi\cos\gamma - \sin\phi\sin\gamma] - 2ay\,[\sin\phi\cos\gamma + \cos\phi\sin\gamma] = 0$$

or
$$\frac{(x - r)^2 + a^2 + y^2 - q^2}{2a} = [(x - r)\cos\gamma + y\sin\gamma]\cos\phi - [\,(x - r)\sin\gamma - y\cos\gamma]\sin\phi$$

Thus the two equations (6.21) and (6.22) simplify to

$$x\cos\phi + y\sin\phi = \frac{x^2 + y^2 + b^2 - p^2}{2b} \qquad (6.23)$$

and, $\quad [(x - r)\cos\gamma + y\sin\gamma]\cos\phi - [(x - r)\sin\gamma - y\cos\gamma]\sin\phi = \dfrac{(x - r)^2 + a^2 + y^2 - q^2}{2a}$

$$(6.24)$$

Solving above equations for $\cos\phi$ and $\sin\phi$ and substituting the values obtained in the trignometric identity $\cos^2\phi + \sin^2\phi = 1$ yields the general equation of coupler curve as

$$\{\sin\alpha[(x - r)\sin\gamma - y\cos\gamma](x^2 + y^2 + b^2 - p^2) + y\sin\beta \ [(x - r)^2 + y^2 + a^2 - q^2]\}^2$$

$$+ \sin\alpha[(x - r)\cos\gamma + y\sin\gamma](x^2 + y^2 + b^2 - p^2) - x\sin\beta \ [(x - r)^2 + y^2 + a^2 - q^2]^2$$

$$= 4k^2\sin^2\alpha \ \sin^2\beta\sin^2\gamma[x(x - r) + y^2 - ry\cot\gamma]^2 \qquad (6.25)$$

where, $$k = \frac{a}{\sin\alpha} = \frac{b}{\sin\beta} = \frac{c}{\sin\gamma}$$

The equation being of sixth order, a straight line will interesect it in more than six points.

6.13 SYNTHESIS FOR PATH GENERATION

In section 6.1, a path generator was described as a mechanism whose tracing point (called coupler point) on the floating link is constrained to describe a prescribed path with reference to the frame link. As mentioned in the above sections, number of coupler curves is infinite. Useful curves can be obtained by choosing proper link proportions and also by selecting suitable locations for coupler point on the floating link. This is again a problem of mechanism synthesis.

Usefulness of a coupler curve depends on particular shape of its segments and also on a peculiar shape of the curve as whole or part of it. Approximate straight line portions or arcs of circle are considered desirable in number of applications. As the interest lies in motion characteristics of the coupler point, the coupler point becomes the output of the linkage. Approximate straight line motion mechanism and parallel motion linkage, devised by Watt in 1784 are the best known examples of coupler-point linkage. The watt linkage is currently used for axle and differential suspensions of some high-performance cars.

As already mentioned in article 6.12. Hrones and Nelson proposed atlas of approximately 7,300 coupler curves that are drawn to large scale. These atlas constitute a very practical tool for the designers. The designers role in this case is reduced to selecting segment of a suitable coupler curve, in accordance with the required path to be traced by the coupler point, and selecting corresponding linkage and coupler point out of these collections of coupler curves. The four-bar mechanisms considered for above purpose are crank-rocker type. We shall use notations for links, same as the ones used for function generators and shown in Fig. 6.16. Thus, if r_0, r_1, r_2 and r_3 be the lengths of frame link, input link, coupler and output link respectively, then conditions to be satisfied for crank-rocker mechanism are:

$$r_0 < (r_2 + r_3) - r_1$$
$$r_0 > |(r_2 - r_3)| + r_1 \qquad (6.26)$$

and, $r_2, r_3, r_1 > r_0$

These conditions imply that r_0 is the smallest link and r_1 is the longest link, satisfying Grashoff's law.

For the sake of convenience, r_1 may be taken as unity.

One such mechanism, along with only six related coupler curves, is reproduced in Fig. 6.23 from the above atlas. The points marked C_1, C_2, C_3, C_4, C_5 and C indicate suitable locations for coupler point necessary to generate the path represented by corresponding coupler curve. It may be noted that the mechanism with coupler point at C generates a curve which has a flat (straight-line) segment between points L and M. This portion of coupler curve, together with corresponding range of rotation of input and output links, may be conveniently used for guiding a suitable point C along an approximate straight line path.

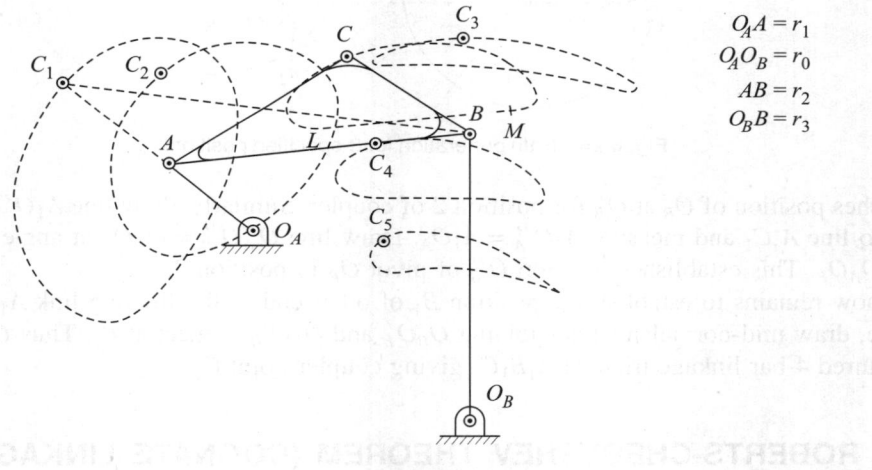

$$O_A A = r_1$$
$$O_A O_B = r_0$$
$$AB = r_2$$
$$O_B B = r_3$$

Fig. 6.23

6.14 GRAPHICAL SYNTHESIS FOR PATH GENERATION (THREE SPECIFIED POSITIONS)

Let us consider the problem of designing a 4-bar linkage whose coupler point C pass through three prescribed positions C_1, C_2 and C_3. The choice of link-length for frame $O_A O_B \equiv r_0$, and its orientation is arbitrary here. Similarly, the length of crank OA_1 and its position for the position of coupler point C_1 is also arbitrary (see Fig. 6.24). The point A then moves along an arc of a circle of radius $O_A A_1$.

The method of designing path generating mechanism involves the principle of inversion used earlier. The procedure used is as follows.

Join $O_A A_1$ and $A_1 C_1$. From coupler points C_2 and C_3 draw arcs of radii $C_2 A_2 = C_3 A_3 = C_1 A_1$ to cut the crank circle at points A_2 and A_3 respectively. Attempt will be made to obtain inversions in positions 2 and 3 of mechanism keeping link $A_1 C_1$ fixed and releasing the frame link. Let $\angle C_1 A_1 O_A = \beta_1$; $\angle C_2 A_2 O_A = \beta_2$ and $\angle C_3 A_3 O_A = \beta_3$.

Fixing the floating link in position $C_1 A_1$, the relative positions of link $A_1 O_A$ and $O_A O_B$ will be obtained for positions 2 and 3. For position 2, draw line $A_1 O'_A$ making angle β_2 with line $A_1 C_1$. Measure $A_1 O'_A = A_1 O_A$ and draw line $O'_A O'_B = O_A O_B$ at an angle $\angle A_1 O'_A O'_B = \angle A_2 O_A O_B$. This

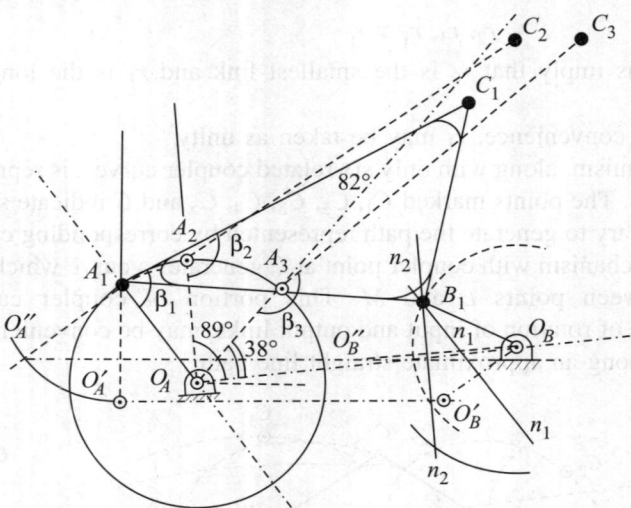

Fig. 6.24 Path generation for 3-specified positions.

establishes position of O_B at O_B' for position 2 of coupler. Similarly, draw line $A_1 O_A''$ at angle β_3 c.c.w. to line $A_1 C_1$ and measure $A_1 O_A'' = A_1 O_A$. Draw line $O_A'' O_B'' = O_A O_B$ at angle $\angle A_1 O_A'' O_B''$ $= \angle A_3 O_A O_B$. This establishes position O_B'' of pivot O_B in position 3.

It now remains to establish the position B_1 of other end of the floating link $A_1 B_1$. For this purpose, draw mid-normals of lines joining $O_B O_B'$ and $O_B O_B''$ to meet at B_1. Thus $O_A A_1 B_1 O_B$ is the required 4-bar linkage triangle $A_1 B_1 C_1$ giving coupler point C_1.

6.15 ROBERTS-CHEBYSHEV THEOREM (COGNATE LINKAGES)

An extremely useful property of planer 4-bar linkages is revealed in a theorem by Roberts and Chebyshev.

The theorem states that the curve generated by a given coupler point of a 4-bar mechanism can also be generated exactly by two other 4-bar mechanisms, coplanar with the first. Stated more clearly, this means that one point common to each of the three different but related 4-bar linkages will trace identical coupler curves. The two additional linkages, which generate identical coupler curves, are called *cognates* or more specifically "Roberts-Chebyshev cognates". Above theorem is attributed to two different scientists for their independent discoveries; Roberts (1875 in U.K.) and Chebyshev (1878 in Russia).

When the major concern is about the curve/path traced by a coupler point of a 4-bar linkage, concept of cognate linkages enables a designer to have two additional alternatives, either of which may prove to be more favorable than the original design in respect of transmission angle or space requirements.

It may be recalled that kinematically equivalent 4-bar linkages are commonly used for velocity and acceleration analysis of planar, direct contact mechanisms. The dimensions of such equivalent mechanisms change with time. As against this, cognate linkages do not look alike and in contrast to equivalent linkages, dimensions of cognate linkages do not change with time. This permits substitution of a given path generator by another one for the entire cycle of motion. The velocity and acceleration characteristics of cognate linkages, however, need not necessarily be identical link for link.

In a 4-bar linkage $OABP$ with coupler point C (as shown in Fig. 6.25), the other two four-bar linkage mechanisms with their coupler points describing the same coupler curve as C as constructed are described below:

Construct a parallelogram $OACE$ on sides OA and AC of Fig. 6.25. Also, construct parallelogram $PBCK$ on adjoining sides PB and BC. On sides CE and CK of these parallelograms, construct triangles, similar to triangle ABC such that $\angle CEF = \angle CAB = \angle KCL = \alpha$.

And
$$\angle ECF = \angle ABC = \angle CKL = \beta.$$

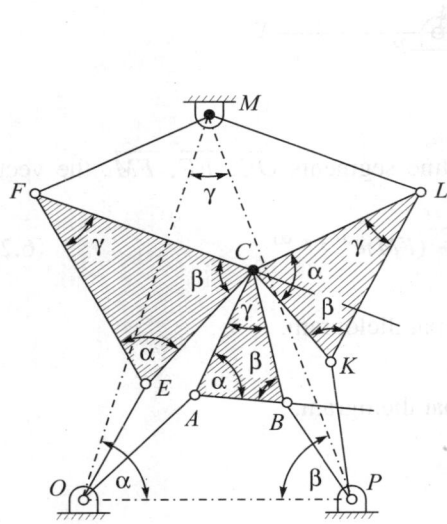

Fig. 6.25 Cognate Linkages.

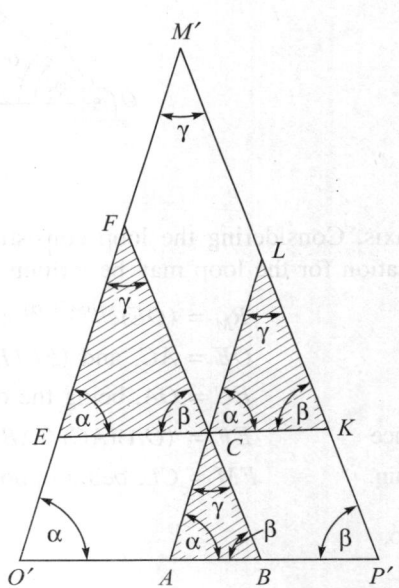

Fig. 6.26 Caley's diagram (Robert's Linkage).

Complete the parallelogram of adjoining sides FC and CL and this gives location M of the third pivot at the intersection points.

For the Roberts-chebyshev theorem to be true, it is necessary to show that the third pivot M is a fixed point and the pivot triangle OPM is similar to the coupler triangle ABC. Geometric proof for this purpose is tedious and what follows below is the proof given by J.B. Schor (1941) which makes use of complex numbers.

Referring to Fig. 6.27, it is clear that the point M is fixed in position, if the position vector

$$R_M = (OM)\,e^{j\delta} = \text{constant} \qquad (6.27)$$

To show this clearly the length OM and the angle δ will have to be expressed in terms of constant parameters (like link-lengths) of the parent chain $OABP$ and must be independent of the angular displacements ϕ_1, ϕ_2 and ϕ_3 of the mechanism. Let us select X-axis along the line joining OP and origin at O, and let the angles be measured positive in c.c.w. sense with respect to

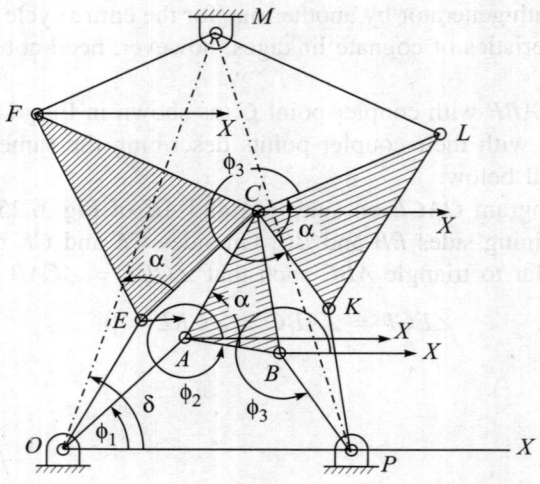

Fig. 6.27

X-axis. Considering the loop consisting of directed line segments $\overrightarrow{OE}$, $\overrightarrow{EF}$, $\overrightarrow{FM}$, the vector equation for the loop may be written as

$$R_M = (OE)e^{j(\phi_2 + \alpha)} + (EF)e^{j(\phi_1 + \alpha)} + (FM)e^{j(\phi_3 + \alpha)} \tag{6.28}$$

But $OE = AC$, and $(EF/EC) = (AC/AB)$

and $EC = OA$, being the opposite side of parallelogram.

Hence $EF = (OA)(AC)/(AB)$

Again, $FM = CL$, being opposite sides of a parallelogram.

Also,
$$\frac{CL}{CK} = \frac{CA}{AB}$$

where $CK = BP,$

It follows that $FM = (BP)(CA)/AB$

It is thus seen that OE, EF and FM are all constant. Substituting for OE, EF and FM in equation (6.28)

$$R_M = (AC)e^{j(\phi_2 + \alpha)} + \frac{(OA)(AC)}{(AB)}e^{j(\phi_1 + \alpha)} + \frac{(BP)(CA)}{(AB)}e^{j(\phi_3 + \alpha)} \tag{6.29}$$

or
$$R_M = \left(\frac{AC}{AB}\right)e^{j\alpha}[ABe^{j\phi_2} + OAe^{j\phi_1} + BPe^{j\phi_3}] \tag{6.30}$$

Readers may note that the summation within squared parenthesis represents vectorial addition of directed line segments $\overrightarrow{OA}$, $\overrightarrow{AB}$ and $\overrightarrow{BP}$ which gives a single directed line segment $\overrightarrow{OP}$. Hence, from equation (6.30)

$$R_M = \left(\frac{AC}{AB}\right)OPe^{j\alpha} \tag{6.31}$$

The magnitude of vector $R_M = \left(\dfrac{AC}{AB}\right)OP$

As $|R_M| = OM$, it follows that $\qquad \dfrac{OM}{OP} = \dfrac{AC}{AB}$

In other words, triangles OPM and ABC are similar. Further as angle $\angle CAB = \alpha$ is constant, right hand side of equation (6.31) is constant. Hence, $\delta = \alpha$ and R_M is constant. Hence proved.

The two desired 4-bar linkages, which can generate identical coupler curves at C, are $OEFM$ and $PKLM$. The construction in Fig. 6.25 gives a redundant chain with $n = 10$, and $1 = 14$ (there being six simple turning pairs and four double joints at O, P, M and C) giving d.o.f. $= 3 \ (10 - 1) - 2 \ (14) = -1$. This suggests that there are two more links than necessary and two links, such as OE and PK, could be removed.

Cayley suggested the plan of Fig. 6.26 and is claimed to be a simple way of determining the link lengths of cognate mechanisms. It also explains rationale behind angular relationship between the three coupler triangles. The side $O'P'$ of the plan in Fig. 6.26 is obtained by pulling out the linkage $OABP$ straight so that cranks OA, PB and the side AB of the coupler triangle ABC lie along a straight line. Thus, length $O'P' = OA + AB + PB$, and is clearly greater than the pivot distance OP. Next draw lines through coupler point C parallel to the three sides of the coupler triangle ABC. Location of third pivot M now becomes essential to establish link lengths of the two cognate mechanisms $OEFM$ and $PKLM$. This is accomplished by constructing triangle $O'P'M'$ similar to the coupler triangle ABC with $O'P'$ as base. Then $O'E$ and $M'F$ give crank lengths and EFC the coupler triangle for cognate mechanism $OEFM$. Similarly, $P'K$ and $M'L$ give the lengths of cranks while LKC gives the required dimensions of the coupler triangle for the second crank lengths and EFC the coupler triangle for cognate mechanism $PKLM$.

Following interesting features of the Fig. 6.26 are potentially useful:

1. The vertex angles at the point C of all the three coupler triangles are all different, *i.e.* no two angles are equal.
2. The included angles between two corresponding sides of any two coupler triangles connected to the same pivot through different cranks, are same. Thus,

$$\angle BAC = \angle CEF = \alpha, \quad \text{and} \quad \angle ABC = \angle CKL = \beta$$

6.16 COUPLER CURVES FROM 5-BAR MECHANISMS

As pointed out earlier Chebyshev in the year 1878 (in Russia) independently arrived at Roberts theorem. One plus point with Chebyshev's findings is that for the first time he pointed out that any coupler curve traced out by a 4-bar mechanism can also be traced out by a 5-bar mechanism, in which the cranks must rotate in the same direction at the same speed.

Thus a given coupler curve is generated by three cognate 4-bar linkages and for each cognate linkage, there is a cognate 5-bar linkage which can generate the same coupler curve. *Hence, as a general rule, any coupler curve can be generated by six cognate linkages; three being four-bar linkages and the other three being 5-bar linkages.*

A cognate 5-bar linkage for a linkage *OABP*, with coupler triangle *ABC*, is obtained by drawing lines *OE* and *EC* parallel to *AC* and *OA* respectively (Fig. 6.28). Similarly lines *PF* and *FC* are drawn parallel to *BC* and *PB* respectively. Linkage *OECFP* is then the required 5-bar cognate of the mechanism *OABP*. Links *OE* and *PF* are required to rotate in the same sense with the same speed. To achieve this, there are two approaches. In the first instance, draw line *PQ* parallel to *OE* and line *EQ* parallel to *OP*. Then, the parallelogram four-bar

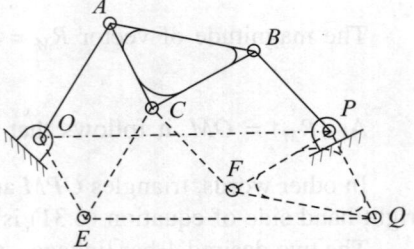

Fig. 6.28 A cognate 5-bar linkage.

mechanism *OEQP* will ensure that the side *PQ* of ternary link *PQF* always remains parallel to *OE* and moves at speed and in sense same as that of *OE*. An another way of achieving this is to connect links *OE* and *PF* through toothed gears.

Concept of cognate linkages is further illustrated through following examples.

EXAMPLE 6.10 Find out a 4-bar cognate linkage of the mechanism shown in Fig. 6.29(a). Given that *OA* = 52 mm; *AB* = 22 mm; *PB* = 52 mm; *BC* = 20 mm; *AC* = 14 mm and base *OP* = 32 mm and is inclined at 45° to the horizontal. The crank *OA* is inclined to the frame link *OP* at an angle of 40° in the given configuration.

(a)

(b)

Fig. 6.29

Solution: The mechanism is thus described in the prescribed position by thick lines *OABP* as shown in Fig. 6.29(a) to the full-scale. Since *OABP* is a crossed 4-bar mechanism pulling out the linkage to ensure that *OA*, *AB* and *BP* lie along a straight line does not produce length *O′P′* = *OA* + *AB* + *BP*. Hence let us consider the position during rotation of cranks when arms *OA* and *BP* do not appear as crossed arm. In such a position, coupler point *C* lies on the lower side of *AB*. The pivot triangle *OPM* corresponds to this configuration of linkage. As discussed earlier, ∠*POM* equals the angle included in coupler triangle at the point where the crank *OA* connects it. Hence ∠*POM* = α and ∠*OPM* = β. Thus the pivot triangle *OPM* is constructed. We now proceed to construct cognate linkage on *P* and *M* as pivots. As *BP* connects pivot *P* to coupler triangle at *B*, side *BC* connecting point *B* to coupler point is of interest to us. Hence, from pivot *P* draw a line parallel to *BC* and complete the parallelogram *PBCK*. Thus, *PK* is the new crank connecting pivot *P* to new coupler triangle, which must again have the coupler point at *C*. Thus, *KC* is one side of new coupler triangle. Since, ∠*OPM* = β, the included angle of new coupler triangle at *K* should also be β. Again as ∠*OMP* = γ, the included angle of new coupler triangle at *L* must be equal to γ, and therefore, ∠*LCK* = α. Hence make ∠*LCK* = α and ∠*LKC* = β to obtain the point *L* of coupler triangle *LCK*. Joint *LM*. The cognate mechanism *PKLM* with coupler point at *C* is shown in dotted lines.

As shown at Fig. 6.29(b) this can be obtained in yet another way. Draw *O′P′* parallel to *OP* to represent stretched out condition of mechanism such that *O′P′* = *OA* + *AB* + *BP*. As argued above, for this to happen, coupler point *C* must lie on the right hand side of *O′P′*. Locate points *A* and *B* on *O′P′* and construct triangle *ABC* similar to the given coupler triangle. From *O′* and *C* draw lines parallel to *AC*. Similarly from *P′* and *C* draw lines parallel to *BC*. Finally from *C* draw line parallel to *O′P′*. This completes the layout shown in Fig. 6.29(b). The new coupler triangle and its dimensions are represented by sides *CK*, *KL* and *CL*. The length *P′K* represents length of the crank connecting pivot *P* to the coupler triangle at *K*. Similarly, length *LM′* measures length of crank connecting new pivot *M* to the new coupler triangle at *L*. These dimensions may be compared with those established in Fig. 6.29(a).

EXAMPLE 6.11 The mechanism shown in Fig. 6.30(a) was designed for use in an automated assembly unit. An insertion tool located at *C* must move in the path determined by the mechanism. Unfortunately the bearing at B_o was found to interfere with something else. Suggest an alternative mechanism relocating the bearing without altering the path of the insertion tool. (AMIE, Summer 1993)

Solution: Referring to Fig. 6.30(a), the given coupler mechanism is A_oABB_o with coupler point *C*. Since the pivot B_o is to be replaced, construct parallelogram with A_oA and *AC* as sides. This locates pivot *E* on new coupler triangle. Thus, *EC* is one side of the new coupler triangle. As reasoned out earlier, the included angle of coupler triangle at *E* must be equal to α and the included angle at *C* for new coupler triangle has to be other then γ (which is the included angle at *C* in original coupler triangle). Hence, the included angle at *C* in new coupler triangle is β. Thus draw *EF* at α to *EC* and *CF* at β to *EC*, the intersection point gives the third point *F* of coupler triangle.

Finally, on A_oB_o as base describe a triangle A_oB_oM similar to coupler triangle *ABC* to locate the third pivot *M*. Join *MF*. Then, A_oEFM is the new mechanism with coupler point at *C* and new pivots at A_o and *M*. This being a cognate mechanism, point *C* will describe identical coupler curve.

Alternatively, after obtaining position M of third pivot as above, construct Cayley's diagram as in Fig. 6.30(b), to obtain dimensions of links A_oE, EF and FM. For this purpose, draw $A'_oB'_o = A_oA + AB + BB_o$, and on AB describe coupler triangle ABC. From C draw lines parallel to the three sides of coupler triangle ABC. Then measured lengths A'_oE, EF and FM' give required link lengths A_oE, EF and FM of Fig. 6.30(a).

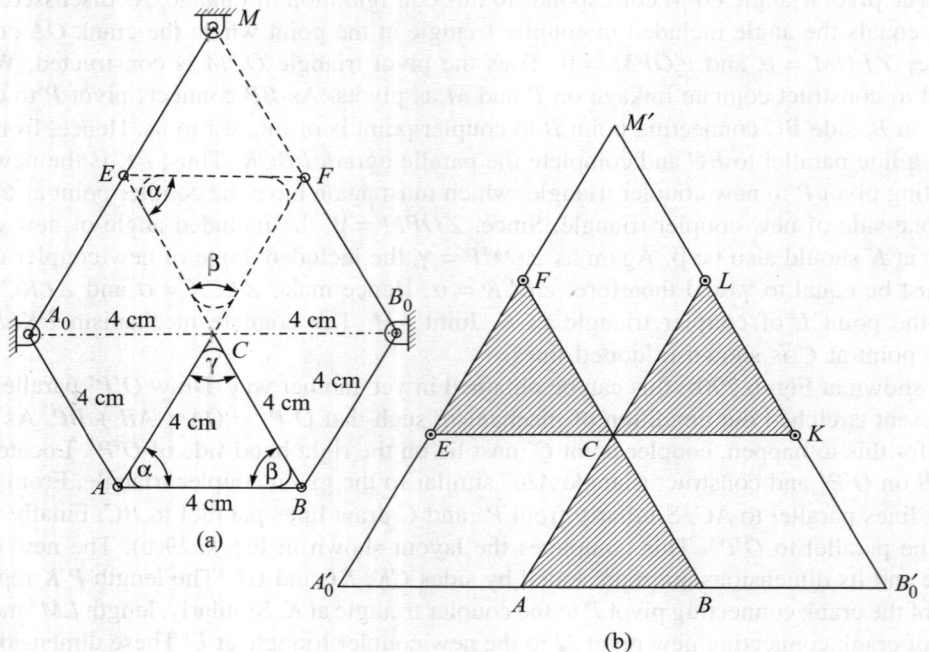

(a)

(b)

Fig. 6.30

REVIEW QUESTIONS

6.1 Synthesise a 4-bar linkage that will satisfy the following relation: $y = x^{1.2}$ for $1 \le x \le 5$ using Chebyshey spacing for three precision points.

Take $\phi_s = 30°$; $\psi_s = 60°$, and $\Delta\phi = \Delta\psi = 90°$.

(*Ans.* $r_0 = 10$ cm; $r_1 = -37.27$ cm; $r_2 = 27.13$ cm; $r_3 = -36.72$ cm)

6.2 Design a 4-bar mechanism to generate function $y = \sin x$ for $0° \le x \le 90°$. The range in ϕ is $120°$ and the range in ψ is $60°$. Solve using 3 accuracy points based on Chebyshev spacing.

(*Ans.* $r_0 = 52.5$ mm; $r_1 = 29.0$ mm; $r_2 = 75.6$ mm; $r_3 = 38.0$ mm)

6.3 A 4-bar mechanism is to be designed to generate the function $y = -\dfrac{x}{8}(x + 2)$ in the range $0 \le x \le 6$. Both the cranks are to rotate $90°$ for this range of x. Solve this problem for $\psi_s = 0$ and $\phi_s = -70.7°$.

(*Ans.* $r_1 = -16.4$ cm; $r_2 = 15.5$ cm; $r_3 = 17.4$ cm; $r_0 = 25.4$ cm)

6.4 Solve example 6.3 for $\psi_s = -45°$ and $\phi_s = 86.9°$.

(**Ans.** $r_1 = 12.4$ cm; $r_2 = 28.1$ cm; $r_3 = 17.0$ cm; $r_0 = 25.4$ cm)

6.5 Determine the lengths of the links of a 4-bar linkage to generate $y = \log_{10} x$ in the interval $1 \le x \le 10$. The length of the smallest link is 5 cm. Use three accuracy points with Chebyshev spacing. Take $\phi_s = 45°$, $\psi_s = 135°$; $\phi_f = 105°$ and $\psi_f = 225°$.

(**Ans.** $r_0 = 10$ cm; $r_2 = 21.85$ cm; $r_3 = -14.20$ cm)

6.6 Design a 4-bar mechanism to guide a rigid body (defined in position by a line PQ in it) through three positions of the input link. Take $PQ = 5$ cm and position of point P be given in terms of polar coordinates (r, α) and inclination of line PQ to X-axis by angles θ as follows:

$$\theta_1 = 40°; \quad r_1 = 90 \text{ mm}; \quad \alpha_1 = 78° \qquad \theta_2 = 55°; \quad r_2 = 40 \text{ mm}; \quad \alpha_2 = 90°$$

$$\theta_3 = 70°; \quad r_3 = 75 \text{ mm}; \quad \alpha_3 = 95°$$

6.7 Lay out a crank rocker mechanism so that output rocker will rotate through 60° and the minimum transmission angle is 40°.

6.8 For the two coupler-link position indicated by lines A_1B_1 and A_2B_2 in Fig. 6.31, locate the pole point. Using points C and D as moving hinge points (shown in dotted lines in Fig. 6.31) design a four-bar mechanism that will move line AB into its two designated positions.

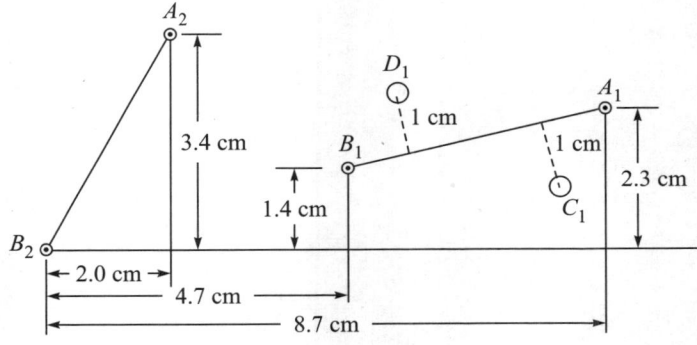

Fig. 6.31

6.9 The pole triangle for three positions of a coupler plane and the position of coupler point C when the coupler is in position 1 are given in Fig. 6.32. Locate the position of C when the coupler is in positions 2 and 3. Use the rotation poles and displacement angles and check using the cardinal position of C.

$P_{23}P_{13} = 2.0$ units; $P_{13}P_{12} = 2.50$ units;

$P_{23}P_{12} = 2.25$ units.

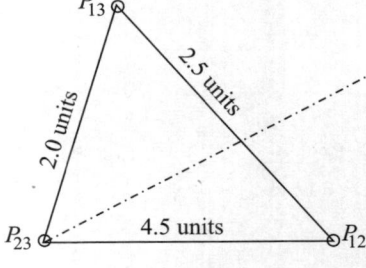

Fig. 6.32

6.10 For the mechanism shown, select a suitable second fixed pivot O_c and locate the associated moving hinge pin C_1. Following data refers to the position of pivot O_A in terms of x and y coordinates as (3.0, 6.5). $O_AA - 1.5$ units; $AB = 3.0$ units

$$\theta_1 = 60°; \quad \theta_2 = 90°; \quad \theta_3 = 135°; \qquad \beta_1 = 85.5°; \quad \beta_2 = 89.0°; \quad \beta_3 = 101°$$

6.11 Lay out a mechanism to satisfy the following displacements as shown in Fig. 6.33.

Input crank	*Output crank*
$\phi_{12} = 25°$	$\psi_{12} = -20°$
$\phi_{13} = 60°$	$\psi_{13} = -40°$

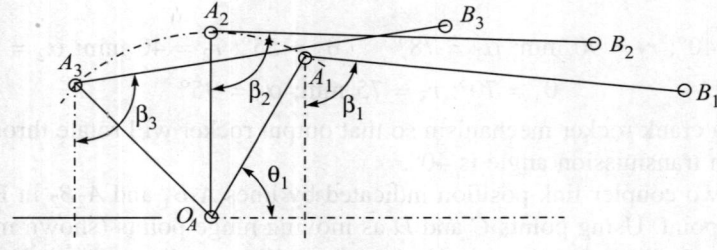

Fig. 6.33

Chapter 7

CAMS

7.1 INTRODUCTION

A cam is a higher paired machine member which is used to impart specified motion, by direct contact, to another machine element, called follower. A cam may also be defined as a convenient device transforming one type of motion into another. Usually the cam either oscillates or rotates while the follower has the motion of either translation or oscillation. The cam has a curved or grooved surface which mates with the follower to impart a specified motion to it.

Cam-follower mechanisms are simple and less expensive. They have fewer moving parts and require a very small working space. Many modern-day automatic machines depend upon cams to provide proper timing of the machine components. Many machines require complicated motion which can be obtained only by means of cam-follower mechanisms. Requirement of complex motion, however, does not complicate the design process for cams. Because of desired features, cam mechanisms are used in printing machines, paper cutting machines, packaging and processing machinery, spinning and weaving textile machinery, and internally expanding shoe-brakes. Cams are also used in automatic screw cutting machines, feed mechanism of automatic lathes, in presses and for operating inlet and exhaust valves of internal combustion engines. Cam mechanisms can also be designed with adjustability, as in duplicating machines. Cam mechanisms often find application where linkages would be suitable. These include instruments, typewriters, computers, and measuring apparatus which often require oscillating motion whose accuracy is not critical.

7.2 COMPARISON BETWEEN CAMS AND LOWER PAIRED MECHANISMS

Like lower paired mechanisms, cam and follower combination may be designed for motion generation (rigid body guidance), path generation or function generation. In majority of applications, however, cam and follower combination is used for function generation. A comparison between the two types of mechanisms may greatly facilitate the type synthesis.

1. Being a higher paired mechanism, wear between cam and follower is greater than that taking place between links connected through lower pairs in mechanisms. As a result of continuous wear, cam-contour is modified until required function is no longer generated.

This may lead to increased value of 'peak-accelerations' and may create excessive inertia forces, which will adversely affect the performance of the machine.

2. In case of force closure, a follower is held in continuous contact with cam through a spring force. Thus, even when there is no load to be driven, there is a fluctuating spring load because of which it may become impossible to connect it to a simple and desirable input. As against this, close-paired/track cam, where the pairing elements are held togather mechanically (see Fig. 7.1), results in increase in the cost of manufacture. Reversals of pressure occur in this type of cam as the roller shifts from one side of the track to the other. This causes accelerated wear and premature failure in certain areas of track. No such problem is anticipated in lower paired mechanism.

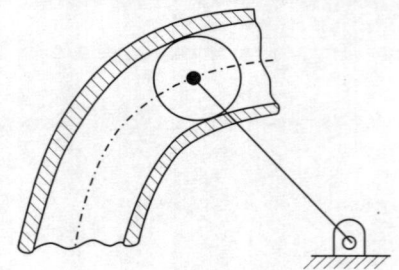

Fig. 7.1 Closed cam pair with track cam and roller.

3. In majority of applications, cam surface remains unprotected from dust and chips. This is in sharp contrast to linkage mechanism where it is much easier to enclose the joint. Linkage mechanisms are also well recognised for smooth and quiet action. They are preferable in applications requiring tight control on noise production.

4. In machines requiring periodic motions, an important aspect is to modify motions produced by various mechanisms. While it is difficult to adjust motion of a cam mechanism, the motion of a linkage mechanism can be modified easily by adjusting link lengths.

5. One of the major plus points of cam mechanism is that it can be designed to generate specified function/motion exactly at every position of the mechanism. Allowance must, of course, be provided in the above statement for deviations on account of manufacturing tolerances and dimensional changes arising out of wear. As against this, a linkage mechanism can generate motion/function which matches exactly with the specified one only at a limited number of precision points.

6. Another plus point of cam mechanisms is that it can produce exact dwell (*i.e.,* stand still position of follower) of finite duration, which is not possible with linkage mechanism.

7. A yet another plus point in favour of a cam mechanism is that unlike cam, the design of linkage is not that easy. Further, in the case of linkage, the design must be reviewed from the point of view of transmission angle. A review of design, from the point of view of space requirement, may also necessiate redesign of cam mechanism.

7.3 CLASSIFICATION OF CAMS AND FOLLOWERS

7.3.1 Classification of Cams

Cams can be classified according to their shapes. Different types of cams are:

1. A plate cam, which is also called a disc cam or a radial cam (see Fig. 7.2a)
2. A wedge cam (see Fig. 7.2b)

3. A cylindrical cam or barrel cam (see Fig. 7.2c)
4. An end or face cam (see Fig. 7.2d)
5. A conical cam (see Fig. 7.2e)
6. A globoidal cam (see Fig. 7.2f).

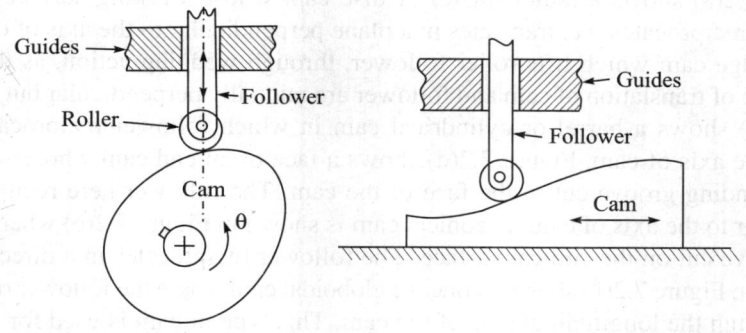

(a) Plate disc or radial cam with translating follower

(b) Wedge cam with translating follower

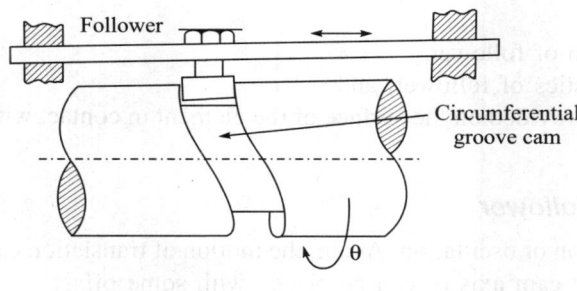

(c) Cylindrical/Barrel cam with translating follower

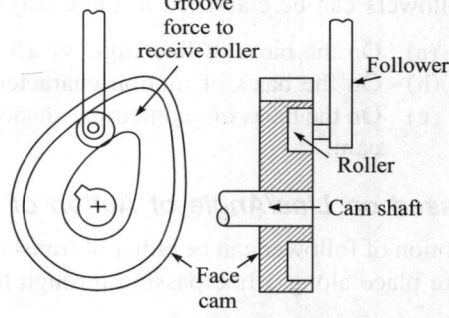

(d) An end or face cam

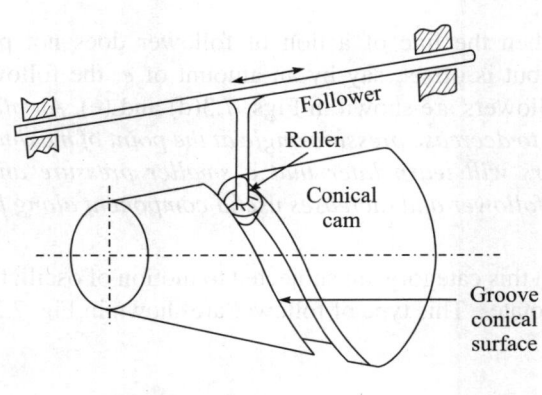

(e) Conical cam

(f) Globoidal concave cam with oscillating follower

Fig. 7.2 Illustration for different types of cams.

The **plate/disc cam** is the one most commonly used, while the least common category is the **wedge cam** as it requires reciprocating rather than continuous rotary motion. In view of this, most of the discussions in this chapter are directed towards the study of plate or disc cam, although the principles involved are equally useful in other types of cams.

Figure 7.2(a) shows a radial (plate) or disc cam whose working surface is so shaped that the follower reciprocates, *i.e.* translates in a plane perpendicular to the axis of cam. Figure 7.2(b) shows a wedge cam which lifts roller follower, through wedging action, as the cam translates. The direction of translation of cam and follower are mutually perpendicular but in the same plane. Figure 7.2(c) shows a barrel or cylindrical cam in which follower reciprocates in a direction parallel to the axis of cam. Figure 7.2(d) shows a face or an end cam whose roller moves along the corresponding groove cut in the face of the cam. The follower here reciprocates in a plane perpendicular to the axis of cam. A conical cam is shown in Figure 7.2(e) where the roller moves along a groove cut on the conical surface. The follower reciprocates in a direction parallel to an end-generator. Figure 7.2(f) shows a concave globoidal cam where the follower oscillates in a plane passing through the longitudinal axis of the cam. This type of cam is used for indexing purpose.

7.3.2 Classification of Followers

Followers can be classified in three ways:

(a) On the basis of line/angle of action of follower,
(b) On the basis of motion characteristics of follower, and
(c) On the basis of geometrical shape provided on the surface of the element in contact with cam.

Based on Line/Angle of Action of Follower

Motion of follower can be either of translation or oscillation. Again, the motion of translation can take place along a line passing through the cam axis or can be placed with some offset.

(i) *Radial Translatory Followers.* When the line of action of follower passes through cam axis of rotation, the follower is called *radial followers*. These followers are shown in Figs. 7.2(a) and 7.3(a, b, c).

(ii) *Offset Translatory Followers.* When the line of action of follower does not pass through the axis of cam rotation but is offset, say by an amount of e, the follower is called *offset follower*. These followers are shown in Figs. 7.3(d) and (e). *An offset is usually provided on a side so as to decrease pressure angle at the point of maximum velocity during outstroke. Readers will learn later that a smaller pressure angle decreases side thrust in guides of follower and increases useful component along line of stroke of follower.*

(iii) *Oscillatory Follower.* Followers in this category are subjected to motion of oscillation about some fixed axis as the cam rotates. This type of follower are shown in Fig. 7.2(f) and 7.3(g, h).

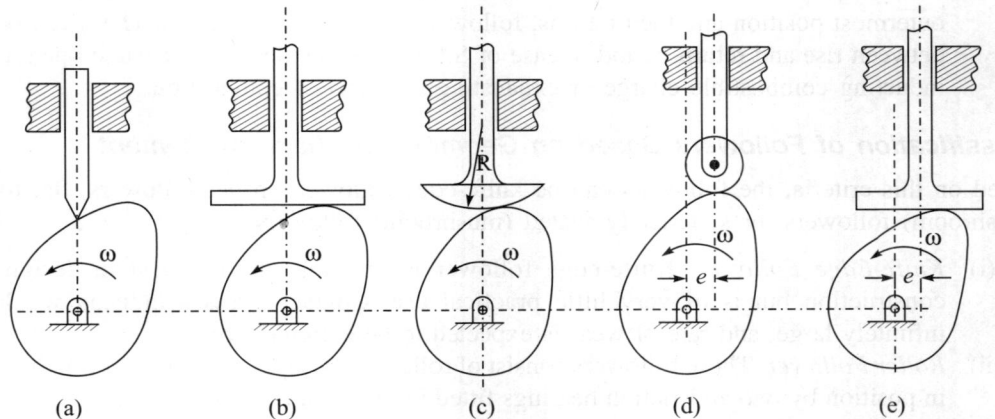

(a) Translating knife edge follower; without offset (*i.e.* radial) (b) Translating flat-footed mushroom follower, without off-set (c) Translating spherically seated mushroom follower, without offset (d) Translating roller follower with offset equal to *e* (e) Translating flat-footed mushroom follower with offset equal to *e*

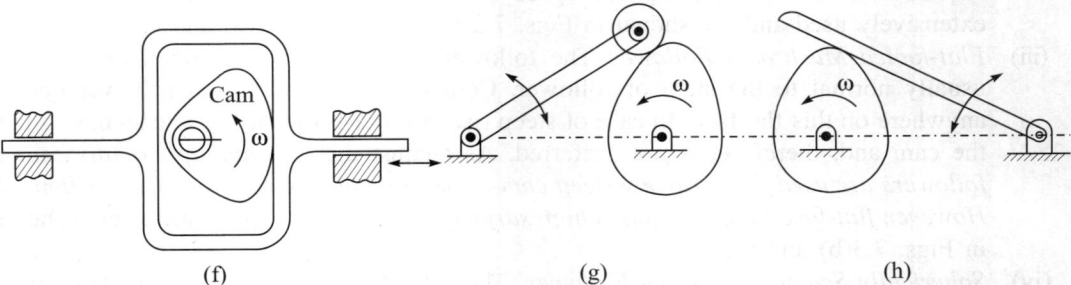

(f) Translating positive acting cam (spring not required) (g) Oscillating roller follower
(h) Oscillating flat-footed mushroom follower

Fig. 7.3 Types of follower motions.

Classification Based on Motion Sequence

During the motion of followers there are three states

Rise period (designated as *R*), when follower rises from inner to outermost position; Return period (again designated as *R*) when the follower returns from outermost to inner most position; and Dwell period (designated as '*D*') when follower remains standstill in a given position. Current classification of followers is based on the sequence of these states of motion.

(i) *Rise-Return-Rise* (R-R-R). Here the follower undergoes rise, return, and rise motion in sequence. There is no dwell period. This is rarely used in industry.

(ii) *Dwell-Rise-Return-Dwell* (D-R-R-D). This type of follower is more commonly used in industry. The follower starts from dwell, rises to the outermost position, return to the innermost position and then follows dwell period again.

(iii) *Dwell-Rise-Dwell-Return-Dwell* (D-R-D-R-D). This type of follower motion is very widely used in industry. Starting with dwell condition, follower rises, dwells in the outermost position and then returns, followed by dwell period again. The dwell period between rise and return period in case of S.I. engine decides time of valve opening for inducting combustible charge or exhausting products of combustion.

Classification of Followers Based on Geometrical Shape of Element

Based on this criteria, the followers can be knife-edge followers, roller followers, flat footed (mushroom) followers, or spherically seated (mushroom) followers.

(i) *Knife-Edge Follower.* Knife-edge follower is probably the simplest in design and construction but is of very little practical use. Contact stresses induced would be infinitely large, and rate of wear is expected to be extremely high.

(ii) *Roller Follower.* These followers consist of roller freely supported on a pin which is held in position by two antifriction bearings fitted in two concentric holes in the forks of the follower-stem. Rolling element replaces sliding friction of knife-edge by rolling friction. It must be noted, however, that the sliding is not entirely eliminated, since the inertia of the roller prevents it from responding instantaneously to the changes of angular velocity required by the varying peripheral speed of the cam. This type of followers are very extensively used and are shown in Figs. 7.2(a) and also in 7.3(d) and (g).

(iii) *Flat-footed Mushroom Follower.* The follower is provided with a flat face which is usually normal to the stem of follower. Contact between cam and follower occurs anywhere on this flat face. In case of steep rise, a roller-follower has a tendency to jam the cam and, therefore, is not preferred. As against this *flat and spherically seated followers are used for relatively steep cam-curves and also*, where *the space is limited. However, flat-face follower causes high surface stresses.* This type of follower is shown in Figs. 7.3(b) and (e).

(iv) *Spherically Seated Mushroom Follower.* The follower in this case is provided with a spherical surface of suitable radius, with centre located on the centre-line of follower. This type of follower is shown in Fig. 7.3(c). As stated above, this type of follower is used for relatively steep cam-curves and is useful where space may not be adequate. *Unlike flat-face follower, the spherical shape here minimises the surface stresses.* It is important to note that in view of limited space and due to weakness of pin of roller (in roller follower), automobile engines use spherically seated mushroom follower. In stationary gas and oil engines, where more space is available, roller follower is preferred.

Both knife-edge and roller-followers produce lot of side thrust between the stem of follower and guides. *As against this, on account of zero pressure-angle, the flat-footed mushroom follower produces side thrust in guides only on account of friction between the contacting surfaces.* In the case of flat-footed follower, wear may be diminished by off-setting the axis of the follower as shown in Fig. 7.4. Such an arrangement causes follower also to rotate about its own axis as the cam rotates.

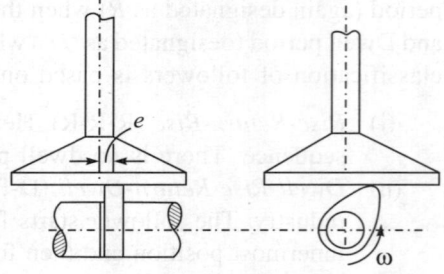

Fig. 7.4

7.4 TERMINOLOGY FOR RADIAL CAM

To understand different terms for Radial cam (Plate/Disc Cam), let us consider the inversion by assuming cam to be fixed and follower to be moving around the cam with same angular velocity co in the opposite sense. In Fig. 7.5, the roller follower is shown moved around the cam in c.c.w. sense so as to occupy positions 1–1, 2–2, 3–3 and 4–4. Note that each one of the above four follower centre lines, pass through the cam axis of rotation. With roller centres corresponding to the above four positions of follower, located at O_1, O_2, O_3 and O_4, let P_1, P_2, P_3 be the respective points of contact of rollers with cam-curve. Then, clearly, *line joining roller centre to the corresponding point of contact with cam denotes the common normal to the roller surface and cam surface*. These common normals for the four positions of rollers, as above, are shown as n_1–n_1, n_2–n_2 and n_3–n_3.

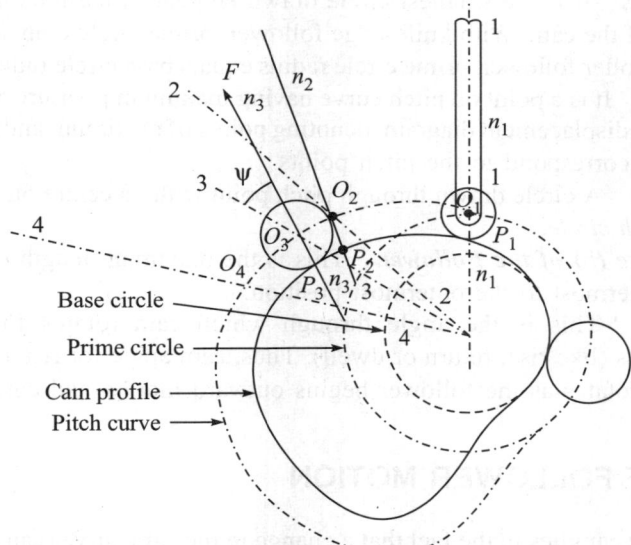

Fig. 7.5 Radial cam with roller follower.

Definitions (*Refer to* Fig. 7.5)

1. **Cam Profile.** Working surface of cam which is physically in contact with follower is called the cam profile. In two-dimensional space, this is represented by the curve which gives outline of the cam.

2. **Base Circle.** Base circle is the smallest circle that can be described from cam axis and tangent to cam profile. Condition of tangency requires that the base circle should merely touch cam profile, but should not cut it. Base circle diameter decides the size of the cam.

3. **Trace Point.** Trace point may be defined as the reference point on the follower and is very useful in generating pitch curve. For knife-edge follower, the knife-edge itself is the reference point; while with roller follower it is represented by the centre of the roller. For flat-faced follower, trace point is the point of contact at the time when follower is in contact with cam on base circle.

4. *Pressure Angle.* For a given position of roller-follower, let its line of action be denoted as 3-3, for which the roller centre is at O_3 and the point of contact with earn profile is at P_3. Then O_3P_3, being the radius of roller, also represents common normal to the cam profile at P_3. The angle ψ included between the follower line of action 3-3 and common normal O_3P_3 is defined as the pressure angle.

Since direct contact is involved, force F from cam to the follower is transmitted along common normal and it follows that only $F \cos\psi$ component is available along follower line of action for driving follower (*see* Fig. 7.5). The other component $F \sin\psi$ produces side thrust in the guides of follower which decides force of friction.

5. *Pitch Curve.* When the cam is held fixed and the follower is moved around the cam in opposite sense for getting inversion, the locus of the trace point is called the *pitch curve*. For knife-edge follower, the pitch curve coincides with the cam profile.

6. *Prime Circle.* It is the smallest circle drawn tangent to the pitch curve from the axis of rotation of the cam. With knife-edge follower, prime circle coincides with base circle while, with roller follower, prime circle radius equals base circle radius plus roller radius.

7. *Pitch Point.* It is a point on pitch curve having maximum pressure angle. The inflection points on the displacement diagram, denoting points of maximum and minimum velocities of follower, correspond to the pitch points.

8. *Pitch Circle.* A circle drawn through pitch point with its centre on cam axis of rotation is called *pitch circle*.

9. *Lift or Stroke (h) of the Follower.* This is the maximum length of travel of follower from the innermost to the outermost position.

10. *Cam Angle.* This is the angle through which cam rotates for specific follower displacements (like rise, return or dwell). Thus, cam angle for rise means angle through which cam rotates as the follower begins outward motion and completes it.

7.5 TYPES OF FOLLOWER MOTION

The versatility of the cam lies in the fact that a change in the cam curve (cam profile) also results in a corresponding change in the characteristics of the follower motion. Thus the number of follower motions possible is indeed very large. *But though a cam can be designed to impart nearly any motion to follower, some of the follower motions produce extreme values of velocity, acceleration, and jerk (which is the time derivative of acceleration), which result in high stresses and vibrations in the associated components.* Satisfactory kinematic design of cams therefore aims at compromising upon a follower motion which will have relatively mild values of velocity, acceleration and jerk.

Most common (standard) follower motions are:

1. Uniform motion,
2. Modified uniform motion,
3. Simple harmonic motion,
4. Parabolic or constant acceleration\deceleration motion, and
5. Cycloidal motion.

7.5.1 Uniform Motion or Constant Velocity Follower Motion

This type of follower motion requires that follower displacement y is linearly proportional to the cam angle θ of rotation. Mathematically,

$$y \propto \theta \quad \text{or} \quad y = C\theta \tag{7.1}$$

where C is the constant of proportionality and θ is the cam angle of rotation in radians.

Let, θ_0 = outstroke angle of cam rotation   and h = stroke length of follower.

Then, from equation (7.1),

$$h = C\theta_0 \quad \text{or,} \quad C = (h/\theta_0)$$

From equation (7.1), therefore,

$$y = \left(\frac{h}{\theta_0}\right)\theta \tag{7.2}$$

For velocity differentiating w.r. to time on both sides of equation (7.2),

$$(dy/dl) = V = \left(\frac{h}{\theta_0}\right)\frac{d\theta}{dt}$$

Therefore
$$V = (h/\theta_0)\omega \tag{7.3}$$

where ω = uniform speed of cam rotation. It follows from equation (7.3) that the velocity of follower remains constant during outstroke. Proceeding in a similar way, it can be shown that velocity of follower during return stroke also remains constant.

From equation (7.1), it follows that the displacement diagram, *i.e.* a plot of y v/s θ, is a straight line from origin with slope $(= h/\theta_0)$. This is shown in Fig. 7.6, which shows displacement, velocity and acceleration diagrams for uniform follower motion. It follows from equation (7.3) that velocity remains constant during outstroke and return stroke.

Since the right hand side of equation (7.3) contain constant terms, it follows that higher time derivative of equation (7.3), is zero during outstroke and return stroke. Thus

$$\text{Acceleration } A = \frac{dv}{dt} = 0$$

This, however, does not give true picture. At the beginning of outstroke, a finite velocity is to be obtained within no time and this requires infinite $(+\infty)$ acceleration. Similarly, at the end of the outstroke, a finite velocity is to be brought down to zero in no time. Thus an infinite retardation $(-\infty)$ is required at the end of outstroke. Similarly, a $(-\infty)$ acceleration is required at the beginning of return stroke and $(+\infty)$ acceleration at the end of return stroke. Infinite acceleration/retardation implies that very large inertia forces $(P = mA)$ will be exerted at the follower bearings and also at the contact point on the cam surface.

7.5.2 Modified Uniform Motion

The uniform motion displacement curve is modified so as to eliminate shock loads at the beginning and end of outstroke and return stroke. A convenient method is to use circular arcs of radius equal to the lift of follower at the beginning and end of outstrokes and return strokes. This is shown for outstroke in Fig. 7.7. Comparing the velocity diagram of Fig. 7.6 with that in Fig. 7.7, the difference is at once clear. In earlier case a finite value of velocity V was required to be achieved in no time. In the present case, the finite value of velocity is to be obtained in time corresponding to cam angle θ' (see Fig. 7.7). Thus, the acceleration involved in the present case is not infinite.

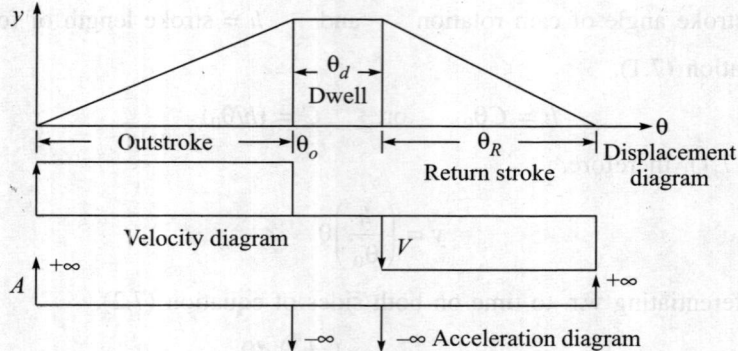

Fig. 7.6 Uniform follower motion.

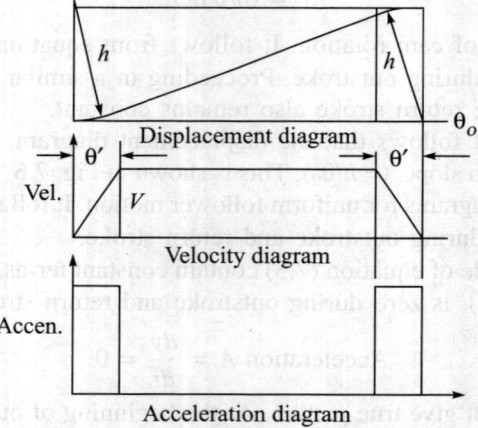

Fig. 7.7 Modified uniform motion.

However, even this modification is not considered adequate, because the accelerations and retardations at the beginning and end of outstroke and return stroke are still quite high.

7.5.3 Simple Harmonic Motion

This follower motion is widely used in designing cam for specific follower motions. It is simple to construct and mark a definite improvement over the two previous motions. Simple harmonic motion is recommended for follower motion when cam rotates at moderate speeds.

Graphically, as shown in Fig. 7.8(a), simple harmonic motion is obtained by constructing a semicircle on the displacement axis of diameter equal to h, the stroke-length. The semicircle is divided into as many equal parts as are used to divide outstroke angle θ_0 and return stroke angle θ_R along θ-axis. Intersection of horizontal projectors from points 1, 2, 3, 4, 5, and 6 with vertical projectors from points 1, 2, 3, 4, 5, and 6 of cam angles θ_0 and θ_R then give the points on simple harmonic displacement diagram. S.H.M. curve is then obtained by joining all these points in correct sequence by a smooth curve.

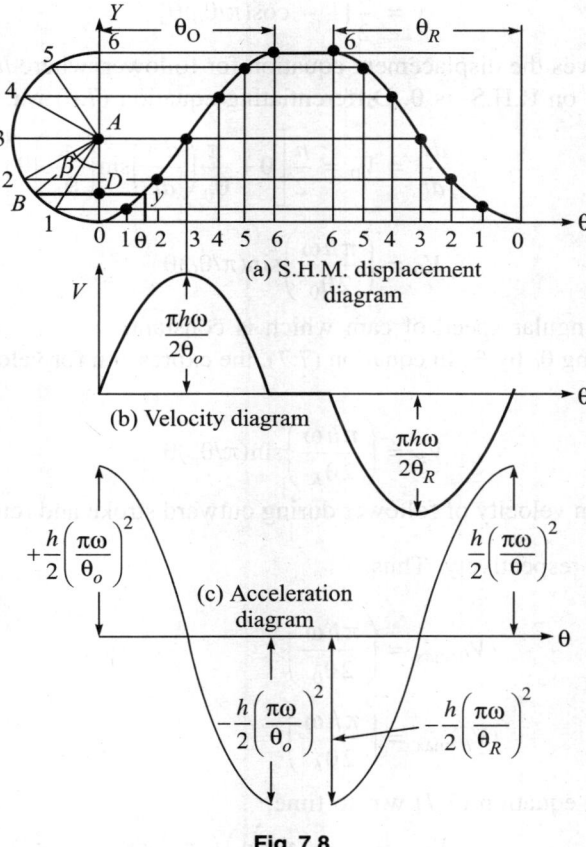

(a) S.H.M. displacement diagram

(b) Velocity diagram

(c) Acceleration diagram

Fig. 7.8

To derive the displacement, velocity and acceleration equation for simple harmonic displacement diagram, let AB be a radius vector at an angle β from vertical diameter AO in Fig. 7.8(a). Further, let BD be the perpendicular dropped from B on vertical diameter. Then, the follower displacement during outstroke.

$$y = OD = AO - AD$$

or

$$y = \frac{h}{2} - AB\cos\beta = \frac{h}{2}(1 - \cos\beta) \qquad (7.4)$$

Now, as the radius AB rotates through π radians, the outstroke is complete and, during this period, cam turns through outstroke angle θ_0. Hence, the angle β can be expressed in terms corresponding cam angle θ as

$$\beta/\pi = \theta/\theta_0$$

Therefore

$$\beta = (\pi/\theta_0)\theta \qquad (7.5)$$

Substituting for β in equation (7.4),

$$y = \frac{h}{2}[1 - \cos(\pi/\theta_0)\theta] \qquad (7.6)$$

Equation (7.6) gives the displacement equation for follower, where h, π and θ_0 are constant and the only variable on R.H.S. is θ. Differentiating equation (7.6) w.r. to time.

$$\frac{dy}{dt} = V_0 = \frac{h}{2}\left[0 + \frac{\pi}{\theta_0}\left(\frac{d\theta}{dt}\right)\sin\left(\frac{\pi}{\theta_0}\right)\theta\right]$$

Therefore

$$V_0 = \left(\frac{\pi h \omega}{2\theta_0}\right)\sin(\pi/\theta_0)\theta \qquad (7.7)$$

where $d\theta/dt \equiv \omega$ = angular speed of cam which is constant.

Similarly, replacing θ_0 by θ_R in equation (7.7), the expression for velocity of follower during return stroke is

$$V_R = \left(\frac{\pi h \omega}{2\theta_R}\right)\sin(\pi/\theta_R)\theta \qquad (7.8)$$

Clearly, maximum velocity of follower during outward stroke and return stroke occurs when $\theta = \dfrac{\theta_0}{2}$ and $\theta = \dfrac{\theta_R}{2}$ respectively. Thus,

$$(V_0)_{max} = \left(\frac{\pi h \omega}{2\theta_0}\right) \qquad (7.9)$$

and

$$(V_R)_{max} = \left(\frac{\pi h \omega}{2\theta_R}\right) \qquad (7.10)$$

Again, differentiating equation (7.7) w.r. to time,

$$\frac{dV_0}{dt} = A_0 = \left(\frac{\pi h \omega}{2\theta_0}\right)\left(\frac{\pi}{\theta_0}\right)\frac{d\theta}{dt}\cos\left(\frac{\pi}{\theta_0}\right)\theta$$

or

$$A_0 = \frac{h}{2}\left(\frac{\pi \omega}{\theta_0}\right)^2\cos(\pi/\theta_0)\theta \qquad (7.11)$$

The acceleration equation for return stroke is obtained by replacing θ_0 by θ_R in equation (7.11). Thus,

$$A_R = \frac{h}{2}\left(\frac{\pi \omega}{\theta_R}\right)^2\cos(\pi/\theta_R)\theta \qquad (7.12)$$

maximum value of accelerations during outstroke and return strokes are obtained when $\theta = 0$ or θ_0 in equation (7.11) and $\theta = 0$ or θ_R in equation (7.12). Thus, maximum acceleration during outstroke and return strokes are

$$(A_0)_{max} = \pm \frac{h}{2} \left(\frac{\pi \omega}{\theta_0} \right)^2 \qquad (7.13)$$

and

$$(A_R)_{max} = \pm \frac{h}{2} \left(\frac{\pi \omega}{\theta_R} \right)^2 \qquad (7.14)$$

The displacement, velocity and acceleration diagrams for the S.H.M. are shown in Fig. 7.8(a), (b) and (c). Remembering that,

$$\frac{dy}{dt} = \frac{dy}{d\theta} \frac{d\theta}{dt},$$

it follows that the slope of the displacement diagram $(dy/d\theta)$ is given by $\left(\dfrac{dy}{dt} \right)/\omega$, and since most of the plate cams are driven at constant speeds ω, it follows that the velocity diagram of Fig. 7.8 also represents slope of the displacement diagram to a scale $(1/\omega)$. Clearly, maximum slope of displacement diagram occurs at $\theta = (\theta_0/2)$ and also at the middle point of return stroke. These are the points of inflection for the curve, where displacement diagram is the steepest. Readers should note that steepness aspect of cam contour is an important aspect in kinematic design of cam.

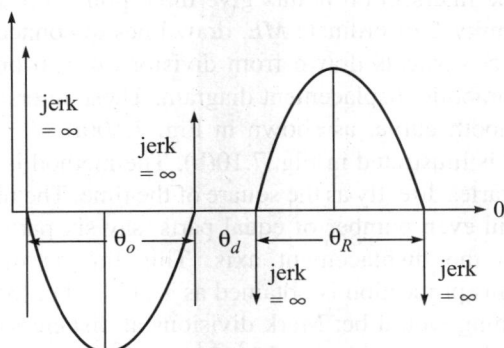

Fig. 7.9 Jerk diagram for simple harmonic motion.

Differentiating equation (7.11) further,

$$\text{jerk} = \frac{dA_0}{dt} = -\frac{h}{2} \left(\frac{\pi \omega}{\theta_0} \right)^3 \sin \left(\frac{\pi}{\theta_0} \right) \theta \qquad (7.15)$$

The jerk at the beginning and end of each of the outstroke and return stroke is thus infinite, because a finite value of acceleration is to be generated in no time (Ref. Fig. 7.9). Hence, the jerk $\left(= \dfrac{dA}{dt} \right)$ is infinite at the beginning and end of the outstroke/return stroke. It follows from the relation

$$F = mA$$

that,

$$\left(\frac{dF}{dt}\right) = m\left(\frac{dA}{dt}\right) = m \times \text{jerk}$$

An infinite jerk therefore indicates a very high rate of force variation with respect to time. Cams with S.H.M. for followers are therefore recommended to be operated at low to moderate speeds, as it denotes dynamic loads of large magnitude.

7.5.4 Uniformly Accelerated and Retarded (Parabolic) Follower Motion

In this type of follower motion, first half of the outstroke is completed with uniform acceleration and the latter half with uniform deceleration. Similarly the first half of return stroke is completed with uniform deceleration and the latter half with uniform acceleration. *Note that actually these are accelerations and decelerations in negative directions of displacements, during return stroke.*

The displacement diagram for parabolic follower motion can be constructed using two methods. In the first instance, *refer to* Fig. 7.10(a). Let lengths *OM, MN* and *NQ* represent respectively the cam angles for rise, dwell and return periods and let ordinate *OB* represent lift or stroke *h* of the follower. Complete rectangle with sides *OB* and *OQ*. Let a line *LD* be drawn perpendicular to *OM* at its mid-point and let *LD = h*. Let *OM* be divided into, say six number of equal parts. Mark the divisions as 0, 1, 2, 3, 4, 5 and 6. Similarly, also divide *LD* into same number of equal parts and mark the divisions as 0, 1, 2, 3, 4, 5 and 6 as shown. Join the origin *O* to divisions 1, 2 and 3 on *LD* by straight lines to cut verticals from divisions 1, 2, 3 respectively on θ axis. The intersection points give three points on the displacement diagram. Also, from the other extremity *E* of ordinate *ME*, draw lines to connect points 4, 5 and 6 on *LD* and let these lines intersect verticals drawn from divisions 4, 5, 6 on θ axis. The intersection points are the points on parabolic displacement diagram. These intersection points are joined in correct sequence by a smooth curve, as shown in Fig. 7.10(a).

A yet another method is illustrated in Fig. 7.10(b). The method is based on the fact that the displacement of follower varies directly as the square of the time. The method consists in dividing the angle of outstroke θ_0 in even number of equal parts, say six parts. Next, a line is drawn at any convenient angle to the displacement axis. This line is divided in proportions of 1, 3, 5, 5, 3, 1. Basically, this proportion is obtained as $1 : (2^2 - 1) : (3^2 - 3 - 1)$, etc. An another way of stating the same thing would be: Mark divisions at distances of $1^2, 2^2, 3^2$ as measured from *O* and other three points at distances of $1^2, 2^3$ and 3^2 as measured from other reference point *O'* on this line.

In short, the line *OO'* should be drawn 18 units long and should be divided in proportion 1 : 3 : 5 : 5 : 3 : 1. The last division *O'* on this line is to be joined to point *B* by a straight line and from all divisions on *OO'* lines are to be drawn parallel to *O'B*. Intersection of these lines with the displacement axis gives points 1, 2, 3, 4, 5 and 6. Intersection of lines, drawn from these points parallel to θ-axis, with verticals drawn from divisions 1, 2, 3, 4, 5 and 6 on θ-axis, gives points on parabolic displacement diagram. This is illustrated in Fig. 7.10(b).

Equation: Two parabolas are required meeting at an inflection point to describe the displacement diagram during outstroke. Thus, an attempt is made to establish displacement equation for both the halves. Let us choose general equation of parabola for the first half of outstroke, as

$$y = a\theta^2 + b\theta + c \tag{7.16}$$

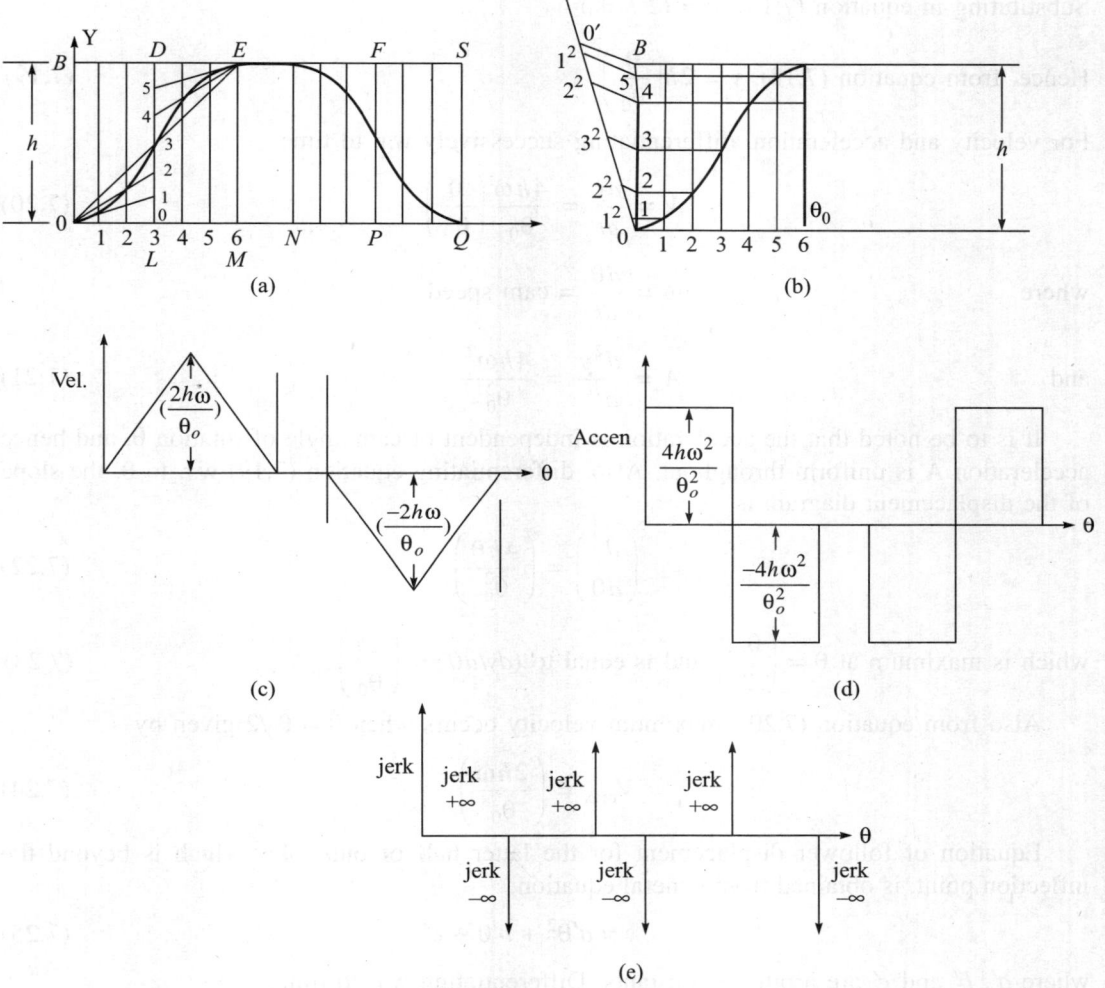

Fig. 7.10 Motion characteristics from parabolic follows motion.

a, b, c are the arbitrary constants.
Differentiating w.r. to θ,

$$\frac{dy}{d\theta} = 2a\theta + b \tag{7.17}$$

Applying initial conditions, $y = \dfrac{dy}{d\theta} = 0$ at $\theta = 0$, We have $b = c = 0$

Hence, the equation of parabolic motion is:

$$y = a\theta^2 \tag{7.18}$$

Again, at the point of inflection, where

$$\theta = (\theta_0/2), \text{ we require } y = h/2$$

Substituting in equation (7.18), $a = (2 h/\theta_0^2)$

Hence, from equation (7.18), $y = 2h\left(\dfrac{\theta}{\theta_0}\right)^2$ (7.19)

For velocity and acceleration, differentiating successively w.r. to time,

$$V = \frac{dy}{dt} = \frac{4h\omega}{\theta_0}\left(\frac{\theta}{\theta_0}\right)$$ (7.20)

where $\omega = \dfrac{d\theta}{dt} = $ cam speed

and $A = \dfrac{d^2y}{dt^2} = \dfrac{4h\omega^2}{\theta_0^2}$ (7.21)

It is to be noted that the acceleration is independent of cam angle of rotation θ, and hence acceleration **A** is uniform throughout. Also, differentiating equation (7.19) w.r. to θ, the slope of the displacement diagram is

$$\left(\frac{dy}{d\theta}\right) = \left(\frac{4h\theta}{\theta_0^2}\right)$$ (7.22)

which is maximum at $\theta = \left(\dfrac{\theta_0}{2}\right)$ and is equal to $(dy/d\theta) = \left(\dfrac{2h}{\theta_0}\right)$ (7.23)

Also from equation (7.20), maximum velocity occurs when $\theta = \theta_0/2$ given by

$$V_{max} = \left(\frac{2h\omega}{\theta_0}\right)$$ (7.24)

Equation of follower displacement for the latter half of outstroke, which is beyond the inflection point, is obtained from general equation:

$$y = a'\theta^2 + b'\theta + c'$$ (7.25)

where a', b' and c' are arbitrary constants. Differentiating w.r. to time,

$$V = (2a'\theta + b')\omega$$ (7.26)

where $\omega = \dfrac{d\theta}{dt} = $ cam speed which is assumed constant

Applying end conditions that at $\theta = \theta_0$, $V = 0$ and at $\theta = \dfrac{\theta_0}{2}$, $V = \left(\dfrac{2h\omega}{\theta_0}\right)$, i.e. same as in first half of outstroke.

Thus, for $V = 0$, we have $\qquad 0 = (2a'\theta_0 + b')\omega$ (7.27)

and, at $\theta = \dfrac{\theta_0}{2}$, $\qquad \left(\dfrac{2h\omega}{\theta_0}\right) = (a'\theta_0 + b')\omega$

Cancelling ω on either side of this equation, $\quad a'\theta_0 + b' = (2 h/\theta_0)$ (7.28)

Also, from equation (7.27) as $\omega \neq 0$ at all times,

$$2a'\theta_0 + b' = 0 \tag{7.29}$$

Solving equation (7.28) and (7.29) simultaneously,

$$a' = -2(h/\theta_0^2) \qquad \text{and} \qquad b' = (4 h/\theta_0)$$

Hence, from equation (7.25),

$$y = -\left(\frac{2h}{\theta_0^2}\right)\theta^2 + \left(\frac{4h}{\theta_0}\right)\theta + c' \tag{7.30}$$

Applying the end condition $y = h$ at $\theta = \theta_0$,

$$h = -\left(\frac{2h}{\theta_0^2}\right)\theta_0^2 + \left(\frac{4h}{\theta_0}\right)\theta_0 + c'$$

or

$$c' = h + 2h - 4h = -h$$

Substituting for c' in equation (7.30), the displacement equation is

$$y = h[-1 + (4/\theta_0)\theta - (2/\theta_0^2)\theta^2]$$

$$= h\left[+1 - 2\left(1 - \frac{\theta}{\theta_0}\right)^2\right] \tag{7.31}$$

Differentiating equation (7.31) w.r. to time for velocity,

$$V = \frac{dy}{dt} = h\left[0 - 4\left(1 - \frac{\theta}{\theta_0}\right)\left(-\frac{1}{\theta_0}\frac{d\theta}{dt}\right)\right]$$

$$= \frac{4h\omega}{\theta_0}\left(1 - \frac{\theta}{\theta_0}\right) \tag{7.32}$$

Differentiating further w.r. to time,

$$A = \frac{dy}{dt} = \frac{4h\omega}{\theta_0}\left(0 - \frac{1}{\theta_0}\right)\frac{d\theta}{dt}$$

$$A = \left(\frac{-4h\omega^2}{\theta_0^2}\right) \tag{7.33}$$

Thus, acceleration is uniform and independent of time. The velocity and acceleration diagrams are shown in Figs. 7.10(c) and (d). Further, since right-hand side of equations (7.21) and (7.33) do not involve cam angle θ, it follows that the next time derivative of acceleration,

$$\text{jerk} = \frac{dA}{dt} = 0$$

Thus, the first impression carried about parabolic follower motion is that probably, it is the best follower motion. No other displacement curve will produce given velocity, starting from rest, in a given time with so small a 'maximum acceleration'. It is for this reason that parabolic follower motion is erroneously known to be the best cam curve. In reality, in many respects, this is the worst of all cam curves.

A little consideration shows that uniform acceleration itself is a source of weakness. As can be seen by comparing jerk diagram (Fig. 7.10e) with acceleration diagram (Fig. 7.10d), finite accelerations are required to be produced in no time at the beginning and end of each of the outstroke and return stroke. Also, at inflection point complete reversals from acceleration to retardation and from retardation to acceleration is required in no time. Hence infinite amount of jerk is produced at all the above points. This is considered to be a highly objectionable feature of parabolic follower motion. Such a jerk characteristic produces abruptly changing contact stresses at the bearings and on the cam surface and leads to noise, surface wear and eventual failure. Infinite jerk encountered in this type of follower motion also necessitates the use of large spring size. Parabolic motion should, therefore, be used for low and moderate speeds.

7.5.5 Cycloidal Follower Motion

It is clear from the discussion of Parabolic follower motion that it is not enough for a good follower motion to have a uniform velocity diagram. A good follower motion should also have uniform acceleration diagram, avoiding any abrupt change in velocity and acceleration curves at all the points in the displacement diagram. Cycloidal follower motion was intended to meet this requirement.

As the name suggests, the displacement curve is generated from a cycloid. It may be recalled that cycloid is the locus of a point on the periphery of a rolling circle, which rolls without slipping on a straight line called directrix. In the present case, displacement axis (i.e., y-axis) with length equal to the stroke length h, becomes the directrix. In order to traverse this length h in one revolution of rolling circle, the radius r of rolling circle should be (= $h/2\pi$). A point P, on the periphery of rolling circle, is assumed to be located at the origin O and traces a cycloidal curve as shown in Fig. 7.11. The displacement curve is then obtained as a graph of the point's vertical position vs. time.

A convenient approach to draw displacement diagram, however, is to draw the rolling circle only once, with point Q as the centre at the right hand top corner. Divide this circle into even number of parts, say six. The division points should be so located that point Q becomes foot of perpendicular from at least one division on the vertical diameter. Join PQ. Also divide the outstroke angle in the same number of equal parts and erect lines from these points perpendicular to θ axis. From foots of perpendiculars M and N, drawn on vertical diameter from peripheral points of circle, draw lines parallel to PQ to cut the vertical lines from points 1, 2, 3, 4, 5, 6 on θ axis. The intersection points are the points on cycloidal displacement curve. This is shown in Fig. 7.11.

Mathematically, a cycloidal curve is represented by

$$y = h \left[\frac{\theta}{\theta_0} - \frac{1}{2\pi} \sin \left(\frac{2\pi}{\theta_0} \theta \right) \right] \tag{7.34}$$

where h = stroke length of follower, θ = cam angle of rotation and, θ_0 = outstroke angle.

Differentiating equation (7.34) for velocity and acceleration, we have

$$\frac{dy}{dt} = V = h \left[\left(\frac{1}{\theta_0} \right) - \frac{1}{2\pi} \left(\frac{2\pi}{\theta_0} \right) \cos \left(\frac{2\pi}{\theta_0} \theta \right) \right] \frac{d\theta}{dt}$$

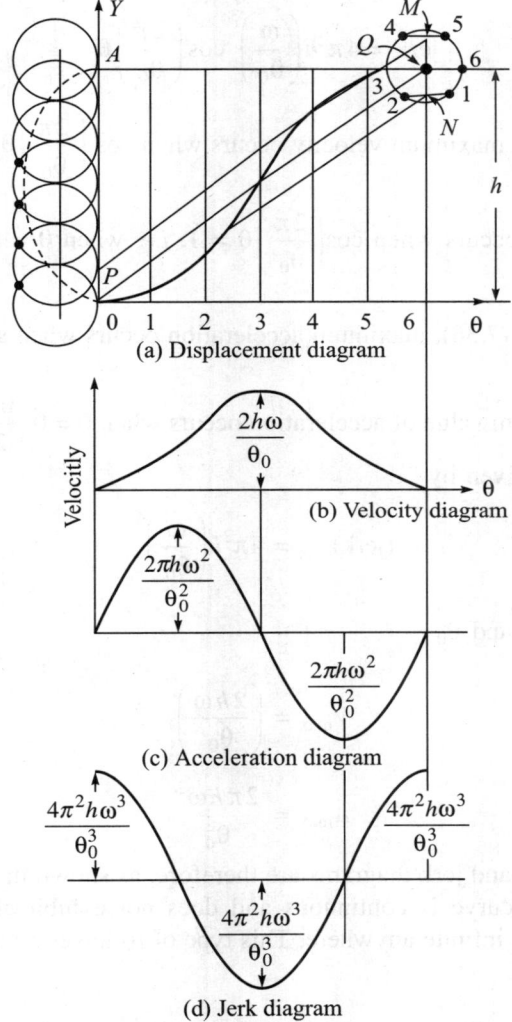

(a) Displacement diagram

(b) Velocity diagram

(c) Acceleration diagram

(d) Jerk diagram

Fig. 7.11 Motion characteristics of cycloidal follower motion.

Therefore

$$V = \frac{h\omega}{\theta_0}\left[1 - \cos\left(\frac{2\pi}{\theta_0}\right)\theta\right] \qquad (7.35)$$

And acceleration

$$A = \frac{dV}{dt} = 2\pi h\left(\frac{\omega}{\theta_0}\right)^2 \sin\left(\frac{2\pi}{\theta_0}\right)\theta \qquad (7.36)$$

Also, jerk is given by

$$\text{jerk} = \frac{dA}{dt} = 2\pi h\left(\frac{\omega}{\theta_0}\right)^2\left(\frac{2\pi}{\theta_0}\right)\frac{d\theta}{dt}\cos\left(\frac{2\pi}{\theta_0}\right)\theta$$

Therefore

$$\text{jerk} = 4\pi^2 h \left(\frac{\omega}{\theta_0}\right)^3 \cos\left(\frac{2\pi}{\theta_0}\right)\theta \tag{7.37}$$

From equation (7.35), maximum velocity occurs when $\cos\left(\frac{2\pi}{\theta_0}\right)\theta = -1$, *i.e.* when $\theta = \frac{\theta_0}{2}$.

Also, minimum velocity occurs when $\cos\left(\frac{2\pi}{\theta_0}\right)\theta = 1$, *i.e.* when $\theta = 0°$ and $\theta_0°$.

Again, from equation (7.36), maximum acceleration occurs when $\sin\left(\frac{2\pi}{\theta_0}\right)\theta = 1$, *i.e.* when $\theta = \frac{\theta_0}{4}$ and $\frac{3\theta_0}{4}$. Minimum value of acceleration occurs when $\theta = 0$, $\frac{\theta_0}{2}$ and θ_0. From equation (7.37) maximum jerk is given by

$$(\text{jerk})_{\text{max}} = 4\pi^2 h \left(\frac{\omega}{\theta_0}\right)^3 \tag{7.38}$$

and occurs at $\theta = 0$, $\frac{\theta_0}{2}$ and θ_0

Thus

$$V_{\text{max}} = \left(\frac{2h\omega}{\theta_0}\right) \tag{7.39}$$

$$A_{\text{max}} = \frac{2\pi h\omega^2}{\theta_0^2} \tag{7.40}$$

Velocity, acceleration and jerk diagrams are therefore, as shown in Fig. 7.11. It is important to note that acceleration curve is continuous and does not exhibit abrupt changes. The jerk therefore does not become infinite anywhere. This type of follower motion is, therefore, equally suitable at high speeds.

7.6 PRESSURE ANGLE (ψ)

As explained in section 2.10, pressure angle ψ (also known as the deviation angle) is the compliment of transmission angle. It was indicated that in mechanisms like cams and gears, it is an index of merit. The pressure angle ψ is defined as an angle included between the line of action of follower and corresponding normal to the pitch curve through trace point. Readers may recall that neglecting friction, force by direct contact is always transmitted from one member to the other along the common normal. It follows therefore that the line of action of contact force F from cam to follower will be inclined at an angle ψ, to the line of action of follower. As in a 4-bar mechanism, the pressure angle varies with cam angle of rotation and is a measure of effectiveness of cam to transfer driving force (effort) to the follower.

As shown in Fig. 7.12, the driving force F from cam, which acts along common normal, is inclined to the line of action of follower at angle 'ψ'. Resolving along the line of action of follower, the driving force on follower is,

$$F_y = F\cos\psi$$

and the component, normal to the line of action of follower is,

$$F_x = F\sin\psi$$

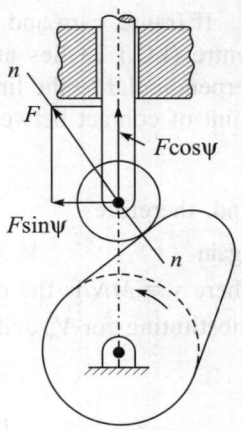

This component, which acts perpendicular to the line of stroke, tends to bend the follower stem and produces friction and wear in guides. When a cam is used to generate a function $y = f(\theta)$, a graph between derivative y' ($\equiv dy/d\theta$) and θ represents the slope of displacement diagram at each value of θ. The derivative $dy/d\theta$ then measures 'steepness' of the displacement diagram and is closely related to the mechanical advantage of the cam mechanism. It will be

Fig. 7.12 Significance of pressure angle.

shown subsequently that the pressure angle is closely related to the slope ($dy/d\theta$), the follower displacement y and the prime/base circle radius. Since $(dy/dt) = (dy/d\theta)(d\theta/dt)$, and for a uniformly rotating cam $\omega = (d\theta/dt)$ is constant, it follows that velocity diagram for a follower motion also represents 'steepness' of the follower displacement diagram to a scale $1/\omega$. Thus the point of maximum velocity, which usually coincides with the inflection point, also corresponds to the point of maximum steepness of displacement diagram.

Clearly, the side thrust on the follower can be reduced by decreasing pressure angle. For satisfactory performance a generally accepted value of maximum permissible pressure angle is $\psi_{max} \le 30°$. In special cases however, where the forces are small and the bearings are accurate, larger values of pressure angle are permissible.

7.7 PARAMETERS AFFECTING PRESSURE ANGLE

In order to establish parameters which decide pressure angle in a cam follower combination, consider a radial cam operating a knife-edge follower, as shown in Fig. 7.13. *It must be noted, however, that by choosing a knife-edge follower, the discussions are not restricted to a single form of follower. A knife-edge follower can be taken to be a special case of roller-follower, in which the radius of the roller is reduced to zero.*

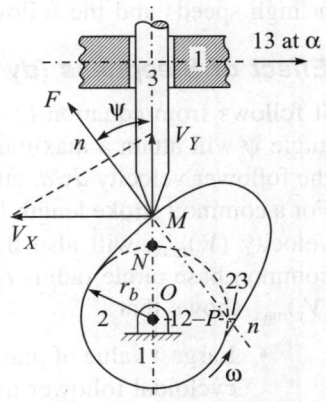

Let r_b be the base-circle radius of cam and y be the follower displacement for the given position. Further, let the cam rotate c.c.w. with an angular speed ω and let $n - n$ denote the direction of common normal at the point of contact. Let V_y and V_x be the velocities of the tip of follower along and perpendicular to the line of stroke respectively. Clearly, the angle made with line of stroke by common normal $n - n$ is the pressure angle 'ψ'.

Fig. 7.13 Radial cam with knife edge follower.

If frame, cam and follower be numbered as links 1, 2 and 3 respectively, the instantaneous centre (I.C.) 23 lies at P, the point of intersection of normal $n - n$ with another line drawn perpendicular to the line of stroke through cam centre O, joining 12 with 13. Thus, if M be the point of contact between cam and follower then, for the given configuration in Fig 7.13,

$$V_y = (OP)\omega \quad \text{and} \quad V_x = (OM)\omega$$

And, therefore

$$(V_y/V_x) = (OP/OM) = \tan\psi \tag{7.41}$$

Again $\quad V_y = \text{Velocity of follower} = (dy/dt)$ and, $V_x = (R_b + y)\omega \tag{7.42}$

where $y \equiv MN$ is the displacement of follower during outstroke

Substituting for V_x and V_y in equation (7.41),

$$\tan\psi = \frac{(dy/dt)}{(r_b + y)\,\omega} \tag{7.43}$$

Again, substituting $\omega(dy/d\theta)$ for (dy/dt) in equation (7.43), we have

$$\tan\psi = \frac{dy/d\theta}{(r_b + y)} \tag{7.44}$$

It follows from equation (7.44) that the pressure angle ψ varies:

- directly with steepness $(dy/d\theta)$ of displacement diagram,
- inversely with the base circle radius r_b, and
- inversely with the follower displacement y.

It follows that pressure angle of a cam goes on changing as the follower displacement changes during rise and return period. Recall that smaller ψ reduces wear & chances of jamming of follower in guides.

Effect of Base Circle radius on ψ

It is clear from equation (7.44) that for a given follower lift y and a given steepness of follower motion, the pressure angle can be reduced by increasing base circle radius r_b. This is, however, not advisable as the resulting large cam requires more space, and it can produce more unbalance at high speeds and the follower is required to traverse a longer path.

Effect of Steepness (dy/dθ) of Displacement Diagram on ψ

It follows from equation (7.44) that for a given base circle radius r_b and constant ω, pressure angle ψ will attain a maximum value when the slope $dy/d\theta$ of displacement diagram and hence the follower velocity dy/dt attains a maximum value and the follower displacement is minimum. For a common stroke length h, it follows that a follower motion having a larger value of follower velocity $(V_y)_{\text{max}}$, will also have a larger pressure angle. For standard follower motions, with common base circle radius r_b and follower lift h, a comparison of maximum follower velocity $(V_y)_{\text{max}}$ shows that

- Largest value of maximum pressure angle (at pitch point) is obtained with parabolic and cycloidal follower motion,
- Intermediate (moderate) value of maximum pressure angle ψ_{max} is obtained with simple harmonic follower motion, and
- Smallest value of maximum pressure angle is obtained with uniform follower motion.

Conversely to offset the effect of different values of $(V_y)_{max}$ and to obtain same value of maximum pressure angle (for a given follower lift), parabolic and cycloidal follower motion require largest size (r_b) of cam while the cam with uniform follower motion requires smallest cam size (r_b). The simple harmonic follower motion requires moderate cam size. Readers should note further that even for a uniform follower motion where velocity is constant, the pressure angle ψ changes with follower displacement.

7.8 EFFECT OF OFFSET FOLLOWER MOTION

Sometimes the line of action of follower is offset so as to clear another part of machine. However, the main reason in opting for an offset follower motion is to reduce the side thrust on the guide. Direction of offset however is very important. This is because *an offset follower arrangement reduces maximum pressure angle in one of the strokes (outward or return) but increases the same in the other stroke. As a rule, when the line of action of follower is offset to the right of cam-shaft centre, the cam must rotate in counterclockwise direction. Similarly, when the offset is provided to the left cam-shaft centre, the cam must rotate in clockwise direction.* This arrangement result in smaller value of maximum pressure angle for outstroke during which larger forces act on the follower.

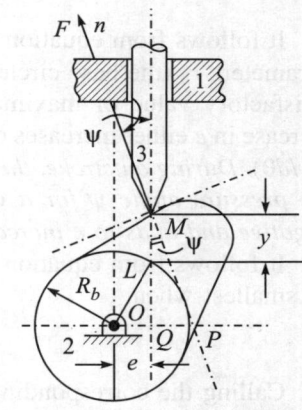

Fig. 7.14 Radial cam with off-set knife edge follower.

Figure 7.14 shows again a radial cam with an offset knife-edge follower. Let the amount of offset be e. Further, let $(V_y)_{max}$ be the maximum velocity of the follower and ψ_{max} be the maximum pressure angle. As in the previous case, the instantaneous centre 23 lies at the point P at the intersection of normal n-n to the cam profile at M and a line from cam centre O drawn perpendicular to the line of stroke. Thus, P being a point common to links 2 and 3, we have

$$V_P = V_y = (OP)\omega$$

which gives

$$\left(\frac{V_y}{\omega}\right) = \frac{dy}{d\theta} = OP \tag{7.45}$$

Again from Fig. 7.14

$$OP = e + QP \tag{7.46}$$

Further, as $\angle PMQ = \psi$, the pressure angle,

$$QP = MQ \ \tan\psi$$

or

$$QP = (LM + QL)\tan\psi = (y + \sqrt{r_b^2 - e^2}\,)\tan\psi \tag{7.47}$$

Substituting for OP and QP from equations (7.45) and (7.47) in equation (7.46),

$$(dy/d\theta) = e + (y + \sqrt{r_b^2 - e^2}\,)\tan\psi$$

Rearranging

$$\tan\psi = \frac{(dy/d\theta) - e}{(\sqrt{r_b^2 - e^2} + y)} \tag{7.48}$$

Multiplying numerator and denominator of equation (7.48) by ω and remembering that $\omega(dy/d\theta) = dy/dt$, we have

$$\tan\psi = \frac{(dy/dt) - e\omega}{(\sqrt{r_b^2 - e^2} + y)\omega} \tag{7.49}$$

When the follower is at pitch point, $\psi = \psi_{max}$ and $dy/dt = (V_y)_{max}$ and therefore,

$$\tan(\psi_{max}) = \frac{(V_y)_{max} - e\omega}{(\sqrt{r_b^2 - e^2} + y)\omega} \tag{7.50}$$

It follows from equation (7.48) that having chosen a displacement equation for a follower, parameters r_b (the base circle radius) and offset amount e may be selected suitably to ensure a satisfactory value of maximum pressure angle. It also follows from equation (7.48) that an increase in e either increases or decreases the magnitude of numerator depending on sign of slope $(dy/d\theta)$. *During outstroke, the slope of $(dy/d\theta)$ is positive and the numerator decreases, reducing the pressure angle ψ for a given value of e. As against this, during return stroke $(dy/d\theta)$ is negative and adds to e increasing pressure angle ψ.*

It follows from equation (7.50) that maximum pressure angle (ψ_{max}) during outstroke will be smallest when,

$$(V_y)_{max} - e\omega \approx 0 \tag{7.51}$$

Calling the corresponding value of offset as the critical value e_c, it follows from equation (7.51) that,

$$e_c \approx (V_y)_{max}/\omega \tag{7.52}$$

Equations (7.50) and (7.52) also suggest that it is possible to reduce pressure angle ψ to zero value, resulting in substantial gain in efficiency of transmission by offsetting follower by an amount $e = e_c$. Practical considerations, however, restrict achievable offset e to a maximum value of 50 per cent of the critical value e_c. Mathematically,

$$e \leq (0.5)e_c \tag{7.53}$$

It may be noted that equations similar to (7.48) and (7.50) can be obtained for roller follower by replacing the term r_b by prime circle radius r_0. Thus, for roller follower,

$$\tan\psi = \frac{(dy/d\theta) - e}{(\sqrt{r_0^2 - e^2} + y)} \tag{7.54}$$

and

$$\tan(\psi_{max}) = \frac{(V_y)_{max} - e\omega}{(\sqrt{r_0^2 - e^2} + y)} \tag{7.55}$$

Reduction in Pressure Angle by Reduction in Steepness dy/dθ

Equation (7.44) suggests that the pressure angle ψ can be decreased for outstroke by decreasing the slope $dy/d\theta$ of the displacement diagram. In other words, steepness of displacement diagram can be decreased to reduce pressure angle ψ. It follows from Fig. 7.15 that any reduction in slope $dy/d\theta$, for same follower lift h requires a larger cam angle. Any step to reduce slope $(dy/d\theta)$ without increasing base circle radius r_b, amount to extending cam angle of rise in the next cam angle. This naturally results in reducing cam angle next in the cycle. In many practical situations this is considered undesirable.

To illustrate this point, consider Fig. 7.15 in which uniform follower motion is assumed for simplicity. It is clear from the figure that any step to reduce pressure angle by reducing steepness of displacement diagram, shifts point Q on displacement diagram to a position Q' increasing cam angle of rise from θ_0 to θ'_0. For a given angle of return and the subsequent dwell angle θ_w, the above step amounts to reducing dwell angle θ_d to θ'_d. In petrol/gas engine, this necessarily implies smaller time for inducting charge, which is undesirable. Instead of decreasing steepness of displacement diagram, in such cases, use of a flat faced follower is a better option.

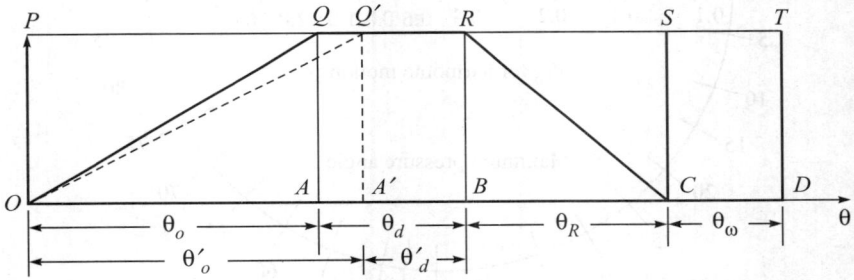

Fig. 7.15 Effect of reducing steepness on pressure angle.

Because common normal to the cam follower interface has to be normal to the flat face of follower, the angle between common normal n-n and follower axis is zero when the flat face is normal to follower axis. Thus, pressure angle for such flat faced translating follower is always zero. It is for this reason that a flat faced follower is used for relatively steep cam curves, and also where space is limited.

Nomogram Relating ψmax to Base Circle Radius

Since a knife-edge follower is a special case of roller follower, with radius of roller reduced to zero, it follows that by replacing r_b by the prime circle radius r_o, equation (7.44) leads to the following equation for roller follower.

$$\tan\psi = \frac{dy/d\theta}{(r_o + y)} \tag{7.56}$$

It follows from equation (7.44) that maximum value of ψ occurs when $(dy/d\theta)$ is maximum and y is minimum. To find out maximum values of ψ, equation (7.44) can be differentiated with respect to θ and equated to zero. Such a procedure then yields values of θ at which maximum and minimum values of ψ are obtained. This is, however, not an easy proposition.

A more convenient approach is to construct nomogram by searching on digital computer ψ_{max} using equation (7.44) for each of the standard follower motions. One such nomogram is shown in Fig. 7.16. With such a nomogram it is possible to obtain maximum pressure angle if the follower lift h and base circle radius r_b are known. A line drawn joining given value of (h/r_b) to 'total-cam angle θ', then give ψ_{max} in the other segment.

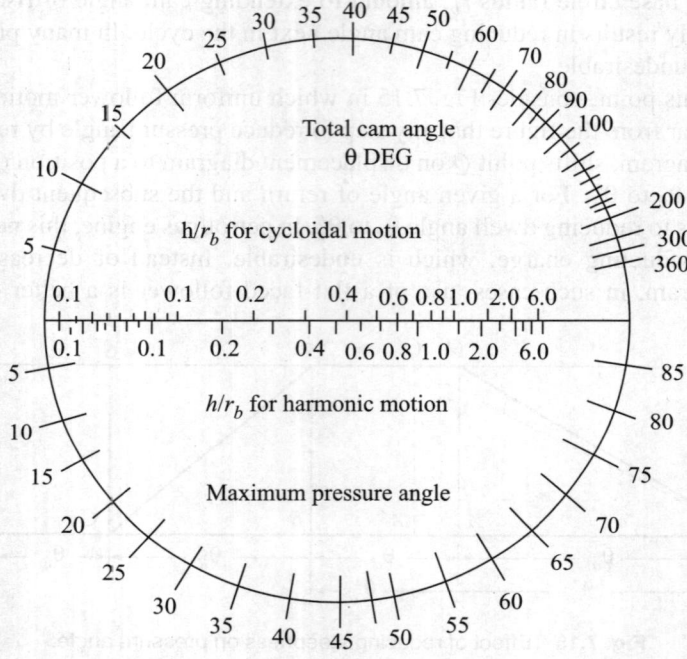

Fig. 7.16 Nomogram relating ψ_{max} to cam angle.

7.9 MAIN CONSIDERATION INFLUENCING CHOICE OF CAM

For lifting follower through a stroke length h, for a given cam angle of rotation, innumerable cam profiles are possible. An answer to this question, namely which one of the cam profile to choose, must be based on the following consideration. An attempt will be made to show that majority of these considerations are related in some way to the derivatives of follower displacement diagram.

These considerations are:
(a) Smaller lateral pressure on guides.
(b) Force required to accelerate the follower should be smaller.
(c) Smooth (jerkless) follower motion and long-life.
(d) Small base circle, consistent with space requirement.

7.9.1 Smaller Lateral Pressure on Guides

For smaller lateral pressure on guides, one must ensure a smaller value for 'maximum-pressure angle' during outstroke. A number of steps to reduce 'maximum pressure angle' during outstroke have been discussed in section (7.7). It is clear from equation (7.44) that pressure angle ψ is not necessarily maximum at inflection point where $dy/d\theta$, and hence the follower velocity, is maximum. To demonstrate the effect of terms in denominator of equation (7.44) on ψ_{max} let us consider an example.

EXAMPLE 7.1 Find out ψ_{max} during outstroke for radial cam of $r_b = 5$ cm, $h = 3$ cm, driving a radial follower with S.H.M. over outstroke angle $0 \leq \theta \leq 90$ degrees.

Solution: For S.H.M., $y = \dfrac{3}{2}\left[1 - \cos\left(\dfrac{\pi}{\pi/2}\right)\theta\right]$

and $dy/d\theta = \dfrac{3\pi}{2\,(\pi/2)}\sin\left(\dfrac{\pi}{\pi/2}\right)\theta$

Thus $y = 1.5(1 - \cos 2\theta)$

and $(dy/d\theta) = 3\sin 2\theta$

Table below lists values *of y, dy/dθ* and ψ for various cam angles θ in the range $35^\circ \leq \theta \leq 45^\circ$ (where ψ_{max} value is likely to occur) using equation (7.44).

Table 7.1   Pressure angle ψ of cam for S.H.M.

S.No.	Cam Angle θ	tan ψ	Pressure Angle ψ	S.No.	Cam Angle θ	tan ψ	Pressure Angle ψ
1	35°	0.47086	25.214°	7	41°	0.4722	25.277°
2	36°	0.47265	25.2980°	8	42°	0.47035	25.190°
3	37°	0.47379	25.3514°	9	43°	0.46795	25.07°
4	38°	0.4743	25.3753°	10	44°	0.4650	24.938°
5	39°	0.4742	25.370°	11	45°	0.4615	24.775°
6	40°	0.4735	25.337°				

clearly, $\psi_{max} \approx 25.3753^\circ$ and occurs at $\theta \approx 38^\circ$ while the inflection point occurs at $\theta = 45^\circ$. **Ans.**

Although point of maximum pressure angle is different from inflection point, and though maximum pressure angle is not equal to $(dy/d\theta)_{max}$, maximum velocity of follower during outstroke can still provide a primitive measure to compare standard follower motions. Thus on the basis of maximum follower velocity, it can be said that, other things remaining the same, uniform motion cam have the smallest value of ψ_{max} while parabolic and cycloidal motion cam has the largest value of ψ_{max}. Simple harmonic motion cam has intermediate value of ψ_{max}.

7.9.2   Smaller Force Required to Accelerate Follower

Required driving force to accelerate follower will be minimum if (a) follower pressure angle ψ_{max} has small value, (b) mass of follower assembly is small, and (c) if, the required 'maximum acceleration is itself small. Factors at (a) and (b) do not require additional comments.

Required acceleration for completing outstroke depends on the type of follower motion. Maximum acceleration values for different (standard) follower motions are as under:

Follower motion		*Maximum acceleration during outstroke*
a.	Uniform follower motion	zero
b.	Simple harmonic motion	$4.935\left(\dfrac{h\omega^2}{\theta_0^2}\right)$
c.	Parabolic follower motion	$4.0\left(\dfrac{h\omega^2}{\theta_0^2}\right)$
d.	Cycloidal follower motion	$6.283\left(\dfrac{h\omega^2}{\theta_0^2}\right)$

Thus, leaving aside uniform follower motion, parabolic motion requires smallest value of *maximum acceleration*, while cycloidal motion requires largest value of *maximum acceleration*. Stated in other words, every other thing remaining the same, maximum inertia force required to be overcome in parabolic motion is least, while in cycloidal motion it is maximum. Uniform follower motion is excluded from above discussions for obvious reasons.

7.9.3 Smooth Jerkless Motion

Except for cycloidal follower motion, all other follower motions produce infinite jerk. Simple harmonic motion involves infinite jerk at the beginning and end of outstroke, while parabolic follower motion involves infinite jerk at the beginning, end as well as at the point of inflection during outstroke. Thus, in this respect, parabolic follower motion is inferior to S.H.M. *Besides jerky motion, presence of infinite jerk introduces abruptly changing contact stresses at the bearings and on the cam surface and leads to noise, surface wear and eventual failure. This is especially true for cams running at high speeds.* Needless to say that in cycloidal motion, jerk is not infinite anywhere in the outstroke. However, even in cycloidal motion jerk is present in the third power of cam speed ω. Thus as the speed increases, this component increases at a very fast rate. In general, however, a cycloidal follower motion is preferable to any of the standard follower motions.

7.9.4 Smaller Base-circle

Besides occupying small space, cams with smaller base-circle produce less of unbalance at higher speeds and the follower has a smaller path to follow. Smaller cams, however, tend to have larger pressure angle for the same lift h. As the cam profile is steeper, it tends to bend the follower sideways. It has been shown that for a common value of maximum pressure angle of 30° (at pitch point), circumference of pitch circles required for standard follower motion are:

(a) Uniform follower motion = 1.73 h
(b) Modified uniform motion = 2.27 h
(c) Simple harmonic motion = 2.72 h
(d) Parabolic and cycloidal motion = 3.46 h

Clearly, uniform follower motion requires smallest size of cam while parabolic and cycloidal motion of follower requires largest size of cam.

7.10 RADIUS OF CURVATURE AND UNDERCUTTING

Even if a cam is proportioned for giving satisfactory pressure angle, the follower still may not produce specified follower motion if the pitch curve is too sharp. Knife-edge follower has the advantage in that there is no restriction on the radius of curvature of the cam profile. *With a flat-footed mushroom follower, the cam profile must not have a concave shape. When using a roller follower, the roller must be free to touch all the points on cam profile. This requires that concave portion, occurring anywhere on the cam profile, must have a radius of curvature greater than that of the roller.*

If r_o be the prime circle radius, r_r be the roller radius and s_ρ be the radius of curvature of pitch curve in convex portion, then,

$$s_\rho = \rho + r_r \qquad (7.57)$$

where ρ = radius of curvature of cam profile

Now if roller radius r_r be increased so that $r_r \to s_\rho$, then under this situation $\rho \to 0$. In other words, the cam profile becomes pointed. Figure 7.17(a) envisages such a situation where roller radius is equal to the radius of curvature of pitch curve. A still more interesting situation arises if r_r exceeds the value of s_ρ. In such a case, as shown in Fig. 7.17(b), the cam-curve loops over itself which is impossible. This situation is known as *undercutting*. Presence of undercutting leads

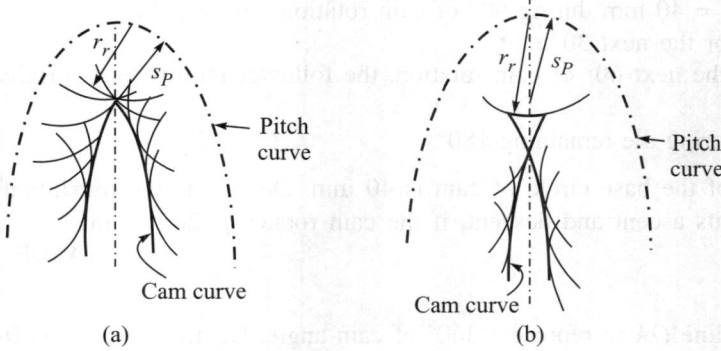

Fig. 7.17   Illustrations depicting conditions (a) $\rho = 0$ and (b) $r_r > s_\rho$.

to deviation of roller centre from pitch curve. Thus, for satisfactory operation (avoiding both the conditions cited),

$$(S_\rho)_{min} > r_r \qquad (7.58)$$

Thus, the minimum radius of curvature of pitch curve must be greater than the roller radius.

7.11 CONSTRUCTION OF CAM PROFILES

Having learnt standard follower motions and their peculiarities, it remains now to generate (*i.e.* lay out) the cam curves based on specified follower motions. *A true cam profile must ensure that every point on it is accessible to the follower. Method of constructing cam profile for a specified follower motion, is based on the principle of inversion.*

Consider a cam rotating with uniform angular speed ω in clockwise direction. In order to obtain the inversion, the cam is held stationary and follower is moved around it with angular speed ω in c.c.w. direction. Further, for ensuring complete inversion the orientation of the line of action of follower with respect to cam and net displacement of follower in respect of cam axis of rotation must be kept same.

As the cam rotates uniformly, equal angular displacements of cam take place in equal time intervals. Thus when the outstroke and return stroke angles are divided into a given number of equal parts, the ordinate at a particular division indicates displacement of trace point with respect to starting position of that cam angle. Hence the general procedure is to obtain displacement diagram for specified follower motion during outstroke angle θ_0, and return stroke angle θ_R with cam angles θ_o and θ_R divided into a fixed number of equal parts. Next, either a base circle (in case of knife-edge, flat-footed and spherically seated follower) or a prime circle (if it is a roller follower) is drawn and follower displacements are plotted on respective divisions beyond base/prime circle, as the case may be. This is illustrated through a number of solved problems below. Readers are advised to divide the outstroke and return stroke angles in six or eight equal parts. This is generally sufficient to produce cam contours with acceptable level of accuracy. Knife-edge follower is basically of academic importance. In view of simplicity of problems however, they are very convenient in illustrating the procedure of cam layout.

EXAMPLE 7.2   Draw the cam profile for a knife-edge follower with the following data:

(i)   Cam lift = 40 mm during 90° of cam rotation with S.H.M.,
(ii)   Dwell for the next 30°,
(iii)   During the next 60° of cam rotation, the follower returns to its original position with S.H.M.,
(iv)   Dwell during the remaining 180°.

The radius of the base circle of cam is 40 mm. Determine the maximum velocity of the follower during its ascent and descent, if the cam rotates at 240 r.p.m.

(AMIE, Summer 1981)

Construction

1.   Draw a line *OA* to represent 360° of cam angle. Let the scale along θ-axis be 1 cm = 30°. Along this line, mark *OP* = 90° to represent outstroke; *PQ* = 30° to represent dwell period and *QR* = 60° to represent return period. Remaining portion *RA* = 180° represents dwell period.

2.   Draw vertical line *OK* = 40 mm (full scale) to represent the stroke length. Construct semicircle on *OK* as diameter. Divide this semicircle in six equal parts and mark the divisions 1, 2, 3, 4, 5, 6 as shown. Divide the length *OP* in six equal parts and number the divisions 1, 2, 3, 4, 5, 6 as shown. Since the return stroke is also executed with S.H.M., divide *QR* also in six equal parts and number them as shown.

3.   From the divisions 1, 2, 3 ......, 6 on semicircle of Fig. 7.18, draw horizontal lines to meet the vertical lines from divisions on *OP* and *QR*. Intersection points of horizontal lines from 1, 2, 3, 4, 5, and 6 with vertical lines from divisions 1, 2, 3, 4, 5, 6 respectively on *OP* and *QR* give the points on displacement diagram shown in Fig. 7.18. Join these points by smooth curves in correct sequence. After drawing the displacement

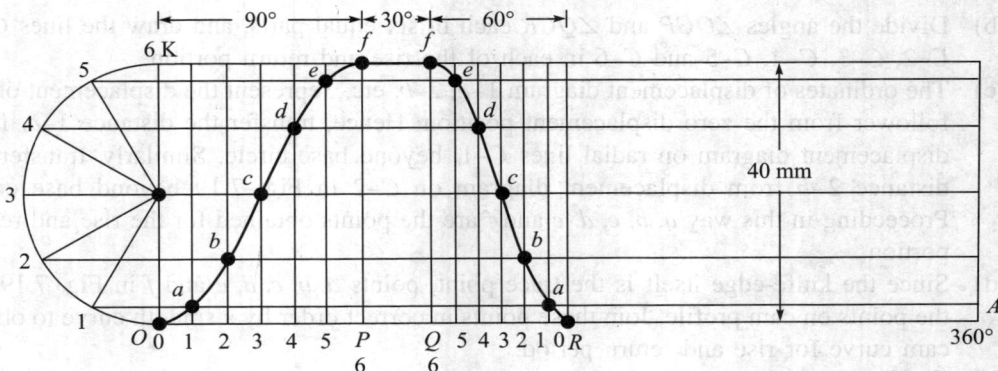

Fig. 7.18 Displacement diagram.

diagram, the cam profile of Fig. 7.19 may be drawn as detailed below. Though the follower is not stated to be a radial one, in the absence of any information about offset, the line of action of follower is assumed to pass through cam-shaft axis. During the dwell period PQ the follower does not move. The displacement diagram during the dwell period is therefore represented by straight line ff.

(a) Draw the base circle of 40 mm diameter with C as the centre. With CO as the assumed initial position of follower, mark angle $\angle OCP = 90°$ for cam angle of rise. Also mark angle $\angle PCQ = 30°$ as angle of dwell and angle $\angle QCR = 60°$ as the angle of return.

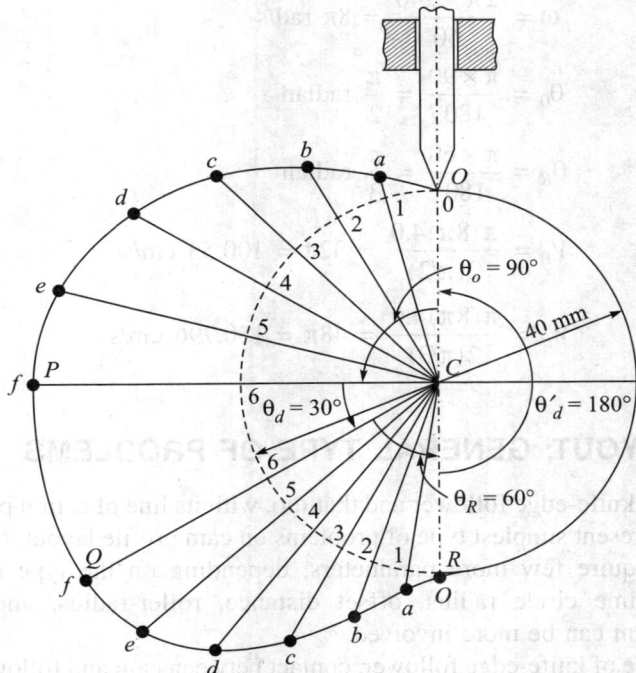

Fig. 7.19 Radial cam with knife edge follower.

(b) Divide the angles $\angle OCP$ and $\angle QCR$ each in six equal parts, and draw the lines C–1, C–2, C–3, C–4, C–5 and C–6 in each of the rise and return portion.

(c) The ordinates of displacement diagram 1–a, 2–b, etc., represent the displacement of the follower from the zero displacement position. Hence, transfer the distance 1–a, from displacement diagram on radial lines C–1, beyond base circle. Similarly, transfer the distance 2–b from displacement diagram on C–2 in Fig. 7.19 beyond base circle Proceeding in this way a, b, c, d, e and f are the points obtained for the rise and return portion.

(d) Since the knife-edge itself is the trace point, points a, b, c, d, e and f in Fig. 7.19 are the points on cam profile. Join these points in correct order by a smooth curve to obtain cam curve for rise and return period.

(e) For the dwell period PCQ, the follower does not move along line of stroke and therefore, cam profile for dwell segment PCQ is given by an arc of circle with Cf as radius. Similarly, for the angle of dwell $\angle RCO = 180°$, the cam profile is given by arc of a circle with CR as radius, (*i.e.* the base circle itself).

Students are advised to ensure that same scale is used for drawing displacement diagram and the cam profile.

For simple harmonic motion, maximum velocity during ascent (Rise) and descent (Return) is given for S.H.M. by

$$V_o = \frac{\pi \omega h}{2\theta_0} \quad \text{and} \quad V_R = \frac{\pi \omega h}{2\theta_R}$$

Now as
$$\omega = \frac{2\pi \times 240}{60} = 8\pi \text{ rad/s}$$

and
$$\theta_0 = \frac{\pi \times 90}{180} = \frac{\pi}{2} \text{ radian}$$

with
$$\theta_R = \frac{\pi \times 60}{180} = \frac{\pi}{3} \text{ radian}$$

Therefore
$$V_0 = \frac{\pi(8\pi)4.0}{2(\pi/2)} = 32\pi = 100.53 \text{ cm/s} \qquad \textbf{Ans.}$$

and,
$$V_R = \frac{\pi(8\pi)4.0}{2(\pi/3)} = 48\pi = 150.796 \text{ cm/s} \qquad \textbf{Ans.}$$

7.12 CAM LAYOUT: GENERAL TYPE OF PROBLEMS

Problems involving knife-edge follower and that too, with its line of action passing through cam-axis of rotation, represent simplest type of problems on cam profile layout. General type of cam-layout problems require few more parameters, depending on the type of problems. These parameters are: prime circle radius, offset distance, roller-radius, and so on. Even the displacement diagram can be more involved.

Again in the case of knife-edge follower, contact between cam and follower must necessarily occur at the knife-edge, which falls on the line of action of the follower. It was therefore possible to obtain points lying on cam profile directly by scaling displacement away from base circle along corresponding lines of action of follower. As against this, with other followers, there is no fixed point on roller or on flat/spherical face where the cam and follower may always have physical contact. It is because of this reason that a reference point (called a trace point) is always needed with other types of followers. *By definition, the trace point location must be related to base circle/prime circle through corresponding follower displacement. For instance, in the case of roller follower, the centre of roller is a trace point. Thus, location of roller centre in any position, with respect to the path of roller centres representing zero displacement (i.e. the prime circle) measures the displacement of follower at the corresponding follower position.* This makes it necessary to draw prime circle and mark displacements of roller centre with respect to the prime circle. In the case of flat-faced follower, the trace point is the point of contact between the cam and flat face of the follower, when the follower is in contact with the base circle portion of cam profile. Here, the emphasis in definition, is on the point of contact when the follower is in the position of zero displacement. For redial cam this position is identical with the point of intersection of flat face of follower and its line of action. Thus any point of contact between cam and follower, at any other position of flat face, can not be taken to be a trace point. In drawing cam profile, for flat-footed follower therefore, amount of follower displacement must be plotted with reference to the zero displacement position of follower.

In both the cases (roller follower and flat-faced follower), since contact between cam and follower may occur anywhere on the follower surface, the best strategy is to locate the trace point for every displaced position; describe the follower surface for each displaced position; and then draw a smooth curve touching each position of the follower surfaces. This is explained through worked examples to follow.

Important Note

In the case of knife-edge follower, contact between cam and follower necessarily occurs at the knife-edge which also lie on the line of action of the follower. *As against this, in case of roller follower and flat-footed follower, contact occurs between cam and follower at a point which may or may not lie on the line of action of follower.* This is clearly visible in position 4-*d* of roller follower in Fig. 7.20(b). Thus, in each of the two cases, surface of follower is required to be described for each displaced position and a smooth curve touching all these follower surfaces, to obtain the required cam curve.

It must be noted that if the cam curve does not touch a follower surface in any displaced position, corresponding follower position is not obtainable. This point must be kept in mind while drawing cam profile.

EXAMPLE 7.3   Design a cam for operating the exhaust valve of an engine. It is required to give equal uniform acceleration and retardation during opening and closing of the valve, each of which corresponds to 60° of cam rotation. The valve must remain in the fully open position for 20 degrees of cam rotation.

The lift of the valve is 37.5 mm and the least radius of the cam is 50 mm. The follower is provided with a roller of 50 mm diameter and its line of stroke passes through the axis of the cam.

(Kerala Univ., 1979; DAV Indore, March 1991)

Construction

1. Draw a line *OR* and extend it as shown in Fig 7.20(a). Select a suitable scale along *x*-axis to represent cam angles (1 cm = 30°). Thus *OP* = 60° represents cam angle of rise; *PQ* = 20° represents dwell period and *QR* = 60° represents cam angle of return. The remaining dwell period of 220° is represented by line beyond point *R*.

2. Divide rise period and return period in six equal parts and from the divisions marked 1, 2, 3, 4, 5, and 6 erect vertical lines. Since the stroke length involves measurement upto half milimeter, full scale has been used for displacement axis of Fig 7.20(a).

3. Divide the middle ordinate line from division 3 of rise and return periods in six equal parts, and number them as 1, 2, 3, ..., 6 as shown. Join point *O* of rise period of divisions 1, 2 and 3 of this line by straight lines cutting vertical lines from divisions, 1, 2 and 3 respectively on θ-axis at points *a, b* and *c*. Similarly, join point *f* of rise period to divisions 4, 5 and 6 to intersect vertical lines from divisions 4, 5, 6 respectively on θ-axis at points *d, e* and *f*. Join points *a, b, c, d, e and f* by a smooth curve. In a similar way construct displacement curve for return period as shown.

4. In view of large space requirement, the cam profile will be constructed to half the full scale, and therefore, only half the values of ordinates from the displacement diagram will be considered. A better practice is to use same scale for Figs. 7.20(a) and (b).

5. With a convenient location *C* for centre of base circle, describe a base circle of radius 50 mm to half scale. Describe also the prime circle with centre at *C* and radius as 75 mm to half scale.

6. Considering vertical diameter *CO* as the reference line, measure angles ∠*OCP* = 60°; ∠*PCQ* = 20° and ∠*QCR* = 60° in c.c.w. direction. Divide the outstroke angle ∠*OCP* and return stroke angle ∠*QCR* in six equal parts by lines *C*–1, *C*–2, *C*–3, *C*–4, *C*–5, and *C*–6, and plot half of the displacements 1–*a*, 2–*b*, 3–*c*, 4–*d*, 5–*e* and 6–*f* respectively from displacement diagram along these lines. (Note the scales of the two diagrams)

7. Join the points *a, b, c, d, e* and *f* in the outstroke and return stroke angles by smooth curves. The two curves are joined together by an arc of circle *f–f* through *C*. This gives the pitch curve, along which the centre of roller must move for satisfactory follower motion.

8. From each of the points *a, b, c, d, e* and *f* as centre, describe arcs of the circle of radius 25 mm to represent roller surfaces. It is necessary that the cam profile must touch each and every arc of these circles representing roller, but it is not necessary for the cam curve to touch roller surface along the line of action of follower only.

9. Describe a smooth curve, emerging from base circle, to touch each and every arc of circle, representing roller surface. For the dwell periods, draw arc of circle from cam axis *C*, joining two open ends of cam curve between rise period and return period. Similarly, draw arc of circle with centre at *C*, joining two open ends of pitch curve between the rise and return periods.

10. The pitch curve and cam profile are as shown in Fig 7.20(b).

Note: Student must note carefully that in this problem only half of the displacements from Fig. 7.20(a) are to be transferred to Fig 7.20(b). If this is not done a wrong cam curve will result. This is necessary on account of different scales used in Figs. 7.20(a) and (b).

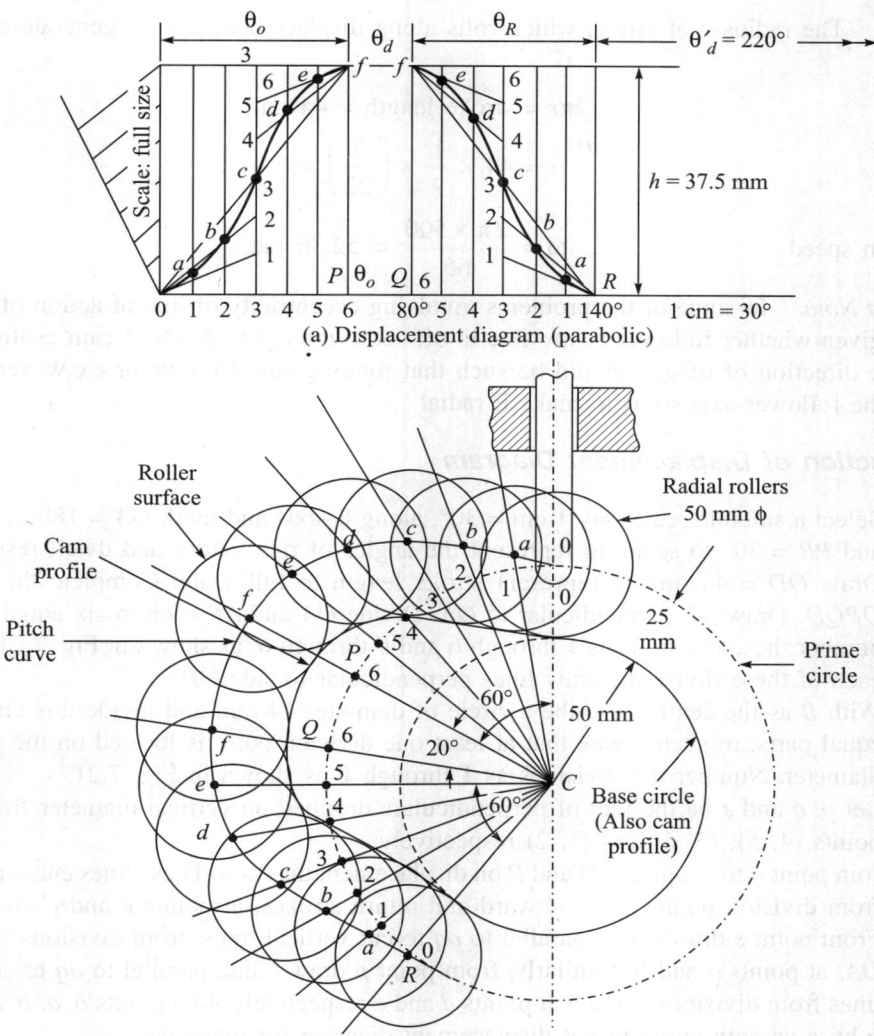

(a) Displacement diagram (parabolic)

(b) Cam profile for Radial cam with Roller follower

Fig. 7.20

EXAMPLE 7.4 Draw the profile of a cam when the roller follower moves with cycloidal motion during outstroke and return stroke, as detailed below:

(a) Outstroke with maximum displacement of 44 mm during 180° of cam rotation,

(b) Return stroke for next 150° of cam rotation, and

(c) Dwell for the remaining 30° of cam rotation.

Minimum radius of the cam is 1.5 cm and the roller diameter of the follower is 1.0 cm. The axis of the follower is offset by 1 cm towards the right from the axis of cam shaft.

Also find out maximum velocity and acceleration during the outstroke and return stroke, if the cam rotates uniformly at 500 r.p.m.

Solution: The radius r of circle, which rolls along displacement axis to generate cycloid, is given by

$$2\pi r = \text{stroke length} = 44 \text{ mm}$$

or,

$$r = 44 \times \frac{1}{2} \times \left(\frac{7}{22}\right) = 7 \text{ mm}$$

Also, cam speed

$$\omega = \frac{2\pi \times 500}{60} = 52.36 \text{ rad/s}$$

Important Note: In some of the problems involving eccentricity of line of action of follower, it is not given whether follower centre-line in off-set to the right on left of cam centre. In such cases, the direction of off-set should be such that rotating cam (in c.w. or c.c.w. sense) tends to push the follower-axis so as to make it radial.

Construction of Displacement Diagram

1. Select a suitable scale, say 1 cm = 30°, along θ–axis and mark $OA = 180°$; $AP = 150°$ and $PR = 30°$ to scale, to represent the angles of rise, return and dwell respectively.
2. Draw $OD = 44$ mm to represent stroke length to full scale. Complete the rectangle $OPQD$. Draw AB perpendicular to OA. Divide OA and AP each in six equal parts and number these divisions as 1 through 6 and 1′ through 6′ as shown in Fig. 7.21(a). From each of these divisions, draw lines perpendicular to side OP.
3. With B as the centre describe a circle of diameter 14 mm and divide this circle in six equal parts, in such a way that at least one division point is located on the horizontal diameter. Number the divisions as 1 through 6 as shown in Fig. 7.21.
   Let p, q and s be the feet of perpendiculars dropped on vertical diameter from pair of points (4, 5); (3, 6) and (1, 2) respectively.
4. Join point q to the points O and P on displacement diagram. These lines cut vertical lines from division points 3, in outward and return strokes, at points c and c' respectively. From point s draw a line parallel to oq to cut vertical lines, from divisions 1 and 2 on OA, at points a and b. Similarly, from point p draw a line parallel to oq to cut vertical lines from divisions 4 and 5 at points d and e respectively. Join points o, a, b, c, d, e and q by a smooth curve to get displacement diagram for outstroke.
5. From point q draw a line qp to cut vertical from point 3′, on line AP at point c'. Again from points p and s draw lines parallel to line pq to cut vertical lines from divisions 5′, 4′ and 2′, 1′ at points e', d', b' and a' respectively, as shown in Fig. 7.10(a). Join points P, a', b', c', d', e' and q by a smooth curve to complete the displacement diagram.

To draw the cam profile, draw from a suitable point C as centre, following circles of radii as indicated below:

(i) Offset circle of radius 1 cm
(ii) Base circle of radius 1.5 cm
(iii) Prime circle of radius 2.0 cm, and then proceed as under:

1. Draw a vertical line tangent to offset circle to the right. Consider this as the reference position of zero displacement of the follower. Let L be the point of tangency and let line of action of follower for this position cut the prime circle at O.

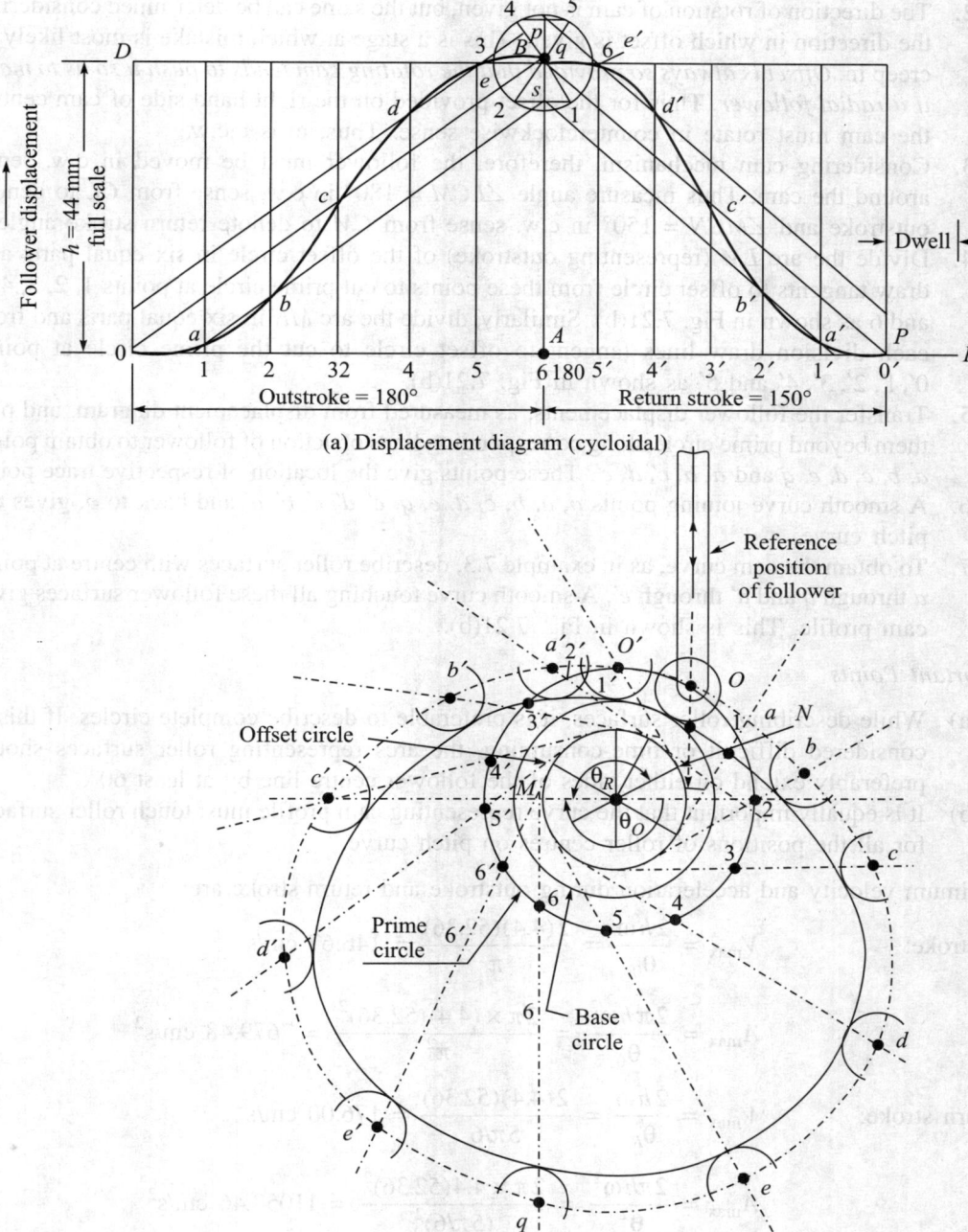

(a) Displacement diagram (cycloidal)

(b) Cam profile for off-centred roller follower

Fig. 7.21(a) and (b)

2. The direction of rotation of cam is not given, but the same can be determined considering the direction in which offset is given. This is a stage at which mistake is most likely to creep in. *Offset is always so provided that the rotating cam tends to push it so as to make it a radial follower.* Thus for the offset provided on the right hand side of cam centre, the cam must rotate in counterclockwise sense. Thus, ω is c.c.w.

3. Considering cam mechanism, therefore, the follower must be moved in c.w. sense around the cam. Thus measure angle $\angle LCM = 180°$ in c.w. sense from CL to denote outstroke and $\angle MCN = 150°$ in c.w. sense from CM to denote return stroke angle.

4. Divide the arc LM (representing outstroke) of the offset circle in six equal parts and draw tangents to offset circle from these points to cut prime circle at points 1, 2, 3, 4, 5 and 6 as shown in Fig. 7.21(b). Similarly, divide the arc MN in six equal parts and from each division draw lines tangent to offset circle to cut the prime circle at points $0', 1', 2', 3', 4'$ and $5'$ as shown in Fig. 7.21(b).

5. Transfer the follower displacements, as measured from displacement diagram, and plot them beyond prime circle along corresponding lines of action of follower to obtain points a, b, c, d, e, q and a', b', c', d', e'. These points give the location of respective trace point.

6. A smooth curve joining points $o, a, b, c, d, e, q, e', d', c', b', a'$ and back to o, gives the pitch curve.

7. To obtain the cam curve, as in example 7.3, describe roller surfaces with centre at points a through q and a' through e'. A smooth curve touching all these follower surfaces gives cam profile. This is shown in Fig. 7.21(b).

Important Points

(a) While describing roller surfaces, it is preferable to describe complete circles. If this is considered difficult or time-consuming, the arcs representing roller surfaces should preferably extend on either sides of the follower centre line by at least 60°.

(b) It is equally important that the curve representing cam profile must touch roller surfaces for all the positions of roller centres on pitch curve.

Maximum velocity and acceleration during outstroke and return stroke are:

Outstroke:
$$V_{max} = \frac{2h\omega}{\theta_0} = \frac{2(4.4)(52.36)}{\pi} = 146.67 \text{ cm/s}$$

$$A_{max} = \frac{2\pi h\omega^2}{\theta_0^2} = \frac{2\pi \times (4.4)(52.36)^2}{\pi^2} = 7679.48 \text{ cm/s}^2$$

Return stroke:
$$V_{max} = \frac{2h\omega}{\theta_R} = \frac{2(4.4)(52.36)}{5\pi/6} = 176.00 \text{ cm/s}$$

$$A_{max} = \frac{2\pi h\omega^2}{\theta_R^2} = \frac{2\pi \times 4.4(52.36)^2}{(5\pi/6)^2} = 11058.46 \text{ cm/s}^2$$

EXAMPLE 7.5 (a) A radial cam, operating a flat mushroom follower, rotates at 200 r.p.m. The follower rises through 20 mm with S.H.M. during 120° of cam rotation. It dwells for 30 degrees of cam rotation and returns to the initial position by S.H.M. in next 150° of cam rotation. Assuming a minimum radius for cam to be 25 mm, draw the cam profile. (DAVV Indore, Nov. 1987)

(b) Determine V_{max} & A_{max} during outstroke. Also determine minimum face width of follower required.

Solution: *Construction* As explained in example 7.2 construct displacement diagram for S.H.M. follower motion during outstroke and return stroke with $h = 20$ mm; $\theta_o = 120°$; $\theta_d = 30°$; $\theta_R = 150°$; and $\theta'_d = 60°$. This is illustrated in Fig. 7.22(a).

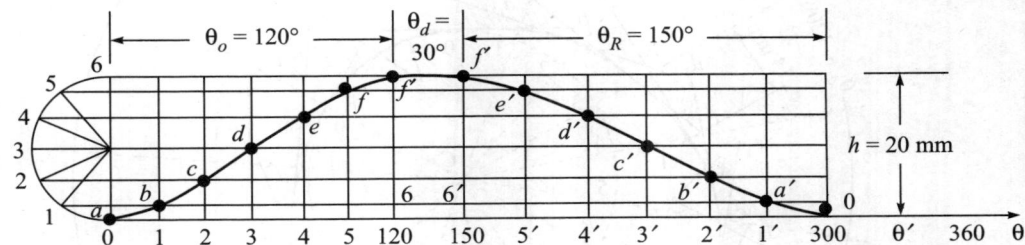

Fig. 7.22(a)   Displacement diagram for S.H.M. follower motion.

Cam Layout

1. Draw the base circle of 25 mm radius with C as the centre. Assume the rotation of cam in c.w. sense.
2. Take reference position, corresponding to zero displacement, of follower at O, the extremity of vertical diameter. Divide the circle in angles $\angle OCP = 120°$ for outstroke $\angle PCQ = 30°$ for dwell, $\angle QCR = 150°$ for return stroke and $\angle RCO = 60°$ for dwell again.
3. Divide angles $\angle OCP$ and $\angle QCR$ in six equal parts. Let the corresponding division marks for outstroke by 1, 2, 3, 4, 5 and 6 and for return strokes by $1'$, $2'$, $3'$, $4'$, $5'$ and $6'$ as shown in Fig. 7.22(b).
4. Draw radial lines from divisions 1 from 6 and also through divisions $1'$ through $6'$ to denote lines of action of follower. Note that cam is assumed to rotate in c.w. direction and therefore, follower is being moved around cam in c.c.w. direction.
5. Along the radial lines C–1, C–2, C–3, C–4, C–5 and C–6 transfer amount of follower displacement away from base circle, to obtain points a, b, c, d, e and f. Similarly plot points a', b', c', d', e' and f' for return stroke.
6. From points a, b, c,.....f and a', b', c',.......f draw straight lines perpendicular to follower centre lines to represent flat faces of follower.
7. Draw a smooth curve touching each and every follower face so described. The resulting cam curve is shown in Fig. 7.22(b). Note that if any follower face does not touch cam, corresponding follower position is not obtainable and the resulting cam profile is rendered incorrect.

Readers must note that if the follower's flat-face is not sufficiently wide, it cannot rest on cam in the corresponding follower position. Thus, consider follower position along C–3, in which the point of contact between cam and flat-face is farthest in c.c.w. sense at a distance of 14 mm. If flat face of follower on this side does not extend by a distance greater than or equal to 14 mm the prescribed follower position is not obtainable. Similarly, in position $C - 3'$ during return stroke of follower, the follower cannot rest on cam unless on the other side flat face does not extend

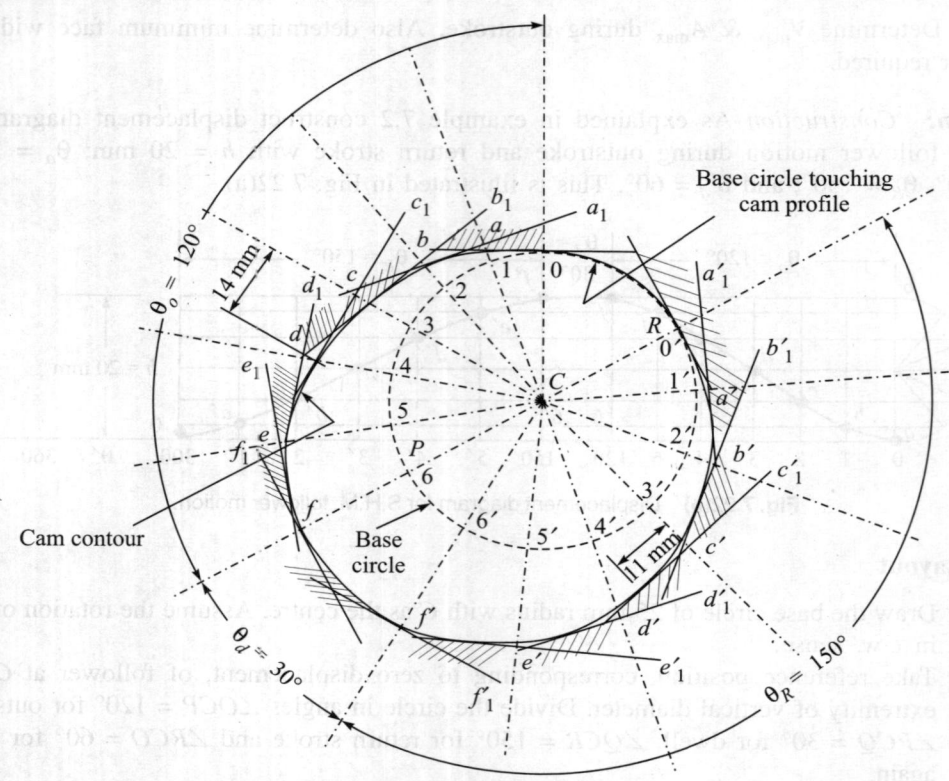

Fig. 7.22(b) Cam profile layout for flat faced follower.

by an amount greater than or equal to 11 mm. In every other position this requirement is smaller than 11 mm. Thus the follower must have a width (14 mm + margin) on left hand side and (11 mm + margin) on the right hand side, as shown in Fig. 7.23. Maximum velocity and acceleration during outstroke for S.H.M. are given by

$$V_{max} = \frac{\pi h \omega}{2\theta_0} = \frac{\pi(2.0)}{2(2\pi/3)} \times \left(\frac{2\pi \times 200}{60}\right) = 31.4159 \text{ cm/s}$$

and, $$A_{max} = \frac{h}{2}\left(\frac{\pi \omega}{\theta_0}\right)^2$$

or, $$A_{max} = \frac{2.0}{2}\left(\frac{\pi}{2\pi/3}\right)^2 \omega^2 = 1.0\left(\frac{3}{2}\right)^2 \omega^2 = 987 \text{ cm/s}^2$$

Fig. 7.23 Facewidth of follower.

Margin ⊢ 14 mm ¦ 11 mm ⊣ Margin

7.13 TRANSLATING FLAT FACE FOLLOWER: ANALYTICAL DESIGN

Whenever a large follower lift is attempted in too little cam rotation with too small a cam, in all probability the cam profile crosses over itself. This phenomenon was described earlier in section

7.10 as undercutting. *In machining of such a cam, during production, this part of the cam shape is lost and hence the intended follower motion is not obtained. Solution to this problem lies in (i) reducing required lift h, or (ii) increasing corresponding cam angle of rotation.* In many design situations, however, this may not be permissible. A yet another alternative is to increase the base circle radius r_b, which results in increasing cam size of course. The pivotal question, therefore, would be: By what amount should r_b be increased? A trial and error approach is obviously out of question. The problem is therefore, approached analytically.

Consider a radial cam operating a radial follower with motion $y = f(\theta)$. Let r_b be the base circle radius and let (x, y) be the coordinates of point of contact Q between cam and follower for the position $A'A'$ shown in Fig. 7.24.

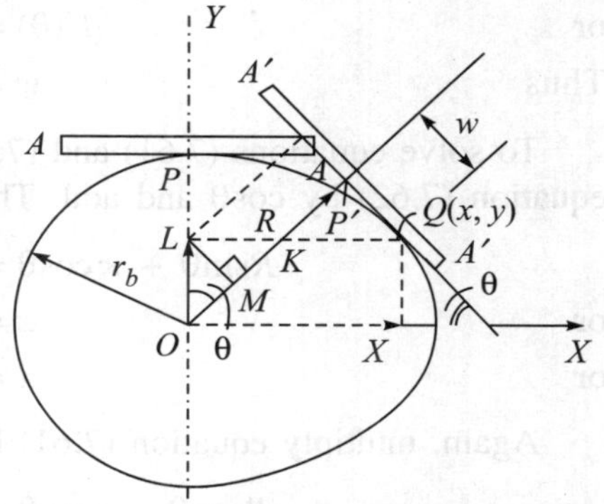

In the position AA, the flat face rests on the circular arc of least radius r_b. Position of AA' of follower corresponds to the position when cam was rotated c.c.w. through angle θ. If P be the trace point on follower, then angle $\angle POP' = \theta$. This also means that the flat face of follower is inclined at θ to the negative side of x-axis. Thus, denoting vector OP' by R, we have

Fig. 7.24 Flat faced follower in contact with cam.

$$R = r_b + \text{follower displacement at}$$
cam angle θ

or
$$R = r_b + f(\theta) \tag{7.59}$$

Taking y-axis along original position of follower axis and x-axis perpendicular to it through O, the slope of the face $A'A'$ is given by

$$(dy/dx) = -\tan\theta \tag{7.60}$$

Let L be the foot of perpendicular QKL dropped from Q, the point of contact, on y-axis. Then $OL = y$. Also let M be the foot of perpendicular dropped from L on the line OP'. As AA' and LM are perpendicular to line OP' and $\angle P'KQ = (90 - \theta)$, we have,

$$OP' = OM + MP'$$

or
$$OP' = (OL)\cos\theta + (LQ)\cos(90 - \theta)$$

or
$$R = y\cos\theta + x\sin\theta \tag{7.61}$$

Also, the distance w of point of contact Q from P' is given by,

$$w = P'Q = (LQ)\cos\angle LQP' - LM$$

or
$$w = x\cos\theta - (OL)\sin\theta$$

or
$$w = x\cos\theta - y\sin\theta \tag{7.62}$$

Again, as $dy/dx = (dy/d\theta)(d\theta/dx)$, from equation (7.60)

$$(dy/d\theta)(d\theta/dx) = -\frac{(\sin\theta)}{(\cos\theta)}$$

or
$$(\sin\theta)\frac{dx}{d\theta} = -(\cos\theta)\frac{dy}{d\theta} \tag{7.63}$$

The rate of follower lift with reference to cam angle of rotation is given by

$$f'(\theta) = \frac{dR}{d\theta}$$

And from equation (7.61), substituting for $dR/d\theta$,

$$f'(\theta) = \frac{dR}{d\theta} = -y\sin\theta + x\cos\theta$$

or

$$f'(\theta) = x\cos\theta - y\sin\theta = w \text{ (from Eq. 7.62)}$$

Thus

$$w = f'(\theta) \tag{7.64}$$

To solve equations (7.61) and (7.62) simultaneously, multiply equation (7.61) by $\sin\theta$ and equation (7.62) by $\cos\theta$ and add. Thus,

$$R\sin\theta + w\cos\theta = x\,(\sin^2\theta + \cos^2\theta)$$

or

$$x = R\sin\theta + w\cos\theta$$

or

$$x = R\sin\theta + f'(\theta)\cos\theta \tag{7.65}$$

Again, multiply equation (7.61) by $\cos\theta$ and equation (7.62) by $\sin\theta$ and subtract. Thus,

$$R\cos\theta - w\sin\theta = y(\cos^2\theta + \sin^2\theta)$$

or

$$y = R\cos\theta - f'(\theta)\sin\theta \tag{7.66}$$

Substituting for R in equations (7.65) and (7.66) from equation (7.59),

$$x = [r_b + f(\theta)]\sin\theta + f'(\theta)\cos\theta \tag{7.67}$$

and

$$y = [r_b + f(\theta)]\cos\theta - f'(\theta)\sin\theta \tag{7.68}$$

Equations (7.67) and (7.68) provide analytical expressions with the help of which *x and y coordinates of any point on cam profile can be obtained for a given value of cam angle θ, if analytical expressions for follower displacement f(θ) and its slope f′(θ) are known in terms of θ. Thus we have a method to plot cam profile (either manually or using computer-aided graphics technique) much more accurately.*

Further, from equation (7.64),

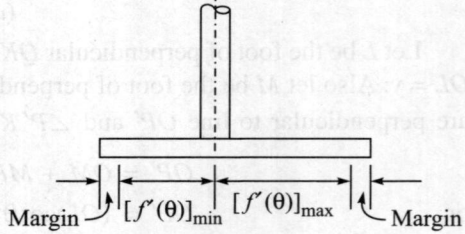

Fig. 7.25 Minimum face width of flat footed follower.

$$w = f'(\theta)$$

it follows that total width of follower required

$$W = [f'(\theta)]_{max} + [f'(\theta)]_{min}$$

Here, maximum and minimum values of $f'(\theta)$ refers to the maximum values of $f'(\theta)$ on the positive (outstroke) and negative (return stroke) side of the trace point over which follower must extend. In addition to the above, some margin is also required to be provided on flat face. Thus, as shown at Fig. 7.25 total width of flat face of follower

$$W'' = [f'(\theta)]_{max} + [f'(\theta)]_{min} + 2(\text{margin}) \tag{7.69}$$

In order to derive expression for minimum radius of cam r_b, let us proceed as follows. Consider the follower to rest on flat face at the point of cusp, say, shown in Fig. 7.26.

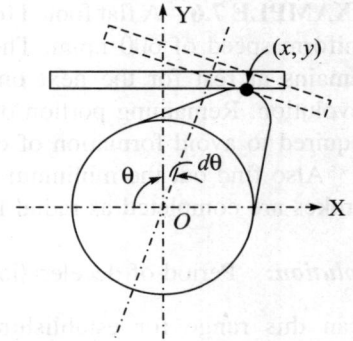

Equations (7.67) and (7.68) are the parametric equations of the envelope, which is the cam profile itself. Again a cusp occurs when anywhere on the cam curve transition from convex to concave shape occurs. Ideally, of course, the cam profile must be convex everywhere. Let us now establish expression for finding minimum permissible value of r_b for avoiding cusp formation.

When the contact occurs at the point of cusp, the follower does not move even though cam rotates through a small angle $d\theta$. Thus, follower displacement components along x-axis and y-axis are zero. In other words,

Fig. 7.26 Follower in contact with flat face at the cusp point.

$$(dx/d\theta) = (dy/d\theta) = 0 \qquad (7.70)$$

To establish expression for minimum permissible value of r_b, for avoiding cusp, differentiate equation (7.67) w.r. to θ, remembering that r_b is constant. Thus,

$$\frac{dx}{d\theta} = [r_b + f(\theta)]\cos\theta + [0 + f'(\theta)]\sin\theta + f''(\theta)\cos\theta - f'(\theta)\sin\theta$$

Simplifying

$$\frac{dx}{d\theta} = [r_b + f(\theta) + f''(\theta)]\cos\theta \qquad (7.71)$$

Also, differentiating equation (7.68) w.r. to θ,

$$\frac{dy}{d\theta} = -[r_b + f(\theta)]\sin\theta + [0 + f'(\theta)]\cos\theta - f''(\theta)\sin\theta - f'(\theta)\cos\theta$$

Simplifying

$$(dy/d\theta) = -[r_b + f(\theta) + f''(\theta)]\sin\theta \qquad (7.72)$$

For cusp to occur, $dy/d\theta = dx/d\theta = 0$, and this is possible at all time t, when

$$r_b + f(\theta) + f''(\theta) = 0 \qquad (7.73)$$

Hence to avoid formation of cusp,

$$r_b + f(\theta) + f''(\theta) > 0 \qquad (7.74)$$

Since the follower displacement $f(\theta)$ is always positive, it follows that a cusp can possibly form only when $f''(\theta)$ is negative, *i.e.* during the period when follower decelerates.

Also, equation (7.69) gives required follower width for radial follower with flat face. For a flat-faced follower with an offset of e for its line of action,

$$W = \{[f'(\theta)]_{max} - e\} + \{[f'(\theta)]_{min} + e\} + 2(\text{margin})$$

or $\qquad\qquad W = [f'(\theta)]_{max} + [f'(\theta)]_{min} + 2(\text{margin}) \qquad (7.75)$

Thus, the final expression giving width of flat face of follower is same for radial as well as for offset follower motion.

EXAMPLE 7.6 A flat footed follower is actuated by a radial cam keyed to a shaft, rotating with uniform speed of 600 r.p.m. The follower is lifted through 3.75 cm in one-fourth revolution; remains at rest for the next one-fourth revolution and returns back in the next one-fourth revolution. Remaining portion of cycle is the dwell period. Determine the least radius of cam required to avoid formation of cusp point.

Also find out the minimum width of flat face of follower, if both the outstroke and return strokes are completed as radial follower with parabolic motion.

Solution: Period of deceleration extends from $\frac{\theta_0}{2} \leq \theta \leq \theta_o$, and as $\theta_0 = \theta_R$, it is sufficient to scan this range for establishing minimum radius r_b. Since, deceleration is constant for $\frac{\theta_o}{2} \leq \theta \leq \theta_o$ we are interested in that point of parabolic motion at which $f(\theta)$ is least. This is clearly the point $\theta = (\theta_o/2)$ in the latter half of outstroke. At this point,

$$y = f(\theta) = \frac{h}{2} = \frac{3.75}{2}$$

$$f''(\theta) = (\text{Acceleration})/\omega^2 = -\frac{4h}{\theta_0^2}$$

Hence

$$r_b + \frac{3.75}{2} - \frac{4(3.75)}{(\pi/2)^2} > 0$$

or
$$r_b > \frac{16(3.75)}{\pi^2} - \frac{3.75}{2} \qquad \text{or} \qquad r_b > 4.204 \text{ cm.}$$

Thus the base circle radius should be greater than or equal to 4.3 cm. Minimum face width of follower is obtained as under. During rise period,

$$[f'(\theta)]_{max} = (V_{max})/\omega = \frac{1}{\omega} \frac{2h\omega}{\theta_0} = \left(\frac{2h}{\theta_0}\right)$$

and, $[f'(\theta)]_{min} = \frac{1}{\omega}$(maximum velocity during return stroke)

$$= -\frac{1}{\omega}\left(\frac{2h\omega}{\theta_R}\right) = -(2h/\theta_R)$$

since, $\theta_o = \theta_R = \pi/2$, numerical addition of the two values with margin on either side gives maximum face width. Thus, taking a margin of 3 mm

$$W = \frac{2h}{\theta_0} + \frac{2h}{\theta_R} + 2(\text{margin}) = \frac{2(3.75)}{(\pi/2)} + \frac{2(3.75)}{(\pi/2)} + 2 \times (0.3)$$

or $W = 10.15$ cm. (*i.e.* roughly 5.1 cm wide long on either side of the follower stem). **Ans.**

7.14 CAM WITH OSCILLATING ROLLER FOLLOWER

Cam with oscillating roller follower is shown in Figs 7.3(g) and (h). In the case of oscillating roller follower, the ordinate of displacement diagram should actually correspond to the arcular distances moved through by the centre of roller. However, if the total angular displacement of follower is reasonably small, the chords of circle are very nearly equal to the corresponding arcular segments. Hence it is reasonably correct to measure ordinate distances from displacement diagram at each station, and transfer directly to the corresponding arcs travelled by the roller centre using dividers (which amounts to measuring chords). Further, one must rotate the fixed pivot of follower in a direction opposite to that of cam rotation, keeping its distance from cam centre the same. This step is necessary to obtain true inversion.

The procedure is best explained with the help of the following example problem.

EXAMPLE 7.7   Draw profile of a cam to give the following motion to an oscillating roller follower.

(i)   Follower to move outwards through an angular displacement of 17° during 90° of cam rotation;

(ii)   Follower to dwell for 60° of cam rotation;

(iii)   Follower to return to its initial position during 90° of cam rotation; and

(iv)   Follower to dwell during remaining 120° of cam rotation.

The pivot of the oscillating follower is 11.5 cm from the axis of rotation of the cam. The distance between the pivot centre and the roller centre is 10 cm, the roller is 3.8 cm diameter and the minimum radius of the cam is 5 cm. The outward stroke of the follower is completed with S.H.M., while the return stroke is completed with uniform acceleration and retardation, the retardation being double the acceleration.

Solution: Let t_1, A_1 and t_2, A_2 be the time required and acceleration/retardation during the two parts of return stroke. Then, starting from zero velocity at the outer most position, the follower achieves a velocity say V_1 in time t_1 such that

$$V_1 = O + (A_1 t_1)$$

and during the latter part of return stroke, the follower velocity is reduced from V_1 to zero in time t_2. Thus,

$$O = V_1 + (A_2 t_2)$$

But $A_2 = 2A_1$ and thus neglecting sign, for same V_1, the above two equations give

$$A_1 t_1 = (2A_1) t_2 \text{ or } t_2 = \frac{t_1}{2}$$

In other words $\theta_1 : \theta_2 = 2 : 1$. The displacement diagram for outstroke and return stroke can be constructed by plotting the chord of the arcular movement of roller centre along y-axis. For small angle of oscillation, variations in the lengths of arc and corresponding chord of circle may be neglected. For instance, the length of chord in the given case is $2 \times 10 \times \sin 8.5 = 2.956$ cm, while the length of arc of circle is $2 \times 10 \times \dfrac{8.5\,\pi}{180} = 2.967$ cm.

The displacement diagram is constructed with stroke length as 2.95 cm and cam angles plotted along x-axis. The displacement diagram for outstroke is constructed as S.H.M. as described previously. For return stroke, as cam angle required for acceleration is twice that required for retardation, let the ordinate *FG* divide the outstroke angle in proportion 2 : 1. Thus

$DF = 60°$ and $FH = 30°$. The displacement curve for return stroke is completed by dividing DF in four equal parts and the length FH in three equal parts and then following the usual procedure. Since cam rotates uniformly and time of acceleration is twice that of retardation during outstroke, follower will cover two-third of the stroke length during the period of acceleration, as shown in Fig. 7.27(a).

Note: Since Fig. 7.27(b) is drawn to half scale, only half of the displacement from Fig. 7.27(a) is to be transferred to Fig. 7.27(b).

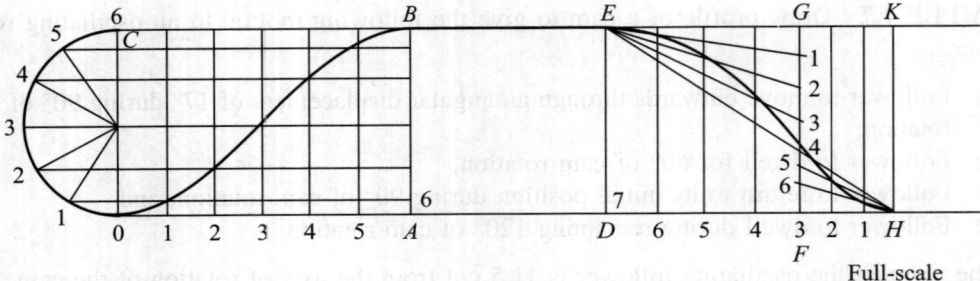

Fig. 7.27(a) Displacement diagram.

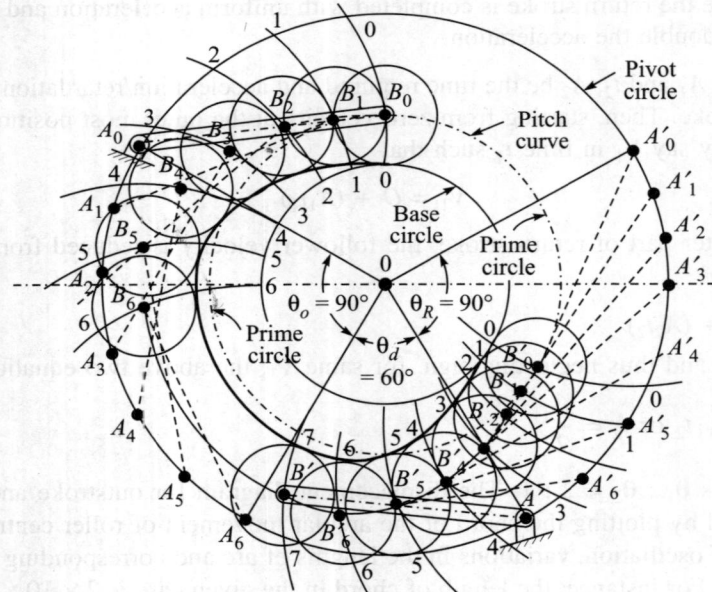

Cam profile for oscillating roller follower.

Fig. 7.27(b) Half-scale.

Construction: Cam Profile

With O as the centre, describe circles of radii 5 cm and 6.9 cm to represent base circle and prime circle. With O as the centre also describe a circle of radius 11.5 cm to represent the path, along which pivot of follower will have to be moved around the cam in a sense opposite to that of cam rotation. Let any point A_o on this circle denote the reference position of pivot for zero displacement of follower. With A_o as centre and radius equal to 10 cm, draw an arc to cut prime circle at B_o. Then A_oB_o gives the reference position of roller follower. Measure angles $\angle A_0OA_6$ = θ_0 = 90°; $\angle A_6OA'$ = θ_d = 60° and $\angle A'OA'_0$ = θ_R = 90°. Divide the arc A_0A_6 into six equal parts at points A_1, A_2, A_3, A_4, A_5 and A_6. From points A_1, A_2, A_3 etc. as centre describe arcs of circle of radius 10 cm to cut prime circle at points 1, 2, 3, 4, 5 and 6 respectively.

Since Fig. 7.27(b) is drawn to half-scale and Fig 7.27(a) is drawn to full scale, only half of the displacements, as measured in Fig 7.27(a), are to be transferred to Fig 7.27(b), For a change, the displacements, are marked with divider on arcs from points 1, 2, 3..., 6 on prime circle. These points are shown as B_1, B_2, B_3,...,B_6 and, when joined by a smooth curve, give pitch curve. In a similar way, points B'_1, B'_2, B'_3,..., B'_6 are to be obtained on pitch curve for return stroke by referring to the displacement diagram. Thus angle $\angle A'OA'_0$ = 90° is divided in proportion 2 : 1. The portion $\angle A'OA'_3$ = 60° and $\angle A'_3OA'_0$ = 30°. Angle $\angle A'O'A'_3$ is divided into four equal parts at A'_3, A'_4, A'_5 and A'_6 while $\angle A'_3 O A'_0$ is divided into three equal parts at A'_0, A'_1, A'_2 and A'_3. As in the case of rise period, position of roller centres are obtained at B'_0, B'_1, B'_2,..., B'_6.

Lastly, with B_1, B_2, B_3,...B_6 and B'_0, B'_1, B'_2..., B'_6 as centres, roller surfaces are described and a smooth curve touching all these roller surfaces then defines the cam profile.

7.15 CAMS WITH SPECIFIED CONTOURS

Cams considered thus far had specified follower motion and, as such, the motion characteristics (*e.g.*, displacement, velocity, acceleration and jerk) are known even before the cam is actually produced on machines. *On the negative side, however, such cams are generally difficult and costly to manufacture. A master cam must be prepared first, mostly by hand, for its subsequent use as template in the production of other cams of identical shape.*

To achieve greater accuracy in the production of cam profile, reducing cost of production simultaneously, it may be advisable to make use of cams whose profile consists of circular arcs and straight lines. Such cams must be analysed for determining their motion characteristics.

7.15.1 Circular Arc Cam with Tangent Follower

The contour of such cam consists of three arcular portions:

    (a)   Arc of base circle,
    (b)   Arc of nose circle, and
    (c)   Circular flanks joining nose circle to the base circle.

Figures 7.28(a) and (b) show one such circular arc cam with flat-faced reciprocating follower. Let r_b and r be the base circle radius and nose circle radius, and let R be the radius of the circular flanks described with I_1, I_2 as centres. Also, let C and O to be the centres of base circle and nose circle respectively. Let us choose y-axis along follower centre line with origin at cam centre C so that as follower is moved around cam from position 1 to 2, the y-axis also moves with it.

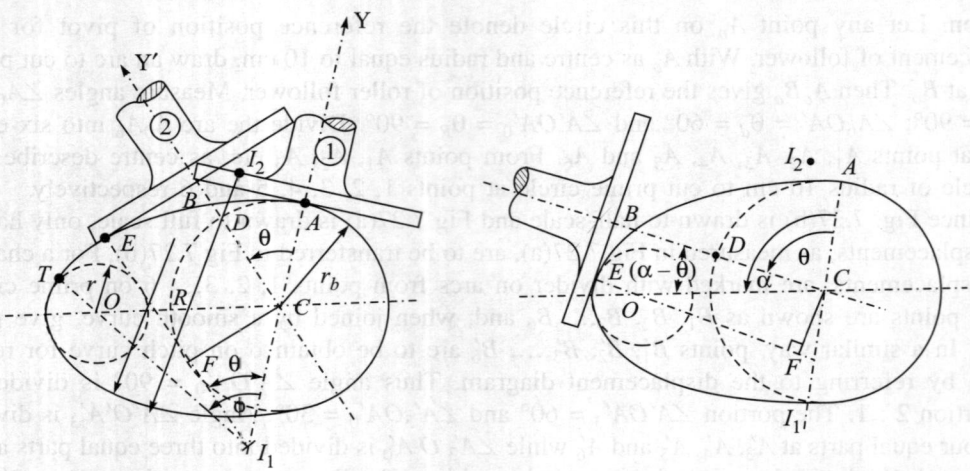

Fig. 7.28(a) Flat footed follower in contact with cam on convex flank.

Fig. 7.28(b) Flat footed follower in contact with cam on nose circle.

In the position 1, the flat face of follower is in contact with the cam at point A on line joining I_1C. In other words, A is the point from which flank curve springs out. Similarly, point T where line joining I_1O meets the nose circle is the point at which flank curve merges into nose circle.

Contact on Flank

Let the angle of rotation of cam in Fig. 7.28(a) be θ, so that centre lines CY and CY' of follower subtend an angle θ at cam centre. Let E be the point of contact of flat face with cam in position 2 of follower. Join I_1E and let F be the foot of perpendicular from C on line I_1E. For the position 2 of follower, the trace point is at B, the point of intersection with follower centre line. Then, follower displacement in position 2 is,

$$y = BD = BC - DC = EF - AC \tag{7.76}$$

But, $EF = I_1E - I_1F = I_1A - I_1C \cos\theta = I_1C + AC - I_1C \cos\theta = AC + I_1C (1 - \cos\theta)$

Note that I_1E and I_1A are radii of same arc. Substituting for EF in equation (7.76),

$$y = I_1C (1 - \cos\theta) \tag{7.77}$$

Velocity of follower is obtained by differentiating w.r. to time (note that I_1C = constant),

$$V = \frac{dy}{dt} = (I_1C)\left[0 + (\sin\theta)\frac{d\theta}{dt}\right]$$

For cam rotating at a uniform speed, $\dfrac{d\theta}{dt} = \omega$ and therefore,

$$V = (I_1C)\omega\sin\theta \tag{7.78}$$

Acceleration is obtained as a next time derivative,

$$A = \frac{dv}{dt} = I_1C\omega^2\cos\theta \tag{7.79}$$

Equations (7.78) and (7.79) are valid for points of contact on flank portion. Thus from equation (7.78), maximum velocity is obtained when $\theta = \phi$, for which condition

$$V_{max} = (I_1 C)\,\omega\sin\phi \qquad (7.80)$$

Also, from equation (7.79) it follows that maximum acceleration occurs when contact occurs at A, *i.e.* when $\theta = 0$, and hence,

$$A_{max} = (I_1 C)\,\omega^2 \qquad (7.81)$$

When $\theta > \phi$, the contact between cam and follower occurs on nose circle portion.

Contact on Nose Circle

Refer to Fig. 7.28(b) for point of contact on nose portion. E is the point of contact of follower with the cam and B is the corresponding trace point of flat-faced follower. Join BC to cut the base circle at D. The normal to the flat face at E passes through nose centre O and F and let CF be the perpendicular drawn from C on this normal.

Let $\angle ACO = \alpha$ and $\angle ACB = \theta$, the cam angle of rotation. Then the follower displacement, when the flat face touches cam at E, is

$$y = BD = BC - CD = EF - CD \qquad (7.82)$$

But

$$EF = EO + OF \qquad (7.83)$$

As OF and BC are parallel and OC cuts them $\angle FOC = \angle OCB = (\alpha - \theta)$. Also, as $\angle OFC = 90°$, from right angled triangle OFC,

$$OF = OC\cos(\alpha - \theta)$$

Substituting in equation (7.83),

$$EF = EO + OC\cos(\alpha - \theta)$$

Substituting for EF in equation (7.82),

$$y = EO - CD + OC\cos(\alpha - \theta) \qquad (7.84)$$

Note that EO, the nose circle radius and CD, the base circle radius are constant and therefore, differentiating equation (7.84) w.r. to time,

$$V = \frac{dy}{dt} = OC\left(\frac{d\theta}{dt}\right)\sin(\alpha - \theta) = (OC)\omega\sin(\alpha - \theta) \qquad (7.85)$$

Differentiating once more w.r. to time for acceleration,

$$A = \frac{dv}{dt} = -(OC)\omega^2\cos(\alpha - \theta) \qquad (7.86)$$

Negative sign on R.H.S. of equation (7.86) indicates that when the contact takes place on nose circle, the follower motion is retarded.

It follows from equation (7.85) that velocity of follower is zero when $\theta = \alpha$, *i.e.* when follower axis is located along line of centres CO. Again from equation (7.86), it follows that retardation is maximum when $\theta = \alpha$, and its maximum value is given by $(OC)\omega^2$.

EXAMPLE 7.8 A symmetrical cam with convex flanks operates a flat-footed follower (see Fig. 7.29). The lift is 8 mm, base circle radius is 25 mm and the nose radius is 12 mm. If the total angle of cam action is 120° find the radius of the convex flanks. Determine the maximum velocity and the maximum acceleration when the cam shaft rotates at 500 r.p.m. (AMIE, Winter 1984)

Solution: Total angle of action $2\alpha = 120°$

Therefore $\qquad\qquad \alpha = 60°$

Base circle radius $r_b = 25$ mm and radius of flank $R = OA$. As $N = 500$ r.p.m.,

$$\omega = \frac{2\pi \times 500}{60} = 52.36 \text{ rad/s}$$

Total lift $h = CM - DC$ or $CM = 8 + r_b = 33$ mm

But $\quad OM =$ nose circle radius = 12 mm;

Therefore $\quad OC = CM - OM = 33 - 12 = 21$ mm

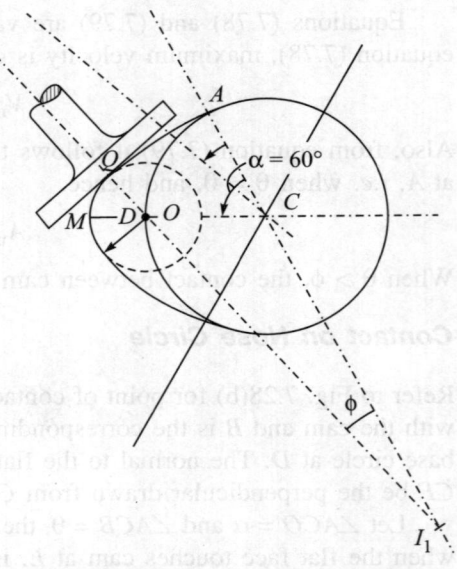

Fig. 7.29 Symmetrical cam with convex flank operating flat faced follower.

Maximum velocity V_{max} occurs where the convex flank merges with nose circle. Angle $\angle AI_1Q = \phi$ at which maximum velocity occurs can be established only if the radius of convex flank $R = I_1A = I_1Q$ and hence, OI_1 and CI_1 are known. To establish R, let us consider obtuse angled triangle OCI_1, in which,

$$OI_1 = QI_1 - OQ = R - \text{nose circle radius} = (R - 12)$$

and $\qquad\qquad CI_1 = AI_1 - AC = R - \text{base circle radius} = (R - 25)$

Hence, from $\qquad (OI_1)^2 = (OC)^2 + (CI_1)^2 - 2(OC)(CI_1)\cos(180 - \alpha)$

or $\qquad (R - 12)^2 = (21)^2 + (R - 25)^2 + 2(21)(R - 25)\cos 60$

or $\qquad (R^2 - 24R + 144) = 441 + (R^2 - 50R + 625) + 21R - 525$

or $\qquad 5R = 397$ which gives $R = 79.4$ mm $\qquad\qquad$ **Ans.**

Hence $\qquad OI_1 = 79.4 - 12 = 67.4$ mm and $CI_1 = 79.4 - 25 = 54.4$ mm

Hence, from triangle OCI_1,

$$\frac{OC}{\sin\phi} = \frac{OI_1}{\sin(180 - 60)}$$

or $\qquad\qquad \sin\phi = \frac{OC}{OI_1} \times \sin 60 = \frac{21}{67.4} \times \frac{\sqrt{3}}{2} = 0.2698$

Hence $\qquad\qquad V_{max} = (I_1C)\omega \sin\phi$

$$= (54.4)\frac{2\pi \times 500}{60} \times 0.2698 = 768.49 \text{ mm/s} = 0.7685 \text{ m/s} \quad \textbf{Ans.}$$

And maximum acceleration occurs at $\theta = 0°$ given by

$$A_{max} = (I_1C)\omega^2 = (54.4)\left(\frac{2\pi N}{60}\right)^2 = 149.14 \text{ m/s}^2 \qquad \textbf{Ans.}$$

EXAMPLE 7.9   A cam having a lift of 1 cm operates the suction valve of a 4-stroke cycle petrol engine. The least radius of the cam is 2 cm and nose radius is 0.25 cm. The crank angle of the engine, when suction valve opens, is 4° after t.d.c. and it is 50° after b.d.c. when the suction valve closes. The cam shaft has a speed of 1000 r.p.m. The cam is of circular type with circular nose and flanks. It is integral with cam shaft and operates as a flat-faced follower. Estimate: (a) the maximum velocity of valve; (b) the maximum acceleration and retardation of the valve; (c) the minimum force to be exerted by the spring to overcome inertia of the valve parts which weigh 2 Newton. (DAV Indore, 1987 June)

Solution: Total crank angle of cam rotation, during which valve remains open

$$= (180 - 4 + 50) = 226°$$

For a 4-stroke cycle engine, cam shaft rotates at half the speed of crank shaft. Hence, the cam angle of rotation during which valve remains open $= \dfrac{226}{2} = 113$.

Cams of specified contour are usually symmetric and as such cam angle of ascent equals cam angle of descent. Hence,

$$\alpha = \frac{113}{2} = 56.5°$$

With a lift of 1.0 cm; $r_b = 2.0$ cm and nose circle radius $= 0.25$ cm; the centre distance between nose centre O and base circle C.

$$OC = (r_b + h) - 0.25 = 2 + 1 - 0.25 = 2.75 \text{ cm}$$

Again, if R be the radius of convex flank, then as in example 7.8 from triangle OCI_1,

$$(OI_1)^2 = OC^2 + (CI_1)^2 - 2(OC)(CI_1)\cos(180 - 56.5)$$

or $(R^2 - 0.5R - 0.0625) = 7.5625 + (R^2 - 4R + 4) + 3.0356(R - 2.0)$

But, $(OI_1)^2 = (R - 0.25)^2$

or $(4 - 3.0356 - 0.5)R = 7.5625 + 0.0625 + 4 - 6.0712$

or $0.4644R = 5.5538$

Therefore $R = 11.959$ or $R = 11.96$ cm

Hence $CI_1 = (R - 2.0) = 9.96$ cm,

Again, referring to Fig. 7.28 from triangle COI_1.

$$\frac{OC}{\sin\phi} = \frac{OI_1}{\sin(180 - 56.5)}$$

Therefore $\sin\phi = \dfrac{OC}{OI_1} \times \sin 56.5 = 0.1958$

Therefore $\phi = 11.293°$

For maximum velocity of follower, $\theta = \phi$, and the maximum velocity is then,

$$V_{max} = (CI_1)\omega\sin\phi$$

$$= (9.96)\left(\frac{2\pi \times 1000}{60}\right) \times 0.1958 = 204.22 \text{ cm/s} = 2.04 \text{ m/s} \qquad \textbf{Ans.}$$

Maximum acceleration

$$A_{max} = (I_1 C)\omega^2 = (9.96)\left(\frac{2\pi \times 1000}{60}\right)^2 = 109223.62 \text{ cm/s}^2 = 1092.23 \text{ m/s}^2 \quad \textbf{Ans.}$$

Maximum retardation occurs at $\theta = \alpha$

$$(A_r)_{max} = -(OC)\,\omega^2 \cos(\alpha - \theta)$$

Thus

$$(A_r)_{max} = -(2.75)\left(\frac{2\pi \times 1000}{60}\right)^2 = 30157.12 \text{ cm/s}^2 = 301.571 \text{ m/s}^2 \quad \textbf{Ans.}$$

Spring force required,

$$F_{min} = \text{mass} \times \text{maximum retardation} = \frac{2}{9.81} \times 301.571 = 61.482 \text{ newtons.} \quad \textbf{Ans.}$$

7.15.2 Tangent Cam with Roller Follower

Instead of joining the base circle and nose circle by circular arc, here the two arcs are connected by a straight flank which is a common tangent to both.

Let r be the roller radius, r' the nose circle radius, r_b the base circle radius and d the distance between base circle centre and nose circle centre. Further, let α be the cam angle corresponding to ascent.

When Roller Contacts Cam on Straight Flank

Let CA and CB be the prime circle radius and base circle radius, at the point of commencement of ascent respectively. Consider a position of cam rotation through θ during rise period, for which centre of roller P is as shown in Fig. 7.30 with point of contact with flank at point Q.

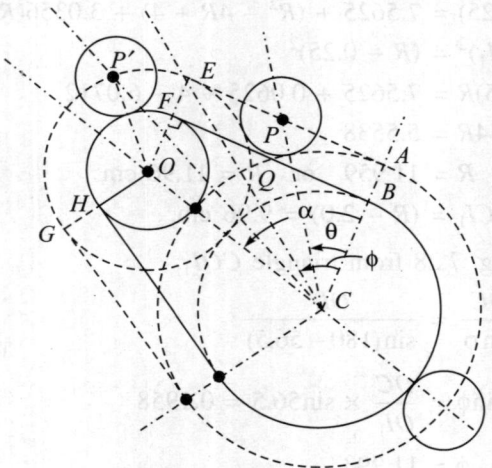

Fig. 7.30 Tangent cam with Roller follower.

For this position of roller, lift of follower,

$$y = CP - CA = \frac{CA}{\cos\theta} - CA = CA\left(\frac{1}{\cos\theta} - 1\right) = (r_b + r)\left(\frac{1}{\cos\theta} - 1\right) \quad (7.87)$$

For velocity, differentiating equation (7.87) w.r. to time,

$$V = dy/dt = (r_b + r)\left[\frac{0 + \sin\theta}{\cos^2\theta} - 0\right]\frac{d\theta}{dt} = (r_b + r)\left(\frac{\sin\theta}{\cos^2\theta}\right)\omega \qquad (7.88)$$

For $\theta < 90°$ as θ increases, $\sin\theta$ increases and $\cos\theta$ decreases and therefore it follows from equation (7.88) that velocity is maximum when θ is maximum. Thus, the velocity is maximum when the straight flank just merges in the nose circle, *i.e.* when $\theta = \phi$. Therefore,

$$V_{max} = (r_b + r)\omega \times \left(\frac{\sin\phi}{\cos^2\phi}\right) \qquad (7.89)$$

Again, differentiating equation (7.88) w.r. to time for acceleration,

$$A = (r_b + r)\omega^2\left[\frac{\cos^2\theta\cos\theta - \sin\theta\,(2\cos\theta)(-\sin\theta)}{\cos^4\theta}\right]$$

$$= (r_b + r)\omega^2\frac{\cos^2\theta + 2\sin^2\theta)}{\cos^3\theta} = (r_b + r)\omega^2\frac{(2 - \cos^2\theta)}{\cos^3\theta} \qquad (7.90)$$

As $\cos\theta$ decreases with increasing θ (for $\theta < 90°$), the numerator $(2 - \cos^2\theta)$ increases as θ increases. Thus minimum value of A occurs when $\theta = 0$, and maximum value occurs when $\theta = \phi$ Thus,

$$A_{min} = (r_b + r)\omega^2 \text{ and } A_{max} = \left(\frac{2 - \cos^2\phi}{\cos^3\phi}\right) \times (r_b + r)\omega^2 \qquad (7.91)$$

Contact between Cam and Roller on Nose Circle

As the roller moves along circular nose, its centre is always placed at a distance of $(r' + r)$ from nose centre. Again, in the uninverted condition the centre of roller always move along the centre line of follower. Hence the roller centre can be thought of as slider of a slider-crank mechanism in which line CP' (for instance) represents frame, CO as crank and OP' as a connecting rod. Thus, by replacing crank angle θ of rotation by $(\theta - \phi)$ in the present case, the expression for slider displacement of the equation (3.68) can be used for follower displacement also. Thus, replacing θ by $(\theta - \phi)$ and r by d,

$$y = \{1 - \cos(\theta - \phi) + n - \sqrt{n^2 - \sin^2(\theta - \phi)}\}$$

where

$$n = \frac{OP'}{OC} = \frac{(r' + r)}{d}$$

Substituting above

$$y = d[1 - \cos(\theta - \phi)] + (r' + r) - \sqrt{(r' + r)^2 - d^2\sin^2(\theta - \phi)} \qquad (7.92)$$

Also, from equation (3.69) by replacing r by d and θ by $(\theta - \phi)$,

$$V = d\omega\left[\sin(\theta - \phi) + \frac{\sin 2(\theta - \phi)}{2\sqrt{n^2 - \sin^2(\theta - \phi)}}\right] = \omega\left[d\sin(\theta - \phi) + \frac{d^2\sin 2(\theta - \phi)}{2\sqrt{(r' + r)^2 - d^2\sin^2(\theta - \phi)}}\right]$$

Finally, from equation (3.70),

$$A = d\omega^2 \left\{ \frac{\cos(\theta - \phi) + n^2\cos^2(\theta - \phi) + \sin^4(\theta - \phi)}{[n^2 - \sin^2(\theta - \phi)]^{3/2}} \right\}$$

EXAMPLE 7.10 Particulars of a symmetric tangent cam operating a roller follower are as under:

Least radius of cam: 3 cm; Roller radius: 1.5 cm; Angle of ascent: 75°;
Total lift: 1.5 cm; Speed of cam shaft: 600 r.p.m.

Calculate the principal dimensions and the equations of displacement curve when the follower is in contact with straight flank and circular nose.

Solution: d = distance between nose centre and base circle centre = CO

r' = nose circle radius; r = roller radius = 1.5 cm

From Fig. 7.31,

$$(d + r') = r_b + \text{lift}$$

Therefore $d \equiv CO = (3 + 1.5) - r' = (4.5 - r')$

Also, $(CQ) + r' = r_b$ or $CQ = (3 - r')$

Hence, from right angled triangle OCQ,

$$\cos\alpha = \cos 75 = \frac{CQ}{OC} \quad \text{or,} \quad 0.2585 = \frac{(3 - r')}{(4.5 - r')}$$

Solving $r' = 2.477$ cm **Ans.**

Also from $d = CO = (4.5 - r') = (4.5 - 2.477) = 2.023$ **Ans.**

From right angled triangle GKC,

$$\tan\phi = \frac{GK}{CK} = \frac{OQ}{CK}$$

$$= \frac{(CO)\sin\alpha}{(r_b + \text{roller rad})} = \frac{(2.023)\sin 75}{(3 + 1.5)} = 0.434$$

Therefore $\phi = 23.46°$ **Ans.**

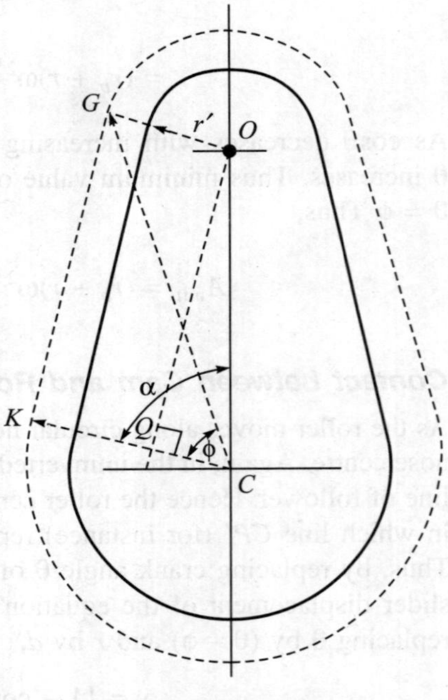

Equation of displacement

Fig. 7.31 Symmetric tangent cam.

(i) When roller is in contact on straight flank, the displacement equation is

$$y = (r_b + r)\left(\frac{1}{\cos\theta} - 1\right) = (3 + 1.5)\left(\frac{1}{\cos\theta} - 1\right) = 4.5\left(\frac{1}{\cos\theta} - 1\right) \quad \textbf{Ans.}$$

(ii) When the roller is in contact on circular nose, the displacement equation is

$$y = d[1 - \cos(\theta - \phi)] + (r' + r)\sqrt{(r' + r)^2 - d^2\sin^2(\theta - \phi)}$$

or $$y = 4.5[1 - \cos(\theta - \phi)] + (2.477 + 1.5)\sqrt{(2.477 + 1.5)^2 - (4.5)^2\sin^2(\theta - \phi)}$$

$$= 4.5[1 - \cos(\theta - \phi)] + (3.977)\sqrt{15.8 - 20.25\sin^2(\theta - \phi)} \quad \textbf{Ans.}$$

EXAMPLE 7.11   A cam consists of a circular disc of diameter 7.5 cm with its centre displaced 2.5 cm from the cam shaft axis. The follower has a flat face (horizontal) in contact with the cam and the line of action of the follower is vertical and passes through the shaft axis as shown in Fig. 7.32. The follower weights 23 Newton and is pressed downwards by a spring which has a stiffness of 35 N/cm. In the lowest position, the spring force is 45 Newton.

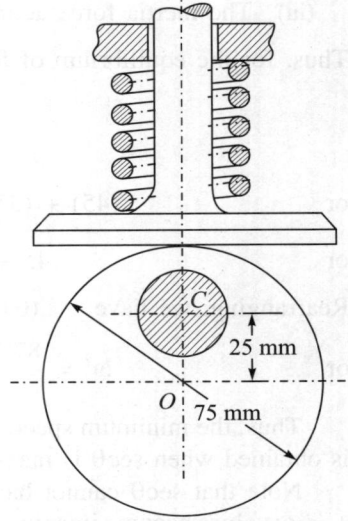

(a) Derive the expression for the acceleration of the follower in terms of the angle of rotation from the beginning of the lift,

(b) As the cam shaft speed is gradually increased, a value is reached at which the follower begins to lift from the cam surface. Determine the cam shaft speed for the condition.   (Univ. of Roorkee, 1980)

Fig. 7.32 Circular disc cam with flat faced follower.

Solution: With centre of rotation at C rather than at O, the geometric centre of circular disc, the centre of disc O rotates in circular path of radius $CO = 25$ mm. The distance y through which the centre of the disc O is lifted in vertical direction, also represents the displacement of flat-faced follower in vertical direction. For the position of circular disc, shown in dotted line in Fig. 7.33, the centre of disc rotates to O',

$$y = EF = OP = OC - PC = OC (1 - \cos\theta) = 2.5 (1 - \cos\theta).$$

For velocity and acceleration of follower, we differentiate this expression successively w.r. to time. Thus

$$V = \frac{dy}{dt} = \left(\frac{dy}{d\theta}\right)\left(\frac{d\theta}{dt}\right) = + (2.5)\omega\sin\theta \text{ cm/s}$$

where ω = speed of rotation of cam, which is uniform

$$= 2\pi \times \frac{N}{60}$$

And acceleration

$$a = \frac{dv}{dt} = \left(\frac{dv}{d\theta}\right)\left(\frac{d\theta}{dt}\right) = (2.5)\omega^2\cos\theta \text{ cm/s}^2$$

The force analysis of follower can be carried out by considering the following forces:

(i) Downwards force due to initial compression of spring = F

(ii) Downwards force due to additional compression of spring = ky, where k is the stiffness of spring

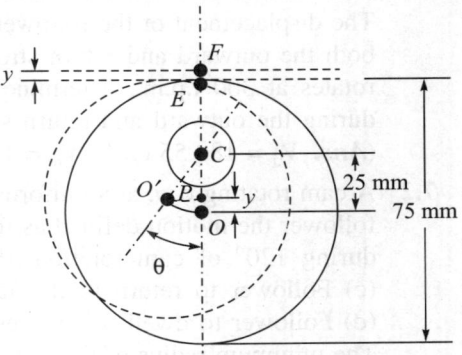

Fig. 7.33 Illustration for displacement of follower.

(iii) The inertia force acting upwards due to follower mass and its acceleration = $(w/g)a$.

Thus, for the equilibrium of follower (*Refer* section 14.2),

$$F + ky + \frac{W}{g}a = 0$$

or $\qquad (45) + (35)(2.5)(1 - \cos\theta) + \left(\dfrac{23}{981}\right)(2.5)\omega^2\cos\theta = 0$

or $\qquad 45 + 87.5(1 \cos\theta) + (0.0586)\omega^2\cos\theta = 0$

Rearranging, we have $\quad (0.0586\omega^2 - 87.5)\cos\theta = (-45 - 87.5)$

or $\qquad \omega^2 = \dfrac{87.5 - (132.5)\sec\theta}{0.0586} = 1493.17 - (2261.09)\sec\theta$

Thus, the minimum speed at which the follower begins to lift away from the cam (*i.e.* jump) is obtained when $\sec\theta$ is maximum.

Note that $\sec\theta$ cannot have values between -1 to $+1$. Further with values $\sec\theta = +1$ to $+\infty$, ω values become imaginary. Hence smallest feasible value of ω is obtained at $\theta = 180$, where $\sec\theta = -1$.

Hence, smallest possible value of ω^2 is

$$\omega^2 = 1493.17 + 2261.09 = 3754.26$$

Therefore $\qquad \omega = 61.27$ rad/s

Therefore $\qquad N_{min} = \dfrac{61.27 \times 60}{2\pi} = 585.1$ r.p.m. **Ans.**

REVIEW QUESTIONS

7.1 Draw the cam profile of a cam operating a knife-edge follower from the following data:
(a) Follower to move outward through 4 cm during 60° of cam rotation. (b) Follower to dwell for the next 45°. (c) Follower to return to its original position during next 90°. (d) Follower to dwell for the rest of the cam rotation.
The displacement of the follower is to take place with simple harmonic motion during both the outward and return strokes. The least radius of the cam is 5 cm. If the cam rotates at 300 r.p.m., determine maximum velocity and acceleration of the follower during the outward and return strokes. (AMIE, Summer 1976.)
(*Ans.* $V_0 = 188.55$ cm/s; $V_R = 125.66$ cm/s; $A_o = 17776.6$ cm/s^2; $A_R = 7895.68$ cm/s^2)

7.2 A cam rotating c.w. at a uniform speed of 1000 r.p.m. is required to give a knife-edge follower the motion defined as follows: (a) Follower to move outward through 2.5 cm during 120° of cam rotation. (b) Follower to dwell for next 60° of cam rotation. (c) Follower to return to its starting position during the next 90° of cam rotation. (d) Follower to dwell for the rest of the cam rotation.
The minimum radius of the cam is 5 cm and the line of stroke of the follower is offset by 2 cm from the axis of the cam shaft. If the displacement of the follower takes place

with uniform and equal acceleration and retardation on both the outward and return strokes, draw profile of the cam and find the maximum velocity and acceleration during outstroke and return stroke. (Nagpur Univ., 1979)

(***Ans.*** V_o = 249.28 cm/s; V_R = 333.44 cm/s; A_o = 24857.3 cm/s^2; A_R = 44472.7 cm/s^2)

7.3 Draw cam profile for the follower motion specified in problem 7.1, if the follower is fitted with a roller of radius 1.0 cm.

7.4 Draw cam profile for the follower motion prescribed in problem 7.2 if a roller of radius 12 mm is fitted to the follower.

7.5 Draw the displacement, velocity and acceleration diagram for a follower moving with S.H.M. during outstroke and return strokes from the following details:

Follower to complete outstroke of 5 cm in 1/3rd revolution of cam; to dwell for a period of 1/12th of the revolution; to return to its initial position in 1/6th of the revolution and then to dwell for the rest of the revolution. The cam shaft runs at a speed of 360 r.p.m.

7.6 A follower is to move outwards 5 cm with S.H.M. while the cam turns through 180° with angular velocity ω. The follower is to return with S.H.M. during the next 150° of rotation and dwell for 30°. Sketch and dimension the displacements velocity and acceleration diagrams of the follower. (AMIE, Summer 1993)

7.7 Draw the profile of a cam for the follower motion prescribed below: (a) Roller follower of 3 cm diameter; roller moves outward with S.H.M. during 160° of cam rotation. (b) It rests for 20 degrees of cam rotation. (c) Returns with uniform equal acceleration and retardation during 160° of cam rotation. (d) Dwells for the remaining 20° of cam rotation.

Assume a minimum radius of cam to be 5 cm and maximum lift of follower to be 4 cm. The cam rotates at a uniform speed in c.w. direction and the line of stroke of follower passes through the cam shaft-centre.

7.8 A disc cam rotating in a c.w. direction is used to move a reciprocating roller follower with S.H.M. in a radial path as given below: (i) Outstroke with maximum displacement of 25 mm during 120° of cam rotation. (ii) Dwell for 60° of cam rotation. (iii) Return stroke with maximum displacement of 25 mm during 90° of cam rotation, and (iv) Dwell during remaining 90° of cam rotation.

The line of action of reciprocating follower passes through the cam shaft axis. The minimum radius of cam is 20 mm. If the cam rotates at a uniform speed of 300 r.p.m. find the maximum velocity and acceleration during outstroke and return stroke. The roller diameter is 8 mm.

(***Ans.*** V_o = 58.9 cm/s; V_R = 78.54 cm/s; A_o = 2775.82 cm/s^2; A_R = 4934.8 cm/s^2)

7.9 Draw the profile of the cam in example problem 7.8 when the line of action of follower is offset by 20 mm to the left from cam shaft axis.

7.10 Find out analytically maximum values of pressure angles occurring during outstroke and return stroke of example problem 7.9. Show that when the follower is offset by same amount of the right of cam axis, value of maximum pressure angle increases during outstroke.

7.11 Draw the profile of a cam which will give a lift of 37.5 mm to a roller follower. The diameter of the roller is 25 mm and the line of stroke is offset by 20 mm from the axis of the cam. The outstroke of the follower takes place with S.H.M. during 72° of cam rotation followed by a period of rest during 18° of cam rotation. The follower then returns with equal uniform acceleration and retardation during 54° of cam rotation. The minimum radius of the cam is 50 mm.

If the cam rotates at a uniform speed of 240 r.p.m., find the maximum acceleration during outstroke and return stroke. (Pune Univ.)

$$(\textbf{\textit{Ans.}}\ A_0 = 74.022\ \text{m/s}^2,\ A_R = 106.667\ \text{m/s}^2)$$

7.12 A flat-faced mushroom follower is operated by a uniformly rotating cam. The follower raised through a distance of 25 mm in 120° rotation of cam, remains at rest for the next 30° and is lowered during further 120° rotation of the cam. The raising of follower takes place with cycloidal motion and the lowering with uniform acceleration and deceleration. However, the uniform acceleration is 2/3rd of the uniform deceleration. The least radius of cam is 25 mm which rotates at 300 r.p.m.

Draw the cam profile and determine the values of the maximum velocity and maximum acceleration during rising, and maximum velocity and uniform acceleration and deceleration during lowering of the follower.

$$[\textbf{\textit{Ans.}}\ V_o = 0.75\ \text{m/s};\ A_o = 35.31\ \text{m/s}^2;\ V_R = 0.075\ \text{m/s};\ A_R = 18.75\ \text{m/s}^2\ \text{and}$$
$$A_R = 28.13\ \text{m/s}^2\ (\text{retardation})]$$

Hint: Uniform accen. $A = \dfrac{2}{3}A'$, where A' is the uniform deceleration. From $s = ut + \dfrac{1}{2}Atr^2$,

as $u = 0$ the distance moved during acceleration $s = \dfrac{1}{2}\left(\dfrac{2}{3}A'\right)\left(\dfrac{3}{2}t'\right)^2 = 1.5\left(\dfrac{1}{2}A't'^2\right)$

$= (1.5)s'$

where, $s' = $ distance travelled during retardation).

7.13 A radial translating flat face follower has a lift of 3 cm. The rise takes place with S.H.M. for 180° cam rotation, followed by a dwell for 30° and return motion with S.H.M. for next 120°. The balance of 30° of cam rotation is dwell. The base circle radius of the cam is 3 cm. Obtain the cam profile and establish the minimum width of the follower face with a clearance of 0.3 cm at both the ends. The cam rotates in c.c.w. direction.

$$(\textbf{\textit{Ans.}}\ 4.37\ \text{cm})$$

7.14 Find out minimum width of follower required for flat faced follower of example 7.13 analytically.

7.15 The following data relates to a cam operating on oscillating roller follower:

Minimum diameter of the cam = 44 mm; Diameter of roller = 14 mm; Length of follower arm = 40 mm; Distance of fulcrum centre from cam centre = 50 mm; Angles of rise, dwell and return in sequence are 75°, 60° and 105° respectively.

Take angle of oscillation of follower = 28° and draw the cam profile if follower completes outward and return strokes both with S.H.M.

7.16 A circular cam with a radius of 25 cm is driving a flat-faced translating follower. The follower has a lift of 20 cm. Determine the minimum required face width of the follower.

(AMIE, Winter 1993)

7.17 A cam profile consists of two circular arcs of radii 24 mm and 12 mm joined by straight lines, giving the follower a lift of 12 mm. The follower is a roller of 24 mm radius and its line of action is a straight line passing through the cam shaft axis. When the cam shaft has a uniform speed of 500 r.p.m., find the maximum velocity and acceleration of the follower while in contact with straight flank of the cam. (Jodhpur Univ.)

$$(\textbf{\textit{Ans.}}\ 1.2\ \text{m/s};\ 197.7\ \text{m/s}^2)$$

7.18 Following particulars relate to a symmetrical tangent cam operating a roller follower: Least radius = 30 mm; nose radius = 24 mm; roller radius = 17.5 mm; distance between cam shaft and nose circle centres = 23.5 mm; angle of action of cam = 150, 198; cam shaft speed = 600 r.p.m. Assuming that there is no dwell between ascent and descent, determine the lift of valve and the acceleration of follower at a point where straight flank merges into the circular nose. (Agra Univ.) (**_Ans._** 17.5 mm; 304 m/s^2)

7.19 Following particulars relate to a symmetrical tangent cam having a roller follower: Minimum radius of the cam = 40 mm; lift = 20 mm; speed = 360 r.p.m.; roller diameter = 44 mm; angle of ascent = 60°.
Calculate the acceleration of follower at the beginning of lift. Also find its values when the roller just touches the nose and is at the apex of the circular nose. Sketch the variation of displacement, velocity and acceleration during ascent.

$$(\textbf{\textit{Ans.}}\ 88.1\ \text{m/s}^2;\ 165\ \text{m/s}^2;\ -92.5\ \text{m/s}^2;\ -111.1\ \text{m/s}^2)$$

7.20 A flat-ended valve tappet is operated by a symmetrical cam with circular arcs for flank and nose profiles. The total angle of action is 150°, base circle diameter is 125 mm and the lift is 25 mm. During the lift, the period of acceleration is half that of the period of deceleration. Speed of camshaft is 1250 r.p.m. The straight line path of the tappet passes through the cam axis. Find the radii of nose and the flank. Also find maximum acceleration and deceleration during the life.

$$(\textbf{\textit{Ans.}}\ 40\ \text{mm};\ 149\ \text{mm};\ 1475\ \text{m/s}^2;\ 814\ \text{m/s}^2)$$

7.21 Find out the maximum value of pressure angle during outstroke for a radial cam with base circle radius of 5 cm; $h = 3$ cm. The cam drives a radial follower with parabolic follower motion for outstroke angle from 0° to 120°.

Chapter 8

GEARS

8.1 INTRODUCTION

A gear is a toothed member (link) which is commonly used for transmitting motion by means of successively engaging teeth from a rotating shaft to another or from a rotating shaft to a body which translates. It may be recalled that a motion of translation can be considered to be that of rotation about an axis at infinity. Like belt/rope drive, a gear drive also involves a higher pair between driving and driven member. Mating gear teeth, acting against each other to produce rotary motion, may be likened to a lobe of a cam and follower. For this reason, it is also possible to define a gear as multi-lobed cam. A study of rolling contact is basic to the development of the theory of gears, and therefore a brief study of rolling contact may be quite useful for a better understanding of theory of gearing.

8.2 ROLLING CONTACT AND POSITIVE DRIVE

A pure rolling contact exists on a direct contact mechanism only if the linear velocities of the bodies, at their point of contact, are identical. A necessary condition for this requires that the point of contact lies on the line of centres. As against this, sliding exists whenever the bodies have relative motion along the tangent through their point of contact. Thus, if V_{p2} and V_{p3} are the tangential velocities of the contacting points P_2 and P_3 on bodies 2 and 3, relative velocity $(V_{p_2} - V_{p_3})$ then represents the velocity of sliding.

A rolling contact between two cylindrical bodies can occur either externally (Fig 8.1a) or internally (Fig 8.1b). For having identical velocities of the contacting points P_2 and P_3, the two members in Fig 8.1(a) must rotate in opposite sense while in the case of internal contact in Fig 8.1(b), the two cylinders must rotate in the same sense. **Positive drive exits in a direct contact mechanism, if the motion of the driving member compels the driven member to move**. It can be shown that for *positive drive to exist, the common normal through the point of contact **must not** pass through any or both of the centre of rotation*. In the case of internal and external contacts between rotating cylinders of Figs. 8.1(a) and (b), positive drive does not exist as the common normal at the point of contact passes through both the centres of rotation. In these

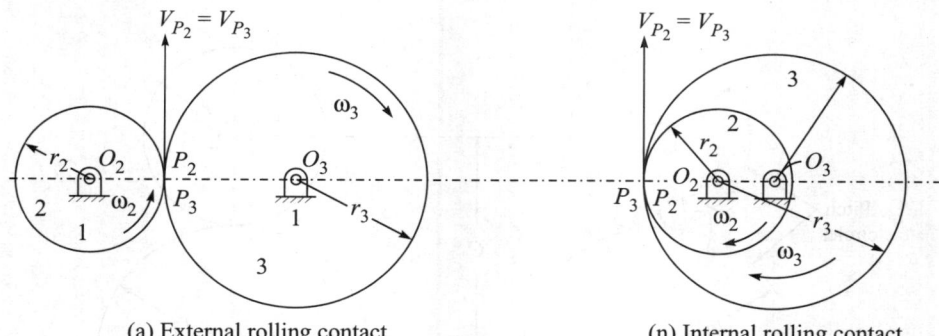

(a) External rolling contact   (n) Internal rolling contact

Fig. 8.1

cases, motion is transmitted from one body to the other only if there is sufficient friction at the contacting surfaces. These mechanisms are therefore called *friction drives*. Friction drives are particularly desirable in certain applications which require overload protection. In such situations when excessive load comes on the shaft of the driven member, the bodies can slip before any of the machine parts are damaged. In general, friction drives are not used to transmit large forces because this necessiates large bearing loads R (Recall that $F = \mu.R$) in radial direction.

The power which can be transmitted by rolling bodies is limited by the friction which can be developed at the surfaces. When excessive load is encountered, slippage occurs. *Hence in order to provide a positive drive, teeth are placed on the contacting members and the resulting machine members are called gears*. In the current chapter, we intend to study only those gears which produce a constant angular velocity ratio. Except for worm gears, all the gears under discussion here are equivalent to rolling bodies. The motion transmitted by gear teeth, whatever be their shape and size, consists partly of rolling and partly of sliding. This is the major reason of considering a gear pair as a higher pair.

8.3 CLASSIFICATION OF GEARS

Gears may be classified on the basis of relative orientation of the two shafts carrying gears.

8.3.1 Gears Mounted on Parallel Axes

Irrespective of the nature of contact, a pair of gears mounted on parallel shaft produces a uniform motion that is equivalent to the rolling motion without slipping between two cylinders. This is shown in Fig. 8.2(b).

Straight spur Gears

Gear pairs, having parallel axes of rotation, are spur gears. Simplest type of spur gears are the straight spur gears. They have straight teeth parallel to the gear axes. A pair of straight spur gear is shown in Fig 8.3(a). As a pair of teeth enters into the engagement, the contact between the two teeth takes place over the entire width along a line parallel to the axes of rotation. Thus, there is a sudden application of load associated with high impact stresses and excessive noise at high speeds. As the gears rotate further, the line of contact between a pair of teeth goes on shifting

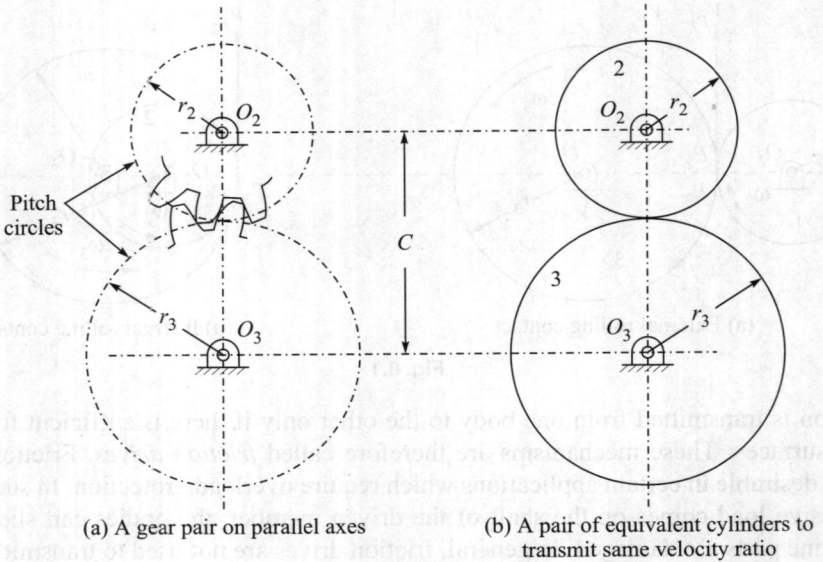

(a) A gear pair on parallel axes (b) A pair of equivalent cylinders to transmit same velocity ratio

Fig. 8.2

parallel to itself as shown by representative successive lines a–a, b–b, c–c, d–d,..., in Fig. 8.3(b). Shape of tooth profile of a straight spur gear is shown in Fig. 8.31(a).

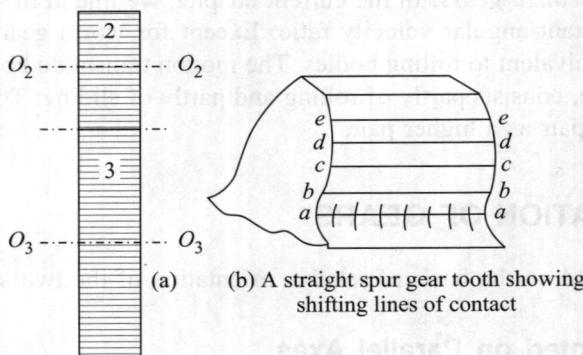

(a) (b) A straight spur gear tooth showing shifting lines of contact

Fig. 8.3 Straight spur gears in mesh.

Helical Spur Gears (or Simply Helical Gears)

In helical gears, the teeth are cut on helices instead of straight across the gear parallel to the axis [(*see* Fig. 8.4(a)]. Two mating toothed wheels of helical type must have the same helix angle. Further the teeth must be of opposite hand as shown in Fig. 8.4(a).

As a pair of teeth comes in contact, the engagement begins only at the point of leading edge of the curved teeth. As the gears rotate further, the contact progresses across the tooth along the diagonal line. This is shown in Fig. 8.4(b) where the straight lines a–a, b–b, c–c, d–d, etc. are the successive lines of contact. These lines are the elements of the involute helicoid. Thus, in helical gears the load application is gradual and as such, impact stresses and noise are also less

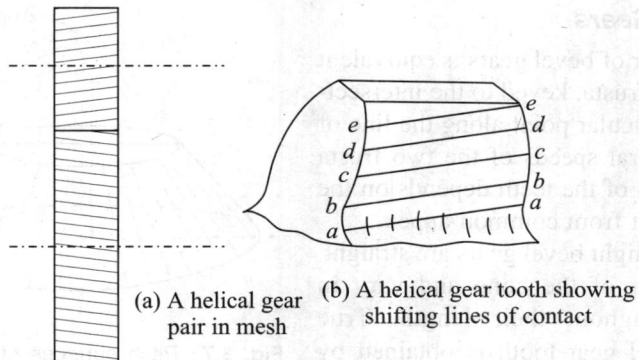

(a) A helical gear pair in mesh

(b) A helical gear tooth showing shifting lines of contact

Fig. 8.4   Helical spur gears.

in comparison to straight spur gears. Helical gears can therefore be used at higher velocities and can carry higher loads compared to straight spur gears. Shape of a helical gear tooth profile is shown in Fig. 8.31(b).

The angle at which the gear teeth are cut in helical gears is known as the helix angle.

Herringbone Gears

Because of helix angle in a single helical gear a component of load, acting on gear tooth, acts in axial direction producing end-thrust vide Fig. 8.34. This is considered to be disadvantage with single helical gear. A herringbone gear is equivalent to two helical gears of same helix angle but of

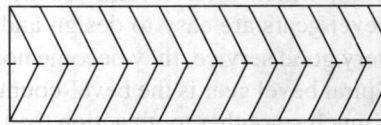

Fig. 8.5   Herringbone gear.

opposite hand which are held together as an integral piece. In the case of herringbone gear, the two rows of teeth are separated by a groove required for tool runout. In view of two rows of teeth of opposite hand, the axial thrusts mutually cancel out. In view of this, herringbone gears can be run at high speeds with less of vibration and noise.

Spur Rack and Pinion

In this case the spur rack can be considered to be a spur gear of infinite pitch radius with its axis of rotation placed at infinity parallel to that of pinion. The pinion rotates while the rack translates. The combination can be used to convert either the rotary motion into translatory motion or vice-versa (*see* Fig. 8.6).

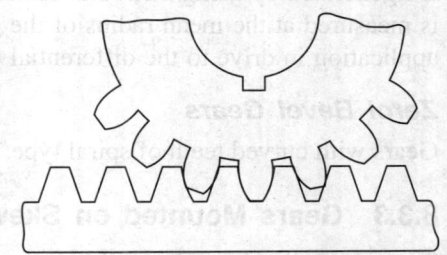

Fig. 8.6   Spur rack and pinion.

8.3.2   Gears Mounted on Intersecting Shaft Axes

Bevel gears are used to transmit power between two shafts whose axes intersect. The pitch surfaces of the wheels are frusta of cones which, by pure rolling motion, will transmit the same motion as the bevel wheels (*see* Fig. 8.7). The point of intersection of the two shafts must coincide with common apex of both the frustums of cones. Any angle can be used as included angle between shaft axes, however an angle of 90° is the most common (*see* Fig. 8.7). The most accurate teeth are obtained by method of generation.

Straight Bevel Gears

Kinematically, a pair of bevel gears is equivalent to a pair of conical frusta, keyed to the intersecting shafts. At a particular point along the line of contact, the peripheral speeds of the two frusta are equal. The pitch of the teeth depends on the distance of the point from common appex.

The teeth of straight bevel gears are straight, along the generator of the cone and vary in cross-section throughout their length. True shape of such bevel gear tooth is obtained by taking a spherical section through the tooth, with

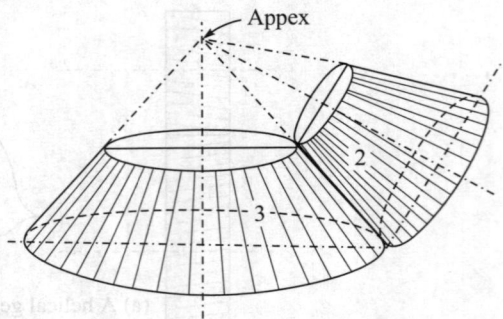

Fig. 8.7 Pitch surfaces of frusta of cone which can transmit same velocity ratio.

the centre of the sphere at common apex. Straight bevel gears usually connect shafts which run at low speeds. When two straight bevel gears of same size connect two shafts at right angles to each other, the gears are called *Mitre Gears*.

Spiral Bevel Gears (Ref Fig. 8.8)

It is also possible to produce bevels with spiral and curved spiral teeth, which offer certain practical advantages. Although straight bevel gears are easy to design and simple to manufacture and give very good service, they become noisy at higher pitch line velocities. Spiral bevel gear is the bevel-counterpart of the helical gear. It gives a much smoother tooth action than straight bevel gears and are very useful for high-speed applications. The tooth elements are theoretically spirals, but in practice they are made of circular arcs because of the ease of manufacture.

Spiral bevel gear teeth are conjugate to a basic crown rack which are generated by using a circular cutter. The spiral or helix angle φ is measured at the mean radius of the gear. Spiral bevel gears find application in drive to the differential of an automobile.

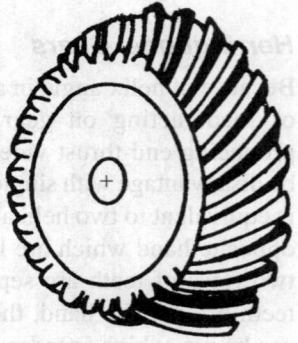

Fig. 8.8 Spiral bevel gear.

Zerol Bevel Gears

Gears with curved teeth of spiral type, but with spiral angle φ = 0° are called Zerol Bevel Gears.

8.3.3 Gears Mounted on Skew Shaft Axes

Several applications demand gears to connect two non-parallel and non-intersecting shafts. Unlike gears with parallel axes or intersecting shaft axes, uniform rotary motion by pure rolling cannot be obtained by gears of this type.

Hypoid Gears

Whenever it is necessary to have a gear similar to a bevel gear but having an off set between shaft-axes, hypoid gear becomes the obvious choice. The discs through which motion is transmitted are frusta of hyperboloides. Hypoid gears are based on hyperboloids of revolution. A necessary condition to connect two non-parallel non-intersecting shafts by hyperboloids, is that the two hyperboloids must be generated by revolving the same line (*i.e.*, their line of contact) about the axis of each of the two connected shafts.

Figure 8.9 shows one such hyperboloid generated through successive positions of such line revolving about the axis of a shaft. Due to manufacturing difficulties, in practice, only portions (frusta) of hyperboloids are used for motion transmission.

Relative motion between gears of this type consist partly of rolling and partly of sliding, along the common line, and has much in common with worm gears. These are used to drive wheel axles.

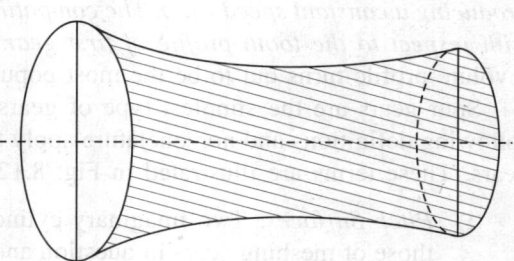

Fig. 8.9 A hyperboloid.

Crossed Helical Gears (Spiral Gears)

The teeth of crossed helical gears have a point contact with each other, which changes to line contact as the gears wear out. For this reason the use of double helical gears or spiral gears is limited to light loads. There is no difference between a crossed helical gear and a simple helical (spur) gear. They are manufactured in the same way. The difference comes to fore when crossed helical gears are mounted in mesh with each other. A pair of meshed crossed-helical gears have teeth which are usually of the same hand. By a suitable choice of helix angle for the mating gears, the two shafts can have any angle of inclinations (Fig. 8.10). These gears find use in the drive for feed mechanisms on machine tools, camshafts and oil pumps on small I.C. engines.

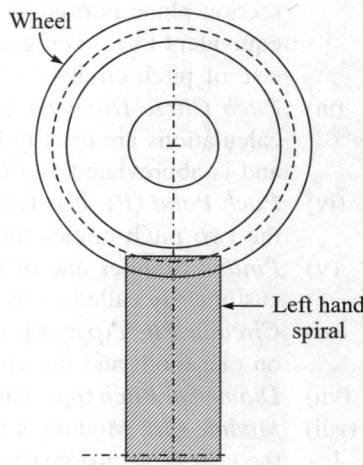

Fig. 8.10 Spiral gears.

Worm Gears

Worm is a member having screwlike thread and it is usual to call worm teeth as threads. A worm and worm-wheel combination is similar to a pair of mating spiral gears except that the larger wheel (usually) has a hollow or concave shape such that it partially encloses the worm. For this reason, they have a line contact and are capable of transmitting more power. The worm-wheel is usually the driven member of the pair and are shown in Fig. 8.11. The two shafts may have any angle between them, but normally it is 90°. The sliding velocity of worm wheel is higher compared to any other type of gear pairs.

8.4 NOMENCLATURE FOR STRAIGHT SPUR GEARS

Almost any tooth profile can be used successfully for a spur gear as long as the mating tooth profile is compatible for

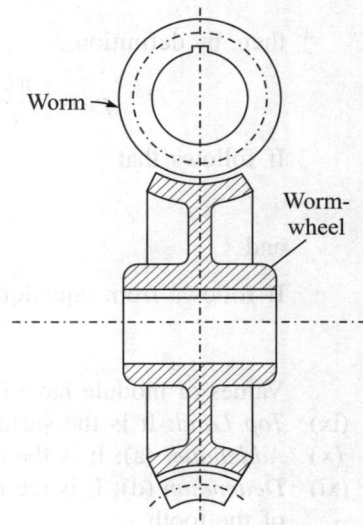

Fig. 8.11 Worm gears in mesh.

producing a constant speed ratio. The compatible tooth profile is technically known as **conjugate** *with respect to the tooth profile of first gear.* Though any tooth profile is possible for gears, involute profile turns out to be the most popular amongst all.

Spur gears are the simplest type of gears and as such may be studied first. Many of the following definitions and nomenclature apply to straight spur gears, and are basic to all types of gears. These terms are illustrated in Fig. 8.12(a) & (b).

(i) *Pitch Surfaces:* Two imaginary cylindrical surfaces having their axes coinciding with those of meshing gears in question and transmitting same velocity ratio through friction alone (without any slip), are called *equivalent pitch surfaces.*

(ii) *Pitch Circle:* Circle obtained by taking transverse section of pitch surface (*i.e.,* using section plane normal to the axis of pitch cylinder) is called *pitch circle.* Pitch circles equivalent to a given gear pair are shown in Fig. 8.2(b). Being imaginary by nature, the size of pitch circles for a given pair of mating gears changes with centre distance.

(iii) *Pitch Circle Diameter* (*PCD*): Pitch circle is a theoretical circle upon which all gear calculations are usually based. Diameter of this circle is called *pitch circle diameter D* and is abbreviated as *PCD.*

(iv) *Pitch Point* (*P*): It is that point on a line joining centres of two mating gears, at which the two pitch circles meet.

(v) *Pinion:* Smaller one of the two mating gears is called *pinion,* while the larger one is customarily called *gear.*

(vi) *Circular Pitch* (p_c): It is the arcular distance measured along pitch circle between a point on one tooth and the corresponding point on the adjacent tooth. (*see* Fig. 8.12a).

(vii) *Diametral Pitch* (p_d): This is defined as the number of teeth per unit pitch circle diameter.

(viii) *Module* (*m*): Module is defined as the ratio of pitch diameter to the number of teeth on the gear. The customary unit of module is millimetre. Module is the index of tooth size.

If T = number of teeth on gear wheel,

D = pitch circle diameter in mm.,

then, by definition

$$p_c = \left(\frac{\pi D}{T}\right); \ p_d = (T/D) \text{ and module } m = (D/T)$$

If follows that

$$(p_c)(p_d) = \pi \quad\quad\quad (8.1)$$

and

$$m = (1/p_d) \quad\quad\quad (8.2)$$

If follows from equations (8.1) and (8.2) that,

$$p_c = m\pi \qu\quad\quad\quad (8.3)$$

Values of module have been standardised by *BIS,* the Bureau of Indian standards.

(ix) *Top Land*: It is the surface at the top of a tooth as shown in Fig. 8.12(b).

(x) *Addendum* (*a*): It is the radial distance between the top land and pitch circle.

(xi) *Dedendum* (*d*): It is the radial distance measured between the pitch circle and the root of the tooth.

(xii) *Total Depth*: It is the radial distance measured between top-land and root of the tooth. In other words, total depth = addendum + dedendum.

(xiii) *Working Depth*: It is that portion of the total depth of gear tooth along which actual contact occurs between mating teeth. In other words, working depth equals sum of the addenda of mating gears. It is thus different from total depth.

(xiv) *Tooth Thickness*: It is the arcular distance measured along pitch circle between involute profiles on the two sides of the tooth (*see* Fig. 8.12b).

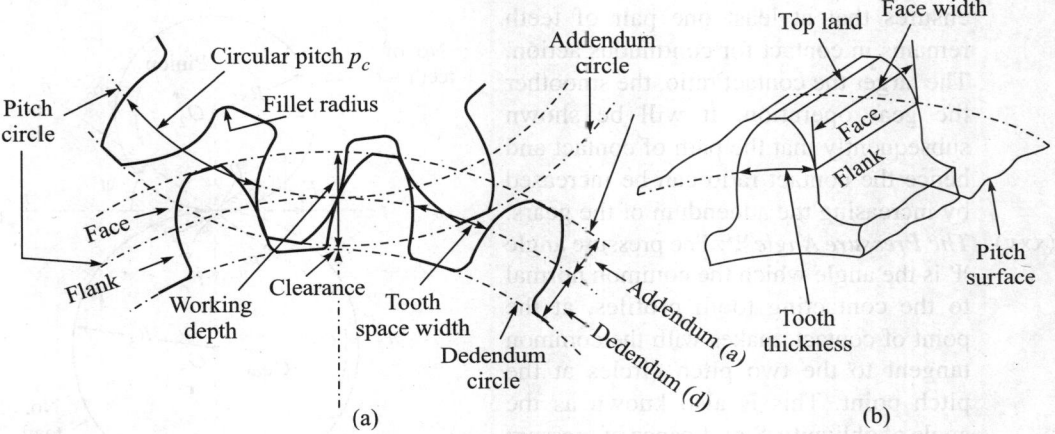

Fig. 8.12 Illustration for gear tooth terminology.

(xv) *Tooth Space Width*: It is the width of the space between two adjacent teeth, measured along pitch circle.

(xvi) *Face:* It is the working surface of tooth profile lying above the pitch surface.

(xvii) *Flank*: This is the portion of the working surface of the tooth profile lying below the pitch surface.

(xviii) *Fillet Radius*: It is the radius of circle (concave curve) that connects the root circle to the flank portion of the tooth.

(xix) *The Clearance* (c): This is the amount by which dedendum of a gear exceeds the addendum of mating gear.

(xx) *Backlash*: This is the amount by which width of tooth space exceeds the tooth thickness of the mating gear.

(xxi) *Gear Ratio*: It is the ratio of larger to the smaller number of teeth in a pair of mating gears.

(xxii) *Path of Contact*: In general, the path of contact is the curve traced by the point of contact of two teeth from the beginning to the end of engagement. In the case of involute gears, the path of contact is along a straight line tangent to base circles.

(xxiii) *Arc of Contact*: The path described by a point on tooth on pitch circle, from beginning to the end of engagement of given tooth pair, is called *arc of contact*.

(xxiv) *Contact Ratio*: In order to reduce average load shared by a pair of teeth, it is desirable that on a time basis, more than one pair should remain in contact at all the time. Contact ratio is defined as the average number of pairs of teeth which are in contact. Mathematically,

$$\text{Contact Ratio} = \frac{\text{length of arc of contact}}{\text{circular pitch}}$$

The contact ratio usually is not a whole number. If the ratio is 1.5, it does not mean that there are always 1.5 pairs of teeth in contact. The real meaning is that there are alternately one pair and two pairs of teeth in contact and on a time average basis the average is 1.5. Theoretical minimum value of contact ratio is 1.0 but the recommended practice is to have a contact ratio of minimum 1.4. A contact ratio of minimum 1.0 ensures that at least one pair of teeth remains in contact for continuous action. The larger the contact ratio, the smoother the gear operation. It will be shown subsequently that the path of contact and hence the contact ratio can be increased by increasing the addendum of the gears.

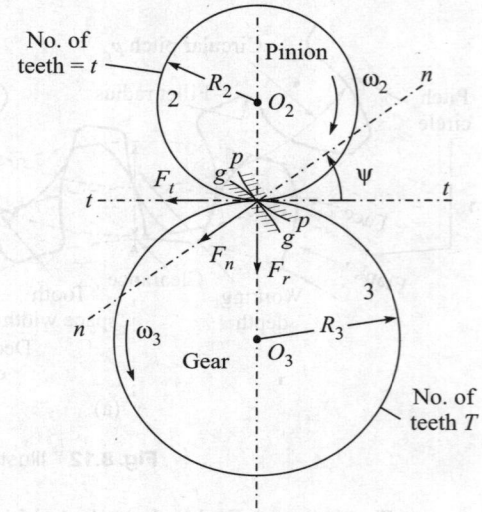

Fig. 8.13 Illustration for significance of pressure angle.

(xxv) *The Pressure Angle* Ψ: The pressure angle Ψ is the angle which the common normal to the contacting tooth profiles, at the point of contact, makes with the common tangent to the two pitch circles at the pitch point. This is also known as the angle of obliquity. Significance of pressure angle can be understood by considering Fig. 8.13 which depicts pitch circles of two mating gears and common normal to the contacting tooth profiles, making angle Ψ with the common tangent.

Remembering that load from one member to the other is transmitted along common, normal to contacting surfaces, let $n - n$ be the common normal to the pinion and gear teeth profiles represented by pp and gg. Let F_n be the force transmitted from pinion to gear along $n - n$.

Resolving this force along the tangent $t - t$ and also along common line of centres (radial component),

$$F_t = F_n \cos\psi \quad \text{and} \quad F_r = F_n \sin\psi \tag{8.4}$$

The component F_t produces driving torque ($= F_t \times R_3$) and is the useful component, while the component F_r tends to separate the shafts at O_2 and O_3 apart. The component F_r is undesirable as it produces bending in the shaft and increases bearing loads. From purely load transmission point of view therefore, it is desirable to have pressure angle Ψ as small as possible (From other considerations however, as will be seen later, a larger value of Ψ is preferable).

Finally, if R_2, ω_2 and t be the pitch circle radius, angular velocity and number of teeth on pinion respectively and R_3, ω_3 and T be the pitch circle radius, angular velocity and number of teeth on gear respectively, then for common module m,

$$\left(\frac{2R_2}{t}\right) = \left(\frac{2R_3}{T}\right) = m \tag{8.5}$$

Further, from the definition of gear ratio,

Gear ratio $\qquad G = \dfrac{T}{t} = \dfrac{R_3}{R_2}$ (from eq. 8.5) $\qquad\qquad$ (8.6)

Again, for same pitch line velocity at pitch-point,

$$V_P = R_2\omega_2 = R_3\omega_3 \qquad\qquad (8.7)$$

Hence, from equations (8.6) and (8.7),

Gear ratio $\qquad G = (T/t) = (R_3/R_2) = (\omega_2/\omega_3) \qquad\qquad$ (8.8)

EXAMPLE 8.1 A spur gear drive transmits a gear ratio of 4.0 over a centre distance of 50 cms. If module is 10 mm, determine the pitch circle diameters of pinion and gear. If pinion transmits a torque of 6000 N·cm, find (i) pressure between the teeth and (ii) load transmitted to bearing if pressure angle is 20°.

Solution: If D and d be the pitch circle diameters and T and t be the number of teeth on gear and pinion respectively, then for a centre distance of $C = 500$ mm,

$$500 = (d + D)/2$$

$$= \frac{1}{2}(mt + mT) = \frac{1}{2}mt\left(1 + \frac{T}{t}\right)$$

$$= \frac{1}{2}(10)t(1 + G)$$

As gear ratio $G = 4.0$, above relation gives,

$$t = \frac{500 \times 2}{10 \times (1 + 4)} = 20$$

And therefore $\qquad T = G \times t = 80$

Therefore $\qquad d = 20 \times m = 200$ mm

and $\qquad D = 80 \times m = 800$ mm. $\qquad$ **Ans.**

Further, assuming 100% transmission efficiency,

$$(\text{Torque} \times \omega)_p = (\text{Torque} \times \omega)_g$$

Hence, Torque on gear $\qquad = \left(\dfrac{\omega_2}{\omega_3}\right) \times 6000 = 24000$ N·cm

If F_t and F_r be the tangential and radial components of tooth load on gear and F_n be the load transmitted by pinion,

$$F_t = \left(\frac{\text{Torque on gear}}{D/2}\right) = \frac{24000}{(80/2)} = 600 \text{ N}$$

Again, from equations (8.4)

$$(F_t/F_r) = \cot \Psi$$

Therefore, $\quad\quad\quad\quad\quad\quad F_r = F_t \tan \Psi = 600 \times \tan 20°$

Hence $\quad\quad\quad\quad\quad\quad\quad F_r = 218.4$ N $\quad\quad\quad\quad\quad\quad$ **Ans.**

Also $\quad\quad\quad\quad\quad\quad\quad\quad F_n = F_t/\cos \Psi = 638.5$ N $\quad\quad\quad\quad$ **Ans.**

8.5 FUNDAMENTAL LAW OF TOOTHED GEARING

A positive drive is the one which transmits an absolute uniform angular velocity ratio during even a small traction of a revolution. In these days of high speed machines, a gear drive must be positive else vibrations and dynamic forces would develop causing immediate failure.

During the period of engagement of two teeth, there is some sliding of one tooth on another. *While the sliding occurs, the gears should continue to transmit a constant velocity ratio. Thus, for a gear drive to be positive, it is not enough to have interlocking teeth between a gear pair. The contacting surfaces of a pair of meshing teeth must be so shaped that during the entire period of engagement between a pair of teeth, angular velocity ratio remains constant.* Such surfaces of mating teeth are called *conjugate*. The law of gearing thus provides a basis to decide conjugate tooth surfaces.

The law of gearing states that for transmitting constant angular velocity vatio, common normal to the contacting surfaces of mating teeth, at every instantaneous point of contact, must pass through a fixed point on the line of centres of the two gears. The fixed point is called the pitch point which divides the line of centres in inverse ratio of the angular velocities of the mating gears.

Consider two rigid bodies 2 and 3, which are pivoted at points O_2 and O_3 respectively. As shown in Fig. 8.14, let the two bodies be in contact along a line through A perpendicular to the plane of paper. The point A is located on the curved surfaces of the two bodies. Further, let the body 2 rotate instantaneously in c.c.w. sense with angular velocity ω_2. The angular velocity ω_3 of body 3 is required to be established.

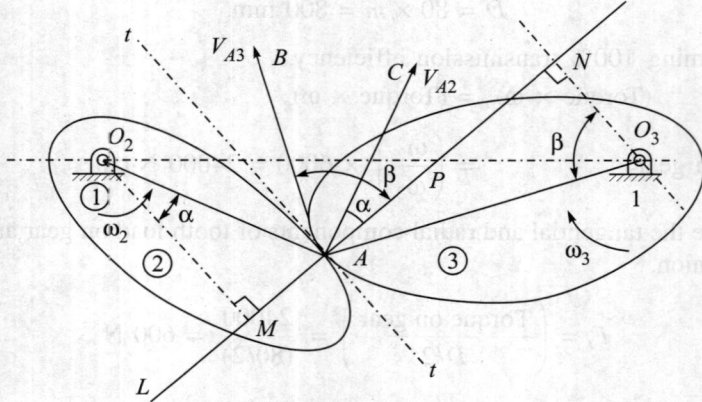

Fig. 8.14 Illustration for law of gearing.

Join O_2 and O_3 and draw line LN normal to the contacting surfaces at A, and let the two lines intersect at P. Draw O_2M and O_3N as perpendiculars to common normal LN from O_2 and O_3 respectively. Join A to O_2 and O_3. Let $\angle AO_2M = \alpha$ and $\angle AO_3N = \beta$.

At the given instant a point A, considered to be point on body 2, is moving with the velocity $V_{A2} = (O_2A)\omega_2$ in a direction perpendicular to line O_2A. Similarly the point A, now considered to be a point on body 3, is moving with velocity $V_{A3} = (O_3A)\omega_3$ in a direction perpendicular to the line O_3A as shown in Fig. 8.14. Note that V_{A2} is perpendicular to O_2A and MN is perpendicular to line O_2M. Hence $\angle CAN = \angle AO_2M = \alpha$. Similarly, it can be shown that $\angle BAN = \angle AO_3N = \beta$.

The component of velocities $V_{A2}\cos\alpha$ and $V_{A3}\cos\beta$ of velocity vectors V_{A2} and V_{A3} respectively, along the common normal LN must be equal. If they were not equal, either body 2 would dig into body 3 or, it would move away from it (which means broken contact). Thus,

$$V_{A2}\cos\alpha = V_{A3}\cos\beta \qquad (8.9)$$

But, $\qquad V_{A2} = (O_2A)\omega_2$ and $V_{A3} = (O_3A)\omega_3$. Substituting in equation (8.9),

$$(O_2A)\omega_2\cos\alpha = (O_3A)\omega_3\cos\beta \qquad (8.10)$$

From right angled triangles O_2MA and O_3NA,

$$(O_2A)\cos\alpha = O_2M; \text{ and } (O_3A)\cos\beta = O_3N$$

Substituting in equation (8.10) and rearranging,

$$(\omega_2/\omega_3) = (O_3N)/(O_2M) \qquad (8.11)$$

The triangles O_2MP and O_3NP are similar as besides right angles at M and N, $\angle O_2PM = \angle O_3PN$. Hence,

$$(O_3N/O_2M) = (O_3P)/(O_2P) = (NP)/(MP) \qquad (8.12)$$

Substituting for (O_3N/O_2M) in equation (8.11), we have

$$(\omega_2/\omega_3) = (O_3P)/(O_2P) \qquad (8.13)$$

Thus the ratio of angular velocities (ω_2/ω_3) varies inversely as the ratio of distances of point P from centres O_2 and O_3. Clearly, the ratio (ω_2/ω_3) will remain constant as long as the position of point P is fixed along the line O_2O_3. This leads us to following law of gearing.

In order that a pair of curved surface (tooth profiles) may transmit a constant angular velocity ratio, the shape of contacting tooth profiles must be such that the common normal passes through a fixed point P on the line of centres. The point P divides the line of centres in an inverse proportion as the ratio of angular velocities. The fixed point P is called the pitch point and the line MN (common normal to contacting surfaces) is called the line of action.

The Velocity of Sliding. Component of velocities V_{A2} and V_{A3} along the common tangent $t - t$ are given by $V_{A2} \sin \alpha$ and $V_{A3} \sin \beta$. The velocity of sliding of surface of body 3 relative to the surface of body 2 at the point of contact is given by

$$\begin{aligned}
\text{Velocity of sliding} &= V_{A3}\sin\beta \sim V_{A2}\sin\alpha \\
&= \omega_3(O_3A)\sin\beta \sim \omega_2(O_2A)\sin\alpha \\
&= \omega_3(AN) \sim \omega_2(AM) \\
&= \omega_3(AP + PN) \sim \omega_2(PM - AP) \\
&= (\omega_2 + \omega_3)AP + \omega_3PN \sim \omega_2PM \qquad (8.14)
\end{aligned}$$

From equations (8.12) and (8.13) it follows that,

$$\frac{\omega_2}{\omega_3} = \frac{O_3P}{O_2P} = \frac{NP}{MP} \text{ or } (\omega_2) MP = \omega_3(NP)$$

Substitution as above in equation (8.14) last two terms cancels out yielding,

$$\text{Velocity of sliding} = V_s = (\omega_2 + \omega_3)AP. \qquad (8.15)$$

This is obvious because point P is an instantaneous centre of rotation.

It follows from equation (8.15) *that velocity of sliding is zero at the pitch point where* $AP \rightarrow O$, *and is maximum at the farthest point along the line of action.*

Readers may verify that angle complimentary to the angle made by the path of contact with the line of centres is the pressure angle ψ.

The Pitch Point as an I.C.

Referring to Fig. 8.14, let frame link be 1. Clearly, the gear centres O_2 and O_3 are the instant centres 12 and 13 respectively. From the theorem of 3 centres, it follows that the third instant centre of rotation of link 2 with respect to link 3 will always lie on the line joining O_2 and O_3. Further, since this instant centre I is a point common to links 2 and 3 at which both bodies have same absolute velocity, it follows that,

$$(O_2I)\,\omega_2 = (O_3I)\,\omega_3$$

A comparison of this equation with (8.13) shows that this instantaneous centre I (23) and the pitch point P are the same. The distances O_2I and O_3I are infact the pitch circle radii. There being no relative motion between gears 2 and 3 at this point, the velocity of sliding is always zero at the pitch point. Also as the pitch point P is the I.C. of rotation of 2 w.r. to 3, the velocity of sliding at any other point of contact A (on the path of contact) is –

$$AP \,(\omega_2 + \omega_3)$$

Note that due to opposite sense of rotation of the two gears the relative rotational speed is $(\omega_2 + \omega_3)$.

8.6 CONJUGATE TEETH

When mating tooth profiles are so shaped that they produce a constant angular velocity ratio during mesh, the tooth surfaces are said to be conjugate with respect to one another. *It is in fact possible to specify any tooth profile and then to find a profile for mating tooth so that the two tooth profiles are conjugate. One of these tooth profiles is that of involute, which is very widely used in gearing. A yet another solution consists of cycloidal profile which was first to be used and is still being used in clocks and watches. Even though a large number of conjugate curves are possible, real problem lies in producing these tooth profiles in large quantities on gear-blanks of steel and other materials using existing machinery. This apart, economics of reproducing these gear tooth profiles restricts the choice mostly in favour of involute tooth profiles.*

We now describe below graphical construction to emphasize the fact that for any assumed gear tooth curve, it is always possible to obtain conjugate tooth curve. For the sake of brevity, construction has been carried out for establishing three points on curve. The procedure is however, general, and can be repeated to establish many more points on conjugate curve so that it can be described accurately.

Fig. 8.15 shows to two gear wheels G and G' in mesh with their axes at O_A and O_B respectively. For simplicity, the tooth profile of gear wheel G is assumed to be a circular arc A_1PA_2 of radius OP. Then, as gear tooth moves around OA with gear rotation, the location of centre of curvature of tooth profile must maintain same relative position vis-a-vis gear tooth curve A_1PA_2. In other words as the gear G rotates, the centre of curvature of tooth profile will move around in a circle with $O_A O$ as radius. It is now required to find out a tooth profile that is conjugate in respect of tooth profile A_1PA_2. Rotate the gear wheel G about wheel axis O_A so that the tooth profile moves from position A_1PA_2 to B_1BB_2. In

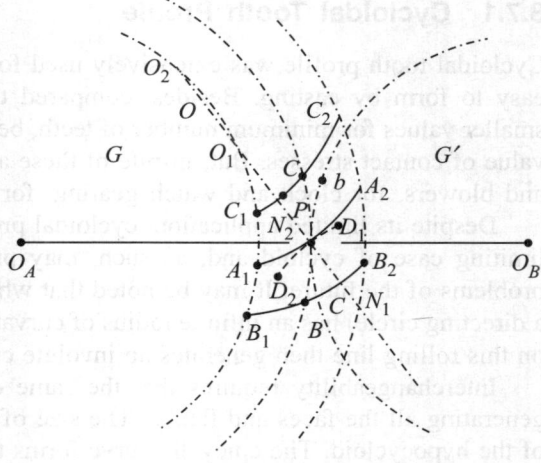

Fig. 8.15

doing so the centre of curve of tooth profile A_1PA_2 shifts from position O to O_1 keeping same distance from OA. In other words point O and O_1 will be located on same circle of radius O_A O. Position of normal to new position is obtained by joining O_1 to P which, when extended, cuts profile $B_1 B B_2$ at point N_1. Since $O_1 P$ is the common normal, it follows that for the given position of profile N_1 is the point of contact. To transfer position of this point of contact to the reference position A_1PA_2, draw arc from N_1 with radius $O_B N_1$. This step is self-explanatory the moment it is realised that N_1, being a point on gear tooth *of* wheel B, must move in an arc of circle with radius $O_B N_1$. As point B is moved back to position P_1 a point coinciding with P on the pitch circle of gear G' moves to position b, such that $Pb = PB$. From b' strike arc of radius $P N_1$ to cut arc of circle $N_1 D_1$ at D_1. This gives a point on conjugate tooth profile besides the point P.

To locate a third point on conjugate tooth profile, consider $C_1 C C_2$ to be third position of tooth profile. Let O_2 be the centre of curvature for this position of tooth profile. Line joining O_2 and P cuts the current tooth profile position at point N_2. $O_2 P$, being a common normal to the two tooth profiles, N_2 gives yet another point of contact between the two tooth profiles. To transfer this position to the reference position, mark point C on pitch circle of G' such that arc $PC = $ arc PC. Draw an arc of circle from N_2 with $O_B N_2$ as radius and strike arc on it of radius $CD_2 = PN_2$. Thus, D_2 is the third point on the conjugate profile, besides points D_1 and P. A more accurate tooth profile can be establish by taking a few more points D_3, D_4, D_5, etc. and passing a smooth curve through them.

8.7 TOOTH PROFILES

It should be evident from discussions of section 8.6 that theoretically any suitable curve can be selected for gear tooth profile and a conjugate curve can be established for mating gear teeth.

But owing to interchangeability aspect, cost and ease of production on large scale, together with many other advantages, the choice practically narrows down to involute curve.

8.7.1 Cycloidal Tooth Profile

Cycloidal tooth profile was extensively used for gear manufacture in distant past because it is easy to form by casting. Besides, compared to involute gearing, cycloidal gearing perrmits smaller values for minimum number of teeth, better contact and wear characterisitics, and lower value of contact stresses. But, inspite of these advantages, its use was restricted to gear pumps and blowers, for clock and watch gearing, for rack jacks, etc.

Despite its limited application, cycloidal profile needs our attention as involute is simply a limiting case of cycloid and, as such, may provide solution to the unsolved and unknown problems of the future. It may be noted that when a generating circle (rolling on the outside of a directing circle) has an infinite radius of curvature, it becomes a straight line, and a fixed point on this rolling line then generates an involute curve.

Interchangeability requires that the same describing (generating) circle must be used in generating all the faces and flanks. The size of the describing circle depends on the properties of the hypocycloid. The epicyclic curve forms the face while the hypocycloid forms the flanks of the teeth. Further, when the describing circle is half the size of directing circle, the flanks will be radial and the resulting tooth will be relatively weak at the root (Fig. 8.16a). If the describing circle is made smaller the hypocycloid curves away from the radial lines giving spreading tooth (Fig. 8.16c) at the root, which is stronger. However, when describing circle is larger the hypocycloid curves the other way, passing inside the radial lines (Fig. 8.16b), producing a weaker tooth that is difficult to shape on milling machine. It follows that the diameter of the describing circle should not be larger than the radius of the directing circle of the smallest gear of the interchangeable set.

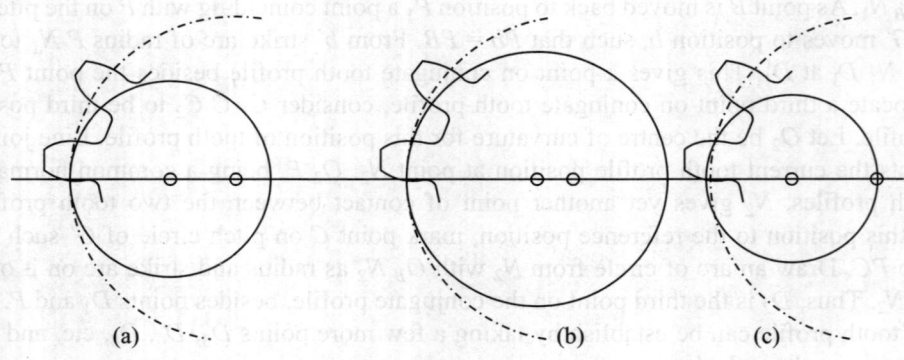

(a)   (b)   (c)

Fig. 8.16

The pressure angle in cycloidal gears varies constantly; it diminishes during the approach, becomes zero at the pitch point, and then increases during the recess. The variation in pressure angle leads to additional noise and wear, and also to changes in bearing reactions at the shaft supports. Further, in order to run properly, cycloidal gears must be operated at exactly the correct centre distance, otherwise the contacting portions of the profiles will not produce conjugate action. In must be noted that in view of deflections produced by load being transmitted, it is virtually impossible to maintain correct centre distance between gears. For gears used in heavy work, experience suggests a maximum permissible pressure angle of 30°, giving a mean value of 15°.

8.7.2 Involute Tooth Profile

It may be recalled that an involute curve is defined in two ways.

(i) It is the locus of a point on a taut string as the string is unwrapped from the circumference of a circle.

(ii) It is the locus of a fixed point on a straight line which rolls without slipping on a circle.

The circle from which the taut string is unwrapped or over which the line rolls without slipping, is called the *base circle. It follows from the definition of involute curve that the involute curve cannot exist inside the base circle. It also follows that involute curves drawn from circles of different diameter (as base circle) are all different.*

Figure 8.17 shows an involute curve generated by rolling a *BC* line without slipping on the base circle. Points C_1, C_2, C_3 ... C_5 are the points on the curve corresponding to the points of tangency P_1, P_2, P_3, ... P_5 on the base circle. It follows that each of the points P_j ($j = 1, 2, 3, ... 5$) is an instantaneous centre of curvature for small arc around corresponding point C_j on the involute curve. Since P_1C_1, P_2C_2, P_3C_3, etc. were drawn tangent to base circle, following important property of involute curve becomes evident:

Tangent to the base circle is also the normal to the involute curve at corresponding point.

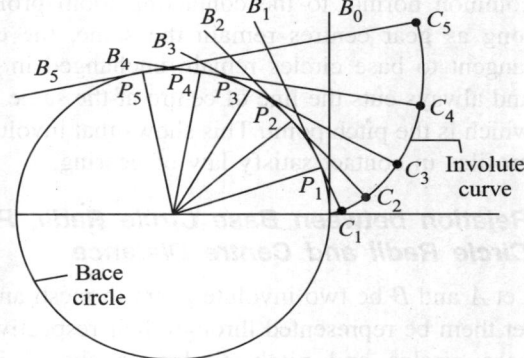

Fig. 8.17 Involute curve as obtained by rolling a straight line *BC* without slipping.

To Show That Involute Gears Satisfy Law of Gearing

To show that involute tooth profile satisfies law of gearing, consider a string whose end points are attached to two rotating cylinders *A* and *B* corresponding to base circles (see Fig. 8.18). Cylinder *A* is free to rotate about axis through O_A and *B* is free to rotate about axis O_B. *N* represents a fixed point or knot on the string through which a pencil point projects out. Keeping cylinder *A* fixed and string taut, if *N* is moved around cylinder *A* (cylinder *B* being free to rotate) in c.c.w. sense, it describes the involute curve $a'a$ on the upper disc. Similarly, keeping wheel *B* fixed (and, releasing cylinder *A*) and string taut, if the point *N* is moved around cylinder *B*, it describes the involute curve $b'be$.

Since motion of point *N* is always perpendicular to corresponding radius of rotation (like *LN* or *MN*), the string *LM* is normal to the two involute curves for all positions of contact.

Finally, consider curved surface $a'a$ to be an integral part of and rotating with cylinder *A*. Similarly, let the curved surface *be* to be an integral part of and moving with cylinder *B*. Now keeping the two curved surfaces in contact, back up both the cylinders so that a point *d* on curved surface $a'a$ comes in contact with point *e* on the curved surface $bb'e$. Note that for this to occur arcular length $a'd$ must be equal to arcular length $b'e$. In order to locate the point N' at which the two points d and e will come in coincidence in Fig. 8.18, draw arcs of circle of radius $O_A d$ and $O_B e$ with O_A and O_B as centres respectively. They meet at N'. It is worth noting that point N' lies on the line (string) LM which is common tangent to both the cylinders. This can be checked further by selecting another pair of points f and g on curved surfaces $a'a$ and be with the condition that arcular length $f a' =$ arcular length gb'.

Above discussions lead us to the conclusion that whenever two involute profiles of mating gear teeth come in contact, the point of contact always lie on the common tangent to the base circles, which is also the common normal to the contacting tooth profiles. So long as gear centres remain the same, the common tangent to base circles remain unchanged in position and always cuts the line of centre at the same point P, which is the pitch point. This shows that involute tooth profiles in contact satisfy law of gearing.

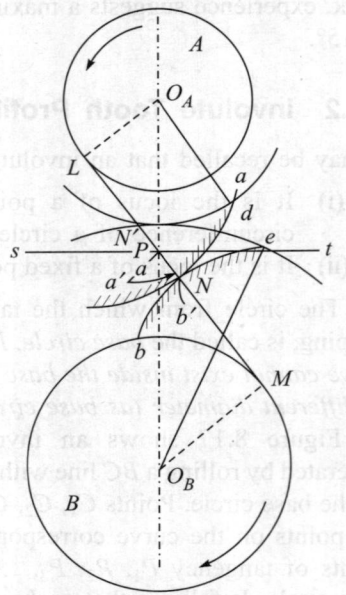

Fig. 8.18 Illustration to show that involute tooth satisfies law of gearing.

Relation between Base Circle Radii, Pitch Circle Radii and Centre Distance

Let A and B be two involute gears in mesh and let them be represented through their respective base circles and pitch circles as shown in Fig. 8.19. Let O_A and O_B be the axis of rotation of gears A and B. Further, let r_p and r_b be the pitch circle radius and base circle radius of gear A and R_p and R_b be the pitch circle radius and base circle radius of gear B. If Ψ is the pressure angle, then $O_A O_B$ being perpendicular to the common tangent $t - t$ and $O_B M$ and $O_A L$ being perpendicular to LM,

$$\angle LO_A P = \angle MO_B P = \Psi$$

From the right angled triangle $O_A LP$,

$$O_A L = (O_A P)\cos\Psi$$

or

$$r_b = r_p \cos\Psi \qquad (8.16)$$

Also, from right angled triangle $O_B MP$,

$$O_B M = (O_B P)\cos\Psi$$

or

$$R_b = R_p \cos\Psi \qquad (8.17)$$

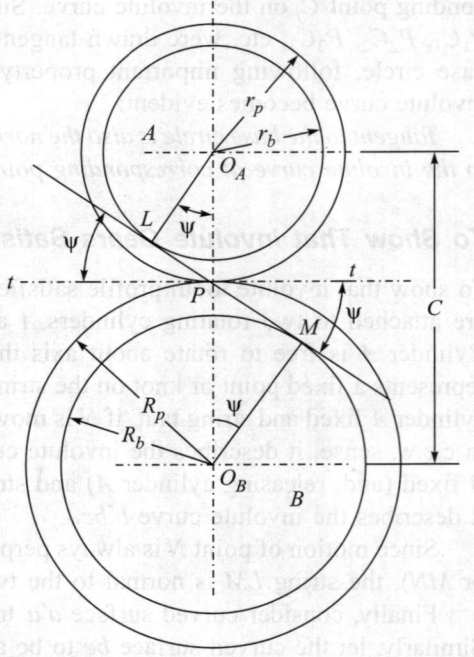

Fig. 8.19 Illustration showing relation of centre distance with base and pitch circle radii.

Adding on the corresponding sides of equations (8.16) and (8.17),

$$(r_b + R_b) = (r_p + R_p)\cos\Psi \tag{8.18}$$

Noting that

$$r_p + R_p = O_A P + O_B P = \text{centre distance} = C,$$

it follows from equation (8.18) that

$$(r_b + R_b) = C\cos\Psi \tag{8.19}$$

or

$$C = \frac{(r_b + R_b)}{\cos\Psi} \tag{8.20}$$

and

$$\cos\psi = \frac{(r_b + R_b)}{C} \tag{8.21}$$

Equation (8.21) indicates that for a given pair of gears in mesh (for which r_b and R_b are constant), pressure angle ψ increases if centre distance C is increased within limits. It must be noted here that as the tooth profile (which is involute here) does not change, the size of base circle remains unaltered.

Effect of Increasing Centre-Distance on Velocity Ratio

Consider a pair of meshing gears having involute teeth with each tooth in contact with mating tooth on either side. Clearly, any attempt to bring the two centres closer will lead to jamming or deforming the teeth. Therefore, consider the case in which the centre distance is slightly increased. This leads to clearance and backlash between the mating teeth.

Base circles are fundamental characteristic of the gears and unless they are changed, tooth profiles remain unchanged. Thus, when the centre distance is increased, we have new pitch circles of larger radius for each gear, keeping the ratio of pitch circle radii the same. This is necessary because the pitch circles of mating gears are imaginary circles and, are always tangent to each other. Following discussions are aimed at elaborating this point.

As shown in Fig. 8.20, let a pinion with centre at O_1 and a gear with centre at O be in mesh with pitch point at P. For this position of gears, the common tangent LM to base circles passes through P. Let

$$O_1 L = r_b \quad O_1 P = r_p \quad OM = R_b \quad \text{and} \quad OP = R_p$$

For this position, the common tangent LM represents line action and $\angle LO_1 P = \angle MOP = \psi$, the pressure angle. Using similarity of triangles $PO_1 L$ and POM, the velocity ratio is given by

$$\frac{\omega_1}{\omega_2} = \frac{OP}{O_1 P} = \frac{OM}{O_1 L} \tag{8.22}$$

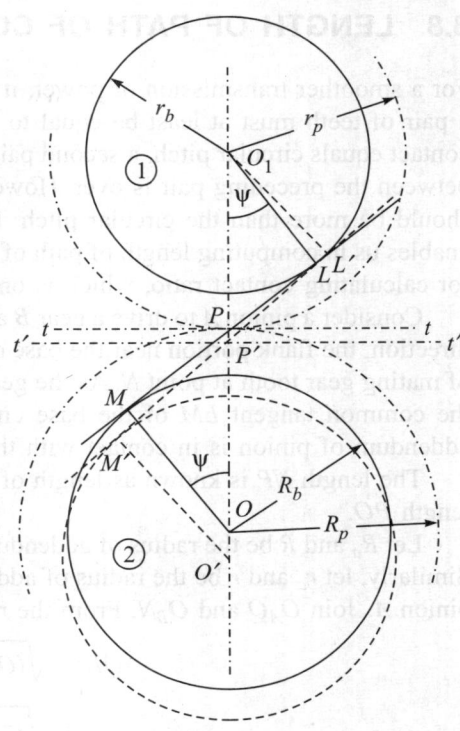

Fig. 8.20

Let the centre of gear 2 be now shifted slightly from O and O'. *The base circle size remains the same, but with changed position of centre of gear 2, the base circle changes its position to the one shown by dotted lines.* As a consequence, common tangent to the two base circles takes up the position $L'M'$ and the pitch point changes to P'. Clearly, the new pressure angle $\angle M'O'P'$ is larger than ψ. This is also clear from the new position of common tangent.

Even in the new position of centre of gear 2, from similar triangles $O'M'P'$ and $O_1L'P'$ we have

$$\frac{O'P'}{O_1P'} = \frac{O'M'}{O_1L'} \tag{8.23}$$

But $\qquad O'M' = OM = R_b$ and $O_1L' = O_1L = r_b$, and

therefore, from equation (8.23), $\quad \dfrac{O'P'}{O_1P'} = \dfrac{OM}{O_1L'} \tag{8.24}$

Comparing equation (8.22) and (8.24),

$$\frac{\omega_1}{\omega_2} = \frac{OM}{O_1L} = \frac{O'P'}{O_1P'} \tag{8.25}$$

This shows that even if the centre distance is changed within limits, the velocity ratio remains unchanged. Pressure angle, however increases as the centre distance is increased.

8.8 LENGTH OF PATH OF CONTACT

For a smoother transmission of power, it is necessary that the length of arc of contact between a pair of teeth must at least be equal to the circular pitch of the teeth. When length of arc of contact equals circular pitch, a second pair of gear teeth begins to engage before the engagement between the preceding pair is over. However, as will be seen later, the length of arc of contact should be more than the circular pitch. It is necessary therefore to establish a relation which enables us in computing length of path of contact. This is important because it will pave the way for calculating contact ratio, which is one of the measures of quality of gear drive.

Consider a pinion A to drive a gear B as shown in Fig. 8.21. As the pinion rotates in clockwise direction, the flank portion near the base circle of pinion comes in contact with addendum circle of mating gear tooth at point N. As the gear pair rotates further, the point of contact shifts, along the common tangent LM of the base circles. The contact continues upto point Q, when the addendum of pinion is in contact with the flank portion of tooth of gear B.

The length NP is known as length of path of approach while length of path of recess is the length PQ.

Let R_a and R be the radius of addendum circle and radius of pitch circle of gear respectively. Similarly, let r_a and r be the radius of addendum circle and radius of pitch circle respectively of pinion A. Join O_AQ and O_BN. From the right angled triangle O_AQL,

$$QL = \sqrt{(O_AQ)^2 - OL^2}$$

$$= \sqrt{r_a^2 - (O_AP\cos\Psi)^2}$$

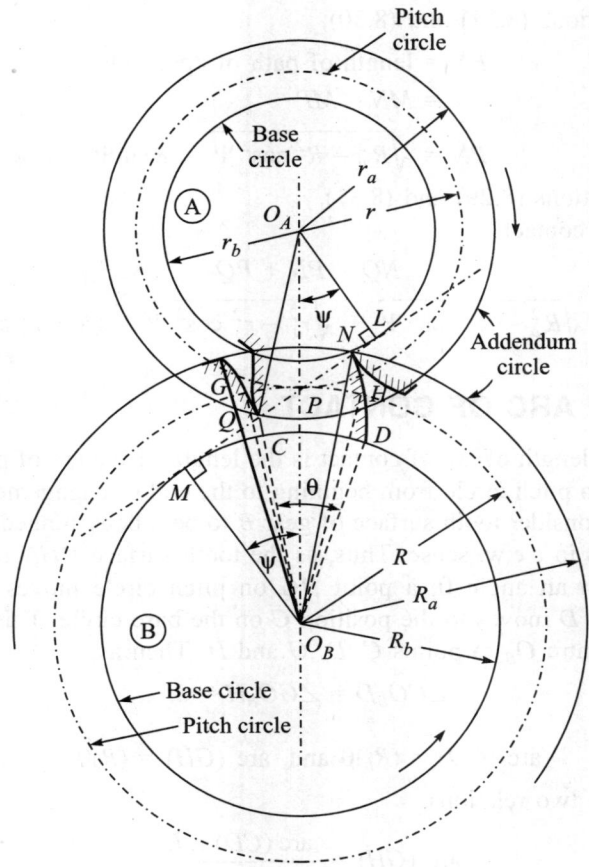

Fig. 8.21

$$= \sqrt{r_a^2 - r^2 \cos^2 \Psi} \qquad (8.26)$$

Also, from right angled triangle $O_B MN$,

$$MN = \sqrt{(O_B N)^2 - (O_B M)^2} = \sqrt{R_a^2 - (R \cos \Psi)^2} \qquad (8.27)$$

And, from right angled triangle $PO_A L$,

$$PL = O_A P \sin \Psi = r \sin \Psi \qquad (8.28)$$

Hence, from equations (8.26) and (8.28), QP = length of path of recess

$$= QL - PL = \sqrt{r_a^2 - r^2 \cos^2 \Psi} - r \sin \Psi \qquad (8.29)$$

Again, from right angled triangle $MO_B P$,

$$MP = (O_B P) \sin \Psi = R \sin \Psi \qquad (8.30)$$

Hence, from equations (8.27) and (8.30),

$$PN = \text{length of path of approach}$$
$$= MN - MP$$

or

$$PN = \sqrt{R_a^2 - R^2 \cos^2 \Psi} - R\sin\Psi \tag{8.31}$$

Hence, using equations (8.29) and (8.31),

Length of path of contact,

$$NQ = PN + PQ$$

$$NQ = \sqrt{R_a^2 - R^2 \cos^2 \Psi} + \sqrt{r_a^2 - r^2 \cos^2 \Psi} - (R + r)\sin\Psi \tag{8.32}$$

8.9 LENGTH OF ARC OF CONTACT

As defined earlier, the length of arc of contact is the length of the arc of pitch circle described by a point of a tooth on pitch circle from begining to the end of engagement between a pair of teeth in mesh. Let us consider tooth surface of gear B to be a body hinged at O_B and revolving about it (*see* Fig. 8.21) in c.c.w. sense. Thus, as the tooth surface DHN revolves c.c.w. to the position CQG, through an angle θ, a point HG on pitch circle moves to the position and simultaneously, a point D moves to the position C on the base circle. This is shown clearly in Fig. 8.21. Join gear centre O_B to points C, D, G and H. Then as,

$$\angle CO_BD = \angle GO_BH = \theta,$$

it follows that,

$$\text{arc } (CD) = (R_b)\theta \text{ and, arc } (GH) = (R)\theta$$

Hence, from the above two relations,

$$\text{arc } (GH) = \frac{\text{arc } (CD) \times R}{R_b} \tag{8.33}$$

Further, from equation (8.17), the pitch circle radius R is related to base circle radius as,

$$R_b = R\cos\Psi.$$

Hence, substituting in equation (8.23) for R_b,

$$\text{arc } (GH) = \frac{\text{arc } (CD)}{\cos\Psi} \tag{8.34}$$

Finally, recall that an involute curve is generated by rolling a straight line over a base circle without slipping. Thus, the involute curves GQC and NHD can be considered to have been generated by rolling segment PQ of stsaight line ML without slipping on the base circle of radius R_b. Clearly, as the line ML is rolled over the base circle, point Q coincides with C and point N coincides with point D on base circle. It follows therefore that,

$$\text{arc length } (CD) = QN = \text{length of path of contact} \tag{8.35}$$

Substituting for arc (CD), in equation (8.34), it follows that

Length of arc of contact $\qquad GH = \dfrac{QN}{\cos\Psi} = \dfrac{\text{Length of path of contact}}{\cos\psi}$ $\qquad$ (8.36)

8.10 CONTACT RATIO

Contact ratio is defined as average number of pairs of teeth which are in contact. Since length of arc of contact of a pair of mating teeth and circular pitch are both measured along pitch circle, it follows that,

$$\text{Contact Ratio} = \frac{\text{Length of Arc of Contact}}{\text{Circular pitch}} \qquad (8.37)$$

The contact ratio usually is not an integer number and its value generally lies between 1 and 2. When cantact ratio is equal to one, it means that one tooth together with adjacent tooth-space will occupy the entire arc of contact *GH*. Stated in other words, when a tooth is just to enter into engagement, the preceding tooth is just to go out of contact from its mating tooth. Thus, all throughout the motion of arc of contact, exactly one pair of teeth remains in contact.

Let us see now as to what happens really when the contact ratio is greater than 1 but less than 2. A contact ratio of 1.4, for instance, implies that when a pair of teeth is just entering into engagement, the preceding tooth-pair continues to remain in contact. Thus, for a short spell of time, two pairs of teeth will be in mesh and as the preceding tooth goes out of engagement only one pair of teeth will remain in engagement for a short spell of time.

EXAMPLE 8.2 The number of teeth on each of the two equal spur gears in mesh is 40. The teeth have 20° involute profile and the module is 6 mm. If the length of arc of contact is 1.75 times the circular pitch, find the addendum. (AMIE, Summer 1980)

Solution: The contact ratio is defined as the ratio of length of arc of contact and circular pitch and is given as 1.75. Thus

$$\text{Length of arc of contact} = 1.75\ p_c = 1.75\ (m\pi) = 1.75\ (6 \times \pi) = 32.99 \text{ mm}$$

Therefore   Length of path of contact = $(\cos\Psi) \times$ length of arc of contact

or   Length of path of contact = $(\cos 20) \times 32.99 = 31.0$ mm

From equation (8.32), as gears are of same proportion,

$$31.0 = 2\sqrt{R_a^2 - (R\cos\Psi)^2} - 2R\sin\Psi \qquad (1)$$

Now $$R = \frac{mT}{2} = \frac{6 \times 40}{2} = 120 \text{ mm}$$

solving equation (1)

$$(31.0) + 2(120)\sin 20 = 2\sqrt{R_a^2 - (120\cos 20°)^2}$$

Simplifying and squaring on both sides,

$$R_a^2 - (112.763)^2 = (113.082/2)^2$$

Therefore $R_a = 126.14$ mm

Hence, required addendum $= R_a - R = 126.14 - 120 = 6.14$ mm   **Ans.**

EXAMPLE 8.3 A pair of 20° full-depth involute spur gears having 30 and 50 teeth respectively of module 4 mm are in mesh, the smaller gear rotate at 1000 r.p.m. Determine (a) sliding velocities at engagement and at disengagement of a pair of teeth and (b) the contact ratio. Take addendum = 1 module.

Solution: Pitch circle radii of pinion and gear are:

$$r = \frac{mt}{2} = \frac{4 \times 30}{2} = 60 \text{ mm}$$

$$R = \frac{mT}{2} = \frac{4 \times 50}{2} = 100 \text{ mm}$$

Angular speed of pinion $$\omega_1 = \frac{2\pi \times 1000}{60} = 104.72 \text{ rad/s}$$

Angular speed of gear $$\omega_2 = \left(\frac{30}{50}\right) \times \omega_1 = 62.832 \text{ rad/s}$$

Now $a = 1 \times$ module $= 4$ mm and, $R_a = R + a = 100 + 4 = 104$ mm

Also, $r_a = r + a = 60 + 4 = 64$ mm

Hence length of path of approach (Fig. 8.21)

$$PN = \sqrt{R_a^2 - (R\cos\Psi)^2} - R\sin\Psi$$

$$= \sqrt{(104)^2 - (100\cos 20°)^2} - (100)\sin 20° = 10.36 \text{ mm}$$

and, length of path of recess,

$$= \sqrt{r_a^2 - (r\cos 20°)^2} - r\sin 20°$$

$$= \sqrt{(64)^2 - (60\cos 20°)^2} - (60)\sin 20° = 9.76 \text{ mm}$$

Since, sliding velocity is proportional to the distance from pitch point, maximum sliding velocity during path of approach occurs at the point of engagement given by

$$= 10.36(\omega_1 + \omega_2) = 1735.84 \text{ mm/s}$$ **Ans.**

Velocity of sliding at the point of disengagement is given by

$$= 9.76(\omega_1 + \omega_2) = 1635.3 \text{ mm/s}$$ **Ans.**

Length of arc of contact $$= \frac{10.36 + 9.76}{\cos 20} = 21.41 \text{ mm}$$

Therefore   circular pitch $p_c = m\pi = 4\pi$

and   contact ratio $$= \frac{10.36 + 9.76}{4\pi\cos 20°} = 1.704$$ **Ans.**

EXAMPLE 8.4 Two equal involute gear wheels of 20° pressure angle have 20 teeth each Calculate length of arc of contact if the addendum was standard and equal to one module. Pitch of teeth is 6 mm of diameter per tooth. What should be the addendum, if the arc of contact is to be maximum possible? What is then the length of arc of contact?

Solution: From the definition of module $\left(m = \dfrac{D}{T} \right)$ it follows that 'pitch of 6 mm of diam. per tooth' is the module itself.

Thus $\qquad\qquad\qquad m = 6$ mm; $t = 20$; $T = 20$

Therefore $\qquad\qquad\qquad r = \dfrac{m \times 20}{2} = 60$ mm; $R = \dfrac{m \times 20}{2} = 60$ mm

For 20° full depth tooth, standard addendum = 1 module = 6 mm

Therefore $\qquad\qquad\qquad\qquad R_a = r_a = 60 + 6 = 66$ mm

Length of path of contact for this case,

$$ L = 2 \sqrt{R_a^2 - (R \cos \Psi)^2} - 2R \sin \Psi $$

or $\qquad\qquad L = 2 \sqrt{(66)^2 - (60 \cos 20°)^2} - 2(60) \sin 20° = 27.56$ mm

Therefore, $\qquad\qquad$ length of arc of contact $= \dfrac{27.56}{\cos 20} = 29.33$ mm

Maximum length of path (and hence of arc) of contact is obtained when addendum circles of pinion and gear pass through points of tangency to base circles. Thus, from Fig. 8.21, maximum length of path of contact $= R \sin \Psi + r \sin \Psi$

or $\qquad\qquad\qquad L' = (60 + 60) \sin 20° = 41.04$ mm

Therefore Maximum length of arc of contact

$$ = \frac{120 \sin \Psi}{\cos \Psi} = 120 \tan 20° = 43.676 \text{ mm} $$

Radius of addendum circle R_a for length of path of contact = 41.04 is,

$$ 41.04 = 2 \sqrt{R_a^2 - (60 \cos 20°)^2} - 2(60) \sin 20 $$

or $\qquad\qquad\qquad \sqrt{R_a^2 - (60 \cos 20°)^2} = \dfrac{82.08}{2}$

Squaring and solving,

$$ R_a = 69.74 \text{ mm} $$

Therefore $\qquad$ required addendum $a = 69.74 - 60 = 9.74$ mm. $\qquad\qquad$ **Ans.**

8.11 INTERFERENCE AND UNDERCUTTING

As explained in section 8.6, a pair of teeth together will transmit a constant angular velocity ratio if and only if the mating tooth profiles are conjugate with respect to one another. One of the major attributes of a gear drive is that it provides a constant velocity ratio. However, when surfaces of mating gear teeth contact over a non-conjugate portion, interference takes place. *All throughout the contact over non-conjugate portion of tooth profile, a non-uniform motion is transmitted, which is undesirable. Interference can therefore be taken to mean 'interference in the transmission of uniform motion'* from pinion to the gear.

To understand this phenomenon, consider a pinion, with centre at O_1 and a gear with centre at O_2 to be in mesh as shown in Fig. 8.22. Let P be the pitch point and AB be the common tangent to the base circles of the two gears. Let the pinion rotate c.w. to drive the gear. The addendum circle of gear is shown to cut the line of action, when extended, at point C which is outside the length BA. Also, let the addendum circle of pinion cut the line of action at point B' between points B and P, as shown in Fig. 8.22.

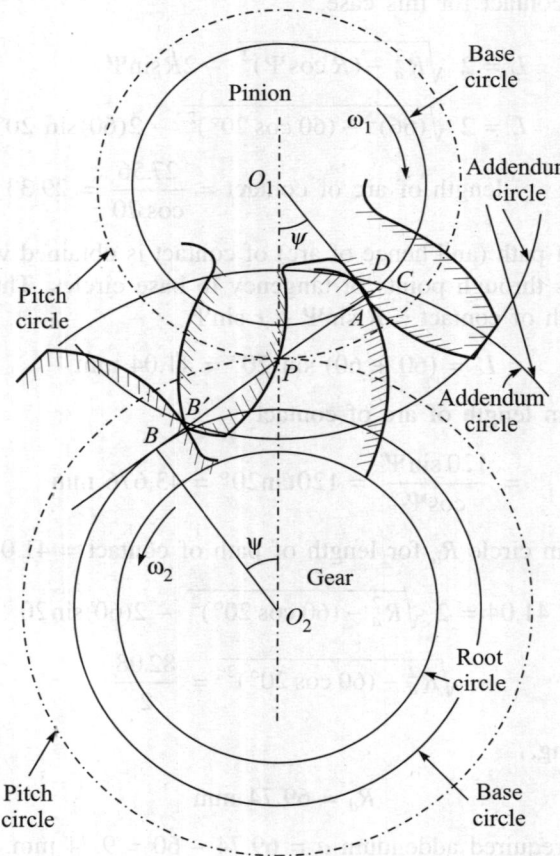

Fig. 8.22 Figure illustrating condition of interference.

Since point C lies outside the length BA, we can not say with certainity if the corresponding point of contact will lie only on line BA produced (As will be shown in the line to follow, relative motion between mating teeth for portion A to C does not correspond to conjugate action). For this reason, we consider the mating teeth at the point of disengagement B' and then consider their relative motion in reverse direction.

As the mating pinion and gear teeth back off, the point of contact between the teeth moves towards P along line $B'P$. During this interval, point of contact slides up the flank portion of gear tooth and slides down the face portion of pinion tooth. As the contact point between mating teeth reach a point P and back off further, the point of contact slides up the face portion of gear tooth and slides down the flank portion of pinion tooth. This continues as the point of contact moves along the line PA. If the back-off motion is continued beyond point A, which is a point of tangency on base circle, the point of contact slides down further on the flank portion of pinion inside the base circle. Clearly, for this part of contact (between points A and C) the profile of pinion tooth is non-involute, and is, therefore, conducive to interference.

It follows that points of tangency A and B to the base circle are the outermost points of contact for conjugate action of teeth in contact. We may conclude therefore that for a pair of mating involute teeth,

(a) maximum possible length of path of contact, avoiding interference, is AB which is the distance between points of tangency to base circles.

(b) interference is possible only when diameter of base circle is larger than the diameter of dedendum circle, and

(c) interference will occur when addendum circle of pinion or gear cuts the line of action outside the points of tangency to base circles.

It should be evident from Fig. 8.22 that BP $(= R \sin \psi)$ is larger then AP $(= r \cdot \sin \psi)$ and therefore, with interchangeable tooth proportions, chances of addendum circle of gear cutting line of contact outside the point of tangency are always better compared to the other case.

To sum up, it must be noted that **interference is possible only when the base circle diameter is larger than the dedendum circle diameter, permitting a non-involute portion to exist on the pinion tooth between base circle and dedendum** circle. *Readers should note that this is a necessary condition rather than sufficiency condition for interference to exist. For, even if pinion tooth profile have a non-involute portion, actual contact between teeth may not extend to this non-involute portion. Thus the sufficiency condition for interference to occur is that the addendum circle of gear must cut the line of common tangent to base circles outside the points of tangency.*

Usual methods of gear cutting include: (1) Forming methods using cutters and (2) methods of generation using generating cutters. In form cutting, the shape of cutter is such that the tooth space takes the exact shape of the cutter. As against this, in method of generation, a tool having a shape different from that of the tooth profile is moved relative to the gear blank so that the desired tooth shape is obtained.

Form cutting can be done using form-milling cutters shaped to conform to the tooth space, and is one of the oldest methods of cutting gear teeth. Generation of tooth profile is possible either by (i) using a pinion-cutter or a rack cutter on shaping machine or (ii) by hobbing process using a cutter tool which is shaped like a worm. *Demand for greater accuracy in the shape of the tooth profiles has led to the development of machines for generating rather than forming the teeth on the gear blanks. Generation methods are particularly well suited to involute teeth.*

Undercutting: A straight sided rack cutter mating with a gear wheel is shown in Fig. 8.23. The rack cutter may be treated as a gear having a pitch circle radius approaching infinity. Since the rack cutter has straight sides, common normal to contacting tooth profiles (which represents line of action) is normal to inclined faces of rack cutter and is shown making angle ψ with the pitch line of rack cutter. Clearly, the straight sides of rack cutter are also inclined at angle ψ to the line of centres *OP*. Addendum *a* has been so chosen in Fig. 8.23 that in the position *BD* (along the radial line from gear centre *O*) of the straight side of the rack, the tip lies at the point of intersection of the line of the action with addendum line of the rack. For this value of addendum *a* of rack cutter, naturally, no interference is possible as *B* itself is the point of tangency on base circle.

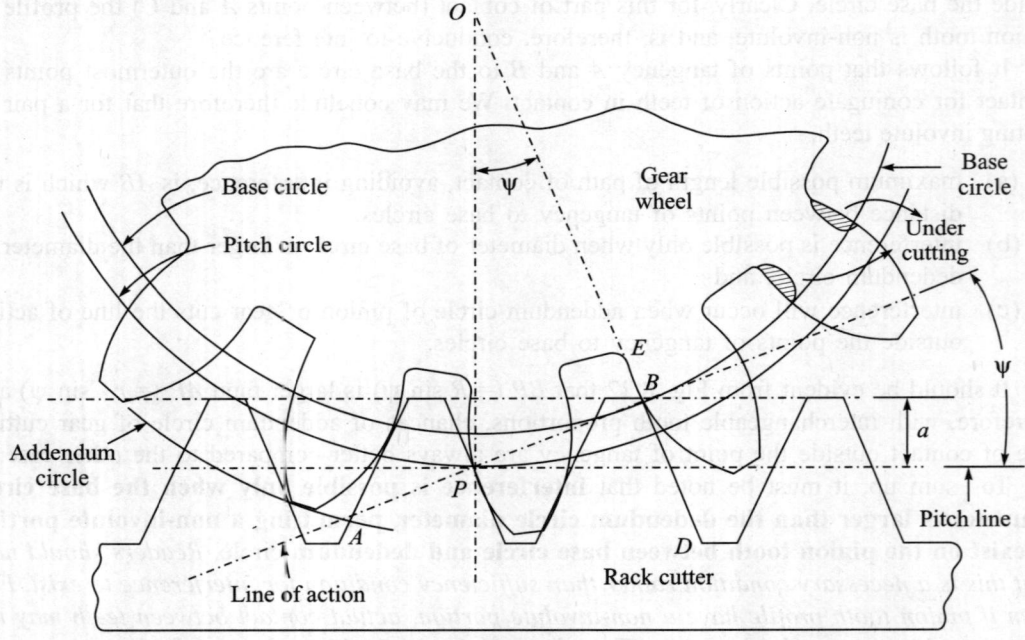

Fig. 8.23 A straight sided rack cutter mating a gear wheel.

Further, as involute curve cannot exist inside base circle, gear tooth profiles are shown radial inside the base circle in the figure. This is represented by the radial portion *BE* of the gear tooth in Fig. 8.23. Since *B* represents a point of tangency to the base circle, segment *BP* of path of contact represents limiting length of path of approach. Hence *a* represents maximum permissible addendum to avoid interference.

Any attempt to increase addendum of rack cutter above the limiting value *a*, amounts to extending rack cutter within the base circle of gear wheel. *Straight portion gear tooth profile overlaps with the straight side of rack cutter producing interference. Such a contact does not permit rolling motion. The tip of rack cutter therefore digs inside flank portion of gear and chiesels out material to clear its way out.*

It may be interesting to note that an increase in addendum of gear wheel will not lead to interference. The rack cutter has an infinite radius of curvature for its base circle and therefore, there is no possibility of addendum circle of gear, cutting line of action beyond point of tangency.

Undercutting weakens the tooth and removes a small portion of involute profile, reducing length of path of contact. Thus, elimination of interference by a generation process leads to undercutting of gear tooth, which is equally undesirable.

8.12 STANDARD PROPORTIONS OF INTERCHANGEABLE GEARS

A set of gears is called interchangeable when any two gears of the set will mesh satisfying the law of gearing. *For a set of gears to be interchangeable, they must have same module (or pitch), pressure angle, addendum and dedendum. Besides the above conditions. the tooth thickness must be ($\pi/2$) times the module.*

Standard tooth forms have been adopted to ensure interchangeability of gears. Non-standard gears are also produced for certain specific applications like automotive and air craft industries. Standard proportions are listed in Table 8.1.

Table 8.1 Standard Tooth Proportions

S.No.	Parameters	$14\frac{1^\circ}{2}$ composite or full depth involute system (m)	20° full depth involute system	20° stub tooth involute system
1.	Addendum	1 m	1 m	0.8 m
2.	Dedendum	1.25 m	1.25 m	1 m
3.	Minimum total depth	2.25 m	2.25 m	1.80 m
4.	Working depth	2 m	2 m	1.60 m
5.	Tooth thickness	1.5708 m	1.5708 m	1.5708 m
6.	Minimum clearance	0.25 m	0.25 m	0.2 m
7.	Fillet radius	0.4 m	0.4 m	0.4 m

Advantage of using interchangeable gears lies in easy availability of production tools and as such, these gears can be produced quickly and economically.

Modules Bureau of Indian Standards (B.I.S.) has recommended the following modules as first, second and third choice:

First Choice (Preferred) : 1, 1.25, 1.5, 2, 2.5, 3, 4, 5, 6, 8, 10, 12, 16 and 20 mm
Second Choice : 1.125, 1.375, 1.75, 2.25, 2.75, 3.5, 4.5, 5.5, 7, 9, 11, 14 and 18 mm
Third Choice : 3.25, 3.75, 6.5

B.I.S. recommends that preference be given to modules listed as first choice. Modulus listed under third choice should be avoided as far as possible.

Full-depth Tooth: A full-depth tooth is one in which addendum is equal to the module and provides a larger working depth.

Stub Tooth: A stub tooth has a working depth smaller than that of a full-depth tooth. This is obtained by cutting short both addendum and dedendum.

Pressure Angles: Standard pressure angles used in practice are $14\frac{1}{2}^{\circ}$ and 20°, but interchangeable gears may be produced with other pressure angles too. For instance, pressure angles of $17\frac{1}{2}^{\circ}$, $22\frac{1}{2}^{\circ}$ and 25° are used some times. Pressure angle of $14\frac{1}{2}^{\circ}$ was used in early designs simply because sine of this angle is very close to $\frac{1}{4}$ and for a cast gear tooth of $14\frac{1}{2}^{\circ}$, layout used to be simpler for pattern makers.

The $14\frac{1}{2}^{\circ}$ composite system is the oldest tooth system. Basic rack combines both involute and cycloidal curves for minimizing interference. This is a full depth tooth produced by form-milling cutters. It uses tools and machinery usually found in even the smallest machine shops.

Effect of Module on size of tooth: It should be clear from Table 8.1, that working depth is twice the module, and tooth thickness is 1.5708 times the module. Thus size of gear tooth increases as the module increases. This is depicted in Figure 8.24 which gives approximate tooth sizes for a few standard modules of choice 1.

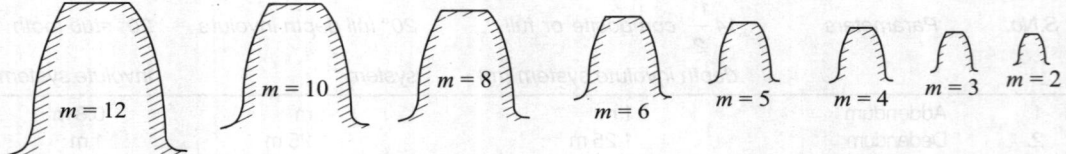

Fig. 8.24 Relative tooth sizes (approximate) for differnt modules *m*.

8.13 MINIMUM NUMBER OF TEETH TO AVOID INTERFERENCE

It was shown in section 8.11 that in order to avoid interference in a pair of involute gears in mesh, the addendum circle of either must not intersect line of contact outside the points of tangency to the two base circles. It will be useful to consider the expressions for length of path of approach and length of path of recess in terms of pitch circle radius r of pinion and R of gear.

$$\text{Length of path of approach} = r\sin\Psi$$

and
$$\text{Length of path of recess} = R\sin\Psi$$

As pinion is smaller of the two gears, it follows that

$$(\text{Length of path of approach}) \leq (\text{Length of path of recess}).$$

Since addenda of pinion and gear of 'standard proportions' are equal, while increasing addenda of the two in steps, limiting condition is reached first by the addendum circle of gear wheel. **Clearly, if conditions are favourable for interference to occur, the interference will occur first between the tip of gear tooth and flank of pinion tooth.** We now proceed to establish a relation for minimum number of teeth on pinion required to avoid interference.

Minimum Number of Teeth on Pinion to Avoid Interference: As explained above, limiting condition, avoiding interference is obtained when the addendum circle of gear cuts the line of action at the point of tangency to the base circle. This is shown in Fig. 8.25 in which A and B are the points at which line of contact is tangent to base circles. O_1P is the pitch circle radius of pinion and O_2P is the pitch circle radius of gear. Thus AP and PB are the limiting paths of approach and recess respectively. Join AO_2 and $B'O_1$ which are the addendum circle radii R_a and r_a for gear and pinion respectively. Note that for equal values of addendum on pinion and gear, the addendum circle of pinion cuts line of action at point B' which lies between points of tangency B and pitch point P.

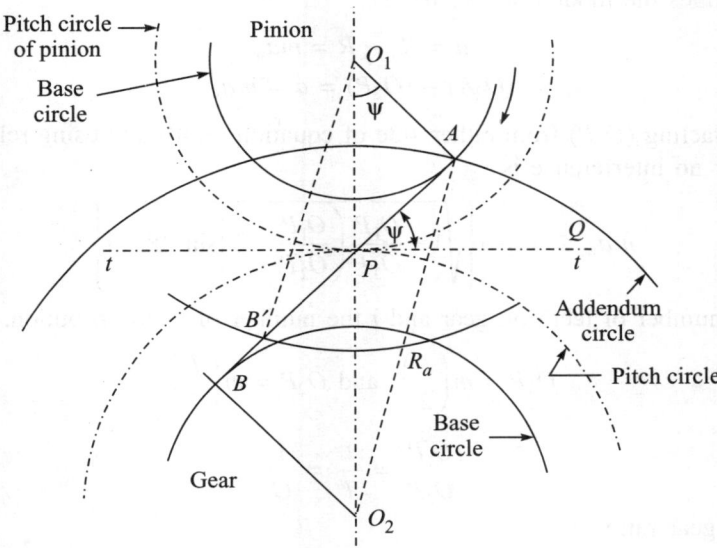

Fig. 8.25 Figure illustrating length of path of contact for no interference.

It follows that if interference occurs, it will occur on pinion tooth flank portion. Consider the limiting condition indicated by point A where a slight increase in addendum circle radius of gear will lead to the condition of interference. From the obtuse angled triangle O_2PA,

$$(AO_2)^2 = (O_2P)^2 + (AP)^2 - 2(O_2P)\,(AP)\cos \angle APO_2 \qquad (8.38)$$

But from the geometry of the Figure 8.25,

$$\angle APO_2 = \angle APQ + \angle QPO_2 = \{\Psi + 90)$$

Substituting in equation (8.38),

$$(AO_2)^2 = (O_2P)^2 + (AP)^2 + 2(O_2P)\,(AP) \sin \Psi$$

Again, from right angled triangle APO_1,

$$AP = (O_1P) \sin\Psi$$

Substituting above for AP

$$(AO_2)^2 = (O_2P)^2 + (O_1P)^2 \sin^2\Psi + 2(O_1P)\,(O_2P) \sin^2\Psi \qquad (8.39)$$

Rearranging therefore, $(AO_2)^2 = (O_2P)^2 \left\{ 1 + \left(\dfrac{O_1P}{O_2P} \right)^2 \sin^2 \Psi + 2 \left(\dfrac{O_1P}{O_2P} \right) \sin^2 \Psi \right\}$

Taking square root on either sides,

$$AO_2 = (O_2P) \sqrt{ 1 + \left(\dfrac{O_1P}{O_2P} \right) \left\{ \left(\dfrac{O_1P}{O_2P} \right) + 2 \right\} \sin^2 \Psi } \qquad (8.40)$$

Since AO_2 is the addendum circle radius of gear, expressing addendum a of gear tooth as a fraction (a_w) times the module m, we have

$$a = R_a - R = ma_w$$

or $$(O_2A) - (O_2P) = a = ma_w \qquad (8.41)$$

Hence, subtracting (O_2P) from either side of equation (8.40) and using relations in (8.41), the condition for no interference is

$$ma_w \le (O_2P) \left\{ \sqrt{ 1 + \dfrac{O_1P}{O_2P} \left(\dfrac{O_1P}{O_2P} + 2 \right) \sin^2 \Psi } - 1 \right\} \qquad (8.42)$$

If T be the number of teeth on gear and t the number of teeth on pinion, so that

$$O_1P = m \left(\dfrac{t}{2} \right); \text{ and } O_2P = m \left(\dfrac{T}{2} \right)$$

we have $$\dfrac{O_1P}{O_2P} = \dfrac{t}{T} = \dfrac{1}{G} \qquad (8.43)$$

where, G is the gear ratio.

Substituting in equation (8.42), condition for no interference becomes,

$$ma_w \le m \left(\dfrac{T}{2} \right) \left\{ \sqrt{ 1 + \dfrac{t}{T} \left(\dfrac{t}{T} + 2 \right) \sin^2 \Psi } - 1 \right\}$$

which simplifies to $$a_w \le \left(\dfrac{T}{2} \right) \left\{ \sqrt{ 1 + \dfrac{t}{T} \left(\dfrac{t}{T} + 2 \right) \sin^2 \Psi } - 1 \right\}$$

Above relation can be put in the following form, $T \ge \dfrac{2(a_w)}{\left\{ \sqrt{ 1 + \dfrac{t}{T} \left(\dfrac{t}{T} + 2 \right) \sin^2 \Psi } - 1 \right\}} \qquad (8.44)$

Stated in terms of gear ratio G, interference will not occur if,

$$T \ge \dfrac{2(a_w)}{\sqrt{ 1 + \dfrac{1}{G} \left(\dfrac{1}{G} + 2 \right) \sin^2 \Psi } - 1} \qquad (8.45)$$

Since T is related to t, the number of teeth on pinion, through gear ratio $G = \dfrac{T}{t}$, substituting for T on L.H.S.,

$$Gt \geq \frac{2\,(a_w)}{\sqrt{1 + \dfrac{t}{T}\left(\dfrac{t}{T} + 2\right)\sin^2 \Psi} - 1}$$

or

$$t \geq \frac{2\,(a_w)\,\dfrac{t}{T}}{\sqrt{1 + \dfrac{t}{T}\left(\dfrac{t}{T} + 2\right)\sin^2 \Psi} - 1} \qquad (8.46)$$

and

$$t \geq \frac{2\left(\dfrac{a_w}{G}\right)}{\sqrt{1 + \dfrac{1}{G}\left(\dfrac{1}{G} + 2\right)\sin^2 \Psi} - 1} \qquad (8.47)$$

for 20° full depth teeth addendum equals module itself and therefore $a_\omega = 1$. Thus for gear ratio of 1, 2, 3, 4, 5, 6, and 7, the values on the basis of equation (8.45) are tabulated as follows:

G	1.0	2.0	3.0	4.0	5.0	6.0	7.0
t_{min}	12.32	14.16	14.98	15.44	15.74	15.947	16.09

minimum number of teeth prescribed in machine design for 20° involute pinion (with full depth teeth) for avoiding interference is 17.

EXAMPLE 8.5 Two mating spur gears with module pitch 6.5 mm have 19 and 47 teeth of 20° pressure angle and 6.5 mm addendum. Determine the number of pairs of teeth in contact and the angle turned through by the larger wheel for one pair of teeth in contact.

Determine also the ratio of the sliding velocity to the rolling velocity at the instant (a) engagement commences, (b) the engagement terminates, and (c) at the pitch point. (AMIE, Winter 1976)

Solution: Given $t = 19$; $T = 47$; $m = 6.5$ mm; $\Psi = 20°$; $a = 6.5$ mm

Therefore $\qquad\qquad\qquad\qquad a_w = 1$

Pitch circle radii of pinion and gear are,

$$r = m(t/2) = 6.5 \times \frac{19}{2} = 61.75 \text{ mm}$$

and

$$R = m(T/2) = 6.5 \times \frac{47}{2} = 152.75 \text{ mm}$$

Hence addendum circle radii of pinion and gear are

$$r_a = r + a = 61.75 + 6.5 = 68.25 \text{ mm}$$

and
$$R_a = R + a = 152.75 + 6.5 = 159.25 \text{ mm}$$

Length of path of approach (Fig. 8.21),

$$PN = \sqrt{R_a^2 - R^2\cos^2\Psi} - R\sin\Psi = 16.73 \text{ mm}$$

Length of path of recess $= PQ = \sqrt{r_a^2 - r^2\cos^2\Psi} - r\sin\Psi$

$$= \sqrt{(68.25)^2 - (61.75\cos 20)^2} - (61.75)\sin 20 = 14.81 \text{ mm}$$

(i) Thus, length of path of contact $= 16.73 + 14.81 = 31.54 \text{ mm}$

Hence, length of arc of contact $= \dfrac{31.54}{\cos 20} = 33.564 \text{ mm}$

Therefore, number of pairs of teeth in contact, namely,

contact ratio $CR = \dfrac{33.564}{p_c} = \dfrac{33.564}{(m\pi)}$

Hence $\qquad CR = \dfrac{33.564}{(6.5)\,\pi} = 1.644$ pairs $\qquad$ **Ans.**

(ii) Angle turned through by larger wheel for one pair of teeth in contact

$$\theta = \dfrac{\text{length of arc of contact} \times 360°}{\text{circumference of gear}}$$

or $\qquad \theta = \dfrac{33.564 \times 360°}{2\,\pi\,(R)} = \dfrac{33.564 \times 360}{2\,\pi \times 152.75} = 12.589° \qquad$ **Ans.**

(iii) Angular speed of none of the mating gears is given. If $\omega_1 =$ speed of pinion, the rolling velocity,

$$V_r = r\omega_1 = (61.75)\,\omega_1 \text{ mm/s}$$

Velocity of sliding at the instant of engagement

$$V_s = (\omega_1 + \omega_2)\,16.73$$

or $\qquad V_s = \left(\omega_1 + \dfrac{19}{47}\omega_1\right) \times 16.73 = (23.493)\omega_1 \text{ mm/s}$

Velocity of sliding at pitch point $= 0$
And, velocity of sliding at the point disengagement

$$V_s' = (\omega_1 + \omega_2) \times 14.81 = \left(\omega_1 + \dfrac{19}{47}\omega_1\right) \times 14.81 = (20.297)\omega_1 \text{ mm/s}$$

Hence the ratio of velocity of sliding to velocity of rolling at

(a) instant of engagement $\quad = \dfrac{(23.493)\,\omega_1}{(61.75)\,\omega_1} = 0.38$ **Ans.**

(b) pitch point $\quad\quad\quad\quad = \dfrac{0}{(61.75)\,\omega_1} = 0$ **Ans.**

(c) instant of disengagement $\quad = \dfrac{(20.797)\,\omega_1}{(61.75)\,\omega_1} = 0.337$ **Ans.**

EXAMPLE 8.6 The following data refer to two mating involute gears of 20° pressure angle shown in Fig. 8.26.

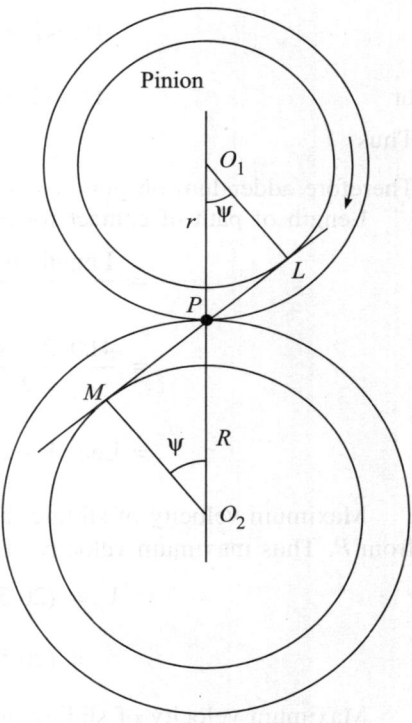

Number of teeth on pinion	20
Gear ratio	2
Speed of pinion (r.p.m.)	250
Module (mm)	12

If the addendum on each wheel is such that the path of approach and path of recess on each side are of half the maximum possible length each, find:

(a) addendum for pinion and gear; (b) the length of arc of contact; and (c) the maximum velocity of sliding during approach and recess. (DAVV Indore, 1975 and AMIE, Winter 1982).

Solution: $r = 12 \times \dfrac{20}{2} = 120$ mm; $R = 12 \times \dfrac{40}{2}$

$= 240$ mm

Maximum length of approach without possibility of interference,

$$PL = r\sin\Psi$$
$$= (120)\sin 20° = 41.042 \text{ mm}$$

Maximum length of recess, without possibility of interference

$$MP = R\sin\Psi = (240)\sin 20° = 82.085 \text{ mm}.$$

Fig. 8.26

Thus, given length of path of approach

$$PL' = \frac{41.042}{2} = 20.521 \text{ mm}$$

and, given length of path of recess

$$PM' = \frac{82.085}{2} = 41.042 \text{ mm}$$

For the length of path of approach = 20.521 mm, we have

$$20.521 = \sqrt{R_a^2 - (240 \cos 20°)^2} - 240 \sin 20°$$

or
$$\sqrt{R_a^2 - (50862)} = 20.521 + 82.0848 = 102.606$$

Therefore
$$R_a = 247.77 \text{ mm}$$

Therefore   addendum on gear = (247.77) − 240 = 7.77 mm   **Ans.**

Similarly for a length of path of recess = 41.042 mm
Hence

$$41.042 = \sqrt{r_a^2 - (120 \cos 20)^2} - (120) \sin 20$$

or
$$41.042 = \sqrt{r_a^2 - (112.763)^2} - 41.0424$$

Thus
$$r_a = 139.47$$

Therefore addendum on pinion = 139.47 − 120 = 19.47 mm   **Ans.**

Length of path of contact for above addenda (actual)

$$= \frac{\text{Length of path of approach}}{2} + \frac{\text{Length of path of recess}}{2}$$

$$= \frac{41.042 + 82.085}{2} = 61.563 \text{ mm}$$

$$= \text{Length of arc of contact} = \frac{61.563}{\cos 20} = 65.51 \text{ mm}$$ **Ans.**

Maximum velocity of sliding during approach occurs when the point of contact is farthest from P. Thus maximum velocity of sliding during approach

$$V_s = (20.521) (\omega_1 + \omega_2)$$

$$= (20.521) (250 + 125) \frac{2\pi}{60} = 805.86 \text{ mm/s}$$ **Ans.**

Maximum velocity of sliding during recess,

$$V_s' = 41.042 (250 + 125) \frac{2\pi}{60} = 1611.72 \text{ mm/s}$$ **Ans.**

EXAMPLE 8.7   Two-gear wheels mesh externally and are to give a velocity ratio of 3 to 1. The teeth are of involute type: module = 6 mm; addendum = one module, pressure angle = 20°. The pinion rotates at 90 r.p.m. Find

(a)   the number of teeth on the pinion to avoid interference on it and the corresponding number on wheel,
(b)   lengths of path and arc of contact,
(c)   the number of pairs of teeth in contact,
(d)   the maximum velocity of sliding. (AMIE, Winter 1980)

Solution: Given : module $m = 6$ mm; addendum $a = 6$ mm

Gear ratio $G = 3 : 1$; $\Psi = 20°$; $N_{pinion} = 90$ r.p.m.

Here, as addendum $= 1 \times$ (module),

$$a_p = a_\omega = 1. \quad \text{and} \quad \omega_1 = \frac{2\pi \times 90}{60} = 9.43 \text{ rad/s}$$

Number of teeth required on gear to avoid interference,

$$T \geq \frac{2\,a_\omega}{\sqrt{1 + \dfrac{1}{G}\left(\dfrac{1}{G} + 2\right)\sin^2\Psi} - 1} \tag{1}$$

$$\geq \frac{2\,(1)}{\sqrt{1 + \dfrac{1}{3}\left(\dfrac{1}{3} + 2\right)\sin^2 20} - 1} ; \ i.e., \geq 44.94 = 45, \text{ say}$$

Therefore $t \geq \dfrac{45}{3} \ i.e., \ t \geq 15$

(***Note***: Since limiting condition of interference with standard module is reached first on the flank of pinion teeth, it is necessary to use relation for limiting number teeth on gear and then to find out limiting number of teeth on pinion.)

(b) length of path of approach

$$= \sqrt{R_a^2 - (R\cos\Psi)^2} - R\sin\Psi$$

But for $t = 15$ and $T = 45$, $r = \dfrac{6 \times 15}{2} = 45$ mm and $R = \dfrac{6 \times 45}{2} = 135$ mm

and $r_a = 45 + 6 = 51$ and $R = 135 + 6 = 141$ mm

Therefore   length of path of approach

$$= \sqrt{(141)^2 - (135\cos 20)^2} - (135)\sin 20 = 15.37 \text{ mm}$$

And, length of path of recess

$$= \sqrt{(51)^2 - (45\cos 20)^2} - (45)\sin 20 = 13.12 \text{ mm.}$$

Therefore maximum velocity of sliding occurs at a point farthest (in the given case at point of engagement) from pitch point. Thus

$$(V_s)_{max} = 15.37 \ (\omega_1 + \omega_2)$$

$$= 15.37\left(9.43 + \frac{9.43}{3}\right) = 193.25 \text{ mm/s} \qquad \textbf{Ans.}$$

Length of path of contact $= 15.37 + 13.12 = 28.49$ mm   **Ans.**

Length of arc of contact $= \dfrac{28.49}{\cos 20} = 30.32$ mm   **Ans.**

Number of pairs of teeth in contact (*i.e.*, the contact ratio) $= \left(\dfrac{30.32}{p_c} \right)$

$$= \frac{30.32}{m \cdot \pi} = \frac{30.32}{6\pi} = 1.608 \qquad\qquad \textbf{Ans.}$$

8.14 MINIMUM NUMBER OF TEETH ON PINION TO AVOID INTERFERENCE WITH RACK

A pinion rotating in c.w. sense and driving a rack is shown in Fig. 8.27. The rack has a pitch line instead of arc of circle. Clearly, the rack can be considered to be a gear of infinite pitch circle radius with its centre located at infinity along the line O_1P.

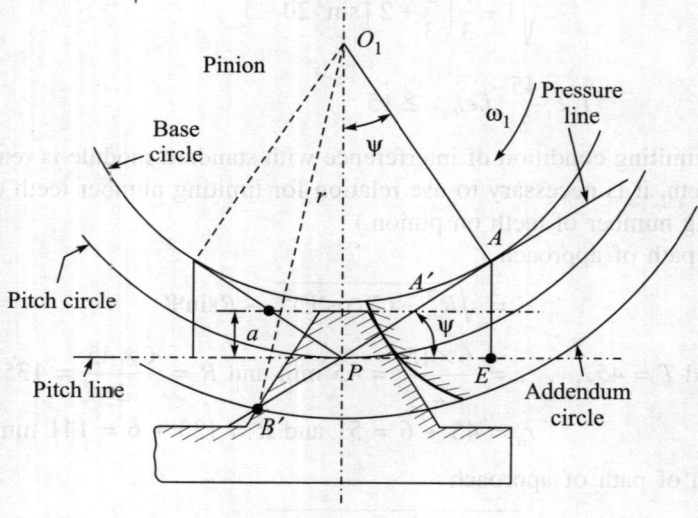

Fig. 8.27

Let PA be the tangent drawn from pitch point P to the base circle. Then, for involute teeth on pinion, PA is also the pressure line (*i.e.* the line of action). Further, let $AE = a'$ be the length of perpendicular dropped from point A on the pitch line of the rack. Clearly, A is the limiting point of contact for avoiding interference. In other words, the length AE gives limiting value of addendum of rack for zero interference. This is because as the rack tooth moves to the right, the point of contact moves along the pressure line AP and the actual point of contact lies always outside the base circle of the pinion. This implies that involute profile is always possible at the point of contact between A and P and hence, interference will not take place.

Hence, the condition for no interference is

$$a \le a'$$

or, $$a \le (AP)\sin\psi \qquad\qquad (8.48)$$

But from right angled triangle O_1AP,

$$AP = (O_1P) \sin\psi = r\sin\psi$$

Substituting for AP in (8.48)

$$a \leq r\sin^2\psi \tag{8.49}$$

But, the pitch circle radius of pinion

$$r = \frac{m\,t}{2}$$

where m is the module and t is the number of teeth on pinion.

Substituting for r in (8.49), the condition for avoiding interference is

$$a \leq \left(\frac{m\,t}{2}\right)\sin^2\psi$$

or

$$t \geq \frac{2a}{(m\sin^2\psi)} \tag{8.50}$$

Expressing addendum a as a fraction a_r times module (*i.e.*, $a = a_r \times m$) we have from (8.50)

$$t \geq \left(\frac{2a_r}{\sin^2\psi}\right) \tag{8.51}$$

For standard proportions of tooth $a_r = 1$ for $20°$ pressure angle and, in that case,

$$t_{min} = \frac{(2)}{\sin^2 20} = 17.1$$

Thus

$$t_{min} = 18$$

For $20°$ stud tooth, $a_r = 0.8$ and as such,

$$t_{min} = \frac{(1.6)}{\sin^2 20} = 13.68$$

Therefore

$$t_{min} = 14$$

EXAMPLE 8.8   Find the minimum number of teeth to avoid undercutting when the addendum for stub-tooth is 0.84 module and when the power component is 0.95 times the normal thrust, if (a) the gear ratio is unity; (b) the gear ratio is 3 to 1; (c) the pinion gears with a rack.

In each case, find the length of arc of contact in terms of module. (DAVV Indore, 1988 July)

Solution: Given, $\dfrac{F_t}{F} = \cos\Psi = 0.95$ or $\cos\Psi = 0.95$

Hence, pressure angle $\Psi = \cos^{-1}(0.95)$ or, $\Psi = 18.195°$

Addendum $\qquad\qquad = (0.84) \times$ module (for stub tooth)

Therefore $\qquad a_r = 0.84 = a_\omega$

(a) While meshing with another gear with $G = 1$, minimum number of teeth on gear

$$T = \frac{2 \times a_\omega}{\sqrt{1 + \dfrac{1}{G}\left(\dfrac{1}{G} + 2\right)\sin^2 \Psi} - 1}$$

or $\qquad T = \dfrac{2 \times 0.84}{\sqrt{1 + 1(1 + 2)\sin^2 18.195} - 1} = \dfrac{2 \times 0.84}{0.13688} = 12.27$

i.e., $\qquad\qquad\qquad\qquad T = 13.$

Length of path of contact (as $r = R$, $t = T = 13$ and $r_a = R_a$) is

$$r = \frac{mt}{2} = (6.5m) \text{ and } r_a = 6.5m + 0.84m = (7.34m)$$

Therefore Length of path of contact $= 2\left[\sqrt{r_a^2 - (r\cos\Psi)^2} - r\sin\Psi\right]$

or $\qquad L = 2\sqrt{(7.34m)^2 - (6.5m \times \cos 18.195)^2} - 2\,(6.5m)\sin 18.195$

or $\qquad\qquad\qquad\qquad L = (3.877m)$ mm

Length of arc of contact $= \dfrac{(3.877m)}{\cos 18.195} = (4.081m)$ mm $\qquad\qquad$ **Ans.**

(b) For a gear ratio $G = 3$,
minimum number of teeth on gear

$$T = \frac{2\,(0.84)}{\sqrt{1 + \dfrac{1}{3}\left(\dfrac{1}{3} + 2\right)\sin^2 18.195} - 1} = \frac{2 \times 0.84}{0.0372}$$

$$= 45.14 = 46 \text{ say}$$

Therefore minimum no. of teeth on pinion $= \dfrac{46}{3} = 15.33 = 16$ say

Therefore revised number of teeth on gear $= 16 \times 3 = 48$

Therefore pitch circle radii $r = \dfrac{16m}{2} = 8m$ mm; $R = \dfrac{48m}{2} = 24\,m$ mm

$$r_a = (8 + 0.84)m = (8.84m)\text{mm}$$

$$R_a = (24 + 0.84)m = (24.84m)\text{mm}.$$

Hence, length of path of contact,

$$= \sqrt{(24.84m)^2 - (24m\cos 18.195)^2} + \sqrt{(8.84m)^2 - (8m\cos 18.195)^2}$$

$$- (24m + 8m)\sin 18.195$$

$$= (9.858m) + (4.515m) - 9.992m = (4.381)m \text{ mm}.$$

Therefore length of arc of contact $= \dfrac{4.381\,m}{\cos 18.195} = (4.61)m$ mm. **Ans.**

(c) When the pinion gears with the rack, minimum number of teeth on pinion,

$$t = \frac{2a_r}{\sin^2 \Psi} = \frac{2\,(0.84)}{\sin^2 18.195} = 17.23 = 18 \text{ say.}$$ **Ans.**

Length of path of contact (due to symmetry, the path of contact will extend by same amount on other side (*see* Fig. 8.27). Thus,

length of path of contact $= 2\sqrt{r_a^2 - (r \cos \Psi)^2} - 2r\sin \Psi$

Here, for pinion $\qquad\qquad r = \dfrac{m \times 18}{2} = 9m$

and $\qquad\qquad\qquad r_a = (9m + 0.84m) = 9.84m$

Therefore length of path of contact $= 2\sqrt{(9.84m)^2 - (9m \cos \Psi)^2} - 2\,(9m)\sin \Psi$
$$= (4.12)m \text{ mm}$$

Therefore length of arc of contact $= \dfrac{4.12\,m}{\cos 18.195} = (4.337m)$ mm. **Ans.**

EXAMPLE 8.9   A pinion of 20 involute teeth and 125 mm pitch circle diameter drives a rack. The addendum of both pinion and rack is 6.25 mm. What is the least pressure angle which can be used to avoid interference? With this pressure angle, find the length of the arc of contact and the minimum number of teeth in contact at a time. (J.U. Gwalior, 1983)

Solution: Module $m = \dfrac{125}{20} = 6.25$ mm

Since, minimum number of teeth to avoid interference is given by

$$t \geq \frac{2\,a}{m \sin^2 \Psi} \quad \text{(see equation 8.50)}$$

it follows that smaller the pressure angle, larger is the required number of teeth on pinion. Hence for $t = 20$, smallest pressure angle Ψ, is given by

$$\sin^2 \Psi = \frac{2\,a}{m\,t} = \frac{2 \times 6.25}{20 \times 6.25} = 0.1$$

Therefore $\qquad\qquad\qquad \Psi = 18.43°$ **Ans.**

Length of the path of contact (as in the previous example),

$$= 2\left(\sqrt{r_a^2 - (r\cos \Psi)^2} - r\sin \Psi\right)$$

However, at the limiting condition with smallest pressure angle, the addendum line of rack passes through the point of tangency A to base circle and, as such, length of path of contact (vide Fig. 8.27) is

$$B'A = \sqrt{r_a^2 - (r\cos\Psi)^2}$$

(This expression can be established by considering right angle traingles $O_1B'A$ and O_1PA and is left as an exercise for students)

For the given problem, length of path of contact

$$= \sqrt{(62.5 + 6.25)^2 - (62.5\cos 18.43)^2}$$

$$= 34.796 \text{ mm} \qquad\qquad \textbf{Ans.}$$

Length of arc of contact $= \dfrac{34.796}{\cos 18.43} = 36.68$ mm

Therefore contact ratio $= \dfrac{36.68}{p_c} = \dfrac{36.68}{m\,\pi} = \dfrac{36.68}{6.25\,\pi} = 1.868$ **Ans.**

EXAMPLE 8.10 A pinion is with 18 teeth meshes with an internally toothed gear with 72 teeth. Find the length of the path of contact when pressure angle is 20°, module = 4 mm and the addenda on pinion and wheel are 8.5 mm and 3.5 mm respectively.

Solution: It is worthwhile to know a few interesting features of internal gearing. Since internal gears have the teeth projecting inward, (towards centre of gears), the face of wheel tooth lies on the inside of pitch circle. Similarly, addendum circle diameter is less than pitch circle diameter. Path of contact, as before, is along common tangent to the two base circles.

Given $t = 18$; $T = 72$ $\qquad\qquad \Psi = 20°$; $m = 4$ mm

addendum on pinion = 8.5 mm $\qquad$ addendum on wheel = 3.5 mm

As shown in Fig. 8.28, the addendum circle of internal gear/cuts common tangent to base circles at point K while addendum circle of pinion cuts the same common tangent at point N.

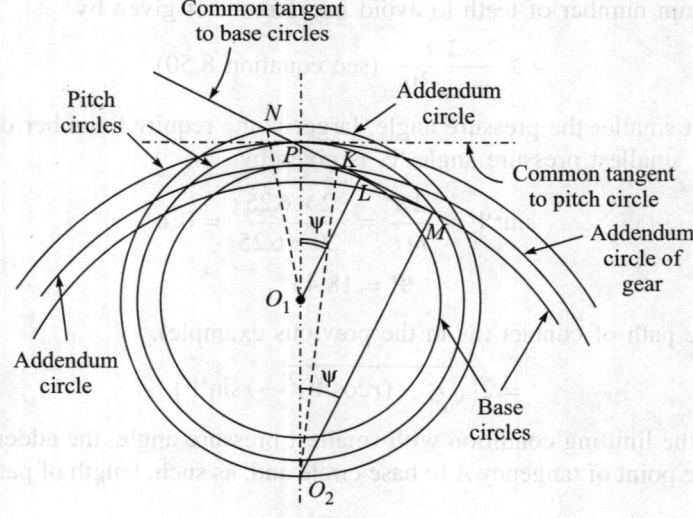

Fig. 8.28

Hence *NPK* is the length of path of contact; *NP* being the length of path of approach and *PK* the length of path of recess. O_1L is the radius of base circle of pinion with *L* as the point of tangency of *NK* with the base circle of pinion. Similarly, O_2M is the radius of base circle of gear with *M* as the point of tangency of *NK* with the base circle of gear.

From right angled triangle O_1NL

$$NL = \sqrt{O_1N^2 - O_1L^2} \tag{1}$$

But
$$O_1L = O_1P\cos\Psi \text{ and } O_1N = r_a$$

Therefore
$$NL = \sqrt{r_a^2 - (r\cos\Psi)^2}$$

Again, *NP* = Length of path of approach = (*NL* – *PL*), $\tag{2}$

where
$$PL = O_1P\sin\Psi$$

Hence from (1) and (2),

$$NP = \sqrt{r_a^2 - r^2\cos^2\Psi} - r\sin\Psi \tag{3}$$

Similarly, from right angled triangle $O_2\,PM$,

$$PM = (O_2P)\sin\Psi = R\sin\Psi \tag{4}$$

and from right angled triangle O_2KM,

$$KM = \sqrt{O_2K^2 - O_2M^2}$$
$$= \sqrt{R_a^2 - (O_2P\cos\Psi)^2} = \sqrt{R_a^2 - R^2\cos^2\Psi} \tag{5}$$

Hence using equations (4) and (5),

$$PK = \text{length of path of recess} = PM - MK$$
$$= R\sin\Psi - \sqrt{R_a^2 - R^2\cos^2\Psi}$$

Hence in the given problem, length of path of contact *NK* = *NP* + *PK*

$$= \left(\sqrt{r_a^2 - r^2\cos^2\Psi} - r\sin\Psi\right) + \left(R\sin\Psi - \sqrt{R_a^2 - R^2\cos^2\Psi}\right)$$

Now
$$R = \frac{mT}{2} \quad i.e., R = \frac{4 \times 72}{2} = 144 \text{ mm}$$

and
$$r = \frac{mt}{2} \quad i.e., r = \frac{4 \times 18}{2} = 36 \text{ mm}$$

And as such,
$$r_a = 36 + 8.5 = 44.5 \text{ mm}$$
and
$$R_a = 144 - 3.5 = 140.5 \text{ mm}$$

Therefore
$$NK = \left(\sqrt{(44.5)^2 - (36\cos 20°)^2} - 36\sin 20°\right)$$

$$+ \left(144\sin 20° - \sqrt{140.5^2 - (144\cos 20°)^2}\right)$$

$$= 16.6 + 11.437 = 28.037 \text{ mm} \qquad \textbf{Ans.}$$

8.15 COMPARISON BETWEEN INVOLUTE AND CYCLOIDAL TOOTH PROFILES

As indicated earlier, involute tooth profile is almost universally used in preference to cycloidal or any other tooth form. This does not mean that involute tooth form is the best from every angle. Table 8.2 will provide a realistic assessment of the positive and negative points about the involute and cycloidal tooth profiles.

Table 8.2

S.No.	Involute tooth profile	Cycloidal tooth profile
1.	Being the angle made by common tangent of base circles with common tangent to pitch circles at pitch point, the pressure angle remains constant throughout the engagement. This ensures smooth running of the gears.	Pressure angle varies continuously; being zero at the pitch point and maximum at the point of engagement and disengagement. This causes continuous variation in power component and also in bearing load. The running is less smooth.
2.	Tooth profile consists of a single (involute) curve and the rack cutter used for generating the profile has straight teeth. The rack cutter is cheaper and the method of manufacture is simpler. This leads to reduction in the cost of manufacture of involute teeth.	The tooth profile consists of two curves—epicycloid and hypocycloid. The method of manufacture, therefore, is more involved and leads to costlier gear teeth.
3.	Perhaps the most desirable feature of involute teeth is that a small variation in centre distance does not change the velocity ratio. Thus distance between shafts need not necessarily be maintained exact as per design specifications. This gives great flexibility during assembly and larger tolerances may be permitted.	Exact centre distance is necessary to transmit constant velocity ratio.
4.	Since involute curve does not exist within base circle, interference is always possible if base circle radius is larger than dedendum circle radius.	Since outside the pitch (Directing) circle epicycloidal curve exists and inside it the hypocycloidal curve exists, cycloidal curve can exist everywhere on tooth profile and no interference exists.
5.	The radius of curvature of involute curve, near the base circle, is quite small and contact stresses are likely to be very high. The tooth profile in flank portion is almost radial. Both the factors together produce a tooth weaker in flank region compared to cycloidal tooth.	Cycloidal curve (hypocycloidal in particular) produces a spreading flank and, for this reason, cycloidal tooth is stronger compared to involute tooth.
6.	Convex surface of pinion tooth comes in contact with convex portion of gear tooth and this leads to more wear.	In cycloidal tooth profile epicycloidal shape of face of gear tooth comes in contact with hypocycloidal flank portion of pinion tooth. Thus a convex flank has a contact with concave face which results in lesser wear.

As a direct consequence of 'no interference' in cycloidal tooth profile these profiles can be used on gear teeth for achieving higher reduction ratio. Further, cycloidal teeth have less sliding action than involute teeth, which results in smaller wear. In view of the above cycloidal gears find applications in clocks, watches and certain instruments because pinions can afford to have low number of teeth without any risk of interference or undercutting. They are also used in some crude purposes where heavy and impact loads act on the machine.

8.16 METHODS OF REDUCING OR ELIMINATING INTERFERENCE

In general, there are three different ways of reducing or eliminating interference and subsequent undercutting at the flank region, when a pinion has less than 'minimum number of teeth' to avoid interference.

By Using Modified Involute or Composite System

When a standard addendum is used for involute pinion with a pressure angle of $14\frac{1}{2}^{\circ}$ the smallest pinion that will gear with a rack without interference has 32 teeth. In many situations, it becomes obligatory to use a pinion which has 12 teeth only. In such cases the shape of basic rack tooth may be modified. *The flank portion of pinion tooth lying inside the base circle and the mating portion of gear tooth face may be made cycloidal in place of involute shape. Remaining portion of the pinion tooth may be of involute profile.*

The cycloidal parts of the profiles are generated by rolling circles which have a diameter equal to the pitch radius of the pinion. This gives rise to what is known as $14\frac{1}{2}^{\circ}$ composite system. Since part of the profile is cycloidal, the gears will have correct gearing only when the centre distance is exact. Another serious disadvantage of such composite system is that the profile of the rack tooth cannot be reproduced with the same degree of accuracy which is obtainable with straight sided rack.

By Modifying Addendum of Gear Tooth

It may be recalled that when conditions favouring interference exist, tip of gear tooth comes in contact with non-involute portion of flank between dedendum circle and base circle. To avoid interference, therefore, an obvious proposition seems to be to 'chop-off' interfering portion of the face of the gear tooth. This is illustrated in Fig. 8.29. The resulting tooth is called 'stub-tooth'

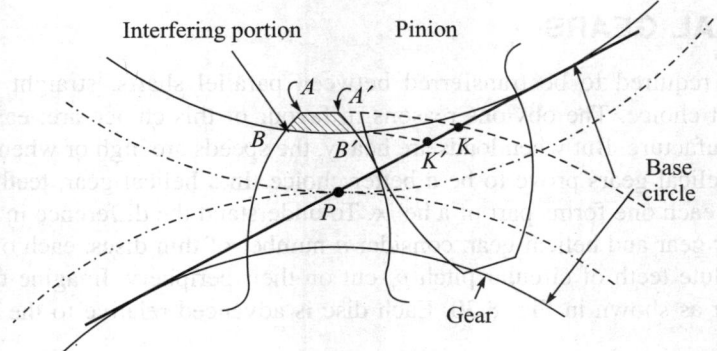

Fig. 8.29 Illustration demonstrating advantage of stub-tooth.

instead of 'full-depth' tooth. As can be seen in the figure, portion $A'B'$ of gear tooth meshes with non-involute portion in the flank of pinion tooth. If portion $AA'B'B$ of gear tooth is chopped off, interference will be eliminated. However, this step amounts to reducing radius of addendum circle of gear, which now cuts line of action at point K' instead of K. In other words, length of path of approach and hence, the contact ratio is reduced.

Increased Centre Distance

It will be relevant and useful for the readers to note that while size of base circle for each of the two mating gears is fixed, their pitch circle size decides gear size only when their centre distance is fixed. When centres O_1 and O_2 of two mating gears are slightly moved apart (the diameters of base circles of the two remaining the same), the common tangent to the base circles cut the line of centres at point P'. O_1P' and O_2P' then become the pitch circle radii of the mating gears. Mathematically (*see* Fig. 8.20)

$$r_b = r\cos\Psi \text{ and } R_b = R\cos\Psi$$

Hence $$(r_b + R_b) = \text{constant} = (r + R)\cos\Psi$$
If $$C \equiv r + R \text{ is the centre distance}$$
$$(r_b + R_b) = \text{constant} = C\cos\Psi$$

This shows that in order to maintain a constant velocity ratio an increase in centre C must be compensated suitably by a decrease in the value of $\cos\Psi$. In other words, an increase in centre distance necessarily implies an increase in pressure angle Ψ.

Thus centre distance between two involute gears may be increased within limits, without disturbing correctness of gearing, and this step will prevent the tip of gear tooth from mating with non-involute flank portion of the pinion, atleast to some extent. There are, however, two adverse consequences:

(a) Actual pitch circle diameter is increased, this tends to reduce effective addenda and hence the arc of contact too.
(b) As the difference in tooth space width and tooth thickness at new pitch circle is further increased, considerable backlash will be introduced between the teeth.

Besides these, undercutting may be permitted to avoid interfernce. However, normally this is not recommended because of reasons stated in section 8.11.

8.17 HELICAL GEARS

When power is required to be transferred between parallel shafts, straight spur gear drive becomes the first choice. The obvious reasons in favour of this choice are: ease in design and economy in manufacture. But when loads are heavy, the speeds are high or when excessive noise is undesirable, helical gears prove to be a better choice. In a helical gear, teeth of spur wheels are cut such that each one forms part of a helix. To understand the difference in the construction of a straight spur gear and helical gear, consider n number of thin discs, each of small width w, and having involute teeth of circular pitch p_c cut on their periphery. Imagine these discs to be fastened together as shown in Fig. 8.30. Each disc is advanced relative to the adjacent one by

an amount equal to $\left(\dfrac{p_c}{w}\right)$.

If the number of discs within a total thickness *b* is increased to infinity (reducing individual disc to a lamina), the stepped teeth will become helicoid in form. With stepped gears the load is applied first to a portion of the face width, later to the next portion of step and so on so forth. Just when the last step of first tooth is to go out of contact, first step of the next tooth comes into contact with driving tooth. As a result the teeth come in contact with less of impact.

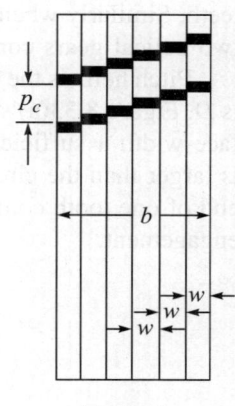

Fig. 8.30

Shapes of gear tooth profile of straight spur gear and helical gear can be compared as shown in Fig. 8.31(a) and (b). Imagine a stiff plane to roll over the cylindrical surface of base cylinder. In Fig. 8.31(a), a line *ab* of the plane, parallel to the axis of cylinder, is shown to generate involute surface of tooth profile. In Fig. 8.31(b) a line *a'b'* of the plane, inclined to the axis of cylinder, is shown to generate a surface called *involute helicoid* and consists of straight line elements as shown. When a pair of gear teeth of helical gears are in contact, the contact takes place between the teeth along one of the successive positions of line *a'b'*.

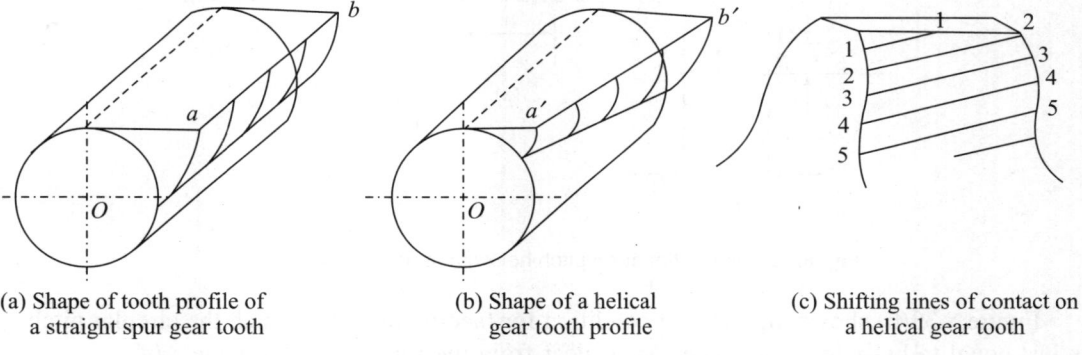

| (a) Shape of tooth profile of a straight spur gear tooth | (b) Shape of a helical gear tooth profile | (c) Shifting lines of contact on a helical gear tooth |

Fig. 8.31   Geometry of gear tooth profiles.

When mating teeth of straight spur gears mesh, the contact takes place across the entire face-width. As against this, in helical gears contact begins at one end of the tooth and progresses across the tooth (Fig. 8.31c). Straight diagonal lines such as 1–1, 2–2, 3–3, etc. indicate successive lines of contact. Because of gradual contact across the teeth, the impact load is reduced and gears turn out to be stronger. They have longer life and generate lesser noise compared to straight spur gears.

The direction in which teeth of a helical gear slope is defined as the hand of the gear. In order to determine the hand of a helical gear, a gear is to be placed on its side as shown in Fig. 8.32(a). Teeth sloping upwards to the right are called right hand

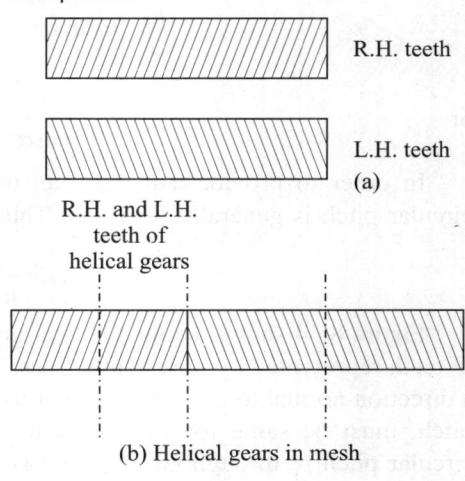

Fig. 8.32

teeth. Similarly when the teeth slope upward to the left they are called left hand teeth. As a rule, two helical gears connecting parallel shafts must have opposite hand (*see* Fig. 8.32b).

Pitch helix is the curve at which the helical tooth intersects the pitch cylinder, whose diameter is *D*. Figure 8.33(a) shows one such curve constituting tooth of a helical gear. It is usual to make face-width *w* sufficiently large so that for a given angle of helix, tooth advance (*see* Fig. 8.33a) is larger than the circular pitch. This step ensures overlapping action which means that leading end of one tooth comes into engagement before the trailing end of preceeding tooth goes out of engagement.

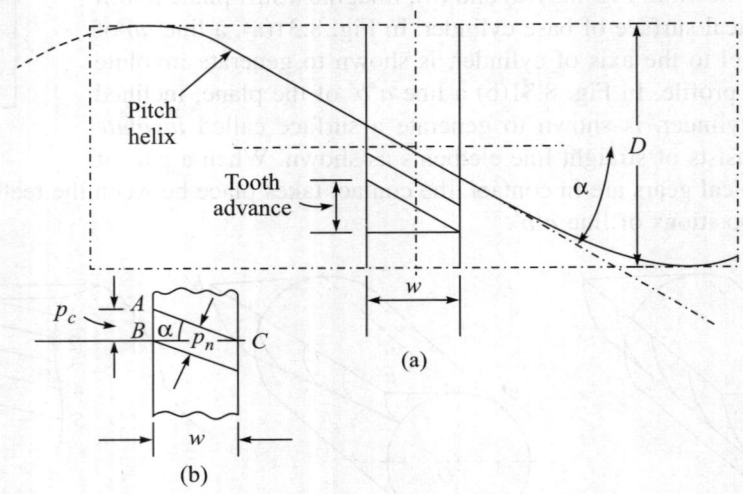

Fig. 8.33 Illustration for a pitch helix curve and tooth advance.

Figure 8.33(b) shows the limiting condition for face-width *w* for which the circular pitch p_c is just equal to helix tooth advance. As is clear from the right angled triangle *ABC*,

$$\frac{p_c}{w} = \tan\alpha$$

or

$$w = \left(\frac{p_c}{\tan\alpha}\right) \qquad (8.52)$$

In order to provide a margin, an overlap of 15% of the circular pitch is generally provided. Thus,

$$w = (1.15)\frac{p_c}{\tan\alpha} \qquad (8.53)$$

Figure 8.34 shows a single helical gear with only a couple of teeth represented by their centre lines. The pitch measured in a direction normal to the centre line of teeth (p_n), called normal pitch, must be same for two gears to mesh. It is related to circular pitch p_c through (*see* Fig. 8.34) the relation

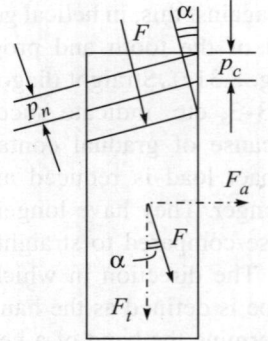

Fig. 8.34 Illustration for relation between p_n and p_c.

$$p_n = p_c \cos\alpha$$

or
$$p_c = \left(\frac{p_n}{\cos\alpha}\right) \tag{8.54}$$

Let m_n be the module in the normal plane, defined as the diameter of an imaginary circle, given by (Tp_n/π), in charge of one tooth. Then,

$$p_n = (m_n)\pi \tag{8.55}$$

Thus, from equations (8.54) and (8.55), as $p_c = m\pi$

$$\pi(m_n) = p_c \cos\alpha = (m\pi)\cos\alpha$$

or
$$m_n = m\cos\alpha \tag{8.56}$$

where m is the module of corresponding straight spur gear of pitch circle diameter D'.

Thus
$$m_n = \left(\frac{D'}{T}\right)\cos\alpha \tag{8.57}$$

For a straight spur gear normal module (identical with module) is given by (D'/T); T being the number of teeth on equivalent spur gear. Thus, substituting for m_n, in equation (8.57),

$$\left(\frac{D}{T}\right) = \left(\frac{D'}{T}\right)\cos\alpha; \quad \text{or} \quad D' = \left(\frac{D}{\cos\alpha}\right) \tag{8.58}$$

We conclude that the pitch circle diameter of a helical gear of given normal pitch p_n, is $(\cos\alpha)$ times the pitch circle diameter of a straight spur gear wheel having same number of teeth and normal pitch. As a general rule, tooth dimensions can be established based on a normal pressure angle of 20°. Most of the proportions, indicated for straight spur gear, can be used for helical gear by replacing module m in the expressions by m_n, the normal module.

8.18 SPIRAL GEARS (SKEW OR SCREW GEARS)

As indicated earlier, spiral gears are used to connect two non-parallel, non-intersecting shafts for transmitting power. The pitch surfaces are cylindrical and the teeth are inclined to the axis of the gear. As in the case of a helical gear, the teeth can be right-handed or left-handed. The teeth of spiral gears have a point contact and are suitable for transmitting small power. Unlike helical gears, however, two spiral gears in mesh may have their teeth either of same or opposite hand.

Centre Distance: The shortest distance between the two skew shafts axes gives the centre distance between a pair of spiral-gears. *Looking in the direction of this centre distance, the angle through which axis of one shaft must be rotated so as to make it parallel to the axis of the other (maintaining same relative directions of rotation), is known as the shaft angle.* This definition of shaft angles shows that a pair of spur-helical gears can be treated as a special case of a pair of spiral gears, in which the shaft angle is zero.

Two pairs of spiral gears having identical relative positions of the shafts are shown in Fig. 8.35. However, for the same sense of rotation of gear 2, the mating gear in left-hand drive rotates in c.c.w. sense and the one in right-hand drive rotates in c.w. sense. The shaft angle for left-hand drive is less than 90°, while for right-hand drive it is greater than 90°. The teeth on wheels with centre of O_1 and C_1, on the mating side, are shown by dashed lines while the teeth on side, facing the observer are shown by full lines in the plan view at Fig. 8.35(a) & (b).

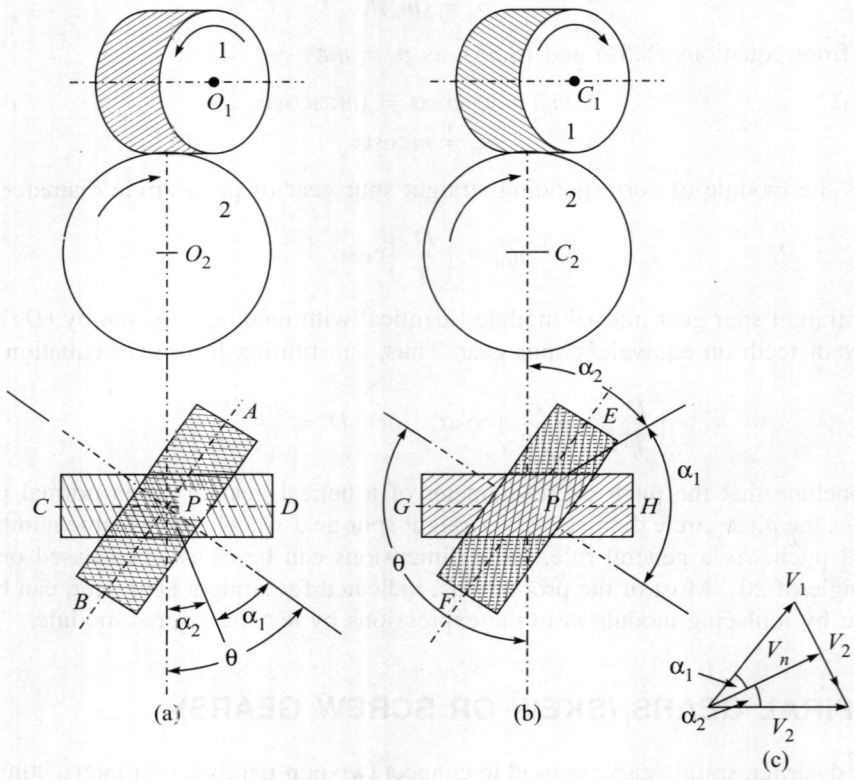

Fig. 8.35

With the pitch point at P, the assumed direction of rotation for wheels with centre at O_2 can be achieved only if the lines of teeth on mating sides lie between AP and PC for the left-hand pair and between EP and PH for the right-hand pair. Direction of movement of teeth on the mating sides of the wheels at the pitch point are shown by arrowheads in Fig. 8.35.

A common feature for both the cases is that the shaft angle equals the sum of spiral angles of the teeth on mating sides. Further, in the left-hand drive both the wheels have left-hand helices (teeth on sides facing observer) while both the wheels in right-hand drive have right-hand helices.

For proper meshing, normal pitch/module for both the mating wheels of a pair must be same, however the circular pitches will be different if the spiral angles of mating wheels are different. Let θ be the shaft angle α_1 and α_2 be the spiral angles of the teeth on the two wheels, so that, $\theta = \alpha_1 + \alpha_2$ with gears of same hand helix and $\theta = \alpha_1 - \alpha_2$ for gears of opposite hand helix.

As shown in Fig. 8.35(c), for common tooth velocity in a direction normal to the line of mating teeth,

$$V_n = V_1\cos\alpha_1 = V_2\cos\alpha_2$$

or
$$(V_2/V_1) = (\cos\alpha_1/\cos\alpha_2)$$

Therefore $\qquad$ Velocity ratio $= \dfrac{\omega_2}{\omega_1} = \dfrac{(V_2/r_2)}{(V_1/r_1)}$ $\qquad$ (8.59)

Substituting $r_2 = d_2/2$, $r_1 = d_1/2$ and also for V_2/V_1 from equation (8.59),

$$(\omega_2/\omega_1) = \left(\frac{d_1}{d_2}\right)\frac{\cos\alpha_1}{\cos\alpha_2} \qquad (8.60)$$

or
$$(\omega_2/\omega_1) = \left(\frac{m_1 T_1}{m_2 T_2}\right)\frac{\cos\alpha_1}{\cos\alpha_2} \qquad (8.61)$$

where m_1 and m_2 are the modules of corresponding spur gears such that (as from equation 8.56).

$$m_n = m_1\cos\alpha_1 = m_2\cos\alpha_2$$

Substituting for m_1 and m_2 in equation (8.61) and cancelling m_n,

$$(\omega_2/\omega_1) = \frac{(T_1/\cos\alpha_1)}{(T_2/\cos\alpha_2)}\frac{\cos\alpha_1}{\cos\alpha_2}$$

Therefore $\qquad$ V. R. $= \left(\dfrac{\omega_2}{\omega_1}\right) = \left(\dfrac{T_1}{T_2}\right)$ $\qquad$ (8.62)

Comparing right hand sides of equations (8.60) and, (8.62), $\dfrac{T_1}{T_2} = \dfrac{d_1\cos\alpha_1}{d_2\cos\alpha_2}$ $\qquad$ (8.63)

Again, the centre distance between mating wheels,

$$C = r_1 + r_2$$

or
$$C = \frac{1}{2}(m_1 T_1 + m_2 T_2) = \frac{m_n}{2}\left(\frac{T_1}{\cos\alpha_1} + \frac{T_2}{\cos\alpha_2}\right) \qquad (8.64)$$

or
$$C = \frac{m_n T_1}{2}\left(\frac{1}{\cos\alpha_1} + \frac{G}{\cos\alpha_2}\right) \qquad (8.65)$$

or
$$C = \frac{D'}{2}\left(\frac{1}{\cos\alpha_1} + \frac{G}{\cos\alpha_2}\right) \qquad (8.66)$$

where D' is the pitch circle diameter of a spur gear with the same (T_1) number of teeth and having module equal to the normal module of the gear 1.

8.19 THE EFFICIENCY OF SPIRAL AND HELICAL GEARS

In helical and spiral gears, teeth are inclined to the respective axis and they can be right handed or left handed. Helical gears have parallel axes of shafts and a line contact. When one of the two shafts is turned through some angle, the axes no longer remain parallel and contact occurs only at a point. In such a skewed position of shafts the gears are called *crossed helical* or *spiral gears*.

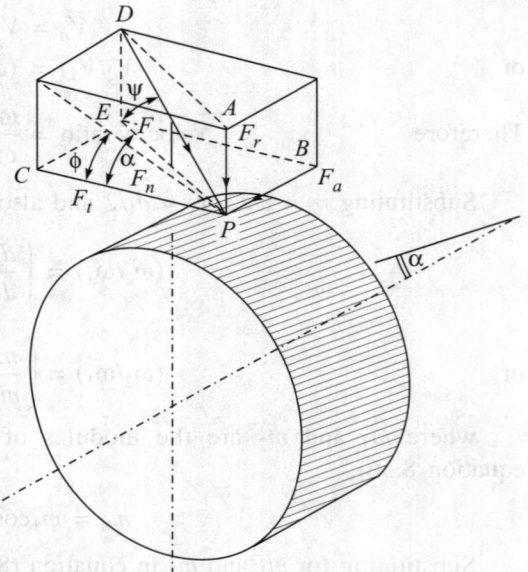

As in the case of spur gear, force F from mating gear tooth is transmitted along line of action which is inclined at pressure angle Ψ to the plane tangent to the pitch surface. When resolved in the diagonal plane *PADE* (*see* Fig. 8.36), the two components are

The radial component $F_r = F\sin\Psi$ (8.67)

Fig. 8.36 Diagram illustrating force analysis of spiral gear.

The other component $F_r = F_n\cos\Psi$ can now be resolved in two mutual perpendicular directions, one parallel to the axis of wheel and second tangent to the pitch circle (power component). Thus, for helix angle the two components are

$$F_a = F_n\sin\alpha = F\cos\Psi \sin \alpha \qquad (8.68)$$
and,
$$F_1 = F_n\cos\alpha = F\cos\alpha \cos\Psi \qquad (8.69)$$

Efficiency: Sliding motion between the surface of teeth of spiral gears takes place mainly along the tangent to pitch helix. If F_n and F_n' be the normal reactions between two mating surfaces of gears 1 and 2 respectively, then corresponding frictional forces (μF_n) and $(\mu F_n')$ act tangential to the pitch helix so as to oppose relative sliding velocities V_{12} and V_{21} respectively as shown in Figs. 8.37(b), (c) and (d). Being normal reaction between mating teeth,

$$F_n = F_n' = R$$

and therefore, corresponding frictional forces must also be equal. Thus,

$$\mu F_n = \mu F_n'$$

As shown in Fig. 8.37(b) the normal force F_n exerted by gear 2 combines with frictional force μF_n to give the resultant force F, which is inclined to the direction of F_n at the friction angle ϕ. The tangential driving force F_t acts in the plane of rotation of gear 1 in the tangential plane and is inclined to F_n at the helix angle α. Thus, resolving F along F_t,

$$F_t = F\cos(\alpha - \phi) \qquad (8.70)$$

Similarly, in Fig. 8.37(c), the normal reaction F_n' combines with frictional force $\mu F_n'$ to give resultant force F'. Resolving along F_t',

$$F_t' = F'\cos(\alpha' + \phi) \qquad (8.71)$$

But as $F_n = F_n'$ and $\mu F_n = \mu F_n'$, it follows that,

$$F = F'$$

and therefore from equations (8.70) and (8.71),

$$F_t = F\cos(\alpha - \phi); \text{ and } F_t' = F\cos(\alpha' + \phi)$$

Hence, efficiency $\qquad \eta = \dfrac{\text{output power}}{\text{input power}} = \dfrac{F_t' V_2}{F_t \times V_1}$

Therefore $\qquad \eta = \dfrac{\cos(\alpha' + \phi)}{\cos(\alpha - \phi)} \dfrac{V_2}{V_1} \qquad (8.72)$

The velocities V_2 and V_1 are the pitch line velocities of gears 2 and 1. As the velocity vectors must lie in respective planes of rotation of gears, the velocities V_1 and V_2 are shown inclined at helix angles α and α' to the normal velocity component V_n in Fig. 8.37(d). For the teeth to remain in contact, the mating teeth must have same velocity V_n in the direction of common normal. Hence, from Fig. 8.37(d),

$$V_1\cos\alpha = V_2\cos\alpha' \quad \text{or} \quad (V_2/V_1) = (\cos\alpha')/\cos\alpha$$

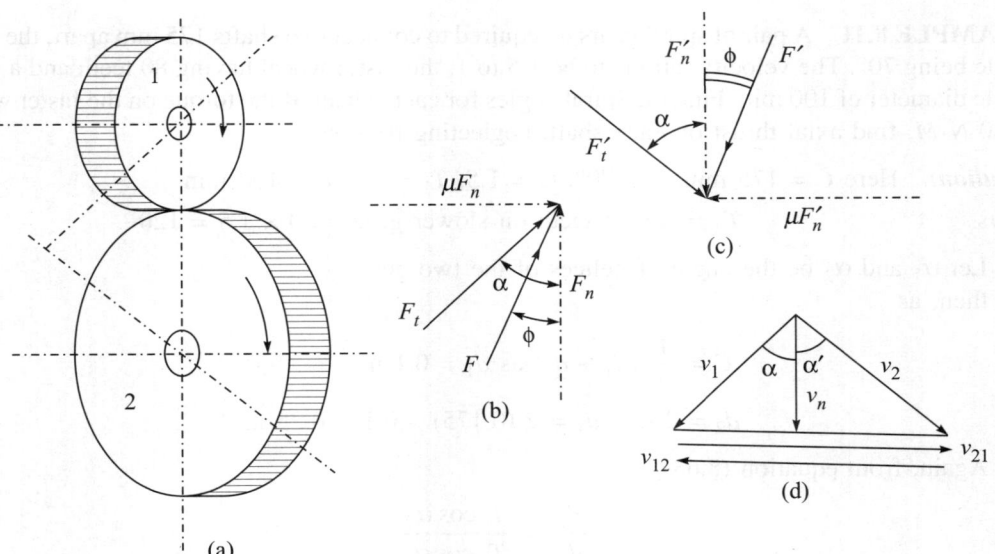

Fig. 8.37 Illustration for efficiency of a spiral gear.

Substituting for (V_2/V_1) in equation (8.72),

$$\eta = \frac{\cos(\alpha' + \phi)}{\cos(\alpha - \phi)} \frac{\cos\alpha}{\cos\alpha'} \qquad (8.73)$$

Using trigonometric identities the R.H.S. reduces to

$$\eta = \frac{\cos(\alpha + \alpha' + \phi) + \cos(\alpha - \alpha' - \phi)}{\cos(\alpha' + \alpha - \phi) + \cos(\alpha' - \alpha + \phi)}$$

Substituting for $\alpha + \alpha' = \theta$ (the shaft angle), as $\cos(-\beta) = \cos\beta$,

$$\eta = \frac{\cos(\theta + \phi) + \cos(\alpha - \alpha' - \phi)}{\cos(\theta - \phi) + \cos(\alpha - \alpha' - \phi)} \qquad (8.74)$$

Again, since the shaft angle θ and angle of friction ϕ is constant, the efficienty η will be maximum when the common term in the numerator and denominator of expression (8.71) is maximum, *i.e.* when

$$\alpha - \alpha' - \phi = 0$$

or   when, $\alpha = \phi + \alpha'$ $= \phi + (\theta - \alpha)$ or $2\alpha = \theta + \phi$

or   when, $$\alpha = \left(\frac{\theta + \phi}{2}\right)$$

Hence $(\eta)_{max} = \dfrac{\cos(\theta + \alpha) + 1}{\cos(\theta - \alpha) + 1}$ (8.75)

EXAMPLE 8.11   A pair of spiral gears is required to connect two shafts 175 mm apart, the shaft angle being 70°. The velocity ratio is to be 1.5 to 1, the faster wheel having 80 teeth and a pitch circle diameter of 100 mm. Find the spiral angles for each wheel. If the torque on the faster wheel is 80 $N \cdot M$, find axial thrust on each shaft, neglecting friction.

Solution: Here $C = 175$ mm; $\theta = 70°$; $G = 1.5$; $T_1 = 80$; $d_1 = 100$ mm

Thus $T_2 = $ No. of teeth on slower gear $= 80 \times 1.5 = 120$

Let α_1 and α_2 be the angle of helices of the two gears, then, as

$$C = \frac{1}{2}(d_1 + d_2) \text{ as } d_1 = 0.1 \text{ m,}$$

$$d_2 = 2C - d_1 = 2(0.175) - 0.1 = 0.25 \text{ m}$$

Again, from equation (8.63),

$$\frac{d_1}{d_2} = \frac{T_1 \cos\alpha_2}{T_2 \cos\alpha_1}$$

Hence $\dfrac{\cos\alpha_1}{\cos\alpha_2} = \left(\dfrac{T_1}{T_2}\right)\left(\dfrac{d_2}{d_1}\right) = \left(\dfrac{80}{120}\right)\left(\dfrac{0.25}{0.1}\right) = \dfrac{5}{3}$

or $\cos\alpha_1 = (5/3)\cos\alpha_2$

Also $\alpha_1 + \alpha_2 = \theta = 70°$ and as such, $\alpha_2 = (70° - \alpha_1)$

Substituting above,

$$\cos\alpha_1 = \frac{5}{3}\cos(70 - \alpha_1)$$

or

$$\cos\alpha_1 = \frac{5}{3}(\cos70\cos\alpha_1 + \sin70\sin\alpha_1)$$

or

$$\cos\alpha_1 = (0.757)\cos\alpha_1 + (1.485)\sin\alpha_1$$

or

$$0.243\cos\alpha_1 = 1.458\sin\alpha_1$$

or

$$\tan\alpha_1 = \frac{0.243}{1.458} = 0.167$$

Therefore $\quad\quad\quad\quad\quad \alpha_1 = 10.51°$ **Ans.**

Therefore $\quad\quad\quad\quad\quad \alpha_2 = 70 - 10.51 = 59.49°$ **Ans.**

Referring to Fig. 8.36, the tangential component of force of faster gear

$$F_t = \frac{\text{Torque}}{\text{pitch circle radius}} = \frac{80 \times 100}{10/2} = 1600 \text{ N}$$

If F_n be the corresponding force normal to tooth line, then

$$F_n\cos\alpha = F_t$$

or

$$F_n = \frac{F_t}{\cos\alpha_1} = \frac{1600.0}{\cos 10.51°} = 1622.05 \text{ N}$$

An equal force $F_n' = F_n$ acts on other gear in the normal direction to tooth line.

Therefore $\quad\quad F_t' = F_n' \cos\alpha_2 = 1622.05 \times \cos 59.49 = 963.899 \text{ N}$

Therefore axial thrust on slow gear $F_a' = F_n' \sin\alpha_2 = 1622.05 \times \sin 59.49 = 1304.586 \text{ N}$ **Ans.**

axial thrust on fast gear, $F_a = F_n \sin\alpha_1 = 16220.5 \times \sin 10.51 = 266.57 \text{ N}$ **Ans.**

EXAMPLE 8.12 In a spiral gear drive, the spiral angle of the teeth on the driving wheel has been fixed at 50°. The normal pitch of the teeth is 12.5 mm. The driving wheel A turns at twice the speed at the driven wheel B; the shafts are at right angles and the shortest distance between them is approximately 17.5 mm. Determine dimensions of suitable gears for the drive, giving for each wheel (a) the number of teeth, (b) the spiral angle of teeth, (c) the circular pitch, and (d) the pitch diameter. Find also the exact centre distance between the axes. If friction angle is 5°, what is the efficiency of the drive? (DAVV Indore, 1977)

Solution: Given $\alpha_1 = 50°$; $\theta = 90°$. Hence, $\alpha_2 = \theta - \alpha_1 = 40°$ **Ans.**

Circular pitch p_c of pinion and p_c' of gear are

$$p_c = \frac{12.5}{\cos 50} = 17.68 \text{ mm and } p_c' = \frac{12.5}{\cos 40} = 15.45 \text{ mm} \quad\quad \textbf{Ans.}$$

Centre distance,

$$C = \frac{D'}{2}\left(\frac{1}{\cos \alpha_1} + \frac{G}{\cos \alpha_2}\right)$$

or

$$17.5 = \frac{D'}{2}\left(\frac{1}{\cos 50} + \frac{2}{\cos 40}\right)$$

Therefore

$$D' = 35/3.886 = 9.00 \text{ cm} = 90 \text{ mm}$$

Here D' is the p.c.d. of spur gear of same number of teeth T. Thus,

$$D' = \frac{p_n T}{\pi} = \frac{12.5 \times T}{\pi}. \quad \text{Thus} \quad T = \frac{\pi \times 90.0}{12.5} = 22.6 \approx 23 \qquad \textbf{Ans.}$$

Thus, for a gear ratio of 2, the number of teeth on driven gear $T = 2 \times 23 = 46$

The diameter of equivalent spur gear $D'' = 23 \times 12.5 = 287.5$ mm

The diameters of spiral gears are,

$$D_1 = \frac{23 \times 12.5}{\pi \times \cos 50} = 129.42 \text{ mm} \qquad \textbf{Ans.}$$

and

$$D_2 = \frac{46 \times 12.5}{\pi \times \cos 40} = 226.235 \text{ mm} \qquad \textbf{Ans.}$$

Therefore exact centre distance

$$C = \frac{D_1 + D_2}{2} = 177.83 \text{ mm} \qquad \textbf{Ans.}$$

Now the efficiency is to be obtained by taking into account the friction angle $\phi = \tan^{-1} \mu = 5°$.

Therefore

$$\eta = \frac{\cos(90+5) + \cos(50-40-5)}{\cos(90-5) + \cos(50-40-5)}$$

or

$$\eta = \frac{\cos 95 + \cos 5}{\cos 85 + \cos 5} = 87.4\% \qquad \textbf{Ans.}$$

EXAMPLE 8.13 The centre distance between two mating spiral gears is 260 mm and the angle between the shafts is 65°. The normal circular pitch is 14 mm and gear ratio is 3.0. The driven gear has a helix angle of 35^c. Find (i) the number of teeth on each wheel, (ii) the exact distance, and (iii) the efficiency assuming the friction angle to be 5.5°.

Solution: Given $C = 260$ mm; $\theta = 65°$; $p_{cn} = 14$ mm ; $G = 3$ $\alpha_2 = 35°$; $\phi = 5.5°$

Hence,

$$\alpha_1 = \theta - \alpha_2 = 65 - 35 = 30°$$

Also,

$$T' = (3.0)\, T_1 \text{ and } m_n = \frac{p_{cn}}{2\pi} = \left(\frac{14}{2\pi}\right)$$

$$260 = C = \frac{14}{2\pi}\left(\frac{T}{\cos 30} + \frac{3T}{\cos 35} \right) = T(10.34)$$

Therefore $\qquad\qquad T = 25.144 \approx 26$ **Ans.**

Therefore $\qquad\qquad T' = 3 \times 26 = 78$ **Ans.**

The exact centre distance,

$$C_{exact} = \frac{14}{2\pi}\left(\frac{26}{\cos 30} + \frac{78}{\cos 35} \right)$$

or $\qquad\qquad C_{exact} = \frac{14}{2\pi}(29.18 + 91.48) = 268.85 \text{ mm}$ **Ans.**

And efficiency,

$$\eta = \frac{\cos(65 + 5.5) + \cos(30 - 35 - 5.5)}{\cos(65 - 5.5) + \cos(30 - 35 - 5.5)}$$

or $\qquad\qquad \eta = \frac{0.446797 + 0.9864}{0.59412 + 0.9864} = \frac{1.4334}{1.5805}$

Thus, $\qquad\qquad \eta = 0.9069 \quad i.e., \quad 90.69\%$ **Ans.**

8.20 WORM AND WORM GEAR

Worm gearing is essentially a form of spiral gearing in which the driving and driven shafts have axes usually at right angles. The pinion or worm has a small number of teeth, usually one to four. Since the teeth completely wrap around the pitch cylinder, they are also called threads. This type of gearing is extensively used to transmit power at high velocity ratios (as high as 300 : 1 or more in a single stage) between non-intersecting shafts. Worm drives may also be used as speed increaser, but generally they are used as speed reducers in which the worm is the driver and the worm wheel is the driven member.

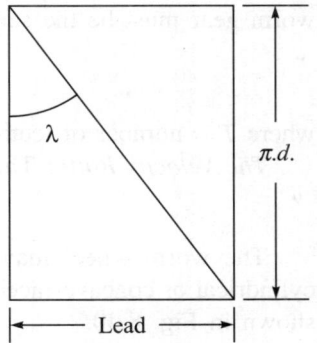

Fig. 8.38 Development of pitch helix.

The worm gear is not a helical gear as its face is made concave so as to fit the curvature of the worm (*see* Fig. 8.40) and provide line contact in place of point contact. This feature is important because the line contact enables worm gearing to transmit high tooth loads. A major disadvantage of worm gearing is that high sliding velocities exist across the teeth similar to the case of crossed helical gears.

The speed ratio of a worm drive is equal to the number of teeth on the worm wheel divided by the number of threads on the worm. Depending on whether one, two, three or four threads are used on worm, it is called single, double, triple, and quadruple-threaded worm respectively.

The axial pitch, or simply the pitch P, is defined as the distance from a point on one thread profile to the same point on an adjacent profile measured along the pitch cylinder. As against this, lead l is the amount the helix advances axially when rotated through one turn about the pitch cylinder. The lead is equal to the number of threads times the pitch, i.e.

$$l = (t_w)p \qquad (8.76)$$

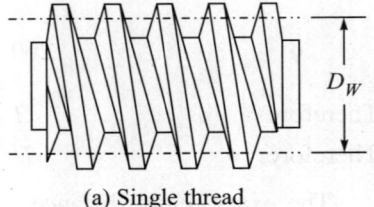

(a) Single thread

This is shown in Fig. 8.39. The slope of the thread is called the lead angle λ. Figure 8.38 shows development of the pitch helix where d is the pitch diameter of worm. From the geometry of figure,

$$\tan\lambda = \frac{1}{\pi d} = \frac{(t_\omega)p}{\pi d} \qquad (8.77)$$

When the lead angle λ is greater than the friction angle of the contacting surfaces, the worm is called reversible or over-running. In the case of over-running worm, the worm can be freely driven by the worm wheel. The lead angle is the compliment of the helix angle, i.e. $\alpha + \lambda = 90°$.

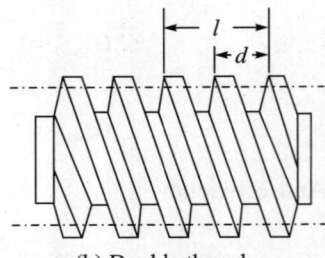

(b) Double thread

Fig. 8.39 Illustration for single & double thread.

Centre Distance: Let the pitch diameters of the worm and worm wheel be d and D respectively. Then the centre distance is mathematically expressed as

$$C = \frac{D + d}{2} \qquad (8.78)$$

In order to ensure proper mating between worm and worm gear, the circular pitch of the worm gear must be the same as the axial pitch of the worm. Thus,

$$p_c = p = \left(\frac{\pi D}{T}\right) \qquad (8.79)$$

where T = number of teeth on worm wheel.

The Velocity Ratio: The velocity ratio for a worm drive is given by

$$\omega/\omega' = (T/t) \qquad (8.80)$$

The worm wheel may have either straight-faced teeth, the outside of the wheel being cylindrical or concave-faced teeth, the outside of the wheel following the curve of the worm as shown in Fig. 8.40.

In the case of worms, the lead angle λ is quite small and therefore helix angle ($\alpha = 90° - \lambda$) approaches 90°.

Again, the axes of worm and wheel are usually at 90°.

Hence if α and α' are the angles of helices for worm and worm wheel respectively, then

$$\theta = \alpha + \alpha' = 90°$$

Thus, substituting $\qquad \alpha = (90 - \lambda)$ for a worm,

$$(90 - \lambda) + \alpha' = 90°$$

or $\qquad\qquad\qquad\qquad \lambda = \alpha' \qquad\qquad\qquad\qquad (8.81)$

and $\qquad\qquad\qquad \alpha = (90 - \alpha') = (90 - \lambda) \qquad (8.82)$

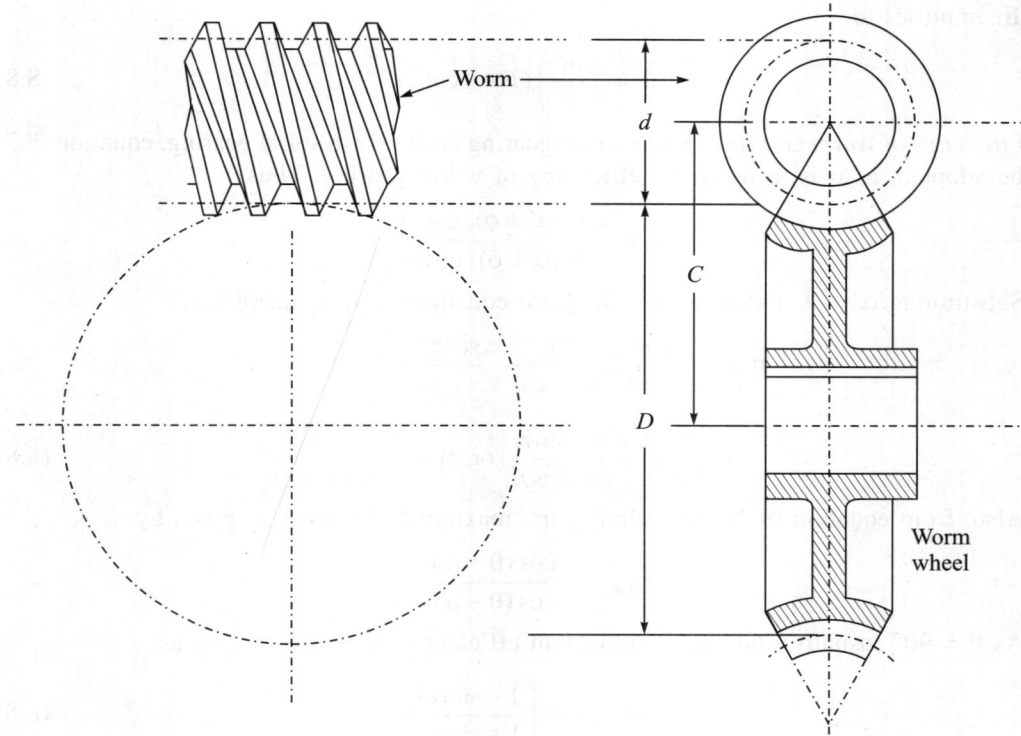

Fig. 8.40 Worm and wheel in mesh.

This shows that the load angle of worm equals helix angle of the worm wheel when $\theta = 90°$.
For proper meshing (as in the case of spiral gears), normal pitch of worm and worm wheel be equal. Thus,

$$(p_n)_{worm} = (p_n)_{wheel}$$

From equation (8.78), the centre distance can be further expressed as

$$C = \frac{1}{2}\ (m'T + mt) = \frac{m_n}{2}\left(\frac{T}{\cos\alpha'} + \frac{t}{\cos\alpha}\right)$$

$$= \frac{m_n}{(2\cos\alpha')}\left(T + t\,\frac{\cos\alpha'}{\cos\alpha}\right)$$

Substituting for α' from equation (8.81) and for α from equation (8.82) and as $\cos(90 - \lambda) = \sin\lambda$

$$C = \left(\frac{m_n}{2\cos\alpha'}\right)\left(T + t\,\frac{\cos\lambda}{\sin\lambda}\right)$$

which simplifies to,

$$C = \left(\frac{m'}{2}\right)(T + t\cot\lambda) \tag{8.83}$$

Efficiency of Worm Gearing: Since worm gearing is similar to spiral gearing, equation (8.73) can be adopted, as it is to represent efficiency of worm gearing. Thus,

$$\eta = \frac{\cos(\alpha' + \phi)}{\cos(\alpha + \phi)}\frac{\cos\alpha}{\cos\alpha'}$$

Substituting $\alpha' = \lambda$ and $\alpha = (90 - \lambda)$ from equations (8.81) and (8.82),

$$\eta = \frac{\cos(\lambda + \phi)}{\cos[90 - (\lambda + \phi)]}\frac{\cos(90 - \lambda)}{\cos\lambda}$$

$$= \frac{\cos(\lambda + \phi)}{\sin(\lambda + \phi)}\frac{\sin\lambda}{\cos\lambda} \text{ or, } \eta = \frac{\tan\lambda}{\tan(\lambda + \phi)} \tag{8.84}$$

Also, from equation (8.75) for spiral gears, maximum efficiency is given by

$$(\eta)_{max} = \frac{\cos(\theta + \alpha) + 1}{\cos(\theta - \alpha) + 1}$$

As $\theta = 90°$ (usually), this gives maximum efficiency for worm gearing as

$$(\eta)_{max} = \left(\frac{1 - \sin\alpha}{1 + \sin\alpha}\right) \tag{8.85}$$

In order to study the effect of coefficient of friction $\mu = \tan\phi$ and angle of helix α on efficiency, Table 8.3 is obtained from equation (8.84), which can be simplified to:

$$(\eta) = \frac{\tan(90 - \alpha)}{\tan[90 - (\alpha - \phi)]}$$

Table 8.3

$\mu = \tan\phi$	$\alpha = 10°$	$20°$	$30°$	$40°$	$50°$	$60°$	$70°$	$80°$	$90°$
$0.24 = \tan 15°$	—	0.24	0.464	0.556	0.587	0.577	0.52	0.378	0
$0.1584 = \tan 10°$	0	0.484	0.63	0.688	0.704	0.688	0.63	0.484	0
$0.0787 = \tan 5°$	0.496	0.736	0.807	0.834	0.839	0.825	0.78	0.658	0
$0.0393 = \tan 2.5°$	0.746	0.866	0.902	0.914	0.916	0.906	0.879	0.795	0

It can be seen that in the useful range of $20 \leq \alpha \leq 80°$, percentage variation in the efficiency for different friction angles are:

$\phi = 15°$; % variation in $\eta = 59.1\%$ $\phi = 10°$; % variation in $\eta = 31.25\%$

$\phi = 5°$; % variation in $\eta = 21.57\%$ $\phi = 2.5°$; % variation in $\eta = 13.2\%$

It is easy to see that per cent variation in efficiency over the working range decreases with μ and is least for $\phi = 2.5°$. Further, coefficient of friction seems to be more important than helix angle in increasing the efficiency of the drive. Though the table shows that maximum efficiency is obtainable for $40° \leq \phi \leq 50°$ highest efficiency in this range can be obtained only with a low coefficient of friction.

EXAMPLE 8.14 A two thread worm drives a worm wheel of 30 teeth. The worm has a pressure angle of $14\dfrac{1°}{2}$, a pitch of 19 mm and a p.c.d of 50 mm. If the worm rotates at 1000 r.p.m., for $\mu = 0.05$ find,

    (i)   the helix angle of the worm,
   (ii) the speed of worm wheel,
  (iii) the centre distance,
  (vi) the efficiency of the set, and
   (v) the lead angle for maximum efficiency.

Solution: The lead angle of worm,

$$\lambda = \tan^{-1}\left(\frac{1}{\pi d}\right) = \tan^{-1}\left(\frac{2 \times 19}{\pi \times 50}\right) = 13.599° \approx 13.6°$$

Hence, the helix angle = $(90 - \lambda) = 76.4°$ **Ans.**

The speed of worm gear = $N \times T/t = 1000 \times \dfrac{2}{30} = 66.666$ r.p.m. ≈ 66.67 r.p.m.   **Ans.**

$$\phi = \tan^{-1}(\mu) = \tan^{-1}(0.05) = 2.862°$$

The centre distance,

$$C = \frac{m'}{2}(T + t \cot \lambda) \text{ where, } m' = \frac{p_c}{\pi} = \frac{19}{\pi} = 6.048 \text{ mm}$$

Therefore $C = \dfrac{19}{2\pi}(30 + 2 \cot 13.6) = 115.72$ mm   **Ans.**

$$\eta = \frac{\tan \lambda}{\tan(\lambda + \phi)} = \frac{\tan 13.6}{\tan(13.6 + 2.862)} = 81.87\% \qquad \textbf{Ans.}$$

For maximum efficiency,

$$\alpha = \frac{\theta + \phi}{2} = \frac{90 + 2.862}{2} = 46.43°$$

Hence, $\alpha' = 90 - 46.43 = 43.57°$

Lead angle for this condition,

$$\lambda = \alpha' = 43.57° \qquad \textbf{Ans.}$$

Maximum efficiency for this case is given by,

$$(\eta)_{max} = \frac{\tan 43.57}{\tan(43.57 + 2.862)} = 90.49\%$$

8.21 BEVEL GEARS

Bevel gears are used to transmit power between shafts whose axes intersect and whose pitch surfaces are rolling cones. The pitch surfaces in this case are defined as the conical surfaces with common apex, which transmit same velocity ratio as the given pair of bevel gears over same centre distance by a pure rolling action. Interlocking projection called teeth are provided on pitch surfaces so as to have a positive drive. As in the case of spur gears, tooth profile of bevel gears may either have a cycloidal or an involute profile.

If the generating cone rolls on the outside of the pitch cone, an element (generator) on generating cone generates tooth profile of face. When the generating cone rolls on the inside of the pitch cone, the element sweeps out the flank portion of the tooth. Figure 8.41 shows involute tooth developed from a base cone. For this, one imagines a piece of paper wrapped around a cone. Let *OP* be a slit cut along a generator. Now let the paper be unwrapped along *OP* from the cone, keeping the paper taut. *The cut edge OC then describes the involute tooth profile. Any point, like point C, on unwrapping slit remains at a constant distance OC from the appex O and therefore, the curve described by it lies on the surface of a sphere of radius OC. A generator curve of this type (say CP) is therefore called a spherical involute.* Readers will appreciate that it is not difficult to conceive that curve *CP* cannot lie in the end-face *ABP* or extension thereof.

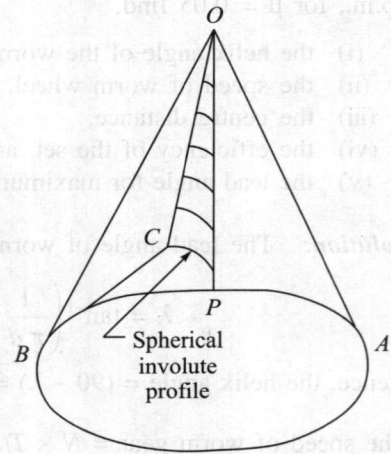

Fig. 8.41 Development of involute tooth from base cone.

As the curve generated by any point like *C*, on the generator *OC*, does not lie on plane surface, it is difficult to understand tooth action using true shape of tooth profile. In practice, Tredgold's approximation is employed, which consists in substituting a conical surface for the actual spherical surface, such that the conical surface is tangential to the actual spherical surface.

Figure 8.42 shows a pair of bevel gears in mesh. The back cone of these gears have elements that are perpendicular to the corresponding elements of pitch cone. The surface of the back cone gives an approximation to the spherical surface. For studying the tooth action of bevel gears, these back cones are laid out flat (not shown in Figure). Each cone forms a portion of a spur gear, whose pitch circle radius equals back cone radius.

As shown in Fig. 8.42, *d* and *D* are the pitch circle diameter at the bigger ends of smaller and bigger gears respectively. The module *m* and circular pitch p_c are defined in the same manner as spur gear at the larger end,

Thus,
$$m = \frac{D}{T} = \frac{d}{t}$$

and
$$p_c = (\pi D/T)$$

where *T* and *t* are the number of teeth on larger and smaller gear respectively. Also,
$$\frac{\omega}{\omega'} = \frac{D}{d} = \frac{T}{t}$$

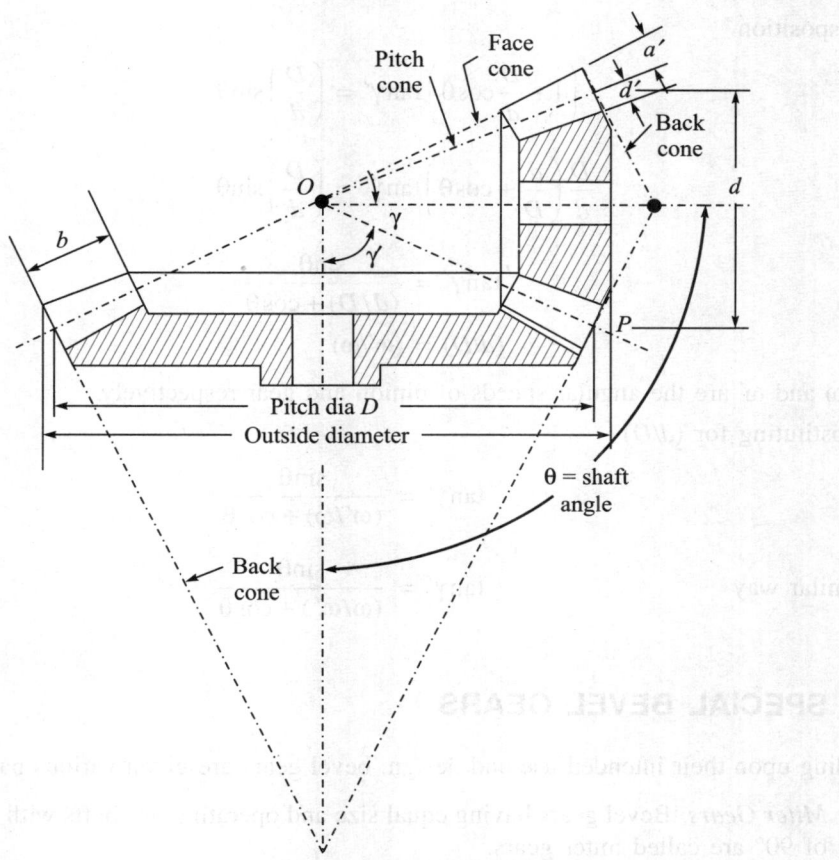

Fig. 8.42 A Pair of Bevel gears in mesh.

Pitch cone angles λ and λ' are infact the semi-cone angles for the smaller and bigger gears.
The tooth elements are straight lines which converge at the apex of cone. Thus the tooth
thickness decrease continuously towards inside; being maximum at the outside.

If γ and γ' be the pitch angles for pinion and gear respectively and d and D be their respective
pitch circle diameters at the larger end, we have

$$\sin\gamma' = \left(\frac{D/2}{OP}\right) = \frac{D/2}{(d/2)/\sin\gamma} = \left(\frac{D}{d}\right)\sin\gamma \qquad (8.86)$$

If θ is the shaft angle, then from θ = γ + γ',

$$\sin\gamma' = \frac{D}{d}\sin(\theta - \gamma')$$

Expanding R.H.S. and dividing both sides by cos λ',

$$\tan\gamma' = \frac{D}{d\cos\gamma}(\sin\theta\cos\gamma' - \cos\theta\sin\gamma') = \left(\frac{D}{d}\right)(\sin\theta - \cos\theta\tan\gamma')$$

By transposition

$$\left(1 + \frac{D}{d}\cos\theta\right)\tan\gamma' = \left(\frac{D}{d}\right)\sin\theta$$

or

$$\frac{D}{d}\left(\frac{d}{D} + \cos\theta\right)\tan\gamma' = \left(\frac{D}{d}\right)\sin\theta$$

Hence

$$\tan\gamma' = \frac{\sin\theta}{(d/D) + \cos\theta} \tag{8.87}$$

But

$$(d/D) = (\omega'/\omega)$$

where ω and ω' are the angular speeds of pinion and gear respectively.

Substituting for (d/D)

$$\tan\gamma' = \frac{\sin\theta}{(\omega'/\omega) + \cos\theta} \tag{8.88}$$

In a similar way

$$\tan\gamma' = \frac{\sin\theta}{(\omega/\omega') + \cos\theta} \tag{8.89}$$

8.22 SPECIAL BEVEL GEARS

Depending upon their intended use and design, bevel gears are given various names:

(i) *Miter Gears:* Bevel gears having equal size and operating on shafts with included angle of 90° are called miter gears.

(ii) *Angular Bevel Gears:* These are the bevel gears of unequal size which generally operate with an included shaft angle θ which is other than 90°.

(iii) *Crown Gears:* When the pitch angle of one of a pair of bevel gears is made equal to 90°, the pitch cone becomes a flat surface and the resulting gear is called a crown gear. The crown gear can be considered to be a counterpart of a rack in spur gearing.

(iv) *Spiral Bevel Gears:* Spiral bevel gears consist of obliquely curved teeth. The spiral angle is such that the face advance of the tooth is greater than the circular pitch of the tooth. The spiral bevel gear thus, is much like a helical gear. The straight tooth bevel gears become noisy at high pitch-line velocities. Spiral bevel gears provide a distinctly better choice in such situations.

(v) *Zerol Bevel Gear:* A bevel gear is a patented gear which has curved teeth such that the spiral angle of the tooth is tangent to a cone element giving a zero degree spiral angle.

(vi) *Hypoid Gears:* As in the case of automobile-differential applications, it is frequently desirable to have a gear similar to bevel gear but with shafts offset. Such gears are called hypoid gears because their pitch surfaces are hyperboloids of revolution. This form of bevel gears was introduced for use with automotive rear-axle drive systems known as the ring gear and pinion.

EXAMPLE 8.15   Two straight bevel gears are assembled with 90° shaft angle. The gears have a module of 10 mm with 42 and 79 teeth. Determine the pitch cone angles.

Solution: Given: $\theta = 90°$; $m = 10$ mm; $t = 42$; $T = 79$

From

$$\tan \gamma' = \frac{\sin\theta}{(\omega'/\omega) + \cos\theta}, \text{ as } \frac{\omega'}{\omega} = \frac{t}{T} = \frac{42}{79} = 0.5316$$

$$\gamma' = \tan^{-1}\left(\frac{\sin 90}{0.5316 + 0}\right) = 62.0° \qquad \textbf{Ans.}$$

Therefore $\gamma = 90 - 62 = 28°$ **Ans.**

REVIEW QUESTIONS

8.1   The following data relate to two meshing gears:
Velocity ratio = 1 : 3; Module: 4mm; Pressure Angle: 20°; Center distance: 200 mm
Determine the number of teeth and the base circle radius of the gear wheel.

(*Ans.* $t = 25$; $T = 75$; $R_b = 141$ mm)

8.2   Two meshing spur gears with 20° pressure angle have a module of 4 mm. The centre distance is 220 mm and the number of teeth on the pinion is 40. What should be the new centre distance so that pressure angle is increased to 22°?

(*Ans.*   221.9 mm).

8.3   State and prove law of gear tooth action for constant velocity ratio and show how the involute teeth profile satisfies the condition. Derive an expression for the velocity of sliding between a pair of involute teeth. State the advantages of involute profile as a gear tooth profile. (AMIE, Winter 1978)

8.4   Derive the formula for the length of the path of contact for two meshing spur gears having involute profile. A pinion having 10 teeth of involute form 20° pressure angle and 6 mm module drives a gear having 40 teeth of addendum = module, find (i) addendum and pitch circle radii of the two years, (ii) The length of path of apprach, (iii) The path of contact and (iv) arc of contact. (AMIE, Winter 1983)

(*Ans.* $r = 6$ cm; $R = 12$ cm; $r_a = 6.6$ cm; $R_a = 12.6$ cm;
1.535 cm; 2.921 cm; 3.11 cm).

8.5   Two 20° involute spur gear mesh externally and give a velocity ratio of 3, module is 3 mm and rotates at 120 r.p.m. determine

(i) The minimum number of teeth on gear and pinion

(ii) Number of pairs of teeth in contact. (Univ. Amravati)

8.6   A pair of gear, having 40 and 20 teeth respectively, are rotating in mesh, the speed of the smaller being 2000 r.p.m. Determine the velocity of sliding between the gear teeth face at the point of engagement, at the pitch point and at the point of disengagements if the smaller gear is the driver. The distances of engagement and disengagement from the pitch point along a common tangent to the base circles are 1.3 cm and 1 cm respectively. (DAV lndore, 1983)

(*Ans.* 408.4 and 314.16 cm/s).

8.7 A pair of involute gears with 16° pressure angle and 6 mm module is in mesh. The number of teeth on pinion is 16 and its rotational speed is 240 r.p.m. When the gear ratio is just 1.75, find, in order that the interference is just avoided,

(a) the addenda on pinion and gear wheel
(b) the length of path of contact, and
(d) the maximum velocity of sliding of teeth on either side of the pitch point.
(DAV lndore, 1989)

(*Ans.* a_p = 9.6 mm; a_g = 4.2 mm; 35.41 mm; V = 95.58 cm/s)

8.8 A pair of gears, having 40 and 30 teeth respectively, are of 25° involute form. The addendum length is 5 mm and the module is 5.5 mm. If smaller wheel is the driver and rotates at 1500 r.p.m. find the velocity of sliding at the point of engagement and at the point of disengagement.

(*Ans.* 3 m/s; 2.4 m/s)

8.9 Two-spur wheels each have 30 teeth of involute shape. The circular pitch is 25 mm. and the pressure angle is 20°. If the arc of contact is to be 2 × pitch, determine the minimum addendum of the teeth.

8.10 Two equal involute gear wheels of 40 teeth each, 3 mm module and 20° pressures angle are in mesh. Determine the minimum addendum if 2 pairs of teeth are always in mesh. The gear wheels rotate at 750 r.p.m. and transmit 4.6 kW. Find the normal force at each pair assuming equal direction of load between the pairs. Neglect friction.

8.11 Determine the number of pairs of teeth in contact at a given instant, if two equal involute gears of 18 teeth of pressure angle $19° \frac{1}{n} 28'$, have addendum of 0.8 module.

8.12 Two wheels with standard involute teeth of 5 mm module are to gear together with a velocity ratio of 4.5, the pressure angle being 15°. Find (i) the least permissible number of teeth in the pinion, if interference is to be avoided; (ii) the pressure between the teeth, when such a pinion is transmitting a torque of 6000 N·cm.

8.13 Determine the minimum number of teeth required on a pinion, in order to avoid interference which is to gear with,

(a) a wheel to give a gear ratio of 3:1; and   (b) an equal wheel
The pressure angle is 20° and a standard module of I module for the wheel may be assumed.

(*Ans.* 15; 13)

8.14 A gear wheel having 24 involute teeth of 5 mm module is generated by a straight sided rack cutter. The addendum of the cutter and the wheel is one module. Determine the minimum pressure angle if interfernce is to be avoided. If such 24 teeth wheel should mesh with a 40 teeth wheel correctly, determine the length of arc of contact and the minimum number of pairs of teeth in contact at any time.

(*Ans.* Ψ = 16.8°; 27.5 mm; 1.75).

8.15 A pair of spiral gears connect two non-intersecting shafts which are inclined at 45°. The speed ratio required is 2.5 to 1 and the least distance between axes of the shafts should be 18 cm. The normal module is 5 mm and pinion should have 18 teeth. Determine the spiral angle for the gear and their pitch circle diameters. The spirals should have the same hand. If the coefficient of friction is 0.05 at the gear teeth determine the power lost in friction if the input into the pinion is 3.7 kW at 1000 r.p.m. Calculate also the sliding velocity at the teeth.

(**Ans.** $\alpha' = 12.3$; $d = 92.3$ mm; $D = 267.7$ mm; 0.20 HP; 4.01 m/s)

8.16   Two non-intersecting shafts are inclined at 60° and the least distance between their axes is approximately 20 cm. They are connected by a pair of spiral gear, whose normal module is 6 mm to give a reduction in speed of 3 to 1. If the spiral angles of the pinion and gear are determined by the condition of maximum efficiency and the limiting angle of friction is 3°, determine the spiral angles of the gears, the pitch circle diameters of the gears, the corresponding numbers of teeth and the efficiency of transmission. Also calculate the axial thrust of both the shafts if the input torque on the pinion is 98 N·m.

(**Ans.** $\alpha = 31.5°$; $\alpha' = 28.5°$; $t = 16$; $T = 48$; $d = 112.7$ mm; $D = 294$ mm; $c = 203.35$ mm; $\eta = 0.942$; $f_a = 946$ N; and 1036 N)

8.17   A quadruple thread worm is driving a 160 teeth worm wheel with the shafts at 90°. Circular pitch of the worm gear is 2.86 cm pitch diameter of the worm is 6.98 cm. Find out the worm lead angle, helix angle of the worm gear, and the distance between shaft centers.

(**Ans.** $\lambda = 27.51°$ = helix = angle of gear; $c = 76.258$ cm; $N/N' = 40$)

Chapter 9

GEAR TRAINS

9.1 INTRODUCTION

An assembly of gear wheels by means of which motion is transmitted from one shaft to another shaft is called gear train. A gear train may include any or all kinds of gears—spur, bevel, spiral, etc. Ordinary (or usual) gear trains are those in which the axes of none of the gears move relative to the frame. Ordinary gear trains comprise simple and compound gear trains. The other gear trains are the epicyclic gear trains in which relative motion between axes of gears is possible.

Gear trains are necessary when—

(a) a large velocity reduction or mechanical advantage is desired,

(b) the distance between two shafts is not too great and, at the same time, is not short enough to permit the use of a single large gear,

(c) when certain specific velocity ratio is desired.

A fundamental requirement of any two meshing gears (involute) is that they must have the same module 'm' and the same pressure angle 'ψ'. This requires that for gears 1 and 2,

$$m = (D_1/T_1) = T_2/T_2)$$

Also, to ensure a positive drive, necessarily the pitch line velocities must be equal. Thus,

$$v = (\pi D_1 N_1) = \pi D_2 N_2)$$

Substituting for pitch circle diameters in terms of m, the pitch line velocity requirement gives

$$v = \pi(mT_1)N_1 = \pi(mT_2)N_2$$

where N_1 and N_2 are r.p.m. of gears 1 and 2 respectively.

Thus, for common module, $(T_1 N_1) = (T_2 N_2)$

This is true for any two meshing (involute) gears.

9.2 CLASSIFICATION

Gear trains can be classified as under:

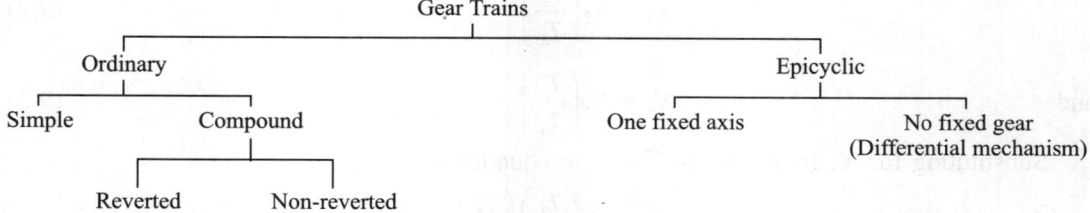

Simple Gear Trains

Effect of number of intermediate gears: As the name implies, this is the simplest type of gear train for transmitting motion from one shaft to the other. Characteristic features of this type of train are: (a) All the gear axes remain position fixed with respect to frame and (b) each gear is mounted on a separate shaft. However, while in Fig. 9.1(a) two intermediote gears connect gears on shaft-axes through O_1 and O_4, in Fig. 9.1(b) a single intermediate gear connects gears on shaft-axes through O_1 and O_4. To understand the role of number of gears separating the connected gears, we proceed as follows:

Let T_1, T_2, T_3, T_4 and T_5 be the number of teeth on gears with shaft axes at O_1, O_2, O_3, O_4 and O_5 respectively. Also let N_1, N_2, N_3, N_4 and N_5 be the r.p.m. of gears mounted on shafts through O_1, O_2, O_3, O_4 and O_5 respectively. Then for Fig. 9.1(a),

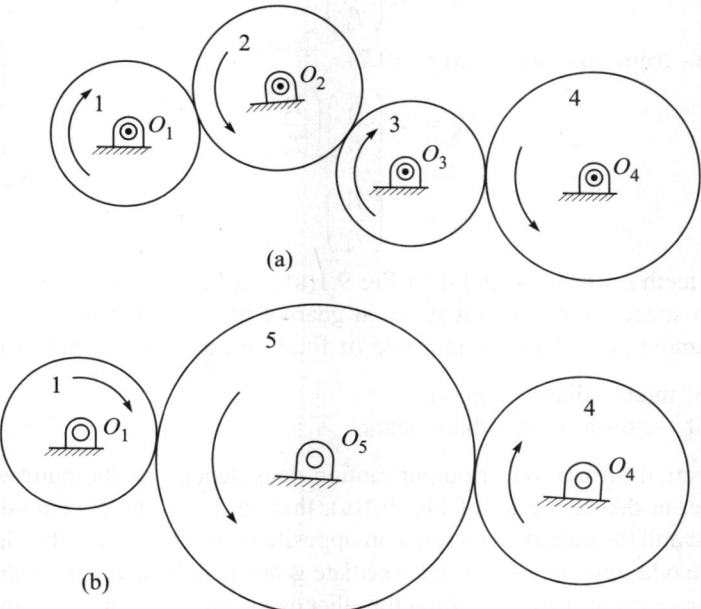

Fig. 9.1 Simple Gear Train: Effect of number of intermediate gears.

$$N_2 = N_1 \left(\frac{T_1}{T_2} \right) \tag{9.1}$$

$$N_3 = N_2 \left(\frac{T_2}{T_3} \right) \tag{9.2}$$

and $$N_4 = N_3 \left(\frac{T_3}{T_4} \right) \tag{9.3}$$

Substituting for N_3 from equation (9.2) in equation (9.3),

$$N_4 = N_2 \left(\frac{T_2}{T_3} \right) \left(\frac{T_3}{T_4} \right) \tag{9.4}$$

Substituting for N_2 from equation (9.1) in (9.4),

$$N_4 = N_1 \left(\frac{T_1}{T_2} \right) \left(\frac{T_2}{T_3} \right) \left(\frac{T_3}{T_4} \right)$$

Therefore $$N_4 = N_1 \left(\frac{T_1}{T_4} \right) \tag{9.5}$$

Also, for Figure 9.1(b),

$$N_5 = N_1 \left(\frac{T_1}{T_5} \right) \tag{9.6}$$

and $$N_4 = N_5 \left(\frac{T_5}{T_4} \right) \tag{9.7}$$

Substituting for N_5 from equation (9.6) in (9.7),

$$N_4 = N_1 \left(\frac{T_1}{T_5} \right) \left(\frac{T_5}{T_4} \right)$$

or $$N_4 = N_1 \left(\frac{T_1}{T_4} \right) \tag{9.8}$$

If number of teeth on gears 1 and 4 in Fig 9.1(a) and Fig. 9.1(b) are equal, it follows that for the same input speed N_1, the output speed at gears 4 in the two trains is the same. Stated in other words, in simple gear trains, magnitude of final output speed is not affected by—

(i)   number of intermediate gears, and
(ii) number of teeth on intermediate gears.

However, direction of rotation of output motion does depend on the number of intermediate gears. For instance, in the gear train of Fig. 9.1(a), there are even number (two) of intermediate gears, and the input and the output gears rotate in opposite sense. As against this, in the gear train of Fig. 9.1(b) there are odd (one) number of intermediate gears, and the input and output gears rotate in the same sense. Above results can be verified for other even and odd numbers of intermediate gears.

Train value.   The ratio of output speed to input speed is called the train value and it varies inversely as the number of teeth on input and output gears.

We see from equations (9.5) and (9.8) that velocity ratio for a simple gear train depends only on the number of teeth on the first and last gears in the train. The intermediate gears do not have any influence on the velocity ratio and become *idler gears*. The idler gears serve the following two purposes:

  (i) **Number of intermediate gears control direction of rotation of output gear**.
 (ii) **They serve the purpose of bridging the gap between the input and output shaft**, when this centre distance is large. A single large gear to bridge this gap requires larger space in transverse direction and is usually uneconomical.

Compound Gear Train

An alternative way of transmitting motion from one shaft to another using gear train is to compound one or more pairs of gears. *A gear pair is compounded if they are mounted on the same shaft and are made into an integral part in some way. A compound gear train is one which consists of one or more compound gears.*

There are many applications in which power is supplied through high speed motor/prime mover. Such motors/prime movers are smaller in size and are usually cheaper than their low speed counterparts. Such applications involve large speed reductions. Simple gear trains in which the input and output gear alone decide the reduction ratio is of little use in such applications. A compound gear train in which each shaft, except the first and last, carries two wheels is more useful in such cases. A higher speed reduction is conveniently obtained with compound gear train. Figure 9.2 shows a compound gear train with two intermediate shafts. Gears 2 and 3 are mounted on shaft through O_2 in such a way that there is no relative rotational motion between them. In other words their r.p.m., N_2 and N_3, is same. Similarly, gears 4 and 5 are mounted on shaft through O_4 such that there is no relative rotational motion between gears 4 and 5. In other words their r.p.m., N_4 and N_5, are same. Let N_j and T_j ($j = 1, 2, 3...,6$) denote r.p.m. and number of teeth respectively of jth gear wheel.

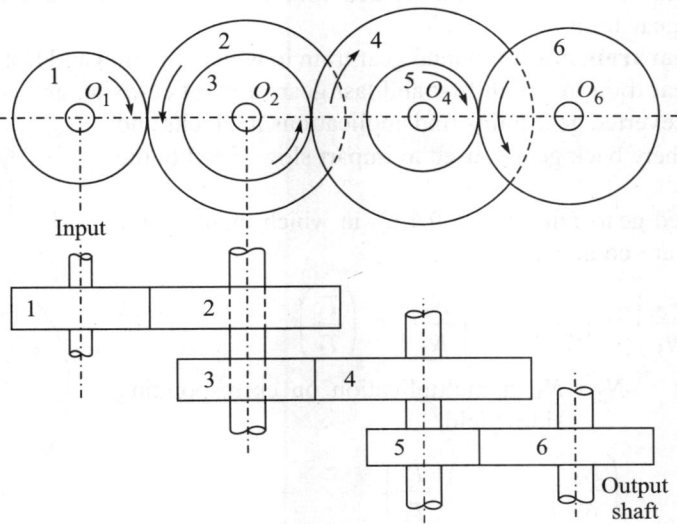

Fig. 9.2 A compound gear train.

Then

$$\left(\frac{N_2}{N_1}\right) = \left(\frac{T_1}{T_2}\right) \tag{9.9}$$

and,

$$\left(\frac{N_4}{N_3}\right) = \left(\frac{T_3}{T_4}\right) \tag{9.10}$$

But gears 2 and 3 are keyed to the same shaft and therefore, $N_2 = N_3$. Thus, multiplying equations (9.9) and (9.10) on corresponding sides,

$$\left(\frac{N_4}{N_1}\right) \equiv \left(\frac{N_4}{N_3}\right)\left(\frac{N_2}{N_1}\right) = \left(\frac{T_3}{T_4}\right)\left(\frac{T_1}{T_2}\right) \tag{9.11}$$

Also,

$$\left(\frac{N_6}{N_5}\right) = \left(\frac{T_5}{T_6}\right) \tag{9.12}$$

Remembering that gears 4 and 5 are keyed to the same shaft, $N_4 = N_5$. Hence multiplying equations (9.11) and (9.12) on corresponding sides, the gear train value is

$$\left(\frac{N_6}{N_1}\right) \equiv \left(\frac{N_6}{N_5}\right)\left(\frac{N_4}{N_1}\right) = \left(\frac{T_5}{T_6}\right)\left(\frac{T_3}{T_4}\right)\left(\frac{T_1}{T_2}\right) \tag{9.13}$$

Thus gear train value equals ratio of product of driving tooth numbers to product of driven tooth numbers. The velocity ratio (overall reduction ratio) is given by

$$\text{V.R} = \left(\frac{N_1}{N_6}\right) = \left(\frac{T_6}{T_5}\right)\left(\frac{T_4}{T_3}\right)\left(\frac{T_2}{T_1}\right) \tag{9.14}$$

Reverted Gear Train

Compound gear trains are further subdivided into two classes: (i) Reverted gear train and (ii) Non-reverted gear train.

A **reverted gear train** is a compound gear train in which, the first and last gears are co-axial. Any compound gear train in which first and last gears are not co-axial, are called **non-reverted type gear train**. Reverted gear trains find applications in clocks and in simple lathes where back gear is used to impart slow speed to the chuck.

For the reverted gear train of Fig. 9.4(a), in which input gear 1 and output gear 4 are co-axial, we have

$$\left(\frac{N_2}{N_1}\right) = \left(\frac{T_1}{T_2}\right) \text{ and } \left(\frac{N_4}{N_3}\right) = \left(\frac{T_3}{T_4}\right)$$

Remembering that $N_2 = N_3$, a multiplication on corresponding sides yield,

$$\left(\frac{N_4}{N_1}\right) = \left(\frac{T_1}{T_2}\right)\left(\frac{T_3}{T_4}\right) \tag{9.15}$$

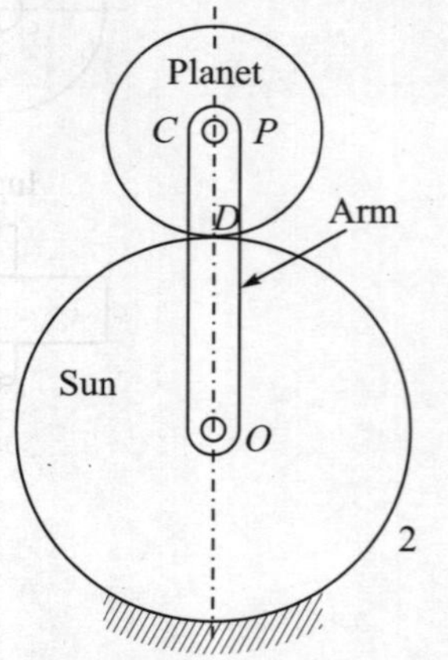

Fig. 9.3 Epicyclic gear train.

Further, if R_j ($j = 1, 2, 3, 4$) be the pitch circle radii of jth gear, then, for common axis of gears 2 and 3,

$$R_1 + R_2 = R_3 + R_4 \qquad (9.16)$$

Assuming common module for all the four wheels, and remembering that $R = m\dfrac{T}{2}$, equation (9.16) yields,

$$\left(m\frac{T_1}{2}\right) + \left(m\frac{T_2}{2}\right) = \left(m\frac{T_3}{2}\right) + \left(m\frac{T_4}{2}\right)$$

Cancelling out common terms, **for reverted gear train**

$$T_1 + T_2 = T_3 + T_4 \qquad (9.17)$$

Planetary or Epicyclic Gear Train

If axis of rotation of one or more gears is allowed to rotate about another axis the gear train is known as Planetary or Epicyclic gear train. A gear whose axis is permitted to move in an arc of a circle about the fixed axis is called a planet wheel. A gear whose axis is fixed in position, and about which axis of planet wheel revolves, is called a Sun wheel. The planet wheel is carried by an arm which is free to revolve about the fixed axis, and revolves with the input shaft. Sun wheel is also free to revolve about this fixed axis and is mounted on driven shaft. The planet gear P is free to revolve with respect to arm on a pin attached to it. A simple epicyclic gear train is shown in Fig. 9.3. The sun and planet wheels are shown to mesh at point D. Usually more than one planet wheel are required for dynamic balancing but, for the sake of simplicity and convenience, only one planet wheel is shown in sketches. A little consideration shows that if the arm in Fig. 9.3 is held stationary and the gear 2 is released, the inversion gives a simple reverted gear train [Fig. 9.4(a)]. In some of the applications, the fixed sun wheel is annular (*note:* an internal gear is called as *Annulus*) wheel and another (planet) wheel rolls on the inside of it. In such a case, the point D on planet wheel traces hypocycloidal path. It is customary, however, to call all such gear trains as epicyclic gear trains. Figure 9.4(b) shows an epicyclic gear train in which three

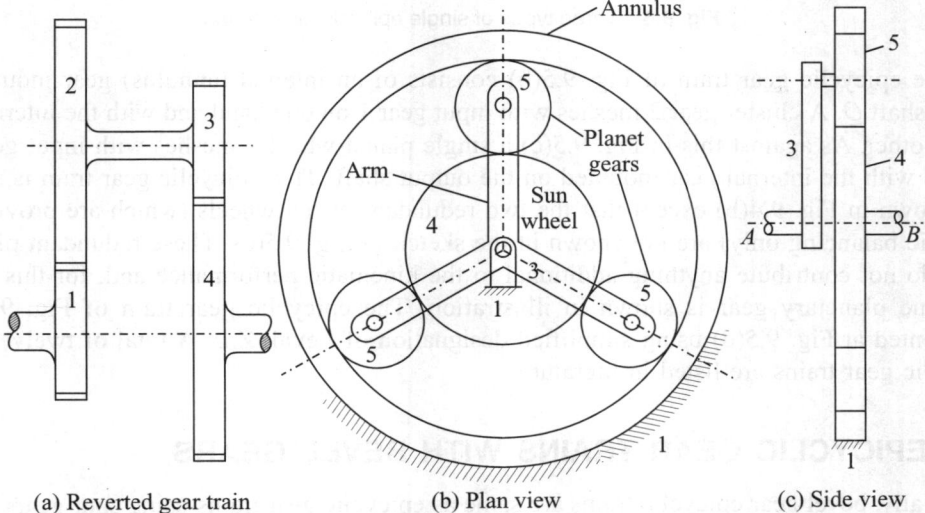

(a) Reverted gear train   (b) Plan view   (c) Side view

Fig. 9.4 Epicyclic gear train with three planet gears.

planet gears, carried on an arm, are in mesh with annulus, which is fixed. The sun gear is free to revolve. Shaft A is the input shaft and shaft B is the output or driven shaft.

In an epicyclic gear train, usually one wheel is fixed, but this is not necessary. There exists epicyclic gear trains in which no wheel is fixed. Epicyclic gear trains permit large speed reduction. An annulus may be used in an epicyclic gear train to make it compact.

A single epicyclic gear train is one in which a single planet gear rolls over another gear, producing epicyclic path. Three common type of single epicyclic gear trains are shown at Figs. 9.5(a), (b) and (c). The one shown in Fig. 9.5(a) is the simplest one where P and Q are the input and output shafts. These are connected by a simple reverted train. The block of gear (called cluster gear) 2 is free to rotate on a shaft which is supported in casing. If the casing is fixed, the epicyclic gear train is reduced to a simple reverted train. When the casing is free to revolve about the axis of shafts P and Q, it becomes an epicyclic gear train. Casing C in this case is called *planet wheel carrier.*

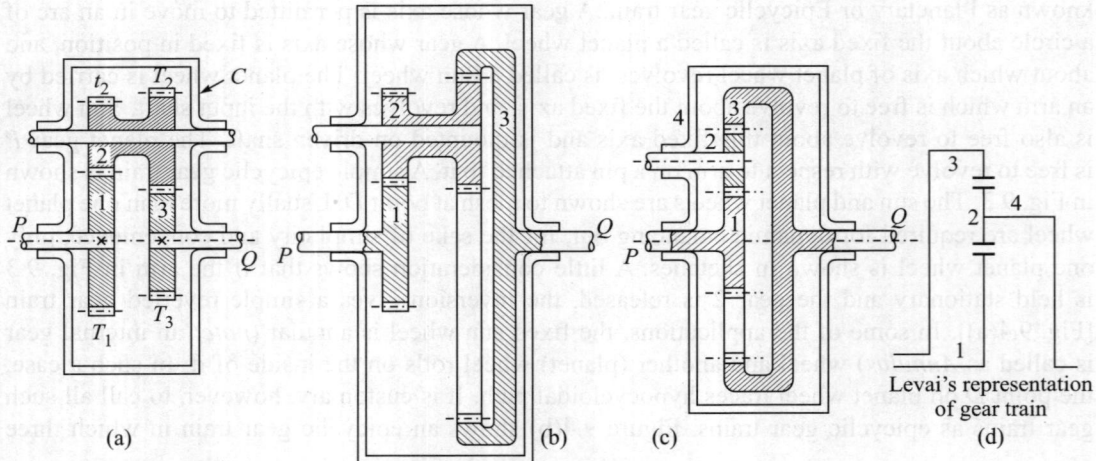

Fig. 9.5 Three types of single epicyclic gear trains.

The epicyclic gear train of Fig. 9.5(b) consists of an internal (annulus) gear mounted on output shaft Q. A cluster gear 2 meshes with input gear 1 on one hand and with the internal gear on the other. As against this in Fig. 9.5(c) a single planet wheel 2 meshes with input gear 1 as well as with the internal gear mounted on the output shaft. This epicyclic gear train is same as that shown in Fig. 9.4(b) except that the two redundant planet wheels (which are provided for dynamic balancing only) are not shown in the sketch in Fig. 9.5(c). These redundant planetary gears do not contribute anything additional to the kinematic performance and, for this reason, only one planetary gear is shown in illustration. The epicyclic gear train of Fig. 9.5(c) is represented at Fig. 9.5(d) using simplified designation of Levai, Z.L. A total of twelve simple epicyclic gear trains are listed in literature.

9.3 EPICYCLIC GEAR TRAINS WITH BEVEL GEARS

Essentially, bevel gear epicyclic trains are same as epicyclic gear trains using spur gears and are used quite frequently. Figure 9.6 shows Humpage's reduction gear and represents a double epicyclic gear train. As will be seen in the next section, the method of analysis of such trains is the same as for spur gear trains. However, when the axes of bevel gears is inclined to the main axis the terms 'clockwise' and 'counterclockwise' lose much of their significance and as such, usual convention of using plus and minus signs for indicating c.w. and c.c.w. sense is not followed in such cases. Instead, it is much more convenient to indicate sense by small arrowheads at the pitch line. Bevel gears are used to provide a compact planetary system that permits high-speed reduction with few gears.

In solving problems on epicyclic gear trains with bevel gears, the most important point to note is that the sense of rotation of a pair of bevel gears, mounted on same/parallel axes and connected by an intermediate bevel gear, is always opposite. For instance, gears A and F in Fig. 9.6 are co-axial and are connected through gear B. *Thus, if gear A rotates c.c.w., gear F will rotate in c.w. sense. Since gears A, B and F have external gearing, gear B will be assumed to have sense of rotation same as that of A.*

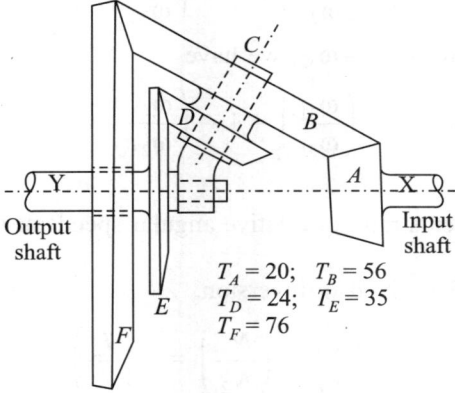

Fig. 9.6 Humpage gear train.

9.4 ALGEBRAIC METHOD OF ANALYSING EPICYCLIC GEAR TRAINS

Consider simple epicyclic gear train shown in Fig. 9.7, where the sun gear 2 is fixed to the ground and arm 3 carries the axis of planet gear 4 along an arc of circle of radius O_2O_4 around O_2.

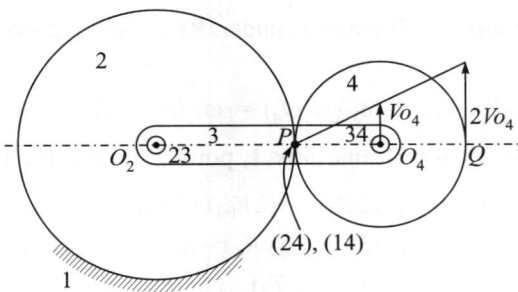

Fig. 9.7 A planetary gear train.

A little consideration shows that this is the inversion of a simple gear train in which arm 3 is fixed and sun gear 2 is released to revolve about O_2. Note that O_2 and O_4 being lower pairs, link 3 can be fixed or released to obtain inversions. Remembering that relative velocity between links does not change with inversion, it follows that relative speeds of wheels 2 and 4 with respect to arm 3 remains same in both the inversions. However, the absolute speed of gears 2 and 4 with respect to the frame does change with inversion.

Hence, the absolute angular velocity ω_{41} of planet wheel 4 w. r. to frame 1, may be obtained using expression

$$\omega_{41} = \omega_{31} + \omega_{43} \qquad (9.18)$$

where ω_{43} is the angular speed of gear 4 w. r. to arm 3.

Dividing by ω_{31} and remembering that $\omega_{32} = \omega_{31}$,

$$\left(\frac{\omega_{41}}{\omega_{31}}\right) = 1 + \left(\frac{\omega_{43}}{\omega_{32}}\right) \qquad (9.19)$$

Again, remembering that $\omega_{23} = -\omega_{32}$, we have

$$\left(\frac{\omega_{41}}{\omega_{31}}\right) = 1 - \left(\frac{\omega_{43}}{\omega_{23}}\right) \qquad (9.20)$$

The ratio $\left(\dfrac{\omega_{43}}{\omega_{23}}\right)$ represents ratio of relative angular speeds of wheels 4 and 2 with respect to arm 3. Thus, when arm 3 is fixed in inversion,

$$\left(\frac{\omega_{43}}{\omega_{23}}\right) = \left(\frac{N_4}{N_2}\right) = -\left(\frac{T_2}{T_4}\right)$$

The negative sign appears on account of external gearing. Substituting in equation (9.20), we get

$$\omega_{41} = \omega_{31}\left(1 + \frac{T_2}{T_4}\right)$$

or $$\omega_{41} = \omega_{31}\left(1 + \frac{R_2}{R_4}\right) \qquad (9.21)$$

Above expression can also be obtained as under. Remembering that P is an I.C. 24, we have $V_q = 2V_{04}$.

But, $$V_{04} = \omega_{31}(O_2O_4) = \omega_{31}(R_2 + R_4)$$

Again as sun gear 2 is fixed to frame link 1, point P is also the I.C. 14. Hence

$$\omega_{41} = V_Q/(2R_4) = (2V_{04})/(2R_4)$$

Substituting for V_{04} $$\omega_{41} = \omega_{31}(R_2 + R_4)/R_4 = \omega_{31}(1 + R_2/R_4)$$

or $$\omega_{41} = \omega_{31}(1 + T_2/T_4) \qquad (9.22)$$

Planetary gear trains are quite useful in making the reduction unit more compact than a compound gear train. They are also useful in applications requiring two degrees of freedom. An epicyclic gear train of 2 degrees of freedom can be obtained from Fig. 9.4(b) by releasing the annulus, or from Fig. 9.7 by releasing sun wheel 2. Figure 9.8 shows the gear train obtained from Fig. 9.7 with Sun wheel 2 released now.

The epicyclic gear train of Fig. 9.8, has two degrees of freedom and, therefore, requires two input motions so as to determine the motion of epicyclic gear train completely. For instance, if the absolute angular speed ω_{21} of sun wheel and ω_{31} of the arm be given, it is possible to derive expression for the absolute speed ω_{41} of the planet wheel.

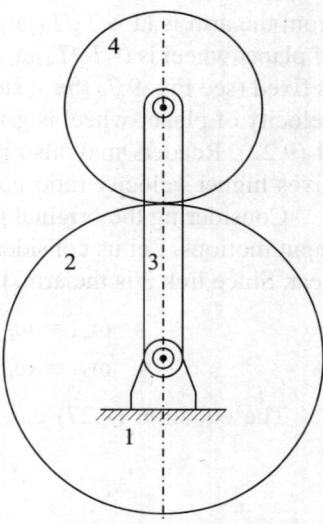

Thus, as before, absolute angular speeds of sun wheel, planet wheel and arm are related as

$$\omega_{41} = \omega_{31} + \omega_{43} \qquad (9.23)$$

and

$$\omega_{21} = \omega_{31} + \omega_{23} \qquad (9.24)$$

Fig. 9.8 Epicyclic Gear Train of 2 d.o.f.

Rearranging above equations,

$$\omega_{43} = \omega_{41} - \omega_{31} \qquad (9.25)$$

$$\omega_{23} = \omega_{21} - \omega_{31} \qquad (9.26)$$

Dividing equation (9.25) by (9.26) on respective sides,

$$\left(\frac{\omega_{43}}{\omega_{23}}\right) = \left(\frac{\omega_{41} - \omega_{31}}{\omega_{21} - \omega_{31}}\right) \qquad (9.27)$$

Noting that L.H.S. of equation (9.27) gives ratio of relative speeds of planet and sun wheel w.r. to arm and remembering that this ratio will remain the same in the inversion, obtained by fixing arm 3, it follows that

$$\left(\frac{\omega_{43}}{\omega_{23}}\right) = \left(-\frac{T_2}{T_4}\right) \qquad (9.28)$$

Hence, from equation (9.27),

$$\left(\frac{\omega_{41} - \omega_{31}}{\omega_{21} - \omega_{31}}\right) = -\left(\frac{T_2}{T_4}\right) \qquad (9.29)$$

Solving for the planetary gear speed ω_{41},

$$\omega_{41} = -\left(\frac{T_2}{T_4}\right)\omega_{21} + \left(1 + \frac{T_2}{T_4}\right)\omega_{31} \qquad (9.30)$$

Equation (9.30) shows that resultant absolute angular velocity of planet gear depends on absolute angular velocities of sun wheel (ω_{21}) and of arm (ω_{31}) and, therefore, represents a 2 degree of freedom system. The contribution from the rotation of sun wheel is $(-T_2/T_4)\omega_{21}$ and from the arm is $(1 + T_2/T_4)\omega_{31}$. If the arm is fixed, ω_{31} is equal to zero and the angular velocity. of planet wheel is $(-T_2/T_4)\omega_{21}$ which corresponds to a simple gear train. Instead, if the sun wheel is fixed (see Fig. 9.7) the system has only 1 degree of freedom and the resultant absolute angular velocity of planet-wheel is given by $(1 + T_2/T_4)\omega_{31}$. Note that this expression agrees with the one at (9.22). Readers may also note that use of epicyclic gear train, with input from rotating arm, gives higher velocity ratio compared to the simple gear train.

Considering the original gear train of 2 d.o.f., with arm 3 of Fig. 9.8 revolving, there are two input motions. Let us consider sun wheel 2 to be the first gear and the planet gear 4 to be the last gear. Since link 3 is the arm, the angular speeds of gears 2 and 4 w.r. to arm can be designated as

$$\omega_{43} \equiv \omega_{LA} = \text{angular velocity of last gear w.r. to arm } A.$$

$$\omega_{23} \equiv \omega_{FA} = \text{angular velocity of first gear w.r. to arm } A.$$

The equation (9.27) can be re-written as,

$$\left(\frac{\omega_{LA}}{\omega_{FA}} \right) = \left(\frac{\omega_L - \omega_A}{\omega_F - \omega_A} \right) \tag{9.31}$$

where

$$\frac{\omega_{LA}}{\omega_{FA}} = -\left(\frac{T_2}{T_4} \right)$$

and

$\omega_L \equiv \omega_4$ = angular velocity of planet wheel (last gear) w.r. to ground link.

$\omega_F \equiv \omega_2$ = angular velocity of sun wheel (first gear) w.r. to ground link.

and

$\omega_A \equiv$ angular velocity of arm link w.r. to ground link.

Before applying equation (9.31) to the problems on epicyclic gear trains, it is necessary to ensure that

(a) first and last gears are mounted on shafts supported on frame link, and
(b) the first and last gears must be in mesh either directly or through intermediate gears whose axis is carried by revolving arm.

The above method is also called 'Formula method'. *The conditions at (a) and (b) above imply that the planet gears must not be taken as the 'first' and the 'last' gears. Further, the first and the last gears must mesh with gears having planetary motion.*

EXAMPLE 9.1 An internal wheel B with 80 teeth is keyed to a shaft F. A fixed internal wheel C with 82 teeth is concentric with B. A compound wheel DE gears with the two internal wheels; D has 28 teeth and gears with C while E gears with B. The compound wheel rotates freely on a pin which projects from disc which is keyed to a shaft A co-axial with F. If all the wheels have the same pitch and the shaft A makes 800 r.p.m.; what is the speed of F? (DAV Indore, 1977, 1987, 1990)

Solution: Refer to Fig. 9.9. Since all the gears have same pitch, for common axis of compound wheel DE,

$$T_B - T_E = T_C - T_D$$

or

$$T_E = (T_B + T_D - T_C) = (80 + 28 - 82) = 26$$

Let B be the first gear and C be the last gear which are co-axial. There are two intermediate gear reductions between first and last gears. Hence the left hand side of equation (9.31) is

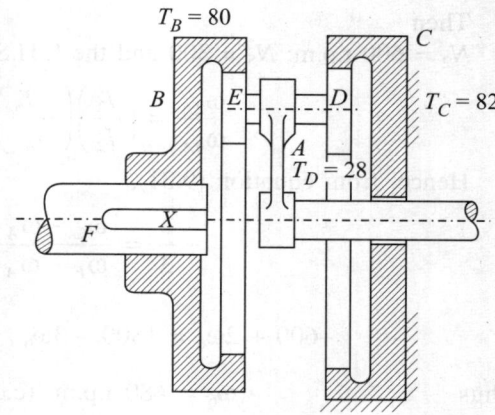

$$\frac{\omega_{LA}}{\omega_{FA}} = \left(\frac{T_D}{T_C}\right)\left(\frac{T_B}{T_E}\right) = \left(\frac{28}{82}\right)\left(\frac{80}{26}\right)$$

The right-hand side of equation (9.31) can be written in terms of corresponding r.p.m. Thus,

$$\frac{(28)\,(80)}{(82)\,(26)} = \frac{N_L - N_A}{N_F - N_A}$$

Fig. 9.9 Planetary Gear Train with internal wheels.

As the gear C is fixed, $N_L = 0$ and as per data $N_A = 800$ r.p.m. Thus,

$$-(82)(26)N_A = (28)(80)(N_F - N_A)$$

or $$N_F = \left[1 - \frac{(82)\,(26)}{(28)\,(80)}\right] N_A = 0.0482 \, N_A = 38.57 \text{ r.p.m.} \qquad \textbf{Ans.}$$

EXAMPLE 9.2 In the gear train of Fig. 9.10 the inputs are the sun gear 5 and the ring gear 2. For the given angular velocities of $N_{51} = 300$ r.p.m. and $N_{21} = 600$ r.p.m. both c.c.w. as seen from the right, find the resulting motion or arm 6.

Solution: This is a 2 d.o.f. system involving two input motions of N_2 and N_5 as shown. Let us consider gear 5 to be the first gear and gear 2 to be the last gear.

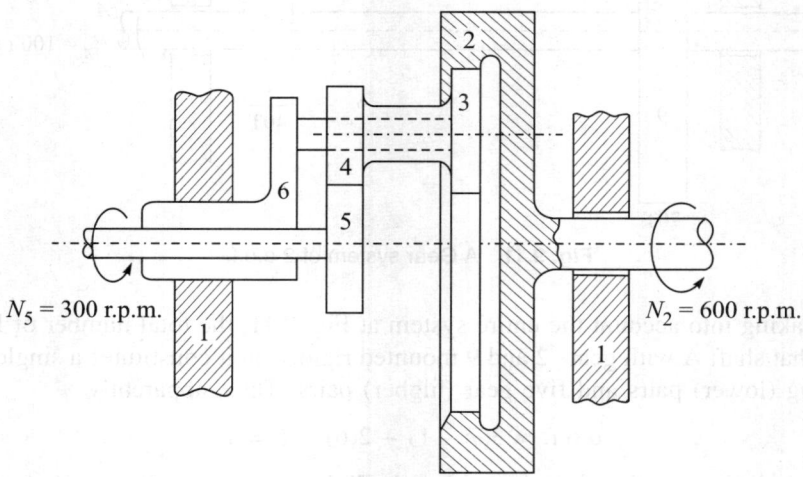

Fig. 9.10 Epicyclic Gear Train of 2 d.o.f.

Then

$N_2 = 600$ r.p.m; $N_5 = 300$ and the L.H.S. of equation (9.31) is

$$\frac{\omega_{2A}}{\omega_{5A}} = \left(\frac{T_3}{T_2}\right)\left(\frac{-T_5}{T_4}\right) = \left(\frac{45}{120}\right)\left(-\frac{48}{27}\right) = -\frac{2}{3}$$

Hence, from equation (9.31),

$$-\frac{2}{3} = \frac{\omega_L - \omega_A}{\omega_F - \omega_A} \quad \text{or} \quad -\frac{2}{3} = \frac{600 - \omega_6}{300 - \omega_6}$$

or

$$-600 + 2\omega_6 = 1800 - 3\omega_6 \quad \text{or} \quad \omega_6 = \frac{2400}{5} = 480 \text{ r.p.m.}$$

Thus

$$\omega_6 = 480 \text{ r.p.m. (c.c.w.), as seen from right.} \qquad \textbf{Ans.}$$

EXAMPLE 9.3 If shaft A of the gear system shown in Fig. 9.11 turns at 100 r.p.m., obtain the speed of shaft B and its direction of rotation. Also find the degrees of freedom of the gear system. (AMIE, Summer 1993)

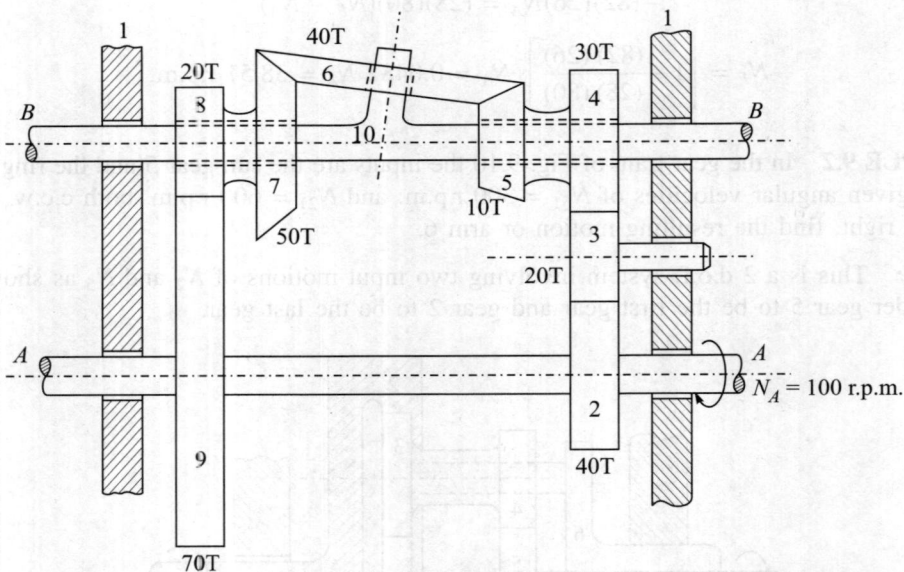

Fig. 9.11 A Gear system of 2 d.o.f.

Solution: Taking into account the entire system at Fig. 9.11, the total number of links is only seven. Note that shaft A with gears 2 and 9 mounted rigidly on it constitutes a single link. There are six turning (lower) pairs and five gear (higher) pairs. Thus, apparently

$$\text{d.o.f.} = 3(7 - 1) - 2(6) - 5 = 1.$$

However, a little consideration shows that the links consisting of input shaft A and gears, 2, 3 and 9 do not constitute part of epicyclic gear train; the purpose of gears 2, 3 and 9 along with shaft A is only to provide two input motions. Thus, considering links 1, 4/5, 6, 7/8 and 10 (total 5 links) only, there are 4 turning pairs (the two bearings of shaft 10 in frame 1 give only one pair) and 2 gear pairs between links 5, 6 and 7. Thus, d.o.f.,

$$F = 3(5 - 1) - 2(4) - 2$$
$$= 12 - 8 - 2 = 2$$

Thus, the degree of freedom of epicyclic gear train is 2. **Ans.**

This is obvious because gears 2 and 9 on input shaft A are meant for providing two different speeds to gears 4 and 8 of epicyclic gear train.

This being a problem of 2 d.o.f., algebraic method is better suited. Considering gear wheel 5 to be the first gear, gear 7 to be the last gear and shaft 10, carrying planet wheel 6, to be the arm, we have from

$$\frac{\omega_L - \omega_{10}}{\omega_F - \omega_{10}} = \frac{\omega_{L10}}{\omega_{F10}}$$

or,

$$\frac{\omega_L - \omega_{10}}{\omega_F - \omega_{10}} = \left(\frac{T_6}{T_7}\frac{T_5}{T_6}\right) = \frac{40 \times 10}{50 \times 40} = \frac{1}{5}$$

Again

$$\omega_5 \equiv \omega_4 = \omega_2\left(\frac{T_2}{T_5}\right)$$

or

$$\omega_5 \equiv \omega_4 = 100\left(\frac{40}{30}\right) = \frac{400}{3} \text{ r.p.m. in clockwise sense, as seen from left}$$

and

$$\omega_7 \equiv \omega_8 = \omega_9\left(\frac{T_9}{T_8}\right)$$

Therefore,

$$\omega_7 \equiv \omega_8 = 100\left(\frac{70}{20}\right) = 350 \text{ r.p.m. (c.c.w. as seen from left)}$$

Thus

$$\omega_F \equiv \omega_5 = \frac{-400}{3} \text{ r.p.m.}$$

and,

$$\omega_L \equiv \omega_7 = 350 \text{ r.p.m.}$$

Hence

$$\frac{350 - \omega_{10}}{-(400/3) - \omega_{10}} = \frac{1}{5} \text{ or } 4\omega_{10} = 5(350) + \frac{400}{3}$$

Therefore

$$\omega_{10} = 470.8 \text{ r.p.m.} \qquad \textbf{Ans.}$$

9.5 TABULATION METHOD FOR ANALYSING EPICYCLIC GEAR TRAIN

As indicated earlier, analysis of epicyclic gear train becomes involved because the arm/carriage carrying planet wheel (s) rotates. This suggests that principle of kinematic inversion can be of great use in analysing the epicyclic gear trains. By fixing the arm and releasing the fixed link (gear), the epicyclic gear train gets converted into a simple reverted train which is simpler to analyse. It must be noted that in kinematic inversion, relative motion between any two links does not change but the absolute motion does change. Since only lower pairs can be inverted [vide section 2.4 B (ii)], and further because arm is the only link in the train having all lower paired connections, it follows that a meaningful inversion results by fixing arm only.

The motion of gear-wheels may be considered in two parts

  (i) Motion in which all the wheels are relatively at rest but are carried round bodily by the arm; and

  (ii) Motion in which the arm is fixed but the gear-wheels are in motion relative to each other. It is this part which corresponds to the inversion of epicyclic gear train.

To understand the first part of motion, which is entirely due to the mobile condition of arm/carriage, consider Fig. 9.12. The figure illustrates four positions of a planet wheels carried by arm 3. Imagine that the sun and planet wheels do not have the teeth and are represented by their pitch cylinders, with contact at pitch point P. Let the planet wheel be rigidly attached to the arm and let it be carried bodily through 1 revolution c.c.w. around the sun wheel. A radius CQ on planet wheel then sweeps through 360° as shown in the figure. Clearly, in a given problem if the arm revolves at y r.p.m. the planet wheel (together with other wheels assumed fixed to the arm) will also undergo y revolutions in one minute. This is contemplated in the third step of tabulation method.

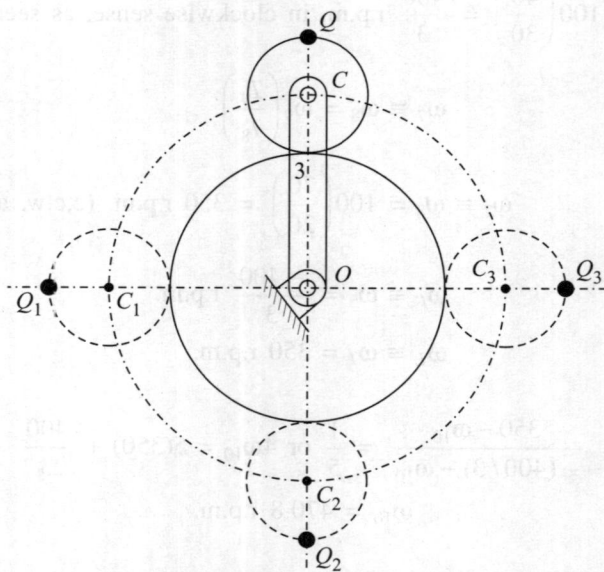

Fig. 9.12   Illustration for Tabulation method.

The step described above ensures position of arm in the 'final orientation'. But some of the other components (like planet wheels and other wheels) are still to assume their proper orientation. In this step therefore, arm is fixed and the inversion of the gear train is considered.

One of the locked wheels is given a full rotation and, treating the gear train as reverted one, corresponding rotations are established in step 1. In step 2 assuming that actual speed of one of the gear wheel in the (inverted) reverted train to be x times the speed so established, we multiply all the entries in the table by x. This step ensures proper orientation of all the gear wheels with respect to one another. Steps described in this para, followed by the steps in the preceding para ensures relative orientation for each member identical to what exists in the original epicyclic gear train.

The procedure can now be summarised in the following steps:

1. Lock the arm so as to obtain the kinematic inversion in the form of a simple reverted train. Except for the arm, all the wheels are now free to rotate.

2. Consider any convenient gear and wheel and turn it in clockwise direction through one revolution. Establish corresponding revolutions of all other gears and enter the information under step 1 and 2 in the first row of the table.

3. Multiply each entry of row 1 of the table by x and enter the values of product in second row of the table. Physically this implies that the chosen wheel is rotated through x revolutions.

4. Add y to all the entries of row 2 of the table and enter the result in row 3 of the table. Physically this implies that all the wheels are locked to the arm in the position corresponding to entries in row 2 and arm (along with the complete assembly) is turned through y revolutions.

In general x and y are unknown and the third row gives r.p.m. of each component in terms of x and y. By equating any two of these expressions for speed, in terms of x and y in row 3, to the given data, variables x and y can be established. With x and y so established, actual speeds of all other components of epicyclic gear train can be determined. This is explained through solved problems below.

EXAMPLE 9.4 An epicyclic speed reduction gear is shown in Fig. 9.13. The driving shaft carries on the arm A a pin, on which the compound wheel B–C is free to revolve. Wheel C

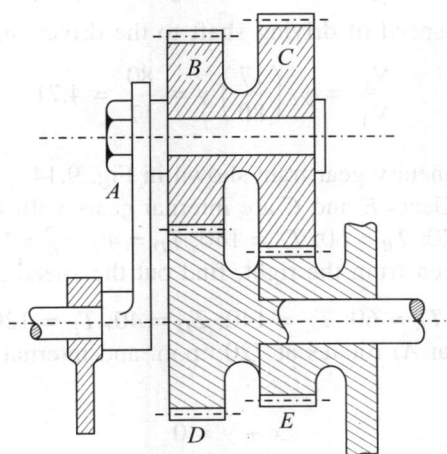

Fig. 9.13 An Epicyclic Speed Reduction Gear train.

meshes with the fixed wheel E and wheel B meshes with a wheel D keyed to the driven shaft. The numbers of teeth on the wheels are:

$T_b = 27$; $T_c = 30$; $T_d = 24$; and $T_e = 21$. Find the ratio of the speed of the driving shaft to the speed of the driven shaft. (DAVV Indore, 1989)

Solution: Given: $T_b = 27$; $T_c = 30$; $T_d = 24$; and $T_e = 21$ and, wheel E is fixed. The table is prepared as follows:

Table 9.1 Solution by Tabulation Method.

S.No.	Operations	Arm A	Wheel E $T_e = 21$	Wheel B–C (compound) $T_b = 27\ T_c = 30$	Wheel D (driven shaft) $T_d = 24$
1.	Fix the arm and rotate wheel E by +1 revolution	O	+1	$\dfrac{-T_e}{T_c}$	$\dfrac{T_e}{T_c}\dfrac{T_b}{T_d}$
2.	Multiply by x	O	x	$\dfrac{-T_e}{T_c}x$	$\dfrac{T_e}{T_c}\dfrac{T_b}{T_d}x$
3.	Add y to all	y	$x + y$	$-\left(\dfrac{T_e}{T_c}\right)x + y$	$\dfrac{T_e}{T_c}\dfrac{T_b}{T_d}x + y$

Since the gear wheel E is fixed,

$$x + y = 0 \qquad \text{or} \qquad x = -y$$

Hence, the output speed, as given by speed of D

$$= \left(\frac{T_e}{T_c}\right)\left(\frac{T_b}{T_d}\right)(-y) + y$$

$$= \left[-\left(\frac{21}{30}\right)\left(\frac{27}{24}\right) + 1\right]y = \left(\frac{17}{80}\right)y$$

Hence, the ratio of the speed of driving shaft to the driven shaft

$$\frac{N_A}{N_D} = y \bigg/ \left(\frac{17}{80}\right)y = \frac{80}{17} = 4.71 \qquad \qquad \textbf{Ans.}$$

EXAMPLE 9.5 In the planetary gear train shown in Fig. 9.14, gear A is the driver and gears B and D are compounded. Gears E and C are internal gears with C as the fixed gear. Number of teeth on gears are: $T_A = 20$; $T_B = 60$; $T_C = 140$; $T_D = 40$; $T_E = 120$. If the driver gear rotates at 720 r.p.m. c.c.w. when seen from the right, find out the speed at which driven gear rotates.

Solution: Given $T_A = 20$; $T_B = 60$; $T_C = 140$; $T_D = 40$; $T_E = 120$

But the input shaft (Gear A) rotates at 720 r.p.m. and internal gear is fixed, *i.e.* $N_A = 720$ and $N_C = 0$

Thus
$$x + y = 0$$

and
$$-7x + y = 720$$

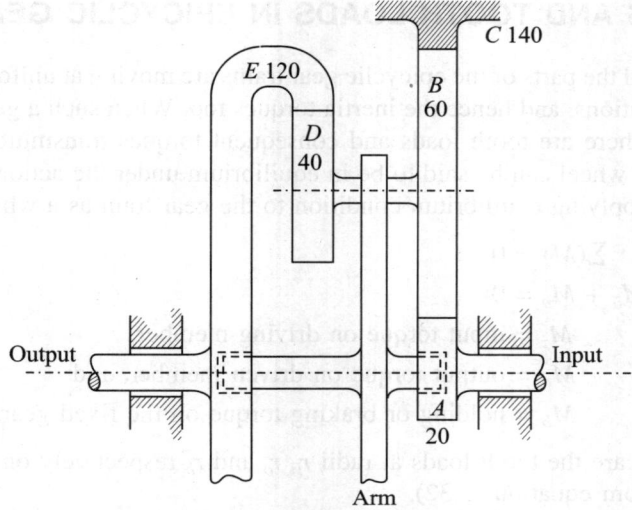

Fig. 9.14 Planetary gear train.

Subtracting

$$8x = -720,$$

Therefore

$$x = -90$$

Substituting back, $y = 90$

Table 9.2 Solution by Tabulation Method.

S.No.	Operations	Arm	Gear A $T_A = 20$	Compound Gear B–D $T_B = 60$ $T_D = 40$	Gear C $T_C = 140$	Gear E $T_E = 120$
1.	Fix the Arm and give 1 revolution to C c.c.w.	O	$-\dfrac{140}{60}\dfrac{60}{20}$	$+\dfrac{140}{60}$	$+1$	$+\dfrac{140}{60}\dfrac{40}{120}$
2.	Multiply by x	O	$-7x$	$\dfrac{14}{6}x$	x	$\dfrac{14}{18}x$
3.	Add y to all	y	$-7x + y$	$\dfrac{14}{6}x + y$	$x + y$	$\dfrac{14}{18}x + y$

Hence, output shaft speed

$$N_E = \frac{14}{18}x + y$$

$$= \frac{14}{18}(-90) + y$$

$$= -70 + 90 = 20 \text{ r.p.m.} \quad \textbf{Ans.}$$

Thus, there is a speed reduction by a ratio $= \dfrac{720}{20} = 36$

9.6 TORQUES AND TOOTH LOADS IN EPICYCLIC GEAR TRAINS

Let us assume that all the parts of the epicyclic gear trains are moving at uniform speeds, avoiding any angular accelerations, and hence the inertia torques too. When such a geared system is used to transmit power, there are tooth loads and consequent torques transmitted from one gear to the other. Each gear wheel can be said to be in equilibrium under the action of external torques acting on it. Thus applying equilibrium condition to the gear train as a whole,

$$\Sigma(M) = 0$$

or

$$M_i + M_o + M_b = 0 \tag{9.32}$$

where

$$M_i = \text{input torque on driving member,}$$

$$M_o = \text{output torque on driven member, and}$$

$$M_b = \text{holding or braking torque on the fixed gear.}$$

If F_i, F_o and F_b are the tooth loads at radii r_i, r_o and r_b respectively on driving, driven and fixed gears then, from equation (9.32),

$$(F_i)r_i + (F_o)r_o + F_b(r_b) \tag{9.33}$$

Finally, if there are no frictional losses within the gear train, either at the bearings or at the contact surfaces of teeth, the energy balance dictates,

$$\Sigma M(\omega) = 0$$

or

$$M_i\omega_i + M_o\omega_o + M_b\omega_b = 0 \tag{9.34}$$

Again, as the angular speed $\omega_b = 0$ for fixed gear, equation (9.34) simplifies to

$$(M_i)\omega_i + (M_o)\omega_o = 0 \tag{9.35}$$

Substituting for M_o from equation (9.35) in equation (9.32),

$$M_b = M_i\left(\frac{\omega_i}{\omega_o} - 1\right) \tag{9.36}$$

Equation (9.36) enables to obtain braking or fixing torque required to be applied to the fixed gear. Equation (9.33) can be used to evaluate tooth loads.

EXAMPLE 9.6   In an epicyclic gear train of the sun and planet type, shown in Fig. 9.4, the pitch circle diameter of the annular wheel is to be nearly 216 mm and module 4 mm. When the annular wheel is stationary, the spider which carries three planet wheels of equal size, is to make one revolution for every five revolutions of the driving spindle carrying the sun wheel.

Determine suitable number of teeth for all the wheels and the exact diameter of the pitch circle of the annular wheel. If an input torque of 19.6 N·m is applied to the spindle carying the sun wheel; determine the fixing torque on the annular ring. (AMIE, Summer 1983, DAVV Indore, 1983; J. Univ. Gwalior, 1983)

Solution: Referring to Fig. 9.4,

$$T_2 = \frac{P.C.D.}{\text{module}} = \frac{216}{4} = 54$$

For kinematic calculations, it is sufficient to consider one planet wheel instead of three. As given in the problem,

$$N_5/N_A = 5,$$

where N_5 and N_A are speeds in **r.p.m.** of sun wheel and spider.

Table 9.3 Solution by Tabulation Method.

S.No.	Operations	Spider/Arm A	Sun wheel (Input) T_4	Planet wheel T_5	Annulus $T_2 = 54$
1.	Fix the spider/arm and give +1 revolution to the annulus	O	$-\left(\dfrac{T_2}{T_5}\right)\left(\dfrac{T_5}{T_4}\right)$	$+\left(\dfrac{T_2}{T_5}\right)$	+1
2.	Multiply by x	O	$-\dfrac{T_2}{T_4}x$	$+\left(\dfrac{T_2}{T_5}\right)x$	$+x$
3.	Add y to all	y	$-\left(\dfrac{T_2}{T_4}\right)x + y$	$+\left(\dfrac{T_2}{T_4}\right)x + y$	$x + y$

When the annulus is stationary, sun wheel makes 5 revolutions for 1 revolutions of spider.

Hence

$$y = 1 \qquad \text{(i)}$$

$$x + y = 0 \qquad \text{(ii)}$$

and

$$-\left(\frac{T_2}{T_4}\right)x + y = 5 \qquad \text{(iii)}$$

From (i) and (ii), $x = -1$ and $y = 1$
Hence from (iii)

$$\left(\frac{T_2}{T_4}\right) + 1 = 5$$

or

$$\left(\frac{T_2}{T_4}\right) = 4 \qquad \text{(iv)}$$

Since

$$T_2 = 54, \qquad T_4 = \frac{54}{4} = 13.5$$

Selecting next higher integer value, $T_4 = 14$ **Ans.**

And therefore

$$T_2 = 14 \times 4 = 56$$

Hence, pitch circle diameter of annulus, $mT = 4 \times 56 = 224$ mm **Ans.**
Again from the geometry of Fig. 9.4,

$$\frac{D}{2} = \left(\frac{d_4}{2}\right) + d_5 \text{ and, dividing by common module,}$$

$$\frac{T_2}{2} = \left(\frac{T_4}{2}\right) + T_5$$

Hence, substituting values of T_4 and T_2,

$$T_5 = \left(\frac{56}{2}\right) - \left(\frac{14}{2}\right) = 21 \qquad \textbf{Ans.}$$

The fixing torque as given by equation (9.36) is

$$M_b = 19.6\left(\frac{\omega_4}{\omega_3} - 1\right)$$

Now

$$\frac{\omega_4}{\omega_3} = \left[-\left(\frac{T_2}{T_4}\right)x + y\right]\bigg/ y = \frac{(4+1)}{1} = 5$$

Hence

$$M_b = 19.6(5 - 1) = 78.4 \text{ N·m} \qquad \textbf{Ans.}$$

EXAMPLE 9.7 Figure 9.15 shows an epicyclic gear train known as Ferguson's paradox. Gear A is fixed to the frame and is, therefore stationary. The arm B and gears C and D are free to rotate on shaft S. Gears A, C, D have 100, 101 and 99 teeth respectively, all cut to same pitch circle diameter from gear blanks of the same diameter so that same planet wheel of 20 teeth meshes with all of them. Determine the revolutions of gears C and D for one revolution of the arm B. (AMIE, Winter 1979)

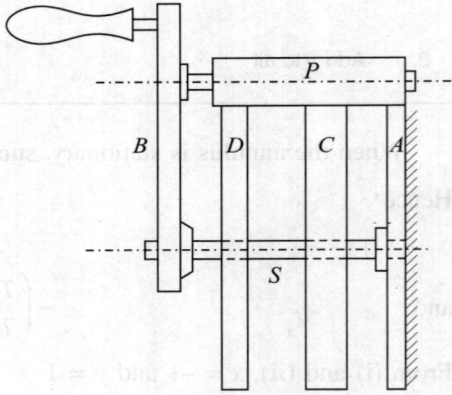

Fig. 9.15 Ferguson's paradox.

Solution: Given $T_A = 100$; $T_C = 101$;

$$T_D = 99; \ T_P = 20$$

Table 9.4 Solution by Tabulation Method.

S.No.	Operations	Arm B	Gear A	Gear C	Gear D
1.	Fix the arm and rotate gear A by + 1 turn	O	+ 1	$\left(\dfrac{T_A}{T_P}\dfrac{T_P}{T_C}\right)$	$\left(\dfrac{T_A}{T_P}\dfrac{T_P}{T_D}\right)$
2.	Multiply by x	O	x	$\left(\dfrac{T_A}{T_C}\right)x$	$\left(\dfrac{T_A}{T_D}\right)x$
3.	Add y to all	y	$x + y$	$\left(\dfrac{T_A}{T_C}\right)x + y$	$\left(\dfrac{T_A}{T_D}\right)x + y$

But gear A is fixed and $N_B = +1$

i.e.,

$$y = 1$$

and,

$$x + y = 0$$

or,

$$x = -1$$

Hence, for 1 revolution of B, number of revolutions of C

$$N_C = \left(\frac{T_A}{T_C}\right)x + y = \left(\frac{100}{101}\right) \times (-1) + 1 = \left(\frac{1}{101}\right) \text{ revolutions}$$ **Ans.**

And, for 1 revolution of B, number of revolutions of D,

$$N_D = \left(\frac{T_A}{T_D}\right)x + y = \left(\frac{100}{99}\right) \times (-1) + 1 = -\left(\frac{1}{99}\right) \text{ revolutions}$$ **Ans.**

It follows that as the arm B is turned, gear C rotates very slowly in the same sense, while gear D rotates very slowly in the opposite sense.

EXAMPLE 9.8 The number of teeth in the gear train shown in Fig. 9.16, are as follows:

$$T_s = 18;\ T_p = 24;\ T_c = 12;\ T_A = 72.$$

P and C from a compound gear carried by the arm D and the annular gear A is held stationary. Determine the speed of the output D. Also find the holding torque required on A if 5 kW is delivered to S at 800 r.p.m. with efficiency of 94%.

In case the annulus A rotates at 100 r.p.m. in the same direction as S, what will be the new speed of D? (DAV Indore, 1993) (Amravati Univ.)

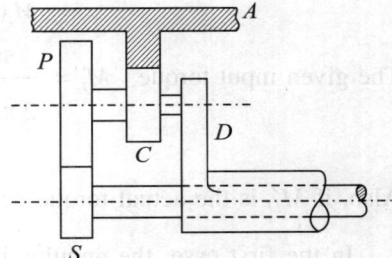

Fig. 9.16 Epicyclic Gear Train.

Solution: Refer Table 9.5

Table 9.5 Solution by Tabulation Method.

S.No.	Operations	Arm D	Sun gear $T_s = 18$	Compound gear P–C $T_p = 24$; $T_c = 12$	Annulus A $T_A = 72$
1.	Fix the arm D and give +1 revolution to sun gear	O	+1	$-\dfrac{T_s}{T_p}$	$-\left(\dfrac{T_s}{T_p}\right)\left(\dfrac{T_c}{T_A}\right)$
2.	Multiply by x	O	x	$-\left(\dfrac{18}{24}\right)x$	$-\left(\dfrac{18}{24}\right)\left(\dfrac{12}{72}\right)x$
3.	Add y to all	y	x + y	$-\dfrac{3}{4}x + y$	$-\dfrac{1}{8}x + y$

The annulus is fixed and, as such, $-\dfrac{1}{8}x + y = 0$ (i)

or $y = (x/8)$

Also, sun gear rotates at 800 r.p.m. and hence,

$$x + y = 800$$ (ii)

Substituting for y, $\left(1 + \dfrac{1}{8}\right)x = 800$

or
$$x = (6400/9)$$

and therefore,
$$y = \frac{x}{8} = \frac{800}{9}$$

The speed of output shaft = speed of arm D = $\dfrac{800}{9}$ = 88.9 r.p.m.   **Ans.**

For energy equilibrium, from equation (9.34),

$$M_i(N_i) + M_o(N_o) + M_b(N_b) = 0 \qquad \text{(iii)}$$

The given input torque $M_i = \dfrac{5000}{2\pi \times \dfrac{N}{60}} = \dfrac{5000 \times 60}{2\pi \times 800} = 59.68$ N·m

Also, if M_D' is the actual torque delivered to arm D, the theoretical torque on arm = $\left(\dfrac{M_D'}{0.94}\right)$

In the first case, the annulus is fixed, and as such,

$$N_A \equiv N_b = 0$$

Hence from equation (iii), $M_o = -M_i\left(\dfrac{N_i}{N_o}\right)$

i.e.,
$$\left(\frac{M_D'}{0.94}\right) = -59.69\left(\frac{800}{88.9}\right)$$

Hence
$$M_D' = -59.68\left(\frac{800}{88.9}\right)0.94$$

or
$$M_D' = -504.83 \text{ N·m}$$

Now, from equation (9.32),

$$M_s + M_D' + M_A = 0 \qquad \text{or} \qquad (59.68) - (504.83) - M_A = 0$$

or
$$M_A = 445.15 \text{ N·m} \qquad \textbf{Ans.}$$

In the second case, as the annulus rotates at 100 r.p.m.,

$$-\frac{x}{8} + y = 100 \qquad \text{(iv)}$$

or
$$y = \left(100 + \frac{x}{8}\right)$$

Substituting for y in equation (ii) above,

$$x + \left(100 + \frac{x}{8}\right) = 800$$

or
$$x = \frac{8}{9}(700) = 622.28$$

and, therefore
$$y = 177.78$$

Therefore, new speed of arm
$$D = 177.8 \text{ r.p.m.}$$
Ans.

EXAMPLE 9.9 An epicyclic gear train has a fixed annular wheel C concentric with sun wheel A. A planet wheel B gears with A and C and can rotate freely on a pin carried by an arm D which rotates about an axis co-axial with that of A and C. If T_1 and T_2 are the numbers of teeth on A and C respectively, show that the ratio of the speeds of D to A is $\dfrac{T_1}{(T_1 + T_2)}$. If the least number of teeth on any wheel is 18 and $T_1 + T_2 = 120$, find the greatest and least speeds of D when wheel A rotates at 500 r.p.m. (DAVV Indore, 1986, 1988)

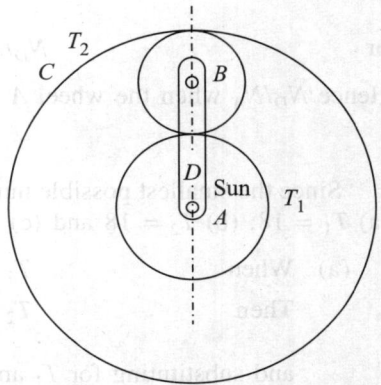

Fig. 9.17 Epicyclic Gear Train with Fixed Annular wheel.

Solution: From the geometry of Fig. 9.17, the number of teeth on planet wheel B,

$$T_B = \left(\frac{T_2}{2}\right) - \left(\frac{T_1}{2}\right) = \frac{(T_2 - T_1)}{2}$$

Table 9.6 For Tabulation Method.

S.No.	Operations	Arm D	Sun wheel A (T_1)	Planet wheel B $\dfrac{(T_2 - T_1)}{2}$	Annulus (T_2)
1.	Fix the arm and rotate sun wheel by *l* twin	O	+1	$-\dfrac{2T_1}{(T_2 - T_1)}$	$\left(-\dfrac{T_1}{T_2}\right)$
2.	Multiply by x	O	x	$-\left(\dfrac{2T_1}{T_2 - T_1}\right)x$	$\left(-\dfrac{T_1}{T_2}\right)x$
3.	Add y to all	y	x + y	$-\left(\dfrac{2T_1}{T_2 - T_1}\right)x + y$	$\left(-\dfrac{T_1}{T_2}\right)x + y$

The annulus C is fixed and therefore,

$$\left(-\frac{T_1}{T_2}\right)x + y = 0 \qquad\qquad (i)$$

or
$$y = \left(\frac{T_1}{T_2}\right)x \qquad \text{(ii)}$$

Thus the ratio of speeds of D to A is,

$$\frac{N_D}{N_A} = \frac{y}{x+y} = \frac{\left(\dfrac{T_1}{T_2}\right)x}{\left(1+\dfrac{T_1}{T_2}\right)x}$$

or
$$N_D/N_A = T_1/(T_1 + T_2) \qquad \textbf{Ans.}$$

Hence N_D/N_A when the wheel A rotates at 500 r.p.m.,

$$x + y = 500 \qquad \text{(iii)}$$

Since the smallest possible number of teeth on any gear is 18, let us examine the cases when (a) $T_1 = 18$; (b) $T_2 = 18$ and (c) $T_b = 18$.

(a) When $\qquad\qquad\qquad T_1 = 18$

Then $\qquad\qquad\qquad\quad T_2 = 120 - 18 = 102$

and substituting for T_1 and T_2 in (i), $-\left(\dfrac{18}{102}\right)x + y = 0 \qquad\qquad$ (iv)

Solving (iii) and (iv) simultaneously,

$$x = \left(\frac{102}{120}\right)500 = 425$$

Therefore $\qquad y = $ speed of $D = 500 - 425 = 75$ r.p.m.

(b) When $\qquad\qquad\qquad T_2 = 18,$

Then $\qquad\qquad\qquad\quad T_1 = 120 - 18 = 102$

Hence from (i), $-\left(\dfrac{102}{18}\right)x + y = 0 \quad$ or $\quad x + y = 500$

Solving simultaneously,

$$x = 500\left(\frac{18}{120}\right) = 75$$

Hence, speed of $D = y = 500 - 75 = 425$ r.p.m.

(c) When $\qquad\qquad\qquad T_b = 18,$

$$\frac{T_2 - T_1}{2} = T_b = 18$$

or $\qquad\qquad T_2 + T_1 = 36 \qquad$ and $\qquad T_1 + T_2 = 120$

Solving simultaneously,

$$T_2 = \frac{156}{2} = 78 \therefore T_1 = 42$$

Hence, from equation (i),

$$-\left(\frac{42}{78}\right)x + y = 0 \text{ and } x + y = 500$$

Solving simultaneously,

$$x = 500\left(\frac{78}{120}\right) = 325$$

Hence, speed of D $\qquad = y = 500 - 325 = 175$ r.p.m.

Thus maximum and minimum possible speeds of D are 425 r.p.m. and 75 r.p.m. **Ans.**

EXAMPLE 9.10 Figure 9.18 shows the arrangement of wheels in a compound epicyclic gear train. The sun wheel S_2 is integral with the annular wheel A_1. The two arms are also integral with each other. The number of teeth on the wheels are as follows:

$$S_1 = 24; A_1 = 96; S_2 = 30; A_2 = 90$$

(a) If the shaft P rotates at 1980 r.p.m., find the speed of the shaft Q, when A_2 is fixed.
(b) At what speed does Q rotate, when A_2 instead of being fixed rotates at 198 r.p.m. in the same direction as S_1, which is rotating at 1980 r.p.m.

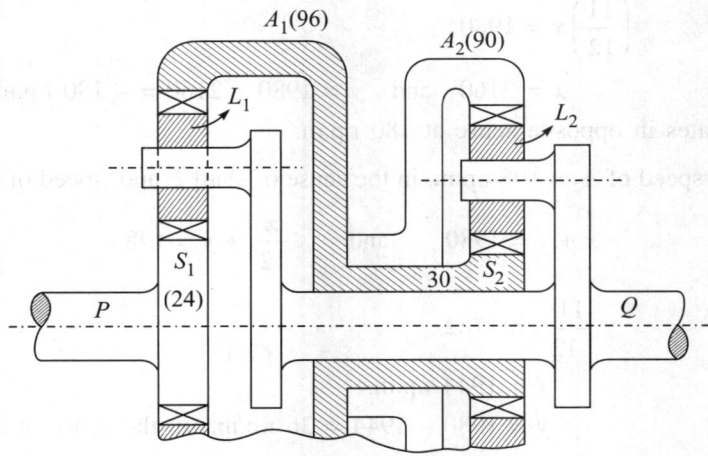

Fig. 9.18 A compound Epicyclic Gear Train.

Solution: The compound epicyclic gear train consists of an epicyclic gear train comprising S_1, A_1, planet wheel L_1 and planet carrier. The other gear train consists of S_2, A_2, etc.

Consider the first epicyclic train of the compound planetary system.

Table 9.7   Tabulation Method.

S.No.	Operations	Arm Q	Sun wheel S_1 (24)	Planet wheel L_1 (36)	Annulus A_1 (96)	Sun wheel S_2 (30)	Planet wheel L_2 (30)	Annulus A_2 (90)
1.	Fix the arm and rotate S_1 by + 1 rev.	O	+1	$-\dfrac{24}{36} = \dfrac{-2}{3}$	$\dfrac{-2}{3} \times \dfrac{36}{96}$	$-\dfrac{1}{4}$	$+\dfrac{1}{4} \times \dfrac{30}{30}$	$\dfrac{1}{4} \times \dfrac{30}{90}$
2.	Multiply by x	O	$+x$	$-\dfrac{2x}{3}$	$-\dfrac{x}{4}$	$-\dfrac{x}{4}$	$+\dfrac{x}{4}$	$\dfrac{x}{12}$
3.	Add y to all	y	$x + y$	$-\dfrac{2x}{3} + y$	$-\dfrac{x}{4} + y$	$-\dfrac{x}{4} + y$	$\dfrac{x}{4} + y$	$\dfrac{x}{12} + y$

Number of teeth on the planet wheels is given by,

$$T_{L1} = \frac{T_{A1} - T_{S1}}{2} = \frac{96 - 24}{2} = 36$$

$$T_{L2} = \frac{T_{A2} - T_{S2}}{2} = \frac{90 - 30}{2} = 30$$

Case (a) Annulus A_2 is fixed and speed of $S_1 = 1980$ r.p.m.

Therefore $\qquad\qquad x + y = 1980 \qquad$ and $\qquad \dfrac{x}{12} + y = 0$

Subtraction on both sides, eliminating y, gives

$$\left(\frac{11}{12}\right)x = 1980$$

or $\qquad\qquad x = 2160 \quad$ and $\quad y = 1980 - 2160 = -180$ r.p.m.

Thus shaft Q rotates in opposite sense at 180 r.p.m. **Ans.**

Case (b) The speed of $A_2 = 198$ r.p.m. in the sense of shaft P and, speed of $S_1 = 1980$ r.p.m.

Thus $\qquad\qquad x + y = 1980 \qquad$ and $\qquad \dfrac{x}{12} + y = 198$

Subtracting $\qquad\qquad \dfrac{11x}{12} = 1782$

Thus $\qquad\qquad x = 1944$ r.p.m.

and, $\qquad\qquad y = 1980 - 1944 = 36$ r.p.m. (in the sense of S_1). **Ans.**

EXAMPLE 9.11   Figure 9.19 shows an epicyclic gear train with the following particulars:
   A has 40 teeth external; B has 80 teeth external
   C has 20 teeth external and Compound wheel D has 50 teeth external (Compound wheel)
   E has 20 teeth external and F has 40 teeth external (Compound wheel)
   G has 90 teeth external

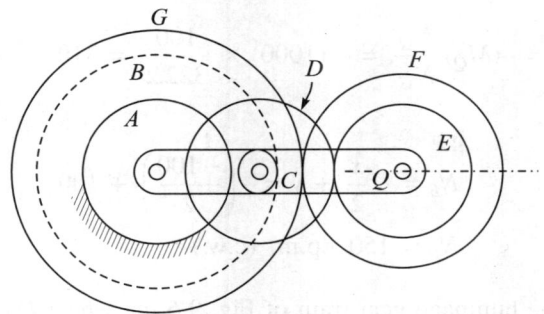

Fig. 9.19 Epicyclic Gear Trains with compounded wheels.

Gear wheel C gears with A, B and wheel D gears with E. Gear F gears with G. Wheel A is fixed and the arm runs at 100 r.p.m. in clockwise direction. Determine the torque exerted on gear wheel G, if the input torque to the arm is 1000 N-m. Also find the speed of wheel B. (Amravati Univ.)

Table 9.8 Tabulation Method.

S. No.	Operations	Arm Q	Wheel A (40)	Compound wheel C-D	Wheel B (80)	Wheel E (20)	Wheel F (40)	Wheel G (90)
1.	Arm fixed, wheel A rotated through +1 rev.	O	+1	$-\dfrac{40}{20} = -2$	$-2 \times \dfrac{20}{80}$ $= -\dfrac{1}{2}$	$+2 \times \dfrac{50}{20}$ $= +5$	$+5$	$-5 \times \dfrac{40}{90}$ $= -\dfrac{20}{9}$
2.	Multiply by x	O	$+x$	$-2x$	$-\dfrac{x}{2}$	$5x$	$+5x$	$-20x/9$
3.	Add y to all	y	$x + y$	$-2x + y$	$-\dfrac{x}{2} + y$	$5x + y$	$5x + y$	$-\dfrac{20x}{9} + y$

Solution:

The arm Q rotates at 100 r.p.m. (c.w.) and wheel A is fixed. Hence, from Table 9.8

$$y = 100 \text{ r.p.m. (c.w.)}$$

and

$$x + y = 0$$

Hence

$$x = -100 \text{ r.p.m. (c.c.w.)}$$

Thus, speed of wheel G,

$$N_G = -\frac{20x}{9} + y = \frac{20}{9}(100) + 100$$

$$= 222.2 + 100 = 322.2 \text{ r.p.m. (c.w.)} \qquad \textbf{Ans.}$$

From equation (9.34),

$$(M_Q)N_Q + M_G N_G + M_A N_A = 0$$

and   as

$$N_A = 0,$$

$$M_G = -(M_Q)\frac{N_Q}{N_G} = -(1000) \times \frac{100}{322.2} = 310.37 \text{ N·m} \qquad \textbf{Ans.}$$

Speed of wheel B,

$$N_B = -\frac{x}{2} + y = -\left(\frac{-100}{2}\right) + 100$$

or $\qquad N_B = 150$ r.p.m. (c.w.) $\qquad\qquad$ **Ans.**

EXAMPLE 9.12 In the humpage gear train of Fig. 9.6 the wheel F is fixed; X and Y are the driving and driven shafts respectively; the compound wheel B–D can revolve on a spindle C which can turn freely about the axes of X and Y.

If the number of teeth on the wheels A, B, F, D and E are 17, 60, 75, 19 and 25 respectively and if 9 k.w. is put into the shaft X at 500 r.p.m., what is the output torque on the shaft Y, and what are the tangential forces at the contact points between the wheels D and E and between the wheels B and F, if the mean diameters of B and D are 375 mm and 118.65 mm.

Solution: The input torque $= \dfrac{9000 \times 60}{2\pi \times 500}$

or $\qquad M_i = 171.89$ N·m

In problems involving bevel gears sign of revolution is difficult to decide. A golden rule to follow is that a pair of bevel gears, mounted on same axis and connected through another bevel gear, will rotate in opposite directions.

Table 9.9 Tabulation Method.

S.No.	Operations	Arm C	Wheel A (17)	Wheel B–D compound (B = 60; D = 19)	Wheel F (75)	Wheel E (25)
1.	Fix the arm and turn A through +1 rev.	O	+1	$\frac{17}{60}$	$-\frac{17}{60}\times\frac{60}{75}=-\frac{17}{75}$	$-\frac{17}{60}\times\frac{19}{25}$
2.	Multiply by x	O	$+x$	$\frac{17}{60}x$	$\frac{-17}{75}x$	$\frac{-17\times19}{60\times25}x$
3.	Add y to all	y	$x+y$	$\frac{17}{60}x+y$	$\frac{-17}{75}x+y$	$-\frac{17\times19}{60\times25}x+y$

Now gear A along with driving shaft rotates at 500 r.p.m. while gear F is fixed. Hence,

$$x + y = 500 \quad \text{and} \quad \frac{-17}{75}x + y = 0$$

Subtracting $\qquad \dfrac{92}{75}x = 500$

Therefore $\qquad\qquad x = 500 \times \dfrac{75}{92} = 407.6$ and, $y = 92.39$

Hence, the output speed on gear E

$$N_E = -\frac{17 \times 19}{60 \times 25}(407.6) + 92.39 = 4.62 \text{ r.p.m.}$$

From equation (9.34), the output torque is,

$$M_o = M_i\left(\frac{N_i}{N_o}\right)$$

or $\qquad\qquad M_o = 171.89\left(\dfrac{500}{4.62}\right) = 18602.8 \text{ N·m}$ **Ans.**

From equation (9.33)

$$(F_i)R_i + (F_o)r_o + F_b(r_b) = 0 \qquad\qquad\text{(i)}$$

The module of wheels B and $D = m = \dfrac{375}{60} = 6.25$ mm

Therefore pitch circle radius of wheel $E = \dfrac{6.25 \times 25}{2} = 78.125$ mm

and, pitch circle radius of wheel $F = \dfrac{6.25 \times 75}{2} = 234.375$ mm

Tangential force between wheels D and E,

$$F_o = \frac{\text{output torque}}{\text{pitch circle radius of } E}$$

or $\qquad\qquad F_o = \dfrac{18602.8 \times 1000}{78.125}$

Thus, $\qquad\qquad F_o = 238115.84 \text{ N} = 238.116 \text{ kN}$ **Ans.**

Tangential force between B and F,

$$F' = \frac{\text{Torque on } F}{\text{Pitch circle radius of } F}$$

From equation (i) above,

$$F_b = \frac{M_i + M_o}{r_F}$$

or $\qquad\qquad F_c = \dfrac{(171.89 + 18602.8)1000}{234.375}$

or $\qquad\qquad F_c = 80105.344 \text{ N} = 80.105 \text{ kN}$ **Ans.**

EXAMPLE 9.13 An epicyclic bevel gear train, as shown in Fig. 9.20, has a fixed wheel B meshing with pinion C. The wheel E on the driven shaft meshes with the pinion D. The pinions C and D are keyed to a shaft, which revolves in bearings on the arm A. The arm A is keyed to the driving shaft. The numbers of teeth are: $T_B = 75$; $T_C = 20$; $T_D = 18$ and $T_E = 70$. Find the speed of the driven shaft if

(1) the driving shaft makes 1000 r.p.m. and
(2) the wheel B turns in the same sense as the driving shaft at 400 r.p.m., the driving shaft still making 1000 r.p.m.

Solution: Data $T_B = 75$; $T_C = 20$; $T_D = 18$, and $T_E = 70$.

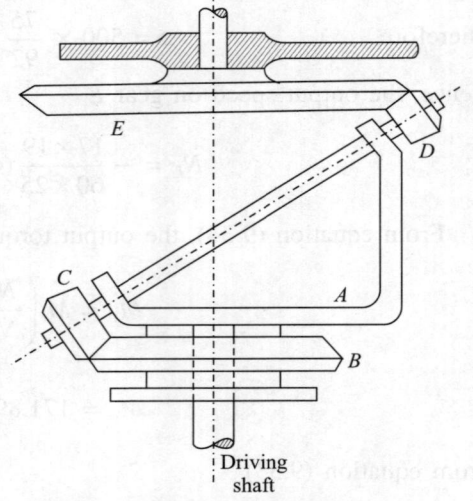

Fig. 9.20 An Epicyclic Bevel Gear Train.

Table 9.10   Tabulation Method.

S.No.	Operations	Arm A	Wheel B; $T_B = 75$	Wheels: C–D T_C = 20; $T_D = 18$	Wheel E; $T_E = 70$
1.	Fix the arm and rotate B through +1 rev.	O	+1	$\dfrac{75}{20} = \dfrac{15}{4}$	$-\dfrac{75}{20} \times \dfrac{18}{70} = -\dfrac{27}{28}$
2.	Multiply by x	O	x	$\dfrac{15}{4}x$	$-\dfrac{27}{28}x$
3.	Add y revolutions to all	$+y$	$x + y$	$\dfrac{15}{4}x + y$	$-\dfrac{27}{28}x + y$

Case (1) $x + y = 0$,   and $y = 1000$ r.p.m. which gives $x = -1000$

Hence, from table, $N_E = -\dfrac{27}{28} \times (-1000) + 1000$

or, $N_E = 1000\left(1 + \dfrac{27}{28}\right) = 1964.3$ r.p.m.   **Ans.**

Case (2)   When $N_B = 400$ r.p.m. & N_A 1000 r.p.m.

$x + y = 400$ and $y = 1000$,

Therefore $x = -600$

Hence, from table, $N_E = -\dfrac{27}{28} \times (-600) + 1000 = +578.57 + 1000 = 1578.6$ r.p.m.   **Ans.**

9.7 BEVEL GEAR DIFFERENTIAL

Bevel gear differentials represent a class of planetary gear trains which are used very widely and requires our special attention. Basically, a bevel gear differential is a mechanism of 2 degrees of freedom, used for adding and subtracting two variables.

An application of epicyclic bevel gear train can be seen in the differential gear used for the rear drive of a motor car. The back axle is made in two separate parts one for each rear wheel. When the vehicle moves along a straight path, both the rear wheels rotates at same speed. While negotiating a curved path, however, the outer wheel must move faster than the inner wheel. This is necessary for avoiding skidding of the wheels. This is achieved through the use of differential gear (*see* Fig. 9.21).

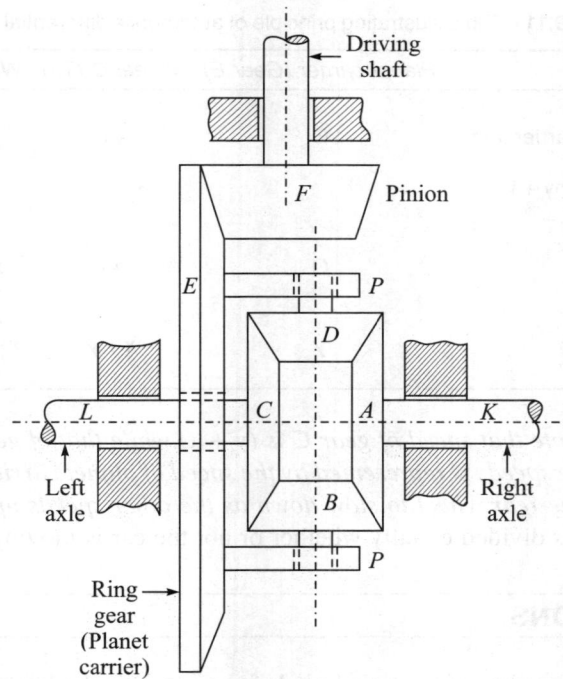

Fig. 9.21 A Bevel Gear Differential of Automobile.

The bevel gear *F* is keyed to the propeller shaft and meshes with the ring gear *E*, which runs freely on the left hand axle. The ring gear *E* carries projections *P–P* for supporting a spindle on which two equal bevel gears *B* and *D* can rotate freely. These bevel gears mesh with two equal bevel gears *A* and *C* which are keyed to the right hand and left hand axles respectively. The bearing brackets *P–P* revolve about the common axis of axles *L* and *K*, as the ring gear *E* rotates. *Thus, the ring gear E which enables planet gears B and D to revolve bodily about the common axis of L and K, can be considered to be a planet carrier.*

When the 4-wheeler moves straight, the ring gear *E* receives motion from driving pinion *F*. As both the rear wheels have to rotate at same speed, all the wheels *A*, *B*, *C* and *D* revolve as a single unit with ring gear *E*. There is thus no relative motion between gear *A*, *C* and planet wheels B, D. In this condition the planet gears play a role similar to the keys, and serve to connect the gears A and C to the ring gear E. Note that in this condition, the planet gears B and D do not rotate about their own axes.

When the automobile takes a turn, the road-wheel on the inside of the curvature makes a fewer rotations than the outer road-wheel moving along a path of larger radius of curvature. In order to avoid sliding of one or both the road-wheels, this difference in speeds of outer and inner road-wheels is to be accommodated in some manner. The differential mechanism, owing to its two degrees of freedom, permits each wheel to rotate at different velocities, providing power simultaneously, to both. The planet gears rotate about their own axes and permit the gears (and hence, the corresponding road wheels) A and C to rotate at different velocities. This becomes evident from Table 9.11 which relates speeds of different gears.

Table 9.11   Table illustrating principle of automobile differential gear.

S.No.	Operations	Planet carrier (Gear E)	Wheel C (T_C)	Wheel D-B	Wheel A (T_A)
1.	Fix the Planet carrier and rotate wheel C by +1 revolution	O	+1	$\left(\dfrac{T_C}{T_D}\right)$	$\dfrac{T_C}{T_D} \times \left(\dfrac{-T_D}{T_C}\right)$ = -1
2.	Multiply by x	O	$+x$	$x\left(\dfrac{T_C}{T_D}\right)$	$-x$
3.	Add y	y	$x + y$	$x\left(\dfrac{T_C}{T_D}\right) + y$	$-x + y$

It is interesting to note that speed of gear C is $(y + x)$ while that of gear A is $(y - x)$ and the arithmatic mean of these speeds is represented by the speed of planet carrier. Thus the differential automatically allows one rear wheel to slow down as the other speeds up. In a usual motor car differential, the torque is divided equally whether or not the car is moving along a straight path.

REVIEW QUESTIONS

9.1 Determine the number of teeth and pitch for two-toothed wheels to transmit a velocity ratio of 4 to 1 between two shafts whose centres are 67.6 cm approximately. The drive must satisfy the following conditions.

   (i) The module in mm must be chosen from amongst the following—24, 22, 20, 18, 16, 15, 14, 13, 12 and 11;

   (ii) The actual distance between shaft centres must not vary by more than one per cent from that given above;

   (iii) The number of teeth must be as small as possible. (Banaras Hindu University)

   (*Ans.*   18 mm; 15 and 60 teeth)

9.2 A four speed sliding gear box of a motor car is required to give speed ratios from the driving shaft driven shaft of 4 : 1, 2.5 : 1, 1.5 : 1, and 1 : 1 approximately in the first, second, third and top gears respectively. The pitch of gears in module is 3.25 mm and centre to centre distance between mating gears is 7 cm. Find suitable number, of teeth on various gears, if the minimum number of teeth on pinion of 14. (M.S. Un. Baroda)

(**Ans.** $T_A = 14$; $T_B = 29$; $T_C = 18$; $T_D = 25$; $T_E = 20$; $T_F = 23$)

9.3 The gearing of a machine tool is shown in Fig. 9.22. The motor shaft is connected to A and rotates at 975 r.p.m. The gear wheels B, C, D and E are fixed to parallel shafts rotating together. The final gear F is fixed on the output shaft G. What is the speed of F? The number of teeth on each wheel are as given below:

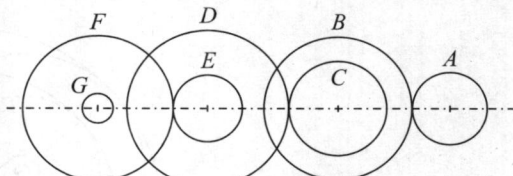

Fig. 9.22 Compound Gear Train.

Gears:	A	B	C	D	E	F
No. of teeth:	20	50	25	75	26	6

(**Ans.** 52 r.p.m.)

9.4 Two shafts A and B in the same line are geared together through an intermediate parallel shaft C. The wheels connecting A and C have a module of 4 mm and those connecting C and B a module of 9 mm, the least number of teeth in any wheel being not less than 15. The speed of B is to be about but not greater than 1/12 the speed of A, and the ratio at each reduction is the same. Find suitable wheels, the actual reduction and the distance of shaft C from A and B. (Hint): for reverted train: $\left(\dfrac{T_A + T_C}{2} \right) 4 = \left(\dfrac{T_C' + T_B}{2} \right) 9$. Further,

for same reduction in each stage $\left(\dfrac{N_A}{N_C} \right) = \left(\dfrac{N_C}{N_B} \right) > \sqrt{12} \approx 3.5$

(**Ans.** $C = 324$ mm)

9.5 In the epicyclic gear shown in Fig. 9.23, the internal wheels A and F and a compound wheel $C - D$ rotate independently about the axis O. The wheels B and E rotate on pins fixed to the arm L. The wheels all have the same module and the number of teeth are: B and E 18, C 28, and D 26. If L makes 150 r.p.m. (c.w.), find the speed of F when: (a) the wheel A is fixed; (b) the wheel A makes 15 r.p.m. c.c.w. (I. mech. E.) (AMIE, Summer 1982)

(**Ans.** (a) $N_F = 6.22$ r.p.m. (c.w.); (b) $N_F = 8.16$ r.p.m. (c.c.w.))

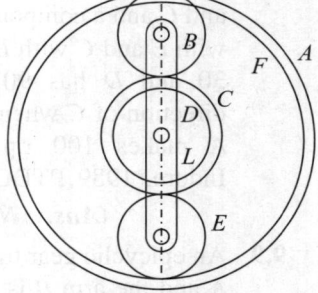

Fig. 9.23 Epicyclic Gear Train.

9.6 A mechanism for recording distance covered by the bicycle is as shown in Fig. 9.24. There is a fixed annular wheel A of 22 teeth and another annular wheel B of 23 teeth which rotates loosely on a common axis O. An arm driven by the bicycle wheel through gearing not described, also revolves freely on the axis through O and carries on a pin at its extremity two wheels C and D which are integral with one another. The wheel C

has 19 teeth and meshes with A and the wheel D with 20 teeth meshes with B. The diameter of the bicycle wheel is 70 cm. What must be the velocity ratio between the bicycle wheel and the arm, if B makes one revolution per 1.6 kilometres covered? (Banaras Hindu University)   (*Ans.*   V.R. = 4.99)

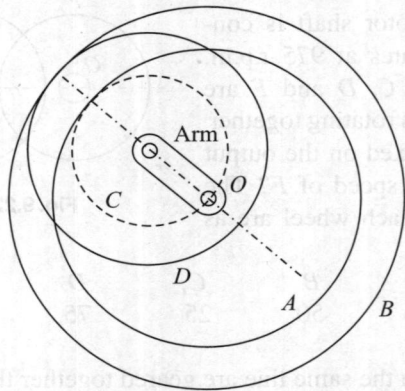

Fig. 9.24   Distance recording Bicycle mechanism.

9.7   An epicyclic gear train, similar to the one shown in Fig. 9.17, consists of three wheels A, B and C. Wheel C has 72 internal teeth, A has 32 external teeth. The wheel B gears with both A and C and is carried on an arm which rotates about the centre of A at 18 r.p.m.

If the wheel C is fixed, determine the speeds of wheels B and A.

(*Ans.* N_A = 58.5 r.p.m. in the sense of rotation of arm; N_B = 46.8 r.p.m. in opposite sense)

9.8   In a reverted train Fig. 9.25 an arm A carries two concentric separate wheels B and C and a compound wheel D–E. B gears with E and C with D. B has 75 teeth, C has 30 and D has 90. Find the speed and direction of C when B is fixed and the arm A makes 100 r.p.m. clockwise. (DAV Indore, 1989 PTDC)

(*Ans.* N_C = 400 r.p.m. c.c.w.)

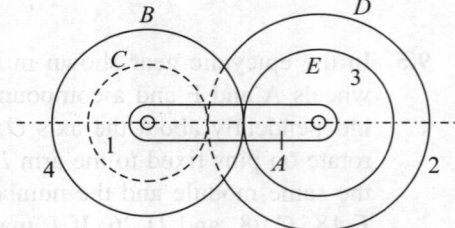

Fig. 9.25   A Reverted Gear Train.

9.9   An epicyclic gear train is shown in Fig. 9.26. The wheel D is held stationary by the shaft A and the arm B is rotated at 200 r.p.m. The wheels E (20 teeth) and F (40 teeth) are fixed together and rotate freely on the pin carried by the arm. The wheel G (30 teeth) is rigidly attached to the shaft C. Find the speed of shaft C, stating direction of rotation relative to that of B.

If the gearing transmits 7.5 kw, what will be the torque required to hold the shaft A stationary, neglecting all frictional losses. (DAV Indore, PTDC-1987)

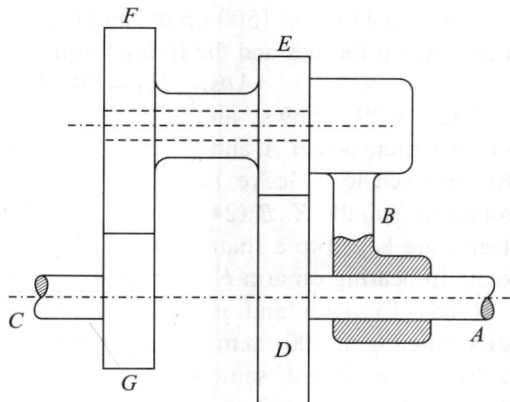

Fig. 9.26 An Epicyclic Gear Train.

(**Ans.** $N_C = 466.7$ opposite to B; $M_b = 511.5$ N·m)

9.10 An aircraft engine drives a propeller through a reduction drive shown in Fig. 9.27. The gears 1, 2, 3 and 4 have 48, 27, 45 and 120 teeth respectively. Find the propeller speed in magnitude and direction if the engine makes 2500 r.p.m.

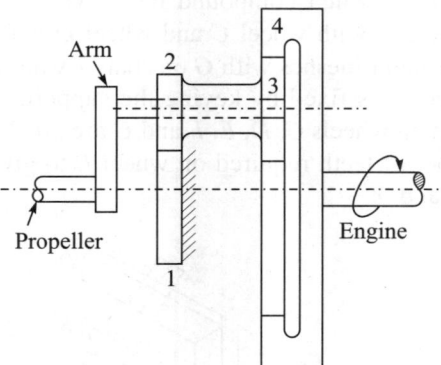

Fig. 9.27 A Reduction drive for propeller.

(**Ans.** 1500 r.p.m., same direction as that of engine shaft)

9.11 In the epicyclic gear unit, shown in Fig. 9.28, the input shaft is attached to the planet carrier Z and the output shaft to the sun wheel A. Sun wheel A_1 is fixed to the casing and does not rotate while the planet wheels C and C_1 are compounded and rotate together. If all the teeth are of same module and A and C each have 20 teeth, find the teeth on A_1 and C_1 so that the output shaft runs in the reverse direction and at half the speed of the input shaft.

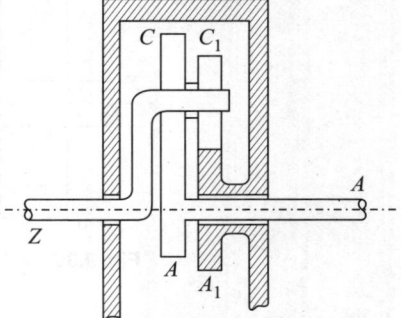

Fig. 9.28 An Epicyclic Gear Train.

If the input shaft transmit 10.45 kw at 1500 r.p.m., and the efficiency of the unit is 96%, calculate the input and output torques and the fixing torques on the casing. (I. Mech. E.)

(**Ans.** $T_{A1} = 24$; $T_{C1} = 16$; $M_b = 194.26$ N·m)

9.12 The diagram of Fig. 9.29 shows an epicyclic gear train in which wheel A and wheel E (30 teeth) are fixed to a sleeve Y which is free to rotate on spindle X. B (24 teeth) and C (22 teeth) are keyed to a shaft which is free to rotate in bearing on arm F. D has 70 teeth. H has 15 teeth and is mounted on a shaft V rotating at 100 r.p.m. Spindle X makes 300 r.p.m. in the same direction as V. All the teeth are of the same module. Find the speed and direction of rotation of Z.

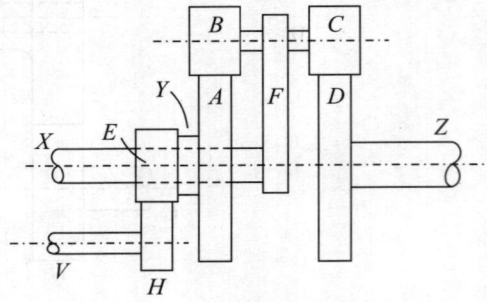

Fig. 9.29   An Epicyclic Gear Train.

(*Note:* H is not a part of the epicyclic gear train, as it turns on a fixed shaft, not in line with the axis of arm. Use algebraic method)

(**Ans.** $N_Z = 11.7$ r.p.m. opposite to V)

9.13 A gear train is shown in Fig. 9.30 in which the shaft X rotates at 500 r.p.m. Solid with shaft X is the arm A on which compound bevel wheels B and D can revolve freely together; wheel B meshes with wheel C and wheel D with E, which is also solid with the spur wheel F; the latter meshes with G on shaft Y which rotates in opposite direction to X. The bevel wheel C is fixed by keying the support.

The numbers of teeth on wheels C, D, E, F and G are 20, 27, 32, 24 and 30 respectively. Determine the number of teeth required on wheel B to give a speed reduction of 20 to 1 between shafts X and Y. (**Ans.** $T_B = 18$)

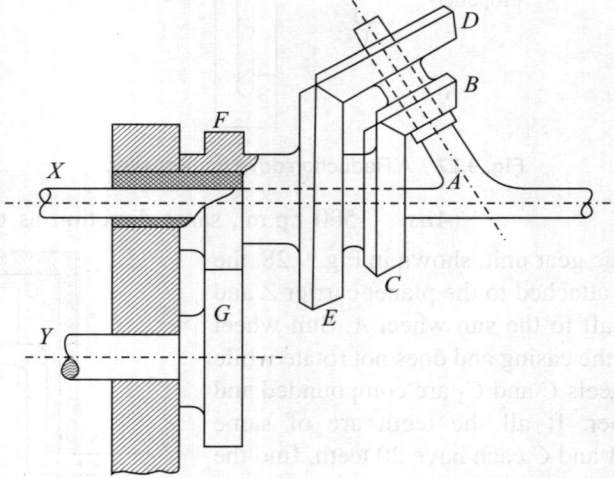

Fig. 9.30   A Gear Train Incorporating Compound Bevel.

9.14 With reference to Fig. 9.31, suppose the input I turns at 120 r.p.m. (c.c.w.), input II turns at 360 r.p.m. (c.w.) determine the speed and direction of rotation of the output shaft.

(*Ans.* N = 50 r.p.m. c.c.w.)

9.15 In the gear drive indicated in Fig. 9.32, A is the driving shaft rotating at 300 r.p.m. in the direction shown and B is the driven shaft. The casing C is held stationary. E and H are keyed to the central vertical spindle; F can rotate freely on this spindle. K and L are rigidly fixed to each other and rotate together freely on a pin fitted to the underside of F. L meshes with internal teeth on the casing C. Take number of teeth on

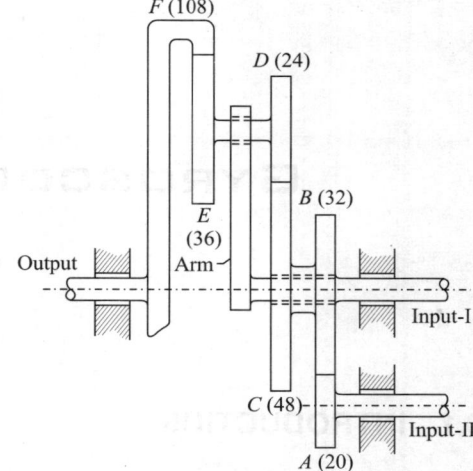

Fig. 9.31

wheels as: D = 40; E = 30; F = 50; G = 80; H = 40; K = 20; L = 30. Find the number of teeth on C and the speed and direction of rotation of B.

(*Ans.* T_C = 90; N_B = 100 r.p.m., opposite direction)

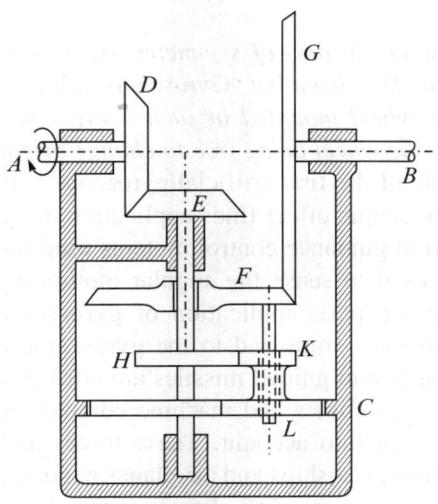

Fig. 9.32

Chapter 10

GYROSCOPIC EFFECTS

10.1 INTRODUCTION

'Gyre' (pronounced as *jir*) is a Greek word meaning 'circular motion'. Thus 'Gyration' means a whirling motion. A very interesting manifestation of inertia is the 'gyroscopic effect' of a rotor whose axis of spin is made to precess, *i.e.*, change its position in space. Historically, Newton's time (1642–1727) marks the period during which earliest studies on gyroscopic phenomena were carried out. Motion of earth, which is a massive gyroscope, was the focal point of such studies. Mathematical foundation of the principles of gyroscopic motion, was however laid down by Euler (1707–1783).

A heavy body rotating about an axis of symmetry, offers a resistance to a change in the orientation of axis of rotation. This is called 'Gyroscopic Effect'. A gyroscope may be defined as a heavy spinning mass or wheel mounted in such a way that only one point, its centre of gravity, is in a fixed position; the wheel being free to change orientation of its axis around C.G.

The most popular example of the first artificial gyroscope is the spinning top which has its origin probably in China. Gyroscopic effect finds application in gyrocompass used on airplanes and ships and also in the inertial guidance control systems used for missiles and space vehicles. In these cases, gyroscope is used to sense the angular motion of a body.

For a very long time in the past, application of gyroscopic principles were limited to gyroscopes for stabilizing seaborne ships and to the gyroscopic compasses. Applications like bomb-sights, control of airplanes and guided missiles are attributed to the developments during world war II. While designing machines and machine components, forces and couples due to gyroscopic effects must be taken into account. These forces and couples are encountered in 2-wheeler and 4-wheeler vehicles, sea-ships and air-planes negotiating curved paths, or in marine turbines as the ship pitches in heavy sea. The characteristic of a spinning body, namely that of retaining the direction of axis of rotation in space, is used in stabilizing, a projectile when a high angular velocity of spin about the longitudinal axis in given to the projectile by riffling the barrel of a gun.

Some times these gyroscopic effects are desirable, but more generally they are undesirable. A designer must account for these forces and couples in selection of bearings and rotating parts. With the present trend of increasing machine speeds and decreasing factor of safety, designers must consider gyroscopic forces and couples in machine design calculations.

10.2 ANGULAR MOTION AND CONVENTIONAL VECTOR REPRESENTATION

In problems involving gyroscopic effects, it is very important to determine correctly the sense of gyro-couples and gyro-reaction couples. This is possible only if one understands the convention used in representing angular motion characteristics like displacement, velocity and acceleration.

Angular velocity ω, like angular acceleration, is a vector quantity and requires following parameters to be specified:

(a) magnitude of angular velocity,

(b) direction of axis of spin (being normal to the plane of spin. This automatically fixes plane of spin), and

(c) the sense of angular velocity ω *i.e.,* whether clockwise or counterclockwise.

A straight line drawn normal to the plane of spin (*i.e.,* parallel to the axis of spin) represents direction while the length of the line represents to some scale the magnitude of angular velocity vector. A conventional way of representing sense of such vector is to use right-handed screw rule. In this convention the arrowhead is placed along the vector in the same direction as a right-handed screw would move, relative to a fixed nut, if rotated in the sense of angular velocity. The working rule of right-handed screw can be stated thus:

Stretch out your right hand keeping forefinger at right angles to the thumb. Keep forefinger along the vector and rotate the thumb in the sense of angular velocity. The direction of arrowhead would then be indicated by the direction in which the tip of the screw moves. This is illustrated in Fig. 10.1 in which a line *OA* rotates in counterclockwise direction at (a) and in clockwise direction at (b). In each of these cases, right-handed screw rule gives direction indicated along normal *n–0–n* in respective figures.

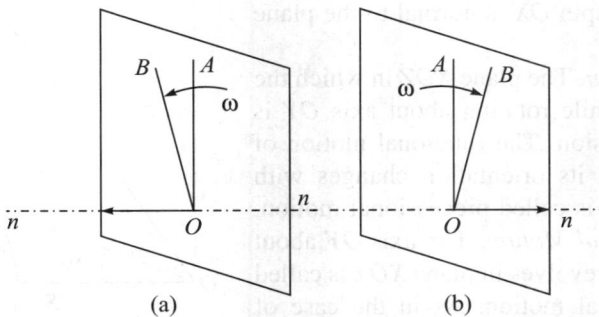

Fig. 10.1 Illustration for vectorial representation of Angular motion.

Although the convention adopted for representing angular acceleration as a vector is same as that for angular velocity, it is important to note that angular acceleration may or may not act in the same sense as that of angular displacement. Hence, the thumb must be rotated in the direction of angular acceleration.

An alterative way of expressing right-hand screw rule, for determining the sense of angular velocity vector, is as follows:

Stretch out the right hand keeping fingers curving on the inside, pointing in the sense of rotation, and the thumb stretched outwards perpendicular to fingers, as shown in Fig. 10.2. The thumb will then point in the sense of arrowhead of the vector. The figure shows a line OA to rotate through $\Delta\theta$ to the position OB in c.c.w. sense as seen from top. The fingers are shown curving and pointing in the direction of rotation as seen from the top. The thumb then indicates direction of arrowhead for angular speed ω. A rotation in reverse direction from OB to OA is in clockwise sense and requires right hand in Fig. 10.2 to be placed in inverted position with thumb pointing downwards.

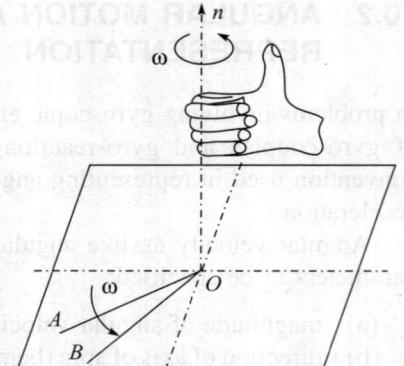

Fig. 10.2 Illustration for Right Handed Screw Rule.

10.3 PRECESSIONAL MOTION AND ANGULAR ACCELERATION

Consider a disc of mass moment of inertia I to be spinning with angular velocity ω about the axis OX in clockwise direction as seen in the direction from O to X. The plane of the disc is normal to the X-axis. As shown in Fig. 10.3, let the axis of disc rotate about Y-axis in the plane XOZ so as to occupy a new position OX' after a short interval of time Δt. Let the angle XOX' be equal to $\Delta\theta$.

Before proceeding for discussion on angular acceleration, let us define a few relevant terms.

Plane of Spin. Referring to Fig. 10.3, the plane like $PQRS$, in which the disc spins with angular velocity ω, is called the plane of spin.

Axis of Spin. The line OX about which the disc rotates in its plane, is called axis of spin. It must be noted that the axis of spin OX is normal to the plane of spin.

Plane of Precession. The plane XOZ in which the axis of spin moves while rotating about axis OY is called plane of precession. The rotational motion of axis of spin in which its orientation changes with respect to the axis OX is called precessional motion.

Axis of Precessional Motion. The axis OY about which the axis of spin revolves in plane XOZ is called the axis of precessional motion. As in the case of spinning motion, it is easy to see that axis of precessional motion is normal to the plane of precession.

Gyroscopic Effect. A body rotating about an axis of symmetry offers resistance to change in the direction of axis of spin. This tendency to oppose precessional motion is called *gyroscopic effect*.

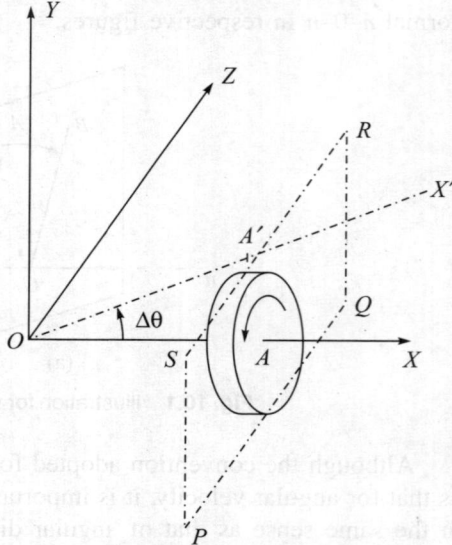

Fig. 10.3 Illustration for Precessional motion.

Angular momentum. As is known, linear momentum of a body of mass m moving with a velocity V is given by (mV), which is a vector having direction and sense same as that of velocity V. Angular momentum of a particle is then defined as the moment of linear momentum. Thus, a mass m rotating about a centre O with angular velocity ω at a radius of R will have an angular momentum of (mVR), where $V = R\omega$.

Referring to Fig. 10.4, a disc of mass m and rotating with angular velocity ω behaves as if the entire mass of disc were concentrated in a thin ring of radius k (the radius of gyration) and of small radial thickness dr. The tangential velocity at this radius is then given by

$$V = k\omega. \qquad (10.1)$$

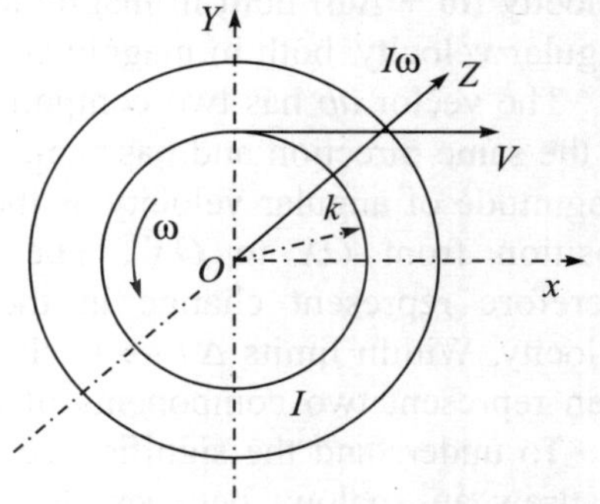

and the angular momentum is given by $= mVk$.

Thus, substituting for V from equation (10.1), the angular momentum is given by

Fig. 10.4 Illustration for Angular momentum.

$$= (m\,k^2)\omega$$

And as mass moment of inertia is given by $I = mk^2$, we have

$$\text{Angular momentum} = I\omega \qquad (10.2)$$

Since angular velocity ω is a vector and I is a scalar quantity, the angular momentum is also a vector quantity and is represented by a vector along the axis of spin. The sense of this vector is again decided using right-hand screw rule. Thus, with fingers curving and pointing in the direction of ω (c.c.w.), the angular momentum vector $I\omega$ will act in the direction of thumb, as shown in Fig. 10.2.

Accelerations. With a reference to Fig. 10.5(a) the disc is revolving with angular speed ω (c.w. as seen from O to X) about axis OX. Let the vector Oa represent the angular velocity ω in Fig. 10.5(b). A short time interval Δt later, assume that the disc revolves with angular velocity $(\omega + \Delta\omega)$ in the same sense (c.w.) about OX'. Thus the axis of spin has precessed **through an angle $\Delta\theta = \angle XOX'$**.

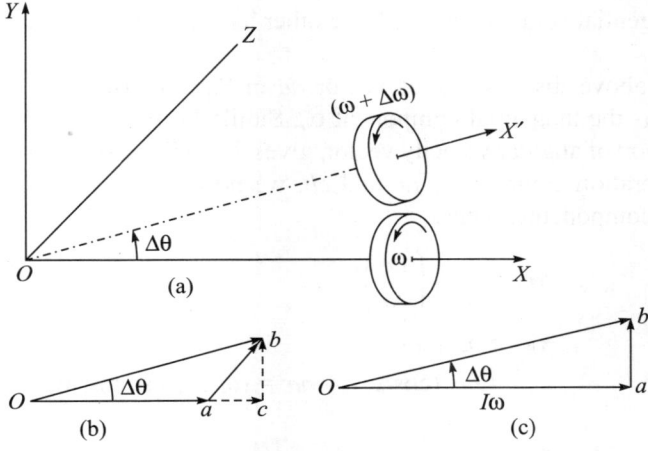

Fig. 10.5 A general case of acceleration of rotating disc involving precessional motion.

The sense and direction of angular velocity vectors, as obtained by applying right-hand screw rule to the disc in the two positions, are shown in Fig. 10.5(b). In the figure, Oab is the vector triangle whose side oa represents angular velocity ω while the side ob represents angular velocity $(\omega + \Delta\omega)$ both in magnitude and direction. Vector ab then represents total change in angular velocity, both in magnitude and direction.

The vector ab has two components in mutually perpendicular directions. Component ac is in the same direction and has same sense as vector oa and therefore, represents increase in the magnitude of angular velocity as the rotor axis changes its position from OX to OX'. The component vector cb therefore represent change in the direction of angular velocity. Within limits $\Delta t \to 0$, the ratios $ac/\Delta t$ and $cb/\Delta t$ then represent two components of acceleration.

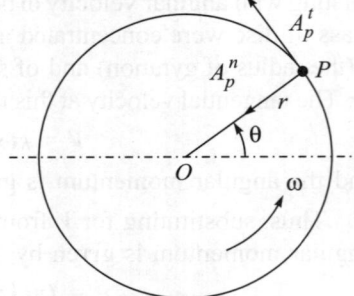

To understand the significance of each component, let us draw an analogy between the precessional motion of rotating disc as above, and the planar motion of rotation of a mass particle P in an arc of a circle (Fig. 10.6). The tangential and radial (normal) components of acceleration of the mass particle P are given by

Fig. 10.6 Accelerations due to motion of a particle along arc of circle.

$$A_p^t = \frac{dV}{dt}$$

$$= \frac{d}{dt}(r\omega) = r\frac{d}{dt}(\omega) = r\alpha \tag{10.3}$$

and

$$A_p^n = \frac{V^2}{r} = V\left(\frac{V}{r}\right) = V\omega$$

$$= V\left(\frac{d\theta}{dt}\right) \tag{10.4}$$

Thus, normal or radial (centripetal in this case) component of acceleration comes into picture because of change in θ w.r. to time, *i.e.,* the component exists because of change in the direction of velocity. The tangential component A_p^t, on the other hand, depends on change in the magnitude of angular speed ω.

In the light of above discussions, as vector ac in Fig. 10.5(b) corresponds to increase in magnitude of ω, it is the tangential component α_t. Similarly the vector cb, as it corresponds to change in the direction of angular velocity vector, gives the radial component α_r. After identifying the nature of acceleration components ac and cb, we now proceed to establish magnitudes of these acceleration components. Thus,

$$\alpha_t = \lim_{\Delta t \to 0}\left(\frac{ac}{\Delta t}\right) \tag{10.5}$$

But

$$ac = oc - oa$$

$$= ob\,(\cos\Delta\theta) - oa = (\omega + \Delta\omega)\cos\Delta\theta - \omega$$

For small values of angle $\Delta\theta$, $\cos\Delta\theta \approx 1$, and therefore, $ac = (\Delta\omega)$ and therefore, substituting for ac in equation (10.5),

$$\alpha_t = \lim_{\Delta t \to 0}\left(\frac{\Delta\omega}{\Delta t}\right) = \left(\frac{d\omega}{dt}\right) \tag{10.6}$$

Also, the second acceleration component,

$$\alpha_r = \lim_{\Delta t \to 0}\left(\frac{cb}{\Delta t}\right) \tag{10.7}$$

But $\qquad\qquad\qquad cb = ob\sin\Delta\theta$

or $\qquad\qquad\qquad cb = (\omega + \Delta\omega)\sin\Delta\theta$

For small angle $\Delta\theta$,

$$\sin\Delta\theta \approx \Delta\theta$$

Therefore $\qquad\qquad cb = (\omega + \Delta\omega)\Delta\theta$

Neglecting product of two small quantities in comparison to others,

$$cb = \omega\Delta\theta$$

Substituting for cb in equation (10.7),

$$\alpha_r = \lim_{\Delta t \to 0}\left(\frac{\omega\Delta\theta}{\Delta t}\right)$$

$$= \omega\frac{d\theta}{dt}$$

or $\qquad\qquad\qquad \alpha_r = \omega\omega_p \tag{10.8}$

where $\qquad\qquad \omega_p \equiv \dfrac{d\theta}{dt}$ is the angular velocity of precession motion.

Thus, the total angular acceleration of spinning disc subjected to precessional motion is given by vector equation

$$\alpha = \alpha_t + \alpha_r \equiv \left(\frac{d\omega}{dt}\right) + \omega\omega_p \tag{10.9}$$

Particular cases
It follows from equation (10.9) that—

(a) When ω is uniform (*i.e.,* when angular acceleration is absent), $\dfrac{d\omega}{dt} = 0$ and then,

$$\alpha = \omega\omega_P \tag{10.10}$$

This is the case in which we are mainly interested in this chapter. It will be shown in section to follow that $\alpha_r (= \omega\omega_p)$ is the gyroscopic acceleration.

(b) When ω is not uniform but $\omega_p = 0$, there is no precessional motion and θ is constant. Total acceleration in this case is given by

$$\alpha = (d\omega/dt) \tag{10.11}$$

which is commonly known as angular acceleration of rotating disc. It follows from above discussion that for gyroscopic acceleration to exist, there must be a precessional motion of the axis of spin.

10.4 GYROSCOPIC COUPLE

Just as a force is needed to produce linear acceleration, a couple or torque is necessary for producing angular accelerations in a rotating disc. Clearly, a couple, like gyroscopic couple must be applied so as to produce precessional motion and gyroscopic acceleration. We now proceed to establish plane of gyro-couple and its magnitude.

Referring to Figs. 10.5(a) and (b), instead of considering ω and $(\omega + \Delta\omega)$ respectively along OX and OX', consider oa and ob to represent angular momentum vectors in the original and precessed conditions respectively. Note that mass moment of inertia 'I' being scalar, the vector diagram at Fig. 10.5(b) now gives vector diagram for angular momentum to a different scale.

If angular velocity is assumed to remain constant, the vector ac in Fig. 10.5(b) becomes zero and line ab, tends to become perpendicular to vector oa at a. The line ab in Fig. 10.5(c) then represents change in angular momentum, due entirely to the precessional motion. For the present case, as ω is uniform, vectors oa and ob represent angular momentum $(I\omega)$ in different orientations of axis of spin. Note that for small angle $\Delta\theta$, even if $oa = ob$, the line ab can be assumed to be approximately perpendicular to oa. Thus, the change in angular momentum (Fig. 10.5c),

$$ab = \Delta(I\omega)$$

And so, the rate of change of angular momentum produced by gyroscopic couple C is given by

$$C = \lim_{\Delta t \to 0}\left(\frac{ab}{\Delta t}\right) \tag{10.12}$$

But, $ab \approx (oa)\,\Delta\theta$ for small angle $\Delta\theta$.
Hence, from equation (10.12),

$$C = (oa)\frac{d\theta}{dt}$$

Hence, $C = (oa)\omega_p$

Again $oa = I\omega$

Hence $C = I\omega\omega_p \tag{10.13}$

Equation (10.13) gives gyroscopic couple required to produce precessional motion in a disc of mass moment of inertia I, rotating at an angular speed of ω. The sense of gyro-couple is same as that of gyroscopic acceleration $\omega\omega_p$ (note that I is a scalar). The vector ab of Fig. 10.5(c), gives direction of gyroscopic acceleration. Applying right-handed screw rule to vector ab, the plane of rotation of the thumb is normal to ab. Hence, gyro-couple in Fig. 10.5(c) acts in a plane normal to the line ab, the sense of the couple being clockwise, as seen from a to b. Evidently, the line ab represents the axis of gyro-couple.

Note that the axis of gyro-couple *ab* in Fig. 10.5(c) is perpendicular to *OX*, the axis of spin. Further the axis of precession *OY* is normal to the plane *XOZ* containing axes of spin and gyro-couple, we conclude that the three axes are mutually perpendicular. Again, the plane *XOZ* is the plane of precessional motion, plane *YOZ* is the plane of spin (being parallel to the plane of disc) and plane *XOY* is the plane of gyro-couple. It follows that all the three planes are perpendicular to one another. This leads to the following rule:

Rule 1:   *Axis of spin, axis of gyro-couple and axis of precessional motion are mutually at right angles to each other. Similarly, the plane of spin, plane of precessional motion and the plane of gyro-couple are normal to each other.*

10.5   GYRO-COUPLE AND GYRO-REACTION COUPLE

In most of the problems on stability analysis, due to the presence of centripetal acceleration and gyroscopic acceleration, the problem ceases to be one of static equilibrium. *Unless corresponding inertia force and inertia couple (i.e., centrifugal force and gyro-reaction couple) are introduced, the problem in dynamics cannot be converted into an equivalent problem of static equilbrium. It is for this reason that equilibrium analysis of two and four-wheeler vehicles considers static equilibrium under the action of overturning moment due to centrifugal force and gyro-reaction couple.* This is perfectly in accordance with *D' Alembert's* principle (see Section 14.2).

In all the cases of stability analysis, it is very important that the difference between gyro-couple and gyro-reaction couple, in respect of their direction and effect, be clearly understood. Gyro-couple, which is also known as active couple, must be applied to a rotating disc for obtaining desirable precessional motion of the axis of the spin. This couple is usually applied through the bearings (which support the shaft) to the shaft. The shaft, in turn, exerts an equal and opposite (reaction) couple on the bearings. This is perfectly in accordance with Newton's third law of motion, which states that action and reaction are equal and opposite. Thus, the precessional motion of the axis of spin causes a gyroscopic reaction couple to act on the frame to which bearings are fixed. Being equal and opposite to gyro-couple, magnitude of gyro-reaction couple is also given by

$$C = I\omega\omega_p.$$

The plane of this reaction couple is same as that of gyro-couple, only the sense is opposite.

10.6   ANALOGY WITH MOTION OF A PARTICLE IN CIRCULAR PATH

When a particle of mass *m* moves along a circular path of radius *r* with constant angular velocity ω, the tangential velocity ($r\omega$) remains constant in magnitude but changes direction continuously. In order to enable the particle to change its direction of motion continuously, a centripetal force must be applied radially either in the form of a tension in a string or by some other means. The centripetal force is given by

$$F_c = mass \times centripetal\ acceleration$$

Thus,
$$F_c = m(V^2/r)$$

or
$$F_c = mV\left(\frac{V}{r}\right)$$

or
$$F_c = (mV)\omega$$

Hence,
$$F_c = \text{(linear momentum)} \times \omega \qquad (10.14)$$

where ω is the rate at which the radius is changing its position continuously.

As against this, centrifugal force is a force of reaction and acts radially outwards. This is an inertia force which resists a change in the orientation of radius vector and hence direction of tangential velocity both. In short, centripetal force is an active force and must be applied externally whereas the centrifugal force is a reactive force and represents an inertia force.

Similarly, the gyroscopic couple, which is given by
$$C = \text{(angular momentum)} \times \omega_p \qquad (10.15)$$

is an active couple and must be applied externally to ensure precessional motion of the axis of spin. Like centripetal force, it causes angular momentum vector to change its direction continuously.

Gyro-reaction couple, like centrifugal force, is a reaction couple and by way of inertia effect offers resistance to a change in the direction of angular momentum vector. Thus in Fig. 10.5. if the sense *ab* represents gyro-couple, the same vector in opposite sense, *i.e. b* to *a* will represent gyro-reaction couple. We formulate second rule as under:

Rule 2: *Like centripetal force, a gyro-couple is an active couple and must be applied externally to produce precessional motion of the axis of spin. Gyro-reaction couple, on the other hand, is comparable to centrifugal force. Being an inertia couple, it opposes any change in the orientation of axis of spin.*

As we proceed for stability analysis of a body (either moving or rotating) students are advised to remember the following two points.

1. Consider effect due to reactive forces and reactive couples only, and
2. *Units:* The mass moment of inertia I in S.I. units should be kg·m². With above units the couple C would have a unit of N·m.

EXAMPLE 10.1 A uniform disc of 150 mm diameter has a mass of 5 kg. It is mounted centrally in bearings which maintain its axle in a horizontal plane. The disc spins about its axle with a constant speed of 1000 r.p.m. while the axle precesses uniformly about the vertical axis at a speed of 60 r.p.m., c.w. as seen from top. Figure 10.7 shows the direction of rotation of the disc. If the distance between the bearings is 100 mm, find the resultant reaction at each bearing due to the weight of mass and gyroscopic effect.

Compare reactions as obtained above, with the ones obtained when the c.g. of the disc is away from axis of rotation by a distance of 2 mm, neglecting gyroscopic effect due to precessional motion.

Solution: Considered from the point of view of gravitational effect alone, the weight (5×9.81) N of the disc is shared equally at the bearings. The vertical reactions R_1 to share this load are:

$$R_1 = \frac{5 \times 9.81}{2} = 24.525 \text{ N (upwards)}$$

Figure at 10.7(b) shows angular momentum vector in position *oa* and in precessed position *ob* in horizontal plane. When the precessional motion occurs in c.w. sense as seen from top, the new position of angular momentum vector is *ob*. The vector *ab* therefore gives sense of gyroscopic acceleration. Consistent with above sense, right-handed screw rule gives sense of gyro-couple in c.c.w. direction as shown at Fig. 10.7(a). This couple acts in longitudinal vertical plane of shaft.

The gyro-couple necessary to induce precessional motion is applied through the bearing to the axle. The axle in turn, exerts gyro-reaction couple on bearings. The reactions in the two bearings are induced in such a way as to constitute reaction couple. Thus, reaction R on left-hand support should act up and reaction R on right-hand support should act down to constitute reaction couple. Hence

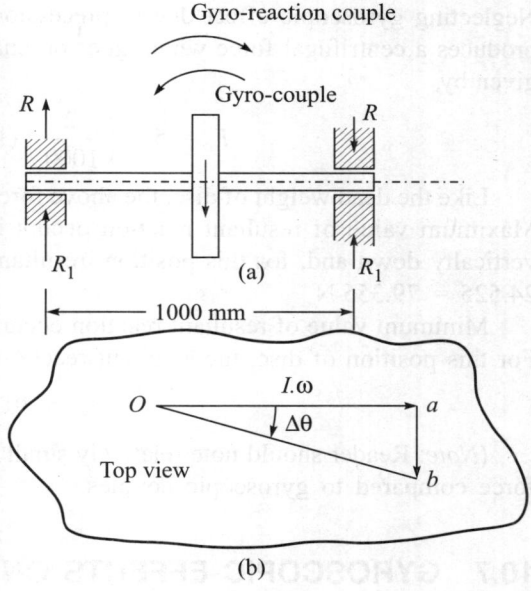

Fig. 10.7

$$R \times \left(\frac{100}{1000}\right) = I\omega\omega_p$$

i.e.
$$R = 10\,(I\omega\omega_p)$$

Assuming radius of gyration to be equal to the radius of disc, the mass moments of inertia of disc,

$$I = mk^2 = 5\left(\frac{75}{1000}\right)^2$$

Thus
$$I = (0.028125)\ \text{kg·m}^2$$

$$\omega = \frac{2\pi \times 1000}{60} = 104.719\ \text{rad/s; and}\ \omega_p = \frac{2\pi \times 60}{60} = 6.283$$

Therefore $C = I\omega\omega_p = 18.505\ \text{N·m}$

Therefore $R = 10\,(C) = 185.05\ \text{N}$

This reaction is upwards on left-hand bearing and downwards on right-hand bearings, giving net reactions as,

$$R_L = R + R_1 = 185.05 + 24.525$$

Therefore $R_L = 209.575\ \text{N (upwards)}$ **Ans.**

Thus $R_R = R - R_1 = 185.05 - 24.525$

or $R_R = 160.525\ \text{N (downwards)}$ **Ans.**

Neglecting gyroscopic effect due to precessional motion, an eccentricity of 2 mm in the disc produces a centrifugal force which goes on changing direction continuously. The magnitude is given by,

$$F_c = 5 \times \left(\frac{2}{1000}\right)(104.719)^2 = 109.66 \text{ N}$$

Like the dead weight of disc, the above force will also be shared equally by the two bearings. Maximum value of resultant reaction occurs in each bearing when the centrifugal force acts vertically down and, for this position, resultant reaction in each bearing = $R' + R_1$ = 54.83 + 24.525 = 79.355 N

Minimum value of resultant reaction occurs when centrifugal force acts vertically upwards. For this position of disc, the resultant reaction in each bearing = 54.83 – 24.525

$$= 30.305 \text{ N.} \qquad \textbf{Ans.}$$

(*Note*: Reader should note relatively small contribution to bearing reaction from centrifugal force compared to gyroscopic couples.

10.7 GYROSCOPIC EFFECTS ON AN AEROPLANE

Figure 10.8 shows elevation and top view of an aeroplane. Let,

m = mass of the engine and propeller in kg
ω = angular velocity of the engine in rad/s
k = radius of gyration in m
I = mass moment of inertia of the rotating parts of the engine and propeller in kg·m^2.
 = (mk^2)

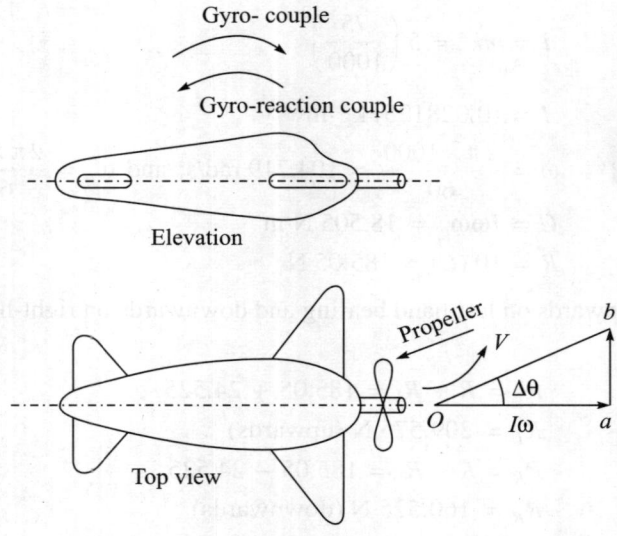

Fig. 10.8 Elevation and top view of an aeroplane.

V = linear velocity of the aeroplane in m/s

R = radius of curvature of the curved path being negotiated (m)

$$\omega_p = \text{angular velocity of precession} \left(= \frac{V}{R} \right) \text{rad/s}$$

Let the engine and propeller rotate in clockwise sense as seen from rear side, and let us assume that the plane is taking a turn to left. As seen in the top view, $oa \equiv I\omega$ represents angular momentum vector when the plane moves straight and ob represents new angular momentum vector when the plane has turned through a small angle $\Delta\theta$ to the left. The vector ab drawn perpendicular to oa now represents sense of gyroscopic couple.

The plane of gyro-couple is the longitudinal vertical plane, shown in, the elevational view. The gyro-couple is clockwise while gyro-reaction couple is c.c.w. Their magnitude is given by

$$C = I\omega\omega_p.$$

Thus,
$$C = (mk^2)\,\omega\left(\frac{V}{R}\right) \tag{10.16}$$

As can be seen in elevational view, the gyro-reaction couple has a tendency to lift the nose and depress the tail. A little consideration will show that when the plane takes a turn to the right, reverse will occur *i.e.*, reaction couple will tend to lift the tail and depress the nose.

EXAMPLE 10.2 An aircraft makes a half circle of 100 m radius towards left when flying at 400 km/hr. The engine and the propeller of the plane weighs 4.9 kN having a radius of gyration of 50 cm. The engine rotates at 3000 r.p.m. clockwise when viewed from the rear. Find the gyroscopic couple and its effect on the aircraft. (AMIE, Summer 1985)

Solution
$$\omega_p = \frac{V}{R} = \frac{400 \times 1000}{3600 \times 100} = 1.11 \text{ rad/s}$$

$$\omega = \frac{2\pi \times 3000}{60} = 314.16 \text{ rad/s}$$

$$I = \frac{4.9 \times 1000}{9.81} \times (0.50)^2 = 124.87 = \text{kg·m}^2$$

Therefore   gyroscopic couple $C = I\,\omega\,\omega_p$

or $C = (124.87)\,(314.16)\,(1.11)$

or $C = 43584.5 \text{ N} - \text{m} = 43.584 \text{ kN·m}$ **Ans.**

When the plane takes a turn to the left, the gyroscopic acceleration and hence the gyro-couple will be given by vector ab in Fig. 10.8. The gyro-reaction couple, which acts c.c.w. in the longitudinal vertical plane, tends to lift the nose and depress the tail.   **Ans.**

10.8 STABILITY ANALYSIS OF 4-WHEELER VEHICLE

A 4-wheeler vehicle negotiating a curve of radius R with a uniform speed V is associated with centripetal acceleration and gyroscopic acceleration is a problem of dynamic equilibrium. This may be converted into an equivalent problem of static equilibrium, (using D'Alembert's principle) by introducing centrifugal force (inertia force) and gyro-reaction couple (inertia couple) in force analysis in addition to reaction due to self weight of vehicle.

Stability considerations require that no wheel of a 4-wheeler is lifted off the ground, which is generally feared when negotiating a curve. This condition of stability is satisfied when ground reaction on any of the four wheels becomes zero or negative. Note that as load is considered positive, when acting downwards corresponding reaction, acting upwards is also considered positive. Conversely, reactions acting downwards, are considered negative.

Because of the placement of engine assembly, the centre of gravity of a self-propelled vehicle does not coincide with the geometrical centre of the four wheels. However, for the sake of simplicity, let us assume that the c.g. lies on a vertical line through the geometric centre G' as shown in Fig. 10.9.

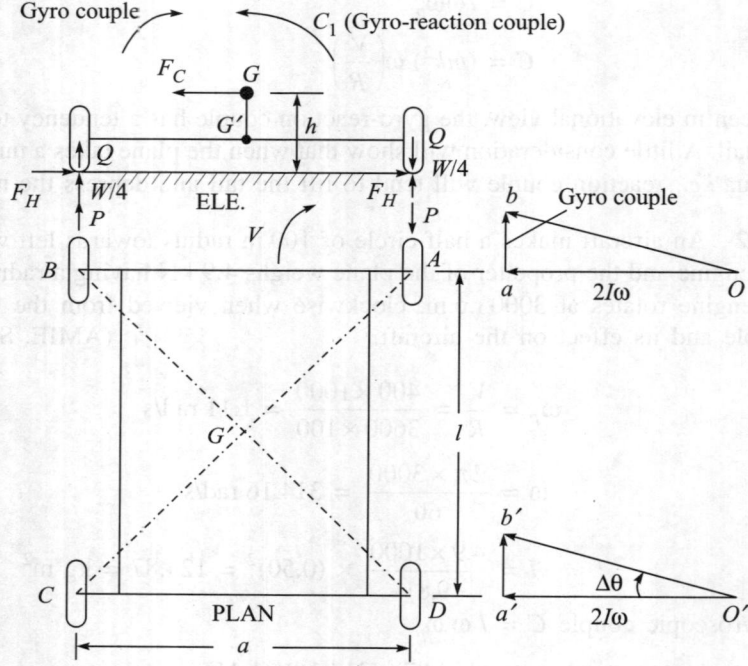

Fig. 10.9 Stability analysis of 4-wheeler vehicle.

Let W = weight of vehicle, assumed to be equally shared by all wheels
        h = height of c.g. above ground level
        V = linear velocity of vehicle in m/s
        r = radius of road wheels in m
        ω = angular speed of wheels ($= V/r$) in rad/s.
        I = mass moment of inertia of a wheel about its own axis
        R = radius of curvature of path, which is assumed to be large in comparison to
              dimensions of the vehicle

ω_p = angular velocity of precessional motion = (V/R) rad/s.

$CD = a$ = wheel track in m

$AD = l$ = wheel base in m

Figure 10.9 shows a simplified elevational and plan view of the chasis, showing only the relevant details. Let the vehicle negotiate a turn to the right as shown in Fig. 10.9.

Neglecting effects of rotating parts of engine and transmission system, there are following types of forces and couples to which the vehicle is subjected:

(A) Reaction forces of the ground on four wheels.

Assuming uniform distribution of dead weight on four wheels, reaction due to gravitational force,

$$R_1 = W/4 \text{ (vertically upwards) on each wheel}$$

(B) Assuming the entire mass to be concentrated at G, the centre of gravity, the motion in curvilinear path produces a centrifugal force F_c (which is an inertia force) at G, given by

$$F_c = \frac{W}{g}\left(\frac{V^2}{R}\right)$$

To restore equilibrium of forces in horizontal direction (i.e., $\Sigma F_H = 0$), ground develops centripetal force through horizontal ground reaction, equal to $F_c/4$ on each wheel, acting towards centre of curvature. The resultant of horizontal reactions F_H, acting on each wheel, can be assumed to act at ground level in transverse vertical plane through the centre of gravity. Thus, $4 F_H = F_c$. But the resultant reaction $4 F_H$ and centrifugal force F_c, being equal and opposite forces (which are not colinear), constitute an overturning couple C_1 which is counterclockwise, as seen from rear side in transverse vertical plane and is shown in elevational view of Fig. 10.9. Thus

$$C_1 = F_c \times h$$

$$= \frac{W}{g}\left(\frac{V^2}{R}\right) h \tag{10.17}$$

To prevent overturning in outward direction due to F_c, ground develops reactions P on each wheel. These reactions act upwards on two outer wheels and downwards on two inner wheels so that a balancing couple is introduced such that,

$$2 (P \times a) = C_1 = \frac{W}{g}\left(\frac{V^2}{R}\right) h$$

or

$$P = \frac{W}{2g}\left(\frac{V^2}{R}\right)\left(\frac{h}{a}\right) \tag{10.18}$$

The centripetal force $\left(\dfrac{W}{g}\dfrac{V^2}{R}\right)$, necessary to keep the car moving in its circular track is supplied by the radial force of friction from ground which appears in the form of reaction F_H at the wheels.

(C) As the 4-wheeler vehicle negotiates the curve, the angular momentum vector $I\omega$ moves through angle $\Delta\theta$ from position oa to ob and from $o'a'$ to $o'b'$ for the front and rear wheel pairs respectively. For the pair of wheels on front side, the gyro-couple C' acts in sense ab and is given in magnitude by

$$C' = (2I\omega)\omega_p$$

For the wheel pairs on rear side, the gyro-couple acts in the sense $a'b'$ and is given in magnitude by

$$C'' = (2I\omega)\omega_p$$

The gyro-couples C' and C'' act in the same sense in parallel 'transverse-vertical' planes, and are therefore additive. The resultant gyro-couple C can be assumed to act in transverse vertical plane through c.g. as shown in elevational view. Thus,

$$C = C' + C'' = 4(I\omega)\omega_p \tag{10.19}$$

The Fig. 10.9 shows the Gyro-reaction couple acting in transverse vertical plane in c.c.w. sense, as seen from rear side. This is an overturning couple. To balance this couple ground reaction Q should be upwards on outer wheels and downwards on inner wheels. Thus,

$$2(Q \times a) = 4 I\omega\omega_p$$

or

$$Q = 2I\left(\frac{V}{r}\right)\left(\frac{V}{R}\right)\frac{1}{a}$$

or

$$Q = 2I\left(\frac{V^2}{rRa}\right) \tag{10.20}$$

Pressure on Wheels. Let P_0 be the pressure on each of the outer and inner wheels. Then

$$P_0 = \text{Pressure on each of the outer wheel}$$

$$= \left(\frac{W}{4} + P + Q\right) \tag{10.21}$$

and

$$P_i = \text{Pressure on each of the inner wheel}$$

$$= \left(\frac{W}{4} - P - Q\right) \tag{10.22}$$

It follows from Eqs. (10.18) and (10.20) that the reactions P and Q are functions of velocity square. Thus, for certain critical velocity $V = V_c$, the reaction on inner wheels $P_i = 0$. At this stage the inner wheels are just at the point of lifting from the ground. A slight increase in velocity ($V > V_c$) will cause the inner wheels to lift from the ground and the stability of the vehicle will be at stake.

Particular cases

(1) There may be rotating masses, other than the wheels, which rotate in planes parallel to those of the wheels. If these masses rotate in same sense as that of wheels, their gyroscopic effect add up, increasing magnitude of corresponding reaction Q. But if the masses rotate in a sense opposite to that of wheels, the net gyroscopic couple and reaction Q is decreased.

In 4-wheelers, therefore, effect of precession, and hence that of gyro reaction couple, may be decreased or even eliminated by arranging flywheel to rotate in a sense opposite to that of the driving wheels.

(2) In yet another category of problems, masses may rotate in transverse vertical planes [see Fig. 10.10(a)]. When the vehicle negotiates a turn, the precessional motion of such rotating masses experience gyroscopic reaction couple in longitudinal vertical plane. In order to balance this reaction couple in c.c.w. same, as seen from L.H.S, ground will have to develop reactions L upwards on front wheels and downwards on rear wheels.

EXAMPLE 10.3   The rotor of a motor used for electric traction weighs 4900 N and has a radius of gyration of 20 cm. The centre of mass of the rotor is midway between the bearings. The speed of the motor and train are 1500 r.p.m. and 75 km/hr respectively. The motor shaft is parallel to the axle of the track wheels which rotate in two bearings 80 cm apart. Find the force on each bearing when the train turns in a curve of 150 m radius. (DAV Indore, 1981; 1987)

Solution: The M.I. of rotor $= \dfrac{4900}{9.81} \times (0.2)^2 = 19.98 \text{ kg·m}^2$

$$\text{Angular speed of motor-rotor} = \frac{2\pi \times 1500}{60} = 157.079 \text{ rad/s}$$

$$\text{Angular speed of precession } \omega_p = \frac{(75 \times 1000)}{(3600 \times 150)} = (5/36) \text{ rad/s}$$

$$\text{Hence, gyroscopic couple} = I\omega\omega_p$$
$$= 19.98 \,(157.079)\,(5/36) = 435.9 \text{ N·m}$$

The gyro-reaction couple has an equal magnitude and as bearings are 80 cm apart, force in each bearing is given by

$$F \times 80 = C = 435.9$$

Therefore $F = \dfrac{435.9 \times 100}{80} = 544.9 \text{ N}$ **Ans.**

EXAMPLE 10.4   A rear engine automobile is travelling along a track of 100 m mean radius. Each of the four road-wheels has a moment of inertia of 2 kg.m² and an effective diameter of 60 cm. The rotating parts of the engine have a moment of inertia of 1 kg.m². The engine axis is parallel to the rear axle and the crank shaft rotates in the same sense as the road wheels. The gear ratio, engine to back axle, is 3 : 1. The vehicle weighs 14.71 kN and has its centre of gravity 50 cm above the road level. The width of the track of the vehicle is 1.5 m. Determine the limiting speed of the vehicle around the curve for all the four wheels to maintain contact with the road surface if this is not cambered. (DAV Indore, 1988; 1989)

Solution: M.I. of each road wheel

Therefore $I_\omega = 2 \text{ kg.m}^2$

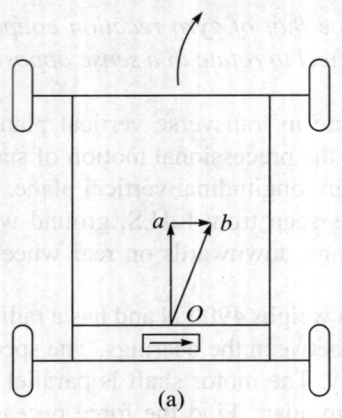

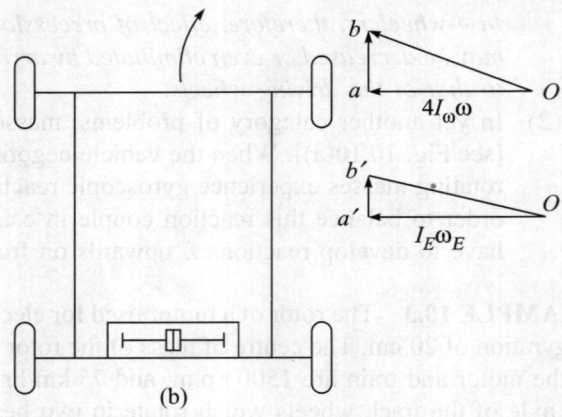

Fig. 10.10

M.I. of rotating parts of engine $I_E = 1$ kg.m^2

$$R = 100 \text{ m};$$

$$r = \frac{0.6}{2} = 0.3 \text{ m}$$

Let V be the limiting speed in m/s ensuring all the wheels to maintain contact with ground.

Then, corresponding wheel speed = $(V/0.3)$ rad/s and the giving angular speed of precessional motion,

$$\omega_p = V/R = (V/100) \text{ rad/s}$$

Hence, gyro-couple due to wheels at limiting condition,

$$C' = 4 (I_\omega \omega) \omega_p$$

or

$$C' = 4 (2) \left(\frac{V}{0.3}\right)\left(\frac{V}{100}\right)$$

Thus

$$C' = (0.2667) V^2 \text{ N·m}$$

Gyro-couple due to engine parts,

$$C'' = I_E \omega_E \omega_p$$

or

$$C'' = I_E (G\omega) \omega_p$$

Thus,

$$C'' = 1 \times 3 \times \left(\frac{V}{0.3}\right)\left(\frac{V}{100}\right)$$

or

$$C'' = (0.1)V^2 \text{ N·m}$$

As the crank shaft and the road wheels rotate in same sense in parallel planes, total gyroscopic couple

$$C = C' + C''$$

or

$$C = (0.2667 + 0.1)V^2$$

or

$$C = (0.3667)V^2 \text{ c.w. as seen from rear}$$

The centrifugal force F_c acting through c.g.,

$$F_c = \frac{W}{g} \frac{V^2}{R} = \frac{14710}{9.81} \times \frac{V^2}{100}$$

or
$$F_c = (14.995)V^2 \text{ N}$$

The overturning couple due to centrifugal force,

$$C_1 = F_c \times h = (14.995)V^2 \times 0.5$$

Hence vertical reaction on each wheel due to C_1

$$P = \frac{(14.995 \times 0.5)V^2}{2 \times 1.5} = (2.5)\, V^2 \text{ N}$$

(upwards on outer wheels and downwards on inner wheels)
Also, the vertical reaction due to gyro-couple

$$Q = \frac{(0.3667)V^2}{2(1.5)} = (0.122)\, V^2 \text{ N}$$

Limiting condition is reached on inner wheels at limiting speed, when

$$\frac{W}{4} - P - Q = 0$$

or
$$(P + Q) = \frac{W}{4}$$

or
$$(2.5 + 0.122)\, V^2 = \frac{14710}{4}$$

Therefore
$$V^2 = \frac{14710}{4 \times 2.622} = 1402.55$$

Therefore
$$V = 37.45 \text{ m/s}$$

Maximum limiting speed

$$V = \frac{37.45 \times 3600}{1000} = 134.82 \text{ km/hr} \qquad \textbf{Ans.}$$

EXAMPLE 10.5 A rail car has a total weight of 39240 N. There are two axles, each of which together with its wheels and gearing has a total moment of inertia of 30 kg·m². The centre distance between the two wheels on an axle is 1.5 m and each wheel is of 37.5 cm radius, Each axle is driven by a motor, the speed ratio between the two being 1 : 3. Each motor with its gear has moment of inertia of 15 kg·m² and runs in a direction opposite to that of axle. The centre of gravity of the car is 105 cm above the rails. Determine the limiting speed for this car when it is rounding a curve of 240 m radius such that no wheel leaves the rail. Consider the centrifugal and gyroscopic effects completely. Assume that no cant is provided for the outer rail,

(DAV Indore, 1981)

Solution: Given: $W = 39240$ N; $I_1 = 30$ kg.m^2; $I_2 = 30$ kg.m^2

Let N_m = RPM of motor; N = RPM of axle

$$a = 1.5 \text{ m}; \qquad r = 37.5 \text{ cm}; \qquad G = \frac{N_m}{N} = 3$$

$$I_m = 15 \text{ kg.m}^2; \qquad N_m = 3 \text{ N (opposite sense)}$$

$$h = 105 \text{ cm}; \qquad R = 240 \text{ m}$$

The load shared per wheel $= \dfrac{W}{4} = \dfrac{39240}{4} = 9810$ N

Let V be the limiting speed in m/s at which wheels leave the rail. Then in this condition,

$$\left(\frac{W}{4} - P - Q \right) \le 0 \text{ at inner wheels}$$

The centrifugal force at the running speed of V

$$F_c = \frac{W}{g}\frac{V^2}{R} = \frac{39240}{9.81} \times \frac{V^2}{240} = 16.67 \, V^2 \text{ N}$$

This produces an overturning couple ($F_c \times h$) which is to be balanced by generating ground reactions P upward on outer wheels and P downward on inner wheels. Thus,

$$2 P \times a = F_c \times h$$

Therefore
$$P = \frac{(16.67 V^2) \times 1.05}{2 \times 1.5} = (5.835) \, V^2 \text{ N}$$

The gyroscopic couples are contributed by rotating parts of the two motors and also by rotating parts of two axles. Assume that the rail car takes a turn to the right, while moving as shown in Fig. 10.10(b). The rotating parts of motor then rotate in planes parallel to those of wheels but in opposite sense. Hence the gyroscopic couples due to two axles along with wheels and gearing is represented by vector *ab* in Fig. 10.11. The gyro-couple due to rotating parts of two motors and gearing is represented by *a′b′*. As can be seen, the two gyro-couple act in same plane but in opposite sense.

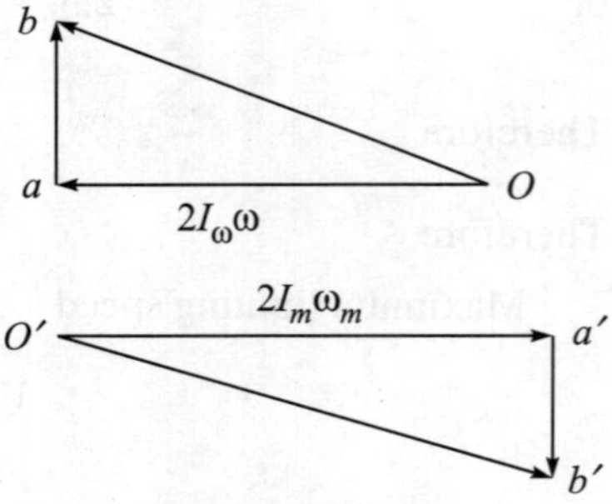

Fig. 10.11

$$\text{For wheels and axle, } \omega = \left(\frac{V}{0.375} \right) \text{ rad/s}$$

$$\text{Precessional velocity, } \omega_p = \left(\frac{V}{240} \right) \text{ rad/s}$$

$$\text{And, for motor parts } \omega_m = 3 \, \omega = \left(\frac{3V}{0.375} \right) \text{ rad/s}$$

The precessional angular velocity for rotating parts of motor is same as that for wheel and axles. Thus gyro-couple due to wheels, axles and gearing is

$$C_1 = 2(30)\left(\frac{V}{0.375}\right)\left(\frac{V}{240}\right)$$

Therefore $\qquad\qquad C_1 = (0.667)\ V^2\ \text{N.m}$

The gyro-couple due to rotating parts of two motors, gearing etc. is

$$C_2 = 2(15)\left(\frac{3V}{0.375}\right)\left(\frac{V}{240}\right)$$

$$= (1.0) \cdot V^2\ \text{N.m}$$

As can be seen from Fig. 10.11, gyro-couple C_1 due to rotating parts of axles contribute to overturning while C_2 opposes it. The net gyro-couple is

$$C = C_2 - C_1$$

$$= (0.333)\ V^2,\ \text{which opposes overturning}$$

If Q be the reactions producing gyro-couple upwards on inner wheels and downwards on outer wheels, then for the two axles,

$$2Q \times a = 0.333\ V^2$$

Therefore $\qquad\qquad Q = \dfrac{0.333\,V^2}{2 \times 1.5} = (0.111)\ V^2 \qquad\qquad$ (Upwards on inner wheel)

Thus, for limiting condition on inner wheel,

$$\frac{W}{4} - P + Q = 0$$

or $\qquad\qquad\qquad\qquad P - Q = \dfrac{W}{4}$

or $\qquad\qquad (5.835 - 0.111)\ V^2 = \dfrac{39240}{4}$

or $\qquad\qquad\qquad\qquad V^2 = 1713.84$

Therefore $\qquad\qquad\qquad V = 41.4\ \text{m/s}$

or $\qquad\qquad\qquad\qquad V = \dfrac{41.4 \times 3600}{1000} = 149\ \text{km/hr} \qquad\qquad$ **Ans.**

10.9 STABILITY ANALYSIS OF A TWO-WHEEL VEHICLE

The case of a two-wheel vehicle, such as bicycle, a scooter or a motorcycle, is different from that of a 4-wheeled vehicle. However, D' Alemebert's principle still plays important role. Taking the case of a motor cycle, let

I_E = moment of inertia of the rotating parts of the engine in kg·m^2
I_w = moment of inertia of rotating parts of wheel in kg·m^2
r_ω = radius of road-wheel in m
R = radius of curvature of curved path in m
G = gear ratio ω_E/ω_w

h = height of c.g. rider and machine together above the ground ($h > r_w$) in m

ω_E = angular velocity of engine parts in rad/s

ω_w = angular velocity of wheels in rad/s

V = linear velocity of motor cycle on road in m/s

W = total weight of machine and rider in N

θ = angle of heel, *i.e.*, the angle of inclination of the plane of rider and motorcycle with vertical plane

Now $\omega_w = (V/r_w)$

and therefore $\omega_E = (\omega_w)\, G$

or $\omega_E = (V/r_w)\, G$ (10.23)

As the planes of rotation of wheels and engine parts are parallel and the sense of rotation is same, the angular momentum vectors are collinear and additive. Thus, if I be the combined equivalent moment of inertia reduced at the wheel speed, then

$$I\,\omega_w = (I_E\,\omega_E) + (I_w\,\omega_w)$$ (10.24)

Simplifying

$$I\,\omega_w = \omega_w\,(GI_E + I_w)$$

Cancelling out ω_w

$$I = (I_w + GI_E)$$ (10.25)

Let Fig. 10.12(a) be the plan view of a motorcycle moving at a speed of V m/s along a curved path of radius R. Figure 10.12(b) shows a wheel in inclined position while negotiating a curve. The figure shows an imaginary cone whose generators represent positions of axis of spin of wheel, as the motorcycle moves along a circular path of radius R keeping its plane inclined at θ to vertical always.

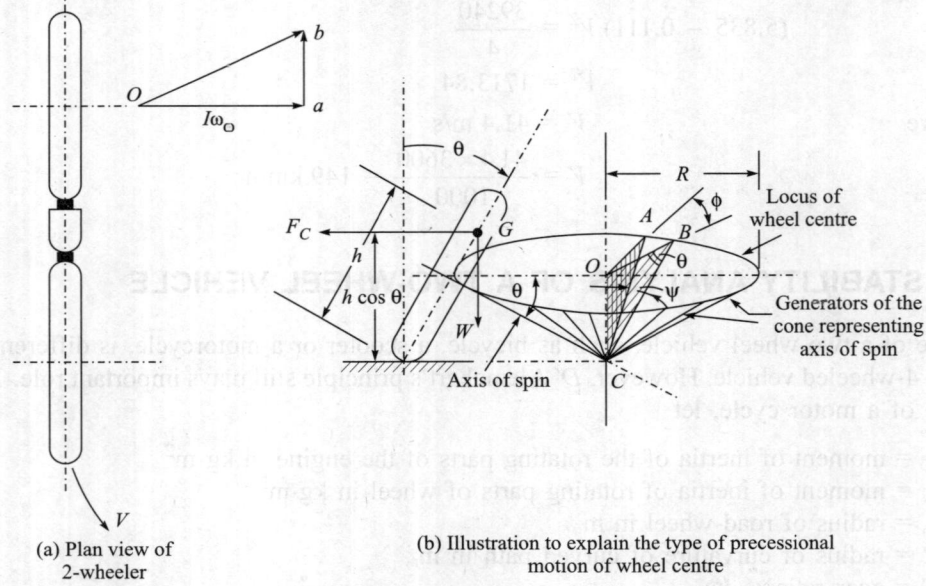

(a) Plan view of 2-wheeler

(b) Illustration to explain the type of precessional motion of wheel centre

Fig. 10.12

Consider the axis of spin to move along the slant surface of cone from position given by generator CB to CA through small angle Ψ. The corresponding shift in centre of the wheel from B to A is indicated by a small angular shift ϕ of the radius OB to OA as shown. Thus, if l be the length of generator and R the radius of the base of cone, then

$$\text{arc } AB \approx l\,\Psi = R\phi$$

Thus, differentiating w.r. to time, $\quad \dfrac{d\psi}{dt} = \dfrac{R}{l}\dfrac{d\phi}{dt}$

For the spinning wheel the term $d\Psi/dt$ gives the rate at which the axis of spin precesses, i.e., $d\Psi/dt = \omega'_p$. Thus ω'_p is the actual angular velocity of precession, while $d\phi/dt = \omega_p$ is a more convenient measure of angular velocity of precessional motion $= (V/R)$. We therefore consider $\omega_p = d\phi/dt$ as angular velocity of precession in the plane in which wheel centre moves. Thus, for a two-wheeler negotiating a curve, angular velocity of precession

$$\omega'_p = \omega_p\left(\frac{R}{l}\right)$$

And, from right angled triangle COB, as $\angle OBC = \theta$ and $\angle BOC = 90°$,

$$\frac{R}{l} = \cos\theta$$

Hence, angular velocity of precessional motion,

$$\omega'_p = \omega_p\cos\theta = \left(\frac{V}{R}\right)\cos\theta \tag{10.26}$$

While negotiating curved path, the plane of motorcycle is inclined to the vertical at an angle θ, for balancing the overturning couples. The balancing couple is provided by the weight W, and corresponding moment arm $= h\sin\theta$. Thus, the restoring couple is $(Wh\sin\theta)$.

The overturning couple is due to

(i) centrifugal force, given by

$$C_1 = \left(\frac{W}{g}\frac{V^2}{R}\right)h\cos\theta$$

(ii) gyro-reaction couple $(I\omega_w\omega_p\cos\theta)$, given by vector ba in the Fig. 10.12(a) and tends to assist couple due to centrifugal force to overturn the vehicle. Thus the total overturning couple

$$C = \left(\frac{W}{g}\frac{V^2}{R}\right)h\cos\theta + I\omega_w\omega_p\cos\theta \tag{10.27}$$

Equating this to balancing couple due to weight W,

$$Wh\sin\theta = \left(\frac{W}{g}\frac{V^2}{R}\right)h\cos\theta + I\omega_w\omega_p\cos\theta$$

or
$$\tan\theta = \frac{V^2}{gR} + \left(\frac{I}{Wh}\right)\left(\frac{V}{r_w}\right)\left(\frac{V}{R}\right)$$

or
$$\tan\theta = \frac{V^2}{WhR}\left[\frac{Wh}{g} + \frac{I_w + GI_E}{r_w}\right] \qquad (10.28)$$

Equation (10.28) specifies required value of angle of wheel θ to avoid skidding.

EXAMPLE 10.6 A solo motorcycle rider and passenger together have a mass of 320 kg, the combined centre of gravity being 525 mm above ground level. The wheels of machine each have a mass of 9 kg., a radius of gyration of 225 mm and an effective rolling radius of 300 mm. The rotating parts of the engine have a mass of 12 kg, a radius of gyration 75 mm and rotate in the same sense as the road wheels. The gear ratio, engine to back-wheel is 3.5 : 1. The machine is travelling around a banked curve. Determine the angle of banking necessary for the machine to ride normal to the track on a bend of 60 m radius at a speed of 160 km/hr, allowing for gyroscopic effects.   (DAVV Indore, 1989)

Solution: M = 320 kg; h = 0.525 m

$$I_w = 9 \times (0.225)^2 = 0.4556 \text{ kg·m}^2 \text{ (per wheel)}$$

$$r_w = 0.3 \text{ m}; I_E = 12 \times (0.075)^2 = 0.0675 \text{ kg·m}^2$$

Engine parts rotate in same sense as wheels, and gear ratio

$$G = \frac{\omega_E}{\omega_w} = 3.5$$

$$R = 60 \text{ m}; V = 160 \text{ km/hr} = 44.44 \text{ m/s}$$

Therefore $\omega_w = \dfrac{160 \times 1000}{3600 \times 0.3} = 148.15$ rad/s and $\omega_p = \dfrac{160 \times 1000}{3600 \times 60} = 0.7407$ rad/s

Angle of banking reqd. = Angle of heel required. Figure 10.13(a) shows position of a road wheel on banked road, while Fig. 10.13(b) shows plan view of motorcycle turning of the left. The equivalent M.I., reduced to wheel speed, is

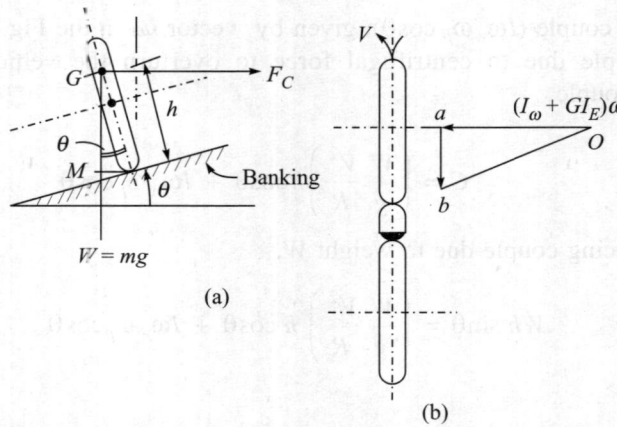

Fig. 10.13 Illustration for effect of banking on roads in Gyroscopic Analysis of motorcycle.

$$I = 2I_w + GI_E = 0.9112 + (3.5) \times 0.0675 = 1.14745 \text{ kg·m}^2$$

Therefore   Gyro-couple $C = I\omega_w\omega_p$

The angular momentum vector $I\omega_w$ shifts to the position ab due to precession, and vector ab gives sense of gyro-couple. The gyro-reaction couple acts in the sense ba and tends to overturn the vehicle in outward direction. Total overturning couple is

$$C' = \frac{MV^2}{R}(h\cos\theta) + I\omega_w(\omega_p\cos\theta)$$

or $$C' = \frac{320 \times (44.44)^2}{60} \times (0.525)\cos\theta + (1.14745)(148.15)(0.7407) \times \cos\theta$$

or $$C' = (5529.76 + 125.915)\cos\theta = (5655.67)\cos\theta$$

This is to be balanced by moment due to self weight.
Thus

$$(mg) \, h \sin\theta = (5655.67) \cos\theta$$

or $$\tan\theta = \frac{5655.67}{320 \times 9.81 \times 0.525} = 3.43$$

$$\theta = 73.75° \qquad \qquad \textbf{Ans.}$$

(*Note:* The angle θ of banking appears to be quite large in comparison to normal values. However, in view of a very large speed of 160 km/hr at the bend, this is necessary).

10.10 GYROSCOPIC EFFECTS ON NAVAL SHIPS

Some of the important terms associated with naval ships are explained below with the help of Fig. 10.14, which shows front view and top view of a sea vessel.

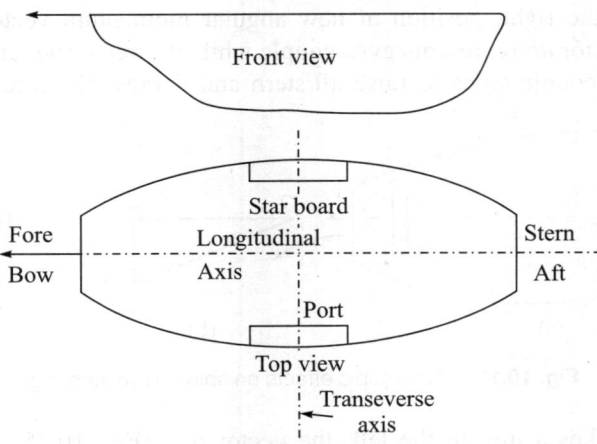

Fig. 10.14 Front and top view of a naval ship.

Bow (Fore):   This is the front end of the ship in the direction of motion.

Stern (*Aft*):   This is the rear end of the ship.

Starboard: This is the right hand side of ship, looking in the direction of the motion.

Port: This is the left hand side of the ship, looking in the direction of motion.

A naval ship has generally three distinct types of motion.

1. *Steering Motion*:   This is the turning of a ship along a curved path, while the ship moves forward.
2. *Pitching Motion*:   Pitching is the oscillatory motion of the ship about a transverse axis. During this motion, fore and aft go up and down alternately.
3. *Rolling Motion*:   Rolling is the side wise oscillatory motion of ship about the longitudinal axis. During rolling motion, port and starboard sides go up and down alternately.

A boat/ship propelled by a turbine has a propeller shaft running parallel to the longitudinal axis. This shaft carries rotating parts that include propeller. As the ship is subjected to any of the three motions, we need to consider the resulting gyroscopic effects.

Gyroscopic Effect Due to Rolling

As the axis of propeller shaft is parallel to longitudinal axis, during rolling motion the axis of rotating parts remains parallel to itself. In other words, there is no precessional motion and therefore, the ship is not subjected to any gyroscopic effect.

Gyroscopic Effect Due to Steering

Figure 10.15 shows a plan view of a ship with turbine rotor mounted on longitudinal shaft. Let I be the mass moment of inertia and ω be the angular speed at which turbine rotor is rotating in clockwise sense as seen from aft/stern side. The angular momentum vector *oa*, as given by right-hand screw rule therefore acts in a direction shown in Figs. 10.15(b) and (c). When the ship takes a turn to the right, position of new angular momentum vector is *ob* as shown in Fig. 10.15(b). The vector *ab* represents gyro-couple while the vector *ba* represents gyro-reaction couple. The reaction couple tends to raise aft/stern and depress the fore or bow.

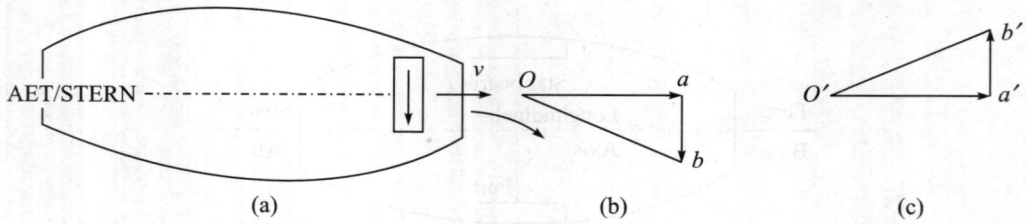

(a)   (b)   (c)

Fig. 10.15   Gyroscopic effects on ship due to steering.

When the ship takes a turn to the left, the vector *b'a'* (Fig. 10.15(c)) gives the sense of reaction couple. This reaction couple tends to raise the bow and depress the stern.

Gyroscopic Effect Due to Pitching

Figure 10.16(a) shows the front view of a ship in three positions during pitching. With the rotor rotating clockwise, as seen from stern side, vector a_1b_1 in 10.16(b) shows the gyro-couple for position A_1OB_1 of ship. Similarly, vector a_2b_2 in Fig. 10.16(c) shows gyro-couple for position A_2OB_2 of the ship.

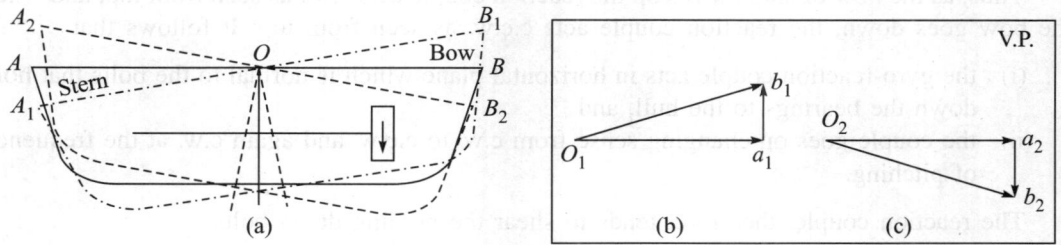

Fig. 10.16 Gyroscopic effects due to Pitching motion.

Thus, when bow goes up, the reaction couple is given by sense b_1a_1 and exerts a couple in horizontal plane in clockwise sense as seen from top. Similarly, when the bow goes down gyro-reaction couple is given by vector b_2a_2 and by right-hand screw rule, the reaction couple acts counterclockwise in horizontal plane as seen from the top.

Let us assume that the pitching motion takes place with simple harmonic motion. Thus let the axis of spin be subjected to precessional motion given by

$$\theta = \phi \sin \omega_0 t \qquad (10.29)$$

where

ϕ = Amplitude of angular oscillations in rad.

$$= (\text{maximum value of } \theta) \times \frac{\pi}{180} \text{ rad.}$$

ω_0 = Angular velocity of pitching in simple harmonic motion, called frequency. (rad/s)

$$= \frac{2\pi}{\text{Time period of S.H.M. in s}}$$

Thus

$$\frac{d\theta}{dt} = \text{Rate at which axis of spin changes position w.r. to time}$$

Thus

$$\frac{d\theta}{dt} = \text{angular velocity of precessional motion } \omega_p$$

or

$$\omega_p = (\phi\,\omega_0)\cos \omega_0 t \qquad (10.30)$$

Thus angular velocity of precessional motion is not constant. From design considerations, we are more interested in maximum value of gyroscopic couple and hence, maximum value of precessional velocity = $\phi\omega_0$

Thus

$$(\omega_p)_{\text{max}} = \phi\,\omega_0 \qquad (10.31)$$

or

$$(\omega_p)_{\text{max}} = \phi\left(\frac{2\pi}{\tau}\right) \qquad (10.32)$$

where

τ = time period of pitching in S.H.M.

The maximum gyroscopic couple is given in magnitude by

$$C_{max} = I\omega(\omega_p)_{max} \tag{10.33}$$

where $\qquad\qquad\qquad\qquad I$ = mass moment of inertia in kg·m^2

and $\qquad\qquad\qquad\qquad \omega$ = angular velocity of rotor in rad/s

Thus, as the bow of ship moves up the reaction couple acts c.w., as seen from top, and when the bow goes down, the reaction couple acts c.c.w. as seen from top. It follows that

(i) the gyro-reaction couple acts in horizontal plane which is normal to the bolts that hold down the bearings to the hull, and

(ii) the couple goes on changing sense from c.w. to c.c.w. and again c.w. at the frequency of pitching.

The reaction couple, therefore, tends to shear the holding down bolts.

EXAMPLE 10.7   The propeller of a steamer weighs 14710 N and has a radius of gyration of 1.2 m. The steamer turns left in a circle of 160 m radius at 25 km/hr, the speed of the propeller being 90 r.p.m. in a clockwise sense when viewed from the rear. Determine the magnitude and effect of gyroscopic couple on the steamer: (AMIE, Summer 1984)

Solution: Given W = 14710 N; R = 160 m; k = 1.2 m; V = 25 km/hr

$$\omega = \frac{2\pi \times 90}{60} = 9.4248 \text{ rad/s}$$

$$I = \frac{W}{g} \times k^2 = \frac{14710}{9.81} \times (1.2)^2 = 2159.3 \text{ kg·m}^2$$

Angular velocity of precession $\omega_p = V/R = \dfrac{25 \times 1000}{3600 \times 160} = 0.0434$ rad/s

Figure 10.17 shows the elevation and plan view of the steamer. *oa* is the angular momentum vector for c.w. sense of rotation of propeller, as seen from rear, when steamer moves straight. The vector *ob* gives new angular momentum vector as the steamer turns to the left. The vector

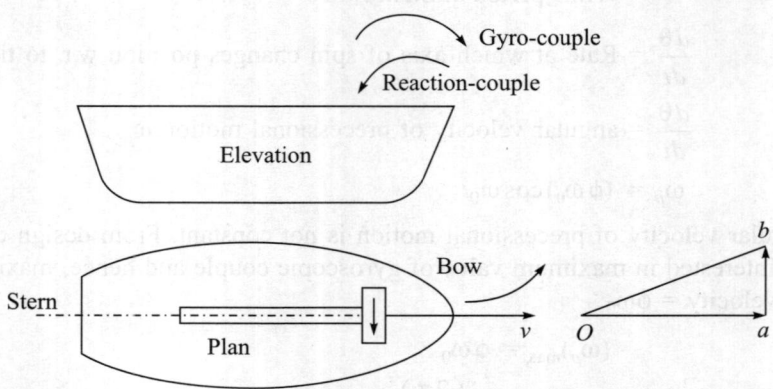

Fig. 10.17   Gyroscopic effects due to steering.

ab, which is perpendicular to *oa*, represents sense of gyro-couple which is clockwise, as shown in elevation. The reaction couple is therefore counterclockwise and tends to lift the bow and depress the stern. The magnitude of gyro-couple and reaction couple is

$$= C = I\omega\omega_p$$
$$= (2159.3)\ (9.4248)\ (0.0434) = 883.2\ \text{N·m.} \qquad \textbf{Ans.}$$

EXAMPLE 10.8   The turbine rotor weighing 9.8 kN rotates at 2000 r.p.m. clockwise when looking from stern. The vessel pitches with an angular velocity of 0.5 rad/s. Calculate the gyroscopic couple during the rise of bow. Assume radius of gyration of the rotor as 25.4 cm.

(DAV Indore, 1991)

Solution: Given $W = 9.8$ kN; $\omega = \dfrac{2\pi \times 2000}{60} = 209.44$ rad/s

$$k = 25.4 \text{ cm}$$

Angular velocity of pitching $\omega_p = 0.5$ rad/s

Moment of inertia
$$I = \frac{9.8 \times 1000}{9.81} \times (0.254)^2$$
$$= 64.516 \text{ kg·m}^2$$

Therefore   Gyroscopic couple $= I\omega\omega_p$

or
$$C = (64.516)\ (209.44)\ (0.5)$$

or
$$C = 6756.11 \text{ N.m} \qquad \textbf{Ans.}$$

As seen in Fig. 10.18, the angular momentum vector *oa* moves to position *ob* as the bow moves up during pitching. Thus sense of gyro-couple is given by vector *ab* which is c.c.w. as seen from top. The gyro-reaction couple acts c.w. and tends to rotate bow towards starboard side, as shown in top view.

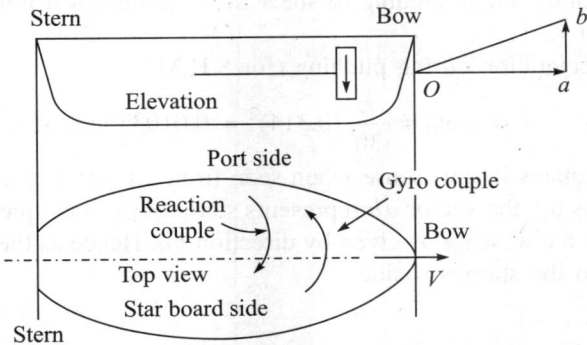

Fig. 10.18   Gyroscopic effect due to pitching.

EXAMPLE 10.9   A turbine rotor of a ship weighs 196.0 kN and has a radius of gyration of 75 cm. Its speed is 2000 r.p.m. The ship pitches 6° above and below the mean position. A complete oscillation takes place in 20 seconds and the motion is simple harmonic. Determine:

(a) the maximum couple tending to shear the holding down bolts of the turbine;

(b) the maximum acceleration of the ship during pitching;

(c) the direction in which the bow will tend to turn while rising, if the rotation of the rotor is clockwise, when looking from the aft. (DAV Indore, 1983, 87, 90; Jeevaji Univ. Gwalior, 1983).

Solution: Given $\quad W = 196000$ N

$$k = 75 \text{ cm}; \ \omega = \frac{2\pi \times 2000}{60} = 209.439 \text{ rad/s}$$

Amplitude of pitching motion $\phi = 6° = \dfrac{\pi}{180} \times 6 = \dfrac{\pi}{30}$ radians

Time period of oscillation = 20 s

Therefore $\quad$ angular frequency of oscillation in pitching $= \omega_0 = \dfrac{2\pi}{20} = \dfrac{\pi}{10}$

and $\quad$ maximum precessional speed

$$(\omega_p)_{max} = \frac{\pi}{30} \times \frac{\pi}{10} = \frac{\pi^2}{300} = 0.0329 \text{ rad/s}$$

Also $$I = \frac{W}{g} k^2 = \frac{196000}{9.81} \times (0.75)^2 = 11238.5 \text{ kg·m}^2$$

(a) Thus maximum gyroscopic couple and reaction couple

$$C = I\omega \, (\omega_p)_{max}$$

or $$C = 11238.5 \, (209.439) \times \frac{\pi^2}{300} = 77436.5 \text{ N·m}$$

Hence, maximum couple tending to shear the holding down bolts = 77436.5 N·m

Ans.

(b) Maximum acceleration during pitching (for S.H.M.),

$$\alpha = \phi \, \omega_0^2 = \frac{\pi}{30} \, (0.314)^2 = 0.010335 \text{ rad/s}^2 \qquad \textbf{Ans.}$$

(c) As the rotor rotates in c.w. sense when seen from aft, we may refer to Fig. 10.18. As the bow moves up, the vector ab represents sense of gyro-couple. Reaction couple will therefore have a c.w. sense as given by direction ba. Hence as the bow moves up it will tend to turn to the starboard side. **Ans.**

10.11 GYROSCOPIC SHIP STABILIZATION

Principle Fundamental requirement of ship stabilization is that the rotor of gyroscope should be made to precess in such a way that the reaction couple exerted by the rotor of gyroscope shall oppose any disturbing couple from sea which may act on the frame of the ship. If at every instant of time, the reaction couple from gyroscope equals the disturbing couple (applied by sea waves) then complete stabilization would be obtained.

For instance, consider the stabilization of ship for reducing amplitude of roll. The disturbing (applied) couple arises from the effect of the difference in buoyancy on the two sides of the centre line (longitudinal vertical plane) of the ship, when on the wave slope. Figure 10.19 shows transverse section of a ship. Clearly the volume of sea water *AOCD* displaced by left half portion is less than the volume of sea water *OBEC* displaced by right half portion of the ship. The difference in buoyancy forces, on the two sides of longitudinal vertical plane, produces a disturbing couple which causes rolling motion of the ship.

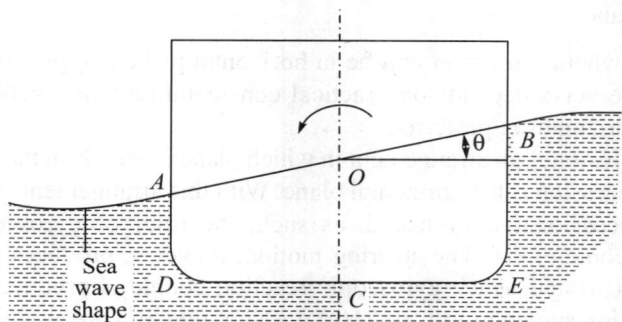

Fig. 10.19   Illustration for Disturbing couple on ship.

This disturbing couple is a periodic couple which has its maximum value in opposite sense when the ship is on the either side of the wave with its centre at the point of maximum slope of the wave. It has zero value when the ship centre is at the crest (Fig. 10.20 a) or at the trough (Fig. 10.20 b). Due to symmetrical shape of wave form about ship centre line, the two halves of the ship displace equal volumes of sea water and the buoyancy forces on the two halves are therefore equal.

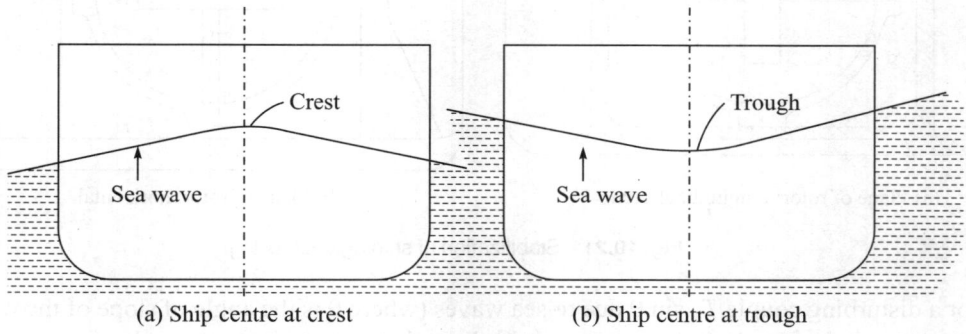

(a) Ship centre at crest   (b) Ship centre at trough

Fig. 10.20

It is important to note that the law according to which the magnitude of couple varies, depends on the shape of wave profile. For an assumed sinusoidal wave profile, the disturbing couple due to waves will also be sinusoidal (say, $T_w \sin \theta$). Hence, in order to maintain the ship at even keel, the gyroscope must, at every instant, provide a reaction couple which is equal and opposite to the applied couple due to the waves.

In order to locate the plane of spin for the rotor of gyroscope and also, the plane of precession, we proceed as under. It may be recalled that plane of spin, plane of precession and plane of gyro-reaction couple are mutually perpendicular to one another. The disturbing couple from sea waves producing rolling motion acts in transverse vertical plane. Therefore, for stabilization, gyroscope must produce reaction couple in transverse vertical plane. This leaves choice of two planes for providing spin and precession.

    (a)   Longitudinal vertical plane, and

    (b)   Horizontal plane

The choice as to whether the spin will be in horizontal plane and precession in longitudinal vertical plane, or vice versa depends on practical considerations. In practice, the arrangement with plane of spin horizontal is preferred.

Figure 10.21(a) illustrates an arrangement in which plane of spin is in the longitudinal vertical plane and plane of precession is the horizontal plane. With this arrangement, pitching motion does not produce any precessional motion and, as such, the rotor of gyroscope is incapable of producing any gyroscopic effect. The steering motion, however, produces precessional motion of rotor in Fig. 10.21(a) and hence gyroscopic reaction couple is produced.

The rotor in position shown in Fig. 10.21(b) does not undergo precessional motion when the ship steers and as such no gyroscopic effect is involved. During pitching motion however, the axis of spin undergoes precessional motion and gyroscopic reaction couple is introduced.

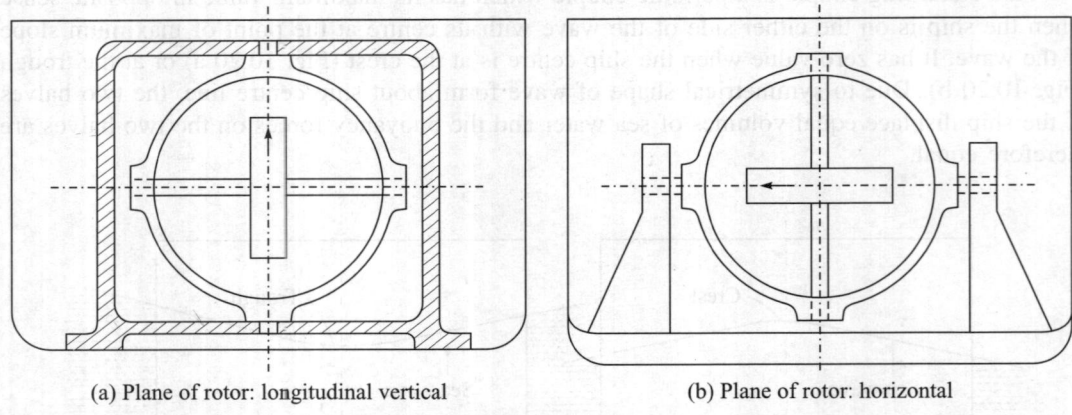

    (a) Plane of rotor: longitudinal vertical   (b) Plane of rotor: horizontal

Fig. 10.21   Stabilization of ship against rolling.

For a disturbing couple $T_w \sin \theta$ due to sea waves (where θ is the angle of slope of the wave), let the rotor axis in Fig. 10.21(b) be precessed through an angle ϕ. As a result of precessional motion, the plane of spin does not remain horizontal and, instead of gyro-reaction couple of $C = I\omega\omega_p$, only the component $I\omega\omega_p \cos \phi$ is available in transverse vertical plane. For complete stabilization, therefore,

$$C \cos\phi = T_w \sin\theta, \text{ at all time.} \tag{10.34}$$

or $$(I\omega\omega_p)\cos\phi = T_w \sin\theta \tag{10.35}$$

From right-hand side of equation (10.34) it is clear that maximum disturbing couple occurs when θ is maximum, *i.e.* when the ship is at the point of maximum slope. At this position, the left-hand side of equation (10.34) must give largest possible contribution. In other words, at this position, φ must be zero which, means that the plane of spin must be horizontal. Again, when the ship rides on trough or crest the applied wave couple is zero (as θ = 0) and for these positions, for producing zero reaction couple, the axis of spin must be precessed through 90°. Thus, over the quarter period of wave (between point of maximum slope of the point of crest/trough), the rotor of gyroscope must be precessed through 90°. This is very difficult indeed.

In practice, parameters I and ω are kept constant and axis of rotor of gyroscope is made to precess by 60° on either sides of the vertical position. This amounts to saying that equation (10.34) is not satisfied fully at all the time, and only partial stabilisation is aimed at. The vertical shaft of gyroscopic rotor is carried in a casing which in supported in bearings fixed to the frame of the ship (see Fig. 10.22), so as to permit the desired precessional motion. The casing is driven by an electrical motor at an angular velocity ω_p, which is practically constant over most of the arc of precession and it is brought to rest and accelerated in the opposite direction over a small arc at the end of outward swing. The drive from electrical motor to the casing is through a gear which is rigidly connected to the casing. This arrangement aims at reducing amplitude of roll.

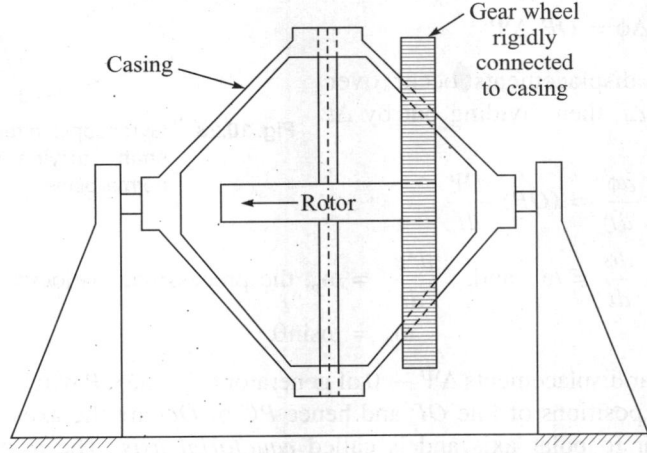

Fig. 10.22 Gyroscope for ship stabilization in Rolling.

In sperry stabilizer the rotor of gyroscope weighed 997 kN or more and was motor-driven. The speed of the rotor was upto 1000 r.p.m. Three such units were carried on a large Italian liner, which was sunk during the war. It is used to produce a stabilizing torque of nearly 23920 kN·m and reduced the average roll amplitude from $7\frac{1}{2}$ to 1 or $1\frac{1}{4}$ degree.

10.12 GYROSCOPIC ANALYSIS OF A DISC FIXED RIGIDLY TO A ROTATING SHAFT AT CERTAIN ANGLE

Consider a disc of uniform thickness to be fixed rigidly to a rotating shaft. Let the polar axis of disc be inclined to the axis of shaft at an angle of θ, as shown in Fig. 10.23(a). Let the shaft rotate with angular velocity ω rad/s in clockwise direction as seen along the axis of shaft from the right hand side. The disc, being rigidly fixed to shaft, also rotates at ω in the same sense. Thus, component of angular velocity of disc about OP, the polar axis = $\omega\cos\theta$.

The rotation of shaft along with disc, causes, the polar axis OP to rotate so as to generate a conical surface with appex at O and point P moving in a plane perpendicular to the plane of paper. This plane is represented in Fig. 10.23(b) by its trace PM. Let the generator OP move through a small angle $\Delta\Psi$ along conical surface and, simultaneously, let PM move through a small angle $\Delta\phi$ in circle with centre at P. Then,

$$(PM)\ \Delta\phi = OP\ \Delta\Psi$$

or, $\quad (OP\sin\theta)\ \Delta\phi = OP\ \Delta\Psi$

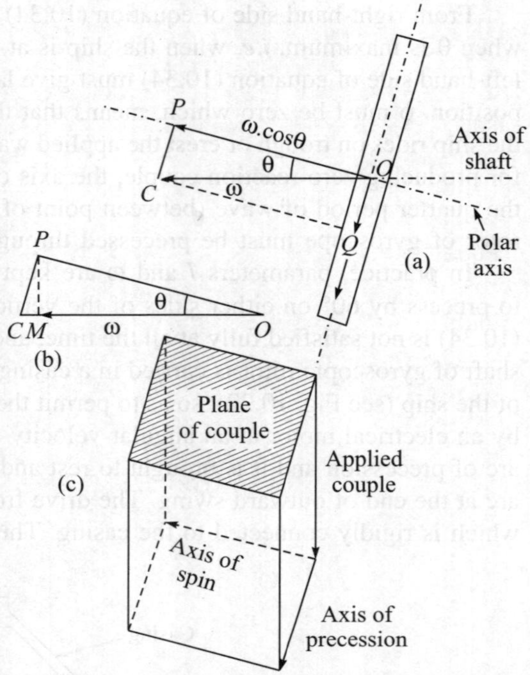

Thus, if these displacements occur over same time interval Δt, then dividing out by Δt and taking limits,

$$(OP\sin\theta)\ \frac{d\phi}{dt} = (OP)\ \frac{d\Psi}{dt}$$

Fig. 10.23 Gyroscopic effects due to a rotating shaft carrying uniform disc not in normal plane.

Noting that, $\quad \dfrac{d\phi}{dt} = \omega$ and, $\dfrac{d\Psi}{dt} = \omega_p$, the precessional velocity,

$$\omega_p = \omega\sin\theta$$

For small angular displacements $\Delta\Psi \to 0$ of generator OP, line CP will be normal to the plane containing adjacent positions of line OP and hence PC or OQ are the axes of precession. Line OQ is perpendicular to polar axis and is called *equatorial axis*. The axis of gyro-couple is therefore perpendicular to OP and OQ both. The gyro-couple producing precessional motion is

$$C = I_p\ (\omega\cos\theta)\ (\omega\sin\theta)$$
$$= I_p\ (\omega^2/2)\ \sin 2\theta \qquad (10.36)$$

The reaction couple tries to turn the disc is c.c.w. direction as seen from top.

Next, let I' be the mass moment of inertia of disc about equatorial axis. The motion of equatorial axis produces precession of axis OP. Angular velocity of disc about equatorial axis is $\omega\cos\theta$. Gyroscopic couple is therefore

$$C' = I'\ (\omega\cos\theta)\ (\omega\sin\theta)$$
$$= I'\ (\omega^2/2)\ \sin 2\theta \text{ opposite to couple } C.$$

Hence, resultant gyroscopic couple $= (I - I')\ (\omega^2/2)\ \sin 2\theta \qquad (10.37)$

10.13 GYROSCOPIC ANALYSIS OF GRINDING MILL

In some of the applications, gyroscopic effects are successfully utilised to enhance functional aspects of the system. In all such applications, a designer aims at introducing gyroscopic effects purposefully. Grinding mill is one such application in which gyroscopic effects boost the crushing force. Figure 10.24 shows a grinding mill in which two conical rollers are placed symmetrically in a pan. The conical grinders are free to rotate on shafts which are hinged to the central driving shaft. As the driving shaft rotates, the conical rollers are carried round the pan crushing the matter that is in the pan. Crushing action of rollers (due to self weight) is supplemented by the additional force due to gyroscopic action.

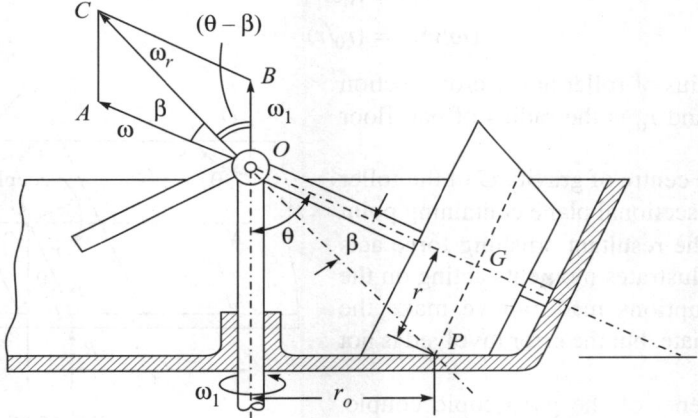

Fig. 10.24 Illustration for Gyroscopic effects in Grinding Mill.

Let ω = angular velocity of roller on shaft represented by vector OA,
  ω_1 = angular velocity of vertical driven shaft
  ω_r = resultant velocity vector
  I = mass moment of inertia of roller about shaft axis OG
  I_1 = equatorial moment of inertia, *i.e.* the mass moment of inertia of roller about any axis perpendicular to the axis of spin.

In Fig. 10.24, vector OA represents vector ω while vector OB represents angular velocity vector ω_1 and therefore, $\omega_r \equiv OC$ represents resultant velocity vector. Line of action of this vector, when extended, cuts the interface of conical roller and pan floor surface at point P. The position of point P depends on the frictional characteristics of the roller and the pan.

From the vector polygon $OACB$, as shaft OG is inclined at θ to vertical and $\angle POG = \beta$, by sine rule

$$\frac{OA}{\sin(\theta - \beta)} = \frac{AC}{\sin \beta} \qquad (10.38)$$

or

$$\frac{OA}{AC} = \frac{\sin(\theta - \beta)}{\sin \beta} \qquad (10.39)$$

But, from velocity polygon,

$$\frac{OA}{AC} = \left(\frac{\omega}{\omega_1}\right)$$

Substituting in equation (10.39) for OA/AC,

$$\frac{OA}{AC} = \frac{\omega}{\omega_1} = \frac{\sin(\theta - \beta)}{\sin\beta} \qquad (10.40)$$

Also at point P; where the resultant velocity vector meets the floor of pan, the relative velocity between the pan and the roller must be zero. Hence,

$$r\omega = r_0\omega_1$$

or
$$(\omega/\omega_1) = (r_0/r) \qquad (10.41)$$

where r is the radius of roller at the cross section through point P and r_0 is the radius of pan-floor at point P.

Assuming the centre of gravity G of the roller to lie in the cross-sectional plane containing point P, and also that the resultant crushing force acts at P, Fig. 10.25 illustrates moments acting on the roller. The assumptions made above make the analysis approximate, but the error involved is not large.

The components of the gyroscopic couple, applied externally is given by

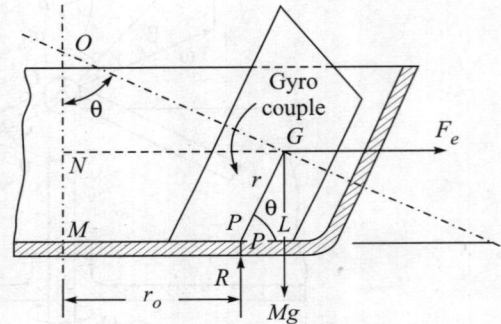

Fig. 10.25

$$C = I\omega\omega_1\sin\theta + (I - I_1)(w_1^2/2)\sin 2\theta$$

$$= I\omega_1^2\left(\frac{r_0}{r}\right)\sin\theta - (I + I_1)\left(\frac{\omega_1^2}{2}\right)\sin 2\theta$$

$$= \left[(I - I_1)\cos\theta + I\left(\frac{r_0}{r}\right)\right]\omega_1^2\sin\theta \qquad (10.42)$$

From Fig. 10.25
$$\angle GOM = \angle GPL = \theta$$
and therefore,
$$NG = ML = (r_0 + r\cos\theta)$$

Hence, from right angled triangle GON

$$ON = NG\cot\theta = (r_0 + r\cos\theta)\cot\theta$$

Considering moments of all forces about O,

$$C = F_c(r_0 + r\cos\theta)\cot\theta - Mg(r_0 + r\cos\theta) + R(r_0) \qquad (10.43)$$

Comparing equations (10.42) and (10.43) on R.H.S.

$$\left[(I - I_1)\cos\theta + I\left(\frac{r_0}{r}\right)\right]\omega_1^2\sin\theta = F_c(r_0 + r\cos\theta)\cot\theta - M(r_0 + r\cos\theta) + R(r_0)$$

Dividing out by ($Mg\, r_0$) and rearranging,

$$\frac{R}{Mg} = \left(1 + \frac{r}{r_0}\cos\theta\right) + \frac{\omega_1^2 \sin\theta}{Mgr_0}\left[(I - I_1)\cos\theta + I\left(\frac{r_0}{r}\right)\right] - \frac{F_c \cot\theta}{Mg}\left(1 + \frac{r}{r_0}\cos\theta\right) \quad (10.44)$$

Clearly if the right-hand side is greater than 1, reaction R will be greater than dead-weight Mg of the roller. In other words, gyroscopic action will contribute to crushing force if $\dfrac{R}{Mg} > 1$.

For suitable values of angle θ, the quantity on R.H.S. of expression will be greater than 1.
For instance, at $\theta = \pi/2$, the conical roller reduces to a cylindrical roller for which,

$$\frac{R}{Mg} = (1 + 0) + \frac{\omega_1^2}{Mg\,r_0}\left[(0) + I\left(\frac{r_0}{r}\right)\right] - 0$$

$$= 1 + \left(\frac{I\omega_1^2}{Mgr}\right) \quad (10.45)$$

which is clearly greater than 1.

EXAMPLE 10.10 Figure 10.26 shows a roller type crushing mill. Find out the percentage increase in the crushing force due to gyroscopic effect. (AMIE, Winter 1993)

Solution:

Given

$$\omega_1 = \frac{2\pi \times 300}{60} = 31.4 \text{ rad/s}$$

$$M = 500 \text{ kg}, \ r = 20 \text{ cm}$$

and

$$I = \frac{1}{2} Mr^2$$

$$= \frac{1}{2} \times 500 \times (0.20)^2 = 10 \text{ kg·m}^2$$

Therefore

$$\frac{I\omega^2{}_1}{Mgr} = \frac{10 \times (31.4)^2}{500(9.81)(0.5)} = 4.02$$

Therefore

$$\frac{R}{Mg} = 1 + \left(\frac{I\omega_1^2}{mgr}\right) = 5.02$$

Thus, percentage increase in crushing force

$$= \left(\frac{R - mg}{Mg}\right) \times 100$$

$$= \left(\frac{R}{Mg} - 1\right) \times 100 = (5.02 - 1) \times 100 = 402\% \qquad \textbf{Ans.}$$

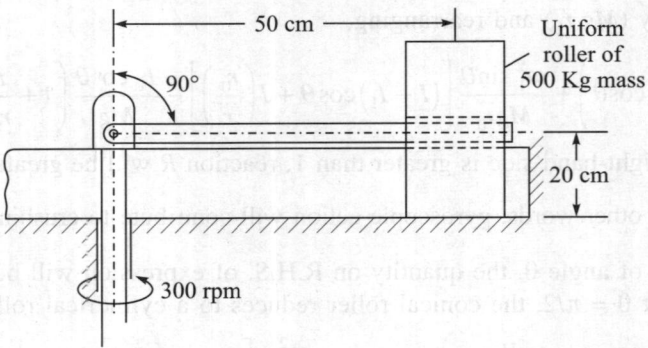

Fig. 10.26 Roller type crushing mill.

REVIEW QUESTIONS

10.1 A pair of locomotive driving wheels with the axles have a moment of inertia of 500 kg·m². The diameter of the wheel treads is 1.9 m and the distance between the wheel centres is 1.525 m. When the locomotive is travelling on a level track at 112 km/hr, defective blasting causes the wheel to fall 9 mm and rise again in a total time of 0.12 seconds. If displacement of wheel takes place with S.H.M., find gyroscopic reaction on locomotive rails. **(Ans.** 2526 N·m in horizontal plane)

10.2 A four-wheeled trolley car weighing 24.52 k·N runs on rails, which are 1.5 m apart and travel around a curve of 30 m radius at 24 km/hr. the rails are at the same level. Each wheel of the trolley is 75 cm in diameter and each of the two axles is driven by a motor running in a direction opposite to that of the wheels at five times the speed of rotation of the wheels. The moment of inertia of each axle with gear and wheel is 18 kg·m². Each motor with shaft and gear pinion has a moment of inertia of 12 kg·m². The centre of gravity of the car is 90 cm above rail level. Determine the vertical forces exerted by each wheel on the rails taking into consideration the centrifugal and gyroscopic effects. State the centrifugal and gyroscopic effects on the trolley. (DAV Indore, July 1982) **(Ans.** 7131.5 N, 5128.5 N)

10.3 A rear engine automobile is travelling around a track of l00 m mean radius. Each of the four road wheels has a moment of inertia of 1.6 kg·m² and an effective diameter of 600 mm. The rotating parts of the engine have a moment of inertia of 0.85 kg.m², the engine axis is parallel to the rear axle and the crankshaft rotates in the same sense as the road wheels. The gears ratio, engine to back axle is 3 : 1. The vehicle has a mass of 1400 kg and its centre of gravity is 450 mm above the road level. The width of the track of the vehicle is 1.5 m. Determine the limiting speed of the vehicle round the curve for all four wheels to maintain contact with the road surface if this is not cambered.
(DAV Indore, March 1990) **(Ans.** 39 m/s)

10.4 A motor car takes a bend of 30 m radius at a speed of 60 km/hr. Determine the magnitudes of centrifugal and gyroscopic couples acting on the vehicle and state the effect each of these has on the road reactions to the road wheels. Assume that each road wheel has a moment of inertia of 3 kg·m² and an effective road radius of 40 cm.

The rotating parts of engine and transmission are equivalent to a flywheel weighing 736 N with a radius of gyration of 10 cm. The engine turns in a clockwise direction when viewed from the front.

The back axle ratio is 4 : 1, the drive through the gear box being direct. The gyroscopic effects of the half shafts at the back axle are to be ignored.

The car weighs 11.768 N and has its centre of gravity 60 cm above the road wheels. The turn is in a right-hand direction.

If the turn has been in a left-hand direction, all other details being unaltered, which answer, if any need modification. (*Ans.* 6664 N·m; 280 N·m)

10.5 A motor cycle with its rider has a mass of 200 kg, the centre of gravity of the machine and rider combined being 60 cm above the ground with machine in vertical position. Moment of inertia of each road wheel is 0.525 kg·m² and the rolling diameter is 60 cm. The engine rotates at six times the speed of the road wheels and in the same sense. The engine rotating parts have a moment of inertia of 0.1685 kg·m². Determine the angle of the heel necessary if the unit is speeding at 60 km/hr round a curve of 30 m. If the road and tyre friction allow for the angle of heel not to exceed 50° what is the maximum road velocity of the motorcycle? (*Ans.* 18.3 m/s)

10.6 A uniform disc of 15 cm diameter and mass of 5 kg mounted centrally on bearings which maintain its axle in a horizontal plane as shown in Fig. 10.27. Speed of spin is 1000 r.p.m. while the uniform precessional speed is 60 r.p.m. about the vertical. If the distance between the bearings is 10 cm, find the resultant reaction at each bearing due to weight and gyroscopic effects and show clearly on sketch, the points of contact between the axle & the bearing.

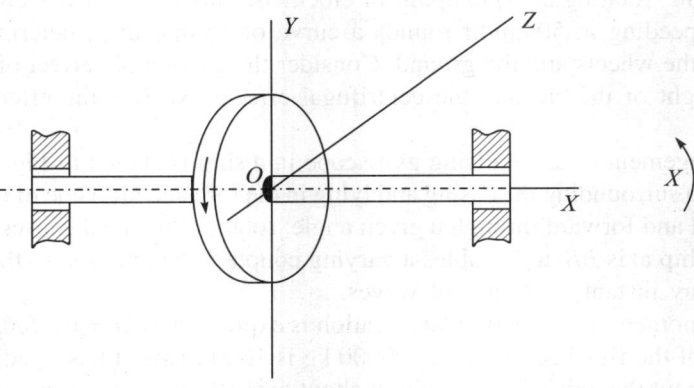

Fig. 10.27

10.7 A solo motorcycle, complete with rider, weighs 1962 N. The centre of gravity of the machine and rider combined being 60 cm above the ground with machine in vertical position. The moment of inertia of each road wheel is 10500 kg·cm² and the rolling diameter is 60 cm. The engine rotates at 6 times the speed of the road wheels and in same sense. The engine rotating parts have a moment of inertia of 1685 kg·cm^2. Determine the angle of the heel necessary if the unit is speeding at 60 km/hr round a curve of 30 m.

(**Ans.** 44.2°)

10.8 The wheel of a motorcycle and the engine parts have a moment of inertia 2.5 kg·m^2 and 0.15 kg·m^2 respectively. The axis of rotation of the engine crank shaft is parallel to that of the road wheels. If the gear ratio is 5 : 1 and the diameter of road wheels is 0.65 m, find the magnitude and direction of the gyroscopic couple when the motorcycle rounds a curve of 30 m radius at 16 m/s. (**Ans.** 85 N·m)

10.9 The moment of inertia of an aeroplane air screw is 6.75 kg·m^2 and rotates at 1200 r.p.m. The air craft makes a complete half turn in 10 seconds. Calculate the gyroscopic couple on the aircraft and state its effect on the aircraft. The air-screw rotates clockwise when viewed from the rear. (**Ans.** 270 N·m)

10.10 The rotor of a turbine installed in boat with its axis along the longitudal axis of the boat makes 1500 r.p.m. clockwise when viewed from the stern. The rotor weighs 7160 N & has a radius of gyration 30 cm. If at an instant the boat pitches in the longitudinal vertical plane so that the bow rises from the horizontal plane with an angular velocity of 1 rad/s determine the torque acting on the boat and the direction in which it tends to turn the boat at the instant. (**Ans.** 1052 kg·m)

10.11 A steam turbine, mounted in a boat makes 3000 r.p.m. The effective moment of inertia of the rotor, shaft and propeller is 47.25 kg·m^2. Find the magnitude and direction of couple acting on the hull when the boat describes a circular path, making a complete turn in 17 seconds. (DAV Indore, April. 1976)

10.12 A racing car of mass 2500 kg has a wheel base of 2 m and track width of 105 cm. The centre of gravity lies midway between the front and the rear axles and is 0.4 m above the ground. The engine rotating parts are equivalent of a flywheel of moment of inertia of 50 kg·m^2 rotating at 6000 r.p.m. in clockwise direction when viewed from front. If the car speeding at 50 km/hr rounds a curve of 15 m radius, determine the reactions between the wheels and the ground. Consider the gyroscopic effect of the flywheel the dead weight of the car and the centrifugal effects. Neglect the effect due to rotating wheels.

10.13 The arrangement of a stabilising gyroscope in a ship is shown in Fig. 10.22. A toothed gear when surrounding the casing and lying in the fore and aft plane of the ship, oscillates backward and forward through a given angle; rotation of wheel causes precession about othwart ship axis *BB* and enables a varying couple to be opposed to the rolling moment due, at any instant, to a turn of waves.

Heeling moment in one particular situation is expressed as $M = 10^7$ (cost) N·m. The spin velocity of the flywheel of mass 100000 kg is 1000 r.p.m. It has a radius of gyration of 1.25 m about the axis of spin and 1 m about axis *BB*. If the motion of the wheel is such that the hull is maintained upright, determine:

(i) Angular velocity of gear wheel at an instant when $t = 0$ and flywheel is horizontal.

(ii) Couple necessary on the wheel to cause precessional angular acceleration of 0.75 rad/s^2 when $t = \pi/2$, if at this instant the flywheel is at the end of oscillation.

(iii) Couple causing pitching if the ship rolls through the upright position with angular velocity 0.1 rad/s.

(**Ans.** (i) 0.6112 rad/s (ii) 117187.5 N·m (iii) 1.636×10^6 N·m)

FRICTION

11.1 INTRODUCTION

When a body slides or is made to slide relative to a second solid body with which it is in contact, there is a resistance to relative motion. The resistance so encountered is called *friction*. The force resisting relative motion is called the force of friction. Thus, by definition, *force of friction acts in a direction opposite to that of relative motion and is tangential to the contacting surfaces of the two bodies in contact*. There are two aspects of friction as a phenomenon. At every joint in a machine, there is a loss of energy due to the force of friction and relative motion. A proper understanding of friction as a phenomenon enables us to reduce frictional forces. Some of the situations permit replacement of sliding friction by rolling friction. In others, use of suitable lubricant may reduce frictional losses significantly. In a number of applications, on the other hand, friction is considered to be quite useful. Friction drives like belt and rope drives, friction clutches, variable speed friction drives, as also friction brakes, are just some of such useful applications. While it is true that the nature of these resisting forces is imperfectly understood, they can be recognised through their effects. Experimental investigations have taught us to account for them with acceptable engineering accuracy.

11.2 TYPES OF FRICTION

Depending upon the type and condition of contacting surfaces, friction can be categorised in three different types:

1. ***Dry Friction.*** This type of friction exists between two bodies having relative motion and whose contacting surfaces are dry and are not separated by any lubricant. It is further sub-divided into two types.
   (a) *Solid Friction:* Dry friction in which the contacting surfaces have a sliding motion relative to each other is called *solid friction*.
   (b) *Rolling Friction:* Rolling friction is said to occur between two bodies in direct contact when they have a relative motion of pure rolling. Such friction exists in ball and roller bearings.

2. *Skin or Greasy Friction.* When contacting surfaces of two bodies in relative motion are separated by a film of lubricant of infinitesimally small thickness, skin or greasy friction is said to exist between them. This type of friction is also known as *boundary friction*.

3. *Film or Viscous Friction.* When contacting surfaces of two bodies in relative motion are completely separated by a relatively thick film of lubricant, film or viscous friction is said to exist between the two.

A detailed discussion on boundary friction and film friction appears in Section 11.21.

11.3 DRY FRICTION

Dry friction of sliding type is also known as solid friction and it exists because of surface roughness. When viewed under a microscope, even a smooth surface is found to have roughness and irregularities which cannot be detected by naked eye or ordinary touch. If a block of one material is placed over the level surface of the same or different material, a certain degree of interlocking of the minute projecting particles takes place. Whenever one block moves or tends to move tangentially with respect to the surface on which it rests, the interlocking property of the projecting particles opposes the relative motion.

Laws of Solid Friction

Based on numerous experiments, following laws have been formulated for dry friction with sliding (solid friction) motion:

(1) Force of friction is directly proportional to the normal load between the surfaces for a given pair of materials.

(2) Force of friction depends upon the materials of the contacting surfaces.

(3) Force of friction is independent of the area of contact surfaces for a given normal load. In other words, force of friction is independent of specific pressure between contacting surfaces.

(4) Force of friction is independent of the velocity of sliding of one body relative to the other body in contact.

Comments

(a) When a small force P is applied parallel to the contacting surfaces, such that P is less than the force necessary to cause relative motion, a tangential frictional force F' is developed due to roughness of contacting surfaces such that,

$$F' = P$$

The induced force F' thus opposes relative motion and the block in Fig. 11.1 remains in static equilibrium condition with

$$\sum F_V \equiv R_n - W = 0 \text{ and} \tag{11.1}$$
$$\sum F_H \equiv F' - P = 0 \tag{11.2}$$

Let the applied force be incremented by small amount ΔP each time. The block and contacting surfaces develop additional frictional force equal to ΔP each time to ensure static equilibrium and as a result, the block does not move. However, there is a limit upto which the contacting surfaces can develop additional frictional force. At this stage, any increment in force P does not lead to corresponding increment in F and a small increase in the magnitude of P will cause the block to slide in the direction of P. This is a limiting condition upto which $P = F'$ and beyond which F' cannot increase. In other words, $F' = F$ is the

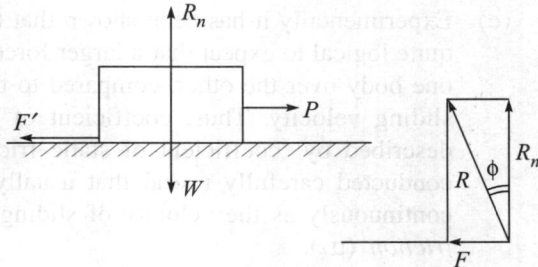

Fig. 11.1 A block under static equilibrium.

largest possible value of frictional force which the bodies can develop. Hence, $F' = F$ is called the *limiting force of friction*. It may be interesting to note that the friction experienced by block during motion is called kinetic friction and the corresponding force of friction is always less than the limiting force of friction.

According to the first law of solid friction, under limiting conditions of friction, the force of friction F is directly proportional to normal reaction R_n, *i.e.*

$$F \propto R_n$$

The ratio of F to R_n is called *limiting coefficient of friction* (μ) however, for brevity, many a time the word 'limiting' is omitted. It must be remembered however that, for $F' < F$ *i.e.*, when limiting condition is not reached, $F'/R_n = \mu'$, *i.e.*, coefficient of friction, which is less than the limiting coefficient of friction μ. Standard values listed in tables and design data books refer to limiting values of coefficient of friction. For limiting condition, it follows that

$$F = \mu R_n \qquad (11.3)$$

It follows from vector diagram of Fig. 11.1 that the ratio (F/R_n) represents tangent of angle ϕ made by resultant reaction R with normal reaction R_n. Thus,

$$(F/R_n) = \tan\phi \qquad (11.4)$$

It follows from equations (11.3) and (11.4) that

$$\mu = \tan\phi \qquad (11.5)$$

the angle ϕ is called *limiting angle of friction*.
The value of limiting coefficient of friction μ varies from material to material.

(b) The third law is only approximately correct. For a given normal load, area of contact determines the intensity of pressure ($p = W/A$). For excessively small area of contact, the intensity of pressure p is quite large and there is a limit upto which pressure intensity can be allowed to increase. When limiting value of pressure intensity, for a given pair of materials, is exceeded a phenomenon called 'seizing' occurs; when particles from one or both contacting materials are torn away and get welded to the other surface. For intensity of pressure below that at which seizing takes place, the value of μ is practically independent of the area of contact surfaces.

(c) Experimentally it has been shown that the fourth law is again approximately true. It is quite logical to expect that a larger force would be required to initiate sliding motion of one body over the other, compared to the one which is necessary to maintain uniform sliding velocity. Thus, coefficient of friction at limiting condition could better be described by 'coefficient of static friction' or 'coefficient of stiction'. Experiments conducted carefully reveal that usually coefficient of friction diminishes slowly but continuously as the velocity of sliding increases. This is called *kinetic coefficient of friction* (μ_k).

Although laws of solid friction stated above are only approximately correct, many practical problems can be solved with sufficient accuracy using these laws. As will be explained later, many revolute joints and sliding pairs have greasy or skin-friction condition and for such condition, laws of solid friction are still applicable with modified (reduced) value of coefficient of friction.

11.4 ANGLE OF REPOSE AND ANGLE OF FRICTION

Angle of repose is the limiting angle of inclination of a plane when a body, placed on the inclined plane, just starts sliding down the plane. This angle equals the limiting angle of friction.

As shown in Fig. 11.2, a body rests on a plane which can be tilted about one of the edges. Let the angle of inclination α of the plane with horizontal be increased gradually until the body starts sliding down the plane. For a value of α, approaching limiting angle of friction, we have for static equilibrium,

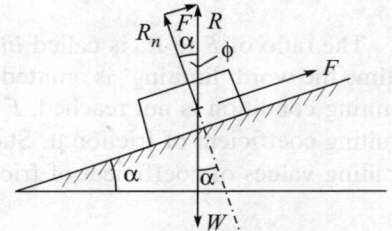

Fig. 11.2 Illustration for Angle of Repose.

$$F = W\sin\alpha \quad \text{and} \quad R_n = W\cos\alpha$$

Dividing out on respective sides,

$$(F/R_n) = \tan\alpha \quad \text{or} \quad F = (\tan\alpha)R_n$$

Thus, as $\alpha \to \phi$ (the limiting angle of friction), $\tan\alpha$ approaches μ, the limiting value of coefficient of solid friction.

Further, as the angle of inclination α of inclined plane is increased, so as to approach liming angle of friction, the body starts sliding down. For equilibrium, the resultant reaction R of the plane on the body must be equal and opposite to the weight W and is inclined at the angle of friction to the normal of the plane.

11.5 MOTION ALONG INCLINED PLANE

There are a number of applications in which inclined plane or its concept is used in one or the other form. We now proceed to establish relationship between different forces under the action of which a body slides either up or down an inclined plane.

Body Having Motion up the Inclined Plane

As shown in Fig. 11.3, let the angle of inclination of the inclined plane to the horizontal be α and let a body of weight W be moved up the plane under the action of a force F inclined at θ to the direction of weight W.

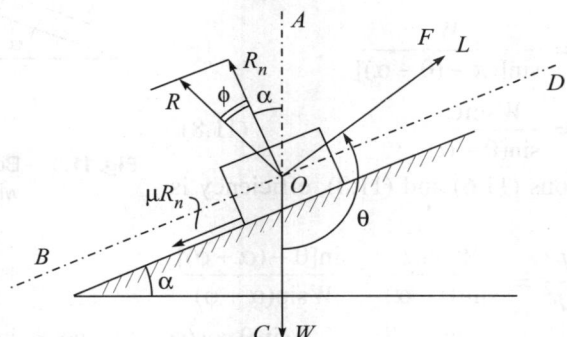

Fig. 11.3 Body having motion up the plane.

As the body moves up the plane, the friction force μR_n acts down the plane. Let R be the resultant of forces R_n and frictional force μR_n, so that the equilibrium of the body can be considered under the action of forces F, W and R.

Applying Lame's theorem,

$$\frac{F}{\sin[\pi - (\alpha + \phi)]} = \frac{W}{\sin[(\pi - \theta) + (\alpha + \phi)]}$$

or $\quad \dfrac{F}{\sin(\alpha + \phi)} = \dfrac{W}{\sin(\theta - \alpha - \phi)]}$ Thus, $\quad F = \dfrac{W\sin(\alpha + \phi)}{\sin[\theta - (\alpha + \phi)]}$ (11.6)

Thus for moving the body up the plane, minimum value of force F is required if the denominator on right-hand side of equation (11.6) is maximum. This is so when $\theta - (\alpha + \phi) = \pi/2$

or when, $\qquad\qquad\qquad \theta - \left(\dfrac{\pi}{2} + \alpha\right) = \phi$ (11.7)

It is clear from Fig. 11.3 that $\angle AOB = (90 + \alpha)$.

Thus, $\qquad\qquad\qquad \angle COD = \angle AOB = \left(\dfrac{\pi}{2} + \alpha\right)$

And hence,

$$\theta - \left(\frac{\pi}{2} + \alpha\right) = \angle LOD, \text{ i.e. the angle at which } F \text{ is inclined to the plane.}$$

Hence, a minimum force F can make the body slide up the plane when it is inclined to the inclined plane at friction angle ϕ.

Efficiency

For the motion of body up the plane, the efficiency of an inclined plane is defined as the ratio of force required to move the body without friction to the force required to move the body with friction.

Thus for the motion of the body up the plane, when frictional force μR_n is neglected, the reaction R coincides with R_n and the various forces acting on body are as shown in Fig. 11.4. Applying Lame's theorem to the forces acting on the body,

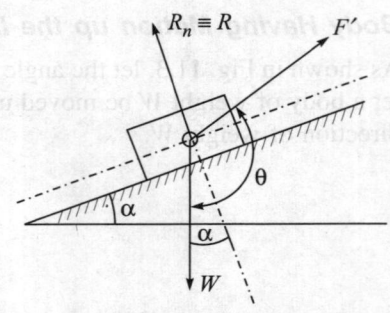

$$\frac{F'}{\sin(\pi - \alpha)} = \frac{W}{\sin[(\pi - (\theta - \alpha)]}$$

or
$$F' = \frac{W\sin\alpha}{\sin(\theta - \alpha)} \qquad (11.8)$$

Fig. 11.4 Body motion up the plane,- without friction.

Hence, from equations (11.6) and (11.8), efficiency is given by

$$\eta_{up} = \frac{F'}{F} = \frac{W\sin\alpha}{\sin(\theta - \alpha)} \times \frac{\sin[\theta - (\alpha + \phi)]}{W\sin(\alpha + \phi)}$$

or
$$\eta_{up} = \frac{\sin\alpha}{\sin\theta\cos\alpha - \cos\theta\sin\alpha} \times \frac{\sin\theta\cos(\alpha + \phi) - \cos\theta\sin(\alpha + \phi)}{\sin(\alpha + \phi)}$$

or
$$\eta_{up} = \frac{\sin\alpha}{\sin\alpha\sin\theta\,(\cot\alpha - \cot\theta)} \times \frac{\sin(\alpha + \phi)\sin\theta[\cot(\alpha + \phi) - \cot\theta]}{\sin(\alpha + \phi)}$$

Therefore
$$\eta_{up} = \frac{\cot(\alpha + \phi) - \cot\theta}{\cot\alpha - \cot\theta} \qquad (11.9)$$

Body Having Motion down the Plane

As the body slides down the plane, the frictional force μR_n acts upwards and therefore, the resultant reaction R is inclined to R_n at ϕ in clockwise sense (see Fig. 11.5). Assuming the force F to act downwards at θ to the line of action of W, Lame's theorem gives

$$\frac{F}{\sin[\pi - (\phi - \alpha)]} = \frac{W}{\sin[\theta + (\phi - \alpha)]}$$

or
$$F = \frac{W\sin(\phi - \alpha)}{\sin[\theta + (\phi - \alpha)]} \qquad (11.10)$$

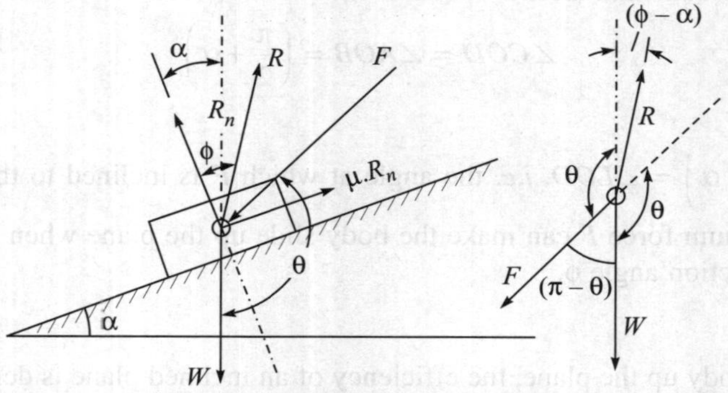

Fig. 11.5 Illustration for Body motion down the plane, with friction.

Above equation suggests that applied force F is positive (*i.e.*, downward force is required to be applied) only when $\phi > \alpha$.

When $\phi = \alpha$, even without applying any force (effort), the body starts sliding down under its own weight.

Again, when $\phi < \alpha$, *i.e.* when angle of friction is less than the angle of inclination of plane, the effort F is negative. In other words, when $\phi < \alpha$, a force equal to F must be applied in the opposite direction to resist the motion.

The condition $\phi > \alpha$ is desirable in applications like screw-jacks where after lifting to a certain height, the load must not come down of its own accord.

It follows from equation (11.10) that for a given value of α, smallest force F is needed when the denominator is maximum. In other words for F to be minimum,

$$\sin[\theta + (\phi - \alpha)] = 1 \quad \text{or} \quad (\theta + \phi - \alpha) = \pi/2$$

giving,
$$F_{min} = W\sin(\phi - \alpha) \tag{11.11}$$

When friction is neglected, $\phi = 0$ and equation (11.10) gives,

$$F' = -\frac{(W\sin\alpha)}{\sin(\theta - \alpha)} \tag{11.12}$$

Negative sign indicates that when friction is absent, a force in opposite direction must be applied to oppose the motion down the plane.

Efficiency. Efficiency of the inclined plane, as in the earlier case, is given by

$$\eta_{down} = \frac{F}{F'} = \frac{W\sin(\phi - \alpha)}{\sin[\theta + (\phi - \alpha)]} \times \frac{\sin(\theta - \alpha)}{W\sin\alpha}$$

or,
$$\eta_{down} = \frac{\sin(\phi - \alpha)}{\sin\theta\cos(\phi - \alpha) + \cos\theta\sin(\phi - \alpha)} \times \frac{\sin\theta\cos\alpha - \cos\theta\sin\alpha}{\sin\alpha}$$

or
$$\eta_{down} = \frac{\sin(\phi - \alpha)}{\sin\theta\sin(\phi - \alpha)[\cot(\phi - \alpha) + \cot\theta]} \times \frac{\sin\theta\sin\alpha[\cot\alpha - \cot\theta]}{\sin\alpha}$$

or
$$\eta_{down} = \frac{(\cot\alpha - \cot\theta)}{\cot(\phi - \alpha) + \cot\theta} \tag{11.13}$$

In special applications (e.g., in screw jacks), $\theta = 90°$ and for such applications from equation (11.9),

$$\eta_{up} = \frac{\cot(\alpha + \phi) - \cot90°}{\cot\alpha - \cot90°}$$

or
$$\eta_{up} = \frac{\cot(\alpha + \phi)}{\cot\alpha} = \frac{\tan\alpha}{\tan(\alpha + \phi)} \tag{11.14}$$

And from equation (11.13),

$$\eta_{down} = \frac{\cot\alpha - \cot90°}{\cot(\phi - \alpha) + \cot90°}$$

or
$$\eta_{down} = \frac{\cot\alpha}{\cot(\phi - \alpha)} = \frac{\tan(\phi - \alpha)}{\tan\alpha} \tag{11.15}$$

11.6 FRICTION OF NUT AND SCREW

A screw, when developed, represents an inclined plane. A turn of a helical curve unwound from a cylinder is shown in Fig. 11.6.

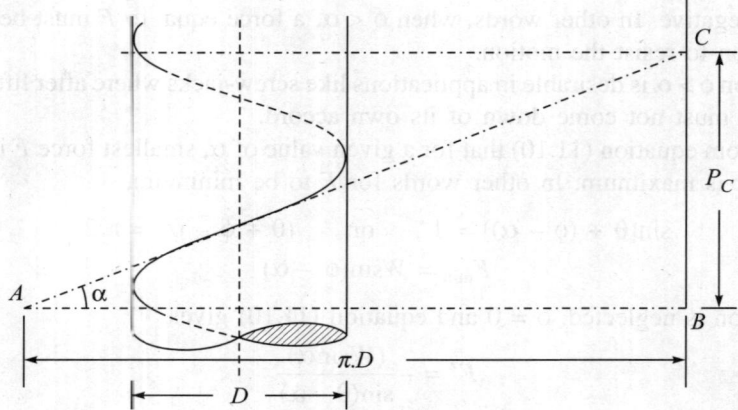

Fig. 11.6 A cylinder with helical curve.

The development is a right angled triangle with base equal to the circumference and height equal to the circumferential pitch of helix p_c. The angle of inclination of line AC with BC is equal to α, the helix angle. In the case of multi-start thread, p_c is to be replaced by the lead l of the screw. Thus, $\tan\alpha = (l/\pi D)$.

The motion of a nut on the given screw may now be considered equivalent to the motion of a body of weight W on the inclined plane such as AC (Fig. 11.7)

(a) *Screw Jack with Square Threads:* A square threaded screw is used for screw jack to lift the load. Consider that the nut of the screw is moved against an axial load W by the rotation of the screw. The load on the nut is transferred to the screw as a distributed load on the surface of the threads in contact. For the purpose of analysis however, this load is assumed to be concentrated at a point on the mean circumference of the thread. Strictly speaking, the helix angle decreases slightly from the root to the tip of the thread. However the depth of the thread is small in comparison with the radius of the screw and, for all practical purposes, it is sufficient if the helix angle at the mean radius of the thread is used.

Referring to the equivalent case of inclined plane depicted in Fig. 11.7(a), force F acting horizontally

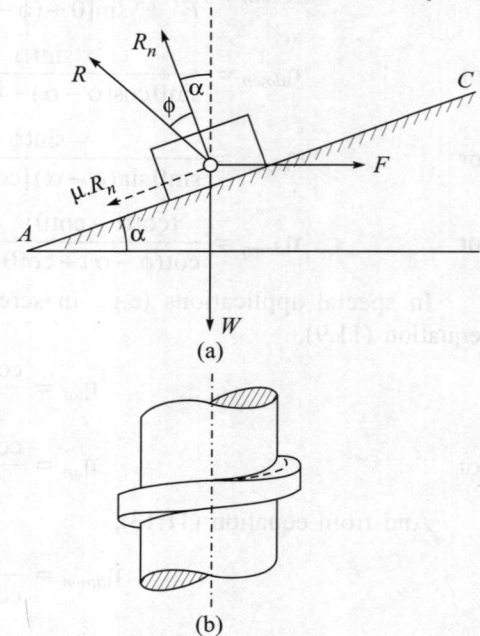

(a)

(b)

Fig. 11.7 Body motion equivalent to that of a Nut on a Screw.

on the sliding mass is the force at the screw-thread required to slide the load W up the inclined plane. For the present case $\theta = 90°$ and from equation (11.6),

$$F = \frac{W\sin(\alpha + \phi)}{\sin[90 - (\alpha + \phi)]}$$

or

$$F = W\tan(\alpha + \phi) \qquad (11.16)$$

or

$$F = W\frac{\tan\alpha + \tan\phi}{1 - \tan\alpha\tan\phi}$$

But

$$\tan\alpha = (l/\pi D) \text{ and}$$

and,

$$\tan\phi = \mu.$$

Hence

$$F = \frac{(l/\pi D) + \mu}{1 - \left(\dfrac{l}{\pi D}\right)\mu}$$

Therefore,

$$F = W\left(\frac{l + \mu\pi D}{\pi D - \mu l}\right) \qquad (11.17)$$

It is usual to provide a lever for applying the force. The lever is fixed to the screw-head. Thus, if L be the length of lever and P be the effort applied at the end of lever then,

$$P(L) = F(D/2) \quad \text{or} \quad P = F\left(\frac{D}{2L}\right) \qquad (11.18)$$

Substituting for F from equation (11.17),

$$P = \frac{WD}{2L}\left(\frac{l + \mu\pi D}{\pi D - \mu l}\right) \qquad (11.19)$$

And substituting for F from equation (11.16) in (11.18),

$$P = \frac{WD}{2L}\tan(\alpha + \phi) \qquad (11.20)$$

Equations (11.19) and (11.20) give effort required at the end of lever for lifting the axial load W. Now consider the case in which the load is to be lowered. From equation (11.10) for $\theta = 90°$,

$$F = \frac{W\sin(\phi - \alpha)}{\sin[90 + (\phi - \alpha)]}$$

or

$$F = W\tan(\phi - \alpha) \qquad (11.21)$$

Substituting for F in equation (11.18),

$$P = \frac{WD}{2L}\tan(\phi - \alpha) \qquad (11.22)$$

or

$$P = \left(\frac{WD}{2L}\right)\frac{\tan\phi - \tan\alpha}{1 + \tan\phi\tan\alpha}$$

Substituting $\tan\phi = \mu$ and, $\tan\alpha = (l/\pi D)$

$$P = \frac{WD}{2L}\left(\frac{\mu\pi D - l}{\pi D + \mu l}\right) \qquad (11.23)$$

Equations (11.22) and (11.23) give effort P required to lower the axial load W. The screw efficiency is given by

$$\text{Screw Efficiency} = \frac{\text{Work done per revolution in lifting load}}{\text{Work done per revolution by the effort}}$$

or

$$\eta = \frac{Wl}{F \times \pi D}$$

Substituting $\dfrac{l}{(\pi D)} = \tan\alpha,$ and $F = W\tan(\alpha + \phi),$

we have

$$\eta = \frac{W\tan\alpha}{W\tan(\alpha + \phi)}$$

Therefore

$$\eta = \frac{\tan\alpha}{\tan(\alpha + \phi)} \qquad (11.24)$$

The screw efficiency for lifting the load is maximum when,

$$\frac{d\eta}{d\alpha} = 0$$

Differentiating equation (11.24) w.r. to α,

$$\frac{\sec^2\alpha\tan(\alpha + \phi) - \tan\alpha\sec^2(\alpha + \phi)}{\tan^2(\alpha + \phi)} = 0$$

or $\sec^2\alpha\tan(\alpha + \phi) - \tan\alpha\sec^2(\alpha + \phi) = 0$

or $\dfrac{\tan(\alpha + \phi)}{\sec^2(\alpha + \phi)} = \dfrac{\tan\alpha}{\sec^2\alpha}$

or $\sin(\alpha + \phi)\cos(\alpha + \phi) = \sin\alpha\cos\alpha$

or $\sin2(\alpha + \phi) = \sin2\alpha$

This is possible either when $\phi = 0$, *i.e.,* when there is no friction (*i.e.,* $\mu = 0$) or when,

$$\sin2(\alpha + \phi) = \sin(\pi - 2\alpha)$$

i.e., when $2(\alpha + \phi) = \pi - 2\alpha$

or $\alpha = \dfrac{\pi - 2\phi}{4}$

or $\alpha = \left(45° - \dfrac{\phi}{2}\right) \qquad (11.25)$

Thus, maximum efficiency during lifting, as obtained by substituting for α from equation (11.25) in (11.24) is

$$\eta_{max} = \frac{\tan\left(45° - \dfrac{\phi}{2}\right)}{\tan\left(45° - \dfrac{\phi}{2} + \phi\right)} = \frac{\tan\left(45° - \dfrac{\phi}{2}\right)}{\tan\left(45° + \dfrac{\phi}{2}\right)}$$

or,
$$\eta_{max} = \left(\frac{\tan 45° - \tan\dfrac{\phi}{2}}{1 + \tan 45° \tan\dfrac{\phi}{2}}\right)\left(\frac{1 - \tan 45° \tan\dfrac{\phi}{2}}{\tan 45° + \tan\dfrac{\phi}{2}}\right)$$

or
$$\eta_{max} = \frac{\left(1 - \tan\dfrac{\phi}{2}\right)\left(1 - \tan\dfrac{\phi}{2}\right)}{\left(1 + \tan\dfrac{\phi}{2}\right)\left(1 + \tan\dfrac{\phi}{2}\right)} = \frac{\left(1 - \tan\dfrac{\phi}{2}\right)^2}{\left(1 + \tan\dfrac{\phi}{2}\right)^2}$$

Substituting for $\qquad \tan\dfrac{\phi}{2} = \dfrac{\sin(\phi/2)}{\cos(\phi/2)}$ and simplifying,

$$\eta_{max} = \frac{\cos^2\dfrac{\phi}{2} + \sin^2\dfrac{\phi}{2} - 2\sin\dfrac{\phi}{2}\cos\dfrac{\phi}{2}}{\cos^2\dfrac{\phi}{2} + \sin^2\dfrac{\phi}{2} + 2\sin\dfrac{\phi}{2}\cos\dfrac{\phi}{2}}$$

or
$$\eta_{max} = \left(\frac{1 - \sin\phi}{1 + \sin\phi}\right) \qquad (11.26)$$

Again, mechanical advantage
$$= \frac{W}{P} = \frac{W}{\dfrac{WD}{2L}\tan(\alpha + \phi)}$$

or
$$\text{M.A.} = \frac{2L}{D\tan(\alpha + \phi)} = \left(\frac{2L}{D}\right)\cot(\alpha + \phi) \qquad (11.27)$$

The velocity ratio is given by V.R. $= \dfrac{\text{Distance moved through by effort } P/\text{revolution}}{\text{Distance moved through load/revolution}}$

Therefore
$$VR = \frac{2\pi L}{l} = \frac{L}{l/2\pi}$$

But from
$$\tan\alpha = (l/\pi D), \qquad (11.28)$$

and,
$$(l/\pi) = D\tan\alpha$$

substituting from equation (11.28), in the expression for V.R.

$$VR = \left(\frac{2L}{D\tan\alpha}\right) \qquad (11.29)$$

Self Locking Screws

It follows from equation (11.24) that efficiency of power screw, while lifting the load, is

$$\eta = \frac{\tan\alpha}{\tan(\alpha + \phi)}$$

It was seen during discussions on equation (11.10) that when $\phi = \alpha$, no extra force is required to cause downward motion of the mass and when $\phi < \alpha$, a force F must be applied in opposite direction to resist downwards sliding of mass. Thus, in order to ensure that in the elevated position the load does not slide down, $\phi > \alpha$.

Stated in other words the screw will not overhaul or there will be self-locking during raised position of mass (i.e., mass will not slide down of its own accord) when friction angle ϕ is greater than angle of helix. Thus, for a limiting condition, when $\phi \rightarrow \alpha$.

$$\eta = \frac{\tan\phi}{\tan(\phi + \phi)} = \frac{\tan\phi}{\tan 2\phi}$$

or

$$\eta = \frac{\tan\phi}{2\tan\phi}\,(1 - \tan^2\phi) = \frac{1}{2}\,(1 - \tan^2\phi)$$

With common values of $\mu = 0.1$ to 0.25, $\mu^2\ (\equiv \tan^2\phi)$ is quite small compared to 1 and can therefore be neglected in comparison to 1. Hence,

$$\eta \approx \frac{1}{2} = 50\%$$

Thus reversal of the nut is avoided and the screw becomes self locking, only if the efficiency is less than 50% approximately.

(b) **V– Threads:** Even if the helix angle is neglected, the faces of V-threads have an included angle of 2β and the faces are inclined to the axis of the spindle. Figure 11.8 shows a section through a V-thread. The axial component of R_n must be equal to the axial load W. Thus

$$W = R_n\cos\beta \qquad (11.30)$$

Hence,

$$R_n = \left(\frac{W}{\cos\beta}\right)$$

Then the friction force acting tangentially to the thread surface is,

$$\mu R_n = \frac{\mu W}{\cos\beta} = \mu_1 W \qquad (11.31)$$

With $\mu_1 = \mu/\cos\beta$ as the equivalent coefficient of friction, all the expressions derived for square threads are applicable to V-threads. This implies that $\tan\phi$ term is to be always replaced by $(\tan\phi/\cos\beta)$.

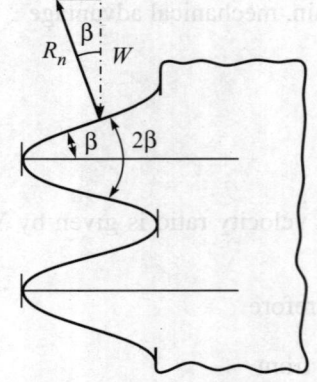

Fig. 11.8 A section through V-threads.

EXAMPLE 11.1 A load of 10 k.N is raised by means of a screw jack, having a square threaded screw of 12.0 mm pitch and of mean diameter 50 mm. If a force of 100 N is applied at the end of a lever to raise the load, what should be the length of the lever used? Take coefficient of friction = 0.15. What is the mechanical advantage obtained? (AMIE, Summer 1984)

Solution: The helix angle α is given by

$$\tan\alpha = \frac{p_c}{\pi D} = \frac{12.00}{\pi(50)} = 0.07639 \text{(single start thread assumed)}$$

Therefore $\alpha = \tan^{-1}(0.07639) = 4.3686$

The motion of nut is analogous to the sliding motion of a body along the inclined plane. As the force F acts horizontally, $\theta = \pi/2$, and is given by

$$F = W\tan(\alpha + \phi)$$

Now $\phi = \tan^{-1}(0.15) = 8.531°$

Therefore $F = 10000 \tan(4.3686 + 8.531)$

or $F = 2290.23$ N

Also, if P be the effort at the end of lever of length L,

$$P \times L = F \times \frac{D}{2}$$

Therefore $L = \frac{F}{P} \times \frac{D}{2} = \frac{2290.23}{100} \times \frac{50}{2} = 572.56 \text{ mm} = 57.25 \text{ cm}$ **Ans.**

$$\text{Mechanical Advantage} = \frac{W}{P} = \frac{10000}{100} = 100 \qquad \textbf{Ans.}$$

EXAMPLE 11.2 (SI)   The mean diameter of a square threaded screw jack, shown in Fig. 11.9, is 50 mm. The pitch of the thread is 10 mm. The coefficient of friction is 0.15. What force must be applied at the end of a 0.7 m long lever which is perpendicular to the longitudinal axis of the screw to raise a load of 20 kN and to lower it.
(Osmania Univ.)

Solution: Given $D = 500$ mm; $p_c = 10$ mm; $\mu = \tan \phi = 0.15$

$$W = 20000 \text{ N}; \phi = \tan^{-1}(0.15) = 8.531°$$

$$\tan\alpha = \frac{10}{\pi \times 50}$$

Therefore $\alpha = \tan^{-1}\left(\dfrac{1}{5\pi}\right) = 3.643°$

Also from

$$F = W\tan(\alpha + \phi) = 20000 \tan(3.643 + 8.531) = 4314.65 \text{ N}$$

The effort P required at the end of the lever is given by

$$P \times L = F \times \frac{D}{2}$$

or $P = F \times \dfrac{D}{2L}$

or $P = 4314.65 \times \dfrac{50}{700 \times 2} = 154.09 \text{ N}$ **Ans.**

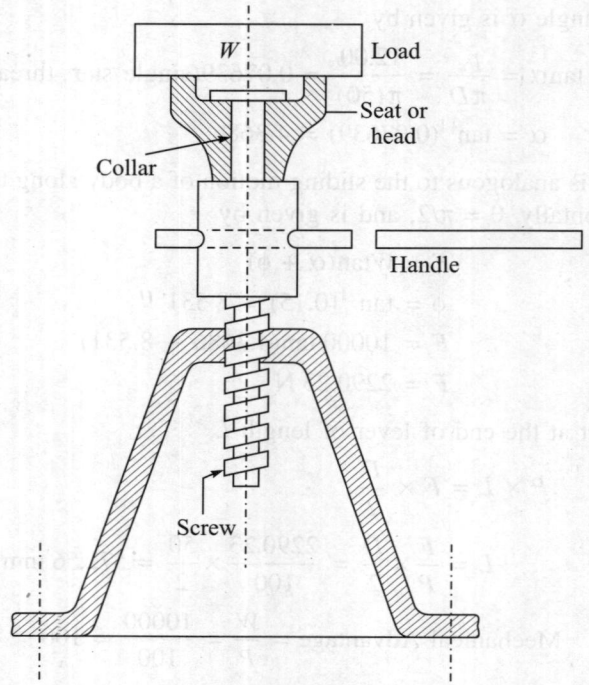

Fig. 11.9 A Screw Jack.

Force required at mean radius to lower the load F'

$$F' = W \tan(\phi - \alpha)$$

or
$$F' = 20000 \tan(8.531 - 3.643)$$

Hence,
$$F' = 1710.385 \text{ N}$$

Hence, for lowering, the effort P' required is given by

$$P' \times L = F' \times \frac{D}{2}$$

Therefore
$$P' = F' \times \frac{D}{2L} = 1710.385 \times \frac{50}{2 \times 700}$$

or
$$P' = 61.085 \text{ N} \qquad \textbf{Ans.}$$

EXAMPLE 11.3 Two tie rods are connected by a turn-buckle having right and left-handed threads. The threads are V-type (metric) and have a pitch of $\frac{1}{2}$ cm (*i.e.*, 2 threads per cm) on a mean diameter of 3 cm and a thread angle of 60°. Assuming $\mu = 0.18$, find the torque required to produce a pull of 39 kN

  (a) When the rods are tightened,
  (b) When the rods are loosened.

Solution: $W = 39$ kN

$$\mu' = \text{equivalent coeff. of friction} = \frac{0.18}{\cos\left(\dfrac{60}{2}\right)} = 0.2078$$

or, $$\phi = \tan^{-1}(0.2078) = 11.739°$$

Also $$\tan\alpha = \frac{0.5}{\pi \times 3} = 0.05305$$

or, $$\alpha = \tan^{-1}(0.05305) = 3.037°$$

(a) Force required at the mean diameter of threads for tightening

$$F = W\tan(\alpha + \phi)$$

or $$F = 39\tan(3.037° + 11.739°) = 10.2868 \text{ kN}$$

$$\text{The torque required} = F \times \frac{D}{2}$$

or $$10.2868 \times \frac{3}{2} = 15.4302 \text{ kN.m} \qquad \textbf{Ans.}$$

(b) The force required at pitch radius when the rods are slackend,

$$F' = W\tan(\phi - \alpha)$$

or $$F' = 39\tan(11.739° - 3.037°) = 5.9692 \text{ kN}$$

Total torque taking into account either side

$$= F' \times \frac{D}{2} = 5.9692 \times \frac{3}{2} = 8.954 \text{ kN.cm} \qquad \textbf{Ans.}$$

11.7 WEDGE

Like a screw jack, a wedge is also used to lift loads. It is a good example of three links forming a mechanism. It consists of three sliding pairs. The first link is the frame link 1, the second link is the wedge 2 itself, and the third link is the slider 3 which carries the load W.

Case (1) *When Friction is Neglected*

Considering the equilibrium of wedge, it is acted upon by

(a) the horizontal driving force F,

(b) the reaction R_{21} of the horizontal face of frame link, in vertical direction, and

(c) the reaction R_{23} of the slider link 3 in a direction normal to the inclined face of wedge (*Refer* to Fig. 11.10)

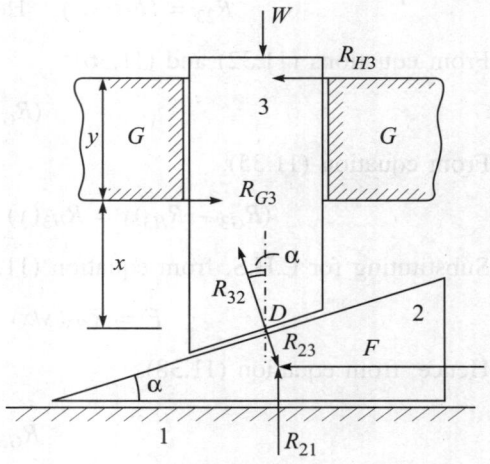

Fig. 11.10 Force analysis in a wedge,-without friction.

For equilibrium of wedge,

$$R_{23}\sin\alpha = F \tag{11.32}$$

and

$$R_{23}\cos\alpha = R_{21} \tag{11.33}$$

Dividing out

$$F = R_{21}\tan\alpha$$

or

$$R_{21} = F\cot\alpha \tag{11.34}$$

Considering equilibrium of slider (3), it is acted upon by

(a) the weight W vertically downwards,
(b) the reaction R_{32} from wedge, equal and opposite to R_{23}, and
(c) the reaction from the guides G.

The line of action of weight W and reaction R_{32} meets at D. Hence, for complete equilibrium of the slider, the reaction from guide G must pass through D. However from practical considerations, guide G is generally placed above D. Thus the reaction R_{G3} from guide is equivalent to a force R'_{G3} ($=R_{G3}$) at D together with a couple ($R_{G3} \times x$) in c.w. direction on the slider. To balance this couple, a reaction R_{H3} acts at the upper end of guides. The effect of this reaction is to apply a force R'_{H3} ($=R_{H3}$) at D, together with a couple R_{H3} ($x + y$) in c.c.w. sense and this couple must balance the earlier couple. Thus,

$$(R_{G3})x = R_{H3}(x + y) \tag{11.35}$$

Also, for force equilibrium at D,

$$(R_{G3} - R_{H3}) = R_{32}\sin\alpha \tag{11.36}$$

and

$$W = R_{32}\cos\alpha$$

But from equations (11.33) and (11.34),

$$R_{23} = (F/\sin\alpha). \quad \text{Hence from above,} \quad W = F\cot\alpha \tag{11.37}$$

From equations (11.32) and (11.36),

$$(R_{G3} - R_{H3}) = F \tag{11.38}$$

From equation (11.35),

$$(R_{G3} - R_{H3})x = R_{H3}(y) \quad \text{or} \quad R_{G3} - R_{H3} = R_{H3}(y/x) \tag{11.39}$$

Substituting for L.H.S. from equation (11.38),

$$F = R_{H3}(y/x) \quad \text{or} \quad R_{H3} = (x/y)F \tag{11.40}$$

Hence, from equation (11.38),

$$R_{G3} = \left(\frac{x+y}{y}\right)F \tag{11.41}$$

Case (2) When Friction is Considered

Let ϕ = the limiting angle of friction between the frame and wedge and also between wedge and slider, and

ϕ' = the limiting angle of friction between guides and slider.

Referring to Fig. 11.11(a), the forces acting on wedge are as shown in Fig. 11.11(b). The reactions R_{21} and R_{23} are the resultant reactions. From Lame's theorem,

$$\frac{F}{\sin[\pi - (\alpha + 2\phi)]} = \frac{R_{21}}{\sin\left[\dfrac{\pi}{2} + (\alpha + \phi)\right]} = \frac{R_{23}}{\sin\left[\dfrac{\pi}{2} + \phi\right]}$$

or

$$\frac{F}{\sin(\alpha + 2\phi)} = \frac{R_{21}}{\cos(\alpha + \phi)} = \frac{R_{23}}{\cos\phi}$$

Therefore

$$R_{21} = F\frac{\cos(\alpha + \phi)}{\sin(\alpha + 2\phi)} \tag{11.42}$$

and

$$R_{23} = F\frac{\cos\phi}{\sin(\alpha + 2\phi)} \tag{11.43}$$

Consider the equilibrium of slider. The slider moves up under the wedging action. The resultant reactions R_{G3} and R_{H3} are therefore inclined to respective normals at ϕ' on lower side. The slider is thus in equilibrium under the action of forces, R_{G3}, R_{H3}, W' and R_{32}. Let O be the point of intersection of R_{G3} and R_{H3}.

Taking moments of components of R_{32} and W' about O,

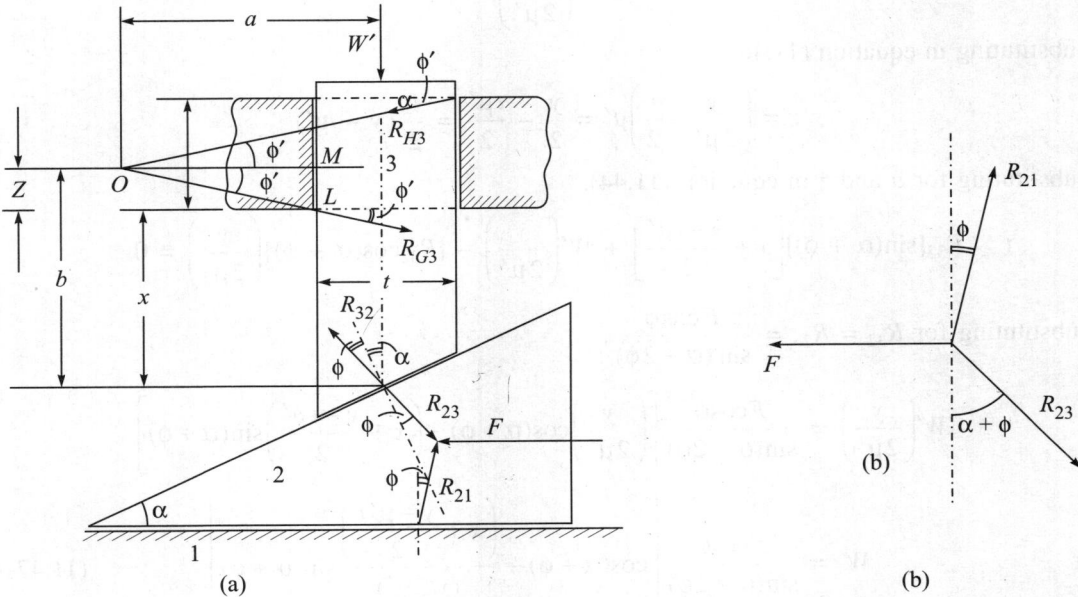

Fig. 11.11 Force analysis of a wedge,-with friction.

$$R_{32}[\sin(\alpha + \phi)](x + z) + W'(a) - R_{32}\cos(\alpha + \phi)a = 0 \qquad (11.44)$$

where
$$R_{32} = R_{23} = \frac{F\cos\phi}{\sin(\alpha + 2\phi)}$$

Also, from the geometry of Fig. (11.11),

$$\tan\phi' = \frac{y - z}{a + \dfrac{t}{2}}$$

Therefore
$$\mu' = \frac{y - z}{a + \dfrac{t}{2}}$$

or
$$\left(a + \frac{t}{2}\right)\mu' = y - z$$

or
$$z = y - \left(a + \frac{t}{2}\right)\mu' \qquad (11.45)$$

Also from right angled triangle OLM,

$$z = \left(a - \frac{t}{2}\right)\tan\phi' = \left(a - \frac{t}{2}\right)\mu' \qquad (11.46)$$

Comparing right-hand sides of equations (11.45) and (11.46),

$$y - \left(a + \frac{t}{2}\right)\mu' = \left(a - \frac{t}{2}\right)\mu'$$

or
$$a = \left(\frac{y}{2\mu'}\right)$$

Substituting in equation (11.46),

$$z = \left(\frac{y}{2\mu'} - \frac{t}{2}\right)\mu' = \frac{y}{2} - \frac{t\mu'}{2} = \frac{1}{2}\,(y - t\mu')$$

Substituting for a and z in equation (11.44),

$$R_{32}[\sin(\alpha + \phi)]\left[x + \frac{y - t\mu'}{2}\right] + W'\left(\frac{y}{2\mu'}\right) - [R_{32}\cos(\alpha + \phi)]\left(\frac{y}{2\mu'}\right) = 0$$

Substituting for $R_{32} = R_{23} = \dfrac{F\cos\phi}{\sin(\alpha + 2\phi)}$,

$$W'\left(\frac{y}{2\mu'}\right) = \frac{F\cos\phi}{\sin(\alpha + 2\phi)}\left[\left(\frac{y}{2\mu'}\right)\cos(\alpha + \phi) - \left(x + \frac{y - t\mu'}{2}\right)\sin(\alpha + \phi)\right]$$

or
$$W' = \frac{F\cos\phi}{\sin(\alpha + 2\phi)}\left[\cos(\alpha + \phi) - \frac{\left(x + \dfrac{y - \mu't}{2}\right)}{(y/2\mu')}\sin(\alpha + \phi)\right] \qquad (11.47)$$

Efficiency of a wedge is the ratio of load lifted when friction is considered, to the load being lifted when the friction is neglected, the force applied remaining the same. Thus, dividing on respective sides of equation (11.47) by equation (11.37)

$$\eta = \frac{W'}{W} = \frac{\cos\phi\tan\alpha}{\sin(\alpha + 2\phi)}\left[\cos(\alpha + \phi) - \left(\frac{2\mu'}{y}\right)\left(x + \frac{y - \mu't}{2}\right)\sin(\alpha + \phi)\right] \qquad (11.48)$$

When x and t are small in comparison to other dimensions, neglecting x and t,

$$\eta = \frac{W'}{W} = \frac{\cos\phi\tan\alpha}{\sin(\alpha + 2\phi)}\left[\cos(\alpha + \phi) - \left(\frac{2\mu'}{y}\right)\left(\frac{y}{2}\right)\sin(\alpha + \phi)\right]$$

$$= \frac{\cos\phi\tan\alpha}{\sin(\alpha + 2\phi)}\left[\cos(\alpha + \phi) - \tan\phi'\sin(\alpha + \phi)\right] \qquad (11.49)$$

$$= \frac{\cos\phi\tan\alpha}{\sin(\alpha + 2\phi)}\left[\frac{\cos(\alpha + \phi)\cos\phi' - \sin(\alpha + \phi)\sin\phi'}{\cos\phi'}\right]$$

$$= \frac{\cos\phi\tan\alpha}{\sin(\alpha + 2\phi)}\left[\frac{\cos(\alpha + \phi + \phi')}{\cos\phi'}\right] \qquad (11.50)$$

And, if $\phi = \phi'$,
$$\eta = \frac{\tan\alpha}{\tan(\alpha + 2\phi)} \qquad (11.51)$$

11.8 ROLLING FRICTION

Friction exists only when one body slides or tends to slide relative to a second body with which it is in contact. If the relative motion is that of pure rolling, ideally there is no friction and the two bodies make line or point contact, parallel to the axis of rotations. It is usual to assume that for a given pair of rolling surfaces, force of rolling friction is proportional to the load and is independent of relative curvatures.

When a cylinder under vertical load W rolls over a flat surface, deformation can occur either in the surface, in the rolling cylinder or both depending on their relative resistance to deformation. These cases are shown at Figs. 11.12(b) and (c) while the one at Fig. 11.12(a) shows a line/point contact.

When both the surfaces are quite hard, as at Fig. 11.12(a), the line of action of W passes through the line of contact, and there is no resisting couple acting on the cylinder. For cylinders at Figs. 11.12(b) and (c) because of depression in ground or wheel the entire cylinder must be rotated bodily about leading edge A, for any further motion. As the line of action of W does not pass through the leading edge, but at a distance of x from it, the wheel is subjected to a resisting couple = Wx.

When rolling occurs with uniform velocity, the couple required to cause rolling–

$$F\,h = W\,x. \qquad (11.52)$$

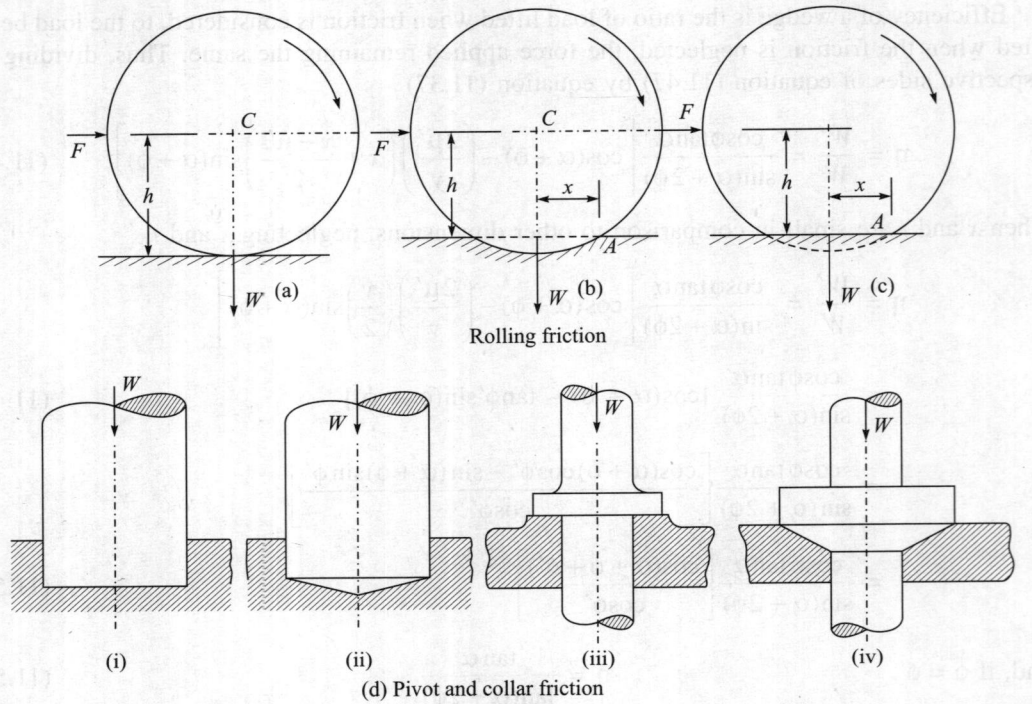

Rolling friction

(d) Pivot and collar friction

Fig. 11.12

In equation (11.52), x is known as coefficient of rolling friction and has linear units like milimetre, centimetre, etc. In this regard, it is different from coefficient of sliding friction μ, which is dimensionless.

Let

$\qquad F_r$ = force required to cause rolling motion

$\qquad F_s$ = force required to cause sliding motion

Then, from equation (11.52),

$$F_r = \left(\frac{x}{h}\right)W \text{ and, for sliding motion,}$$

$$F_s = \mu W.$$

Then as the applied force F is increased gradually, body will roll without sliding if,

$$F_s > F_r$$

or, substituting for F_r & F_s, this condition reduces to $\mu > \left(\dfrac{x}{h}\right)$

Similarly, body will have pure sliding motion when

$$F_r > F_s$$

or, when

$$\left(\frac{x}{h}\right) > \mu$$

Finally body will roll and slide if,

$$F_s = F_r. \quad \text{That is when, } \mu = \left(\frac{x}{h}\right)$$

11.9 PIVOT AND COLLAR FRICTION

Unlike journal bearings, which support transverse (radial) loads, pivot and collar bearings are required to take up axial loads on shafts. Again a collar bearing differs from a pivot bearing in the sense that a shaft extends through and beyond the collared bearing while it is terminated at pivot in pivot bearings. When placed at the end of a vertical shaft, a pivot bearing is called *a foot-step bearing.*

Examples of axial thrust can be found in

(a) the propeller shafts of naval ships,
(b) the shafts of steam turbines,
(c) the vertical shafts of machines and spindles of drilling machines
(d) shaft carrying helical gear singly, etc.

The surface or surfaces, on which thrust is carried, are usually plane surfaces at right angles to the axis of shaft. Occasionally conical surfaces, with the cone axis coinciding with the axis of rotation, are also used for above purpose. The collars may have flat bearing surface or conical bearing surface (as shown in Fig. 11.12(d)), however, the flat surfaces are more common. Several collars may be provided on shaft to reduce the intensity of pressure.

Friction in Pivot and Collar

In order to preserve the shafts in correct axial position, bearing surfaces are provided. Relative motion between the contact surfaces of a thrust bearing is resisted by the friction between the rubbing surfaces. Thus, before a relative motion can occur between the rubbing surfaces, a torque or a couple must be applied to overcome the friction torque.

Present knowledge of friction is inadequate to state the exact physical changes taking place between rubbing surfaces. There is a considerable uncertainty in regards to the distribution of the axial load over the entire contacting surface area. However, if the bearings are in brand new condition, the fit between the two rubbing surfaces can be assumed to be perfect. In such cases the axial load can be assumed to be distributed more or less uniformly and resulting normal intensity of pressure can be assumed to be constant. It is usual to assume that coefficient of friction μ is constant for all the points on the surface.

When the bearings become old and worn out, the uneven bearing surfaces are incapable of ensuring a uniform pressure distribution. The rate of wear of bearing surfaces, on the other hand, depends not only on the intensity of pressure but also on rubbing velocity i.e., rate wear = φ (v, p). Exact relationship between pressure, rubbing velocity and rate of wear is not known.

In one sense however, wear appears to be a self adjusting phenomenon. If at one spot the wear is less than the rest of the surface, the increased pressure on high spots on this area will produce increased wear at the same spot. It is therefore reasonable to assume that the rate of wear in unlubricated bearings is uniform. The rate of wear is approximately proportional to the pressure and velocity of rubbing both. For uniform rate of wear, therefore, *pv* = constant. For a given relative speed of rotation ω, the velocity of rubbing v is directly proportional to the radius r. It follows that

$$pr = \text{constant (for uniform rate of wear)} \qquad (11.53)$$

Above discussions thus lead to the following two common assumptions in problems involving pivot and collar friction:

1. *Assumption of Uniform Pressure Intensity,* which assumes axial load to be uniformly distributed over the entire bearing surface *i.e.,*

$$p = \text{constant} = C$$

2. *Assumption of Uniform Rate of Wear,* which implies

$$pr = \text{constant} = C$$

Thus if p_o and p_i are the pressures at outer and inner radii r_o and r_i respectively, then

$$p_o r_o = p_i r_i = C \qquad (11.54)$$

Note: The law $p \propto 1/r$ is obtained on the basis of assumption that the friction occurs between two solid surfaces, *i.e.,* no lubricant exists in between the two bearing surfaces. Further, the relation $p \propto 1/r$ leads to an absurd conclusion that at the centre, where $r = 0$, the pressure intensity is infinite. *Thus, none of the two assumptions stated above can give correct value of friction moment and the exact value may lie somewhere between the values obtained on the basis of the two assumptions.*

11.10 AXIAL FORCE AND FRICTION MOMENT IN PIVOTS AND COLLARS

Relative motion between the contacting surfaces in a pivot and collar bearing is resisted by the friction. Thus, before any relative motion takes place between contacting surfaces a driving torque must be applied to overcome this friction.

To arrive at a most generalised expression for friction torques, let us consider a conical bearing surface as shown in Fig. 11.13. Let W be the axial load supported by the conical bearing surface with an apex angle of 2α. Let the minimum and maximum radii of actual contact be r_2 and r_1 respectively. Consider a ring of elemental bearing surface of radius r and radial width δr, which has a width δl along the conical surface.

Let, p = intensity of normal, pressure over bearing surface at radius r,

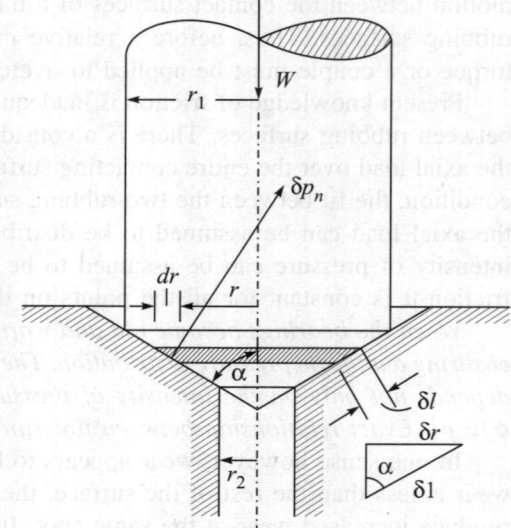

Fig. 11.13 Truncated Conical Pivot Bearing.

$$\delta A = \text{elemental ring area} = (2\pi r)\delta l$$

or $$\delta A = (2\pi r)\frac{\delta r}{\sin\alpha}.$$

Then, the elemental normal force on the ring area of radial width δr,

$$\delta P_n = p\delta A = p(2\pi r)\frac{\delta r}{\sin\alpha} \qquad (11.55)$$

Component of this force in a direction normal to axis cancels out due to symmetry. The axial component, along the axis of the shaft, is given by

$$\delta P = (p\delta A)\cos(90° - \alpha) = (p\delta A)\sin\alpha = \left(p2\pi r\frac{\delta r}{\sin\alpha}\right)\sin\alpha$$

Therefore $$\delta P = (p2\pi)r\delta r$$

The entire bearing surface can be assumed to consist of a large number of such elemental rings and the total force contribution in axial direction is

$$P = \int_{r_2}^{r_1} (p2\pi)rdr \qquad (11.56)$$

Again, the frictional force on the elemental ring area (in the peripheral direction),

$$\delta F = \mu\delta P_n = \mu p(2\pi r)\frac{\delta r}{\sin\alpha} \qquad (11.57)$$

The elemental frictional moment due to ring area about the axis of shaft is

$$\delta M_f = (\mu\delta P_n) \times r$$

$$= \left(\mu\frac{2\pi r\delta r p}{\sin\alpha}\right) \times r$$

or $$\delta M_f = \mu r^2\frac{2\pi p}{\sin\alpha}\delta_r$$

Hence, the total frictional moment in bearing

$$M_f = \int_{r_2}^{r_1} \mu r^2\left(\frac{2\pi p}{\sin\alpha}\right)dr \qquad (11.58)$$

Resultant force P in the axial direction must be equal to weight W for equilibrium. Hence, from equation (11.56),

$$W = P = \int_{r_2}^{r_1} (p2\pi)rdr$$

or $$W = 2\pi \int_{r_2}^{r_1} (pr)dr \qquad (11.59)$$

Also, frictional moment from equation (11.58),

$$M_f = \frac{2\pi\mu}{\sin\alpha} \int_{r_2}^{r_1} (pr^2)dr \qquad (11.60)$$

Flat pivot and collar bearings can be considered to be a particular case of conical surfaces with semicone angle $\alpha = 90°$ and $\sin \alpha = 1$.

Hence from equations (11.59) and (11.60), for flat pivot & collar bearing

$$W = 2\pi \int_{r_2}^{r_1} (pr)dr \tag{11.61}$$

and

$$M_f = 2\pi\mu \int_{r_2}^{r_1} (pr^2)dr \tag{11.62}$$

Assumption of Uniform Pressure Intensity

With pressure intensity $p = $ constant all over the bearing surfaces, equation (11.59) simplifies to

$$W = 2\pi p \int_{r_2}^{r_1} rdr = (2\pi p)\left(\frac{r^2}{2}\right)_{r_2}^{r_1}$$

$$W = \pi p(r_1^2 - r_2^2) \tag{11.63}$$

and from equation (11.60)

$$M_f = \frac{2\pi\mu p}{\sin\alpha} \int_{r_2}^{r_1} r^2 dr$$

or,

$$M_f = \frac{2\pi\mu p}{\sin\alpha} \left(\frac{r^3}{3}\right)_{r_2}^{r_1} = \frac{2}{3}\frac{\mu p\pi}{\sin\alpha} (r_1^3 - r_2^3) \tag{11.64}$$

In a number of problems, intensity of pressure p is not given but axial load W is specified. For such problems, it is more convenient to express equation (11.64) in terms of axial load W. Hence, dividing equation (11.64) by (11.63) on respective sides,

$$\frac{M_f}{W} = \frac{2}{3}\frac{\mu}{\sin\alpha}\left(\frac{r_1^3 - r_2^3}{r_1^2 - r_2^2}\right)$$

or

$$M_f = \frac{2}{3}\frac{\mu W}{\sin\alpha}\left(\frac{r_1^3 - r_2^3}{r_1^2 - r_2^2}\right) \tag{11.65}$$

Equations (11.63) and (11.65) therefore give expressions for required axial load W and frictional torque for conical bearing surfaces.

For Flat Bearing Surfaces (Particular Case). As in earlier case, substituting $\alpha = 90°$ in equations (11.63) and (11.65), the respective expressions for flat pivot/collar bearings are

$$W = \pi p(r_1^2 - r_2^2) \tag{11.66}$$

and

$$M_f = \frac{2}{3}\mu W\left(\frac{r_1^3 - r_2^3}{r_1^2 - r_2^2}\right) \tag{11.67}$$

11.10.2 Assumption of Uniform Rate of Wear

With this assumption, $(pr) = C = $ constant. Substituting for (pr) in the integrand of equation (11.59),

$$W = 2\pi \int_{r_2}^{r_1} C dr \text{ or, } W = 2\pi C (r_1 - r_2) \tag{11.68}$$

Similarly, substituting for product (pr) in equation (11.60),

$$M_f = \frac{2\pi\mu}{\sin\alpha} \int_{r_2}^{r_1} C r dr$$

$$M_f = \frac{2\pi\mu C}{\sin\alpha} \left(\frac{r_1^2 - r_2^2}{2} \right) \tag{11.69}$$

Thus, dividing equation (11.69) by (11.68) on each side,

$$\frac{M_f}{W} = \frac{\mu}{2\sin\alpha} (r_1 + r_2)$$

or

$$M_f = \frac{1}{2} \frac{\mu W}{\sin\alpha} (r_1 + r_2) \tag{11.70}$$

Equations (11.68) and (11.70) give the required axial force and frictional torque generated for conical surfaces.

For Flat Surfaces taking $\alpha = 90°$, equations (11.68) and (11.70) yield

$$W = 2\pi C (r_1 - r_2) \tag{11.71}$$

and

$$M_f = \frac{1}{2} \mu W (r_1 + r_2) \tag{11.72}$$

EXAMPLE 11.4 A pair of dry friction surfaces is required to transmit power at 3600 r.p.m., with friction surfaces having coefficient of friction as 0.3 and an outer radius of 12 cm. The axial pressure is not to exceed 14.5 N/cm². Calculate the axial force required to develop the required pressure as also the frictional moment transmitted using assumptions of

(a) uniform pressure intensity, and
(b) uniform rate of wear

for ratio of radii $(r_o/r_i) = $ 2, 1.8, 1.7, 1.6, 1.5, 1.4, 1.3, 1.25, 1.2, 1.15, 1.1. Compare the results in both the cases.

Solution: *For uniform intensity of pressure, let* $k = (r_1/r_2) \equiv (r_o/r_i)$.
Hence, from equation (11.63)

$$W = \pi p(r_1^2 - r_2^2) = \pi p r_1^2 \left(1 - \frac{r_2^2}{r_1^2} \right)$$

or

$$W = \pi(14.5)(12)^2 \left(1 - \frac{1}{k^2} \right) = \pi (14.5)(12)^2 \left(\frac{k^2 - 1}{k^2} \right)$$

or,

$$W = (6559.64)\left(\frac{k^2 - 1}{k^2}\right) \qquad \text{(i)}$$

Also

$$M_f = \frac{2}{3}\mu p\pi\,(r_1^3 - r_2^3)$$

or

$$M_f = \frac{2}{3}(0.3)(14.5)(\pi)(12)^3\left(1 - \frac{1}{k^3}\right)$$

or

$$M_f = 15743.149\left(\frac{k^3 - 1}{k^3}\right) \qquad \text{(ii)}$$

Values W and M_f, for uniform pressure assumption are listed in Table 11.1 for various values of k as calculated using expressions (i) and (ii).

For uniform rate of wear, as the limiting (maximum) pressure occurs at inner radius r_2 (as $p \propto 1/r$), and is given by

$$p_i \equiv p_2 = 14.5 \text{ N/cm}^2; \ (r_1/r_2) = k \qquad \text{and} \qquad r_1 = 12 \text{ cm}$$

Then

$$C = p_1 r_1 = p_2 r_2 = p_2(r_1/k)$$

Thus,

$$W = 2\pi C\,(r_1 - r_2)$$

or

$$W = 2\pi(p_2 r_2)(r_1 - r_2) = 2\pi p_2\left(\frac{r_1}{k}\right)r_1\left(1 - \frac{r_2}{r_1}\right)$$

or

$$W = 2\pi p_2 r_1^2 \frac{1}{k}\left(1 - \frac{1}{k}\right) = 2\pi p_2 r_1^2\left(\frac{k - 1}{k^2}\right)$$

Hence,

$$= 2\pi \times 14.5 \times (12)^2\left(\frac{k - 1}{k^2}\right) = 13119.29\left(\frac{k - 1}{k^2}\right) \qquad \text{(iii)}$$

And

$$M_f = \pi\mu C(r_1^2 - r_2^2)$$

$$= \pi\mu(p_2)(r_1/k)\,r_1^2\left(1 - \frac{1}{k^2}\right) = \pi\mu p_2 r_1^3\left(\frac{k^2 - 1}{k^3}\right)$$

$$= \pi(0.3)(14.5)(12)^3\left(\frac{k^2 - 1}{k^3}\right) = 23614.723\left(\frac{k^2 - 1}{k^3}\right) \qquad \text{(iv)}$$

Values of W and M_f, as calculated for various values of ratio $k \ (= r_1/r_2)$ for uniform wear rate assumption, are listed in rows 3 and 4 of Table 11.1.

Table 11.1

		$k = 2.0$	1.8	1.7	1.6	1.5	1.4	1.3	1.25	1.2	1.15	1.1
Uniform Pressure Assumption	W: N	4919.7	4535.0	4289.8	3997.3	3644.2	3212.9	2678.2	2361.5	2004.3	1599.6	1138.5
	M_f: N·cm	13775.3	13043.7	12538.7	11899.6	11078.5	10005.8	8577.4	7682.65	6632.5	5391.77	3915.0
Uniform Wear Assumption	W: N	3279.8	3239.3	3177.7	3074.8	2915.4	2677.4	2328.86	2099.08	1822.1	1488	1084.2
	M_f: N·cm	8855.5	9070.12	9084.4	8993.9	8746.2	8261.7	7416.5	6801	6013	5007	3725.8

We thus conclude that for same limiting pressure,

(1) The axial load required for transmitting friction torque decreases, (for both assumptions) as the ratio of radii (r_1/r_2) goes on decreasing.

The required axial load as estimated using "Uniform-Wear assumption" model, is always less than that anticipated using "Uniform pressure" model for all values of $k = (r_1/r_2)$ under consideration.

(2) Frictional torque transmitted, as predicted by "Uniform-pressure assumption", is higher for all ratios of radii (k), compared to that predicted using uniform wear assumption.

(3) However, 'frictional torque' predicted by uniform pressure assumption, goes on increasing continuously as k increases. It is not so for other assumption. With uniform wear assumption, the friction torque first increases upto $k = 1.7$ but decreases thereafter. This is illustrated graphically in Fig. 11.14.

11.11 DESIGN CONSIDERATIONS IN THE CHOICE OF ASSUMPTION

As indicated earlier, there exists considerable uncertainty regarding the nature of distribution of axial load over the area of the contact surfaces. In the absence of exact relation describing mechanism of friction and wear, assumptions of uniform pressure intensity and uniform rate of wear should be looked upon as mathematical models to predict amount of frictional torque as closely as possible.

It is clear from the graph that assumption of uniform pressure intensity gives an estimate on higher side while assumption of uniform rate of wear gives an estimate of frictional torque on the lower side. Since an accurate estimate is not possible, a designer must opt for that assumption which leads to a safer result or a conservative design. For instance, while calculating H.P. lost in friction for a bearing, an assumption of uniform pressure intensity leads to an estimate of H.P. lost on higher side. Such an assumption thus compels a designer to provide for more effective provisions for carrying away heat generated or even to provide for a more effective lubrication.

Similarly, while designing a clutch for power transmission, frictional torque is calculated based on the assumption of uniform rate of wear. This assumption leads to an estimate of

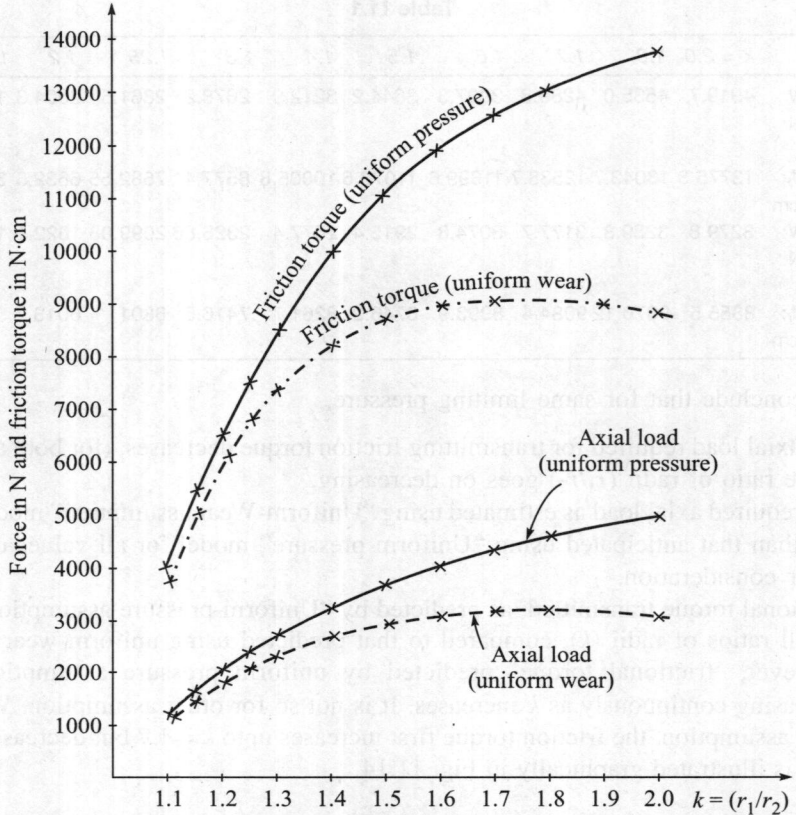

Fig. 11.14 Assumption of uniform Pressure and uniform Rate of Wear—compared.

frictional torque on the lower side. Such an estimate compels a designer to increase torque carrying capacity by increasing friction area or by using better lining material or even by using more than one effective pairs of friction surfaces. All these steps amount to providing friction torque capacity more than what is needed.

Therefore, as a rule, if the problem is concerned with the power absorption by friction (as in bearings), one should opt for assumption of uniform pressure intensity. As against this, when problem is concerned with power transmission capacity, one should opt for the assumption of uniform rate of wear.

11.12 THRUST BEARING

Figure 11.15 shows general arrangement of a horse-shoe shaped ordinary thrust bearing diagramatically. The bearing consists of a number of collars C which are turned integrally with the shaft.

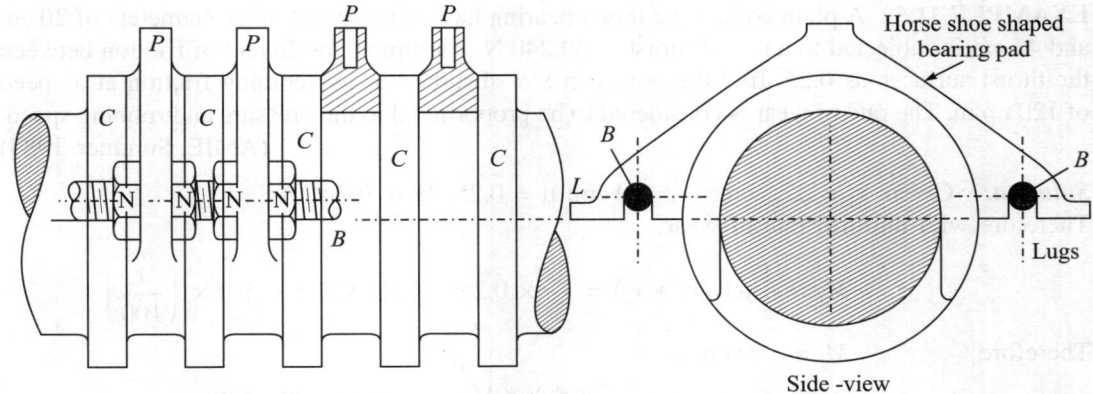

Fig. 11.15 Horse shoe shaped Thrust Bearing.

As will be seen later, the number of collars does not have any effect on frictional moment. However they do help in distributing total axial load over larger surface area, thereby reducing the average bearing pressure.

A horse-shoe shaped bearing pad P is interposed between each pair of adjacent collars. As shown in side-view, the bearing pads ride on the shaft and are held in position by two or more lugs. Two long screwed bolts B run parallel to the axis of the shaft and pass through lugs. Nuts N or either side of lugs locate the bearing pads in the axial direction. When pads are correctly positioned, all the n bearing pads share the total thrust load W equally.

Thus, let W = total axial thrust

           n = total number of bearing pads

           r_o = outer radius of bearing surfaces

           r_i = inner radius of bearing surfaces

           μ = coefficient of friction

In view of prevailing boundary friction condition, laws of solid friction are still applicable, though with reduced coefficient of friction. This being a case of power absorption, pressure can be assumed to be uniform across bearing surfaces. Thus with load shared by each collar as W/n, the frictional moment for each collar is

$$= \frac{2}{3}\mu\left(\frac{W}{n}\right)\left(\frac{r_o^3 - r_i^3}{r_o^2 - r_i^2}\right) \tag{11.73}$$

Hence total frictional moment with n collars,

$$M_f = \frac{2}{3}\mu W\left(\frac{r_o^3 - r_i^3}{r_o^2 - r_i^2}\right) \tag{11.74}$$

Thus, number of collars does not have any effect on frictional moment absorbed but it does reduce load per collar and hence the intensity of pressure to $W/[n\pi(r_o^2 - r_i^2)]$. With reduced intensity of pressure, the bearing surfaces can be lubricated more effectively.

EXAMPLE 11.5 A plain collar type thrust bearing having inner and outer diameters of 20 cm and 45 cm is subjected to an axial thrust of 39,240 N. Assuming coefficient of friction between the thrust surfaces as 0.25. find the power in kW absorbed in overcoming friction at a speed of 120 r.p.m. The rate of wear is considered to be proportional to the pressure and rubbing speed.

(AMIE, Summer 1985)

Solution: Given: $r_1 = 22.5$ cm; $r_2 = 10$ cm; $\mu = 0.25$; $W = 39240$ N

Therefore with uniform rate of wear,

$$M_f = \frac{1}{2}\mu W(r_1 + r_2) = \frac{1}{2} \times 0.25 \times 39240(22.5 + 10) \times \left(\frac{1}{100}\right)$$

Therefore $\qquad M_f = 1594$ N·m

Hence power absorbed in friction kW $= \dfrac{2\pi \times N \times M_f}{1000} = \dfrac{2 \times \pi \times 120 \times 1594}{60 \times 1000} = 20$ kW **Ans.**

EXAMPLE 11.6 Determine the power lost in overcoming friction in the bearing and number of collars required for a bearing whose contact surfaces are 200 mm external radius and 120 mm internal radius. The coefficient of friction is 0.08. The total axial load is 29450 N. Intensity of pressure is 34.3 N/cm². Shaft speed is 420 r.p.m. Assume uniform pressure intensity at contacting surfaces.

(AMIE, Winter 1981)

Solution: Given: $r_1 = 20$ cm; $r_2 = 12.0$ cm

$$\mu = 0.08; \ W = 29450 \text{ N}; \ N = 420 \text{ r.p.m.}$$

For uniform pressure intensity assumption,

$$M_f = \frac{2}{3}\mu W\left(\frac{r_1^3 - r_2^3}{r_1^2 - r_2^2}\right)$$

or $\qquad M_f = \dfrac{2}{3} \times 0.08 \times 29450\left(\dfrac{20^3 - 12^3}{20^2 - 12^2}\right)$

or $\qquad M_f = 38493.6$ N·cm $= 384.94$ N·m

Therefore power lost in friction $= \dfrac{2\pi \times N \times M_f}{(1000 \times 60)} = \dfrac{2\pi \times 420 \times 384.94}{(1000 \times 60)} = 16.93$ kW.

Let n be the number of collars required to ensure a uniform pressure intensity of 34.3 N/cm². Hence, from

$$W = n\pi p(r_1^2 - r_2^2)$$
$$29450 = n \times \pi \times 3.5\,(20^2 - 12^2)$$

Therefore $\qquad n = \dfrac{29450}{\pi \times 34.3(20^2 - 12^2)} = 1.0675$

Hence $\qquad n = 2$, say $\qquad\qquad\qquad\qquad\qquad$ **Ans.**

EXAMPLE 11.7   The thrust of a propeller shaft in a marine engine is taken up by a number of collars integral with the shaft which is 30 cm in diameter. The thrust on the shaft is 196.2 kN and the speed is 75 r.p.m. Taking μ to be constant and equal to 0.055 and assuming intensity of pressure as uniform and equal to 29.4 N/cm², find the external diameter of the collars and the number of collars required; if the power lost in friction is not to exceed 16.4 kW.

(SGSITS, April 2006)

Solution: $r_2 = \dfrac{30}{2} = 15$ cm; $W = 196200$ N

$$\mu = 0.055;\ N = 75 \text{ r.p.m.};\ p = 29.4 \text{ N/cm}^2$$

The frictional torque M_f is given by

$$16.4 = \frac{2\pi \times 75 \times M_f}{60 \times 1000}$$

Therefore

$$M_f = \frac{16.4 \times 60 \times 1000}{2\pi \times 75} = 2088.1 \text{ N.m}$$

Again, for uniform pressure assumption, total frictional torque is also given by

$$M_f = \left(\frac{2}{3}\right)\mu W(r_1^3 - r_2^3)/(r_1^2 - r_2^2)$$

$$= \left(\frac{2}{3}\right)(0.055)(196200)(r_1^2 + r_1 r_2 + r_2^2)/(r_1 + r_2)$$

Therefore

$$\frac{(r_1^2 + r_1 r_2 + r_2^2)}{(r_1 + r_2)} = \frac{2088.1 \times 3}{2(0.055)(196200)} = 0.29$$

Substituting value $r_2 = 0.15$ m,

we have $r_1^2 + 0.15 r_1 + (0.15)^2 = 0.29(r_1 + 0.15)$

or $r_1^2 - 0.14 r_1 - 0.021 = 0$

Solving quadratic in r_1, $r_1 = \dfrac{0.14 \pm \sqrt{(0.14)^2 + 4(0.021)}}{2}$

or $r_1 = \dfrac{0.14 \pm 0.322}{2} = 0.231$ m (Neglecting negative solution)

Therefore $r_1 = 23.1$ cm

The number collars required is calculated on the basis of limiting pressure p. Thus, from

$$(W/n) = \pi p(r_1^2 - r_2^2)$$

we have $n = \dfrac{W}{\pi p(r_1^2 - r_2^2)}$

$$= \frac{196200}{\pi \times 29.4[(23.1)^2 - (15)^2]} = 6.88$$

Therefore   number of collars reqd. = 7   **Ans.**

EXAMPLE 11.8   A conical pivot supports a load of 20 kN. The angle of cone is 120° and intensity of pressure is not to exceed 3.5 bar. The external radius is 3 times the internal radius. Find the diameter of the bearings surface. If μ = 0.065 and r.p.m. of shaft is 120, what power in kW. is absorbed by friction?

Solution: Given: $W = 20000 \ N$; $2\alpha = 120°$; $p = 3.5$ bar (1 bar = 10^5 pascal = $10^5 \ N/m^2$);

$$r_1 = 3 \ r_2; \ \mu = 0.65 \text{ and } N = 120 \text{ r.p.m.}$$

Let r_1 and r_2 be the outer and inner radii in metres. Then, for uniform pressure assumption,

$$W = p\pi (r_1^2 - r_2^2)$$

or $20000 = 3.5 \times 10^5 \times \pi (9r_2^2 - r_2^2)$

or $r_2^2 = \dfrac{20,000}{3.5 \times 10^5 \times \pi \times 8}$

or $r_2^2 = 2.2736 \times 10^{-3}$

Thus, $r_2 = 0.0477 \text{ m or } r_2 = 4.77 \text{ cm}$

and, $r_1 = 14.3 \text{ cm}$

Hence, the frictional torque is,

$$M_f = \frac{2}{3} \frac{\mu W}{\sin\alpha} \left(\frac{r_1^3 - r_2^3}{r_1^2 - r_2^2} \right)$$

$$= \frac{2 \times 0.065 \times 20000}{3 \times \sin 60} \left(\frac{14.3^3 - 4.77^3}{14.3^2 - 4.77^2} \right)$$

$$= 15504.6 \text{ N·cm}$$

$$= 155.046 \text{ N·m}$$

Therefore   power absorbed in friction $P = \dfrac{2\pi NM_f}{60}$

or $P = \dfrac{2\pi \times 120 \times 155.046}{60} = 1948.4 \text{ W} = 1.948 \text{ kW}$ **Ans.**

11.13   FRICTION CLUTCHES

Friction clutches are used in machines where frequent engagement or disengagement of driving shafts, with connected loads, is required. Such condition is frequently encountered in automobiles where for a short interval of time, power supply may have to be cut off either partially or completely, without stopping the engine. A friction clutch is also quite useful in transmission of power where machines must be started and stopped frequently.

Positive or jaw clutches are used to connect driven shaft to the prime mover when it is not objectionable to start the drive suddenly and where the connected load (including inertia resistance) is not so great as to produce injurious shock when the clutch is engaged suddenly. *Positive jaw clutch is also used when sudden initial load (in the form of inertia and other loads) is not considered harmful to the prime mover itself.*

Friction clutches are particularly important when engagement must be made under load and prime mover is to be protected against sudden loads. It is also important when two shafts, running at different speeds, are to be connected, without stopping them. Above features make friction (*i.e.,* non-positive type) clutches more suitable for vehicular and process machinery as also in the drive unit of machine tools. Like friction clutches, hydraulic (fluid) couplings also perform well in the above applications. The slip occurring between the friction surfaces, to a certain extent, is thus desirable in providing cushion against sudden transference of power.

Types of Clutches

Clutches are classified as under:

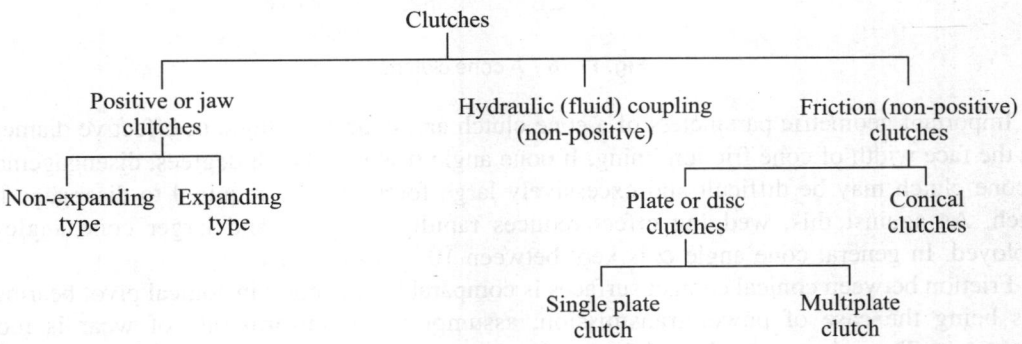

11.14 CONE CLUTCH

As illustrated in Fig. 11.16, a cone clutch consists of a cup keyed with a sunk key to driving shaft and a corresponding mating cone lined on the outside with friction lining. This cone is either mounted on splined shaft or is connected through a feather key to the driven shaft. Inner conical surface of the cup portion fits exactly onto the outer conical surface of the mating cone mounted on driven shaft. A helical compression spring, placed around the driven shaft, holds the two friction surfaces of clutch in contact. The initial compression in the spring maintains necessary pressure between mating surfaces and produces frictional torque.

The clutch is disengaged by shifting cone portion to the right against additional spring compression by a fork which fits into the shifting groove on the hub portion on the cone. The hub is free to slide axially over splines.

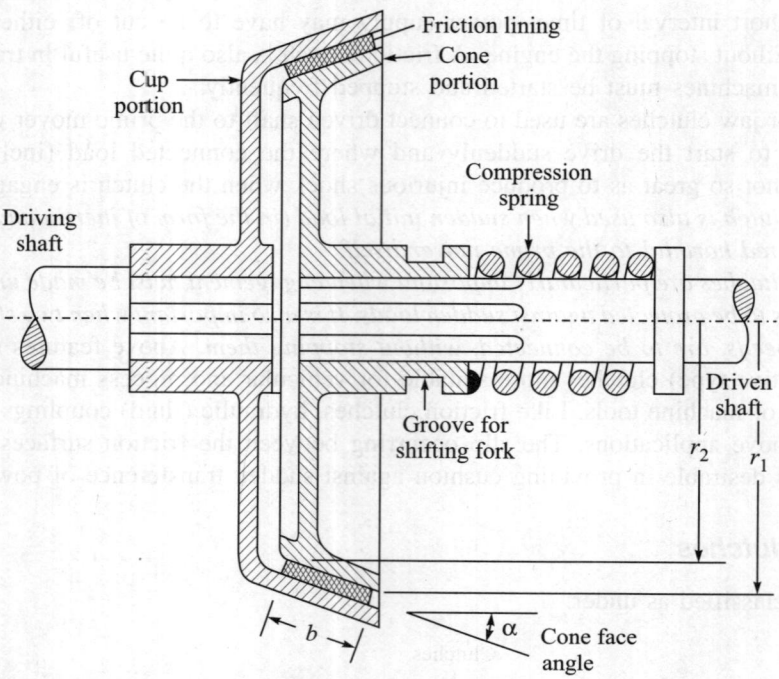

Fig. 11.16 A cone clutch.

Important geometric parameters of a cone clutch are: cone-face angle α, effective diameter and the face width of cone friction lining. If cone angle α is less than 8 degrees, disengagement of cone clutch may be difficult and excessively large force may be required to disengage the clutch. As against this, wedging effect reduces rapidly as larger and larger cone angle is employed. In general cone angle α is kept between 10 to 15 degrees.

Friction between conical contact surfaces is comparable to friction in conical pivot bearings. This being the case of power transmission, assumption of uniform rate of wear is more appropriate. Thus, the required axial force W and the transmitted frictional torque are given by

$$W = 2\pi C(r_1 - r_2) \tag{11.75}$$

where $\qquad C = p_1 r_1 = p_2 r_2 \quad$ and, $\quad T = \dfrac{1}{2}\dfrac{\mu W}{\sin\alpha}(r_1 + r_2) \tag{11.76}$

In the earlier stages of development of automobile, clutches of conical type with leather face, were employed. This design has the limitation in that loss of roughness or hardening of the clutch lining leads to violent engagement of clutch.

EXAMPLE 11.9 A cone clutch with a semi-cone angle α of 15°, transmits 10 kW at 600 r.p.m. The normal pressure between the surfaces of contact is not to exceed 100 kN/m². The axial width of the friction surfaces is half the mean diameter. Assuming coeff. of friction as 0.25, determine:

(a) dimensions of friction plate, and

(b) axial force

(Amravati Un.)

Solution: Given $\alpha = 15°$; $\mu = 0.25$; $p_{max} = 100$ kN/m²

$$N = 600 \text{ r.p.m. and } b = \frac{d_m}{2} = (R_m)$$

It follows from Fig. 11.14 that

$$b = \frac{r_1 - r_2}{\sin\alpha}$$

Also from data

$$b = \frac{d_m}{2} = \frac{1}{2}\left(\frac{d_1 + d_2}{2}\right) = \frac{1}{2}(r_1 + r_2)$$

Equating R.H.S.

$$\left(\frac{r_1 - r_2}{\sin\alpha}\right) = \left(\frac{r_1 + r_2}{2}\right)$$

or

$$\left(\frac{r_1 - r_2}{r_1 + r_2}\right) = \frac{\sin\alpha}{2} = 0.1294$$

Hence

$$\frac{r_1}{r_2} = \frac{1 + 0.1294}{1 - 0.1294} = 1.2973$$

Again

$$(10 \times 1000) = \frac{2\pi NT}{60}$$

Therefore

$$T = \frac{60 \times 10000}{2\pi \times 600} = 159.15 \text{ N·m}$$

Again, for uniform wear assumption,

$$\text{as } p \propto 1/r; \quad p_2 > p_1$$

Let

$$p_2 = p_{max} = (100 \times 1000) \text{ N/m}^2$$

Therefore from the relation for frictional moment,

$$T = \frac{1}{2}\frac{\mu W}{\sin\alpha}(r_1 + r_2)$$

Also because

$$W = 2\pi c(r_1 - r_2) = 2\pi (p_2 r_2)(r_1 - r_2)$$

or

$$W = (2\pi \times 100 \times 1000) r_2 (r_1 - r_2)$$

Substituting above

$$T = \frac{1}{2}\frac{\mu}{\sin\alpha}(2\pi \times 100 \times 1000) r_2 (r_1 - r_2)(r_1 + r_2)$$

or

$$T = \frac{1}{2}\frac{(0.25)(2\pi \times 100 \times 1000)}{\sin 15} r_2 (r_1^2 - r_2^2)$$

or $\qquad T = (303454.54)\, r_2(r_1^2 - r_2^2)$

But $\qquad T = 159.15$ N.cm.

Therefore $\qquad 303454.54\, r_2\, (r_1^2 - r_2^2) = 159.15$

or $\qquad r_2\, [(1.2973 r_2)^2 - r_2^2] = 0.00052446$

or $\qquad r_2^3 = 0.0007678$ Hence, $r_2 = 0.09156$ m $= 91.6$ mm $\qquad$ **Ans.**

And therefore $\qquad r_1 = (1.2973) r_2 = 118.8$ mm $\qquad$ **Ans.**

Also $\qquad b = \dfrac{r_1 - r_2}{\sin 15} = \dfrac{118.8 - 91.6}{\sin 15} = 105$ mm $\qquad$ **Ans.**

The axial force required $\qquad W = 2\pi C\,(r_1 - r_2)$

or $\qquad W = 2\pi(100 \times 1000)(0.0916)(0.1188 - 0.0916)$

or $\qquad W = 1565.5$ N $\qquad$ **Ans.**

EXAMPLE 11.10 The semicone angle of a cone-clutch is 12.5° and the contact surfaces have a mean diameter of 80 mm. Coefficient of friction is 0.32. What is the minimum torque required to produce slipping of the clutch for an axial force of 250 N? If the clutch is used to connect an electric motor with a stationary flywheel, what is the time needed to attain the full speed and the energy lost during slipping? Motor speed is 900 r.p.m. and the moment of inertia of the flywheel is 0.45 kg.m^2.

Solution: Given $\alpha = 12.5°$; $d_m = \dfrac{d_1 + d_2}{2} = 80$ mm; $\mu = 0.32$

Let $T_{\min}$ be the time required for slipping; $W = 250$ N;
Also let t be the time reqd. to attain full speed

$$(N)_{motor} = 900 \text{ r.p.m.}; \quad (I)_{fly\ wheel} = 0.45 \text{ kg·m}^2$$

For uniform rate of wear assumption,

$$T = \frac{1}{2}\,\frac{\mu W}{\sin\alpha}\,(r_1 + r_2)$$

or $\qquad T = \dfrac{\mu W}{\sin\alpha}(r_m)$

where $\qquad r_m = \text{mean radius} = \dfrac{r_1 + r_2}{2}$

Hence $\qquad T = 0.32 \times \dfrac{250}{\sin 12.5} \times \left(\dfrac{80}{2}\right) = 14784.7$ N·mm

Thus $\qquad T = 14.785$ N·m. $\qquad$ **Ans.**

This is the limiting torque beyond which slipping will occur as μ is the limiting coeff. of friction.

The driven shaft is to be accelerated from a speed $\omega = 0$ to a speed of motor $\omega' = \dfrac{2\pi \times 900}{60}$

= 94.25 rad/s. The angular, acceleration required for this purpose, is supplied by the friction torque in the clutch. Thus, torque

$$M_f = I\alpha' \qquad \text{or,} \qquad \alpha' = \frac{14.785}{0.45} = 32.856 \text{ rad/s}^2$$

Assuming a uniform angular acceleration α', we have

$$\omega' = \omega + \alpha' \cdot t$$

And as $\omega = 0$, time required to attain full speed

$$t = (\omega'/\alpha') = (94.25/32.856) = 2.868 \text{ s.} \qquad \textbf{Ans.}$$

Angle turned by driving shaft, slipping (*i.e.*, during the period when $\omega < \omega'$),

$$\theta = t \times \frac{2\pi \times 900}{60}$$

or

$$\theta = 2.868 \times \frac{2\pi \times 900}{60} = 270.3 \text{ rad}$$

And during this time, angle turned through by driven shaft,

$$\theta' = \omega t + \frac{1}{2}\alpha' t^2$$

or

$$\theta' = 0 + \frac{1}{2}(32.856)(2.868)^2 = 135.13 \text{ rad.}$$

Hence, work of friction $= T(\theta - \theta')$

or

$$W_f = 14.785(270.3 - 135.13) = 1998.5 \text{ m.N} \qquad \textbf{Ans.}$$

11.15 SINGLE PLATE CLUTCH

A disc clutch and a plate clutch are different only in respect of the thickness of the disc and details of construction. A disc clutch has one or more discs that engage with a second set of discs so as to connect a source of power to a machine or a machine part. A single plate clutch and multi-disc clutch operate on the same principle and represent a category of clutches most common in practice.

As can be seen in example 11.10, clutch-plate wear is a function of work of friction and depends on slipping time. The other factors, which have a direct influence on clutch-plate wear, are:

(i) thermal load,
(ii) friction material, and
(iii) surface structure.

Wear remains negligible if the friction surface temperature during engagements is kept within permissible limits. Oil as a coolant (*e.g.,* in multi-disc clutch) reduces wear. For maximum effect, the cooling oil should pass as close as possible to the friction face and ideally through internal oilways. Wet running clutches therefore undergo less wear than dry running clutches, where only the design and specific construction (involving design of the clutch, type and method of installation, forced air cooling, etc.) can ensure good heat dissipation and restrict the faster rate of wear to a tolerable limit. As against this, the friction coefficient of dry running plates is much higher than that of wet running plates. This is one of the reasons as to why in certain scooters, oil-immersed multi-disc clutches are used while in certain passenger cars a single plate dry clutch is used.

The clutch shown in Fig. 11.17 is representative of single plate friction clutches used frequently in automobiles. It operates on the principle that no frictional torque can be transmitted unless the contacting friction surfaces are pressed together under a normal load W. In the single plate clutch, as shown in the figure, this force is provided through a pressure plate P by a number of helical compression springs S which are spaced uniformly around the circumference and subjected to certain value of initial compression. The initial compression in the springs is sufficient to produce axial force W, required to generate given frictional torque. Needless to say that when clutch plate wears out, the initial compression in the springs decreases and as a result a smaller axial force W is available for generating frictional torque. This explains loss in torque transmission capacity of a worn out clutch, which could possibly be restored only by tightening the screws and producing same amount of initial compression in the springs.

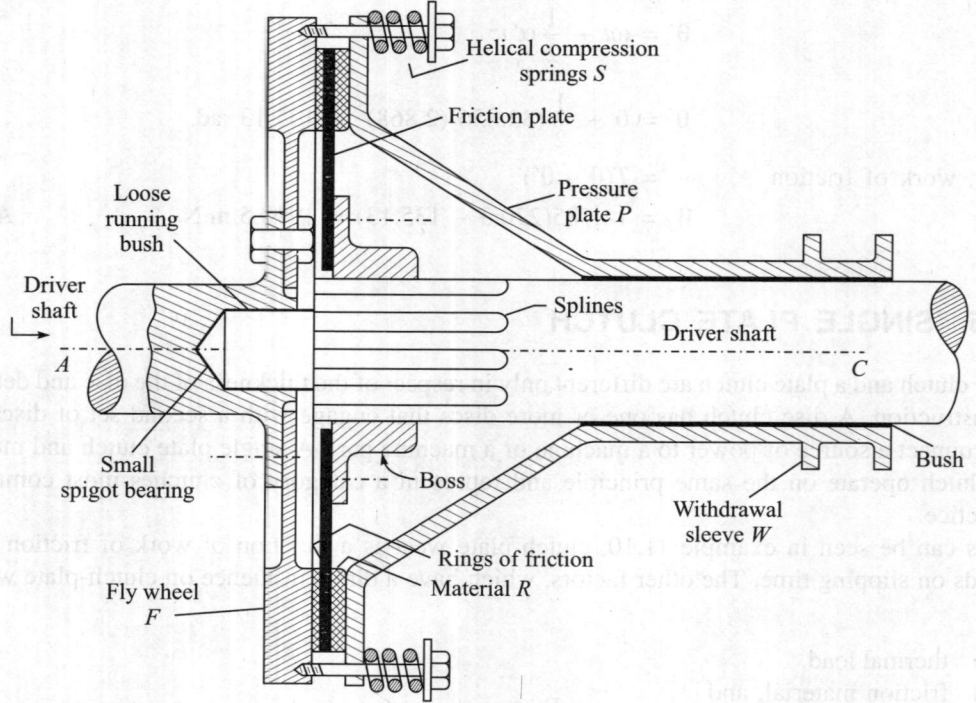

Fig. 11.17 Single plate friction clutch.

The friction plate D is fixed to a boss which is free to slide axially on splines provided on the driven shaft. Provision of splines permit a relative motion in axial direction between the boss and driven shaft, but no relative motion of rotation is permitted between the two. Two rings of friction material are riveted to the friction plate, one on each side. Alternatively, the rings of friction material can be riveted, one to the flywheel F and other to the pressure plate. The flywheel F is bolted to the flange which is formed integrally on the driving shaft. The flywheel is also connected to the pressure plate P through a number of equispaced screws that permit pressure plate to slide axially with respect to flywheel but does not permit relative motion of rotation between the two. The pressure plate along with the withdrawal sleeve thus rotates at the speed of rotation of flywheel. In the engaged condition, the pressure plate is pressed against friction plate under the action of force of initial compression in springs S. The pressure plate is bushed internally, so as to revolve freely with respect to the driven shaft. A withdrawal sleeve formed integrally on the pressure plate permits the same to be pulled out of engagement to the right side against additional compression in springs. The withdrawal is facilitated by a roller carried at the forked end of a lever, which is pivoted to the frame.

Even in the disengaged condition, the pressure plate continues to rotate at the same speed as flywheel but no torque is transmitted to the friction plate, and hence to the driven shaft, in the absence of any axial force. A small spigot bearing provided in the end of driving shaft enables to secure and maintain correct alignment of driven shaft.

11.16 MULTI-DISC CLUTCH

A multi-disc clutch represents an improvement in the design of a single plate clutch. The outer casing C is bolted down to the flywheel F and rotates with it. The flywheel, in turn, is bolted to the driving shaft A. The main feature which distinguishes a multi-disc clutch from a single plate clutch is the provision of a number of axial grooves on the inner periphery of casing and also on the outer periphery of the driven member B, which is made to rotate with driven shaft G. Alternate discs marked O have tongues or projections on the outer edge as shown at Fig. 11.18(b). These discs have a sliding fit in the internal axial grooves provided in the casing. A number of discs, marked I, have tongues or projections at the inner edge as shown at Fig. 11.18(b). These tonques on discs I have a sliding fit with the axial grooves provided on the outside periphery of driven member B. Thus, while discs O are free to slide in axial direction with respect to casing, they do not have relative motion of rotation with respect to casing. Same is true about relative motion between discs I and the driven member.

The inner radius of outer disc O is larger than the outer radius of the driven member while the outer radius of inner disc I should be less than the inner most radius of the outer casing. Fig. 11.18(b) shows only half of the inner and outer discs about a vertical diameter. While assembling the clutch plates, the outer and inner discs are slid in their respective axial grooves alternately. The outer discs O are arranged to be the first and last discs to avoid any wear on flywheel and pressure plate which rotate at the same r.p.m. together with the outer discs.

In the engaged condition, the initial compression in the spring exerts an axial force W with which the friction plates are pressed together. This force, normal to the friction surfaces, is responsible for transmitting friction torque from outer plates to the inner plates and thence to the driven member. To disengage the clutch, the withdrawal sleeve, together with the pressure plate, is pulled to the right against additional compressive force in the spring. This movement relieves the friction discs of the force W and therefore, no friction torque is transmitted from outer discs to the inner discs and thence to the driven member as well.

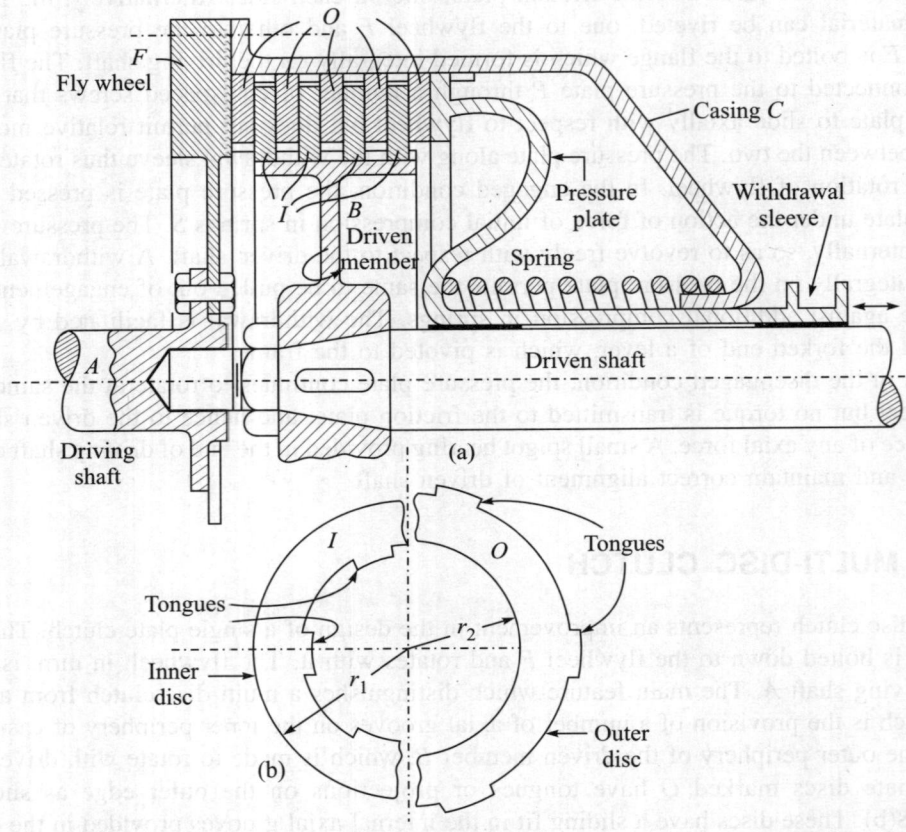

Fig. 11.18 A multi-disc clutch.

11.17 EFFECT OF NUMBER OF PAIRS OF ACTIVE SURFACES

It must be noted that in case of a multi-disc clutch, discs are free to slide in their respective axial grooves under the action of axial force W. And therefore, every disc, and hence every pair of friction surfaces is subjected to the same axial force W.

A pair of active surfaces (also called friction surfaces) consists of two friction surfaces in contact and having a relative motion of angular sliding between them. Clearly, every interface between outer and inner discs, at which relative motion exists, constitutes a pair of active surfaces. The number of pairs of active surfaces n_a is then counted as follows:

Let there be a total of 7 (seven) friction discs as shown in Fig. 11.19. Discs 1, 3, 5 and 7 (total 4) are the outer discs while

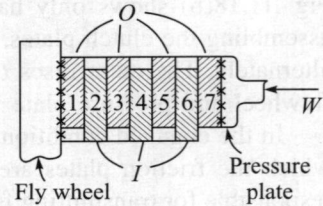

Fig. 11.19

discs 2, 4, 6 (total 3) are the inner discs. The friction discs are alternately arranged as outer and inner discs. To avoid friction and subsequent wear on flywheel as well as pressure plate, the first and last discs are arranged as outer discs. This required that

$$n_o = n_i + 1 \qquad (11.77)$$

where $\quad n_i$ = number of inner discs $I \quad$ and, $\quad n_o$ = number of outer discs O

Each friction disc offers two friction surfaces, one on each side. Hence, as a general rule, number of pairs of active surfaces equals n, the total number of discs.

With discs 1 and 7 as outer discs, the friction surfaces of discs 1 and 7, in contact with flywheel and pressure plate, become inactive; there being no relative motion. Thus in all, there is a loss of one pair of friction surfaces. Here actual number of pairs of active surfaces,

$$n_a = (n_o + n_i) - 1 \equiv (n - 1) \qquad (11.78)$$

In the case of a single plate clutch, there are two pairs of friction surfaces, one on either side. When both the pairs of friction surfaces are active, $n_a = 2$.

Thus, based on the assumption of uniform rate of wear, expression for friction torque for multi disc clutch can be written as:

$$T = \frac{1}{2}\mu W(n_a)(r_1 + r_2) \qquad (11.79)$$

or $$T = \mu W(n_a) r_m \qquad (11.80)$$

where $\qquad n_a$ = number of pairs of active surfaces

$\qquad\qquad$ = 2 for a single plate clutch with both the surfaces active,

$\qquad\qquad$ = $(n_o + n_i - 1)$ for multi-disc clutch

$\qquad W$ = axial force on clutch plates,

$\qquad r_1, r_2$ = outer and inner radii respectively of friction surfaces

$\qquad \mu$ = coefficient of friction

$\qquad r_m$ = $(r_1 + r_2)/2$ = mean radius.

Similarly, the axial force for uniform rate of wear,

$$W = 2\pi C(r_1 - r_2) \qquad (11.81)$$

where $\qquad C = p_1 r_1 = p_2 r_2$ = constant

EXAMPLE 11.11 A single plate clutch (both sides effective) is required to transmit 27 kW at 1600 r.p.m. The outer diameter of the plate is limited to 30 cm, and the intensity of pressure between the plates is not to exceed 6.87 N/cm². Assuming uniform wear and a coefficient of friction of 0.3, find the required inner diameter of the plates, and the axial force necessary to engage the clutch. (Punjab University)

Solution: Given $\quad n_a = 2$; $P = 27$ kW; $N = 1600$ r.p.m. $d_1 = 30$ cm; $p_{max} = 6.87$ N/cm²; $\mu = 0.3$

Friction torque required to transmit 27 kW, is

$$T = \frac{P \times 60 \times 1000}{2\pi N}$$

or
$$T = \frac{27 \times 60 \times 1000}{2\pi \times 1600} = 161.14 \text{ N·m}$$

With uniform rate of wear, $p_1 r_1 = p_2 r_2$

and as $r_2 < r_1$, the pressure $p_2 > p_1$

Hence $$p_2 = 6.87 \text{ N/cm}^2$$

and hence, $$(6.87) r_2 = C$$

Thus, axial force,

$$W = 2\pi C(r_1 - r_2) = 2\pi(6.87 \, r_2)(15 - r_2) = 13.74\pi \, (15 - r_2)r_2$$

Also, the friction torque

$$T = \frac{1}{2} \mu W(n_a)(r_1 + r_2)$$

Substituting for W $$T = \frac{1}{2}(0.3)(2)(13.74)\pi r_2(15 - r_2)(15 + r_2)$$

or $$16114 = 12.95 \, (15^2 - r_2^2) \, r_2$$

or $$225 \, r_2 - r_2^3 = 1244.3$$

By trial and error, value of r_2 that satisfies the above equation is,

$$r_2 \approx 10.077 \text{ cm}$$

Thus, $$r_2 = 10.1 \text{ cm, say}$$

Therefore $$d_2 = \text{inner diameter} = 20.2 \text{ cm} \qquad \textbf{Ans.}$$

The axial force required, $$W = 2\pi C(r_1 - r_2)$$

$$= 2\pi(10.1 \times 6.87) \, (15 - 10.1) = 2136.26 \text{ N} \qquad \textbf{Ans.}$$

EXAMPLE 11.12 A car engine rated at 9 kW gives a maximum torque of 88 N·m. The clutch is of the single plate type both sides of the plates being effective. If the coefficient of friction 0.3, the mean axial pressure is 8 N/cm^2 and external radius of the friction surface is 1.25 times the internal radius, find the dimensions of the clutch plate and the total axial pressure that must be exerted by the springs.

Solution: Given $P = 9$ kW; $T_{\text{max}} = 88$ N·m; $n_a = 2$

$$\mu = 0.3; \text{ mean pressure } p_m = 8 \text{ N/cm}^2 \quad \text{and} \quad \frac{r_1}{r_2} = 1.25$$

Note: Normally the axial force W is computed for a clutch using assumption of wear. In the present case however, as mean pressure is given, it follows that this pressure can be assumed to be uniformly distributed. In other words,

$$W = p_m \pi(r_1^2 - r_2^2)$$

or $$W = 8\,(\pi)r_2^2 \, (1.25^2 - 1)$$

or, $$W = (14.14)\, r_2^2$$

For 2 pairs of active surfaces (one on each side) $n_a = 2$ and hence, from equation (11.79),

$$T = \frac{1}{2}\mu W(n_a)(r_1 + r_2)$$

or

$$T = \frac{1}{2}(0.3)(14.14)r_2^2(2)r_2(1.25 + 1) = (9.545)r_2^2$$

This must be equal to the required maximum torque 88 N·m. Hence,

$$(9.545)r_2^3 = 88$$

Therefore $\qquad\qquad\qquad r_2 = 9.22$ cm $\qquad\qquad$ **Ans.**

Hence $\qquad\qquad\qquad r_1 = 1.25 \times 9.22 = 11.5$ cm $\qquad\qquad$ **Ans.**

EXAMPLE 11.13 A single plate friction clutch, with both sides of the plate being effective, is used to transmit power of an engine at 2400 r.p.m. It has outer and inner radii 4.5 and 3.5 cm respectively. The pressure is applied axially by means of springs and the maximum intensity of pressure is 7.85 N/cm^2. If the coeff. of friction is 0.3, find

   (i) the total axial pressure exerted by the springs, and
  (ii) power transmitted

Solution: Given $n_a = 2$; $N = 2400$ r.p.m.; $\mu = 0.3$

$$p_{max} = 7.85 \text{ N/cm}^2; \ r_1 = 4.5 \text{ cm}; \ r_2 = 3.5 \text{ cm}$$

With the assumption of uniform wear rate, as $p_1 r_1 = p_2 r_2$, maximum pressure occurs at inner radius r_2. Hence,

$$p_2 = 7.85 \text{ N/cm}^2$$

and, axial force $\qquad\qquad W = 2\pi C(r_1 - r_2)$

or $\qquad\qquad\qquad W = 2\pi(7.85 \times 3.5)(4.5 - 3.5) = 172.6$ N

Again, the frictional torque transmitted

$$T = \frac{1}{2}\mu W(n_a)(r_1 + r_2)$$

Thus, $\qquad\qquad\qquad T = \frac{1}{2}(0.3)(172.6)(2)(4.5 + 3.5)$

or $\qquad\qquad\qquad T = 414.2$ N·cm

or $\qquad\qquad\qquad T = 4.14$ N·m

Therefore power transmitted $\qquad = \dfrac{2\pi \times 2400 \times 4.14}{(60 \times 1000)} = 1.04$ kW $\qquad$ **Ans.**

EXAMPLE 11.14 A rotor is driven by a co-axial motor through a single plate clutch, both sides of the plate being effective. The external and internal diameters of the plate are respectively 22 cm and 16 cm and the total spring load pressing the plates together is 559 N. The motor armature and shaft weighs 7.846 kN with an effective radius of gyration of 20 cm. The rotor weighs 11.768 kN with an effective radius of gyration of 18 cm. The coefficient of friction for the clutch = 0.35.

The driving motor is brought upto a speed of 1250 r.p.m. when the current is switched off and the clutch is suddenly engaged. Determine,

  (a) the final speed of motor and rotor,
  (b) the time to reach this speed, and
  (c) the kinetic energy lost during the period of slipping.

How long would slipping continue if it is assumed that a constant resisting torque of 58.8 N.m were present? If instead of a resisting torque, it is assumed that a constant driving torque of 58.8 N.m is maintained on the armature shaft, what would then be slipping time?

Solution: Given: $n_a = 2$; $r_1 = \dfrac{22}{2} = 11$ cm; $r_2 = \dfrac{16}{2} = 8$ cm;

$$W = 559 \text{ N; weight of shaft and motor armature} = 7846 \text{ N}$$
$$k = 20 \text{ cm; rotor weight} = 11768 \text{ N; } k' = 18 \text{ cm; } \mu = 0.35$$

With uniform wear assumption

$$T = \frac{1}{2}\mu W(n_a)(r_1 + r_2)$$

or

$$T = \frac{1}{2}(0.35)(559)(2)(11 + 8) = 3717.3 \text{ N·cm}$$

Also, M.I. of motor armature and shaft $I = \left(\dfrac{7846}{9.81}\right) \times (0.2)^2 = 32 \text{ kg·m}^2$

and, M.I. of rotor, $I' = \left(\dfrac{11768}{9.81}\right) \times (0.18)^2 = 38.9 \text{ kg·m}^2$

  (a) Let the final speed of both motor and rotor be ω.

Then, as angular momentum before and after slipping are same,

$$I\omega_1 + I'\omega' = (I + I')\omega$$

As the rotor was initially standstill, $\omega' = 0$. Hence, from momentum equation,

$$32\left(\frac{2\pi \times 1250}{9.81}\right) + 0 = (32 + 38.9)\omega$$

Therefore $\omega = \dfrac{32}{(32 + 38.9)} \times \dfrac{2\pi \times 1250}{60} = 59.08 \text{ rad/s}$

Therefore, $N = 564.2 \text{ r.p.m.}$ **Ans.**

(b) Let t be the time required by rotor to reach this speed of $\omega = 59.08$ rad/s. If α' be the angular acceleration produced in rotor by friction torque from clutch, then

$$T = I'\alpha \qquad \text{or,} \qquad \alpha = \frac{T}{I'} = \left(\frac{37.17}{38.9}\right) = 0.956 \text{ rad/s}^2$$

Assuming this to be the uniform acceleration for rotor, with initial velocity $\omega' = 0$, we have

$$\omega = 0 + \alpha't$$

Therefore $$t = (\omega/\alpha') = \frac{59.08}{0.956} = 61.8 \text{ s} \qquad \qquad \textbf{Ans.}$$

(c) The kinetic energy of driver and driven components together before and after engagement are different and K.E. lost during slipping is given by

$$E = E_1 - E_2$$

or $$E = \left[\frac{1}{2}I\omega_1^2 + \frac{1}{2}I'(0)^2\right] - \frac{1}{2}(I + I')\omega^2$$

or $$E = \frac{1}{2}(32)\left(\frac{2\pi \times 1250}{60}\right)^2 - \frac{1}{2}(32 + 38.9)(59.08)^2$$

Thus $$E = (274155.7 - 123736.3)$$

or $$E = 150419.4 \text{ N·m.} \qquad \qquad \textbf{Ans.}$$

Time of slipping with a constant resisting torque:

If additional torque of 58.8 N·m is available as resisting torque B on driven side, the total torque acting as resisting torque (to retard the driving shaft) is

$$T' = 5880 + 3717.3 = 9597.3 \text{ N·cm}$$

Hence, angular retardation caused in driving shaft α' is given by

$$T' = I\alpha', \text{ or } \alpha' = \left(\frac{T'}{I}\right) = \frac{95.973}{32} = 3 \text{ rad/s}^2$$

Thus, if t' be the new time of slipping, for a final speed of ω'' rad/s we have

$$\omega'' = \omega_1 - \alpha't'$$

or $$\omega'' = \left(\frac{2\pi \times 1250}{60}\right) - (3)t'$$

or $$\omega'' = (130.899) - (3)t' \qquad \qquad \text{(i)}$$

Also for driven shaft, the final speed is given by,

$$\omega'' = 0 + \alpha t' = \left(\frac{37.17}{38.9}\right)t' \qquad \qquad \text{(ii)}$$

Equating R.H.S. of (i), (ii),

$$\left(\frac{37.17}{38.9}\right)t' = (130.899) - (3)t'$$

Therefore $\qquad\qquad\qquad t' = \dfrac{130.899}{3.9628} = 33.09 \text{ s}$ **Ans.**

Time of slipping for constant driving torque of 58.8 N·m

The torque available for acceleration of driving shaft = 5880 – 3717.3 = 2162.7 N·cm
Thus if ω_3 be the final speed of driver and driven shaft, then for common ω_3,

$$\frac{2\pi \times 1250}{60} + \left(\frac{21.63}{32}\right)t'' = 0 + \left(\frac{37.17}{38.9}\right)t''$$

or $\qquad\qquad (0.2796)t'' = 130.899.$ Hence, $t'' = 468.2$ s $\qquad$ **Ans.**

EXAMPLE 11.15 A multiple disc clutch has five plates having four pairs of active friction surfaces. If the intensity of pressure is not to exceed 12.46 N/cm^2, find the power in k.w. transmitted at 500 r.p.m. if the outer and inner radii of friction surfaces are 125 and 75 mm, respectively. Assume uniform wear and take coefficient of friction = 0.3 (AMIE, Summer 1984)

Solution: Given $\quad n_a = 4; \; n = 5; \; p_{max} = 12.46$ N/cm^2

$\qquad\qquad N = 500$ r.p.m; $r_1 = 125$ mm; $r_2 = 75$ mm, $\mu = 0.3$

Since maximum pressure occurs at inner radius,

$$p_2 = 12.46 \text{ N/cm}^2 \text{ and } r_2 = 7.5 \text{ cm}$$

Therefore $\qquad\qquad C = p_2 r_2 = (12.46)(7.5)$

Therefore, Axial force $\qquad W = 2\pi C(r_1 - r_2)$

or $\qquad\qquad W = 2\pi \times 12.46 \times 7.5(12.5 - 7.5) = 2935.82$ N

For uniform wear, friction torque,

$$T = \frac{1}{2}\mu W(n_a)(r_1 + r_2)$$

or $\qquad\qquad T = \frac{1}{2}(0.3)(2935.82)(4)(12.5 + 7.5) = 35229.8$ N·cm

or $\qquad\qquad T = \frac{35229.8}{100} = 352.3$ N·m

Therefore power transmitted = $T\omega$

$$P = 352.3 \times \frac{2\pi \times 500}{60} \text{ Watts}$$

$$P = 18446.4 \text{ watts or } P = 18.45 \text{ kW} \qquad\qquad \textbf{Ans.}$$

EXAMPLE 11.16 A multi-plate clutch of alternate bronze and steel plates having effective diameters of 17.5 cm and 7.25 cm has to transmit 30 H.P. at 2000 r.p.m. The end thrust is

1670 N and the coefficient of friction is 0.11. Calculate from the first principles the number of plates necessary assuming uniform pressure distribution on the plates.

(DAVV Indore, 1986, 1990, 1994).

Solution: Given $r_1 = \left(\dfrac{17.5}{2}\right) = 8.75$ cm; $r_2 = \dfrac{7.25}{2} = 3.625$ cm

$$N = 2000 \text{ r.p.m.; H.P.} = 30; W = 1670 \text{ N}; \mu = 0.1$$

The mean torque T transmitted is given by

$$\text{H.P.} = \frac{2\pi N T_m}{4500}$$

$$T_m = \frac{30 \times 4500}{2\pi \times 2000} = 10.743 \text{ kgf·m} = (10.743 \times 9.81) \text{ N·m}$$

Equating this to frictional torque based on the assumption of uniform pressure,

$$T_f = \frac{2}{3}\mu W(n_a)\left(\frac{r_1^3 - r_2^3}{r_1^2 - r_2^2}\right)$$

or $(10.743 \times 981) = \dfrac{2}{3}(0.11)(1670)(n_a)\dfrac{(8.75^3 - 3.625^3)}{(8.75^2 - 3.625^2)}$

Therefore $(10.743 \times 981) = (1201.6)\, n_a$

$$n_a = \frac{10.743 \times 981}{1201.6} = 8.77 \approx 9$$

Thus $n_a = n - 1$

Therefore $n = $ Total number of plates required $= 9 + 1 = 10$ **Ans.**

Thus, if $n_o = $ number of outer discs and

$$n_i = \text{number of inner discs,}$$

$$n_o + n_i = 10 \quad \text{and,} \quad n_i = n_o - 1$$

Solving $2n_o - 1 = 10$

As $n_o = 5.5$, we adopt $n_o = 6$ and $n_i = 5$

Assuming uniform pressure intensity p, the axial force is given by,

$$W = \pi p(r_1^2 - r_2^2)$$

Therefore $p = \dfrac{W}{\pi(r_1^2 - r_2^2)} = \dfrac{1670}{\pi(8.75^2 - 3.625^2)} = 8.3816 \text{ N/cm}^2$ **Ans.**

EXAMPLE 11.17 A multi-disc clutch has three discs on the driving shaft and two on the driven shaft. The outside diameter of the contact surfaces is 24 cm and inside diameter 12 cm. Assuming uniform wear and coefficient of friction as 0.3, find the maximum axial intensity of pressure between the discs for transmitting 33.5 HP (25 kw) at 1575 r.p.m.

Solution: Given: $n_o = 3$; $n_I = 2$; $r_1 = \dfrac{24}{2} = 12$ cm; $r_2 = \dfrac{12}{2} = 6$ cm

$$\mu = 0.3; \ N = 1575 \text{ r.p.m.}; \ \text{Power} = 25 \text{ kw.}$$

Number of pairs of active surfaces $= (n_o + n_I) - 1 = 3 + 2 - 1 = 4$.

The friction torque T_f is given by

$$25000 = \frac{2\pi N T_f}{60}$$

$$T_f = \frac{(25 \times 1000) \times 60}{2\pi \times 1575} = 151.6 \text{ N·m.}$$

For uniform wear rate,

$$T_f = \frac{1}{2}\mu W (n_a)(r_1 + r_2)$$

$$151.6 \times 10^2 = \frac{1}{2}(0.3)W(4)(12 + 6)$$

Therefore $\qquad W = \dfrac{151.6 \times 100 \times 2}{0.3 \times 4 \times 18} = 1403.7$ N

Again $\qquad\qquad W = 2\pi c(r_1 - r_2)$

where $\qquad\qquad c = p_2 r_2 = (6p_2)$

Hence $\qquad 1403.7 = 2\pi(6p_2)(12 - 6)$

Hence $\qquad\qquad p_2 = \dfrac{1403.7}{2\pi \times 6 \times 6}$ 6.206 N/cm^2

The maximum pressure intensity $p_2 = 6.206$ N/cm^2 **Ans.**

EXAMPLE 11.18 A plate clutch has three discs on the driving shaft and two discs on the driven shaft, providing four pairs of contact surfaces. The outside diameter of the contact surfaces is 240 mm and inside diameter 120 mm. Assuming uniform pressure and $\mu = 0.3$, find the total spring load pressing the plates together to transmit 23 kw at 1575 r.p.m.

If there are 6 springs each of stiffness 13 kN/m and each of the contact surfaces has worn away by 1.25 mm, find the maximum power that can be transmitted, assuming uniform wear.

(London Univ.)

Solution: Given $\ n_o = 3$; $n_i = 2$; and hence $n = 3 + 2 = 5$

and $\qquad\qquad n_a = 5 - 1 = 4$; $r_1 = 12$ cm; $r_2 = 6$ cm; $\mu = 0.3$

Power transmitted,

$$23 \times 1000 = \frac{2\pi \times 1575 \times T_f}{60}$$

Therefore $\qquad\qquad\qquad\qquad\qquad\qquad T_f = 139.45$ N.m

And angular speed $\qquad\qquad\qquad\qquad \omega = \dfrac{2\pi \times 1575}{60} = 164.93$ rad/s

As required in problem, based on uniform pressure assumption, the axial load W to be supplied by compression springs for transmitting above torque, is given by

$$T_f = \frac{2}{3}\mu W(n_a)\left(\frac{r_1^3 - r_2^3}{r_1^2 - r_2^2}\right)$$

or $\qquad\qquad (139.45 \times 100) = \dfrac{2}{3}(0.3)W(4)\left(\dfrac{12^3 - 6^3}{12^2 - 6^2}\right)$

or $\qquad\qquad\qquad\qquad\qquad W = 1245.1$ N $\qquad\qquad\qquad\qquad\qquad$ **Ans.**

As wear takes place uniformly on each of the 4 pairs of friction surfaces and these are 4×2 friction surfaces, total amount of wear on all the discs in axial direction

$$\Delta = 8 \times 1.25 \text{ mm} = 1.0 \text{ cm}$$

As the total width of all the 5 friction discs decreases by 1.0 cm, the compression in all the springs decreases by an amount of 1.0 cm. As the springs are in parallel, their joint stiffness K is

$$K = 6 \times 13 = 78 \text{ kN/m} = \frac{78 \times 1000}{100} = 780 \text{ N/cm}$$

Total reduction of 1 cm in the total width of clutch plates in axial direction, causes a reduction of 1 cm in the amount of initial compression of all the springs and, as such, the compressive force producing axial load W drops by

$$= 1 \times 780 = 780 \text{ N}$$

Hence, in worn out condition of clutch, new axial force,

$$W' = W - \Delta (k) = 1245.1 - 780 = 465.1 \text{ N}$$

Hence, new friction torque, based on assumption of uniform wear rate, is

$$T_f' = \frac{1}{2}\mu W' (n_a)(r_1 + r_2)$$

$$= \frac{1}{2}(0.3)(465.1)(12 + 6) \times 4$$

$$= 5023.08 \text{ N·cm} = 50.2308 \text{ N·m}$$

Hence power transmitted by worn out clutch,

$$= T_f'\omega$$

$$= 50.2308 \times 164.93 = 8284.56 \text{ watts} = 8.285 \text{ kw} \qquad\qquad \textbf{Ans.}$$

Note: Wear of clutch plates in this case has caused a drop in power transmission capacity by

$$= \frac{23 - 8.285}{23} \times 100 = 63.98\%$$

11.18 CENTRIFUGAL CLUTCH

Centrifugal clutches are normally located inside the motor pulleys. One such clutch is shown in Fig. 11.20(a) and (b). It consists of a number of shoes which slide in the radial direction inside the rim of the pulleys. The outer surfaces of shoes are lined with friction material and make contact with the inside of the rim over these friction surfaces. The shoes are constrained to move radially in guides and are held in position by springs which prevent contact of shoes with rim until a predetermined rotational speed is reached. The radially inward force of spring increases as the speed increases and the shoe moves out. However, once the shoe touches rim at speed ω_1, the spring force can be assumed to remain constant thereafter. As against this, centrifugal force F_c is proportional to ω^2. Thus, at any moment the shoe is under the action of force ($F_c - P$) in radial direction; P being the spring force and F_c being the centrifugal force. When F_c is greater than P the shoe moves radially outwards, while if F_c be less than P, the shoe moves radially inwards.

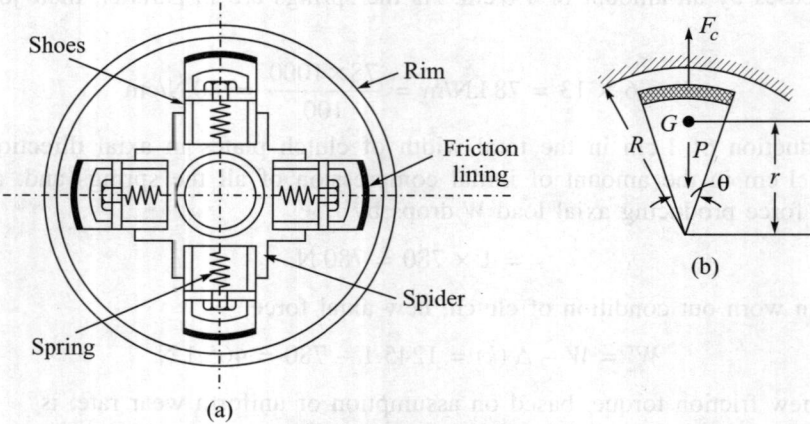

Fig. 11.20 A centrifugal clutch.

Let ω_1 be the angular speed at which the shoe makes contact with the rim and let P_1 be the spring force at this position. For any speed equal to or higher than ω_1, the shoe occupies the same position and therefore the spring force remains constant at $P = P_1$. When the shoe comes in contact with rim, any further increase in centrifugal force F_c is not associated with any increase in spring force and the entire increase in centrifugal force results in increasing the normal force and hence the frictional force between shoe-lining and rim.
Thus, let

$\qquad W$ = weight of each shoe
$\qquad n$ = number of shoes
$\qquad r$ = radial distance of c.g. of shoe from the centre of the spider

R = inside rim radius for pulley

N = running speed of pulley in r.p.m.

ω = angular speed of pulley corresponding to the r.p.m. of N

ω_1 = angular speed at which the engagement occurs

and, μ = coefficient of friction between shoe and rim

When rotating at a speed ω $(\omega > \omega_1)$, the centrifugal force F_c and spring force acting on each shoe are:

$$F_c = \frac{W}{g} r\omega^2$$

The clearance c between shoe and rim being small, r can be assumed to remain constant.

and

$$P = \frac{W}{g} r\omega_1^2$$

Thus, the net outward force, pressing the shoe against living is therefore

$$F = (F_c - P) \tag{11.82}$$

or

$$F = \left(\frac{W}{g} r\omega^2 - \frac{W}{g} r\omega_1^2\right) = \frac{W}{g}(r)(\omega^2 - \omega_1^2) \tag{11.83}$$

Therefore, the frictional force acting on each shoe,

$$\mu F \equiv \mu(F_c - P) = \mu\left(\frac{W}{g} r\right)(\omega^2 - \omega_1^2)$$

The frictional torque transmitted by n such shoes,

$$T_f = n\left(\mu \frac{W}{g} r\right) R\,(\omega^2 - \omega_1^2) \tag{11.84}$$

Thus, if frictional torque T_f, n, μ, r and R are known, the weight W can be obtained for given values of ω and ω_1.

To evaluate size of the shoe, let

l = arcular length of contact of shoe surface

b = face width of shoes

R = contact radius of shoes and equals inside rim radius

θ = angle, subtended at the centre of spider by shoe

p = intensity of shoe-pressure on rim. (For reasonable shoe life, this may be restricted to $9.8\ \mathrm{N/cm^2}$)

Since, $l = R\theta$ and, area of contact for each shoe = lb, the force with which shoe is pressed against rim = $pb = p(R\theta)b$

Comparing with equations (11.82) and (11.83),

$$(F_c - P) = p(R\theta)b$$

or

$$\frac{W}{g} r(\omega^2 - \omega_1^2) = p(R\theta)b$$

or

$$b = \frac{Wr}{(pR\theta)}(\omega^2 - \omega_1^2) \tag{11.85}$$

EXAMPLE 11.19 A centrifugal clutch has 4 shoes which slide radially in a spider keyed to the driving shaft and make contact with internal cylindrical surface of a rim keyed to the driven shaft. When the clutch is at rest, each shoe is pulled against a stop by a spring so as to leave a radial clearance of 5 mm between the shoe and the rim. The pull exerted by the spring is then 500 N. The mass centre of shoe is 160 mm from the axis of the clutch.

If the internal diameter of the rim is 400 mm, the mass of each shoe is 8 kg., the stiffness of each spring is 50 N/mm and the coefficient of friction between the shoe and the rim is 0.3, find the power transmitted by the clutch at 500 r.p.m. (Cambridge Univ.)

Solution: Given $n = 4$; $c = 0.5$ cm; $s = 500$ N; $r = 16.0$ cm

Rim radius $\quad\quad\quad\quad R = 20.0$ cm; $W/g = $ mass of shoe $= 8$ kg

$k = $ spring stiffness $= 500$ N/cm

$\mu = 0.3$; $N = 500$ r.p.m.

Therefore $\quad\quad\quad\quad \omega = \dfrac{2\pi \times 500}{60} = 52.36$ rad/s

Since clearance c is specified, considering it

$r_1 = $ radius of shoe mass centre

or $\quad\quad\quad\quad r_1 = r + c = 16 + 0.5 = 16.5$ cm

Hence, the centrifugal force acting radially outwards,

$$F_{c_1} = \frac{w}{g} r_1 \omega^2$$

or,

$$F_{c_1} = (8)\frac{(16.5)}{100}(52.36)^2 = 3618.872 \text{ N}$$

Radially inward pull due to spring force

$F = $ given spring force $+ k \times$ clearance

or $\quad\quad\quad\quad F = 500 + 500(0.5) = 750$ N

Hence, frictional force, tangential to rim, for one shoe

$$= \mu(3618.87 - 750) = 860.66 \text{ N}$$

Hence, frictional torque due to all the 4 shoes

$$T_f = 4 \times R \times 860.66$$
$$= 68852.88 \text{ N·cm} = 688.53 \text{ N·m}$$

Therefore power transmitted $= \dfrac{T\omega}{1000}$ kW $= \dfrac{688.53 \times 52.36}{1000} = 36.05$ kW   **Ans.**

11.19 FRICTION CIRCLE AND FRICTION AXIS

A journal bearing forms a turning pair and the relative motion between the journal and the bearing is that of sliding. Many bearings or turning pairs are not perfectly lubricated and this is particularly true for many of the turning joints found in linkages. With imperfect lubrication (*i.e.*, when thick film lubrication is nor possible) a skin, greasy, or boundary friction conditions prevail and laws of solid friction are still applicable. For the purpose of analysis thus, the frictional force is assumed to be dependent on normal force of reaction.

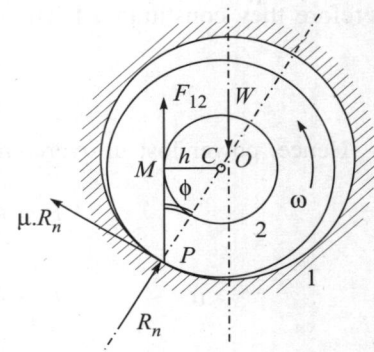

Figure 11.21 shows two members of a turning pair with the mechanical clearance between the contact surfaces much exaggerated. A journal/pin rotating in its bearings, rolls or climbs into the contact position as at P. For this position the axis of journal is at C, which lies on the radial line OP of the bearing. For assumed c.c.w. direction of rotation of the journal, the frictional force μR_n acts on journal in a direction shown in the figure. The resultant F_{12} (exerted by the bearing 1 on the journal 2) of normal reactin R_n and frictional force μR_n acts at friction angle ϕ to the normal direction. For equilibrium in vertical direction, this force F_{12} must be equal and parallel to external load W.

Fig. 11.21 Illustration for friction circle.

The line action of force F_{12} can be considered to be tangent to an imaginary circle of radius h, drawn through the centre C of journal. If R be the radius of the journal, then from the right angled triangle CMP,

$$MC = h = R\sin\phi \qquad (11.86)$$

where ϕ = limiting angle of wet friction

The angle ϕ is usually small and as such,

$$\sin\phi \approx \tan\phi$$

Hence,

$$\sin\phi = \mu$$

Hence from equation (11.86).

$$h = \mu R \qquad (11.87)$$

Note that pin climbs up in a direction opposite to that of rotation of the pin. The imaginary circle of radius $h = \mu R$, drawn from pin centre is called *friction circle*. The friction circle thus depends only on the radius R of the journal/pin and the coefficient of journal friction. Evidently the force F_{21}, exerted by the journal on the bearing, is equal and opposite to F_{12} and must act tangent to the same friction circle.

The concept of friction circle is of considerable importance in graphical analysis as it helps in locating line of action of force (thrust or pull) transmitted along a link. When a link of a mechanism is connected to adjacent links through two frictionless pin joints, the line of action (pull or thrust) is transmitted along the line of centres of these pins. But when friction is taken into account at both the pin joints, the line of action of pull or thrust does not pass through the pin centres but is placed tangent to friction circles at the two pin joints. In other words, the line of action of the thrust or pull is placed along one of the four common tangents to the friction circles at the two pin centres. This common tangent to the friction circles, along which the line of thrust or pull in the link lies, is called the *friction axis* of the link. Thus, the next important step is to identify the particular common tangent which determines the friction axis. But before we do that, let us see the significance of force F_{12} acting tangent to the friction circle.

The resultant force F_{12} as exerted by bearing 1 on journal 2 is equal and opposite to W and therefore they constitute a friction couple,

$$C = F_{12} \times h = W \times h$$
$$C = W \times \mu R \text{ N·m} \tag{11.88}$$

Hence, power lost in overcoming friction,

$$P = C\omega \text{ watts} = \frac{C \times 2\pi N}{60 \times 1000} \text{ kw}$$

or

$$P = \frac{(\mu W R) \times 2\pi N}{60 \times 1000} \text{ kw} \tag{11.89}$$

Basic Principle Involved in Locating Friction Axis is as Under

Force of friction at each pin must oppose relative motion between the pin and the bearing. Also the thrust or pull along the friction axis, in turn, must exert a moment about pin centres which opposes (overcomes) this friction moment.

Steps to Locate Friction Axis

(1) Determine whether a given force P acting along a given link is a thrust (compressive load) or a pull (tensile force). The convention of representing thrust or pull along a link is as shown in Fig. 11.22.

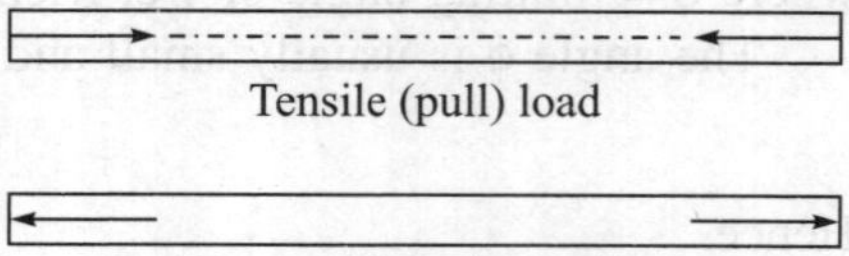

Tensile (pull) load

Compressive (thrust) load

Fig. 11.22

(2) Determine the nature of relative motion of the given link with respect to the connected link at the pin joint under consideration. This helps to identify friction couple which tends to oppose the relative motion.

(3) For the direction of force obtained in (1) above, determine the side on which the force P in (1) should be tangent to friction circle so as to oppose friction couple determined in step (2) above.

(4) Location of sides on which friction axis should become tangent to friction circles at both the pin joints fixes the common tangent along which the friction axis lies.

EXAMPLE 11.20 Figures 11.23(a), (b), (c) and (d) show four positions of I.C. engine mechanism operating under gas load P. For diameters of gudgeon pin and crank-pins as 10 mm and 15 mm respectively and coefficients of friction of 0.06 for both the pins, draw the friction axis for connecting rod in each configuration.

Solution: The mechanism with force P acting on piston is shown in four configurations at Figs. 11.23(a), (b), (c) and (d). The radius of friction circle

at gudgeon pm $\quad r = \dfrac{10}{2} \times 0.06 = 0.3$ mm; at crank pin $\quad r' = \dfrac{15}{2} \times 0.06 = 0.45$ mm

In the figures these friction circles have been drawn in an exaggerated way for increasing clarity of diagrams. We now proceed to locate friction axis position for each case out of the four possible positions of common tangents.

Mechanism at Fig. 11.23(a)

Step 1: The force P on piston pushes the connecting rod against the inertia load connected to the crank shaft. This produces a compression in the connecting rod, and the compressive force is indicated through arrowheads using convention, as shown in figure.

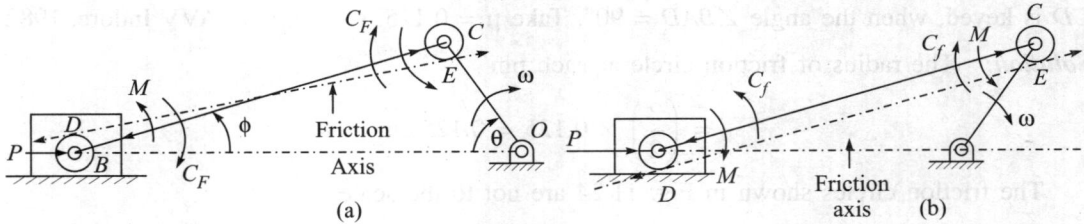

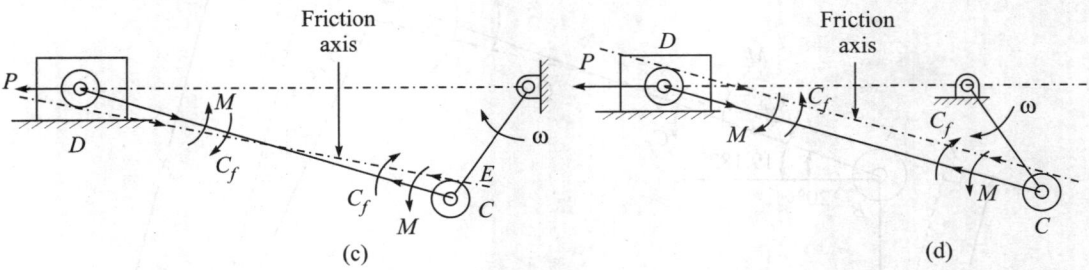

Fig. 11.23 Location of friction axis.

Step 2: For the position shown, any displacement of piston to the right increases angle ϕ which the rod makes with line of stroke and decreases angle α which it makes with the crank. This means that at the piston' end, the relative motion between piston and connecting rod is such that the rod must swing in c.c.w. sense to increase ϕ. This is shown by arrow M. Similarly, at the crank end, for maintaining same relative motion w.r. to crank, the connecting rod must swing in c.c.w. since (as shown by M) for increasing α.

The friction opposes relative motion and therefore the friction couples are shown by C_f at both the ends, opposite to the relative motion of connecting rod.

Step 3: The line of action of force (*i.e.*, friction axis) should now be placed such that with the direction of arrowheads as shown at each end, the driving couple should act in a sense opposite to that of C_f. Hence the friction axis is shown tangent to friction circle on top side at D and on lower side of friction circle at E at the other end of connecting rod.

Mechanisms at Fig. 11.23(b), (c) and (d)

Proceeding in steps as described in (a) above, and with direction of relative motion indicated by M and friction couple by, c_f in each case at the two ends, the friction axes are indicated by DE in Figs. 11.23 (b), (c) and (d).

EXAMPLE 11.21 *ABCD* is a 4-bar chain with *AB* as the driving link and *AD* as the fixed link. The lengths of the links are as follows:

$$AB = 7.5 \text{ cm}; \quad BC = 17.5; \quad CD = 15 \text{ cm}; \quad AD = 22.5 \text{ cm}$$

The diameters of the pins at *A*, *B*, *C* and *D* are all 2 cm. What torque must be applied to the driving shaft in order to overcome a resisting torque of 29.4 N.m applied to the shaft, to which *CD* is keyed, when the angle $\angle BAD = 90°$. Take $\mu = 0.125$. (DAVV Indore, 1982)

Solution: The radius of friction circle at each pin

$$= \left(\frac{2}{2}\right) \times 0.125 = 0.125 \text{ cm}$$

The friction circles shown in Fig. 11.24 are not to the scale.

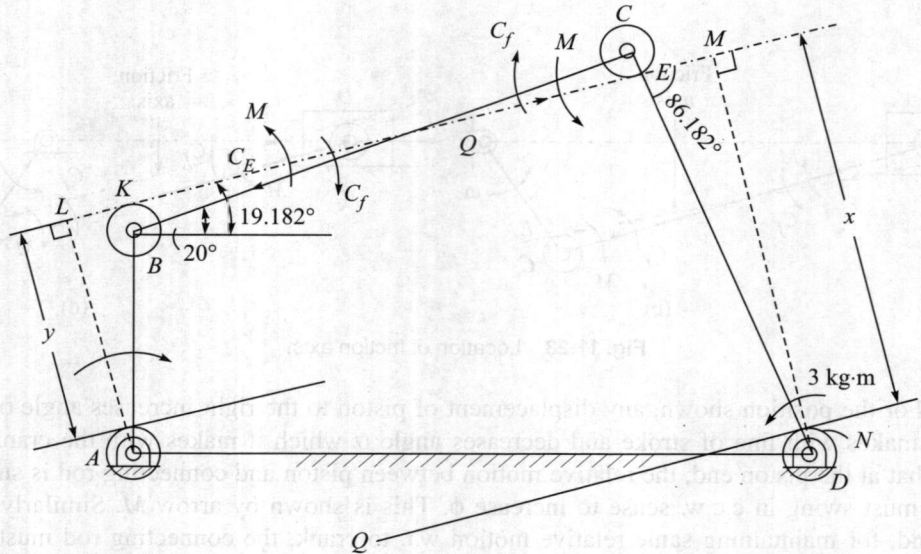

Fig. 11.24 Friction force in coupler.

Location of Friction Axis

Step 1: For assumed c.c.w. sense of resisting torque of 29.4 N·m on shaft at *D*, the driving torque on link *AB* must be applied in c.w. sense. For the assumed sense of driving and resisting torque, the link *BC* is in compression and this is shown by arrowheads along the link.

Step 2: Any further displacement in c.w. sense of link *AB*, tends to increase the angle *ABC* and decrease the angle *BCD*. Hence link *BC* has to swing c.c.w. at *B* with respect to link *AB* and also swing in c.c.w. sense at *C* w.r. to link *DC* to maintain the same relative motion.

Hence the frictional couple C_f at joints *B* and *C* act in clockwise sense.

Step 3: The line of action of thrust in coupler *BC*, with direction of arrowheads as indicated in Fig. 11.24, should be placed along common tangent to friction circles, so as to oppose friction couple at both the ends. Hence, the friction axis must be placed tangent to friction circle on the top side at *B* and tangent to the friction circle on the lower side at *C*. Hence *KE* is the required friction axis along which thrust *Q* in coupler is transmitted.

Let r_B and r_C be the radii of friction circles at *B* and *C* respectively. Then the inclination of friction axis of coupler with coupler *BC* is,

$$\sin^{-1}\left(\frac{r_B + r_C}{BC}\right) = \sin^{-1}\left(\frac{0.125 + 0.125}{17.5}\right) = 0.818°$$

By measurement, for the given configuration

$$\angle BCD = 93°$$

Hence, angle of inclination of the friction axis with the link *CD*,

$$\angle KED = 93° + 0.818° = 93.818°$$

Let *DM* be the perpendicular dropped from *D* on the friction axis. Then,

$$\angle DEM = 180° - \angle KED = 180 - 93.818° = 86.182°.$$

Hence, from right angled triangle,

$$DM = DE\sin86.182$$
$$= (15 - 0.125)\sin86.182° = 14.842 \text{ cm}$$

Since the thrust in coupler is to be obtained only to overcome the resisting torque, the frictional resistance at *D* may be neglected. Hence, let *NQ* be drawn parallel to friction axis from *N*, the intersection point of friction circle with *DM*. Then,

$$x = NM = DM - DN = 14.842 - 0.125 = 14.717 \text{ cm}$$

Hence, the thrust in coupler along friction axis,

$$T = \frac{29.4 \times 100}{14.717} = 199.8 \text{ N}$$

Again, by measurement, the coupler *BC* is inclined at an angle 20° to frame *AD*. Hence friction axis is inclined to frame *AD* at,

$$= 20° - 0.818° = 19.182°$$

Let AL be the perpendicular dropped from A on friction axis KE extended. Hence,

$$\angle LKA = 90 - 19.182 = 70.818$$

Hence, from right angled triangle AKL,

$$AL = AK\sin\angle AKL$$

$$= (AB + r_B)\sin 70.818 = 7.625\sin 70.818 = 7.2016\ \text{cm}$$

Given resisting torque on driven shaft may be assumed to include friction couple at D, but torque to be applied to link AB should take care of friction couple at A. Hence, total torque C required to be applied to link AB should be such that,

$$C = T \times AL = 199.8 \times 7.2016 = 1438.6\ \text{N·cm} = 14.386\ \text{N·m} \qquad \textbf{Ans.}$$

11.20 LUBRICATED SURFACES

Since the inception of machines, it has been the continual endeavour of its makers to reduce the friction between the parts of machines. Probably the earliest and the simplest approach in this direction was to introduce lubricant between the rubbing surfaces. A lubricant is usually a fluid having two essential properties:

    (i) Viscosity, and   (ii) Oiliness

Viscosity. Ability to sustain shear stress and resistance offered to the sliding of one layer of lubricant over an adjacent layer is measured in terms of viscosity of a lubricant. The absolute viscosity of a lubricant is defined as the force required to cause a plate of unit area to slide with unit velocity relative to a parallel plate when the two plates are separated by a layer of lubricant of unit thickness.

Consider a fluid lubricant to consist of a number of parallel layers (Fig. 11.25). The layers of lubricant immediately in contact with a friction surface is assumed to have no motion relative to those surfaces.

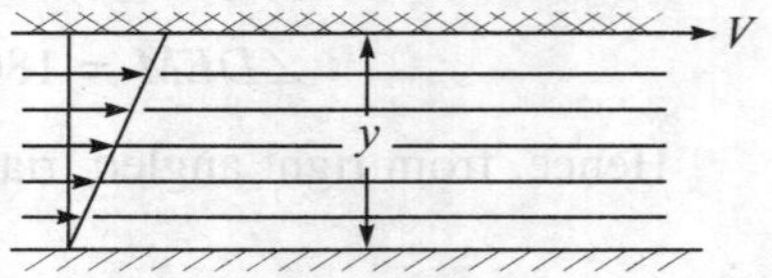

Thus, if one of the two friction surfaces is fixed, the layer of lubricant in contact with it will also be stationary while the layer adjacent to the other friction surface, moving at say velocity V, also moves at velocity V.

Fig. 11.25 Layers of a fluid lubricant.

It may be assumed that the intermediate layers of lubricant move with velocities which are directly proportional to their distances from the surface of the fixed plate. The above assumption leads to a linear velocity gradiant.

Thus, viscous force needed to cause a plate of area A to slide with a velocity V relative to a parallel plate, which is separated from it by a layer of lubricant of thickness y, is expressed as-

$$F = \eta A(V/y) \qquad (11.90)$$

where η = absolute viscosity of the lubricant in N-s/mm^2

              V = relative velocity mm/s

              y = distance in mm

              A = Area of plate mm^2

The viscosity of a liquid diminishes rapidly with an increase of temperature, but is only slightly influenced by an increase of pressure.

Oiliness: Compared to the term viscosity, oiliness is more difficult to define. If two lubricants having same viscosity are used at same temperature and under similar conditions, then the coefficient of friction of one lubricant in general will not be equal to the coefficient of friction of the other lubricant. The difference in coefficient of friction of the two lubricants is attributed to the property of oiliness. Other things remaining the same, a lubricant which is less oily than the other, will have higher coefficient of friction than the other.

Oiliness and viscosity are entirely independent properties. The difference between the two classes of oils namely, the animal and vegetable oils (which give higher oiliness) and mineral oil, is thought to be due to the fact that animal and vegetable oils spread over any surface in contact, whilest mineral oils do not have so high a power of spreading. The spreading power also seems to depend upon the material of the surface. In a specific situation, which one of the two properties (namely, viscosity and oiliness) play more significant role, in deciding coefficient of friction, depends on the thickness of the layer of the lubricant. When the thickness of the lubricating oil film, separating friction surfaces, is finite a thick film is said to exist and it is the property of viscosity which determines coefficient of friction. As against this, when thickness of oil film is extremely small (of molecular size) skin, greasy or boundary friction condition is said to exist, and it is the property of oiliness which determines the value of coefficient of friction.

11.21 FRICTION BETWEEN LUBRICATED SURFACES

Boundary Friction (Skin and Greasy Friction)

When seen under a microscope, even carefully machined surfaces of bearings do not appear to be perfectly smooth. In view of the size of molecular thickness of lubricating oil, such surfaces must be considered as rough. Thus, when an oil film of molecular thickness separates the two friction surfaces, chances of metal to metal contact at high spots do exist. In situations like this, relative sliding between layers of lubricating oil (and consequently the property of viscosity) does not play any significant role in determining friction between surfaces. Therefore, in situations like this, it is the property of oiliness which plays an important role in determining coefficient of friction. Oiliness property varies not only with lubricants, but also with the material of the friction surfaces. Intra-molecular forces are generally considered to play an important role in the entire mechanism of boundary friction. According to Hardy and others, there is certain bond between the molecules of lubricant and the metallic surfaces in contact. This is comparable to a chemical bond and the strength of this bond depends partly on the lubricant and partly on the metal. In boundary lubrication, the lubricant gets absorbed in the metal surface. The absorbed film is extremely thin and is very difficult to be removed from the surface.

With extremely thin layer of lubricant between friction surfaces, boundary friction condition resembles that of solid friction with coefficient of friction considerably smaller than the limiting coefficient of solid friction. Further, presence of a thin film also reduces the chances of seizure between rubbing surfaces significantly. An important feature of boundary lubrication is that once a thin film is established between friction surfaces, a lubricant of higher oiliness can be replaced by an another lubricant of lower oiliness. Usual lubricants of higher oiliness are vegetable oils like castor oil, rape oil and olive oil which have a greater affinity for the metallic surfaces and are readily absorbed on friction surfaces. These oils are more susceptible to degeneration and oxidation which is considered undesirable in lubrication practices. As against this, mineral oils are generally deficient in respect of property of oiliness. An introduction of mineral oil, subsequent to formation of thin film using vegetable oils, thus ensures thin film lubrication which is practically free of chances of degeneration and oxidation.

Chemically pure graphite in extremely finely divided powder form is called colloidal form. When mixed with lubricant in colloidal form, the graphite particles remain in suspension for an indefinite period. A thin layer of graphite deposited on friction surfaces is known to increase absorption of oil on friction surfaces and bearing is rendered almost immune to seizure. Boundary condition is likely to occur in heavily loaded slow running bearings, in bearings where the relative motion is slow and reciprocating, and in all such cases where supply of oil is not sufficient for maintaining a film.

Film Lubrication (also Called Thick Film or Viscous Lubrication)

Thick film or viscous lubrication condition is said to have been achieved when the friction surfaces are completely separated by a thick film/layer of lubricant, so that metal to metal contact is completely replaced by relative sliding movement between the layers of oil. Friction forces thus arise from relative sliding of the layers of the oil and not from the rubbing of actual friction surfaces. This condition is considered the ideal form of lubrication for, as a rough guidance, the limiting coefficient of friction for dry condition can be three times as high as that for boundary friction and can be as high as 15 times that of viscous friction.

Unfortunately, bearings must meet certain special conditions of operation for forming and maintaining a thick film of lubricant between the friction surfaces. These conditions will be listed after the discussions on Towers's experiment.

The present theory of hydro-dynamic lubrication has its origin in the laboratory of *B*. Tower, who was employed to study the friction in rail-road journal bearings and to recommend the best methods of lubricating them. It was an accident or error, during the course of his investigations, which prompted Tower to look at the problem in more details and this resulted in a discovery which eventually led to the development of the *Theory of Hydrodynamic Lubrication* in the year 1885.

He used a railway carriage bearing (see Fig. 11.26) where a load was applied through a bearing brass step *B* and was supported by a shaft or a journal *A*. The bearing pads embraced rather less than one-half of the circumference of the journal and bottom portion of shaft dipped into a bath of oil. The oil adhered to the surface of rotating shaft and was carried round to the clearance space between journal and bearing pad at *P*.

A number of holes were radially drilled in the bearing at different locations along a line parallel to the axis and also along the periphery. The tappings from these holes were taken to a Bourdon pressure gauge. He plugged all but one hole at a time so as to measure the pressure at each hole. Tower found that there was a fluid pressure at all parts of

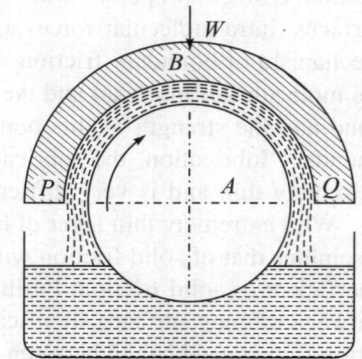

Fig. 11.26 Railway carriage bearing.

a well lubricated bearing. Presence of oil pressure at all points established for the first time that there exists a pressurised oil film between friction surfaces.

Tower also found that the coefficient of friction was neither constant nor dependent on load W, as it should have been if laws of solid friction were valid. Infact, the coefficient of friction was found to be a function of bearing pressure p and surface speed v of the journal. Thus,

$$\mu = k \frac{\sqrt{v}}{p}$$

(11.91)

i.e., the coefficient of friction μ decreases with an increase in bearing pressure p and increases with increase in surface speed v. Further, the actual value of coefficient of friction was very small, being of the order of 0.001 to 0.002 with mineral oil as the lubricant and the temperature of oil bath maintained at 90° F. The most interesting aspect of Tower's experiment was the variation in pressure of oil measured at different points in the bearing while the journal was still rotating. The reported pressure distribution in the oil film was very similar to the one shown in Fig. 11.27(a) and (b). Across the diametral width of the bearing, the pressure was found to increase from zero at the oil inlet P to a maximum value at a point little to the outlet side of the centre line. The pressure thereafter decreased to zero at the outlet, where the clearance space was open to the atmosphere. Along the length of the bearing pad l, the pressure remained very nearly constant except near the two ends, where it rapidly fell down to zero.

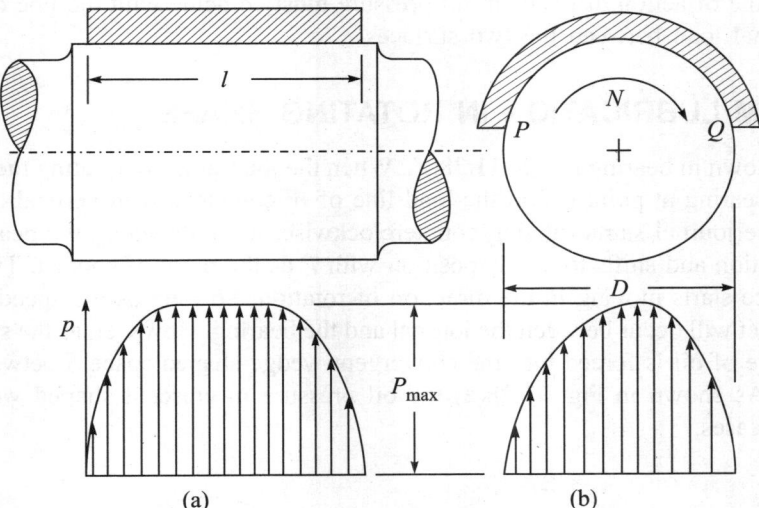

Fig. 11.27   Oil Pressure Distribution in Journal Bearing.

The maximum pressure p_{max} of the lubricating oil was found to be much greater than the nominal bearing pressure. However, the total load, as calculated on the basis of recorded bearing pressures, was almost equal to the load W coming through bearing pad. The product of 'projected area' of bearing and effective pressure equalled the total load on bearing. This proved conclusively that the film of oil in viscous lubrication carries the entire load and shaft virtually floats in the oil.

The other results of Tower's experiment can be listed as under:

(i) The tangential frictional force per unit area of bearing surface is almost independent of the load;

(ii) The tangential frictional force diminishes with an increase in oil bath temperature;

(iii) The tangential frictional force is independent of material used for friction surfaces;

(iv) Tangential frictional force is different for different lubricants;

(v) The thickness of the oil film is greater at the inlet side of the bearing pad compared to the outlet side.

In a subsequent paper in the year 1886 A.D., Reynold established conditions essential for the formation of hydro-dynamic film.

Conditions Essential for the Formation of Hydro-dynamic Film

(1) There must exist a relative motion between the two friction surfaces, in a direction approximately tangential to the surfaces.

(2) A continuous supply of lubricating oil should be ensured.

(3) One of the two bearing surfaces should be capable of taking up some inclination with respect to the other surface in the direction of relative motion, so that a wedge of lubricant is formed.

(4) The line of action of resultant oil pressure must coincide with the line of action of the external load between the two surfaces.

11.22 FILM LUBRICATION IN ROTATING SHAFS

A journal is shown in bearing in Fig. 11.28(a). When the journal is not rotating the journal makes contact with bearing at point C and the load line of W coincides with vertical diameter of the bearing. As the journal starts rotating counterclockwise, it climbs along the bearing surface in opposite direction and shifts to a new position with P as the point of contact. The oil adhering to shaft surface starts moving in the direction of rotation. So long as the speed is low, metal-to-metal contact will occur between the journal and the bearing. However, as the speed increases, more and more of oil is forced into the convergent wedge shaped space S between the journal and bearing. As shown in Fig. 11.28(a) the oil pressure in crescent shaped wedge portion S therefore increases.

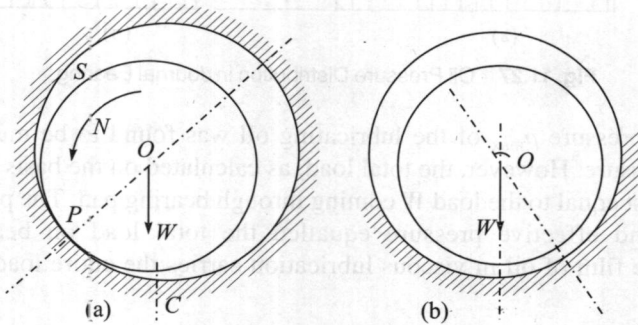

Fig. 11.28 Mechanism of Oil Film Formation.

As the speed increases further the pressure in wedge portion is enough to lift the journal up to the right as shown in Fig. 11.28(b). Metal to metal contact between journal and bearing is replaced by a film of oil that is formed between the journal and bearing, which carries the entire load.

Figure 11.29 shows a plot of graph between the coefficient of friction μ and the speed of rotation N in r.p.m. The graph shows that in low speed region, the coefficient of friction drops down as the speed increases from P to Q. This is because, as more and more oil is forced into the small convergent space S (Fig. 11.28 a), a component of oil-pressure in vertically upward direction pushes the shaft up and more of oil enters beneath the shaft. Thus, as we move from P to Q the conditions change from greasy (skin or boundary) friction to fluid friction at Q, which represents condition of complete film formation.

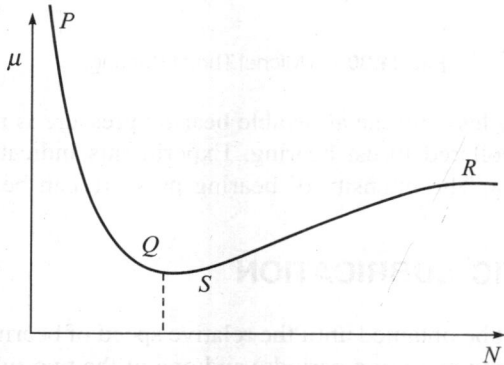

Fig. 11.29 Variation of coefficient of friction with speed of rotation in r.p.m.

Beyond a certain speed, the coefficient of friction starts rising again. This is represented by portion SR of the curve. In practice, a bearing is designed to ensure shaft rotation in the region SR, which promises stable working condition under film friction.

11.23 MICHEL THRUST BEARING

A.G. Michel in U.K. and A. Kingsbury in U.S.A. aimed at artificially producing conditions essential for the formation of hydro-dynamic film. Michel thrust bearing is illustrated diagramatically in Fig. 11.30. The thrust from the shaft is transmitted to the casing through a single collar C. Actual bearing surface consists of a number of metallic bearing pads B, which are sector shaped and are supported in casings. The pads are housed in casings in such a way that they can take up small inclination to the flat surface of collars but are prevented from motion in circumferential direction. In the case of large bearings, the lubricating oil is supplied under P pressure but in the case of small bearings the collar dips into a sump of oil. As the oil is carried round by the collar, flat surface of collar and the inclined surface of tilted pad provides a wedge shaped oil film. The axial load is then transmitted from the collar and through the wedge shaped film to the casing.

Figure 11.30(b) shows two alternative arrangements for supporting the pads in tilted position. In the first case a step provided underneath the bearing pad permits it to tilt slightly on the edge. In the second case a screw point acting off the centre tilts the pad.

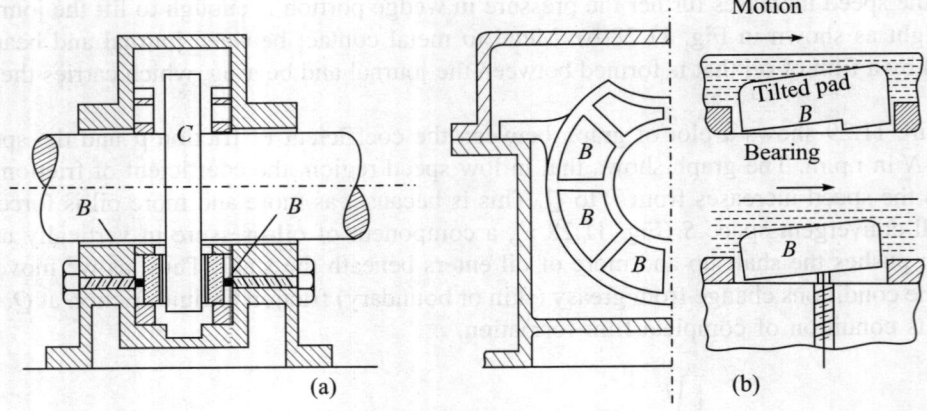

Fig. 11.30 A Michel Thrust Bearing.

Friction is significantly less and the allowable bearing pressure is much higher than that for horse-shoe shaped multi collared thrust bearing. Experiments indicate that the coefficient of friction is only about 0.003. The intensity of bearing pressure can be as high as 4670 N/cm².

11.24 HYDROSTATIC LUBRICATION

Hydro-dynamic film cannot be obtained until the relative speed of bearing surfaces is sufficiently high (and is never zero during operating periods) and one of the two rubbing surfaces is capable of taking some inclination with respect to the other. There are good number of applications in which this is not possible. In reciprocating components and in particular in machine tool units like tables, saddle, slide, etc., hydro-dynamic film cannot be formed. To reduce friction and wear in such applications, therefore, hydrostatic guideways have been developed.

In hydrostatic bearings and guideways, the lubricating oil is pressurised externally and the bearings/guideways have provisions for supplying oil at considerable pressure to several pockets in the bearings/guideways. One such hydrostatic bearing is shown in Fig. 11.31. The oil flows out through the clearance between a shoulder of the spindle and the end of the bearing, such hydrostatic bearings can operate under fluid friction conditions even at the slowest speeds of rotation.

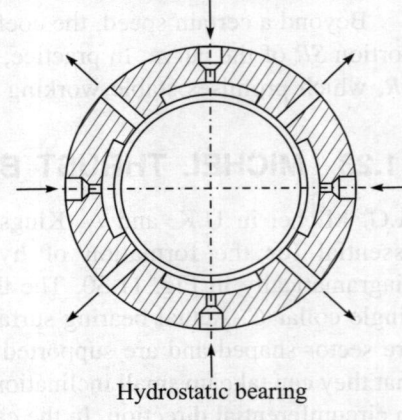

Hydrostatic bearing

Fig. 11.31

11.25 ROLLING CONTACT BEARINGS

Ball and roller bearings are used in almost every kind of machine and mechanical device with rotating parts. However, such bearings cannot be used indiscriminately without taking into account types of loads and operating conditions. As described earlier, substitution of rolling friction in place of sliding friction results in a reduction of frictional resistance in a significant way. The surfaces of the two elements of the turning pairs are separated by a number of rolling elements. Before we proceed in detail, a word of caution is required regarding the term 'antifriction bearings'. The popular meaning conveyed through the term is 'a bearing with zero friction'—which is incorrect. Of course the introduction of rolling friction cuts down frictional resistance significantly. To reduce the coefficient of rolling friction further, the balls or the rollers are made of chromium stated or chrome-nickel steel, which are heat treated to increase their hardness. These rolling elements are precisely ground and polished. The starting friction is thus cut down significantly.

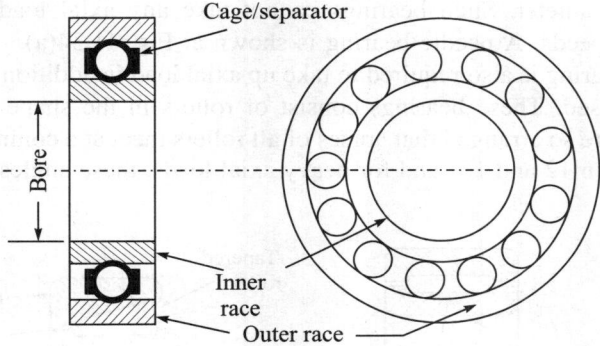

Fig. 11.32 An antifriction Ball Bearing.

1. **Ball Bearings:** (see Fig. 11.32) A ball bearing usually consists of four parts—an inner ring or race, an outer ring or race, the steel balls and the cage or separator. To increase the contact area and permit larger loads to be carried, the balls run in curvilinear grooves. The inner race is usually fitted to the shaft with tight press-fit and rotates with it. The outer race has a press-fit with fixed housing and is therefore stationary. There are shallow grooves in both the raceways having radius slightly larger than that of the balls. A separator or cage keeps the balls apart at a fixed distance. Raceways are hardened and a small amount of lubricant (usually grease) is used to prevent rust formation.

Such bearings are intended to take up radial loads (*i.e.,* loads at right angles to the axis of rotation) basically, although it can also sustain small amount of axial load. Each ball makes a point contact with both the inner race and outer race. With double row ball bearings, the load carrying capacity is increased but not by the same amount. When both axial and radial loads are to be taken up, angular contact ball bearings take care of both the loads. As against this, self aligning bearings permit rotation of shaft axis in the longitudinal plane and take care of large amounts of angular misalignment. This arrangement eliminates bending stresses in the neck of the shaft on which the inner race is mounted.

2. **Roller Bearings:** Constructionwise, a roller bearing is more or less similar to a ball bearing except that balls are replaced by cylindrical rollers which increases load carrying capacity for a given overall size of the bearing. When the rollers are solid, the length is

equal to the diameter. The inner race of the bearing is grooved while the outer race is either plane or convex in some designs. In some of the designs, outer race is spherical so as to allow self-alignment of the shaft. The rollers in this case are slightly barreled. Roller bearings carry only the radial loads (*see* Fig. 11.33).

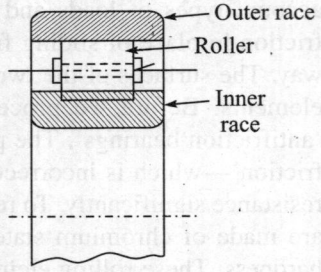

Fig. 11.33 A Roller Bearing.

Needle bearings fall in the category of roller bearings with rollers of very small diameters. These needles are used without cage and are capable of carrying heavy loads. The diameter of needles is of the order of 2 to 4 mm and having length between 3 to 10 times the diameter. Such bearings cannot take any axial load and are suitable for comparatively low speeds. A needle bearing is shown at Fig. 11.34(a).

When a roller bearing is also required to take up axial load in addition to radial load, tapered roller bearings are used. These bearings consist of rollers in the shape of frustum of a cone (Fig. 11.34b). They are so arranged that apices of all rollers meet at a common point. The contact angle is taken between 12 and 16° and for heavy axial loads, these angles are taken between 28 and 30°.

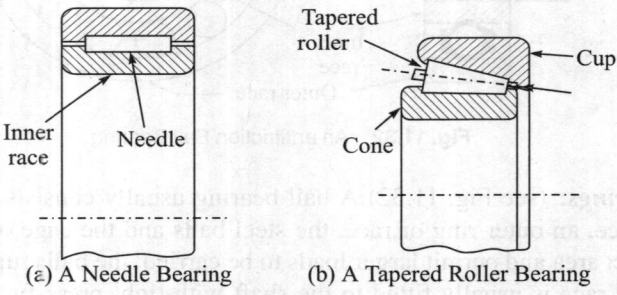

(a) A Needle Bearing   (b) A Tapered Roller Bearing

Fig. 11.34

11.26 ADVANTAGES AND DISADVANTAGES OF ROLLING CONTACT BEARINGS

(1) Antifriction bearings have a smaller axial length compared to plain hydro-dynamic bearing for the same load. For instance, length of antifriction bearing is generally less than the bore size while the length of journal bearing is usually between 2–3 times the bore diameter. However, antifriction bearings usually occupy more radial space.

(2) Rolling contact bearings require less of starting torque. This is because in hydrodynamic bearing, initially there is no film formation and as such frictional forces are large. As against this, rolling friction is quite less when motion is about to begin. This results in saving almost 40% of power.

(3) More accurate centering of shaft in housing is possible using antifriction bearings. This is obvious as no clearance is required for accomodating lubricating oil required for film formation.

(4) Rolling contact bearings require precision in their manufacture and are therefore more costly.
(5) The hardened raceways of antifriction bearings are liable to surface failure. When bearings are overloaded, the raceways flake off.
(6) At high speeds, antifriction bearings tend to become more noisy than the journal bearings.
(7) Rolling contact bearings are convenient from the point of view of maintenance, as no lubrication is required.

REVIEW QUESTIONS

11.1 A block of weight 12 N is resting on a flat surfaces with a coefficient of friction $\mu = 0.25$. It is acted upon by a force of 2 N applied parallel to the surface of contact. What is the frictional force acting on the block? (**Ans.** 2 N)

11.2 A flat footstep bearing 20 cm in diameter, supports a load of 2 tonnes. If the coefficient of friction is 0.05 and r.p.m. is 100, calculate the H.P. lost in friction for:

(a) uniform intensity of pressure and
(b) uniform rate. (**Ans.** 936 and 0.698 H.P.)

11.3 A wedge *B* is being raised against the action of a resistive force *R* by applying a force *F* to another wedge *A* as shown in Fig. 11.35. Derive the expression for the mechanical efficiency as a function of θ and μ. Neglect all masses. (AMIE, Winter 1993)

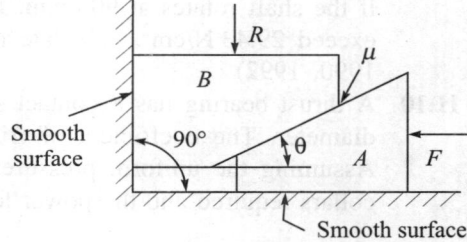

Fig. 11.35 Lifting through wedge action.

11.4 A vertical screw with a single start square thread, 5 cm mean diameter and 1.25 cm pitch, is raised against an axial load of 9810 N by means of a hand-wheel, the boss of which is threaded to act as a nut. The axial load is taken by a thrust collar which supports the wheel boss and has a mean diameter of 6 cm. If the coefficient of friction is 0.15 for the screw and 0.18 for the collar and the tangential force applied by each hand to the wheel is 98 N, find the diameter of the hand-wheel. Derive the formula used for computing the screw torque. Comment on the result (AMIE, Summer 1981)
(**Ans.** Dia of hand-wheel: 112.2 cm)

11.5 The pitch of the 50 mm mean diameter threaded screw of a screw jack is 12.5 mm. The coefficient of friction between the screw and the nut is 0.13. Determine the torque required on the screw to raise a load of 24.525 kN, assuming the load to rotate with the screw. Determine also the ratio of the torque required to raise the load to the torque required to lower the load and the efficiency of the machine. (AMIE, Summer 1983)
(**Ans.** $T = 12986$ N·m; Ratio = 4.25; $\eta = 37.5\%$)

11.6 In a screw jack, the vertical screw of mean diameter 50 mm has single start square threads of 10 mm pitch. It support a load of 19.62 kN. The axial thrust is taken by a collar bearing of inner diameter 70 mm and outer diameter 120 mm. The coefficient of friction for the collar bearing is 0.1 and that for the screw and nut 0.12. Find (i) the torque in N·cm units required to be applied to the nut to raise the load and (ii) efficiency of the screw lack. (***Ans.*** 9535 N·cm; μ = 68.8%)

11.7 The spindle of a screw jack has single start square threads with an outside diameter of 4.5 cm and a pitch of 1 cm. The spindle moves in a fixed nut. The load is carried on a swivel head but is not free to rotate. The bearing surface of the swivel head has a mean diameter of 6 cm. The coefficient of friction between the nut and screw is 0.12 and that between the swivel head and the spindle is 0.10. Calculate the load which can be raised by efforts of 98 N each applied at the end of two levers each of effective length of 35 cm. Also determine the velocity ratio and the efficiency of the lifting arrangement. (Jodhpur Univ.) (***Ans.*** 9761 N; 220; 39.4%)

11.8 A pivot bearing of a shaft consists of a frustum of a cone. The diameters of the frustum are 20 cm and 40 cm and its semi-cone angle is 60°. The shaft carries a load of 39240 N and rotates at 240 r.p.m. Determine (i) the magnitude of the pressure intensity on the bearing surface, and (ii) the power lost in friction. Given μ = 0.025. Make suitable assumption if required giving reasons for the same. Prove the formulae used. (DAV Indore, 1983) (***Ans.*** p = 41.63 N/cm²; 4.488 kw)

11.9 A thrust bearing supports a load of 235.44 kN on collars 40 cm external diameter and 30 cm internal diameter. If μ = 0.07, calculate the power lost in friction at the bearings if the shaft rotates at 90 r.p.m. If the permissible bearing pressure intensity is not to exceed 29.43 N/cm², calculate the number of collars required. (DAV Indore, 1982, 1990, 1992) (***Ans.*** n = 15; p = 27.74 kw)

11.10 A thrust bearing has 8 contact surfaces, 20 cm external diameter and 15 cm internal diameter. The coefficient of friction is 0.08 and the total axial thrust is 22072 N. Assuming the uniform pressure intensity of 34.34 N/cm², determine the number of collars required and the power lost in friction at 450 r.p.m. (Delhi Univ.) (***Ans.*** 5; 7.43 kw)

11.11 The thrust of the propeller shaft of a marine engine is taken up by 8 collars whose external and internal diameters are 660 mm and 420 mm respectively. The thrust pressure intensity is 0.4 MN/m² and may be assumed uniform. The coefficient of friction between the shaft and the collars is 0.4. If the shaft rotates at 90 r.p.m. find (i) total thrust on the collars and (ii) power absorbed by friction at the bearing. (London Univ.) (***Ans.*** 651.4 kN; 68 kw)

11.12 A shaft has a number of collars integral with it. The external diameter of the collars is 40 cm and the shaft diameter is 25 cm. If the uniform intensity of pressure is 34.33 N/cm² and its coeff. of friction is 0.05, estimate (i) power absorbed in overcoming friction when the shaft runs at 105 r.p.m. and carries a load of 147.15 kN, (ii) number of collars required? (Annamalai Univ.) (***Ans.*** 13.57 kw; n = 6)

11.13 Following data refer to a cone-clutch: max. contact surface rad. = 15 cm; minimum contact surface rad. = 12.5 cm; semi-cone angle = 20; coeff. of friction = 0.25; Allowable normal pressure = 13.73 N/cm². Estimate: (i) the axial load and (ii) power transmitted at 1000 r.p.m. (AMIE, Summer 1929) (***Ans.*** 2965 N; 31.62 kw)

11.14 The thrust of a propeller shaft in a marine engine is taken up by a number of collars, solid with the shaft, which is 30 cm in diameter. The total thrust on the shaft is 176.58 kN and its speed is 75 r.p.m. coefficient of friction = 0.05 and pressure = 34.33 N/cm^2, constant over the surface. Find (a) the external diameter of collars, and (b) the number of collars required, if the power lost in friction is 14.9 kW (AMIE, Summer 1980)   (*Ans.* D = 52.55 cm; n = 4)

11.15 A leather faced conical clutch as a cone angle of 30°. If the intensity of pressure between the contact surfaces is limited to 34.33 N/cm^2 and the breadth of the conical surface is not to exceed one-third of the mean radius, find the dimensions of the contact surfaces to transmit 22.37 kW at 2000 r.p.m. Assume uniform rate of wear and take coefficient of friction as 0.15. (AMIE, Winter 1982)
(*Ans.* b = 3:36 cm; r_1 = 10.514 cm; r_2 = 9.646 cm)

11.16 A conical friction clutch is used to transmit 93.2 kW at 1500 r.p.m. The semi-cone angle is 20° and the coefficient of friction is 0.2. If the mean diameter of the clutch is 37.5 cm and the intensity of normal pressure is not to exceed 24.53 N/cm^2, find the dimensions of the conical clutch and the axial force required.
(*Ans.* r_1 = 19.726 cm; b = 5.364 cm)

11.17 A single plate friction clutch, with both sides of the plate being effective, is used to transmit power of an engine at 2400 r.p.m. It has outer and inner radii 4.5 and 3.5 cm respectively. The pressure is applied axially by means of springs and the maximum intensity of pressure is 7.85 N/cm^2. If the coefficient of friction is 0.3, find (i) total axial pressure exerted by the springs and (ii) power transmitted. (Bombay Univ.)
(*Ans.* 172.56 N; 1.05 kw)

11.18 A motor car clutch is required to transmit 8.95 kw at 3000 r.p.m. It is of single plate type, both sides of plate being effective. If μ = 0.25 and the axial pressure is limited to 833.9 N/cm^2 of plate area and the external diameter of the plate is 1.4 times the internal diameter, determine the dimensions of the plate.   (*Ans.* r_1 = 8.5 cm)

11.19 A multiple disc clutch, steel and bronze discs, is to transmit 7.46 kw at 450 r.p.m. The inner radius of the contact is 4 cm and the outer radius 9 cm, μ = 0.1. The maximum allowable pressure 2.61 N/cm^2. Determine the following, (i) total number of discs (ii) the average pressure; (iii) the axial force regd. and (iv) maximum pressure.
(*Ans.* n = 7; p = 19.62 N/cm^2; W = 4002.5 N; p_{max} = 31.88 N/cm^2)

11.20 A machine is driven from a constant speed shaft rotating at 300 r.p.m. by means of a friction clutch. The moment of inertia of the rotating parts of the machine is 5 kg·m^2. The clutch is of the disc type, both sides of the disc being effective in producing driving friction. The diameters of the friction plate are 20 cm and 12.5 cm. The intensity of axial pressure applied to the disc is 0.7 bar (1 bar = 10^5 n/m^2). Assuming uniform pressure and coefficient of friction to be 0.25, determine the time required for the machine to attain full speed when the clutch is suddenly engaged. Determine also the energy supplied during clutch slip. Determine the ratio of power transmitted with uniform wear to that with uniform pressure, and the maximum intensity of pressure in case of uniform wear. (DAV Indore, 1992)
(*Ans.* t = 2.8456 sec.; E = 4934.8 N·M; Power transmitted ratio = 0.984; p = 0.7 bar)

11.21 74.57 kw power is transmitted at 300 r.p.m. by a multiplate disc friction clutch. The plates run in oil and have the friction surfaces of steel and phosphor bronze alternately. $\mu = 0.07$, and the axial intensity of pressure is not to exceed 14.7 N/cm². External radius is 1.25 times the internal radius. The external radius is 12.5 cm. Determine the number of plates needed to transmit the required torque. Make suitable assumptions in solving the problem and derive the relations used.

(*Ans.* Required number = 14; However as $n_i = n_i = n_0 - 1$ Take $n = 15$;
$n_o = 8$: $n_i = 7$)

11.22 Explain what is meant by friction circle. In a direct acting steam engine, the stroke is 60 cm and the diameter of the piston is 30 cm, the length of the connecting rod is 4 times the length of the crank. The diameter of the cross-head pin, crank pin, crank-shaft are 8.7 cm, 11.2 cm and 12.5 cm respectively. The coefficient of friction between the cross-head and guides is 0.08 and for the two pins and crank-shaft it is 0.005. When the crank has moved through an angle of 45°, from the i.d.c., the effective steam pressure on piston is 39.24 N/cm².
Draw a diagram to show the direction of forces acting in the link-work, and calculate the turning moment on the crank-shaft. (DAV Indore, 1917)

11.23 A multi-plate clutch is to transmit 3.73 kw at 1500 r.p.m. The inner and outer radii of the plates are to be 5 cm and 10 cm respectively. The maximum axial spring force is restricted to 981 N. calculate the necessary number of pairs of surfaces if $\mu = 0.35$, assuming constant wear. What will be the necessary axial force?

(*Ans.* Calculated value of $n_a = 1$)

Chapter 12

BELT, ROPE AND CHAIN DRIVES

12.1 INTRODUCTION

Belt, rope and chain drives are the examples of higher pair. As explained earlier, in Chapter 2, pairs of elements having relative motions of sliding, turning or screw motion are known as lower pairs. Relative motion between pairing elements of a higher pair is more complicated compared to that in lower pairs. In the case of belts if we neglect slip, the relative motion is that of rolling; the belt winding on to the pulley at one end and leaving the pulley at the other. The belt drive thus is not a positive drive due to slip associated with the rolling motion. Power is transmitted from one shaft to another through friction between belt and pulleys.

Ordinarily, a belt drive is used to transmit power between two parallel shafts separated by a certain minimum distance; the minimum distance being dependent on the type of belt used. The above condition of minimum distance is a desirable condition for the belts to work most efficiently. Flat belts are used to connect shafts with centre distance as large as 10 m, while *V*-belts are used to connect shafts with centre distance less than or equal to 5 m. V-belts permit a speed ratio upto a maximum of 7. As against this, a rope drive is used for connecting grooved pulleys or drums at a centre distance upto a maximum of 30 m.

When using flat belts, clutch action may be obtained by shifting the belt from a loose to a tight pulley. Step pulleys may be used to obtain a desired change in velocity ratio economically. Belts and ropes transmit power due to friction between them and the pulleys. *When power being transmitted exceeds the torque capacity, corresponding to limiting coefficient friction, the belt/ rope slips over the pulley. A positive aspect of this is that a belt/rope drive may also be used as an overload protection device.*

12.2 VELOCITY RATIO

A set of parallel shafts may be connected by a flat belt so as to produce either—

    (i) an open belt drive or   (ii) a crossed belt drive

These arrangements are shown in Figs. 12.1(a) and (b). One of the two connected pulleys is called driver while the other one is called the driven pulley. Note that unlike gears, it is not necessary for the driver pulley to be smaller than driven pulley as the belt drive may be used either as a speed-reducer or for increasing the speed.

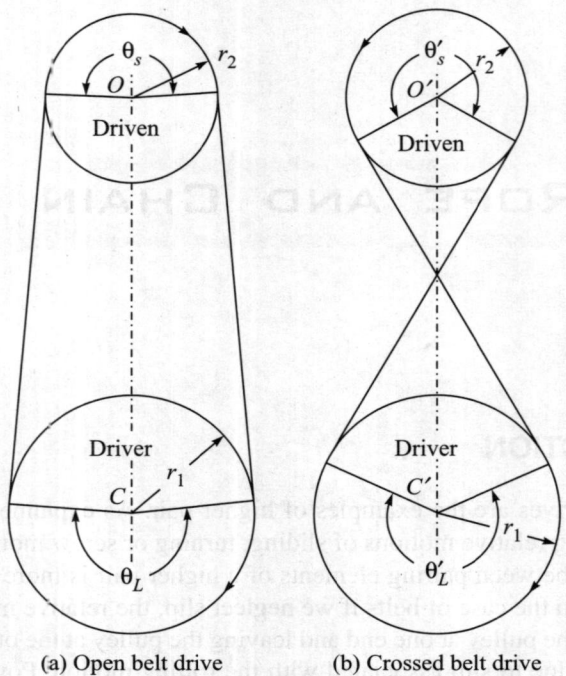

(a) Open belt drive   (b) Crossed belt drive

Fig. 12.1

Assuming the driver pulleys to rotate in the clockwise sense, it is easy to see that driver and driven pulleys rotate in the same sense in open belt drive, but in crossed belt drive, they rotate in opposite sense. Let us first neglect belt thickness.

Let r_1 = the radius of driver pulley in either case,

     r_2 = the radius of driven pulley in both the cases,

     ω_1 = angular speed of driver pulley, and

     ω_2 = angular speed of driven pulley.

Assuming no slip to occur between belt and pulleys, the peripheral speed of rotation of pulleys must equal the belt speed. Thus

$$V = r_1 \omega_1 = r_2 \omega_2$$

or
$$\frac{\omega_2}{\omega_1} = \frac{r_1}{r_2} \qquad\qquad (12.1)$$

Thus, if N_2 be the speed in r.p.m. of driven pulley and N_1 be the speed in r.p.m. of driver pulley, we have

$$\frac{\omega_2}{\omega_1} \equiv \frac{N_2}{N_1}$$

Hence, from equation (12.1),

$$\frac{N_2}{N_1} = \frac{r_1}{r_2} \qquad (12.2)$$

In other words, the velocity ratio of belt drive varies inversely as their radii.

Effect of Belt Thickness t

Assume the belt thickness to be uniform. The effective radii at the neutral surface of belt, which is neither compressed nor elongated as the belt is subjected to flexure (while passing over the pulleys), are

$$R_1 = \left(r_1 + \frac{t}{2}\right) \quad \text{and} \quad R_2 = \left(r_2 + \frac{t}{2}\right)$$

Hence, for no sleep,

$$V = R_1\omega_1 = R_2\omega_2 \qquad \text{or} \qquad (\omega_2/\omega_1) = R_1/R_2$$

Substituting for R_1 and R_2,

$$(\omega_2/\omega_1) = (r_1 + t/2)/(r_2 + t/2) \qquad (12.3)$$

And in terms of speeds in r.p.m.,

$$(N_2/N_1) = (r_1 + t/2)/(r_2 + t/2) \qquad (12.4)$$

As $\qquad r_1 = d_1/2 \qquad$ and $\qquad r_2 = d_2/2$, this leads to the expression

$$(N_2/N_1) = (d_1 + t)/(d_2 + t) \qquad (12.5)$$

Effect of Slip on Velocity Ratio

On the driver pulley, the pulley drives the belt and on driven pulley, the belt drives the pulley. When there is no slip between belt and pulleys, the peripheral speed of pulleys and the belt speed are equal. The slip is thus measured by relative velocity between the belt and the pulley.

Let $\quad s_1$ = percentage slip between driver pulley and belt, and
$\quad s_2$ = percentage slip between belt and driven pulley.

Clearly for a peripheral speed $r_1\omega_1$ of the driver pulley, the belt speed will be less than the peripheral speed of pulley and is given by

$$\text{belt speed} = (r_1\omega_1)\left(\frac{100 - s_1}{100}\right) \qquad (12.6)$$

As the belt drives driven pulley, when slip occurs, the belt speed must be larger than the peripheral speed $(r_2\omega_2)$ of the driven pulley. Thus

$$r_2\omega_2 = (\text{belt speed}) \times \left(\frac{100 - s_2}{100}\right)$$

Substituting for belt speed from equation (12.6),

$$r_2\omega_2 = r_1\omega_1\left(\frac{100 - s_1}{100}\right)\left(\frac{100 - s_2}{100}\right) \qquad (12.7)$$

If s be the total percentage slip between the driver and the driven pulleys, then

$$r_2\omega_2 = r_1\omega_1\left(\frac{100-s}{100}\right) \qquad (12.8)$$

Equating R.H.S. of (12.7) and (12.8),

$$\left(\frac{100-s}{100}\right) = \left(\frac{100-s_1}{100}\right)\left(\frac{100-s_2}{100}\right)$$

or $\qquad (1 - 0.01s) = (1 - 0.01s_1)(1 - 0.01s_2)$

or $\qquad (1 - 0.01s) = 1 - (s_1 + s_2)0.01 + (0.0001)s_1 s_2$

or $\qquad s = (s_1 + s_2) - (0.01)s_1 s_2 \qquad (12.9)$

The last term in equation 12.9 is generally quite small compared to other terms, and can therefore be neglected. Hence,

$$s \approx s_1 + s_2 \qquad (12.10)$$

And therefore,

$$\omega_2 r_2 = (\omega_1 r_1)\left[1 - \frac{s_1 + s_2}{100}\right] \qquad (12.11)$$

EXAMPLE 12.1 A prime mover running at 400 r.p.m. drives a D.C. generator at 600 r.p.m. through a belt drive. Diameter of pulley on the output shaft of the prime mover is 60 cm. Assuming a total slip of 3%, determine the diameter of the generator pulley, if the belt thickness is 5 mm.

Solution: Given $s_1 + s_2 = 3$; $d_1 = 60$ cm; $t = 0.5$ cm and $N_1 = 400$; $N_2' = 600$ r.p.m.

Let d_2 and N_2 be the diameter and speed in r.p.m. of driven pulley and d_1, N_1 be the diameter and speed of driver pulley. Then considering effect of belt thickness alone,

$$\frac{N_2}{N_1} = \left(\frac{d_1 + t}{d_2 + t}\right) \quad \text{or} \quad N_2 = N_1\left(\frac{d_1 + t}{d_2 + t}\right)$$

Incorporating effect of slip,

$$N_2' = N_2\left(1 - \frac{s}{100}\right)$$

Therefore

$$N_2' = N_1\left(\frac{d_1 + t}{d_2 + t}\right)\left(1 - \frac{s}{100}\right)$$

or,

$$600 = 400\left(\frac{60 + 0.5}{d_2 + 0.5}\right)\left(1 - \frac{3}{100}\right)$$

or

$$\frac{60.5}{(d_2 + 0.5)} = \left(\frac{600}{400}\right)\frac{100}{97}$$

Therefore $\qquad d_2 + 0.5 = \dfrac{400}{600} \times \dfrac{97}{100} \times 60.5 = 39.123$

Therefore $\qquad d_2 = 39.123 - 0.5 = 38.623$ cm ≈ 38.62 cm $\qquad$ **Ans.**

12.3 BELT LENGTH

For Open Belt Drive

The angle of lap θ (also called angle of wrap or contact) is the angle subtended by segment of belt, in contact with smaller pulley, at the centre of that pulley. Let P, Q and M, N be the points at which belt is tangent to driver and driven pulley respectively (*see* Fig. 12.2). Let O and C be the centres of the driver and driven pulleys.

Join OP, OQ and CM, CN. Then, as MP is a common tangent to the two pulleys,

$$\angle OPM = \angle CMP = 90° \text{ and thus, } PO \text{ is parallel to } MC.$$

From O drop a perpendicular OL on CM and let α be the angle made by OL with the line of centres. Then, from right angled triangle OCL,

$$\sin\alpha = (LC/OC)$$

Thus, if $L' = OC$ be the centre distance, then with $LC = (r_2 - r_1)$

$$\sin\alpha = \dfrac{(r_2 - r_1)}{L'}$$

Further, from right angled triangle OLC,

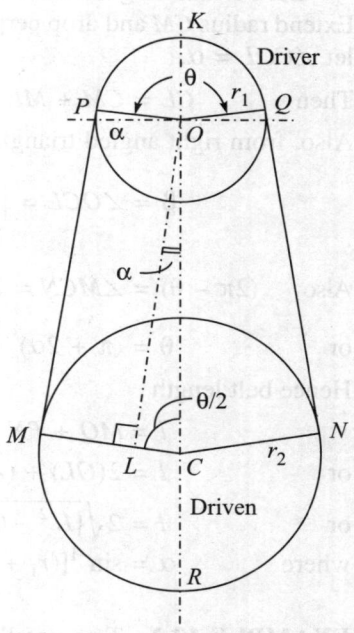

Fig. 12.2 Flat-Belt length.

$$OL = OC\cos\alpha = L'\cos\alpha$$

Hence, by symmetry of Figure 12.2 about centre lines OC,

$$NQ = PM = OL = L'\cos\alpha$$

Again $\qquad\qquad \theta = (\pi - 2\alpha)$

The belt length l is given by,

$$l = 2PM + \text{arc } PKQ + \text{arc } MRN$$

or $\qquad l = 2L'\cos\alpha + r_1(\pi - 2\alpha) + r_2[2\pi - (\pi - 2\alpha)]$

or $\qquad l = 2L'\cos\alpha + r_1(\pi - 2\alpha) + r_2(\pi + 2\alpha)$

or $\qquad l = 2L'\cos\alpha + (r_1 + r_2)\pi + (r_2 - r_1)2\alpha \qquad\qquad (12.12)$

For Crossed Belt Drive

For the crossed belt drive shown in Fig. 12.3, let

r_1 = radius of driver pulley

r_2 = radius of driven pulley

$OC = L'$ = centre distance of pulleys

θ = angle of lap which can be shown to be equal on both the pulleys

Let PN and MQ be the common tangents to the circles. Extend radius CM and drop perpendicular OL on it from O and let $\angle COL = \alpha$,

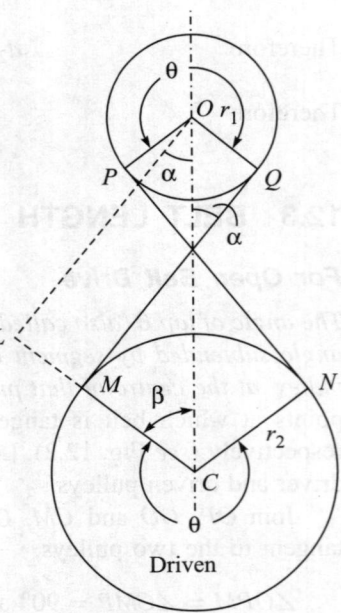

Then $CL = CM + ML = CM + OQ = (r_1 + r_2)$

Also, from right angled triangle OLC,

$$\beta = \angle OCL \doteq \left(\frac{\pi}{2} - \alpha\right)$$

Also $(2\pi - \theta) = \angle MCN = 2\beta = 2\left(\frac{\pi}{2} - \alpha\right)$

or $\theta = (\pi + 2\alpha)$ (12.13)

Fig. 12.3 Crossed-Belt length.

Hence belt length

$l = MQ + PN + r_1\theta + r_2\theta$

or $l = 2(OL) + (r_1 + r_2)\theta$

or $l = 2\sqrt{\{L'^2 - (r_1 + r_2)^2\}} + (r_1 + r_2)(\pi + 2\alpha)$ (12.14)

where $\alpha = \sin^{-1}[(r_1 + r_2)/L']$

EXAMPLE 12.2 Two parallel shafts are to be connected by a suitable drive with pulleys of 320 mm and 640 mm diameters. Determine the required length of belt if the centre distance is 3.2 m, and

(a) the pulleys must rotate in same sense, and

(b) when the pulleys must rotate in opposite sense.

Solution: Given $r_1 = 16$ cm; $r_2 = 32$ cm; $L = 320$ cm

If the pulleys are to rotate in same sense, they must be connected through an open belt drive and for securing rotations in opposite sense, they must be connected through a crossed belt drive.

(a) Belt length for open belt drive

$$\sin\alpha = \frac{r_2 - r_1}{L} = (32 - 16)/320 = 0.05 \quad \text{Hence } \alpha = 2.867 \text{ degrees}$$

Belt length $l = 2(320)\cos 2.866 + (32 + 16)\pi + (32 - 16)\dfrac{2.866 \times \pi}{180} \times 2$

$= 639.199 + 150.796 + 1.6$

$= 791.595 = 791.6$ cm   **Ans.**

(b) Belt length for crossed belt drive

$$\sin\alpha = (16 + 32)/320 = 0.15 \quad \text{Hence } \alpha = 8.627°$$

Hence belt length
$$l = 2\sqrt{(320)^2 - (16+32)^2} + (16+32)[\pi + 2\alpha]$$

Therefore
$$l = 632.759 + 48\left[\pi + 2\left(\frac{8.627 \times \pi}{180}\right)\right] = 798.01 \text{ cm} \qquad \textbf{Ans.}$$

Analysis for Belt Tensions, A Preamble.   Even when a pulley rotates uniformly, the belt mass in contact with the pulley is subjected to centrifugal acceleration. *Thus, a pulley rotating uniformly can be likened to a stationary pulley (in accordance with Newton's First Law) only when the centrifugal force and its effects are considered separately.* Theory in Sections 12.4 and 12.6 is developed on this philosophy.

12.4 LIMITING RATIO OF BELT-TENSIONS

Flat Belts

Consider a flat belt wound round a fixed pulley, as shown in Fig. 12.4. Because of flexible material, the belt can function as a link only when it is in tension and as such, at all times the belt must be subjected to finite tensile loads.

Let T_2 be the tension at one end of the belt and let the tension T_1 at the other end of the belt be increased gradually, until the belt starts slipping bodily round the pulley rim. This is equivalent to increasing torque, by way of increasing tension difference $(T_1 - T_2)$ until slip occurs between the belt and pulley. The value of tension T_1 at which slip occurs depends on the tension T_2 at the other end, the angle of lap θ and coefficient of friction μ between the belt and the pulley.

Consider an elemental length pq of the belt as shown in Fig. 12.4. Let $\delta\theta$ be the angle subtended

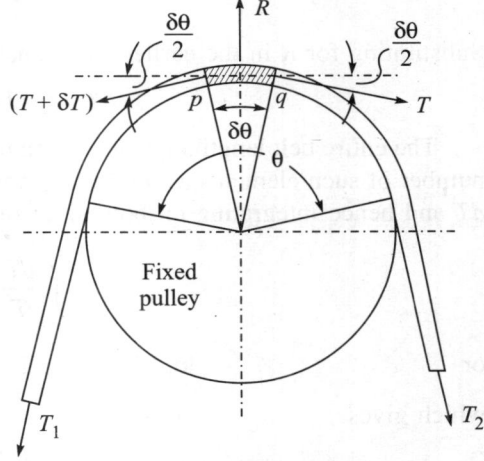

Fig. 12.4 Belt Tensions.

by belt length pq at the centre of the pulley. Further, let T be the tension in the belt at q and $(T + \delta T)$ be the tension in the belt at p. The difference in tensions at p and q in belt is due to the friction between belt length pq and pulley. The friction force depends on normal reaction R between belt and pulley.

Thus, considering equilibrium of the belt length pq, the various forces are:

(i)   tension T at q
(ii) tension $T + \delta T$ and p
(iii) normal reaction R normal to pulley rim at the mid point of length pq
(iv) the force of friction μR, tangential to the elemental length pq at the point of symmetry.

Resolving forces along and perpendicular to the tangent at the point of symmetry of pq.

$$(T + \delta T)\cos\frac{\delta\theta}{2} - T\cos\frac{\delta\theta}{2} = \mu R \tag{12.15}$$

and

$$(T + \delta T)\sin\frac{\delta\theta}{2} + T\sin\frac{\delta\theta}{2} = R \tag{12.16}$$

For small angles $\delta\theta/2$,

$$\sin\frac{\delta\theta}{2} \approx \frac{\delta\theta}{2} \quad \text{and} \quad \cos\frac{\delta\theta}{2} \approx 1$$

Equations (12.15) and (12.16) therefore simplify to

$$(T + \delta T) - (T) = \mu R \tag{12.17}$$

and

$$(T + \delta T)\frac{\delta\theta}{2} + T\left(\frac{\delta\theta}{2}\right) = R \tag{12.18}$$

Neglecting product of two small quantities (*e.g.*, δT and $\delta\theta/2$) in comparison to other terms, these equations reduce to

$$\delta T = \mu R \quad \text{and} \quad 2T\left(\frac{\delta\theta}{2}\right) = R$$

Substituting for R in the earlier equation,

$$\delta T = \mu(T\delta\theta) \quad \text{or,} \quad (\delta T/T) = \mu\delta\theta \tag{12.19}$$

The entire belt length in contact with the pulley over the angle of contact θ, consists of large number of such elements as pq. In the limiting condition, $\delta\theta$ approaches $d\theta$ and δT approaches dT and hence integrating on both sides of equation (12.19),

$$\int_{T_2}^{T_1} \frac{dT}{T} = \int_{0}^{\theta} \mu d\theta$$

or

$$\log(T_1) - \log(T_2) = \mu\theta \quad \text{or,} \quad \log(T_1/T_2) = \mu\theta \tag{12.20}$$

which gives

$$(T_1/T_2) = e^{\mu\theta} \tag{12.21}$$

When μ is the limiting coefficient of friction, the ratio (T_1/T_2) is called limiting ratio of tensions. While increasing the tension T_1 (and hence the torque), if the above limiting ratio is exceeded, the belt will slip over the pulley. *Note that θ in the above expression is the angle of lap in radians on smaller one of the two pulleys. It must be remembered that equation (12.21) is also valid with $\mu < \mu_{lim}$, even when belt is not at the point slipping.*
Following conclusions are drawn from equation (12.21):

1. Limiting ratio of tensions is independent of the radius of the pulley,
2. Larger the angle of lap, larger is the tension T_1 (for a given value of tension T_2) which can be sustained avoiding any slip. This means that for an open belt drive, as smaller pulley has smaller angle of lap, slip will occur first on the smaller pulley. Hence the limiting ratio of tension must be computed for smaller of the two pulleys.

In the case of crossed belt drive, angle of lap is equal on both the pulleys and therefore, limiting ratio of tension can be established for any of the two pulleys.

V-Belt and Rope Drives

V-belt and rope drive arrangements are shown in Fig. 12.5. Figures at (a) and (b) show grooved cross-sections used for wedge section (V-section) and circular section belts or ropes. The most striking feature being that the lower side of belt and rope do not touch bottom of the V-groove, but are wedged between the side faces of the groove. The moment belt or rope touches bottom of the groove, the wedge action become inoperative, and the drive is considered to have failed. Flat rim section may be treated as the limiting case of grooved section with groove angle 2α approaching $180°$.

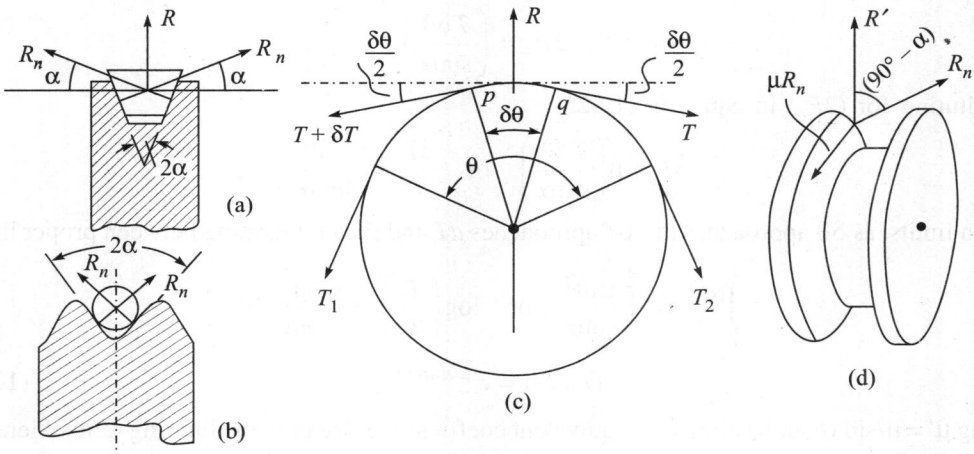

(a)

(b)

(c)

(d)

Fig. 12.5 Tension in Belts of V-and circular cross-section.

Let the groove angle of the wedge-portion of belt/pulley be 2α. Also, let T_1 and T_2 be the tensions on tight and slack side respectively of the belt and rope.

The resultant reaction R in this case is contributed by normal reactions R_n acting on side faces of the V-groove (Figs. 12.5a, b and d) and belt/rope. The frictional forces act tangential to the periphery of circle on the side faces as shown at Fig. 12.5(d). Thus, from Fig. 12.5(a),

$$R = 2R_n\sin\alpha \tag{12.22}$$

And for equilibrium of element pq from Fig. 12.5(c),

$$(T + \delta T)\cos\frac{\delta\theta}{2} - T\cos\frac{\delta\theta}{2} = 2(\mu R_n) \tag{12.23}$$

and

$$(T + \delta T)\sin\frac{\delta\theta}{2} + T\sin\frac{\delta\theta}{2} = R \tag{12.24}$$

For small angle $\delta\theta/2$, therefore, from equations (12.23), and (12.24)

$$\delta T = 2\mu R_n$$

and

$$T(\delta\theta/2) + (\delta T)(\delta\theta/2) = R$$

Neglecting product of two small quantities, above equations may be rewritten as

$$\delta T = 2\mu R_n \tag{12.25}$$

and

$$T\delta\theta = R \tag{12.26}$$

Substituting for R in equation (12.26) from (12.22),

$$T\delta\theta = 2R_n\sin\alpha$$

This yields

$$2R_n = \left(\frac{T\delta\theta}{\sin\alpha}\right).$$

Substituting for $(2R_n)$ in equation (12.25)

$$\delta T = \mu\left(\frac{T\delta\theta}{\sin\alpha}\right) \quad \text{or} \quad \frac{\delta T}{T} = \frac{\mu\delta\theta}{\sin\alpha}$$

Within limits, as $\delta\theta$ approaches $d\theta$, δT approaches dT and then integrating between proper limits,

$$\int_{T_2}^{T_1} \frac{dT}{T} = \int_0^\theta \frac{\mu d\theta}{\sin\alpha} \quad \text{or} \quad \log\left(\frac{T_1}{T_2}\right) = \left(\frac{\mu}{\sin\alpha}\right)\theta$$

or

$$(T_1/T_2) = e^{(\mu/\sin\alpha)\theta} \tag{12.27}$$

Letting $\mu' = \mu/\sin\alpha$ and calling it as equivalent coefficient of friction, the limiting ratio of tensions is

$$(T_1/T_2) = e^{\mu'\theta} \tag{12.28}$$

where

$$\mu' = (\mu/\sin\alpha)$$

Clearly, for $2\alpha = 180°$; $\sin\alpha = 1$ and $\mu' = \mu$. Thus the wedge action in V-belt and ropes apparently tends to increase the coefficient of friction. The wedge angle 2α is usually of the order of $34°$–$38°$. With a wedge angle $2\alpha = 38°$ and assumed values of $\theta = 165°$ and $\mu = 0.2$, the limiting ratio of tensions for V-belt is 5.865 as against bare 1.778 for flat belts.

12.5 MAXIMUM EFFECTIVE TENSION AND H.P. TRANSMITTED

The maximum effective tension (*i.e.*, tangential pull) transmitted by a belt or a rope is given by the difference in tension on tight and slack side *i.e.*, $(T_1 - T_2)$. Thus,

Maximum effective Pull

$$T = (T_1 - T_2) = T_1\left(1 - \frac{T_2}{T_1}\right)$$

or

$$(T_1 - T_2) \equiv T = T_1(1 - e^{-\mu\theta}) \tag{12.29}$$

The tight and slack sides of a belt is decided on the basis of sense of rotation of driving pulley. The side from which the belt is pulled (received) is the tight side while the side on which belt is delivered by the pulley is the slack side.

Let V be the peripheral speed of the belt in m/s. Then work done per second is

$$\text{W.D.} = (T_1 - T_2)V, \text{ in N} \cdot \text{m/s when } (T_1 - T_2) \text{ is in N}$$

Then, H.P. transmitted $= (T_1 - T_2)V/75$; where $(T_1 - T_2)$ is in kgf. $\hspace{2em}$ (12.30)

Also, power transmitted in kW $= \dfrac{(T_1 - T_2)V}{1000}$ kW; where $(T_1 - T_2)$ is in N. $\hspace{1em}$ (12.31)

and $(T_1 - T_2) =$ Maximum effective tension.

Also, substituting for $(T_1 - T_2)$ from equation (12.29)

$$\text{H.P.} = \frac{T_1(1 - e^{-\mu\theta})V}{75} \text{ for flat belts; where } T_1 \text{ is in kgf.} \quad (12.32)$$

and $\hspace{4em}$ $\text{H.P.} = \dfrac{T_1(1 - e^{-\mu'\theta})V}{75}$ for V-belts; where T_1 is in kgf. $\hspace{1em}$ (12.33)

Also, power transmitted in kW,

$$P = \frac{T_1(1 - e^{-\mu\theta})V}{1000} \text{ for flat belts; where } T_1 \text{ is in N.} \quad (12.34)$$

and $\hspace{4em}$ $P = \dfrac{T_1(1 - e^{-\mu'\theta})V}{1000}$ for V-belts; where T_1 is in N. $\hspace{1em}$ (12.35)

EXAMPLE 12.3 A flat belt runs on a pulley 1 m in diameter and transmits 7.5 kw at a speed of 200 r.p.m. Taking angle of lap as 170° and coefficient of friction as 0.2, find the necessary width of the belt, if pull is not to exceed 196 N/cm width of belt. Neglect centrifugal tension.

Solution: Given $r_1 = 100$ cm; power $= 7.5$ kW; $N = 200$ r.p.m. Let $b =$ width of belt.

Also, $\hspace{6em}$ $\theta = 170°$; $\mu = 0.2$; $(T_1)/b = 196$ N

Because, $\hspace{5em}$ Power $= (7.5 \times 1000)$W

$$(T_1 - T_2) = \frac{7.5 \times 1000}{V} = \frac{7.5 \times 1000 \times 60}{\pi \times 1 \times 200} = 716.197 \text{ N} \quad (a)$$

Also $\hspace{6em}$ $T_1/T_2 = e^{0.2 \times \frac{170 \times \pi}{180}} = 1.81$

Therefore $\hspace{5em}$ $T_1 = (1.81)T_2$

Hence from (a),

$$(1.81 - 1)T_2 = 716.197$$

Therefore $\hspace{5em}$ $T_2 = 884.194$ and $T_1 = 1600.39$ N

Now $\hspace{6em}$ $(T_1/b) = 196$ N/cm

Therefore $\hspace{5em}$ $b = \dfrac{T_1}{196} = \dfrac{1600.39}{196} = 8.165$ cm $\hspace{4em}$ **Ans.**

EXAMPLE 12.4 A shaft running at 100 r.p.m. is to drive another shaft at 250 r.p.m. and transmit 11 kw. The belt is 11.5 cm wide and 12 mm thick and the coefficient of friction between the belt and the pulley is 0.25. The distance between the shafts is 2.8 m and the smaller pulley is 60 cm in diameter. Calculate the stress in belt, connecting the two pulleys, when the belt is arranged as (a) open belt and (b) crossed belt.

Solution: Given: $N_1 = 100$ r.p.m.; $N_2 = 250$ r.p.m.; $P = 11$ kw; $r_2 = 30$ cm;

$$L = 280 \text{ cm}; \ b = 11.5 \text{ cm}; \ t = 1.2 \text{ cm}; \ \mu = 0.25$$

Since

$$P = \frac{(T_1 - T_2)V}{1000} \text{ in kw,}$$

we have

$$11 = \frac{(T_1 - T_2)2\pi N_2\left(r_2 + \dfrac{t}{2}\right)}{60 \times 1000}$$

or

$$(T_1 - T_2) = \frac{11 \times 60 \times 1000}{2\pi \times 250 \times (30 + 0.6)/100}$$

or

$$(T_1 - T_2) = \frac{11 \times 60 \times 1000}{2\pi \times 250 \times 0.306} = 1373.1 \text{ N} \tag{a}$$

For an open belt drive,

$$\alpha = \sin^{-1}\left(\frac{r_1 - r_2}{L}\right) = \sin^{-1}\left[\frac{\left(\dfrac{250}{100} \times 30\right) - 30}{280}\right]$$

or $\alpha = \sin^{-1}(0.1607)$ or $\alpha = 9.248°$

Therefore $\theta = (\pi - 2\alpha) = 161.5° = \dfrac{161.5}{180} \times \pi = 2.819$ rad,

Therefore $\dfrac{T_1}{T_2} = $ limiting ratio of tensions $= e^{0.25 \times 2.819}$

or $$\frac{T_1}{T_2} = 2.023 \tag{b}$$

From equations (a) and (b),

$$(2.023 T_2 - T_2) = 1373.1$$

Therefore $T_2 = 1342.2$ N   and $T_1 = 2715.33$ N

Hence stress in belt on tight side $\sigma = \dfrac{T_1}{11.5 \times 1.2} = \dfrac{2715.33}{11.5 \times 1.2} = 196.76 \text{ N/cm}^2$ **Ans.**

For crossed belt drive,

$$\alpha = \sin^{-1}\left(\frac{r_1 + r_2}{L}\right) = \sin^{-1}\frac{\left(\dfrac{250}{100} \times 30 + 30\right)}{280}$$

or $\alpha = \sin^{-1}(0.375) = 22° = 0.384$ rad.

Therefore $\qquad\qquad \theta = (\pi + 2\alpha) = 3.14 + 2(0.384) = 3.908$ radians

Hence $\qquad\qquad T_1/T_2 =$ limiting ratio of tensions $= e^{\mu\theta}$

or $\qquad\qquad T_1/T_2 = e^{0.25 \times 3.908} = e^{0.977} = 2.656$

Substituting in equation (a),

$$(2.656 T_2 - T_2) = 1373.1$$

Therefore $\qquad\qquad T_2 = 828.9$ N and $T_1 = 2.657 \times T_2 = 2202.5$ N

Hence $\qquad$ Max. stress in belt $= \dfrac{T_1}{b \times t} = \dfrac{2201.76}{11.5 \times 1.2} = 159.6$ N/cm^2 $\qquad$ **Ans.**

12.6 CENTRIFUGAL TENSION AND STRESSES IN BELTS OR ROPE

Let r be the radius of pulley, V the speed of belt or rope, a the area of cross-section of belt/rope and w be the weight of belt or rope per unit length.

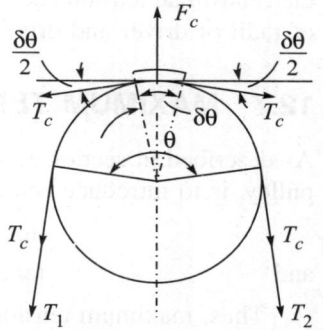

As an elemental length Δl of belt/rope passes over a pulley, it is subjected to centrifugal force F_c and a centripetal force must be applied to ensure that it maintains a radius of curvature r of the path. This centripetal force needed by the belt/rope length is supplied by belt/rope which develops a centrifugal tension T_c both on tight and slack side.

To establish the expression for T_c, consider the elemental length Δl of belt/rope to move along the pulley. Let this elemental length subtend an angle $\delta\theta$ at the centre of pulley (Fig. 12.6). Thus, the weight of this elemental length

Fig. 12.6 Centrifugal Tension.

$$\delta w = w(r\delta\theta)$$

In moving with peripheral velocity V along pulley curvature, this elemental length is subjected to centrifugal force given by

$$F_c = \frac{w(r\delta\theta)}{g}\frac{V^2}{r}$$

or $\qquad\qquad\qquad F_c = \dfrac{wV^2}{g}\delta\theta \qquad\qquad\qquad (12.36)$

The element is in equilibrium under the action of centrifugal tension T_c on either side and centrifugal force F_c. Due to symmetry the tangential components of T_c mutually cancel out while the radial components add up to balance centrifugal force on the element. Thus,

$$F_c = 2T_c \sin\frac{\delta\theta}{2} \qquad\qquad\qquad (12.37)$$

Equating right hand sides of equations (12.36) and (12.37),

$$(2T_c)\sin\frac{\delta\theta}{2} = \frac{wV^2}{g}\delta\theta \qquad\qquad\qquad (12.38)$$

For small elements, $\delta\theta/2$ is quite small and as such $\sin\left(\dfrac{\delta\theta}{2}\right) \approx \dfrac{\delta\theta}{2}$. Hence, from equation (12.38),

$$2(T_c)\frac{\delta\theta}{2} = \frac{wV^2}{g}\delta\theta \quad \text{or} \quad T_c = \frac{wV^2}{g} \tag{12.39}$$

The additional stress in belt/rope material due the inertia effect is thus,

$$\sigma_c = \frac{T_c}{a} = \frac{wV^2}{ag} \tag{12.40}$$

where a = area of cross-section of belt or rope.

Since equations (12.39) and (12.40) do not involve any variable, it follows that centrifugal tension and centrifugal stress is uniform over the entire arc. It is equally important to note that the centrifugal tension and centrifugal stress do not depend on radius of pulley. Thus, irrespective of radii of driver and driven pulleys, the centrifugal tension is same at both the pulleys.

12.7 MAXIMUM TENSION IN BELT/ROPE

As described in section 12.6 effect of centrifugal force acting on belt segment, moving along a pulley, is to introduce centrifugal tension T_c on both sides of the pulley. Thus,

$$\text{total tension on tight side, } T_t = T_1 + T_c \tag{12.41}$$

and

$$\text{total tension on slack side, } T_s = T_2 + T_c \tag{12.42}$$

Thus, maximum tension to which the belt or rope is subjected is T_t. And therefore, if σ be the maximum stress induced in the belt/rope,

$$T_t = \sigma \times b \times t \tag{12.43}$$

where b = width of belt in cm

and t = thickness of belt in cm

and σ = maximum stress in belt/rope in N/cm^2

12.8 INITIAL TENSION AND ITS ROLE IN POWER TRANSMISSION

Belt/rope drive being a friction drive, frictional forces must exist between belt/rope and pulley when power transmission begins. This requires that belt/rope must be installed with some initial tension T_o. As shown in Fig. 12.7, component of initial tension T_o in radial direction produces an inward force ($2T_o\sin\delta\theta/2$) over a belt element and, in turn, the pulley produces a normal reaction R_n acting radially outward. This reaction R_n helps in producing frictional force μR_n. Thus, an initial T_o is necessary for power transmission.

As the power transmission takes place, the driver pulley pulls the belt from one side (increasing tension on that side) and delivers the same to the other side (decreasing the tension thereby on that

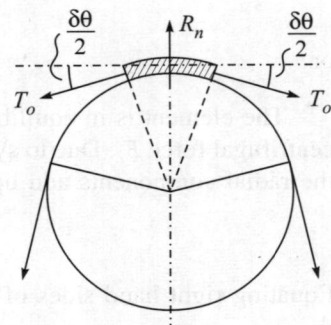

Fig. 12.7 Initial Tension in Belt.

side). Thus, when power is transmitted, tension on tight side increases from T_o to T_1 while on slack side it decreases from T_o to T_2 (note that T_2 must be tensile for power transmission).

Assuming the belt material to be linearly elastic and α to be the coefficient of elongation/contraction in length per unit force, increase in belt length on tight side = $\alpha(T_1 - T_0)$ and, decrease in belt length on slack side = $\alpha(T_0 - T_2)$.

Assuming further that the total belt length remains the same, if follows that

$$\alpha(T_1 - T_0) = \alpha(T_0 - T_2)$$

or
$$(T_1 + T_2) = 2T_0 \tag{12.44}$$

or
$$T_0 = \left(\frac{T_1 + T_2}{2}\right) \tag{12.45}$$

When the belt is at the point of slipping,

$$T_1/T_2 = e^{\mu\theta}$$

Substituting for T_2 in equation (12.44),

$$(T_1 + T_1/e^{\mu\theta}) = 2T_0$$

or
$$T_1(1 + e^{-\mu\theta}) = 2T_0$$

or
$$T_1 = 2T_0/(1 + e^{-\mu\theta}) \tag{12.46}$$

Therefore
$$(T_1 - T_2) = T_1(1 - e^{-\mu\theta})$$

Substituting for T,
$$(T_1 - T_2) = 2T_0(1 - e^{-\mu\theta})(1 + e^{-\mu\theta})$$

When centrifugal tension is also considered, T_1 and T_2 can be replaced by T_t and T_s respectively in equations (12.44) and (12.45), giving

$$(T_t + T_s) = 2T_0$$

and
$$T_0 = \left(\frac{T_t + T_s}{2}\right)$$

Thus
$$T_0 = \frac{(T_1 + T_c) + (T_2 + T_c)}{2} = \frac{(T_1 + T_2 + 2T_c)}{2} \tag{12.47}$$

Substituting for T_1 in equation (12.32) and (12.34) the power transmission for the flat belts is given by

$$\text{H.P.} = \frac{2T_0(1 - e^{-\mu\theta})V}{75(1 + e^{-\mu\theta})} \quad \text{(MKS)} \tag{12.48}$$

and
$$P = \frac{2T_0(1 - e^{-\mu\theta})V}{1000(1 + e^{-\mu\theta})} \quad \text{kw (SI Units)} \tag{12.49}$$

This shows that tension T_1, and hence the power transmitted through a belt, depends on initial tension T_0.

EXAMPLE 12.5 (MKS) A belt running on a pulley 105 cm in diameter, rotating at 230 r.p.m., is to transmit 8.95 kW. Taking $\mu = 0.27$ and the angle of lapping = 165°, find (a) the necessary width of the belt if the pull is not to exceed 157 N per cm of width (b) the necessary initial tension in the belt.

Solution: The peripheral belt speed $V = \dfrac{\pi \times 105 \times 230}{60} = 1264.49$ cm/s = 12.645 m/s

Also
$$\vartheta = \frac{165 \times \pi}{180} = 2.8798 \text{ rad}$$

Therefore
$$T_1/T_2 = e^{0.27 \times 2.8798} = 2.176 \quad \text{or} \quad T_2 = (2.176)T_1$$

And, therefore
$$T_1 - T_2 = T_1(1 - e^{-\mu\theta}) = T_1(1 - 0.4595) = (0.54)T_1$$

Also
$$P = \frac{(T_1 - T_2) \times V}{1000} \text{ kw}$$

Hence,
$$(T_1 - T_2) = \frac{8.95 \times 1000}{12.645} = 707.8 \text{ N}$$

Thus, equating
$$0.54 T_1 = 707.8$$

Therefore
$$T_1 = 1310.7 \text{ N}$$

And
$$T_2 = T_1/e^{\mu\theta} = 1310.7/2.176 = 602.34 \text{ N}$$

For a belt width of b cm, the permissible tension on tight side = $(157)b$ N. Hence,
$$(157)b = 1310.7$$

or
$$b = 8.348 \text{ cm} \approx 8.35 \text{ cm} \qquad \qquad \textbf{Ans.}$$

Also, the initial tension,
$$T_0 = \frac{T_1 + T_2}{2} = \frac{1310.7 + 602.34}{2} = 956.52 \text{ N} \approx 956.5 \text{ N} \qquad \textbf{Ans.}$$

EXAMPLE 12.6 (SI) An open belt drive connects two parallel shafts 1.2 m apart. The driving and driven shafts rotate at 375 r.p.m. and 150 r.p.m. respectively and the driven pulley is 45 cm in diameter. The belt is 5 mm thick and 8 cm wide. Coefficient of friction between belt and pulley is 0.3 and maximum permissible tension in the belting is 137 N/cm². Determine

   (i) the diameter of driving pulley
  (ii) maximum power in kw that may be transmitted by the belting, and
 (iii) the required initial tension in the belt.

Solution: Given: $N_1 = 375$ r.p.m.; $N_2 = 150$ r.p.m.; $\mu = 0.3$, $d_2 = 45$ cm; $L = 120$ cm
belt section: $\qquad\qquad b = 8$ cm; $t = 0.5$ cm; $\sigma_t = 137$ N/cm²

Diameter of driving pulley,

$$d_1 = d_2 \left(\frac{N_2}{N_1} \right) = 45 \left(\frac{150}{375} \right) = 18 \text{ cm}$$ **Ans.**

The angle of lap for an open belt drive is given by

$$\theta = (\pi - 2\alpha)$$

where $$\sin\alpha = \frac{r_1 - r_2}{L} = \frac{22.5 - 9}{120} \text{ Hence } \alpha = \sin^{-1}(0.1125) = 6.46°$$

Therefore $$\theta = \pi - 2(6.46) = 167.08° = \frac{167.08}{180} \times \pi = 2.916 \text{ radians}$$

The limiting ratio of tensions,

$$T_1/T_2 = e^{(0.3) \times 2.916} = 2.3984$$

Also for transmitting power, maximum permissible tension in belt (neglecting centrifugal tension)

$$T_1 = 137 \times (8 \times 0.5) = 548 \text{ N}$$

Therefore $$T_2 = T_1/2.2984 = 228.49 \text{ N}$$

Also the velocity of belt $$= \frac{\pi \times d_2 \times N_2}{60} = \frac{\pi \times 45 \times 150}{60}$$

$$= 353.43 \text{ cm/sec} = 3.5343 \text{ m/s}$$

Hence the power transmitted in kW,

$$P = \frac{(T_1 - T_2)V}{1000} = \frac{(548 - 228.49) \times 3.5343}{1000} = 1.129 \text{ kw}$$

or $$P \approx 1.13 \text{ kw}$$ **Ans.**

Also, the initial tension required,

$$T_0 = \frac{T_1 + T_2}{2} = \frac{548 + 228.49}{2} = 388.25 \text{ N}$$ **Ans.**

EXAMPLE 12.7 An open belt connects two flat pulleys. The smaller pulley is 30 cm in diameter and runs at 200 r.p.m. The angle of lap on this pulley is 160 degrees and the coefficient of friction between the belt and pulley face is 0.25. The belt is on the point of slipping when 2.61 kw is being transmitted. Which of the following alternatives would be more effective in increasing the power which could be transmitted.

(i) Increasing the initial tension in the belt by 10%.
(ii) Increasing the coefficient of friction by 10% by the application of a suitable dressing to the belt

(DAV Indore, 1982)

Solution: Given: $d_1 = 30$ cm; $N_1 = 200$ r.p.m.; $\theta = 160°$; $\mu = 0.25$; Power = 2.61 kw

Limiting ratio of tensions

$$T_1/T_2 = e^{0.25(160\pi/180)} = e^{0.698} = 2.01$$

Belt speed

$$V = \frac{\pi \times 30 \times 200}{60 \times 100} = 3.142 \text{ m/s}$$

Thus power transmitted

$$= \frac{T_1(1 - e^{-\mu\theta})V}{1000} \text{ kW}$$

Therefore

$$2.61 = \frac{T_1(1 - 1/2.01)\, 3.142}{1000}$$

or

$$T_1 = \frac{1000 \times 2.61}{3.142 \times 0.5025} = 1653.1 \text{ N}$$

Therefore

$$T_2 = 1653.1/2.01 = 822.44 \text{ N}$$

Therefore

$$T_0 = \frac{T_1 + T_2}{2} = \frac{1653.1 + 822.44}{2} = 1237.8 \text{ N}$$

Also

$$e^{-\mu\theta} = \frac{1}{2.01} = 0.4975$$

First Alternative:

Increasing initial tension by 10% means new value of initial tension $T_o' = 1237.8 \times 1.10$

or

$$T_0 = 1361.58 \text{ N}$$

New value of power transmitted

$$= \frac{2T_0'(1 - e^{-\mu\theta})V}{75(1 + e^{-\mu\theta})}$$

or

$$P' = \frac{2 \times 1361.58\,(1 - 0.4975) \times 3.142}{1000\,(1 + 0.4975)} = 2.871 \text{ kW}$$

Therefore, percentage of increase in power $= \dfrac{2.871 - 2.61}{2.61} \times 100 = 10.0\%$ **Ans.**

Second Alternative:

Increasing coeff. of friction μ by 10% means,

New value of coeff. of friction

$$\mu' = 1.10 \times 0.25 = 0.275$$

Therefore

$$e^{+\mu'\theta} = e^{0.275(160\pi/180)} = 2.155$$

Hence

$$e^{-\mu'\theta} = \frac{1}{2.155} = 0.464$$

Therefore power transmitted

$$P'' = \frac{2T_0(1 - e^{-\mu'\theta})V}{1000(1 + e^{-\mu'\theta})}$$

or

$$P'' = \frac{2 \times 1237.8\,(1 - 0.464) \times 3.142}{1000\,(1 + 0.464)} = 2.8478$$

Therefore, percentage of increase in power transmitted = $\dfrac{2.8478 - 2.61}{2.61} \times 100 = 9.11\%$ **Ans.**

Hence the first alternative of increasing initial tension by 10% is better, as it gives larger percentage increase in h.p. transmitted. **Ans.**

EXAMPLE 12.8   Following data refer to a flat belt drive:
   Power transmitted: 18.6 kw; Pulley diameter: 180 cm; Angle of contact: 175°; Speed of pulley: 300 r.p.m; Coeff. of friction between belt and pulley surface: 0.3
   Permissible stress in belt = 294 N/sq.cm; Thickness of belt = 8 mm; Density of belt material = 0.00932 N/cu.cm.
   Determine the width of the belt required taking into account the centrifugal tension.

Solution: Weight of 1 metre length of belt

$$w = (b \times t \times 100) \times 0.00932$$

or

$$w = (b \times 0.8 \times 100) \times 0.00932 = (0.7456)b \text{ N}$$

$$\text{Belt speed} = \frac{\pi \times d \times N}{60} = \frac{\pi \times 1.8 \times 300}{60} = 9\pi = 28.26 \text{ m/s}$$

Again, limiting ratio of tensions,

$$\frac{T_1}{T_2} = e^{0.3 \times (175\pi/180)} = e^{0.9163} = 2.5$$

Therefore

$$e^{-\mu\theta} = \frac{1}{2.5} = 0.4$$

Again

$$T_c = \frac{wV^2}{g} = \frac{0.7456 \times b \times (28.26)^2}{9.81} = (60.7)b \text{ N}$$

Maximum stress in belt occurs in tight side due to tension $T_t = T_1 + T_c$.

Thus,

$$T_t = (T_1 + 60.7b)$$

This should also be equal to tension corresponding to permissible stress

$$T' = (294 \times b \times 0.8) = (235.2)b \text{ N}$$

Hence

$$T_1 + (60.7)b = (235.2)b$$

or

$$T_1 = 174.5 \text{ N}$$

Again, power transmitted,

$$18.6 = \frac{T_1(1 - e^{-\mu\theta})V}{1000}$$

Therefore

$$T_1 = \frac{18.6 \times 1000}{(1 - 0.4) \times 28.26} = 1096.96$$

Hence

$$b = \frac{1096.96}{174.5} = 6.286 \approx 6.29 \text{ cm}$$ **Ans.**

EXAMPLE 12.9 An open belt drive connects two pulleys 1.2 and 0.5 m diameter on parallel shafts 3.6 m apart. The belt has a mass of 0.9 kg/m length, and the maximum tension in it is not to exceed 2.0 kN.

The 1.2 m pulley, which is the driver, runs at 200 r.p.m. Due to belt slip on one of the pulleys, the velocity of the driven shaft is only 450 r.p.m. Calculate the torque on each of the two shafts, the power transmitted, and the power lost in friction. $\mu = 0.3$.

What is the efficiency of the drive?

(AMIE, Winter 1979)

Solution: The angle of lap is given by $\theta = (\pi - 2\alpha)$

where,
$$\alpha = \sin^{-1}\left[\frac{(1.2/2) - (0.5/2)}{3.6}\right]$$

$$= \sin^{-1}(0.09722) = 5.579°$$

$$\theta = 180 - (5.579)\cdot 2 = 168.84°$$

Hence
$$T_1/T_2 = e^{\mu\theta} = e^{0.3(168.84\pi/180)}$$

Therefore
$$\frac{T_1}{T_2} = e^{0.884} = 2.42 \text{ Therefore } e^{-\mu\theta} = 0.413$$

Also, belt speed
$$= \frac{\pi d N}{60} = \frac{\pi \times 1.2 \times 200}{60} = 12.566 \text{ m/s}$$

Hence the centrifugal tension,

$$T_c = \frac{w}{g}V^2 = mv^2 = (0.9) \times (12.566)^2 = 142.11 \text{ N}$$

The maximum tension in the belt

$$T_t = T_1 + T_c = 2000 \text{ N}$$

Also

$$\left(\frac{T_t - T_c}{T_s - T_c}\right) = \frac{T_1}{T_2} = e^{\mu\theta} = 2.42$$

Therefore
$$\frac{2000 - 142.11}{T_s - 142.11} = 2.42$$

Therefore
$$T_s = 142.11 + \frac{2000 - 142.11}{2.420} = 909.83 \text{ N}$$

Hence torque on driving pulley $= (2000 - 909.83) \times 0.6 = 654.1$ N·m   **Ans.**

Torque on the driven pulley,

$$M = (T_t - T_s) \times r'$$

or
$$M = (2000 - 909.83) \times 0.25 = 272.54 \text{ N·m}$$ **Ans.**

The power transmitted by driver pulley,

$$= \frac{2\pi \times 200 \times 654.1}{60 \times 1000} = 13.7 \text{ kw}$$ **Ans.**

When there is no slip, the r.p.m. of driven pulley

$$= \frac{1.2}{0.5} \times 200 = 480 \text{ r.p.m.}$$

However, due to slip the actual speed of driven shaft = 450 r.p.m. Hence, with torque of · 272.54 N·m on driven shaft, the power actually transmitted to follower

$$= \frac{2\pi \times 450 \times 272.54}{60 \times 1000} = 12.84 \text{ kw}$$ **Ans.**

Therefore power lost in friction = 13.69 – 12.84 = 0.85 kw **Ans.**

The efficiency of drive $\quad \eta = \dfrac{12.84}{13.69} \times 100 = 93.79\%$ **Ans.**

EXAMPLE 12.10 Two parallel horizontal shafts, whose centre lines are 4.8 m apart, one being vertically above the other, are connected by an open belt drive. The pulley on the upper shaft is 1.05 m diameter while, that on the lower shaft is 1.5 m diameter. The belt is 150 mm wide and the initial tension in it when stationary and when no torque is being transmitted is 3 kN. The belt has a mass of 1.5 kg/m length; the gravitational force on it may be neglected but centrifugal force must be taken into account. The material of the belt may be assumed to obey Hooke's law and the free lengths of the belt between the pulleys may be assumed to be straight. The coefficient of friction between the belt and either pulley is 0.3. Calculate

(a) the pressure in N/m^2 between the belt and the upper pulley when the belt and pulleys are stationary and no torque is being transmitted;

(b) the tension in the belt and the pressure between the belt and the upper pulley if the upper shaft rotates at 400 r.p.m. and there is no resisting torque on the lower shaft, hence no power being transmitted;

(c) the greatest tension in the belt if the upper shaft rotates at 400 r.p.m. and the maximum possible power is being transmitted to the lower shaft. (Univ. London)

Solution: *Case* (a): As the belt and pulley are stationary, tension in belt is the initial tension T_0. The downward force due to initial tension in belt

$$= 2T_0 \sin \frac{\delta\theta}{2} \approx 2T_0\left(\frac{\delta\theta}{2}\right) \approx T_0(\delta\theta) \qquad \text{(i)}$$

Let p be the pressure exerted by pulley on belt. The upward force on a segment of belt, subtending angle $\delta\theta$ at the pulley centre (Fig. 12.8),

$$R = p(r\delta\theta)b = p\left(\frac{1.05}{2}\right)0.15 \times \delta\theta$$

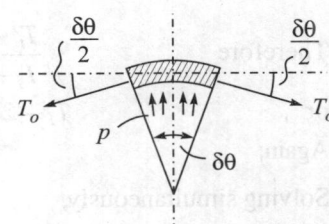

Fig. 12.8

Also, as $\qquad\qquad T_0 = 3000$ N, equating (i) & (ii),

$$p\left(\frac{1.05}{2}\right) \times 0.15 \times \delta\theta = 3000 \times \delta\theta$$

Therefore $\qquad\qquad\qquad\qquad p = 38095$ N/m^2 $\qquad\qquad\qquad\qquad$ **Ans.**

Case (b): When the shaft rotates at 400 r.p.m., the belt is subjected to centrifugal force as it passes over the pulley. This is balanced by a part of initial tension which we call as centrifugal tension. Note that there is no resisting torque and as such the belt length remains the same as in case (a) above.

Thus,

$$T_0' + T_c = 3000 \text{ N} \quad \text{But, } T_c = mv^2$$

$$= (1.5)\left(\frac{\pi \times 1.05 \times 400}{60}\right)^2 = 725.4 \text{ N}$$

Therefore $\qquad\qquad\qquad\qquad T_0' = 3000 - 725.4 = 2274.6$ N

Hence by proportion, the pressure between belt and pulley

$$p' = \frac{2274.6}{3000} \times 38095 = 28883.6 \text{ N/m}^2 \qquad\qquad\qquad \textbf{Ans.}$$

Case (c): With upper shaft rotating at speed same as that in case (b)

$$T_c = 725.4 \text{ N}$$

When the power is being transmitted, let

$\qquad\qquad T_t$ = tension on tight side $\qquad$ and $\qquad T_s$ = tension on slack side

Then

$$\frac{T_1}{T_2} = \frac{T_t - 725.4}{T_s - 725.4} = e^{0.3(\theta)}$$

Again, angle of lap, $\qquad\qquad \theta = \pi - 2\alpha \quad$ where $\quad \alpha = \sin^{-1}\left(\frac{15 - 1.05}{2 \times 4.8}\right) = 2.687°$

Therefore $\qquad\qquad\qquad\qquad \theta = 180 - 2(2.687) = 174.6°;$

But $\qquad\qquad\qquad\qquad \frac{T_1}{T_2} = e^{0.3(174.6\pi/180)} = 2.495$

Therefore $\qquad\qquad\qquad\qquad \frac{T_t - 725.4}{T_s - 725.4} = 2.495$

or $\qquad\qquad\qquad\qquad T_t - 2.495\, T_s = 1084.47$ N

Again, $\qquad\qquad\qquad\qquad T_t + T_s = 2T_0 = 6000$ N

Solving simultaneously,

$$(3.495)T_s = 4915.53$$

Therefore $\qquad\qquad T_s = 1406.45$ N $\qquad$ and $\qquad T_t = 4593.55$ N $\qquad\qquad$ **Ans.**

12.9 CONDITION FOR MAXIMUM POWER TRANSMISSION

A belt drive transmits maximum power when

(a) the maximum tension on tight side equals maximum permissible value of the tension for the belt, and

(b) condition of limiting ratio of tension is reached, *i.e.*, when the limiting frictional force acts between belt and pulley surface.

Based on expressions (12.32) and (12.34) for belt drive, power transmitted by a flat belt drive can be expressed as

$$P = kT_1(1 - e^{-\mu\theta})v \tag{12.50}$$

where k is a constant, value of which depends on units being followed. Thus, let

$$T_1 = \text{Tension on tight side}$$

and
$$v = \text{velocity of belt in m/s}$$

Now
$$T_1 = (T_t - T_c) = \left(T_t - \frac{w}{g}v^2\right) \tag{12.51}$$

Substituting for T_1 from above in equation (12.50),

$$P = k\left(T_t - \frac{wv^2}{g}\right)(1 - e^{-\mu\theta})v$$

Note that in the above expression, value of maximum permissible (total) tension T_t on tight side depends on material properties of belt and can, therefore, be treated as constant for a given belt. Again factors k and $e^{-\mu\theta}$ are assumed invariant with velocity v. Writing above expression as

$$P = k(1 - e^{-\mu\theta})\left(T_t v - \frac{wv^3}{g}\right) \tag{12.52}$$

Differentiating w.r. to velocity for maximum power and equating to zero,

$$\frac{dP}{dv} = 0 = k(1 - e^{-\mu\theta})\frac{d}{dv}\left[T_t.v - \frac{wv^3}{g}\right]$$

or
$$k(1 - e^{-\mu\theta})\left[T_t - \frac{3wv^2}{g}\right] = 0 \tag{12.53}$$

Since $k \neq 0$ and also as

$$(1 - e^{-\mu\theta}) \neq 0$$

it follows that,

$$T_t - \frac{3wv^2}{g} = 0 \tag{12.54}$$

or
$$T_t = \frac{3wv^2}{g} \equiv 3T_c \tag{12.55}$$

It follows that for transmitting maximum h.p., the tension on tight side

$$T_t = 3T_c \tag{12.56}$$

or

$$T_c = \frac{T_t}{3}$$

Also, from equation (12.55),

$$v = \sqrt{\frac{T_t g}{3w}}$$

This gives optimum value of velocity at which maximum h.p. is transmitted. Thus,

$$v_{opt} = \sqrt{\left\{\frac{T_t g}{3w}\right\}} \tag{12.57}$$

It follows from equation (12.56) that

$$T_1 = 3T_c - T_c \equiv 2T_c$$

Also,

$$T_1 = T_t - T_c = \left(T_t - \frac{T_t}{3}\right) = \left(\frac{2}{3}\right)T_t \tag{12.58}$$

Hence for transmitting maximum power, the centrifugal tension must be equal to a value one-third that of maximum allowable tension in belt and simultaneously, the belt should be at the point of slipping.

It follows from equation (12.52) that the power transmitted is zero when,

(i) either $v = 0$ or (ii) $T_t - \dfrac{wv^2}{g} = 0$ *i.e.*, when $v = \sqrt{\left\{\dfrac{gT_t}{w}\right\}}$ \hfill (12.59)

At velocity $v = \sqrt{(T_t g / w)}$ the centrifugal force is probably so high that the elasticity in belt fails to produce enough centrifugal tension to balance the effect of centrifugal force on belt passing over pulleys. Denoting this velocity by v', the optimum value of velocity v_{opt} at which maximum power is transmitted, is given (from a comparison of equations 12.57 and 12.59) by

$$v_{opt} = \frac{1}{\sqrt{3}} v'$$

or

$$v_{opt} = (0.577)v' \tag{12.60}$$

Thus v_{opt} does not coincide with arithmetic mean of the velocities producing zero power, and given by expression (12.59). Maximum power transmitted by a belt drive at operating velocity $v = v_{opt}$, is obtained by substituting for $v = v_{opt}$ in equation (12.52). Thus

$$P_{max} = k(1 - e^{-\mu\theta})\left[T_t - \frac{w}{g}\left(\frac{T_t g}{3w}\right)\right]\sqrt{\frac{T_t g}{3w}}$$

$$= k(1 - e^{-\mu\theta})\left(T_t - \frac{T_t}{3}\right)\sqrt{\frac{T_t g}{3w}}$$

$$= k(1 - e^{-\mu\theta})\frac{2}{3}T_t\sqrt{\frac{T_t g}{3w}}$$

which leads to the following expressions for maximum power transmission in MKS and SI units. Thus

$$(\text{H.P.})_{max} = \frac{2}{3}\,(T_t)\sqrt{\left\{\frac{T_t g}{3w}\right\}}\left(\frac{1 - e^{-\mu\theta}}{75}\right) \tag{12.61}$$

and

$$(\text{P})_{max} = \frac{2}{3}\,(T_t)\sqrt{\left\{\frac{T_t g}{3w}\right\}}\left(\frac{1 - e^{-\mu\theta}}{1000}\right)\text{kw} \tag{12.62}$$

12.10 POWER TRANSMITTED BY BELT: FURTHER COMMENTS

It is repeatedly demonstrated experimentally that the sum of tensions T_1 on tight side and T_2 on slack side is always greater than twice the initial tension T_0. Researchers have tried to reason out this deviation from the established relation

$$T_1 + T_2 = 2T_0$$

It was first argued that the belt material is not perfectly elastic and the stress-strain curve is not a straight line. For the type of stress-strain curve shown at Fig. 12.9 for instance, stress increases at a greater rate than the strain. As a result, for producing a given strain Δ, increase in stress $\Delta\sigma_2$ required at B, in higher stress range, is greater than $\Delta\sigma_1$ the increase is stress required at A in the lower stress range. This means that the total belt length remaining unchanged, increase in tension $(T_1 - T_0)$ required on tight side is much larger than the required decrease in tension $(T_0 - T_2)$ on slack side for producing equal elongation and contraction in the belt. Hence

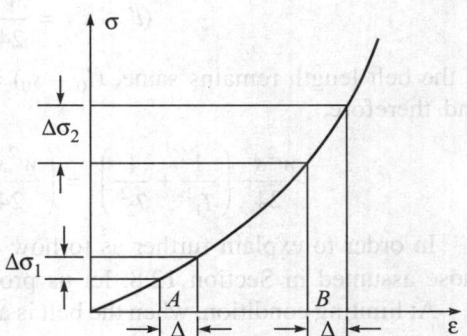

Fig. 12.9 Nonlinear Belt Behaviour.

$$(T_1 - T_0) > (T_0 - T_2)$$

or

$$T_1 + T_2 > 2T_0 \tag{12.63}$$

Although this is a fairly convincing argument, it is considered inadequate to explain large amount of difference between the sum of tensions $T_1 + T_2$ and twice the value of initial tension. Another reason has been given to explain the large difference. Thus the free-lengths of the belt between the two pulleys, when the drive is horizontal, is considered to be a catenary curve. For a given length of span s (*i.e.*, the centre distance between the pulleys), the length of the catenary curve l varies with tension in the belt at supports.

Let w be the weight per unit span of belt, s the length of span and T the tension at the support. The length of catenary in such a case exceeds the length of span by an amount (Fig. 12.10),

$$(l - s) = \frac{(w^2 s^3)}{24 T^2} \tag{12.64}$$

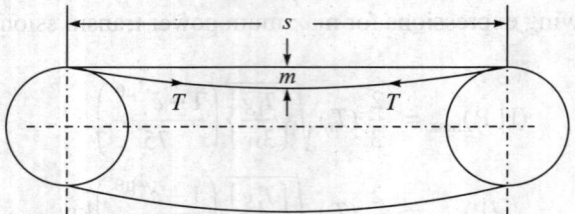

Fig. 12.10 Catenary effect.

When the belt does not transmit any torque, the belt tension is T_0 everywhere and

$$(l_0 - s_0) = \frac{w^2 s^3}{24 T_0^2} + \frac{w^2 s^3}{24 T_0^2} \equiv \frac{2(w^2 s^3)}{24 T_0^2}$$

When the belt transmits torque and tensions on tight and slack side are T_1 and T_2 respectively, the length of catenary exceeds length of span by an amount

$$(l' - s') = \frac{w^2 s^3}{24 T_1^2} + \frac{w^2 s^3}{24 T_2^2}$$

If the belt length remains same, $(l_0 - s_0) = (l' - s')$
and therefore,

$$\frac{w^2 s^3}{24}\left(\frac{1}{T_1^2} + \frac{1}{T_2^2}\right) = \left(\frac{w^2 s^3}{24}\right)\frac{2}{T_0^2} \quad \text{or} \quad \frac{1}{T_1^2} + \frac{1}{T_2^2} = \frac{2}{T_0^2} \qquad (12.65)$$

In order to explain further as to how conditions in actual practice differ very widely from those assumed in Section 12.8, let us proceed as follows.

At limiting condition, when the belt is at the point of slipping, it follows from equation (12.65) that

$$\frac{1}{T_1^2}\left(1 + \frac{T_1^2}{T_2^2}\right) = \frac{2}{T_0^2}$$

or

$$2(T_1/T_0)^2 = 1 + e^{2\mu\theta}$$

or

$$e^{2\mu\theta} = 2\left(\frac{T_1}{T_0}\right)^2 - 1$$

or

$$e^{\mu\theta} \equiv T_1/T_2 = \sqrt{\{2(T_1/T_0)^2 - 1\}} \qquad (12.66)$$

Again,

$$\frac{T_1 + T_2}{2T_0} = \frac{T_1(1 + e^{-\mu\theta})}{2T_0} \qquad (12.67)$$

and

$$\frac{T_1 - T_2}{2T_0} = \frac{T_1(1 - e^{-\mu\theta})}{2T_0} \qquad (12.68)$$

Based on assumed values of ratio of tension (T_1/T_2), corresponding values of $(T_1 + T_2)/2T_0$ and $(T_1 - T_2)/T_0$ are calculated using equations (12.66) through (12.68) and are tabulated below:

Table 12.1

(T_1/T_2)	1.1	1.3	1.5	1.7	1.9	2.0
$e^{\mu\theta}$	1.1916	1.5427	1.8708	2.1863	2.494	2.6457
$(T_1 + T_2)/2T_0$	1.01156	1.0713	1.1509	1.2388	1.3309	1.378
$(T_1 - T_2)/T_0$	0.1768	0.45732	0.6982	0.9224	1.1382	1.244

It follows from above that as power is transmitted, not only the ratio $(T_1 - T_2)/2T_0$ but also $(T_1 + T_2)/2T_0$ increases with (T_1/T_2). Firstly, this negates the basis of equation (12.44) and secondly it shows that as power transmission increases, the sum of tensions $(T_1 + T_2)$ increases and hence the belt length also increases. Thus, even equation (12.65) is not strictly correct.

Creep: This is yet another feature of belt transmission which goes to show that actual conditions in practice differ from those assumed in deriving equation (12.65). The phenomenon of creep arises because there exists a need for a difference in tension on the two sides of a belt drive, and because all materials stretch in more or less direct proportion to the force applied to them. It exists in all belt drives even below the limiting ratio of tensions.

Figure 12.11 portrays graphically as to what happens when the belt approaches driver pulley from tight side in stretched (and hence, in laterally contracted) condition and leaves the pulley on slack side in comparatively unstretched (and hence, laterally expanded) condition. This is shown in an exaggerated way in the figure. As the driver pulley receives belt on tight side, an elongated segment Δl of belt received by driver pulley, contracts in length to $\Delta l'$ as the segment passes over the pulley from tight to slack side. In other words, driving pulley receives longer length and delivers a shorter length on slack side and therefore the belt slips back relative to pulley. This is called *creep*.

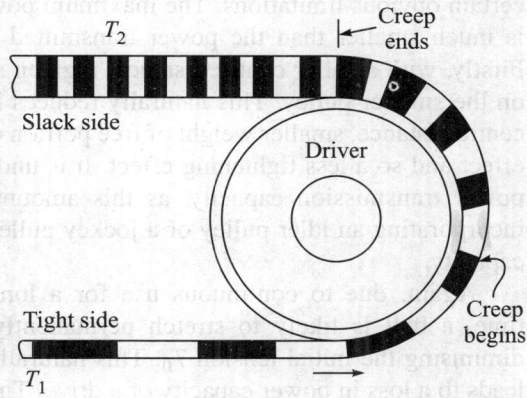

Fig. 12.11 Illustration for creep.

On the driven pulley, on the other hand, the driven pulley receives smaller belt length on slack side and delivers longer length as the belt segment passes over from unstretched condition on slack side to stretched condition on tight side. The belt, therefore, slips forward relative to the pulley. Clearly, creep occurs because of elastic stretch and the recovery of the belt as it passes over the pulleys. This results in a speed loss; the belt speed being smaller than peripheral speed of driver pulley and greater than peripheral speed of driven pulley. Presence of creep leads to loss in power which is dissipated out as heat energy. In practice, amount of creep varies from belt to belt, depending on their capacity to stretch and varies between 0.1 and 1 per cent.

Thus the slip and creep both contribute to a reduction in velocity ratio and represent a relative motion between belt and pulley. However, slip occurs when limiting coefficient of friction (and hence, limiting ratio of tension) is exceeded, the creep occurs even below the limiting ratio of tension and is due entirely to the elastic nature of belt material

Additional Features of Open Belt Drive

While connecting two pulleys of a horizontal drive, open belt drive is to be so arranged that the upper portion of the belt drive should be the slack side. The belt sag, being greater on the slack side, enables to obtain a larger angle of lap. The tight side in lower portion reduces sag and reduces the loss in angle of lap on that side. This arrangement, therefore, leads to a better grip between belt and pulley.

In drives having fluctuating sources of power (e.g. a gas engine), it is preferable to have tight side on the top side. A sudden shock, if any, in this case, causes the belt to vibrate and slip. Such an arrangement, therefore, reduces shock on the driven pulley or machine.

12.11 IDLER AND JOCKEY PULLEYS

An open belt drive for a short centre distance, particularly when the velocity ratio is high, has certain obvious limitations. The maximum power that can be transmitted using such a belt drive is much smaller than the power transmitted using same pulleys over longer centre distance. Firstly, with smaller centre distance, a given set of pulleys will produce a smaller angle of lap on the smaller pulley. This naturally reduces limiting ratio of tension. Secondly, due to shorter centre distance, smaller weight of free portion of belt between pulleys produce a smaller catenary effect and so a less tightening effect. It is undesirable to use higher initial tension for restoring power transmission capacity, as this amounts to increasing bearing loads. An arrangement incorporating an idler pulley or a jockey pulley enables to minimise loss in power transmission capacity.

Again, due to continuous use for a long time, a belt is likely to stretch permanently, diminising the initial tension T_0. This naturally leads to a loss in power capacity of a drive. The initial tension can be restored in a belt drive either by increasing centre distance or by using an idler or jockey pulley on the slack side of the belt, as shown in Fig. 12.12. A jockey pulley or idler is supported on an arm which is free to turn about pivots in a fixed frame. The jockey/idler pulley is supported on an arm while pressure of this pulley on the belt is produced using a spring or a hanging weight. The idler pulley is pressed against the slack side of the belt and generally results in increasing the angle of lap on smaller pulley. The idler pulley also tends to increase the tension T_2 on slack side.

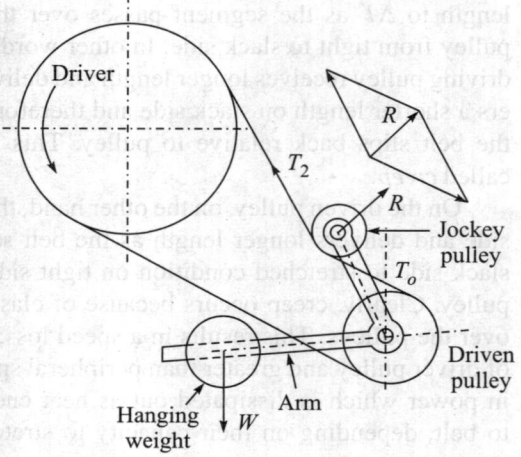

Fig. 12.12 An Idler/Jockey Pulley.

12.12 TIMING BELT

Moulded, endless and flat belts with regularly spaced teeth, formed on one side, have a comparatively recent origin. They incorporate the positive action of a chain drive with most of the advantages of the other types of belt drives (Fig. 12.13).

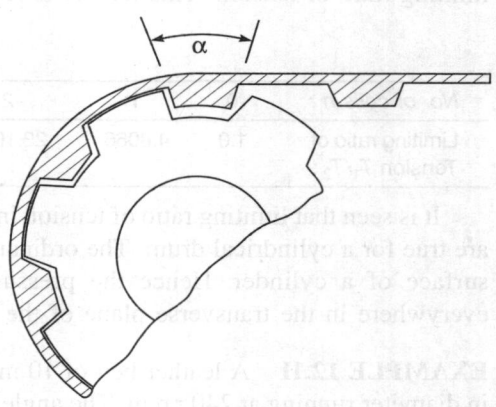

Fig. 12.13

Actual constructional details vary among manufacturers, but the tooth shear strength always exceeds the tensile strength of belt, provided that six or more teeth are in mesh. There is a positive tooth and groove engagement of the belt without slip, creep or speed variations. Using this drive a precise synchronization between the driver and driven elements is possible. This can be used for belt speeds of 4800 m/min and for h.p. from sub-fraction to 60 h.p.

This drive is deliberately designed as a positive drive and will not slip in conditions of sudden overloads or starts. Applications include business machines, sewing machines, timing drives, portable wood working equipment and power transmission units. Besides being positive, the drive is compact, light, quiet, versatile and low in maintenance.

12.13 ROPE DRIVE

Expression for limiting ratio of tension, derived in equation (12.27) for circular cross-section and V-shaped groove namely,

$$T_1/T_2 = e^{\mu'\theta}$$

holds good in the case of a rope drive when slip is just to begin. By coiling rope for number of turns n round a rotating cylinder or capstan, a large value of limiting ratio T_1/T_2 can be obtained. Physically this means that by applying a small force T_2 at one end of the rope, leaving the capstan, comparatively much larger resistance (equal to T_1) can be overcome at the other end of rope approaching the capstan.

If n = number of coils/turns (integral or fraction) of the rope around cylinder,

T_1 = tension on tight side, corresponding to limiting ratio of tension,

T_2 = tension on slack side at the point of slipping,

Then the total angle of lap,

$$\theta = (2\pi n)\,\text{rad} \qquad (12.69)$$

Thus, assuming $\mu' = 0.25$, the limiting ratio of tension is,

$$T_1/T_2 = e^{0.25(2\pi)n} = e^{(1.57)n} \qquad (12.70)$$

It is interesting to note the effect of number of turns/coils of rope around capstan on the limiting ratio of tension. This feature is reflected in Table 12.2

Table 12.2

No. of coils n :	0	1	2	3	4	5	6
Limiting ratio of Tension T_1/T_2 :	1.0	4.8066	23.1038	111.0522	533.788	2565.734	12332.58

It is seen that limiting ratio of tension increases rapidly as n increases. Note that above results are true for a cylindrical drum. The ordinary capstan has a concave profile rather than the plane surface of a cylinder. Hence the pressure exerted by capstan surface on the rope is not everywhere in the transverse plane of the rope.

EXAMPLE 12.11 A leather belt of 10 mm thickness transmits 37.3 kw from a pulley 120 cm in diameter running at 240 r.p.m. The angle of lap is 170° and coeff. of friction is 0.3. The weight of 1 cu cm leather is 9.316×10^{-3} N. What width of belt is required if the stress is limited to 245.2 N per centimetre of width?

Solution: Given $P = 37.3$ kw; $d_2 = 120$ cm; $N_2 = 240$ r.p.m.; $\theta = 170°$; $\mu = 0.3$;

weight density = 9.316×10^{-3} N/cm³; $T_t = 245.2$ N/cm of width

Let b be the width of belt in cm

Then weight per m length of belt,

$$w = (b \times 1.0) \times 100 \times 9.316 \times 10^{-3} = (0.9316)b \text{ N/m}$$

The maximum tension on tight side,

$$T_t = (245.2)b \text{ N}$$

Peripheral belt speed assuming no slip,

$$v = \frac{\pi \times 1.2 \times 240}{60} = 15.0796 \text{ m/s}$$

Therefore from, $P = \dfrac{(T_1 - T_2)V}{1000}$, we have

$$(T_1 - T_2) = \frac{37.3 \times 1000}{15.0796} = 2473.5 \text{ N} \tag{a}$$

Also, for limiting ratio of tension,

$$T_1/T_2 = e^{0.3 \times (170\pi/180)} = e^{0.890} = 2.435$$

But, $(T_1 - T_2) \equiv T_1(1 - e^{-\mu\theta})$

Therefore from (a), $T_1 = \dfrac{2473.5}{(1 - 0.41067)} = 4197.1 \text{ N}$

Also
$$T_c = \frac{w}{g} V^2 = \frac{(0.9316)b}{9.81} \times (15.0796)^2 = (21.594)b\,\text{N}$$

Hence from
$$T_t = T_1 + T_c$$

We have
$$(245.2)b = 4197.2 + (21.594)b$$

or
$$b = 18.77 \text{ cm} \qquad \textbf{Ans.}$$

EXAMPLE 12.12 A belt transmitting power from a motor to a machine weighs 23.54 N/m and the maximum permissible tension for it is 5395 N. The angle of contact of belt with pulley is 200° and coeff. of friction is 0.28. If the belt runs under maximum power conditions, determine the maximum power transmitted and initial tension in the belt so that drive can fulfil the above conditions.

Solution: Given: $w = 23.54$ N/m; $T_t = 5395$ N; $\theta = 220°$; $\mu = 0.28$;
Limiting ratio of tensions,
$$(T_1/T_2) = e^{\mu\theta} = e^{(0.28)(200\pi/180)} = e^{0.97738} = 2.6575$$

For maximum h.p. transmitted,
$$T_c = \frac{1}{3}(T_t) = \frac{5395}{3} = 1798.33 \text{ N}$$

Hence
$$T_1 = T_t - T_c = 5395 - 1798.33 = 3596.67 \text{ N}$$

Therefore the tension on slack side,
$$T_2 = T_1/e^{\mu\theta} = 3596.67/2.6575 = 1353.4 \text{ N}$$

Hence initial tension,
$$T_0 = \frac{T_1 + T_2}{2} = 2475 \text{ N} \qquad \textbf{Ans.}$$

Max power transmitted
$$P = \frac{(T_1 - T_2)V}{1000}$$

where
$$V = v_{\text{opt}} = \sqrt{\frac{gT_t}{3w}} = \sqrt{\frac{9.81 \times 5395}{3 \times 2354}} = 27.38 \text{ m/s}$$

Hence, power transmitted
$$P = \frac{(3596.67 - 1353.4) \times 27.38}{1000} = 61.42 \text{ kw} \qquad \textbf{Ans.}$$

EXAMPLE 12.13 A belt embraces the shorter pulley by an angle of 165° and runs at a speed of 1700 m/min. Dimensions of the belt are: width = 20 cm and thickness = 8 mm. It weighs 0.00981 N/cm³. Determine the maximum power that can be transmitted at the above speed if the maximum permissible stress in the belt is not to exceed 245.25 N/cm^2 and $\mu = 0.025$ (AMIE, Winter 1982)

Solution: Given: $\theta = 165°$; $\mu = 0.25$; $V = 1700$ m/min; $b = 20$ cm; $t = 0.8$ cm;

$$w' = 0.00981 \text{ N/cm}^3; \sigma_{per} = 245.25 \text{ N/cm}^2$$

Maximum tension in the belt,

$$T_t = \sigma_{per} \times b \times t = 245.25 \times 20 \times 0.8 = 3924 \text{ N}$$

Weight of belt per metre length

$$w = (100 \times 20 \times 0.8) \times \frac{0.00981}{1} = 15.696 \text{ N/m}$$

The centrifugal tension in the belt

$$T_c = \frac{wv^2}{g} = \frac{15.696}{9.81} \times \left(\frac{1700}{60}\right)^2 = 1284.4 \text{ N}$$

For optimum power, $\qquad T_c = \frac{1}{3}T_t = \frac{3924}{3} = 1308 \text{ N}$

It follows that the belt drive is not operating at optimum condition. Now, at limiting condition,

$$(T_1/T_2) = e^{0.25(165\pi/180)} = e^{0.7199} = 2.054$$

From above, $\qquad T_1 = T_t - T_c = 3924 - 1284.4 = 2639.6 \text{ N}$

Therefore $\qquad T_2 = T_1/e^{\mu\theta} = \frac{2639.6}{2.054} = 1285.1 \text{ N}$

Max. power $\qquad = \frac{(2639.6 - 1285.1) \times 1700}{60 \times 1000} = 38.38 \text{ kw}$ **Ans.**

EXAMPLE 12.14 Power is transmitted using a V-belt drive. The included angle of v-groove is 30°. The belt is 2 cm deep and its maximum width is 2 cm. If the mass of the belt is 0.03434 N/cm length and the maximum allowable stress is 137.34 N/cm^2, determine the maximum h.p. transmitted when the angle of lap is 140°, $\mu = 0.15$. (AMIE, Winter 1984)

Solution: Given $2\alpha = 30°$; $t = 2$ cm; $b = 2$ cm; $S = 0.03434$ N/cm length

$$\sigma_{per} = 137.34 \text{ N/cm}^2; \theta = 140°; \mu = 0.15$$

Let w be the belt width in the inner side.
The limiting ratio of tension is,

$$(T_1/T_2) = e^{\mu\theta/\sin\alpha}$$

or $\qquad (T_1/T_2) = e^{0.15(140\pi)/(180\times\sin15°)} = e^{1.416} = 4.12$

As seen from the geometry of V-belt cross-section, shown in Fig. 12.14,

$$h = \frac{1}{\tan15°} \approx 3.8637 \text{ cm}$$

Therefore $\quad w/2 = (h - 2)\tan 15°$. Therefore $w \approx 0.965$ cm

Therefore area of c/s of belt $= \dfrac{(b + w)}{2} \times t$

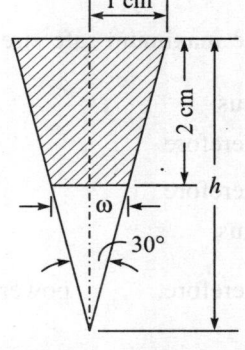

or $\qquad$ area of c/s $= \left(\dfrac{2 + 0.965}{2}\right) \times 2 = 2.965$ cm^2

Therefore, maximum tension in the belt,

$$T_t = \sigma_{per} \times a = 137.34 \times 2.965 = 407.2 \text{ N}$$

For transmitting maximum power,

$$T_c = \frac{1}{3}(T_t) = \frac{407.2}{3} = 135.73 \text{ N}$$

Fig. 12.14 Geometry of V-belt.

Hence from,

$$T_c = mv^2$$

$$v = \sqrt{\frac{135.73}{m}} = \sqrt{\frac{135.73}{0.03434 \times 100}} = 6.287 \text{ m/s}$$

Therefore $\qquad T_1 = T_t - T_c = \dfrac{2}{3}T_t = 271.47$ N

and, $\qquad$ Power $= \dfrac{T_1(1 - e^{-\mu'\theta})v}{1000} = \dfrac{271.47\left(1 - \dfrac{1}{4.12}\right) \times 6.287}{1000} = 1.292 \text{ kw}$ **Ans.**

EXAMPLE 12.15 A belt drive consists of two V-belts in parallel, on grooved pulleys of the same size. The angle of the groove is 30°. The cross-sectional area of the belt is 750 mm^2 and $\mu = 0.12$. The density of the belt material is 1.2 mega-gm/m^3 and the maximum safe stress in the material is 7 MN/m^2. Calculate the power that can be transmitted between pulleys 300 mm diameter rotating at 1500 r.p.m. Find also the shaft speed in rev/min at which the power transmitted would be maximum (London Univ.)

Solution: Limiting ratio of tensions, ($\theta = \pi$ for equal size pulleys)

$$T_1/T_2 = e^{0.12\pi/\sin 15} = e^{1.45658} = 4.29$$

Also, as $d_1 = d_2 = 300$ mm and $N = 1500$ r.p.m.,

$$v = \frac{\pi \times 300 \times 10^{-3} \times 1500}{60} = 23.562 \text{ m/s}$$

The belt cross-sectional area $= 750 \times 10^{-2} = 7.5$ cm^2

Therefore, mass of belt per metre length $m = (1.2 \times 10^6) \times \dfrac{7.5}{(100)^2} \times 1 = 900$ gm/m

or $\qquad m = \dfrac{900}{1000} = 0.9$ kg/m

Therefore $\qquad T_c = (0.9) \times (23.562)^2 = 499.65$ N

The maximum safe stress in belt $= \dfrac{7 \times 10^6}{(100)^2} = 700$ N/cm^2

Thus $\qquad\qquad\qquad\qquad T_t = 700 \times 7.5 = 5250$ N

Therefore $\qquad\qquad\qquad T_1 = T_t - T_c = 5250 - 499.65 = 4750.35$

Therefore $\qquad\qquad\qquad T_2 = T_1/e^{\mu'\theta} = 4750.35/4.29 = 1107.3$ N

Thus $\qquad\qquad\qquad\qquad T_s = T_2 + T_c = 1107.3 + 499.65 = 1606.95$

Therefore, $\qquad$ power transmitted $= \dfrac{(T_t - T_s)V}{1000}$ for one belt

Thus $\qquad\qquad$ Power/belt $= \dfrac{(5250 - 1606.95) \times 23.562}{1000} = 85.837$ kw

Therefore $\quad$ total power transmitted by 2 belts $= 85.867 \times 2 = 171.67$ kw $\qquad$ **Ans.**

Speed at which power transmitted is maximum, is given by

$$V = \sqrt{\frac{gT_t}{3w}} = \sqrt{\frac{T_t}{3m}} = \sqrt{\frac{5250}{3 \times 0.9}} = 44.095 \text{ m/s}$$

The shaft speed in r.p.m. for this speed,

$$N = \frac{44.095}{\pi d} = \frac{44.095 \times 100 \times 60}{\pi \times 30} = 2807 \text{ r.p.m.} \qquad \textbf{Ans.}$$

EXAMPLE 12.16 An electric motor drives a compressor through V-belt. The following data are known:

	motor pulley	compressor pulley
Pitch diameters cm	27.5	120
Angle of contact radians	2.0	4.28
Coeff. of friction	0.3	0.3
Groove angle	34°	34°
Power transmitted kw	26.82 kw	
Speed r.p.m.	1450	

Each belt weighs 2.943 N/m and the maximum allowable tension is 441.5 N/belt. Calculate the number of belts required for the drive.

Solution: The small pulley is the motor pulley. Hence ratio of limiting tensions,

$$(T_1/T_2) = e^{0.3(2)/\sin 17} = e^{2.052} = 7.785$$

Therefore $\qquad\qquad e^{-\mu'\theta} = 0.12845$

Also belt speed $\qquad\qquad V = \dfrac{\pi \times 27.5 \times 1450}{60 \times 100} = 20.8785$ m/s

Also $\qquad\qquad\qquad$ Power $= \dfrac{(T_1 - T_2)V \times n}{1000}$ kw

$$n(T_1 - T_2) = \frac{1000 \times 26.82}{20.8785} = 1284.6 \text{ N}$$

where n = number of belts operating the compressor. Maximum tension in each belt,

$$T_t = 441.5 \text{ N}$$

$$T_c = \frac{2.943 \times (20.8785)^2}{9.81} = 130.77 \text{ N}$$

Therefore $T_1 = 441.5 - 130.77 = 310.73 \text{ N}$

Therefore $T_2 = T_1/e^{\mu'\theta} = \dfrac{310.73}{7.785} = 39.91 \text{ N}$

Substituting for $(T_1 - T_2)$ above,

$$n = \frac{1284.6}{(310.77 - 39.91)} = 4.74 \simeq 5 \qquad \textbf{Ans.}$$

EXAMPLE 12.17   The maximum power that can be transmitted by a belt is 55.93 kw. The belt is 25 cm wide and 10.5 mm in thickness and weights 0.00883 N/cm^3. If the ratio of tension is 2, find value of maximum stress induced in the belt at this condition, for the belt.

Solution: Given belt cross-sectional area = a = 25 × 1.05 = 26.25 cm^2

And $T_1/T_2 = e^{\mu\theta} = 2.0$

Therefore,   weight of belt per metre length = (26.25 × 100 × 0.00883) N

or $w = \dfrac{26.25 \times 100 \times 0.00883}{1} = 23.18 \text{ N/m}$

For maximum power,

$$T_c = \frac{1}{3}T_t$$

Also $v_{opt} = \sqrt{\dfrac{gT_t}{3w}} = \sqrt{\dfrac{9.81 \times T_t}{3 \times 23.18}} = 0.376\sqrt{T_t}$

Again, from

$$\text{Power} = \frac{T_1(1 - e^{-\mu\theta})v}{1000} \text{ kw}$$

or $\text{Power} = \dfrac{(T_t - T_c)(1 - e^{-\mu\theta})v}{1000} = \dfrac{(2/3)\,T_t(1 - 1/2)v_{opt}}{1000}$

Substituting for V_{opt},

$$55.93 = \frac{\left(\dfrac{2}{3}\right)T_t\left(1 - \dfrac{1}{2}\right)0.376\sqrt{T_t}}{1000}$$

Therefore $(T_t)^{\frac{3}{2}} = \dfrac{55.93 \times 1000}{\left(\dfrac{2}{3}\right)\left(\dfrac{1}{2}\right)(0.376)} = 446250$ Thus, $T_t = 5840$ N

Thus if σ be the permissible stress in the belt, then

$$\sigma(b \times t) = T_t$$

Therefore $\qquad \sigma = \dfrac{5840}{(25 \times 1.05)} = 222.5 \text{ N/cm}^2$ **Ans.**

EXAMPLE 12.18 A rope, 4 cm in diameter, drives a grooved pulley at a speed of 20 m/s. The angle lap of rope on the pulley is 200°, the angle of groove for the pulley is 60°, $\mu = 0.3$ and weight per m length of rope = 8.83 N. If permissible tension in the rope is 147.15 N/cm², calculate power that can be transmitted at this speed.

Solution: Given $\sigma_{per} = 147.15$ N/cm²; d = rope dia. = 4 cm;

$$v = 20 \text{ m/s}; \ \theta = 200°; \ \mu = 0.3; \ \alpha = \dfrac{60}{2} = 30°; \ w = 8.83 \text{ N/m}$$

We have, $\qquad T_t = \sigma_{per} \times \dfrac{\pi}{4} d^2 = 147.15 \times \dfrac{\pi}{4} (4)^2 = 1849.14$

Also, $\qquad T_c = \dfrac{8.83}{9.81} \times (20)^2 = 360.0 \text{ N}$

Therefore $\qquad T_1 = 1849.14 - 360 = 1489.14 \text{ N}$

Also $\qquad \mu'\theta = \dfrac{0.3}{\sin 30} \times \dfrac{200\,\pi}{180} = 2.094$

Therefore $\qquad T_1/T_2 = e^{2.094} = 8.12$

Therefore $\qquad e^{-\mu'\theta} = \dfrac{1}{8.12} = 0.12315$

Therefore power transmitted $= \dfrac{T_1(1 - e^{-\mu'\theta})\,v}{1000}$

$$= \dfrac{1489.14\,(1 - 0.12315)\,20}{1000} = 26.11 \text{ kw} \qquad \textbf{Ans.}$$

EXAMPLE 12.19 A pulley used to transmit power by means of ropes has a diameter of 3.6 m and has 15 grooves of 45° angle. The angle of contact is 170° and the coefficient of friction between the ropes and the groove sides is 0.28. The maximum possible tension in the ropes is 941.8 N and the weight of the rope is 14.72 N per m length. What is the speed of the pulley in r.p.m. and the power transmitted if the condition of maximum power prevail?

(AMIE, Summer 1983)

Solution: Given $D = 3.6$ m; $n_g = 15$; $2\alpha = 45°$; $\mu = 0.28$; $T_t = 941.8$ N; $\theta = 170°$; $w = 14.72$ N/m

For maximum h.p. condition,

$$T_c = \frac{1}{3} T_t = \frac{941.8}{3} = 313.9 \text{ N}$$

and

$$v_{\text{opt}} = \sqrt{\frac{gT_t}{3w}} = \sqrt{\frac{9.81 \times 941.8}{3 \times 14.72}} = 14.464 \text{ m/s}$$

Also, limiting ratio of tensions,

$$T_1/T_2 = e^{\mu'\theta}$$

or

$$T_1/T_2 = e^{(0.28 \times 170\pi)/(180\sin 22.5)} = e^{2.1709} = 8.766$$

Therefore

$$e^{-\mu'\theta} = 0.1141$$

Also

$$T_1 = T_t - T_c = \frac{2}{3} T_t = \frac{2}{3}(941.8) = 627.87 \text{ N}$$

Therefore

$$T_2 = (T_1/e^{\mu'\theta}) = 71.63 \text{ N}$$

The r.p.m. of pulley $= \dfrac{60 \times v}{\pi D}$

or

$$\text{r.p.m.} = \frac{60 \times 14.464}{\pi \times 3.6} = 76.73 \text{ r.p.m.}$$

Max power transmitted per rope $= \dfrac{(T_1 - T_2)v}{1000} = \dfrac{(627.87 - 71.63) \times 14.464}{1000}$

$$= 8.045 \text{ kw/rope}$$

Therefore, total max power transmitted using 15 ropes $= 8.045 \times 15 = 120.62$ kw   **Ans.**

EXAMPLE 12.20 A capstan and a rope are used in a railway goods yard for moving trucks. The capstan runs at 50 r.p.m. The rope from the line of trucks makes 2.75 turns around the capstan at a radius of 20 cm and the free end of the rope is pulled with a force of 147.15 N. Determine the pull on the trucks, the power taken by the trucks, and the power supplied by the capstan. Take $\mu = 0.25$.

Solution: Given $N = 50$ r.p.m.; $n = 2.75$ turns; $r = 20$ cm;

$$F = 147.15 \text{ N}; \mu = 0.25$$

Here, total angle of lap $\theta = 2.75 \times 2\pi = (5.5)\pi$

Therefore limiting ratio of tensions,

$$T_1/T_2 = e^{0.25 \times 5.5\pi} = e^{4.3197} = 75.165$$

(Note: The grooves in capstan permit contact with ropes over the entire groove surface and as such, groove angle 2α loses its significance)

For deriving mechanical advantage,

$$T_2 = 147.15 \text{ N}$$

Therefore $T_1 = 147.15 \times 75.165 = 11060.53 \text{ N}$ and, $v = \dfrac{2\pi r N}{60}$

or $V = \dfrac{2\pi \times 0.20 \times 50}{60} = 1.047 \text{ m/s}$

Also power transmitted $= \dfrac{(T_1 - T_2)v}{1000}$

or power transmitted $= \dfrac{(11060.53 - 147.15) \times 1.047}{1000} = 11.428 \text{ kw}$

Again, power taken by trucks $= \dfrac{11060.53 \times 1.047}{1000} = 11.58 \text{ kw}$ **Ans.**

The pull on trucks $T_1 = 11060.5 \text{ N}$ **Ans.**

EXAMPLE 12.21 The motion of a vessel drifting away from a dockside is retarded by a rope secured to the vessel and given 3 complete turns around a bollard on the dockside. A pull of 392.4 N is applied to the free end of the rope at an instant when the speed of the vessel, which weighs 39.24 MN is 10 cm/s. After 10 seconds the rope begins to slip. Assuming that the rope stretches elastically, calculate the stretch 's' in the rope between the bollard and the vessel and the speed of the vessel when the rope slips (μ between rope and bollard = 0.25).

Solution: Given $\theta = 3 \times 2\pi = 6\pi$; $T_2 = 392.4 \text{ N}$; $v = 10 \text{ cm/s}$; $\mu = 0.25$; $t = 10 \text{ s}$.

The limiting ratio of tension,

$$(T_1/T_2) = e^{0.25 \times 6\pi} = 111.318$$

Therefore $T_1 = 392.4 \times 111.318 = 43681.2 \text{ N}$

As the force $T_2 = 392.4 \text{ N}$ is applied at the free end, tension $T_1 = 43681.2 \text{ N}$ is developed at the other end in 10 seconds (as the vessel continues to drift, stretching the rope). Thus the tension rises from 0 to T_1 in 10 seconds as the velocity of vessel drops down from u to v. During this period, the distance moved through by the vessel,

$$s = \frac{1}{2}(u + v) \times t$$

Hence, from work done consideration,

$$\frac{1}{2}(T_1)s = \frac{W}{2g}(u^2 - v^2)$$

or $\dfrac{1}{2}(43681.2) \times \dfrac{1}{2}(0.1 + v) \times 10 = \dfrac{39240000}{2 \times 9.81}(0.1^2 - v^2)$

Therefore $(0.1 + v) = 18.315(0.1^2 - v^2)$

or $(0.1 + v) = 18.315(0.1 - v)(0.1 + v)$

Cancelling $(0.1 + v)$ terms on either side, we have

$$(0.1 - v) = \frac{1}{18.315}$$

Therefore

$$v = \left(0.1 - \frac{1}{18.315}\right) = 0.0454 \text{ m/s}$$

Therefore, stretch in the rope,

$$s = \frac{1}{2}(u + v) \times t$$

or

$$s = \frac{1}{2}(0.1 + 0.0454) \times 10 = 0.727 \text{ m} = 72.7 \text{ cm}$$ **Ans.**

EXAMPLE 12.22 Following data is given for a rope pulley transmitting 23.49 kw

Diameter of pulley = 40 cm   Speed of pulley = 110 r.p.m.

Angle of groove = 45°   Angle of lap (smaller pulley) = 160°

Coeff. of friction μ = 0.28   Number of ropes = 10

Weight of rope in N/m length = $0.05199C^2$

Working tension is limited to $11.968C^2$ N

When C = girth of rope in cm

Find initial tension and diameter of each rope (AMIE, Summer 1979)

Solution: Given: N = 110 r.p.m.; power = 23.49 kw; D = 40 cm; μ = 0.28; 2α = 45°;

$$\theta = 160°; \ w = 0.05199 \ C^2$$

$$T_{max} = 11.968C^2 \text{ N}.$$

The peripheral rope speed $v = \dfrac{\pi \times 0.4 \times 110}{60} = 2.304$ m/s

The centrifugal tension in rope,

$$T_c = \frac{wv^2}{g} = \frac{(0.05199C^2)(2.304)^2}{9.81} = (0.02813)C^2$$

Limiting ratio of tensions,

$$T_1/T_2 = e^{0.28(160\pi/180)\,cosec\,22.5°} = e^{2.043} = 7.715$$

Also

$$T_1 = T_{max} - T_c$$

or

$$T_1 = (11.968 - 0.02813)C^2 = (11.93987)C^2$$

Therefore

$$T_2 = T_1/e^{\mu'\theta} = \left(\frac{11.93987}{7.715}\right)C^2 = 1.5476\,C^2$$

Again, power transmitted by each rope,

$$= \frac{23.49}{10} = 2.349 \text{ kw}$$

Therefore,   from power considerations,

$$2.349 = \frac{(T_1 - T_2)V}{1000}$$

Therefore

$$(T_1 - T_2) = \frac{1000 \times 2.349}{2.304} = 1019.53 \text{ N}$$

Equating

$$(11.93987 - 1.5476)C^2 = 1019.53$$

We have

$$C^2 = \frac{1019.53}{10.39227} = 98.1$$

Hence

$$C = 9.9$$

Therefore,   initial tension

$$T_0 = \frac{T_1 + T_2}{2}$$

or

$$T_0 = \frac{(11.93987 + 1.5476)C^2}{2} = 661.6 \text{ N} \qquad \textbf{Ans.}$$

If d be the diameter of the rope, then girth (perimeter) $C = \pi d$

Therefore

$$d = \frac{9.9}{\pi} = 3.15 \text{ cm} \qquad \textbf{Ans.}$$

12.14 MATERIALS OF BELT AND ROPE

Primarily, materials used for belt and rope must be strong, flexible, durable and have high coefficient of friction. Following materials are most commonly used.

Flat Belt: Oak-tanned leather is a standard material but chrome leather is used where pliable materials are required.

Wherever relatively cheap material is required for light power transmission, fabric such as canvas or cotton duck impregnated with a filler material is commonly used. Such fabric may be vulcanised to form rubber belts in applications involving exposure to oil or sunlight. They can be used to connect shafts with centre distance upto 8–10 metre.

V-Belts: V-belts consist of fabric and cords moulded in rubber, generally covered with fabric. Figure 12.15 shows a typical cross-section of a V-belt. As shown in the figure, the portion that carries direct tensile (power) load is situated near the neutral axis of the section where stresses due to bending are insignificant as the belt

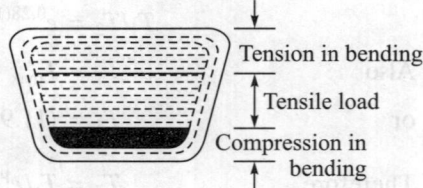

Fig. 12.15 A *V*-belt structure.

moves around the pulley. The fabric near the top surface carries the tensile stress due to bending while the rubberised portion at the bottom carries compressive stresses due to bending. The V-belts are vulcanized to form a loop of exact length and are available in standard lengths to suit requirements in installations. They can connect shafts with centre distance upto 4 metre.

A recent practice is to provide a load (tensile) carrying element, consisting of small endless steel cables, which increases load carrying capacity with less of stretching. Velocity ratios upto 7:1 can be transmitted using V-belts.

Rope: Ropes are available in sizes varying from 18 mm to 50 mm in diameter. Common materials are: manila, hemp and cotton. These can be used to connect shafts with centre distance upto 30 metres.

Wire ropes are used where power is to be transmitted over long centre distance as in mining, hauling, winding, etc. The wire rope is constructed of strands each of which is made of small wires twisted together.

Table 12.3 provides a comparative study of various mechanical drives and should be quite useful as a general guide for making the first choice.

Table 12.3

Type of drive	Relative cost	Space requirement	Factor that limits power capacity	Design criteria
1. Flat Belts	1	Quite large	Available space	Stretching and slip in belts
2. V-Belts	3	Large	Available space and cost	Wear of side faces and stretching
3. Roller chains	5	Medium	Speed and width	Breakage of links and wear of pins
4. Silent chains	7	Medium	Available space and speed	Strength
5. Straight spur gears	15	Small	Available space	Tooth strength and wear-rate
6. Helical gears	25	Small	Available space	Tooth-strength and wear-rate

12.15 CHAINS

In the case of belt and rope a constant velocity ratio can never be transmitted, mainly due to slip arising out of overloads and also because of creep. Timer belt was shown to be an exception to the above rule. As against this, a chain drive is a positive drive and is used to transmit a constant velocity ratio.

A chain consists of rigid links, which are hinged together providing flexibility that is necessary for the wrapping action around driving and driven wheels, called sprockets. These wheels have projecting teeth, which fit into suitable recesses in the links of the chain, ensuring a positive drive thereby. However, a chain drive is not suitable for precise timing applications. The reason for this is to be seen in the kinematics of the chain drive. In some cases, idlers are used as tensioning devices. The links, pins and bushings used in chain drive are fabricated from high grade steel and the pin and bushings are ground to ensure accuracy of pitch. The rollers turn on bushings that are press fitted to the inner link-plates. The pins are prevented from rotating in the outer link plates by the press-fit assembly. Chain drives are used for transmitting speed ratios upto 8 and can have efficiency of transmission as high as 98 per cent. Figure 12.16 shows a typical roller chain, where the side plates are blanked from cold rolled steel.

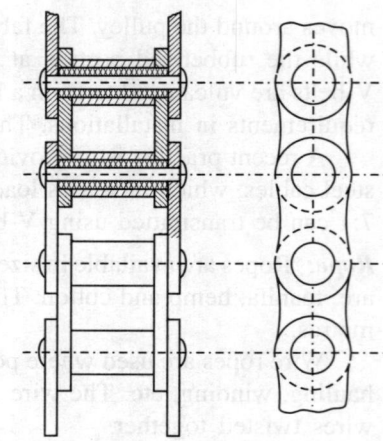

Fig. 12.16   A roller chain.

Pitch: The distance between centres of adjacent hinges/pins is called pitch.

The *pitch circle diameter* of the chain-sprocket is the diameter of an imaginary circle on which centres of pins lie, when the chain is wrapped around the sprocket.

With reference to Fig. 12.17, let D be the pitch circle diameter, p the pitch and θ angle subtended at the centre by chordal distance between the centres of adjacent pins. Then,

$$p = 2\left(\frac{D}{2}\right)\sin\frac{\theta}{2}$$

or $D = \dfrac{p}{\left(\sin\dfrac{\theta}{2}\right)}$ But $\theta = \left(\dfrac{360}{T}\right)$

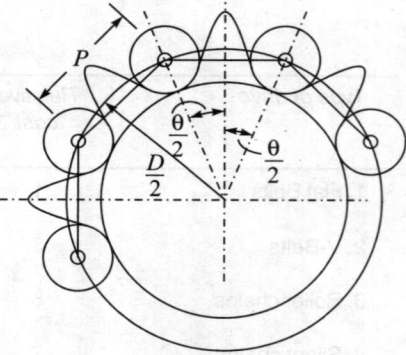

Fig. 12.17   Chain over a sprocket.

where T = number of teeth on the sprocket.
Hence

$$D = \frac{p}{\sin(180/T)} \tag{12.71}$$

Let us assume that the sprocket rotates at angular velocity ω in c.c.w. direction. In Fig. 12.18(a), the chain is received at A on sprocket and the chain velocity is maximum possible

$$v = (OA)\omega \tag{12.72}$$

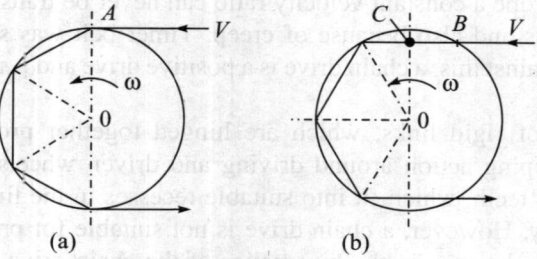

Fig. 12.18   Chain velocity.

As against this, minimum chain velocity v' is obtained at Fig. 12.18(b), where chain is received at C with smallest distance OC from the centre. Thus,

$$v' = (OC)\omega = \left(OA\cos\frac{\theta}{2}\right)\omega$$

$$v' = (OA)\omega \cos\frac{\theta}{2} \tag{12.73}$$

Thus, as each link of the chain is rigid and forms a chord of pitch circle in all positions, the speed of the chain varies between limits

$$(OA)\omega\cos\frac{\theta}{2} \le v \le (OA)\omega$$

The speed ratio of the shafts, connected by chain, depends upon the number of teeth on sprockets.
Thus if

N_1, N_2 = speeds in r.p.m. of driven and driver sprockets respectively, and

n_1, n_2 = the number of teeth on driven and driver sprockets.

The average velocity v of chain is established on the basis of pitch p. Thus, as

$$(N_2/N_1) = n_1/n_2$$

pn_1 gives sum of sides of regular polygon formed by chain links on driven sprocket while pn_2 gives sum of sides of regular polygon formed by chain links on driver sprocket.
Then

$$v = \frac{(pn_1)N_1}{100} = \frac{(pn_2)N_2}{100}\text{m/s} \tag{12.74}$$

12.16 INVERTED TOOTH CHAIN (SILENT CHAIN)

The chain consists of a series of flat plates, each provided with two projections or teeth. The projections are ground on outer faces to give an included angle of 60° or, in some cases, 75°. These ground surfaces bear against the working faces of the sprocket teeth, however the inner faces do not participate in load transmission and are given shape so as to clear the sprocket teeth. A number of such plates are placed alternately to obtain required width of chain and by connecting them together by hardened steel pins. The hardened steel pins pass through hardened steel bushes inserted in the ends of the links. The pins are riveted over the outside plates. One such silent chain is shown in Fig. 12.19. Each link, as it enters the sprocket, pivots about the pin on the adjacent link which is in contact with the sprocket. The working faces of the links thus come into contact with working surface of the sprocket teeth gradually. A similar action occurs at the time of leaving the sprocket.

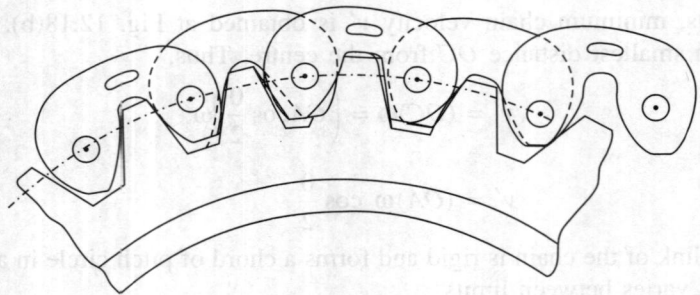

Fig. 12.19 Inverted tooth chain.

REVIEW QUESTIONS

12.1 Two pulleys of diameters 450 mm and 200 mm are on parallel shafts 1.95 m apart, and are connected by a crossed belt. Find the belt length required and the angle of contact between the belt and each pulley.

What power can be transmitted by the belt when the larger pulley rotates at 200 r.p.m., if the maximum permissible tension in the belt is 1 kN, and the coeff. of friction between the belt and the pulley is 0.25? (Univ. London)   (*Ans.* $L = 4.980$ m; $p = 2.74$ kw)

12.2 The following data refer to an open belt drive: Diameter of larger pulley: 400 mm, Dia. of smaller pulley: 250 mm; Distance between pulleys: 2 m; Coeff. of friction between belt and smaller pulley surface: 0.4; Maximum tension when the belt is at the point of slipping = 1200 N.

Find the power transmitted at a speed of 10 m/s. It is desired to increase the power. Which of the following two methods will you select?

1. Increasing the initial tension in the belt by 10%
2. Increasing the coeff. of friction between the smaller pulley and belt by 10% by the application of suitable dressing of the belt.

(*Ans.* 8.48 kw; option of increasing initial tension is preferable)

12.3 A shaft which rotates at a constant speed of 160 r.p.m. is connected by belting to a parallel shaft 72 cm apart which has to run at 60, 80 and 100 r.p.m. The smallest pulley on the driver shaft is 4 cm in radius. Determine the remaining radii of the two stepped pulleys for (a) a crossed belt, (b) an open belt. Neglect belt thickness and slip (AMIE, Summer 1982)

(*Ans.* $r_2 = 10.67$ cm; $r_3 = 4.89$ cm; $r_4 = 9.78$ cm; $r_5 = 5.64$ cm; $r_6 = 9.02$ (crossed belt); $r_2' = 6.0$ cm; $r_6' = 9.6$ cm. (open belt)

12.4 A leather belt 9×250 mm is used to drive a cast iron pulley 90 cm in diameter at 336 r.p.m. If the active arc on the smaller pulley is 120° and the stress in the tight side is 196.2 N/cm², find the h.p. capacity of the belt which weighs 0.009614 N/cm³, coefficient of friction of leather on cast iron is 0.35.   (*Ans.* 36.76 kw)

12.5 An open belt running over two pulleys 24 cm and 60 cm diameter connects two parallel shafts 3 metres apart and transmits 3.73 kw from the smaller pulley that rotates at

300 r.p.m., coefficient of friction between the belt and pulley is 0.3 and the safe working tension is 98.1 N per cm width. Determine (a) minimum width of belt (b) initial tension in belt and (c) length of belt required.

(**Ans.** $b = 16.69$ cm; $T_0 = 1148.4$ N; $L = 7.33$ m)

12.6 A flat belt is required to transmit 35 kw from a pulley of 1.5 m effective diameter running at 300 r.p.m. The angle of contact is spread over 11/24 of the circumference and the coefficient of friction between belt and pulley surface is 0.3. Determine, taking centrifugal tension into account, width of the belt required. It is given that the belt thickness is 9.5 mm, density of its material is 1.1 Mg/m^3 and the related permissible working stress is 2.5 N/mm^2. (Note: 1 Mg = 1×10^6 gm = 1000 kg)

(**Ans.** $b = 14.3$ cm)

12.7 A compressor, requiring 89.5 kw, is to run at about 250 r.p.m. The drive is by V-belts from an electric motor running at 750 r.p.m. The diameter of the pulley on the compressor shaft must not be greater than 1 m while the centre distance between the pulleys is limited to 1.75 m. The belt speed should not exceed 1600 m/min.

Determine the number of V-belts required to transmit the power if each belt has a cross-sectional area of 3.75 cm^2 and weighs 0.00981 N/cm^3 and has an allowable tensile stress of 245.2 N/cm^2. The groove angle of the pulley is 35°. The coefficient of friction between the belt and the pulley is 0.25. Calculate also the length required of each belt.

(**Ans.** No. of belts = 6; $L = 5.655$ cm)

12.8 As a rule the difference between the tight and slack side tensions for a leather belting should not exceed 98.1 N/cm of width for a belt 0.5 cm thick. Find maximum stress in the belt. Given,

Angle of lap = 170°; $\mu = 0.25$ and belt speed = 1000 m/min.

Density of leather may be taken as 1×10^{-3} kg/cm^3 (**Ans.** $\sigma_{max} = 402.3$ N/cm^2)

12.9 A V-belt having a lap of angle 180° has a cross-sectional area of 2.5 sq. cm and runs in a groove of included angle 45°. The density of the belt is 0.0015 kg/cm^3, the maximum stress is limited to 392.3 N/cm^2 the coefficient of friction being 0.15.

Find the maximum power that can be transmitted, if the wheel has a mean diameter of 30 cm and runs at 1000 r.p.m. (**Ans.** 9.42 kw)

12.10 A small generator is driven by means of a V-belt which has a total angle of 60° between the faces of the *V*. The angle of lap of the pulley is 120° and the mean radius of the belt as it passes round the pulley is 50 mm. If $\mu = 0.2$ and the mass of the belt is 0.45 kg/m, find the tension in each side of the belt when 750 W is being transmitted at a pulley speed of 1800 r.p.m. (**Ans.** 180 N; 100.5 N)

12.11 A belt drive consists of a V-belt working on a grooved pulley, with an angle of lap of 160°. The cross-sectional area of the belt is 650 mm^2, the groove angle is 30° and $\mu = 0.1$. The density of the belt material is 1 mg/m^2 and its maximum safe stress is 8 MN/m^2 of cross-section.

Derive an expression for the ratio of the tensions on the two sides of the drive when the belt is about to slip.

Calculate the power that can be transmitted at a belt speed of 25 m/s.

(**Ans.** 79 kw)

12.12 A pulley is driven by a flat belt, the angle of the lap being 120°. The belt is 100 mm wide by 6 mm thick and has a mass of 1 Mg/m². If μ = 0.3 and the maximum stress in the belt is not to exceed 1.5 MN/m², find the greatest power which can be transmitted and the corresponding speed of the belt. (*Ans.* 6.265 kw, 22.36 m/s)

12.13 Following particulars apply to one pulley of a rope drive between two parallel shafts. Effective pulley dia.: 1.5 m; Total angle of groove: 45°. Minimum angle of lap: 180°; Maximum permitted load per rope: 650 N

Mass of rope per m run: 0.45 kg; coeff. of friction: 0.25

(a) find the power transmitted per rope at a pulley speed of 200 r.p.m, if centrifugal tension is neglected.

(b) find the pulley speed when centrifugal tension accounts for half the permitted load in the rope, and the power which can be transmitted at that speed.
(*Ans.* 8.9 kw; 342 r.p.m; 7.62 kw)

12.14 A capstan and rope are used in railway goods yard for moving trucks. The capstan runs at 50 r.p.m. The rope from the line of trucks makes 2.75 turns round the capstan at a radius of 20 cm, and the free end of the rope is pulled with the force of 150 N. Calculate the pull exerted on the trucks, the power taken by the trucks and the power supplied by the capstan. Take μ = 0.25. (*Ans.* 11240 N; 11.78 kw; and 11.6 kw)

12.15 Five turns of a rope of 2 cm diameter are wound round the post. If the weight of 266983 N is to be supported at one end of the rope, calculate the required pull at the other end. The coefficient of friction between the rope and post surface is 0.285.
(*Ans.* 38.25 N)

12.16 Derive the conditions for maximum power to be transmitted by a belt drive. What maximum power can be transmitted per sq. cm of cross-section, if the tension in the belt is not to exceed 245.17 N/cm² and the ratio of tensions in the tight and slack side 1.8. Assume weight of 1 cu.cm of belt as 0.01079 N.
(*Ans.* Power = 2.0 kw; v = 27.26 m/s) (DAV Indore, 1987; 1992)

12.17 Deduce an expression for the tension in a belt due to centrifugal force and prove that the velocity at which the maximum power can be transmitted, is given by $\sqrt{\dfrac{Tg}{3w}}$, where

T is the maximum tension on tight side and w is the weight of belt per metre length. A rough rule for leather belting is that the difference between tight and slack side tensions should not exceed 98 N/cm of the width of belt for a belt of 4.75 rnm thickness. If this rule is applied under following conditions, what is the maximum stress on the tight side of the belt? Angle of lap = 170 degrees, coeff. of friction = 0.25; belt speed = 200 m/min; density of leather = 0.0011 kg/cu.cm.

(*Ans.* 408 N/cm²) (DAV Indore, 1982; 1994)

Chapter 13

BRAKES AND DYNAMOMETERS

13.1 INTRODUCTION

A brake is a machine element whose primary purpose is to absorb kinetic energy by way of frictional resistance. While in majority of cases, brakes are used to absorb kinetic energy by stopping or slowing down some moving part, there are brakes which are used to withhold conversion from potential to kinetic energy of the objects as they are lowered by hoists, elevators, etc. Other applications of brakes include holding a body in the state of rest or of uniform motion against accelerating forces/couples and preventing an unwanted reversal of the direction of rotation. The major functional difference between a clutch and a brake is that unlike a brake, which connects a moving member to a stationary frame, a clutch connects a rotating member to another member that is free to rotate. It must be noted, however, that clutch is also required to connect a rotating member to a stationary one, when the latter is to be set into motion. However, dissipation of heat due to energy absorbed is always a more serious problem with brakes rather than with clutches.

The purpose of dynamometers, on the other hand, is to measure forces or couples which tend to change the state of rest or uniform motion. It is thus the purpose, rather than the principles of operation, which distinguishes one from other and a brake arrangement may be frequently used to serve as dynamometer by incorporating some force/couple measuring device.

13.2 CLASSIFICATION OF BRAKES

Brakes can be classified as under:

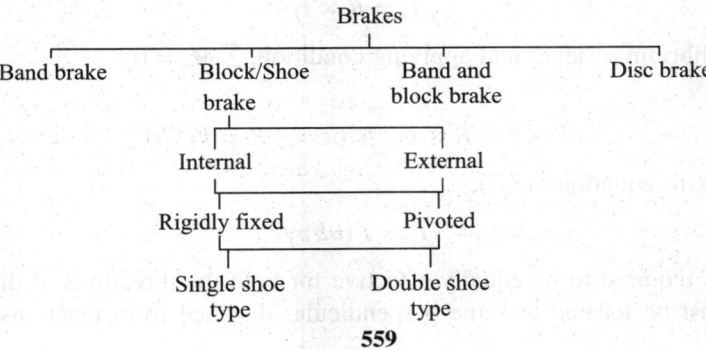

559

13.3 THE SIMPLE BLOCK OR SHOE BRAKE

This arrangement consists of a shoe or a block, either attached rigidly or pivoted to a lever with the help of which it is pressed against a rotating cylinder called brake drum. The shoe or block is shown pivoted in Fig. 13.1(a) and fixed in Fig. 13.1(b), to the brake lever. The block is made up of wood or metal that is sometimes lined with special friction material to ensure a higher coefficient of friction. The friction between the shoe and drum surfaces produces a friction torque that retards rotation.

(a) *Pivoted Shoe:* As the shoe is free to rotate about the pin C (Fig. 13.1a), resultant of normal reaction from drum and the frictional force on shoe must pass through the pin. This is shown by reaction R for counterclockwise drum rotation, and by R' for clockwise drum rotation. This feature enables more accurate determination of relation between applied force P and tangential friction force between the drum and the shoe. This arrangement, however, is mechanically very complicated.

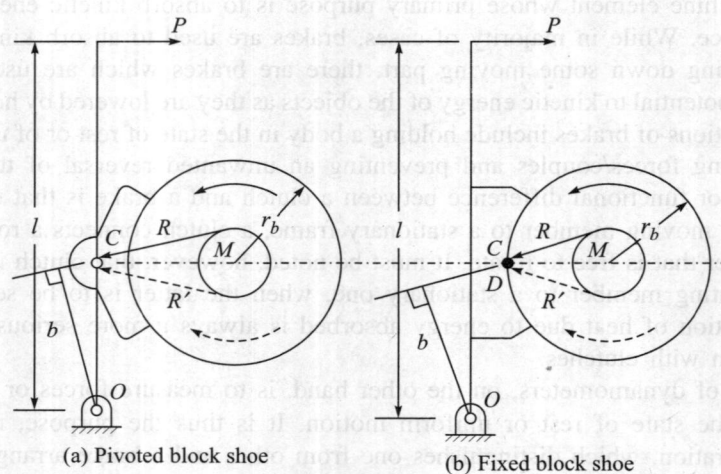

(a) Pivoted block shoe   (b) Fixed block shoe

Fig. 13.1

The resultant reaction R is tangent to the friction circle drawn with drum centre as the centre. Let r_f be the radius of friction circle. Then, friction torque is given by

$$T_f = R \times r_f \tag{13.1}$$

Considering equilibrium of lever and applying condition, $\sum M_o = 0$ we have

$$P \times l = R \times b \qquad \text{or} \qquad R = P(l/b) \tag{13.2}$$

Substituting for R in equation (13.1),

$$T_f = P(l/b)r_f \tag{13.3}$$

If the shoe is required to be equally effective for both the directions of drum rotation, the lever pivot O must be located at same perpendicular distance from reactions R and R'. This necessarily requires that the line joining pin centre C to lever pivot O must be perpendicular to the line joining pin centre C to drum centre M.

The radius of friction circle in equation (13.3) is given by

$$r_f = (\mu r_b) \qquad\qquad (13.4)$$

where $\qquad\qquad r_b$ = effective brake drum radius

Hence $\qquad\qquad T_f = \mu P r_b (l/b) \qquad\qquad (13.5)$

The free body diagram of pivoted shoe indicates that it is subjected to two forces: drum reaction R and the force applied by lever through pin C. Clearly, for preventing rotation of pivoted shoe, the reaction R must pass through pin so that forces are balanced. This condition is particularly more difficult to satisfy as the shoe continues to wear thus reducing the distance CM continuously.

(b) Block/Shoe Rigidly Attached to Brake Lever: A block/shoe attached rigidly to the brake lever is shown in Fig. 13.1(b). The friction torque is developed and its magnitude depends on the assumption in respect of pressure distribution between shoe and the drum, which ultimately decides the line of action of resultant reaction R. *When shoe is short, the pressure distribution between drum and shoe may be assumed to be uniform. With this assumption, the line of thrust R may be assumed to pass through point D—the point of symmetry of the shoe contact surface.* Thus, as shown in the figure, the resultant reaction R is placed tangent to friction circle on top side for c.c.w. rotation of drum and on the lower side for c.w. rotation of drum. In both the positions, the resultant reaction passes through the point D. With the above assumption therefore, the friction torque is again given by equation (13.1).

13.4 SHORT-SHOE BRAKES (CONDITION OF SELF ENERGIZATION)

When the contacting surface of a shoe subtends a small angle, say 45° of less, at the drum centre, the shoe is considered to be a short shoe. Conversely, when contacting surface of a shoe subtends on angle $\theta > 45°$ at the drum centre, the shoe is considered to be a long shoe. The line of demarcation of $\theta = 45°$ between the two categories is rough and is meant only for general guidance. For $\theta \leq 45°$ assumption of uniform pressure intensity between contacting surfaces of drum and shoe introduces a smaller error in computing friction torque which can be neglected. Assumption of 'small shoe' simplifies the problem of computing friction torque to a great extent.

In order to understand the effect of position of lever pivot in relation to drum centre and also the effect of direction of drum rotation on friction torque, consider Fig. 13.2. The figure shows three distinct locations for lever pivots that are spaced at a distance of c from one another. The pivot O_2 lies on tangent drawn to the periphery of drum at the mid-point of contacting surface of shoe. The pivot point O_1 lies exactly above O_2 while pivot position O_3 lies exactly below O_2.

Let P = effort applied at the end of lever for applying brake,

$\qquad \mu$ = coefficient of friction between the shoe and the drum,

$\qquad R$ = normal reaction from drum in radial direction on shoe, and hence on lever,

$\qquad r_b$ = radius of brake drum,

$\qquad l$ = distance of line of action of effort P from lever pivot, and

$\qquad a$ = distance of the line of action of normal reaction R from the lever pivot.

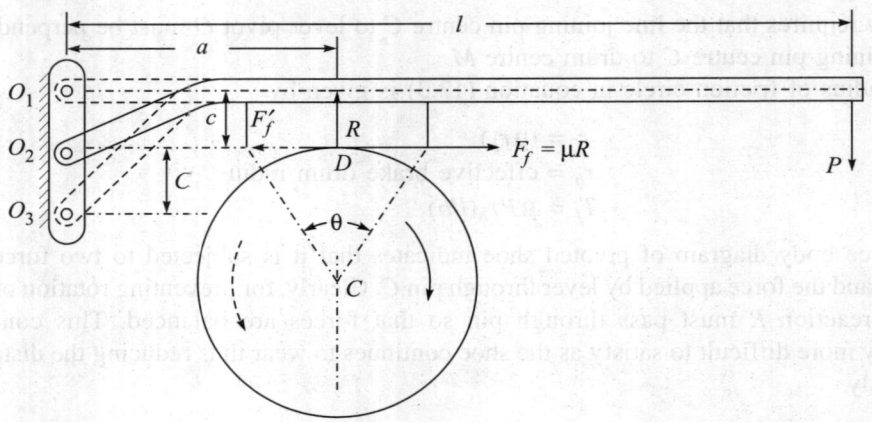

Fig. 13.2 Effect of lever pivot position & direction of drum rotation.

For Clockwise Rotation of Drum

For rotation of brake drum in c.w. direction, the normal reaction R and friction force μR are shown by the solid line in Fig. 13.2. Considering free body diagram of brake lever, the equilibrium of lever requires that equation $\sum M_o = 0$ must be satisfied for each of the three pivot positions. Thus, for pivot position at O_1,

$$Pl = Ra + (\mu R)c \qquad \text{or} \qquad P = (a + \mu c)R/l \qquad (13.6)$$

For pivot position at O_2, as $c = O$,

$$Pl = Ra + \mu R(c) \qquad \text{or} \qquad P = R(a/l) \qquad (13.7)$$

and for pivot position at O_3,

$$Pl = Ra - \mu Rc \qquad \text{or} \qquad P = (a - \mu c)R/l \qquad (13.8)$$

A comparison of equations (13.6), (13.7) and (13.8) shows that pivot position O_3 is best for c.w. drum rotation, as it requires smallest effort P to generate required frictional force F_f and required braking torque therefore.

Note that in the corresponding equilibrium equation, the frictional force μR and effort P produce moments about O_3 in the same sense and are therefore additive. Such a condition in which frictional force helps in applying the brake is termed condition of self energization and the brake is called self-energizing or self-actuating. A limiting condition occurs in equation (13.8) when $a = \mu c$. When $a = \mu c$ the effort P required to apply the brake is zero. This may be erroneously considered to be an advantage but in reality the brake may 'grab' and result in uncontrolled braking. With such brakes, once contact is established between shoe and drum, the brake applies of its own accord. Such condition goes under the name *self locking* and the brake is called *self-locking brake*.

For Counterclockwise Rotation of Drum

For rotation of brake drum in counterclockwise direction, the direction of normal reaction remains same but the friction force μR now acts in the direction indicated by dashed line in

Fig. 13.2. As before, considering free body diagram of brake lever and applying equilibrium condition $\sum M_o = 0$,

for pivot position at O_1,

$$Pl = Ra - (\mu R)c$$

or
$$P = (a - \mu c)R/l \qquad (13.9)$$

for lever-pivot position at O_2, as $c = 0$,

$$Pl = Ra + \mu R(O)$$

or
$$P = R(a)/l \qquad (13.10)$$

and, for pivot position at O_3,

$$Pl = Ra + (\mu R)c$$

or
$$P = (a + \mu c)R/l \qquad (13.11)$$

As can be easily seen from the above equations, self-energization exists for lever pivot position O_1.

Conclusion: Conclusion of the above discussions may be summarized as under:

1. The brake is equally effective for both the directions of drum rotation, requiring same actuating force for a given frictional torque, when the lever-pivot is at O_2.
2. Pivot position O_1 ensures condition of self-energization for counterclockwise sense of drum rotation. The actuating force required for producing a given frictional torque for clockwise sense of drum rotation is larger than that required in (1) above.
3. The pivot position O_3 ensures self-energization condition for clockwise sense of drum rotation, while the actuating force P required for operating brake for c.c.w. sense of drum rotation is larger than that required in (1) above.

Thus the pivot positions O_1 and O_3 have the advantage in that though they require considerable actuating force for one direction of drum rotation, they can be designed to be self-locking for the reversed direction of rotation. Such an arrangement goes under the name '*back-stop*', which is useful in preventing reversal in the direction of movement of load, being lifted, in the event of sudden power-failure. In the event of accidental power failure when gravitational forces cause reversal in direction of motion and self locking becomes effective and possible accidents can be prevented.

Equations (13.8) and (13.9) indicate that brakes can be designed as self-energized brakes, avoiding self-locking, for one direction of rotation. Such applications where unidirectional rotations are expected such brakes provide larger frictional torque for a given actuating force capacity. Needless to say that increased braking capacity of such brakes for one direction of rotation is at the cost of braking capacity in the reversed direction of rotation.

13.5 DOUBLE BLOCK BRAKES

A double shoe brake is used for—

   (i) decreasing radial thrust on bearings,

(ii) increasing braking capacity, and

(iii) reducing heat generated per unit area.

In general, both the shoes are not self energizing for a given direction of drum rotation. The normal reactions of drum on left and right shoe are also not equal. Usually, the method of applying brakes ensure same effort P on the two levers as shown in Fig. 13.3. For equilibrium of the two levers, we must have

$$\sum M_{OL} = 0 \quad \text{and} \quad \sum M_{OR} = 0$$

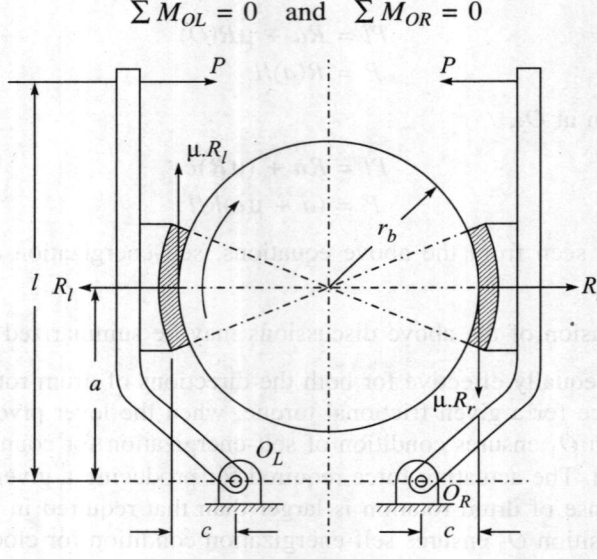

Fig. 13.3

Here, equations (13.9) and (13.10) can be simplified as,

$$R_l = \frac{P(l)}{(a - \mu c)}$$

and

$$R_r = \frac{P(l)}{(a + \mu c)} \qquad (13.12)$$

Thus the total frictional torque,

$$T_f = (\mu R_l)r_b + (\mu R_r)r_b$$

or

$$T_f = (R_l + R_r)\mu r_b \qquad (13.13)$$

Substituting for R_l and R_r from equations (13.11) and (13.12),

$$T_f = \left[\frac{P(l)}{(a - \mu c)} + \frac{P(l)}{a + \mu c} \right] \mu r_b$$

or

$$T_f = \left[\frac{2a}{(a^2) - (\mu c)^2} \right] (\mu P l) r_b \qquad (13.14)$$

EXAMPLE 13.1 (SL) A brake drum of 35 cm radius with a single fixed shoe arrangement is as shown in Fig. 13.2 with lever pivot at O_1 and distance $c = 36$ mm. The effort P is applied at a distance of $l = 90$ cm while the thrust on the shoe acts at a distance of $a = 35$ cm from the lever pivot. The brake drum sustains a torque of 245 N·m at 480 r.p.m. If coefficient of friction between shoe and brake drum is 0.28, find (i) the total normal force on the shoe and the frictional force F, (ii) the effort P required to apply the brake for clockwise and anticlockwise directions of rotation of the braking drum, (iii) condition to make the brake self-locking, and (iv) rate of heat generation in J/min.

Solution: (i) Assuming the brake to be short-shoe type, the frictional torque on drum,

$$T_f = 245 = (\mu R) \times r_b = (0.28)R \times 0.35$$

Therefore $$R = \frac{245}{0.28 \times 0.35} = 2500 \text{ N}$$ **Ans.**

The frictional force $= \mu R = 0.28 \times 2500 = 700$ N **Ans.**

   (ii) Considering the free body diagram of the lever for (c.c.w.) sense of rotation and using equilibrium equation: $\sum M_{o_1} = 0$

$$P(90) = R(a) - (\mu R)c$$

Therefore $$P = \frac{2500(a - \mu c)}{90}$$

or $$P = \frac{2500(35 - 3.6 \times 0.28)}{90} = 944.22 \text{ N}$$ **Ans.**

For clockwise rotation of drum,

$$P = \frac{2500(35 + 3.6 \times 0.28)}{90} = 1000.22 \text{ N}$$ **Ans.**

(Note that the condition of self-energization here has resulted in reducing actuating force by 56 N)

   (iii) For self-locking, which is possible here only when drum rotates clockwise, is obtained when

$$a - \mu c \leq 0 \quad \text{or} \quad c \geq \frac{35}{0.28} \quad i.e., \ c \geq 125 \text{ cm}$$

Work lost in friction resulting in heat generation, = Torque $\times 2\pi N$

$$= (\mu R)r_b \times 2\pi N = 0.28 \times 2500 \times 0.35 \times 2\pi \times 480$$
$$= 738,902.6 \text{ J/min}$$ **Ans.**

EXAMPLE 13.2 (SI) A bicycle and rider of mass 100 kg are travelling at a speed of 16 km/hr on a level road. A brake is applied to the rear wheel which is 0.9 m in diameter and this is the only resistance acting. How far will the bicycle travel and how many turns will it make before it comes to rest? The pressure applied on the brake is 100 N and $\mu = 0.05$ (Oxford Univ.).

Solution: The peripheral speed $v = \dfrac{1600}{3600} = 4.444$ m/s

The frictional force on rim $= \mu R = 0.05 \times 100 = 5$ N

Let s be the distance moved through before stopping as the brake is applied. The work lost in friction,

$$= (\mu R) \times s = 5s$$

The K.E. of the bicycle and rider to be absorbed,

$$= \frac{1}{2}mv^2 = \frac{1}{2}(100)(4.444)^2 = 987.65 \text{ N·m}$$

Equating the work lost in friction to K.E. destroyed,

$$5s = 987.65 \quad \text{Hence } s = 197.53 \text{ m} \qquad \textbf{Ans.}$$

Number of revolutions made by bicycle wheel before stopping,

$$= \frac{197.53}{\pi \times 0.9} = 69.86 \text{ revolutions} \qquad \textbf{Ans.}$$

13.6 LONG SHOE BRAKES

Consider Fig. 13.4 which shows an external shoe brake. The shoe is rigidly attached to the brake lever, which is pivoted at C to the frame. The actuating force P acts to rotate the lever in c.c.w. direction about the lever pivot C. If the drum were not there, a rotation of lever through an angle α would mean a displacement of shoe from a position, shown by solid lines to the one shown by dotted lines. Since the drum is there, the movement from solid to the broken line position (shown exaggerated) would be because of deformation in lining. Everything, except brake lining, is assumed to have infinite rigidity, and therefore the new position of brake shoe is obtained by entire deformation in the lining material. Figure 13.4 illustrates this deformation by a point A on shoe which moves to position A', by way of rotation about pivot C. Thus the compression of

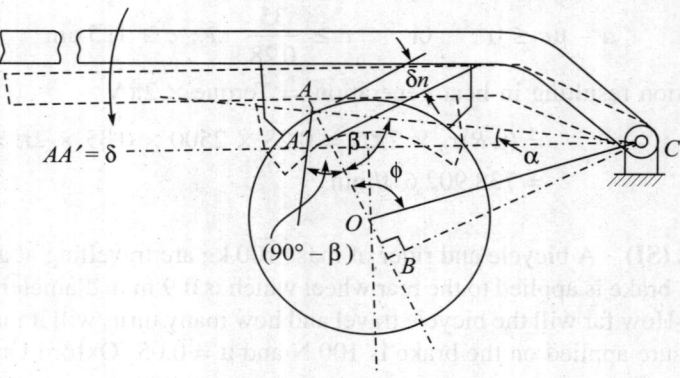

Fig. 13.4 Long Shoe Brake.

the lining $AA' = \delta$. Let δn be the component of δ when resolved along radial direction AOB. Thus as seen in figure, for small angles of rotation α,

$$\delta n = \delta \cos(90 - \beta)$$

or
$$\delta n = \delta \sin\beta \qquad (13.15)$$

Also
$$\delta = AA' = (CA)\alpha \qquad (13.16)$$

Therefore from equations (13.15) and (13.16),

$$\delta n = (CA)\alpha \sin\beta \qquad (13.17)$$

To express the distance CA in terms of other known (fixed) parameters, we note from right angled triangle CAB that

$$CA = (CB/\sin\beta) \qquad (13.18)$$

where B is the foot of the perpendicular from pivot C on radius AOB.

Join the drum centre O to C and let $\angle AOC = \phi$. Also let $OC = L$.

Then, from right angled triangle COB,

$$CB = L\sin(180 - \phi)$$
$$CB = L\sin\phi \qquad (13.19)$$

Substituting for CA in equation (13.17) from equations (13.18) and (13.19),

$$\delta n = \frac{L\sin\phi}{\sin\beta} \times \alpha \sin\beta$$

or
$$\delta n = L\alpha\sin\phi \qquad (13.20)$$

Just as in clutches, in brakes too the major consideration is wear. The assumption of uniform rate of wear implies that wear is proportional to the product (pv), which represents work of friction. As the drum is cylindrical, velocity v is constant everywhere on drum surface. Hence, pressure distribution can be established solely by considering motion of the shoe relative to the drum.

Brake linings, in general, do not have linear stress-strain characteristic. However, the error introduced by assuming a linear relationship is small in comparison to other uncertainties, particularly the value of coefficient of friction. Thus, based on linear relation between pressure p and normal deflections, we have

$$p = K\delta n \qquad (13.21)$$

where
$$K = \text{constant of proportionality}$$

Substituting for deflection δn from equation (13.20) in (13.21),

$$p = KL\alpha\sin\phi \qquad (13.22)$$

Equation (13.22) shows that normal pressure p is maximum when $\sin\phi$ attains its maximum value $(\sin\phi)_{max}$. Thus, from equation (13.22),

$$p_{max} = KL\alpha(\sin\phi)_{max} \qquad (13.23)$$

or
$$KL\alpha = \frac{p_{max}}{(\sin\phi)_{max}} \qquad (13.24)$$

Substituting for $KL\alpha$ from equation (13.24) in (13.22),

$$p = p_{max}\left[\frac{\sin\phi}{(\sin\phi)_{max}}\right] \tag{13.25}$$

In majority of practical cases, the range of included angles $\phi_1 \le \phi \le \phi_2$, includes angle $90°$, and therefore $(\sin\phi)_{max} = 1$. Thus

$$p = p_{max}\sin\phi \tag{13.26}$$

However, when the range $\phi_1 \le \phi \le \phi_2$ does not include angle $90°$, $(\sin\phi)_{max} \ne 1$ and relation in equation (12.25) must be used.

Braking Torque: Consider free body diagram of the brake lever of the long shoe brake shown in Fig. 13.5. Consider an elemental shoe area at A, which is located at an angle ϕ w.r. to reference line CO and which subtends an angle $d\phi$ at the centre of the drum. For c.c.w. rotation of drum, the normal force of reaction and the frictional force are as shown at Fig. 13.5. For the equilibrium of brake lever, the algebraic summation of moments of all forces about lever pivot C is

$$M_p + M_f + M_n = 0 \tag{13.27}$$

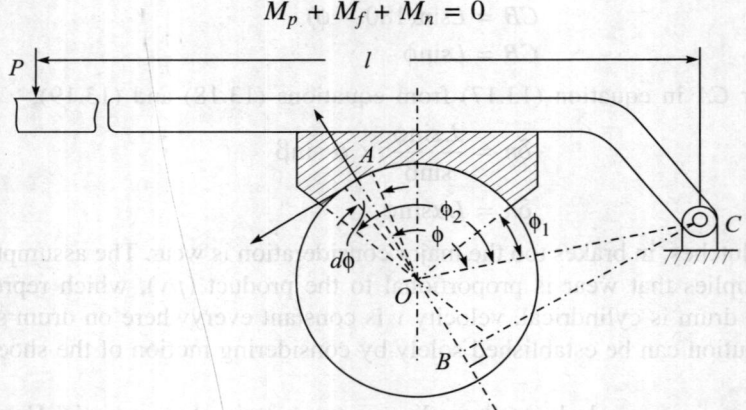

Fig. 13.5 Force analysis in Long shoe brake.

where M_p is the moment due to actuating force P at C, M_n is the summation of moments at C due to normal forces of reaction at each elemental area, and M_f is the summation of moments at C due to frictional forces at each elemental area.

Now

$$M_p = P(l) \tag{13.28}$$

and

$$M_n = -\int_{\phi_1}^{\phi_2} (pbr_b d\phi)L\sin\phi \tag{13.29}$$

Substituting for p from equation (13.26), and realising that like maximum normal pressure p_{max}, the drum parameters b, r_b and distance L are constant, equation (13.29) leads to

$$M_n = -\frac{(p_{max})br_bL}{(\sin\phi)_{max}}\int_{\phi_1}^{\phi_2}\sin^2\phi\,d\phi$$

which, upon integration, simplifies to

$$M_n = -\frac{(p_{max})br_bL}{4(\sin\phi)_{max}}[2(\phi_2 - \phi_1) - \sin2\phi_2 + \sin2\phi_1] \qquad (13.30)$$

Similarly

$$M_f = \int_{\phi_1}^{\phi_2} \mu p(br_b\,d\phi)[r_b + L\cos(180 - \phi)]$$

Substituting for p from equation (13.26) and simplifying,

$$M_f = \frac{\mu p_{max}br_b}{(\sin\phi)_{max}} \int_{\phi_1}^{\phi_2} [r_b\sin\phi - L\sin\phi\cos\phi]d\phi$$

or

$$M_f = \frac{\mu p_{max}br_b}{4(\sin\phi)_{max}}[4r_b(\cos\phi_1 - \cos\phi_2) + L(\cos2\phi_2 - \cos2\phi_1)] \qquad (13.31)$$

Self Energization

Above derivations for M_p, M_n and M_f are not completely general. The quantities M_p and M_n offer no difficulties in judging direction of moments about C. As a rule the moments M_p and M_n will always be in opposite direction, and their direction can be determined by inspection. The direction of frictional moment M_f depends not only on the geometry of the brake but also on the direction of rotation. For instance equation (13.31) is derived based on elemental frictional moment for c.c.w. drum rotation. Thus a positive value of M_f from eqn. (13.31) means that M_f acts in a c.c.w. direction and a negative value for M_f means that it acts in c.w. direction. If the direction of rotation of the drum is reversed, the right-hand side of equation (13.31) should be multiplied by 1, to maintain a consistent sign convention.

Brake Torque: The frictional torque on drum due to frictional force on the element at A,

$$dT_f = \mu(br_b)\,pd\phi r_b$$

The entire shoe area may be considered to consist of a large number of such elements and as such,

$$T = \mu b(r_b)^2 \int_{\phi_1}^{\phi_2} pd\phi$$

Substituting, for p from equation (13.26)

$$T_f = \frac{\mu b(r_b)^2 p_{max}}{(\sin\phi)_{max}} \int_{\phi_1}^{\phi_2} \sin\phi d\phi$$

$$T_f = -\frac{\mu br_b^2 p_{max}}{(\sin\phi)_{max}}(\cos\phi_2 - \cos\phi_1) \qquad (13.32)$$

Conclusion

It is clear from equation (13.26) that as $\sin\phi$ approaches $(\sin\phi)_{max}$, the pressure intensity p approaches (p_{max}) value. When $\phi_1 \leq 90°$ and $\phi_2 \geq 90°$, $(\sin\phi)_{max} = 1$ and the maximum pressure occurs at $\phi = 90°$. But when the range ϕ_2 to ϕ_1 does not include angle $\phi = 90°$, and $\phi_2 < 90°$, $\sin\phi_2$ equals $(\sin\phi)_{max}$.

It follows that friction material located at and near the point, where (sinϕ) has its largest possible value, is most effectively utilised in producing friction torque on drum. Similarly, friction lining provided at $\phi = 0$ or in regions where ϕ is very small, is used most uneconomically.

13.7 LONG SHOE BRAKES (SHORTER METHOD)

Figure 13.6 shows a cylindrical drum rotating with angular velocity ω radians per second in c.c.w. direction. When the shoe is pressed against the drum, the wear mainly takes place on the brake lining. The wear takes place on shoe lining in such a way that they retain their cylindrical shape. This necessarily requires that all points of contact on shoe are displaced by a constant distance $BA = \delta$ parallel to the direction of force R. Hence, wear in a direction normal to the drum, *i.e.*, $CA = \delta n$, is given by

$$\delta n = \delta \cos\phi \tag{13.33}$$

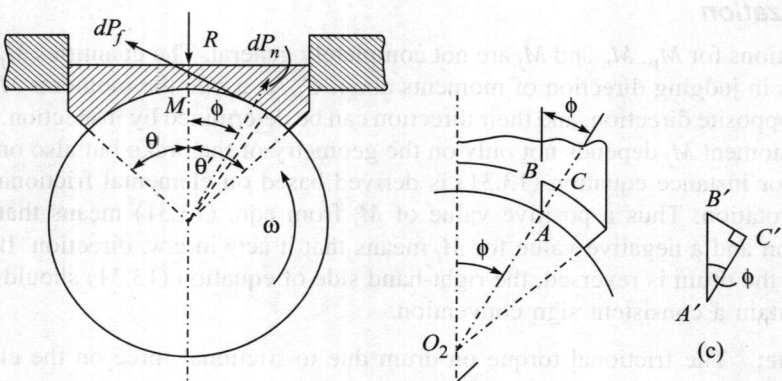

Fig. 13.6 A Long Shoe Subjected to Normal Force *R*.

where amount of wear $BA = \delta$ is constant. Since pressure is not uniform over the contacting surface, a suitable alternative assumption must be made to establish an expression for pressure distribution across shoe surface. Thus, let us assume normal wear δn to be proportional to the work of friction. Again, as peripheral drum velocity is constant everywhere, the work of friction is proportional to the normal pressure p. Hence,

$$p \propto \delta n$$
or
$$p \propto \delta \cos\phi$$
or
$$p = C \cos\phi \tag{13.34}$$

The wear component δ, being constant, has been merged with the constant of proportionality. Thus,

$$C = \delta \times \text{constant of proportionality}$$

It follows from equation (13.34) that at the point of symmetry of shoe, where $\phi = 0$, the pressure is maximum and the pressure decreases towards outer edges of shoe as ϕ increases.

For an included angle of $2\phi = 45°$, as $\cos 22.5 = 0.92$, it follows that the pressure variation in a short shoe can be neglected.

In order to derive expressions for friction torque and effective frictional force, we proceed as follows.

Let

b = width of brake drum

2θ = angle subtended at the drum centre by shoe

μ = coefficient of friction

dA = elemental area on shoe surface, located at ϕ from the line of symmetry, which subtends angle $d\phi$ at the drum centre

p = normal pressure on elemental area dA

r_b = drum radius

Then
$$dA = b(r_b d\phi)$$

The elemental normal force on this element,
$$dP_n = pdA = pb(r_b)d\phi \tag{13.35}$$

Hence, the elemental frictional force,
$$dF_f = \mu dP_n = \mu pb(r_b)d\phi \tag{13.36}$$

The contribution from the element to the braking torque,
$$dT_b = r_b \times dF_f = \mu pb(r_b^2)d\phi \tag{13.37}$$

Substituting for p from equation (13.34) and integrating, the braking torque is
$$T_b = \int_{-\theta}^{\theta} dT_b = \mu br_b^2 \int_{-\theta}^{\theta} C\cos\phi d\phi$$

or
$$T_b = (\mu br_b^2)C(\sin\theta + \sin\theta)$$

or
$$T_b = (2\mu br_b^2)C\sin\theta \tag{13.38}$$

Again, resolving the elemental normal force of equation (13.35) along the direction of force R and equating R to it after integration,
$$R = \int_{-\theta}^{\theta} (dP_n)\cos\phi = br_b \int_{-\theta}^{\theta} p\cos\phi d\phi$$

Substituting for p from equation (13.34),
$$R = br_b C \int_{-\theta}^{\theta} \cos^2\phi d\phi$$

or
$$R = C(br_b)\left(\theta + \frac{1}{2}\sin 2\theta\right) \tag{13.39}$$

or
$$C = \frac{2R}{br_b(2\theta + \sin 2\theta)} \tag{13.40}$$

The resultant frictional force need not necessarily act at the point of symmetry M of the shoe. However, to simplify solution of problems on long shoes, a single equivalent force F_f acting at the point of symmetry on shoe will be assumed to exist and a suitable expression for the same will be established now.

Substituting for C from equation (13.40) in equation (13.38),

$$T_b = (\mu R r_b)\frac{4\sin\theta}{(2\theta + \sin2\theta)} \tag{13.41}$$

Assuming that a single equivalent frictional force F_f acting at the point of symmetry M produces this torque, we have

$$F_f \times r_b = T_b$$

Using this expression to eliminate T_b from equation (13.41),

$$F_f = R\left(\frac{4\mu\sin\theta}{2\theta + \sin2\theta}\right) \tag{13.42}$$

The term $(4\mu\sin\theta)/(2\theta + \sin2\theta)$ in equation (13.42) may now be looked upon as 'equivalent coefficient of friction' and designated as μ'.

Hence $$F_f = \mu' R \tag{13.43}$$

where $$\mu' = \mu\left(\frac{4\sin\theta}{2\theta + \sin2\theta}\right) \tag{13.44}$$

Thus, in the case of a long block shoe as shown in Fig. 13.7, we have for the equilibrium of lever,

or $$P(l) = R(a) + F_f(c)$$
$$P(l) = R(a + c\mu')$$

Therefore $$R = \frac{P(l)}{(a + \mu'c)} \tag{13.45}$$

Therefore Frictional torque $= \mu' R r_b$ is given using equation (13.45) as

$$T_b = (\mu' r_b)\frac{Pl}{(a + \mu'c)}$$

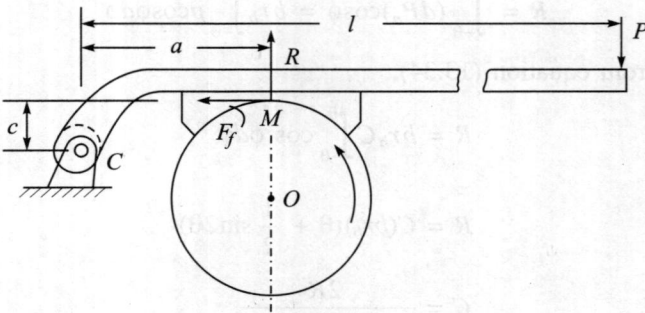

Fig. 13.7 Force Analysis in a Long Shoe Brake.

Therefore   Frictional (braking) torque $T_b = \dfrac{\mu' P r_b l}{(a + \mu' c)}$ (13.46)

Note that if the brake is self energizing, the term $(a + \mu' c)$ will be replaced by $(a - \mu' c)$ in equation (13.46). As in evident from above expression this will increase braking capacity.

13.8 INTERNALLY EXPANDING SHOES

Self propelled vehicles nowadays invariably use internally expanding shoe-brakes. These brakes have replaced band brakes, which were used in earlier designs of automobiles. The band brakes were discarded as it was difficult to operate them efficiently in environment exposed to dirt and water. A typical internally expanding shoe-brake arrangement is shown in Fig. 13.8. It consists of two semi-circular segments of shoes which are lined externally with friction material. These shoes are pivoted at convenient points like O_L and O_R and for applying the brakes, the ends A and B are moved away from each other mechanically or hydraulically, against a spring force. The friction lining, usually of Ferodo, is pressed against inner flange of the drum when the brake-shoes are expanded. The force F can be applied using either a cam or a piston moving inside a hydraulic cylinder, under pressurised oil received from a master cylinder.

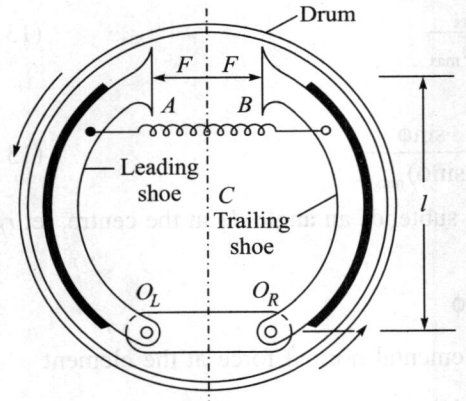

Fig. 13.8   An Internally Expanding Shoe Brake.

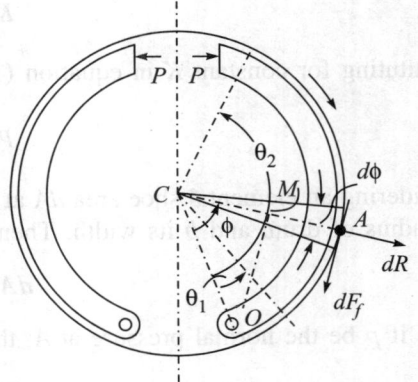

Fig. 13.9   Force Analysis in Internal Shoe Brake.

For the type of the arrangement shown in figure, only one of the two shoes will be self-energizing for a given direction of rotation, which increases effectiveness. When not in operation, the outer diameter of friction lining is less than the inner flange diameter and the wheel is free to rotate. The portion of shoe nearer to pivot is called *heel* and the usual practice is to omit the friction material for a short distance away from the heel. This avoids interference and leads to a more economical and effective utilization of friction lining since material near heel contributes very little to the friction torque. Discussions below refer to Fig. 13.9.

As in the case of external shoes, here too we assume that normal wear is proportional to the work of friction (pv). And as the peripheral velocity v is constant everywhere on drum surface, we have

$$\delta n \propto P$$ (13.47)

Again, normal wear at any element A on the shoe is proportional to the displacement in radial direction CA at A. Let M be the foot of perpendicular from O on CA. As the brake lever is pivoted at O, it follows that wear in normal direction at A will be proportional to OM. Let $OC = L$, Then

$$\delta n \propto OM$$

or
$$\delta n \propto L\sin\phi \qquad (13.48)$$

where ϕ is the angle at which element A is located w.r. to reference line CO. From equations (13.47) and (13.48) it follows that,

$$p \propto L\sin\phi$$

And as distance OC is constant, we have

$$p = K\sin\phi \qquad (13.49)$$

where K is constant which takes into account distance OC and also the constant of proportionality.

Thus, if (p_{max}) be the maximum value of pressure occuring at the maximum value of $\sin\phi$,

$$p_{max} = K(\sin\phi)_{max}$$

or
$$K = \frac{p_{max}}{(\sin\phi)_{max}} \qquad (13.50)$$

Substituting for constant K in equation (13.49),

$$p = p_{max}\frac{\sin\phi}{(\sin\phi)_{max}} \qquad (13.51)$$

Considering an elemental shoe area dA at A, which subtends an angle $d\phi$ at the centre, let r_b be the radius of drum and b its width. Then,

$$dA = b(r_b)d\phi$$

Thus, if p be the normal pressure at A, then the elemental normal force at the element

$$dR = pb(r_b)d\phi \qquad (13.52)$$

Hence, the elemental frictional force at A,

$$dF_f = \mu dR = \mu pbr_d d\phi \qquad (13.53)$$

Substituting for pressure p in equations (13.52) and (13.53) from equation (13.51) and taking moments about fulcrum O,

$$M_n = \int_{\theta_1}^{\theta_2} (OM)dR = (br_b)L\left[\frac{p_{max}}{(\sin\phi)_{max}}\right]\int_{\theta_1}^{\theta_2}\sin^2\phi d\phi$$

or
$$M_n = \frac{br_bLp_{max}}{(\sin\phi)_{max}}\left[\frac{1}{2}(\theta_2 - \theta_1) - \frac{1}{4}(\sin2\theta_2 - \sin2\theta_1)\right]$$

Thus
$$M_n = \frac{br_b L p_{max}}{4(\sin\phi)_{max}} [2(\theta_2 - \theta_1) - (\sin 2\theta_2 - \sin 2\theta_1)] \qquad (13.54)$$

And from equation (13.53),

$$M_f = \int_{\theta_1}^{\theta_2} dF_f (r_b - L\cos\phi)$$

or,
$$M_f = (\mu b r_b) \frac{p_{max}}{(\sin\phi)_{max}} \int_{\theta_1}^{\theta_2} (r_b \sin\phi - L\sin\phi\cos\phi) d\phi$$

Thus
$$M_f = \frac{(\mu b r_b) p_{max}}{4(\sin\phi)_{max}} [4 r_b (\cos\theta_1 - \cos\theta_2) + L(\cos 2\theta_2 - \cos 2\theta_1) \qquad (13.55)$$

Readers may verify that equations (13.54) and (13.55) are identical to equations (13.30) and (13.31) for external shoes.

The braking torque for internal shoes is given by,

$$T_b = \int_{\theta_1}^{\theta_2} (dF_f) r_b = (\mu b r_b^2) \int_{\theta_1}^{\theta_2} p \, d\phi$$

And substituting for p from equation (13.51) and integrating

$$T_b = \frac{(\mu b r_b^2) p_{max}}{(\sin\phi)_{max}} (\cos\theta_1 - \cos\theta_2) \qquad (13.56)$$

With expressions for M_n and M_f so established, following equilibrium equations for the two shoes assist in solving problems on brakes.

For left-hand shoe brake (c.w. drum rotation), $Fl = M_n - M_f$

and, *for right-hand shoe*, $\quad Fl = M_n + M_f$

EXAMPLE 13.3 Assuming the block brake of example 13.1 to be a long shoe with included angle of 120°, find out (i) the total normal force on the shoe and the frictional force, (ii) effort P required to apply the brake for c.w. and c.c.w. direction of drum rotation, (iii) condition for the brake to become self locking.

Solution: For long shoe brake, braking torque

$$T_b = 245 \text{ N·m} = (\mu r_b) R \left(\frac{4\sin\theta}{2\theta + \sin 2\theta} \right)$$

As
$$2\theta = 120° = \frac{2}{3}\pi \text{ radians,}$$

We have
$$\frac{4\sin\theta}{2\theta + \sin 2\theta} = 1.17$$

Therefore
$$R = \text{normal force on shoe, is given by}$$

$$R = \frac{245}{(0.28)(0.35) \times 1.17} = 2136.75 \text{ N} \qquad \textbf{Ans.}$$

Therefore Equivalent single friction force $F_f = \mu'R$

where $\qquad \mu' = (0.28) \times \dfrac{4\sin\theta}{2\theta + \sin 2\theta} = 0.3276$

Therefore $\qquad F_f = \mu' \times 2136.75 = 699.999 \approx 700 \text{ N}$ **Ans.**

(ii) Considering Fig. 13.2, with lever pivot at O_1, c.c.w. rotation of drum makes the shoe self energizing and equilibrium equation for lever becomes,

$$Pl = M_n - M_f \quad \text{Therefore} \quad Pl = (R)a - (F_f)c$$

Therefore $\qquad P = \dfrac{(2136.75)(0.35) - (700)(0.036)}{0.9} = 802.96 \text{ N}$ **Ans.**

And for c.w. rotation of drum, when the shoe is not self-energized, equilibrium equation for lever is,

$$Pl = M_n + M_f$$

Therefore $\qquad P = \dfrac{(2136.75)(0.35) + (700)(0.036)}{0.9} = 858.96$ **Ans.**

(iii) For long shoe brake to become self-locking for c.c.w. direction of drum rotation,

$$M_n = M_f \quad \text{or} \quad a - \mu'c \le 0$$

i.e., $\qquad c \ge \dfrac{35}{(0.28)(1.17)} \quad \text{or} \quad c \ge 106.84 \text{ cm}$ **Ans.**

EXAMPLE 13.4 Calculate the value of maximum pressure p_{max}, the braking torque T_b and the power in kw for the brake shown in Fig. 13.10 The coefficient of friction is equal to 0.2. Assume the brake to have long shoe. Solve the problem using the method discussed in Section 13.6.

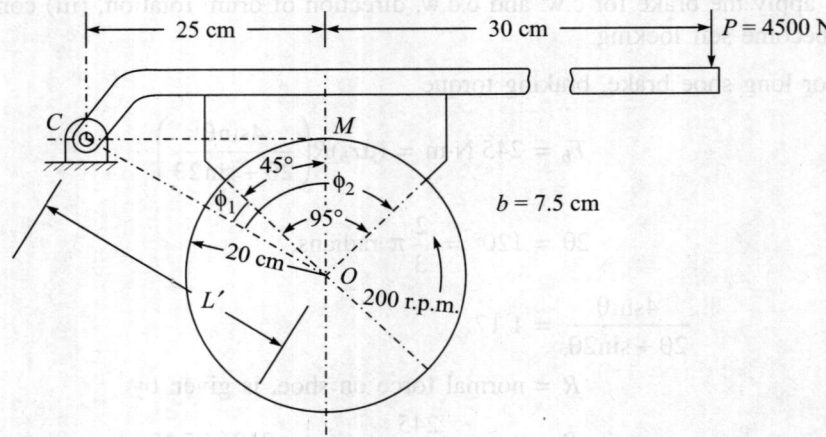

Fig. 13.10 A long shoe brake (External).

Solution: From the right angled triangle *OMC* of figure,

$$OC = \sqrt{(OM)^2 + (CM)^2}$$

or $$OC = \sqrt{(20)^2 + (25)^2}$$

Thus $$OC = 32.02 \text{ cm}$$

Also $$\angle COM = \tan^{-1}\left(\frac{25}{20}\right)$$

or $$\angle COM = 51.34°$$

Therefore $$\phi_1 = 51.34 - 45° = 6.34°$$

and $$\phi_2 = 6.34 + 95° = 101.34°$$

Therefore $$2(\phi_2 - \phi_1) = 2(101.34 - 6.34) = 190° = 3.316 \text{ rad}$$

Range of angle ϕ is $\qquad 6.34° < \phi < 101.34$ which includes angle $90°$

and hence $$(\sin\phi)_{max} = 1$$

From equation (13.30),

$$M_n = -\frac{(7.5)(20)(32.02)p_{max}}{4(1)}[(3.316) - \sin 202.68° + \sin 12.68]$$

$$M_n = (4708.25)p_{max}$$

Similarly, from equation (13.31),

$$M_f = \frac{(0.2)\,p_{max}\,(7.5)\,(20)}{4(1)}[4 \times 20\,(\cos 6.34° - \cos 101.34) + 32.02\,(\cos 202.68 - \cos 12.68)]$$

or $$M_f = (258.44)\,p_{max}$$

Positive value of M_f indicates that it acts c.c.w. Hence the moments M_n and M_f are additive in the moment equation. Thus

$$Pl = (4708.25 + 258.44)\,p_{max}$$

Therefore $$p_{max} = \frac{4500 \times 55}{(4708.25 + 258.44)} = 49.83 \text{ N/cm}^2 \qquad \textbf{Ans.}$$

The braking torque from equation (13.32)

$$T_b = \frac{(0.2)(7.5)(20)^2 \times 49.83}{1}[\cos 101.34 - \cos 6.34]$$

or $$T_b = 35594.0 \text{ N·cm} = 355.94 \text{ N·m}$$

Frictional power absorbed $= \dfrac{2\pi TN}{60 \times 1000} = \dfrac{355.94 \times 200 \times 2\pi}{60 \times 1000} = 7.455 \text{ kw}$ $\qquad \textbf{Ans.}$

EXAMPLE 13.5 Brake drum of a long shoe (external) brake is 17.5 cm in radius and 6.25 cm wide. The shoe is rigidly attached to a lever which is pivoted at C and the shoe subtends an angle of 90° at the drum centre O as shown in Fig. 13.11. The actuating force is located at a distance of 50 cm from lever pivot as measured along lever and 27.5 cm from the line of action of normal reaction. The lever pivot itself is located at a height of 15.0 cm above the horizontal centre line of drum. If the permissible value of $p_{max} = 0.7$ M.Pa, find corresponding actuating force and the friction power that the brake will absorb. Take $\mu = 0.25$.

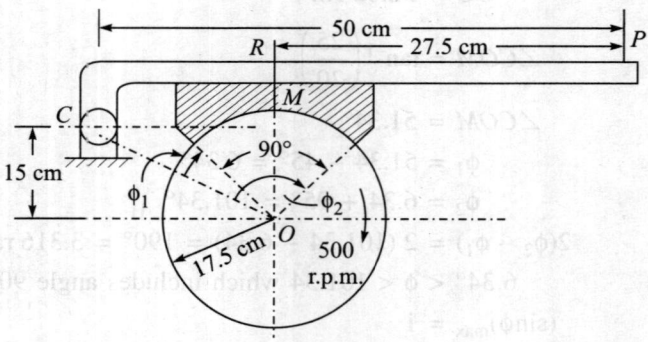

Fig. 13.11 An external long shoe brake.

Solution: From the geometry of figure,

$$OC = \sqrt{(50 - 27.5)^2 + (15)^2}$$

$$= 27.04 \text{ cm}$$

Also,

$$\angle COM = \tan^{-1}\left(\frac{50 - 27.5}{15}\right) = 56.3°$$

Therefore

$$\phi_1 = 56.3° - 45° = 11.3°$$

$$\phi_2 = 56.3 + 45 = 101.3°$$

Hence

$$2(\phi_2 - \phi_1) = 2(101.3 - 11.3) = 180° = \pi \text{ rad}$$

As the range of ϕ is $11.3° < \phi < 101.3°$; $(\sin\phi)_{max} = 1.0$

$$p_{max} = 0.7 \text{ MPa} = 70 \text{ N/cm}^2$$

Hence, from equation (13.30),

$$M_n = -\frac{(70)(6.25)(17.5)(27.04)}{4 \times 1.0} [\pi - \sin 202.6 + \sin 22.6]$$

or

$$M_n = 202376.42 \text{ N·cm} = 2023.76 \text{ N·m}$$

And from equation (13.31),

$$M_f = \frac{0.25 \times 70 \times 6.25 \times 17.5}{4 \times 1.0} [4 \times 17.5(\cos 11.3 - \cos 101.3) + 27.04(\cos 202.6 - \cos 22.6)]$$

Therefore

$$M_f = 15519.24 \text{ N·cm} = 155.19 \text{ N·m}$$

Since M_f turns out to be positive, the couple M_f acts c.c.w. about pivot. Thus the moment M_f opposes the moment M_n in equilibrium equation. Thus,

$$P \times 50.0 = M_n - M_f$$

Therefore $P = \dfrac{202376.4 - 15519.24}{50} = 3737.1 \text{ N}$ **Ans.**

And from equation (13.32), the braking torque,

$$T_b = -\frac{0.25 \times 6.25 \times (17.5)^2 \times 70}{1} \times [\cos 101.3 - \cos 11.3]$$

$$= 39410.2 \text{ N·cm} = 394.1 \text{ N·m}$$

Therefore   Power lost in friction $= \dfrac{2\pi \times 500 \times 394.1}{60 \times 1000} = 20.635 \text{ kw}$ **Ans.**

EXAMPLE 13.6 A double-shoe brake, as shown in Fig. 13.12, is capable of absorbing a torque of 1400 N·m. The diameter of the brake drum is 350 mm and the angle of contact for each shoe is 100°. If $\mu = 0.4$, find:

(a) spring force necessary to set brake
(b) the width of shoes if bearing pressure on lining material is not to exceed 0.3 N/mm^2

Solution: Assume the drum to rotate c.w. For this sense of rotation, only the left-hand shoe will be self energizing and hence for a given spring force S on both the levers (as shown in Fig. 13.12), the normal reactions R_L and R_R will not be equal.

Now, following the shorter method for long shoes, equivalent coefficient of friction

$$\mu' = \mu\left(\frac{4\sin\theta}{2\theta + \sin 2\theta}\right)$$

where $2\theta = 100° = 1.745$ rad and $\mu = 0.4$

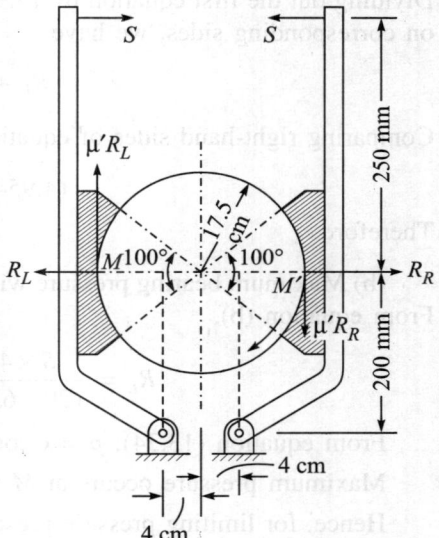

Fig. 13.12 A Double Shoe Brake.

Therefore $\mu' = 0.4\left(\dfrac{4\sin 50}{1.745 + \sin 100}\right) = 0.449$

The braking torque for each shoe is then given by

$$T_b = (\mu' R) \times r_b$$

where the frictional forces are assumed to act at points M and M', the mid-points of shoe surfaces. Thus joint braking torque from left-hand and right-hand shoe is

$$1400 \times 100 = \mu' \times r_b(R_L + R_R)$$

or $(R_L + R_R) = \dfrac{1400 \times 100}{0.449} = 17817.4 \text{ N}$ (a)

Again, for each shoe, the friction force at each shoe is related to the corresponding normal reaction as

$$F_L = \mu' R_L \qquad \text{and} \qquad F_R = \mu' R_R$$

And for equilibrium, moment equations for levers on left and right hand side, are

$$S \times 45 = R_L \times 20 - (\mu' R_L)(17.5 - 4.0)$$

and

$$S \times 45 = R_R \times 20 + (\mu' R_R)(17.5 - 4.0)$$

Simplifying above equations, we have

$$S \times 45 = (20 - 6.06)R_L \qquad \qquad \text{(b)}$$

and

$$S \times 45 = (20 + 6.06)R_R \qquad \qquad \text{(c)}$$

Dividing out the first equation by 13.94 and second one by 26.06 and adding the two equations on corresponding sides, we have

$$R_L + R_R = S(3.228 + 1.7268) \qquad \qquad \text{(d)}$$

Comparing right-hand sides of equations (a) and (d),

$$(4.9548)S = 17817.4$$

Therefore

$$S = 3596 \text{ N} \qquad \qquad \textbf{Ans.}$$

(b) Maximum bearing pressure will occur on left-hand shoe where normal reaction is larger. From equation (b),

$$R_L = \frac{S \times 45}{(20 - 6.06)} = \frac{3596 \times 45}{13.94} = 11608.3 \text{ N}$$

From equation (13.34), $p = C\cos\phi$

Maximum pressure occurs at M where $\cos\phi = 1$, and therefore, $p_{max} = C$

Hence, for limiting pressure prescribed in problem, $p_{max} = C = (0.3 \times 100) = 30 \text{ N/cm}^2$

Hence, from equation (13.40),

$$b = \frac{2R}{Cr_b(2\theta + \sin2\theta)}$$

$$= \frac{2 \times 11608.3}{30 \times 17.5 \, (1.745 + \sin100)} = 16.2 \text{ cm} \qquad \qquad \textbf{Ans.}$$

EXAMPLE 13.7 (MKS) The layout and dimensions of a block brake are shown in Fig. 13.13. Assuming $\mu = 0.4$ and that the angle of contact for each block is 110°, determine the force P on the operating arm required to set the brake for counterclockwise direction of rotation of the wheel which is 30 cm in diameter. The torque on the wheel is 3237.3 N·cm. (Note that the operating arm is mounted so that the tensions on the two sides of chain are equal).

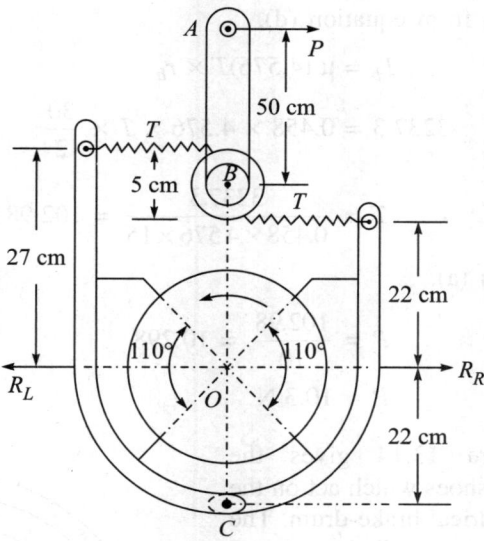

Fig. 13.13 A Block Brake.

Solution: For the equilibrium of operating arm AB, taking moments about B,

$$P \times 50 = (T \times 2.5) + (T \times 2.5)$$

or $\hspace{6cm} T = 10P \hspace{5cm}$ (a)

Again, the equivalent coeff. of friction,

$$\mu' = (0.4)\frac{4 \times \sin 55}{\left(\dfrac{110 \times \pi}{180}\right) + \sin 110°} = 0.458$$

For c.c.w. rotation of wheel, the right-hand shoe becomes self-energizing. Thus, let R_L and R_R be the normal reactions of wheel on the left-hand and right-hand shoe respectively. The equilibrium equation for left-hand shoe ($\sum M_c = 0$) is,

$$T \times 49 = R_L \times (22) + (\mu' R_L) 15$$

or $\hspace{4cm} R_L = \dfrac{T \times 49}{(22 + 0.458 \times 15)} = (1.697)T \hspace{3cm}$ (b)

And for right-hand shoe,

$$T \times 44 = R_R \times (22) - (\mu' R_R) \times 15$$

or $\hspace{4cm} R_R = \dfrac{T(44)}{(22 - 0.448 \times 15)}$

or $\hspace{4cm} R_R = (2.879)T \hspace{5cm}$ (c)

From equations (b) and (c), $\hspace{2cm} (R_L + R_R) = (4.576)T \hspace{3cm}$ (d)

Again, braking torque, $\hspace{4cm} T_b = \mu'(R_L + R_R) \times r_b$

Substituting for $(R_L + R_R)$ from equation (d),

$$T_b = \mu'(4.576)T \times r_b$$

or

$$3237.3 = 0.458 \times 4.576 \times T \times \frac{30}{2}$$

Therefore

$$T = \frac{3237.3}{0.458 \times 4.576 \times 15} = 102.98 \text{ N}$$

Therefore From equation (a),

$$P = \frac{102.98}{10} = 10.298$$

or

$$P \approx 10.3 \text{ N}$$ **Ans.**

EXAMPLE 13.8 Figure 13.14 gives the arrangement of two brake shoes which act on the internal surface of a cylindrical brake-drum. The braking forces Q_1 and Q_2 are applied as shown and each shoe pivots on its fulcrum C_L and C_R. The width of shoe lining is 4 cm and the intensity of pressure as any point K is $44.1 \sin\phi$ N/cm² where ϕ is measured as shown from either pivot. Take $\mu = 0.35$ and determine the braking torque and magnitude of forces Q_1 and Q_2. Given that $d = 30$ cm; $L = 20$ cm and angle OC_L or OC_R make with vertical = 25°.

The brake lining on each shoe is 60° above and 40° below the horizontal centre-line.

(London Un.)

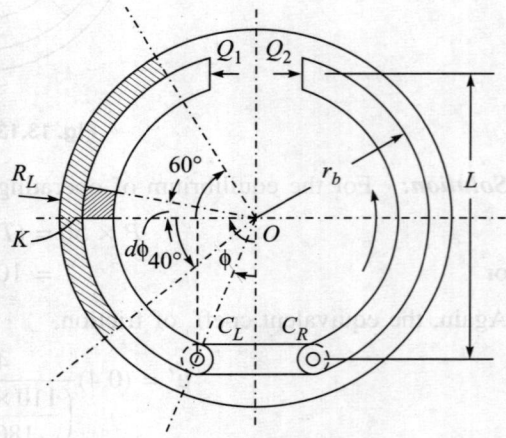

Fig. 13.14 An internal shoe brake.

Solution: The normal pressure at K

$$p_n = 44.1 \sin\phi; \text{ included angle at centre by the brake lining, } \theta = 100°$$
$$\theta_1 = 90 - (40°) - (25°) = 25°$$
$$\theta_2 = (90 + 60) - 25 = 125°$$

Total braking torque due to one shoe,

$$T_L = \mu b r_b^2 \int_{\theta_1}^{\theta_2} p \, d\phi$$

$$= 0.4 \times 4 \times (15)^2 44.1 \int_{\theta_1}^{\theta_2} \sin^2\phi \, d\phi$$

or

$$T_L = 15876 \left[\frac{1}{2}(\theta_2 - \theta_1) - \frac{1}{4}(\sin 2\theta_2 - \sin 2\theta_1) \right]$$

or

$$T_L = \frac{15876}{2} \times \left[(1.745) - \frac{1}{2}(\sin 250 - \sin 50) \right] = 20621.9 \text{ N·cm}$$

Therefore   Total braking torque = $2 \times 20{,}621.9 = 41243.8$ N·cm   **Ans.**

Magnitude of forces Q_1 and Q_2:

p_{max} *i.e.*, the maximum pressure is achieved at $\phi = 90°$, and as per statement of problem,

$$p_{max} = 44.1\sin\phi \ \text{N/cm}^2$$

Also, as seen in Fig. 13.14,

$$OC_L = OC_R = \frac{L/2}{\cos 25} = \frac{10}{\cos 25} = 11.034 \ \text{cm}$$

Hence, from equation (13.54),

$$M_n = \frac{(4)(15)(11.034)\times 44.1}{4\,(1)}\left[2\left(\frac{\pi}{180}\times 100\right) - (\sin 250 - \sin 50)\right]$$

or $M_n = 37928.4$ N·cm

And from equation (13.55),

$$M_f = \frac{(0.35)\times 4 \times 15 \times 44.1}{4\times (1)}\,[4\times 15(\cos 25 - \cos 125) + 11.034\,(\cos 250 - \cos 50)]$$

or $M_f = 18042$ N·cm

As M_f is positive, it acts c.c.w. about C_L and equilibrium eqn. for lever is,

$$Q_1 \times 20 = M_n - M_f$$

Therefore $$Q_1 = \frac{37928.4 - 18042}{20} = 994.32 \ \text{N}$$ **Ans.**

For R.H. shoe, $Q_2 \times 20 = M_n + M_f$

and therefore,

$$Q_2 = \frac{37928.4 + 18042}{20} = 2798.5 \ \text{N}$$ **Ans.**

13.9   BAND BRAKES

A band brake consists of a rope, belt or flexible steel band, lined with friction material, which is wrapped partly around a drum or sheave. Such brakes find applications in power excavators, hoists and other machines. One end of the band is fastened to one end of a straight or curved lever, while the other end is connected either (i) to an intermediate point on the lever on the other side of fulcrum, or (ii) to the fulcrum pin.

With former arrangement the band brake is called '*a differential band brake*' while with the latter arrangement, it is called '*a simple band brake*'.

In order to apply brake, effort is applied at the end of a lever in such way as to tighten the band around the drum. The friction between the band and the drum provides the tangential braking force. If T_1 and T_2 be the tensions on the tight and slack side of bands at the time of slipping, then

$$T_1/T_2 = e^{\mu\theta}$$

where θ is the angle of lap in rad, and μ = coeff. of friction.
Also, the tangential braking force,

$$F = (T_1 - T_2) \tag{13.57}$$

The braking torque of the band brake is given by

$$T_b = (T_1 - T_2) \times r_b \tag{13.58}$$

Simple Band Brake

It is evident that when effort P acts upwards, end A moves nearer to the drum and the band around drum is slackned. For the simple band brake shown in Fig. 13.15, the effort P must act downwards at B for applying the brake. In the given figure drum rotates in clockwise sense. When tight end of the band (with tension T_1) is connected to fulcrum pin, the moment contribution of T_1 about C is zero. The braking torque is still the same as given by equation (13.58) but the advantage lies in a smaller effort P required to balance moment due to T_2 at C. Thus

$$P(l) = T_2(l_2)$$

or

$$P = T_2(l_2/l) \tag{13.59}$$

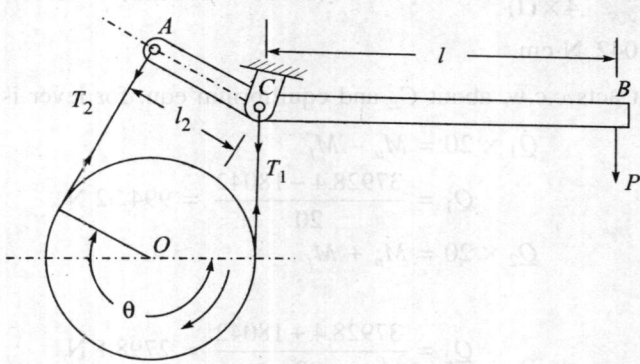

Fig. 13.15 A Simple Band Brake.

Thus above arrangement results in smaller effort for the same braking torque as given by equation (13.58). Net braking force $Q = (T_1 - T_2)$ acts along tension T_1.

Differential Band Brake

As shown in Fig. 13.16, the two ends of the band are connected to the lever on the opposite sides of fulcrum. If $l_1 < l_2$, an effort P acting downwards apply the brake. Similarly when $l_2 < l_1$, an upward force P will be required to tighten the band around drum.

As with simple band brake arrangement, the effective braking force is $Q = (T_1 - T_2)$ and acts in the direction of T_1. The braking torque is therefore $(T_1 - T_2) \times r_b$ and the equilibrium equation for lever is given by

$$P(l) + T_1(l_1) - T_2(l_2) = 0$$

Therefore

$$P = \frac{T_2(l_2) - T_1(l_1)}{l} \tag{13.60}$$

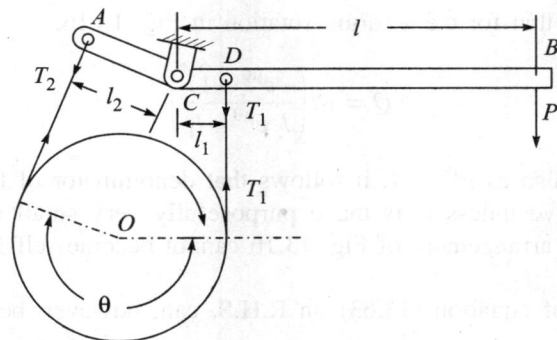

Fig. 13.16 A Differential Band Brake.

In both the cases (*i.e.*, simple and differential band brakes), the braking torque on the drum is,

$$T_B = \text{(Effective pull in band)} \times r_b$$

or $$\qquad T_B = (T_1 - T_2) \times r_b \qquad\qquad (13.61)$$

where r_b is the brake drum radius. Maximum braking torque is generated when the band is at the point of slipping on drum, the condition for which is,

$$(T_1/T_2) = e^{\mu\theta}$$

where μ = limiting coefficient of friction between band and drum, and

$$\theta = \text{angle of lap in radians.}$$

Taking T_1 common on R.H.S. in equation (13.61) and substituting for ratio $(T_1/T_2) = e^{\mu\theta}$ we have

$$T_B = T_1(1 - e^{-\mu\theta}) \times r_b$$

Equation (13.60) may be rewritten as

$$P = \frac{T_2(l_2 - l_1 e^{\mu\theta})}{l}$$

A comparison of numerator of above expression with that in equation (13.59) shows that for given breaking torque, brake drum radius and limiting ratio of tension, a smaller effort P is required in differential band brake.

For simple band brake arrangement shown in Fig. 13.15, net braking force,

$$Q = T_1 - T_2 = T_2(e^{\mu\theta} - 1) = Pl(e^{\mu\theta} - 1)/l_2 \qquad\qquad (13.62)$$

Also, for clockwise drum rotation of differential band brake arrangement (Fig. 13.16),

$$Q = T_1 - T_2 = T_2(e^{\mu\theta} - 1)$$

and $$\qquad Pl = T_2 l_2 - T_1 l_1 = T_2(l_2 - l_1 e^{\mu\theta})$$

or $$\qquad T_2 = Pl/(l_2 - l_1 e^{\mu\theta})$$

Substituting for T_2 in the above equation for Q,

$$Q = Pl\left(\frac{e^{\mu\theta} - 1}{l_2 - l_1 e^{\mu\theta}}\right) \qquad\qquad (13.63)$$

It may be verified that for c.c.w. drum rotation in Fig. 13.16,

$$Q = Pl\left(\frac{e^{\mu\theta} - 1}{l_2 e^{\mu\theta} - l_1}\right)$$

Since $l_2 > l_1$ and also as $e^{\mu\theta} > 1$, it follows that denominator of this equation can never become zero or negative unless l_2 is made purposefully very small compared to l_1. Hence differential band brake arrangement of Fig. 13.16 cannot become self locking for c.c.w. drum rotation.

The denominator of equation (13.63) on R.H.S. can, however, become negative or zero when,

$$l_1 e^{\mu\theta} \geq l2 \qquad \text{or} \qquad e^{\mu\theta} \geq (l_2/l_1)$$

Under this condition, for c.w. rotation of drum, brake is called *self-supporting* or *self-locking*. When the band is tightened around the drum, the effort P may be reduced to zero under self-locking condition. Such a condition is undesirable as it renders 'gradual application of brake' impossible.

Equation (13.62) indicates that a simple band brake can never become self-locking. This is obvious also because the net braking force Q acts through lever pivot. As against this, for c.w. rotation of drum, differential band brake of Fig. 13.16 is self-energizing because net braking force Q produces a couple about C in a direction same as that due to P.

Students may verify that in Fig. 13.15, instead of tight side of band, if slack side is connected to fulcrum, larger effort is required to apply the brake.

Effects of Distances l₁ and l₂ And Direction of Drum Rotation

Let us now consider effect of distances l_1 and l_2 and direction of drum rotation on the magnitude and direction of actuating force P. Necessarily, an actuating force P must act in a direction so as to tighten the band around the brake drum. In reference to Fig. 13.16, it follows that

  (i)   when $l_1 = l_2$, actuating force can neither tighten nor slacken the band around the drum.
 (ii)   when $l_1 < l_2$, a downward force P causes a larger movement of end A away from drum compared to that of D towards the drum. This is evident as $l_2\theta > l_1\theta$; θ being the angle of rotation of lever. Thus increased band length requirement $(l_2\theta - l_1\theta)$ produces tightening effect around the drum and brake applies. An upward force P, on the other hand, causes end A to move closer to the drum by a distance $l_2\theta$ and causes end D to move away from drum by an amount $l_1\theta$. Excess band length $(l_2\theta - l_1\theta)$ available at drum tends to slacken the band around the drum.
(iii)   when $l_1 > l_2$, a downward force P causes end D to move closer to drum by a larger distance $l_1\theta$ and causes the end A to move away from drum by a smaller distance $l_2\theta$. This results in slackening the band around the drum. On the other hand, a force P acting upwards causes end D to move away from the drum by larger distance $l_1\theta$ and causes end A to move closer to drum by a smaller distance $l_2\theta$. These movements tighten the band around the drum and brake applies.

For differential band brake arrangement, shown in Fig. 13.16, we therefore conclude as under:

For $l_1 < l_2$, the actuating force P must act downwards to apply the brake.

For $l_1 > l_2$, the actuating force P must act upwards to apply the brake.

The direction of drum rotation does not decide the direction of actuating force P for applying brake. But the direction of drum rotation does decide whether or not the brake is self-energizing. For instance in the arrangement shown in Fig. 13.16, the effective braking force $(T_1 - T_2)$ which acts in the direction of T_1 at D, produces a clockwise moment at fulcrum C and adds to the moment due to P. Thus drum rotation in c.w. sense, arrangement in Fig. 13.16 produces self-energization. But for c.c.w. drum rotation tight and slack sides exchange their position. With the tight side now connected to lever at A, the effective braking force $(T_1 - T_2)$ acting along T_1 and A now opposes couple due to P and the brake does not remain self-energizing.

13.10 BAND AND BLOCK BRAKE

A band and block brake consists of a flexible band—either in the form of a leather strap (one or more ropes) or a thin strip of steel, which are lined on the inside with blocks of wood or any other suitable material. The friction between the blocks and the drum provides friction force necessary for braking. Each block embraces equal arc on the drum. Blocks of wood have the advantage in that they can be replaced easily when in worn out condition. With reference to Fig. 13.17,

let n = number of blocks in contact with drum,

  μ = coefficient of friction between the drum and block,

  2θ = angle subtended by each of the blocks at the centre of drum, and

T_n, T_O = tension on tight and slack side respectively of the band and block brake.

Let $T_1, T_2, T_3, ...,$ be the tension in the band after first, second, third,------- block respectively, so that, tension in band before n^{th} block is T_{n-1}.

Consider the first block. It is in equilibrium under the action of tension T_0 and T_1 on the two sides of block, radially outward reaction R of drum and friction force μR as shown in Fig. 13.17(b), to an enlarged scale. The tension T_0 and T_1 are inclined at θ to the tangent line to drum periphery at the point of symmetry.

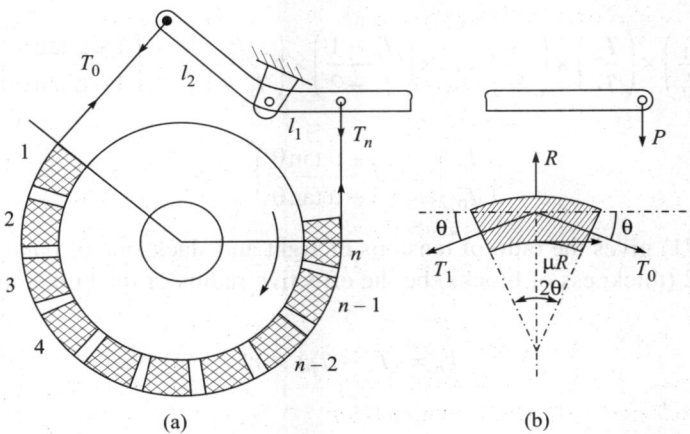

Fig. 13.17 A Band and Block Brake.

Resolving along the tangent and normal direction, for equilibrium

$$(T_0 + T_1)\sin\theta = R \tag{13.64}$$

and

$$(T_1 - T_0)\cos\theta = \mu R \tag{13.65}$$

Dividing equation (13.64) by (13.65) on corresponding sides,

$$\left(\frac{T_1 + T_0}{T_1 - T_0}\right)\tan\theta = \frac{1}{\mu}$$

or

$$\left(\frac{T_1 - T_0}{T_1 + T_0}\right) = \mu\tan\theta \tag{13.66}$$

Adding 1 on both sides of equation (13.66),

$$\left(\frac{2T_1}{T_1 + T_0}\right) = 1 + \mu\tan\theta \tag{13.67}$$

and, subtracting from 1, either side of equation (13.66),

$$\left(\frac{2T_0}{T_1 + T_0}\right) = 1 - \mu\tan\theta \tag{13.68}$$

Dividing equation (13.67) by (13.68) on respective sides,

$$\left(\frac{T_1}{T_0}\right) = \left(\frac{1 + \mu\tan\theta}{1 - \mu\tan\theta}\right) \tag{13.69}$$

A similar analysis for block 2,3,4,..., n will show that

$$\left(\frac{T_1}{T_0}\right) = \left(\frac{T_2}{T_1}\right) = \left(\frac{T_3}{T_2}\right) = ... = \left(\frac{T_n - 1}{T_n - 2}\right) = \left(\frac{T_n}{T_n - 1}\right) = \frac{1 + \mu\tan\theta}{1 - \mu\tan\theta} \tag{13.70}$$

Hence, multiplying out

$$\left(\frac{T_1}{T_0}\right) \times \left(\frac{T_2}{T_1}\right) \times \left(\frac{T_3}{T_2}\right) \times ... \times \left(\frac{T_n - 1}{T_n - 2}\right) \times \left(\frac{T_n}{T_n - 1}\right) = \left(\frac{1 + \mu\tan\theta}{1 - \mu\tan\theta}\right)^n$$

or

$$\left(\frac{T_n}{T_0}\right) = \left(\frac{1 + \mu\tan\theta}{1 - \mu\tan\theta}\right)^n \tag{13.71}$$

Equation (13.71) gives the ratio of tensions on tight and slack side of band and block brake. If $r_b' = r_b + 1/2$ (thickness of blocks) be the effective radius of the brake, the braking torque is given by

$$T_b = (T_n - T_0) \times r_b' \tag{13.72}$$

EXAMPLE 13.9 A simple band brake is fitted on a crane having the diameter of barrel as 50 cm. The band embrace 3/4th of the circumference of the brake drum of diameter 70 cm. The tight end of the band is attached to the fulcrum of the brake-lever while the slack end of the band is attached to a pin which is 10 cm away from the fulcrum. Calculate the braking torque acting on the drum shaft if the operating force of 392.4 N is acting on the lever at a distance of 65 cm from the fulcrum. The coefficient of friction is 0.3. Instead of attaching the tight end of the band to the fulcrum of the lever, if it is attached to a pin 2 cm away from the fulcrum of the lever and on opposite side of the pin to which the slack end of the band is attached, what will be the increase of braking torque acting on the drum shaft?   (DAV Indore, 1982)

Solution: Given $\mu = 0.3$; $l = 65$ cm; $l_2 = 10$ cm

Also, $\theta = \dfrac{3}{4} \times 2\pi = (1.5\pi)$ radians, and

drum radius $r_b = \dfrac{70}{2} = 35$ cm; Actuating force $P = 392.4$ N

Referring to Fig. 13.15,

$$T_2 \times 10 = 392.4 \times 65. \text{ Thus, } T_2 = 2550.6 \text{ N}$$

Also, the limiting ratio of tensions is

$$\frac{T_1}{T_2} = e^{0.3 \times 1.5\pi} = 4.11$$

Therefore $T_1 = T_2 \times e^{\mu\theta} = 10{,}483$ N

Therefore   Braking torque $T_b = (10483 - 2550.6) \times 35 = 277634.0$ N·cm

or $T_b = 2776.34$ N·m   **Ans.**

If the tight end of the band is attached to a pin 3 cm from the fulcrum on the effort side, the arrangement gives a differential band brake. Referring to Fig. 13.16, assume that angle of lap does not change.

$$l_1 = 2 \text{ cm}; l_2 = 10 \text{ cm}; l = 65 \text{ cm}$$

The equilibrium equation of lever is then,

$$392.4 \times 65 = T_2 \times 10 - T_1 \times 2$$

or $392.4 \times 65 = T_2 \times 10 - (4.11 T_2) \times 2$

or $T_2 = \dfrac{392.4 \times 65}{(10) - (4.11 \times 2)} = 14329.2$ N

$T_1 = 14329.2 \times e^{\mu\theta} = 58893$ N

$T_b = (58893 - 14329.2) \times 35$ N·cm

or $T_b = 1559733$ N·cm $= 15597.33$ N·m

Increase in braking torque $= 1556956.6$ N·cm

i.e.,   461.79 percent increase   **Ans.**

EXAMPLE 13.10 (SI) A crane is required to hold a load of 100 kN. This load is attached to a rope wound round the crane barrel which is 40 cm in diameter. The brake drum which is fixed to the barrel shaft is 55 cm in diameter. The band embraces three-quarter of the circumference of the drum and the coefficient of friction between the band and the drum is 0.3. The brake is to be applied by a hand-lever above the drum and the operating force acting vertically downwards must not exceed 500 N.

Find suitable length of the lever on both sides of the fulcrum. Neglect effect of rope-diameter and the thickness of the band and assume that one of the ends of the band is attached to the fulcrum pin directly.

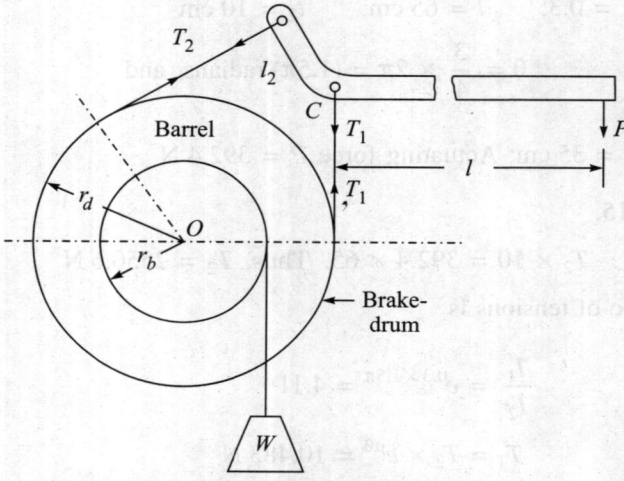

Fig. 13.18 A simple band brake to hold a load W.

Solution: $W = 100,000$ N; $r_b = \left(\dfrac{40}{2}\right)$ m;

$$r_d = \left(\frac{55}{2}\right) \text{ cm;}$$

$$\theta = \frac{3}{4}(2\pi) = 1.5\pi \text{ rad;}$$

$$\mu = 0.3; \ P \le 500 \text{ N}$$

Under limiting conditions,

$$\left(\frac{T_1}{T_2}\right) = e^{0.3 \times 1.5\pi} = 4.11$$

The effecting braking force is $(T_1 - T_2)$ and it produces a braking torque,

$$T_b = (T_1 - T_2) \times r_d$$

and this must balance moment due to supported load.
Thus

$$W \times r_b = (T_1 - T_2)r_d$$

Hence

$$(T_1 - T_2) = W\left(\frac{r_b}{r_d}\right)$$

or

$$T_2(e^{\mu\theta} - 1) = 100,000 \left(\frac{20}{27.5}\right) \quad \text{or,} \quad T_2 (4.11 - 1) = 72727.27$$

Therefore

$$T_2 = 23384.975 \text{ N} \simeq 23385 \text{ N}$$

Therefore For equilibrium of lever,

$$P \times l = T_2 \times l_2 \quad \text{or,} \quad \frac{l}{l_2} = \frac{23385}{500} = 46.77$$

Assuming a reasonable value of $l_2 = 3$ cm;

$$l = 46.77 \times 3 = 140.31 \text{ cm} \qquad \textbf{Ans.}$$

EXAMPLE 13.11 A simple band brake, as shown in Fig. 13.19, is applied to shaft carrying a flywheel of 3924 N. The radius of gyration of the flywheel is 450 mm and runs at 300 r.p.m. If the coefficient of friction is 0.2 and the brake drum diameter is 240 mm, find

1. the torque applied due to hand load of 98.1 N
2. the number of turns of the wheel before it is brought to rest, and
3. the time required to bring it to rest, from the moment of the application of the brake.
   (Saurashtra Univ., 1980)

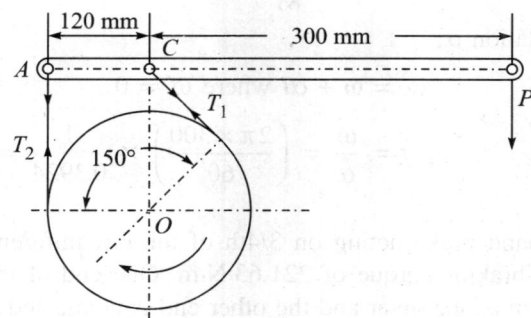

Fig. 13.19 A simple band brake.

Solution: Given: $r_b = 12.0$ cm; $\theta = \left(\dfrac{210}{180}\right) \pi$ rad; $\mu = 0.2$

Therefore Limiting ratio of tensions = $\dfrac{T_1}{T_2} = e^{(7\pi/6) \times 0.2} = 2.08$

(1) For equilibrium of brake lever, moment equation is

$$P \times 30.0 = T_2 \times 12.0$$

or $\qquad T_2 = (98.1 \times 30)/12 = 245.25 \text{ N}$

Hence, $\qquad T_1 = 245.25 \times 2.08 = 510.12 \text{ N}$

Therefore $\qquad$ Braking torque $T_b = (T_1 - T_2) \times r_b$

Therefore $\qquad T_b = (510.12 - 245.25) \times 12 = 3178.44 \text{ N·cm}$ $\qquad$ **Ans.**

(2) Moment of inertia of flywheel $\quad = \dfrac{3924}{9.81} \times (0.45)^2 = 81.0 \text{ kg·m}^2$

The kinetic energy of flywheel, $\quad E = \dfrac{1}{2} I \omega^2$

Therefore $\qquad E = \dfrac{1}{2}(81.0)\left(\dfrac{2\pi \times 300}{60}\right)^2 = 39971.9 \text{ N·m}$

This is required to be absorbed by the band brake. Let n be the number of revolutions completed by flywheel before stopping. Then,

$$T_b \times 2\pi n = 39971.9$$

Therefore $\qquad n = \dfrac{39971.9}{31.7844 \times 2\pi} = 200.15 \text{ revolutions.}$ $\qquad$ **Ans.**

(3) Let α be the angular retardation produced in flywheel by brake. Then, from

$$I\alpha = T_b = 3178.44$$

We have $\qquad \alpha = \dfrac{3178.44/100}{81} = 0.3924 \text{ rad/s}^2$

Assuming uniform retardation α,

$$\omega' = \omega + \alpha t \text{ where } \omega' = 0$$

It follows that, $\qquad t = \dfrac{\omega}{\alpha} = \left(\dfrac{2\pi \times 300}{60}\right) \times \dfrac{1}{0.3924} = 80.06 \text{ s}$ $\qquad$ **Ans.**

EXAMPLE 13.12 A band brake acting on 3/4th of the circumference of a drum of 45 cm diameter is to provide a braking torque of 221.63 N·m. One end of the band is attached to a pin A, 10 cm from fulcrum of the lever and the other end is connected to a pin D, 2.5 cm from the fulcrum on the other side. If the operating force is applied at a distance of 50 cm from the fulcrum and the coeff. of friction is 0.30, find the two values of operating force corresponding to the two directions of the rotation of drum.

If the coeff. of friction is likely to change during operation, at which value of μ will the brake become self locking. (DAV Indore, 1990)

Solution: Given $\theta = \left(\dfrac{3}{4}\right) 2\pi = 1.5\pi \text{ rad}$; $\quad \mu = 0.30$

$$r_b = \left(\frac{45}{2}\right) \text{cm}; \qquad T_b = 22163 \text{ N·cm}$$

$$l_2 = 10 \text{ cm}; \qquad l_2 = 2.5 \text{ cm}; \quad l = 50 \text{ cm}$$

For an arrangement similar to the one shown in Fig. 13.16, for c.w. sense of drum rotation.

$$P \times 50 - T_2(l_2) + T_1(l_1) = 0$$

and $$(T_1 - T_2) \times r_b = 22163$$

or $$(T_1 - T_2) = \frac{22163 \times 2}{45} = 985.0 \text{ N} \qquad \text{(ii)}$$

Also $$\frac{T_1}{T_2} = e^{1.5\pi \times 0.3} = 4.11$$

Hence, from (ii),

$$T_2(e^{\mu\theta} - 1) = 985$$

or $$T_2 = \frac{985}{(4.11 - 1)} = 316.72 \text{ N}$$

Therefore $$T_1 = 4.11 \times 316.72 = 1301.72 \text{ N}$$

Hence, from equation (i),

$$P = \frac{T_2 l_2 - T_1 l_1}{l}$$

$$= \frac{(316.72) \times 10 - (1301.72)\,(2.5)}{50}$$

$$= -1.742 \text{ N} \qquad \textbf{Ans.}$$

A negative value of P results because, $l_2 < l_1 e^{\mu\theta}$ and indicates a tendency for self-locking. For c.c.w. sense of drum rotation, the moment equation for brake lever is

$$P \times 50 = T_1 l_2 - T_2 l_1$$

or $$P = \frac{(1301.72) \times 10 - (316.72) \times (2.5)}{50} = 244.5 \text{ N} \qquad \textbf{Ans.}$$

The brake cannot become self locking for c.c.w. drum rotation. For clockwise drum rotation, the drum is already self-locking, however a limiting condition of $P = 0$ is achieved when,

$$l_2 - l_1 e^{\mu'\theta} = 0$$

or $$e^{\mu'\theta} = \left(\frac{l_2}{l_1}\right) = (10/2.5) = 4.0$$

Therefore $$\mu'\theta = 1.38629 \quad \text{Hence, } \mu' = \frac{1.38629}{1.5\pi} = 0.294 \qquad \textbf{Ans.}$$

EXAMPLE 13.13 In a crane, the rope supports a load 'W' and is wound round a barrel 45 cm in diameter. A differential band brake acts on a drum of 75 cm diameter, which is keyed to the same shaft as the crane barrel. The two ends of bands are attached to pins on opposite sides of the fulcrum of the brake lever at distances of 3.0 and 15 cm from fulcrum. The angle of lap of the brake is 270° and $\mu = 0.3$. What is the maximum load 'W' which can be supported by the brake, when a force of 392.2 N is applied to the lever at a distance of 90 cm from the fulcrum? (DAV Indore, 1990)

Solution: Given $r_d = \dfrac{75}{2}$ cm; $r_b = \dfrac{45}{2}$ cm; $l_2 = 15.0$ cm;

$$l_1 = 3.0 \text{ cm}; \ \theta = \pi 270/180 = 1.5\pi \text{ rad}$$
$$\mu = 0.3; \ P = 392.2 \text{ N}; \ l = 90 \text{ cm}$$

For limiting condition,

$$(T_1/T_2) = e^{0.3 \times 1.5\pi} = 4.11$$

Also

$$W \times r_b = (T_1 - T_2) \times r_d$$

or

$$W = \left(\frac{75}{45}\right)(T_2)(e^{\mu\theta} - 1)$$

Also, for equilibrium of lever

$$Pl = T_2 l_2 - T_1 l_1 = T_2(l_2 - l_1 e^{\mu\theta})$$

or

$$T_2 = \frac{Pl}{(l_2 - l_1 e^{\mu\theta})} = \frac{392.2 \times 90}{(15 - 3 \times 4.11)} = 13220.224 \text{ N}$$

Hence, substituting for T_2 above.

$$W = 5.183 \times 13220.224 = 68520.4 \text{ N} \qquad \textbf{Ans.}$$

EXAMPLE 13.14 A band and block brake having 12 blocks, each of which subtends an angle of 16° at the centre, is applied to a rotating drum of 600 mm diameter. The blocks are 75 mm thick. The two ends of the band are attached to pins on opposite sides of the fulcrum at distances of 40 mm and 150 mm. Determine the maximum braking torque, if a force of 250 N is applied to the lever at a distance of 900 mm from the fulcrum. Take $\mu = 0.3$. (Amaravati Un.)

Solution: $2\theta = 16°$; $n = 12$; $r_b = \dfrac{60}{2} = 30$ cm; . $t = 7.5$ cm

$$l_2 = 15 \text{ cm}; \quad l_1 = 4.0 \text{ cm.}; \quad \mu = 0.3; \quad P = 250 \text{ N}; \quad l = 90 \text{ cm.}$$

For the limiting condition

$$\frac{T_n}{T_o} = \left(\frac{1 + 0.3 \tan 8.0}{1 - 0.3 \tan 8.0}\right)^{12} = 2.7524$$

The arrangement being similar to a differential band brake for equilibrium of lever (Fig. 13.16),

$$P \times l = T_2 l_2 - T_1 l_1$$
$$250 \times 90 = T_2 (l_2 - l_1 e^{\mu\theta})$$

Therefore

$$T_2 = \frac{250 \times 90}{15 - 4.0 \times 2.7524} = 5638.53 \text{ N}$$

Therefore Braking Torque $= (T_1 - T_2)(r_b + t) = T_2(e^{\mu\theta} - 1) \times (30 + 7.5)$

$$= 370535.99 \text{ N·cm} = 3.70536 \text{ kN·m.} \qquad \textbf{Ans.}$$

(*Note:* When the direction of drum rotation w.r. to lever position is not given, it is advisable to assume condition that ensures self-energization).

EXAMPLE 13.15 A band and block brake having 14 blocks, each of which subtends an angle of 15° at the centre, is applied to a drum of 1 m effective diameter. The drum and flywheel mounted on the same shaft weigh 19620 N and a combined radius of gyration of 50 cm. The two ends of the band are attached to pins on opposite sides of the brake lever at distances of 3 cm and 12 cm from the fulcrum. If a force of 196.2 N is applied at a distance of 75 cm from the fulcrum, find

(1) maximum braking torque,
(2) angular retardation of drum, and
(3) time taken by the system to come to rest from the rated speed of 360 r.p.m.

The coeff. of friction between blocks and drum may be taken as 0.25. (Calcutta Un.)

Solution: Given $n = 14$; $2\theta = 15°$; $(r_b)_{effective} = \dfrac{100}{2} = 50$ cm

$$W = 19620 \text{ N}; \qquad k = 50 \text{ cm}; \qquad l_1 = 3 \text{ cm}; \qquad l_2 = 12 \text{ cm};$$
$$P = 196.2 \text{ N}; \qquad l = 75 \text{ cm}; \qquad \mu = 0.25$$

(Ref. Fig. 13.17)

(1) M.I. of flywheel and drum $= \dfrac{w}{g} \times k^2 = \dfrac{19620}{9.81} \times (0.5)^2 = 500 \text{ kg·m}^2$

For the limiting condition,

$$\frac{T_n}{T_o} = \left(\frac{1 + 0.25 \tan 7.5}{1 - 0.25 \tan 7.5} \right)^{14} = 2.514$$

From moment equation of lever,

$$P \times l = T_2 \times l_2 - T_1 \times l_1$$

or

$$196.2 \times 75 = T_2(l_2 - l_1 e^{\mu\theta})$$

Therefore

$$T_2 = \frac{196.2 \times 75}{(12 - 3 \times 2.514)} = 3300.8 \text{ N}$$

Therefore $\qquad\qquad T_1 = T_2 e^{\mu\theta} = 8298.2$ N

Therefore $\qquad$ Braking torque $T_b = T_2(e^{\mu\theta} - 1) \times 50 = 249870.56$ N·cm

(2) Let α be the angular retardation of drum. Then

$$I\alpha = T_b$$

or $\qquad\qquad \alpha = \dfrac{2498.7056}{500} = 4.9974 \text{ rad/s}^2$ $\qquad$ **Ans.**

(3) Let t be the time taken in seconds to bring the drum to rest from a speed of 360 r.p.m. Then as,

$$\omega_1 = \frac{2\pi \times 360}{60} = 12\pi$$

$$\omega_2 = 0$$

Therefore From, $\qquad \omega_2 = \omega_1 + \alpha t$

or $\qquad\qquad t = \dfrac{-\omega_1}{\alpha} = \dfrac{12\pi}{4.9974}$

$$t = 7.54 \text{ s} \qquad\qquad\qquad \textbf{Ans.}$$

13.11 THE BRAKING OF A VEHICLE

Let us now consider retardation of a 4-wheeler vehicle caused by the application of brakes. Figure 13.20 shows diagramatically a car moving up an inclined surface having an inclination of α to the horizontal.

Let $\quad W$ = weight of the car

$\qquad G$ = centre of gravity of car

$\qquad h$ = height of c.g. above road surface

$\qquad x$ = distance of c.g. from rear axle along the chasis

$\qquad l$ = wheel base of the car

R_a and R_b = Reactions of ground on front wheels and rear wheels respectively.

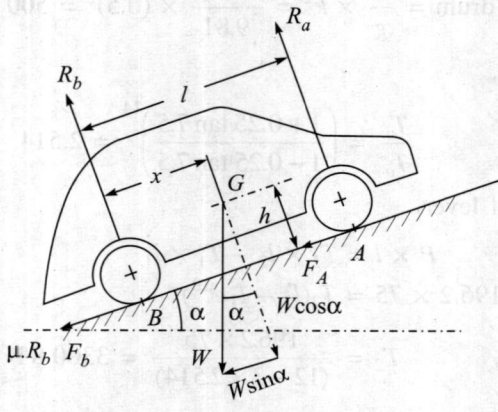

Fig. 13.20 A 4-wheeler vehicle moving up the gradient.

The brakes can be applied to the car in three ways:

    (a) to the rear wheels only,
    (b) to the front wheels only,
    (c) to all the four wheels.

This is basically a problem of dynamics and can be reduced to an equivalent problem in statics by d'Alembert's principle. According to this principle, if P is the resultant of a system of forces applied to a rigid body of mass m, the magnitude of P is given by

$$P = ma$$

This may be written as
$$P - ma = 0$$

The term $(-ma)$ is then termed as reversed effective force. In the case of a car moving up the plane, the reversed effective force (*i.e.*, the inertia force) is included in the system of forces which are actually applied to the car. With the introduction of inertia force, the forces can now be considered in equilibrium.

Brakes Applied to the Rear Wheels

Let F_b = total braking force applied at the rims of rear wheels by brakes = (μR_b)
        f = retardation of the car
        μ = coeff. of adhesion between the tyres and the road surface
        W = weight of the car
   $R_a R_b$ = normal reaction between ground on front and rear wheels respectively.

Reducing the forces, acting on the car to coplanar system, consider equilibrium of the car. Resolving along and perpendicular to the inclined plane,

$$F_b + W\sin\alpha = \left(\frac{W}{g}\right)f \tag{13.73}$$

and
$$R_a + R_b = W\cos\alpha \tag{13.74}$$

Also, considering moments of forces about, say G and applying equilibrium condition $\Sigma M_G = 0$,

$$(F_b)h + (R_b)x = R_a(l - x)$$

or
$$(F_b)h = (R_a)l - (R_a + R_b)x \tag{13.75}$$

Substituting for $(R_a + R_b)$ from equation (13.74),

$$(F_b)h = (R_a)l - (W\cos\alpha)x$$

which leads to

$$R_a = \frac{(F_b)h + (Wx)\cos\alpha}{l} \tag{13.76}$$

Also, rearranging equation (13.74)

$$R_b = (W\cos\alpha) - R_a$$

or
$$R_b = (W\cos\alpha) - \frac{(F_b)h + (Wx)\cos\alpha}{l}$$

or
$$R_b = \frac{W(l - x)\cos\alpha - (F_b)h}{l} \qquad (13.77)$$

From equation (13.73),

$$\left(\frac{f}{g}\right) = \frac{F_b + W\sin\alpha}{W} \qquad (13.78)$$

The braking force F_b depends on the effort applied by driver on brake-pedal, but in any case the maximum value of force F_b cannot exceed the limiting force of friction $= \mu R_b$. Any additional effort by the driver on the pedal results in the skidding of wheels on road. Hence maximum value of retardation in equation (13.78) is given by

$$\left(\frac{f}{g}\right) = \frac{\mu R_b + W\sin\alpha}{W} = \mu\left(\frac{R_b}{W}\right) + \sin\alpha \qquad (13.79)$$

Also, substituting μR_b for F_b in equation (13.77),

$$R_b = \frac{W(l - x)\cos\alpha - (\mu R_b)h}{l}$$

Rearranging,

$$R_b = \frac{W(l - x)\cos\alpha}{(l + \mu h)} \qquad (13.80)$$

Substituting for (R_b/W) from equation (13.80) in (13.79),

$$\left(\frac{f}{g}\right) = \mu\frac{(l - x)\cos\alpha}{(l + \mu h)} + \sin\alpha \qquad (13.81)$$

Equation (13.81) gives retardation f produced in the car when moving along the inclined plane and brakes are applied to rear wheels.

Also, from equation (13.76), substituting for F_b and R_b from equation (13.74)

$$R_a = \frac{\mu h(W\cos\alpha - R_a) + Wx\cos\alpha}{l}$$

or, rearranging,
$$R_a = \frac{(x + \mu h)W\cos\alpha}{(l + \mu h)} \qquad (13.82)$$

Case (a) When $\alpha = 0$, *i.e.*, when car move on level track, from equation (13.81),

$$\left(\frac{f}{g}\right) = \frac{\mu(l - x)}{(l + \mu h)} \qquad \text{and,} \qquad R_a = \frac{(x + \mu h)W}{(l + \mu h)}$$

Case (b) When car moves down the plane, equation (13.79) modifies to

$$\left(\frac{f}{g}\right) = \mu(R_b/W) - \sin\alpha$$

and equation (13.81) modifies to,

$$\left(\frac{f}{g}\right) = \frac{\mu(l - x)\cos\alpha}{(l + \mu h)} - \sin\alpha \qquad (13.83)$$

Brakes Applied to the Front Wheels

When brakes are applied to the front wheels rather than to rear wheels, the braking force F_a (with a maximum limiting value μR_a) acts through the point A. Except for this change, all other forces remain the same. Hence resolving parallel and normal to the plane,

$$R_a + R_b = W\cos\alpha \qquad \text{(a)}$$

and

$$F_a + W\sin\alpha = \left(\frac{w}{g}\right)f \qquad \text{(b)}$$

Also, from moment equation $\Sigma M_o = 0$,

$$F_a(h) + R_b x = R_a(l - x)$$

or

$$(\mu R_a)h + R_b x = R_a(l - x) \qquad \text{(c)}$$

or

$$(R_a + R_b)x = R_a(l - \mu h)$$

or

$$R_a = \frac{(R_a + R_b)x}{(l - \mu h)} = \frac{Wx\cos\alpha}{(l - \mu h)} \quad \text{from equation (a)} \qquad \text{(13.84)}$$

and

$$R_b = W\cos\alpha - R_a = W\cos\alpha - \frac{Wx\cos\alpha}{(l - \mu h)}$$

or,

$$R_b = \frac{(l - \mu h - x)\,W\cos\alpha}{(l - \mu h)} \qquad \text{(13.85)}$$

Also, from equation (b),

$$(f/g) = (F_a/W) + \sin\alpha$$

Substituting $F_a = \mu R_a$ and then substituting for R_a from equation (13.84),

$$(f/g) = \frac{\mu x\cos\alpha}{(l - \mu h)} + \sin\alpha \qquad \text{(13.86)}$$

Case (a) When travelling on level track (*i.e.*, $\alpha = 0$)

$$\left(\frac{f}{g}\right) = \frac{\mu x}{(l - \mu h)}$$

Case (b) For car moving down the plane, equation (b) modifies to

$$F_a - W\sin\alpha = (W/g)f$$

or

$$\mu R_a - W\sin\alpha = (W/g)f$$

so that,

$$\frac{W\mu x\cos\alpha}{(l - \mu h)} - W\sin\alpha = (W/g)f$$

or

$$\left(\frac{f}{g}\right) = \frac{\mu(x\cos\alpha + h\sin\alpha) - l\sin\alpha}{(l - \mu h)} \qquad \text{(13.87)}$$

Brakes Applied to All the Four Wheels

As before, resolving along the plane and normal to it vehicle moving up the plane,

$$R_a + R_b = W\cos\alpha \tag{13.88}$$

and

$$\mu R_a + \mu R_b + W\sin\alpha = \frac{Wf}{g}$$

so that

$$\mu(R_a + R_b) + W\sin\alpha = \frac{Wf}{g}$$

and substituting for $(R_a + R_b)$ from equation (13.88),

$$(f/g) = (\mu\cos\alpha + \sin\alpha) \tag{13.89}$$

Also from condition of equilibrium $\Sigma M_G = 0$,

$$(F_a + F_b)h + R_b(x) = R_a(l - x)$$

which, for a limiting condition $F = \mu R$, through equation (13.88), reduces to

$$\mu(W\cos\alpha)h + (R_b + R_a)x = (R_a)l,$$

and as such,

$$R_a = \frac{(x + \mu h)W\cos\alpha}{l} \tag{13.90}$$

Finally, from equations (13.88) and (13.90)

$$R_b = W\cos\alpha - R_a$$

or

$$R_b = \frac{(l - x - \mu h)W\cos\alpha}{l} \tag{13.91}$$

Case (a) When moving on level track ($\alpha = 0$),

$$R_A = (x + \mu h)W/l$$
$$R_B = (l - x - \mu h)W/l$$
$$(f/g) = (\mu)$$

Case (b) When car moves down the plane,

$$\mu(R_a + R_b) - W\sin\alpha = \frac{W}{g}f$$

or

$$(f/g) = (\mu\cos\alpha - \sin\alpha) \tag{13.92}$$

When brakes are applied in a moving car, couples are introduced which tend to rotate the car about its c.g. in vertical longitudinal plane. The couples tend to increase pressure between road and front wheels and decrease the pressure between road and rear wheels. This is reflected in the magnitudes of reactions R_a and R_b on front wheels and rear wheels. This can be verified in the following example problem.

EXAMPLE 13.16 A lorry weighing 29430 N has its centre of gravity 0.6 m above the ground level. The axle centres are 2.4 m apart, the centre of gravity being 0.9 m in front of the back axle and 1.5 metre behind the front axle. The lorry is running at a speed of 36 km/hr. Find the minimum distance in which the car may be slopped when

(a) the rear wheels are braked,
(b) the front wheels are braked, and
(c) all the wheels are braked and the coeff. of friction between tyre and road is (i) 0.15, and (ii) 0.45

What is the required ratio of R_a/R_b for case (i) and (ii) when four wheel-brakes are used? (DAV Indore, 1978)

Solution: $W = 29430$ N; $l = 2.4$ m; $h = 0.6$ m; $x = 0.9$ m; $v = 36$ km/hr

While moving up the gradient and rear brakes are applied, the equilibrium equations are:

$$R_a + R_b = W\cos\alpha$$

and, $$(\mu R_b)h + R_b x = R_a(l - x)$$

Also, $$\left(\frac{W}{g}\right)f = F_b + W\sin\alpha$$

Putting $\alpha = 0$, the three equations are

$$R_a + R_b = W,$$
$$\mu R_b h + (R_a + R_b)x = (R_a)l$$

and $$\left(\frac{W}{g}\right)f = F_b + 0$$

(a) Substituting values above,

$$R_a + R_b = 29430 \tag{a}$$
$$\mu(R_b)h + (29430)x = (R_a)l \tag{b}$$
$$29430\left(\frac{f}{g}\right) = F_b \tag{c}$$

From equation (b) with $\mu = 0.15$,

$$(0.09)R_b + 26487 = 2.4(R_a)$$

Solving this and equation (a) simultaneously,

$$(0.09)R_b + 26487 = 2.4(29430 - R_b)$$

or $$(2.49)R_b = (70632 - 26487)$$

Therefore $$R_b = 17728.9 \text{ N and } R_a = 11701 \text{ N}$$

Therefore $$F_b = \mu R_b = 0.15 \times 17728.9 = 2659.3 \text{ N}$$

Hence from equation (c),

$$f = \frac{2659.3 \times 9.81}{29430} = 0.8864 \text{ m/s}$$

Therefore If s be the distance required to stop the lorry, then from $V^2 = 0 = u^2 + 2fs$

$$s = \frac{-(36 \times 1000/3600)^2}{2(0.8864)} = 56.4 \text{ m}$$ **Ans.**

When $\mu = 0.45$ from equation (b),

$$(0.6)(0.45)R_b + (29430)x = 2.4(R_a)$$

Solving this and equation (a) simultaneously

$$(0.27)R_b + 29430(0.9) = (29430 - R_b)(2.4)$$

or, $$(2.67)R_b = (70632 - 26487)$$

Therefore $$R_b = 16533.7 \text{ N}, \ R_a = 12896.3 \text{ N}$$

Therefore $$F_b = 0.45 \times 1653.7 = 7440.2 \text{ N}$$

Hence, from equation (c),

$$f = \frac{7440.2 \times 9.81}{29430} = 2.48 \text{ m/s}^2$$

Therefore $$s = \frac{-(36000/3600)}{2(2.48)} = 20.16 \text{ m}$$ **Ans.**

Note: In normal course, when the vehicle is standstill off-centred location of c.g. of lorry

calls for reactions: $R_b = \dfrac{29430 \times 1.5}{2.4} = 18393.75 \text{ N}$ and $R_a = \dfrac{29430 \times 0.9}{2.4} = 11036.25 \text{ N}$ and the

braking in above problem has resulted in increasing reaction in front wheel from 11036.25 to 11701 N and decreasing reaction in rear wheels from 18393.75 N to 17728.9 N

(b) When the front wheels are braked,

The equilibrium equations are

$$R_a + R_b = W \quad\quad\quad\quad (d)$$

$$\mu R_a h + R_b x = R_a(l - x) \quad\quad\quad\quad (e)$$

and $$\left(\frac{W}{g}\right)f = (\mu R_a) \quad\quad\quad\quad (f)$$

(i) Putting values, (for $\mu = 0.15$),

$$R_a + R_b = 29430$$

$$0.09 \, R_a + 0.9 \, R_b = 1.5 \, R_a$$

or $$1.41 \, R_a - 0.9 \, R_b = 0$$

[Solving equation simultaneously]

$$1.41(R_a) - 0.9(29430 - R_a) = 0$$

$$2.31R_a = 26487$$

Therefore $$R_a = 11466.2 \text{ N}; \ R_b = 17963.8 \text{ N}$$

And $$f = \frac{0.15 \times 11466.2}{29430} \times 9.81 = 0.5733 \text{ m/s}^2$$

Therefore $$s = \frac{-(36000/3600)^2}{2 \times f} = 87.21 \text{ m}$$

Again, when $\mu = 0.45$, equations d, e and f become

$$R_a + R_b = 29430$$

$$0.27 R_a + 0.9 R_b = 1.5 R_a$$

Solving simultaneously, $$R_a = \frac{29430}{2.367} = 12433.5 \text{ N}; \quad R_b = 16996.5 \text{ N}$$

Hence, from $$\left(\frac{w}{g}\right) f = \mu R_a$$

$$f = \frac{0.15 \times 12433.5}{29430} \times 9.81 = 0.6217 \text{ m/s}^2$$

Therefore $$s = \frac{(36000/3600)^2}{2 \times f} = 80.417 \text{ m}$$ **Ans.**

(c) Lastly, when all the brakes are applied, equilibrium equations are:

$$R_a + R_b = 29430$$

$$\mu(R_a + R_b)h + R_b x = R_a(l - x)$$

Putting values & simplifying,

$$\mu(29430 \times 0.6) + 0.9 R_b = 1.5 R_a$$

or $$(1.5) R_a - 0.9 R_b = (17658)\mu$$

Thus, solving for $\mu = 0.15$,

$$2.4 R_a = 0.9(29430) + (17658)0.15$$

Therefore $$R_a = 12139.875 \text{ N}; \quad R_b = 17290.125 \text{ N}$$

Therefore   From $$\left(\frac{W}{g}\right) f = \mu(R_a + R_b)$$

or $$f = \frac{29430 \times \mu}{29430} \times 9.81 = (9.81)\mu$$

Therefore $$s = \frac{(36000/3600)^2}{2 \times (9.81)\mu} = \frac{50}{(9.81)\mu}$$

Therefore   For $\mu = 0.15$, $s = 33.98 \text{ m}$ **Ans.**

   For $\mu = 0.45$, $s = 11.326 \text{ m}$ **Ans.**

Reqd. ratio for $\mu = 0.15$,

$$\frac{R_a}{R_b} = \frac{12139.875}{17290.125} = 0.702 \qquad \textbf{Ans.}$$

and for $\mu = 0.45$, the equations are,

$$R_a + R_b = 29430 \text{ and, } 1.5R_a - 0.9R_b = 7946.1$$

Solving $\qquad\qquad R_a = 14347.1; \ R_b = 15082.9 \text{ N}$

Therefore $\qquad\qquad \dfrac{R_a}{R_b} = \dfrac{14347.1}{15082.9} = 0.9512 \qquad \textbf{Ans.}$

13.12 TYPES OF DYNAMOMETERS

As indicated earlier, a dynamometer is a brake which incorporates suitable arrangement to measure forces or couples which tend to change the state of rest or of uniform motion of a body. Broadly speaking, dynamometers can be grouped into two categories:

(1) Absorption Dynamometers and
(2) Transmission Dynamometers.

As the name suggests an *absorption dynamometer* works on the principle of absorbing available power in doing work usually against frictional resistance. In contrast, a *transmission dynamometer* transmits all the available power unchanged, except perhaps for the power lost in overcoming inherent friction at various joints of the dynamometer.

While absorption type of dynamometers are suitable for measuring power output of machines of moderate power capacity, the transmission dynamometers are suitable for measuring power output of machines with large power capacity.

Absorption type dynamometers are further subdivided as:

(a) *Mechanical Friction Dynamometers,* which include Prony brake dynamometers and Rope brake dynamometers.
(b) *Hydraulic Friction Dynamometers,* which include Tresla fluid friction dynamometer and Froude water vortex dynamometer.
(c) *Electrical Dynamometer,* which is commonly known as Swinging Field dynamometer.

Transmission type dynamometers can also be subdivided into three categories:

(a) *Belt Transmission Dynamometer,*
(b) *Epicyclic Train Dynamometer,* and
(c) *Torsion Dynamometer,* which includes Bevis Gibson Flash Light dynamometer; the Fotinger torsion dynamometer and Amsler torsion dynamometer.

Only a few representative types of dynamometers are proposed to be discussed in the following sections.

ABSORPTION DYNAMOMETERS

13.13 PRONY BRAKE DYNAMOMETER

This is a simple type of absorption dynamometer in which two wooden blocks, each embracing less than half the circumference, press against a motor which is keyed to the engine or motor shaft as shown in Fig. 13.21. The wooden blocks are held tight against a pulley from opposite sides with the help of bolts and nuts. Closed coil helical springs compressed between nuts and upper block, ensure a constant force between the blocks and the pulley.

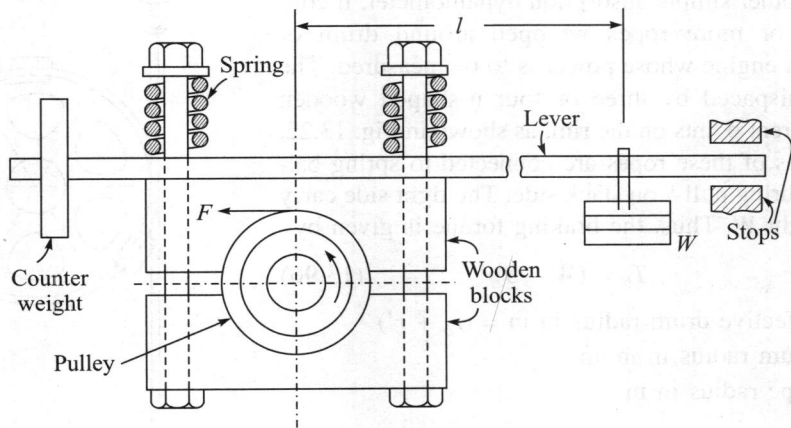

Fig. 13.21 A Simple Absorption Dynamometer.

A weight W (adjustable) is attached to a lever which in turn is attached to either of the two blocks. A counter weight is provided at the other end of the lever for balancing the brake when unloaded. Compression in the springs is controlled by screwing nuts up and down and this controls the friction torque between the blocks and the pulley. Frictional force F between block and pulley tends to rotate the block in c.c.w. sense, which is prevented by moment (Wl) of the dead weight W on the lever. On the other hand, the frictional force F is increased by increasing compression in the spring so as to absorb all kinetic energy of pulley. Thus, for equilibrium of lever,

$$Wl = F \times r \qquad (13.93)$$

where r = effective radius of pulley.

Also, if N be the speed of shaft in r.p.m., the H.P. absorbed

$$\text{H.P.} = \frac{2\pi N T_f}{4500} \quad \text{and} \quad \text{Power in kW} = \frac{2\pi N M_f}{60 \times 1000}$$

where, T_f = frictional torque in kgf·m, $\quad M_f$ = frictional torque in N·m

and, $\quad T_f = M_f = Fr = Wl$

Hence, brake horse-power = $\dfrac{2\pi N \times Wl}{4500}$ H.P. $\qquad$ (13.94)

and brake power in kw = $\dfrac{2\pi N \times W \times l}{60 \times 1000}$ kw $\qquad$ (13.95)

Great care must be exercised to ensure that the lever always floats between the stops.

13.14 ROPE BRAKE DYNAMOMETER

This is yet another simple absorption dynamometer. It consists of two or more ropes wrapped around drum or flywheel of an engine whose power is to be measured. The ropes are equispaced by three or four u-shaped wooden blocks at different points on the rim, as shown in Fig. 13.22. The slack sides of these ropes are connected to spring balance for measuring pull S on slack side. The tight side carry the dead weight W. Thus, the braking torque is given by

$$T_b = (W - S)r \qquad (13.96)$$

where r = effective drum radius in m = $(r_b + r')$

r_b = drum radius in m and

r' = rope radius in m

Hence, brake horse power is given by

$$\text{B.H.P.} = \frac{2\pi N(W - S)r}{4500} \text{ H.P.} \quad (13.97)$$

where W & S are in kgf.

And brake power on engine shaft = $\dfrac{2\pi N(W - S)r}{60 \times 1000}$ kw $\qquad$ (13.98)

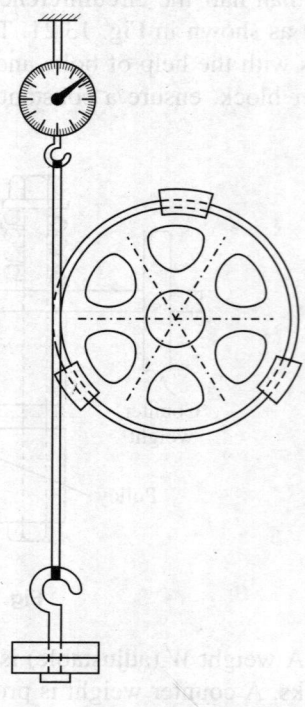

Fig. 13.22 Rope Brake Dynamometer.

where W and S are in N.

When heat generated is considered excessive, cooling arrangement may be provided by using a pulley of channel section with flanges turned on the inside (see Fig. 13.23). Cooling water is supplied in this channel section by a pipe and is discharged by an outlet pipe with a flattened end which enables it to scoop the running water.

Critical speed: In order to ensure that cooling water remains in contact with the channel section

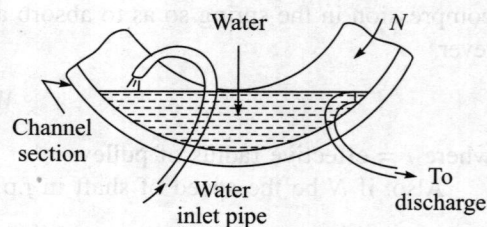

Fig. 13.23 Cooling arrangement.

in any position during rotation, it is necessary that the centrifugal force on water particles must be equal to or greater than the gravitational force acting on it. Thus,

$$F_c \geq w \qquad (13.99)$$

where w = weight of water particle

or
$$\frac{w}{g} r\omega^2 \geq w$$

where r = mean radius at which water is revolving with channel section

ω = angular speed of drum/pulley

or
$$\omega^2 \geq (w)\left(\frac{g}{wr}\right)$$

or
$$\omega \geq \sqrt{(g/r)} \qquad\qquad (13.100)$$

The critical speed of drum/pulley, above which water particles remain in contact with, is,

$$\omega_{critical} = \sqrt{(g/r)}$$

Therefore
$$(N)_{critical} = \frac{60}{2\pi} \sqrt{(g/r)} \ \text{r.p.m.} \qquad\qquad (13.101)$$

EXAMPLE 13.17   Following data refer to a rope-brake dynamometer:

Radius of the brake drum = 125 cm, Diameter of rope = 25 mm,

Dead load in pan = 2451.7 N, Spring balance reading = 392.3 N, and r.p.m. = 125.

Find the B.H.P. and I.H.P. of the engine if mechanical efficiency is 80%.

(DAV Indore, 1990)

Solution: Effective radius of drum = $125 + \dfrac{2.5}{2}$ cm = 126.25 cm

$(W-S)$ = Effective tension in rope = $(2451.7 - 392.3)$ = 2059.4 N

Therefore   Brake power

$$P_b = \frac{(W-S) \times r \times 2\pi N}{60 \times 1000} \ \text{kw}$$

or
$$P_b = \frac{125 \times (2059.4) \times 2\pi \times 126.25}{60 \times 1000 \times 100} = 34.03 \ \text{kw} \qquad\qquad \textbf{Ans.}$$

Therefore   Indicated power $P_i = \dfrac{\text{Brake power}}{\text{Mechanical efficiency}}$ **Ans.**

or
$$P_i = \frac{34.03}{0.80} = 42.54 \ \text{kw} \qquad\qquad \textbf{Ans.}$$

EXAMPLE 13.18   The rim of the brake wheel of a turbine is of channel section and its internal diameter is 45 cm. Find the minimum speed in r.p.m. at which the wheel will hold a layer of water 2.5 cm deep at the top of the rim. (DAV Indore, 1992)

Solution: Effective radius of water layer which can be held in channel $r = \left(\dfrac{45}{2}\right) - 2.5$

$$= 22.5 - 2.5 = 20.0 \text{ cm}$$

The critical speed below which water cannot be held in channel section,

$$(N)_{critical} = \frac{60}{2\pi} \sqrt{\frac{g}{r}}$$

$$= \frac{60}{2\pi} \sqrt{\frac{9.81}{(20/100)}} = 66.88 \text{ r.p.m.} \qquad \textbf{Ans.}$$

TRANSMISSION DYNAMOMETERS

13.15 EPICYCLIC TRAIN DYNAMOMETERS

This is a transmission dynamometer and enables measurement of power which is usefully consumed by a machine. Figure 13.24 illustrates a simple epicyclic train of spur or bevel wheels and is used between the source of power and the machine. The spur gear S is keyed to the driving shaft and revolves with it. The annular wheel Q is keyed to the driven shaft and is driven through the wheel P by driving wheel S. The wheel P revolves freely on a pin fixed to the arm L. The lever L itself can revolve freely about the common axis of the driving wheel S and driven wheel Q.

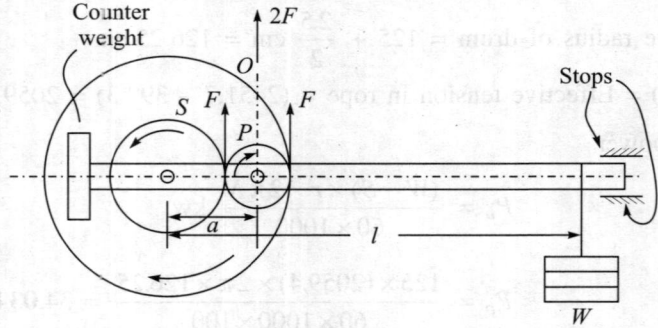

Fig. 13.24 Epicyclic Train Dynamometer.

A close look at the free body diagram of the wheel P shows that it is subjected to a tangential upward effort F as exerted by *annular wheel* Q. Note that the tangential force exerted by S on P and by Q on P will be equal only if friction in the pin of wheel P is neglected. The two forces are equivalent to a single equivalent force $= 2F$ at the pin of wheel P. Hence taking moments at the fulcrum of lever and considering its equilibrium,

$$W \times l = 2F \times a$$

or

$$F = \left(\frac{Wl}{2a}\right) \qquad (13.102)$$

With F so established, a knowledge about the speed of rotation N of the wheel S and its radius r enables one to compute power transmitted by the wheel S. Thus,

$$P = \frac{F \times 2\pi rN}{4500} \text{ H.P.}$$

where F = tangential force in kgf

r = radius of wheel S in m

and N = speed of rotation of S in r.p.m.

Also

$$P = \frac{F \times 2\pi rN}{60 \times 1000} \text{ kw}$$

where F = tangential force in N

r = radius of wheel S in m

N = speed of rotation of S in r.p.m.

Thus, substituting for F,

$$P = \frac{Wl}{2} \times \frac{2\pi rN}{4500} \text{ H.P.} \tag{13.103}$$

$$P = \frac{Wl}{2} \times \frac{2\pi rN}{60 \times 1000} \text{ kw} \tag{13.104}$$

13.16 BELT TRANSMISSION DYNAMOMETER

Figure 13.25 illustrates one design of a belt transmission dynamometer which determines power by measuring difference in tensions on the two sides of the belt, as the belt transmits power. The

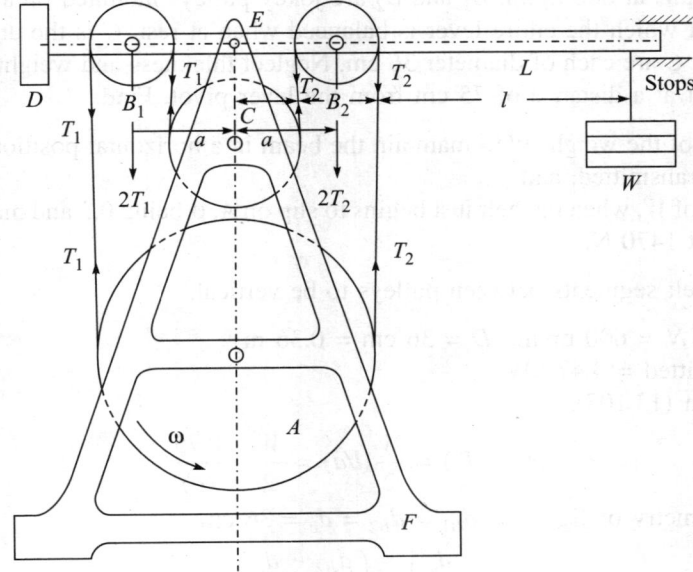

Fig. 13.25   A Belt Transmission Dynamometer.

shaft of prime mover is connected to the driver pulley A. An endless belt passes over the driving pulley A, the intermediate pulleys B_1, B_2 and driven pulley C. The driving and driven pulleys revolve about fixed axis on pins mounted on frame F.

As can be seen in Fig. 13.25, side of the belt connecting pulley B_1 to A becomes tight side with tension T_1 and side of the belt between pulleys A and B_2 becomes slack side with tension T_2. The pulleys B_1 and B_2 are free to revolve on pins mounted on lever L and do not transmit any power. The tensions on the two sides of these pulleys are, therefore, same. In other words, these pulleys serve the purpose of measuring difference of tensions only.

The lever L is pivoted at fulcrum E and the pulleys B_1 and B_2 are mounted at same distance a w.r. to this fulcrum. The total downward force $2T_1$ at the pin centre of pulley B_1 and $2T_2$ at the centre of pulley B_2 produce net c.c.w. moments at the fulcrum $E = 2 (T_1 - T_2) \times a$. This moment is balanced by the dead load W mounted at a distance l from fulcrum E. Thus,

$$2(T_1 - T_2) \times a = W \times l$$

or

$$(T_1 - T_2) = \frac{W}{2}(l/a)$$

With difference in tension on the two sides of the driving wheel A so established, only speed of rotation N of this pulley and its radius r are required to .calculate its power P. Thus,

$$P = \frac{2\pi Nr}{4500}\left(\frac{W}{2}\frac{l}{a}\right) \text{ kw}$$

where r = radius of pulley A in m

l = moment arm of dead weight W in m and

a = centre to centre distance of pulleys B_1 and B_2 from fulcrum of lever in m.

EXAMPLE 13.19 In a belt transmission dynamometer of the type shown in Fig. 13.25, the driving pulley A runs at 600 r.p.m. B_1 and B_2 are jokey pulleys mounted on a horizontal lever pivoted at E about which the entire lever is balanced when at rest. C is the driven pulley. The pulleys B_1, B_2 and C are each of diameter 36 cm. Neglect thickness and weight of the belt. The dead weight is put at a distance of 75 cm from the lever pivot. Find,

(a) the value of the weight W to maintain the beam in a horizontal position when 4.47 kW is being transmitted, and

(b) the value of W, when the belt just begins to slip on A, μ being 0.2 and maximum tensions in the belt 1470 N.

Assume the belt segments between pulleys to be vertical.

Solution: Given N = 600 r.p.m.; D = 36 cm = 0.36 m

Power transmitted = 4.47 kW

From equation (13.105),

$$(T_1 - T_2) = \frac{W}{2}(l/a) = \frac{W}{2} \times \frac{0.75}{(a)}$$

From the geometry of figure as $d_{B1} = d_{B2} = d_c$ = 36 cm

$$a = \left(\frac{d_{B1}}{2} + \frac{d_c}{2}\right) = \left(\frac{d_{B2}}{2} + \frac{d_c}{2}\right) = 36 \text{ cm}$$

and $d_A = d_{B1} + d_c + d_{B2} = 3(36) = 108$ cm Hence, $r_A = 0.54$ m

Also, $$(T_1 - T_2) = \frac{W}{2}\left(\frac{0.75}{0.36}\right)$$

(a) From $$P = \frac{2\pi N(T_1 - T_2)r}{1000 \times 60}$$

Therefore $$(T_1 - T_2) = \frac{4.47 \times 1000 \times 60}{2\pi Nr}$$

or $$(T_1 - T_2) = \frac{4.47 \times 1000 \times 60}{2\pi \times 600 \times 0.54} = 131.74 \text{ N}$$

Hence, from equation (i),

$$W = \frac{2(T_1 - T_2)a}{0.75}$$

or $$W = \frac{2(131.74) \times 0.36}{0.75} = 126.47 \text{ N}$$ **Ans.**

(b) When slip occurs on pulley,

$$\frac{T_1}{T_2} = e^{\mu\theta} = e^{(0.2)\pi} = 1.87445$$

Therefore $$T_1 - T_2 \equiv T_1(1 - e^{-\mu\theta})$$

or $$(T_1 - T_2) = 1470\left(1 - \frac{1}{1.87445}\right) = 685.77 \text{ N}$$

Hence, from equation (i),

$$W = \frac{2a}{l}(T_1 - T_2) = \frac{2 \times 0.36}{0.75} \times 685.77 = 685.34 \text{ N}$$ **Ans.**

13.17 TORSION DYNAMOMETERS

A need to measure large power that is transmitted along the propeller shaft of a turbine or motor vessel, has led to the introduction of a number of torsion dynamometers. These dynamometers make use of elastic deformation in steel shafts for determining the power that is transmitted.

It follows from the torsion formula that within elastic limit,

$$\frac{T}{J} = \frac{G\theta}{l} \quad \text{so that,} \quad (T/\theta) = \frac{GJ}{l}$$

is constant for a given section of the shaft and is called torsional stiffness K_t. Thus, the torque transmitted by a shaft is proportional to the angle of twist. Mathematically,

$$T = (K_t)\theta$$

Hence, by measuring angle of twist θ in radians, the torque transmitted by a shaft can be determined. Remembering that modulus of rigidity for a steel shaft is 80×10^9 N/m², the problem of measuring torque on small lengths of shaft does not appear to be a simple affair. For instance, with a reasonable value of maximum shear stress, $\tau = 3740$ N·cm², we have from torsion formula,

$$T = \left(\frac{GJ}{l}\right)\theta = \tau\left(\frac{2}{d}\right)J$$

which, for a solid circular shaft, becomes

$$\theta = 2\left(\frac{\tau}{G}\right)\left(\frac{l}{d}\right)$$

It follows that even 1 degree of angle of twist in shaft requires (l/d) ratio of the order,

$$(l/d) = \frac{\pi}{180} \times \frac{G}{2\tau} = 18.7$$

Usually, shafts are available in lengths shorter than required by ratio $(l/d) = 18.7$, which means that angle of twists required to be measured are less than 1 degree and therefore, some arrangement for amplifying angle of twist is necessary. Two such arrangements to achieve this objective are described as follows:

Bevis Gibson Flash-light Dynamometer

The principle of operation assumes that light travels in straight line at infinite speed. The set-up consists of two discs A and B fixed to the shaft at a maximum possible distance 1 apart. Each of the two discs A and B carries a narrow radial slit and the discs are so adjusted that in zero load condition these slits face each other. A bright light source L, which is masked all around except for a narrow radial slit, is mounted on a bearing cap to the left of the disc A. When all the three slits are in line, a narrow pencil of light is passed parallel to the axis of the shaft. An eye-piece E is supported on a fixed bracket to the right of the disc B, which can be moved along concentric circular path (with axis coinciding with that of shaft) by means of a vernier arrangement. This is shown at Figs. 13.26(a), (b) and (c). In Fig. 13.27(a), the eye-piece opening falls in line with slits in lamp L, and discs A and B and the flash is visible.

Thus, when the shaft does not transmit any torque, a flash of light may be seen after every revolution of the shaft. But when the torque is transmitted (Fig. 13.26 b), twisting of shaft results in misalignment of slits in discs A and B and the eye-piece is not capable of detecting a flash. However a circular displacement of vernier and eye-piece brings the slits in lamp, discs A, B and eye-piece in one line and a flash is seen. The difference in the two readings of vernier gives the angle of twist. This arrangement can measure angle of twist upto 1/100 degree.

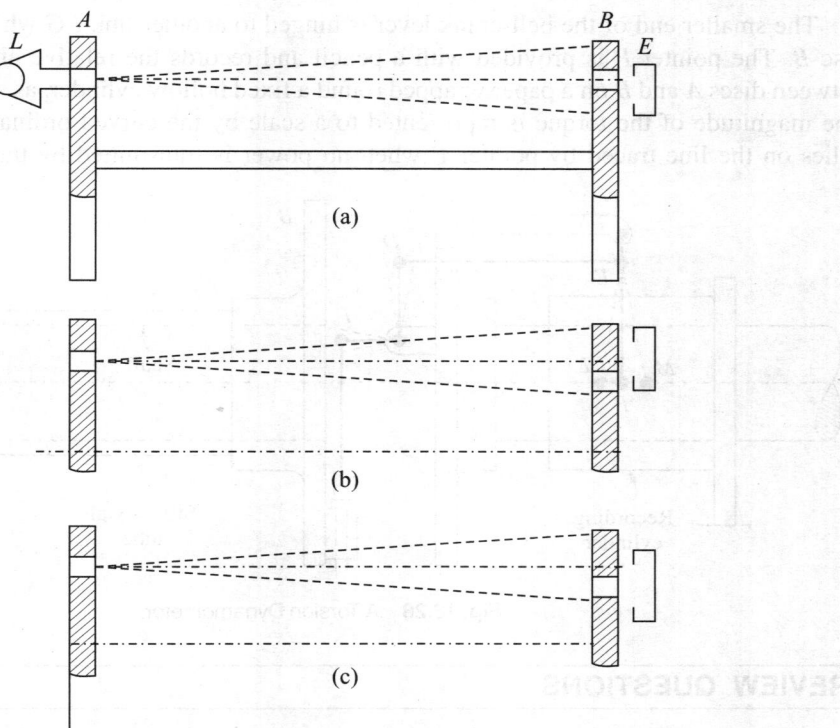

Fig. 13.26 A Flash-light dynamometer.

When the shaft is subjected to a uniform torque, angle of twist may be measured, as above, at a single angular position. However, when the torque does not remain uniform, angle of twist must be measured at several different angular positions of the shaft. For this purpose, discs A and B are provided with short radial slots arranged along a spiral cuve as shown in Fig. 13.27. The lamp and eye-piece in such a case is also required to be moved in radial direction.

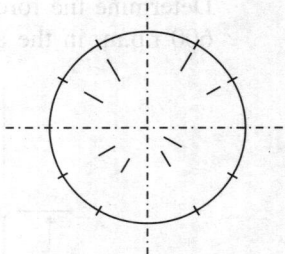

Fig. 13.27 Disc with radial slot.

Foetinger Torsion Dynamometer

This dynamometer makes use of mechanical lever principle to magnify readings of angle of twist. It consists of disc A fixed directly to the shaft and disc B which is fixed to a stiff tube, co-axial with the shaft and fixed to the shaft at a distance of 1 from the flange of disc A. The relative angular displacements between discs A and B then measure angle of twist of the shaft over a length l.

Magnification for angle of twist is obtained by using a bell-crank lever having its fulcrum on disc A. The longer end of the bell-crank lever is connected to a bar DE which is hinged to a pointer PE pivoted to an extenssion of disc A.

The smaller end of the bell-crank lever is hinged to another link *FG* which is pivoted to the disc *B*. The pointer *P* is provided with a pencil and records the relative angular displacement between discs *A* and *B* on a paper wrapped round a fixed hollow cylinder, as shown in Fig. 13.28. The magnitude of the torque is represented to a scale by the curved ordinate *LM*, where point *L* lies on the line traced by pointer *P* when no power is transmitted by the shaft.

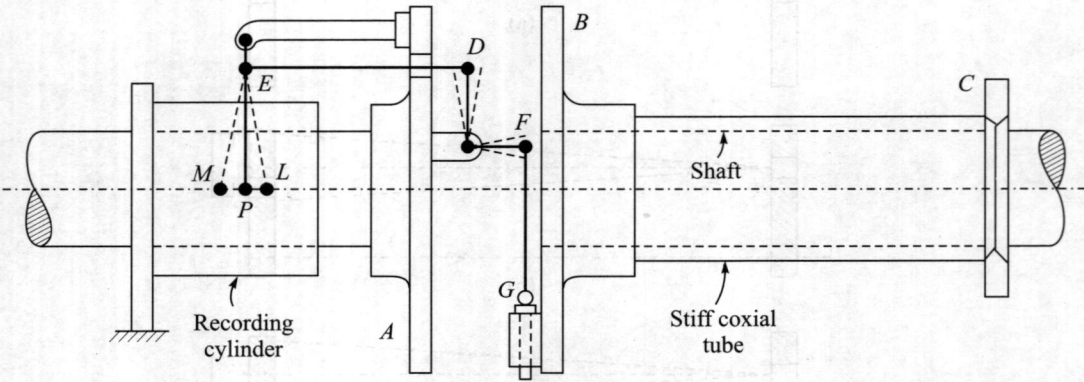

Fig. 13.28   A Torsion Dynamometer.

REVIEW QUESTIONS

13.1 A double block brake arrangement is shown in Fig. 13.29. The coefficient of friction between the cast iron drum and the wooden blocks lined with friction material is 0.3. Determine the force required at the brake lever to absorb 52.2 kw at a drum speed of 600 r.p.m. in the anticlockwise sense.   (***Ans.***   460 N) (DAV Indore, 1979)

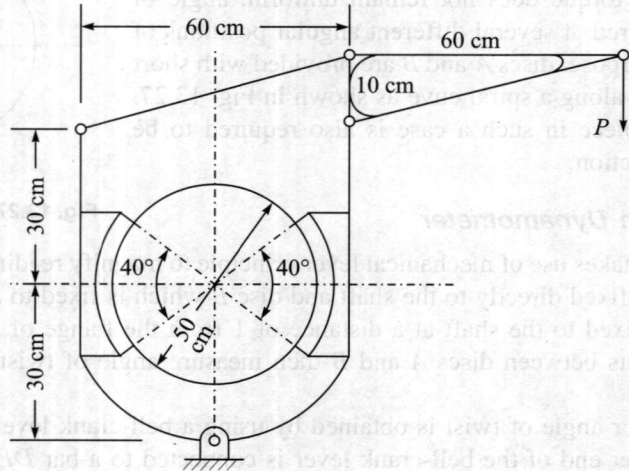

Fig. 13.29   A double block brake.

13.2 Figure 13.30 shows the arrangement and dimensions for the preliminary design of a double block brake to be set by spring providing a force S on the bell crank. The brake is to have a torque rating 11768.4 cm·N at 250 r.p.m. Assuming that angle of contact for each shoe is 120 degrees and that the coefficient of friction for the materials in contact is 0.35, determine the following: (1) The direction of rotation that gives the largest spring force for the rated torque and the value of the spring force S for that direction; (2) Width of the shoe required assuming p.v. = 19614 cm·N per min. per square cm of projected area; (3) The ratio of the shoe width to drum diameter

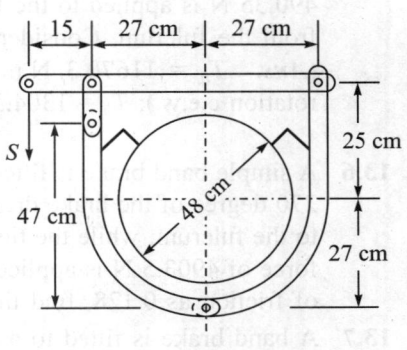

Fig. 13.30

(**Ans.** c.w. direction gives largest $S = 98.1$ N; $b = 12.96$ cm; $\dfrac{b}{d} = 6.27$)

13.3 Determine the torque that may be resisted by the single block brake shown in Fig. 13.31. The brake drum diameter is 25 cm, the contact angle of brake shoe is 90° and the effort applied $P = 627.65$ N. The coefficient of friction between the drum and the shoe-lining $\mu = 0.35$. (**Ans.** $T_b = 6605$ cm·N)

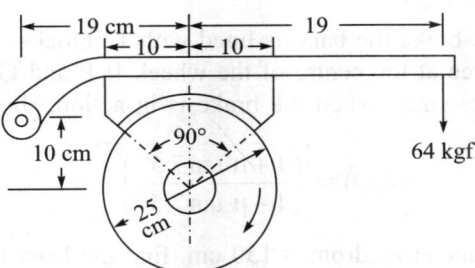

Fig. 13.31

13.4 A block brake shown in Fig. 13.32 is set by a spring that produces a force S on each arm equal to 3432.45 N. The wheel diameter 35 cm and the angle of contact is 100° with each shoe. If the coefficient of friction between the drum and the lining is 0.4, find the maximum torque that the brake is capable of absorbing. (**Ans.** 1336.1 N·m)

13.5 A band brake used for a winch is wound around a drum of 75 cm diameter, keyed to the shaft. The two ends of the band are attached to the pins on the opposite sides of the fulcrum of the brake lever at distances of 2.5 cm and 10 cm from the fulcrum. The angle of lap on the drum is 240 degree. The coeff. of friction is 0.25. Find the torque which can be applied by the brake when a force of

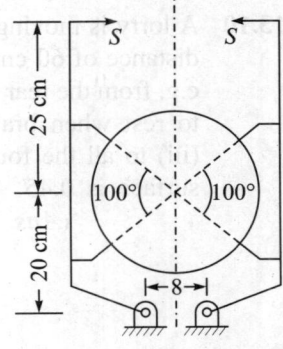

Fig. 13.32

490.35 N is applied to the lever (a) upwards and (b) downwards, at a distance of 1 m from the fulcrum. Consider both directions of rotation.

(*Ans.* T_b = 11670.3 N.m, the force must act upwards to apply the brake (drum rotation c.c.w.); T_b = 1304.3 N·m, the force must act upwards for c.w. drum rotation).

(DAV Indore, 1990 March)

13.6 A simple band brake is fitted on a crane of 60 cm diameter. The band is in contact over 270 degree of the brake drum of 80 cm diameter. The slack end of the band is attached to the fulcrum, while the tight end is attached to the fulcrum of the lever. An operating force of 4903.5 N is applied at a distance of 70 cm from the fulcrum. Assuming coeff. of friction as 0.128, find the braking torque. (*Ans.* T_b = 947.36 N·m)

13.7 A band brake is fitted to a circumference of a wheel of 90 cm diameter, the angle of contact being 220 degree. The band is fixed at one end and a pull of 176.53 N can be applied at the other end. The weight of the wheel is 5344.8 N and its radius of gyration is 37.5 cm and μ = 0.2. If the wheel is rotating at 600 r.p.m. and the brake is fully applied, find the time required for the wheel to come to rest. (*Ans.* Time reqd. = 52.45s)

13.8 A band and block brake having 12 blocks each subtending an angle of 14° at the centre, is applied to a rotating drum having 105 cm diameter. The blocks are 5 cm thick. The two ends of the band are attached to pins on the opposite sides of the brake lever at distances of 2 cm and 8 cm from the fulcrum. What is the minimum force required to be applied at the end of a lever one meter long for the brake to absorb 223.7 kW at 250 r.p.m. Take μ = 0.35. (*Ans.* P = 193.59 N)

13.9 In a band and block-brake the band is lined with 12 blocks of each of which subtends an angle of 15 degree at the centre of the wheel. If P and Q be the greatest and least tensions in the brake strap, when the brake is in action, show that

$$(P/Q) = \left(\frac{1 + \mu \tan 7.5°}{1 - \mu \tan 7.5°} \right)^{12}$$

For an effective diameter of drum = 150 cm, find the least force F required at 56 cm from the fulcrum when the two ends of the band are connected to pins on the opposite sides of the lever at the distances of 3 cm and 15 cm from the fulcrum. Power to be absorbed may be taken as 182 kw at 300 r.p.m. Take μ = 0.4. (*Ans.* P = 392.3 N)

13.10 A lorry is moving on a level road at a speed of 36 km/hr. Its centre of gravity lies at a distance of 60 cm from the ground level. The wheel base is 2.4 m and the distance of c.g. from the rear wheels is 90 cm. Find the distance travelled by the lorry before coming to rest when brakes are applied (i) to the rear wheels, (ii) to the front wheels and (iii) to all the four wheels. The coefficient of friction between the tyres and the road surface is 0.45.

(*Ans.* 21.55 m; 26.82 m; 11.36 m) (Rajasthan Univ., DAV Indore, 1986)

Chapter 14

DYNAMICS OF MACHINES, TURNING MOMENT, FLYWHEEL

14.1 ROLE OF FORCE ANALYSIS IN DESIGN CALCULATIONS

In the preceding chapters on kinematic analysis and synthesis, we were concerned more with the geometry of motion and the lengthwise dimensions of links for obtaining desired motion. The forces that produce such motions or the motions which would result as a consequence of applying such force-system were completely disregarded. Specifications about machine components are, however, incomplete if shape and size is not specified. To determine the size of machine members, knowledge of the magnitude and direction of forces is essential.

Transmission of forces from one member to the other takes place through mating surfaces. Distribution of forces at these boundaries or mating surfaces must be reasonable and attempts must be made to keep their intensities within working limits of the materials constituting the elements of the pair. A too high force operating on sleeve bearing will squeeze out the oil film causing metal-to-metal contact, overheating, and rapid failure of the bearing. Similarly, when the forces between gear teeth are too large, the oil film between the teeth may be squeezed out. A large unit force, in excess of the surface strength of the material, produces flaking and spalling of the metal—culminating in noise, rough motion and eventual failure.

Forces on machine members, if underestimated, lead to a design of insufficient strength and failure would occur rapidly. On the other hand, if these forces are overestimated, designed machine component would have strength more than what is required. This amounts to over-design and the resulting machine is heavier, costlier and may not be competitive with other similar machines. Even if a single machine component turns out to be heavier, the excessive weight results in larger inertia force which is transmitted to the connected members.

Forces in machine members may be due to weight of parts, forces of assembly, forces from applied loads and forces from energy transmitted. Besides, forces of friction, inertia forces, spring forces, impact forces and forces due to changes of temperature do exist. All these forces are required to be considered in the final design of machine components. Forces due to assembly, impact and temperature changes are not intended to be considered in this textbook. Inertia forces (caused by acceleration) are neglected when machine components are analysed for static forces. When inertia forces are also taken into account, the force analysis is called 'Dynamic Force

617

Analysis.' Weights of machine components in many cases are small compared with other static forces and they are therefore neglected in static force analysis.

14.2 LAWS OF MOTION AND D'ALEMBERT'S PRINCIPLE

The fundamental principles of dynamics are Newton's laws of motion. They are not new to most readers but in view of their importance in developing proper understanding, they are reproduced here.

1st Law: A body will remain in state of rest or of uniform motion in a straight line, unless it is acted upon by an external force.

2nd Law: The rate of change of momentum of a body, acted upon by a force or forces, is proportional to the resultant external force and acts in the direction of that force.

3rd Law: For every action of a force, there is an equal and opposite reaction.

The first law can be treated as the 'law of inertia' and states that body starts moving or its velocity changes when it is acted upon by a force. However, no quantitative relation is offered between change in velocity and force. *This law is significant in the sense that it draws line of demarcation between statics and dynamics and lays down that like state of rest, state of uniform motion in a straight line, also belongs to the realm of statics.* This law can be treated as a special case of 2nd law when force **F** and acceleration **A** are both zero.

The second law quantitatively relates change in velocity to the force producing the change. Our everyday experience tells us that material bodies resist any change in their state of motion. The property of matter by virtue of which it resists changes in motion is called *inertia*. Second law of motion, relating rate of change of momentum to the resultant external force **F**, can be expressed mathematically as:

$$\mathbf{F} = \frac{d}{dt}(mV) \tag{14.1}$$

$$= V\left(\frac{dm}{dt}\right) + m\left(\frac{dV}{dt}\right)$$

For constant mass *m*, this becomes

$$\mathbf{F} = m\left(\frac{dV}{dt}\right)$$

or $$\mathbf{F} = mA \tag{14.2}$$

As against this, the third law of motion is quite useful in studying the forces which act upon bodies at rest or upon the bodies in uniform motion (*i.e.* bodies in static equilibrium).

D'Alembert's Principle: Static force analysis serves as a basis for a tentative proportioning of the members. *Final design, particularly in high speed applications, must however take into consideration combined effect of both the static and dynamic force systems. D'Alembert's principle provides a method of converting a problem in dynamics to one in statics.*

Consider a moving body of mass m which is acted upon by a system of coplanar forces $\mathbf{F_1}$, $\mathbf{F_2}$, $\mathbf{F_3}$,..., as shown in Fig. 14.1. Let G be the centre of gravity of the body and $\mathbf{F}$ be the single resultant force. If the resultant force $\mathbf{F}$ is zero and summation of moments of these forces about any single axis is also zero, the body is said to be in static equilibrium. In general, however, the resultant force $\mathbf{F}$ will not be zero, and further, it will not pass through the centre of gravity. Such resultant force $\mathbf{F}$ thus has two effects:

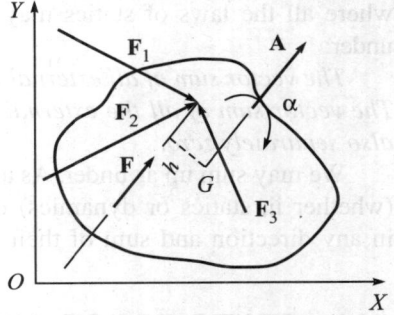

Fig. 14.1 Body in motion under the action of forces.

(i) It produces a linear acceleration $\mathbf{A}$ in the direction of $\mathbf{F}$, given by

$$\mathbf{F} = m\mathbf{A} \tag{14.3}$$

(ii) The line of action of $\mathbf{F}$, which is placed at a perpendicular distance of h from G, also produces an angular acceleration α, given by

$$\mathbf{T} = I\alpha \tag{14.4}$$

where I is the mass moment of inertia of the body about G and T is summation of all external torques and the couple (Fh).

However, when the resultant force $\mathbf{F}$ passes through centre of gravity G, it produces only the linear acceleration $\mathbf{A}$ and the angular acceleration α does not exist.

Effective Force: The single resultant force $\mathbf{F}$ of a system of planer forces $\mathbf{F_1}$, $\mathbf{F_2}$, $\mathbf{F_3}$, ..., which produces a linear acceleration $\mathbf{A}$, given by equation (14.3), is called effective force. The equation (14.3) may also be written as,

$$\mathbf{F} + (-m\mathbf{A}) = 0 \tag{14.5}$$

The quantity $(-m\mathbf{A})$ represents a force that opposes acceleration $\mathbf{A}$ of the body and is called 'Reversed effective force' or inertia force and can be expressed by a term $\mathbf{F_i}$. Hence, equation (14.5) may be rewritten as

$$\mathbf{F} + \mathbf{F_i} = 0 \tag{14.6}$$

Thus, the body may be considered to be in static equilibrium under the combined action of all external forces $\mathbf{F_1}$, $\mathbf{F_2}$, $\mathbf{F_3}$,..., and a reversed effective (*i.e.*, inertia) force $\mathbf{F_i}$. *This concept is known as D' Alembert's principle.*

Proceeding exactly in the same way, equation (14.4) may be rewritten as,

$$\mathbf{T} + (-I\alpha) = 0$$

or

$$\mathbf{T} + \mathbf{T_i} = 0 \tag{14.7}$$

where $\mathbf{T_i}$ is called *reversed effective (or inertia) torque.*

Thus, by adding inertia force and inertia torque to a body which is acted upon by a resultant (effective) force and a resultant torque, the body is brought to equilibrium. Stated in other words, the inclusion of the force $(-m\mathbf{A})$, in the system of applied forces of a dynamic system, converts the problem in dynamics to that in statics. D' Alembert's principle thus aids in the solution of problems in dynamics by permitting them to be solved as problem of 'kinetostatic equilibrium' where all the laws of statics may be applied. D' Alembert's principle may be summarised as under:

The vector sum of all external forces and the inertia forces, acting upon a rigid body, is zero. The vector sum of all the external moments and the inertia torque, acting upon a rigid body, is also separately zero.

We may sum up as under. As a direct consequence of D' Alembert's principle, force analysis (whether in statics or dynamics) can be based on principles of statics, namely, sum of forces in any direction and sum of their moments about any point must be zero.

14.3 STATIC FORCE ANALYSIS

Readers may recall that conditions for a rigid body to be in static equilibrium are:

(i)   Vector sum of all the forces acting on it must be zero; and
(ii) The sum of moments of all the forces acting on the body, about any single axis must be zero.

Mathematically,

$$\Sum F = 0 \tag{14.8}$$

$$\text{and} \qquad \Sum T = 0 \tag{14.9}$$

Free Body Diagram: Implied meaning of term 'body' here is either a single machine component, several components of machine, a portion of a machine or the entire machine itself.

If a mechanical system is in equilibrium as a whole, then each of its component parts is also in equilibrium. It is therefore possible to assume any link in a mechanism to be in equilibrium under the combined influence of all active and passive forces directly applied to it and its own inertia force. *A free body diagram of a member is a complete diagrammatical sketch of the member, physically disconnected from other links and showing all of the external forces acting at their respective points of application. The external forces acting on the member include reactive forces caused by supports and weight of the body.* Strictly speaking, like inertia forces due to acceleration (*e.g.*, centrifugal force), the weight of a body is a 'body force' and acts throughout the body. However, in most cases these body forces can be considered as external loads, acting through body's centre of gravity.

It may be interesting to note that lower pairs are inherently associated with two unknown quantities: either the 'magnitude and direction' of transmitted force (as in the case of turning and rolling pairs), or 'magnitude and the position of line of action' (as in sliding pairs). On the other hand, in higher pairs (i.e., those involving slip-rolling) the 'line of action of transmitted force' is fixed and only magnitude is unknown.

Free body diagram is an essential tool in the force analysis of mechanisms. If the free body diagram is that of a complete machine part, then forces shown on it are the external forces together with moments exerted by adjacent connected parts. If the diagram is for a portion of a machine part, then the forces and moments acting on the cut portion are the internal forces and moments exerted by the part which has been cut away. Figure 14.2 illustrates free body diagrams for three different cases and are self-explanatory.

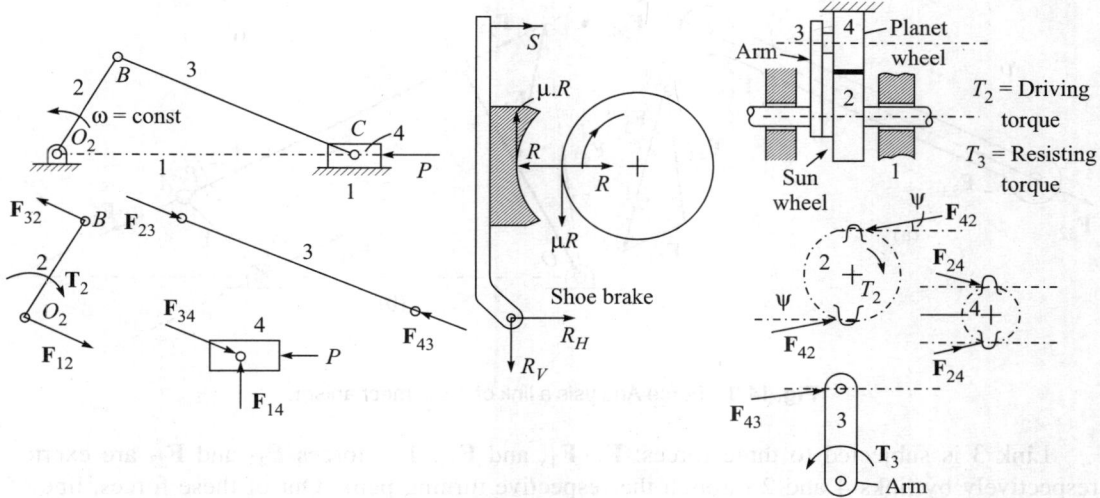

Fig. 14.2 Examples of free-body diagrams.

14.4 STATIC FORCE ANALYSIS FOR MECHANISMS

As a general rule, any link of a mechanism which is connected to two other links through two revolute pairs constitutes a two-force system. Since such links are free to rotate with respect to connected links about the axis of corresponding revolute joint, force exerted by connected links on the 'two-force' link (in the absence of friction) must act along the centre line of that link. In other words, there can be no moment at the pin connection. In the case of a ternary link, which is subjected to forces from three connected links (*i.e.*, three-force member), the direction of resultant force is found by considering summation of forces or moments. As seen under discussions on friction axis, the effect of friction at a turning pair is to shift line of action of force to a position that is tangent to the friction circles. In the analysis to follow, friction forces, together with their effects, have been neglected.

Notation for Denoting Forces: Readers are advised to give special attention to the subscripts used for denoting a force. *A force associated with a single subscript will be taken to mean an external force acting on the link indicated by the subscript.* For instance, external force acting on link 3 of Fig. 14.3(b) is indicated by $\mathbf{F}_3$. *When a force is exerted by one link (say, 3) on another (say, link 4), the symbol is associated with two subscripts the first one indicating the link that exerts the force and the second one indicating link on which the force is exerted.* Thus, in the case cited here, force $\mathbf{F}_{34}$ denotes force exerted by link 3 on link 4, while $\mathbf{F}_{43}$ is the force exerted by link 4 on 3. Needless to say that $\mathbf{F}_{43} = \mathbf{F}_{34}$; their directions are however opposite. This is shown in Fig. 14.3(a).

Static Force Analysis: Consider the four-bar machanism shown in Fig. 14.3(b), in which an external force $\mathbf{F}_3$ is shown to act on link 3. Let us begin the force analysis for link 3.

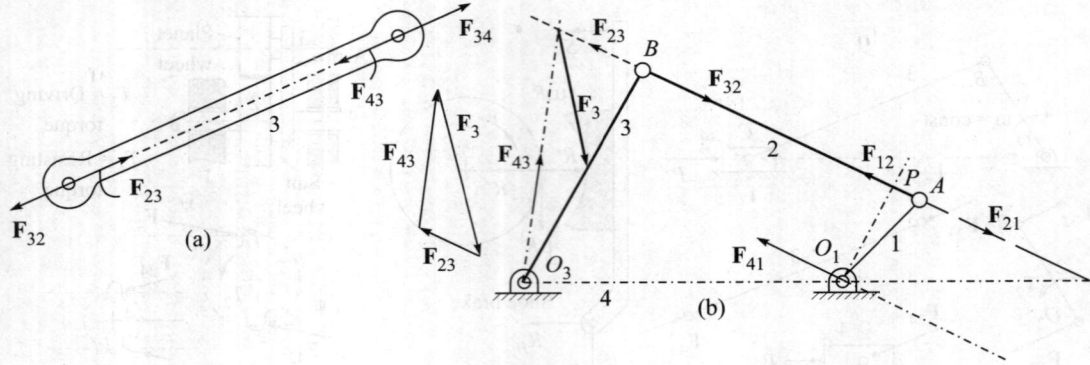

Fig. 14.3 Force Analysis a link of 4-bar mechanism.

Link 3 is subjected to three forces: $\mathbf{F}_3$, $\mathbf{F}_{43}$ and $\mathbf{F}_{23}$. The forces $\mathbf{F}_{43}$ and $\mathbf{F}_{23}$ are exerted respectively by links 4 and 2 through the respective turning pairs. Out of these forces, line of action of force $\mathbf{F}_3$ is known while the force $\mathbf{F}_{23}$ must be equal and opposite to $\mathbf{F}_{32}$ and therefore must be placed along the line of centres of link 2. For the static equilibrium of link 3 all the three forces $\mathbf{F}_3$, $\mathbf{F}_{23}$ and $\mathbf{F}_{43}$ must meet at a common point. Hence for static equilibrium of link 3, force $\mathbf{F}_{43}$ must pass through the point of intersection of forces $\mathbf{F}_3$ and $\mathbf{F}_{23}$ as shown in Fig. 14.3(b). With the magnitude and direction of force $\mathbf{F}_3$ known completely and lines of action of forces $\mathbf{F}_{23}$ and $\mathbf{F}_{43}$ so established, force polygon can be conveniently drawn to a scale so as to obtain magnitude of forces $\mathbf{F}_{43}$ and $\mathbf{F}_{23}$. This is shown in Fig. 14.3.

There are only two forces acting on link 2: $\mathbf{F}_{32}$ exterted by link 3 and $\mathbf{F}_{12}$ exerted by link 1. At a pin connection there can be no moment. Thus at each end of the link 2 there can only be a force. For static equilibrium both these forces must be equal and opposite. Hence, $\mathbf{F}_{12} = \mathbf{F}_{32}$ and acts along the line of centres of link 2.

Link 1 is again a two-force link, subjected to forces $\mathbf{F}_{21}$ and $\mathbf{F}_{41}$. As the link 1 is in static equilibrium, force $\mathbf{F}_{41}$ must be equal and opposite to $\mathbf{F}_{21}$ and act through O_1. For static equilibrium of link 1, the applied (driving) torque T on link 1 must be equal to the torque due to non-collinear forces $\mathbf{F}_{21}$ and $\mathbf{F}_{41}$. Thus,

$$T = \mathbf{F}_{21} \times O_1 P \qquad (14.10)$$

where P is the foot of perpendicular dropped from O_1 on the line of action of force $\mathbf{F}_{21}$.

14.5 MASS MOMENT OF INERTIA AND INERTIA TORQUES

A problem of frequent occurrence is regarding moment of forces about a specified axis, when the forces are distributed over an area. This is particularly true in calculating bending stresses. Magnitude of such forces usually vary with the distance from the moment axis and mathematical

analysis of such a problem usually leads to a form $\int (\text{distance})^2 \times$ elemental area. This integral

is usually referred to as '*area moment of inertia*'. Few others prefer to call this integral as '*second moment of area*' simply because an area cannot possess inertia and the term *moment of inertia* for such an integral, therefore, appears to be a misnomer. The term 'area moment of inertia' is, however, widely used to describe the above integral and we shall continue to use it in the present text.

The second moment of areas in z-y plane about the x and y axes are, respectively,

$$I_x = \int y^2 dA \qquad \text{and} \qquad I_y = \int x^2 dA \qquad (14.11)$$

Calling I_x and I_y as rectangular moments of inertia, the polar moment of inertia is defined as

$$J = I_x + I_y$$

or
$$J = \int (x^2 + y^2) dA$$

or
$$J = \int r^2 dA \qquad (14.12)$$

Sometimes, it is more convenient to express moment of inertia in the form

$$I = Ak^2 \qquad (14.13)$$

Expression (14.13) amounts to assuming the entire area A to be concentrated in a thin ring at a radius of k.

The term k in expression (14.13) is called the *radius of gyration* and is a quantitative measure of the distribution of the area from the moment axis. The moment of inertia of area A about an axis distant d from the centroidal axis is given by (parallel axes theorem),

$$I_x' = I_x + Ad^2_x; \qquad I_y' = I_y + Ad^2_y$$
and
$$I_z' = I_z + Ad^2 \qquad (14.14)$$

Moment of inertia of a volume is a true moment of inertia because a volume is associated with mass, and to distinguish this from second moment of area, it is frequently called *mass moment of inertia*. The inertia integrals for volume are

$$I_x = \int (y^2 + z^2) dm; \qquad I_y = \int (x^2 + z^2) dm$$
and,
$$I_z = \int (y^2 + x^2) dm \qquad (14.15)$$

Products of inertia may also exist and are given by the following set of integrals:

$$I_{xy} = \int xy\, dm; \qquad I_{yz} = \int yz\, dm$$

$$I_{zx} = \int zx\, dm \qquad (14.16)$$

Expressions in equation (14.16) for products of inertia are useful in the sense that when these integrals become zero, they define the three coordinate axes of a body called the principal axes. Corresponding values of integrals in equation (14.15) then define the principal mass moments of inertia. Table 14.1 provides a list of mass moments of inertia (about principal axes) for different geometric shapes of more common occurrences, obtained by solving equation (14.15).

Using parallel axes theorem, mass moment of inertia of a solid, about an axis distant d and parallel to centroidal axis, is given by

$$I = I_G + md^2 \qquad (14.17)$$

where I_G = mass moment of inertia about centroidal axis.

The term radius of gyration used in conjuction with the mass moment of inertia, is given by

$$I_G = mk^2$$

or

$$k = \sqrt{I_G/m} \qquad (14.18)$$

Table 14.1 Mass Moments of Inertia about Principal Axes

Geometrical Shapes		Mass Moments of Inertia
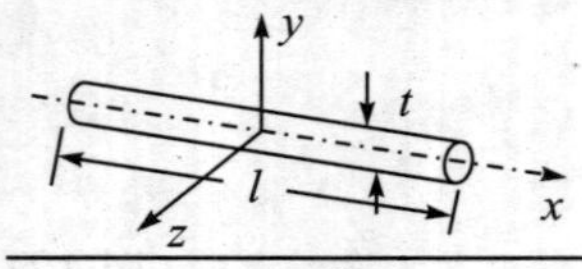	Thin rod ($l > 10t$)	$I_y = I_z = \dfrac{ml^2}{12}$ $I_x = \dfrac{1}{3}\, ml^2$
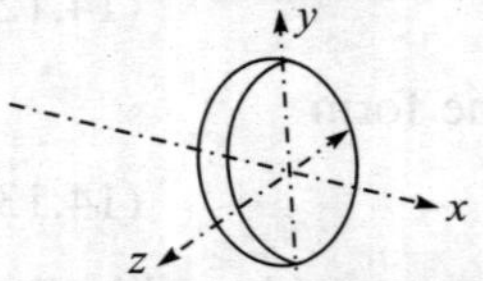	Round disk	$I_x = I_y = \dfrac{mr^2}{4}$ $I_z = \dfrac{mr^2}{2}$
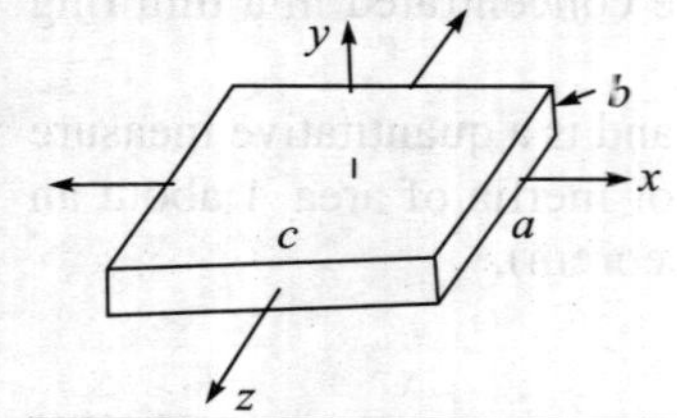	Rectangular Prism	$I_x = \dfrac{m(a^2+b^2)}{12}$; $I_y = \dfrac{m(a^2+c^2)}{12}$ and, $I_z = \dfrac{m(b^2+c^2)}{12}$
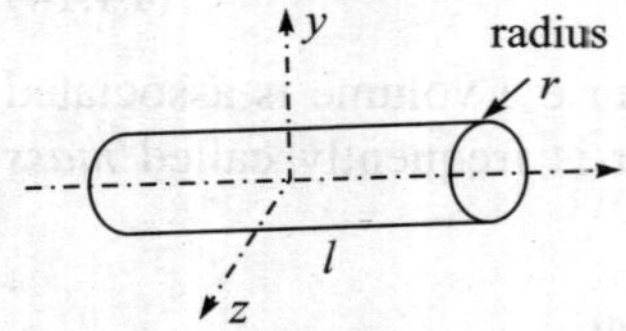	Cylinder (Solid)	$I_x = \dfrac{mr^2}{2}$ $I_y = I_z = \dfrac{m(3r^2+l^2)}{12}$
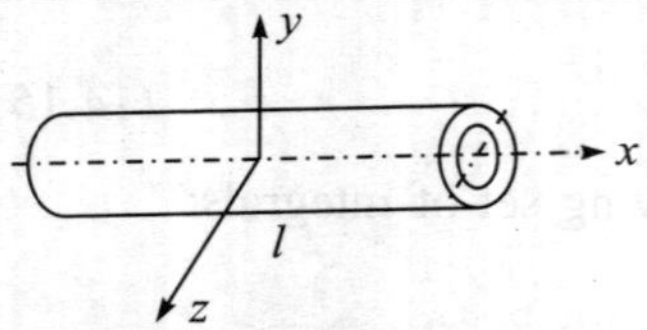	Hollow cylinder (inner rad. = a, outer rad. = b)	$I_y = I_z = \dfrac{m(3a^2+b^2+l^2)}{12}$ $I_x = \dfrac{m(a^2+b^2)}{2}$
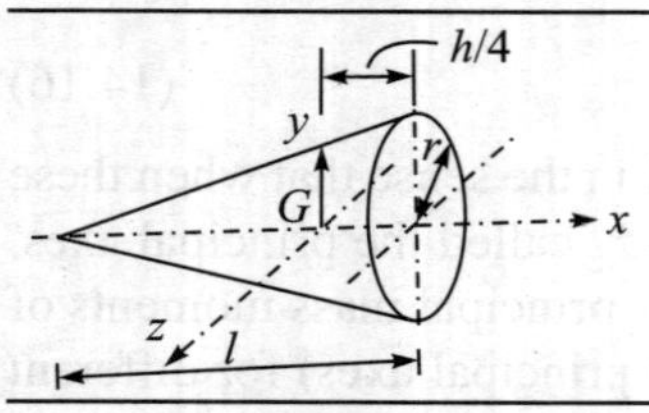	cone	$I_x = \dfrac{3mr^2}{10}$ $I_y = I_z = \dfrac{m(12r^2+3h^2)}{80}$
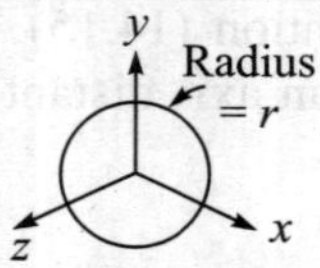	sphere	$I_x = I_y = I_z = \left(\dfrac{2mr^2}{5}\right)$

Units: Since unit of mass in **MKS** units, is kgf.s^2/m, the unit of moment of inertia (= mk^2) is kgf.m.s^2. Also, as the unit of mass in SI units is kg, the unit of moment of inertia in SI units will be kg·m^2.

Couple: According to Newton's second law, rate of change of angular momentum is equal to the external torque or couple and is given by

$$C = \frac{d}{dt}(I\omega)$$

or

$$C = I\frac{d\omega}{dt} = I\alpha \tag{14.18a}$$

Clearly, when a resultant torque T is acting on a body of mass moment of inertia I, it will produce an angular acceleration α in the body, given by

$$\alpha = (T/I) \tag{14.19}$$

14.6 SIMPLE HARMONIC MOTION

Simple experimental methods of determining moments of inertia of objects are based on the principles of simple harmonic motion. A body is said to execute simple harmonic motion (abbreviated as SHM) when it oscillates about a mean equilibrium position (abbreviated as m.e.p.) in such a way that its acceleration is always directed towards m.e.p. and is proportional to the displacement from it.

Referring to Fig. 14.4(a), when a point P moves with uniform speed along the circumference of a circle of radius r, the projection M of point P on horizontal diameter AB describes SHM. The amplitude of oscillations of point M is equal to OB, the radius r of the circle.

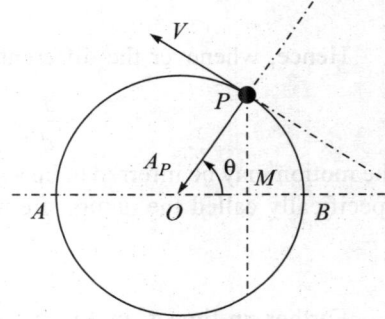

Fig. 14.4(a)   Simple Harmonic Motion.

Let the radius OP rotate uniformly in c.c.w. sense at ω radians per second. Then,

$$V = (r\omega) \tag{14.20}$$

Further, the only acceleration to which point P is subjected is the centripetal acceleration directed towards centre of rotation O and is given by

$$A_P = \left(\frac{V^2}{OP}\right) = \frac{(r\omega)^2}{r} = r\omega^2 \tag{14.21}$$

Taking accelerations directed towards O as negative, the acceleration of point M is given by component of acceleration A_P along the line of stroke. Thus, the acceleration of point M is

$$A_M = -A_P\cos\theta = -(r\omega^2)\cos\theta \tag{14.22}$$

However, from Fig. 14.4(a), the displacement of point M from the m.e.p. is given by

$$x_M = OM = OP\cos\theta = r\cos\theta \tag{14.23}$$

Hence, from equations (14.22) and (14.23),

$$\frac{|\text{Acceleration of } M|}{|\text{Displacement of } M|} = \frac{(r\omega^2)\cos\theta}{(r\cos\theta)} = (\omega^2) = \text{constant} \qquad (14.24)$$

Therefore

$$\omega = \sqrt{\frac{(\text{Amplitude of acceleration of } M)}{(\text{Displacement amplitude of } M)}} \qquad (14.25)$$

Now, as x_M is the displacement of point M from O at time t, then the acceleration of M is given by

$$\frac{d^2}{dt^2}(x_M) = -(r\omega^2)\cos\theta = -\omega^2(r\cos\theta)$$

or

$$A_M = -\omega^2(x_M) \qquad (14.26)$$

which follows from equations (14.22) and (14.23). Rearranging equation (14.26) and deleting the subscript M from x,

$$\left(\frac{d^2x}{dt^2}\right) + \omega^2(x) = 0 \qquad (14.27)$$

Hence, whenever the differential equation of motion is of the form

$$\frac{d^2x}{dt^2} + bx = 0, \text{ where } b \text{ is constant,}$$

the motion may be inferred to be a simple harmonic motion, and the frequency of vibration (more specifically called the natural frequency ω_n) is given by

$$\omega_n \equiv \omega = \sqrt{b} \qquad (14.28)$$

Further, in time τ, in which the radius OP rotates through 2π radian, point M moves from position M to A, followed by A to B and back from B to M. Thus a complete cycle of motion of M is completed in time τ.

Hence, the time period of oscillatory motion of point M is given by

$$\tau = \left(\frac{2\pi}{\omega}\right) = \frac{2\pi}{\sqrt{b}} \qquad (14.29)$$

And frequency in cycles per second,

$$f = \frac{1}{\tau} = \frac{\sqrt{b}}{2\pi} \qquad (14.30)$$

Examples of Simple Harmonic Motion (Determination of M.I.)

(a) *Simple Pendulum:* A simple pendulum consists of a heavy bob of negligible dimensions suspended vertically by means of an inextensible weightless cord. Referring to Fig. 14.4(b), let

W = weight of the pendulum bob,
L = length of the cord from the point of suspension to the centre of the bob.

Let the cord carrying the bob be displaced from the vertical centre line through a small angle θ, and then released to execute free oscillations.

Then the restoring couple on bob, which tends to bring it back to the vertical equilibrium position, is

$$C = W(L\sin\theta) \qquad (14.31)$$

Considering expansion of function sinθ in terms of angle θ in radians, we have

$$\sin\theta \simeq \theta - \frac{\theta^3}{3!} + \frac{\theta^5}{5!} - \frac{\theta^7}{7!} + \dots \text{ (for } -\infty < \theta < \infty) \qquad (14.32)$$

For an angle of oscillation of $\theta = \dfrac{\pi}{9} = 0.349$ radians

(*i.e.*, 20°), the first, second and third terms in the above expansion are respectively 0.349, 0.0070888, 0.00004319. Thus the first term is approximately 49 times that of second term and 8080.5 times that of third term. Again for π/18 =0.1745 radian (10°), the second and third terms are respectively 0.005075 and 0.0000077268 times the first term. Thus for small angles θ (say ≤ 10°),

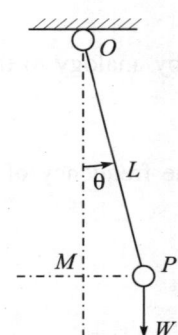

$$\sin\theta \approx \theta \text{ (in radians)} \qquad (14.33)$$

Hence, the restoring couple from equation (14.31),

Fig. 14.4(b) A simple pendulum.

$$C = (WL)\theta \qquad (14.34)$$

Again, using parallel axis theorem, moment of inertia of bob about the pivot O,

$$I_O = I_G + \left(\frac{W}{g}\right)L^2 \qquad (14.35)$$

Neglecting the moment of inertia of bob about its own c.g. (*i.e.*, I_G) in comparison to the term $\left(\dfrac{W}{g}\right)L^2$ we have from equation (14.35),

$$I_O \approx \left(\frac{W}{g}\right)L^2$$

Hence, from Newton's second law,

$$C = \text{Rate of change of angular momentum}$$

$$= \frac{d}{dt}(I_O\omega), \text{ where } \omega = \text{angular velocity of pendulum}$$

Since the moment of inertia I_O is generally constant,

$$C = I_O \frac{d\omega}{dt}$$

or

$$C = I_O \alpha \qquad (14.36)$$

where α = angular acceleration of the bob and the cord of pendulum.

Hence, from a comparison of right-hand sides of equations (14.34) and (14.36),

$$I_o \alpha = (WL)\theta$$

or

$$\alpha = \frac{(WL)\theta}{(W/g)L^2} = g(\theta/L) \qquad (14.37)$$

Hence by analogy to linear S.H.M., $b = \dfrac{\text{(Angular acceleration of bob)}}{\text{(Angular displacement of bob)}} \equiv \dfrac{g(\theta/L)}{\theta}$

or

$$b = (g/L) \qquad (14.38)$$

Thus the frequency of oscillations,

$$\omega_n = \sqrt{b} = \sqrt{g/L} \qquad (14.39)$$

and

$$f_n = \frac{1}{2\pi}\sqrt{b} = \frac{1}{2\pi}\sqrt{g/L}$$

And the time period $\qquad \tau = \dfrac{1}{f_n} = 2\pi\sqrt{(L/g)} \qquad (14.40)$

(b) The Compound Pendulum: A rigid body suspended vertically so as to oscillate freely in a plane with small amplitude, under the action of gravity, constitutes a compound pendulum. In contrast to the case of a simple pendulum, where the entire mass is actually concentrated in bob, the mass here is distributed all over the body. The analysis problem can be simplified by assuming that the entire mass m of the body of weight W is concentrated at its centre of gravity G (see Fig. 14.5).

Let k = radius of gyration of the rigid body about an axis through the c.g. perpendicular to the plane of oscillation, and a = distance of c.g. from the point of suspension.

Then, the mass moment of inertia of the body about its centroidal axis,

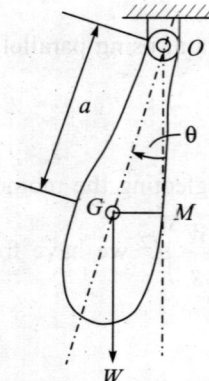

Fig. 14.5 A compound pendulum.

$$I_G = (mk^2)$$

and the mass moment of inertia about axis through O parallel to the centroidal axis is

$$I_O = (mk^2) + ma^2$$

or

$$I_O = m(k^2 + a^2) \qquad (14.41)$$

The restoring couple

$$C = Wa\sin\theta \approx Wa\theta \quad \text{(for small angles } \theta) \tag{14.42}$$

The angular acceleration α of the pendulum, is such that $C = I_o \alpha$

or,
$$\alpha = \frac{C}{I_O} \approx \frac{Wa\theta}{m(k^2 + a^2)}$$

or
$$\alpha \approx \frac{ga\theta}{(k^2 + a^2)} \tag{14.43}$$

Hence the ratio,

$$b = \frac{\text{(Angular acceleration of pendulum)}}{\text{(Angular displacement of pendulum)}}$$

or
$$b = \frac{ga\theta}{(k^2 + a^2)} \times \frac{1}{\theta} = \frac{ga}{(k^2 + a^2)} \tag{14.44}$$

Hence natural frequency of oscillations,

$$\omega_n = \sqrt{b} = \sqrt{\frac{ga}{(k^2 + a^2)}}$$

and
$$f_n = \frac{1}{2\pi}\sqrt{b} = \frac{1}{2\pi}\sqrt{\frac{ga}{(k^2 + a^2)}} \tag{14.45}$$

The equivalent length of a simple pendulum, which would have the same frequency of vibration [compare with equation (14.39)]

$$L_E = \frac{(k^2 + a^2)}{a} = \left(\frac{k^2}{a} + a\right) \tag{14.46}$$

(c) Torsional Pendulum (Trifiliar Suspension): A disc, a flywheel or an equilateral triangular plate suspended with its axis vertical from a ceiling by three long flexible and parallel wires of equal length l, constitute what is popularly known as a trifiliar suspension system. The wires are attached to the plate/disc/flywheel in such a way that the three points form corners of an equilateral triangle—keeping a constant distance a of each point from the axis of disc or plate as shown in Fig. 14.6.

If the disc/plate is twisted about its axis through a small angle and then released, it will oscillate with frequency f_n, which may be related to the dimensions of the system. Since the three wires are initially vertical and attached symmetrically with respect to the centroidal axis of the plate, the initial tension in each wire is

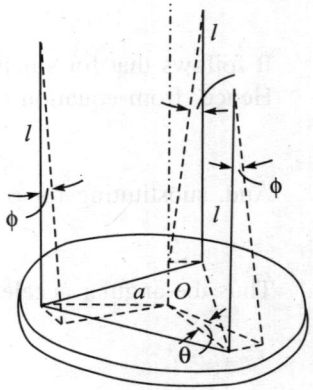

Fig. 14.6 Trifiliar suspension.

$T = W/3$ where W is the weight of the disc/plate.

For an angular displacement θ of the disc/plate, as shown in the figure, each wire undergoes an angular displacement of ϕ and for small displacements,

$$l\phi \approx a\theta$$

or

$$\phi \approx \left(\frac{a}{l}\right)\theta \qquad (14.47)$$

If T' be the tension in each wire subsequent to their angular displacements, then

$$T'\cos\phi = W/3$$

Hence

$$T' = \frac{W}{(3\cos\phi)} \qquad (14.48)$$

In the twisted condition of the plate or disc, the wires become inclined to the surface of the plate at an angle of $\left(\frac{\pi}{2}-\phi\right)$. Hence, the moment at O due to the horizontal component of T' in each wire, perpendicular to the radius 'a' is

$$T'_H = T'\cos\left(\frac{\pi}{2}-\phi\right)\times a$$

Hence total restoring couple about O due to tension T' in each of the three wires is given by—

$$C = 3T'\cos\left(\frac{\pi}{2}-\phi\right)\times a = 3\left[\frac{W}{3\cos\phi}\right]\sin\phi\times a$$

or

$$C = Wa\tan\phi \qquad (14.49)$$

Again, from the expanded series for $\tan\phi$, namely,

$$\tan\phi = \phi + \frac{\phi^3}{3} + \frac{2\phi^5}{15} + \frac{17\phi^7}{315} + \dots$$

It follows that for small angles ϕ, $\tan\phi \approx \phi$
Hence, from equation (14.49),

$$C = Wa\phi$$

And, substituting for ϕ from equation (14.47),

$$C = Wa^2(\theta/l) \qquad (14.50)$$

Thus the angular acceleration towards the equilibrium position,

$$\alpha = \frac{(\text{Restoring torque})}{L}$$

or

$$\alpha = \frac{(Wa^2\theta/l)}{\left(\frac{W}{g}k^2\right)} = \left(\frac{g}{l}\right)\left(\frac{a^2}{k^2}\right)\theta \qquad (14.51)$$

Hence
$$b = \frac{(\text{angular acceleration})}{(\text{angular displacement})} = \left(\frac{g}{l}\right)\left(\frac{a^2}{k^2}\right)\frac{\theta}{\theta} = \left(\frac{g}{l}\right)\left(\frac{a^2}{k^2}\right)$$

Hence natural frequency

$$\omega_n = \sqrt{b} = \left(\frac{a}{k}\right)\sqrt{\frac{g}{l}} \text{ rad/s}$$

and

$$f_n = \frac{1}{2\pi}\left(\frac{a}{k}\right)\sqrt{\frac{g}{l}} \tag{14.52}$$

The trifiliar suspension is used particularly for finding out moment of inertia of bodies which are not bar type, but have a broad base. As against this in bifiliar suspension, members like connecting rod are suspended by two vertical wires placed symmetrically with respect to the c.g.

EXAMPLE 14.1 The connecting rod of a diesel engine weighs 556 N, the distance between the bearing centres is 90 cm and the diameter of the crank pin bearing 13 cm and that of the wrist pin bearing is 7.0 cm. When suspended vertically on a knife-edge through the crank pin bearing, it performs 61 oscillations in 100 s and with knife-edge through the wrist pin bearing, it makes 55 oscillations in the same time interval. Determine the moment of inertia about the axis through the centre of gravity (Note that the position of c.g. is unknown).

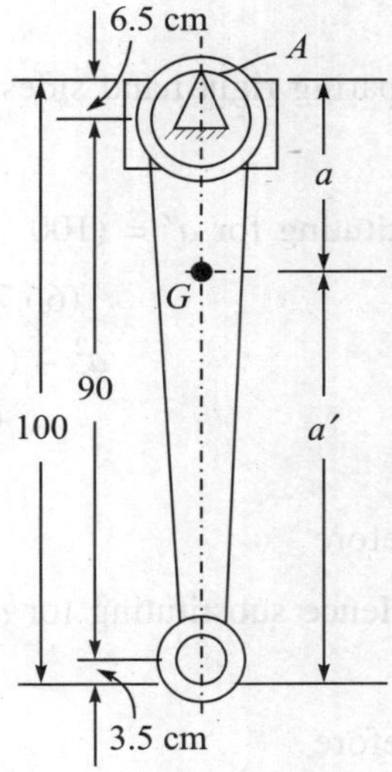

Fig. 14.7 Connecting rod suspended at big end bearing.

Solution: Refer to Fig. 14.7.

For a compound pendulum, the natural frequency is given by (equation 14.45)

$$f_n = \frac{1}{2\pi}\sqrt{\frac{ga}{(k^2 + a^2)}}$$

and the time period by,

$$\tau = 2\pi\sqrt{\frac{(k^2 + a^2)}{ga}}$$

Thus, when the connecting rod is suspended on knife-edge at the crank-pin end,

$$\left(\frac{100}{61}\right) = 2\pi\sqrt{\frac{(k^2 + a^2)}{ga}}$$

or
$$\frac{k^2 + a^2}{a} = \left(\frac{100}{61}\right)^2 \times \frac{981}{4\pi^2} = 66.78$$

Therefore
$$k^2 = (66.78)\,a - a^2 \tag{i}$$

Also, when the rod is suspended from a knife-edge at gudgeon pin (Note that $aa' \neq k^2$)

$$\tau' = 2\pi\sqrt{\frac{(k^2 + a'^2)}{a'g}}$$

or

$$\left(\frac{100}{55}\right) = 2\pi\sqrt{\frac{(k^2 + a'^2)}{a'g}}$$

or

$$\frac{(k^2 + a'^2)}{a'} = \left(\frac{100}{55}\right)^2 \times 981/4\pi^2 = 82.145$$

or

$$k^2 = (82.145)a' - a'^2 \qquad \text{(ii)}$$

Comparing right hand sides of equations (i) & (ii),

$$(66.78)a - a^2 = (82.145)a' - a'^2$$

Substituting for $a' = (100 - a)$ in above equation,

$$(66.78)a - a^2 = (82.145)(100 - a) - (100 - a)^2$$

or

$$a^2 - (148.92)a + 8214.5 - (100)^2 - a^2 + 200a = 0$$

or

$$(51.08)a = 1785.5$$

or

$$a = 34.95 \text{ cm}$$

Therefore

$$b = (100 - 34.95) = 65.05 \text{ cm}$$

Hence substituting for a in equation (i),

$$k^2 = (66.78)(34.95) - (34.95)^2 = 1112.4585$$

Therefore

$$k = 33.35 \text{ cm}$$

Therefore

$$\text{M.I.} = \left(\frac{w}{g}\right)k^2$$

or

$$I = \left(\frac{556}{9.81}\right)(0.3335)^2 = (6.304) \text{ kg·m}^2 \qquad \textbf{Ans.}$$

EXAMPLE 14.2 A connecting rod of mass 5.5 kg is placed on a horizontal platform whose mass is 1.5 kg. It is suspended by three equal wires, each 1.25 m long from a rigid support. The wires are equally spaced round the circumference of a circle of 125 mm radius. When the c.g. of the connecting rod coincides with the axis of the circle, the platform makes 10 angular oscillations in 30 s. Determine the mass moment of inertia about an axis through its c.g. (Oxford Univ.)

Solution: a = distance of wire suppport from centroidal axis = 12.5 cm

l = length of wire = 125 cm

m = mass of connect rod = 5.5 kg

m' = mass of platform = 1.5 kg

Time period of oscillations = $\tau = \dfrac{30}{10} = 3s$

Therefore

$$f_n = \dfrac{1}{\tau} = \dfrac{10}{30} = (1/3) \text{ c.p.s.}$$

Therefore From $f_n = \dfrac{1}{2\pi}\left(\dfrac{a}{k}\right)\sqrt{g/l}$ Hence, $f_n = \dfrac{1}{2\pi}\left(\dfrac{12.5}{k}\right)\sqrt{g/(125)}$

Therefore

$$k^2 = \left(\dfrac{12.5 \times 3}{2\pi}\right)^2 \times \dfrac{981}{125} = 279.55 \text{ cm}^2$$

Therefore

$$k = 16.72 \text{ cm}$$

Hence, moment of inertia about an axis through c.g.,

$$I = mk^2$$
$$= (5.5 + 1.5) \times 279.55 = 1956.85 \text{ kg·cm}^2 \qquad \textbf{Ans.}$$

(*Note:* Value of I so calculated above is a joint M.I. of the platform and rod. To obtain M.I. of the rod, it is necessary to determine time period when the platform is unloaded. This will enable to determine M.I. of the platform alone. The difference then gives moment of inertia of connecting rod alone).

Measurement of Mass Moment of Inertia

In general, shape of a body is so complex that it is impossible to compute the moment of inertia using geometry. For such bodies, it is usually possible to obtain moment of inertia by observing certain dynamic behaviour (e.g., time period of oscillatory motion) for a known input.

Connecting rods and cranks are representative of a class of problems in which the bodies are so shaped that the mass can be assumed to lie in a single plane. If these bodies can be weighed and their centres of gravity located, they can be suspended like a pendulum to execute oscillatory motion. Measurement of time period of oscillations then enables us to compute mass moment of inertia of the machine part. A spoked wheel, a gear, a pulley and even a connecting rod can be suspended on a knife-edge at the rim or at the small/big end bearing. Principle of compound pendulum then enables to establish the mass moment of inertia.

A torsional pendulum (Fig. 14.8) can also be used to establish mass moment of inertia, without actually weighing the body. The rotor of inertia I is connected to a wire or slender rod at the mass centre of inertia. Torsional stiffness k_t of the rod or wire is the torque required per unit twist angle. The torsional stiffness is either known or is computed from the size and material of rod/

wire. The time period in this case is given by $\tau = 2\pi\sqrt{I/k_t}$.

Thus, a determination of time period, also yields the moment of inertia I.

A trifiliar pendulum (Fig. 14.6) provides a very accurate method of measuring mass moment of inertia. The platform

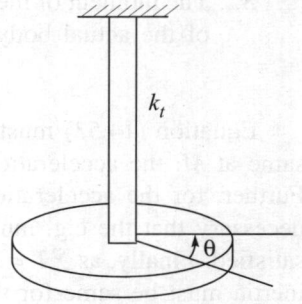

Fig. 14.8 Torsional Pendulum.

should be light and the wires/strings must be of equal length. The part, whose moment of inertia is required, is placed carefully on the platform ensuring that its centre of mass coincides with the centre of platform. The platform is then made to oscillate in such a way that the centroidal axis does not change position, and time period is determined, once when platform is empty and next when platform carries the given object.

Thus, from equation (14.52), when platform is empty,

$$\tau = \frac{1}{f_n} = 2\pi\left(\frac{k}{a}\right)\sqrt{l/g} \qquad \text{or} \qquad k = \frac{a\tau}{2\pi}\sqrt{g/l}$$

and when platform carries the given object,

$$\tau' = \frac{1}{f_n'} = 2\pi\left(\frac{k'}{a}\right)\sqrt{l/g} \qquad \text{or} \qquad k' = \frac{a\tau'}{2\pi}\sqrt{g/l}$$

If M be the mass of object and m the mass of platform, then $(M + m)k'^2 - mk^2 = I$, the M.I. of the object.

14.7 DYNAMICALLY EQUIVALENT TWO MASS SYSTEM

Occasionally, problems of dynamics may be simplified using the concept of dynamic equivalence. Thus, a rigid body of mass M and weight W, and having a radius of gyration about centroidal axis as k, can be replaced by an equivalent system of two rigidly connected masses m_A and m_B, situated respectively at distances a and b from the c.g. In order to be equivalent dynamically, the two-mass system must satisfy following conditions:

1. The sum of the two masses must be equal to the mass of the rigid body *i.e.*,

$$m_A + m_B = M \tag{14.53}$$

or

$$w_A + w_B = W$$

2. The centre of gravity of the two masses must be in the same relative position as the centre of gravity of the body, *i.e.*,

$$(m_A)a = (m_B)b \tag{14.54}$$

or

$$(w_A)a = (w_B)b$$

3. The moment of inertia of the two mass system must be the same as the moment of inertia of the actual body, *i.e.*,

$$(m_A)a^2 + (m_B)b^2 = Mk^2 \tag{14.55}$$

Equation (14.53) must be satisfied because $\Sigma F = MA_G$ and, unless the total mass remains same at M, the acceleration of c.g. *i.e.*, A_G will not remain same for the equivalent systems. Further, for the acceleration of c.g. of the two equivalent systems to remain same, it is also necessary that the c.g. must have the same relative location, requiring equation (14.54) to be satisfied. Finally, as $\Sigma T = I\alpha$ and as α is to be the same for equivalent systems, the moment of inertia must be same for the two equivalent systems.

As against three equations (14.53) through (14.55), there are four unknowns: m_A, m_B, a and b. Any one of these unknowns may be assumed, and the other three will be determined from above equations. To establish a compact relation, substitute for m_B from equation (14.54) in (14.53). Thus,

$$m_A + (a/b)m_A = M$$

or

$$m_A = Mb/(a + b) \qquad (14.56)$$

Again, substituting for m_A in equation (14.54).

$$m_B = Ma/(a + b) \qquad (14.57)$$

Finally substituting for m_A and m_B in equation (14.55),

$$\left(\frac{a^2 b}{a + b}\right) M + \left(\frac{ab^2}{a + b}\right) M = Mk^2$$

or

$$ab = k^2 \qquad (14.58)$$

It follows from equation (14.58) that while selecting locations A and B for masses m_A and m_B, the choice of both the distances a and b cannot be arbitrary. In other words, position of only one of the two masses can be prescribed arbitrarily, the position of the other mass is then automatically decided by equation (14.58).

There are situations when the radius of gyration k of a body is not known and it is necessary to find out location B of second mass for a dynamically equivalent two-mass system. A simple and convenient approach is to suspend the body vertically so that the body is free to swing about an axis through the assumed location A of the first mass. The location B of the second mass then coincides with the centre of the percussion. The distance AB corresponds to equivalent length of simple pendulum.

Figure 14.9(a) illustrates a rigid body of mass M with centre of mass at G. The figure also shows dynamically equivalent two-mass system of masses m_A and m_B connected together by a rigid weightless rod. Figure 14.9(b) is an illustration of a rigid body that is allowed to swing freely in the vertical plane with knife-edge support at one of the two equivalent mass locations, say A. Corresponding dynamically equivalent two-mass system AB is shown in Fig. 14.9(c).

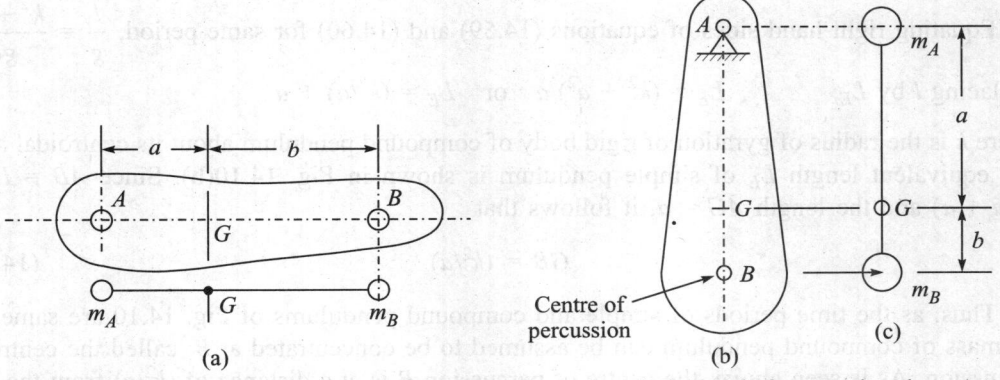

(a)	(b) (c)
Dynamically equivalent two mass system	A rigid body free to swing about A

Centre of percussion

Fig. 14.9

14.8 CENTRE OF PERCUSSION

A body, which is free to rotate about a point A, is always associated with an another point P such that, when stuck with a force at P, there is no force of reaction at A. The right point on a cricket bat or a baseball bat to hit the ball is the centre of percussion associated with the point of hand grip on the bat (centre of rotation). Hitting the ball at any point other than the centre of percussion results in a stinging force delivered to the hand.

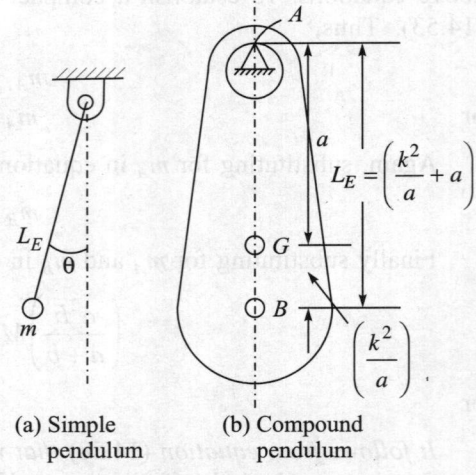

$$L_E = \left(\frac{k^2}{a} + a \right)$$

$$\left(\frac{k^2}{a} \right)$$

(a) Simple pendulum (b) Compound pendulum

Fig. 14.10 Compound pendulum.

Concept of centre of percussion is quite useful in the study of dynamics. To develop the concept, consider a simple and compound pendulum as shown in Figs. 14.10(a) and (b). The time period of oscillation of a simple pendulum of length l is given by

$$\tau = 2\pi \sqrt{l/g} \qquad (14.59)$$

Compound pendulum is a body that oscillates in a manner similar to that of a simple pendulum, but whose mass is distributed and not concentrated as in the case of a simple pendulum. *The mass of compound pendulum may be considered concentrated at a single point so as to produce same time period of oscillation. When this is possible, the point is called centre of percussion.* The length of resulting simple pendulum is called equivalent length L_E of simple pendulum. Readers may recall that the time period of oscillation of a compound pendulum, as derived from equation (14.45) is,

$$\tau' = 2\pi \sqrt{\frac{k^2 + a^2}{ga}} \qquad (14.60)$$

Equating right hand sides of equations (14.59) and (14.60) for same period, $\dfrac{l}{g} = \dfrac{k^2 + a^2}{ga}$

Replacing l by L_E, $\qquad L_E = (k^2 + a^2)/a \quad$ or $\quad L_E = (k^2/a) + a$

where k is the radius of gyration of rigid body of compound pendulum about its centroidal axis. The equivalent length L_E of simple pendulum is shown in Fig. 14.10(b). Since $AB = L_E = (k^2/a + a)$ and the length $AG = a$, it follows that

$$GB = (k^2/a) \qquad (14.61)$$

Thus, as the time periods of simple and compound pendulums of Fig. 14.10 are same, all the mass of compound pendulum can be assumed to be concentrated at B, called the centre of percussion. As is seen above, the centre of percussion B is at a distance of (k^2/a) from the c.g.

It is important to note that point B is the centre of percussion relative to point of suspension A. We cannot speak merely of the centre of percussion of a body without any reference to point of suspension on the body. Thus if the body is suspended from some other point, the centre of percussion would be at some other point B'.

If a pendulum is given an acceleration α about the point of suspension, a single force at the centre of percussion can be shown to replace the inertia force and inertia couple. For instance, in Fig. 14.11(a), the compound pendulum is subjected to an angular acceleration α in c.c.w. sense. Hence an inertia force $F_i = MA_G$ acts at G in a sense opposite to that of linear acceleration $A_G = a\alpha$ and this is also accompanied by inertia couple $C = I\alpha$ which acts in a sense opposite to that of α. In Fig. 14.11(b), a single force F_i at the centre of percussion in shown to produce a couple, $C = (F_i)h$

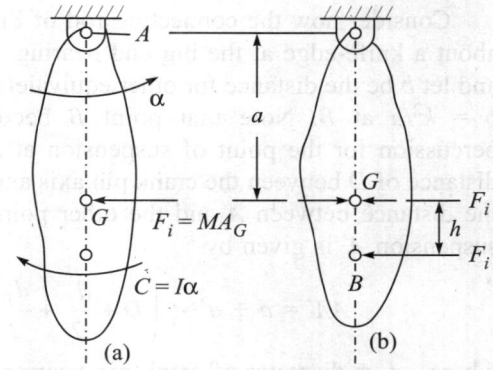

Fig. 14.11 Inertia Force Equivalent in Compound Pendulum.

or
$$I\alpha = (F_i)h = (MA_G)h$$

Hence
$$h = \frac{I\alpha}{MA_G} = \frac{(Mk^2)\alpha}{(M)A_G} = k^2\left(\frac{\alpha}{A_G}\right)$$

or,
$$h = k^2\left(\frac{\alpha}{a\alpha}\right) = \left(\frac{k^2}{a}\right) \qquad (14.62)$$

Thus h turns out to be equal to distance GB of centre of percussion from c.g. The equal and opposite forces F_i at G and B mutually cancel out in Fig. 14.11(b). This means that if the pendulum is struck with a blow at the centre of percussion, no reaction will be produced at the point of suspension.

Significance of the concept of centre of percussion can be experienced by a batsman in the form of stinging effect on hands when the cricket ball or base ball hits the bat at any point other than the centre of percussion. The concept is also of importance in the design of impact testing machines, tilt-hammers, etc.

Location of Point B When Radius of Gyration k and 'C. G.' are Unknown

As the systems in Figs. 14.10(a) and (b) are dynamically equivalent, the time period of oscillations of the physical (compound) and mathematical (simple) pendulum are equal. This forms the basis of a method for obtaining moment of inertia of bodies like a connecting rod of I.C. engines. Referring to Fig. 14.10(b), if a be the distance of c.g. from the point of suspension, the equivalent length L_E of a simple pendulum is given by

$$L_E = \left(\frac{k^2}{a} + a\right)$$

where k = radius of gyration of compound pendulum about the centroidal axis, and the time period of oscillation is given by

$$\tau = 2\pi\sqrt{L_E/g}$$

Consider now the connecting rod of Fig. 14.12, which is free to oscillate in vertical plane about a knife-edge at the big end bearing. Let a be the distance of point of support from c.g. and let b be the distance for other equivalent mass given by $b = k^2/a$ at B. Note that point B becomes centre of percussion for the point of suspension at A. For a centre distance of D between the crank pin axis and wrist pin axis, the distance between A and the other point of knife-edge suspension A' is given by

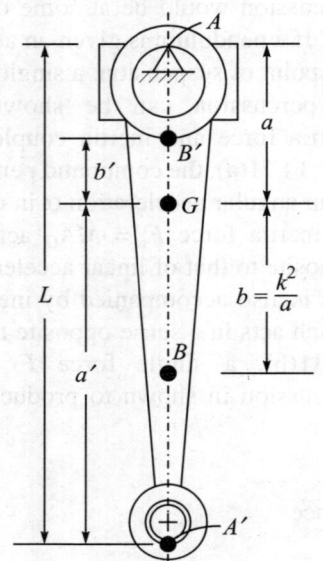

$$AA' = a + a' = \left(D + \frac{d_c}{2} + \frac{d_p}{2} \right) = L, \text{ say.}$$

where d_c = diameter of crank pin bearing, and
d_p = diameter of gudgeon pin bearing.

Again, for a point of suspension A' at the knife-edge in gudgeon pin bearing, let B' be the centre of percussion such that $A'G = a'$ and $B'G = b'$.

Equivalent length of simple pendulum,

(a) with knife-edge at A is $L_E = a + b$ and (b) with knife-edge at A' is $L_E' = a' + b'$

Thus the corresponding time periods are,

Fig. 14.12 Location of second mass, k and CG.

$$\tau_A = 2\pi \sqrt{\frac{(a+b)}{g}} \qquad \text{and} \qquad \tau_A' = 2\pi \sqrt{\frac{(a'+b')}{g}}$$

Hence
$$(a + b) = \frac{g\tau_A^2}{4\pi^2} \qquad \text{and} \qquad (a' + b') = \frac{g\tau_{A'}^2}{4\pi^2} \tag{14.63}$$

Thus
$$b = \left(\frac{g\tau_A^2}{4\pi^2} \right) - a \qquad \text{and} \qquad b' = \left(\frac{g\tau_{A'}^2}{4\pi^2} \right) - a'$$

Hence from
$$k^2 = ab = a\left[\left(\frac{g\tau_A^2}{4\pi^2} \right) - a \right] \tag{14.64}$$

and from
$$k^2 = a'b' = a'\left[\frac{g\tau_{A'}^2}{4\pi^2} - a' \right] \tag{14.65}$$

For the same radius of gyration k at the centroidal axis, equating right hand side of equations (14.64) and (14.65),

$$a\left(\frac{g\tau_A^2}{4\pi^2} \right) - a^2 = a'\left(\frac{g\tau_{A'}^2}{4\pi^2} \right) - a'^2 \tag{14.66}$$

As seen in Fig. 14.12,

$$a' = (L - a)$$

it follows that equation (14.66) can be solved for a using experimentally obtained values of τ_A and τ_A' and hence the position c.g. and k can be obtained using equation (14.64). With radius of gyration 'k' and 'a' known the location of centre of percussion can be obtained as $BG = k^2/a$.

14.9 SIGNIFICANCE OF KINETIC EQUIVALENCE

There is no advantage of 'kinetically (Dynamically) equivalent two-mass system' if the inertia force determination is to be made for only one crank-position. The advantage of using concept of dynamically equivalent system becomes evident only when inertia forces are required to be determined for a number of crank positions. In I.C. engines, for instance, force and torque analysis may be required at as many as 36 to 48 crank positions of a 4-stroke cycle engine. The inertia force due to the mass of reciprocating parts falls along the line of stroke. *For a general case of angular acceleration, the inertia force of an unbalanced crank always passes through a fixed point on the crank i.e., through the centre of percussion. As the crank revolves the centre of percussion is located at a fixed distance from the crank shaft. This is, however, not possible for a connecting rod, which is a floating link.* The distance $h = (I_G \alpha)/F$ (*see* Fig. 14.11), which decides the line of action of inertia force of the connecting rod, is naturally different for different crank positions.

When a connecting rod is replaced by a dynamically equivalent two mass-system, the inertia forces on connecting rod must pass through each of mass positions. Thus, all the inertia forces in the mechanism will have their lines of action through the fixed points on links. Evidently, this is quite convenient for inertia force analysis for a number of crank positions.

14.10 EQUIVALENT DYNAMIC SYSTEM: GRAPHICAL DETERMINATION

Consider a body of mass M and having its centre of mass at G as shown in Fig. 14.13. It is required to locate position B of mass M_B for a given position A of the mass M_A of a dynamically equivalent two-mass system. For locating the required position B, join the points A and G by a line and extend it beyond G. From the point G erect a line perpendicular to AG and measure $GP = k$ (the radians of gyration of the connecting rod) along it. Jon PA. From the point P now draw a line perpendicular to PA to meet the line AG extended at point B. The point B then defines suitable location for the second mass M_B.

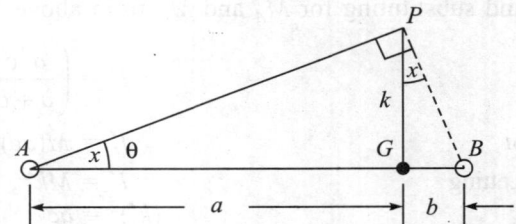

Fig. 14.13 Location of second mass (graphical method).

It is clear from the construction of Fig. 14.13 that if $\angle BAP = \theta$, then from right angled triangle AGP, $\angle APG = (90 - \theta)$. Again since $\angle APB = 90°$, it follows that $\angle BPG = \theta$. Thus the right angled triangles APG and BPG are similar, and hence

$$\left(\frac{PG}{AG}\right) = \left(\frac{BG}{PG}\right) \quad \text{or} \quad \left(\frac{k}{a}\right) = \left(\frac{b}{k}\right)$$

or
$$k^2 = ab$$

This shows that the locations A and B, as obtained above for masses M_A and M_B, meet the requirements of dynamic equivalence.

14.11 CORRECTION COUPLE REQUIRED FOR ARBITRARY CHOICE OF BOTH THE MASS LOCATIONS

Situations arise in practice when, instead of having a dynamically equivalent system, the locations for both the masses M_A and M_B are to be selected arbitrarily. For instance, while considering inertia effect of the connecting rod, on the turning moments at the crank shaft, it is quite convenient to assume masses M_A and M_c to be placed respectively at the gudgeon pin and the crank pin in such a way as to satisfy the first two conditions of dynamic equivalence. In other words, the magnitude of the mass M_A at the wrist pin and M_c at the crank pin may be so chosen that they add up to the mass M of connecting rod.

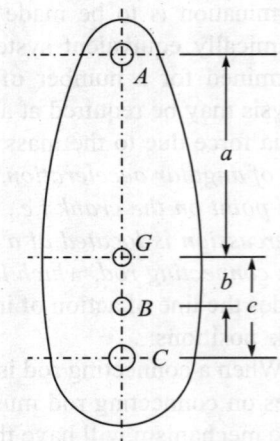

Similarly for keeping the c.g. unchanged, the moment of the two masses at G must be equal. Thus, referring to Fig. 14.14,

Fig. 14.14 The correction couple.

$$M_A + M_C = M \quad \text{and} \quad (M_A)a = (M_C)c$$

Hence

$$M_A = Mc/(a + c) \quad \text{and} \quad M_C = Ma/(a + c)$$

Thus, moment of inertia about centroidal axis due to these masses at A and C is,

$$I' = (M_A)a^2 + (M_C)c^2$$

and substituting for M_A and M_C from above,

$$I' = \left(\frac{a^2c}{a+c}\right)M + \left(\frac{ac^2}{a+c}\right)M = \frac{ac(a+c)}{(a+c)}M$$

or

$$I' = M(ac) \tag{14.67}$$

Letting

$$I' = Mk'^2, \text{ it follows that,}$$

$$(k')^2 = ac \tag{14.68}$$

Thus, the mass moment of inertia of an arbitrarily chosen two-mass system differs from that of a dynamically equivalent two-mass system by an amount:

$$I' - I = M(k'^2 - k^2) \tag{14.69}$$

or

$$I' - I = Ma(c - b)$$

But

$$(c - b) \equiv (a + c) - (a + b) = (l - L_E)$$

where $(a + b) \equiv L_E$ = equivalent length of simple pendulum

and $(a + c) \equiv l$ = length of rod between pin centres.

Thus,

$$Ma(c - b) \equiv Ma(l - L_E) \tag{14.70}$$

EXAMPLE 14.3 (MKS) Explain what is meant by 'equivalent dynamic system'. The following data relate to a connecting rod of a reciprocating engine:

Weight = 539.4 N; Distance between bearing centres = 85 cm; Diameter of small-end bearing = 7.5 cm; Diameter of big-end bearing = 10 cm; Time of oscillations when the connecting rod is suspended from small end = 1.83 s.

Time of oscillation when the connecting rod is suspended from big-end bearing: 1.68 s.

Determine: (a) the radius of gyration of the rod about an axis through the centre of gravity perpendicular to plane of oscillation;

(b) the moment of inertia about the same axis;

(c) the dynamically equivalent system for the connecting rod, constituted of two masses, one of which is situated at the small-end centre.

(AMIE, Summer 1976)

Solution: Time period of a compound pendulum having radius of gyration as k and distance of c.g. from point of suspension as 'a' is given by

$$\tau = 2\pi \sqrt{\frac{k^2 + a^2}{ga}}$$

(a) Substituting the value for $\tau = 1.68$ s for knife-edge support at A in big-end (Fig. 14.12),

$$\left(\frac{k^2 + a^2}{a}\right) = \left(\frac{1.68}{2\pi}\right)^2 \times 981 = 70.134 \text{ cm}$$

Thus, from Fig. 14.12 $\left(\text{as } \frac{k^2}{a} = b\right)$,

$$a + b = 70.134 \text{ cm} \tag{i}$$

Also, when the knife-edge support is at A' in the small-end bearing,

$$1.83 = 2\pi \sqrt{\frac{k^2 + a'^2}{ga'}}$$

or

$$\left(\frac{k^2}{a'} + a'\right) = \left(\frac{1.83}{2\pi}\right)^2 \times 981 = 83.217 \text{ cm}$$

or

$$(a' + b') = 83.217 \tag{ii}$$

Also, for knife-edge pivots at A and A', we have

$$k^2 \equiv ab = a(70.134 - a) \quad \text{and,} \quad k^2 \equiv a'b' = a'(83.217 - a')$$

Equating right hand sides, for a common value of k^2,

$$a(70.134 - a) = a'(83.217 - a') \tag{iii}$$

Again, as is seen from Fig. (14.12),

$$AA' = (85) + \frac{7.5}{2} + \frac{10}{2} = 93.75 \text{ cm}$$

Hence

$$a' = (AA' - a) = (93.75 - a)$$

Substituting for a' in equation (iii),

$$(70.134a - a^2) = (93.75 - a)(a - 10.533)$$

or

$$(70.134a - a^2) = -987.47 + 104.28a - a^2$$

or

$$(34.146)a = 987.47$$

or $\qquad a = \dfrac{987.47}{34.146} = 28.92$ cm

Therefore $\qquad a' = 93.75 - 28.92 = 64.83$ and $b = (70.134 - 28.92) = 41.214$ cm

Therefore $\qquad k^2 = ab = (28.92)(41.214) = 1191.9$

Therefore $\qquad k = 34.52$ cm **Ans.**

(b) The moment of inertia about centroidal axis,

$$I = \frac{539.4}{9.81} \times (0.3452)^2 = 6.552 \text{ kg·m}^2 \qquad \textbf{Ans.}$$

(c) Distance of c.g. from the small-end centre

$$= a' - \frac{7.5}{2} = (93.75 - 28.92) - \frac{7.5}{2} = 61.08 \text{ cm}$$

Hence, the distance of second mass from c.g. of a dynamically equivalent system,

$$b' = \frac{k^2}{61.08} = \frac{1191.9}{61.08} = 19.51 \text{ cm} \qquad \textbf{Ans.}$$

Hence, mass at small-end $= \dfrac{19.51}{(61.08 + 19.51)} \times 539.4 = 130.58$ N $\qquad$ **Ans.**

and mass at the other point, distant 19.51 cm from c.g. $= (539.4 - 130.58) = 408.82$ N **Ans.**

EXAMPLE 14.4 (SI) A connecting rod is suspended from a point 2.5 cm above the centre of small-end, and 65 cm above its centre of gravity, its mass being 37.5 kg. When permitted to oscillate, the time period is 1.87 second. Find the dynamically equivalent system constituted by two masses, one of which is located at the small-end centres. (SGSITS Indore, Feb. 1994)

Solution: Let a' = distance of point of suspension at small-end, from c.g. = 65 cm

$\qquad\qquad\quad a$ = distance of small-end centre from c.g.

and $\qquad\qquad b$ = distance of other mass of equivalent system from c.g.

Then $\qquad\qquad ab = k^2$, where k = radius of gyration about centroidal axis

Also, equivalent length of simple pendulum

$$L_E = \left(\frac{k^2 + a'^2}{a'} \right) \qquad\qquad \text{(i)}$$

Hence, from the expression for time period of simple pendulum,

$$\tau = 2\pi \sqrt{L_E / g}$$

Therefore $\qquad L_E = g\left(\dfrac{\tau}{2\pi}\right)^2 = 981\left(\dfrac{1.87}{2\pi}\right)^2 = 86.89$

Hence from (i), $\qquad 86.89 = \dfrac{k^2}{a'} + a'$

Therefore $\qquad\qquad\qquad k^2 = (86.89 - a')a' = (86.89 - 65)65 = 1422.85$

Therefore $\qquad\qquad\qquad k = 37.72$ cm

Now for required dynamically equivalent two mass system, with one mass at the centre of small end bearing, $a = 65 - 2.5 = 62.5$ cm **Ans.**

Hence, location for other mass of equivalent system,

$$b = \frac{k^2}{a} = \frac{1422.85}{62.5} = 22.765 \text{ cm} \qquad\qquad \textbf{Ans.}$$

Also, $\qquad\qquad M_A = \left(\frac{22.765}{62.5 + 22.765}\right) \times 37.5 = 10.01 \text{ kg} \qquad\qquad \textbf{Ans.}$

$$M_B = \frac{62.5}{(62.5 + 22.765)} \times 37.5 = 27.49 \text{ kg} \qquad\qquad \textbf{Ans.}$$

EXAMPLE 14.5 A connecting rod weighs 20 N and the distance between the centres of crank pin and gudgeon pin is 250 mm. The c.g. falls at a point 100 mm from the crank pin along the line of centres. The radius of gyration about an axis through the c.g. perpendicular to the plane of rotation is 110 mm. Find the equivalent dynamical system if only one of the masses is located at the gudgeon pin.

If the connecting rod is replaced by two masses, one at the gudgeon pin and other at the crank pin and the angular acceleration of the rod is 23000 rad/s^2. clockwise, determine the correction couple applied to the system to reduce it to a dynamically equivalent system.

Solution: $\quad k = 11.0$ cm

$\qquad\qquad a = 15.0$ cm, the distance of mass at gudgeon pin centre from c.g.

$\qquad\qquad b' = 10.0$ cm, the distance of crank pin centre from c.g.

Therefore   For dynamically equivalent two-mass system, location of other mass from c.g. is

$$b \equiv \frac{k^2}{a} = \frac{(11)^2}{15} = 8.067 \text{ cm} \qquad\qquad \textbf{Ans.}$$

Hence $\qquad\qquad W_A = \frac{8.067}{(15 + 8.067)} \times 20 = 6.994 \text{ N} \qquad\qquad \textbf{Ans.}$

$$W_B = 20 - 6.9994 = 13.006 \text{ N} \qquad\qquad \textbf{Ans.}$$

Correction Couple: For a mass placed at gudgeon pin and the other mass placed at crank pin,

$$k'^2 = a' \times b'$$
$$= 10 \times 15 = 150 \text{ cm}^2$$

Therefore   Correction couple $= \dfrac{W}{g}(k'^2 - k^2) \times \alpha$

$$= \frac{20}{981}(150 - 11^2) \times 23000 = 13{,}598.37 \text{ N·cm} \qquad\qquad \textbf{Ans.}$$

14.12 THE EFFECTIVE FORCE AND THE INERTIA FORCE

In a planar motion, all the elements of a rigid body move in parallel planes. Hence, the lines of action of inertia forces of these elements are also in parallel planes. Thus, the inertia forces and inertia couples on various links of a planar mechanism can be assumed to form a coplanar system. *As a consequence we will neglect certain couples that are usually unimportant, but may be determined by means of a simple secondary analysis, if desired.* An assumption of a coplanar system for inertia forces is equivalent to assuming that the mass of the body is concentrated in the plane of motion.

Again, all the forces acting on a machine are considered static in nature except those due to accelerations. Forces due to accelerations are called inertia forces or dynamic forces. A knowledge of acceleration analysis is an essential pre-requisite to understand force analysis involving inertia forces. Students are therefore advised to ensure that they are well acquainted with the acceleration analysis.

Links of a mechanism are subjected to static as well as inertia forces. In high speed machines, the accelerations and resulting inertia forces can be very large in comparison to static forces which do useful work. For instance, in a reciprocating engine of an automobile the inertia forces (at high speeds) can be greater than the gas load. Similarly, inertia forces due to a small unbalance in the rotor of a gas-turbine can produce forces on the bearings far in excess of the weight of the rotor. Thus, in the case of high speed machines, the inertia forces must be considered in their design. In slow speed machines, this is not necessary.

A rigid body is acted upon by a system of planar forces, which may be reduced to a single resultant force called the *effective force*. Magnitude, direction and the line of action of this resultant force may be established either using methods of 'graphic statics' or analytical methods. Line of action of resultant force F, in general, does not pass through the centre of gravity but is placed at a distance of h from it, (*see* Fig. 14.15). To understand complete effect of this resultant force F, introduce two equal and opposite forces F' at G, parallel to the line of action of the resultant force such that $F' = F$. Introduction of equal and opposite forces F' at G does not have any influence on the resultant force F acting on the body. The force F' at G, acting in a direction opposite to F together with the effective force F constitutes a couple Fh producing

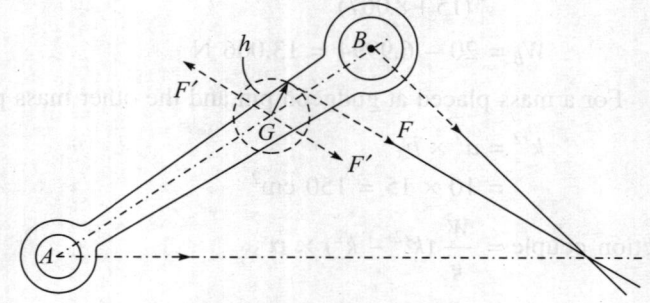

Fig. 14.15 The Effective Force on a Link.

angular acceleration α, in a c.w. sense. The other force F' at G causes linear acceleration A_G. Thus, if m be the mass of the body and k its radius of gyration about an axis through c.g. perpendicular to the plane of oscillation, then,

$$I = mk^2$$

Thus, with A_G as linear acceleration of c.g. and α as the angular acceleration of the body,

$$F = mA_G \quad \text{and} \quad Fh = mk^2\alpha \tag{14.71}$$

From these equations, with F, h, m and k given, the acceleration A_G and α can be worked out. Thus a comprehensive definition of effective force is as under.

A link in a mechanism moves in a definite way owing to the type of constraints imposed by the connected links through the common joints (pairs). The forces exerted on the link by the connected links, together with any other external force (*e.g.*, a gravitational force) can be compounded to give a resultant force F, which must be equal to the force required to produce necessary acceleration in the link. The force F, which is called the *effective force*, has a magnitude given by mass times the linear acceleration of the centre of gravity of the link. Thus,

$$F = mA_G$$

This force may or may not pass through the c.g. of the link under consideration.

Instead of finding resultant of all external forces acting on a link, a direct way of finding the magnitude and the line of action of the effective force, for a given link, is to determine linear acceleration A_G of the c.g. as also the angular acceleration α of the link through velocity and acceleration analysis. When A_G and α are so established, the magnitude of the effective force F is given by,

$$F = \frac{W}{g} A_G$$

where W is the weight of the link. The line of action and direction of the effective force is parallel to the direction of acceleration A_G. The distance h of the line of action of F from the c.g. and the side of c.g., on which the line of action lies, is decided by the magnitude and the sense of angular acceleration α. Thus,

$$h = \left(\frac{W}{g}\right) k^2 \frac{\alpha}{F} \tag{14.72}$$

where k is the radius of gyration of the link about the centroidal axis, perpendicular to the plane of oscillations. For the position of slider-crank mechanism, shown in Fig. 14.16(a), the direction of acceleration A_{pc}^t shows that the point P has an acceleration component acting upwards w.r. to C, and hence the angular acceleration of link $\alpha = (A_{pc}^t)/CP$ must act in c.w. direction as shown in Fig. 14.16(a). Further, with the known direction of acceleration A_G as indicated by line the o_ag in the acceleration polygon $o_app'c$ of Fig. 14.16(c), the line of action of the resultant linear acceleration A_G must lie tangent to the circle drawn with radius h from the centre G, on the right hand side. Note that the total acceleration line pc in the acceleration polygon represents 'acceleration image' of the link PC and the acceleration image g of point G on PC is obtained using the relation $cg/cp = CG/CP$.

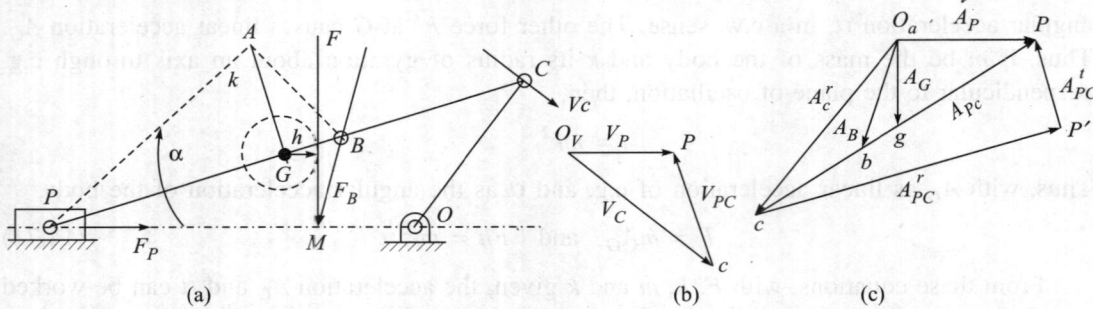

Fig. 14.16 Dynamic Force Analysis of a Slider Crank Mechanism.

To Obtain Line of Action of F Using Concept of Dynamic Equivalence: Line of action of the resultant accelerating (effective) force can be established using a complete graphical approach based on the concept of dynamically equivalent two-mass system. Consider the connecting rod *PC* of the engine mechanism shown in Fig. 14.16(a). Let *W* be the total weight of connecting rod acting through its c.g. *G* and let *k* be its radius of gyration about the centroidal axis. Let the connecting rod *PC* be now replaced by a dynamically equivalent two-mass system; one mass being placed at *P* and the other placed at a point *B* on the rod such that

$$W_P + W_B = W \quad \text{and} \quad pb = k^2$$

where W_P and W_B are the weights of masses placed respectively at *B* and *P* at the distances of *p* and *b* from the c.g. Figure 14.16(a) illustrates the method of obtaining position *B* of second equivalent mass W_B for the assumed position at *P* of the first mass W_P. For the dynamically equivalent two-mass system, the accelerations of the two masses W_P and W_B are given by lines $o_a p$ (=A_P) and $o_a b$ (=A_B) in the acceleration polygon. The lines of action of corresponding accelerating forces for the two masses, namely $\mathbf{F}_P$ and $\mathbf{F}_B$, intersect at point *M* suggesting that the resultant accelerating force $\mathbf{F}$ (*i.e.*, the effective force) for the connecting rod must pass through point *M*. Further, in order that the c.g. of the rod is accelerated in the direction $o_a g$, the direction of the accelerating force $\mathbf{F}$ must be parallel to that of A_G. Hence, a line drawn from *M* parallel to the line $o_a g$ of acceleration polygon (Fig. 14.16c), gives the direction of the effective force $\mathbf{F}$. It may be verified that this line of action of the effective force $\mathbf{F}$ is tangent to the circle drawn from c.g. *G* of radius *h* given by expression (14.72).

14.13 REVERSED EFFECTIVE (INERTIA) FORCE AND FORCE ANALYSIS

Using the procedure described above, it is possible to determine magnitude and direction of the effective force causing accelerations in each link of a mechanism. *A force, equal and opposite to the effective force, is called reversed effective (inertia) force.* According to D' Alembert's principle, inclusion of the reversed effective force amounts to converting the problem in dynamics to an equivalent problem in statics. Reversed effective force can now be included in the force analysis of a link, in addition to all external forces which produce a single resultant accelerating (effective) force. The conditions of static equilibrium namely, $\Sigma F = 0$ and $\Sigma M = 0$ can then be used to complete the force analysis.

Let *OABC* be a four-bar mechanism and let the coupling link have a weight *W* acting through its c.g. *G* as shown in Fig. 14.17(a). Let the points *A* and *B* move in arcs of circle of radii *OA* and *CB* respectively, and let F_i be the inertia force for the link *AB* through point *L*, which we intend to consider in force analysis as per D'Alembert's principle. Let us assume that the magnitude and direction of the inertia force F_i is established using any of the methods described earlier. The points *A* and *B* of the coupler link are constrained to move along respective paths because of radial components of forces *F* and *P* exerted by adjacent links *OA* and *CB* respectively. The tangential components F_t and P_t of the force *F* and *P* exerted by links *OA* and *CB* do useful work on the link *AB*. If the tangential component F_t is given, then F_r and both the components of force *P* may be obtained from the necessary conditions of static equilibrium. Thus, referring to Fig. 14.17(a) taking moments of all the forces about the instantaneous centre *I*,

$$(P_t)IB = (F_t)AI + W(IY) - F_i(IX) \tag{14.73}$$

where *IX* and *IY* are the perpendicular distances of the lines of action of F_i and *W* respectively from the instantaneous centre *I*.

With component P_t so established from known F_t, the components P_r and F_r may be obtained by drawing force polygon to a scale as shown at Fig. 14.17(b).

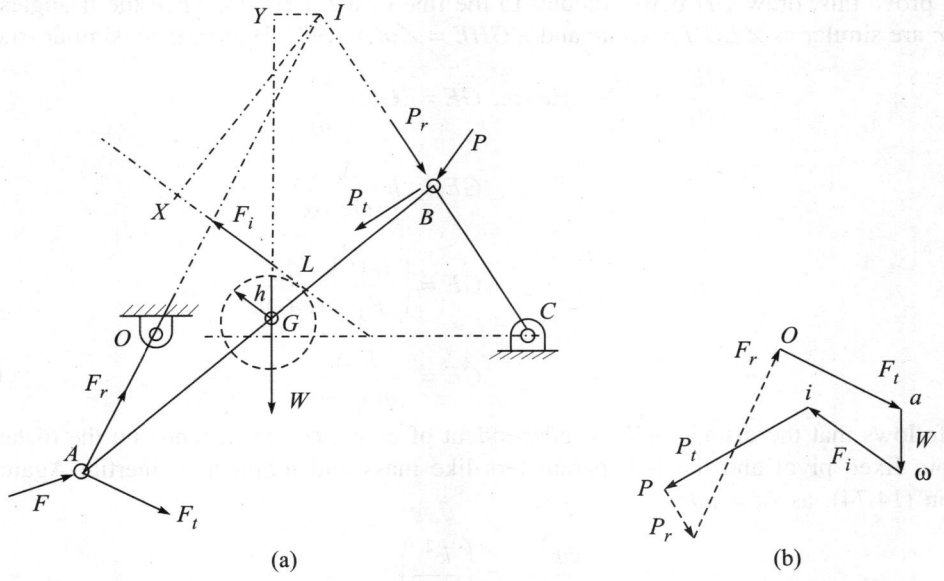

Fig. 14.17 Dynamic Force Analysis of a 4-bar Mechanism.

Inertia Force of a Link Rotating about a Fixed Axis

Figure 14.18 shows a link rotating about a fixed axis through *O* with an angular velocity ω and an angular acceleration α in c.w. sense. Let *m* be the mass of the link and I_G be the mass moment of inertia about an axis through c.g., *G*, perpendicular to the plane of the paper.

Let A_G^r and A_G^t be the radial and tangential components of acceleration of the c.g. which are compounded to give the resultant acceleration A_G. The magnitudes of the two components are given by,

$$A_G^r = (OG)\omega^2 \quad \text{and} \quad A_G^t = (OG)\alpha$$

The resultant accelerating force $F_G \ (= mA_G)$, therefore, acts in the direction of acceleration A_G. The line of action of this accelerating (effective) force should be placed at a distance h from the c.g., such that,

$$h = (I_G)\alpha/F_G$$

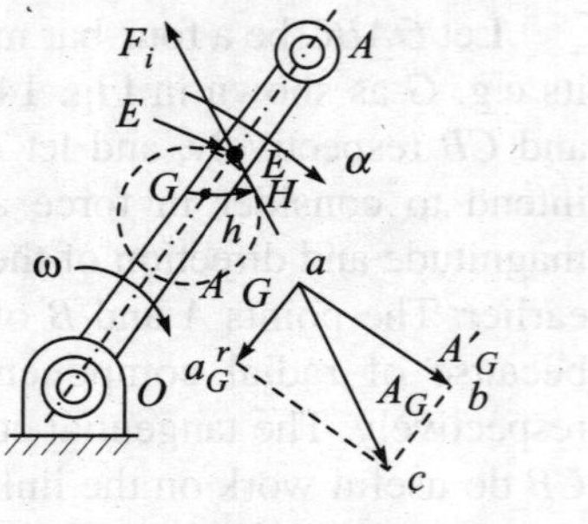

Fig. 14.18 Dynamic Force Analysis: A Rotating Link.

The side of c.g., where line of action of reversed effective force $F_i \ (= -F_G)$ is tangent to the circle of radius h drawn from G as centre, should be such that it opposes angular acceleration α and is as shown in Fig. 14.18.

The inertia force F_i in this case always passes through the same point E on the centre line of the link. It can be proved that E is the centre of percussion and for the fixed axis of rotation of the link, its position is fixed and does not depend on ω or α.

To prove this, draw GH perpendicular to the line of action of F_i. Then the triangles GEH and abc are similar as $\angle EGH = \angle bac$ and $\angle GHE = \angle abc = 90°$. Hence, from similar triangles,

$$\frac{GE}{GH} = \frac{A_G}{A_G^t}, \text{ Hence, } GE = (GH)\frac{A_G}{A_G^t}$$

or

$$GE = h\frac{A_G}{(OG)\alpha}$$

or

$$GE = \left(\frac{I_G\alpha}{F_G}\right)\frac{A_G}{(OG)\alpha}$$

or

$$GE = \frac{I_G A_G}{(mA_G)OG} = \frac{I_G}{mOG} \tag{14.74}$$

It follows that the distance GE is independent of ω and α, but depends on the distance of c.g. from fixed pivot and the link parameters like mass and moment of inertia. Again from equation (14.74), as $I_G = mk^2$,

$$GE = \frac{mk^2}{m(OG)} = \left(\frac{k^2}{OG}\right)$$

Hence E is the centre of percussion.

14.14 DYNAMIC FORCE ANALYSIS OF A FOUR-LINK MECHANISM

Consider the four-bar mechanism $OABC$, shown in Fig. 14:19 in which the crank OA is assumed to rotate uniformly at angular velocity ω_2 c.c.w. Let the three mobile links 2, 3 and 4 have their centres of gravity at G_2, G_3, G_4 and masses M_2, M_3 and M_4 respectively. We now proceed to determine the torque which the shaft at O must exert on link 2 to give the desired motion.

The first step invariably is to draw the velocity and acceleration polygon for determining linear accelerations of points G_2, G_3 and G_4. This is shown in Fig. 14.19(b). From the magnitude and sense of tangential components of accelerations of G_3 and G_4, the magnitude and sense

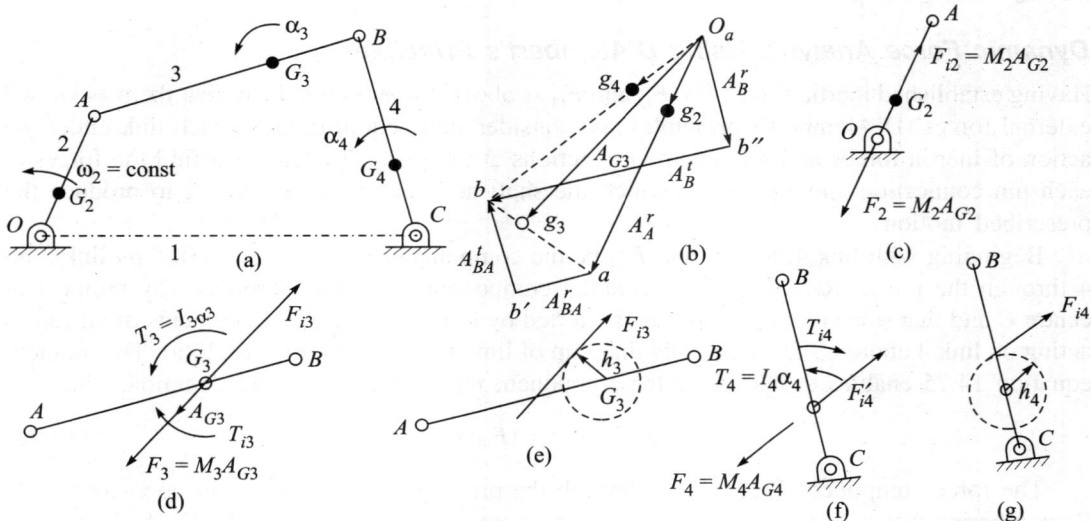

Fig. 14.19 Dynamic Force Analysis: A 4-bar Mechanism.

of angular accelerations α_3 and α_4 can be determined. The sense of α_3 and α_4 so obtained is indicated in Fig. 14.19(a).

Inertia Forces and Couples

Link 2 is shown in Fig. 14.19(c), where the linear acceleration A_{G2} (represented by vector $o_a g_2$) and hence the accelerating force $F_2 = M_2 A_{G2}$, acts towards the centre of rotation. The inertia force F_{i2} therefore acts along the link away from the centre of rotation. There being no angular acceleration, no inertia torque is involved.

Link 3 is shown in Fig. 14.19(d) with linear acceleration of c.g. G_3 in the direction of acceleration vector $o_a g_3$ of the acceleration polygon. This is also the direction of accelerating force $F_3 = M_3 A_{G3}$. Correspondingly, the inertia force F_{i3}' acts in opposite direction. However, in view of angular acceleration α_3, there must be a resultant torque $T_3 = I_3 \alpha_3$ acting in the sense of α_3, where I_3 is the mass moment of inertia of link about an axis through G_3 perpendicular to the plane of paper. The inertia torque T_{i3} is equal and opposite to T_3, as shown in Fig. 14.19(d). Figure 14.19(e) shows a single inertia force F_{i3} which can replace the inertia force F_{i3}' and inertia torque T_{i3}. Note that inertia force F_{i3} is tangent to circle of radius h_3, drawn from G_3, on the top side of it so as to oppose the angular acceleration α_3. Note that $h_3 = (I_3 \alpha_3)/M_3 A_{G3}$.

For link 4, the resultant acceleration A_{G4} acts through G_4 in the direction $o_a g_4$, shown in the acceleration polygon. As seen in Fig. 14.19(f), the accelerating force $M_4 A_{G4}$ acts through G_4 in the direction of A_{G4}. Corresponding inertia force F_{i4}', which is equal and opposite to the accelerating force M_4A_{G4}, is as shown in Fig. 14.19(f). The link 4 is also subjected to angular acceleration α_4 and hence an accelerating torque $T_4 = I_4\alpha_4$ and an equal and opposite inertia torque T_{i4} also exists. Figure 14.19(g) shows a single inertia force F_{i4} which replaces the inertia force F_{i4} and inertia torque T_{i4}. In order to oppose angular acceleration α_4 this inertia force must be tangent to a circle of radius h_4 drawn from G_4 on the top side, as shown in the figure. Note that $h_4 = (I_4\alpha_4)/(M_4A_{G4})$

Dynamic Force Analysis Using D'Alembert's Principle

Having established inertia forces F_{i2}, F_{i3} and F_{i4} as above, the next step is to treat them as known external forces (D'Alembert's principle) and consider static equilibrium of each link under the action of inertia forces and the unknown reactions at the pins. The aim is to find the forces at each pin connection and the torque which the shaft at O exerts on the link 2 to produce the prescribed motion.

Beginning with link 4, let F_{34}^t and F_{34}^r be the components of force F_{34} exerted by link 3 on 4 through the pin B. Realising that the radial component F_{34}^r cannot produce any moment at centre C and that same is true about forces exerted by link 1 on 4, we take moments of all forces acting on link 4 about C. The free body diagram of link 4 is shown at Fig. 14.20(a). The moment equation 14.75 enables to determine the component F_{34}^t in magnitude and direction. Thus,

$$(F_{34}^t)BC = (F_{i4})x_4 \tag{14.75}$$

The force component F_{43}^t, exerted through the pin on link 3, is equal and opposite to F_{34}^t. Remembering that moments due to unknown force components F_{23} and F_{23}^r at pin A will vanish, we now take moments of all forces acting on link 3 about A. The free body F_{23}^t diagram of link 3 is shown in Fig. 14.20(b). Taking moment about A of all the forces acting on link 3,

$$(F_{43}^r)AB\cos\phi = (F_{i3})x_3 - (F_{43}^t)AB\sin\phi \tag{14.76}$$

The equation may be solved to obtain magnitude and direction of force component F_{43}^r. This automatically fixes magnitude and direction of F_{34}^r which is equal and opposite to F_{43}^r. With forces F_{43}^r, F_{43}^t, F_{i3} known in magnitude and direction, the force F_{23} is given by the closing side of the force polygon shown in Fig. 14.20(c).

In the case of link 2, the external force F_{32} is equal and opposite to the force F_{23} and is, therefore, known. This force when compounded with the inertia force F_{i2}, gives the resultant force R which must be equal and opposite to F_{12}. The shaft at O must apply a torque T_2 in c.c.w. direction on link 2 to overcome resisting torque given by:

$$T_2 = (R) \times p \tag{14.77}$$

This is shown in Fig. 14.20(d).

The force F_{14} is obtained by drawing force polygon for link 4. Thus, with forces F_{34}^t, F_{34}^r and F_{i4} known, the closing side of the polygon gives the required force F_{14}. This is shown in Fig. 14.20(e).

Shaking Force is the resultant of all forces acting on the frame of the mechanism due to inertia forces only. This concept is important in practice because shaking force sets up troublesome vibrations in the trame. Frame should be strong enough to withstand shaking forces.

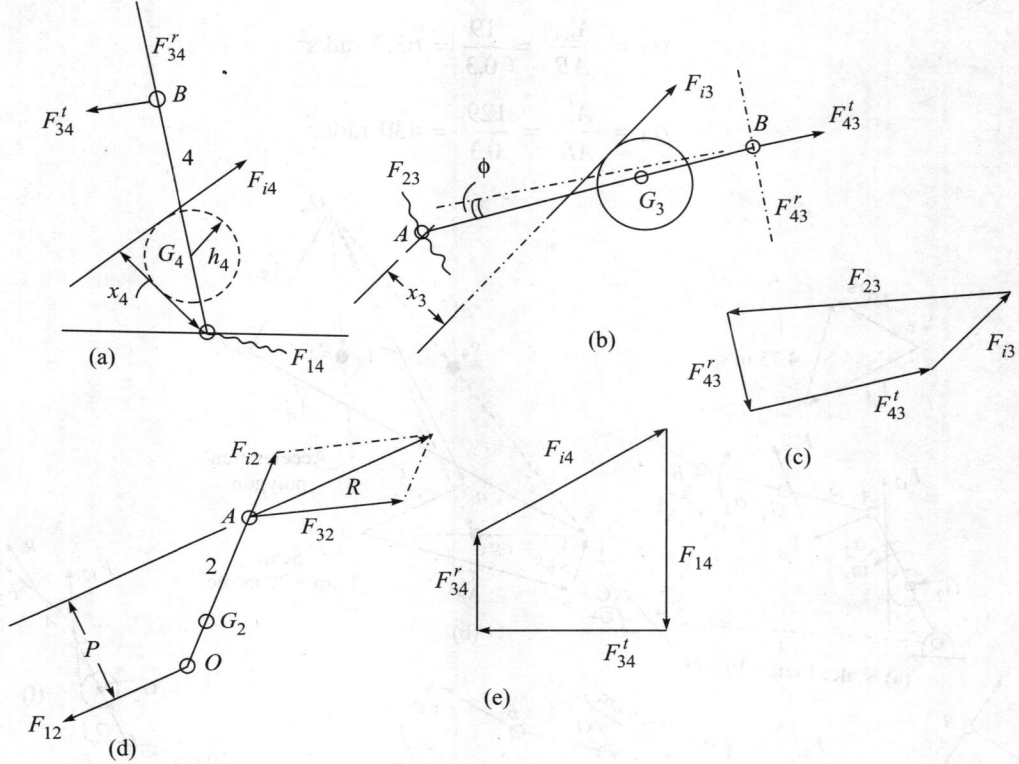

Fig. 14.20

EXAMPLE 14.6 It is required to carry out dynamic force analysis of the four-bar mechanism shown in Fig. 14.21. The angular velocity and angular acceleration of link 2 are given as: $\omega_2 = 20$ rad/s and $\alpha_2 = 160$ rad/s^2. The necessary linear dimensions are:

$OA = 25$ cm; $OG_2 = 11$ cm; $AB = 30$ cm; $AG_3 = 15$ cm; $BC = 30$ cm; $CG_4 = 14$ cm; $OC = 55$ cm and $\angle AOC = 60°$.

The masses and moments of inertia of various members are $\dfrac{1}{m}$

Link	mass m	moment of inertia I_G kg·m^2
2	20.7 kg	0.01872
3	9.66 kg	0.01105
4	23.47 kg	0.0277

Determine (1) the inertia forces of the moving members and (2) the torque which must be applied to link 2.

$$O_a g_2 = A_{G2} = 48 \text{ m/s}^2$$

$$O_a g_3 = 120 \text{ m/s}^2 = A_{G3}$$

$$O_a g_4 = A_{G4} = 65.4 \text{ m/s}^2$$

$$\alpha_3 = \frac{A_{BA}^t}{AB} = \frac{19}{0.3} = 63.3 \text{ rad/s}^2$$

$$\alpha_4 = \frac{A_B^t}{AB} = \frac{129}{0.3} = 430 \text{ rad/s}^2$$

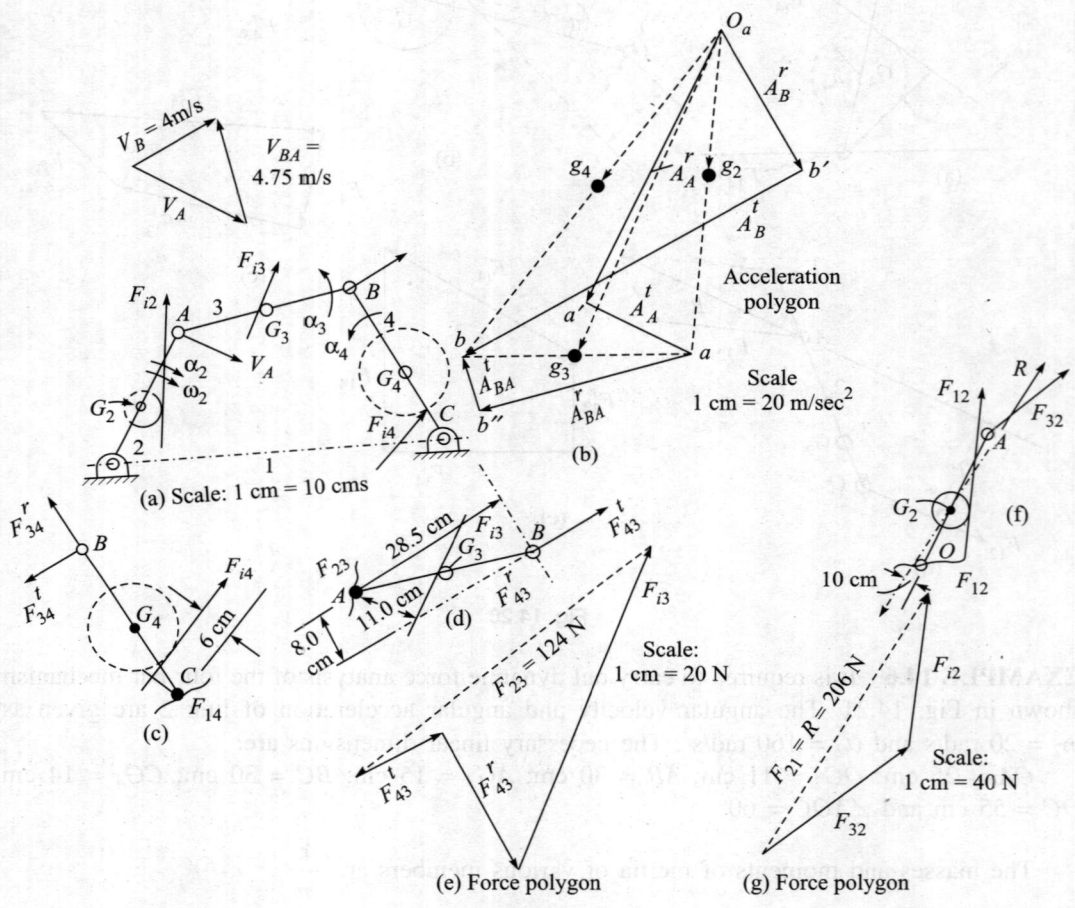

Fig. 14.21

Solution:

(1) **Inertia Forces.** For plotting the velocity and acceleration diagrams, the known velocities and accelerations are:

$$V_A = (25) \times 20 = 500 \text{ cm/s} = 5 \text{ m/s}$$
$$A_A^r = 25 \times 20^2 = 10{,}000 \text{ cm/s}^2 = 100 \text{ m/s}^2$$
$$A_A^t = 25 \times 160 = 4000 \text{ cm/s}^2 = 40 \text{ m/s}^2$$

From velocity polygon,

$$V_B = 4 \text{ m/s}; \ V_{BA} = 4.75 \text{ m/s}$$

Therefore
$$A_B^r = \frac{V_B^2}{CB} = \frac{(4)^2}{0.3} = 53.33 \text{ m/s}^2$$

$$A_{BA}^r = \frac{V_{BA}^2}{BA} = \frac{(4.75)^2}{0.3} = 75.21 \text{ m/s}^2$$

$$o_a g_2 = A_{G2} = 48 \text{ m/s}^2$$

$$o_a g_3 = 120 \text{ m/s}^2 = A_{G3}$$

$$o_a g_4 = A_{G4} = 65.4 \text{ m/s}^2$$

$$\alpha_3 = \frac{A_{BA}^t}{AB} = \frac{19}{0.3} = 63.3 \text{ rad/s}^2$$

$$\alpha_4 = \frac{A_B^t}{CB} = \frac{129}{0.3} = 430 \text{ rad/s}^2$$

The magnitudes of the angular acceleration α_3 and α_4 as obtained from acceleration polygon are listed above. The sense of α_3 and α_4 is indicated in Fig. 14.21(a). The magnitudes of linear accelerations of c.gs. of links 2, 3, 4 as indicated by vectors $o_a g_2$, $o_a g_3$ and $o_a g_4$ in acceleration polygon are also listed below:

$$A_{G2} = 48 \text{ m/s}^2; \qquad A_{G3} = 120 \text{ m/s}^2; \qquad A_{G4} = 65.4 \text{ m/s}^2$$
and, $$\alpha_2 = 160 \text{ rad/s}^2; \qquad \alpha_3 = 63.3 \text{ rad/s}^2; \qquad \alpha_4 = 430 \text{ rad/s}^2.$$

The radii of circles h_2, h_3, h_4 drawn from centres G_2, G_3 and G_4 are obtained after determining accelerating (and inertia) forces as under:

$$F_{G2} = m_2 \times A_{G2} = \frac{20.7}{9.81} \times 48 = 101.28 \text{ N} \qquad \text{(in the direction } o_a g_2)$$

$$F_{G3} = m_3 \times A_{G3} = \frac{9.66}{9.81} \times 120 = 118.17 \text{ N} \qquad \text{(in the direction } o_a g_3)$$

$$F_{G4} = m_4 \times A_{G4} = \frac{23.47}{9.81} \times 65.4 = 156.47 \text{ N} \qquad \text{(in the direction } o_a g_4)$$

Also

$$h_2 = \frac{I_{G2}(\alpha_2)}{F_2} = \frac{(0.01872)(160)}{101.28} = 0.0295 \text{ m} = 2.95 \text{ cm}$$

$$h_3 = \frac{I_{G3}(\alpha_3)}{F_3} = \frac{(0.01105)(63.3)}{118.17} = 0.005919 \text{ m} = 0.59 \text{ cm}$$

$$h_4 = \frac{I_{G4}(\alpha_4)}{F_4} = \frac{(0.0277)(430)}{156.47} = 0.0761 \text{ m} = 7.61 \text{ cm}$$

The inertia forces F_{i2}, F_{i3} and F_{i4} have magnitudes equal and opposite to the respective accelerating forces and will be tangent to circles of radii h_2, h_3, h_4 drawn from centres G_2, G_3, G_4 so as to oppose accelerations α_2, α_3 and α_4. The sense of angular accelerations and locations of inertia forces F_{i2}, F_{i3} and F_{i4} are as indicated in Fig. 14.21(a). Their magnitudes are:

$$F_{i2} = 101.28 \text{ N} \qquad F_{i3} = 118.17 \text{ N} \qquad F_{i4} = 156.47 \text{ N} \qquad \textbf{Ans.}$$

(2) **The Torque Required to be Applied to Link 2 by Shaft at O.** Each of the link is to be analysed for static equilibrium under the action of all external forces on that link plus the inertia force.

Considering the free body diagram of link 4 (scale as in Fig. 14.21a), and taking moments of F_{34}^r, F_{34}^t, F_{i4} and F_{14} about C,

$$(F_{34}^t) \times 0.3 = F_{i4} \times \frac{6}{100}$$

Therefore $(F_{34}^t) = 156.47 \times \dfrac{0.06}{0.3} = 31.294$ N (Acting to the left).

Figure 14.21(d) shows the free body diagram of link 3 to the same scale. Perpendicular distances of the lines of action of F_{i3}, F_{43}^r and F_{43}^t as measured from the construction are respectively 11 cm, 28.5 cm and 7.5 cm.

Taking moments of all the forces about pin A,

$$(F_{43}^r) \times 28.5 = (F_{i3}) \times 11 + (F_{43}^t) \times 8.0$$

Therefore $F_{43}^r = \dfrac{(118.17)\,11 + (31.294) \times 8.0}{28.5} = 54.39$ N

The direction F_{43}^r should be such that it balances the moments (combined) due to F_{43}^t and F_{i3} about A. Hence, the force F_{43}^r is as shown in Fig. 14.21(d). A force polygon of all the forces (including inertia force F_{i3}) acting on link 3 is drawn to scale in Fig. 14.21(e). The closing line then gives the force exerted by link 2 on 3. Thus,

$$F_{23} = 124 \text{ N}$$

The force F_{32} exerted by link 3 on 2 is equal and opposite to F_{23} determined above. Considering free body diagram of link 2 at Fig. 14.21(f), the forces acting on this link are F_{32}, F_{i2} and F_{12}. These forces F_{32} and F_{i2} are assumed to act at G_2. Known forces F_{32} and F_{i2} have been compounded in Fig. 14.21(g) so as to obtain the resultant $R = 206$ N. The ground will provide equal and opposite force,

$$F_{12} = R = 206 \text{ N}$$

Thus, with a moment arm of 10 cm for the resultant force R, the moment of the couple.

$$= P \times L$$
$$= 206 \times 10 = 2060 \text{ N·cm} = 20.6 \text{ N·m} \qquad \textbf{Ans.}$$

This couple is exactly equal (but opposite) to the couple formed by R and F_{12}. It is interesting to note that in this case, accelerating couple (effort) is opposing the motion of the mechanism. The actual driving force in this case is largely supplied by the kinetic energy of the link 4.

14.15 ANALYTICAL EXPRESSIONS FOR VELOCITY AND ACCELERATION OF SLIDER IN SLIDER-CRANK MECHANISM

In many practical problems on engine mechanisms, it becomes desirable to calculate the velocity and acceleration of the piston. This is particularly important where effect of the inertia of the reciprocating parts on the turning moment diagram of the engine is to be studied.

For instance, let

r = crank radius

ω = angular velocity of crank

l = length of the connecting rod = nr.

n = ratio of connecting rod length to crank radius = (l/r)

θ = angle of inclination of the crank to the i.d.c.

ϕ = angle of inclination of the connecting rod to the line of stroke

x = displacement of the piston from the commencement of its stroke

V_p = velocity of piston

A_p = acceleration of piston

ω' = angular velocity of connecting rod

α' = angular acceleration of connecting rod

Referring to Fig. 14.22, let the crank be at i.d.c. in position C_1, when the piston pin occupies the position P_1. In this position, the crank and connecting rod coincide with the line of stroke so that

$$P_1O = P_1C_1 + C_1O = (l + r)$$

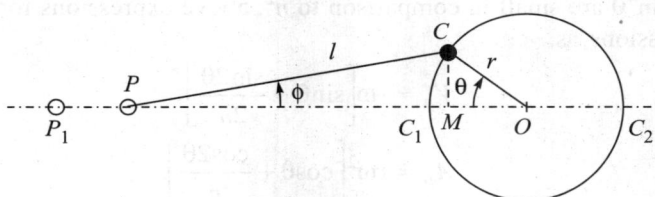

Fig. 14.22

Now, let the crank OC rotate in c.w. direction through θ so that crank occupies position OC and the connecting rod takes up position CP. Let CM be the perpendicular dropped from C on the diameter C_1OC_2. Then piston displacement,

$$x = P_1P = (P_1O - PO)$$
$$= (l + r) - (l\cos\phi + r\cos\theta)$$
$$= r(1 - \cos\theta) + l(1 - \cos\phi) \qquad (14.78)$$

Again, for the common side CM of the right angled triangles COM and PCM,

$$CM = l\sin\phi = r\sin\theta$$

Therefore

$$\sin\phi = (r/l)\sin\theta = \frac{1}{n}\sin\theta \qquad (14.79)$$

Hence

$$\cos\phi \equiv \sqrt{1 - \sin^2\phi} = \left(\frac{1}{n}\right)\sqrt{(n^2 - \sin^2\theta)} \qquad (14.80)$$

Substituting for $\cos\phi$ in equation (14.78), and noting that $l = nr$,

$$x = r(1 - \cos\theta) + nr\left(1 - \frac{1}{n}\sqrt{n^2 - \sin^2\theta}\right) \tag{14.81}$$

or
$$x = r[1 + n - \cos\theta - \sqrt{n^2 - \sin^2\theta}\,]$$

Differentiating w.r. to time,

$$V_p = dx/dt = \left(\frac{dx}{d\theta}\right)\left(\frac{d\theta}{dt}\right) = \omega\frac{dx}{d\theta}$$

or
$$V_p = \omega r\left[\sin\theta + \frac{\sin 2\theta}{2\sqrt{(n^2 - \sin^2\theta)}}\right] \tag{14.82}$$

For uniform ω, differentiating V_p w.r. to time, the acceleration of piston is,

$$A_p = \frac{dV_p}{dt} = \left(\frac{dV_p}{d\theta}\right)\left(\frac{d\theta}{dt}\right)$$

Which can be shown equal to $A_p = r\omega^2\left[\cos\theta + \dfrac{n^2\cos 2\theta + \sin^4\theta}{(n^2 - \sin^2\theta)^{3/2}}\right]$ \hfill (14.83)

As $\sin^2\theta$ and $\sin^4\theta$ are small in comparison to n^2, above expressions for V_p and A_p leads to approximate expressions as,

$$V_p \approx r\omega\left[\sin\theta + \frac{\sin 2\theta}{2n}\right] \tag{14.84}$$

$$A_p \approx r\omega^2\left[\cos\theta + \frac{\cos 2\theta}{n}\right] \tag{14.85}$$

For determining angular velocity and angular acceleration of the connecting rod, let us differentiate both sides of equation (14.79) w.r. to time. Thus,

$$(\cos\phi)\frac{d\phi}{dt} = \frac{1}{n}(\cos\theta)\frac{d\theta}{dt}$$

As
$$\omega' = \frac{d\phi}{dt} \qquad \text{and} \qquad \omega = \frac{d\theta}{dt},$$

$$\omega' = \frac{1}{n}\frac{\cos\theta}{\cos\phi}\omega = \omega\frac{\cos\theta}{\sqrt{n^2 - \sin^2\theta}} \tag{14.86}$$

and
$$\alpha' = \frac{d\omega'}{dt} = \frac{d\omega'}{d\theta}\frac{d\theta}{dt} = \frac{-\omega^2(n^2 - 1)\sin\theta}{(n^2 - \sin^2\theta)^{3/2}} \tag{14.87}$$

As before, since n^2 is large in comparison to both unity and $\sin^2\theta$, the above expressions leads to approximate expressions as

$$\omega' \simeq \left\{\frac{\omega}{n}\right\}\cos\theta \tag{14.88}$$

and
$$\alpha' \approx \left(-\frac{\omega^2}{n}\right)\sin\theta \tag{14.89}$$

It follows from equation (14.89) that α' is always negative, *i.e.*, the sense of α' is always such that it tends to reduce the inclination of the connecting rod w.r. to the line of stroke.

Equation (14.81) can be rewritten as,

$$x = r(1 - \cos\theta) + l(1 - \sqrt{1 - (\sin^2\theta)/n^2})$$

Expanding $\sqrt{1 - (\sin^2\theta)/n^2}$ by binomial theorem, above equation yields,

$$x = r(1 - \cos\theta) + l\left[\frac{\sin^2\theta}{2n^2} + \frac{\sin^4\theta}{8n^2} + \frac{\sin^6\theta}{16n^6} + ...\right]$$

Substituting $l = nr$ in the above equation, we have

$$x = r\left[1 - \cos\theta + \frac{\sin^2\theta}{2n} + \frac{\sin^4\theta}{8n^3} + \frac{\sin^6\theta}{16n^5} + ...\right] \qquad (14.90)$$

It follows that equation (14.90) deviates from an expression for simple harmonic motion due to third term and onwards. It can be seen that if n is large, third term and onwards have a relatively insignificant contribution to displacement and motion approaches as S.H.M.

14.16 PISTON EFFORT, CRANK PIN EFFORT AND CRANK EFFORT

Starting from net force acting on piston, we shall first establish expression for the driving turning moment (T), called crank effort. In section 14.17 later, we shall establish expression for T.M. at crank due to inertia effect.

Piston Effort

Net or effective force applied to the piston is called *Piston effort*. In the case of reciprocating engines, reciprocating masses are subject to acceleration during the first half of the stroke. Hence, inertia forces of the reciprocating masses tend to oppose the motion. This results in reduced net force on the piston. As against this, in the next half stroke, reciprocating masses are subject to retardation and therefore, inertia force tends to increase the net force on piston. This apart, even the pressure of gaseous fluid does not remain constant during the stroke. Thus, even the gas load on piston is subject to variation throughout the working stroke.

In the case of a double acting engine mechanism, the indicator diagram shows that the pressure on piston from one end is p_1 and from the other side, it is p_2. The net force on piston due to steam/gas load only in that case is,

$$F_g = (p_1)a_1 - (p_2)a_2$$

where a_1 = piston area on the cover end

  a_2 = piston area on the other side

Net (effective) force on piston is then

$$F = F_g - F_i = (p_1a_1 - p_2a_2) \pm \frac{R}{g} A_p$$

where R = weight of reciprocating parts

A_p = acceleration/retardation of reciprocating parts

F_i = inertia force due to reciprocating parts = $\dfrac{R}{g} r\omega^2 \left(\cos\theta + \dfrac{\cos 2\theta}{n}\right)$

In the case of vertical reciprocating engines, the weight of reciprocating parts adds to the gas load when piston goes down but subtracts from gas load when piston moves up. Thus net (effective) force on piston,

$$F_p = (p_1 a_1 - p_2 a_2) \pm \frac{R}{g} A_p \pm W \qquad (14.91)$$

If frictional force F_f is also considered, it will always subtract from the right hand side.

Thrust in Connecting Rod

Figure 14.23 shows a slider-crank mechanism with piston subjected to piston effort F_P. Then the gudgeon pin (cross-head pin in steam engine) is in equilibrium under the action of forces: (i) piston effort F_P, (ii) thrust Q in connecting rod, and (iii) side reaction F_N from cylinder wall or cross-head guide bars as the case may be. Using Lame's theorem therefore,

$$\frac{F_P}{\sin(90° + \phi)} = \frac{F_N}{\sin(180° - \phi)} = \frac{Q}{\sin(90°)} \qquad (14.92)$$

or

$$Q = \frac{F_P}{\cos\phi} = \frac{F_N}{\sin\phi}$$

Thus, substituting for $\cos\phi$,

$$Q = \frac{F_P}{\cos\phi} = \frac{nF_P}{\sqrt{n^2 - \sin^2\theta}} \qquad (14.93)$$

and, substituting for $\sin\phi$,

$$F_N = F_P \frac{\sin\phi}{\cos\phi} = \frac{F_P \sin\theta}{\sqrt{n^2 - \sin^2\theta}} \qquad (14.94)$$

The thrust in the connecting rod Q at the other end (crank end) can be resolved along and perpendicular to the tangent to crank circle. As seen in Fig. 14.23, tangent to the crank circle at B meets line of stroke at point L making angle $\angle BLO = (90° - \theta)$.

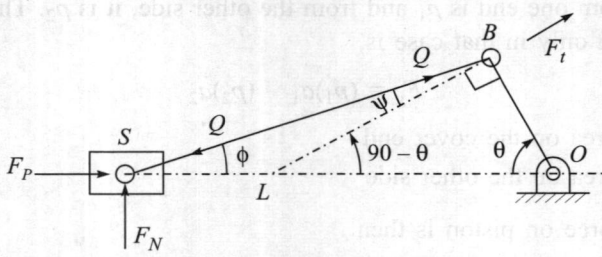

Fig. 14.23

Thus, from the triangle BSL, the exterior angle,

$$\angle BLO = \angle BSL + \angle SBL$$

or $$(90° - \theta) = (\phi + \psi)$$

or $$\psi = 90° - (\theta + \phi)$$

Component of the thrust in connecting rod along the tangent to the crank circle is called crank pin effort. It follows that crank pin effort

$$F_t = Q\cos\psi$$

or $$F_t = Q\cos[90° - (\theta + \phi)] = Q\sin(\theta + \phi)$$

The crank effort, on the other hand, is the turning moment produced on the crank shaft and is given by—

$$T = F_t \times \text{crank radius} = Qr\sin(\theta + \phi) \qquad (14.95)$$

Expanding $\sin(\theta + \phi)$ and substituting for $Q = (F_P/\cos\phi)$,

$$T = (F_P r/\cos\phi)\{\sin\theta\cos\phi + \cos\theta\sin\phi\}$$

or $$T = F_P r\left\{\sin\theta + \cos\theta\,\frac{\sin\phi}{\cos\phi}\right\}$$

or $$T = F_P r\left\{\sin\theta + \frac{1}{2}\frac{\sin2\theta}{\sqrt{n^2 - \sin^2\theta}}\right\} \qquad (14.96)$$

14.17 INERTIA FORCES AND TORQUES IN SLIDER-CRANK MECHANISM

It is often necessary to know the torque which must be exerted on the crank shaft so as to overcome the inertia of the connecting rod and of the reciprocating parts.

(a) Graphical Method: This method is based on the concept of dynamically equivalent two mass system for the floating link. The concept is used to determine the inertia force for connecting rod in magnitude and direction. As in the earlier section, the first step is to draw velocity and acceleration polygon. Klein's construction may also be used alternatively. Figure 14.24(b) shows the acceleration polygon in which the line bp represents acceleration image of the connecting rod and point 'g' gives the acceleration image of G the c.g. of the rod. Vector $O_a g$ in the acceleration polygon, therefore, gives acceleration A_G in magnitude and direction.

For the purpose of determining inertia force F_{i3} of the connecting rod in magnitude, direction and line of action, the weight of the connecting rod is first assumed to be split into two masses one placed at piston pin P and the other at D such that $(PG)(GD) = k^2$, k being the radius of gyration. *Note that placing portion of connecting rod weight at P is only for the purpose of determining inertia force F_{i3} of the rod and is not to be clubbed with the weight of reciprocating parts for finding inertia force of reciprocating parts.*

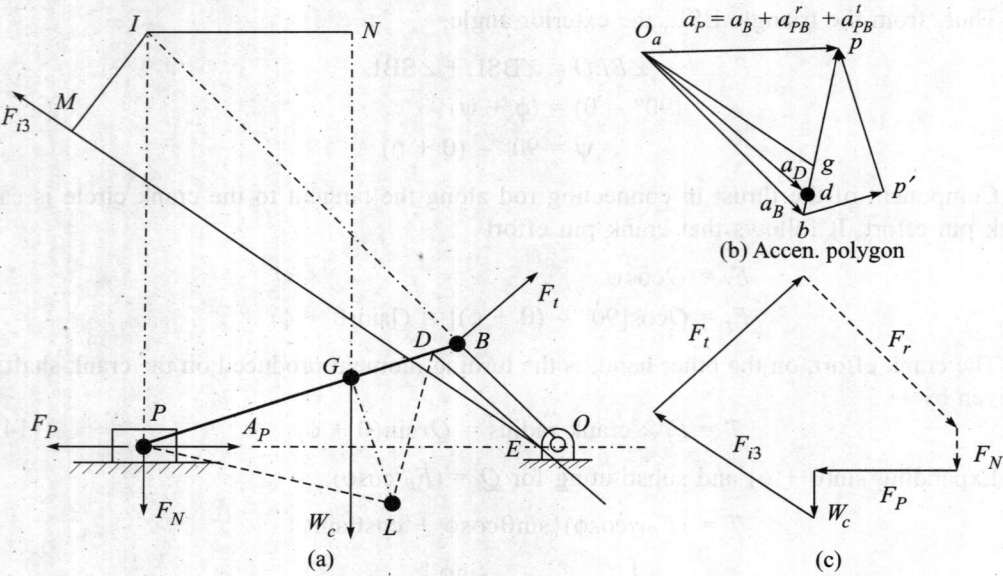

Fig. 14.24

In so far as the connecting rod is concerned, the point D has acceleration image at point d, and its acceleration is given by the vector $O_a d$. Thus the accelerating force on mass at P acts along the line of stroke, while the accelerating force on mass at D acts along DE parallel to the vector $O_a d$. The two accelerating forces intersect at E, suggesting that the resultant accelerating force (and hence the inertia force) on the connecting rod must pass through E. *Having located point E through which the resultant inertia force of the rod passes, we shall disregard the two-mass system now.* Since the accelerating force on the rod must act in the direction of A_G and the inertia force must oppose it, it follows that the inertia force of the rod passes through E and acts in a direction opposite to that of A_G. *The inertia force F_{i3} of the rod is parallel to $O_a g$ and therefore, is as shown in Fig. 14.24(a).* **Considering free body diagram of the connecting rod, external forces acting on it are:**

(i) The inertia force due to reciprocating parts F_p, (ii) the side thrust F_N due to walls of cylinder or guide-bars of cross-head on piston/cross-head, at P normal to the line of stroke, (iii) the inertia force F_{i3} of the rod, (iv) the weight W_c of the rod, vertically through c.g., (v) the forces F_r and F_t which act through the crank-pin B, parallel and perpendicular to the crank OB. Taking moments of these forces about instantaneous centre I, the point of intersection of the lines of action of the forces F_N and F_r (*see* Fig. 14.24a),

$$(F_t)IB = (F_p)IP + (F_{i3})IM + (W_c)IN \qquad (14.97)$$

where IM and IN are the length of perpendiculars dropped from I on the extended lines of action of F_{i3} and W_c respectively. Note that $F_{i3} = (W/g) A_G$, is completely known.

Effort F_t required at the crank to overcome inertia resistance can thus be obtained using equation (14.97). Finally a force polygon for all forces acting on the rod can be drawn as in Fig. 14.24(c) to establish forces F_N and F_r. The torque to be applied to the crank-shaft in order to overcome inertia of moving parts is thus,

$$T = F_t \times OB \qquad (14.98)$$

(b) *Analytical Method:* Alternatively the inertia errect of the connecting rod on the crank-shaft torque may be taken into account by assuming that the total mass of the rod is divided into two parts; one at the crank-pin B and the other at the piston pin P in such a way that c.g. 'remains unchanged. Since this distribution of mass is not in confirmity with the condition $k^2 = ab$, a correction couple is required to be applied to the crank shaft. For maintaining c.g. of the system unchanged, the weight components at piston pin and crank pin should be,

$$\text{Weight at } P, \; W_P = W_c \left(\frac{l-a}{l} \right)$$

$$\text{Weight at } B, \; W_B = W_c \left(\frac{a}{l} \right)$$

Note that the dimensions corresponding to moment arms are: $(l-a)\cos\phi$; $l\cos\phi$; and $a.\cos\phi$. The term $\cos\phi$ cancels out from moment equation, – being the common term.

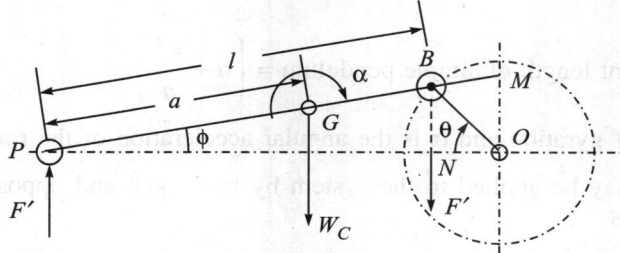

Fig. 14.25

The component of connecting rod weight at B does not contribute to crank-shaft torque because a uniformly rotating crank is subjected to radial component of acceleration only and the only inertia force is the centrifugal force. The component of weight at P can be clubbed with the reciprocating masses (of weight R) producing an inertia force given by

$$F_P = -\left(R + \frac{l-a}{l} W_c \right) \frac{A_p}{g} \tag{14.99a}$$

where A_P = acceleration of reciprocating parts.
When friction and other losses are neglected,

$$F_P V_P = F_B V_B \quad \text{or} \quad F_P = F_B \left(\frac{V_B}{V_P} \right)$$

But, from Klein's construction, $\left(\dfrac{V_B}{V_P} \right) = \left(\dfrac{OB}{OM} \right)$ and substituting for the velocity ratio above,

$$(F_P)OM = (F_B)OB = T_P, \text{ the turning moment} \tag{14.99b}$$

This suggests a way to obtain T_P in graphical method. Substituting for A_P, the acceleration of reciprocating parts from equation (14.85) in (14.99a), we get

$$A_P = r\omega^2 \left(\cos\theta + \frac{\cos 2\theta}{n} \right)$$

Hence

$$F_P = -\left(R + \frac{l-a}{l}W_c\right)\frac{r\omega^2}{g}\left(\cos\theta + \frac{\cos 2\theta}{n}\right)$$

Finally, from equation (14.96), corresponding inertia torque,

$$T = F_P r\left\{\sin\theta + \frac{1}{2}\frac{\sin 2\theta}{\sqrt{n^2 - \sin^2\theta}}\right\} \quad (14.100)$$

Since the weight component at B does not contribute to inertia torque, equation (14.100) represents inertia torque contribution from connecting rod and also from reciprocating masses.

Equation (14.100) however gives approximate inertia torque as the assumed two-mass system is not true in dynamic equivalence. For this reason, as shown in equation (14.70), a correction couple must be applied to the two-mass system. This is given by

$$T' = \left(\frac{W_c}{g}\right)a(l - L_E)\alpha \quad (14.101)$$

where L_E = equivalent length of simple pendulum = $\left(a + \dfrac{k^2}{a}\right)$

k = radius of gyration and α is the angular acceleration of the rod.

The couple T' may be applied to the system by two equal and opposite forces F' acting through pins P and B.

$$F' \times PN = T' = (W_c/g)a(l - L_E)\alpha \quad (14.102)$$

Corresponding torque on crank shaft produced by F' at B,

$$T_c = F' \times NO$$

Substituting for F' from equation (14.102),

$$T_c = \left(\frac{W_c}{g}\right)a(l - L_E)\alpha\left(\frac{NO}{PN}\right)$$

Again, $NO = r\cos\theta$ and, $PN = l\cos\phi$

$$= (l/n)\sqrt{n^2 - \sin^2\theta} = r\sqrt{n^2 - \sin^2\theta}$$

Hence

$$T_c = \left(\frac{W_c}{g}\right)a(l - L_E)\alpha\frac{\cos\theta}{\sqrt{n^2 - \sin^2\theta}} \quad (14.103)$$

Substituting for the angular acceleration of connnecting rod α from equation (14.87),

$$T_c = -\left(\frac{W_c}{g}\right)a(l - L_E)\frac{\omega^2(n^2 - 1)\sin\theta}{(n^2 - \sin^2\theta)^{3/2}}\frac{\cos\theta}{\sqrt{n^2 - \sin^2\theta}}$$

$$= -\left(\frac{W_c}{g}\right)a(l - L_E)\frac{\omega^2(n^2 - 1)\sin 2\theta}{2(n^2 - \sin^2\theta)^2} \quad (14.104)$$

Usually n^2 is large in comparison to $\sin^2\theta$ and unity both and therefore equation (14.104) may be rewritten as,

$$T_c \approx -\left(\frac{W_c}{g}\right)a(l - L_E)\frac{\omega^2}{2\,n^2}\sin 2\theta \qquad (14.105)$$

Besides, there is also a torque due to the force of gravity on connecting rod. The component of the weight of rod at crank pin B is $\left(\dfrac{PG}{PB}\right)W_c$. Torque exerted by this force at B on crank shaft is

$$T_w = -\left(\frac{PG}{PB}\right)W_c(NO)$$

or

$$T_w = -\left(\frac{a}{l}\right)W_c r\cos\theta$$

or

$$T_w = -W_c(a/n)\cos\theta \qquad (14.106)$$

Total torque exerted on the crank shaft by the inertia of moving parts is the algebraic sum of T_p, T_c and T_w.

14.18 SHAKING FORCE ANALYSIS OF A SLIDER-CRANK MECHANISM

Consider a slider-crank engine mechanism shown in Fig. 14.26(a) in which the piston is subjected to a known gas load F_g. Let m_2 and I_2 be the mass and moment of inertia respectively of crank 2 and m_3 and I_3 be the mass and moment of inertia respectively of the connecting rod 3. Finally let m_4 be the mass of reciprocating parts. Note that I_2 and I_3 are the moments of inertia about respective centroidal axes perpendicular to the plane of motion. The crank speed ω_2 is assumed to be constant. This is, however, not necessary and analysis can be carried out by considering resultant acceleration of crank pin as $AB = A_B^r + A_B^t$. Let G_2 and G_3 represent centres of gravity of links 2 and 3 respectively. Suppose that the torque T_2' is exerted by crank 2 on crank shaft and that the shaking forces are required to be obtained.

The velocity and acceleration diagrams are constructed as a first step; the latter is shown in Fig. 14.26(b). The problem is greatly simplified if the links 3 and 4 are grouped together and considered a single free body diagram. This eliminates need to consider force F_{34} which exists at the cnnnection between links 3 and 4 and does not remain an external force for the combined free body diagram. Unless otherwise stated specifically the weights of links will be neglected, being small in comparison to the inertia forces.

The accelerating forces F_2, F_3 and F_4 act in the direction of respective acceleration vectors, o_ag_2, o_ag_3 and o_ap. Their magnitudes are given by

$$F_2 = m_2 A_{G2}; \quad F_3 = m_3 A_{G3}; \quad \text{and} \quad F_4 = m_4 A_p$$

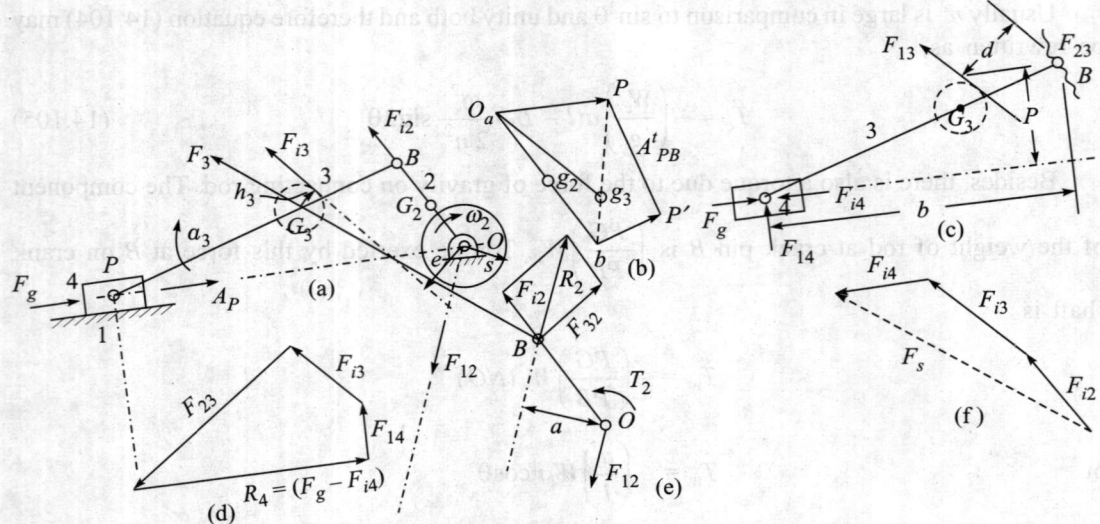

Fig. 14.26

Inertia forces F_{i2} and F_{i4} are equal but opposite to the corresponding accelerating forces. The inertia force F_{i4} acts along line of stroke as shown, while the inertia force F_{i2} acts along the centre line of link 2 radially outwards. The inertia force F_{i3} numerically equals F_3, but is tangent to a circle of radius h_3 drawn from G_3. The radius is given by $h_3 = (I_3\alpha_3)/(m_3 A_{G3})$. The force F_{i3} is tangent to the circle of radius h_3 on the right side so as to oppose α_3.

Consider the combined free body diagram of Fig. 14.26(c) the external forces are: F_g, F_{i4}, F_{14}, F_{i3} and F_{23}, out of which the unknown forces are F_{14} and F_{23}. Taking moments of these forces about pin B,

$$(F_{14})b - (F_g - F_{i4})p + (F_{i3})d = 0$$

or
$$F_{14} = \frac{(F_g - F_{i4})p - (F_{i3})d}{b} \qquad (14.107)$$

Force polygon can then be constructed as in Fig. 14.26(d), so as to obtain force F_{23} as the closing side of the polygon.

Figure 14.26(e) shows the free body diagram of link 2. Force R_2 is the resultant of forces F_{32} (equal and opposite to F_{23}) and F_{i2} equal the force R_2 which is equal and opposite to F_{12}. Torque T_2' which the crank exerts on crank shaft is equal and opposite to the resisting torque T_2. Thus,

$$T_2' = R_2 \times a \qquad (14.108)$$

Magnitude and direction of shaking force is obtained by finding resultant of all the inertia forces for the links 2, 3 and 4. This is shown to be the vector F_s in Fig. 14.26(f), which represents the closing side of the polygon. If simultaneously, line of action of the shaking force is also required, we may take moments of these inertia forces about any convenient point and equate the same to the moment due to F_s. Taking moments of these inertia forces about O (so that moments due to F_{i2} and F_{i4} vanish),

$$(F_s)s = (F_{i3})e \qquad \text{(vide Fig. 14.26a)}$$

or
$$s = e\left(\frac{F_{i3}}{F_s}\right)$$
(14.109)

Thus, the shaking force is completely known.

EXAMPLE 14.7 The connecting rod of a vertical reciprocating engine is 2.5 m long between centres and has a mass of 400 kg. Its mass centre is 1 m from big-end bearing. When suspended from cross-head pin and allowed to swing, the period of oscillation is 2.93s. The crank is 0.5 m long and rotates at 240 r.p.m.

When the crank has turned through 45° from top-dead centre, find, due to inertia of connecting rod,

(a) Magnitude and line of action of the resultant force, acting upon the connecting rod
(b) Reaction at the cross-head guide
(c) Force on the main bearing
(d) Torque on the crank shaft   (DAV Indore, 1975)

Solution:

Given: Distance between the centres   = 2.5 m
               Distance of mass centre from big-end bearing = 1.0 m
Therefore,   distance of c.g. from small-end bearing   = 2.5 – 1.0 = 1.5 m

and
$$n = \frac{l}{r} = \frac{2.5}{0.5} = 5$$

Time period of oscillation $\tau = 2.93$ s

If L_E is the length of equivalent simple pendulum, then from,
$$\tau = 2\pi\sqrt{L_E/g},$$

$$L_E = \left(\frac{\tau}{2\pi}\right)^2 g = \left(\frac{2.93}{2\pi}\right)^2 \times 9.81 = 2.13 \text{ m}$$

Thus $L_E = a + b = 2.13$ m   (Ref. equation 14.63)

Therefore $b = 2.13 – 1.5$ m $= 0.63$ m

Therefore $k_2 = ab = (1.5)(0.63) = 0.945$ m^2

Total inertia torque at the crank shaft consists of

(i)   torque due to mass of reciprocating parts together with that due to component of connecting rod mass at P, i.e., T_P
(ii) the correction torque T_c'
(iii) the torque due to weight of rod i.e., T_w.

The acceleration of reciprocating parts,

$$A_P = r\omega^2\left(\cos\theta + \frac{\cos2\theta}{n}\right)$$

or
$$A_P = (0.5)\left(\frac{2\pi \times 240}{60}\right)^2\left(\cos 45 + \frac{\cos 90}{n}\right) = 223.32 \text{ m/s}^2$$

Hence
$$T_P = -\left(R + \frac{l-a}{l} W_c\right)\frac{A_p}{g}$$

The weight of reciprocating parts is not given, hence the torque contribution from reciprocating masses is neglected.

(a) the weight component of connecting rod at cross-head pin

$$W_P' = \frac{l-a}{l} W_c = \left(\frac{2.5-1.5}{2.5}\right)400 \times 9.81$$

$$W_P' = 1569.6$$

Being a vertical engine this component weight itself gives an inertia force along line of stroke and its torque contribution is given by

$$T_P = W_P' \times r\left(\sin 45 + \frac{\sin 90}{2 \times 5}\right)$$

or
$$T_P = 1569.6 \times 0.5\left(0.707 + \frac{1}{10}\right) = 633.33 \text{ N·m (clockwise)}$$

(b) The weight component of connecting rod at crank pin,

$$W_B = \left(\frac{a}{l}\right)W_c = \frac{1.5}{2.5} \times 400 \times 9.81 = 2354.4 \text{ N}$$

Therefore Torque on crank shaft due to this weight component (*refer* Fig. 14.27),

$$T_B = (2354.4) \times NB = (2354.4) \times r\sin 45$$

or
$$T_B = 2354.4 \times 0.5 \times 0.707 = 832.4 \text{ N·m} \qquad \text{(clockwise)}$$

(c) The weight of connecting rod transferred to P also undergoes acceleration A_p and the corresponding inertia force,

$$F_p' = A_p \times r\left(\sin\theta + \frac{\sin 2\theta}{2n}\right) \times \frac{W_p'}{g}$$

or
$$F_p' = \frac{1569.6}{9.81} \times 223.32 \times 0.5\left(\sin 45° + \frac{\sin 90}{2 \times 5}\right) = 14{,}419.4 \text{ N·m (c. c. w)}$$

(d) The correction couple to be applied

$$T_c' \approx -\left(\frac{W_c}{g}\right)a\,(l - L)\left(\frac{\omega^2}{2\,n^2}\right)\sin 2\theta$$

Therefore
$$T_c' \approx -(400)(1.5)(2.5 - 2.13) \times \left(\frac{2\pi \times 240}{60}\right)^2 \frac{1}{2 \times 5^2}\sin 90$$

or
$$T_c' = 2804.54 \text{ N·m (c.c.w.)}$$

Hence total torque on shaft = 14,419.4 + 2804.54 − 633.3 − 832.4

or $$T = 15,758.2 \text{ N·m} \qquad \textbf{Ans.}$$

Hence $$F_t = \frac{T}{r} = \frac{15,758.2}{0.5} = 31,516.4 \text{ N}$$

(Note that to balance moments due to F_p, F_{i3}, etc. about I.C. F_t must act as shown). The inertia force of connecting rod is given by,

$$F_{i3} = m_c \times A_G$$

By measurement from Klein's construction,

$$A_G = \frac{1}{2.4} \times \omega^2$$

$$= \frac{1}{2.4} \left(\frac{2\pi \times 240}{60} \right)^2$$

$$= 263.189 \text{ m/s}^2$$

Therefore $$F_{i3} = 400 \times 263.189 = 10,5275.6 \text{ N}$$

The direction of F_{i3} is exactly opposite to that of acceleration. A_G indicated by line gO in Klein's construction. Out of the four forces F_t, F_r, F_{i3} and F_N, two are completely known. These are F_t and F_{i3} while for the remaining two forces i.e., F_r and F_N, only direction is known.

The force polygon is completed as shown in Fig. 14.27. The forces F_N and F_r measure,

$$F_N = 7500 \text{ N} \qquad \text{and} \qquad F_r = 96,000 \text{ N}$$

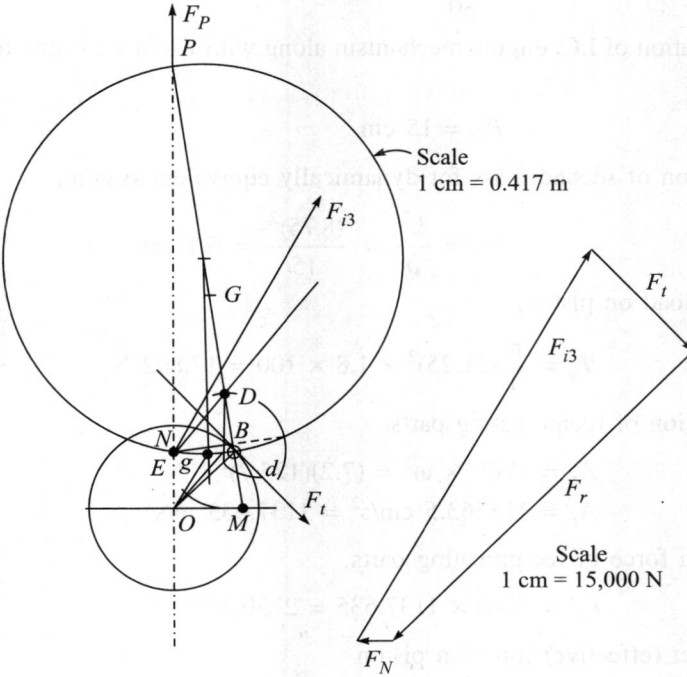

Fig. 14.27

Thus

(a) the magnitude and line of action of resultant (inertia) force acting on rod is given by
$F_{i3} = 1,05,275.6$ N

(b) The reaction at cross-head guide $= F_N = 7500$ N

(c) The force on the main bearing $= F_r = 96,000$ N

(d) The torque on the crank shaft, $T = 15,758.2$ N·m **Ans.**

EXAMPLE 14.8 The connecting rod of an I.C. engine is 22.5 cm long and has a mass of 1.6 kg. The mass of the piston and the gudgeon pin is 2.4 kg and the stroke is 15 cm. The cylinder bore is 11.25 cm. The c.g. of the connecting rod is 15 cm from the small end. Its radius of gyration about the c.g. for oscillations in the plane of swing of the connecting rod is 8.75 cm.

Determine the magnitude and direction of the resultant force on the crank pin when the crank is at 40° and the piston is moving away from the i.d.c. under an effective pressure of gas of 1.8 N/mm². The engine speed is 1200 r.p.m.

Solution: Given: d = cyl. bore = 11.25 cm; $\dfrac{W_c}{g}$ = 1.6 kg; l = 22.5 cm;

$$\frac{R}{g} = 2.4 \text{ kg}; \quad r = \frac{15}{2} = 7.5 \text{ cm};$$

$$a = PG = 15 \text{ cm}; \quad k = 8.75 \text{ cm}; \quad p = 18 \text{ N/mm}^2$$

$$\omega = \frac{2\pi \times 1200}{60} = (40\pi) = 125.7 \text{ rad/s}$$

The configuration of I.C. engine mechanism along with Klein's construction is as shown in Fig. 14.28.

$$P_G = 15 \text{ cm}$$

Hence, location of second mass for dynamically equivalent system,

$$D_G = \frac{k^2}{a} = \frac{(8.75)^2}{15} = 5.1 \text{ cm}$$

(a) The gas load on piston,

$$F_g = \frac{\pi}{4}(11.25)^2 \times 1.8 \times 100 = 17,892 \text{ N}$$

(b) Acceleration of reciprocating parts,

$$A_p = (NO) \times \omega^2 = (7.2)(125.7)^2$$

or $$A_p = 113763.5 \text{ cm/s}^2 = 1137.635 \text{ m/s}^2$$

Hence, inertia force of reciprocating parts,

$$F_p' = (2.4) \times 1137.635 = 2730.3 \text{ N}$$

Hence, the net (effective) force on piston

$$F_p = F_g - F_p' = (17,892 - 2730.3) = 15,161.7 \text{ N}$$

The lines of action of accelerating forces for the dynamically equivalent masses at P and D for the rod intersect at E as shown in Fig. 14.28. The inertia force, therefore, acts from E parallel to acceleration $A_G \equiv$ 'gO', but opposite in direction.

From the Klein's construction,

$$A_G = (go) \times \omega^2$$
$$= (6.9) \times (125.7)^2 = 1090.234 \text{ m/s}^2$$

Hence the total inertia force of the connecting rod $= \dfrac{W_c}{g} \times A_G = 1.6 \times 1090.234$

or, $\qquad\qquad\qquad\qquad F_i = 1744.37 \text{ N}$

Taking the moments of all forces acting on the rod about I, the I.C.,

$$F_t \times IB = (F_i)IX + (W_c)IY + (F_p)IP$$

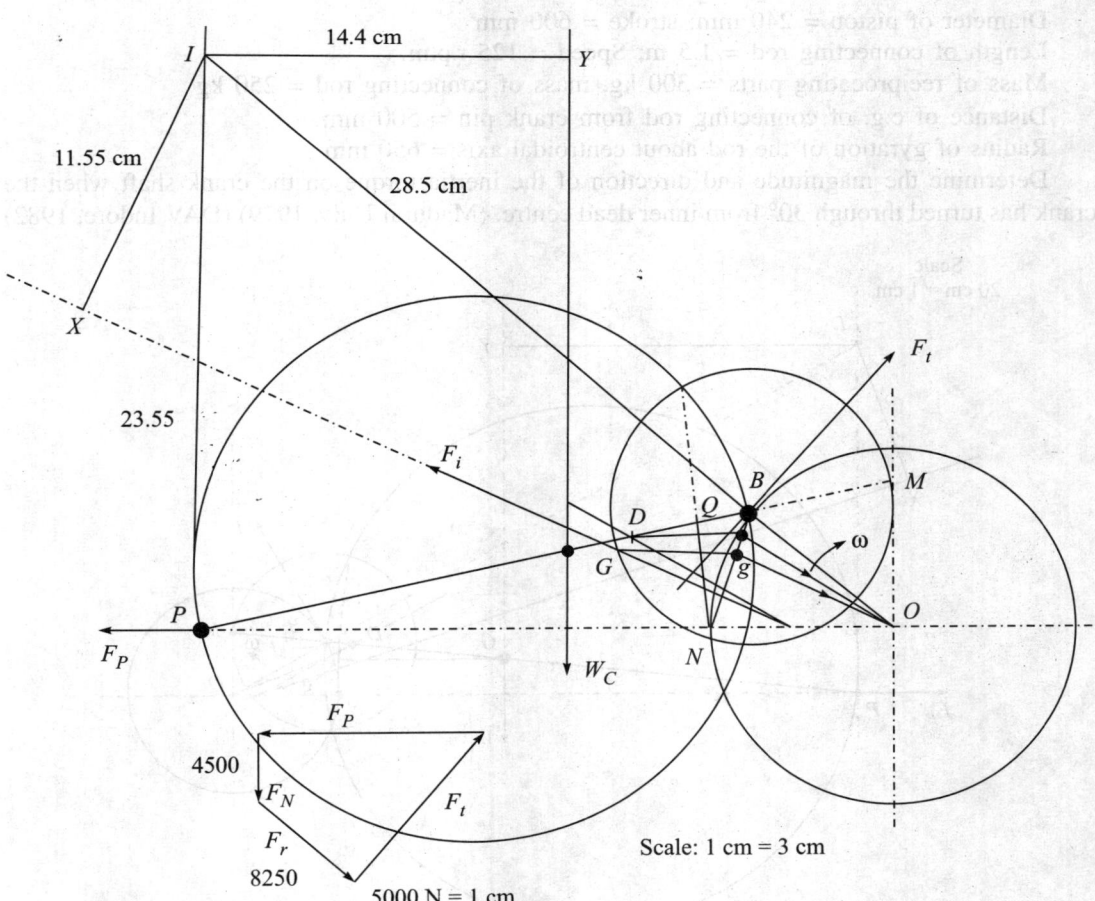

Fig. 14.28

By direct measurement from the figure, using scale,

$$(F_t) \times 28.5 = (F_i) \times 11.55 + (1.6 \times 9.81) \times 14.4 + (F_p) \times 23.55$$

or $\quad F_t = \dfrac{(1744.37) \times 11.55 + (1.6 \times 9.81) \times 14.4 + (15,161.7) \times (23.55)}{28.5} = 13{,}240.15 \text{ N}$

For finding the resultant force on crank pin, we must also know F_r. To determine F_N and F_r, a force polygon is constructed as in Fig. 14.28. By measurement from force polygon, using scale,

$$F_N = 4500 \text{ N} \quad \text{and} \quad F_r = 8250 \text{ N}$$

Therefore Resultant force on crank pin

$$= \sqrt{F_t^2 + F_r^2} = \sqrt{(13240.15)^2 + (8250)^2} = 15{,}600 \text{ N} \qquad \textbf{Ans.}$$

EXAMPLE 14.9   Following data refers to a steam engine—
Diameter of piston = 240 mm; stroke = 600 mm
Length of connecting rod = 1.5 m; Speed = 125 r.p.m.
Mass of reciprocating parts = 300 kg; mass of connecting rod = 250 kg
Distance of c.g. of connecting rod from crank pin = 500 mm
Radius of gyration of the rod about centroidal axis = 650 mm
Determine the magnitude and direction of the inertia torque on the crank shaft when the crank has turned through 30° from inner dead centre. (Madurai Univ. 1979) (DAV Indore, 1982)

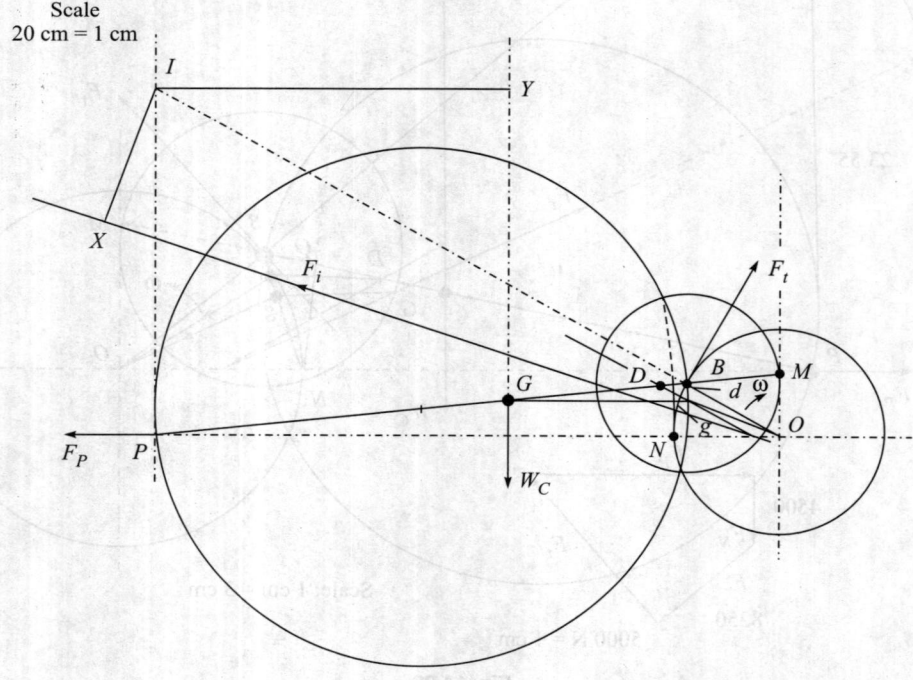

Fig. 14.29

Solution: Given: $d = 24.0$ cm; Rad. of crank $r = \dfrac{60}{2} = 30$ cm

$$l = 150 \text{ cm}; \quad \frac{R}{g} = 300 \text{ kg};$$

$$\frac{W_c}{g} = 250 \text{ kg}; \quad k = 65 \text{ cm}; \quad \theta = 30°; \quad l - a = 50 \text{ cm}$$

$$a = 150 - 50 = 100 \text{ cm}; \quad \omega = \frac{2\pi \times 125}{60} = 13.09 \text{ rad/s}$$

Draw the mechanism to the scale and construct Kleins diagram as in Fig. 14.29.

$$PG = 100 \text{ cm} = a$$

$$GD = b = \frac{k^2}{PG} = \frac{(65)^2}{100} = 42.25 \text{ cm}$$

By measurement from Kleins' construction,

$$A_G = gO = 28(13.09)^2$$

or
$$A_G = 4797.7 \text{ cm/s}^2 = 47.977 \text{ m/s}^2$$

$$A_p = NO = 28(13.09)^2 = 4797.7 \text{ cm/s}^2 = 47.977 \text{ m/s}^2$$

Hence, the inertia force of connecting rod,

$$F_i = (250) \times 47.9770 = 11{,}994.25 \text{ N}$$

The inertia force of reciprocating parts,

$$F_p = 300 \times 47.977 = 14{,}393.1 \text{ N}$$

Let F_t be the equivalent inertia force at crank pin. Then, taking moments of all forces about I,

$$(F_t)IB = (F_p)IP + (F_i)IX + (W_c)IY$$

or
$$F_t = \frac{(14393.1)(100) + (11994.25)(40) + (250 \times 9.81)(100)}{172}$$

or
$$F_t = 12{,}583.3 \text{ N}$$

Therefore Inertia torque to be overcome at crank shaft

$$T_i = F_t \times r = 12{,}583.3 \times 0.3 = 3774.99 \text{ N·m} \qquad \textbf{Ans.}$$

EXAMPLE 14.10 Solve Example 14.9 analytically.

Solution: $l = 150$ cm, $a = 150 - 50 = 100$ cm, $n = \dfrac{150}{30} = 5$, $\omega = 13.09$ rad/s

Equivalent length of simple pendulum, $L_E = a + \dfrac{k^2}{a}$

or
$$L_E = 100 + \frac{65^2}{100} = 142.25 \text{ cm}$$

Equivalent weight at P (weight of reciprocating parts plus component of connecting rod weight)

$$W_P = R + \frac{(l - a)}{l} W_c$$

or

$$W_P = 9.81 \times \left[300 + \left(\frac{150 - 100}{150} \right) \times 250 \right] = 3760.47 \text{ N}$$

The inertia force due to equivalent weight at P,

$$F_P = m_p r\omega^2 \left(\cos 30 + \frac{\cos 60}{5} \right)$$

or

$$F_P = \frac{3760.47}{9.81} \times 0.3 \times (13.09)^2 \left(\cos 30 + \frac{\cos 60}{5} \right) = 9035.41 \text{ N}$$

Corresponding inertia torque,

$$T_P = F_p r \left(\sin\theta + \frac{\sin 2\theta}{2 \sqrt{(n^2 - \sin^2\theta)}} \right)$$

or

$$T_P = (19035.41) \times 0.3 \left(\sin 30 + \frac{\sin 60}{2 \cdot \sqrt{(5^2 - \sin^2 30)}} \right) = 3352.36 \text{ N·m}$$

The correction couple at crank shaft,

$$T_c = \left(\frac{W_c}{g} \right) a \ (l - L_E) \frac{\omega^2}{2n^2} \sin 2\theta$$

or

$$T_c = (250) \left(\frac{100}{100} \right) \left(\frac{150 - 142.25}{100} \right) \times \frac{13.09^2}{2 \times 5^2} \times \sin 60 = 57.50 \text{ N·m}$$

Inertia torque due to weight of connecting rod,

$$T_w = W_c \left(\frac{a}{n} \right) \cos\theta$$

or

$$T_w = (250 \times 9.81) \left(\frac{100}{100} \times \frac{1}{5} \right) \cos 30 = 424.78 \text{ N·m}$$

Total inertia torque to be overcome,

$$T_i = T_p + T_c + T_w$$

or

$$T_i = 3352.36 + 57.5 + 424.78 = 3834.64 \text{ N·m} \qquad \textbf{Ans.}$$

EXAMPLE 14.11 Figure 14.30 shows a vertical single-cylinder compressor whose crank revolves at 955 r.p.m. c.c.w. The stroke of the piston is 150 mm, and the length of the connecting rod is 300 mm. The piston weighs 37.8 N, and the connecting rod 20 N and the crank shaft

88.97 N. The c.g. of the rod is 78 mm from the crank-pin centre A, and the corresponding moment of inertia is 0.0294 kg·m^2. The c.g. of the crank is located at 12.5 mm from the axis of rotation so that its inertia force (centrifugal force) F_{i2} balances the centrifugal force of the rotating part of the connecting rod. In the phase depicted, the gas force F_4 on the piston amounts to 5338 N. Determine the driving torque T_2, the slide reaction F_{14}, the pin force F_O, F_A, F_P and the forces exerted on the engine block by the foundation. Weights of members may be neglected. Consider the crank to have rotated 45° from i.d.c.

Solution: Figure 14.30(a) shows the inertia forces of the moving links in their respective positions. Figure 14.30(b) shows the acceleration diagram where g_2 and g_3 denote acceleration images respectively of the c.g.s. of links 2 and 3. Hence $o_a g_2$ and $o_a g_3$ gives respective accelerations in magnitude and direction.

The acceleration values measured from acceleration diagram of Fig. 14.30(b) are: $A_{G2} = 125$ m/s^2; $A_{G3} = 640$ m/s^2 and $A_p = 505$ m/s^2. Hence inertia forces are

$$F_{i4} = \frac{37.8}{9.81} \times 505 = 1945.87 \text{ N}$$

$$F_{i3} = \frac{20}{9.81} \times 640 = 1304.39 \text{ N}$$

$$F_{i2} = \frac{88.97}{9.81} \times 125 = 1133.66 \text{ N}$$

Gas force on piston, $\qquad F_g = 5338 \text{ N}$

Hence, the resultant of gas force F_g and inertia force F_{i4}, is

$$R_4 = F_g - F_{i4} = 5338 - 1945.87 = 3392.13 \text{ N}$$

The angular acceleration α_3 of the rod is established from $A_{PA}^t = 560$ m/s^2 (from measurement).

Thus $\qquad \alpha_3 = (A_{PA}^t)/AP = \dfrac{560}{0.3} = 1866.67 \text{ rad/s}^2$

The inertia force F_{i3} of the connecting rod, will have line of action tangential to a circle of radius h_3 drawn from G_3, such that

$$h_3 = (I_3 \alpha_3)/F_{i3} = (0.0294)(1866.67)/(1304.79) = 0.042 \text{ m}$$

or $\qquad\qquad\qquad\qquad h_3 = 4.2 \text{ cm}$

The line of action of F_{i3} will be tangent to this circle on right hand side so as to oppose α_3. This is shown in Fig. 14.30(c). Thus taking moments of all external forces about A (moment arms by measurement),

$$(F_{14}) \times 29.5 = (R_4) \times 5.25 - (F_{13}) \times 1.25$$

or $\qquad\qquad F_{14} = \dfrac{(3392.13) \times 5.25 - (1304.79) \times 1.25}{29.5} = 548.4 \text{ N} \qquad$ **Ans.**

A force polygon is drawn next in Fig. 14.30(d) for forces acting on link 3 and the closing line of the polygon then establishes the force F_{23} exerted by crank 2 on link 3. By measurement, $F_{23} = 2725$·N.

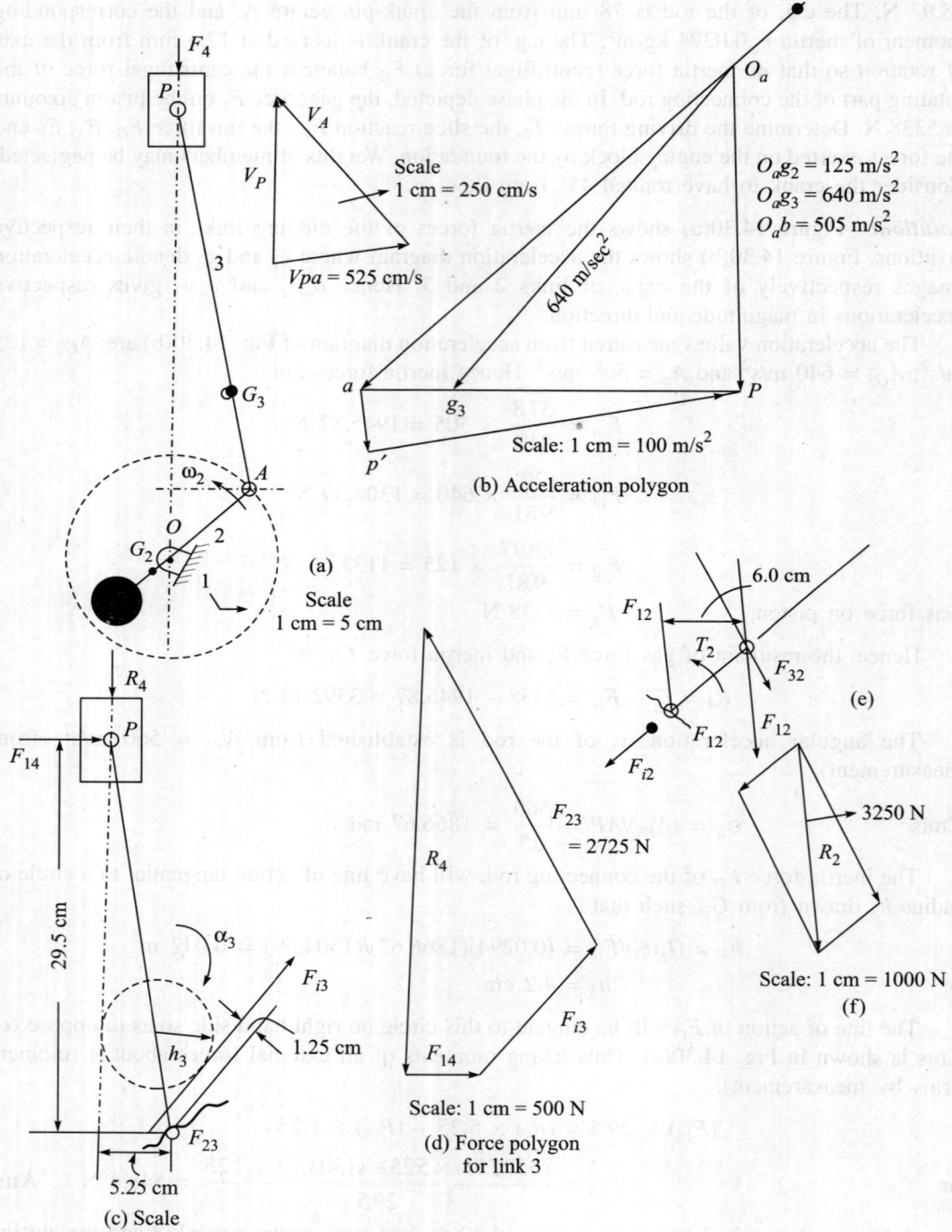

F_4

P

V_A

Scale
1 cm = 250 cm/s

V_P

3

$Vpa = 525$ cm/s

O_a

$O_ag_2 = 125$ m/s^2
$O_ag_3 = 640$ m/s^2
$O_ab = 505$ m/s^2

640 m/sec^2

a

g_3

P

p'

Scale: 1 cm = 100 m/s^2

(b) Acceleration polygon

G_3

A

ω_2

O

G_2

2

1

(a)

Scale
1 cm = 5 cm

6.0 cm

F_{12}

T_2

F_{32}

F_{12}

F_{12}

(e)

F_{i2}

R_4

P

F_{14}

29.5 cm

α_3

F_{i3}

h_3

1.25 cm

F_{23}

5.25 cm

(c) Scale
1 cm = 5 cm

R_4

F_{23}
= 2725 N

F_{i3}

F_{14}

Scale: 1 cm = 500 N

(d) Force polygon
for link 3

3250 N

R_2

Scale: 1 cm = 1000 N

(f)

Fig. 14.30

Considering the free body diagram of crank now (Fig. 14.30e), the force F_{32} exerted by link 3 on 2 is equal and opposite to F_{23}, as shown. The inertia force

$$F_{i2} = \frac{88.97}{9.81} \times A_{G2} = \frac{88.97}{9.81} \times 125 = 1133.66 \text{ N}$$

Since direction of acceleration of G_2 is og_2 as shown in acceleration polygon, the inertia force F_{i2} will act in the direction A to O. Resultant of forces F_{32} and F_{i2} is obtained in Fig. 14.30(f) as the vector

$$R_2 = 3250 \text{ N}$$

An equal and parallel force F_{12} in a direction opposite to that of R_2 is provided by frame at O. This constitutes an inertia torque,

$$T_2' = (F_{12}) \times d$$
$$= (3250) \times 6.0 = 19{,}500 \text{ N·cm (c.w.)} = 195 \text{ N·m (c.w.)}$$

Hence, driving torque required at the crank shafts,

$$T_2 = T' = 195 \text{ N·m (c.c.w.) as shown in Fig. 14.30(e)} \qquad \textbf{Ans.}$$

Thus

(a) Driving torque required $T_2 = 195$ N·m (c.c.w.)
(b) The slide reaction $F_{14} = 548.4$ N (to the right)
(c) The pin force at O $FO = R_2 = 3250$ N
(d) The pin force at A $FA = F_{32} = 2725$ N
(e) The pin force at P $FP = R_4 = 3392.13$ N **Ans.**

14.19 TURNING MOMENT DIAGRAM (CRANK EFFORT DIAGRAM)

Indicator Diagram

A turning moment diagram gives distribution of crank effort (*i.e.*, turning moment) against the crank angle θ of rotation. For plotting turning moment diagram (abbreviated as T.M. diagram) gas/steam pressure must be known for various crank positions. The values of pressure p for various crank angles θ can be obtained from the indicator diagram. An indicator mechanism usually enables to plot pressure values along ordinate for crank angles (along base) at an interval of say 15° for expansion and compression strokes. Let us take the case of a gasoline engine. The ordinates of the indicator diagram represent absolute pressures and therefore the effective pressures on piston are obtained by deducting atmospheric pressure from the measured pressure and multiplying by piston area. A representative indicator diagram is shown in Fig. 14.31(a) for a 4-stroke cycle engine.

Inertia Forces and Resultant Forces

Having established from the indicator diagram the gas pressures acting on the piston, throughout the cycle, the next step in the analysis is to determine contribution from inertia forces of reciprocating parts, connecting rod, and the crank. Analytical methods are available for this purpose, but a lot of effort is saved by graphical methods discussed in Sections 14.17 and 14.18.

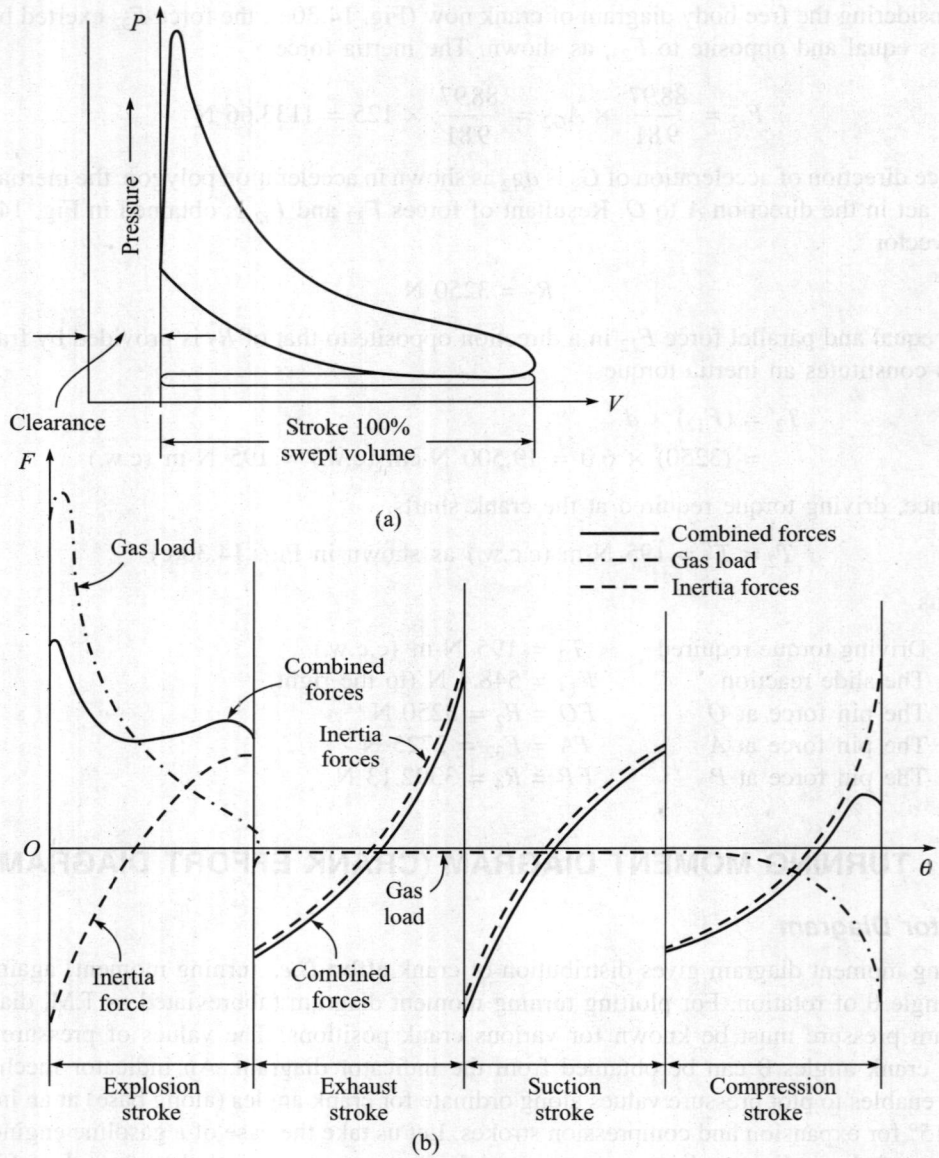

Fig. 14.31(a–b)

In order to obtain resultant forces on different links, it is necessary to combine static forces, like gas forces, with inertia forces. Forces on various links on account of gas pressure alone can be established using static force analysis discussed in Section 14.4. These forces can be recorded in tabular form. The table can then be completed by combining static forces with inertia forces discussed previously. The characteristics of these forces and their effects can be seen from the study of Figs. 14.31(a) and (b). Note that as indicated in equation 14.99(b), the ordinate OM

(Fig. 14.25) is subject to cyclic variations and turns out to be an additional factor for variations in T.M. diagram. It follows from equation 14.99(b) that,

$$\text{T.M.} = (F_g \pm F_i) \times \text{OM}$$

Figure 14.31(b) shows piston force variation for one cylinder. The forces assisting the motion of piston are considered positive and have been plotted above the base line. Gas torque diagram is shown by dotted lines in Fig. 14.31(c). During the suction stroke, the pressure of the gases on the working side of the piston is slightly less than the pressure of the atmosphere on the crank side of the piston and there is a small negative gas-torque loop. During the compression stroke, work is done by the piston on the gases, and as such a large negative gas-torque loop results. During expansion stroke, however, expanding gases do work on the piston and a large positive gas-torque loop is obtained. Finally during exhaust stroke, work is done by the piston on the gases. This is because pressure inside the cylinder is slightly greater than the external atmospheric pressure. This again results in small negative gas-torque loop. The negative gas-torque loops during suction and exhaust stroke are too small and therefore not shown in Fig. 14.31.

Turning Moment Diagram

The torque required on the crank shaft in order to overcome the inertia and accelerate the reciprocating parts and connecting rod, is shown in Fig. 14.31(d). For each half revolution of the crank, the diagram has a positive and negative loop. The two loops may have different shape due to the obliquity effect of connecting rod, but their areas are equal. This is obvious because energy absorbed in accelerating the parts must equal the energy released during subsequent retardation of these parts.

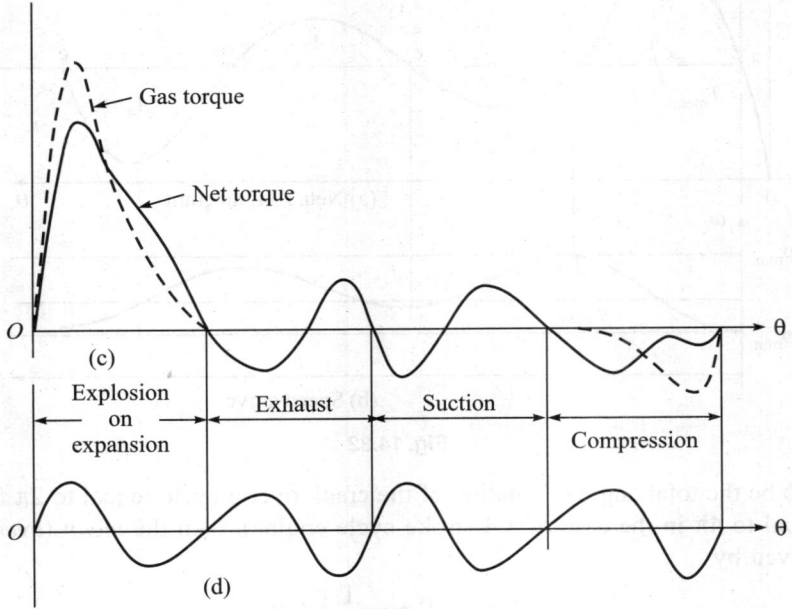

Fig. 14.31(c–d)

The net crank effort (*i.e.*, T.M.) for a given crank position is given by the difference between the gas-torque and the torque required to over come inertia and accelerate the moving parts, and is shown by full lines in Fig. 14.31(c). In the case of multi-cylinder engine, the resultant turning moment is obtained by the superposition of the turning moments of each cylinder, with the starting point suitably shifted depending on the crank offset.

14.20 FLUCTUATION OF CRANK SHAFT SPEED

The couple or torque (called load) that resists rotation of the crank shaft remains constant over a cycle for most engines. And as we have seen in the earlier sections, the driving turning moment exerted on the crank shaft, fluctuates substantially. This means that except for a few crank positions, an unbalanced torque always exists which has a tendency either to increase or to decrease the angular speed of the crank shaft.

Let T_r be the resisting torque, which is assumed to be constant over a cycle while the turning moment T is a function of crank angle θ, and is developed by an engine. Let Fig. 14.32(a) represent a netturning moment developed by an engine, over one cycle. *The area between turning moment curve and the θ-axis represents energy delivered by the engine over the cycle under consideration.* This is expressed mathematically as

$$E = \int_{\theta} T d\theta \qquad (14.110)$$

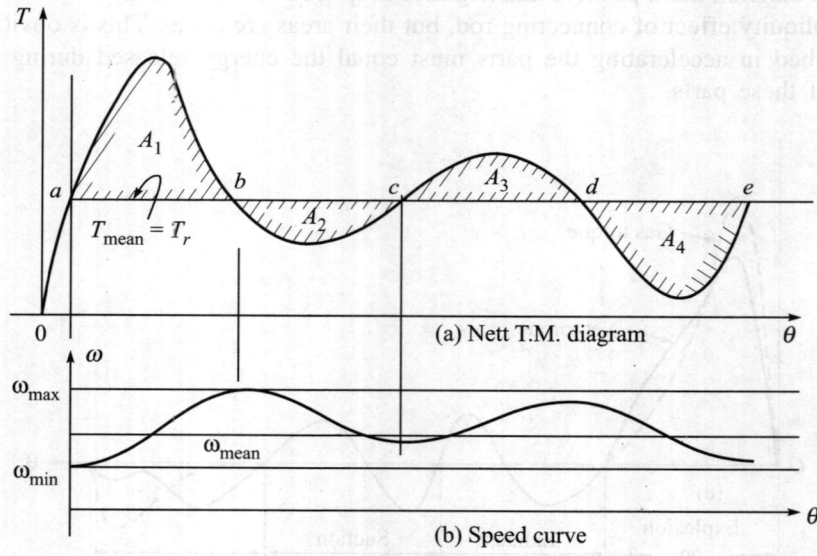

(a) Nett T.M. diagram

(b) Speed curve

Fig. 14.32

Thus, if ϕ be the total angle of rotation of the crank over a cycle (equal to 2π for a 2-stroke cycle and equal to 4π in the case of a 4-stroke cycle engine), then the *mean (average)* turning moment is given by

$$T_{\text{mean}} = E/\phi = \frac{1}{\phi} \int T d\theta \qquad (14.111)$$

Mean (Average) Turning Moment (T_m)

The mean T.M. is then defined as a uniform turning moment, which when acting over the complete cycle, does the same work as the work produced by the actual turning moment. In other words, a rectangular area under the straight line $T = T_m$ over the period of cycle ϕ and the area under the actual T.M. curve, over the same period, are equal.

Assuming the resistance to rotation of the crank shaft to be uniform, the resisting moment line may be assumed to coincide with T_m line. Thus, as shown in Fig. 14.32(a) the resisting moment T_r line cuts the net (actual) turning moment curve at the points a, b, c, d and e. The net turning moment and resisting turning moment are equal at these points and therefore the slope of speed curve in Fig. 14.32(b) at corresponding crank-angles is zero. At all other points between a and b, the actual T.M. is greater than the resisting moment and the excess torque tends to accelerate the rotating parts. Between points b and c, on the other hand, actual T.M. is less than the resisting T.M. and the unbalanced torque decelerates the rotating parts. Thus, while the angular speed at b is greater than that at a, the angular speed at c is less than that at b. Similarly, at points between c and d, the actual turning moment is greater than the resisting T.M. and the unbalanced torque tends to accelerate the rotating parts. Thus, angular speed of rotating parts at d is more than that at c. Finally, between points d and e, actual T.M. is less than the resisting T.M. and the unbalanced torque decelerate the rotating parts, so that angular speed at e is less than that at d. Finally, since mean T.M. equals uniform resisting torque, the angular speeds at a and e are equal.

We conclude therefore that so long as resisting torque T_r is uniform and equals mean T.M., the speed of rotating parts at the beginning and the end of a cycle is just the same. However, when $T_r > T_m$, the speed drops down from cycle to cycle (this can be experienced when a running machine is to be stopped). Also, when $T_r < T_m$ the speed goes up from cycle to cycle. It is also important to note that maximum and minimum speeds of rotation of the crank shaft need not necessarily occur at the opposite ends of the same shaded loop. Also, it is not necessary that maximum speed of crank shaft will occur at the leading end of the shaded loop of maximum area. This may be confirmed by taking an arbitrary case where areas of loops are $A_1 = +60$ units; $A_2 = -20$ units; $A_3 = +40$ units and $A_4 = -80$ units.

Confining our discussions to the case where $T_r = T_{mean}$ let E_a be the kinetic energy at point a in Fig. 14.32(a), the expressions for kinetic energy at points a, b, c, d and e can be written as follows.

$$E_b = E_a + \int_{\theta_a}^{\theta_b} (T - T_r)\, d\theta \equiv E_a + \text{(hatched loop area } A_1)$$

or
$$E_b = (E_a + 60)$$

$$E_c = E_b + \int_{\theta_b}^{\theta_c} (T - T_r)\, d\theta \equiv E_b - \text{(hatched loop area } A_2)$$

or
$$E_c = E_a + 60 - 20 = E_a + 40$$

Substituting for E_b

$$E_c = E_a + \text{(hatched loop area } A_1 - \text{hatched loop Area } A_2)$$

Also

$$E_d = E_c + \int_{\theta_c}^{\theta_d} (T - T_r)\,d\theta = E_c + \text{(hatched loop area } A_3)$$

and substituting for E_c,

$$E_d = E_a + \text{(hatched loop area } A_1) - \text{(hatched loop area } A_2)$$
$$+ \text{(hatched loop area } A_3)$$

or

$$E_d = E_a + 60 - 20 + 40 \equiv E_a + 80$$

Finally

$$E_e = E_d + \int_{\theta_d}^{\theta_e} (T - T_r)\,d\theta = E_d - \text{(hatched loop area } A_4)$$

and substituting for E_d,

$$E_e = (E_a + 80 - 80) = E_a$$

Expressing hatched loop areas by corresponding terms A_1 through A_4, above expressions may also be written in terms of corresponding sums and differences of kinetic energies. Thus let K.E. at a be E_a.

Then, K.E. at point b:
$$(E_a + A_1) = E_a + \frac{1}{2}I(\omega_b^2 - \omega_a^2)$$

K.E. at point c:
$$(E_a + A_1 - A_2) = E_a + \frac{1}{2}I(\omega_c^2 - \omega_a^2)$$

K.E. at point d:
$$(E_a + A_1 - A_2 + A_3) = E_a + \frac{1}{2}I(\omega_d^2 - \omega_a^2)$$

K.E. at point e:
$$(E_a + A_1 - A_2 + A_3 - A_4) = E_a + \frac{1}{2}I(\omega_e^2 - \omega_a^2)$$

Now, if K.E. is to be the same at the beginning and end of cycle then $E_e = E_a$, which implies $A_1 - A_2 + A_3 - A_4 = 0$

or

$$\frac{1}{2}I(\omega_e^2 - \omega_a^2) = 0 \qquad \text{or} \qquad \omega_e = \omega_a$$

Above discussions enable us to identify points at which *kinetic energy is maximum and minimum. Evidently, these are the points at which angular speed is maximum and minimum respectively. The difference in the kinetic energies at these points is called the maximum fluctuation of energy (E_f)*. The maximum fluctuation of energy E_f may be expressed as a fraction of the indicated work done by the engine during one cycle. The fraction is known as the **coefficient of fluctuation of energy**, denoted by K_e. Thus

$$K_e = (E_f)/E \tag{14.112}$$

where E is the indicated work done per cycle of the engine. The values of K_e are representative of type of prime mover and provides a rough guidance in fly-wheel design, particularly when actual T.M. diagrams are not known. It must be noted however that even with engines of the same category there can be a considerable deviation in the values of K_e. This is because (i) indicator diagrams differ considerably, and (ii) inertia torque influences the shape of T.M. diagram to a varied extent. A few representative values of K_e established on the basis of respective T.M. diagrams are as under.

	Type of prime moves	K_e
1.	Single cylinder, double acting steam engine:	0.21
2.	Cross-compound steam engine:	0.096
3.	Single cylinder, single acting 4-stroke gas engine:	1.93
4.	Four-cylinder, single acting 4-stroke gas engine:	0.066
5.	Six-cylinder, single acting 4 stroke gas engine:	0.031

Readers should note improvement achieved in the value of K_e by way of increasing number of cylinders at S.NO. 3 through 5.

The Coefficient of Fluctuation of Speed (K_S)

Another term widely used in flywheel calculations is the 'coefficient of fluctuation of speed'. This is defined as the ratio of the difference between the maximum and minimum angular speeds during a cycle to the mean speed of rotation of the crank shaft. Thus, if ω_{max} and ω_{min} are the values of maximum and minimum angular speeds and ω the mean angular speed of the crank shaft in a cycle, then

$$K_s = (\omega_{max} - \omega_{min})/\omega \qquad (14.113)$$

When the fluctuation in the angular speed of the crank shaft is not very large,

$$\omega \approx (\omega_{max} + \omega_{min})/2 \qquad (14.114)$$

For a given fluctuation of energy E_f, the range of angular speed $(\omega_{max} - \omega_{min})$ may be made as small as possible by increasing the mass moment of inertia of a rotating mass, meant for absorbing kinetic energy. It is, however, unwise to reduce the range $(\omega_{max} - \omega_{min})$ more than what is necessary. Permissible range of variation of angular speeds, and hence the value of K_s is dependent on the purpose for which the given engine is required. Thus, when driving line-shafts for machine tools, a much larger value for K_s and hence for $(\omega_{max} - \omega_{min})$ may be permitted, while when the engine is coupled directly to an electric generator, a much smaller value of K_s may be allowed. A few representative values of K_s are as under:

	Type of application	K_s
1.	For engines delivering power to agricultural machinery	0.05
2.	For engines driving shaft of a workshop	0.03
3.	Engines delivering power to weaving and spinning machines	0.02 – 0.01
4.	Engines driving d.c. generator	0.006

14.21 THE FLYWHEEL

A flywheel is a rotating mass that is used as an energy reservoir in a machine. It absorbs energy in the form of kinetic energy, during those periods of crank rotation when actual turning moment is greater than the resisting moment and releases energy, by way of parting with some of its K.E., when the actual turning moment is less than the resisting moment. A flywheel thus controls fluctuations in angular speeds of crank shaft over a cycle, in a passive way. Absorption of energy is necessarily accompanied by an increase in the speed of rotation of the flywheel while, in releasing the energy, the speed of rotation of the flywheel drops down. Central theme in the design of flywheel, therefore, is to proportion the flywheel in such a way that the fluctuations in the speed do not exceed the permissible limits. It is customary to neglect effect of rotary inertia of other rotating parts while calculating the size of flywheel.

Let I = mass moment of inertia of the flywheel = $\dfrac{W}{g} k^2$; k being the radius of gyration

and, ω_1, ω_2 = maximum and minimum speeds of rotation respectively

Also, let ω = mean (average) angular speed of the crank shaft,

K_e, K_s = coefficient of fluctuation of energy and speed respectively

and E = indicated work per cycle

If E_1 and E_2 be the kinetic energy of the flywheel at maximum and minimum angular speeds respectively, then maximum fluctuation of energy,

$$E_f = E_1 - E_2 = \frac{1}{2} I(\omega_1^2 - \omega_2^2) \qquad (14.115)$$

or

$$E_f = I\left(\frac{\omega_1 + \omega_2}{2}\right)(\omega_1 - \omega_2)$$

Multiplying and dividing on right hand side by $(100)\omega$

$$E_f = I\omega\left(\frac{\omega_1 + \omega_2}{2}\right)\left(\frac{\omega_1 - \omega_2}{\omega} \times 100\right) \times \frac{1}{100}$$

or

$$E_f = \frac{I\omega^2}{100} K_s \qquad (14.116)$$

where K_s is the percentage value of coeff. of fluctuation of speed

Also, substituting for $E_f = K_e E$, from equation (14.112), we have

$$K_e E = \frac{I\omega^2}{100} K_s$$

or

$$I = \frac{100E}{\omega^2}\left(\frac{K_e}{K_s}\right) \qquad (14.117)$$

Any of the equations (14.115), (14.116) and (14.117) may be used to establish required moment of inertia of the flywheel, depending on type of data.

14.22 FLYWHEEL FOR PUNCHING PRESS

The flywheels used for prime movers constitute a class of problems in which the resisting torque (load) is assumed to be constant and the driving torque varies. Flywheels used in punching, rivetting and similar machines constitute another class of problems in which the actual (driving) turning moment provided by an electric motor is more or less constant but the resisting torque (load) varies.

In the case of punching press, the crank is driven by a motor which supplies a uniform torque. If the speed of rotation remains nearly constant, this implies that the energy is transmitted almost at a steady rate. Figure 14.33 shows such punching press schematically. Let us assume that the cycle of punching extends over 0 to 2π radian and that the actual punching occurs during $\theta_1 \leq \theta \leq \theta_2$. The resisting load (due to punching operation) is encountered only during the rotating of the crank from $\theta_1 \leq \theta \leq \theta_2$. For the rest of the rotation of the crank, there is no resisting torque. Thus, when there is no flywheel to store the excess energy in the periods $\theta_2 \leq \theta \leq 2\pi$ and $0 \leq \theta \leq \theta_1$, the speed of the crank shaft will rise excessively. Conversely, when $\theta_1 \leq \theta \leq \theta_2$ (*i.e.*, during actual punching/rivetting operation), the speed of the crank shaft will drop down drastically, as the rate of energy supply is far below the punching requirement.

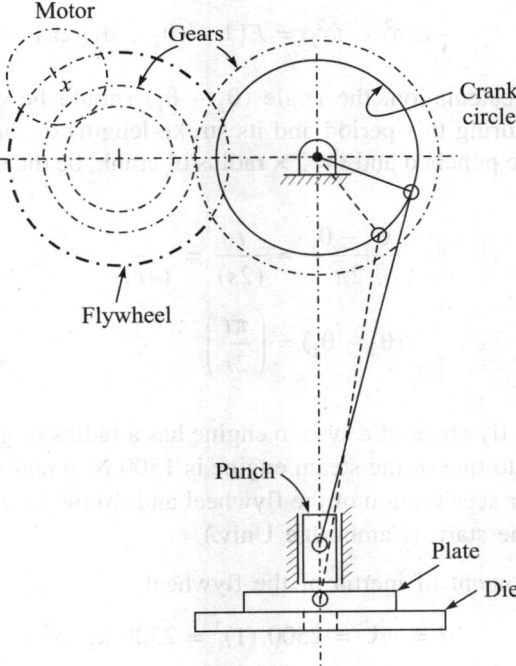

Fig. 14.33 Flywheel for a Punching Press.

An economical proposition is therefore to provide a flywheel which absorbs excess of energy being supplied in the form of kinetic energy during the period $\theta_2 \leq \theta \leq 2\pi$ and $0 \leq \theta \leq \theta_1$, and to release the same during the period of heavy demand *i.e.*, $\theta_1 \leq \theta \leq \theta_2$. This helps to keep fluctuation of speed within permissible limit by way of choosing a flywheel of adequate mass moment of inertia.

Let E be the energy of total work required for punching a single hole. The value of E is dependent on thickness of the plate, size of the hole and the mechanical properties of the material. Let us assume that a single hole is punched per revolution of the crank. Hence theoretically (neglecting other losses) energy supplied as input to the crank shaft should also be E per revolution. Thus the uniform rate of work supplied per radian of crank revolution is $(E/2\pi)$. Hence the actual work supplied by motor during punching operation,

$$E_1 = \frac{E}{2\pi}(\theta_2 - \theta_1) \qquad (14.118)$$

But as the total work required for punching a hole is E, the rest of the work has to be supplied by the flywheel, by way of releasing stored energy. This is given by

$$E_f = E - E_1 = E\{1 - (\theta_2 - \theta_1)/2\pi\} \qquad (14.119)$$

and this must be equal to $I(\omega_1^2 - \omega_2^2)/2$, where I is the mass moment of inertia of the flywheel while ω_1 and ω_2 are respectively the maximum and minimum speeds of rotation of the flywheel.

Thus,

$$\frac{1}{2}I(\omega_1^2 - \omega_2^2) = E\{1 - (\theta_2 - \theta_1)/2\pi\} \qquad (14.120)$$

For a more precise calculation, the angle $(\theta_2 - \theta_1)$ should be expressed in terms of the displacement of punch during this period and its stroke-lengths or radius of r crank. Thus, if t be thickness of plate to be punched and $s = 2 \times$ radius of crank, be the stroke length of the punch, then

$$\frac{\theta_2 - \theta_1}{2\pi} \approx \frac{t}{(2s)} = \frac{t}{(4r)}$$

or

$$(\theta_2 - \theta_1) = \left(\frac{\pi t}{2r}\right) \qquad (14.121)$$

EXAMPLE 14.12   The flywheel of a system engine has a radius of gyration of 1 m and a mass of 2500 kg. The starting torque of the steam engine is 1500 N·m and may be assumed constant. Determine (a) the angular acceleration of the flywheel and (b) the kinetic energy of the flywheel after 10 seconds from the start. (Cambridge Univ.)

Solution: The mass moment of inertia of the flywheel,

$$I = mk^2 = 2500 \ (1)^2 = 2500 \ \text{kg·m}^2$$

For the starting torque of 1500 N·m, the angular acceleration α produced is given by,

$$T = I\alpha$$

or $$\alpha = (T/I) = (1500/2500) = 0.6 \ \text{rad/s}^2 \qquad \textbf{Ans.}$$

The angular speed ω achieved after 10 seconds with $\alpha = 0.6$ rad/s, starting from rest,

$$\omega = 0 + \alpha t = (0.6)10 = 6 \text{ rad/s}$$

Hence, kinetic energy 10 seconds after starting

$$\text{k.E.} = \frac{1}{2} I \omega^2 = \frac{1}{2} (2500)(6.0)^2$$

$$= 45,000 \text{ N·m} = 45.0 \text{ kN·m} \qquad \textbf{Ans.}$$

EXAMPLE 14.13 Turning moment curve for one revolution of a multi-cylinder engine above and below the line of mean resisting torque are given by -0.32, $+4.06$, -2.71, $+3.29$, -3.24, $+24$, $-3.74 + 2.71$, -2.45 sq. cm. The vertical and horizontal scales are 1 cm = 588420 N·cm and 1 cm = 24° respectively. The fluctuation of speed is limited to ± 1.5 percent of mean speed which is 250 r.p.m. The hoop stress in rim material is limited to 549.2 N/cm^2. Neglecting effect of boss and arms determine suitable diameter and cross-section of flywheel rim. Weight density of rim material is 0.0706 N/cu. cm. Assume width of rim equal to four times its thickness.

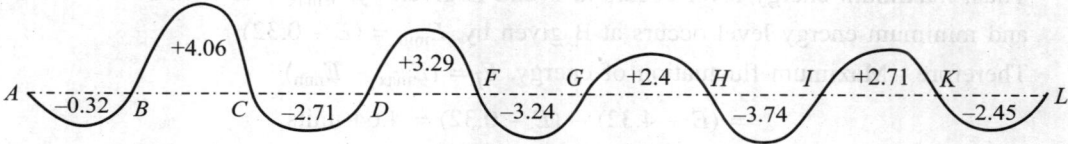

Fig. 14.34 Turning Moment Diagram.

Solution: The hoop stress in flywheel rim,

$$\sigma_{hoop} = \frac{\rho V^2}{g} = \frac{(0.0706) V^2}{981}$$

Hence, maximum velocity V_{max}, corresponding to maximum hoop stress $\sigma_{hoop} = 549.4$ N/cm^2 is,

$$V_{max} = \sqrt{\frac{g (\sigma_{hoop})_{max}}{\rho}} = \sqrt{\frac{981 \times 549.4}{0.0706}}$$

or $\qquad V_{max} = 2763 \text{ cm/s} = 27.63 \text{ m/s}$

The peripheral velocity V_{max} is related to corresponding r.p.m. N_{max} through,

$$V_{max} = \frac{2\pi N_{max} \times r}{60}$$

Also $\qquad \omega = \frac{2\pi N}{60} = \frac{2\pi \times 250}{60} = 26.18 \text{ rad/s}$

But $\qquad N_{max} = N + \frac{1.5}{100} \times N$

or $\qquad N_{max} = (1.015) \text{ N} = 1.015 \times 250 = 253.75 \text{ r.p.m.}$

Hence $\qquad r = \frac{60 V_{max}}{2\pi N_{max}} = \frac{60 \times 27.63}{2\pi \times 253.75} = 1.04$

Therefore $\qquad$ $r = 1.04$ m say,

Therefore $\quad$ Diameter of rim $\qquad$ $d = 2.08$ m $\qquad\qquad$ **Ans.**

Let E be the energy level at point A, then referring to Fig. 14.34,

Energy level at B	$= E - 0.32$	
Energy level at C	$= E - 0.32 + 4.06$	$= E + 3.74$
Energy level at D	$= E + 3.74 - 2.71$	$= E + 1.03$
Energy level at F	$= E + 1.03 + 3.29$	$= E + 4.32$
Energy level at G	$= E + 4.32 - 3.24$	$= E + 1.08$
Energy level at H	$= E + 1.08 + 2.4$	$= E + 3.48$
Energy level at J	$= E + 3.48 - 3.74$	$= E - 0.26$
Energy level at K	$= E - 0.26 + 2.71$	$= E + 2.45$
Energy level at L	$= E + 2.45 - 2.45$	$= E$

Thus, maximum energy level occurs at F and is given by, $E_{max} = E + 4.32$

and minimum energy level occurs at B given by, $E_{min} = (E - 0.32)$

Therefore $\quad$ Maximum fluctuation of energy, $E_f = (E_{max} - E_{min})$

$$= (E + 4.32) - (E - 0.32) = 4.64 \text{ units}$$

Hence $\qquad\qquad$ $E_f = (4.64) \times \left(\dfrac{\pi}{180} \times 24° \right) \times (1 \times 588600)$

$$= 1144002 \text{ N·cm} = 11440 \text{ N·m}$$

Hence, from $\qquad$ $\dfrac{I\omega^2 \, k_s}{100} = 11440$

Therefore $\qquad$ $I = \dfrac{11440 \times 100}{(26.18)^2 \times 2\,(1.5)} = 556.4 \text{ kg·m}^2$

Therefore $\quad$ If W be the weight of the flywheel, and $k \equiv r = 1.04$ m be the radius of gyration, then from

$$I = \frac{W}{g} k^2$$

$$W = 556.4 \times \frac{9.81}{(1.04)^2} = 5046.5 \text{ N}$$

Assuming all the mass to be concentrated in rim and further assuming rim width $b = 4t$, the thickness of rim, the rim cross-sectional area,

$$a = bt = 4t^2$$

$\qquad$ Volume of material in rim $= (2\pi r)a$

Therefore Mass of material in rim $= (2\pi r \cdot a)0.0706 = 5046.5$

Therefore $\qquad\qquad$ $5046.5 = (2\pi \times 104 \times 0.0706) \times 4t^2$

or $\qquad t^2 = \dfrac{5046.5}{2\pi \times 104 \times 0.0706 \times 4} = 27.35$

Therefore $\qquad\qquad t = 5.23$ cm

Therefore $\qquad\qquad b = 4 \times t = 20.9$ cm

Therefore $\qquad\qquad b = 20.9$ cm and $t = 5.23$ cm $\qquad\qquad$ **Ans.**

EXAMPLE 14.14 The turning moment diagram for an engine consists of a curve represented by the equation

$$T = (19614 + 9316.7\sin2\theta - 5590\cos2\theta) \text{ N·m}$$

where θ is the angle moved by the crank from inner dead centre. If the resisting torque is constant, determine—

(a) power developed by the engine.

(b) moment of inertia of flywheel in kg·m^2, if the total fluctuation of speed is not to exceed one per cent of mean speed which is 180 r.p.m. and

(c) angular acceleration of the flywheel when the crank has·turned through 45° from the inner dead centre. $\qquad$ (DAV Indore, Jan. 1989, Nov. 1992) (AMIE, Summer 1979)

Solution: $\omega = 6\pi$ rad/s, $K_s = 1.0$

(a) Since T–θ diagram involves trigonometric functions of angle 2θ it follows that the cycle repeats itself after a crank angle of rotation of 180°. Hence,

$$T_{mean} = \frac{1}{\pi} \int_0^\pi Td\theta$$

or $\qquad T_m = \dfrac{1}{\pi} \displaystyle\int_0^\pi [19614 + 9316.7\sin2\theta - 5590\cos2\theta]d\theta$

or $\qquad T_m = \dfrac{1}{\pi} \left[19614\theta - \dfrac{9316.7}{2} \cos 2\theta - \dfrac{5590}{2} \sin 2\theta \right]_0^\pi$

or $\qquad T_m = \dfrac{1}{\pi} [19614\pi - 4658.35 \cdot (1 - 1) - 2795(0 - 0)] = 19614$ N·m

Therefore $\quad$ Power developed by engine $= \dfrac{2\pi NT_{mean}}{60 \times 1000}$

$$= \frac{2\pi \times 180 \times 19614}{60 \times 1000} = 369.7 \text{ kW}$$

(b) Deviation of actual torque w.r. to mean (or uniform resisting) torque, as a function of crank angle θ,

$\qquad\qquad T' = T - T_{mean}$

or $\qquad\qquad T' = (19614 + 9316.7\sin2\theta - 5590\cos2\theta) - 19614$

or $\qquad\qquad T' = (9316.7\sin2\theta - 5590\cos2\theta)$

The points of intersection of T_m line and actual torque curve is obtained by equating T' to zero.

Thus $(9316.7\sin2\theta - 5590\cos2\theta) = 0$

or $$\tan2\theta = \frac{5590}{9316} = 0.606$$

Therefore $$2\theta = 31.2°; \; 211.2°$$

Hence the fluctuation of energy represented by loop area between T_m line and actual torque curve between intersection points $2\theta = 31.2°$ and $2\theta = 211.2°$ is given by

Therefore $$E_f = \int_{15.6}^{105.6} (9316.7\sin2\theta - 5590\cos2\theta)d\theta$$

or $$E_f = \left(-\frac{9316.7}{2}\cos2\theta - \frac{5590}{2}\sin2\theta\right)_{15.6}^{105.6}$$

or $$E_f = -4658.35 \, (-1.711) - 2795 \, (-1.036)$$

or $$E_f = 7970.4 + 2875.72 = 10866$$

From $$E_f = \frac{I\omega^2 k_s}{100}$$

We have, $$10866 = \frac{I (6\pi)^2}{100} 1$$

Therefore $$I = \frac{10866 \times 100}{(6\pi)^2} = 3058.2 \text{ kg·m}^2$$

Hence $$I = 3058.2 \text{ kg·m}^2 \qquad \textbf{Ans.}$$

(c) The acceleration is produced by the torque which is in excess of mean torque. The excess torque at any instant is given by

$$T' = T - T_m$$

or $$T' = (9316.7\sin2\theta - 5590\cos2\theta)$$

When crank has turned through $\theta = 45°$ from i.d.c., as $2\theta = 90°$;

or, $$T' = 9316.7(1) - 5590(0) = 9316.7$$

Therefore $$\alpha = T'/I = 9316.7/3058.2 = 3.046 \text{ rad/s}^2 \qquad \textbf{Ans.}$$

EXAMPLE 14.15 The equation of the turning moment curve of a three–crank engine is $(4905 + 1471.5\sin3\theta)$ N·m where θ radians is the crank angle. The moment of inertia of the flywheel is one tonne-metre^2, and the mean engine speed is 300 r.p.m. Calculate:

(a) the power of the engine.
(b) the total fluctuation of speed of the flywheel in percentage under the following conditions:

(i) when the resisting torque is constant,

(ii) when the resisting torque is $(4905 + 588.6\sin\theta)$ N·m   (AMIE, Winter 1979)

Solution:

Given
$$I = (1 \times 1000) \text{ kg·m}^2$$
$$\omega = \frac{2\pi \times 300}{60} = 31.42 \text{ rad/s}$$

(a) Referring to Fig. 14.35, for a complete revolution of crank, we have

$$2\pi \times T_m = \int_0^{2\pi} (4905 + 1471.5\sin3\theta)d\theta$$

or
$$T_m = \frac{1}{2\pi}\left[4905\theta - \frac{1471.5}{3}\cos3\theta\right]_0^{2\pi}$$

or
$$T_m = \frac{1}{2\pi}[(4905 \times 2\pi) - 490.5(1-1)] = 4905 \text{ N·m}$$

Thus, the power of the engine,

$$\text{Power} = \frac{2\pi N T_m}{60 \times 1000}$$

$$= \frac{2\pi \times 300 \times 4905}{60 \times 1000} = 154.09 \text{ kw}$$ **Ans.**

(b) When the resisting torque is constant and equal to T_m, the point (*i.e.*, crank angles) at which the actual torque equals T_m are given by,

$$T = T_{mean}$$

or
$$4905 + 1471.5\sin3\theta = 4905$$

or
$$\sin3\theta = 0$$

or
$$3\theta = 0 \quad \text{or} \quad 180° \quad \text{Thus,} \quad \theta = 0° \text{ or } 60°$$

The largest loop area between T_m line and actual T.M. curve decides the maximum fluctuation of energy E_f. Because of symmetric distribution of actual torque curve about $T_m = 4905$ N·m line, however, all these loop areas are equal. Hence,

$$E_f = \int_0^{60} (T - T_{mean})d\theta$$

or
$$E_f = \int_0^{60} (4905 + 1471.5\sin3\theta - 4905)d\theta$$

or
$$E_f = \int_0^{60} (1471.5\sin3\theta)d\theta = \frac{-1471.5}{3}(\cos3\theta)_0^{60} = 981 \text{ N·m}$$

Hence, from

$$E_f = \frac{IK_s\omega^2}{100}$$

$$K_s = \frac{100E_f}{I\omega^2} = \frac{100 \times 981}{1000 \times (31.42)^2} = 0.099\%$$ **Ans.**

(ii) When the resisting moment is given by $(4905 + 588.65 \sin\theta)$ N·m, referring to Fig. 14.35, the points of intersection of T_m line with actual T.M. line (where, the two are equal) are obtained by equating the two turning moments. Thus,

$$(4905 + 1471.5 \sin 3\theta) = 4905 + 588.6 \sin\theta$$

Simplifying, we have $2.5 \sin 3\theta = \sin\theta$ (a)

Now as, $\sin 3\theta = (3 \sin\theta - 4 \sin^3\theta)$

We have $\sin 3\theta = (3 - 4 \sin^2\theta) \sin\theta$

We have, from equation (a) above, substituting for $\sin 3\theta$, we have

$$2.5(3 - 4 \sin^2\theta) \sin\theta = \sin\theta$$

or $\sin^2\theta = 0.65$

Therefore $\sin\theta = 0.8062$

or $\theta = 53.729$ and 126.271

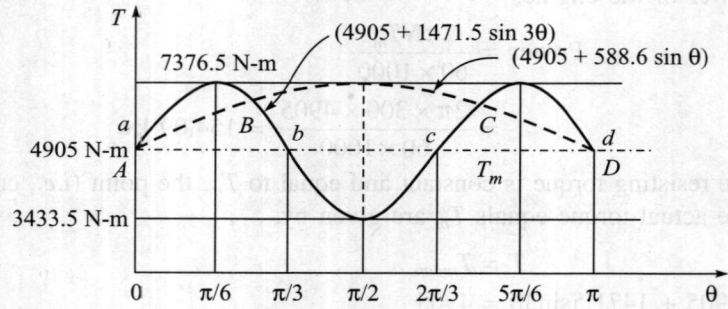

Fig. 14.35 Turning Moment Diagram.

Maximum fluctuation of energy E_f will be given by the largest of the three-loop areas enclosed between the resisting moment line and the actual T.M. line. A little observation reveals that the central loop gives largest area and hence,

$$E_f = \int_{53.729}^{126.271} [(4905 + 1471.5 \sin 3\theta) - (4905 + 588.6 \sin\theta)]d\theta$$

or $$E_f = \int_{53.729}^{126.271} [(1471.5 \sin 3\theta - 588.6 \sin\theta)]d\theta$$

or $$E_f = \left[-\frac{1471.5}{3} \cos 3\theta + 588.6 \cos\theta \right]_{53.729}^{126.271}$$

or $E_f = (-490.5)(+1.89315) + (588.6)(-1.1832)$

or $E_f = -928.59 - 696.43 = -1625.03$ N·m

Hence, from $$E_f = \frac{Ik_s\omega^2}{100}$$

$$k_s = \frac{100 \times 1625.03}{1000 \times (31.42)^2} = 0.1646 \text{ per cent}$$ **Ans.**

EXAMPLE 14.16 A shaft fitted with a flywheel rotates at 250 r.p.m. and drives a machine. The torque of the machine varies in a cyclic manner over a period of 3 revolutions. The torque rises from 735.75 N·m to 2943 N·m in half revolution, remains constant during next one revolution and drops down to 735.75 N·m in the next half revolution. It remains constant at 735.75 N·m during next revolution, the cycle being repeated thereafter.

Determine the power required to drive the machine and percentage fluctuation in speed if the driving torque applied to the shaft is constant and the flywheel weighs 5787.9 N with radius of gyration of 600 mm.

(DAV Indore, March 1990, AMIE, Winter 1981)

Solution: Mean speed of flywheel

$$\omega = \frac{2\pi \times 250}{60} = 26.18 \text{ rad/s}$$

Weight of flywheel = 5787.9 N

Radius of gyration $k = 0.6$ m

Therefore Mass moment of inertia $I = \dfrac{5787.9}{9.81} \times 0.6^2 = 212.4$ kg·m^2

Referring to Fig. 14.36, the area under the T.M. diagram is

$$A = \text{(area of rectangle } AEFO) + \text{(area of trapezium } ADCB)$$

or

$$A = 735.75(6\pi) + \frac{BC + AD}{2} \times (2943 - 735.75)$$

$$= 4414.5\pi + \frac{2\pi + 4\pi}{2} \times 2207.25 = (11036.25)\pi \text{ N·m}$$

Thus, if T_m is the mean tuning moment, then

$$(T_m)6\pi = (11036.25)\pi$$

Therefore

$$T_m = \frac{(11036.25)\pi}{6\pi} = 1839.4 \text{ N·m}$$

Note that excess energy available between points L and P accelerate the rotating parts.

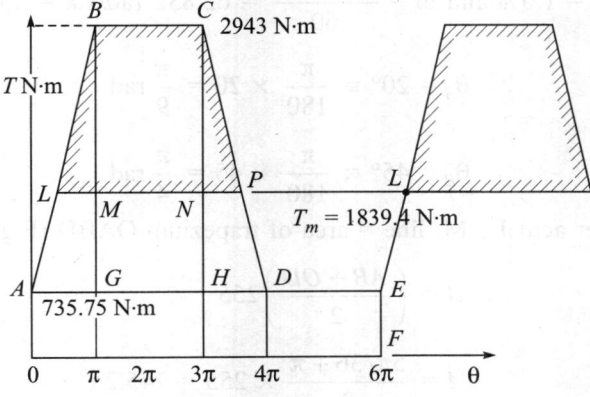

Fig. 14.36 Turning moment diagram.

Let the T_m line cut the actual T.M. line at L and P. Then the maximum fluctuation energy

$$E_f = \text{shaded area } LBCP$$

or $$E_f = \left(\frac{BC + LP}{2}\right) \times BM$$

But $$LP = BC + \frac{(AD - BC)}{(2943 - 735.75)}(2943 - 1839.4)$$

or $$LP = 2\pi + \frac{4\pi - 2\pi}{2207.25} \times 1103.6 = 9.425 \text{ radians}$$

And $$BM = (2943 - 1839.4) = 1103.6 \text{ N·m}$$

Therefore $$E_F = \left(\frac{2\pi + 3\pi}{2}\right) \times 1103.6 = 8667.78 \text{ N·m}$$

Power transmitted $$= \frac{2\pi \times 250 \times T_m}{60 \times 1000} = \frac{2\pi \times 250 \times 1839.4}{60 \times 1000} = 48.16 \text{ kw}$$

Also $$E_f = \frac{I K_s \omega^2}{100}$$

or $$K_s = \frac{100 \times E_f}{I\omega^2} = \frac{100 \times 8667.65}{212.4 \times (26.18)^2} = 5.954\% \qquad \textbf{Ans.}$$

EXAMPLE 14.17 The variation of crank-shaft torque of a four-cylinder petrol engine may be approximately represented by taking the torque as zero at crank angle 0° and 180° and as 255 N·m for crank angles 20° and 45°, the intermediate portions of the torque graph being straight lines. The cycle is repeated in every half revolution. The average speed is 600 r.p.m. Supposing that the engine drives a machine requiring a constant torque, determine the weight of the flywheel of radius of gyration 25 cm which must be provided so that the total variation of the speed shall be one per cent. (DAV Indore, Sept. 1990, AMIE, Winter 1980; SGSICS 2006)

Solution: Given $K_s = 1.0\%$ and $\omega = \frac{2\pi \times 600}{60} = 62.832$ rad/s $k = 25$ cm $= 0.25$ m

$$\theta_A = 20° \equiv \frac{\pi}{180} \times 20 = \frac{\pi}{9} \text{ rad}$$

$$\theta_B = 45° \equiv \frac{\pi}{180} \times 45 = \frac{\pi}{4} \text{ rad}$$

Therefore, Area under actual T.M. line = area of trapezium OABD (Fig. 14.37)

$$A = \left(\frac{AB + OD}{2}\right) 255$$

or $$A = \frac{5\pi/36 + \pi}{2} \times 255 = 145.2\pi$$

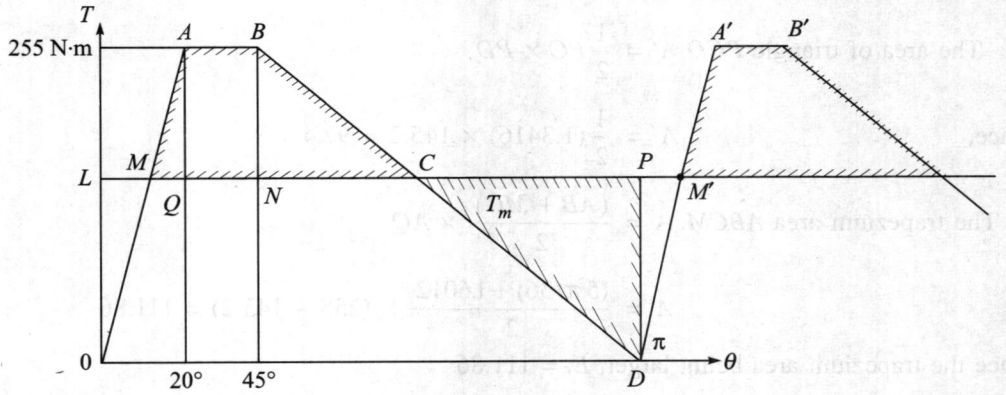

Fig. 14.37 Turning moment diagram.

Therefore, as the cycle covers half revolution only,

$$\pi \times T_m = 145.2\pi$$

or $$T_m = 145.2 \text{ N·m}$$

Referring to Fig. 14.37, consider areas $ABCM$ and PCD,

As, $$BN = 255 - 145.2 = 109.8$$

and $$PD = 145.2$$

Therefore $$\frac{NC}{PC} = \frac{BN}{PD}$$

or $$\frac{NC}{PC} = \frac{109.8}{145.2} = 0.7562$$

Therefore $$\frac{NC}{PC} + 1 = 1.7562$$

or $$PC = \frac{NC + PC}{1.7562} = \frac{(180 - 45)\,\pi/180}{1.7562} = 1.3416$$

Therefore $$NC = NP - PC$$

$$= \left(\pi - \frac{\pi}{180}\,45\right) - 1.3416 = 1.0146$$

Therefore $$MC = MQ + QN + NC$$

$$= \frac{(\pi/9)}{255}(255 - 145.2) + (25)\frac{\pi}{180} + 1.0146$$

$$= 0.1503 + 0.4363 + 1.0146 = 1.6012 \text{ rad}$$

The largest of the areas $ABCM$ and PCD decides E_f, the maximum fluctuation of energy.

The area of triangle PCD $A' = \dfrac{1}{2} PC \times PD$

Hence, $A' = \dfrac{1}{2}(1.3416) \times 145.2 = 97.4$

The trapezium area $ABCM$, $A = \dfrac{(AB + MC)}{2} \times AQ$

or $A = \dfrac{(5\pi/36) + 1.6012}{2} \times (255 - 145.2) = 111.86$

Hence the trapezium area being larger, $E_f = 111.86$

Hence from, $E_f = \dfrac{IK_s\omega^2}{100}$

Hence, $I = \dfrac{W}{g}k^2 = \dfrac{100E_f}{K_s\omega^2} = \dfrac{100 \times 111.86}{1.0 \times (62.832)^2} = 2.833$

Therefore $W = \dfrac{9.81}{(0.25)^2} \times 2.833 = 444.67$ N   **Ans.**

EXAMPLE 14.18   A certain engine develops an output torque of $(1600 + 300\sin2\theta)$ N·m where θ is the crank angle measured from some datum. This engine drives a machine which requires a driving torque of $(1600 + 170\sin\theta)$ N·m. The rotating parts have a mass of 240 kg with a radius of gyration of 0.5 m. The maximum speed of rotation is observed to be 200 r.p.m. Determine:

   (i)   the minimum speed of rotation,
   (ii) the coefficient of fluctuation of speed,
   (iii) power developed by the engine at its mean speed. (Amravati University)

Solution: W/g = 240 kg; k = 0.5 m

$$\omega_1 \approx \omega_{max} = \frac{2\pi N_{max}}{60} = \frac{2\pi \times 200}{60}$$

or $\omega_1 = 20.944$ rad/s

As is clear from the actual T.M., $(1600 + 300\sin2\theta)$ N·m, the cycle is completed in π radians. The mean turning moment T_m is given by (*see* Fig. 14.38),

$$(T_m)\pi = \int_0^\pi (1600 + 300\sin2\theta)d\theta$$

Therefore $T_m = \dfrac{1}{\pi}\left[1600\,\theta - \dfrac{300}{2}\cos2\theta\right]_0^\pi$

or $T_m = \dfrac{1}{\pi}[1600\pi - 150(1 - 1)] = 1600$ N·m

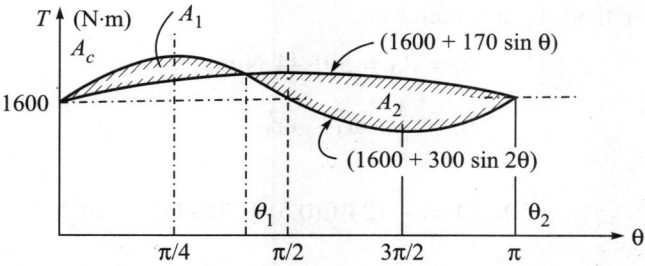

Fig. 14.38 Turning Moment Diagram.

Points of intersection of actual T.M. line with resisting moment line are obtained by equating

$$(1600 + 300 \sin 2\theta) = (1600 + 170 \sin \theta)$$

or
$$\sin 2\theta = \frac{170}{300} \sin \theta$$

or
$$2 \sin \theta \cos \theta = \left(\frac{170}{300}\right) \sin \theta$$

or
$$\sin \theta \left(\cos \theta - \frac{85}{300}\right) = 0$$

Thus either $\sin \theta = 0$ *i.e.*, $\theta = 0°$ and $180°$

or
$$\cos \theta = \frac{85}{300} = 0.2833$$

i.e.,
$$\theta_1 = 73.54$$

The two loop areas A_1 and A_2 between the actual T.M. line and resisting T.M. line are obtained as under.

$$A_1 = \int_0^{73.54°} [(1600 + 300 \sin 2\theta) - (1600 + 170 \sin \theta)] d\theta$$

$$= \left[-\frac{300}{2} \cos 2\theta + 170 \cos \theta\right]_0^{73.54°}$$

$$= (-150)(-1.839) + (170)(-0.7166) = 154.08 \text{ units}$$

$$A_2 = \int_{73.54°}^{180°} [(1600 + 300 \sin 2\theta) - (1600 + 170 \sin \theta)] d\theta$$

$$= \left[-\frac{300}{2} \cos 2\theta + 170 \cos \theta\right]_{73.54°}^{180°}$$

$$= (-154)(1 + 0.8394) + (170)(-1 - 0.2833)$$

$$= -283.27 - 218.1688 = 501.44 \text{ units}.$$

Thus A_2 is larger than A_1 and therefore,

$$E_f = A_2 = 501.44 \text{ N·m}$$

Thus, from
$$E_f = \frac{1}{2}I(\omega_1^2 - \omega_2^2)$$

$$501.44 = \frac{1}{2}(240)(0.5)^2[(20.944)^2 - \omega_2^2]$$

or
$$(20.944)^2 - \omega_2^2 = \frac{2 \times 494.08}{240 \times 0.5^2}$$

or,
$$\omega_2^2 = (20.944)^2 - 16.469 = 421.94$$

or,
$$\omega_2 = 20.54 \text{ rad/s}$$

(i) Thus, the minimum speed,

Therefore
$$N_2 = \frac{60 \times \omega_2}{2\pi} = 196.14 \text{ r.p.m.} \qquad \textbf{Ans.}$$

Average speed,
$$\omega = \frac{1}{2}(\omega_1 + \omega_2) = \frac{20.944 + 20.54}{2} = 20.74 \text{ rad/s}$$

(ii) Coeff. of fluctuation of speed,

$$K_s = \frac{20.944 - 20.54}{20.74} \times 100 = 1.95\% \qquad \textbf{Ans.}$$

(iii) Power developed by the engine at mean speed

$$= \frac{T_m \times \omega}{1000} \text{ kw}$$

$$= \frac{1600 \times 20.74}{1000} = 33.18 \text{ kw} \qquad \textbf{Ans.}$$

EXAMPLE 14.19 A press has to punch 30 holes per minute and the motor delivers 1.5 kW. to the press uniformly. The mechanical efficiency is 100%. If the actual punching of each hole is accomplished during 30° of crank rotation of the punching machine (which is a simple slider-crank mechanism, where punch is connected to the slider) and the fluctuation of speed is not to exceed ± 10%, find out the minimum required moment of inertia of the flywheel which is directly mounted on the crank-shaft. Neglect the inertia of the rest of the system.

(AMIE, Winter 1993)

Solution: With ±10% permissible speed fluctuation, if ω be the average angular speed, then maximum angular speed $\omega_1 = 1.1 \omega$

and minimum angular speed $\omega_2 = 0.9 \omega$

Number of holes punched per minute = 30

Since a slider-crank mechanism has one useful stroke per revolution of crank, the r.p.m. of crank = 30.

Hence average speed $\qquad \omega = \dfrac{2\pi(30)}{60} = \pi$ rad/s

Let T_m be the uniform torque supplied by the motor, while the actual torque required during punching is say T', (where T' is uniform but large) which is supplied during punching *i.e.*, when crank rotates from θ_1 to θ_2 through $30°$.

Thus, if E be the work done per cycle supplied by the motor, then the work supplied purely by motor during punching,

$$W = \left(\frac{\theta_2 - \theta_1}{2\pi}\right)E = \frac{30}{360} \times E = \left(\frac{E}{12}\right)$$

Taken over a complete cycle, the remaining part of cycle requires no work and, as such, the remaining work supplied $\left(E - \dfrac{E}{12}\right)$, tends to increase K.E. of the flywheel. Thus,

$$E\left(1 - \frac{1}{12}\right) = E_f$$

or $\qquad E_f \equiv \dfrac{1}{2} I(\omega_1^2 - \omega_2^2) = \dfrac{1}{2} I(1.1^2 - 0.9^2)\omega^2$

$$= \frac{1}{2} I(1.1^2 - 0.9^2)(\pi)^2$$

or $\qquad \left(\dfrac{11}{12}\right)E = (1.9739)I$

Again power in kilo-watt,

$$P = (\text{W.D. per cycle}) \times (\text{No. of cycles per second})$$

Therefore $\qquad$ W.D./cycle, $E = \dfrac{1.5 \times 1000}{1/2} = 3000$ N·m

Hence $\qquad \dfrac{11}{12}(3000) = (1.9739)I$

Therefore $\qquad I = 1393.18$ kg·m² $\qquad\qquad$ **Ans.**

EXAMPLE 14.20 A punching machine punches 3 cm holes in a 4 cm plate. It does 540 N·m of work per sq. cm. of sheared area. The punch has a stroke of 10 cm and punches one hole every 10 second. The maximum speed of the flywheel at the radius of gyration is 28 m/s. Find the weight of the wheel if the speed at this radius is not to fall below 25 m/s during each punch. (DAV Indore, 1990)

Solution: $\quad$ Area (cylindrical) sheared off $= \pi dt$

$$= \pi \times 3 \times 4 = 12\pi \text{ sq. cm.}$$

Therefore W.D. during punching $= 540 \times 12\pi = 20357.52$ N·m

Let W = W.D. supplied by the motor for one complete cycle of punching.

Since total movement up and down of the punch is 20 cm in one cycle, the W.D. supplied (assumed uniform) per cm of punch movement = $\dfrac{W}{20}$. Hence W.D. supplied by motor during

actual punching = $4 \times \dfrac{W}{20} = \dfrac{W}{5}$

The remaining work = $\dfrac{16\,W}{20}$ supplied by motor during remaining part of the cycle and gets absorbed in flywheel in the form of K.E. This represents maximum fluctuation of energy. Thus,

$$E_f = \frac{16\,W}{20}$$

or $$E_f = \frac{4\,W}{5} = 0.8\ W$$

or $$E_f = 0.8 \times \text{work required for punching one hole}$$

or $$E_f = 0.8 \times 20357.52\ \text{N·m.}$$

Hence, from $$\frac{1}{2} I(\omega_1^2 - \omega_2^2) = E_f$$

$$\frac{1}{2} mk^2(\omega_1^2 - \omega_2^2) = 0.8 \times 20357.52$$

or $$\frac{1}{2} m[(k_1\omega_1)^2 - (k_2\omega_2)^2] = 0.8 \times 20357.52$$

But $$k_1\omega_1 = 28\ \text{m/s} \quad \text{and} \quad k_2\omega_2 = 25\ \text{m/s}$$
Hence

$$m = \frac{2 \times 0.8 \times 20357.52}{(28^2 - 25^2)} = 204.855\ \text{kg}$$

Therefore   Weight of flywheel   = $204.855 \times 9.81 = 2009.6$ N   **Ans.**

EXAMPLE 14.21 A punching press is driven by a constant torque electric motor. The press is provided with a flywheel that rotates at a maximum speed of 225 r.p.m. Radius of gyration of the flywheel is 50 cm. The press punches 720 holes per hour, each punching operation takes 2 seconds and requires 14715 N·m of energy. Find the minimum weight of the flywheel if speed of the same is not to fall below 200 r.p.m. Also determine the required power of the motor.

Solution: $$\omega_1 = \frac{2\pi \times 225}{60} = 23.562\ \text{rad/s}$$

$$\omega_2 = \frac{2\pi \times 200}{60} = 20.944\ \text{rad/s}$$

Number of holes punched per hour = 720

Number of holes punched per min = 12

Hence, number of strokes/min = 12

Power reqd. = (work done per cycle) × No. of cycles per second

Therefore $P = 14715 \times \dfrac{12}{60} = 2943$ watts = 2.943 kw **Ans.**

Total time required for completing punching cycle = $\dfrac{3600}{720} = 5$ s

Time required by actual punching through plate = 2 s. Thus, for the remaining 3 seconds flywheel absorbs kinetic energy which represents maximum fluctuation of energy. Thus,

$$E_f = \frac{3}{5} \times 14715 \text{ N·m} = 8829 \text{ N·m}$$

Hence from, $$E_f = \frac{1}{2} \frac{W}{g} k^2 (\omega_1^2 - \omega_2^2)$$

$$W = \frac{8829 \times 9.81 \times 2}{(0.5)^2 (23.562^2 - 20.944^2)} = 5946.8 \text{ N} \qquad \textbf{Ans.}$$

EXAMPLE 14.22 The torque exerted on the crank shaft of a two-stroke engine is given by the equation:

$$T = (14224.5 + 2256.3\sin2\theta - 1864\cos2\theta) \text{ N·m}$$

where θ is the crank angle of displacement from the i.d.c. Assuming the resisting torque to be uniform, determine–

(a) the power of the engine when the speed is 150 r.p.m.

(b) the moment of inertia of the flywheel if the speed variation is not to exceed ± 0.5.% of the mean speed

(c) the angular acceleration of the flywheel when the crank has turned through 30° from the i.d.c. (AMIE, Summer 1982, Summer 1979)

Solution: As is seen from the expression for T, the torque at $\theta = 0$ is

$$T = 14224.5 - 1864 = 12360.5 \text{ and, } T = 14224.5 \text{ at } \theta \text{ given by,}$$

$$14224.5 = 14224.5 + 2256.3\sin2\theta - 1864\cos2\theta$$

or when, $\tan2\theta = \dfrac{1864}{2256.3} = 0.826$ or when, $\theta = 19.78°$

Hence the $T - \theta$ diagram is as shown in Fig. 14.39.

$$\omega = \frac{2\pi \times 150}{60} = 5\pi \text{ rad/s}$$

$$K_s = 0.5 + (0.5) = 1.0$$

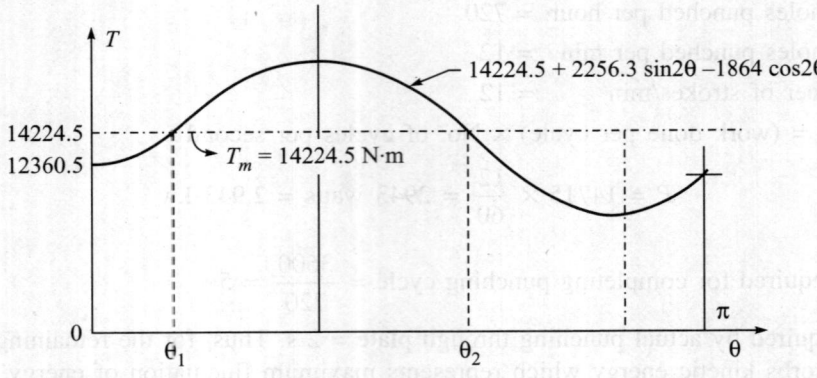

Fig. 14.39

Since the trigonometric functions in the equation for T involves angle 2θ, the cycle is completed in π radians. Thus, if T_m is the mean turning moment, then

$$T_m \times \pi = \int_0^\pi (14224.5 + 2256.3\sin 2\theta - 1864\cos 2\theta)d\theta$$

Therefore

$$T_m = \frac{1}{\pi}\left[14224.5\theta - \frac{2256.3}{2}\cos 2\theta - \frac{1864}{2}\sin 2\theta\right]_0^\pi$$

$$= \frac{1}{\pi}[14224.5\pi - 1128.15\,(1-1) - 932\,(0-0)] = 14224.5$$

(a) Hence the power of the engine $= \dfrac{2\pi \times 150 \times 14224.5}{60 \times 1000} = 223.4$ kw **Ans.**

(b) The points of intersection $(\theta_1$ & $\theta_2)$ or T_m line with actual T.M. line are obtained by equating, as under

$$14224.5 = 14224.5 + 2256.3\sin 2\theta - 1864\cos 2\theta$$

or

$$\tan 2\theta = \frac{1864}{2256.3} = 0.826$$

Therefore $\qquad 2\theta = 39.56$ and $(180 + 39.56)$ degrees

Therefore $\qquad \theta = 19.78°$ and $109.78°$

Since $\sin 2\theta$ and $\cos 2\theta$ terms are symmetric about $T_m = 14224.5$ N·m line, the two loop areas between T_m line and actual T.M. curve are equal. Hence,

$$E_f = \int_{19.78}^{109.78} (T - T_m)d\theta = \int_{19.78}^{109.78} (2256.3\sin 2\theta - 1864\cos 2\theta)d\theta$$

or

$$E_f = \left[\frac{-2256.3}{2}\cos 2\theta - \frac{1864}{2}\sin 2\theta\right]_{19.78}^{109.78}$$

Therefore
$$E_f = (-1128.15)(-1.5419) - (+932)(-1.274)$$
$$= 1739.51 + 1187.4 = 2926.9 \text{ N·m}$$

Hence from
$$E_f = \frac{I\omega^2 K_s}{100}$$

$$I = \frac{100 \times 2926.9}{(5\pi)^2 \times 1.0} = 1186.2 \text{ kg·m}^2 \qquad \textbf{Ans.}$$

(c) Angular acceleration is caused by excess torque given by

$$T - T_m = (14224.5 + 2256.3\sin2\theta - 1864\cos2\theta) - 14224.5$$
$$= (2256.3\sin2\theta - 1864\cos2\theta)$$

At $\theta = 30°$, this excess of torque is,

$$(T - T_m)_{\theta=30} = 2256.3\left(\frac{\sqrt{3}}{2}\right) - 1864\left(\frac{1}{2}\right) = 1022 \text{ N·m}$$

Therefore
$$\alpha = \frac{(T - T_m)}{I} = \frac{1022}{1186.2} = 0.8616 \text{ rad/s}^2 \qquad \textbf{Ans.}$$

REVIEW QUESTIONS

14.1 The crank pin circle radius of a horizontal engine is 30 cm. The weight of the reciprocating parts is 2452.5 N. When the crank has travelled 60° from i.d.c., the difference between the driving and the back pressure is 34.34 N/cm^2. The connecting rod length between centres is 120 cm and the cylinder bore is 50 cm. If the engine runs at 250 r.p.m. and the effect of piston rod diameter is neglected, calculate (i) pressure on slide bar, (ii) thrust in the connecting rod, (iii) tangential force on the crank pin (iv) turning moment on the crank shaft.

(*Ans.* 10675 N, 49308, 47026 and 14107.8 N·m)

14.2 During a trial on a steam engine, it is found that the acceleration of the piston is 36 m/s^2 when the crank has moved 30° from the inner dead centre position. The net effective steam pressure on the piston is 49.05 N/cm^2 and the frictional resistance is equivalent to a force of 588.6 N. The area of the piston is 610 cm^2 and the weight of the reciprocating parts is 1765.8 N. If the length of the crank is 30 cm and the ratio of connecting rod length to the crank length is 4.5, find: (i) reaction on the guide bars, (ii) thrust on the crank shaft bearings, and (iii) turning moment on the crank shaft.

(Roorkee Univ. 1980)   (*Ans.* 2554.5 N, 18512.5 N and 4091.5 N)

14.3 The crank and connecting rod of a petrol engine, running at 1800 r.p.m. are 50 mm and 200 mm respectively. The diameter of the piston is 80 mm and the mass of the reciprocating parts is 1 kg. At a point during the power stroke, the pressure on the piston is 0.7 N/mm^2, when it has moved 10 mm from the i.d.c. Determine (i) net load on the gudgeon pin, (ii) thrust in the connecting rod, (iii) reaction between the piston and cylinder, and (iv) the engine speed at which the above values become zero.

(*Ans.* 1847.9 N, 1865.3 N, 253.8 N and 273.6 rad/s)

14.4 A vertical petrol engine 100 mm diameter of piston and 120 mm stroke has a connecting rod 250 mm long. The mass of the piston is 1.1 kg. The speed is 2000 r.p.m. On the expansion stroke with a crank 20° from top dead centre, the gas pressure is 700 kN/m². Determine: (1) net force on the piston, (2) resultant load on the gudgeon pin, (3) thrust on the cylinder walls, and (4) speed above which other things remaining same, the gudgeon pin load would be reversed in direction.

(*Ans.* 2258 N, 2266 N, 186.3 N, and 2603 r.p.m.)

14.5 A vertical, single cylinder, single acting diesel engine has a cylinder diameter 300 mm, stroke length 500 mm and connecting rod length 4.5 times the crank length. The engine runs at 180 r.p.m. The mass of the reciprocating parts is 280 kg. The compression ratio is 14 and the pressure remains constant during the injection of oil for 1/10th of the stroke. If the compression and expansion follows the law $pv^{1.25}$ = constant, find

   (i)   crank pin effort

   (ii) thrust on the bearings, and

  (iii) turning moment on the crank shaft, when the crank displacement is 45° from the i.d.c. position during expansion stroke. The suction pressure may be taken as 0.1 N/mm².   (*Ans.* 109.520 kN, 79.455 kN, and 27,380 N·m)

14.6 The length of connecting rod of an engine is 50 cm measured between the centres, and it weighs 157 N. The centre of gravity is 12.5 cm from the crank pin centre and the crank radius is 10 cm. Determine the dynamically equivalent system keeping one mass at the small end. The frequency of oscillation of the rod, when suspended from the centre of the small end, is 43 vibrations per minute.

(*Ans.* m_1 = 3.629 kg and m_2 = 12.371 kg)

14.7 It is required to determine the total turning effort and the various reactions of the engine mechanism covered by the following description:

Lengh of crank = 10 cm, length of connecting rod = 30 cm;

Distance of centre of percussion E of the connecting rod from the centre of wrist pin B = 27.3 cm; speed of the crank (line 2), which is constant = 1800 r.p.m.; total mass of the connecting rod (link 3) = 1.63 kg; mass assumed concentrated at centre of percussion E = 1.29 kg; mass assumed concentrated at wrist pin = 0.34 kg; mass of piston (link 4) = 0.91 kg; force on piston due to gas pressure (45° crank position) = 8006.5 N.

Assume the crank to be fully counterbalanced by adding weights to the crank shaft, opposite to the crank pin such that the c.g. of resulting system is situated at the centre line of the main bearing. As such, the inertia force of the crank may be assumed zero.

(*Ans.* F_{32} = 3959 N, F_t = 3603 N, F_{34} = 5782.4 N, F_{12} = 3959 N, and F_{41}; 712 N and F_4 = 5649 N)

14.8 The dimensions of a four-link mechanism are: crank AB = 500 mm; couple 660 mm; output link CD = 560 mm, and frame AD = 1000 mm. The link AB has an angular velocity of 10.5 rad/s c.c.w. and an angular retardation of 26 rad/s² at the instant when it makes a angle of 60° with AD.

The mass of the link BC and CD is 4.2 kg/m length. The link AB has a mass of 3.54 kg and the centre of gravity lies at 200 mm from A and a moment of inertia of

88,500 kg·mm^2. Take c.g.s of links *BC* and *CD* to coincide with their respective mid-points, and their radii of gyration to be $k_3 = 190.5$ mm and $k_4 = 161.7$ mm respectively. Neglecting gravity and frictional effects, determine the instantaneous value of required driving torque on *AB* to overcome inertia forces. (***Ans.*** 23.5 N·m c.c.w.)

14.9 A slotted-lever quick return motion mechanism *OABCD*, in which *AB* is the crank, *OC* is the slotted lever and *B* is the slider on slotted-lever to which crank *AB* is pin-connected, is to be analysed for static equilibrium. A couple link *CD* connects the slotted lever to ram through pin connections. Take *OA* = 400 mm, *AB* = 200 mm, *OC* = 800 mm, *CD* = 300 mm and perpendicular distance of line of action of ram from point pivot *A* = 400 mm. Find required input torque T_2 on crank *AB* for a force of 300 N on the ram. Assume crank *AB* to make angle $\angle BAO = 105°$ and assume $\mu = 0.15$ for each sliding pair. The ram is to move to the right. (***Ans.*** $T_2 = 64.239$ N·m c.c.w.)

14.10 Following data relate to a four-link mechanism:

Link	Length	Mass	MOI about centroidal axis
AB	60 mm	0.2 kg	80 kg·mm^2
BC	200 mm	0.4 kg	1600 kg·mm^2
CD	100 mm	0.6 kg	400 kg·mm^2
AD	140 mm		

AD is the fixed link. The centres of mass for the links BC and CD lie at their mid-points whereas the centre of mass for the link AB lies at A. Find the driving torque on the link AB at the instant when it rotates at an angular velocity of 4.75 rad/s c.c.w. and $\angle DAB = 135°$. Neglect gravity effects. (***Ans.*** 1.675 N·m c.w.)

14.11 A horizontal cross-compound steam engine develops 298.3 kW at 105 r.p.m. The coefficient of fluctuation of energy is found to be 1/12, and the speed is limited to $\pm 0.5\%$ of the normal speed. Find the weight of the flywheel required, if the radius of gyration is 125 cms. (***Ans.*** $W = 81717.3$ N)

14.12 A single cylinder, single acting four-stroke I.C. engine develops 29.83 kW at 300 r.p.m. The turning moment diagram for the expansion and compression stroke may be taken as two isosceles triangles on the bases 2π to 3π and π to 2π radians respectively and the net work done during exhaust and inlet strokes is zero. The work done during compression stroke is negative and is 1/4th of that during expansion. Sketch the T.M. diagram for one cycle and find the maximum value of the turning moment during expansion. If the load remains constant, mark on the diagram the points of maximum and minimum speed. Also find the moment of inertia of flywheel to keep the speed fluctuation within $\pm 1.2.5\%$ of the mean speed.

(***Ans.*** $\theta = 8.25°$ from i.d.c. and o.d.c. I = 578 kg·m^2)

14.13 During the forward stroke of the piston of the double acting steam engine, the turning moment has a maximum value of 196.2 N·m when the crank makes an angle of 80 degrees with the i.d.c. During the backward stroke, the maximum turning moment is 1471.5 N·m when the crank makes 100 degrees with o.d.c. The turning moment diagram for the engine may be assumed for simplicity to be represented by two triangles. Determine the crank angles at which the speed has its maximum and minimum values and the coefficient of fluctuation of energy.

If the crank makes up to 100 r.p.m., the radius of gyration of the flywheel is 175 cm and speed is kept within ± 0.75% of the mean speed, find the weight of the flywheel.

(*Ans.* W = 1909 N, θ = 43.75 degrees before o.d.c. and θ = 35° after i.d.c.)

14.14 The turning moment diagram for a 4-stroke gas engine may be assumed for simplicity to be represented by four triangles, the areas of which from the line of zero pressure are as follows:

Suction stroke: 0.45×10^{-3} m^2, compression stroke = 1.7×10^{-3} m^2, expansion stroke = 6.8×10^{-3} m^2, exhaust stroke = 0.65×10^{-3} m^2. Each square metre of area represents 3 MN·m of energy. Assuming the resisting torque to be uniform find the mass of the rim of a flywheel required to keep the speed between 202 and 198 r.p.m. The mean radius of the rim is 1.2 m. (*Ans.* 1380 kg).

14.15 A certain machine requires a torque of (5000 + 500sinθ) N·m to drive it, where θ is the angle of rotation of shaft measured from certain datum. The machine is directly coupled to an engine which produces a torque of (5000 + 600sin2θ) N·m. The flywheel and the other rotating parts attached to the engine have a mass of 500 kg at a radius of gyration of 0.4 m. If the mean speed is 150 r.p.m., find (a) the fluctuation of energy, (b) the total percentage fluctuation of speed, and (c) the maximum and minimum angular acceleration of the flywheel and the corresponding shaft positions.

(*Ans.* E_f = 1204.2 N·m, K_s = 6.1%, θ = 35° and 127°, α_{min} = 3.46 rad/s^2, and α_{max} = 12.2 rad/s^2)

14.16 A multi-cylinder engine is to run at a speed of 600 r.p.m. On drawing the crank-effort diagram to scale of 1 cm = 2452.5 N·m and 1 cm = 30°, the areas above and below the mean torque line in sq.cm. are as follows: +1.6, −1.72, +1.68, −1.91, +1.97 and −1.62. The speed is to be kept within ±1% of the mean speed of the engine. Calculate the necessary moment of inertia of the flywheel. Determine the suitable dimensions of a rectangular flywheel rim, if the breadth is twice the thickness. The density of the cost-iron is 725 gm/cm^3 and its hoop stress is 588.6 N/cm^2. Assume that the rim contributes 92% of the flywheel effect. (Agra Un. 1980)

(*Ans.* I = 320.3 kg·m^2, t = 5.88 cm and b = 11.76 cm)

14.17 A single cylinder I.C. engine working on the 4-stroke cycle, develops 75 kW at 360 r.p.m. The fluctuations of energy can be assumed to be 0.9 times the energy developed per cycle. If the fluctuation of speed is not to exceed 1 per cent and the maximum centrifugal stress in the flywheel is to be 5.5 MN/m^2, estimate the mean diameter and the cross-sectional area of the rim. The material of the rim has a density of 7.2 mg/m^3. (Vikram Un.) (*Ans.* 1.461 m and 0.09 m^2)

14.18 A 4-stroke engine develops 18.5 kw at 250 r.p.m. The turning moment diagram is rectangular for the expansion and compression strokes. The turning moment for expansion stroke is 2.8 times that of the compression stroke (negative). Assuming constant load, determine the moment of inertia of the flywheel to keep the total fluctuation of the crank shaft speed within 1% of the average speed of 250 r.p.m. (SGSITS Indore, June 1994) (*Ans.* 495 kg·m^2)

Chapter 15

GOVERNORS

15.1 INTRODUCTION

The term 'governing' can be used to describe any process which helps to control the output of a machine (more precisely, the output of a prime mover) in accordance with certain prescribed standards. Prime movers, like steam, gas or diesel engines or steam or gas turbines, may be governed to run at a constant speed irrespective of the power they are producing. In a similar way, a pump may be governed to produce a constant pressure irrespective of its discharge.

A *governor* is a device for automatically regulating the output of a machine. When used in conjunction with prime movers, a governor may be described as a device that regulates the supply of working fluid to the prime movers so as to produce an output in accordance with the changing demand of the load. As the external (resisting) load changes producing a change in the mean speed of rotation, the governor partakes the motion of the machine, producing a change in the moving parts of the governor. Changes in the configuration of the governor parts, in turn, cause suitable alterations in the pressure or quantity of fluid delivered to the machine through a suitable mechanism.

15.2 FUNCTIONS OF A GOVERNOR

Engines are frequently required to run at approximately constant speed, irrespective of the power they produce. It is therefore necessary to alter the energy supplied to the engine by an appropriate amount immediately after a change of load takes place. The control for this purpose could be a valve in the case of a steam engine, a throttle-valve in the case of a petrol engine, a fuel-pump setting for an oil engine, and so on. A governor is expected to set the control automatically at the appropriate level. Following discussions will help to distinguish precisely the function of a governor from that of a flywheel.

Variations in the speed of a prime mover may be produced by

(a) Variations in the driving torque over a complete cycle
(b) Variations in the resisting torque (*i.e.,* the external load) owing to changes in the load on the engine.

As far as the variations in the driving torque are concerned, readers may refer back to Section 14.19. Variations in the driving torque, and hence the variations in the angular speed above, are cyclic in nature. This, in other words, mean that the changes in (driving) turning moment and angular speed occur exactly in same way in each cycle, and can be assumed to be characteristic of the type of the engine. The mean or average driving torque, however, remains constant for successive cycles, and is equal to the average resisting torque. For this condition, the mean speed remains almost constant from cycle to cycle. For a given type of engine, speed fluctuation can be restricted to a desired limit by increasing the inertia of rotating parts (the flywheel effect). The governor exercises no control over the cyclic fluctuation about the mean speed.

Variations in average resisting load (as indicated in point b above), depends on factors external to the prime mover. It may be due to variations in lighting or industrial load (in the case of prime movers, driving an electric generator) or due to fluctuations in power demand on account of different cutting/machining operations being carried out in a shop. In the case of self-propelled vehicles (two and four-wheelers) the average resisting load (i.e., the external load) changes as the vehicle moves up or down the gradient or against a wind load. The role of a governor is thus to ensure that average (mean) turning moment of a prime mover is raised or lowered so as to equal the average resisting torque. The role of a governor is thus to maintain a constant mean speed of rotation by setting mean (average) turning moment on driving shaft equal to the average resisting torque. The flywheel, on the other hand, only controls the range of speed fluctuation about the mean speed, and does not have any influence on the mean turning moment. Thus once the governor has altered fuel/fluid supply to the prime mover so as to make mean turning moment equal to new value of average resisting moment (and assuming that resisting moment remains constant for sufficient time period thereafter), the rest of the task of limiting speed fluctuation, about the mean speed over a cycle, is left to the flywheel.

Discussions above have been summarized in Table 5.1.

Table 15.1 Comparison between Functions of a Flywheel and a Governor

Governor	Flywheel
1. A governor is used only with a prime mover.	A flywheel may be used for any machine (e.g. punching and rivetting machine) including prime movers.
2. Being a part of feedback control system, a governor is essential for a prime mover.	A flywheel is not an essential element of every prime mover.
3. A governor regulates the speed by adjusting fuel-supply to the prime mover and therefore constitutes an active control.	A flywheel regulates the speed by storing and releasing kinetic energy of rotation. It provides a passive control and flywheel acts as a reservoir of energy.
4. A governor is capable of raising or lowering mean turning moment by adjusting fuel supply. The role of governor is not restricted to a cycle. It regulates mean torque and mean speed of rotation over a number of cycles.	A flywheel reduces fluctuation of speed during a cycle above and below the mean speed. It does not control mean speed but controls cyclic fluctuation of speed about a mean speed.

Specific Purpose of Governors

Speed-sensitive governor may find its use in any of the following specific purposes:

1. *Usual Speed Regulation.* A plot of efficiency of an engine against the speed of rotation reveals that there exists an optimum speed of operation at which the efficiency of engine

is maximum. For the efficient operation of an engine it is therefore desirable that irrespective of the load, the speed of rotation of the engine is maintained as close to this optimum speed as possible. A governor provides a regulatory control required for the above purpose. Governors are called upon to automatically regulate the supply of working fluid to the engine in accordance with the variations in power demand.

2. *An Emergency Governor*. A governor may be designed to spring into action when the higher speed of the prime mover exceeds say, 10–12% of the normal operating speed. In such cases, the governor acts as an over-speed shut off safety device. Every steam turbine is provided with an 'over-speed trip', which shuts off the steam supply to the turbine if the r.p.m. exceeds a particular value.

15.3   TYPES OF GOVERNORS

Governors are broadly classified in two groups: centrifugal and inertia governors.

Centrifugal Governors

Governors of this type are more common in practice. Changes in centrifugal forces of rotating masses, due to changes in the speed of rotation of engine, are used in this case for the movement of the sleeve (*see* Fig. 15.1). Two or more revolving masses, called governor balls, revolve about the axis of a shaft (called spindle), which is driven through suitable gearing from the engine crank-shaft. Necessary centripetal force in the radially inward direction is provided either by the components of tensions in upper and lower arms, a spring force or a combination of both. This force is called *controlling force*. The controlling force must increase in magnitude as the distance of the balls from the axis of rotation (called radius of rotation) increases. When rotating at an equilibrium speed, the radius of rotation will be such that the inward controlling force completely balances the outward inertial (centrifugal) force.

Any decrease in the load on engine leads to an increase in the speed of rotation.

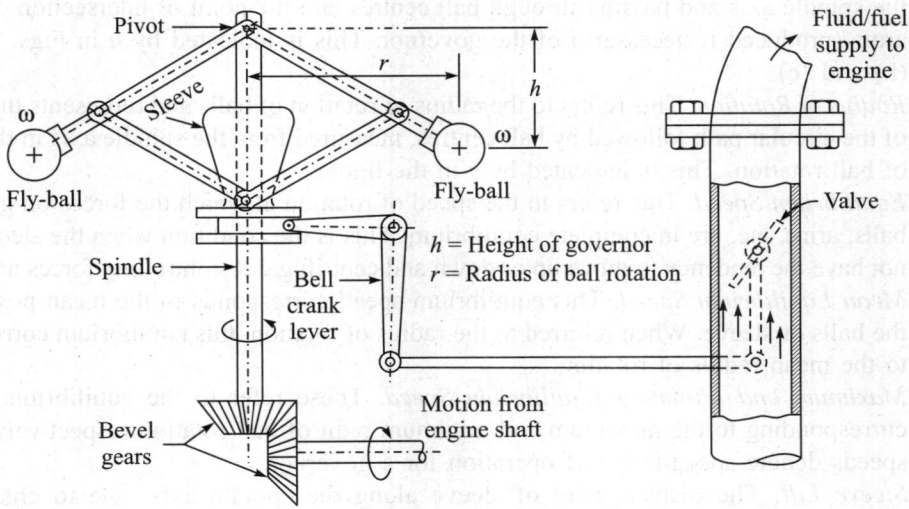

h = Height of governor
r = Radius of ball rotation

Fig. 15.1   A Governor as a Feed-back Mechanism.

This tends to increase the centrifugal force acting on the balls. The governor balls move out until the centrifugal force is again balanced by the controlling force acting inward. Conversely, as the load on the engine increases and the speed of rotation falls down, the governor balls move inward until the centrifugal force is again balanced by the controlling force. The movement of the sleeve, caused by the radially inward and outward movement of balls, is transmitted to the valve through a suitable mechanism which controls the amount of energy supplied to the engine. The outward movement of balls, caused by decreased load on engine, tends to reduce the valve opening.

Centrifugal governors may be either 'gravity controlled' or 'spring force controlled'. In gravity controlled governors, the inward controlling force on balls is provided by force of gravity whereas in the latter case, it is provided by a spring force.

Inertial Type Governors

Inertia governors operate through changes of acceleration (linear or angular), which precede changes in speed and are considered superior to the centrifugal type on account of rapid response to the effect of a change of load. This is because the displacement of the balls is determined by the rate of change of speed of rotation rather than by an actual change of speed of rotation. Above advant--age is offset, however, by difficulty in ensuring complete balance of revolving parts of the governor.

These governors are particularly important for marine applications, especially internal combustion engines where the resisting torque may vary rapidly and continuously, while moving across the sea.

15.4 TERMS USED IN GOVERNORS

Following terms are frequently used in discussions on governors and as such an exposure to these terms at this stage should be helpful for the students.

1. *Height of Governor*. It is the perpendicular distance between the planes, transverse to the spindle axis and passing through ball centres and the point of intersection of upper arms (produced if necessary) of the governor. This is indicated by h in Figs. 15.2(a), (b) and (c).

2. *Radius af Rotation*. This refers to the radius of rotation of balls and represents the radius of the circular path followed by ball centres, measured from the spindle axis in the plane of ball rotation. This is indicated by r in the figure.

3. *Equilibrium Speed*. This refers to the speed of rotation at which the forces on governor balls, arms, etc. are in complete equilibrium. This is the condition when the sleeve does not have the tendency to move up or down and centrifugal & controlling forces are equal.

4. *Mean Equilibrium Speed*. This equilibrium speed corresponds to the mean position of the balls or sleeve. When referred to the radius of rotation, this equilibrium corresponds to the mean radius of rotation.

5. *Maximum and Minimum Equilibrium Speed*. These refer to the equilibrium speeds corresponding to the maximum and minimum radii of ball rotation respectively. These speeds denote speed-range of operation for a governor.

6. *Sleeve Lift*. The displacement of sleeve along the spindle axis, due to changes in equilibrium speed, is called *sleeve lift*.

15.5 THE WATT GOVERNOR

This is the simplest type of centrifugal governor and was used by Watt on some of his early steam engines. Although this governor is now obsolete, it is of interest as it marked the beginning of a class of governors which fall in the category (conical) *pendulum governors*. The other governors of this category are porter governor and Proell governor. An important limitation of all these governors is that they depend heavily on force of gravity, and the axis of spindle for these governors must be kept vertical.

In the Watt governor shown in Fig. 15.2(a), each ball is attached to the lower end of arms A-A, which in turn are pivoted on the axis of rotation at O. The balls therefore, rotate at the same r.p.m. as the spindle C. The sleeve S is attached to the balls through arms B-B which are pin-joined at both the ends. There are two other versions shown in Fig. 15.2(b) and (c), which differ from version at (a) in respect of connecting upper arms A-A to the spindle. In Figs. 15.2(b) and (c), there is an offset between the axis of rotating spindle and points P_1 and P_2 at which upper arms are actually pivoted to the spindle. In version shown in Fig. 15.2(c), the upper arms cross at an intermediate point O. The rotating balls and upper arms move along imaginary conical surface, the appex point of which is always the point of intersection O of the upper arms on the axis. Hence, the height of governor is defined always with respect to point O and not with respect to pivot points P_1 and P_2. This is illustrated in Figs. 15.2(b) and (c). While the lower ends of lower arms are free to slide with sleeve along the spindle, the end O of the upper arms is position fixed with respect to the spindle, and therefore, the length of conical pendulum is 'L' as shown in Fig. 15.2(a).

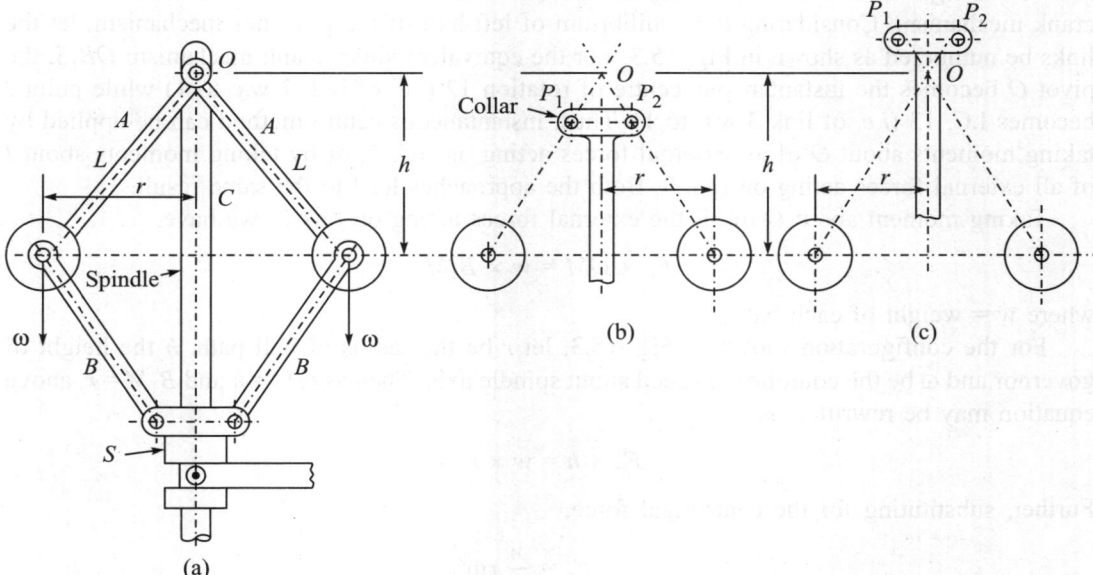

Fig. 15.2 A Watt governor with three versions.

Force Analysis using D'Alembert's Principle: Refer Figs. 15.2 and 15.3.

The arms A and B, as also the sleeve S of the Watt governor, are assumed to have negligible weight. This is in sharp contrast to Porter and Proell governor in which sleeve weight plays a significant role in force analysis.

In the force analysis of all the pendulum types of governor (namely, Watt, Porter and Proell governors), note that for governor balls to move in circular path of radius r, a centripetal force has to be applied through tensions T_1 and T_2 in arms 2 and 3 respectively. Thus, so long as there is a rotational motion, a centripetal acceleration always exist and therefore, the 'force-analysis' problem is basically a problem in dynamics rather than that of statics. *To convert the problem into an equivalent problem in statics, using D'Alembert's*

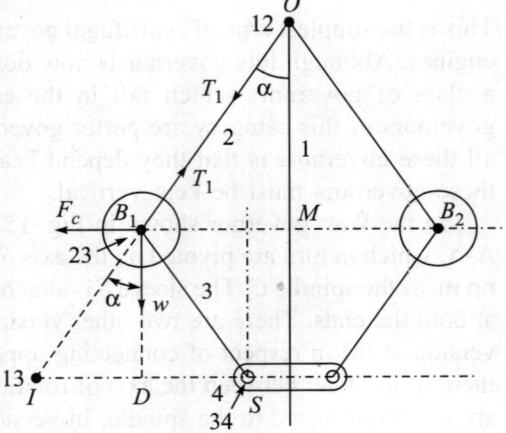

Fig. 15.3 I.Cs. of a Watt Governor.

principle, the rotational motion of the spindle and arms must be suppressed and an inertia force F_c (which is a centrifugal force in this case) must be applied through the centre of balls. With the rotational motion suppressed, the governor mechanism behaves as a crank-slider mechanism, with the sleeve working as a slider. For the difference, however, the link OB (unlike usual crank) is not allowed to have complete rotation in the plane of paper. With angular motion of spindle and arms around governor axis suppressed, the spindle can be likened to the frame-link of a slider-crank mechanism. Considering the equilibrium of left-half of the governor mechanism, let the links be numbered as shown in Fig. 15.3. For the equivalent slider-crank mechanism OB_1S, the pivot O becomes the instantaneous centre of rotation 12 (*i.e.* of link 2 w.r. to 1) while point I becomes I.C. 13 (*i.e.* of link 3 w.r. to 1). Thus, instantaneous centre method can be applied by taking moments about O of all external forces acting on link 2, or by taking moments about I of all external forces acting on link 3. Both the approaches lead to the same result.

Taking moment about O of all the external forces acting on link 2, we have,

$$F_c \times OM = w \times B_1M$$

where w = weight of each ball.

For the configuration shown in Fig. 15.3, let r be the radius of ball path, h the height of governor and ω be the equilibrium speed about spindle axis. Then as $OM = h$ and $B_1M = r$, above equation may be rewritten as

$$F_c \times h = w \times r$$

Further, substituting for the centrifugal force,

$$F_c = \frac{w}{g} r\omega^2,$$

we have,

$$\left(\frac{w}{g} r\omega^2 \right) h = w \times r \tag{15.1}$$

Cancelling common terms on both the sides,

$$\omega^2 = (g/h) \tag{15.2}$$

or

$$h = (g/\omega^2) \tag{15.3}$$

Expressing r and h in centimetres, as $g = 981$ cm/s^2 and $\omega = (2\pi N/60)$

$$h = \frac{981}{(2\pi N/60)^2} = \frac{(89456)}{N^2} = K/N^2 \tag{15.4}$$

Differentiating equation (15.4) w.r. to N,

$$\frac{dh}{dN} = \frac{-2K}{N^3} = -\left(\frac{2K}{N^2}\right)\frac{1}{N}$$

which, within limits may be written as,

$$\delta h = (2K/N^2)(\delta N/N) \tag{15.5}$$

Important Features of Watt Governor

(i) It follows from equation (15.2) that equilibrium speed ω of the governor is independent of ball weight w.

(ii) It also follows from equation (15.3) that the height of the governor is inversely proportional to the square of angular speed.

(iii) Another conclusion from equation (15.5), which is shared by all other pendulum governors, is that change in the height of governor and change in speed have opposite signs. Simply said, an increase in speed causes a decrease in the height of governor.

(iv) Probably, the most important conclusion, which follows from equation (15.5), is that *for a given percentage change of speed (i.e., for a given value of ratio $\delta N/N$), the change in height of governor δh is inversely proportional to the square of the speed N. For a governer to be effective at high speeds, the quantity δh (which decides amount of valve opening) should not drop down rapidly at high speeds for a given percentage of speed fluctuation.* Values of δh for a value of 7% speed fluctuation (*i.e.,* $\delta N/N = 0.07$) for different mean speed N r.p.m. are given in the following table.

N (r.p.m.)	40	60	80	100	120	140	160	180	200
δh cm:	7.827	3.479	1.957	1.252	0.87	0.639	0.489	0.386	0.313

It is seen that the change in the height of governor, for a given percentage change of speed, drops down rapidly and may not be sufficient to change the energy supplied to the engine, by required amount, at high speeds. For this reason the Watt governor is rarely used for speeds beyond say, 75 r.p.m. This limitation is offset significantly by adding sufficient weight at the sleeve in Porter governor.

15.6 EFFECT OF MASS OF ARMS IN WATT GOVERNOR

Since only the effect of forces acting on upper arm is considered in taking moments about pivot O, let W_1 be the total weight of upper arm of length l. Referring to Fig. 15.4, consider an element of small length δx at a distance of x from the pivot O. Let r be the radius of ball rotation and

h be the height of the governor. Let the arm be inclined at θ to the spindle axis in the assumed configuration.

Assuming mass to be uniformly distributed along the length of the arm OB, the weight of the element

$$dw = \left(\frac{W_1}{l}\right)\delta x; \text{ and the moment due to this weight about } O,$$

$$dM = (dw)x\sin\theta = \left(\frac{W_1}{l}\right)(\delta x)x \sin \theta$$

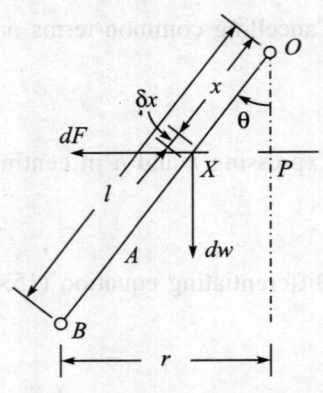

For a given configuration, θ is constant but x is variable.

Assuming the arm to consist of a large number of such elements, moments due to the weight of arm about O,

Fig. 15.4 Effect of mass of arm.

$$M = \int_0^l \left(\frac{W_1}{l}\right)x \sin \theta \, dx$$

or $\qquad M = \left(\frac{W_1}{l}\sin \theta\right)\left(\frac{l^2}{2}\right) = \frac{1}{2} W_1(l \sin \theta) = \frac{1}{2} W_1 r$ $\qquad$ (i)

Again, the inertia force (centrifugal) due to elemental mass is given by,

$$dF = \left(\frac{W_1}{gl}\delta x\right)(x \sin \theta)\omega^2$$

and moment due to this force about O is

$$dM_f = (dF) \times x \cos \theta$$

$$= \left(\frac{W_1}{gl}\delta x\right)(x \sin \theta)\omega^2 \times x \cos \theta$$

Hence, moment about O due to inertia force acting on arm OB,

$$M_f = \int_0^l \left(\frac{W_1\omega^2}{lg}\sin \theta \cos \theta\right)x^2 dx$$

or $\qquad M_f = \left(\frac{W_1}{gl}\right)\omega^2\sin \theta \cos \theta\left(\frac{l^3}{3}\right)$

or $\qquad M_f = \left(\frac{W_1}{3g}\right)(l \sin \theta)(l \cos \theta)\omega^2$

Replacing $l \sin \theta$ by r and $l \cos \theta$ by h, we have

$$M_f = \left(\frac{W_1}{3g}\right)r\omega^2h$$

Combining above moments at (i) and (ii) on corresponding sides with equation (15.1),

$$\left(\frac{w}{g}r\omega^2\right)h + \left(\frac{W_1}{3g}\right)r\omega^2 h = w \times r + \frac{1}{2}W_1 r$$

Dividing out by r
$$\left(w + \frac{W_1}{3}\right)\frac{\omega^2}{g}\, h = (w + W_1/2)$$

or
$$\omega^2 = \left(\frac{w + W_1/2}{w + W_1/3}\right) \times g/h \qquad (15.6)$$

and,
$$h = \left(\frac{w + W_1/2}{w + W_1/3}\right) \times g/\omega^2 \qquad (15.7)$$

15.7 THE PORTER GOVERNOR

A Porter governor differs from a Watt governor only in respect of a heavily weighted sleeve. Inward controlling force required to ensure that balls move in circular path, is provided by tensions T_1 and T_2 in upper and lower arms respectively. As in Watt governor, an increase in the speed of rotation results in an increase in the radius of rotation and also a lift in the sleeve by corresponding amount. Conversely, a drop in equilibrium speed results in decreased radius of ball rotation and the sleeve is lowered by corresponding amount. In both the cases, movement of sleeve is communicated through a bell-crank lever and a suitable mechanism to the valve which controls energy input to the engine.

Figure 15.5 shows left-half of a Porter governor diagrammatically. Weights of upper and lower arms are neglected in comparison to the weight of balls. As explained in the case of a Watt governor, the rotational motion of balls about spindle axis involves centripetal acceleration and the problem must be treated as a problem in dynamics. Introducing inertia force F_c equal to the centripetal (controlling) force and by suppressing rotational motion of arms and spindle, we convert problem of dynamics into an equivalent problem in statics.

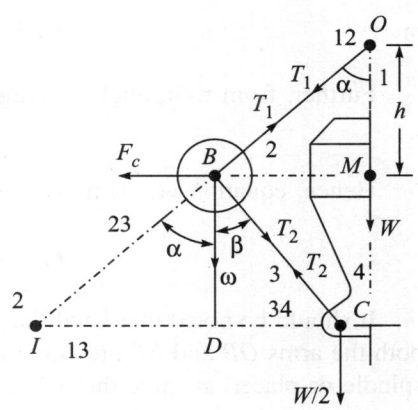

Fig. 15.5 Left half of Porter governor.

The left-half portion of the Porter governor in Fig. 15.5 then represents a slider-crank mechanism. Point of intersection I of the normal to the line of stroke of sleeve (*i.e.*, spindle axis) and the extended line of crank OB then represents instantaneous centre of rotation of link 3 w.r. to spindle. The force analysis can then be carried out by taking moments about I of all the forces acting externally on link 3.

Let, $\quad w$ = weight of each ball
$\qquad W$ = weight of sleeve
$\quad T_1, T_2$ = tension in upper and lower arms respectively
$\qquad F_c$ = the centrifugal force acting at the ball centre

Let BD be the perpendicular from ball centre on the line IC, and let $\angle IBD = \angle IOM = \alpha$ and $\angle CBD = \beta$. Assume that half of the sleeve weight, in charge of left-half portion, acts through C. Taking moments of forces acting externally on link 3, about the I.C., I, we have

$$F_c(BD) = w(ID) + \frac{W}{2}(IC)$$

or

$$F_c = w\left(\frac{ID}{BD}\right) + \frac{W}{2}\left(\frac{IC}{BD}\right) \tag{15.8}$$

But from the right angled triangles IBD and BCD,

$$(ID/BD) = \tan\alpha, \text{ and}$$
$$IC/BD = ID/BD + CD/BD$$

or

$$IC/BD = \tan\alpha + \tan\beta$$

Hence, substituting in equation (15.8),

$$F_c = w\tan\alpha + \frac{W}{2}(\tan\alpha + \tan\beta)$$

Defining ratio $\left(\dfrac{\tan\beta}{\tan\alpha}\right) = k$, above equation can be expressed in terms of variable $\tan\alpha$. Thus,

$$F_c = w\tan\alpha + \frac{W}{2}(1 + k)\tan\alpha$$

or

$$F_c = \left\{w + \frac{W}{2}(1+k)\right\}\tan\alpha \tag{15.9}$$

Further, from right angled triangles OBM,

$$\tan\alpha = r/h$$

Hence, equation (15.9) may be written as,

$$F_c = \left\{w + \frac{W}{2}(1+k)\right\}\left(\frac{r}{h}\right) \tag{15.10}$$

It should be noted that k will have different values for different radii of ball rotation. When both the arms OB and BC are equal in length and their pivots O and C are either on the axis of spindle or placed at same distance from the axis, $k = 1$ as α and β are equal.

Replacing the centrifugal force F_c by its value $F_c = \dfrac{w}{g}r\omega^2$ in equation (15.10), we have

$$\frac{w}{g}r\omega^2 = \left\{w + \frac{W}{2}(1+k)\right\}\left(\frac{r}{h}\right)$$

or

$$\omega^2 = \left\{\frac{w + (1+k)\dfrac{W}{2}}{w}\right\}\left(\frac{g}{h}\right) \tag{15.11}$$

Also,
$$h = \left\{ \frac{w + (1+k)\dfrac{W}{2}}{w} \right\} \left(\frac{g}{\omega^2} \right) \tag{15.12}$$

With the conditions indicated above, when $k = 1$, above expressions reduce to

$$\omega^2 = \left(\frac{w + W}{w} \right) \frac{g}{h} \tag{15.13}$$

and
$$h = \left(\frac{w + W}{w} \right) \frac{g}{\omega^2} \tag{15.14}$$

It is more convenient to express height of governors in cm. Hence taking $g = 981$ cm/s^2 and, $\omega = 2\pi \dfrac{N}{60}$, as in the case of Watt governor, equation (15.14) reduces to

$$h = \left(\frac{w + W}{w} \right) \frac{89456}{N^2} \tag{15.14a}$$

It is interesting to note that when weight of the sleeve W becomes negligible in comparison to the weight of ball w, equation (15.14a) reduces to equation (15.4) for Watt governor.

It follows from equation (15.14) that height of the governor can be increased substantially by choosing a suitable value for the ratio W/w. For instance by choosing a moderate value of $(W/w) = 10$, the factor $(w + W)/w$ takes a value of 11 and values of 'change of height δh' for Porter and Watt governor then compared as indicated in Table 15.2.

As for Watt governor, differentiating equation (15.14a) w.r. to N,

$$\delta h = \left(\frac{W + w}{w} \right) \frac{89456}{N^2} \left(\frac{-\delta N}{N} \right) 2$$

or,
$$\delta h = -\left(\frac{W + w}{w} \right) \frac{2\,(89456)}{N^2} \left(\frac{\delta N}{N} \right) \tag{15.15}$$

For an assumed speed fluctuation of 7% for instance (*i.e.*, $\delta N/N = 0.07$), the δh values for the two governors compare as indicated in Table 15.2.

Table 15.2 (For $\delta N/N = 0.07$ and $W/w = 10$)

N (r.p.m.)	40	60	80	100	120	140	160	180	200
δh cm Watt governor	7.827	3.479	1.957	1.252	0.87	0.639	0.489	0.386	0.313
δh cm: Porter governor	86.1	38.27	21.53	13.78	8.57	7.03	5.38	4.25	3.44

It is clear from the table that the Porter governor can be made more effective at higher speeds by increasing the weight ratio (W/w).

Alternative Approach

Equations (15.11) and (15.12) can also be obtained by taking free body diagrams and considering equilibrium of forces acting on the governor ball and sleeve separately.

The free body diagram of governor ball and sleeve is as shown in Fig. 15.6(a) and (b) while the corresponding force polygons are as at Fig. 15.6(c) and (d) respectively. Neglecting friction, the four forces acting at the pin joint of the ball are:

(i)   tension T_1 in upper arm
(ii) tension T_2 in lower arm
(iii) weight w of the ball
(iv) the centrifugal force F_c acting at the ball centre.

A force polygon, consistent with the lines of action of these forces is shown in Fig. 15.6(c).

Similarly, the three forces acting on the sleeve are: (i) tension T_2 in the two lower arms on the two sides and (ii) the sleeve weight W. Corresponding force polygon, consistent with the respective lines of action, is as in Fig. 15.6(d).

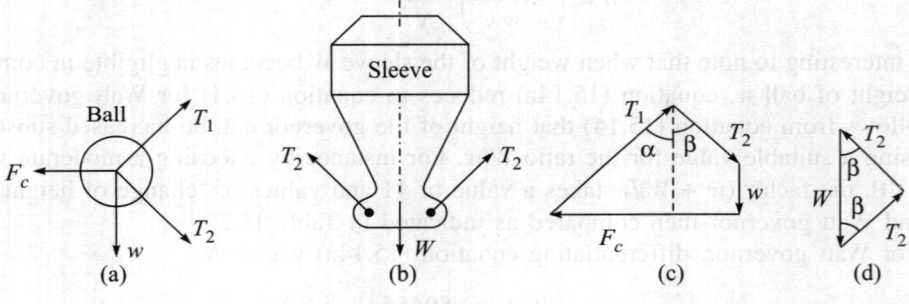

Fig. 15.6

Resolving the forces in Fig. 15.6(d) along the direction of sleeve weight W, it follows that,

$$2T_2\cos\beta = W \quad \text{or} \quad T_2 = \frac{W}{(2\cos\beta)}$$

Also, resolving forces in Fig. 15.6(c) along the direction of w

$$T_1\cos\alpha = T_2\cos\beta + w$$

Substituting for T_2 from above

$$T_1\cos\alpha = \left(\frac{W}{2\cos\beta}\right)\cos\beta + w$$

or

$$T_1 = \left(w + \frac{W}{2}\right)\Big/\cos\alpha$$

Also, resolving forces in Fig. 15.6(c) in a direction normal to that of w, we have

$$F_c = T_1\sin\alpha + T_2\sin\beta$$

Substituting for T_1 and T_2 in the above equation,

$$F_c = \left(w + \frac{W}{2}\right)\sin\alpha/\cos\alpha + W\sin\beta/(2\cos\beta)$$

or

$$F_c = \left(w + \frac{W}{2}\right)\tan\alpha + \left(\frac{W}{2}\right)\tan\beta$$

Taking out $\tan\alpha$ common, $\quad F_c = \left\{w + (1+k)\,\frac{W}{2}\right\}\tan\alpha$

Substituting for $\dot{F}_c$ $\quad \dfrac{w}{g}r\omega^2 = w + \left\{(1+k)\,\dfrac{W}{2}\right\}(r/h)$

or

$$\omega^2 = \left\{\frac{w + (1+k)\,\dfrac{W}{2}}{w}\right\}g/h, \text{ as before.}$$

15.8 EFFECT OF FRICTION

Since the force of friction always opposes motion, when the sleeve moves up the force of gravity on sleeve W and the force of friction f, both act downward. Similarly, when the sleeve moves down the spindle, the force of friction f moves up the spindle while the gravitational pull continues to act down. Thus, when the sleeve moves up, the net force on the sleeve is $(W + f)$ but when it moves down the spindle, the net force on the sleeve is $(W - f)$.

Replacing the term W in equation (15.12) by $(W \pm f)$ the expression for the height of governor, taking into account the effect of friction, is

$$h = \left\{\frac{w + (1+k)\,(W \pm f)/2}{w}\right\}\left(\frac{g}{\omega^2}\right) \tag{15.16}$$

The expression indicates that a given height of governor is obtainable at two distinct speeds of rotation, namely at ω_1 when the speed is rising and at ω_2 when the speed is decreasing. This can be explained as follows.

Assume that the initial speed lies somewhere between ω_1 and ω_2 and that the sleeve has just moved down by a small amount so that the net force downwards on sleeve is $(W - f)$. Let the speed of the engine increase now. The force resisting movement of sleeve is now $(W + f)$ and until speed increases to ω_1 so as to produce corresponding centrifugal force on the balls, the sleeve will not move up. Again with some speed between ω_1 and ω_2, let us assume that the sleeve has just moved up so that the force acting downwards on sleeve is $(W + f)$. Let the speed of engine decrease now. *The sleeve will not move down until the speed of rotation does not fall down so as to reduce centrifugal force on balls to a value that corresponds to a force $(W - f)$ at the sleeve. The governor thus becomes insensitive over the speed range ω_1 to ω_2.*

EXAMPLE 15.1 A loaded governor of the Porter type has equal links 25 cm long pivoted at the axis. The weight of each ball is 29.4 N and the weight of the central load is 137.34 N. The ball radius is 15 cm when the governor begins to lift and 20 cm at the maximum speed. Determine the maximum and minimum speeds and the range of speed.

If the friction at the sleeve is equivalent to 14.7 N, find the maximum and the minimum speeds and the range of speed. (AMIE, Summer 1982)

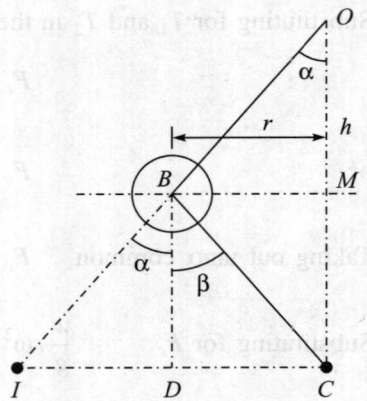

Fig. 15.7 Porter governor: general configuration.

Solution: Referring to Fig. 15.7,

$OB = BC = 25$ cm; $r_1 \equiv (r)_{\max} = 20$ cm
$W = 137.34$ N; $w = 29.4$ N
$r_2 \equiv (r)_{\min} = 15$ cm
$f = 14.7$ N

Case (i) Neglecting friction,

$$\omega^2 = \left\{ \frac{w + (1 + k) \, W/2}{w} \right\} \frac{g}{h}$$

where $k = \dfrac{\tan\beta}{\tan\alpha} = 1$ Thus, $\omega^2 = \left(\dfrac{29.4 + 137.34}{29.4} \right) \dfrac{981}{h}$

Therefore $\omega_1^2 = \left(\dfrac{166.74 \times 981}{29.4} \right) \times \dfrac{1}{h_1}$ and $\omega_2^2 = \left(\dfrac{166.74 \times 981}{29.4} \right) \times \dfrac{1}{h_2}$

Again, from right angled triangle, at maximum and minimum speeds,

$$h_1 = \sqrt{(OB)^2 - r_1^2}; \quad \text{and} \quad h_2 = \sqrt{(OB)^2 - r_2^2}$$

or, $h_1 = \sqrt{25^2 - 20^2}$ and $h_2 = \sqrt{25^2 - 15^2}$

Therefore $h_1 = 15$ cm $h_2 = 20$ cm

Hence $\omega_1 = \left(\dfrac{166.74 \times 981}{29.4 \times 15} \right)^{1/2} = 19.26$ rad/s.

$\omega_2 = \left\{ \dfrac{166.74 \times 981}{29.4 \times 20} \right\}^{1/2} = 16.679$ rad/s.

Therefore $\left. \begin{array}{l} N_1 = 183.92 \text{ r.p.m} \\ N_2 = 159.27 \text{ r.p.m} \end{array} \right\}$ **Ans.**

Range of speed = $183.92 - 159.27 = 24.65$ r.p.m.

Case (ii) When frictional force $f_1 = 14.7$ N exists, with h_1 and h_2 remaining the same,

$$\omega_1^2 = \left\{ \frac{W + (1 + k)(W_1 + f_1)/2}{w} \right\} \frac{g}{h_1}$$

or $\omega_1^2 = \left\{ \dfrac{29.4 + (2)(137.34 + 14.7)/2}{29.4} \right\} \dfrac{981}{15} = 403.6$

Therefore $\qquad \omega_1 = 20.09$ and $N_1 = \dfrac{60 \times 20.09}{2\pi} = 191.85$ r.p.m. **Ans.**

and, $\qquad \omega_2^2 = \left\{ \dfrac{29.4 + (2)\,(137.34 - 14.7)/2}{29.4} \right\} \dfrac{981}{20} = 253.66$

Therefore $\qquad \omega_2 = 15.92$ rad/s; and $N_2 = 152.0$ r.p.m. **Ans.**

Hence $\qquad$ range of speed $= 39.85$ r.p.m. **Ans.**

(Note: For each height of governor, there exists two equilibrium speeds. By selecting higher of the two at $h_1 = 15$ cm & lower of the two at $h_2 = 20$ cm, a maximum possible range of speeds is reported above).

EXAMPLE 15.2 A Porter governor has all four arms 300 mm long. The upper arms are pivoted on the axis of rotation and the lower arms are attached to the sleeve at a distance of 35 mm from the axis. Each ball has a mass of 7 kg and mass of sleeve is 54 kg. If the extreme radii of rotation of the balls are 200 mm and 250 mm, determine the range of speed of the governor. Also find the sleeve lift between the extreme radii of rotation. (Osmania University)

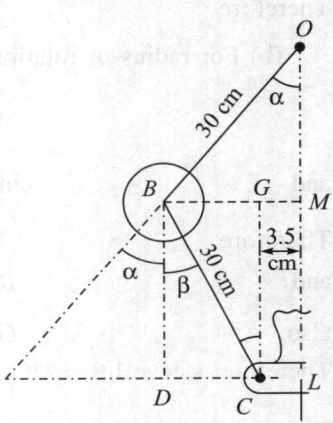

Fig. 15.8 Porter Governor.

Solution: Referring to Fig. 15.8, $OB = BC = 30$ cm

$$CL = 3.5 \text{ cm} \quad M = 7 \text{ kg} \quad M = 54 \text{ kg}$$
$$r_2 = 20.0 \text{ cm} \quad r_1 = 25.0 \text{ cm}$$

(a) For radius of rotation $r_2 = 20$ cm, from right angled triangle OBM,

$$\sin\alpha = \left(\dfrac{20}{30} \right). \text{ Hence, } \alpha = 41.81°$$

and $\qquad h_2 = OM = \sqrt{30^2 - 20^2} = 22.36$ cm

and $\qquad \tan\alpha = 0.8944$

Also $\qquad BG = 20 - 3.5 = 16.5$ cm

and $\qquad CG = \sqrt{30^2 - 16.5^2} = 25.06$ cm; $OL = 22.31 + 25.06 = 47.42$ cm.

Hence from the right angled triangle BCG,

$$\sin\beta = \dfrac{BG}{BC} = \dfrac{16.5}{30} = 0.55$$

Therefore $\qquad \beta = 33.367°$ and $\tan\beta = 0.6585$

Therefore $\qquad k = \dfrac{\tan\beta}{\tan\alpha} = 0.7363$

Hence $\qquad \omega_2^2 = \left\{ \dfrac{w + (1 + k)\,W/2}{w} \right\} \dfrac{g}{h_2}$

or,
$$\omega_2^2 = \left\{ \frac{mg + (1 + k)\, Mg/2}{mg} \right\} \frac{g}{h_2}$$

or,
$$\omega_2^2 = \left\{ \frac{m + (1 + k)M/2}{m} \right\} \frac{g}{h_2}$$

Thus
$$\omega_2^2 = \left\{ \frac{7 + (1 + 0.7363)54/2}{7} \right\} \frac{981}{22.36} = 337.697$$

Therefore $\omega_2 = 18.376$ rad/s   and $N_2 = 175.5$ r.p.m.   **Ans.**

(b) For radius of rotation $r_1 = 25.0$ cm, from right angled triangle OBM,

$$h_1 = OM = \sqrt{30^2 - 25^2} = 16.583$$

and
$$\sin\alpha = \frac{BM}{BO} = \frac{25}{30} = 0.833$$

Therefore $\alpha = \sin^{-1}(0.833) = 56.44°$, or $\tan\alpha = 1.507$

and $BG = BM - 3.5 = 25 - 3.5 = 21.5$ cm

also, $GC = \sqrt{30^2 - 21.5^2} = 20.9$ cm; $OL' = 16.583 + 20.9 = 37.483$ cm

Therefore, sleeve lift $= OL - OL' = 47.42 - 37.48 = 9.94$ cm   **Ans.**

Hence
$$\sin\beta = \frac{BG}{BC} = \frac{21.5}{30} = 0.7167$$

Therefore $\beta = \sin^{-1}(0.7167) = 45.78°$ and $\tan\beta = 1.0276$

Therefore
$$k = \frac{\tan\beta}{\tan\alpha} = \frac{1.0276}{1.507} = 0.682$$

Thus
$$\omega_1^2 = \left\{ \frac{7 + (1 + 0.682)\dfrac{54}{2}}{7} \right\} \times \frac{981}{16.583} = 442.95$$

Therefore $\omega_1 = 21.046$ rad/s   and $N_1 = \dfrac{21.046 \times 60}{2\pi} = 200.978$

Therefore $N_1 = 201.0$ r.p.m.   **Ans.**

Therefore   Speed range $= 201 - 175.5 = 25.5$ r.p.m.   **Ans.**

EXAMPLE 15.3   Figure 15.9 shows a Porter governor for which the speed range can be varied by means of the auxiliary spring S. The spring force is transmitted to the sleeve by the arm AB which is pivoted at A. The two balls each of mass 0.36 kg are supported by four links C_1, C_2, C_3, and C_4, each 90 mm in length. The sleeve carries a mass of 0.9 kg.

The sleeve begins to rise when the balls revolve at 200 r.p.m. in a circle of 75 mm radius. The speed of the governor is not to exceed 220 r.p.m. when the sleeve has risen 10 mm from its original position.

Determine (a) the necessary stiffness of the spring S and (b) the tension in the link C_1 when the sleeve begins to rise. (University of London)

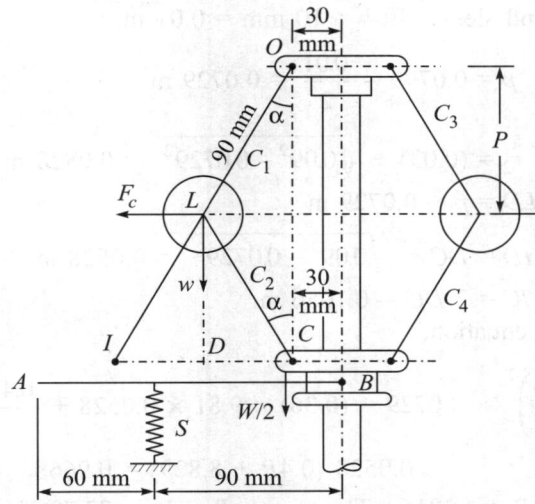

Fig. 15.9 Porter Governor with Auxiliary Spring.

Solution: Let P be the force in N in the spring S and R be the equivalent force exerted at the sleeve. Then taking moments about A, for equivalence,

$$R \times (90 + 60) = P \times 60 \quad \text{or} \quad R = \left(\frac{60}{150}\right)P = (0.4)P$$

This, together with the mass of 0.9 kg at sleeve, produces a total sleeve load of,

$$W = (0.4P) + (0.9) \times 9.81 = (0.4P) + 8.829 \text{ N}$$

For the equilibrium of the link C_2, the moment of external forces acting on link C_2 about the instantaneous centre I must vanish. Thus,

$$F_c \times LD = w \times ID + \frac{W}{2} \times IC \tag{i}$$

(a) For $N = 200$ r.p.m. and $r = \dfrac{75}{1000} = 0.075$ m,

$$LD = p = \sqrt{(0.09)^2 - (0.075 - 0.03)^2} = 0.0779 \text{ m}$$

also
$$ID = DC = (r - 0.03) = 0.045 \text{ m}, \quad \text{and} \quad IC = 2DC = 0.09 \text{ m}$$

Therefore

$$(0.36)(0.075)\left(\frac{2\pi \times 200}{60}\right)^2 \times 0.0779 = (0.36 \times 9.81) \times 0.045 + \left(\frac{0.4P + 8.829}{2}\right) \times 2 \times 0.045$$

or
$$(0.4P + 8.829) = 16.97$$

Therefore
$$0.4P = 8.142 \quad \text{Thus} \quad P = +20.35 \text{ N}$$

(b) For $N = 220$ r.p.m. and sleeve lift $s = 10$ mm $= 0.01$ m,

Therefore
$$p = 0.0799 - \frac{0.01}{2} = 0.0729 \text{ m}$$

Therefore
$$r_2 = (0.03) + \sqrt{0.09^2 - 0.0729^2} = 0.0828 \text{ m}$$

Thus,
$$LD = p = 0.0729 \text{ m}$$

and
$$ID = DC = \sqrt{0.09^2 - 0.0729^2} = 0.0528 \text{ m}$$

Also,
$$IC = 2\, DC = 0.1056 \text{ m}$$

Hence from moment equation,

$$(0.36)(0.0828)\left(\frac{2\pi \times 220}{60}\right)^2 \times 0.0729 = (0.36) \times 9.81 \times 0.0528 + \frac{0.4\,P + 8.829}{2} \times 2(0.0528)$$

or
$$0.0528\,(0.4P + 8.829) = 0.9668$$

or
$$0.4P = 4.4816 \quad \text{Therefore} \quad P = P_2 = 23.70 \text{ N}$$

Therefore, spring stiffness $= \dfrac{(P_2 - P_1)}{\text{Amount of spring compression/tension}}$

or, spring stiffness $= \dfrac{(P_2 - P_1)}{\left(\dfrac{60}{150}\right)(\text{sleeve movement})}$

or spring stiffness $= \dfrac{(P_2 - P_1)}{(0.01) \times \dfrac{2}{5}} = \dfrac{5\,(23.70 - 20.35)}{(0.01) \times 2} = 837.5$ N/m   **Ans.**

Vertical load at pin $C = \dfrac{1}{2}\,(0.4P_1 + 8.829)$ N $= 0.2\,(20.35) + 4.414 = 8.484$ N

Therefore, vertical reaction at pin O, at 200 r.p.m.,

$$R_V = \text{Vertical load at } C + \text{ball weight}$$

or
$$R_V = (8.484) + 0.36 \times 9.81$$

or
$$R_V = 8.484 + 3.5316 = 12.016 \text{ N}$$

This must be equal to component of tension T_1 in arm C_1 in vertical direction. Thus,

$$T_1 \cos\alpha = 12.016$$

or
$$T_1 = (12.016) \times \frac{0.09}{p} = 12.016 \times \frac{0.09}{0.0779} = 13.88 \text{ N}$$ **Ans.**

EXAMPLE 15.4 In an engine governor of the Porter type, the upper and lower arms are, respectively 20 cm and 25 cm long and pivoted on the axis of rotation. The central load is 147.2 N. The weight of each ball is 19.6 N and friction of sleeve together with the resistance of

the operating gear is equal to the weight of 24.5 N at the sleeve. If the limiting inclinations of the upper arms to the vertical are 30° and 40° find, taking friction into account, range of speeds of the governor.
(Allahabad Univ., 1980)

Solution: Given $W = 147.2$ N; $w = 19.6$ N
$$f = 24.5 \text{ N}, \ \alpha_1 = 40°; \ \alpha_2 = 30°$$

(a) For minimum speed of rotation (*see* Fig. 15.10),

$$\alpha_2 = 30°; \ r_2 = 20 \sin30 = 10 \text{ cm}.$$
and, $$h_2 = 20 \cos30 = 17.32 \text{ cm}$$
Also $$BM = r_2 = 25 \sin\beta_2$$

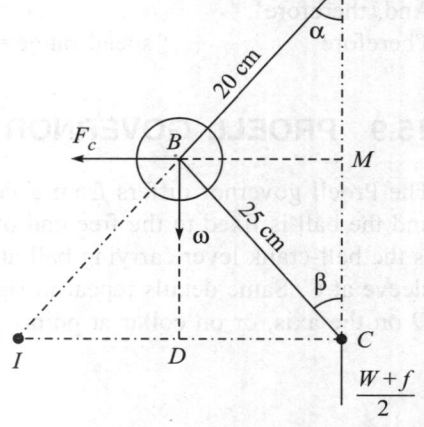

Therefore $\sin\beta_2 = \dfrac{10}{25} = 0.4$ Hence, $\beta_2 = 23.578°$

Fig. 15.10 Engine Governor: Porter Type.

Therefore $\tan\beta_2 = 0.436$ and $\tan\alpha_2 = 0.5773$
Thus $k_2 = 0.7559$

Taking net load $= (w - f)/2$ on sleeve for obtaining lower most speed, we have

$$\omega_2^2 = \frac{w + (1 + k_2)(W - f)/2}{w} \times \frac{g}{h_2}$$

or $$\omega_2^2 = \frac{19.6 + (1 + 0.7559)(147.2 - 24.5\,N)/2}{19.6} \times \frac{981}{17.32} = 367.9$$

Therefore $\omega_2 = 19.18$ rad/s   or, $N_2 = 183.2$ r.p.m.

(b) Similarly, for maximum speed of rotation N_1,

$$\alpha_1 = 40° \quad \text{and} \quad r_1 = 20 \sin40 = 12.856 \text{ cm}$$
$$h_1 = 20 \cos40 = 15.32 \text{ cm}$$

Therefore $\tan\alpha_1 = 0.8391$
Also, from $BM = r_1 = 25 \sin\beta_1$
Therefore $\sin\beta_1 = 12.856/25 = 0.5142$
Thus $\beta_1 = 30.947°$ and $\tan\beta_1 = 0.5996$
Therefore $k_1 = \tan\beta_1/\tan\alpha_1 = 0.7146$

For maximum speed of rotation, taking sleeve load is $(W + f) = (147.2 + 24.5)$N

or, $$\omega_1^2 = \frac{w + (1 + k_1)(W + f)/2}{w} \times \frac{g}{h_1}$$

or $$\omega_1^2 = \frac{19.6 + (1 + 0.7146)(147.2 + 24.5)/2}{19.6} \times \frac{981}{15.32} = 544.94$$

Therefore $\omega_1 = 23.34$ rad/s

And, therefore $N_1 = 222.9$ r.p.m.

Therefore   speed range $= (222.9 - 183.2) = 39.7$ r.p.m.   **Ans.**

15.9 PROELL GOVERNOR

The Proell governor differs from a Porter governor in that the lower arm is a bell-crank lever and the ball is fixed to the free end of this bell-crank lever. Thus, referring to Fig. 15.11, *CBG* is the bell-crank lever carrying ball at *G*, and hinged to upper arm at *B*. The lever is pivoted to sleeve at *C*. Same details repeat in right half. The upper arms may be pivoted to the spindle at *O* on the axis, or on collar at points a small distance away from the axis.

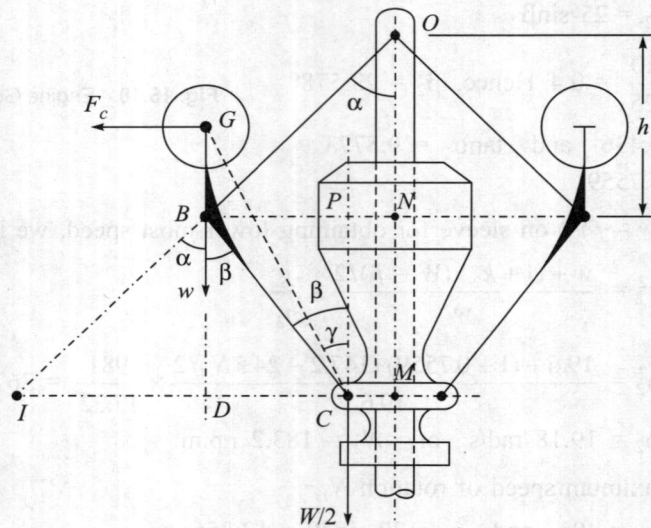

Fig. 15.11   A Proell governor.

The instantaneous centre *I* and point *D* are located as in the case of Porter governor. The portion BG of the bell-crank lever is shown to be vertical in the configurations of the governor in figure.

Considering equilibrium of link *CBG* and taking moment of all external forces on it about *I*,

$$F_c(DG) = w(DI) + \frac{W}{2}(IC)$$

or, $$F_c(DG) = w(DI) + \frac{W}{2}(ID + DC)$$

Dividing by *BD*,

$$F_c\left(\frac{DG}{BD}\right) = w\left(\frac{DI}{BD}\right) + \frac{W}{2}\left(\frac{DI}{BD} + \frac{DC}{BD}\right)$$

or,
$$F_c\left(\frac{DG}{BD}\right) = w\tan\alpha + \frac{W}{2}(\tan\alpha + \tan\beta)$$

or,
$$F_c\left(\frac{DG}{BD}\right) = \left[w + \frac{W}{2}(1+k)\right]\tan\alpha$$

where,
$$k = (\tan\beta/\tan\alpha)$$

Therefore
$$F_c = \left(\frac{BD}{DG}\right)\left[w + \frac{W}{2}(1+k)\right]\tan\alpha \tag{15.17}$$

As
$$F_c = (w/g)\, r\omega^2 \text{ and } \tan\alpha = r/h, \; h = \left(\frac{BD}{DG}\right)\left[\frac{w + \frac{w}{2}(1+k)}{\omega}\right]\frac{G}{\omega^2}$$

Comparing equation (15.17) to equation (15.9) for the Porter governor, it follows that corresponding equation for a Proell governor is obtained by multiplying equation (15.9) for Porter governor by a ratio (BD/DG).

EXAMPLE 15.5 A Proell governor has all the four arms of length 25 cm. The upper and lower ends of the arms are pivoted on the axis of rotation of the governor. The extension arms of the lower links are each 10 cm long and parallel to the axis when the radius of the ball path is 15 cm. The weight of each ball is 44.15 N and the central weight of 353.2 N. Determine the equilibrium speed of the governor. (Bangalore Univ.)

Solution: Referring to Fig. 15.11,
$$OB = BC = 25 \text{ cm}; \; CM = 0; \; BG = 10 \text{ cm}$$

and,
$$w = 44.15 \text{ N}; \; W = 353.2 \text{ N}; \; r = 15 \text{ cm}$$

As
$$BN = r = 15 \text{ cm}$$

We have,
$$OB = \sqrt{25^2 - 15^2} = 20$$

Therefore
$$\tan\alpha = \frac{BN}{NO} = \frac{15}{20} = 0.75$$

and
$$\alpha = 36.87°$$

Again, as the arms are equal and pivoted on spindle axis, $\alpha = \beta = 36.86°$

Therefore
$$BD = BC \cos\beta$$

or
$$BD = (25)\cos 36.87° = 20 \text{ cm}$$

Now
$$F_c = \frac{BD}{DG}\left[w + \frac{W}{2}(1+k)\right]\tan\alpha$$

But as
$$\alpha = \beta, \; k = 1$$

Therefore
$$F_c = \frac{20}{(20+10)}\left[44.15 + \frac{353.2}{2}(2)\right] \times 0.75 = 198.67 \text{ N}$$

Thus
$$\frac{w}{g}r\omega^2 = 198.67$$

Therefore
$$\omega^2 = \frac{198.67 \times 9.81}{44.15 \times (0.15)} = 294.3$$

Therefore $\quad\quad\quad\quad\quad\quad\quad\quad \omega = 17.155$ rad/s

and, $\quad\quad\quad\quad\quad\quad\quad N = \dfrac{17.155 \times 60}{2\pi} = 163.8$ r.p.m. $\quad\quad\quad$ **Ans.**

EXAMPLE 15.6 Each arm of a Proell governor is 25 cm long. The pivots of upper and lower arms are 25 mm from the axis. The central load acting on the sleeve has a mass of 25 kg and each of the rotating balls has a mass of 3.0 kg. When the governor sleeve is in the mid-position, the extension link of the lower arm is vertical and the radius of the path of rotation of the masses is 175 mm. The vertical height of the governor is 200 mm. If the governor speed is 160 r.p.m. when in mid-position, find

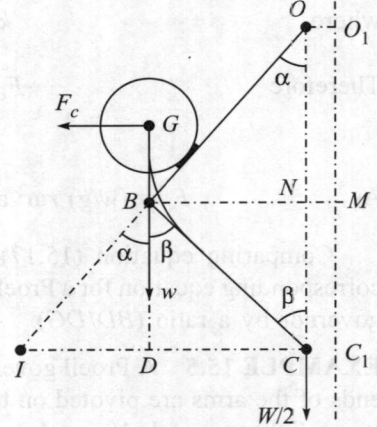

Fig. 15.12 Proell governor.

    (i)   the length of the extension link

    (ii) tension in the upper arm

Solution: Referring to Fig. 15.12,

$$OB = BC = 0.25 \text{ m and } r = MB = 0.175 \text{ m}$$

$$OO_1 = CC_1 = 25 \text{ mm} = 0.025 \text{ m}$$

and, $\quad\quad\quad\quad W/g = 25 \text{ kg}; \; \dfrac{w}{g} = 3.0 \text{ kg}$

Hence, $\quad\quad\quad\quad w = 3 \times 9.81 = 29.43 \text{ N}$

As, $\quad\quad\quad O_1M = h = 0.2 \text{ m}; \; N = 160 \text{ r.p.m.}; \; \sin\alpha = \dfrac{0.175 - 0.025}{0.25} = 0.6$

Therefore $\quad\quad\quad\quad \alpha = 36.87°$

Therefore $\quad\quad\quad\quad \omega = \dfrac{2\pi \times 160}{60} = 16.755 \text{ rad/s}$

Thus by symmetry of geometry, $O'O_1 = \dfrac{0.025}{\tan 36.87} = 0.033 \text{ m}$

$$BD = O_1M = (0.2 - 0.033) = 0.167 \text{ m}$$

Also $\quad\quad\quad\quad DC = BM - 0.025 = 0.175 - 0.025 = 0.15 \text{ m}$

Therefore $\quad\quad\quad IC = 2DC = 0.30 \text{ m} \quad \text{and} \quad ID = DC = 0.15 \text{ m}$

    Thus taking moments of external forces on link BC about I,

$$F_c \times DG = w \times ID + \dfrac{W}{2} + IC$$

or $\quad\quad (3.0) \times 0.175 \times (16.755)^2 \times (BD + BG) = 29.43 \times 0.15 + \dfrac{(25 \times 9.81)}{2} \times 0.3$

Therefore $\quad\quad\quad\quad\quad\quad (BD + BG) = \dfrac{41.262}{3 \times 0.175 \times 16.755^2}$

or $\qquad\qquad (0.167 + BG) \approx 0.28$ m

Therefore $\qquad\qquad BG = 0.113$ m $= 11.3$ cm $\qquad\qquad$ **Ans.**

Tension in the upper arm:

Total reaction at O in vertical direction $\dfrac{W}{2} + w = \left(\dfrac{25}{2} + 3\right) \times 9.81 = 152.06$ N

If T_1 be the tension in the upper arm. Resolving the forces along the axis of spindle, we have

$$T_1 \cos \alpha = 152.06$$

Substituting for cos α from triangle OBN

$$T_1 \left(\frac{0.2}{0.25}\right) = \frac{15.5 \times 9.81}{0.81}$$

Therefore $\qquad\qquad T_1 = \dfrac{15.5 \times 9.81}{0.8} = 190.1$ N $\qquad\qquad$ **Ans.**

EXAMPLE 15.7 A Proell governor has all four arms of length 30 cm long. The upper arms are pivoted on the axis of rotation and the lower arms are attached to the sleeve at a distance of 3.5 cm from the axis. The weight of each ball is 78.5 N and are attached to the extension of the lower links which are 10 cm long. The weight on the sleeve is 588.6 N. The minimum and maximum radii of the governor are 20 cm and 25 cm. Assuming that the extension of the links are parallel to the governor axis at the minimum radius, find the corresponding equilibrium speeds.

(Agra Univ. 1979)

Solution: Referring to Fig. 15.13(a), at minimum equilibrium speed,

$\qquad\qquad OB = BC = 30$ cm; $C_1M_1 = 3.5$ cm; $w = 78.5$ N; $BG = 10$ cm

$\qquad\qquad W = 588.6$ N; $r_1 = 25$ cm; $r_2 = 20$ cm

Thus, $\qquad$ sin$\alpha = BN/OB = 20/30 = 0.6667$

Therefore $\qquad$ $\alpha = 41.81°$; and $\tan \alpha = 0.8944$

Also $\qquad$ sin $\beta = BK/BC = (BN - KN)/BC = (20 - 3.5)/30 = 0.55$

Therefore $\qquad$ $\beta = 33.367°$ and $\tan \beta = 0.65855$

Therefore $\qquad$ $k = \tan \beta/\tan \alpha = 0.736$

From equation (15.17), at minimum speed ω_1, taking moments about I, we have

$$\frac{78.5}{9.81} \times 0.25 \times \omega_1^2 = \frac{BD}{DG}\left[78.5 + \frac{588.6}{2}(1 + 0.736)\right]0.8944$$

where $\qquad\qquad BD = 30 \cos \beta = 25.055$ cm

As the extension arm is vertical at minimum speed,

$\qquad\qquad DG = BD + 10$ cm $= 35.055$ cm

Therefore $\qquad$ $\omega_1^2 = \left(\dfrac{25.055}{35.055}\right)\dfrac{9.81}{78.5 \times 0.25}[78.5 + 294.3(1 + 0.736)]0.8944 = 188.34$

Thus $\qquad$ $\omega_1 = 13.72$ rad/s $\quad$ Hence, $\quad N_1 = 131.05$ r.p.m.

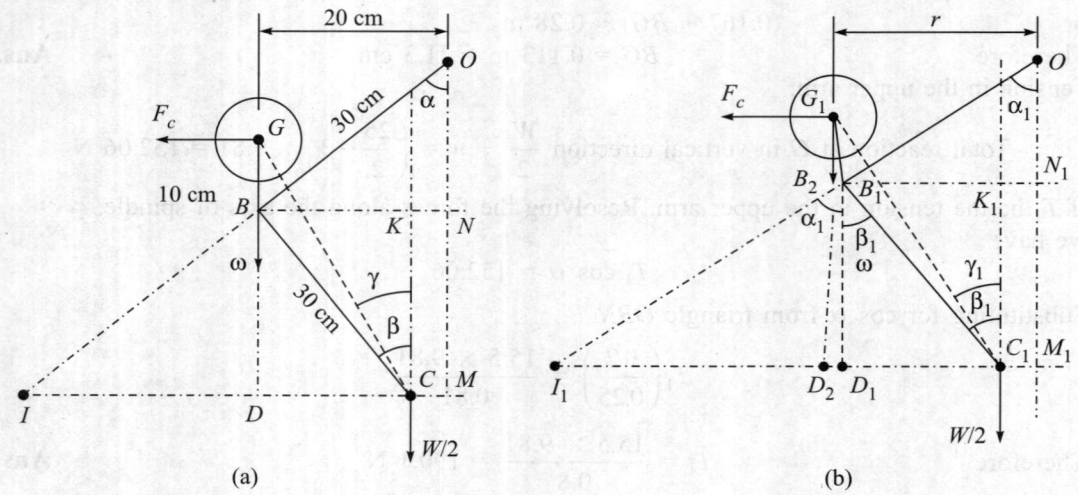

Fig. 15.13 The Proell Governor Geometry.

Let N_2 be the maximum equilibrium speed and let the configuration in Fig. 15.13(b) correspond to this speed. Then

$$B_1N_1 = 25 \text{ cm}$$

and,

$$GC = G_1C_1 = \sqrt{GD^2 + DC^2} = \sqrt{(35.055)^2 + (20 - 3.5)^2}$$

or

$$GC = 38.74 \text{ cm}$$

Also,

$$\tan\gamma = \frac{DC}{DG} = \frac{20 - 3.5}{35.055} = 0.4707$$

Therefore

$$\gamma = 25.2°$$

Also

$$\sin\alpha_1 = \frac{B_1N_1}{OB_1} = \frac{25}{30} = 0.8333$$

Therefore

$$\alpha_1 = 56.44° \text{ and, therefore, } \tan\alpha_1 = 1.507$$

Also

$$\sin\beta_1 = \frac{B_1K_1}{B_1C_1} = \frac{25 - 2.5}{30} = 0.7167$$

Therefore

$$\beta_1 = 45.78°; \quad \text{and} \quad \tan\beta_1 = 1.0276$$

Therefore

$$k_1 = \tan\beta_1/\tan\alpha_1 = 0.682$$

Since angle $(\gamma - \beta)$ remains unchanged from one configuration to the other,

$$\gamma - \beta = \gamma_1 - \beta_1$$

Therefore

$$\gamma_1 = \gamma - \beta + \beta_1 = 25.2 - 33.367 + 45.78 = 37.61°$$

Hence, radius of ball rotation,

$$r = C_1G_1\sin\gamma_1 + 3.5$$

or

$$r = (38.74 \sin 37.61) + 3.5 = 27.14 \text{ cm}$$

Also $\qquad\qquad G_1D_2 = G_1C_1\sin(90 - \gamma_1)$

or $\qquad\qquad\quad G_1D_2 = 38.74 \sin(90 - 37.61) = 30.689$ cm

Also $\qquad\qquad\quad I_1C_1 = I_1D_1 + D_1C_1$

or $\qquad\qquad\qquad I_1C_1 = (B_1D_1 \tan\alpha_1) + B_1C_1 \sin\beta_1$

But $\qquad\qquad\quad B_1D_1 = B_1C_1\cos\beta_1 = 30\cos45.78° = 20.92$

Therefore $\qquad\quad I_1C_1 = 20.92\tan 56.44° + 30 \sin45.78° = 53.03$ cm

Also $\qquad\qquad\quad C_1D_2 = G_1C_1\cos(90 - \gamma_1) = 38.74\sin37.61 = 23.64$ cm

Taking moments of all forces acting on link B_1C_1 at I, we have

$$\left(\frac{W}{g} \times r_2 \times \omega_2^2\right)G_1D_2 = w(I_1D_2) + \frac{W}{2} \times I_1C_1$$

or $\qquad \left(\frac{78.5}{9.81}\right)(0.20)\omega_2^2 \times (0.3069) = 78.5(I_1C_1 - C_1D_2) + \frac{588.6}{2}(0.5303)$

or $\qquad\qquad (0.491)\omega_2^2 = 78.5(0.5303 - 0.2364) + 155.54$

or $\qquad\qquad\qquad \omega_2^2 = \frac{(78.5 \times 0.2939) + 155.54}{0.491} = 363.77$

Therefore, $\qquad\qquad \omega_2 = 19.07$ rad/s

Therefore $\qquad\qquad N_2 = \frac{19.07 \times 60}{2\pi} = 182.1$ r.p.m. $\qquad\qquad$ **Ans.**

15.10 SPRING CONTROLLED GOVERNORS

Rather than using controls through a gravitational pull in gravity controlled governors, a spring force is used, either wholly or partly, to provide controls in a spring controlled governor. This feature becomes interesting particularly when the governor shaft is either horizontal or is inclined to the vertical at some angle. In any of the above cited positions of the governor shaft, only a component of controlling force from gravitational pull is available for controls. Due to this governor loses some of its effectiveness. As against this, a spring force may be used with equal effectiveness for all the positions of governor-shaft axis.

An additional plus point with a spring-controlled governor lies in adjustable initial compression in the spring. The initial compression in spring can be adjusted to ensure any required equilibrium speed for a given ball radius.

15.11 HARTNELL GOVERNOR

A spring controlled governor of the Hartnell type is represented diagrammatically in Fig. 15.14. The sleeve S, along with its attachments, slides on the rotating spindle of the governor, to which a framework, carrying the fulcrum pins of the bell crank levers, are attached. The frame is rigidly attached to the governor spindle and therefore rotates with it at the same speed. Each bell crank lever carries a ball at the end of vertical arm and a roller at the end of horizontal arm. A helical

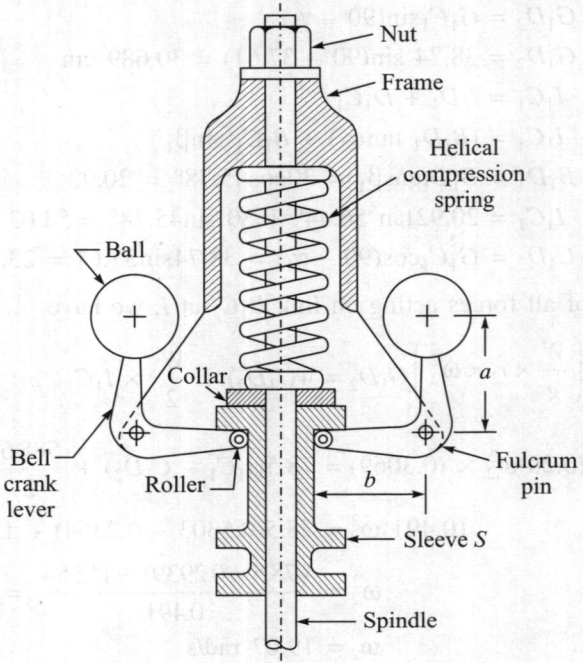

Fig. 15.14 A Hartnell Governor.

compression spring, around governor spindle, provides equal downward force on the two rollers through the collar and sleeve. This downward force on rollers, together with the sleeve weight provides the controlling force, required for rotation of balls in a circular path. Lifting force on the rollers is due to the centrifugal force at the centres of governor balls.

Let S = downward force on sleeve due to the compression of the spring

     W = weight of sleeve

     S_o = downward spring force when angle of rotation α of the vertical arm of bell crank lever with spindle axis is zero degree

     s = stiffness of the spring *i.e.*, the force required to compress the spring by 1 cm

     a = length of vertical arm *i.e.*, the length of the arm of bell crank lever carrying ball

     b = length of the horizontal arm *i.e.*, the arm of the bell crank lever carrying roller

     w = weight of each ball

   r_1, r_2 = maximum and minimum radii of ball rotation

   ω_1, ω_2 = maximum and minimum speeds corresponding to maximum and minimum radii of rotation respectively

     F_1 = centrifugal force on ball at maximum speed = $\dfrac{w}{g} r_1 \omega_1^2$

     F_2 = centrifugal force on ball at minimum speed = $\dfrac{w}{g} r_2 \omega_2^2$

For the configuration shown in Fig. 15.15, taking moments of all external forces, acting on the bell crank lever, about the fulcrum pin O, we have

$$F(a\cos\alpha) + w(a\sin\alpha) = \left(\frac{W+S}{2}\right)b\cos\alpha$$

or

$$\left(\frac{w}{g}r\omega^2\right)(a\cos\alpha) + w(a\sin\alpha) = \left(\frac{W+S}{2}\right)b\cos\alpha \quad (15.18)$$

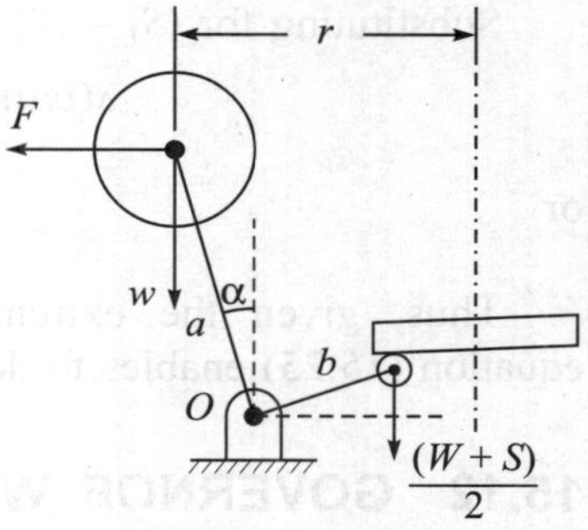

Fig. 15.15

In practice contribution from the weight of governor balls in equation (15.18), namely the term ($wa\sin\alpha$), is quite small in comparison to other components and is neglected. Thus, equation (15.18) reduces to,

$$\left(\frac{w}{g}r\omega^2\right)a\cos\alpha = \left(\frac{W+S}{2}\right)b\cos\alpha$$

Cancelling $\cos\alpha$ term on either side, we have

$$\left(\frac{w}{g}r\omega^2\right)a = \left(\frac{W+S}{2}\right)b \quad (15.19)$$

Using the suffixes 1 and 2 to denote the values at maximum and minimum radius respectively, equation (15.19) may be rewritten for maximum and minimum speed as,

$$(W + S_1) = 2\left(\frac{w}{g}r_1\omega_1^2\right)a/b$$

or

$$(W + S_1) = 2F_1(a/b) \quad (15.20)$$

and

$$(W + S_2) = 2\left(\frac{w}{g}r_2\omega_2^2\right)a/b$$

or

$$(W + S_2) = 2F_2(a/b) \quad (15.21)$$

Subtracting equation (15.21) from (15.20),

$$(S_1 - S_2) = 2(a/b)(F_1 - F_2) \quad (15.22)$$

or

$$(S_1 - S_2) = 2(a/b)\frac{w}{g}(r_1\omega_1^2 - r_2\omega_2^2)$$

For small angle of rotation α of the bell crank lever, the ends of the arms of bell crank lever can be assumed to move along straight lines. Thus, if u be the sleeve lift, corresponding to increase in radius of ball rotation from r_2 to r_1, then

$$\alpha \approx \frac{u}{b} = \frac{(r_1 - r_2)}{a} \quad \text{or} \quad u = (r_1 - r_2)\,b/a \quad (15.23)$$

The increased force of compression $(S_1 - S_2)$ is related to sleeve lift u through

$$S_1 - S_2 = (u)s \equiv s(b/a)(r_1 - r_2) \quad (15.24)$$

Substituting for $(S_1 - S_2)$ in equation (15.22)

$$s(b/a)(r_1 - r_2) = 2(a/b)(F_1 - F_2)$$

or

$$s = 2(a/b)^2\left(\frac{F_1 - F_2}{r_1 - r_2}\right) \tag{15.25}$$

Thus, given the extreme radii of rotation and corresponding equilibrium speeds, equation (15.25) enables to determine required spring stiffness s.

15.12 GOVERNOR WITH SPRING CONNECTED BALLS (WILSON–HARTNELL GOVERNOR)

Referring to Fig. 15.16 the governor is assembled with an initially stretched main spring connecting the two balls which are placed symmetrically with respect to spindle axis. The equilibrium speed for a given ball radius is adjusted using an auxiliary spring attached to the sleeve mechanism. The two bell crank levers are hinged to the bracket. The bracket is keyed to the governor spindle and rotates with it. Rollers at the ends of the horizontal arms of bell crank levers, press against the sleeve. When the speed of rotation rises, radius of ball rotation increases against additional spring tension and the rollers tend to lift the sleeve.

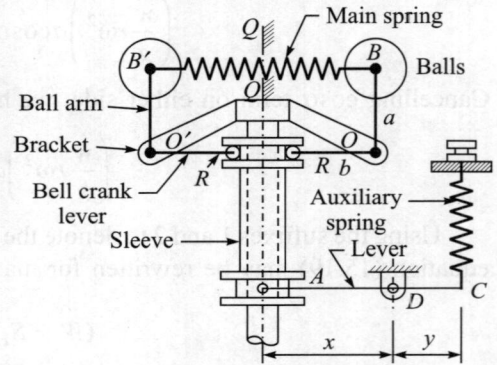

Fig. 15.16 Wilson-Hartnell Governor.

Let w = weight of each governor ball
 W = weight of the sleeve
 P = combined pull of the main spring
 S = pull of the auxiliary spring
 s_a = stiffness of the auxiliary spring
 F = centrifugal force on each ball
 r = radius of the path of ball rotation
 a, b = lengths of ball arm and roller arms respectively of bell crank lever
 s = stiffness of main spring

As the speed increases to ω', each of the ball move out by same amount and the central coil of the main spring therefore does not move either way. The main spring in effect may be considered equivalent to two springs, each fixed at the mid-point in section Q–Q. Thus if s be the stiffness of main spring, then as the number of coils are halved, each of the two equivalent portions of the main spring has a stiffness of $2(s)$.

As the speed increases, the increase in radius of rotation of balls causes further extension in the main spring and the ball movement causes the sleeve to move up and cause elongation in auxiliary spring. If S is the total pull in the auxiliary spring, its equivalent force S' on the sleeve is obtained by taking moments about pivot D. Thus,

$$S \times CD = S' \times AD$$

Therefore

$$S' = S(y/x)$$

Thus, for a pull S in auxiliary spring, total downwards force on the sleeve, $W + S(y/x)$. Again the balls are subjected to centrifugal force F in outward direction and an inward spring force P. Thus, taking moments of all external forces acting on bell crank lever about fulcrum O,

$$(F - P)a = \frac{W + S(y/x)}{2} \times b$$

Using suffixes 1 and 2 to denote quantities at maximum and minimum equilibrium speeds, above equation leads to equations (15.26) and (15.27)

$$(F_1 - P_1)a = \frac{W + S_1(y/x)}{2}b \qquad (15.26)$$

and

$$(F_2 - P_2)a = \frac{W + S_2(y/x)}{2}b \qquad (15.27)$$

Subtracting equation (15.27) from (15.26),

$$\{(F_1 - F_2) - (P_1 - P_2)\}a = \frac{(S_1 - S_2)(y/x)}{2}b \qquad (15.28)$$

Now, as radius of ball path increases from r_2 to r_1, the main spring extends by an amount $2(r_1 - r_2)$. This produces a sleeve movement of

$$u = \frac{(r_1 - r_2)}{a}b$$

and an elongation in auxiliary spring by an amount,

$$v = (u/x)\,y = (r_1 - r_2)(b/a)(y/x)$$

Thus, for an increase in radius of rotation of balls from r_2 to r_1, increase in main spring force and auxiliary spring force are:

$$P_1 - P_2 = (2s)2(r_1 - r_2)$$
$$= 4s(r_1 - r_2)$$

and

$$S_1 - S_2 = s_a(r_1 - r_2)(b/a)(y/x)$$

Substituting for $(P_1 - P_2)$ and $(S_1 - S_2)$ in equation (15.28),

$$(F_1 - F_2)a - 4s(r_1 - r_2)a = \frac{1}{2}s_a(r_1 - r_2)(b^2/a)(y/x)^2$$

Transposing the terms and rearranging,

$$\frac{(F_1 - F_2)}{(r_1 - r_2)} = 4s + \frac{1}{2}s_a(b/a)^2(y/x)^2$$

or

$$4s + \frac{1}{2}(s_a)(b/a)^2(y/x)^2 = \left(\frac{F_1 - F_2}{r_1 - r_2}\right) \qquad (15.29)$$

EXAMPLE 15.8 A governor of the Hartnell type has equal balls of weight 29.43 N set initially at a radius of 20 cm. The arms of the bell crank lever are 12 cm vertically and 15 cm horizontally.

Find: (a) the initial compressive force on the spring, if the speed for an initial ball radius of 20 cm is 240 r.p.m., and (b) the stiffness of the spring required to permit a sleeve movement of 0.4 cm on a fluctuation of 7 per cent in the engine speed.

Solution: Given: $w = 29.43$ N; $r_2 = 20$ cm $a = 12$ cm $b = 15$ cm $N_2 = 240$ r.p.m.
  Sleeve movement $h = 0.4$ cm

Now,
$$\omega_2 = \frac{2\pi \times 240}{60} = 8\pi \text{ rad/s}$$

$$\omega_1 = (1.07)\omega_2 = (8.56\pi) \text{ rad/s}$$

(a) Let S_2 be the initial compressive force in spring. Thus, for minimum speed of rotation ω_2,

$$\frac{(W + S_2)}{2} b = F_2 \times a$$

As weight of sleeve is not mentioned, let $W = 0$. Then from above,

$$S_2 = 2(a/b)\frac{w}{g}r_2\omega_2^2$$

or,
$$S_2 = 2(0.12/0.15)\frac{29.43 \times 0.2}{9.81}(8\pi)^2 = 606.4 \text{ N} \qquad \textbf{Ans.}$$

(b) Let r_1 be the radius of ball rotation when the speed is 7% higher *i.e.*, at (8.56π) rad/s. Then, for a sleeve movement of 0.4 cm,

$$\frac{0.4}{b} = \frac{(r_1 - 20)}{a}$$

or
$$r_1 = 20 + \left(\frac{a}{b}\right) \times 0.4 = 20 + \left(\frac{12}{15}\right) \times 0.4 = 20.32 \text{ cm}$$

Finally, using equation (15.25),

$$s = 2(a/b)^2 \frac{F_1 - F_2}{r_1 - r_2}$$

or,
$$s = 2\left(\frac{12}{15}\right)^2 \frac{w}{g}(r_1\omega_1^2 - r_2\omega_2^2)/(r_1 - r_2)$$

or
$$s = 2\left(\frac{12}{15}\right)^2 \frac{29.43}{981}\{(20)(8.56\pi)^2 - (20.32)(8\pi)^2\}/(20.32 - 20)$$

or
$$s = 195.4 \text{ N/cm} \qquad \textbf{Ans.}$$

EXAMPLE 15.9 A Hartnell governor having a central sleeve spring and two right angled bell crank levers, operates between 290 r.p.m. and 310 r.p.m. for a sleeve lift of 1.5 cm. The sleeve arms and the weight arms are 8 cm and 12 cm respectively. The levers are pivoted at 12 cm from the governor axis and each ball weighs 24.5 N. The weight arms are parallel to governor axis at the lowest equilibrium speed. Determine (a) loads on the spring at the lowest and the highest equilibrium speeds and (b) stiffness of the spring. (DAV Indore, Sept. 1990 and Aug. 1992)

Solution: Given: $w = 24.5$ N, $a = 12$ cm $b = 8$ cm

$$\omega_1 = \frac{2\pi \times 310}{60} = 32.46 \text{ rad/s}$$

$$\omega_2 = \frac{2\pi \times 290}{60} = 30.369 \text{ rad/s}$$

The radius of balls at lowest equilibrium speed ω_2, is $r_2 = 12$ cm
The radius at highest equilibrium speed ω_1,

$$r_1 = r_2 + \frac{1.5}{8} \times 12 = 12 + 2.25 = 14.25 \text{ cm}$$

The centrifugal forces F_1 and F_2 and maximum and minimum equilibrium speeds are:

$$F_1 = \frac{24.5}{981} \times 14.25 \times (32.46)^2 = 374.98 \text{ N}$$

$$F_2 = \frac{24.5}{981} \times (12) \times (30.369)^2 = 276.4 \text{ N}$$

For the lowest equilibrium speed, moment of external forces on bell crank lever about fulcrum gives

$$F_2 \times 12 = \frac{S_2}{2} \times 8$$

or $\qquad\qquad S_2 = \frac{24}{8} \times F_2 = 829.2$ N $\qquad\qquad$ **Ans.**

Similarly taking moments of forces at highest equilibrium speed about fulcrum,

$$F_1 \times a = \frac{S_1}{2} \times b \text{ (neglecting contribution of the weight of ball)}$$

Therefore $\qquad S_1 = 2F_1(a/b) = 3F_1 = 1124.94$ N $\qquad\qquad$ **Ans.**

Also, the stiffness of spring

$$s = 2(a/b)^2 \frac{F_1 - F_2}{r_1 - r_2}$$

or, $\qquad\qquad s = 2\left(\frac{12}{8}\right)^2 \frac{374.98 - 276.4}{14.25 - 12.0} = 197.16$ N/cm $\qquad$ **Ans.**

EXAMPLE 15.10 A spring loaded governor of the Hartnell type has arms of equal lengths. The weights rotate in a circle of 13 cm diameter when the sleeve is in the mid-position and the weight arms are vertical. The equilibrium speed for this position is 450 r.p.m., neglecting friction. The maximum sleeve movement is to be 2.5 cm and the maximum variation of speed, taking friction into account, is to be ± 5% of the mid-position speed. The weight of the sleeve is 39 N and the friction may be considered equivalent to 29 N at the sleeve. The power of the governor must be sufficient to overcome the friction by a one per cent change of speed either way at mid-position. Determine, neglecting obliquity effect of arms:

(a) Weight of each rotating mass
(b) Spring stiffness in N/cm
(c) Initial compression of spring   (AMIE, Winter 1982)

Solution: Mean equilibrium speed $\omega = \dfrac{2\pi \times 450}{60} = 47.12$ rad/s

With ball arm vertical at this speed, $r = \dfrac{13}{2} = 6.5$ cm

Since the ball arm and sleeve arm have equal lengths,
Max. sleeve lift = difference in radii $(r_1 - r_2) = 2.5$ cm

Hence $r_1 = r + \dfrac{2.5}{2} = 6.5 + 1.25 = 7.75$ cm

                                   $r_2 = r - \dfrac{2.5}{2} = 6.5 - 1.25 = 5.25$ cm

Let S be the spring force on sleeve at the mean radius. For being able to overcome friction by one per cent change of speed either way at mid-position,

$$\frac{w}{981} \times 6.5 \times (47.12 \times 1.01)^2 \times a = \left(\frac{S + 39 + 29}{2}\right) \times b$$

and $$\frac{w}{981} \times 6.5 \times (47.12 \times 0.99)^2 \times a = \left(\frac{S + 39 - 29}{2}\right) \times b$$

(a) For equal arms, substituting $a = b$, above equations reduce to

$$(15.0)w = (S + 68)/2$$
and $$(14.419)w = (S + 10)/2$$
Solving simultaneously, $$w = 50 \text{ N}$$ **Ans.**

(b) Maximum equilibrium speed, $\omega_1 = 1.05 \times 47.12 = 49.476$ rad/s
Minimum equilibrium speed, $\omega_2 = 0.95 \times 47.12 = 44.764$ rad/s

The highest equilibrium speed is obtained when frictional force adds the spring force S_1. Thus,

$$\left(\frac{S_1 + 39 + 29}{2}\right) \times b = \frac{50}{981} \times 7.75 \times (49.476)^2 \times a$$

Cancelling a and b on either side,
$$S_1 = 1933.8 - 68 = 1865.8 \text{ N}$$
Also, from

$$\left(\frac{S_2 + 39 - 29}{2}\right) \times b = \frac{50}{981} \times 5.25 \times (44.764)^2 \times a$$

Cancelling a and b on either side,

$$S_2 = 1072.38 - 1 = 1062.38 \text{ N}$$

Therefore spring stiffness, $\quad s = \left(\dfrac{1865.8 - 1062.38}{2.5} \right) = 321.37 \text{ N/cm}$

(c) Initial compression in the spring,

$$P = \frac{S_2}{s} = \frac{1062.38}{321.37} = 3.3 \text{ cm} \qquad \qquad \textbf{Ans.}$$

EXAMPLE 15.11 The Hartnell type spring loaded governor shown in Fig. 15.17 has two balls of 49 N each, which revolve in a circle of 40 cm diameter when the sleeve is at mid-travel. The total movement of the sleeve is 2.5 cm, which is the same as that of the balls. An adjustable load is applied to the sleeve by means of a spring B attached to the centre of a lever moving about a fixed centre C in the manner shown, the tension in B being adjusted by hand. When there is no tension in spring B, the sleeve just commences to rise from its lowest position at a speed of 200 r.p.m. and reaches the upper limit of its travel at 208 r.p.m. Find the stiffness of spring A and also of the spring B, which

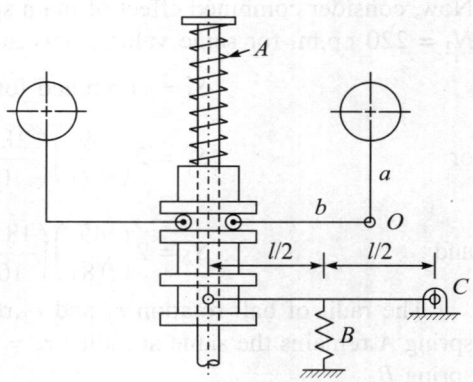

Fig. 15.17 A typical Hartnell Governor.

will make the two limiting speeds 220 and 232 r.p.m. (London University)

Solution: Given, radius of ball rotation at mid-travel $r = 20$ cm
Total sleeve movement $s = 2.5$ cm and $(r_1 - r_2)_{max} = 2.5$ cm

Hence $\qquad\qquad\qquad\qquad\qquad a = b$

Thus, $\qquad\qquad\qquad\qquad\qquad r_1 = 20 + \dfrac{2.5}{2} = 21.25$ cm

and, $\qquad\qquad\qquad\qquad\qquad r_2 = 20 - \dfrac{2.5}{2} = 18.75$ cm

When there is no tension in spring B, taking moments of forces about O, when the sleeve just commences to rise, from lowest position,

$$\left(\frac{S_1}{2} \right) b = F_1 \times a$$

and $\qquad\qquad\qquad\qquad\qquad \left(\dfrac{S_2}{2} \right) b = F_2 \times a$

Therefore, as $a = b$, $\qquad\qquad\qquad S_1 = 2F_1$
and $\qquad\qquad\qquad\qquad\qquad S_2 = 2F_2$
Thus, from $\qquad\qquad\qquad\qquad S_1 = 2F_1,$

$$S_1 = 2 \cdot \left(\frac{49}{9.81}\right)\left(\frac{21.25}{100}\right)\left(\frac{2\pi \times 208}{60}\right)^2 = 1007.16 \text{ N}$$

and $\qquad S_2 = 2F_2$

or, $\qquad S_2 = 2 \times \left(\frac{49}{9.81}\right)\left(\frac{18.75}{100}\right)\left(\frac{2\pi \times 200}{60}\right)^2 = 821.629 \text{ N}$

Therefore stiffness of spring $A = \left(\dfrac{1007.16 - 821.629}{2.5}\right) = 74.21$ N/cm **Ans.**

Now, consider combined effect of main spring A and auxiliary spring B for $N_1 = 232$ r.p.m. and $N_2 = 220$ r.p.m. for same values of r_1 and r_2.

$\qquad S_1' =$ combined force due to main and auxiliary springs

or $\qquad S_1' = 2\left(\frac{49}{9.81}\right)\left(\frac{21.25}{100}\right)\left(\frac{2\pi \times 232}{60}\right)^2 = 1252.99 \approx 1253 \text{ N}$

and $\qquad S_2' = 2\left(\frac{49}{9.81}\right)\left(\frac{18.75}{100}\right)\left(\frac{2\pi \times 220}{60}\right)^2 = 994.17 \text{ N}$

The radii of ball rotation r_1 and r_2 remaining the same, the spring force on sleeve due to spring A remains the same at radius $r_1 = 21.25$ cm and $r_2 = 18.75$ cm. Hence, initial tension in spring B

$$S_0 = (S_2' - S_2) \times \frac{l}{(l/2)}$$

or $\qquad S_0 = (994.17 - 821.629) \times 2 = (172.541 \times 2) = 345.08 \text{ N}$ **Ans.**

Similarly, final tension in spring B,

$$S_0' = (S_1' - S_1) \times \frac{l}{(l/2)} = 2(1253 - 1007.16) = (2 \times 246) = 492 \text{ N}$$

Therefore, stiffness of spring $B = \dfrac{492 - 345}{2.5\left(\dfrac{l/2}{l}\right)} = \dfrac{147 \times 2}{2.5} = 117.6$ N/cm **Ans.**

EXAMPLE 15.12 In a spring loaded governor of the Hartnell type, the weight of each ball is 49 N and the lift of the sleeve is 5 cm. The speed at which the governor begins to float (lowest speed) is 240 r.p.m. and at this speed the radius of the ball path is 11 cm. The mean working speed of the governor is 20 times the range of the speed when friction is neglected. If the length of ball and roller arms of the bell crank lever are 12 cm and 10 cm respectively and if the distance between the centre of pivot of the bell crank lever and axis of governor spindle is 14 cm, determine the initial compression of the spring, taking into account obliquity of the arms.

If friction is equivalent to a force of 29.4 N at the sleeve, find the total alteration in the speed before the sleeve begins to move from mid-position.

(AMIE, Summer 1979); (SGSITS, Indore, Feb. 1994)

Solution: Let ω be the mean equilibrium speed and $\pm\Delta\omega$ represent maximum variations in ω (Fig. 15.18). Then

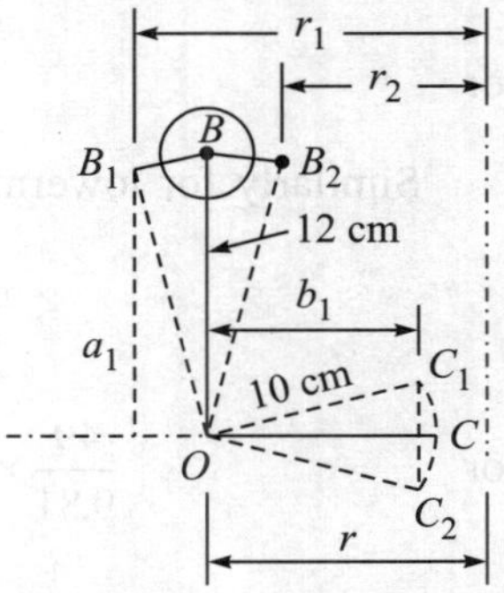

$$\text{maximum speed } \omega_1 = (\omega + \Delta\omega)$$
$$\text{minimum speed } \omega_2 = (\omega - \Delta\omega)$$

Thus, range of speed $(\omega_1 - \omega_2) = 2\Delta\omega$
Then $\omega = 20(2\Delta\omega)$

or

$$\Delta\omega = \left(\frac{\omega}{40}\right)$$

Therefore

$$\omega_1 = \omega + \frac{\omega}{40} = \left(\frac{41}{40}\right)\omega$$

and,

$$\omega_2 = \omega - \frac{\omega}{40} = \left(\frac{39}{40}\right)\omega$$

But

$$N_2 = 240 \text{ r.p.m.}$$

Therefore

$$\omega_2 = \frac{2\pi \times 240}{60} = 8\pi \text{ rad/s}$$

Thus, equating, $(39/40)\omega = 8\pi$

or

$$\omega = \frac{40 \times 8\pi}{39} = 25.78 \text{ rad/s}$$

and

$$\omega_1 = \frac{41}{40}\omega = \frac{41 \times 25.78}{40} = 26.42 \text{ rad/s}$$

At the lowest equilibrium speed ω_2, the radius of ball rotation $r_2 = 11$ cm. Also, assuming the ball arm to be vertical at mean equilibrium speed ω,

$$r = 14 \text{ cm}$$

Therefore

$$r_1 = r_2 + \frac{x}{10} \times 12 = 11 + \frac{5 \times 12}{10} = 17 \text{ cm}$$

For considering obliquity effect of arms, from Fig. 15.18,

$$a_1 = \sqrt{12^2 - (17-14)^2} = 11.62 \text{ cm}$$

and

$$b_1 = \sqrt{10^2 - (5/2)^2} = 9.68 \text{ cm}$$

Thus taking moments about fulcrum pin O of all the forces on lever at the highest equilibrium speed,

$$F_1 \times a_1 + w(r_1 - r) = \left(\frac{S_1}{2}\right)(b_1) \tag{i}$$

or

$$\frac{49}{9.81} \times \left(\frac{17}{100}\right)(26.42)^2 \times \frac{11.62}{100} + 49\left(\frac{3}{100}\right) = \left(\frac{S_1}{2}\right) \times \frac{9.68}{100}$$

Fig. 15.18

or
$$S_1 = \frac{200}{9.68}(70.343) = 1453.4 \text{ N}$$

Similarly for lowermost position (due to symmetry)

$$F_2 \times a_1 - w(r - r_2) = \left(\frac{S_2}{2}\right)b_1$$

or
$$\frac{49}{9.81} \times \left(\frac{11}{100}\right)(8\pi)^2 \times \frac{11.62}{100} - 49\left(\frac{3}{100}\right) = \left(\frac{S_2}{2}\right) \times \frac{9.68}{100}$$

or
$$S_2 = \frac{200}{9.68}(38.858) = 802.85 \text{ N}$$

Hence, stiffness of spring $= \dfrac{S_1 - S_2}{\text{lift}}$

Thus
$$s = \frac{(1453.4 - 802.85)}{5} = 130.1 \text{ N/cm}$$

and initial compression $= \dfrac{S_2}{s} = \dfrac{802.85}{130.1} = 6.17 \text{ cm}$ **Ans.**

Also, spring force at mid-position of sleeve,

$$S' = S_2 + \left(\frac{5}{2}\right) \times s = 802.85 + (2.5) \times 130.1 = 1128.1 \text{ N}$$

Also as $\qquad \Delta\omega = \omega/40$

Therefore $\qquad \Delta N = N/40 = 240/40 = 6 \text{ r.p.m.}$

$\qquad N_1 = 240 + 6 = 246 \text{ r.p.m.}$

and $\qquad N_2 = 240 - 6 = 234 \text{ r.p.m.}$

Finally, neglecting contribution of the weight of balls in equation (i), it follows that

$$\omega^2 = \left(\frac{S}{2}b\right)\frac{g}{w}\,r^2 a_1$$

or $\qquad \omega \propto \sqrt{S} \quad i.e., \quad N \propto \sqrt{S}$

Thus, with a friction force of 29.4 N at the sleeve, the higher speed

$$N' = 246\sqrt{\frac{1128.1 + 29.4}{1128.1}} = 249.2 \text{ r.p.m.}$$ **Ans.**

and
$$N'' = 246\sqrt{\frac{1128.1 - 29.4}{1128.1}} = 242.8 \text{ r.p.m.}$$ **Ans.**

The alternation of speed from (249.2 to 242.8) r.p.m. = 6.4 r.p.m **Ans.**

EXAMPLE 15.13 Following particulars relate to a Wilson-Hartnell governor:

Weight of each ball = 19.0 N; minimum radius of ball rotation = 12.5 cm; maximum radius = 17.5 cm; minimum speed = 240 r.p.m.; maximum speed = 252 r.p.m.; length of the ball arm of each bell crank lever = 15 cm; length of the sleeve arm of each bell crank lever = 10 cm and combined stiffness of the two ball springs = 1.96 N/cm. Find the equivalent stiffness of the auxiliary spring referred to the sleeve.

Solution: Given: $w = 19.0$ N $r_2 = 12.5$ cm $r_1 = 17.5$ cm

$$N_1 = 252 \text{ r.p.m.}   N_2 = 240 \text{ r.p.m.}   a = 15 \text{ cm}$$

$$b = 10 \text{ cm}   s = 1.96 \text{ N/cm}   s_a = ?$$

$$\omega_1 = \frac{2\pi \times 252}{60} = 26.389 \text{ rad/s}$$

$$\omega_2 = \frac{2\pi \times 240}{60} = 8\pi \text{ rad/s}$$

Now

$$F_1 = \frac{19}{9.81} \times \left(\frac{17.5}{100}\right) \times (26.389)^2 = 236.0 \text{ N}$$

and,

$$F_2 = \frac{19}{9.81} \times \left(\frac{12.5}{100}\right)(8\pi)^2 = 152.92 \text{ N}$$

Hence, from equation (15.29),

$$4s + \frac{1}{2}(s_a)(b/a)^2(y/x)^2 = \left(\frac{F_1 - F_2}{r_1 - r_2}\right)$$

or

$$s_a(y/x)^2 = \left\{\left(\frac{236 - 152.92}{17.5 - 12.5}\right) - 4(1.96)\right\} \times 2\left(\frac{15}{10}\right)^2 = 39.492 \text{ N/cm}$$ **Ans.**

EXAMPLE 15.14 In a Wilson-Hartnell type of governor, the two springs attached directly to the balls, each has a stiffness of 7.36 N/cm, and a free length of 10 cm. The weight of each ball is 39 N, the length of the ball arm of each bell crank lever is 8 cm and that of the sleeve arm is 6 cm; the lever *BOR* is pivoted at its mid-point. When the radius of rotation of balls is 8 cm, the equilibrium speed is 240 r.p.m. If the sleeve is to lift 0.75 cm, for an increase of speed of 5 per cent, determine the required stiffness of the auxiliary spring.   (Manchester Un.)

Solution: Given:

$$\omega_2 = \frac{2\pi \times 240}{60} = 8\pi \text{ rad/s}$$

$$\omega_1 = 1.05(\omega_2) = 8.4\pi \text{ rad/s}$$

$$F_2 = \left(\frac{39}{9.81}\right)\left(\frac{8}{100}\right)(8\pi)^2 = 200.89 \text{ N}$$

For a sleeve lift of 0.75 cm, increase in ball radius = $\frac{0.75}{6} \times 8 = 1.0$ cm

Therefore $\qquad$ $r_1 = 8 + 1 = 9$ cm

Therefore $\qquad$ $F_1 = \dfrac{39}{9.81} \times \left(\dfrac{9}{100}\right)(8.4\pi)^2 = 249.17$ N

Using equation (15.29),

$$4s + \frac{1}{2}(s_a)(b/a)^2(y/x)^2 = (F_1 - F_2)/(r_1 - r_2)$$

or $\qquad$ $4(7.36) + \dfrac{1}{2}(s_a)(6/8)^2(y/x)^2 = (249.17 - 200.89)/(9 - 8)$

or $\qquad$ $(s_a)(y/x)^2 \times 0.28125 = 18.84$

or $\qquad$ $(s_a)(y/x)^2 = 66.99$

and as $\qquad$ $(y/x) = 1$

$\qquad\qquad s_a = 67$ N/cm $\qquad\qquad\qquad$ **Ans.**

15.13 GOVERNOR WITH GRAVITY AND SPRING CONTROL

As shown in Fig. 15.19, the governor consists of a close-coiled helical spring, accommodated between the hollow space in the sleeve (around governor spindle) and a cap. The sleeve carries brackets to which two bell crank levers are hinged. One end of the bell crank lever supports governor balls while the other end carries roller which presses against the cap fixed to the spindle. When speed of rotation increases and the radius of ball rotation also increases, the rollers press the cap and lift the sleeve against the force of compression in spring.

The force analysis can be done most conveniently by considering the instantaneous centre I of bell crank lever 3 and taking moments of all the forces acting on link 3.

Let W = weight of the sleeve

$\quad\ S$ = compressive spring force on sleeve together with any other force

$\quad\ w$ = weight of governor ball

$\quad\ r$ = radius of ball rotation

$\quad\ \omega$ = equilibrium speed

The centrifugal force F, ball weight w and the force $(W + S)/2$ on sleeve provide the following moment equation:

$$F(BD) = w(DI) + \left(\frac{W+S}{2}\right)(AI)$$

or $\qquad\qquad F(BD) = w(DA + AI) + \left(\frac{W+S}{2}\right)(AI) \qquad\qquad (15.30)$

EXAMPLE 15.15 A governor of gravity and spring control type, shown in Fig. 15.19(a), has the following data:

Weight of each ball = 15 N, weight of load on sleeves, inclusive of its own weight = 49 N, distance of fulcrum from axis of rotation = 5 cm, length of vertical arm = 10 cm, length of horizontal arm = 4 cm.

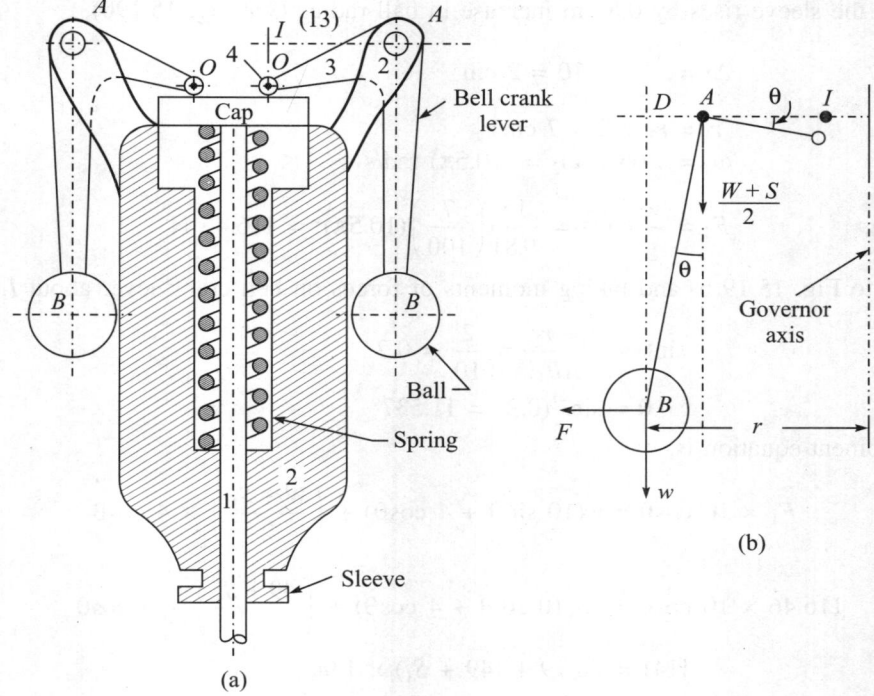

Fig. 15.19 Governor with Gravity and Spring Control.

The speed at which sleeve begins to rise = 300 r.p.m. The rise of sleeve for 5% increase of speed = 0.8 cm

Find the initial thrust in the spring and its stiffness.

Solution: $$\omega_2 = \frac{2\pi \times 300}{60} = 10\pi \text{ rad/s}$$

Therefore $$F_2 = \frac{w}{g} r_2 \omega_2^2 = \frac{15}{9.81} \times \left(\frac{5}{100}\right)(10\pi)^2 = 75.45 \text{ N}$$

Since the ball arm AB is vertical at this position, taking moments of forces about I,

$$F_2 \times 10 = \left(w + \frac{W+S}{2}\right) \times 4$$

or $$75.45 \times 10 = \left(15 + \frac{49 + S_2}{2}\right) \times 4$$

or $$S_2 = 2\left\{\frac{75.45 \times 10}{4} - 15\right\} - 49 = 298.25 \text{ N}$$

Hence, initial thrust in the spring = 298.25 N **Ans.**

When the sleeve rises by 0.8 cm increase in ball radius (*see* Fig. 15.19b),

$$\Delta r = \frac{0.8}{4} \times 10 = 2 \text{ cm}$$

Therefore $\qquad r_1 = r_2 + 2 = 7 \text{ cm}$

Also $\qquad \omega_1 = 1.05 \times \omega_2 = (10.5\pi) \text{ rad/s}$

Therefore $\qquad F_1 = \dfrac{w}{g} r_1 \omega_1^2 = \dfrac{15}{9.81}\left(\dfrac{7}{100}\right)(10.5\pi)^2 = 116.46 \text{ N}$

Referring to Fig. 15.19(b) and taking moments of forces on bell crank lever about I.

$$\sin\theta = \frac{r_1 - r_2}{AB} = \frac{2}{10} = 0.2$$

Therefore $\qquad \theta = \sin^{-1}(0.2) = 11.537°$

Hence, moment equation is,

$$F_1 \times 10 \cos\theta = w(10 \sin\theta + 4 \cos\theta) + \left(\frac{W + S_1}{2}\right) \times 4 \cos\theta$$

or $\qquad 116.46 \times 10 \cos\theta = 15(10 \sin\theta + 4 \cos\theta) + \left(\dfrac{49 + S_1}{2}\right) \times 4 \cos\theta$

or $\qquad 1141 = 88.79 + (49 + S_1) \times 1.96$

Therefore $\qquad S_1 = 487.84 \text{ N}$

Therefore, stiffness of spring $= \dfrac{S_1 - S_2}{0.8}$

or, $\qquad$ Spring stiffness $= \dfrac{487.84 - 298.25}{0.8} = 237 \text{ N/cm}$ $\qquad$ **Ans.**

15.14 HARTUNG GOVERNOR

The ends of vertical arm of the bell crank lever *BAC* (*see* Fig. 15.20) are fitted with springs which are compressed against the frame of the governor, when the rollers at the other end of bell crank lever press against the sleeve.

Taking moments about the hinge *A* of the bell crank lever,

$$(F - S) \times a = W \times b \qquad (15.31)$$

EXAMPLE 15.16 Figure 15.20 shows a spring controlled governor of Hartung type. In this governor the lengths of horizontal and vertical arms of bell crank levers are 10 cm and 8 cm respectively. The distance of the fulcrum of bell crank lever from the axis of the governor is 12 cm. The weight of each revolving mass is 97 N. The stiffness of the spring is 245 N/cm.

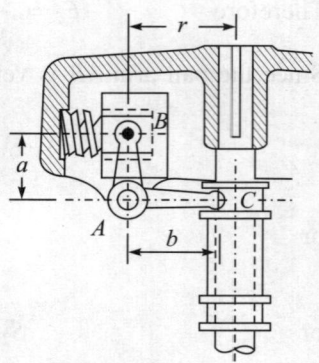

Fig. 15.20

If the length of each spring is 12 cm, when the radius of rotation is 7 cm and the equilibrium speed is 360 r.p.m, find the free length of each spring. Find also the equilibrium speed for the radius of rotation of 12 cm.

Solution: At radius of rotation of 7 cm,

$$\omega = \frac{2\pi \times 360}{60} = 12\pi \text{ rad/s}$$

Therefore centrifugal force $F = \dfrac{w}{g} \times r\omega^2 = \dfrac{97}{9.81} \times \left(\dfrac{7}{100}\right)(12\pi)^2 = 983.7$ N

As the weight of sleeve is not given, it is assumed to be negligible.

With radius of rotation 7 cm and the distance of fulcrum (of bell crank lever) at a distance of 12 cm, the distance of line of action of ball weight from fulcrum

$$m = 12 - 7 = 5 \text{ cm (towards spindle axis).}$$

With the ball arm of the bell crank lever rotated towards governor spindle, the line of action of centrifugal force passes through ball centre, at a distance from fulcrum,

$$p = \sqrt{8^2 - 5^2} = 6.24 \text{ cm}$$

The force of compression in spring acts along spring axis at a height of 8 cm from fulcrum. Taking moments of forces about fulcrum,

$$(983.7) \times 6.24 = 97 \times 5 + S \times 8$$

Therefore

$$S = \frac{(983.7 \times 6.24) - (97 \times 5)}{8} = 706.66 \text{ N}$$

As the spring stiffness = 245 N/cm, the compression in spring,

$$\frac{706.66}{245} = 2.88 \text{ cm}$$

Hence free length of each spring = 12 + 2.88 = 14.88 cm   **Ans.**

Now, when the ball radius becomes 12 cm, additional compression in the spring is

$$12 - 7 = 5 \text{ cm}$$

Total spring force of compression = $S + s(5) = 706.6 + 245(5) = 1931.6$ N

Since fulcrum is already rotating at a radius of 12 cm, it follows that the ball arm is vertical and sleeve arm is horizontal. For this position, weight of ball contributes nothing to the moment at fulcrum. Further, the spring force and centrifugal force act along the same line. Hence, for equilibrium of the ball,

Centrifugal force on ball = spring force

Therefore

$$\frac{w}{g} r\omega^2 = 1931.6$$

or

$$\frac{97}{9.81} \times \left(\frac{12}{100}\right)\omega^2 = 1931.6$$

Therefore $\omega = 40.35$ rad/s

Therefore $N = 385.3$ r.p.m.   **Ans.**

15.15 PICKERING GOVERNOR

Pickering governor is more commonly used in gramophones in clock-work motor device for adjusting the speed of the turntable. It consists of three straight leaf springs each consisting of a single leaf. The three identical leaf springs are symmetrically placed around the governor spindle. Figure 15.21(b) shows the governor with only one leaf spring. A fly-ball, in the form of a disc of weight w, is attached to each leaf spring at the centre. The upper end of each leaf spring is rigidly fixed to the spindle through a hexagonal nut while the lower end of the leaf springs are screwed down to a sleeve through a hexagonal nut and can move up and down the spindle.

When speed of rotation rises, increased centrifugal force causes deflection δ in the leaf spring. The c.g. of the fly-ball moves outward, causing the sleeve to move up as shown in Fig. 15.21(c) through a distance x,

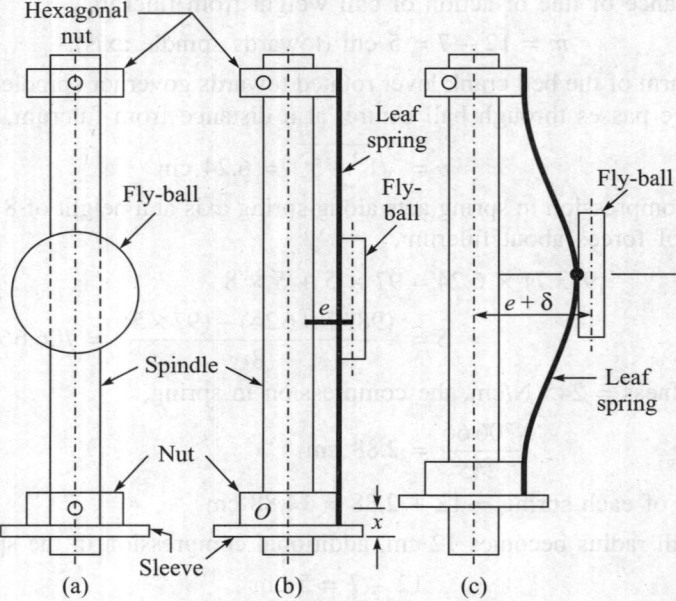

Fig. 15.21 A Pickering Governor.

Let e be the distance of the c.g. of the fly-ball from the spindle axis in rest position, and δ be the deflection of c.g. of the fly-ball at the equilibrium speed ω. Let x be the corresponding sleeve movement at angular speed ω.

The centrifugal force acting at the c.g. of the fly-ball,

$$F = \frac{w}{g}(e + \delta)\omega^2 \text{ N} \tag{15.32}$$

The leaf of the leaf spring can be treated as a fixed beam of length l and of uniform section with flexural moment of inertia I and modulus of elasticity E. Then,

$$\delta = Fl^3/(192EI) \tag{15.33}$$

or

$$\delta = \frac{w}{g}(e + \delta)\ \omega^2 l^3/(192\ EI)$$

Also, the lift x of the sleeve of governor is related to deflection δ and length l by approximate relation,

$$x \approx 2.4\delta^2/l \qquad (15.34)$$

15.16 INERTIA GOVERNORS

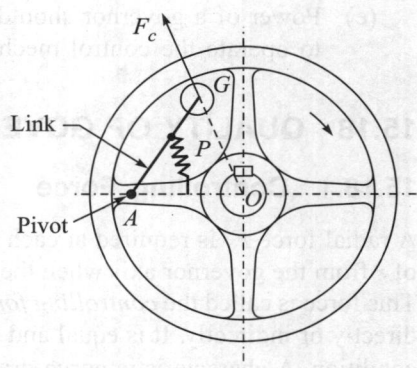

Fig. 15.22 The Inertia Governor.

Elements of an inertia governor are illustrated in Fig. 15.22. The operation of this form of governor is based on the existence of inertia forces which come into effect the moment there is some angular acceleration/retardation. In the case of inertia governors, angular accelerations and retardations provide the necessary inertia couple. As these governors operate by sensing rate of change of angular velocity, rather than change in angular velocity, they are quicker in response. The figure shows a weight w pivoted at point A on the arm of a fly- wheel through a link. As the flywheel rotates, the weight G tends to move out due to centrifugal action. The action of centrifugal force F_c is opposed by the tension in the spring. The pivot A is to be selected such that the three points O, A and G are not collinear. The end of the link is connected to an eccentric, which operates the fuel supply valve of the prime mover.

Let v = velocity of G, the c.g. of the weight w
r = radial distance of G from O
e = perpendicular distance of point A from the line OG

Then, the centrifugal force on the weight,

$$F_c = (w/g)v^2/r$$

and the moment of F_c about A,

$$C = F_c e = (w/g)(v^2/r)e \qquad (15.35)$$

As the prime mover begins to accelerate, the weight w is not accelerated at the same instant, but lags behind owing to inertia. The inertia force acting on the weight in a direction perpendicular to the radial line OG is given by

$$F_1 = (w/g)dv/dt \qquad (15.36)$$

The moments due to centrifugal force and inertia force add up to provide a rapid action to the governor,

15.17 CHARACTERISTICS OF CENTRIFUGAL GOVERNORS

For the governing process to be satisfactory, a centrifugal governor should possess following qualities.

(a) When due to sudden load on prime mover, the speed drops down and sleeve reaches the lowermost position, the prime mover should develop maximum power
(b) When the load on the prime mover drops down suddenly and speed tends to shoot up, the sleeve should at once reach the top most position
(c) Under normal working condition when the load on the engine fluctuates between reasonable limits, the governor sleeve should float at some intermediate position
(d) A governor should respond rapidly to a change of speed
(e) Power of a governor should be enough so as to exert a force on the sleeve, sufficient to operate the control mechanism.

15.18 QUALITY OF GOVERNOR: DEFINITIONS

15.18.1 Controlling Force

A radial force F' is required at each ball so as to maintain the balls at a constant radial distance of r from the governor axis when the governor is either stationary or rotating at a uniform speed. This force is called the *controlling force* and is exerted in radially inward direction on balls, either directly or indirectly. It is equal and opposite to the centrifugal (inertia) force under equilibrium condition. A characteristic curve drawn to show how the pull F' (*i.e.,* controlling force) varies with the radius r of rotation is called controlling force curve. In certain types of governors, controlling force curves become straight lines. Stability and sensitivity aspect of governors can be examined using controlling force curves.

Constant Speed Lines

Being equal and opposite to the centrifugal force F at equilibrium speed ω, the controlling force is numerically given by

$$F' = (w/g)r\omega^2 \tag{15.37}$$

or

$$\omega = \{(g/w)(F'/r)\}^{1/2}$$

or,

$$\omega = \{(g/w)(\tan\phi)\}^{1/2} \tag{15.38}$$

where ϕ is the angle of inclination to the r-axis of a line joining a given point on the controlling force curve to the origin.

It follows from equation (15.37) that,

$$(F'/r) = (F/r) = (w/g)\omega^2 = CN^2 \tag{15.39}$$

Thus, the ratio (F'/r) remains constant for a given equilibrium speed and equation (15.39) therefore represents a constant speed line. In Fig. 15.23, a number of such lines, representing equilibrium speeds ω_0, ω_1, ω_2, etc have been drawn from O, each one cutting controlling force curve PQR at one point. *These constant speed lines also represent the centrifugal-force characteristics at constant speed.* Each such intersection point with controlling force curve represents a radius of ball path for a given equilibrium speed.

As explained in Section 15.5 variations in the controlling force F' with the radii of rotation of the ball path, can be studied under static condition also. **Thus by freezing the rotational motion (thereby eliminating centripetal acceleration and corresponding force), the controlling force can be estimated by measuring the outward radial force required to be applied on**

the balls to keep then at a specified radius of rotation. This can be repeated for all the ball radii in the range, so as to establish controlling force F' as a function of r. Thus

$$F' = f(r)$$

Figure 15.23 illustrates two controlling force curves: the curve PQR corresponding to a stable governor and curve ABC corresponding to an unstable governor. As explained in preceding paragraph, the controlling force curve in Fig. 15.23 is derived from purely static considerations.

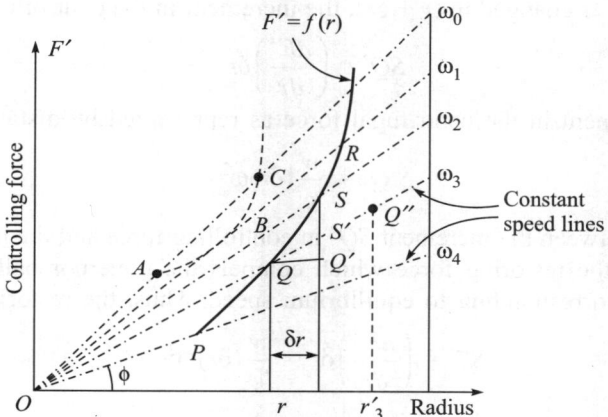

Fig. 15.23 Constant speed lines and controlling force curve.

Suppose now that the governor balls are rotating with a speed of ω at a radius of rotation of r. Then, the centrifugal force exerted on the balls, rotating at radius r, is

$$F = (w/g)r\omega^2$$

Clearly, the plot of centrifugal force v/s radius for a given speed ω_3 is a straight line passing through the origin as shown by line OS' in Fig. 15.23. Point of intersection of this constant speed line with the controlling force curve, namely the point Q, determines the equilibrium radius r. *Only at this point on controlling force curve, the centrifugal force equals the controlling force.* Mathematically,

$$F' = F = (w/g)r\omega_3^2$$

or

$$(F/r) = \left(\frac{w}{g}\right)\omega_3^2 \qquad (15.40)$$

Clearly for a governor to be stable (and consistent in operation) in the working range, it should have only one radius of rotation for a given equilibrium speed. In other words, a centrifugal force line must cut the controlling force curve only at one point. While the curve PQR in Fig. 15.23 satisfies this condition, the controlling force curve ABC does not satisfy the condition.

15.18.2   Stability and Isochronism

A governor is said to be stable over its working range when, for a given equilibrium speed of rotation, say ω_3 in the range, there is only one radius of ball rotation at which the governor is in equilibrium. Physically, this implies that for a given uniform speed of rotation ω_3, if the configuration is disturbed through a small incremental force on sleeve temporarily, the balls may rotate at same speed ω_3 but at a radius say r'_3 instead of r as indicated by point Q''. But the moment this additional force is withdrawn from the sleeve, the balls will start rotating at the same old radius r, as decided by the controlling force curve for equilibrium speed ω_3.

To express the above condition mathematically, let the equilibrium position of governor balls, for equilibrium speed ω_3, be defined by point Q in Fig. 15.23. Now keeping speed the same, if the radius of ball rotation is changed to $(r + \delta r)$, the increment in the controlling force is given by

$$SQ' = \left(\frac{dF'}{dr}\right)\delta r$$

Corresponding increment in the centrifugal force as represented by distance $S'Q'$ is,

$$S'Q' = \frac{w}{g}(\delta r)\omega_3^2$$

The difference between the increment SQ' in controlling force and the increment $S'Q'$, in the centrifugal force, is the restoring force which compels the governor balls to assume original radius of rotation r corresponding to equilibrium speed. Thus, the restoring force,

$$SS' = \left(\frac{dF'}{dr}\right)\delta r - \frac{w}{g}(\delta r)\omega^2$$

or

$$SS' = \left(\frac{dF'}{dr} - \frac{w}{g}\omega^2\right)\delta r \qquad (15.41)$$

For the governor to be stable, evidently, the quantity in the bracket of Eq. (15.41) should be positive.

or,

$$\left(\frac{dF'}{dr}\right) > \frac{w}{g}\omega^2$$

or, using relation (15.40),

$$\left(\frac{dF'}{dr}\right) > (F/r) \qquad (15.42)$$

Thus, the condition of stability requires that the *slope of the 'controlling force curve' should be more than that of the line representing the centrifugal force at the equilibrium speed under consideration.*

Controlling force curve for spring controlled governors is approximately a straight line. Figure 15.24 shows the straight lines representing controlling force curves for spring controlled governors. The one at A is having a positive intercept while the one at C has a negative intercept on controlling force axis. The straight line at B represents controlling force curve which passes through the origin. *By increasing initial tension the line C can be made to move up and the Governor may become Isochronous or unstable.*

Straight line drawn from origin to represent centrifugal force, and to cut the controlling force line at

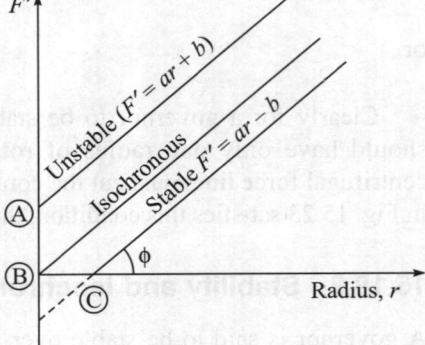

Fig. 15.24 Controlling force curve for spring controlled Governor.

A, will have slope (tanϕ = F/r) greater than the slope (tanθ) of controlling force curve (*see* Fig. 15.25a) and therefore it denotes unstable operation. This can be explained in yet another way. The equation of line at Fig. 15.25(a), with positive intercept on y-axis is,

$$F' = ar + b \qquad (15.43)$$

and dividing out by r,
$$(F'/r) = a + (b/r) \qquad (15.44)$$
or
$$\tan \theta = \tan \theta + (b/r)$$

The term 'a' in equation (15.44) represents tanθ, the slope of controlling force line. The equation shows that as radius r increases the difference between tanϕ and tanθ goes on decreasing but tanϕ > tanθ always. The governor is thus unstable.

When the controlling force curve (line) passes through the origin as at B in Fig. 15.24, only one constant speed line (centripetal force line) overlaps and no other constant speed line can cut it. Thus the governor has one equilibrium speed which is constant for all radii of rotation. Such a governor is called *Isochronous governor*. From the equation of controlling force $F' = ar$ it follows that,

$$(F'/r) = a \qquad (15.45)$$

and the slope of centrifugal force (constant speed) line is constant and is equal to a. *This shows that the isochronous governor has only one equilibrium speed and the slope of controlling force curve equals the slope of centrifugal force line.*

Finally, the controlling force curve at C in Fig. 15.24 has a negative intercept on y-axis. Centrifugal force lines at different equilibrium speeds drawn to cut controlling force line, will have slope always smaller than that of controlling force curve, and the governor is therefore stable. Alternatively the equation for line at C in Fig. 15.24 is

$$F' = ar - b \qquad (15.46)$$

Dividing out by r, we have
$$(F'/r) = a - (b/r) \qquad (15.47)$$

As the radius r increases, the difference between tanθ and tanϕ goes on decreasing but, tanθ > tanϕ always, and the governor is stable.

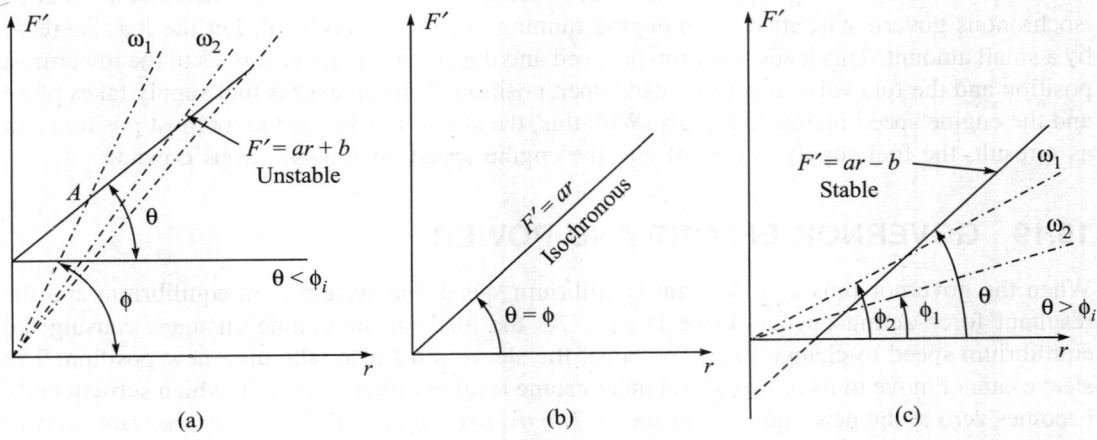

Fig. 15.25 Stability criteria for spring controlled Governor.

15.18.3 Sensitiveness

For maintaining as closely as possible a constant mean speed of rotation, whatever may be the load on engine, it is desirable that the sleeve movement is as large as possible and the resulting change of speed is as small as possible. Thus, taken in a more general sense a governor A is more sensitive compared to the governor B if:

  (i)   governor A requires a smaller fractional change in speed for producing a given sleeve displacement or

  (ii) for a given fractional change of speed, governor A produces larger sleeve displacement.

   When used in conjunction with an engine, a practical requirement will, however, be that the change of equilibrium speed from the full-load to no-load position of sleeve, should be as small a fraction of mean equilibrium speed as possible. This is because the actual amount of sleeve displacement is immaterial so long as it is sufficient to change the energy supplied to the engine by the required amount. A more appropriate definition of sensitiveness is that it is the ratio of the difference between the maximum and minimum equilibrium speed to mean equilibrium speed.

   Thus, if N_1 and N_2 are the maximum and minimum equilibrium speeds, and N be the mean equilibrium speed $= (N_1 + N_2)/2$

$$\text{Sensitiveness} = \left(\frac{N}{N_1 - N_2} \right) \tag{15.48}$$

$$\text{or sensitiveness} = \frac{(N_1 + N_2)}{2(N_1 - N)} \tag{15.49}$$

An isochronous governor is infinitely sensitive as $N_1 = N_2$.

15.18.4 Hunting

A governor is said to hunt if the speed of the engine fluctuates continuously above and below the mean equilibrium speed. This is caused by a governor that is too sensitive and which changes fuel supply in a much larger proportion than is actually needed. As an instance assume that an isochronous governor is fitted to an engine running under a steady load. Let the load increase by a small amount. This leads to a drop in speed and the governor sleeve moves to the lowermost position and the fuel valve is set in a wide open position. Thus an excess fuel supply takes place and the engine speed increases rapidly. With this, the sleeve reaches to the topmost position and as a result, the fuel supply is cut off and the engine speed once again drops down.

15.19 GOVERNOR EFFORT AND POWER

When the governor runs at a constant equilibrium speed, the system is in equilibrium and the resultant force acting on the sleeve is zero. As the load on the engine changes, causing the equilibrium speed to change to a new value, the sleeve must also take up a new position. The sleeve cannot move to its new position unless some resultant force acts on it, which subsequently becomes zero at the new equilibrium speed. *The average force which the governor can exert at the sleeve, producing the sleeve displacement, is called the governor effort.* For the sake of convenience in comparing different types of governors, it is usual to define the effort as that average force which can be exerted at sleeve for one per cent change of speed.

Taking the example of a Porter governor of general configuration and rotating at an equilibrium speed ω. The height of governor is then given by,

$$h = \left[\frac{w + \dfrac{W}{2}(1+k)}{w}\right]g/\omega^2 \tag{15.50}$$

Now let the equilibrium speed change from ω to (qω). The sleeve has a tendency to shift to a new position because of the effort exerted by the governor mechanism at the sleeve. For an indirect assessment of effort, now assume that a force equal and opposite to the 'effort' is applied at the sleeve to prevent its displacement to the new equilibrium position, corresponding to speed (qω). For the same governor height at a different equilibrium speed (qω) let Q be the additional force applied at the sleeve. Then, for new equilibrium speed (qω),

$$h = \left[\frac{w + \dfrac{W+Q}{2}(1+k)}{w}\right]\frac{g}{(q\omega)^2} \tag{15.51}$$

Equating right hand sides of equations (15.50) and (15.51),

$$\left[\frac{w + \dfrac{W}{2}(1+k)}{w}\right]\frac{g}{\omega^2} = \left[\frac{w + \dfrac{W+Q}{2}(1+k)}{w}\right]\frac{g}{(q\omega)^2}$$

or

$$\left[w + \frac{W}{2}(1+k)\right] = \left[w + \frac{W+Q}{2}(1+k)\right]\frac{1}{q^2}$$

or

$$\left[w + \frac{W}{2}(1+k)\right]\left(1 - \frac{1}{q^2}\right) = \frac{Q}{2}(1+k)\frac{1}{q^2}$$

or

$$\text{Effort} = \frac{Q}{2} = \left[\frac{w}{(1+k)} + \frac{W}{2}\right](q^2 - 1) \tag{15.52}$$

For k = 1 this expression reduces to $\dfrac{Q}{2} = \left(\dfrac{w+W}{2}\right)(q^2 - 1)$

In the case of Watt governor as W is negligible, equation (15.52) modifies to,

$$\text{Effort} = \frac{Q}{2} = \frac{1}{2}(q^2 - 1)w \tag{15.53}$$

Clearly, effort required in Porter governor for causing a given displacement of sleeve is larger compared to that required for Watt governor.

Power

The power of a governor is defined as the work done on sleeve (in moving it) for a given percentage change of speed. Thus the power of a governor is given by the product of effort and sleeve displacement. For a Porter governor of the type just described,

lift of the sleeve $x = 2(h_1 - h_2)$

where h_1 and h_2 are heights of the governor at equilibrium speeds ω and $(q\omega)$ respectively.

Also
$$h_1 = \left(\frac{W+w}{w}\right)\frac{g}{\omega^2}\tag{15.54}$$

and
$$h_2 = \left(\frac{W+w}{w}\right)\frac{g}{(q\omega)^2}$$

Hence
$$(h_1 - h_2) = \left\{\frac{1}{\omega^2} - \frac{1}{(q\omega)^2}\right\}\left(\frac{W+w}{w}\right)g = \left(\frac{q^2-1}{q^2}\right)\left(\frac{W+w}{w}\right)\frac{g}{\omega^2}$$

Using equation (15.54), the R.H.S. of above expression reduces to

$$(h_1 - h_2) = \left(\frac{q^2-1}{q^2}\right)h_1$$

Therefore, the lift of the governor is

$$x = 2(h_1 - h_2) = 2\left(\frac{q^2-1}{q^2}\right)h_1$$

and therefore from equation (15.52),

$$\text{Power} = \text{Effort} \times \text{sleeve displacement} = \frac{1}{2}Qx$$

or
$$\text{power} = \frac{1}{2}(q^2-1)(W+w) \times 2\left(\frac{q^2-1}{q^2}\right)h_1$$

Therefore   power of Porter governor $= \{(q^2-1)^2/q^2\}(W+w)h_1$ (15.55)

For Watt governor, where weight of sleeve W is negligible, equation (15.55) reduces to,

$$\text{Power of Watt governor} = \{(q^2-1)^2/q^2\}wh_1\tag{15.56}$$

When $\alpha \neq \beta$ in Porter governor, equations (15.52) and (15.55) become,

$$\text{Effort,}\quad \frac{Q}{2} = \frac{1}{2}(q^2-1)\left\{\frac{w+(1+k)W/2}{w}\right\}\tag{15.57}$$

and
$$\text{Power} = \frac{(q^2-1)^2}{q^2}\left\{\frac{w+(1+k)W/2}{w}\right\}h_1\tag{15.58}$$

15.20 EFFECT OF FRICTION: INSENSITIVENESS

In Section 15.7 it was shown that friction force at the sleeve always acts so as to oppose the motion of the sleeve. As the speed increases and sleeve moves up, the frictional force f acts down and the total force on sleeve $= (W+f)$. Similarly, when speed drops down and the sleeve moves down, the net downward force on sleeve $= (W-f)$. It was shown during the discussions that the effect of friction at sleeve is to render the governor insensitive over a speed range. To understand the effect of friction at sleeve on controlling force diagram and, in order to assess the extent of insensitiveness, we proceed as follows.

Let f = force required at the sleeve to overcome friction of the governor and its mechanism

  f_b = corresponding radial force required at each ball

  W = total load at the sleeve

and F' = controlling force exerted at each ball

The total load required to be overcome at the sleeve is $W + f$ if the governor is speeding up and $W - f$, if the governor is speeding down. Correspondingly, the controlling force exerted on the ball when the speed goes up $(F' + f_b)$ and when the speed goes down, is $(F' - f_b)$. For simple type of governors, it is possible to establish a simple relation between f and f_b .

For a Porter governor, taking moments (*see* Fig. 15.5) about the instantaneous centre *I*,

$$f_b \times BD = (f/2)IC$$
$$= (f/2)(ID + DC)$$

Therefore

$$f_b = (f/2)(\tan\alpha + \tan\beta)$$
$$= (f/2)(1 + k)\tan\alpha$$
$$= (f/2)(1 + k)(r/h) \tag{15.59}$$

And similarly for spring loaded governor (*e.g.,* Hartnell and Wilson Hartnell governors),

$$(f_b)a = (f/2)b \quad \text{Therefore} \quad f_b = \frac{1}{2}(b/a)f \tag{15.60}$$

Figure 15.26 shows clearly the effect of friction on controlling force diagram. As is evident, there are three controlling force curves, the ordinates of which are in the proportion $F' + f_b$, F' and $F' - f_b$. The middle curve corresponds to the case when friction is neglected. The upper curve corresponds to the condition when speed is increasing while the lowermost curve represents the condition when the speed is decreasing. At the radius of rotation R , the controlling force, neglecting friction, is LB and the equilibrium speed on speed scale is N . When the speed increase takes place, the controlling force has a larger value equal to LC and corresponding equilibrium speed on speed scale is N' . Similarly

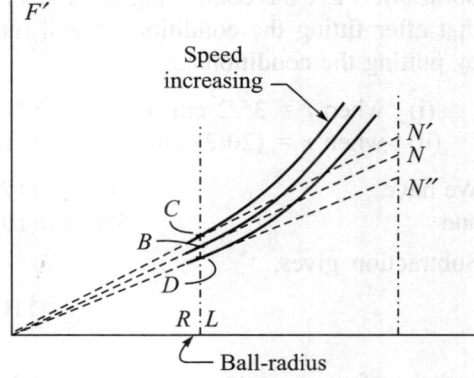

Fig. 15.26

controlling force, when speed decreases, is LD and corresponding equilibrium speed is N'' . Thus, for a radius of rotation R , the speed of rotation varies between N' and N'' . Thus the governor is rendered insensitive over this speed range.

The ratio $(N' - N'')/N$ is called coefficient of insensitiveness of the governor. This coefficient may be expressed in terms of friction force f_b and the controlling force F' as follows. Remembering that controlling force at a given radius is proportional to the square of the ball speed $(F' = wr\omega^2/g)$, the ordinates of the three controlling force curves are expressed as

$$F' = CN^2 \tag{15.61}$$
$$F' + f_b = CN'^2 \tag{15.62}$$
and
$$F' - f_b = C(N'')^2 \tag{15.63}$$

Considering equations (15.62) and (15.63), subtraction gives,

$$2f_b = C\{(N')^2 - (N'')^2\}$$

and dividing now by equation (15.61) on either sides,

$$2f_b/F' = \frac{(N')^2 - (N'')^2}{N^2} = \frac{(N' - N'')}{N} \frac{(N' + N'')}{N}$$

But $N' + N'' = 2N$ and so,

$$2f_b/F' = 2\ \frac{(N' - N'')}{N}$$

Hence the coefficient of insensitiveness,

$$= (N' - N'')/N = (f_b/F') \tag{15.64}$$

EXAMPLE 15.17 In a spring controlled governor, in which the controlling force curve is a straight line, the balls are 35 cm apart when the controlling force is 1160 N, and 20 cm when it is 580 N. At what speed will the governor run, when the balls are 25 cm apart, and each one of them weighs 58 N? By how much should the initial tension be increased to make the governor isochronous and what would then be the speed of rotation?   (DAV Indore, Feb. 1982)

Solution: Let the controlling force curve be represented by general equation $F' = ar + b$, so that after fitting the conditions b will turn out to be either positive, negative or zero. Hence by putting the conditions:

  (i)   when $r = 35/2$ cm, $F' = 1160$ N and
  (ii) when $r = (20/2)$ cm, $F' = 580$ N.

We have $1160 = a(17.5) + b$
and $580 = a(10) + b$

Subtraction gives,

$$a = \frac{(1160 - 580)}{7.5} = 77.33 \text{ N}$$

and therefore, $b = 580 - (10)(77.33) = -193.33$ N

Therefore the equation for controlling force is,

$$F' = (77.33)a - 193.33$$

Since the intercept b is negative, the governor is stable. For balls to run at 25 cm apart, $r = (25/2)$ cm. Hence,

$$F' = (77.33)\left(\frac{25}{2}\right) - 193.33 = 773.295$$

Therefore from $\qquad \dfrac{w}{g} r\omega^2 = F$

$$\left(\dfrac{58}{981}\right)(12.5)\omega^2 = 773.295$$

Hence $\qquad\qquad\qquad \omega = 32.347$ rad/s

And hence required r.p.m. $\quad N = \dfrac{60 \times \omega}{2\pi} = 308.9$ r.p.m. **Ans.**

For governor to become isochronous, the line must be shifted parallel to itself, making intercept on F'-axis equal to zero, as shown in Fig. 15.27. Thus, the slope of the controlling force line remaining the same, the equation of corresponding controlling force curve of isochronous governor is,

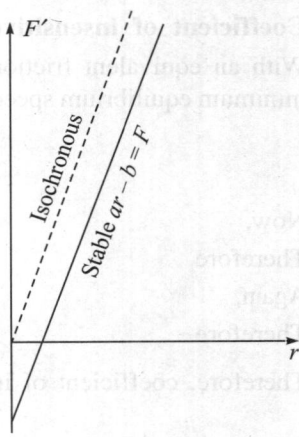

Fig. 15.27

$$F' = (77.33)r$$

Therefore $\qquad \left(\dfrac{58}{981}\right) r\omega^2 = (77.33)r$

or $\qquad\qquad\qquad \omega = 36.165$ rad/s

Therefore $\qquad\qquad N = 345.35$ r.p.m. **Ans.**

EXAMPLE 15.18 The controlling force F' in newtons and the radius of rotation r in cm for a spring loaded governor are related by expression,

$$F' = 27.5r - 75$$

The weight of each ball is 49 N and the extreme radii of ball rotation are 10 cm and 17.5 cm. Find the maximum and minimum equilibrium speeds. If the friction of the governor mechanism is equivalent to a force of 4.9 N at each ball, find the coefficient of insensitiveness of the governor at the extreme radii.

Solution: For the given extreme radii of 10 cm and 17.5 cm, corresponding controlling forces are

$$F_1' = 27.5(17.5) - 75 = 406.25 \text{ N}$$
$$F_2' = 27.5(10) - 75 = 200 \text{ N}$$

Corresponding equilibrium speeds ω_1 and ω_2 are given by,

$$\dfrac{w}{g} r_1 \omega_1^2 = F_1'$$

i.e., $\quad \dfrac{49}{981}(17.5) \times \omega_1^2 = 406.25 \quad$ Hence $\quad \omega_1 = 21.558$ and $N_1 = 205.9$ r.p.m. **Ans.**

Also, $\qquad \dfrac{49}{981}(10)\omega_2^2 = F_2' = 200$ N or $\omega_2 = 20.01$ rad/s Hecne $N_2 = 191.1$. r.p.m. **Ans.**

Coefficient of insensitiveness

With an equivalent frictional force of 4.9 N at the balls, controlling force at maximum and minimum equilibrium speed is:

$$F_1'' = F_1' + 4.9 = 411.15 \text{ N and}$$
$$F_2'' = F_2' - 4.9 = 195.1 \text{ N}$$

Now,
$$\omega_1'^2 = (411.15)/(49 \times 17.5/981)$$

Therefore
$$\omega_1' = 21.688 \text{ rad/s} \quad \text{or} \quad N_1' = 207.1 \text{ r.p.m.}$$

Again,
$$\omega_2'^2 = 195.1/(49 \times 10/981)$$

Therefore
$$\omega_2' = 19.76 \quad \text{Hence} \quad N_2' = 188.7 \text{ r.p.m.}$$

Therefore, coefficient of insensitiveness

$$= \frac{2(N_1' - N_2')}{(N_1' + N_2')} = \frac{2(207.1 - 188.7)}{(207.1 + 188.7)} = 0.093$$

Coefficient of insensitiveness $= 9.3\%$ **Ans.**

EXAMPLE 15.19 The radius of rotation of the balls of a Hartnell governor is 8 cm at the minimum speed of 300 r.p.m. Neglecting gravity effect, determine the speed after the sleeve has lifted by 6 cm. Also determine the initial compression of the spring, the governor effort and the power. The particulars of the governors are given below:

Length of ball arm = 15 cm; length of sleeve arm = 10 cm;
Weight of each ball = 39 N; stiffness of the spring = 245 N/cm.

(DAV Indore, 1990 and AMIE, Winter 1980)

Solution: For a sleeve lift of 6 cm, the radius r_1 is given by,

$$r_1 = r_2 + \frac{x}{b}a = 8 + 6\left(\frac{15}{10}\right) = 17 \text{ cm}$$

Let S_1 and S_2 be the spring force of compression at radius, r_1 and r_2. Then,

$$F_2' = \frac{w}{g}r_2\omega_2^2$$

or
$$F_2' = \frac{39}{981}(8)\left(\frac{2\pi \times 300}{60}\right)^2 = 313.9 \text{ N}$$

And taking moments about lever fulcrum,

$$F_2' \times a = \left(\frac{W + S_2}{2}\right)b$$

Neglecting gravitational effects due to weight of ball and the sleeve,

$$313.9 \times 15 = (S_2/2) \times 10$$

Therefore
$$S_2 = 2\left(\frac{15}{10}\right) \times 313.9 = 941.7 \text{ N}$$

The spring force generated due to additional compression by 6 cm in spring,

$$S' = 6 \times 245 = 1470 \text{ N}$$

Therefore $\qquad S_1 = S_2 + 1470 = 2411.7 \text{ N}$

Taking moments of sleeve load and controlling force at radius r_1 about lever fulcrum,

$$F_1 \times a = \frac{S_1}{2} \times b$$

Therefore $\qquad F_1 = \left(\frac{2411.7}{2}\right)\left(\frac{10}{15}\right) = 803.9$

Hence the speed of governor ω_1 when sleeve is lifted is given by

$$\frac{w}{g} r_1 \omega_1^2 = F_1$$

or $\qquad \omega_1^2 = (803.9) \times \frac{981}{39} \times \frac{1}{17} = 1189.5$

Therefore $\qquad \omega_1 = 34.5 \text{ rad/s and } N_1 = \frac{60 \times \omega_1}{2\pi} = 329.3 \text{ r.p.m.}$ **Ans.**

Initial compression in the spring,

$$\Delta = \left(\frac{S_2}{s}\right) = \frac{941.7}{245} = 3.84 \text{ cm} \qquad\qquad \textbf{Ans.}$$

Force S_2 being the force of initial compression at which there is no sleeve displacement,

$$\text{effort} = \frac{1}{2}(S_1 - S_2) = \frac{1}{2}(2411.7 - 941.7) = 735 \text{ N} \qquad \textbf{Ans.}$$

Governor Power = effort × sleeve displacement

$$= 735 \times 6 = 4410 \text{ N·cm.} = 44.1 \text{ N·m} \qquad\qquad \textbf{Ans.}$$

EXAMPLE 15.20 The arms of a Porter governor are 20 cm each and are pivoted 4 cm from the axis of rotation. The ball and sleeve weights are 20 N and 245 N respectively. Determine the governor effort and power available when the radius of rotation is 12 cm.

Solution: As the arms are of same length and are pivoted at same distance from governor axis, $\alpha = \beta$ and hence $k = 1$. (*see* Fig. 15.28)

Therefore $\qquad\qquad F = \left\{w + (1 + k)\frac{W}{2}\right\}\tan\alpha$

For $k = 1$ $\qquad\qquad F = (w + W)\tan\alpha$

Now $\qquad\qquad \sin\alpha = \frac{(r - 4.0)}{20} = \frac{12 - 4.0}{20} = 0.4$

Therefore $\qquad\qquad \alpha = 23.58°$

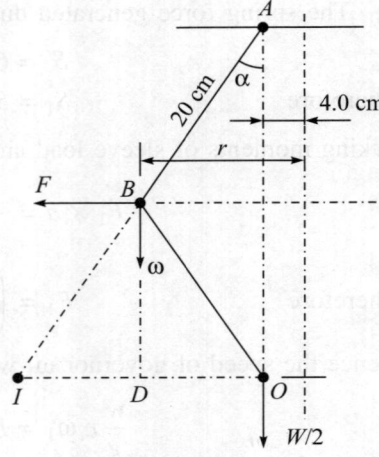

Fig. 15.28

Therefore $\tan\alpha = 0.4364$

Therefore $F = (20 + 245) \times 0.4364 = 115.65$ N

Therefore $\dfrac{w}{g} r\omega^2 = F$

or $\dfrac{20}{981} \times 12 \times \omega^2 = 115.65$

or $\omega = 21.74$ rad/s and $N = 207.6$ r.p.m.

For 1% increase of speed assume that the same radius of rotation could be ensured by changing load at sleeve from W to W'. Then, for new controlling force F',

$$F' = \left\{ w + (1 + k) \dfrac{W'}{2} \right\} \tan\alpha$$

or $F' = \{20 + W'\} \times 0.4364$

Again, force F' for a new speed $\omega' = 1.01\omega$, (as $F \propto \omega^2$),

$$F' = (1.01)^2 F = 1.02 \times 115.65$$

Hence, the new equation is,

$$115.65 \times 1.02 = \{20 + W'\} \times 0.4364$$

or $W' = 250.3$ N

Therefore effort during this change of speed,

$$= \dfrac{1}{2}(W' - W)$$

Therefore $P = \dfrac{1}{2}(250.3 - 245) = 2.65$ N   **Ans.**

For calculating power, sleeve lift is calculated as under. The height of the governor in the two positions,

$$h_1 = r_1/\tan\alpha = \dfrac{12}{0.4364} = 27.497 \text{ cm}$$

$$\omega_2 = 1.01\omega = 21.957$$

Value of h_2 is obtained assuming that governor is now free to take other radius of rotation of ball. Thus,

$$h_2 = \left(\dfrac{w + W}{w} \right) \dfrac{g}{\omega^2}$$

or $h_2 = \left(\dfrac{20 + 245}{20} \right) \dfrac{981}{(21.957)^2} = 26.96 \text{ cm}$

Therefore Lift of sleeve $= 2(h_1 - h_2) = 1.074$ cm

Therefore
$$\text{power} = P \times 1.074$$
$$= 2.65 \times 1.074 = 2.845 \text{ cm·N}$$ **Ans.**

EXAMPLE 15.21 The upper arms of a Porter governor are pivoted on the axis of rotation and the lower arms are pivoted to the sleeve at a distance of 30 mm from the axis of rotation. The length of each arm is 300 mm and weight of each ball is 59 N. If the equilibrium speed is 200 r.p.m., when the radius of rotation is 200 mm, find the required load on the sleeve.

If the friction is equivalent to a force of 39 N at the sleeve, find the coefficient of insensitiveness at 200 mm radius. (Gujarat Un., 1980)

Solution: Referring to Fig. 15.29,

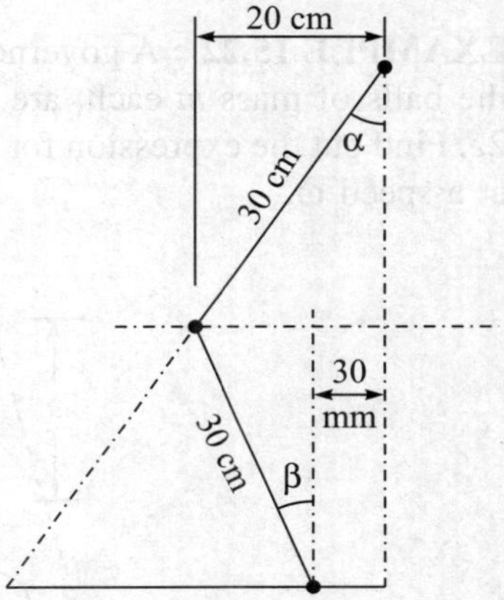

$$\tan\alpha = \frac{20}{\sqrt{(30^2 - 20^2)}} = 0.894$$

$$\tan\beta = \frac{(20 - 3.0)}{\sqrt{\{30^2 - (20-3)^2\}}} = 0.6877$$

Therefore $k = \dfrac{\tan\beta}{\tan\alpha} = (0.769)$

Fig. 15.29

Let W = load on the sleeve, then

$$\omega^2 = \left\{\frac{w + (1+k)\,W/2}{w}\right\}\frac{g}{h}$$

But $h = \sqrt{30^2 - 20^2} = 22.36 \text{ cm}$

Therefore $\omega^2 = \dfrac{59 + (1.769/2)\,W}{59} \times \dfrac{981}{22.36}$

or $\left(\dfrac{2\pi \times 200}{60}\right)^2 = \dfrac{59 + (0.8845)\,W}{59} \times \dfrac{981}{22.36}$

Therefore $W = 600.2 \text{ N}$ **Ans.**

With additional force of 39 N at sleeve due to friction, we have

$$W_1 = W + f = 600.2 + 39 = 639.2 \text{ N}$$
$$W_2 = W - f = 600.2 - 39 = 561.2 \text{ N}$$

Therefore $\omega_1^2 = \dfrac{59 + (0.8845)\,639.2}{59} \times \dfrac{981}{22.36}$

$$\omega_1 = 21.547 \text{ rad/s}$$

Therefore $N_1 = 205.76 \text{ r.p.m.}$

Also $\omega_2^2 = \dfrac{59 + (0.8845)\,561.2}{59} \times \dfrac{981}{22.36}$

Therefore $\qquad \omega_2 = 20.32$ rad/s

Therefore $\qquad N_2 = 194.06$ r.p.m.

Therefore, coefficient of insensitiveness $= \dfrac{2\,(N_1 - N_2)}{(N_1 + N_2)}$

$$= 2\left(\frac{205.76 - 194.06}{205.76 + 194.06}\right) = 0.0585$$

Therefore Coefficient of insensitiveness $= 5.85\%$ **Ans.**

EXAMPLE 15.22   A governor is shown schematically in Fig. 15.30. The two links which carry the balls of mass m each, are connected by a spring of stiffness k and has a natural length of $2e$. Find out the expression for the inclination of the links with vertical when the governor rotates at a speed ω. (AMIE, Winter 1993)

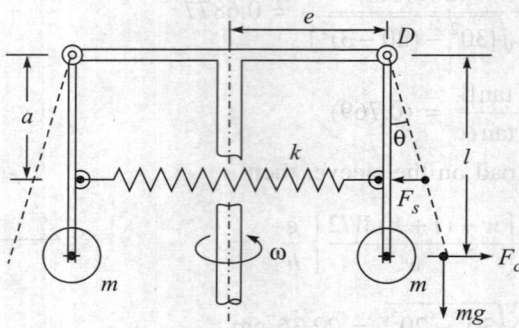

Fig. 15.30

Solution: With the increase and decrease in the speed of rotation of spindle, the central coil of spring connecting the two ball arms does not move. This spring is thus equivalent to two springs, fixed at the central coil plane. Hence, equivalent stiffness of spring $= 2k$. Neglecting mass of arms, various forces acting on arm and ball are as shown in Fig. 15.30 on ball arm on right side.

At the speed of rotation ω_1 let the arm be inclined to the vertical at angle θ as shown. Taking moment of forces acting on ball arm about fulcrum O for equilibrium,

$$F_c \times l\cos\theta = mg(l\sin\theta) + F_s\,(2a\cos\theta) \qquad (i)$$

where F_c and F_s are the centrifugal forces on ball and the spring force respectively. For small angles θ, $\sin\theta \approx \theta$ and $\cos\theta \approx 1$, and therefore above equation becomes,

$$F_c \times l = mgl\theta + (F_s)a$$

But $\qquad F_c = m(e + l\sin\theta)\omega^2$

or $\qquad F_c = m\,(e + l\theta)\omega^2$, for small angle θ

and $\qquad F_s = 2k(a\sin\theta) \approx (2ka\theta)$

Hence the moment equation is,

$$m(e + l\theta)l\omega^2 = mgl\theta + (4ka^2)\theta$$

Therefore By transposing and collecting terms with coefficient θ,

$$(mgl + 4ka^2 - ml^2\omega^2)\theta = mel\omega^2$$

Therefore
$$\theta = \frac{mel\omega^2}{ml(g - l\omega^2) + 4ka^2}$$
Ans.

Similarly, if effect of gravitational pull on ball and its effect is neglected in comparison to other factors, then eliminating corresponding term '*mgl*' from above expression,

$$\theta = \frac{mel\omega^2}{(4ka^2 - ml^2\omega^2)}$$
Ans.

REVIEW QUESTIONS

15.1 A simple Watt governor, in which the arms intersect on the axis, is running at 100 r.p.m. Find the percentage variation of its height if the speed change = ± 5%.
(*Ans.* 20.1%) (DAV/Indore, Nov. 1992)

15.2 In a Porter governor the central load is 176.6 N and the weight of each ball is 19.6 N. The top arms are 25 cm while the bottom arms are each 30 cm long. The friction of the sleeve is 1.4 kgf. If the top arms make 45° with the axis of rotation in the equilibrium position, find the range of speed of the governor in position. (*Ans.* 13 r.p.m.)

15.3 In a Porter governor, all the four arms are 400 mm long. Each ball has a mass of 8 kg and the load on the sleeve is 588.6 N. The upper arms are pivoted on the axis of the spindle whereas the lower arms are attached to the sleeve at a distance of 45 mm from the axis. What will be the equilibrium speeds for the two extreme radii of 250 mm and 300 mm of rotation of the balls. (*Ans.* 147 and 159.0 r.p.m.)

15.4 The upper arms of a Porter governor are pivoted at the axis of rotation and are 30 cm in length. The lower arms are 27.5 cm in length and are attached to the sleeve at a distance of 5 cm from the axis of rotation. The weight of each ball is 58.9 N and the central weight 470.9 N. Calculate the equilibrium speed of the governor when the radius of rotation in 17.5 cm. Determine also the effort and power of the governor for 1 per cent variation of speed. (*Ans.* 170 r.p.m.; 5.4 N and 4.4 N)

15.5 In a Porter governor the upper and lower arms are each 200 mm long and are each inclined at 30° to the vertical when the sleeve is in its lowest position. The points of suspension are each 36 mm from the axis of the spindle. The mass of each rotating ball is 3 kg and that of the central load on the sleeve 196.2 N. If the movement of the sleeve is 36 mm, find the range of speed of the governor. (U.London) and (Jiwaji Univ.)
(*Ans.* 170.5 to 185.5 r.p.m.)

15.6 In a Porter governor, the arms and links are each 25 cm long and intersect on the main axis. Each ball weighs 44.15 N and the central load is 196.2 N. The sleeve is in its lowest position when the arms are inclined at 30 degree to the axis. The lift of the sleeve is 5 cm. What is the force of friction at the sleeve, if the speed at ascent from the lowest position is equal to the speed at the beginning of descent from the highest position? What is then the range of speed of the governor, all other things remaining the same?
(*Ans.* f = 14.7 N, $N_1 - N_2$ = 19.2 r.p.m.)

15.7 The weight of each ball of a Proell governor is 88.3 N, the central load is 1471.5 N and the arms are 25 cm long. The arms are open and are each pivoted at a distance of 5 cm from the axis of rotation. The extension of the lower arm to which each ball is attached is 12.5 cm long and the radius of rotation of the balls is 22.5 cm. When the arms are inclined at 40 degree to the axis of rotation find (i) the equilibrium speed for the above configuration, (ii) the coefficient of insensitiveness if the friction of the governor mechanism is equivalent to a force of 19.6 N at the sleeve.

(Ans. 231 r.p.m. and 1.2%)

15.8 In a Proell governor, the lower arm ABC is formed in one piece. Each ball has a mass of 1.8 kg and the effect of friction is equivalent to a force of 9 N at the sleeve. AB = BD = 150 mm and extension arm BC = 50 mm. Find the magnitude of the central mass, if the sleeve is to rise to its mean position when distance, parallel to the axis between pivots A and D of lower and upper arms, is 225 mm. The speed in this position is 120 r.p.m. and the portion BC of the extension arm is vertical. The pivots A and D of lower and upper arms are both 25 mm away from the governor axis. The gravitational effect of the balls must be taken into account.   (<i>Ans.</i> 3.17 kg) (London Univ.)

15.9 In a spring controlled governor the radial force acting on the balls was 4414.5 N when the centre of the balls was 20 cm from the axis, and 7357.5 N when at 30 cm. Assuming that the force varies directly as the radius, find the radius of the ball path when the governor runs at 270 r.p.m. Also find what alteration in spring load is required in order to make the governor isochronous, and the speed at which it would then run. Weight of each ball = 294.3 N. (<i>Ans.</i> $r = 26.78$ cm, $\Delta F = 5886$ N and N = 299.3 r.p.m.)

15.10 In a Hartnell governor, the ball and sleeve arms are 8.5 cm and 5 cm respectively. The ball arms are parallel to the axis of rotation of the governor when the ball circle radius is 7 cm and the equilibrium speed is 900 r.p.m. Determine:

(i) The weight of each ball when the speed is increased by 1 per cent without any change of radius for the given position and an axial load of 29.43 N is required at the sleeve to maintain equilibrium. (ii) The stiffness and initial extension of each spring if the rate of the sleeve movement, when in the mid-position, is 2 cm for a change of speed by 480 r.p.m.   (<i>Ans.</i> 7.21 N; 166.77 N/cm and 2.74 cm)

15.11 A governor of the Hartnell type has ball arm and sleeve arm of lengths 125 mm and 62.5 mm respectively, the fulcrum of the bell crank lever being 100 mm away from spindle axis. The governor runs at a mean speed of 300 rev/min, each ball has a mass of 2.3 kg, and a 3 per cent reduction in speed causes a sleeve movement of 6 mm. If the ball arm is vertical at the mean speed, and gravitational effects are ignored, find the spring stiffness in N/m. Neglect the mass of the arms. By how much must the adjusting nut be screwed down to render the governor isochronous and what will be the resulting operational speed of the governor. (London Univ.)

(Ans. 26.1 kN/m, 15.17 mm and 359 r.p.m.)

15.12 A Hartnell governor has two rotating balls of mass 2.7 kg each. The ball radius is 125 mm in the mean position when the ball arms are vertical and the speed is 150 rev/min with the sleeve rising. The length of ball arms is 140 mm and the length of the sleeve arms is 90 mm. The stiffness of the spring is 7 kN/m and the total sleeve movement is

± 12 mm from the mean position. Allowing for a constant friction force of 14 N acting at the sleeve, determine the speed range of the governor in the lowest and highest sleeve positions. Neglect the obliquity of the ball arms. (London Univ.)

(*Ans.* 154.3 to 161 r.p.m. and 122.5 to 133.6 r.p.m.)

15.13 Figure 15.31 shows a spring loaded governor. Each ball weighs 49 N. The balls are connected by two horizontal springs A. An auxiliary spring provides an additional force at the sleeve through the medium of lever which is pivoted to a fixed point K on the governor frame. In the mean position, the radius of the ball path is 15 cm and the governor speed is 600 r.p.m. The tension in each of the ball springs A is 1226.3 N. Find the pull exerted by the auxiliary spring for this position. If now the sleeve rises by 2 cm and the corresponding speed is to be 630 r.p.m., find the stiffness of auxiliary spring B if each of the ball springs A has a stiffness of 98.1 N/cm. Neglect the effect of gravity on the balls. (*Ans.* 5194.4 N and 1216.44 N/cm)

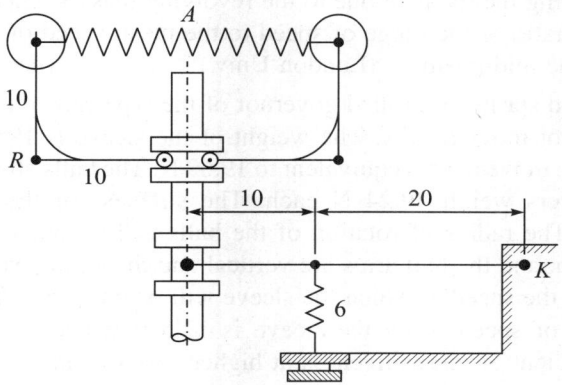

Fig. 15.31

15.14 In Wilson Hartnell governor of the type shown in Fig. 15.16, the two springs attached directly to balls each has a stiffness of 7.36 N/cm and a free length of 10 cm. The weight of each ball is 39.24 N, the length of the ball arm of each bell crank lever is 8 cm, and that of the sleeve arm is 6 cm; the lever is pivoted at its mid-point. When the radius of rotation of the ball is 8 cm, the equilibrium speed is 240 r.p.m. If the sleeve is to lift 0.75 cm for an increase of speed of 5%, determine the required stiffness of the auxiliary spring. (*Ans.* 83.68 N/cm)

15.15 A gramophone is driven by a Pickering governor. The weight of each disc attached to the centre of a leaf spring is 0.196 N. Each spring is 5 mm wide and 0.125 mm thick. The effective length of each spring is 4 cm. The distance from the spindle axis to the centre of gravity of the weight when the governor is at rest is 1 cm. Find the speed of the turn-table when the sleeve has risen 0.8 mm and the ratio of the governor speed to the turn-table speed is 10.5. Take E = 2.06×10^7 N/cm² (Kerala Univ.)

(*Ans.* 76.3 r.p.m.)

15.16 In a spring and gravity controlled governor of the type shown in Fig. 15.19, the mass of each ball is 1.5 kg and the mass of the sleeve is 8 kg. The two arms of the bell crank lever are at right angles and their lengths are $AB = 100$ mm and $AO = 40$ mm.

The distance of the fulcrum A of each bell crank lever from the axis of the rotation is 50 mm and minimum radius of rotation of the governor balls is also 50 mm. The corresponding equilibrium speed is 240 r.p.m. and the sleeve is required to lift 10 mm for an increase in speed of 5 per cent. Find the stiffness and initial compression of the spring. (Cambridge Univ.) **(Ans.** 14.36 N/mm and 9 mm)

15.17 In a spring controlled governor of the Hartung type, the length of the ball and sleeve arms are 80 mm and 120 mm respectively. The total travel of the sleeve is 25 mm. In the mid-position, each spring is compressed by 50 mm and the radius of rotation of the mass centres is 140 mm. Each ball has a mass of 4 kg and the spring has a stiffness of 10 kN/m of compression. The equivalent mass of the governor gear at the sleeve is 16 kg. Neglecting the moment due to the revolving masses when the arms are inclined, determine the ratio of the range of speed to the mean speed of the governor. Find also the speed in the mid-position. (London Univ.) **(Ans.** 317 r.p.m. and 5%)

15.18 In a gravity and spring controlled governor of the type shown in Fig. 15.19, the central spindle does not move axially. The weight of the sleeve is 196.2 N and the frictional resistance to its movement is equivalent to 19.62 N. The balls attached to the right angled bell crank levers weigh 39.24 N each. The stiffness of the spring is 392.4 N/cm compression. The radius of rotation of the balls is 125 mm when the sleeve is in its lowest position, and the ball arms are vertical and the spring exerts a force of 588.6 N. Determine: (i) the speed at which the sleeve will begin to rise from its lowest position, (ii) the range of speed when the sleeve is 12.5 mm above the lowest position and (iii) the coefficient of insensitiveness at higher speed. (Bihar University, 1979)

(Ans. 189.6 r.p.m., 3.1 r.p.m. and 1.44%)

Chapter 16

BALANCING

16.1 INTRODUCTION

Chapter 14 was devoted to the study of inertia forces in slider-crank and four-bar mechanisms. It was pointed out that inertia forces set up shaking forces in the structure. If the parts are not balanced, these shaking forces tend to induce vibrations in the frame. Vibrations are responsible for increasing stresses in the components and subject the bearings to repeated loads, which may lead to premature fatigue failure of parts. *Vibrations, especially when they occur in light weight, high-speed machinery may produce excessive noise, cause undue wear of mating components or lead to faulty performance of the machine.* One of the important applications of the inertia force analysis lies in the balancing of machinery. Balancing of rigid rotors is of great consequence in the operations of many types of machinery. Rotating components such as tyres, flywheels, fans, motors, turbines, etc. also need to be balanced. In a similar way, many machines (*e.g.* I.C. engines, compressors) have reciprocating components and require our attention.

The technique of correcting or eliminating unwanted inertia forces and moments is called *balancing*. Production of machine components with tight tolerances, though desirable, generally leads to a higher cost of production. In general, a more economical proposition is to produce parts with reasonable tolerances, subject them to balancing procedures and apply corrections as and where needed. It is possible to balance either wholly or partly the inertia forces in the machine component by either introducing additional masses to counteract original forces or by removing masses from reference planes from positions diametrically opposite to those of balancing masses.

We shall first discuss the balancing of rotating parts and then consider balancing of the reciprocating components.

16.2 BALANCING OF ROTATING MASSES

Unbalanced rotating masses, mounted on shaft, give rise to centrifugal forces on the shaft. Balancing of such a system requires these masses, forming the system, to be rearranged or introduce additional masses so that the centrifugal forces and couples acting on the shaft are in equilibrium.

The forces acting on rotating masses always act in transverse planes normal to the shaft axis. When the shaft rotates uniformly, the lines of action of all these forces pass through the axis of the shaft, keeping angular relationship same between any two forces.

16.3 STATIC AND DYNAMIC BALANCING

Static Balancing

Machine components in motion may be either in (i) static or standing balance; or (ii) dynamic or running balance. The static imbalance is due to gravity or a centrifugal force. A 'static imbalance' in a rotating system is the effect of an eccentric (or unbalance) mass which can be detected by quasi-static methods in which the weight of out-of-balance mass itself reveals its position. One such arrangement used for this purpose is illustrated in Fig. 16.1(a), which consists of a disc fixed to a straight shaft, resting on rigid rails, which permit the shaft to roll freely without friction. A reference system *XOY* is assumed to be described on the disc which moves with the disc.

To examine the disc for static imbalance, if any, the disc along with the shaft is rolled over rails gently by hand and permitted to assume equilibrium position of its own accord. A chalk mark is placed at a point that is lowermost on the periphery of the disc. This procedure is repeated quite a number of times. If the chalk marks of different trials are scattered at different positions around the disc, the disc is considered to be in static balance. If all the chalk marks coincide at a particular point, the disc is said to have static imbalance.

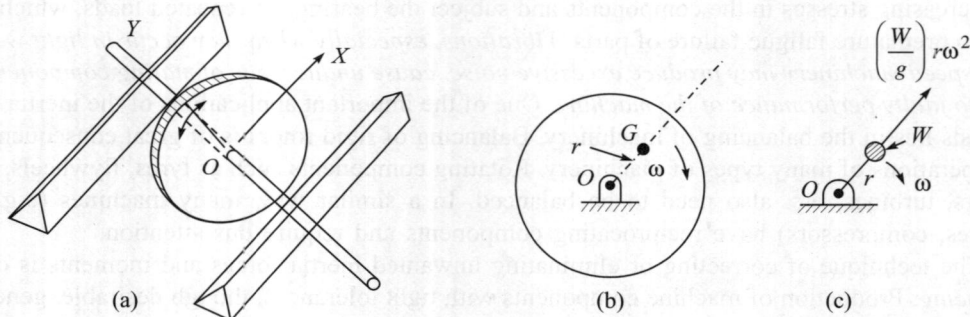

Fig. 16.1

Static balance exists if the forces acting on the parts are in equilibrium among themselves when not running, irrespective of the position in which the parts may be placed. Note that this is possible only when the position of the centre of mass of all moving parts remain unchanged relative to the machine frame. This is true only when the c.g. coincides with the geometric centre. When all the chalk marks in the experimental set-up coincide, the centre of mass of the disc lies on the radial line joining the chalk mark to the geometric centre of the disc. In the lowest peripheral position, the gravitational pull acts downwards along this radial line and therefore, the same point occupies the lowermost position in all trials. Theoretically it is possible to obtain static equilibrium with the unbalance above the shaft axis. However, it is practically impossible to obtain a condition in which two chalk marks are 180 degree apart.

When a static imbalance exists, it is possible to correct the same by drilling out material at the chalk mark. Static balancing can also be achieved by adding weights until the disc remains 'static' when shaft is placed on knife edges as shown in Fig. 16.1(a). *Static balancing process is thus equivalent to bringing the centre of mass of the system to its axis of rotation by adding correction masses or by drilling out material suitably.*

Dynamic Balancing

Even when a rotor is statically balanced, it may exhibit unwanted vibration when allowed to rotate about its axis. *A static balance of machine components is not enough to eliminate inertia forces because these forces may act in various planes (particularly when the disc is not thin) or in various directions depending on the directions and magnitudes of the accelerations (see Fig. 16.2 where equal and opposite inertia forces constitute a couple).* A dynamic imbalance is thus indicated by the presence of a couple with its axis perpendicular to that of rotation for the body. For complete dynamic balancing therefore, in addition to inertia forces, inertia couples must also be balanced.

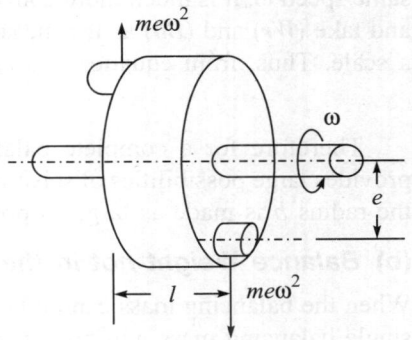

Fig. 16.2

A *dynamically* balanced rotor is statically balanced automatically but, in general, the opposite is not true. As a general rule, a rotor whose thickness is significant in relation to its diameter should be balanced dynamically. For thin discs static balancing is generally enough.

16.4 UNBALANCED ROTATING MASS

Static imbalance in an otherwise symmetric and homogeneous disc can be described by a mass of weight W equal to the weight of the disc, placed at a radius equal to r, the distance between the geometric centre and the c.g. as shown in Figs. 16.1(b) and (c). In some of the situations, it is possible to place balancing mass in the plane of rotation while in others the balance weight and disturbing mass cannot revolve in the same plane. We would discuss both the cases.

(a) *Balance Weight in the Plane of Disturbing Mass*

This is a relatively simpler case. Consider a mass m of weight W to be rigidly attached to a shaft at radius r and rotating with angular velocity ω. Due to inertia effect, a centrifugal force acts radially outwards and the shaft is subjected to a bending moment. To counteract this effect, a balance weight B is introduced at a radius b on the diametrically opposite side as shown in Fig. 16.3. For equilibrium the inertia forces (*i.e.*, centrifugal forces) due to disturbing mass W and balancing mass B must be equal, opposite and collinear. It is easy to see that the two forces are collinear and act in opposite direction. Thus, for complete equilibrium,

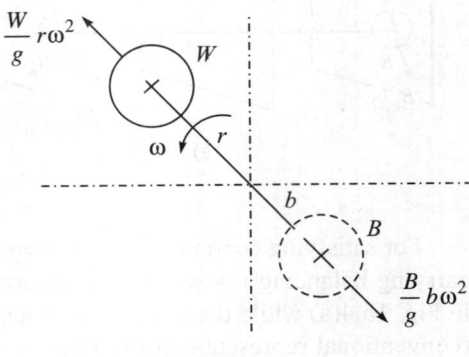

Fig. 16.3

$$(W/g)r\omega^2 = (B/g)b\omega^2 \qquad (16.1)$$

where B = weight of balance mass, and
   b = radius measured from axis of rotation at which the balance mass is placed.

Since in all such cases the disturbing mass and balancing mass rotate with the shaft at the same speed ω, it is much more convenient to cancel out the terms ω^2 and g from equation (16.1) and take (Wr) and (Bb) as the measures of disturbing force and balancing force respectively to a scale. Thus, from equation (16.1)

$$Wr = Bb \qquad (16.2)$$

Therefore for a complete balance the product (Bb) must equal the product $(Wr.)$. This provides large possibilities of selecting any one of the two, namely B and b, arbitrarily. Generally the radius b is made as large as possible so as to restrict B to a lower value.

(b) *Balance Weight not in the Plane of Disturbing Mass*

When the balancing mass cannot be provided in the plane of rotation of disturbing weight W, a single balancing mass will not be enough. This follows from the following discussions.

Referring to Fig. 16.4(a), by arranging a balance weight B_l at radius (b_l) in plane L, in a direction diametrically opposite to that of W, one can ensure the condition $\Sigma F = 0$ for equilibrium. This requires that $B_l b_l = Wr$. This is, however, not enough. As the inertia forces Wr and $B_l b_l$ are equal, opposite but not collinear, they give rise to a couple in longitudinal plane through the shaft axis and passing through the masses W and B_l. This couple tends to rock the shaft in bearings. Evidently, yet another balance mass B_m at radius b_m must be introduced in another reference plane M. For ensuring complete balance, besides satisfying condition $\Sigma F = 0$, it is necessary that Σ couples $= 0$ at any point in the longitudinal (axial) plane passing through the three mass centres. Condition $\Sigma F = 0$ requires that the lines of action of inertia forces Wr, $B_l b_l$, and $B_m b_m$ must be parallel, *i.e.,* the lines of action must be located in the same longitudinal (axial) plane.

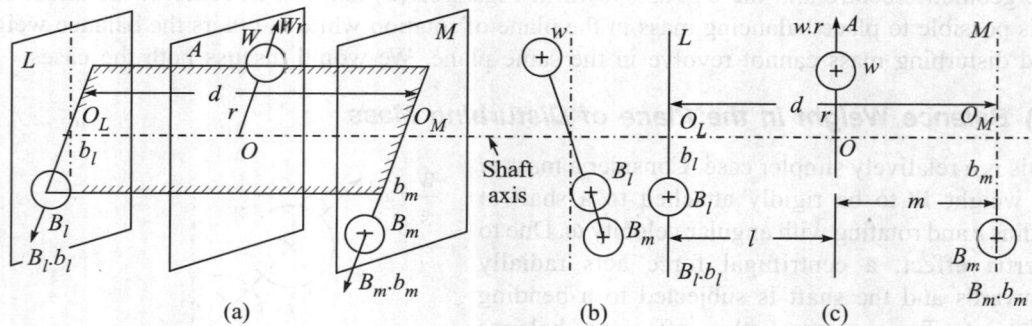

(a)   (b)   (c)

Fig. 16.4

For satisfying condition $\Sigma C = 0$, two arrangements are possible for reference planes L and M carrying balancing masses. One such arrangement is shown by a three-dimensional illustration in Fig. 16.4(a) while the same arrangement is illustrated with the help of side view and elevation (conventional representation) in Figs. 16.4(b) and (c) respectively. In the arrangement shown in Fig. 16.4, the reference planes L and M are located one on each side of the plane of rotation A of the given mass. Taking moments at O_L or O_M, the points of intersection of planes L and M with shaft axis, it becomes evident that both the balancing masses B_l and B_m in planes L and M must be placed on diametrically opposite side of W to ensure that $\Sigma M = 0$ at O_L and O_M. Then for equilibrium,

$$B_l b_l + B_m b_m = Wr \qquad (16.3)$$

and $\qquad (B_l b_l)l = (B_m b_m)m;$ for $\sum C = 0$ at O $\qquad$ (16.4)

Alternatively, $\qquad (B_l b_l)(l + m) = Wr(m);$ for $\sum C = 0$ at O_M $\qquad$ (16.5)

$\qquad (B_m b_m)(l + m) = Wr(l);$ for $\sum C = 0$ at O_L $\qquad$ (16.6)

Any of the equations (16.4) through (16.6) may be used in conjunction with equation (16.3) to evaluate unknown quantities.

Figure 16.5 shows yet another possible arrangement in which both the reference planes L and M are placed on the same side of the plane of rotation A of W. As in the earlier arrangement, for $\sum F = 0$, it is necessary that both the balance masses B_l and B_m be located in the same longitudinal (axial) plane which passes through the disturbing mass centre. Further, for $\sum C = 0$ at all points on shaft axis, the masses B_l and B_m should be placed on the opposite sides of the shaft centre line, as shown in Fig. 16.5(a). The conventional representation using end-view and elevation appears in Fig. 16.5(b) and (c). It follows from Fig. 16.5 that for equilibrium of shaft,

$$Wr + B_m b_m = B_l b_l \qquad (16.7)$$

and $\qquad (B_l b_l)l = (B_m b_m)m,$ for $\sum C = 0$ at O $\qquad$ (16.8)

Alternatively, $\qquad (Wr)l = (B_m b_m)(m - l),$ for $\sum C = 0$ at O_L $\qquad$ (16.9)

and $\qquad (Wr)m = (B_l b_l)(m - l),$ for $\sum C = 0$ at O_M $\qquad$ (16.10)

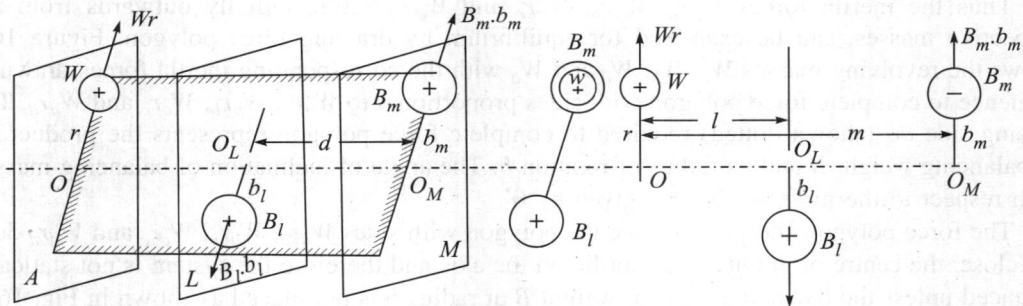

Fig. 16.5

where

$\qquad l$ = distance measured parallel to the axis of the reference plane L from that of the disturbing weight W

and $\quad m$ = distance measured parallel to the axis of the reference plane M from that of disturbing weight W

If $\quad d$ = distance measured parallel to the axis between the reference planes L and M, then replacing term $(m - l)$ by d, equations (16.9) and (16.10) reduce to

$$(B_m)b_m = (Wr)l/d \qquad (16.11)$$

and $$(B_l)b_l = (Wr)m/d \qquad (16.12)$$

In Fig. (16.4) since $(l + m) = d$, it may be verified that equations (16.6) and (16.5) reduce to equations (16.11) and (16.12). Thus equations (16.11) and (16.12) are applicable to both.

16.5 BALANCING OF SEVERAL MASSES REVOLVING IN THE SAME PLANE

Let W_a, W_b, W_c, W_d represent a system of coplanar unbalanced masses that are rigidly attached to the shaft and revolve with it at an angular velocity ω rad/s. Let r_a, r_b, r_c and r_d be the radii of rotation of unbalanced masses and let relative angular positions of masses W_b, W_c and W_d with respect to W_a be θ_b, θ_c and θ_d. When the shaft revolves with masses W_a, W_b, W_c and W_d they maintain same angular relationship.

Unbalanced masses, rigidly attached to and revolving with the shaft and subjected to inertia forces, represent a problem in dynamics on account of centrifugal accelerations. Balancing of such revolving masses can be attempted using laws of static equilibrium only by eliminating rotational motion and consequently centrifugal acceleration thereof. This is attempted by considering product quantities weight times crank radius to represent inertia force to a scale. This is achieved by taking ω^2/g quantity common from each term of vector equation and plotting vector force/couple polygon based on quantities inside the parenthesis shown as follows.

$$(W_a r_a + W_b r_b + W_c r_c + \ldots + W_n r_n = Wr)\omega^2/g$$

and
$$(W_a r_a l_a + W_b r_b l_b = W_c r_c l_c + \ldots + W_n r_n l_n = Wrd)\omega^2/g$$

Thus the inertia forces $W_a r_a$, $W_b r_b$, $W_c r_c$ and $W_d r_d$ acting radially outwards from the respective masses, can be examined for equilibrium by drawing force polygon. Figure 16.6 shows the revolving masses W_a, W_b, W_c and W_d with the corresponding inertia forces drawn in sequence to complete force polygon with sides proportional to $W_a r_a$, $W_b r_b$, $W_c r_c$ and $W_d r_d$. The closing side *do* (shown dotted) required to complete force polygon represents the product Bb of balancing weight B and its radius of rotation b. The angle of inclination of balancing mass B with respect to the mass W_a is then given by θ'.

The force polygon shows that since the polygon with sides $W_a r_a$, $W_b r_b$, $W_c r_c$ and $W_d r_d$ does not close, the centre of gravity does not lie on the axis and therefore the system is not statically balanced unless the balancing mass of weight B at radius b is not placed as shown in Fig. 16.6.

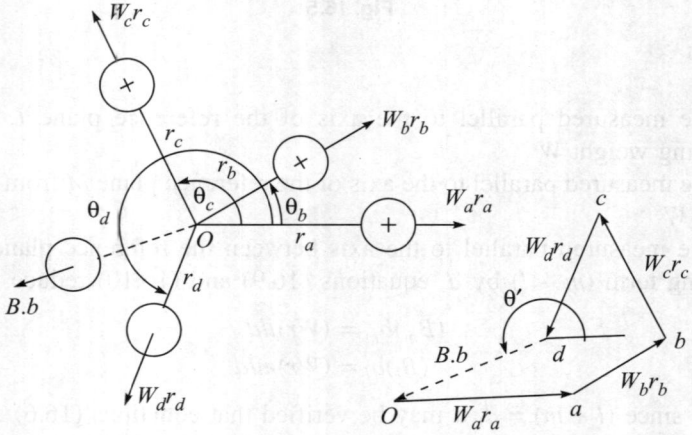

Fig. 16.6 A system of Co-planar forces and their force polygon.

EXAMPLE 16.1 The weights W_1, W_2, W_3 and W_4 are 1962 N, 2943 N, 2354 N and 2550.6 N respectively. The corresponding radii of rotation are 20 cm, 15 cm, 25 cm and 30 cm respectively and the angles between the successive masses are 45°, 75° and 135°. Find the position and magnitude of the balance weight required if its radius of rotation is 20 cm. (Bihar Univ. 1979)

Solution: The inertia (centrifugal) force at the four masses is proportional to:

$$W_1 r_1 = (1962 \times 20) = 39240 \text{ N·cm}$$
$$W_2 r_2 = (2943 \times 15) = 44145 \text{ N·cm}$$
$$W_3 r_3 = (2354 \times 25) = 58860 \text{ N·cm}$$
$$W_4 r_4 = (2550.6 \times 30) = 76518 \text{ N·cm}$$

The force polygon, at Fig. 16.7(b), is completed by drawing vectors *oa*, *ab*, *bc* and *cd* equal to 39240, 44145, 58860 and 76518 N·cm respectively to the scale 1 cm = 19620 N·cm in the respective radially outward directions. The closing side *od* which measures 23544 N·cm represents the resultant of all inertia forces.

The balance weight *B* at radius *b* must be so placed that its inertia force is equal and opposite to the resultant inertia force *od* = 23544 N·cm. The balance weight must therefore produce inertia force equal to 23544 N·cm in the direction *d* to *o*. Hence balance weight *B* is as shown in Fig. 16.7(a) such that

$$B = \frac{23544}{b} = \frac{23544}{20} = 1177.2 \text{ N}$$ **Ans.**

$$\theta_b = 202°$$

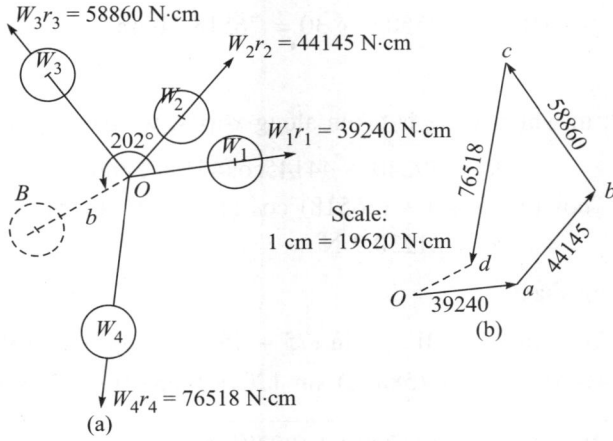

Fig. 16.7 Inertia forces & Force Polygon.

16.6 ANALYTICAL APPROACH FOR SEVERAL ROTATING MASSES IN SAME PLANE

Referring to Fig. 16.6, analytical method simply uses the condition of equilibrium for coplanar forces passing through a common point, namely,

$$\Sigma F_V = 0 \quad \text{and} \quad \Sigma F_H = 0$$

Thus taking the centrifugal direction for revolving weight W_a to be positive X-axis (this is most convenient as all angles are measured wr to this direction), and resolving all forces along X-axis,

$$F_x = (W_a r_a) + (W_b r_b) \cos\theta_b + (W_c r_c) \cos\theta_c + (W_d r_d) \cos\theta_d$$

Similarly,

$$F_Y = (W_b r_b) \sin\theta_b + (W_c r_c) \sin\theta_c + (W_d r_d) \sin\theta_d$$

and then the magnitude of resultant force is given by,

$$R = \sqrt{F_X^2 + F_Y^2}$$

The inclination of this resultant force R to the x-axis is given by, $\tan\theta = \left(\dfrac{F_Y}{F_X}\right)$

The angle of inclination of the balancing weight w.r. to X-axis is thus, $(\theta + 180°)$ and the magnitude of the resultant R then represents the product Bb.

Therefore $\qquad\qquad\qquad\qquad (Bb) = R$

EXAMPLE 16.2 Solve Example 16.1 by analytical method. Resolving along X-axis,

$$\begin{aligned}
(W_a r_a) &= 1962 \times 20 = 39240 \text{ N·cm} \\
(W_b r_b) &= 2943 \times 15 = 44145 \text{ N·cm} \\
(W_c r_c) &= 2354 \times 25 = 58860 \text{ N·cm} \\
(W_d r_d) &= 2550.6 \times 30 = 76518 \text{ N·cm}
\end{aligned}$$

Solution:
Referring to Fig, 16.8, in which $W_a r_a$ is taken along x-axis, we have

$$R_x = 39240 + 44145\cos45° + 58860$$
$$\cos(75 + 45) + (76518)\cos(135 + 75 + 45)$$

Therefore $\qquad R_x = 21220.9$ N

Also resolving along Y-axis,

$$R_Y = (W_b r_b)\sin 45 + (W_c r_c)\sin(75 + 45) + (W_d r_d)\sin(135 + 75 + 45)$$

or $\qquad R_\gamma = (44145)\sin 45 + (58860)\sin 120 + (76518)\sin 255 = 8278.8$ N

Therefore $\quad R = \sqrt{R_X^2 + R_Y^2} = \sqrt{(21220)^2 + (8278.8)^2} = 22777.8$ N

Therefore $\quad Bb = 22778$ N

$$\theta = \tan^{-1}\left(\frac{R_Y}{R_X}\right) = 21.31°$$

Therefore $\quad B = (22778/20) = 1139$ N $\qquad\qquad$ **Ans.**

Therefore $\quad \theta_B = 180 + 21.31 = 201.3°$ $\qquad\qquad$ **Ans.**

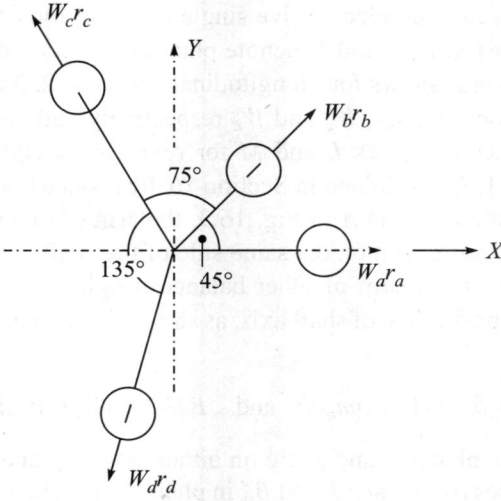

Fig. 16.8

16.7 BALANCING OF SEVERAL MASSES REVOLVING IN DIFFERENT PLANES

A practical example of this type is a long rotor which is required to be mounted in bearings and causing unbalance in different transverse planes along the thickness of the rotor. In all such cases, besides balancing of inertia forces due to revolving masses, one has also to account for the rocking couples in the longitudinal planes through the axis of rotation. This is a problem representing the case of dynamic balancing. The unbalance in different transverse planes can be attributed to improper configurations of the part, errors in machining (as also in casting and forging), improper boring or to keys as well as to assembly problems.

For the sake of simplicity in understanding, let us consider only four unbalanced masses of weights W_a, W_b, W_c and W_d revolving at radii r_a, r_b, r_c and r_d in planes A, B, C and D respectively. The relative angular positions of masses W_b, W_c and W_d relative to W_a is assumed to be θ_b, θ_c and θ_d measured in c.c.w. sense.

There are two methods to solve such a problem. The first one is simple and a direct extension of the method of balancing a single mass by two revolving masses in two different planes. The second method is potentially good and simple although, to begin with, it may not appear to be simpler than the first method.

16.7.1 First Method

The method is based on the findings in Section 16.4(b). Thus, when balancing weights cannot be provided in the plane of disturbing masses, minimum of the two balancing masses must be provided suitably in two different transverse planes. The method consists in selecting two reference planes L and M suitably and finding out the balancing masses required in planes L and M for each disturbing mass taken separately. Various balancing forces in each of the planes L

and M, separately, are then compounded to give single resultant force. As shown in Fig. 16.9(a) let transverse vertical planes A, B, C and D denote planes of rotation of masses W_a, W_b, W_c and W_d respectively. Figure 16.9(a) shows four longitudinal planes 1, 2, 3 and 4 passing through the arms of the revolving masses W_a, W_b, W_c and W_d respectively and also through the axis of the shaft. The balancing masses in planes L and M for revolving weight W_a must be located in longitudinal plane marked 1. As explained in Section 16.4(b), since both the reference planes 'L' and 'M' lie on the same side of plane A in Fig. 16.9, the arm of one of the balance weight B_a' in plane M must be taken parallel and on the same side of the shaft axis as W_a. In that event, for satisfying condition $\Sigma M = O$, the arm of other balance weight B_a in plane L must be located parallel to W_a but on the opposite side of shaft axis, as shown in the figure. Further from equations (16.11) and (16.12),

$$B_a b_a = (W_a r_a)m_a/d \quad \text{and} \quad B_a' b_a' = (W_a r_a)l_a/d$$

Again, as the reference planes L and M lie on either side of plane B of revolving mass W_b, the arms of balancing masses B_b in plane L and B_b' in plane M must be placed along parallel radial lines on the side opposite to that of W_b in relation to shaft axis. The balancing masses are as shown in Fig. 16.9(a) and their magnitudes are given by,

$$B_b b_b = (W_b r_b)m_a/d \quad \text{and} \quad B_b' b_b' = (W_b r_b)l_a/d$$

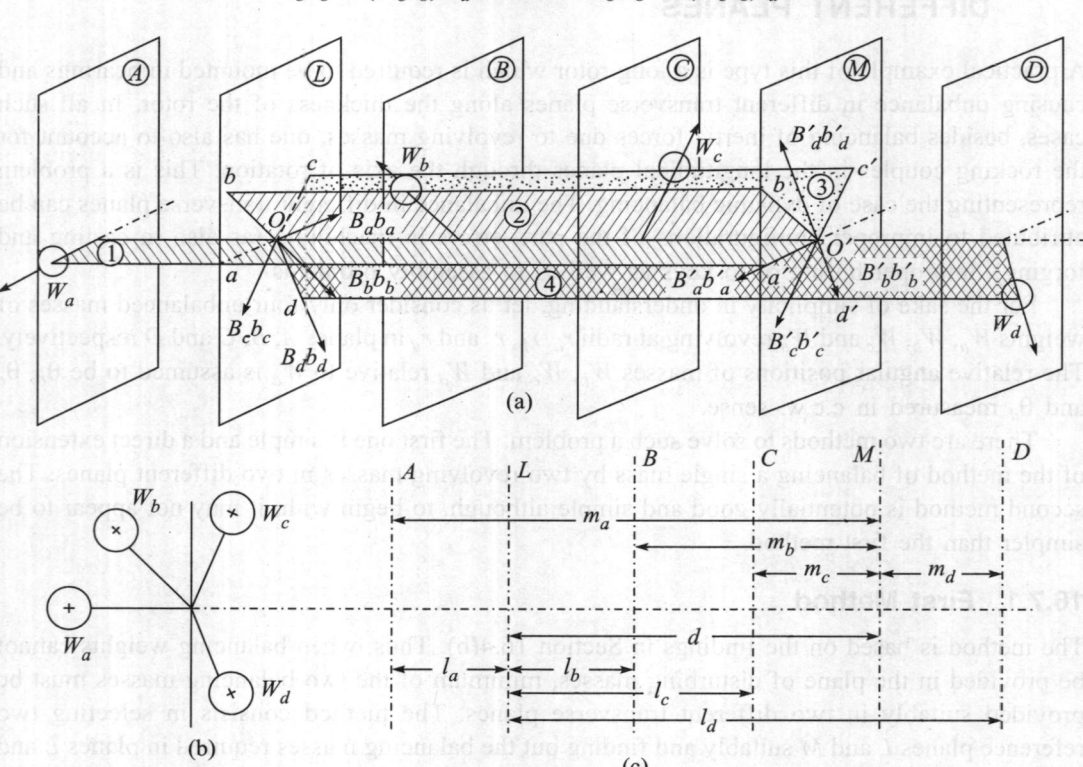

Fig. 16.9

For the disturbing mass W_c in plane C again, the reference planes L and M lie on either side and therefore, as in the case of disturbing mass W_b, the balancing weight B_c in plane L and B_c' in plane M will both be located parallel to W_c but on opposite side of the shaft axis. Their magnitudes are given by,

$$B_c b_c = (W_c r_c) m_c/d \quad \text{and} \quad B_c' b_c' = (W_c r_c) l_c/d$$

For the disturbing mass W_d, lastly, both the reference planes lie on the same side of plane D. Thus balancing weight B_d in plane L must be arranged parallel to and on the same side as W_d, while the balancing weight B_d' must be arranged in plane M parallel to but on opposite side of W_d as shown in Fig. 16.9. The magnitudes of balancing masses are given by,

$$B_d b_d = (W_d r_d) m_d/d \quad \text{and} \quad B_d' b_d' = (W_d r_d) l_d/d$$

Let us adopt a convention that balancing masses in planes L and M lying in the longitudinal plane of disturbing mass, on the side of shaft axis same as that of the disturbing mass, will be taken as positive. But the ones placed on opposite side of disturbing masses will be taken as negative. The data has been represented in Table 16.1.

Table 16.1

Plane	Weight W	Radius r cm	Force ÷ ω²/g = Wr	Distance From		Balancing force ÷ ω²/g	
				Plane L	Plane M	Plane M B'b'	Plane L Bb
A	W_a	r_a	$W_a r_a$	l_a	m_a	$+ (W_a r_a) l_a/d$	$- (W_a r_a) m_a/d$
B	W_b	r_b	$W_b r_b$	l_b	m_b	$- (W_b r_b) l_b/d$	$- (W_b r_b) m_b/d$
C	W_c	r_c	$W_c r_c$	l_c	m_c	$- (W_c r_c) l_c/d$	$- (W_c r_c) m_c/d$
D	W_d	r_d	$W_d r_d$	l_d	m_d	$- (W_d r_d) l_d/d$	$+ (W_d r_d) m_d/d$

Note that in the case of disturbing weight W_a, the reference plane L lies between plane A and M and hence to ensure $\Sigma C = 0$ at the centre of rotation of balance mass B_a, the radii of W_a and B_a' must be placed on the same side of the shaft axis. Similar argument holds good in deciding signs of quantities in the last two columns of Table 16.1 for plane D.

EXAMPLE 16.3 Four masses of mass $m_1 = 100$ kg; $m_2 = 175$ kg; $m_3 = 200$ kg and $m_4 = 25$ kg are fixed to the cranks of 20 cm radius and revolve in planes 1, 2, 3 and 4 respectively. The angular positions of cranks in planes 2, 3 and 4 w.r. to the crank in plane 1 are respectively 75°, 135°, and 200° taken in the same sense. The distances of planes 2, 3 and 4 from plane 1 are 60 cm, 180 cm and 240 cm respectively. Determine the position and magnitude of the balance mass at a radius of 60 cm in plane L and M located midway between planes 1 and 2, and planes 3 and 4 respectively. (DAV Indore, April 1990; SGSITS, April 2006)

Solution: The arrangement of revolving masses in different planes is as shown in Fig. 16.10

Thus, $\quad l_a = 30$ cm; $\quad l_b = 30$ cm; $\quad l_c = 150$ cm; $\quad l_d = 210$ cm

$\quad m_a = 210$ cm; $\quad m_b = 150$ cm; $\quad m_c = 30$ cm; $\quad m_d = 30$ cm

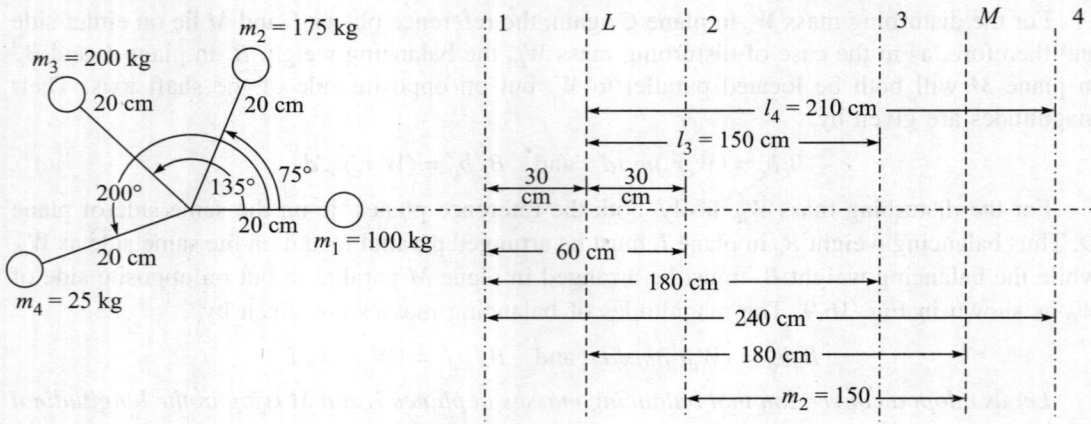

Fig. 16.10 Disturbing masses in several planes.

Plane	Weight $W = mg$ N	Radius r	Force $\div \omega^2/g$ $= W.r$	Distance From cm.		Balancing force	$\div \omega^2/g$
				Plane L	Plane M	Plane M	Plane L
1	100 g	20 cm	2000 g	30	210	$+ (2000)\dfrac{30}{180} g$	$- (2000\ g)\dfrac{210}{180}$
2	175 g	20 cm	3500 g	30	150	$- (3500)\dfrac{30}{180} g$	$- (3500\ g)\dfrac{150}{180}$
3	200 g	20 cm	4000 g	150	30	$- (4000)\dfrac{150}{180} g$	$- (4000\ g)\dfrac{30}{180}$
4	25 g	20 cm	500 g	210	30	$- (500)\dfrac{210}{180} g$	$+ (500\ g)\dfrac{30}{180}$

Thus is plane M the balancing forces are:

$$B_1b_1 = + (333.3)g; \quad B_2b_2 = - (583.3)g;$$
$$B_3b_3 = - (3333.3)g; \quad B_4b_4 = - (583.3)g;$$

and the balancing forces required in plane L are:

$$B_1'b_1' = - (2333.3)g; \quad B_2'b_2' = - (2916.7)g;$$
$$B_3'b_3' = - 666.7g; \quad B_4'b_4' = + (83.3)g$$

Force Polygon at Fig. 16.11(a) is drawn to compound balancing forces required in plane M for balancing centrifugal forces on m_1, m_2, m_3 and m_4. This resultant balancing force is,

$$od = Bb = (3465)g \text{ N·cm}$$

Therefore $(B/g) = \dfrac{3465}{60} = 57.75$ kg at $293°$ to mass m_1 in c.c.w. sense **Ans.**

Similarly, resultant balancing force required in plane L, obtained by compounding individual balancing forces required for masses m_1, m_2, m_3 and m_4, as at Fig. 16.11(b), is

$$o'd' = B'b' = (6480) \text{ g N·cm}$$

Therefore $(B'/g) = \dfrac{6480}{60} = 108$ kg at $308.6°$ to the mass m_1 in c.c.w. sense   **Ans.**

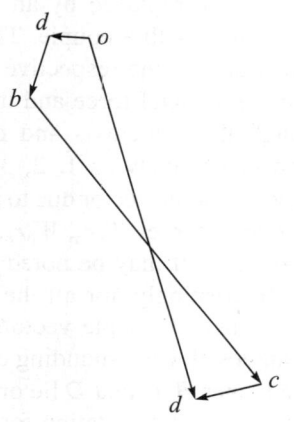

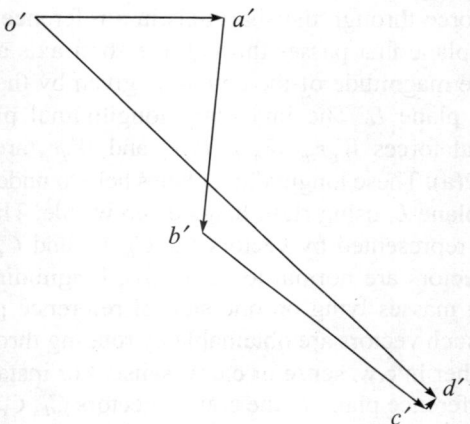

(a) Force polygon in plane (M)
scale:1 cm = 700 g N·cm
$od = (3465 \text{ g}) \text{ N·cm at } 293° \text{ to } m_{1z}$

(b) Force polygon for plane (L)
scale:1 cm = 900 g N·cm $o'd'$
$(6480 \text{ g}) \text{ N·cm at } 0308.6° \text{ to } m_1$

Fig. 16.11

16.7.2 Second Method (Dalby's Method)

This method was proposed by Prof. W.E. Dalby. The basic principle involved is as follows. A force acting on a rigid body, which is fixed at one point, is equivalent to an equal and parallel force acting through the fixed point, together with a couple which tends to rotate the body in the plane containing the line of action of the force and the fixed point. The magnitude of the couple is given by the product of the force and the distance between the plane of rotation and the fixed point.

When a number of centrifugal forces due to unbalanced rotating masses act along the length of the shaft, each such force may be replaced by an equal and parallel force through a fixed point in a reference plane together with a couple. When each of the centrifugal force on the shaft is reduced to a corresponding force and couple in the reference plane, the resultant force and resultant couple must be zero for the equilibrium of the rigid body. The construction of force polygon does not pose any problem but for drawing couple polygon, some understanding is required to be developed.

It is usual to represent a couple using right-handed screw rule. Thus to represent a couple a vector is drawn perpendicular to the plane of couple, with length proportional to the magnitude of couple and direction of arrowhead consistent with the sense of couple as decided by right-handed screw rule. However, when planes of all such couples pass through the axis of rotation, a much simpler convention is possible. This feature makes the proposed method more effective and quite simple. To develop this convention, let us proceed as follows.

Figure 16.12(b) shows angular locations of four revolving unbalanced weights, while Fig. 16.12(a) shows longitudinal planes 1, 2, 3 and 4 passing through the respective cranks of revolving weights and the longitudinal shaft axis. Figure 16.12(c) shows the relative locations transverse of planes A, B, C, D and reference planes L and M along the shaft axis.

Centrifugal force, due to each of the revolving masses, can be replaced by an equal and parallel force through the shaft axis in a reference plane L, together with a couple. This couple acts in a plane that passes through the shaft axis and the crank arm of the respective revolving mass. The magnitude of the couple is given by the product of centrifugal force and its distance from the plane L. The imaginary longitudinal planes through the shaft axis and containing centrifugal forces $W_a r_a$, $W_b r_b$, $W_c r_c$ and $W_d r_d$ are numbered respectively as 1, 2, 3 and 4 in Fig. 16.12(a). These longitudinal planes help to understand sense of couple vector due to respective force in plane L, using right-handed screw rule. The couples due to forces $W_a r_a$, $W_b r_b$, $W_c r_c$ and $W_d r_d$ are represented by vectors C_a, C_b, C_c and C_d in Fig. 16.12(d). It may be noted that these couple vectors are normal to respective longitudinal planes. Interestingly, for all the planes of revolving masses lying on one side of reference plane L, the sense of couple vectors is same. Thus all such vectors are obtainable by rotating through right angles all corresponding centrifugal forces either in c.w. sense or c.c.w. sense. For instance, as the planes B, C and D lie on the same side of reference plane L, the couple vectors C_b, C_c and C_d are obtained by rotating force vectors $W_b r_b$, $W_c r_c$ and $W_d r_d$ through right angles in c.c.w. sense. As against this, the plane of rotation of mass W_a lies on the other side of plane L and the couple vector C_a is obtained by rotating force vector $W_a r_a$ through right angle in the opposite (i.e., c.w. sense) sense as shown in Fig. 16.12(d).

Now if the couple vectors of Fig. 16.12(d) are rotated through right angles in clockwise sense the couple vectors C_b, C_c and C_d for planes lying on one side of reference plane L point in the direction of centrifugal forces $W_b r_b$, $W_c r_c$ and $W_d r_d$ while the couple vector C_a for plane A lying on the opposite side of plane L points in the radially inward direction of the crank carrying W_a. This is shown in Fig. 16.12(e). It is important to note that when all the couple vectors are rotated through right angles in the same sense, their relative positions remain unaffected, but drawing the couple polygon becomes much more easier. At the same time, it becomes relatively easier to work without the aid of three-dimensional visualization as in Fig. 16.12(a). *Thus, the working rule is that the couple vectors for all the rotating masses lying on one side of reference plane may be drawn in radially outward direction along corresponding crank, and in radially inward direction along corresponding crank for masses lying on the other side of the reference plane.* Figure 16.12(f) shows a couple polygon obtained by adding the couple vectors, represented in magnitude and direction (obtained by rotating through 90° in c.w. direction) by C_b, C_c, C_d and C_a. The closing side of the polygon represents balancing couple vector C_m due to unknown balancing mass B_m at radius b_m in the reference plane M. Note that O being the point of intersection of shaft axis with plane L, the moment about O due to centrifugal force on balancing mass B_l in plane L vanishes. Note also that due to rotation of couple vectors through right angles (as per the convention) the direction of couple vector C_m coincides with the direction of corresponding centrifugal force F_m in this case. The magnitude of this force is given by $B_m b_m \equiv F_m = (C_m/d)$. Once the force F_m is established, the balancing force F_l required in plane L is determined by completing force polygon in plane L as shown in Fig. 16.12(g).

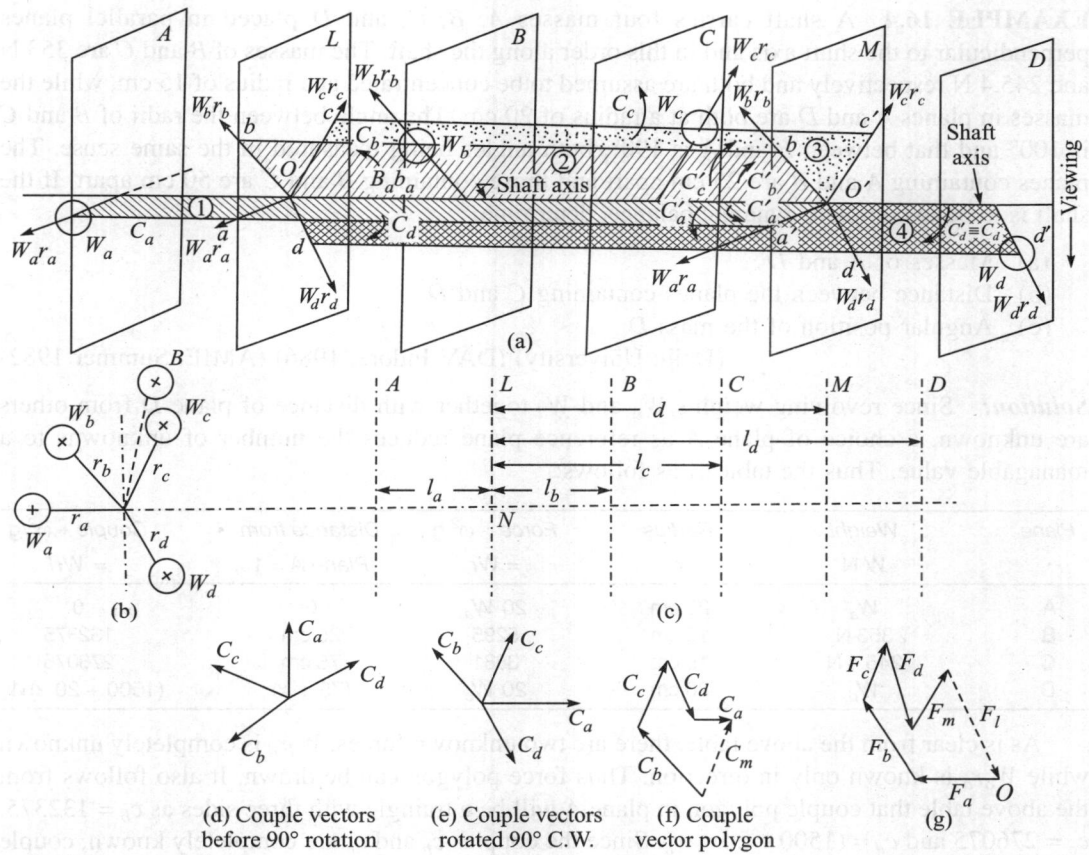

Fig. 16.12   Illustration for Dalby's Method.

It is advisable to tabulate the data with planes indicated in the first column, reading them from left to right. The distances to the left of reference plane may be regarded as negative and the ones to the right as positive. The negative sign of a couple then indicates that corresponding couple vector is to be drawn in radially inward direction. Table 16.2 depicts planes in Fig. 16.12(a) and corresponding forces & couples.

Table 16.2

Plane	Weight W	Radius r	Force $\div \omega^2/g$ $= W.r$	Distance from plane L l	Couple $\div \omega^2/g$ $= W.r.l$
A	W_a	r_a	$W_a r_a$	$-l_a$	$-W_a r_a l_a$
L	B_l	b_l	$B_l b_l$	0	0
B	W_b	r_b	$W_b r_b$	l_b	$W_b r_b l_b$
C	W_c	r_c	$W_c r_c$	l_c	$W_c r_c l_c$
D	W_d	r_d	$W_d r_d$	l_d	$W_d r_d l_d$

EXAMPLE 16.4 A shaft carries four masses A, B, C, and D placed in parallel planes perpendicular to the shaft axis and in this order along the shaft. The masses of B and C are 353 N and 245.4 N respectively and both are assumed to be concentrated at a radius of 15 cm, while the masses in planes A and D are both at a radius of 20 cm. The angle between the radii of B and C is 100° and that between B and A is 190°, both angles being measured in the same sense. The planes containing A and B are 25 cm apart and those containing B and C are 50 cm apart. If the shaft is to be in complete dynamic balance, determine:

(a) Masses of A and D
(b) Distance between the planes containing C and D
(c) Angular position of the mass D

(Delhi University) (DAV Indore, 1986) (AMIE, Summer 1982)

Solution: Since revolving weights W_A and W_d together with distance of plane D from others are unknown, a choice of plane A as reference plane reduces the number of unknowns to a managable value. Thus the table is as follows:

Plane	Weight W N	Radius r	Force ÷ ω^2 g = Wr	Distance from Plane A = 1	Couple ÷ ω^2/g = Wrl
A	W_a	20 cm	20 W_a	0	0
B	353 N	15 cm	5295	25 cm	132375
C	245.4 N	15 cm	3681	75 cm	276075
D	W_d	20 cm	20 W_d	(75 + x)	(1500 + 20 x)W_d

As is clear from the above table, there are two unknown forces. $W_d r_d$ is completely unknown while $W_a r_a$ is known only in direction. Thus force polygon can be drawn. It also follows from the above table that couple polygon in plane A will be a traingle with three sides as c_b = 132375, c_c = 276075 and c_d = (1500 + 20x) w_d. Since the couples c_b and c_c are completely known, couple polygon is completed as at Fig. 16.13(c). By measurement, $c_d = c_o$ = 291357. Thus equating,

$$(1500 + 20x)W_d = 291357 \qquad (i)$$

Further, since the direction of centrifugal force on W_d will be same as that of C_d (= co), the force polygon for plane A is as shown in Fig. 16.13(d). Note that direction of force ($W_a r_a$) is along the radius OA. By measurement from Fig. 16.13(d)

$$cd = W_a r_a = 3825.9$$

Therefore

$$W_a = \frac{3825.9}{20} = 191.3 \text{ N} \qquad \textbf{Ans.}$$

Also,

$$do' \equiv W_d r_d = 3139.2$$

Therefore

$$W_d = \frac{3139.2}{20} = 157 \text{ N} \qquad \textbf{Ans.}$$

Hence from equation (i),

$$(1500 + 20x) \times 157 = 291357$$

Therefore

$$x = \left(\frac{291357}{157} - 1500\right)\Big/20 = 17.8 \text{ cm} \qquad \textbf{Ans.}$$

Angular location of W_d from W_b c.c.w. = θ_d = 251°   **Ans.**

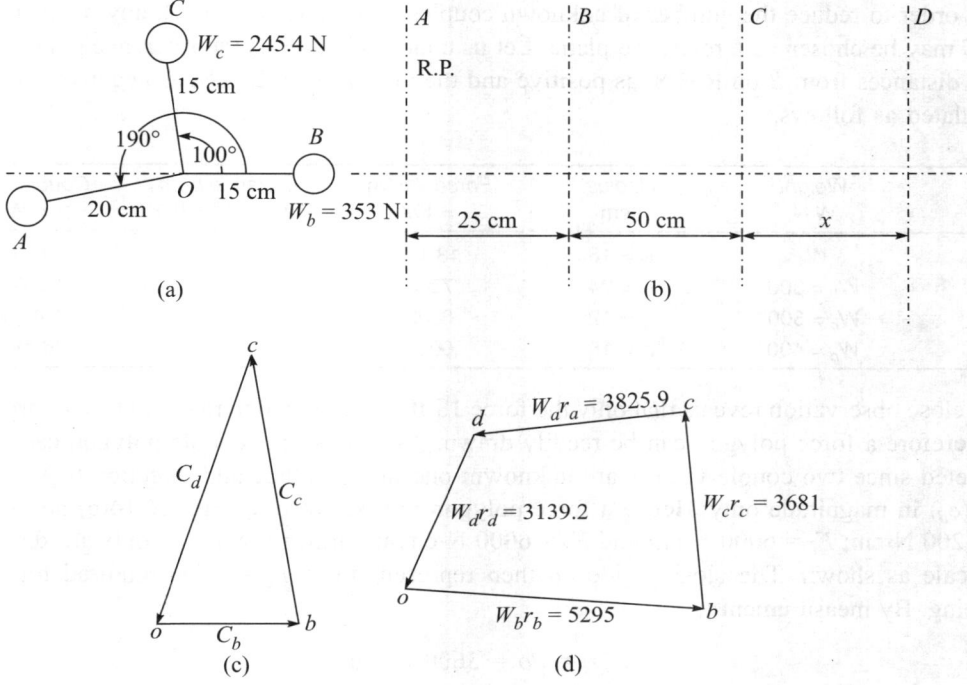

Fig. 16.13

EXAMPLE 16.5 A shaft carries four rotating masses A, B, C, and D in this order along its axis. The mass A may be assumed to be concentrated at a radius of 18 cm; B at 24 cm; C at 12 cm and D at 15 cm. The weights B, C and D are 300 N, 500 N and 400 N respectively. The planes containing B and C are 30 cm apart. The angular spacing of the longitudinal planes containing C and D are 90° and 210° respectively relative to B, measured in the same sense. If the shaft and masses are to be in complete dynamic balance, find (a) the weight and angular position of mass A, (b) the position of planes A and D. (DAV Indore, April 1992)

(AMIE, Summer 1979; Summer 1976 modified)

Solution: The relative position of masses along the axis of shaft and their relative angular spacing is as shown in Figs. 16.14(a) and (b).

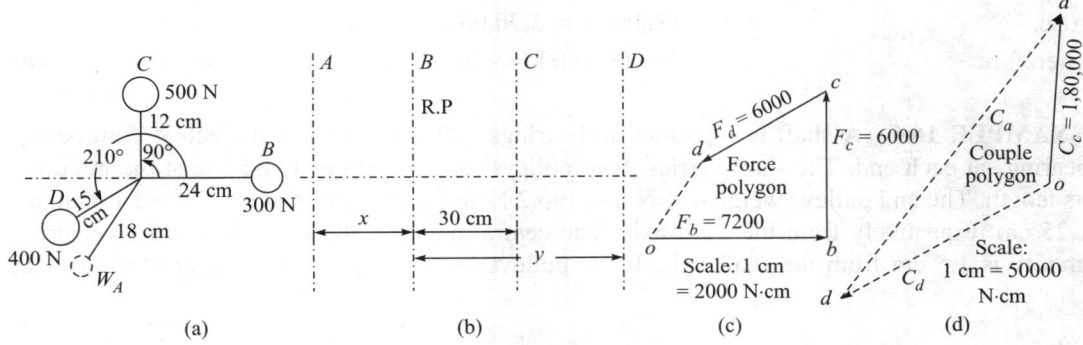

Fig. 16.14

In order to reduce the number of unknown couples to a managable level, any of the planes B or C may be chosen as a reference plane. Let us choose plane B as the reference plane. Then taking distances from B on R.H.S. as positive and the distances on L.H.S. as negative, the data is tabulated as follows.

Plane	Weight W N	Radius r cm	Force ÷ ω^2/g = Wr	Distance from Plane (R.P) B = l	Couple ÷ ω^2/g = Wrl
A	W_a	$r_a = 18$	$18\,W_a$	$-x$	$-18\,W_a x$
B	$W_b = 300$	$r_b = 24$	7200	0	0
C	$W_c = 500$	$r_c = 12$	6000	30	1,80,000
D	$W_d = 400$	$r_d = 15$	6000	y	(6000) y

A close observation reveals that only the force $18\,W_a$ is unknown in magnitude and direction and therefore a force polygon can be readily drawn. As against this couple polygon cannot be completed since two couple vectors are unknown; one in magnitude and direction (c_a) and the other (c_d) in magnitude only. Hence a force polygon is first drawn at Fig. 16.14(c) and forces $F_b = 7200$ N·cm; $F_c = 6000$ N·cm and $F_d = 6000$ N·cm are drawn in radially outward direction to a scale as shown. The closing side do then represent force $F_a = W_a r_a$ required for force balancing. By measurement

$$W_a r_a = do = 3600 \text{ N·cm}$$

Therefore $\qquad\qquad W_a = 3600/18 = 200$ N $\qquad\qquad$ **Ans.**

The angular location of mass W_a from the mass W_b measures

$$\theta_A = 235° \qquad\qquad\qquad\qquad$$ **Ans.**

With location of mass W_a and its weight known, the direction of couple vector $C_a = -18\,W_a x$ (which is radially inward at W_a) is known. From the end points of vector $C_c = 1,80,000$ N·cm^2, which is indicated by line oc in Fig. 16.14(d), line od can be drawn parallel to the directions of couple vector $= (6000)y$ (radially outward direction at W_d) and line ad parallel to $c_a = -18 \times 20 \times x$ N·cm in radially inward direction at A to complete the triangle. Measurement shows,

$$c_d = 6000\ y = 2,45,000$$

Therefore $\qquad\qquad y = 40.83$ cm $\qquad\qquad\qquad\qquad$ **Ans.**

Also, $\qquad\qquad\qquad c_a = -(3600)\ x = 3,70,000$

Therefore $\qquad\qquad x = 3,70,000/3600 = -102.78$ cm $\qquad\qquad$ **Ans.**

EXAMPLE 16.6 A shaft is supported in bearings 180 cm apart and projects 45 cm beyond bearings at each end. The shaft carries three pulleys' one at each end and one at the middle of its length. The end pulleys weigh 471 N and 196.2 N and their centres of gravity are 1.5 cm and 1.25 cm respectively from the shaft axis. The central pulley weighs 549.4 N and its centre of gravity is 1.5 cm from the shaft axis. If the pulleys are arranged so as to give static balance, determine (a) relative angular positions of the pulleys, and (b) dynamic forces produced on the bearings when the shaft rotates at 300 r.p.m.(AMIE, Summer 1978)

Solution: Figure 16.15(b) shows a schematic sketch of shaft and pulleys. Since the pulleys are arranged so as to produce static balance, the c.g. of the system should lie on the axis of rotation. In other words, $W_a r_a$, $W_b r_b$ and $W_c r_c$ for the three pulleys must close a triangle. And this gives a clue to the relative angular positions of the three pulley masses. Thus assuming weight of pulley B to be arranged in vertical diameter, Fig. 16.15(a) determines relative angular positions of pulleys.

Plane	Weight W N	Radius r cm	Force ÷ ω^2/g Wr	Distance from plane L, l cm	Couple ω^2/g wrl
A	471	1.5	706.5	− 45	− 31784.4
L	W_l	r_l	$W_l r_l$	0	0
B	549.4	1.5	824.1	90	74169
M	W_m	r_m	$W_m r_m$	180	180 $W_m r_m$
C	196.2	1.25	245.3	225	55181.3

Thus angle between pulleys A and B = 180 − 16 = 164° **Ans.**
and angle between pulleys B and C = 180 − 52 = 128° **Ans.**
    Referring to the above table, the column for couple vectors involve only one unknown in (180 $W_m r_m$). Hence, couple polygon is first drawn at Fig. 16.15(c) to establish ($W_m r_m$)—the resultant unbalanced force in plane M.
    Representing couple vector *ao* = 31784.4 to scale in the negative direction (*i.e.*, radially inward direction for A) and vector *oc* = 55181.3 and *cb* = 74169 unit in directions radially outward for W_c and W_b respectively gives the closing line for polygon as *ba*. The closing side *ba* represents the unknown quantity 180 $W_m r_m$ and measures 87554.25 to scale. Hence $W_m r_m$ = 486.4 N·cm, which is inclined at 38° to W_b in c.c.w. direction, and represents resultant unbalanced force in bearing M.

$$W_m r_m = 486.4 \text{ N·cm}$$

Therefore   The dynamic force in bearing M

$$= (W_m r_m)(\omega^2/g) = \frac{486.4}{981}\left(\frac{2\pi \times 300}{60}\right)^2 = 489.4 \text{ N} \qquad \textbf{Ans.}$$

With $W_m r_m$ known, a force polygon is completed at Fig. 16.15(d), and the closing side *oa* then represents resultant unbalanced force $W_l r_l$ and measures 471 N to scale. This being inclined at 216° c.c.w. from W_b. The unbalanced dynamic force at bearing L is, then

$$= (W_l r_l)\frac{\omega^2}{g} = (471)\frac{1}{981} \times \left(\frac{2\pi \times 300}{60}\right)^2 = 473.4 \text{ N} \qquad \textbf{Ans.}$$

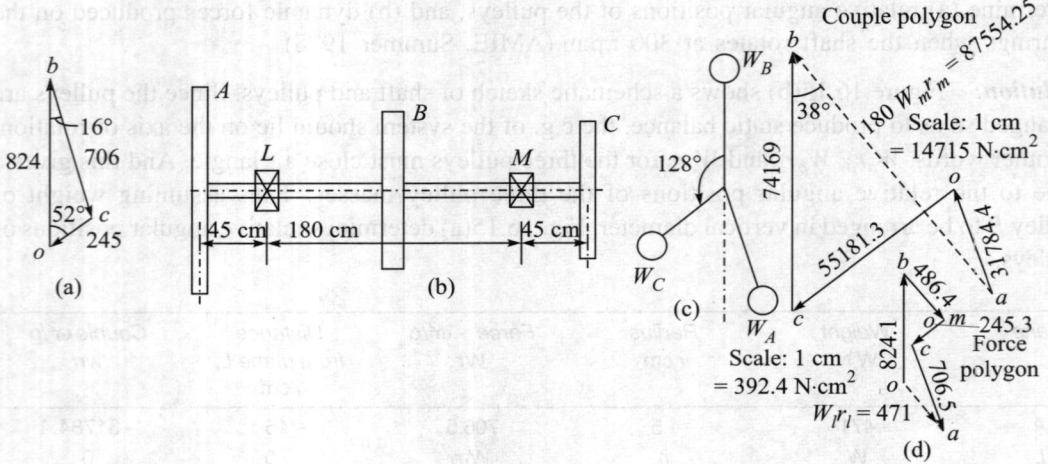

Fig. 16.15

EXAMPLE 16.7 A small three-throw crank shaft has cranks of radii 125 mm, set at 120° to each other, and equally spaced with a pitch of 250 mm. The revolving masses at crank radii are the same for each line and of amount 15 kg. The shaft is supported in two bearings symmetrically arranged with respect to the cranks and 850 mm apart. Determine the dynamical loads on the bearings for a speed of 500 r.p.m.

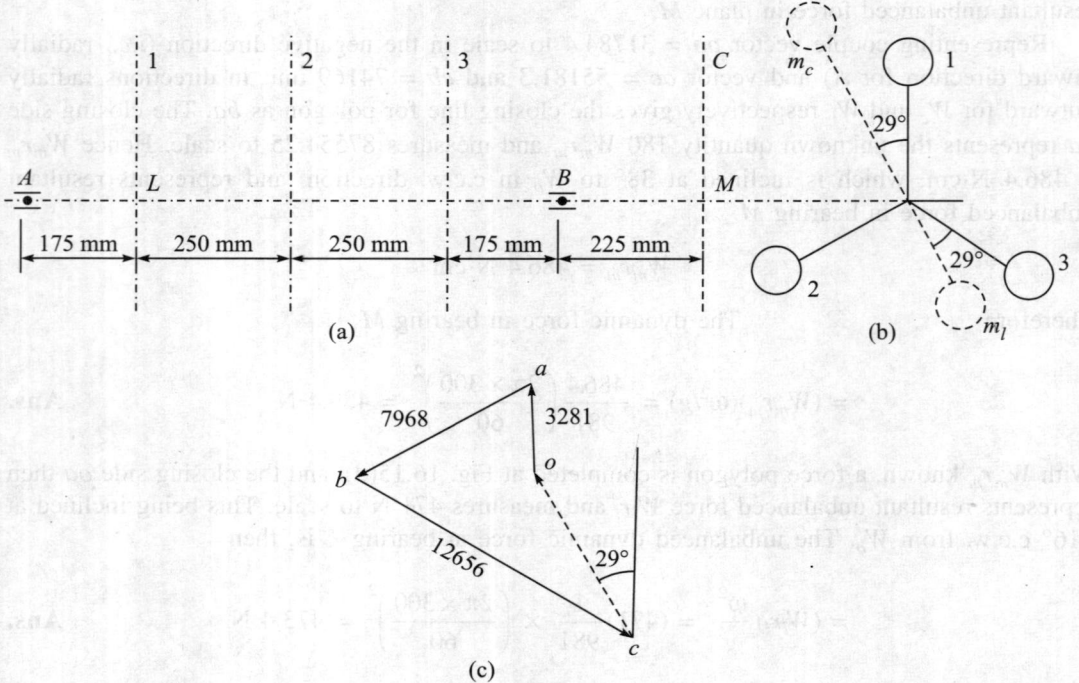

Fig. 16.16

The shaft is to be balanced by means of a mass at a radius of 187.5 mm in the plane of No.1 crank, and a mass at radius 250 mm attached to the flywheel situated 225 mm, beyond the bearing adjacent to No. 3 crank. Determine the magnitude of these balance masses and their angular positions relative to No.1 crank. (Univ. London)

Solution: Referring to Fig. 16.16(a) and (b), as the cranks carrying equal masses at same radii are symmetrically placed, the forces are balanced. Let us choose plane A as the reference plane. The data can be tabulated as follows:

Plane	Mass in kg	Radius in cm	Force ÷ ω^2 mr	Distance from plane $A : l$ cm	Couple ÷ ω^2 = mrl
A	–	–	R	0	0
1	15	12.5	187.5	17.5	3281.25
2	15	12.5	187.5	42.5	7968.75
3	15	12.5	187.5	67.5	12656.25
B	–	–	R	85.0	85 R

Since column of centrifugal force involves two unknown forces in planes A and B, it is preferable to solve couple polygon first.

The closing side of the polygon co shown in Fig. 16.16(c) denotes couple in plane A due to force R in plane B. Thus,

$$85R = 8100$$

Therefore $$R = 8100/85 = 95.29$$

Thus, dynamic load at bearing B

$$= (mr)\omega^2$$

$$= 95.29 \left(\frac{2\pi \times 500}{60} \right)^2 \bigg/ 100 = 2612.43 \text{ N}$$ **Ans.**

The dynamic load at bearing A is equal and opposite to that in bearing B.

Let the shaft be now balanced by a mass m_1 at a radius of 18.75 cm in plane No.1, and by another mass m_c at radius 25 cm attached to the flywheel.

The out of balance couple $mrl = 8100$ kg·cm^2 will be balanced by equal, parallel and opposite mass–arm products ($mr \equiv F$) such that

$$F \times LM = 8100$$

Therefore $$mr = \frac{8100}{(25.0 + 25.0 + 17.5 + 22.5)} = \frac{8100}{90} = 90 \text{ kg·cm}$$

The radius of mass m_1 in plane 1 being 18.75 cm,

$$(m_1) \times 18.75 = 90 \quad \text{Therefore} \quad m_1 = 90/18.75 = 4.8 \text{ kg}$$ **Ans.**

Again, the radius of mass m_c in plane C being 25 cm,

$$(m_c) \times 25 = 90 \quad \text{Therefore} \quad m_c = 3.6 \text{ kg}$$ **Ans.**

The angular position of balance masses are shown by m_1 and m_c in Fig. 16.16(b).

EXAMPLE 16.8   As shown in Fig. 16.17(a) and (b) the out-of-balance of a machine rotor is equivalent to 5 kg at 10 mm radius in one plane A together with an equal mass at 15 mm radius in a second plane B. $AB = 375$ mm and the two radii are 120°. Find the mass required in a third plane C at a radius of 125 mm and its angular position with respect to the given radii so that there is no resultant out-of-balance force. Find also the position of C along the axis for the residual couple to be a minimum and the value of this couple when the speed is 500 r.p.m.

(SGSITS Indore, Dec. 1994) (IMech. Engg. London)

Solution: There being only one unknown force, a force polygon is completed as in Fig. 16.17(c). The closing side *bo* gives direction and magnitude of force in plane C. Thus,

$$(12.5)m_c = 6.625$$

Therefore $m_c = 0.53$ kg   **Ans.**

and is located at $\theta_c = 259°$ from m_a **Ans.**

For minimum residual couple, side '*ed*' of couple polygon should be smallest. The smallest value of the length of side '*ed*' = p (*see* Fig. 16.17d). For this, the value of couple C_c is given by

$$C_c = C_b \cos 41 = (281.25) \times \cos 41$$
$$= 212.26$$

Hence, $(12.5)\ (0.53)\ x$ $\qquad\qquad = 212.26$

$$x = 32.04 \text{ cm from } A$$

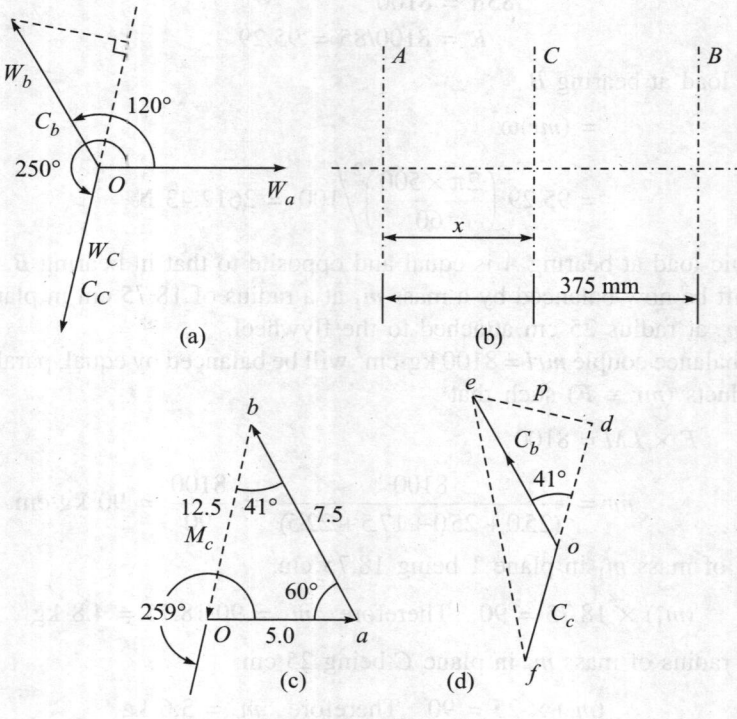

(a)

(b)

(c)   (d)

Fig. 16.17

Unbalance couple: $(C_b \sin 41) \omega^2$

$$= (281.25) \cdot (0.656) \cdot (52.36)^2$$
$$= 505817.2 \text{ cm} \cdot \text{kg} \cdot \text{cm/s}^2 = 50.58172 \text{ m.N} \qquad \textbf{Ans.}$$

Plane	Mass m kg	Radius r cm	Force $\div \omega^2$ $= mr$	Distance l from A cm	Couple $\div \omega^2$ $= mrl$
A	5	1.0	5.0	0	0
B	5	1.5	7.5	37.5	281.25
C	m_c	12.5	$12.5\,m_c$	x	$12.5\,m_c \cdot x$

16.8 BALANCING OF ROTORS

One of the principal causes of vibration in rotating machinery is 'unbalanced rotors'. The amount of unbalance that is permissible for a rotor depends upon the type of application of the machine. Following general guideline may, however, be prescribed:

(a) The higher the speed of rotation, lower is the permissible unbalance. This is obvious since the unbalanced centrifugal force varies as the square of the r.p.m. N (*see* Fig. 16.18).
(b) Nearer the normal running speed of the rotor is to the resonating speed of the 'rotor-mounting', smaller is the permissible unbalance.
(c) The larger the ratio of 'rotor-mass' to 'the total mass', smaller is the permissible unbalance of the rotor.

General Comments

Readers may recall that amount of counter-balance needed for balancing was expressed as a product of weight/mass and radius. Clearly, if one of the two is chosen arbitrarily, the other can be easily determined. In practice it is usual to choose a convenient value for radius and that fixes the weight/mass required for balancing. The balancing weights are either added by attaching a known weight to the rotor or by drilling out and removing mass from the rotor from a position 180° away in the same transverse plane. To reduce the required balancing weight, the counterweight should be placed at the maximum available radius. Further, when the counterweights are placed in two transverse planes, the planes should be arranged as far away from one another as possible. This step leads to a reduction in the magnitude of counterweights.

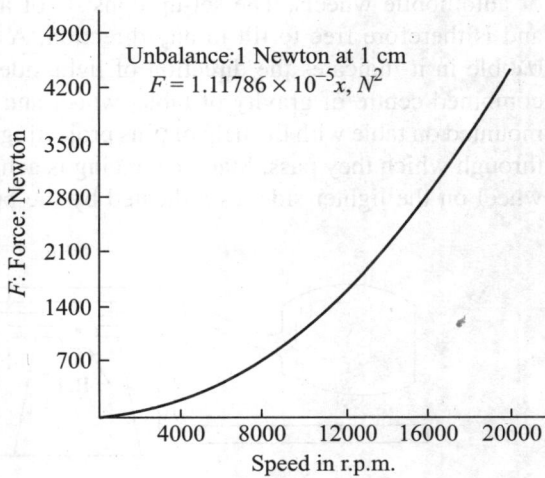

Fig. 16.18 Effect of speed on unbalanced force.

Although any unbalance in a rotor can be corrected by adding counterweights in two transverse planes, this step may leave some bending moment in the shaft. It is advisable therefore that, as far as possible, each unbalance should be balanced by putting counterweights in the same plane. For instance, unbalance due to each crank in an automobile is frequently balanced by adding counterweights opposite to each crank.

16.9 STATIC AND DYNAMIC BALANCING

The type of balancing required depends on the type of machine, its size, speed and economics. The curve in Fig. 16.18 emphasizes the fact that even a small amount of unbalance can produce large centrifugal forces at high speeds. The curve is drawn to indicate centrifugal forces produced by a centimeter-Newton of unbalance at various speeds.

In general, static balancing suffice the purpose when rotating members have short axial length and which rotate at low to moderate speeds. At these speeds the unbalanced couples due to inertia forces are relatively small. Automobile wheels, propellers of airplanes and narrow fans are a few examples of this type.

16.10 STATIC BALANCING MACHINES

Static balancing is carried out by 'trial and error' method. It is essentially a weighing process in which the part is acted upon by either a pull of gravity or a centrifugal force. When machine parts are produced in large quantities, one of the requirements a balancing machine should meet is that it should measure both the amount and location of the unbalance and indicate the required correction directly and quickly. In order to save time, it may be further desired that the part is not required to be rotated. Figure 16.19 illustrates diagrammatically a machine for static balancing of automobile wheels. The set-up consists of a table which is supported on a spherical ball B and is therefore free to tilt in any direction. A round spirit level mounted on the table and the bubble in it indicates the direction of light side. When the wheel is mounted in the set-up, the combined centre of gravity of table, wheel and tyre lies below the centre of ball. The wheel is mounted on table with the help of pins projecting out from the table and the lug-holes in the wheel, through which they pass. Static balancing is achieved by fastening lead-weights to the rim of the wheel on the lighter side, as indicated by the bubble in the spirit level.

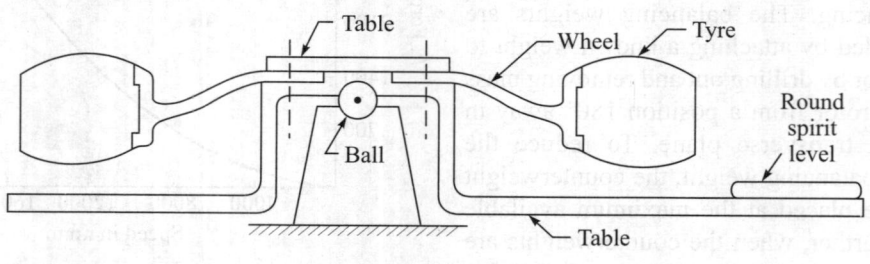

Fig. 16.19

Figure 16.20 shows a yet another static balancing machine. The machine permits oscillation of the cradle, carrying the rotor and electric motor, about an axis of oscillation P-P, which is parallel to the axis of rotation A-A. The part to be balanced is mounted on a cradle C which is held in position by springs S and pivots T. Springs in transverse planes permit angular oscillations of the cradle in transverse plane about axis P-P and allows the system to have its own natural frequency of vibration.

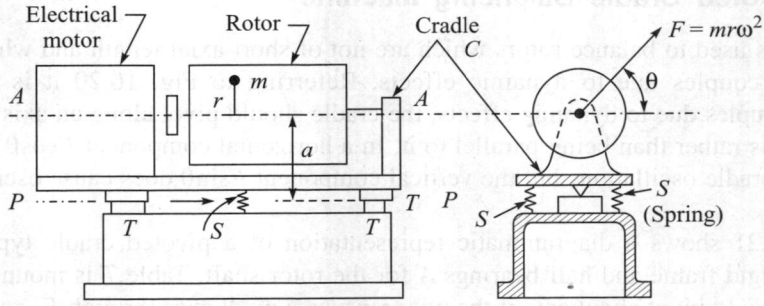

Fig. 16.20 A static balancing machine.

A flexible coupling connects the rotor to an electric motor and the speed is adjusted to coincide with the natural frequency of the system. Under resonating condition (*see* Chapter 17), even a small amount of unbalance produces large amplitude of angular oscillations of the cradle. The centrifugal force or its component in longitudinal vertical plane does not produce any rotation couple, but the component in longitudinal horizontal plane does produce a couple $F_c a\cos\theta$ as shown in the figure. If m is the out-of-balance mass at radius r then, for an angular speed of rotation ω radians per second, this couple is given by $(mr\omega^2)a\cos\theta$, and it acts in transverse vertical plane. This centrifugal force due to static unbalance is maximum (= $mr\omega^2$)a, when mass m occupies position in the horizontal longitudinal plane. The type of pivot support does not permit oscillations (in longitudinal vertical plane) due to dynamic unbalance, if any.

16.11 DYNAMIC BALANCING MACHINES

Customary unit in which unbalance is measured is gram·centimeter. If correct practice is followed in S.I. units, the most appropriate unit of unbalance is miligram·metre (mg·m). It was earlier pointed out that static balancing is sufficient for rotating discs, wheels, gears, etc., where the unbalanced mass can be assumed to exist in a single rotating plane. As against this, in longer machine members like turbine rotors or motor armatures, the unbalance may exist in more than one transverse planes and the unbalanced centrifugal forces give rise to couples in longitudinal planes through shaft axis. The purpose of dynamic balancing is therefore to measure such unbalanced couple, if any, and to add a new couple of equal magnitude but opposite in sense in the same longitudinal plane. The required balancing couple is introduced by adding masses in two pre-selected correction planes. In general, a rotor in question requires static as well as dynamic balancing and as such magnitude as well as the radial location of both the balancing masses need not be the same in the correction planes. Hence, we need to measure not only the angular locations but also the magnitudes of the balancing masses.

There are three methods to measure the required correction for dynamic balancing. These are: (i) the pivoted cradle method, (ii) the nodal point method and (iii) the mechanical-compensation method. In the following text, we will discuss the pivoted cradle method only for brevity.

16.11.1 Pivoted Cradle Balancing Machine

The machine is used to balance rotors which are not of short axial length and which, therefore, must balance couples due to dynamic effects. Referring to Fig. 16.20 it is clear that for considering couples due to dynamic effects, the cradle should pivot along an axis perpendicular to the shaft axis rather than being parallel to it. In a horizontal component $F\cos\theta$ does not have any effect on cradle oscillations but the vertical component $F\sin\theta$ does cause oscillations of the cradle.

Figure 16.21 shows a diagrammatic representation of a pivoted cradle type machine. It consists of a rigid frame and half bearings A for the rotor shaft. Table T is mounted on springs and can be made to pivot about any of the two transverse pivot axes through F_1 and F_2 as shown in the end view. Needless to say that this arrangement permits the cradle, along with rotor, to oscillate in longitudinal vertical plane and have its own natural frequency of oscillations. In the type of machine shown in the figure, a balancing head is solidly coupled to the rotor under test and contains two arms with mass M. Unless otherwise adjusted with push buttons, these arms rotate with the disc and the rotor. The disc B (*i.e.*, the balancing head) incorporates a motor and a system of gears which permit the operator to rotate the masses relative to the disc B. The power is supplied through slip-rings. Out of the two push buttons, the first one causes the two arms to rotate in the same direction, keeping a fixed angle between them, while the other one causes the arms to rotate in mutually opposite directions. In either case, the arms revolve at 12 r.p.m. relative to disc B (*i.e.*, 5 sec. per revolution). As the masses carried on revolving arms constitute the only unbalance in balancing head, the arrangement permits the machine operator to change magnitude as well as the direction of the unbalance. When the two masses rotate in same direction relative to the balancing head (keeping same angle between their arms), the magnitude of resultant centrifugal force remains same but its direction changes. But when the two masses revolve in opposite direction with respect to one another, the direction of resultant inertia force remains fixed but magnitude changes.

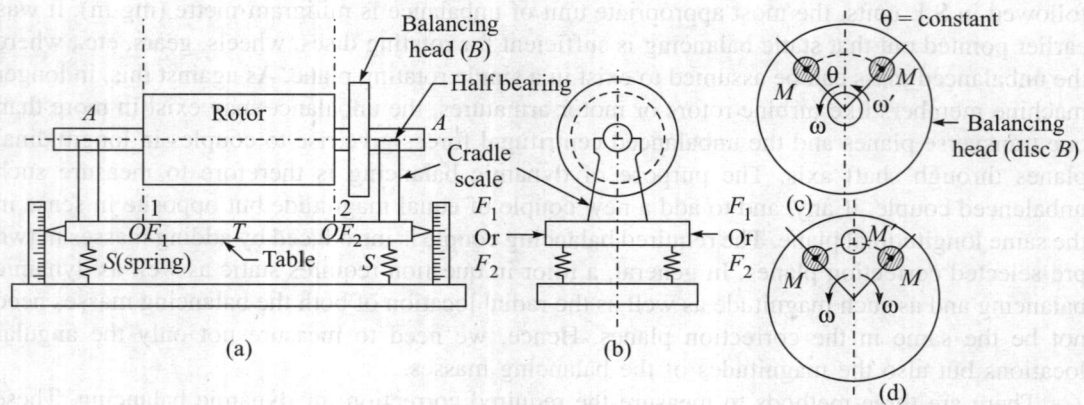

Fig. 16.21 Representation of a pivoted cradle type Dynamic Balancing machine.

The table is first pivoted about pivots $F_1 - F_1$ and the speed of rotor, being driven by a belt or a flexible shaft, is brought up to match with the resonating speed. If now the first button is pressed and the arms are made to rotate in same direction (keeping angle between them fixed), a maximum and minimum amplitude will be observed once in every 5 seconds. Clearly, the maximum amplitude occurs when resultant of the centrifugal forces of these two rotating masses coincides with the radius of unbalanced mass and the minimum amplitude occurs when the resultant centrifugal force due to the two revolving masses is 180° out of phase with the radius of unbalanced mass.

When the minimum amplitude is obtained, the finger is removed from button 1 and the button 2 is now pressed so that the two masses start revolving in mutually opposite directions at the same speed. With above setting, the direction of resultant centrifugal force is kept fixed at the position corresponding to the minimum amplitude and its magnitude is changed. The magnitude of the resultant centrifugal forces due to the two revolving masses is maximum (equal to $2 Mr\omega^2$) when the two masses have zero phase angle and is zero when the two masses are 180° out of phase. The second button is released when the vibration is reduced to zero. The rotor is stopped and from the position of the two mass arms, the magnitude and direction of the required correction (balancing mass) is established.

Referring to Fig. 16.21(d), as a result of steps described above, let the position of masses for zero amplitude be as shown. Then the single equivalent mass M', which produces same centrifugal force as the two masses together is given by,

$$2Mr_1\cos\theta = M'r_1$$

or
$$M' = 2M\cos\theta \qquad (16.13)$$

Having established balancing mass required in the plane of balancing head, let us next find equivalent mass M_2 in plane 2 to be added at any convenient radius r_2 which will have same effect as the pair of masses in balancing head. Consider Fig. 16.22.

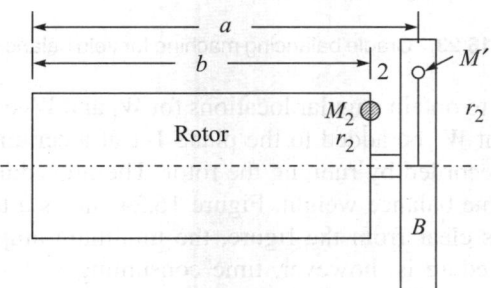

Fig. 16.22 Determination of equivalent mass.

For equivalent effect, taking moments about plane 1-1,

$$(M'r_1)a = (M_2r_2)b$$

Therefore
$$M_2 = M'(r_1/r_2)(a/b) \qquad (16.14)$$

In a similar way, the balancing head B is next coupled rigidly to the shaft on the left hand end of the rotor. The pivot is removed from F_1 and introduced at F_2 instead. The entire procedure is then repeated to establish magnitude and location of balancing mass to be added in plane 2, for balancing moments in plane 2.

16.12 FIELD BALANCING

Field balancing is particularly required for very large rotors for which balancing machines are impractical. Further, even if high-speed rotors are balanced in the shop floor, during production, slight deformations caused during shipping or by creep may need rebalancing of rotors in the field.

Consider cradle balancing machine without balancing head shown in Fig. 16.23. Let W be the unbalanced weight and let W_1 and W_2 be the balancing weights required in planes 1-1 and 2-2. Taking moments in planes of pivots F_1 and F_2 for achieving balance of the rotor,

$$W_1 = W(a - x)/a \tag{16.15}$$

and

$$W_2 = W(x)/a \tag{16.16}$$

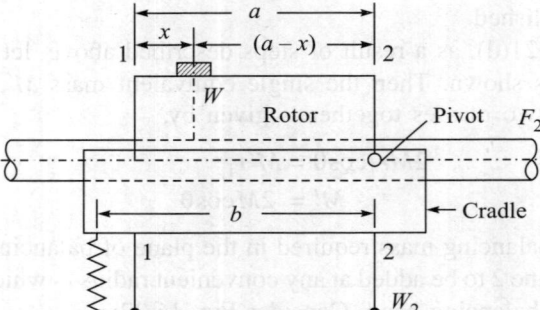

Fig. 16.23 Cradle balancing machine for field balancing.

Next, we shall see how to obtain angular locations for W_1 and W_2 experimentally. With cradle pivot at F_2, let a test weight W' be added to the plane 1-1 at a certain angular position and the vibration of the cradle be recorded by running the rotor. The procedure is repeated for different angular locations of the same balance weight. Figure 16.24 shows a tentative plot with a series of such observations. As is clear from the figure, the minimum amplitude of cradle-vibration occur at $\theta = \theta_1$. This procedure is, however, time consuming.

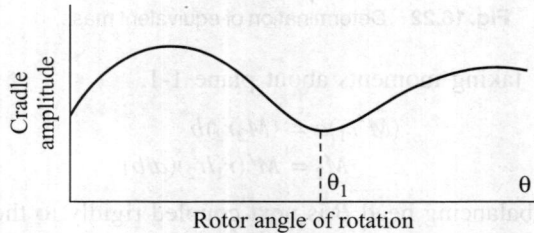

Fig. 16.24 Cradle vibration amplitude and rotor angle of rotation.

16.12.1 Balancing of a Thin Disc

A number of methods are available for field balancing but we will discuss the method developed by E.L. Thearle here. As a first step towards understanding the method, we need to develop an understanding for the 'balancing of a thin disc'.

The method makes use of a vibration pick-up at any bearing to measure vibrations of the unbalanced thin disc, and the output is displayed on the screen of double-beam *CRO*. A sine wave generator, connected to the disc, provides a reference 'sine-wave' through the second beam of the *CRO*.

To begin with, the disc is rotated at a speed ω, which may be selected nearer to resonating frequency for producing measurable vibration. The amplitude of oscillations A_1 together with the phase-difference θ_1 [*see* Fig. 16.25(a)] are measured. The vector *oa* in Fig. 16.25(b), drawn at an angle of θ_1 with respect to positive x-axis then represents the oscillations of amplitude A_1 fully. In the absence of any trial mass, the vector *oa* represents the effect of unbalance alone in the disc.

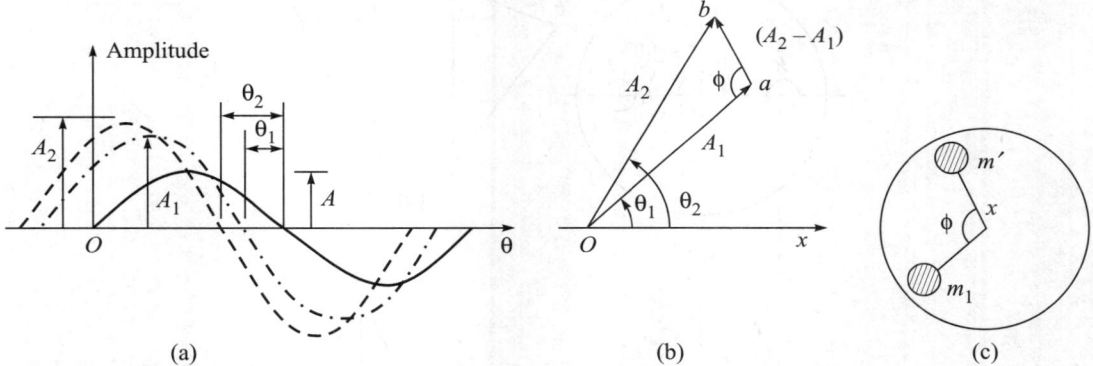

(a)   (b)   (c)

Fig 16.25 Balancing of Thin disc.

A trial mass m' is next attached to the disc at a known radius r and at a known angular position. When the disc is rotated at the same speed ω, let the amplitude of oscillation be A_2 with θ_2 as the phase difference. This is represented by vector *ob* in Fig. 16.25(b). The new amplitude A_2 ($=ob$) is due to the combined effect of original unbalance and the unbalance due to the trial mass m'. The closing side of the vector triangle then represents the vector difference,

$$ab = ob - oa = (A_2 - A_1)$$

which is due entirely to the trial mass m'. If the position of trial mass m' is now advanced by the angle ϕ, shown in the vector diagram of Fig. 16.25 and if the magnitude m' is simultaneously increased to m' (oa/ab), the vector difference ab will become equal and opposite to vector oa. Thus, to balance the disc, we need to add balancing mass m_1 at the same radius r, with the mass adjusted to $m_1 = (oa/ab)$ and located at an angle ϕ, counterclockwise as shown in Fig. 16.25(c), from the location of the trial mass.

In field balancing sometimes it becomes difficult either to couple a sine wave generator or to use a phase meter. In situations like this, the required balancing mass for a single plane rotor can be obtained by measuring vibration amplitudes alone. This, however, calls for two additional test runs, using same trial mass m'. This is described as follows.

16.12.2 Balancing by Four Observations

Let us assume that the four runs of the rotor using same trial mass m', provides four values of amplitudes (namely A_1, A_2, A_3 and A_4) of vibration of the cradle as given in Table 16.3. Note that speed of rotation ω remains same in all the four runs and further, the radial distance of the trial mass m' remains unchanged as indicated in Fig. 16.26(a)

Table 16.3

S. No.	Test condition	Amplitude of cradle
1.	No trial mass ($m' = 0$)	$A_1 \equiv oa$
2.	Trial mass m′ at $P (= 0°)$	$A_2 \equiv ob$
3.	Trial mass m′ at $R (= 180°)$	$A_3 \equiv oc$
4.	Trial mass m′ at $Q (= 90°)$	$A_4 \equiv od$

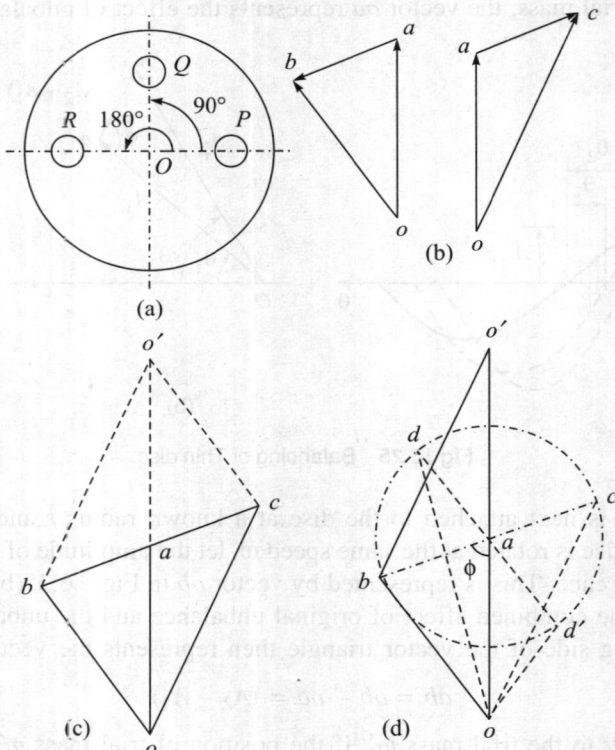

Fig. 16.26 Balancing by Four observations.

From entries at S. No. 1 and 2,

$$ob = oa + ab \tag{16.17}$$

or $\qquad ob = $ (original unbalance) + (unbalance due to m')

and from entries at S. No. 1 and 3,

$$oc = oa + ac \tag{16.18}$$

or $\qquad oc = $ (original unbalance) + (unbalance due to m')

Equations (16.17) and (16.18) are represented in vector diagrams in Figs. 16.26(b) and (c). Sides ab' and ac' in Fig. 16.26 are equal and opposite as m' and the radius remain unchanged and the trial masses are placed 180° apart. Let oa be extended to o', making $oa = ao'$. Then, $obo'c$ represents a parallelogram in Fig. 16.26(c) with $o'c$ equal and parallel to ob and $o'b$ equal and parallel to oc. This leads to the following method of construction.

Referring to Fig. 16.26(d), describe a triangle $oo'b$ with sides $oo' = 2A_1$, $ob = A_2$ and $o'b = A_3$. Let a be the mid-point of line oo'. Join ab, which represents effect of the trial mass at location P in Fig. 16.26(a). Hence the magnitude of required balancing mass at a radius equal to that of the trial mass m' is given by,

$$m_1 = m'(oa/ab) \qquad (16.19)$$

The location of required balancing mass is now obtained at an angle ϕ measured from the reference line, where $\phi = \angle\, oab$. The angle ϕ is to be measured in clockwise or counterclockwise sense depending on whether the magnitude of the fourth amplitude A_4 is given by od or od'; the points d and d' being located by rotating line ab' about a' through a right angle in c.w. sense. The rotation through 90° from position ab follows on account of position of trial mass at Q which is at 90° from the positions at P and R. Note that in all the three positions magnitude and radius of trial mass remain the same. The above construction is illustrated in Fig. 16.26(d).

In the construction explained above, either od or od' must represent the fourth amplitude A_4. This is checked by measuring lengths od and od' to scale, and comparing them with A_4. Since points d and d' correspond to 90° position (namely position Q) of trial mass, while points b and c correspond respectively to 0° and 180° position of trial mass m', the order b–d'–c gives the sense of ϕ if od' represents A_4 truely. Conversely, if od represents A_4 truely, the sense of ϕ is given by the order b–d–c. Since oa is the original unbalance and ab represents effect due to correction (trial) mass m' at 0° position, angle ϕ measured in appropriate sense from the reference line in Fig. 16.26(a) denotes desired location for balancing mass m_1, whose magnitude is given by equation (16.19).

EXAMPLE 16.9 In an experiment to balance a thin disc the following results were obtained. When the disc was run at 2000 r.p.m., vibration velocity level due to original unbalance was $v_0 = 2.6$ mm/s. When a trial mass of 10 gm was fastened in position 1 [see Fig. 16.27(a)], vibration level was $v_1 = 6.5$ mm/s. When the same mass was fastened in position 2, the vibration level was $v_2 = 1.9$ mm/s. When the same trial mass was fastened in position 3, the vibration level was $v_3 = 5.5$ mm/s. Shaft speed was always the same. Find the magnitude of the balancing mass at the radial distance R and its position. (AMIE, Summer 1993)

Solution: The given data may be tabulated as under:

S. No.	Test condition	Velocity of cradle (amplitude) (mm/s)
1	No trial mass (*i.e.*, m′ = 0)	$oa = A_1 = 2.6$
2	Trial mass m′ (= 10 gm) at position 1 (*i.e.*, at 0°)	$ob = A_2 = 6.5$
3	Trial mass m′ at position 2 (*i.e.*, at 180°)	$oc = A_3 = 1.9$
4	Trial mass m′ at position 3 (*i.e.*, at 90°)	$od = A_4 = 5.5$

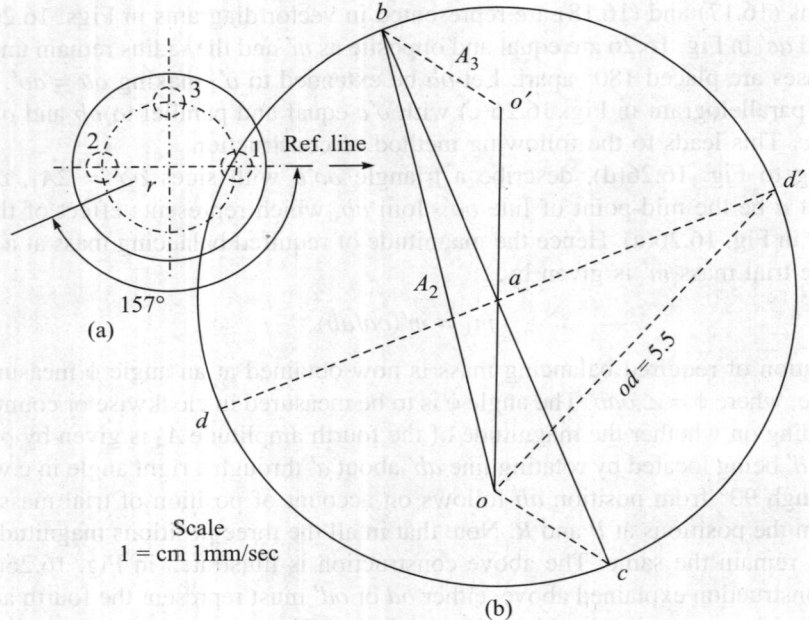

Fig. 16.27 Determination of balancing mass.

Construction: Refer to Fig. 16.27(b). Draw a line $oo' = 2 A_1$ to represent 5.2 mm/s to scale in vertical direction. Let a be its mid-point. From O' draw an arc of radius $ob = A_2 = 6.5$ mm/s to scale and let it be cut at b by an another arc of radius $o'b = A_3 = 1.9$ mm/s to scale from o'. Join ab and extend it to meet the circle, drawn from 'a' as centre and radius equal to ab, at point c. Join oc, which represents $A_3 = 1.9$ mm/s. Angle $\angle oab \, (= \phi)$ measures 157° in the figure while, ab measures 3.9 mm/s. Thus the magnitude of required balancing mass,

$$m_1 = m' \, (oa/ab) = 10 \, (2.6/3.9) = 6.67 \text{ gm} \qquad \textbf{Ans.}$$

To determine the sense of angle ϕ, rotate bc through right angle so as to get the position of points d and d' on the circle. The distance $od' = 5.5$ mm/s which equals A4 while, the distance od measures much less. Thus amplitude A4 in position 3 of trial mass is represented by od'. Hence, the angle ϕ must be measured in the sense $b - d' - c$ (*i.e.*, clockwise) from the reference line in Fig. 16.27(a). This gives the angular location of required balancing mass.

Thus the balancing mass of 6.67 gm must be located at 157° in c.w. sense from the reference line. **Ans.**

16.13 BALANCING OF RECIPROCATING MASSES

The discussions in the preceding articles lead us to the conclusion that it is always possible to balance a system of rotating masses by placing suitable balancing masses at proper angular and radial locations in any two desired planes of rotation. Articles to follow are devoted to a more complex problem of balancing masses which have either motion of translation at variable speed (e.g. piston of engine mechanism) or a combination of motion of translation and rotation (e.g. a swinging connecting rod of an engine).

When the motion of the crank pin is given, corresponding acceleration of the piston in an engine mechanism can be found for any phase of the mechanism. *The convention followed for the sign of acceleration is that all accelerations directed towards mean equilibrium position are considered positive while those directed away from mean equilibrium position are considered negative.* A certain value of force is required to induce this acceleration in the piston. In order to determine the force transmitted along the connecting rod to the crank, the inertia force must either be subtracted from or added to the total steam/gas load acting on the piston, depending on whether the acceleration is positive or negative. Thus, there arises an unbalanced force (called *inertia force*) which is equal and opposite to the accelerating force. This force must be absorbed by the frame and foundation as 'shaking force' and 'shaking-couple'. An attempt must be made to reduce shaking force and shaking couple by adding suitable balancing weights.

16.14 INERTIA EFFECTS OF RECIPROCATING MASSES IN ENGINE MECHANISM

The most general expression for the acceleration of piston is given by

$$A_P = r\omega^2(\cos\theta + A_1\cos2\theta + B_1\cos4\theta + C_1\cos6\theta +...) \qquad (16.20)$$

where

$$A_1 = 2\left[\frac{1}{2n} + \frac{1}{(2n)^3} + \frac{15}{8(2n)^5} + \frac{7.5}{8(2n)^7} + ...\right]$$

$$B_1 = 4\left[-\frac{1}{16n^3} - \frac{3}{4}\frac{1}{16n^5} - ...\right]$$

$$C_1 = 6\left[\frac{3}{16}\frac{1}{16n^5} + \frac{3}{8}\frac{5}{128n^7} + ...\right]$$

$$D_1 = 8\left[-\frac{1}{16}\frac{5}{128n^7} - ...\right]$$

Evidently, the type of slider motion is greatly influenced by the ratio length of connecting rod to crank-radius (*i.e.*, $n = l/r$). When the ratio of connecting rod length to crank radius (*i.e.*, n) is large, terms involving powers of n in the denominator become insignificant and can therefore be neglected. For instance, with a reasonably small value of $n = 4$; it can be verified that coefficients A_1, B_1, C_1, D_1 have the values

$$A_1 = 0.254; \quad B_1 = -0.004088; \quad C_1 = 0.0000738; \quad D_1 = 0.00000119.$$

Even in the coefficient A_1, the contribution of terms other than the first term (for $n = 4$) is of the order of 0.004 and can be neglected in comparison to the term $1/(2n)$. Thus, except for very high speeds of rotation, these terms remain insignificant in respect of their contribution and can be neglected.

Thus a fairly accurate expression for the acceleration of piston, for most practical applications is

$$A_P \approx r\omega^2[\cos\theta + (\cos2\theta)/n] \qquad (16.21)$$

where r is the radius of crank, ω is the angular speed in rad/s and the angle made by crank to the i.d.c. position θ.

With the expression as in equation (16.21), the motion of piston/cross-head is not a simple harmonic motion. The deviation from simple harmonic motion is due to terms involving $1/r$ $(= n)$ ratio in the denominator. In equation (16.21), the second term can be made insignificant by choosing a very long connecting rod in comparison to crank radius and in that case, the motion of sliding member will approach simple harmonic motion. The deviation of actual motion of slider from simple harmonic motion is thus attributed to what is popularly known as *obliquity* (or angularity) effect of connecting rod.

Let R be the weight of reciprocating parts (which includes weight of piston, piston-rings, piston-pin and part of connecting rod weight transferred to gudgeon-pin). Then the force required along the line of stroke to produce above the piston-acceleration is given by

$$F = \text{mass} \times \text{acceleration} = \frac{R}{g}(r\omega^2)[\cos\theta + (\cos2\theta)/n] \qquad (16.22)$$

This force is provided by the pull of the connecting rod. For the position of engine mechanism shown in Fig. 16.28, the connecting rod is in tension and the force Q exerted at the crank pin C is equivalent to an equal and parallel force Q' at the crank shaft bearing O together with a couple Qp. This couple opposes the rotation of crank shaft and must be considered in determining net turning moment exerted on the crank shaft. The force Q and Q' at S and O are equal and each one may be resolved along and perpendicular to the line of stroke. The component along the line of stroke at S ($Q\cos\phi = F$) is used for accelerating the reciprocating parts while the component of Q' $(= Q)$ along the line of stroke at O remains unbalanced. This unbalanced force $Q'\cos\phi = F$ at O acts in a direction opposite to that of accelerating force and represents reversed effective (*i.e.*, inertia) force, and is transmitted to the engine frame at the main bearings. This unbalanced force has a tendency to cause back and forth motion of the engine frame w.r. to the foundation.

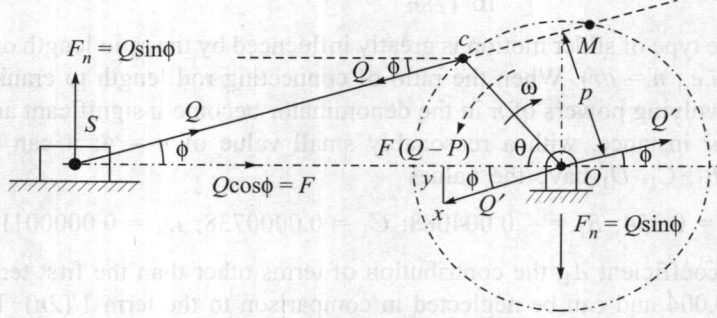

Fig. 16.28 Force analysis in engine mechanism.

The components perpendicular to the line of stroke of Q at S and O constitute a couple given by

$$C = (Q\sin\phi) \times OS = (F_n \times OS)$$

This couple tends to rotate the frame in clockwise sense. Following discussion will show that this couple has a magnitude and direction same as that of the external couple ($= Qp$) applied to the crank shaft.

Complete the force triangle oxy. The triangle oxy and OSM are similar as angles $\angle MOS$ and $\angle OYX$ are right angles and each of the angles $\angle osm$ and $\angle xoy$ is equal to ϕ. Hence, from the similar triangles,

$$OM/OS = xy/oy = F_n/F$$

or
$$F_nOS = FOM \tag{16.23}$$

Again, from the same similar triangles,

$$OM/SM = xy/ox = F_n/Q \tag{16.24}$$

Also
$$p/OM = \sin(90 - \phi)$$

or
$$p/OM = \cos\phi = OS/SM$$

Therefore
$$OM/SM = (p/OS) \tag{16.25}$$

Comparing right hand sides of equations (16.24) and (16.25),

$$F_n/Q = p/OS$$

or
$$F_nOS = Qp \tag{16.26}$$

This shows that the magnitude of the couple which tends to rotate the engine frame with respect to foundation equals the external couple ($=Qp$) applied to the crank shaft through the connecting rod. As is clear from Fig. 16.28, this couple has a sense same as that of the external couple on the crank shaft.

We therefore conclude that the full effect of the inertia of reciprocating masses on the engine frame is equivalent to

(1) an inertia force

$$F' = \frac{R}{g} r\omega^2 (\cos\theta + \cos2\theta/n) \text{ along the line of stroke at } O \text{ the main bearings of the crank shaft, and} \tag{16.27}$$

(2) a couple of direction and magnitude same as that of the external couple on crank shaft ($= Qp$) due to the connecting rod.

16.15 PRIMARY AND SECONDARY UNBALANCED FORCES DUE TO RECIPROCATING MASSES

It was shown in the preceding section that the inertia force of the reciprocating masses of the engine mechanism is given with fair amount of accuracy by the relation

$$F = \left(\frac{R}{g}\right) r\omega^2 (\cos\theta + \cos2\theta/n) \tag{16.28}$$

where R = weight of the reciprocating masses

  ω = angular speed of crank in radians per second

  r = radius of the crank

  l = length of connecting rod

  n = (l/r) ratio

  θ = angle of crank rotation with respect to i.d.c.

When equation (16.28) is positive, the inertia force will be directed away from the main bearing and when it is negative, it will be directed towards the main bearing. Expression (16.28) represents an unbalanced force which may be conveniently split into two parts. Thus,

$$F = \left(\frac{R}{g}r\omega^2\right)\cos\theta + \left(\frac{R}{gn}\right)r\omega^2\cos2\theta = F_p + F_s \qquad (16.29)$$

where primary unbalanced (disturbing) force

$$F_p = (R/g)r\omega^2\cos\theta \qquad (16.30)$$

and secondary unbalanced (disturbing) force

$$F_s = \left(\frac{R}{gn}\right)r\omega^2\cos2\theta \qquad (16.31)$$

Thus the primary disturbing force is the total inertia force if the connecting rod length were infinite, rendering the term in equation (16.31), corresponding to secondary force, insignificant. Thus, the secondary disturbing force of the reciprocating mass represents a correction which is required in order to allow approximately for the effect of the obliquity of the connecting rod.

It is interesting to note that the amplitude (*i.e.*, the maximum value) of the secondary force is only $1/n$ times the amplitude of the primary disturbing force. However, the frequency of the occurance of the maximum value of secondary force is twice that of the primary disturbing force.

16.16 INERTIA EFFECTS OF CRANK AND CONNECTING ROD

Assuming a uniform speed of rotation, a centrifugal force alone represents inertia effect of the crank. The inertia effect in such a case is given by mass times the acceleration of centre of gravity of the crank. While analysing for the shaking forces of the engine however, it is customary to replace the mass of the crank by an equivalent mass at the crank pin. Balancing the crank is relatively simpler since the inertia force of the crank is a pure centrifugal force. The crank can be completely balanced, therefore, by attaching counter balancing mass at suitable radius opposite the crank.

Inertia effects of the connecting rod are best studied using the concept of dynamically equivalent two-mass system (*see* Fig. 16.29). One of the two masses m_p is placed at the piston pin P while the other mass m_E is placed at the centre of percussion E with respect to the piston pin centre at a distance of l_2 from c.g. If l_1 be the distance of c.g. of the connecting rod from gudgeon-pin centre, then

$$l_2 = I_G/(ml_1) = (k_G^2/l_1) \qquad (16.32)$$

where m is the mass of the connecting rod while I_G is the mass moment of inertia at its c.g. The masses assumed to be concentrated at the gudgeon pin P and centre of percussion E are given by,

$$m_E = ml_1/(l_1 + l_2) \quad \text{and} \quad m_p = ml_2/(l_1 + l_2) \tag{16.33}$$

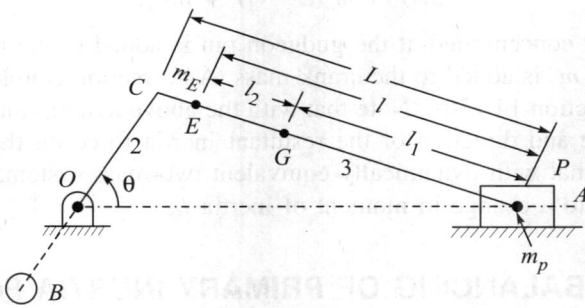

Fig. 16.29 Inertia effects of connecting rod.

The inertia forces acting on component masses m_E and m_p can be obtained either by determining accelerations of P and E using Klein's construction or by analytical expressions. The horizontal and vertical components of the inertia force of mass m_E are expressed as under:

$$(F_E)_H = m_E r\omega^2\left[\cos\theta + \left(1 - \frac{l}{L}\right)\frac{r}{L}\cos 2\theta\right]$$

or

$$(F_E)_H = m\left(\frac{l_1}{l}\right)r\omega^2\left[\cos\theta + \left(1 - \frac{l}{L}\right)\frac{r}{L}\cos 2\theta\right] \tag{16.34}$$

where

$$l = l_1 + l_2$$

and L = length of connecting rod between gudgeon pin and crank pin centres and

$$(F_E)_v = m_E r\omega^2 (l/L)\sin\theta = m(l_1/L)r\omega^2\sin\theta \tag{16.35}$$

The vertical component of the inertia force of the mass m_p concentrated at P is zero. The horizontal component is given by,

$$(F_p)_H = m_p r\omega^2\left(\cos\theta + \frac{\cos 2\theta}{n}\right)$$

or

$$(F_p)_H = m\left(\frac{l_2}{l}\right)r\omega^2\left(\cos\theta + \frac{r}{L}\cos 2\theta\right) \tag{16.36}$$

Departing from an exact dynamically equivalent system of two masses used for connecting rod above a convenient and more popular approach, though approximate, is to assume one mass m_p at the gudgeon pin and another one equal to m_c at crank pin c. Then

$$m = m_p + m_c \quad \text{and} \quad (m_p)l_1 = m_c(L - l_1) \tag{16.37}$$

which yields,

$$m_c = m\left(\frac{l_1}{L}\right) \quad \text{and} \quad m_p = m\left(\frac{L - l_1}{L}\right) \tag{16.38}$$

Such a system of masses can be described at best as 'pseudo dynamically equivalent system' as it does not meet the third condition of dynamic equivalence in respect of identical moment of inertia. Thus,

$$m_p(l_1)^2 + m_c(L - l_1)^2 \neq mk^2_G$$

The mass assumed concentrated at the gudgeon pin is added to the mass of reciprocating masses while the mass m_c is added to the crank mass. A correction couple is needed in such a case as described in Section 14.17(b). Note that with the above assumption of mass distribution at P and C, magnitude and direction of the resultant inertia force on the connecting rod are precisely the same as that with dynamically equivalent two-mass system; the line of action is however different due to a change in moment of inertia I_G.

16.17 PARTIAL BALANCING OF PRIMARY INERTIA FORCES

An unbalanced force due to a revolving mass is constant in magnitude but changes direction continuously. As against this an unbalanced force due to a reciprocating mass remains constant in direction but changes magnitude continuously. For this reason, as a general rule, a single revolving mass cannot serve the purpose of balancing completely a reciprocating mass and vice versa. There exists situations where partial balancing of reciprocating masses, using revolving balance-weights, is considered desirable.

The primary inertia force $F_p = \dfrac{R}{g} r\omega^2\cos\theta$, may be thought of as a component along the line

of stroke of the centrifugal force acting on a revolving mass of weight R placed at radius r and revolving with angular velocity ω. Let us consider, therefore, the possibility of balancing such a reciprocating unbalance by providing a balancing weight B at a radius b diametrically opposite to the crank pin and rotating with crank shaft, as shown at Fig. 16.30. Thus, for complete primary balance, the component along the line of stroke of the inertia force due to balancing mass must be equal and opposite to the primary inertia force. Thus,

$$(B/g)b\omega^2\cos\theta = (R/g)r\omega^2\cos\theta$$

or
$$(Bb) = (Rr) \tag{16.39}$$

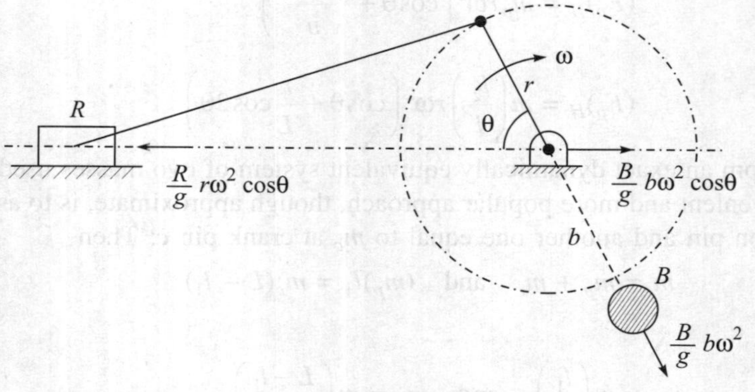

Fig. 16.30 Balancing of primary inertia forces.

This shows that a revolving balance mass may be arranged to balance primary inertia force completely. But this is not all. The centrifugal force acting on the balancing weight B has a component perpendicular to the line of stroke given by $(B/g)\, b\omega^2\sin\theta$ that remains unbalanced. This component has a magnitude equal to that of primary inertia force and goes through same variations of magnitude as the primary force. The only difference being that this component, which is perpendicular to the line of stroke, is maximum when primary inertia force is zero and vice versa. *Thus the introduction of revolving balance weight only serves to change the direction of the primary disturbing force.*

In applications like locomotives, disturbing forces in a direction perpendicular to the line of stroke are considered less dangerous than those along the line of stroke. It is preferable, therefore, to go for partial balancing of primary inertia forces by making $Bb = cRr$, where $c < 1$. With this, the balancing force along the line of stroke is,

$$F_p' = \left(\frac{B}{g}\right) b\omega^2\cos\theta$$

or

$$F_p' = c\left(\frac{R}{g}\right) r\omega^2\cos\theta \qquad (16.40)$$

and therefore, the unbalanced primary force along the line of stroke is

$$(F_p)_H = \{(Rr - Bb)/g\}\omega^2\cos\theta$$

Therefore

$$(F_p)_H = (1 - c)\,(R/g)r\omega^2\cos\theta \qquad (16.41)$$

while, the unbalanced force in a direction perpendicular to the line of stroke is,

$$(F_p)_v = c\left(\frac{R}{g}\right) r\omega^2\sin\theta \qquad (16.42)$$

For ensuring equal values of net disturbing forces in directions along and perpendicular to the line of stroke, obviously $c = 0.5$. But when a disturbing force along the line of stroke is considered more harmful than one perpendicular to it, a larger value of c may be chosen. In the case of two-cylinder steam locomotive, general practice is to use a value of c from 2/3 to 3/4. A larger value of c in such cases is dictated by a need to control swaying couple which is considered more harmful than the oscillating couple in vertical plane about a horizontal axis. In large four-cylinder locomotives with three or more pairs of coupled wheels, the value of c can be as low as 2/5.

Frequently, a balancing weight B is required to balance complete revolving weight W at crank radius r, besides providing a partial balance for the reciprocating parts. In such a case, the requirement is

$$Bb = Wr + cRr = (W + cR)r \qquad (16.43)$$

where

$\quad W =$ weight of unbalanced rotating masses
$\quad r =$ radius of crank
$\quad R =$ weight of reciprocating masses

EXAMPLE 16.10 A single cylinder oil engine has a stroke of 38 cm and the crank makes 360 r.p.m. The reciprocating parts weigh 670 N and the revolving parts are equivalent to 800 N at crank radius. A revolving balance weight is introduced at a radius of 15 cm to balance the whole of the revolving parts and one-half of the reciprocating parts. Find the balancing weight required and the residual unbalanced force on the crank shaft.

Solution: Let B be the required balancing weight at radius b to balance the whole of the revolving weight W (assumed to be concentrated at crank-pin) and half of the reciprocating parts of R. Then, for crank radius = 38/2 = 19 cm,

$$Bb = Wr + cRr$$

or $$Bb = (800)19 + \frac{1}{2}(670) \times 19 = 21565$$

Therefore $$B = 21565/15 = 1437.66 \text{ N (As } b = 15 \text{ cm)}$$

A usual practice is to attach half of this value of balance weight to each of the two crank-webs of a centre crank, as shown in Fig. 16.31.

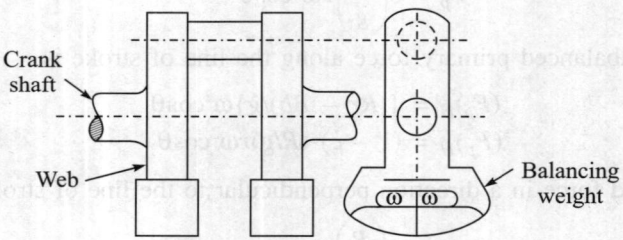

Crank shaft

Web

Balancing weight

Fig. 16.31 Crank with revolving balance weight.

As $c = 1/2$, the unbalanced primary force along the line of stroke,

$$(F_p)_H = c\left(\frac{R}{g}\right)r\omega^2\cos\theta$$

or $$(F_p)_H = \frac{1}{2}\left(\frac{670}{9.81}\right)\left(\frac{19}{100}\right)\left(\frac{2\pi \times 360}{60}\right)^2 \cos\theta = (9221.3)\cos\theta \text{ N}$$

and the unbalanced force, due to balancing mass, perpendicular to the line of stroke,

$$(F_p)_v = \frac{1}{2}\left(\frac{670}{9.81}\right)\left(\frac{19}{100}\right)\left(\frac{2\pi \times 360}{60}\right)^2 \times \sin\theta = (9221.3)\sin\theta \text{ N}.$$

Hence, resultant unbalanced force

$$F_R = \sqrt{(F_p)_H^2 + (F_p)_v^2}$$

or $$F_R = (9221.3)\sqrt{\cos^2\theta + \sin^2\theta} = 9221.3 \text{ N}$$

16.18 PARTIAL BALANCING OF LOCOMOTIVES

Two cylinder locomotives consist of two identical cylinders placed symmetrically either between the planes of the driving wheels as shown in Fig. 16.32(a) or outside the planes of wheels, as shown in Fig. 16.32(b). In the first case the arrangement is known as 'inside cylinder locomotives' and in the latter arrangement as 'outside cylinder locomotives'. In each such arrangements, the two cranks are invariably set at right angles to each other. This ensures that at least one crank will be away from the dead-centre position and it is always possible to start the locomotive. Further, the length of connecting rod is kept large in relation to the crank radius r, for reducing the secondary forces to a small value.

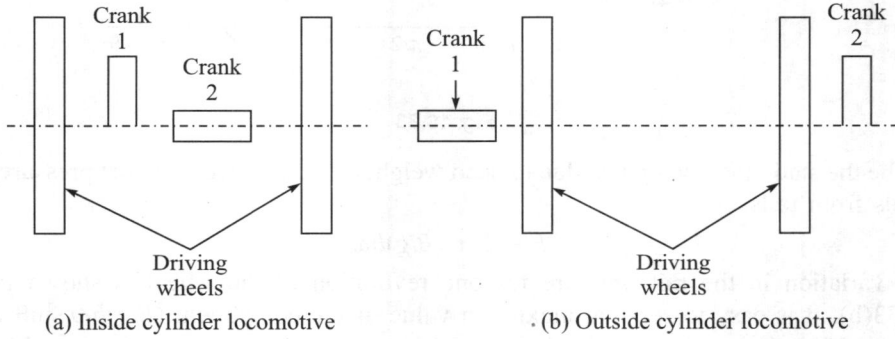

Fig. 16.32 Inside and outside cylinder locomotive.

Locomotives may also be grouped in two categories, namely,

(i) Coupled locomotives
(ii) Uncoupled locomotives

It is usual to have two or three pairs of wheels coupled together so as to increase adhesive weight. Cranks of such coupling rods of one set of coupled wheel are set at 180° to the adjacent driving cranks. Each coupled axle must be treated as separate axle for determining position and magnitude of balancing weights required. Thus, in the case of driving axle, one needs to consider two coupling rod masses besides two sets of cylinder masses and balancing masses in the planes of the wheels. Further, with an intention of controlling hammer blow (to be explained later), it is desired that balancing weights needed for reciprocating parts are not concentrated in the driving wheels. The required balancing weights in all such cases are distributed between all the coupled wheels.

16.19 EFFECT OF PARTIAL BALANCING IN LOCOMOTIVES

Hammer Blow

In steam locomotives the balancing weights are provided on driving wheels (*see* Fig. 16.33) which are the planes other than the planes of cylinder centre lines. If the difference in the planes of wheels and cylinder centre lines is neglected, the plane of primary disturbing force and the plane of centrifugal force, due to balance weight, are the same. The component, perpendicular to the line of stroke, of the inertia (centrifugal) force due to the balance weight B required at radius b to balance the reciprocating parts alone is $(B/g)b\omega^2\sin\theta$. The maximum magnitude (amplitude) of this component is called *hammer blow* and is expressed as.

$$\text{Hammer blow} = (B/g)b\omega^2 \qquad (16.44)$$

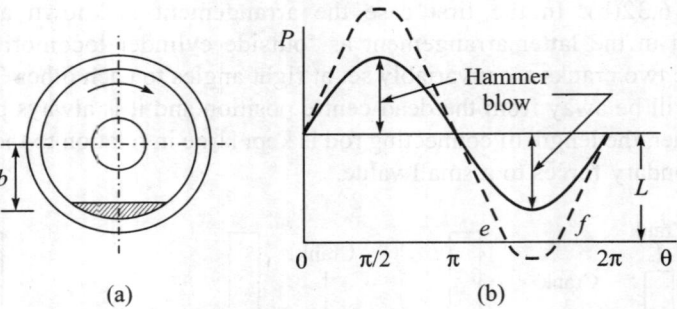

(a)   (b)

Fig. 16.33

If L be the static load per wheel due to dead weight of locomotive, then net pressure/reaction on wheels from rails

$$P = L \pm (B/g)b\omega^2 \qquad (16.45)$$

The variation in the rail pressure for one revolution of the wheel is shown plotted in Fig. 16.33(b). It is easy to see that maximum value of the component $(B/g)b\omega^2\sin\theta$ occurs at $\theta = 90°$ and $270°$; the maximum value being negative at $\theta = 270°$. Since the hammer blow varies as square of the angular speed ω it follows that near $\theta = 270°$, the net rail pressure can be reduced to zero. This limiting condition for the wheels to lift from rails is denoted by,

$$L - (B/g)b\omega^2 = 0 \qquad (16.46)$$

Thus the limiting speed beyond which wheels are lifted from rails is obtained from equation (16.46) as,

$$\omega_{limiting} = \sqrt{(Lg)/(Bb)} \qquad (16.47)$$

At speed corresponding to the curve shown in dotted lines in Fig. 16.33(b), the wheels will clearly lose contact with rails between points e and f.

Let W be the weight of revolving masses and R be the weight of reciprocating masses. If c be the fraction of reciprocating masses to be balanced, then

$$Bb = Wr + cRr$$

which, by transposition of terms, gives

$$c(Rr) = (Bb - Wr) \qquad (16.48)$$

which represents that part of balancing force which balances only the reciprocating parts.

Variation of Tractive Effort

The component along the line of stroke of the inertia force due to the rotating balancing weight B at radius b (such that $Bb = cRr$), leaves an unbalanced part of primary disturbing force for the two cylinders as,

$$(1 - c)(R/g)r\omega^2\{\cos\theta + \cos(90 + \theta)\}$$

The variation of tractive effort, which is caused by unbalanced part of primary disturbing force due to reciprocating masses, is thus given by

Variation in tractive effort $F_t = (1 - c)(R/g)r\omega^2\{\cos\theta + \cos(90 + \theta)\}$

$$= (1 - c)(R/g)r\omega^2\{\cos\theta - \sin\theta\} \qquad (16.49)$$

Variation in the tractive effort will be maximum when,

$$dF_t/d\theta = 0$$

i.e. when, $(1 - c)(R/g)r\omega^2\{-\sin\theta - \cos\theta\} = 0$

or   when, $\sin\theta = -\cos\theta$ or $\tan\theta = -1$

Thus maximum or minimum variation in tractive effort occurs at $\theta = 135°$ or $315°$. Minimum variation of tractive effort occurs when $\theta = 135°$, and is given by

$$F_t = (1 - c)(R/g)r\omega^2(\cos 135° - \sin 135°)$$

$$= (1 - c)(R/g)r\omega^2(-2/\sqrt{2})$$

$$= -\sqrt{2}(1 - c)(R/g)r\omega^2$$

Also when $\theta = 315°$, maximum variation of tractive effort is

$$F_t = (1 - c)(R/g)(r\omega^2)(\cos 315° - \sin 315°)$$

$$= \sqrt{2}(1 - c)(R/g)r\omega^2$$

Hence, maximum and minimum values of variation of tractive effort

$$(F_t)_{max} = \pm\sqrt{2}(1 - c)(R/g)r\omega^2 \qquad (16.50)$$

Swaying Couple

The unbalanced forces in the two cylinders are having a phase angle of 90° and as such, the two unbalanced forces are never equal. Infact when one is maximum/minimum, the other is zero. The force in each cylinder can be reduced to an equivalent force of same magnitude in a plane parallel to and mid-way between the cylinder centre-lines, together with a couple (Fig. 16.34). This couple has a tendency to rotate the frame of the locomotive about a vertical axis, mid-way between cylinder centre lines, and is called

$F'_p = (1 - C)(R/g)r\omega^2\cos\theta$

Cylinder ①

$-a/2$

Centre line

$+a/2$

Cylinder ②

$F_p = (1 - C)(R/g)r\omega^2\cos(90 + \theta)$

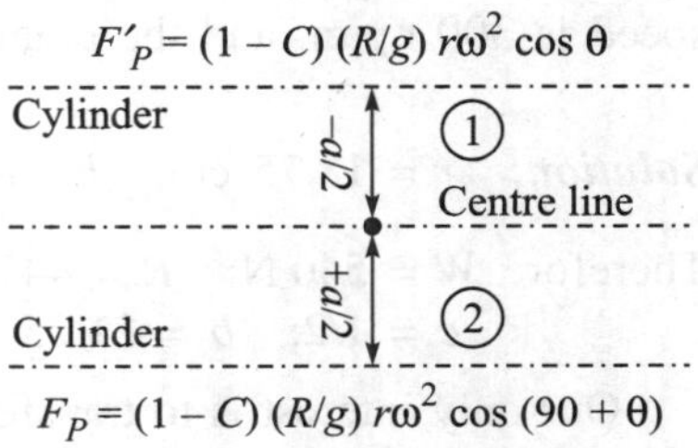

Fig. 16.34 Forces producing swaying couple.

swaying couple. The unbalanced portion of the primary disturbing force for the two cylinders differ in phase by 90°. Treating distances on one side of axis of couple as positive and on the opposite side as negative, the couple is given by

$$C = (1 - c)(R/g)r\omega^2\left\{\left(\frac{a}{2}\right)\cos\theta + \left(-\frac{a}{2}\right)\cos(90 + \theta)\right\}$$

or $C = (1 - c)(R/g)r\omega^2(a/2)[\cos\theta + \sin\theta] \qquad (16.51)$

This couple will have maximum value when

$$dC/d\theta = 0$$

i.e., when, $(1 - c)(R/g)r\omega^2(a/2)[-\sin\theta + \cos\theta] = 0$

i.e., when $-\sin\theta + \cos\theta = 0$

or $\tan\theta = +1$

Thus for maximum swaying couple,

$$\theta = 45° \quad \text{or} \quad 225°$$

Therefore $(C_{max})_{\theta=45} = (1 - c)(R/g)r\omega^2 a/2\left(\dfrac{1}{\sqrt{2}} + \dfrac{1}{\sqrt{2}}\right)$

$$= \sqrt{2}\,(1 - c)(R/g)r\omega^2 a/2$$

and $(C_{max})_{\theta=225} = (1 - c)(R/g)r\omega^2 a/2\left(-\dfrac{1}{\sqrt{2}} - \dfrac{1}{\sqrt{2}}\right)$

$$= \sqrt{2}\,(1 - c)(R/g)r\omega^2 a/2$$

Therefore $C_{max} = \pm\,(1 - c)(R/g)r\omega^2\,(a/\sqrt{2})$ (16.52)

EXAMPLE 16.11 A single cylinder horizontal oil engine has a crank of radius 18.75 cm and a connecting rod of 82.5 cm long. The revolving parts are equivalent to 540 N at crank radius and the weight of piston and gudgeon pin in 441.5 N. The connecting rod has its weight equal to 564 N and its mass centre is located at 26.25 cm from the crank pin centre. The revolving balance weights are fixed to the extensions of the crank webs at a radius of 21.25 cm to balance the revolving parts and half of the reciprocating parts. Neglecting the obliquity of the connecting rod, find the magnitude of the balance weight and the residual unbalanced force when the engine speed is 300 r.p.m. and the crank has rotated 60° from i.d.c. (AMIE, Summer 1985)

Solution: $r = 18.75$ cm; $l = 82.5$ cm; $\omega = \dfrac{2\pi \times 300}{60} = 10\pi$ rad/s

Therefore $W = 540$ N; $R = 441.5$ N; $\theta = 60°$ from i.d.c

$c = 1/2;$ $b = 21.25$ cm

Our only interest is to transfer weight of connecting rod at wrist pin and crank pin and we are not interested in moment of inertia of rod. The weight component at gudgeon pin is,

$$W_p = \frac{564 \times 26.25}{82.5} = 179.5 \text{ N}$$

and the weight component at crank pin

$$W_c = 564 - 179.5 = 384.5 \text{ N}$$

Thus total reciprocating masses = 441.5 + 179.5 = 621 N

Total revolving parts at crank radius = 540 + 384.5 = 924.5 N

For balancing.

$$Bb = (W + cR)r$$

Therefore $B = \left\{924.5 + \dfrac{1}{2}\,621\right\} \times 18.75/21.25 = 1085.7$ N **Ans.**

Hence, the magnitude of residual unbalanced force

$$F' = (1 - c) \frac{R}{g} r\omega^2$$

or $\qquad F' = (1 - 1/2) \dfrac{621}{9.81} \times \dfrac{18.75}{100} (10\pi)^2 = 5851.3 \text{ N}$ **Ans.**

Unbalanced force along the line of stroke

$$F'_h = (1 - c) \frac{R}{g} r\omega^2 \cos 60° = 5851.3\cos 60° = 2925.6 \text{ N} \qquad \textbf{Ans.}$$

Unbalanced force perpendicular to line of stroke

$$F'_v = (c) \frac{R}{g} r\omega^2 \sin 60$$

or $\qquad F'_v = 5851.3 \times \sin 60 = 5067.4 \text{ N}$ **Ans.**

EXAMPLE 16.12 The following data refer to two cylinder locomotive shown at Fig. 16.35(a) and (b) with cranks at 90°. Reciprocating mass per cylinder = 2943 N, crank radius = 30 cm, driving wheel diameter = 180 cm, distance between cylinder centre lines = 65 cm, and distance between driving wheels, centre planes = 155 cm. Determine:

   (i)   Fraction of the reciprocating masses to be balanced if the hammer-blow is not to exceed 45126 N at 90 k.m.p.h.

   (ii) Variation in tractive effort   (DAV Indore, 1986, 1992)

Solution: Given R = 2943 N; r = 30 cm

             Also, $v = r'\omega$

Therefore angular speed of driving wheels

$$\omega = \left(\frac{90 \times 1000}{3600} \right) \times \frac{1}{0.90} = (27.778) \text{ rad/s}$$

Let c = fraction of the reciprocating masses to be balanced and

     B = balancing weights placed at radius b at the driving wheel

Then weight of reciprocating parts to be balanced,

$$cR = 2943c \text{ N}$$

Any of the two planes of driving wheels may be treated as a reference plane so as to eliminate one of the unknown couples in couple polygon. Let us take plane A as the reference plane.

Plane	Weight (W) N	Radius (r) m	Cent. force ÷ ω^2/g = Wr N·M	Distance from Ref. plane A 1 m	Couple ÷ ω^2/g = (Wrl) N·M
A (R.P.)	B	0.9	0.9 B	0	0
B	2943 c	0.3	882.9 c	0.45	397.3 c
C	2943 c	0.3	882.9 c	1.1	971.2 c
D	B	0.9	0.9 B	1.55	1.395 B

Since the couple vectors (397.3c) and (971.2c) are mutually at right angles, [*see* Fig. 16.35(c)], the resultant couple vector is given by,

$$= \sqrt{(397.3\,c)^2 + (971.2\,c)^2} = 1049.3c$$

For complete balance, this must be equal and opposite to the third couple vector 1.55 *Bb*.

Hence, $1.395\,\text{B} = 1049.3c$ Therefore $B = \dfrac{1049.3\,c}{1.395} = 752.19c$

The hammer blow $= \dfrac{B}{g}\,b\omega^2 = \dfrac{752.19 \times 0.9}{9.81} \times (27.778)^2 c = 59164.4c$

Equating this to the limiting value of hammer blow,

$$59164.4c = 45126$$

or

$$c = \frac{45126}{59164.4} = 0.763 \qquad \textbf{Ans.}$$

(ii) **Variation in tractive effort**

$$F_t = \pm \sqrt{2}\,(1 - c)\frac{R}{g}\,r\omega^2$$

or

$$F_t = \pm \sqrt{2}\,(1 - 0.763)\frac{2943 \times 30}{981}(27.778)^2 = \pm 114669.4\ \text{N} \qquad \textbf{Ans.}$$

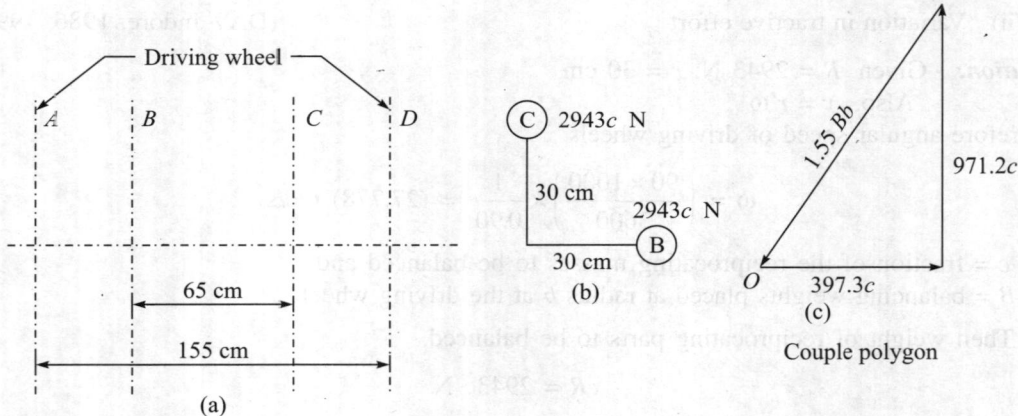

Fig. 16.35 Two cylinder locomotive.

EXAMPLE 16.13 An inside cylinder locomotive has its cylinder centre lines 70 cm apart and has a stroke of 60 cm. The rotating masses per cylinder are equivalent to 1471.5 N at the crank pin, and the reciprocating masses per cylinder to 1765.8 N. The wheel centre lines are 150 cm apart. The cranks are at right angles.

The whole of the rotating and 2/3rd of the reciprocating masses are to be balanced by weights placed at a radius of 60 cm. Find the magnitude and direction of the balancing weights. Also find the fluctuation in rail pressure under one wheel, variation in tractive effort and the magnitude of swaying cople at a crank speed of 300 r.p.m. (Punjab Univ. 1980; DAV Indore, Jan. 1989)

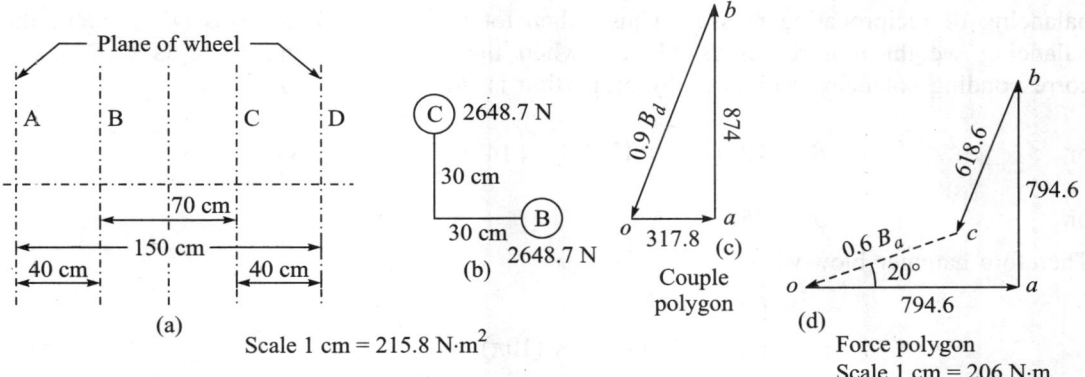

Fig. 16.36

Solution: Given $r = \dfrac{60}{2} = 30$ cm; $W = 1471.5$ N; $R = 1765.8$ N; $a = 70$ cm; $c = 2/3$;

$b = 60$ cm

Equivalent revolving mass, placed at crank pin, which requires balancing $= W + cR$

$$= 1471.5 + \frac{2}{3}(1765.8) = 2648.7 \text{ N}$$

Plane	Weight (W)N	Radius r m	Cent. force $\div \omega^2/g$ $= (Wr)$ N·m	Distance from Ref. plane A l m	Couple $\div \omega^2/g$ $= Wrl$ N·m^2
A	B_a	0.6	0.6 B_a	0	0
B	2648.7	0.3	794.6	0.4	317.8
C	2648.7	0.3	794.6	1.1	874
D	B_d	0.6	0.6 B_d	1.5	0.9 B_d

Since only one unknown exists in the column of couple vectors, the couple polygon is completed first as in Fig. 16.36(c). The closing side *bo* represents the couple vector $(0.9B_d)$ which measures 928 N·m^2 in the polygon when converted using scale.

Thus, $\qquad\qquad\qquad\qquad B_d = (928/0.9) = 1031$ N $\qquad\qquad\qquad$ **Ans.**

and angle at which this is to be placed with respect to crank of mass B, $\theta_D = 180 + 70 = 250°$
$\qquad\qquad\qquad\qquad\qquad\qquad\qquad\qquad\qquad\qquad\qquad\qquad\qquad\qquad\qquad\qquad\qquad$ **Ans.**

The balancing weight B_a is next determined by completing force polygon as in Fig. 16.36(d). By measurement $B_a r_a (= 0.6\ B_a) = 618.6$ N·m. The closing side of the force polygon, namely *co* represents the balancing force 0.6 B_a and is equal to 618.6. Hence

$$B_a = 618.6/0.6 = 1031 \text{ N} \qquad\qquad\qquad \textbf{Ans.}$$
and $\qquad\qquad\qquad \theta_A = $ the angle with mass B $= 200°$ $\qquad\qquad$ **Ans.**

Hammer Blow: Thus a balance weight $B = 1031$ N is required in plane D, at radius b, to completely balance revolving masses and partially balance ($\equiv c.R.r$) the reciprocating masses. However, the hammer blow is caused by that part of balancing mass which is responsible for balancing of reciprocating masses. Thus, when total disturbing force was $(W + c.R)r$, the balancing weight required is B. Hence, when the total disturbing force is $(c.R.r)$, the corresponding balancing weight B_r by proportion is $B(c.R)/(W + c.R)$.

or
$$B_r = 1031\left\{\left(\frac{2}{3}\times1765.8\right)\Big/\left(1471.5 + \frac{2}{3}\times1765.8\right)\right\}$$

or
$$B_r = 458.2 \text{ N}$$

Therefore hammer blow/wheel

$$= (B_r/g)\, b\omega^2$$
$$= (458.2/9.81)\, 0.6 \times (10\pi)^2 = 27659 \text{ N} \qquad \textbf{Ans.}$$

Therefore fluctuation of rail pressure = 27659 N

Variation in Tractive Effort: Maximum variation of tractive effort

$$(F_t)_{\max} = \pm \sqrt{2}\,(1 - c)\frac{R}{g}r\omega^2$$

or
$$(F_t)_{\max} = \pm \sqrt{2}\left(1 - \frac{2}{3}\right)\frac{1765.8}{9.81} \times 0.3 \times (10\pi)^2 = \pm25120 \text{ N} \qquad \textbf{Ans.}$$

Swaying Couple: The maximum swaying couple

$$C_{\max} = \pm(1 - c)\frac{a}{\sqrt{2}}\frac{R}{g}r\omega^2$$

or
$$C_{\max} = \pm\left(1 - \frac{2}{3}\right)\frac{0.70}{\sqrt{2}}\times\frac{1765.8}{9.81} \times 0.3(10\pi)^2 = \pm8794.7 \text{ N·m} \qquad \textbf{Ans.}$$

EXAMPLE 16.14 The three cranks of a 3-cylinder locomotive are all on the same axle and are set at 120°. The pitch of the cylinders is 1 m and the stroke of each piston is 0.6 m. The reciprocating masses are 300 kg for inside cylinder and 260 kg for each outside cylinder and the planes of rotation of the balance weights are 0.8 m from inside crank.

If 40% of the reciprocating parts are to be balanced, find: (a) the magnitude and position of the balancing masses required at a radius of 0.6 m and (b) the hammer blow per wheel when the axle makes 6 r.p.s. (DAV Indore, July 1990) (Allahabad Univ, 1979)

Solution: Crank radii $= \dfrac{0.6}{2} = 0.3$ m

$$\frac{R_1}{g} = \frac{R_3}{g} = 260 \text{ kg}; \quad \frac{R_2}{g} = 300 \text{ kg}$$

The masses to be balanced in cylinders 1, 2 and 3 are $m_1 = m_3 = c\dfrac{R_1}{g}$

or $\qquad\qquad m_1 = m_2 = m_3 = 0.4 \times 260 = 104$ kg

and, $\qquad\qquad m_2 = c\dfrac{R_2}{g} = 0.4 \times 300 = 120$ kg

Plane	Mass m kg	Radius r m	Cent. force ω^2 $= mr$ kg·m	Distance from Ref. plane L l m	Couple $\div\ \omega^2$ $= mrl$ kg·m^2
1	104	0.3	31.2	−0.2	−6.24
L	B_l	0.6	0.6 B_l	0	0
2	120	0.3	36.0	0.8	28.8
M	B_m	0.6	0.6 B_m	1.6	0.96 B_m
3	104	0.3	31.2	1.8	56.16

Let B_l and B_m be the revolving balance masses in planes L and M respectively. Equivalent revolving unbalance (mr) corresponding to a fraction of reciprocating unbalance (cRr) is obtained in accordance with equation (16.39) and are listed in column 4 of the table. Corresponding couple values are listed in the last column, with L as the reference plane.

Since column for couples involve only one unknown, couple polygon is drawn first as in Fig. 16.37(b). The closing side co of couple polygon, which measures 55.5 kg.m^2, represents the quantity $0.96B_m$.

Hence, $\qquad\qquad B_m = \dfrac{55.5}{0.96} = 57.81$ kg $\qquad\qquad$ **Ans.**

and is located at 26° from the crank of mass 1. With this value of B_m known, the unknown quantity $0.6B_m$ in column 4 of the table represents 34.7 kg.m. The force polygon is now completed as in Fig. 16.37(c). $\qquad\qquad$ **Ans.**

The closing side of the force polygon do then represents to scale unknown force $0.6\ B_l$. By measurement,

$$0.6B_l = 35.6 \quad \text{Therefore} \quad B_l = 59.3 \text{ kg} \qquad\qquad \textbf{Ans.}$$

The balance weight B_l is located at an angle of 216° with respect to the crank of first mass. Figure 16.37(a) also shows the angular position of these masses. $\qquad\qquad$ **Ans.**

Hammer blow per wheel

$$\omega = (2\pi \times 6) = 12\pi \text{ rad/s}$$

The difference in the values of B_l and B_m is on account of geometrical errors. Taking an average value,

$$B = \frac{B_l + B_m}{2} = \frac{57.81 + 59.3}{2} = 58.55 \text{ kg}$$

Therefore Hammer blow per wheel $= Br\omega^2$

$$= (58.55)(0.6)(12\pi)^2 = 49927.6 \text{ N} \qquad\qquad \textbf{Ans.}$$

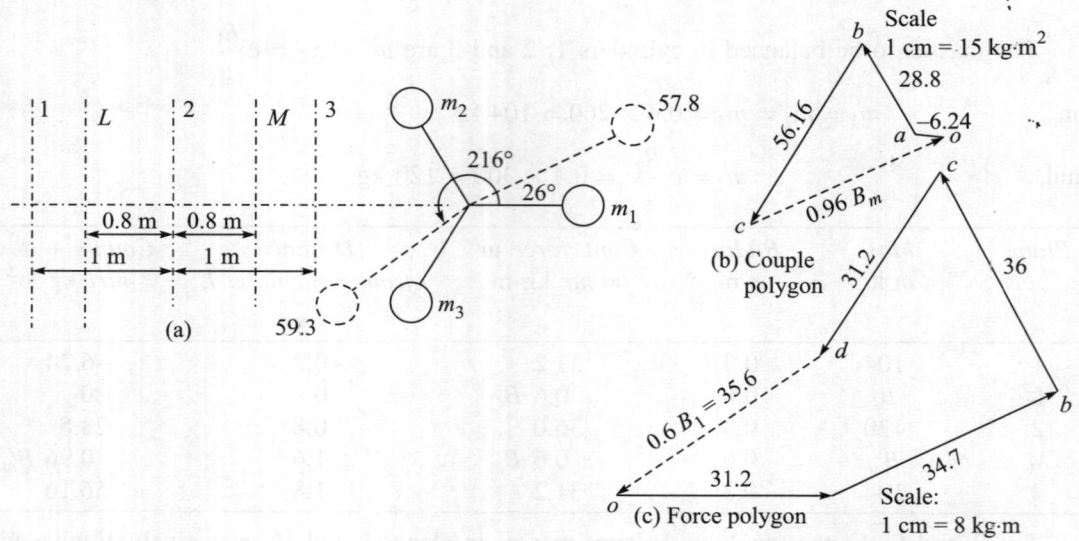

Fig. 16.37 Three cylinder locomotive.

EXAMPLE 16.15 The following data apply to an outside cylinder uncoupled locomotive:

Weight of rotating parts per cylinder = 3433.5 N and weight of reciprocating parts per cylinder = 2943 N.

Angle between the cranks = 90°, crank radius = 30 cm, distance between cylinder centres = 175 cm, radius of balance weights = 75 cm and wheel centres = 145 cm. If whole of the rotating and two-third of the reciprocating parts are to be balanced in plane of the driving wheels, find:

(i) Magnitude and angular position of the balance weights
(ii) Speed in kilometers per hour at which the wheels will lift off the rails when the load on each driving wheel is 26977.5 and the diameter of tread of driving wheels is 180 cm.
(iii) Swaying couple at a speed arrived at in (ii) above.

Solution: $W = 3433.5$ N; $R = 2943$ N, $r = 0.3$ m; $b = 0.75$ m; $c = 2/3$

Plane	Weight to be balanced $(W + cR)$ N	Radius r m	Cent. force $\div \omega^2/g$ $= Wr$ N·m	Distance from Ref. plane B l m	Couple $\div \omega^2/g$ $= Wrl$ N·m^2
A	5395.5	0.3	1618.7	−0.15	−242.8
B	B_b	0.75	0.75 B_b	0	0
C	B_c	0.75	0.75 B_c	1.45	1.0875 B_c
D	5395.5	0.3	1618.7	1.60	2590

Equivalent rotating parts at each crank,

$$= W + cR = 3433.5 + \frac{2}{3}(2943) = 5395.5 \text{ N}$$

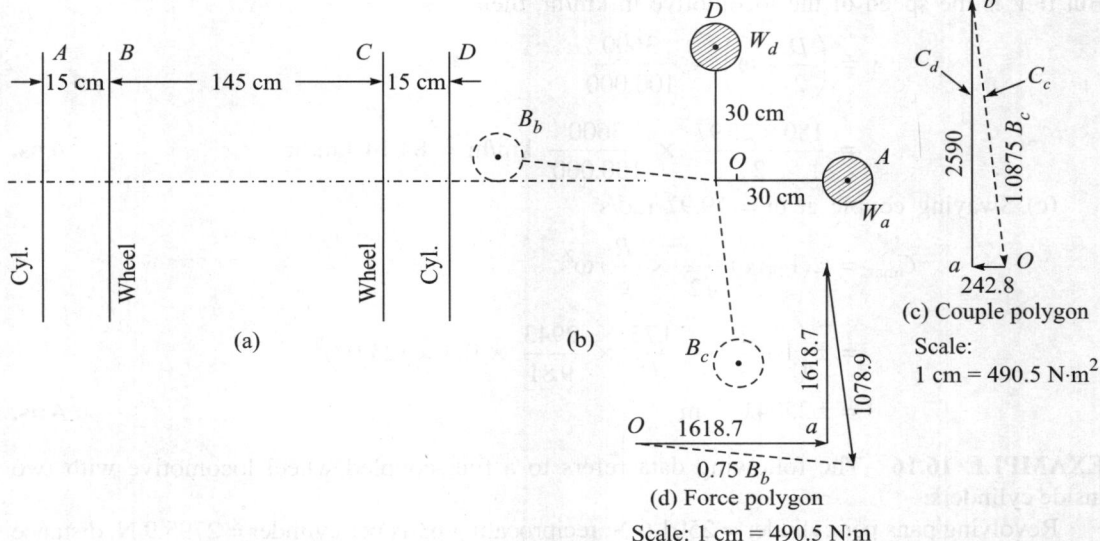

Fig. 16.38 Outside cylinder uncoupled engine.

Figure 16.38(c) shows a couple polygon drawn to scale and the closing side bo', which measures 2600 N, and represents couple $1.0875\,B_c$.

Therefore $\qquad\qquad\qquad B_c = 2600/1.8075 = 1438.5\text{ N}$ **Ans.**

and is located at $\qquad\qquad\qquad \theta_c = 276°\text{ from } W_a$ **Ans.**

$$B_c b = 0.75\,B_c = 1078.9\text{ N}$$

Completing the force polygon at Fig. 16.38(d), the closing side 'co' measures 1078.9 N·m and the side represents force $0.75\,B_b$. Hence from,

$$0.75\,(B_b) = 1078.9$$

Therefore $\qquad\qquad\qquad B_b = 1438.5\text{ N}$ **Ans.**

Angle θ_b at which weight B_b is located from W_A is 174° **Ans.**

The hammer blow is caused by balancing weight required purely to balance the reciprocating mass. Thus,

$$B_r = \left(\frac{cR}{W + cR}\right) \times 1438.5 = \frac{(2/3)\,2943}{53955} \times 1438.5 = 523\text{ N}$$

Therefore, $\qquad$ hammer blow $= \left(\dfrac{B_r}{g}\right) b\omega^2 = \dfrac{523}{9.81} \times 0.75 \times \omega^2$

For preventing the wheels from lifting,

$$26977.5 = \frac{523}{9.81} \times 0.75 \times \omega^2$$

Therefore $\qquad\qquad\qquad \omega^2 = 674.7 \quad\text{Thus,}\quad \omega = 25.97\text{ rad/s}$

But if V is the speed of the locomotive in km/hr, then

$$v = \left(\frac{D}{2} \times \omega\right) \times \frac{3600}{100,000}$$

$$= \frac{180 \times 25.97}{2} \times \frac{3600}{100,000} \text{ km/hr} = 84.14 \text{ km/hr} \qquad \textbf{Ans.}$$

(c) Swaying couple at $\omega = 19.97$ rad/s

$$C_{max} = \pm(1 - c)\frac{a}{\sqrt{2}} \times \frac{R}{g} r\omega^2$$

$$= \pm\left(1 - \frac{2}{3}\right) \times \frac{1.75}{\sqrt{2}} \times \frac{2943}{9.81} \times 0.3 \times (25.97)^2$$

$$= \pm25041 \text{ N·m} \qquad \textbf{Ans.}$$

EXAMPLE 16.16 The following data refers to a four-coupled wheel locomotive with two inside cylinders:

Revolving pans per cylinder = 2501.6 N, reciprocating parts per cylinder = 2795.9 N, distance between planes of driving wheels = 150 cm, pitch of cylinders = 60 cm, diameter of tread of driving wheels = 190 cm, distance between planes of coupling rod cranks = 190 cm, revolving parts for each coupling rod crank = 2452.5 N, angle between engine cranks = 90° N, length of coupling rod cranks = 22 cm, angle made by coupling rod crank with adjacent crank = 180°, distance of c.g. of balance weights in planes of driving wheels from axle centre = 75 cm, crank radius = 32 cm.

Determine: (a) Magnitude and position of balance weights required in leading and trailing wheels to balance two-third of reciprocating and whole of the revolving parts if half of the required reciprocating parts are to be balanced in each pair of coupled wheels, and (b) the maximum variation of tractive force and hammer blow when locomotive speed is 100 km/hr.

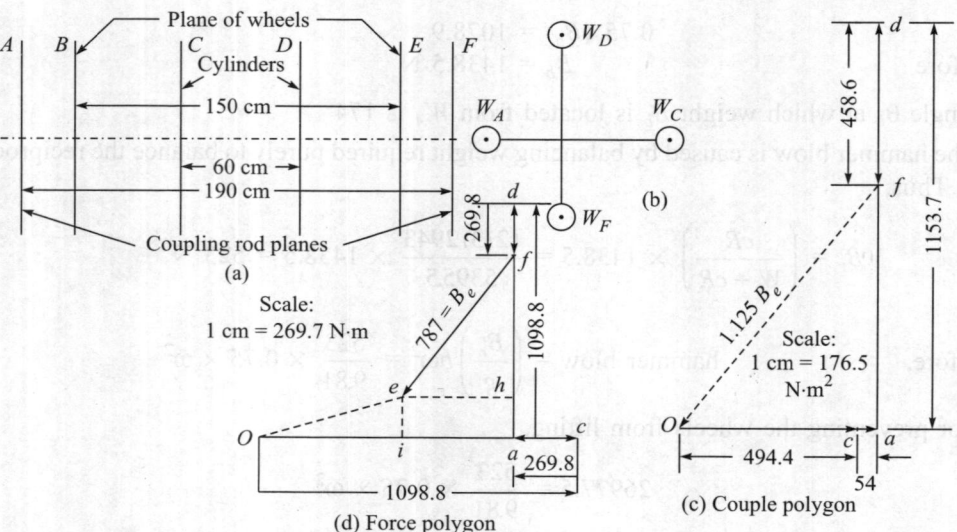

Fig. 16.39(a–d) Inside cylinder coupled locomotive.

Solution: (a) Balancing weight required in each leading wheel: The weight of reciprocating parts per cylinder to be balanced,

$$= cR = \frac{2}{3} \times 2795.9 = 1864 \text{ N}$$

This reciprocating weight is to be divided equally between the driving wheels and the trailing wheels. Thus 932 N is to be taken for driving wheel and 932 N for the trailing wheel. For each driving wheel, balancing weights are required for balancing of

(i) Half of the weight of coupling rod = 2452.5/2 (*i.e.*, 1226.3 N),
(ii) Whole of the revolving weight *i.e.*, 2501.6 N and the share of weight of reciprocating part, which is 932 N. Thus the equivalent rotating weight at crank pin = 2501.6 + 932 = 3433.6 N.

Plane	Weight W N	Radius r m	Cent. force $\div \omega^2/g$ $= Wr$ N·m	Distance from Ref. plane B l m	Couple $\div \omega^2/g$ $= (Wrl)$ N·m^2
A	1226.3	0.22	269.8	−0.2	−54
B (R.P.)	B_b	0.75	0.75 B_b	0	0
C	3433.6	0.32	1098.8	0.45	494.4
D	3433.6	0.32	1098.8	1.05	1153.7
E	B_e	0.75	0.75 B_e	1.5	1.125 B_e
F	1226.3	0.22	269.8	1.7	458.6

The couple polygon is as shown in Fig. 16.39(c). The closing side *fo* can be determined accurately by considering the right angled triangle *ofa*.

Thus, $$fo = \sqrt{(494.4 + 54)^2 + (1153.7 - 458.6)^2} = 885.4$$

But *fo* represents couple 1.125 B_e and therefore, from

$$(1.125) B_e = 885.4$$
$$B_e = 787 \text{ N} \qquad \textbf{Ans.}$$

The angular location of B_e w.r. to W_c is

$$\theta_c = 180 + \tan^{-1}\left(\frac{70.85}{55.9}\right) = 231.7° \qquad \textbf{Ans.}$$

The other balance mass must be equal to 787 due to symmetry. However, a force polygon as in Fig. 16.39(d), may also be completed to determine magnitude and angular location of B_b. The side *fe* (= $B_e b$) = 787 × 0.75 = 590.25 N·m. Hence, from the right angled triangle *feh*,

$$fh = (590.25) \sin 51.7° = 463.2$$
$$eh = (590.25) \cos 51.7° = 365.8$$

Therefore $$oi = (1098.8 - 269.8) - 365.8 = 463.2$$
and $$ei = ah = (1098.8 - 269.8) - 463.2 = 365.8$$

Therefore $$eo = \sqrt{(463.2)^2 + (365.8)^2} = 590.2$$

and $\angle eoi = \tan^{-1}(365.8/463.2) = 38.3°$

Therefore $B_b = 590.2/0.75 = 786.5$ N **Ans.**

and $\theta_B = 180 + 38.3 = 218.3°$ **Ans.**

(b) Balance Weights in Trailing Wheels: The arrangement of cylinders, cranks and wheels remain the same as in Figs. 16.39(a) and (b). For each trailing wheel, the following weights are to be balanced:

(i) Half of the weight of coupling rod $= \dfrac{2452.5}{2} = 1226.3$ N

Therefore $W_A = W_F = 1226.3$ N

(ii) Half of the reciprocating weights being balanced i.e., $W_C = W_D = \dfrac{1864}{2} = 932$ N .

Plane	W N	r m	Wr	l m	Wrl N·m²
A	1226.3	0.22	269.8	−0.2	−54
B (R.P.)	B_b'	0.75	0.75 B_b'	0	0
C	931.7	0.32	298.1	0.45	134.1
D	931.7	0.32	298.1	1.05	313.1
E	B_e'	0.75	0.75 B_e'	1.5	1.125 B_e'
F	1226.3	0.22	269.8	1.7	458.6

In the couple polygon of Fig. 16.39(e),

$oa = -54$; $ac = 134.1$; $cd = 313.1$ and $cf = 458.6$

Therefore, vector od, the closing side represents the unknown vector 1.125 B_e'. From the geometry of the figure,

$$(1.125)B_e' = \sqrt{(54 + 134.1)^2 + (458.6 - 313.1)^2}$$

Therefore $B_e' = (237.8/1.125) = 211.3$ N **Ans.**

The angular location of B_e' w.r. to W_c, is

$$\theta_e = 180 - \tan^{-1}\left(\frac{458.6 - 313.92}{54 + 134.1}\right) = 142.3° \qquad \textbf{Ans.}$$

The force polygon can now be constructed as in Fig. 16.39(f) with, $oc = 298.1$ N·m; $ca = 269.8$ N·m; $ad = 298.1$ N·m; $df = 269.8$ N·m and $fe = 211.3 \times 0.75 = 158.48$ N·m. The closing side eo of the force polygon then represents (0.75) B_b'. By measurement

$$(0.75)B_b' = 164.85$$

Therefore $B_b' = 219.8$ **Ans.**

Angle made by B_b' with $W_c = \theta_b' = 307°$ **Ans.**

(b) The angular speed of driving wheel,

$$\frac{190}{2} \times \omega = \frac{100 \times 1000 \times 100}{3600}$$

Therefore $$\omega = \frac{10^7 \times 2}{190 \times 3600} = 29.24 \text{ rad/s}$$

Therefore, maximum variation of tractive effort

$$(F_t)_{max} = \pm\sqrt{2}\,(1-c)\frac{R}{g}r\omega^2$$

or $$(F_t)_{max} = \pm\sqrt{2}\left(1-\frac{2}{3}\right)\frac{2795.9}{9.81}(0.32)(29.24)^2 = \pm3746.92 \text{ N} \qquad \textbf{Ans.}$$

Hammer blow can be established only when balancing weight B_r ($= B_e''$ or B_b'') is known for balancing reciprocating weights 932 N in planes C and D. This may be done using the following table.

Plane	W N	r in m	Wr N·m	l m from B	Wrl N·m²
B (R.P.)	B_b''	0.75	0.75 B_b''	0	0
C	932	0.32	298.24	0.45	134.2
D	932	0.32	298.24	1.05	313.1
E	B_e''	0.75	0.75 B_e''	1.5	1.125 B_e''

The couple polygon is a right angled triangle [*see* Fig. 13.39(g)].

Then, $$1.125\,B_e'' = \sqrt{(134.2)^2 + (313.1)^2} = 340.6 \text{ N·m}$$

Therefore $$B_e'' = 302.8 \text{ N}$$

Therefore, hammer blow $$= \frac{302.8}{9.81}(0.75)(29.24)^2 = 19792.6 \text{ N} \qquad \textbf{Ans.}$$

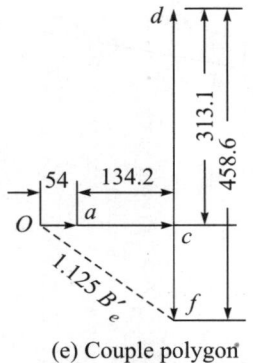

(e) Couple polygon
Scale: 1 cm = 108 N·m²

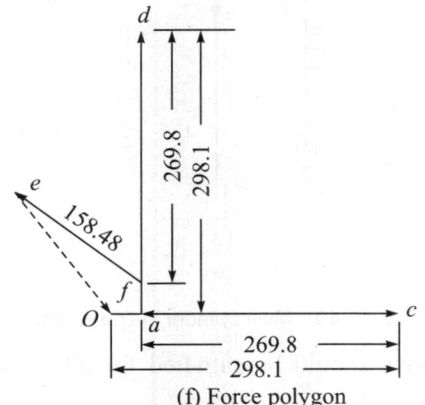

(f) Force polygon
Scale: 1 cm = 78.5 N·m

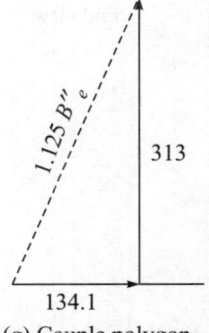

(g) Couple polygon
Scale: 1 cm = 80 N·m

Fig. 16.39(e–g)

16.20 PRIMARY BALANCE OF MULTI-CYLINDER IN-LINE ENGINE

It was shown in the preceding sections that the inertia forces of reciprocating parts of engine mechanisms are transmitted to the crank shaft through the connecting rods and cranks. If these forces on crank shaft of a multi-cylinder engine are not in equilibrium, then either a force, a couple or both are exerted on the bearings and the engine frame is subjected to shaking.

Multi-cylinder in-line engine consists of cylinders having their centre lines in one common plane and all the cylinders lying on the same side of crank shaft centre-line. Multi-cylinder in-line engines are very widely used and are capable of rotating at high speeds. Consider a multi-cylinder in-line engine having n number of identical cylinders of common stroke length and same connecting rod length. The reciprocating parts of all cylinders are identical. Let the cranks of these cylinders be represented in end-view by OA, OB, OC, OD, etc. (Fig. 16.40). Condition for primary balance requires that the algebraic sum of all the primary disturbing forces is zero and that the algebraic sum of all couples about any point in the plane of cylinder centre-lines is also zero. Mathematically,

$$\Sigma\ (R/g)r\omega^2 \cos\ \theta = 0 \tag{16.53}$$

and
$$\Sigma\ (R/g)r\omega^2 a \cos\ \theta = 0 \tag{16.54}$$

where a = distance of a given plane of rotation of the crank from a parallel reference plane. and, θ = generalised angle of inclination of cranks w.r. to i.d.c. position in the plane of cylinder centre-lines.

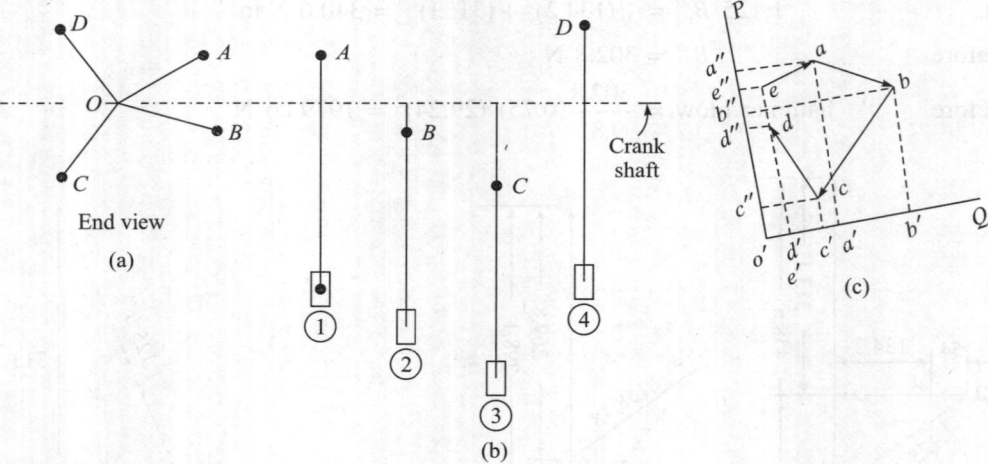

Fig. 16.40   Multi cylinder in-line engine.

Further, the above equations should be satisfied for all the angular positions during the rotation of the crank shaft.

Again as the cranks of individual cylinders are fixed to and rotate with the same crank shaft, the centrifugal forces in different crank positions do not change (in magnitude and direction) relative to one another. It may be said that the force polygon rotates with the cranks. As was shown earlier, the primary inertia force for each cylinder is identical to the component along the line of stroke of the centrifugal force acting on an imaginary mass R placed at the crank pin and rotating with the crank shaft. Such a crank is called **primary crank**.

Thus the balancing problem for primary forces of a multi-cylinder engine can be likened to a shaft carrying a number of rotating masses placed at the corresponding crank-pins. When the corresponding force polygon closes, the primary forces are balanced and when the primary couple polygon closes, the primary couples are balanced. *But when force or couple polygon does not close, an unbalance exists. Since the primary force/couple polygon can be assumed to rotate with crank shaft, the crank position at which unbalanced force/couple reaches maximum value can be obtained by rotating these polygons until the closing side is made parallel to the line of stroke.* One such rotating polygon for primary forces is shown in Fig. 16.40(c) in which sides *ea, ab, bc, cd* represent respectively the primary forces for cranks *OA, OB, OC* and *OD*. In the position of cranks in which the out-of-balance side *de* is perpendicular to the line of stroke *OQ*, the forces will be treated as balanced but in the crank positions identified by the line of stroke *OP* parallel to the out-of-balance force, represented by side *de*, the out-of-balance force will have its maximum value. Thus for all angular positions of crank shaft, the forces are balanced only if point *e″* coincides with *d″* for all crank positions.

16.21   SECONDARY BALANCE OF MULTI-CYLINDER IN-LINE ENGINES

As stated earlier, the secondary disturbing force F_s exists because of the obliquity effect of connecting rod and is given by,

$$F_s = (R/g)(r/n)\omega^2\cos 2\theta$$

which can be rewritten as,

$$F_s = (R/g)\left(\frac{r}{4n}\right)(2\omega)^2\cos 2\theta \qquad (16.55)$$

It is thus more convenient to regard the secondary force as equal to the component along the line of stroke of a centrifugal force acting on an imaginary mass, of weight R placed at the crank pin of a crank of radius $= (r/4n)$ and rotating at twice the speed of the actual crank such a crank is called **Secondary crank**. Thus though the secondary force has a magnitude $1/n$ times that of primary force and can be neglected at low to moderate speeds, it must be considered in multi-cylinder in-line engines where speed is usually large. Further, while at a given instant a primary crank is located at an angle of θ with respect to a dead centre position, the secondary crank will be located at angle of 2θ therefrom. Table 16.4 gives a tabular representation of the comparison between primary and secondary cranks.

Table 16.4   Comparison of Primary and Secondary Cranks

Crank	Radius	Angle of rotation w.r. to a dead centre position	Speed of rotation	Remark
Primary Crank	r	θ	ω	Actual crank
Secondary Crank	$(r/4n)$	2θ	2ω	It is an imaginary crank position

As in the case of primary balancing, the conditions for complete secondary balance of an engine are that:

$$\sum (R/g)(2\omega)^2(r/4n)\cos 2\theta = 0 \qquad (16.56)$$

and

$$\sum (R/g)(2\omega)^2(r/4n)a\cos 2\theta = 0 \qquad (16.57)$$

for all positions of crank shaft, during rotation, relative to dead centres. These conditions are satisfied only if secondary force and couple polygons are closed for the corresponding system of rotating masses.

Above discussions for primary and secondary balance may be summarised as under:

1. For complete primary balance of reciprocating masses, an imaginary mass of weight R, reciprocating masses in weight, is assumed to be concentrated at the crank pin of the actual crank and such a system must be balanced for forces and couples.
2. For complete secondary balance of reciprocating masses, an imaginary mass of weight R, having weight equal to the reciprocating masses, is assumed to be attached to an imaginary secondary crank at a radius of $r/4n$ and rotating at twice the speed of the crank shaft. Such a system must be balanced for forces and couples.

16.22 BALANCING OF 2-STROKE AND 4-STROKE IN-LINE ENGINES

Consider an in-line engine of N number of cylinders of identical reciprocating masses for each cylinder. The cranks of individual cylinder are arranged with an aim to provide a uniform firing interval and balance the reciprocating parts.

In 4-stroke cycle engines the combustible gases are ignited when the piston is nearly at the top dead centre and power is delivered to the crank shaft during 180 degrees of crank rotation until the piston reaches the bottom dead centre position. The products of combustion are exhausted during the next 180° of rotation. Fresh charge consisting of combustible mixture is now inducted into the cylinder during the next 180° of crank rotation. This is finally compressed during next 180° and the cycle is repeated thereafter. Thus a 4-stroke cycle requires two revolutions (4π rad) for completing the cycle, out of which half revolution is utilised for delivering power to the crank shaft. Let the power stroke, exhaust stroke, intake stroke and compression stroke be denoted respectively by labels P, E, I and C. In the case of 2-stroke cycle engine all these events are completed in one revolution (*i.e.*, 2π radian of rotation). Power and exhaust events take place during one stroke while the intake and compression events take place in the next stroke; each event occupying approximately one-fourth of a revolution.

In the case of a 4-stroke cycle engine, thus, uniform firing interval is given by $\alpha = 4\pi/N$. As was shown in preceding articles, total inertia force due to reciprocating masses of first cylinder may be expressed in the form of a Fourier series as,

$$F_1 = (R/g)r\omega^2\{\cos\theta + A\cos 2\theta + B\cos 4\theta + ...\} \qquad (16.58)$$

The series may be expressed in a generalised way as,

$$F_1 = \sum k_j \cos j\theta \qquad (16.59)$$

where $\qquad j = 1, 2, 4, 6,...$etc.

and $\qquad k_j = (R/g)r\omega^2 \times$ (appropriate constants out of 1, A, B, C,...etc.)

With the uniform angular spacing of $\alpha = 4\,\pi/N$, the angular positions of the cranks of cylinders (2, 3, 4, ..., N) may be obtained by replacing crank angle θ of cylinder 1 in eqn. (16.59) by angles $(\theta + \alpha)$, $(\theta + 2\alpha)$, $(\theta + 3\alpha)$,...etc. Hence, from this expression for total inertia forces may be obtained by replacing θ in equation (16.59) by $(\theta + \alpha)$, $(\theta + 2\alpha)$, $(\theta + 3\alpha)$,..., $[\theta + (N-1)\alpha]$.

The combined inertia force therefore is given by summation of a number of cosine series similar to equation (16.59). With each term representing a cosine series as in equation (16.59), the general expression for inertia forces due to all cylinders is,

$$F' = F_1 + F_2 + F_3 + \cdots + F_n$$

Thus, $F' = \sum k_j[\cos j\theta + \cos j(\theta + \alpha) + \cos j(\theta + 2\alpha) + \cdots + \cos j\{\theta + (N-1)\alpha\}]$ (16.60)

with $j = 1, 2, 4,$...etc. and k_j coefficients are as defined above.

The summation of trignometric terms, in which angles are in arithmetic progression (as detailed inside the square bracket of expression (16.60), is given by

$$S = \left[\cos j\left\{\theta + \frac{N-1}{2}\alpha\right\}\right] \times \left\{\sin\left(jN\frac{\alpha}{2}\right)\right\} \Big/ \sin\left(j\frac{\alpha}{2}\right) \qquad (16.61)$$

Further, since for 4-stroke cycle engines, $\alpha = 4\pi/N$, the term $\sin(jN\,\alpha/2)$ modifies to $\sin(2\pi j)$ and the expression (16.61) may be rewritten as

$$S = \left[\cos j\left\{\theta + \frac{N-1}{2}\alpha\right\}\right] \times \sin(2\pi j)/\sin\left(j\frac{\alpha}{2}\right) \qquad (16.62)$$

As j is always an integer the term $\sin(2\pi j)$ in the numerator of equation (16.62) is always zero. Thus, the summation $S = 0$, except when

$$\sin\left(j\frac{\alpha}{2}\right) = 0 \qquad (16.63)$$

In other words, only those harmonics will not be balanced for which $\sin(j\alpha/2) = 0$ *i.e.*, when $(j\alpha)$ is an integer multiple of 2π. For this value of $(j\alpha)$ each of the cosine terms of expression (16.60) reduces to $\cos(j\theta)$. In other words, when $j\alpha$ is an integer multiple of 2π, all the cosine terms inside the square bracket of equation (16.60) are same and equal to $\cos(j\theta)$. The summation is then given by,

$$F' = \sum_j k_j N \cos j\theta \qquad (16.64)$$

The summation in equation (16.64) then expresses unbalanced forces. Again, as $\alpha = 4\pi/N$, the product $(j\alpha)$ can be a multiple of 2π only when j is a multiple of $N/2$.

Thus, it is concluded that in a multi-cylinder four stroke in-line engine, with N number of identical cylinders and with cranks spaced so as to give uniform firing interval, all the harmonics are balanced except those for which 'j' is a multiple of half the number of cylinders (*i.e.*, N/2). Thus for a 2-cylinder, 4-stroke, in-line engine none of the harmonics is balanced and with 3-cylinder engine first, second and fourth harmonics are balanced. Similarly with 4-cylinder engine only first harmonic is balanced but with 6-cylinder engine again, first, second and fourth harmonics are balanced.

The discussion so far was centred around harmonic forces. Couples will be balanced only if the engine is symmetrical with respect to a central plane normal to the axis of crank shaft. *This requires that cranks of cylinders on the two sides of plane of symmetry must be arranged in pairs of parallel cranks which are located at the same distance from this plane.* In short, one side of the plane of symmetry should be a mirror image of the other half. This is possible only with even number of cylinders. Hence in 4-stroke multi-cylinder in-line engine with identical cylinders and reciprocating parts, couples can be balanced only with even number of cylinders.

For a 4-stroke, 4-cylinder engine $\alpha = 4\pi/4\ (= \pi)$ for uniform firing interval. Hence if cranks are arranged in pairs, with crank 1 parallel to say crank 4 and crank 2 parallel to 3 [as shown in Fig. 16.41(a)], the angular spacing between crank 1 and 2 and also between 3 and 4 is π. For primary couples to balance the distance of cranks 2 and 3 must be same from plane L and similarly distance of cranks 1 and 4 from plane L must be same. Further as $N/2 = 2$, only the primary forces will be balanced.

In the case of a 4-stroke, six cylinder engine, $\alpha = 4\pi/N = 120°$. The cranks are to be arranged in three pairs of parallel cranks on both the sides of the plane of symmetry [*see* Fig. 16.41(b)]. Since $N/2 = 3$ only harmonics like 6th and 12th, are unbalanced. Thus primary, secondary and fourth order forces are balanced. For balancing the couples, the parallel cranks on the two sides must be placed at same distance from plane L.

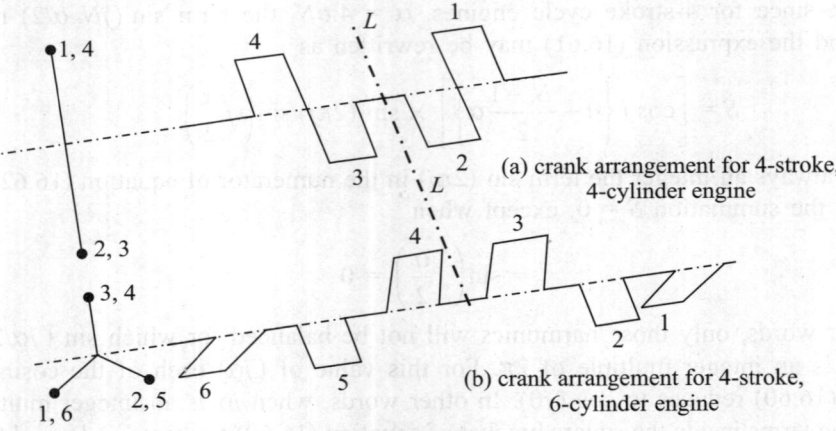

(a) crank arrangement for 4-stroke, 4-cylinder engine

(b) crank arrangement for 4-stroke, 6-cylinder engine

Fig. 16.41

Two-Stroke, Multicylinder In-line Engine

For 2-stroke cycle engine, with identical cylinders and reciprocating parts, $\alpha = 2\pi/N$ for uniform firing interval. It follows from equation 16.61 that as

$$jN\alpha/2 = \frac{jN}{2}\frac{2\pi}{N}$$

the term in numerator in equation (16.61)

$$\sin(jN\,\alpha/2) \equiv \sin(j\pi) = 0 \text{ (for interger } j)$$

Thus, here again, the sum S is always zero except when the denominator $\sin\left(\dfrac{j\alpha}{2}\right)$ is also zero. The term $\sin(j\alpha/2) = 0$, when $j\alpha$ is an integer multiple of 2π, *i.e.* when $j(2\pi/N)$ is an integer multiple of 2π. This means that for $\sin(j\alpha/2) = 0$, j must be an integer multiple of N.

Hence for 2-stroke cycle, multi-cylinder in-line engine all the harmonics are balanced except those for which j is an integer multiple of N, the number of cylinders. Thus with 2-cylinder, 2-stroke engines only primary forces are balanced. Similarly, for a 4-cylinder engine only primary and secondary forces are balanced. The most unfortunate thing about 2-stroke engines is that even after using even number of cylinders, one half of the crank shaft cannot be arranged to provide a mirror image of the other and therefore the couples cannot be balanced [*see* Figs. 16.42 (a) and (b)].

16.23 FIRING ORDER

Balancing is not the only criterion for deciding the arrangement of cranks for a multi-cylinder engine. Probably a more important criterion is that of supplying power impulses from cylinders at equal time intervals, which results in a smoother torque supply from the crank shaft. Firing order implies sequence in which firing takes place in different cylinders of a multi-cylinder engine.

The number of possible firing orders depends upon the number of cylinders and throws of the crank shaft (i.e. angular locations of various cranks and their relative distances from a reference plane O–O). It is desirable to have the power impulses equally spaced and this has led to certain conventional arrangements of crank shaft throws. A greater variations in firing order is possible with the four-stroke cycle engine than with the two-stroke cycle engine which fires every time a piston is at the top dead centre position. Method of determining possible firing order can be illustrated through the following example problem.

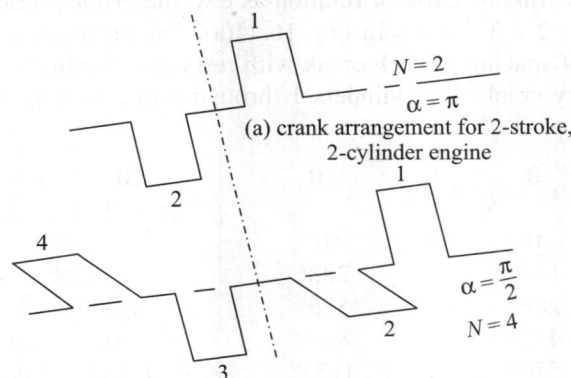

(a) crank arrangement for 2-stroke, 2-cylinder engine

(b) crank arrangement for 2-stroke, 4-cylinder engine

Fig. 16.42

EXAMPLE 16.17 Determine possible firing order for a 4-stroke six-cylinder engine.

Solution: For a 4-stroke cycle engine, firing interval $\alpha = \dfrac{4\pi}{6} = 120°$.

For force and couple balancing, a conventional crank arrangement with the above spacing can be as in Fig. 16.41(b) and can be expressed as in Fig. 16.43. It is left for the reader to verify that with cranks in this fashion, primary and secondary forces as well as primary secondary couples are balanced.

For a clockwise rotation of the crankshaft, cylinder 1 is in the firing position and cylinders 2 or 5 and the cylinders 3 or 4 will follow next. Thus the possible firing orders are (also refer to Fig. 16.41b):

(a) $1 - 2 - 3 - 6 - 5 - 4$
(b) $1 - 2 - 4 - 6 - 5 - 3$
(c) $1 - 5 - 4 - 6 - 2 - 3$
(d) $1 - 5 - 3 - 6 - 2 - 4$

The last firing order is favourable as no two consecutive explosions occur in adjacent cylinders, and therefore, preferred.

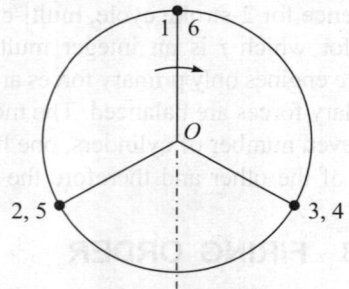

Fig. 16.43 A conventional crank arrangement satisfying firing order.

EXAMPLE 16.18 An engine having five cylinders in line has successive cranks 144° apart, the distance between cylinder centre lines being 450 mm. The reciprocating mass for each cylinder is 16 kg, the crank radius is 135 mm and the connecting rod length is 540 mm. The engine runs at 600 r.p.m. Examine the engine for balance of primary and secondary forces and couples. Determine the maximum values of these and the position of the central crank at which these maximum values occur. (University of London)

Solution: Since the angular spacing of five cranks together is $144 \times 5 = 720°$, it is a 4-stroke cycle engine. Selecting crank number 1 to be along the line of stroke, the relative position of the cylinder centre lines is as in Fig 16.44(a) while the primary and secondary cranks are as shown in Fig. 16.44(b) and (c). Taking sense of rotation as c.w. the primary cranks are shown at 144° interval in the order $1 - 2 - 3 - 4 - 5$ in Fig. 16.44(a). The positions of secondary cranks are obtained by the angle of spacing of each crank with respect to the line of stroke. These angular spacings θ' of secondary cranks for cylinders 1 through 5 in c.w. sense with respect to line of stroke are:

Crank	θ	2θ	$2\theta - k(360°)$ $(k = 0, 1, 2, 3, ...)$	θ'
1	0	0	0	0
2	144°	288°	288°	288°
3	288°	576°	(576 − 360)	216°
4	432°	864°	(864 − 720)	144°
5	576°	1152°	(1152 − 1080)	72

The secondary cranks therefore appear in order $1 - 5 - 4 - 3 - 2$ in Fig. 16.44(c). This leads to the following table.

Plane	Reciprocating mass R kg	Radius of crank r m	Force ÷ ω²/g = Rr	Distance from R.P.3 = l	Couple ÷ ω²/g = Rrl
1	16	0.135	2.16	−0.9	−1.944
2	16	0.135	2.16	−0.45	−0.972
3 (R.P)	16	0.135	2.16	0	0
4	16	0.135	2.16	0.45	0.972
5	16	0.135	2.16	0.90	1.944

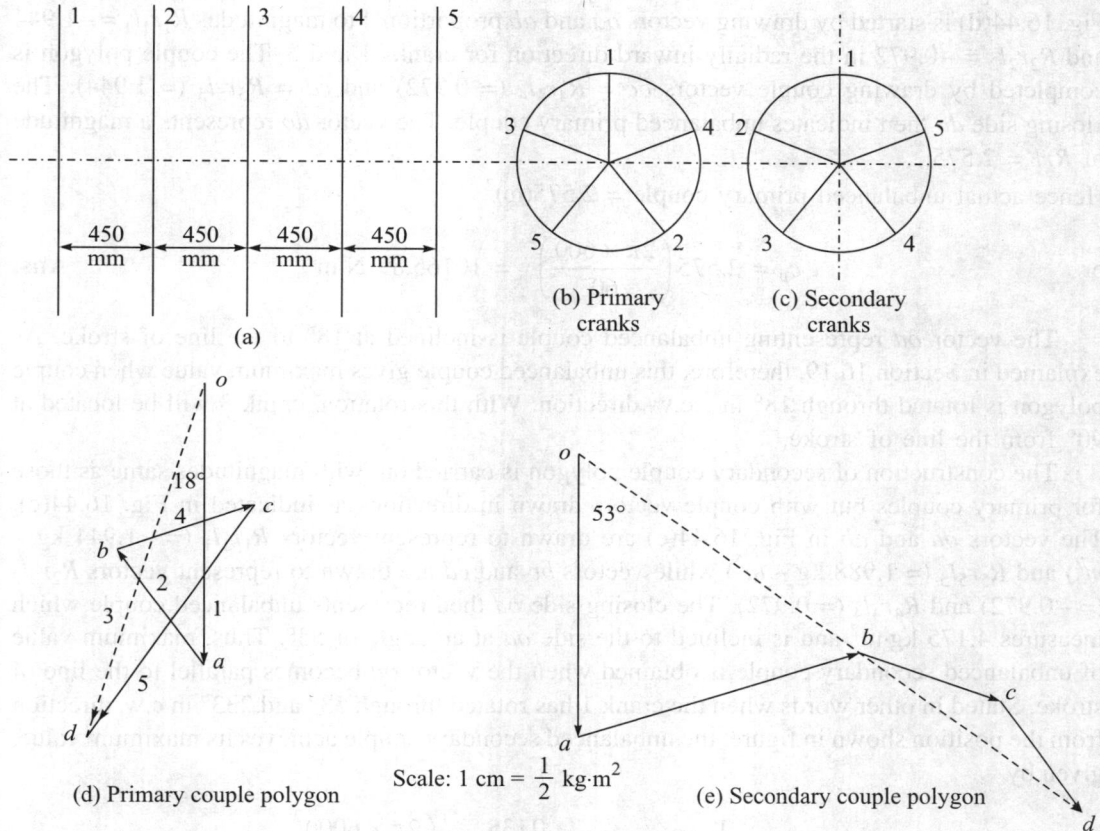

(b) Primary cranks

(c) Secondary cranks

(a)

(d) Primary couple polygon

Scale: 1 cm = $\frac{1}{2}$ kg·m²

(e) Secondary couple polygon

Fig. 16.44 A 5-cylinder inline engine.

Primary and Secondary Forces: Proportional value of primary force for a cylinder = (Rr) and that of a secondary force = $\frac{1}{n}(Rr)$. It follows from Figs. 16.44(b) and (c) that the cranks are symmetrically placed at 72° interval and forces (Rr) and $\left(\dfrac{Rr}{n}\right)$ are same along each crank. Hence primary and secondary forces are in balance.

Primary and Secondary Couples: The primary and secondary couples are proportional to:

$$\text{primary couples: } Rrl \text{ and secondary couple: } \frac{Rr}{n}l.$$

Thus magnitudes of primary couples can represent the corresponding secondary couples to a new scale (as n is a constant for given engine). The construction of couple polygon at

Fig. 16.44(d) is started by drawing vectors oa and ab proportional to magnitudes $R_1 r_1 l_1 = -1.944$ and $R_2 r_2 l_2 = -0.972$ in the radially inward direction for cranks 1 and 2. The couple polygon is completed by drawing couple vectors $bc = R_4 r_4 l_4$ (= 0.972) and $cd = R_5 r_5 l_5$ (= 1.944). The closing side do then indicates unbalanced primary couple. The vector do represents a magnitude of $Rrl = 2.575$.

Hence actual unbalanced primary couple = $2.575(\omega)^2$

or $\qquad\qquad\qquad c_p = 2.575\left(\dfrac{2\pi \times 600}{60}\right)^2 = 10165.89$ N·m $\qquad\qquad$ **Ans.**

The vector od representing unbalanced couple is inclined at 18° to the line of stroke. As explained in Section 16.19, therefore, this unbalanced couple gives maximum value when couple polygon is rotated through 18° in c.c.w. direction. With this rotation, crank 3 will be located at 90° from the line of stroke.

The construction of secondary couple polygon is carried out with magnitudes same as those for primary couples but with couple vectors drawn in directions as indicated in Fig. 16.44(c). The vectors oa and ab in Fig. 16.44(e) are drawn to represent vectors $R_1 r_1 l_1$ (= -1.944 kg – m^2) and $R_5 r_5 l_5$ (= 1.988 kg – m^2) while vectors bc and cd are drawn to represent vectors $R_2 r_2 l_2$ (= -0.972) and $R_4 r_4 l_4$ (= 0.972). The closing side od then represents unbalanced couple which measures 4.175 kg·m² and is inclined to the side oa at an angle of 53°. Thus, maximum value of unbalanced secondary couple is obtained when the vector od becomes parallel to the line of stroke. Stated in other words when the crank 1 has rotated through 53° and 233° in c.w. direction from the position shown in figure, the unbalanced secondary couple achieves its maximum value, given by

$$= (Rrl)\frac{1}{n}\omega^2 = (4.175)\frac{0.135}{0.540} \times \left(\frac{2\pi \times 600}{60}\right)^2$$

Thus $\qquad\qquad (C)_{max} = \dfrac{4.175}{4} \times (20\pi)^2 = 4120.56$ N·m $\qquad\qquad$ **Ans.**

EXAMPLE 16.19 A four-crank engine has the two outer cranks set at 120° to each other, and their reciprocating masses are each 3290 N. The distance between the planes of rotation of adjacent cranks are 45.75 and 60 cm. If the engine is to be in complete primary balance, find the reciprocating mass and the relative angular position for each of the inner cranks. If the length of each crank is 30 cm, the length of each connecting rod is 120 cm and the speed of rotation is 240 r.p.m, what is the maximum secondary unbalanced force? (DAV Indore, 1990)

Solution: $\qquad\qquad\qquad \omega = \dfrac{2\pi \times 240}{60} = 8\pi$ rad/s

$$n = l/r = 120/30 = 4.0$$

Relative positions of cylinder centre-lines is shown in Fig. 16.45(a) and relative angular positions of cranks appear in Fig. 16.45(b).

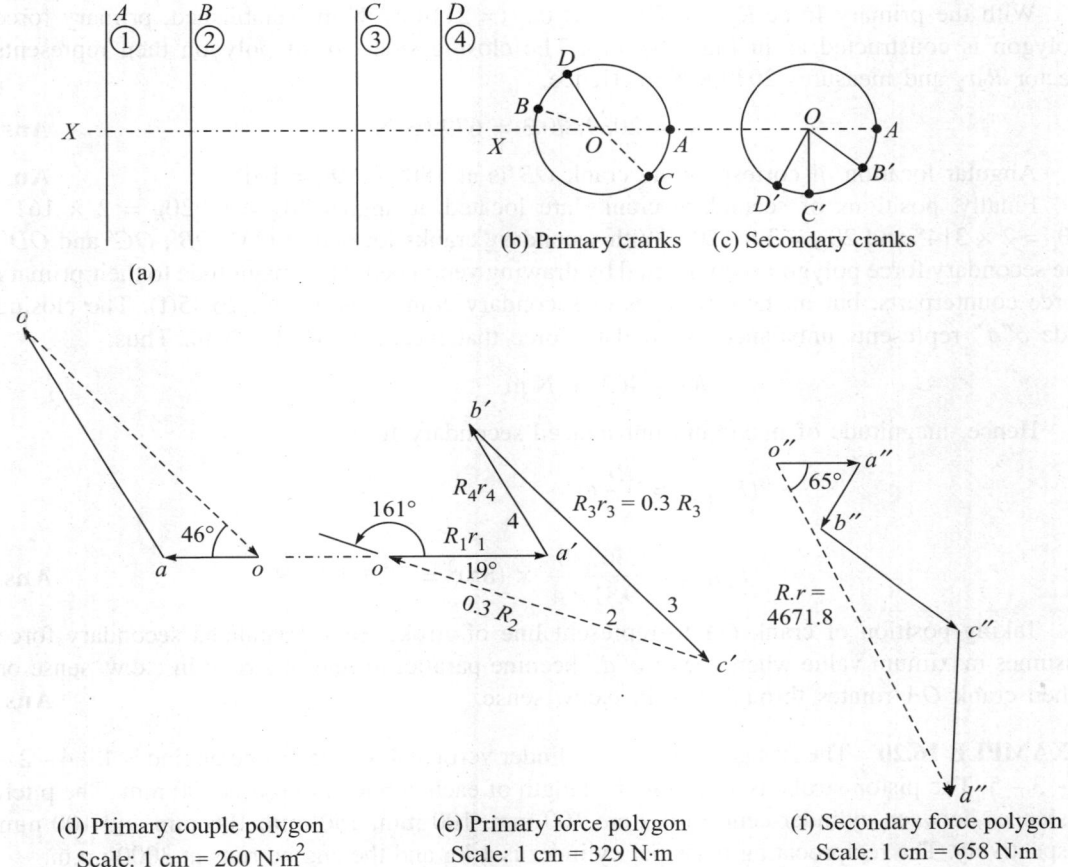

(a)

(b) Primary cranks (c) Secondary cranks

(d) Primary couple polygon
Scale: 1 cm = 260 N·m²

(e) Primary force polygon
Scale: 1 cm = 329 N·m

(f) Secondary force polygon
Scale: 1 cm = 658 N·m

Fig. 16.45 Primary and secondary cranks for 4-crank engine.

Plane	Weight W	Radius r m	Cent. force ÷ ω^2/g = R.r. N·m	Distance l from R.P. B. : lm	couple ÷ ω^2/g = Rrl N·m²
A	3290	0.3	987	−0.45	−444.15
B (R.P.)	R_2	0.3	$0.3 R_2$	0	0
C	R_3	0.3	$0.3 R_3$	+0.75	$0.225 R_3$
D	3290	0.3	987	+1.35	1332.45

Since the last column of the above table contains only one unknown, couple polygon is completed first. After drawing known couple vectors $oa = R_1r_1l_1$ (= −444.15 N·m²) and $ab = R_4r_4l_4$(= 1332.45 N·m²), the closing side bo represents the unknown primary couple $0.225 R_3$. By measurement bo = 1586 N·m². Hence,

$$R_3 = 1586/0.225 = 7048.9 \text{ N} \qquad \textbf{Ans.}$$

The corresponding crank is located at $360° − 46° = 314°$ in c.c.w. direction from crank OA [*see* Figs. 16.45(b) and (c)] **Ans.**

With the primary force $R_3 r_3 = 7048.9 \times 0.3$ (= 2114.67 N·m) established, primary force polygon is constructed as in Fig. 16.45(e). The closing side $c'o'$ of polygon then represents vector $R_2 r_2$ and measures 2039.8 N·m. Hence,

$$R_2 = 2039.8/0.3 = 6799.3 \text{ N} \qquad \textbf{Ans.}$$

Angular location of corresponding crank OB is at $180° - 19° = 161°$ **Ans.**

Finally, positions of secondary cranks are located at angles $2\theta_A = 0$; $2\theta_B = 2 \times 161°$; $2\theta_c = 2 \times 314°$ and $2\theta_D = 2 \times 120°$. With secondary cranks located at OA', OB', OC' and OD', the secondary force polygon is completed by drawing vectors equal in magnitude to their primary force counterparts, but in the directions of secondary cranks, as in Fig. 16.45(f). The closing side $o''d''$ represents unbalanced secondary force that measures 4671.8 N·m. Thus,

$$Rr = 4671.8 \text{ N·m}$$

Hence, magnitude of maximum unbalanced secondary force,

$$(F_s)_{max} = \frac{Rr}{gn} \omega^2$$

or
$$(F_s)_{max} = \frac{4671.8}{9.81 \times 4} \times (8\pi)^2 = 75202.96 \text{ N} \qquad \textbf{Ans.}$$

Taking position of crank OA to represent line of stroke, the unbalanced secondary force assumes maximum value when vector $o''d''$ become parallel to line of stroke in c.c.w. sense or when crank OA rotates through $65°$ in c.c.w. sense. **Ans.**

EXAMPLE 16.20 The firing order in a 6-cylinder vertical 4-stroke in-line engine is $1 - 4 - 2 - 6 - 3 - 5$. The piston stroke is 100 mm and length of each connecting rod is 200 mm. The pitch distances between cylinder centre lines are 100 mm, 100 mm, 150 mm, 100 mm and 100 mm respectively. The reciprocating mass per cylinder is 1 kg and the engine runs at 3000 r.p.m.

Determine the out-of-balance primary and secondary forces and couples on this engine, taking a plane mid-way between the cylinders 3 and 4 as the reference plane.

(DAV Indore, Nov. 1982; April 1990, April 1992)

Solution: $\omega = \dfrac{2\pi \times 3000}{60} = 314$ rad/s

Radius of crank $r = 100/2 = 50$ mm $= 5$ cm

Length of connecting rod, $l = 200$ mm. Hence, $n = l/r = 200/50 = 4.0$

Plane	Mass m kg	Radius r cm	Cent. force $\div \omega^2$ = (mr) kg·cm	Distance from R.P. l cm	Couple $\div \omega^2$ = mrl kg·cm^2
1	1	5.0	5.0	−27.5	−137.5
2	1	5.0	5.0	−17.5	−87.5
3	1	5.0	5.0	−7.5	−37.5
4	1	5.0	5.0	7.5	37.5
5	1	5.0	5.0	17.5	87.5
6	1	5.0	5.0	27.5	137.5

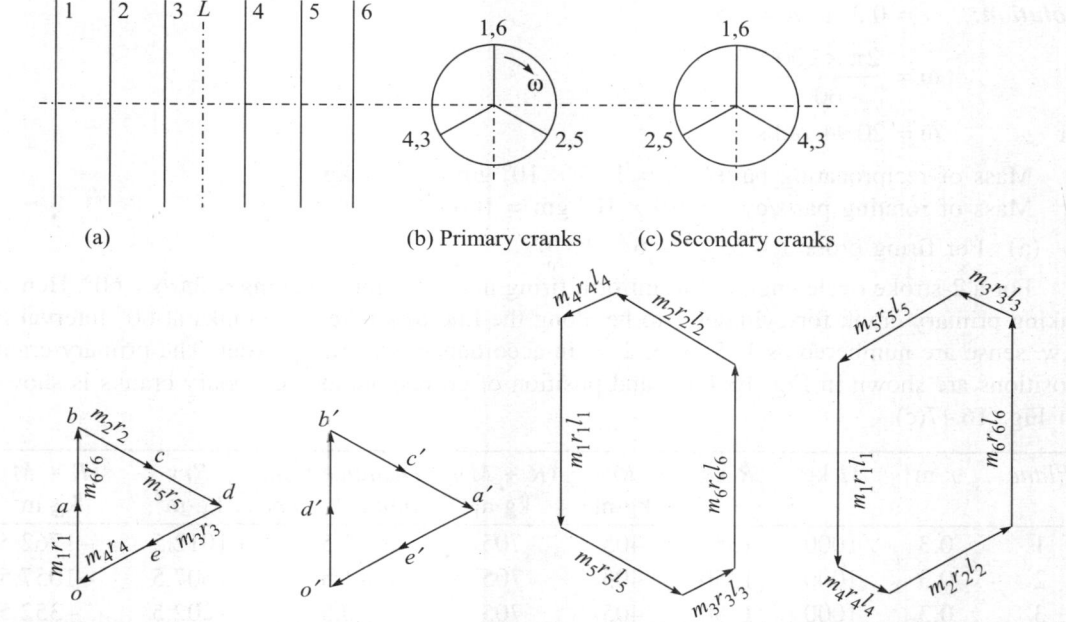

(a)

(b) Primary cranks   (c) Secondary cranks

(d) Primary force polygon (e) Secondary force polygon (f) Primary couple polygon (g) Secondary couple polygon

Scale: 1 cm = $\frac{10}{3}$ kg·cm   Scale: 1 cm = $\frac{10}{3}$ kg·cm   Scale: 1 cm = 35 kg·cm^2 Scale: 1 cm = 35 kg·cm^2

Fig. 16.46 Firing order in 6-cylinder, 4-stroke in-line engine.

Primary and Secondary Forces: The primary force polygon *oabcde* is drawn as in Fig. 16.46(d) using entries in column 4 of the table while the secondary force polygon is drawn as in Fig. 16.46(e) using entries same as above but by drawing vectors in directions indicated by secondary crank positions in Fig. 16.46(c).

The primary force polygon and secondary force polygon close and as such there is no out-of-balance primary and secondary force.

Primary and Secondary Couples: Primary and secondary couple polygons are drawn in Figs. 16.46(f) and (g) using entries of column 6 of the table and drawing vectors in directions indicated respectively by primary and secondary cranks. As the primary and secondary couple polygons close, there is no out-of-balance primary or secondary couple.   **Ans.**

EXAMPLE 16.21 The six cylinders of a single acting, two-stroke cycle diesel engine are pitched 1 m apart and the cranks are spaced at 60° intervals. The crank length is 300 mm and the ratio of connecting rod to crank is 4.5. The reciprocating mass per line is 1.35 Mg and the rotating mass is 1 Mg. The speed is 200 rev/min.

Show with regard to primary and secondary balance, that the firing order 1 – 5 – 3 – 6 – 2 – 4 gives unbalance in primary moment only, and the order 1 – 4 – 5 – 2 – 3 – 6 gives secondary moment unbalance only. Compare the maximum values of these moments, evaluating with respect to the central plane of the engine.   (Univ. of London)

Solution: $r = 0.3$ m; $n = 4.5$

$$\omega = \frac{2\pi \times 200}{60}$$

or $\omega = 20.94$ rad/s

Mass of reciprocating parts/cyl. $= 1.35 \times 10^6$ gm $= 1350$ kg
Mass of rotating parts/cyl. $= 1.0 \times 10^6$ gm $= 1000$ kg

(a)   For firing order $1 - 5 - 3 - 6 - 2 - 4$

For a 2-stroke cycle engine, for uniform firing interval, crank spacing $= 2\pi/6 = 60°$. Hence, taking primary crank for cylinder 1 to be along the line of stroke, the cranks at 60° interval in c.w. sense are numbered as 1, 5, 3, 6, 2, 4, in accordance with firing order. The primary crank positions are shown in Fig. 16.47(a) and position of corresponding secondary cranks is shown in Fig. 16.47(c).

Plane	r m	M kg	R kg	Rr kg·m	$(R + M)r$ kg·m	Distance from plane $L : x$m	Rrx kg·m^2	$(R + M)r$ kg·m^2
1	0.3	1000	1350	405	705	-2.5	-1012.5	-1762.5
2	0.3	1000	1350	405	705	-1.5	-607.5	-1057.5
3	0.3	1000	1350	405	705	-0.5	-202.5	-352.5
4	0.3	1000	1350	405	705	0.5	202.5	352.5
5	0.3	1000	1350	405	705	1.5	607.5	1057.5
6	0.3	1000	1350	405	705	2.5	1012.5	1762.5

The primary couple polygon is drawn in Fig. 16.47(d) and the closing line $O6$ represents unbalanced primary couple of magnitude $R.r = 2415$ kg·m^2 at 30° to the line of stroke. Note that both revolving and reciprocating masses have been considered for primary couples for both the firing orders, while only reciprocating masses will be considered for secondary couples.

The maximum value of unbalanced primary couple

$$(C_p)_{max} = Rr\omega^2$$

or $(C_p)_{max} = 2415 (20.94)^2 = 1058937.9$ N·m $= 1.059$ MN·m   **Ans.**

The secondary couple polygon, drawn in Fig. 16.47(e), shows addition of vectors $R_1 r_1 l_1$ (-1012.5), $R_6 r_6 l_6 (= 1012.5)$, $R_5 r_5 l_5 (= 607.5)$, $R_4 r_4 l_4 (= 202.5)$, $R_3 r_3 l_3 (= -202.5)$ and $R_2 r_2 l_2$ ($= -607.5$). The last vector $R_2 r_2 l_2$ thus terminates at starting point O and the polygon closes. Hence secondary couples are balanced.

(b)   For firing order, $1 - 4 - 5 - 2 - 3 - 6$

Let $R' \equiv (R + M)$. The primary and secondary crank positions for this firing order are shown in Figs. 16.47(f) and (g). As can be seen in Fig. 16.47(h), the additions of vectors $R_1' r_1 l_1$ ($= -1762.5$), $R_2' r_2 l_2 (= -1057.5)$, $R_3' r_3 l_3 (= -352.5)$, $R_4' r_4 l_4 (= +352.5)$, $R_5' r_5 l_5 (= 1057.5)$, and $R_6' r_6 l_6 (= 1762.5)$ finally leads the vector $R_6' r_6 l_6$ to terminate at the starting point. Thus the primary couple polygon closes and there is no unbalanced primary couple.

The secondary couple polygon in Fig. 16.47(i) represents addition of vectors $R_1 r_1 l_1$ ($= -1012.5$), $R_2 r_2 l_2 (= -607.5)$, $R_4 r_4 l_4 (= 202.5)$, $R_3 r_3 l_3 (= -202.5)$, $R_5 r_5 l_5 (= 607.5)$ and $R_6 r_6 l_6$

(= 1012.5). The closing line $O6$ represents secondary unbalanced force and measures 2800 kg·m^2. Hence maximum value of unbalanced secondary couple,

$$(C_s)_{max} = (Rrl) \times \omega^2 \times \frac{1}{\pi}$$

$$(C_s)_{max} = \frac{2800}{4.5} \times (20.94)^2 = 272834.2 \text{ N·m} = 0.2728 \text{ MN·m} \qquad \textbf{Ans.}$$

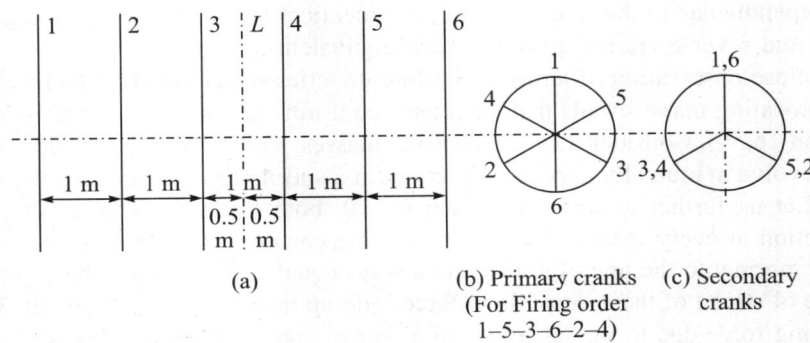

(a)

(b) Primary cranks (For Firing order 1–5–3–6–2–4)

(c) Secondary cranks

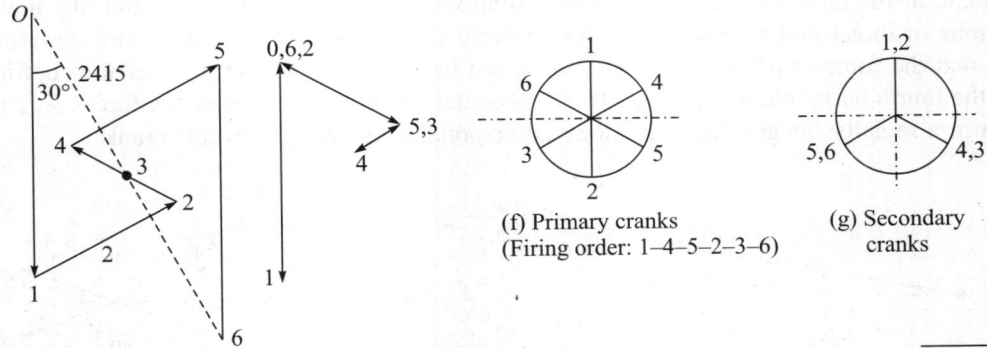

(d) Primary couple polygon
Scale: 1 cm = 350 kg·m^2)
(Firing order: 1–5–3–6–2–4)

(e) Secondary couple polygon
(Scale: 1 cm = 250 kg·m^2)

(f) Primary cranks (Firing order: 1–4–5–2–3–6)

(g) Secondary cranks

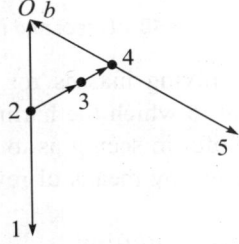

(h) Scale: 1 cm = 435 kg·m^2
Primary couple polygon
(Firing order: 1–4–5–2–3–6)

(i) Scale: 1 cm = 400 kg·m^2
Secondary couple polygon

Fig. 16.47 Balancing of a 6-cylinder, two stroke cycle engine.

16.24 DIRECT AND REVERSE CRANKS

Method of Direct and Reverse Cranks is useful in assessing the extent to which different harmonic forces are in balance.

V-engines and radial engines represent a class of engines in which the cylinders are not in-line. For complete balancing, reciprocating masses in such applications must be replaced by an exactly equivalent system of rotating masses. Such a system of imaginary rotating masses can be treated to be exactly equivalent only when both the components of its centrifugal forces, along and perpendicular to the line of stroke, are identical to the actual disturbing force. Method of 'direct and reverse cranks' provides such equivalent system.

A close observation of terms in the fourier series (equation (16.20)) for the inertia force of a reciprocating mass reveals that each term conforms to the general expression $K_j \cos j\theta$ (vide equation 16.59). Consider a case when two masses, each of which produces an inertia force of $K_j/2$, revolve in mutually opposite directions at a speed j times ω, the angular speed of the crank shaft. Let us further assume that when $\theta = 0$, both the masses lie at the i.d.c. With above assumption at every instant during rotation, the components of the centrifugal forces of these masses, normal to the line of stroke, are always equal and opposite. The components parallel to the line of stroke of these centrifugal forces add up to give a force $K_j \cos j\theta$. Thus, the resultant disturbing force due to inertia forces of a pair of such rotating masses is identical with the jth harmonic of the inertia force due to reciprocating mass. Figures 16.48(b), (c) and (d) show the positions of direct and reverse cranks respectively for the first, second and fourth harmonics. Note that the forces $K_1/2$, $K_2/2$ and $K_4/2$ have not been shown to scale. Further the coefficient B of the fourth harmonic in equation (16.58) is actually negative and hence the forces $K_4/2$ must taken in a radially inward direction along corresponding direct and reverse crank.

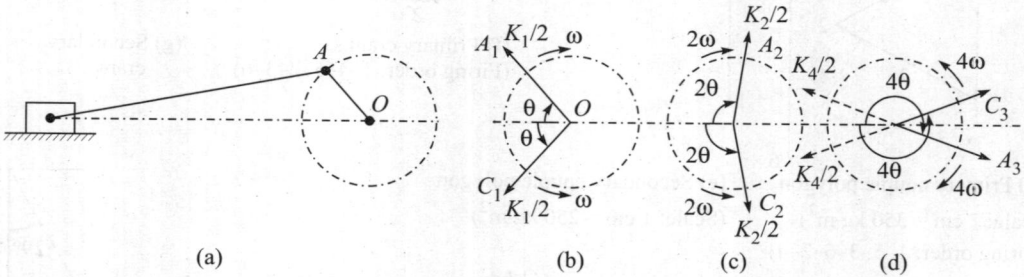

(a) (b) (c) (d)

Fig. 16.48 Direct and reverse cranks.

The substitution of two revolving masses for a single reciprocating mass simplifies the problem of examining the extent to which the harmonic forces are balanced in certain types of engines. The method is also helpful in seeing as to what extent can any unbalanced harmonics be balanced partially or completely by means of revolving balance weights. It may be recalled from equation (16.20) that

$$K_1 = (R/g)r\omega^2 \tag{16.65}$$

$$K_2 = (R/g)(r/4n)(2\omega)^2 \tag{16.66}$$

$$K_4 = -\left(\frac{r}{64n^3}\right)(4\omega)^2$$

i.e., the radius of primary crank is r while that of secondary crank is $(r/4n)$.

EXAMPLE 16.22   A three-cylinder radial engine has the cylinders spaced at angular intervals of 120°. The three connecting rods are coupled directly to a single crank. The stroke is 128 mm, the length of each connecting rod is 230 mm and the weight of the reciprocating parts per cylinder is 18 N. Find the resultant primary and secondary forces at a crank shaft speed of 1500 r.p.m.

Solution: $\omega = \dfrac{2\pi \times 1500}{60} = 157.08$ rad/s

$r = \dfrac{128}{2} = 64$ mm. Hence $n = l/r = 230/64 = 3.59$

For convenience, the common crank OC is assumed to lie along cylinder centre line OA of cylinder 1. Let OA, OB and OD represent respectively the centre lines of cylinders 1, 2 and 3. All the three connecting rods are hinged at the common crank pin. Readers may recall that such connection goes under the name of multiple-joint (a p-tuple joint, with $p = 2$, -vide Section 2.13).

For the position shown in Fig. 16.49(a), the common crank OC is inclined to centre lines of cylinders 1, 2 and 3 respectively at angles $\theta_1 (= 0°)$, $\theta_2 (= 120°)$ and $\theta_3 (= 240°)$ measured in clockwise direction. Direct primary cranks OD_1, OD_2 and OD_3 for cylinders 1, 2, and 3 are therefore located at angles 0°, 120° and 240° in forward (clockwise) direction from the respective centre lines. Thus, in Fig. 16.49(b), all the three direct primary cranks are shown to overlap. The reverse primary cranks OR_1, OR_2 and OR_3 are located respectively at angles $\theta_1 (= 0°)$, $\theta_2 (= 120°)$ and $\theta_3 (= 240°)$ measured in reverse (counterclockwise) direction from respective cylinder centre lines. Thus, as shown in Fig. 16.49(c), the reverse crank of cylinder 1 remains unchanged while the reverse crank of cylinder 2 takes up position along centre line of cylinder 3 at angle 120° c.c.w. The reverse crank of cylinder 3 is located at 240° in c.c.w. sense, *i.e.* along centre line of cylinder 2. Since all the three cylinders and their reciprocating parts are identical, $(K_1)/2$ is same for each crank. Hence, due to symmetry in Fig. 16.49(c), the reverse primary cranks are mutually balanced. But the direct primary cranks are not balanced. The resultant primary unbalanced force is equal to that given by direct primary cranks. Hence,

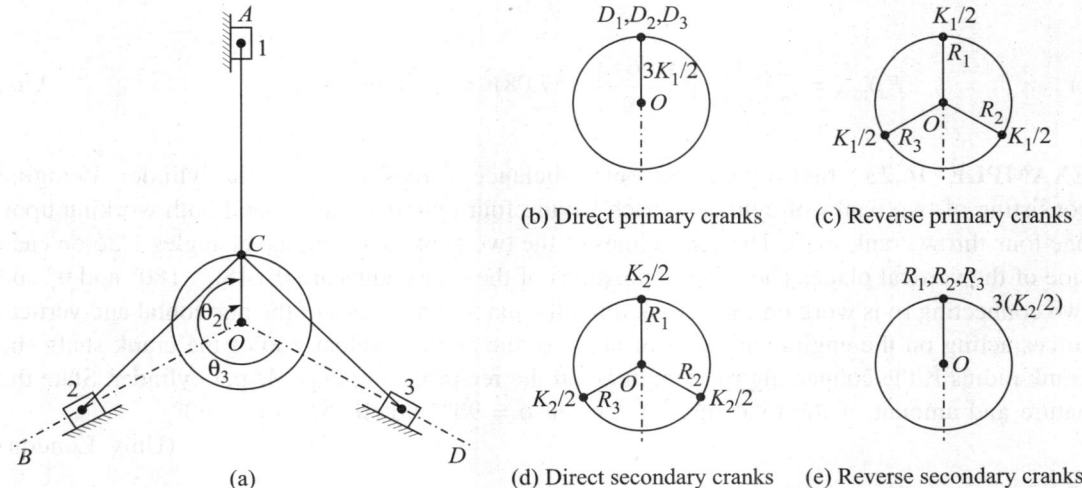

Fig. 16.49   Direct and Reverse Cranks for a 3-cylinder Radial Engine.

$$(F_p)_{max} = 3(K_1)/2 = (3/2)(R/g)r\omega^2$$

or $$(F_p)_{max} = \frac{3}{2}\left(\frac{18}{9.81}\right)(0.064)(157.08)^2 = 4346.27 \text{ N} \qquad \textbf{Ans.}$$

The resultant primary forces may be balanced by attaching a revolving mass B at radius b to the crank shaft at $180°$ to the common crank such that

$$\omega^2 \times (Bb) = 3(K_1)/2$$

or $$Bb = \left(\frac{3}{2}\right)Rr$$

and taking a convenient value of $b = 90$ mm, we have,

$$B = \left(\frac{3}{2}\right) \times 18 \times \left(\frac{0.064}{0.090}\right) = 19.2 \text{ N}$$

Secondary Cranks and Secondary Forces: The angles for secondary cranks are:

$$2\theta_1 = 0; \quad 2\theta_2 = 240° \quad \text{and} \quad 2\theta_3 = 480°$$

Hence measuring these angles in forward (clockwise) direction from respective cylinder centre lines, the direct secondary cranks are as shown in Fig. 16.49(d). It follows from the symmetrical distribution of direct secondary cranks that direct secondary cranks are mutually balanced.

The secondary reversed cranks are located at 2θ, $2\theta_2$ and $2\theta_3$ measured in c.c.w. direction from respective cylinder lines, and are as shown at Fig. 16.49(e). It follows that the reversed secondary cranks are unbalanced and the resultant value of unbalanced secondary force is given by

$$(F_s)_{max} = 3(K_2)/2 = \frac{3}{2}\left(\frac{R}{g}\right)\left(\frac{r}{n}\right)\omega^2$$

or $$(F_s)_{max} = \frac{3}{2}\left(\frac{18}{9.81}\right)\left(\frac{0.064}{3.59}\right)(157.08)^2 = 1210.66 \text{ N} \qquad \textbf{Ans.}$$

EXAMPLE 16.23 Investigate the out-of-balance forces of an eight-cylinder V-engine consisting of two banks of cylinders, each having four cylinders in line and both working upon one four-throw crank shaft. The centre lines of the two banks are inclined at angles $1/2\phi$ on each side of the vertical plane. The relative positions of the four cranks are $0°$, $180°$, $180°$ and $0°$ and two connecting rods work on each crank. Find the maximum values of the horizontal and vertical forces acting on the engine, in terms of angle ϕ, the angular velocity ω of the crank shaft, the crank radius r, the connecting rod length l, and the reciprocating mass M per cylinder. State the nature and amount of the total force, (a) when $\phi = 90°$ and (b) when $\phi = 60°$

(Univ. London)

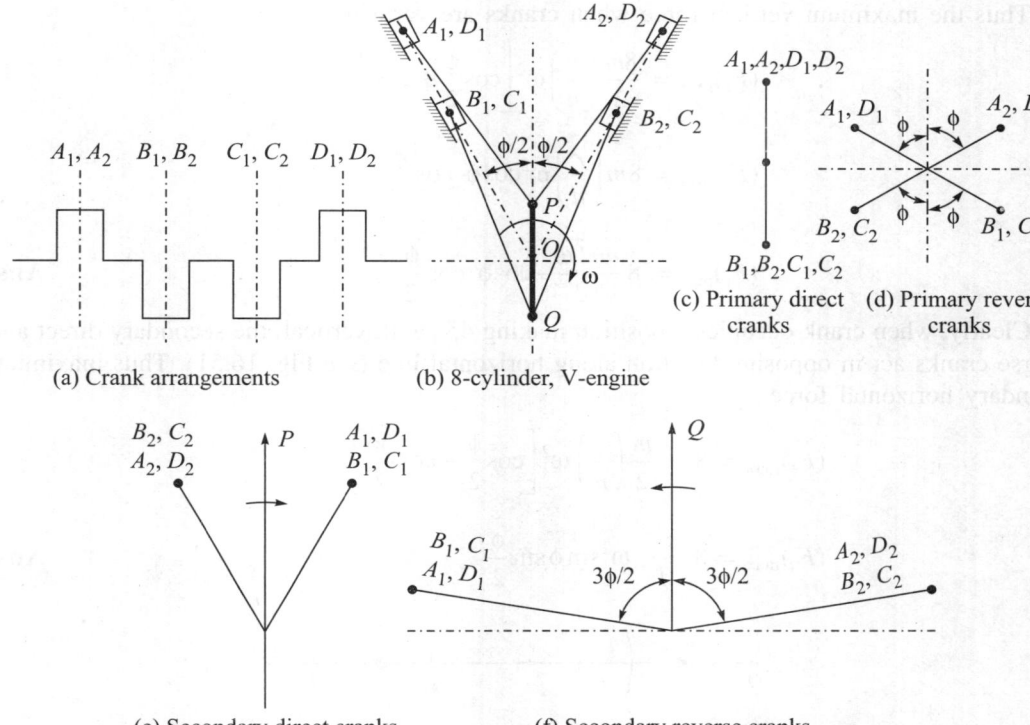

(a) Crank arrangements

(b) 8-cylinder, V-engine

(c) Primary direct cranks

(d) Primary reverse cranks

(e) Secondary direct cranks

(f) Secondary reverse cranks

Fig. 16.50   Direct and Reverse Cranks for 8-cylinder V-engine.

Solution: Figures 16.50(a) and (b) show schematic elevational and end-view of the eight-cylinder V-engine. Suffices 1 and 2 refer to left-hand and right-hand bank respectively. For sliders and B_1, C_1 and B_2, C_2 the common crank is OQ while for sliders A_1, D_1 and A_2, D_2, crank OP is the common crank. Thus in the position shown, the angles made by common crank with the line of strokes are (in c.w. sense):

cylinders A_1, D_1 : $\phi/2$,   cylinders A_2, D_2 : $(2\pi - \phi/2)$
cylinders B_1, C_1 : $(\pi + \phi/2)$,   cylinders B_2, C_2 : $(\pi - \phi/2)$

Hence direct and reverse primary cranks are as shown in Figs. 16.50(c) and (d). It is evident from symmetry of direct-primary and reverse-primary cranks that primary forces are completely balanced.

Secondary Forces: The secondary direct and secondary reverse cranks are shown in Figs. 16.50(e) and (f). It follows that direct secondary cranks give rise to a single resultant force,

$$P = [4K_2\cos\phi/2] \times 2 = 8 \times m\left(\frac{r}{n}\right)\omega^2 \cos\frac{\phi}{2}$$

Similarly, reverse secondary cranks are unbalanced and give rise to a single resultant force,

$$Q = \left[4\frac{K_2}{2}\cos\frac{3\phi}{2}\right] \times 2 = 8\frac{m}{2}\left(\frac{r}{n}\right)\omega^2\cos\frac{3\phi}{2}$$

Thus the maximum vertical force when cranks are vertical,

$$(F_v)_{max} = \frac{8m}{2}\left(\frac{r}{n}\right)\omega^2\left[\cos\frac{\phi}{2} + \cos\frac{3\phi}{2}\right]$$

or

$$(F_v)_{max} = 8m\left(\frac{r}{n}\right)\omega^2\cos\phi\,\cos\frac{\phi}{2}$$

or .

$$(F_v)_{max} = 8\frac{mr^2\omega^2}{l}\cos\phi\,\cos\frac{\phi}{2} \qquad \textbf{Ans.}$$

Clearly, when crank occupies a position making 45° with vertical, the secondary direct and reverse cranks act in opposite direction along horizontal line (*see* Fig. 16.51). Thus maximum secondary horizontal force.

$$(F_s)_{max} = 8 \times \frac{m}{2}\left(\frac{r}{n}\right)\omega^2\left[\cos\frac{\phi}{2} - \cos\frac{3\phi}{2}\right]$$

or

$$(F_s)_{max} = 8\frac{mr^2}{l}\omega^2\sin\phi\sin\frac{\phi}{2} \qquad \textbf{Ans.}$$

Fig. 16.51

Consider a general case when the cranks have rotated through an angle θ. For this case, angle at which resultant direct secondary P and resultant reverse secondary are inclined to vertical = 2θ. Vertical component of this general force,

$$F_v' = (P + Q)\cos2\theta = \left\{\frac{8mr^2\omega^2}{l}\cos\phi\cos\frac{\phi}{2}\right\}\cos2\theta$$

and the horizontal component of the general force,

$$F_H' = (P - Q)\sin2\theta = \left\{\frac{8mr^2\omega^2}{l}\sin\phi\sin\frac{\phi}{2}\right\}\sin2\theta$$

Therefore, resultant force,

$$F' = \sqrt{F_v'^2 + F_H'^2}$$

or,

$$F' = \frac{8mr^2\omega^2}{l}\sqrt{\left(\cos\phi\cos\frac{\phi}{2}\cos2\theta\right)^2 + \left(\sin\phi\sin\frac{\phi}{2}\sin2\phi\right)^2}$$

When $\phi = 90°$,

$$F' = \frac{8mr^2\omega^2}{l}\sqrt{\frac{1}{2}(\sin2\theta)^2}$$

or
$$F' = \frac{4\sqrt{2}\,mr^2\,\omega^2}{l}\sin 2\theta$$ **Ans.**

and when $\phi = 60°$,

$$F' = \frac{8mr^2\omega^2}{l}\sqrt{\frac{3}{16}(\cos 2\theta)^2 + \frac{3}{16}(\sin 2\theta)^2} = \frac{2\sqrt{3}mr^2\omega^2}{l}$$ **Ans.**

16.25 BALANCING V-ENGINES

A V-type engine consists of two in-line engines having a common crank shaft. The two planes in which the piston reciprocates (constituting two banks of cylinders) intersect along the crank shaft axis and form a V of included angle 2α. In multi-cylinder V-type engines, the bank angle (*i.e.*, V-angle) is not arbitrary but is selected in conjunction with the number of cylinders so as to provide uniform firing intervals. The V-8 engine (4 cylinders in each bank) with 90° bank angle is widely used in automotive practice. V-6 or V-12 engines with $\phi = 60$ degree or 90 degree are, however, not uncommon. In the analysis of a V-type engine, each bank of cylinders may be analysed separately and then the unbalanced forces, if any, of the banks can then be combined vectorially.

Consider a V-engine, consisting of 2 cylinders only. With common crank OC as shown in Fig. 16.52, let ϕ be the fixed V-angle so that $\alpha \equiv \phi/2$ is the angle made by each plane of cylinder centre lines with the plane of symmetry indicated by OY. Further, let ω be the angular speed of the crank in radians per second (c.w.) and θ the angle moved by the common crank from the line of symmetry at the given instant. Then, angle made by crank with respect to $OA = (\alpha + \theta)$

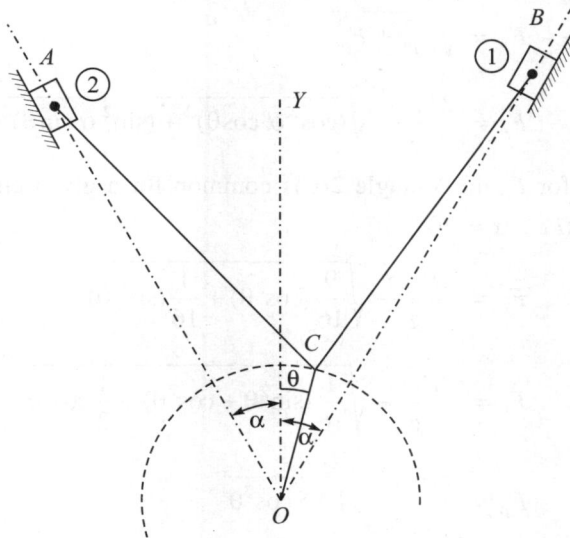

Fig 16.52 A V-engine with two inline engines.

and angle made by crank with respect to $OB = (\alpha - \theta)$. Thus the inertia forces due to reciprocating parts for the two cylinders are

$$F = \frac{R}{g} r\omega^2 \left\{ \cos(\alpha + \theta) + \frac{1}{n} \cos 2(\alpha + \theta) \right\} \qquad (16.67)$$

and

$$F' = \frac{R}{g} r\omega^2 \left\{ \cos(\alpha - \theta) + \frac{1}{n} \cos 2(\alpha - \theta) \right\} \qquad (16.68)$$

where R = weight of reciprocating parts per cylinder
   r = radius of crank.

Considering primary forces first for both the cylinders and resolving them along and perpendicular to the line of symmetry OY,

$$F_{pv} = \frac{R}{g} r\omega^2 \{ \cos(\alpha + \theta) + \cos(\alpha - \theta)\cos\alpha$$

or

$$F_{pv} = \frac{2R}{g} r\omega^2 \cos^2\alpha \cos\theta \qquad (16.69)$$

and,

$$F_{ph} = \frac{R}{g} r\omega^2 \{ \cos(\alpha + \theta) - \cos(\alpha - \theta)\sin\alpha$$

or

$$F_{ph} = \frac{2R}{g} r\omega^2 \sin^2\alpha \sin\theta \qquad (16.70)$$

The resultant primary force, as obtained from equations (16.68) and (16.69) in mutually perpendicular directions, is

$$F_p = \sqrt{F_{pv}^2 + F_{ph}^2}$$

or

$$F_p = \frac{2Rr\omega^2}{g} \sqrt{(\cos^2\alpha\cos\theta)^2 + (\sin^2\alpha\sin\theta)^2} \qquad (16.71)$$

In the expression for F_p the V-angle 2α is common for a given engine. Thus,

(a) For $2\alpha = 60$ (*i.e.*, $\alpha = 30°$),

$$F_p = \frac{2Rr\omega^2}{g} \sqrt{\frac{9}{16}(\cos^2\theta) + \frac{1}{16}(\sin^2\theta)}$$

$$F_p = \frac{2Rr\omega^2}{g} \sqrt{\frac{1}{16}(\sin^2\theta + \cos^2\theta) + \frac{1}{2}\cos^2\theta}$$

or

$$F_p = \frac{2Rr\omega^2}{2g} \sqrt{1 + 8\cos^2\theta} \qquad (16.72)$$

(b) For $2\alpha = 90°$ (*i.e.*, $\alpha = 45°$),

$$F_p = \frac{2Rr\omega^2}{g}\sqrt{\frac{1}{4}\cos^2\theta + \frac{1}{4}\sin^2\theta}$$

or $\qquad\qquad F_p = (Rr\omega^2/g)$ $\qquad\qquad$ (16.73)

(c) For $2\alpha = 120°$ (*i.e.*, $\alpha = 60°$),

$$F_p = \frac{2Rr\omega^2}{g}\sqrt{\frac{1}{16}\cos^2\theta + \frac{9}{16}\sin^2\theta}$$

or $\qquad\qquad F_p = \frac{Rr\omega^2}{2g}\sqrt{1+8\sin^2\theta}$ $\qquad\qquad$ (16.74)

Again, considering components of secondary force only, along and perpendicular to the line of symmetry OY,

$$F_{sv} = \frac{R}{gn}r\omega^2[\cos2(\alpha + \theta) + \cos2(\alpha - \theta)]\cos\alpha$$

or $\qquad\qquad F_{sv} = \frac{R}{gn}r\omega^2[2\cos2\alpha\cos2\theta]\cos\alpha$

and, $\qquad\qquad F_{sh} = \frac{R}{gn}r\omega^2[\cos2(\alpha + \theta) - \cos2(\alpha - \theta)]\sin\alpha$

or $\qquad\qquad F_{sh} = \frac{R}{gn}r\omega^2[2\sin2\alpha\,\sin2\theta]\sin\alpha$

Hence, the resultant secondary force

$$F_s = \sqrt{F_{sv}^2 + F_{sh}^2}$$

or $\qquad F_s = \frac{2R}{gn}r\omega^2\sqrt{(\cos2\alpha\,\cos2\theta\,\cos\alpha)^2 + (\sin2\alpha\,\sin2\theta\,\sin\alpha)^2}$ $\qquad$ (16.75)

(a) For $2\alpha = 60°$ (*i.e.*, $\alpha = 30°$),

$$F_s = \frac{2Rr\omega^2}{gn}\cdot\sqrt{\left(\frac{\sqrt3}{4}\right)^2(\cos^2 2\theta + \sin^2 2\theta)}$$

or $\qquad\qquad F_s = \frac{\sqrt3}{2}\frac{Rr\omega^2}{gn}$ $\qquad\qquad$ (16.76)

(b) When $2\alpha = 90°$ (*i.e.*, when $\alpha = 45°$),

$$F_s = \frac{2Rr\omega^2}{gn}\sqrt{0 + \left(\frac{1}{\sqrt2}\sin2\theta\right)^2}$$

or
$$F_s = \frac{\sqrt{2}\, Rr\omega^2}{gn} \sin 2\theta \qquad (16.77)$$

(c) when $2\alpha = 120°$, (*i.e.*, when $\alpha = 60°$)

$$F_s = \frac{2Rr\omega^2}{gn} \sqrt{\left(-\frac{1}{4}\cos 2\theta\right)^2 + \left(\frac{3}{4}\sin 2\theta\right)^2}$$

or
$$F_s = \frac{1}{2}\frac{Rr\omega^2}{gn} \sqrt{\cos^2 2\theta + 9\sin^2 2\theta}$$

or
$$F_s = \frac{1}{2}\frac{Rr\omega^2}{gn} \sqrt{1 + 8\sin^2 2\theta} \qquad (16.78)$$

EXAMPLE 16.24   The pistons of a 60° V-engine have strokes of 12 cm. The two connecting rods operate on a common crank pin and each is 20 cm long. If the mass of the reciprocating parts is 1 kg per cylinder and the crank shaft speed is 2500 r.p.m., determine the maximum and minimum value of (a) primary force and (b) secondary force.   (SGSITS Indore, Dec. 1994)

Solution: $M = 1$ kg; $r = 12/2 = 6$ cm; $1 = 20$ cm. Therefore $n = 20/6 = 3.33$
$\omega = 2\pi \times 2500/60 = 261.8$ rad/s; $\alpha = 30°$

For $2\alpha = 60°$, from equation (16.72) resultant primary unbalanced force,

$$F_p = \frac{1}{2} (Rr\omega^2/g) \sqrt{1 + 8\cos^2\theta}$$

Maximum value of primary force occurs when $\theta = 0°$. Thus,

$$(F_p)_{max} = \frac{3}{2}\cdot\frac{Rr}{g}\omega^2 = \frac{3}{2}\cdot\frac{Rr}{g}\omega^2$$

or
$$(F_p)_{max} = \frac{3}{2}\cdot(1)(0.06)(261.8)^2 = 6168.5 \text{ N}$$

$$(F_p)_{min} = (F_p)_{max}/3 = 2056.17 \text{ N} \qquad \textbf{Ans.}$$

Also, the resultant secondary force, from equation 16.76,

$$F_s = \frac{\sqrt{3}}{2}\frac{Rr\omega^2}{g\cdot n}$$

or
$$F_s = \frac{\sqrt{3}}{2}(1)\frac{(0.06)}{(3.33)} \times (261.8)^2 = 1068.5 \text{ N} \qquad \textbf{Ans.}$$

REVIEW EXERCISES

16.1 A workman is required to machine a casting, weighing 2060 N on a lathe. He fixes it in such a way that the centre of gravity of the casting is 18 cm from the lathe axis and 32 cm from the face plate. He balances it statically by bolting two weights to the face plate, one of them weighs 588.0 N and is fixed as shown in Fig. 16.53. Its centre of gravity is 16 cm from the face plate. The other weighs 490 N and its c.g. is 10 cm. from the face plate. Determine (a) the radial and the angular position of 490 N weight and (b) the rocking couple at 50 r.p.m.

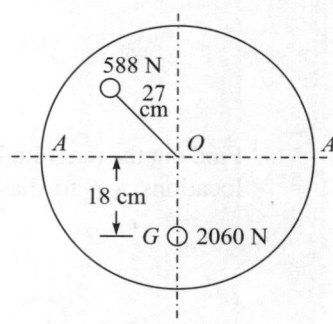

Fig. 16.53

(*Ans.* $r = 57$ cm; θ with $OA = 66.5°$; $C = 210.4$ N·m)

16.2 Four masses A, B, C and D are to be completely balanced. The weights of masses B, C and D are respectively 294.3 N, 490.5 N and 392.4 and they revolve with shaft at radii $r_b = 24$ cm, $r_c = 12$ cm and $r_d = 15$ cm. The mass A revolves at radius $r_a = 18$ cm. The transverse planes containing masses B and C are 30 cm apart. The angle between longitudinal planes containing masses B and C is 90°. B and C make angles of 120° and 210° respectively with D in the same sense. Find

(a) the weight and angular position of mass A,

(b) the position of planes A and D.

(*Ans.* $W_A = 204.5$ N; $\theta_A = 28°$ c.c.w. from D; $l_o = 30.7$ cm from B to the left; $l_A = 81.2$ cm to the right)

16.3 A rotating shaft carries four unbalanced masses 176.6 N, 137.3 N, 157 N and 117.7 N at radii 5 cm, 6 cm, 7 cm and 6 cm respectively. The second, third and fourth masses revolve in planes 8 cm, 16 cm and 28 cm respectively measured from the plane of the first mass and are angularly located at 60°, 135° and 270° respectively measured clockwise from the first mass looking from this mass end of the shaft. The shaft is dynamically balanced by two masses, both located at 5 cm radii and revolving in planes mid-way between those of 1st ang 2nd masses and mid-way between those of 3rd and 4th masses. Determine graphically or otherwise, the magnitude of the masses and their respective angular positions. (AMIE Summer 1977)

(*Ans.* $W_F = 130.5$ N at 25° at c.c.w. from A; $W_E = 294.3$ N at 221° c.c.w. from A)

16.4 A shaft carries three pulleys A, B and C at distances 60 cm, 120 cm apart. The pulleys are out of balance to the extent of 19.6 N, 17.2 N and 24.5 respectively at a radius of 2.5 cm in each case. The angular positions of the out-of-balance masses in the pulley B and C with respect to pulley A are 90° and 210° respectively. Determine in position and magnitude, the balance weights required, in plane L and M mid-way between planes A, B and C respectively. The radius of rotation of balance weights is 12.5 cm. (D.A.V. Indore 1986)

16.5 Referring to Fig. 16.54 the particulars are:

$$w_a = 890 \text{ N}, \; r_a = 22.9 \text{ cm}$$
$$w_b = 1335 \text{ N}, \; r_b = 17.8 \text{ cm}$$
$$w_c = 1068 \text{ N}, \; r_c = 25.4 \text{ cm}$$
$$w_d = 1157 \text{ N}, \; r_d = 30.5 \text{ cm}$$

Find out the magnitudes of balancing masses required in planes L and M and their angular locations w.r. to mass w_a.

(*Ans.* $B_l = 716.5 \text{ N}; \; B_m = 476 \text{ N}, \; \theta_l = 40°$ and $\theta_m = 49°$)

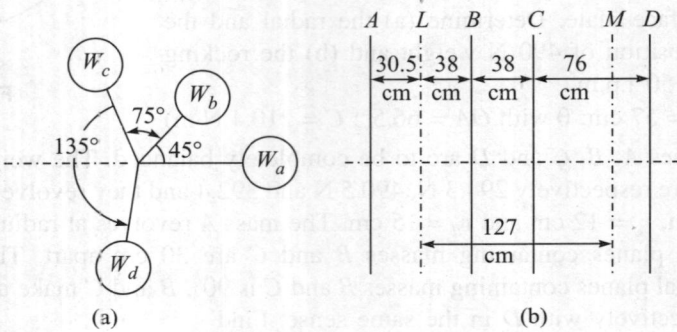

(a) (b)

Fig. 16.54

16.6 The following data refer to two cylinder locomotive with cranks at 90°. Reciprocating mass per cylinder = 2943 N, crank radius = 30 cm, driving wheel diameter = 180 cm, distance between cylinder centre-lines = 65 cm, distance between the driving wheel 'centre-planes' = 155 cm. Determine:

(a) Fraction of the reciprocating masses to be balanced if the hammer blow is not to exceed 45126 N at 96.5 km.p.h.

(b) Variation in tractive force. (D.A.V. Indore: 1982, 1990)

(*Ans.* $c = 0.736; \; ±30420.8$)

16.7 An inside cylinder uncoupled locomotive has the following data:

Revolving weight per cylinder = 3924 N, reciprocating weight per cylinder = 4414 N, crank radius = 0.4, track = 1.5 m, distance between cylinder centre lines = 1.2 m, driving wheel diameter = 1.6 m, radius at which balance weights are located = 0.65 m.

If the hammer blow should not exceed 39240 N when the speed of the locomotive is 75 km.p.h. determine the fraction of reciprocating masses that could be balanced and also the weight of each balancing mass which should also balance completely the revolving masses. (D.A.V. Indore, Feb. 1992) (*Ans.* $c = 0.3; \; 3460 \text{ N}$)

16.8 A two-cylinder uncoupled locomotive has inside cylinders 60 cm apart. The radius of each crank is 30 cm and are at right angles. The weight of the revolving mass per cylinder is 2452.5 N and the weight of the reciprocating mass per cylinder is 2943 N. The whole of the revolving and 2/3rd of the reciprocating masses are to be balanced and the balanced weights are placed, in the planes of rotation of the driving wheels, at a radius of 80 cm. The driving wheels are 2 m in diameter and 1.5 m apart. If the speed of the engine is 80 km.p.h., find the hammer blow, maximum variation of tractive effort and maximum swaying couple. **(Ans.** 18296 N; 16922.3 N; 16186.5 N·cm)

16.9 A two-cylinder uncoupled locomotive with cranks at 90° has the following particulars:

The reciprocating mass per cylinder 3433.5 N, rotating mass per cylinder 2943 N, crank radius 30 cm and pitch of cylinder 0.6 m. The balance weights are placed at a radius of 0.6 m in the planes of driving wheels 1.5 m apart. If the diameter of the driving wheel is 1.8 m and the engine runs at 60 km.p.h., find

(a) Magnitude and angular position of the balancing weights, when whole of the rotating and 2/3rd of the reciprocating masses are to be balanced

(b) Swaying couple

(c) Maximum and minimum pressure on the rails, if pressure due to dead load on each wheel is 29430 N

(d) Maximum speed at which the wheels do not lift from the rails.

(Ans. 1937.5 N; 6131.3 N·m; 62588 N; 87km/hr)

16.10 A four-cylinder engine has cranks arranged symmetrically along the shaft as shown in Fig. 16.55. The distance between the outer cranks A and D is 5.4 metres and that between the inner cranks B and C is 2.4 metres. The weight of the reciprocating parts belonging to each of the outer cylinders is 19620 N and that belonging to each of the inner cylinders is W N.

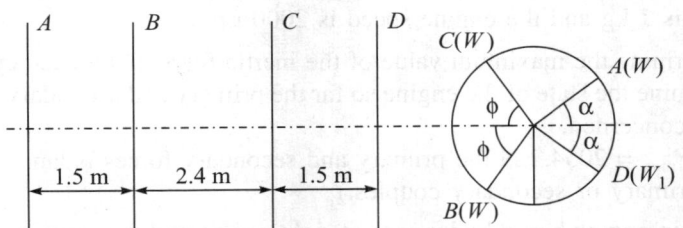

Fig. 16.55 Symmetrical cranks for 4-cylinder engine.

If the primary and secondary forces are to be balanced and also the primary couples, determine the crank angle positions and the weight of the reciprocating parts W corresponding to inner cylinders.

Find also the maximum value of the unbalanced secondary couple, if the stroke is 1 metre, the connecting rod length 2 metres, and the speed of the engine is 110 r.p.m.

(Ans. $\alpha = 31.5°$; $\phi = 54°$; 264870 N·m)

16.11 Figure 16.56 shows the arrangement of the cranks in a four-crank symmetrical engine in which the weight of the reciprocating parts at cranks 1 and 4 are each equal to W_1 and at cranks 2 and 3 are each equal to W_2.

Show that the arrangement is balanced for primary forces and couples and for secondary forces provided that

$$\frac{W_1}{W_2} = \frac{\cos\theta_2}{\cos\theta_1}; \quad \frac{a_1}{a_2} = \frac{\tan\theta_2}{\tan\theta_1} \quad \text{and} \quad \cos\theta_1 \cdot \cos\theta_2 = 1/2.$$

(AMIE, Summer 1980; Engg. Services 1977)

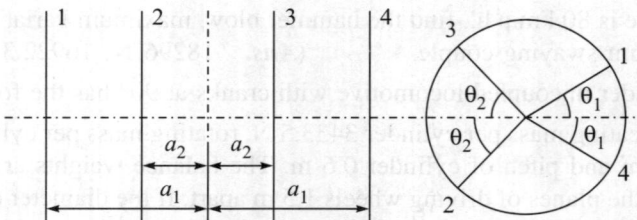

Fig. 16.56 Arrangement of cranks in 4-crank symmetric engine.

16.12 A five-cylinder inline engine running at 750 r.p.m. has successive cranks 144° apart, the distance between the cylinder centre-lines being 37.5 cm. The piston stroke is 22.5 cm and the ratio of the connecting rod to crank is 4. Examine the engine for the balance of primary and secondary forces and couples. Find the maximum values of these and the position of the central crank at which these maximum values occur. The reciprocating mass for each cylinder is 147.2 N. (Bihar University 1980)
(*Ans.* Primary and secondary forces are balanced; maximum primary unbalanced couple: 19996.4 N·m; maximum secondary unbalanced couple: 4159 N·m)

16.13 A six-cylinder inline engine of the 4-stroke internal combustion type, has a firing order 1-4-2-6-3-5. The distance between the centre lines of the cylinder is 10 cm, the stroke of each piston is 7.5 cm, the connecting rod length is 15 cm, the reciprocating mass per cylinder is 1 kg and the engine speed is 2000 r.p.m.

(a) Determine the maximum value of the inertia force along each cylinder centre line.
(b) Examine the state of the engine so far the primary and secondary forces and couples are concerned.
(*Ans.* F_{max} = 2054.2 N the primary and secondary forces balance themselves. There are no primary or secondary couples.)

16.14 A vee-twin engine has cylinder axes at right angles and the connecting rods operate on a common crank. The reciprocating mass per cylinder is 12 kg, the crank is 7 cm, connecting rods 35 cm. Show that the engine may be balanced for primary effect by means of a revolving balance weight.
   If the speed is 500 r.p.m., what is the maximum value of the resultant secondary force and in which direction does it act? (*Ans.* 650.4 N occurs at θ = 0; 90, 180; 270°)

16.15 The cranks of a 4-cylinder marine oil engine are arranged at angular intervals of 90°. The engine speed is 70 r.p.m. and the reciprocating mass per cylinder is 800 kg. The inner cranks are 1 m apart and are symmetrically arranged between the outer cranks which are 2.6 m apart. Each crank is 400 mm long. Determine the firing order of the cylinders for the best balance of reciprocating masses and also the magnitude of the unbalanced primary couple for that arrangement.
   (*Ans.* Firing orders: 1, 4, 2, 3 and 1, 3, 2, 4; c_{max} = 19450 N·m)

16.16 The three cylinders of an air-compressor have their axes 120° to one another, and their connecting rods are coupled to a single crank. The stroke is 10 cm and the length of each connecting rod is 15 cm. The weight of the reciprocating parts per cylinder is 14.7 N. Find the maximum primary and secondary forces acting on the frame of the compressor when running at 3000 r.p.m. Describe clearly a method by which such forces may be balanced.

(*Ans.* F_p = 11105 N; $B_1 b_1$ = 110.4 N-cm; F_s = 3702.3 N; $B_2 b_2$ = 9.19 N·cm) (DAV Indore: Dec. 1982)

16.17 A 3-cylinder radial engine, driven by a common crank has the cylinders spaced at 120°. The stroke is 125 mm, length of the connecting rod is 225 mm and the weight of the reciprocating mass per cylinder is 19.6 Newton. Calculate the primary and secondary forces at crank-shaft speed of 1200 r.p.m. (D.A.V. Indore Sept. 1990 and Mysore Un.)

(*Ans.* 2945 N; 817 N)

16.18 A twin-cylinder V-engine has the cylinders set at an angle of 45°, with both pistons connected to the same crank. The crank radius is 6.25 cm and the connecting rods are 27.5 cm long. The reciprocating weight per line is 14.7 N and the total rotating weight is equivalent to 2 kgf at the crank radius. A balance weight fitted opposite to the crank, is equivalent to 22 N at radius of 8.75 cm. Determine for an engine speed of 1800 r.p.m.; the maximum and minimum values of the primary and secondary forces due to the inertia of reciprocating and rotating masses. (Utkal University)

(*Ans.* Max. F_p = 3178 N; min. F_p = 1805 N; max. F_s = 1000.6 N; min. F_s = 471 N)

16.19 The six cylinders of a single acting two stroke diesel engine are in line and are symmetrically spaced on either side of the central plane. The cranks are spaced at 60° intervals and the centre lines of cylinders 1 and 6 are 4.8 m apart. Cylinders 2 and 5 are 3 m apart and cylinders 3 and 4 are 1.2 m apart. The reciprocating mass per cylinder is 800 kg, the crank radius is 0.3 m, the speed is 180 r.p.m. and the connecting rod is 1.35 m long.

Show that primary and secondary forces are in balance for any order of firing in the cylinders and investigate the out-of-balance couple effects (primary and secondary) when the firing order is 1, 4, 5, 2, 3, 6, giving maximum values. (Univ. London)

(*Ans.* Primary couple = 25.4 kN·m; secondary couple = 127.5 kN·m)

Chapter 17

VIBRATION ANALYSIS

17.1 INTRODUCTION

The topic *vibration* deals with oscillatory behaviour of dynamic systems. Vibratory motion in machines and structures are of frequent concern in engineering practice. In fact all the bodies, possessing mass and elasticity, are capable of vibration. In studying mechanical vibrations, it is therefore desirable to modify the assumption of a rigid body followed in this book so far. Machine parts vibrate because they are elastic bodies.

Vibratory motion is usually objectionable as it produces unwanted noise, high stresses and wear. Frequently vibration leads to premature failure of one or more parts. In certain other cases however, vibratory motion may be desirable. In few other situations, its very presence reveals the inner working of complex machinery in operation which is of great importance in diagnostic maintenance.

When a body has mass, it can possess kinetic energy by virtue of its velocity. Similarly, when a body is elastic, it is capable of storing elastic strain energy which is comparable to the potential energy of a body. In a conservative system it is usually the continuous conversion of potential (elastic strain) energy into kinetic energy and vice versa that keeps the body vibrating without external excitation.

In a non-conservative system however, energy must be supplied from outside to ensure that the body keeps on vibrating.

17.2 DEFINITIONS

1. ***A Periodic Motion:*** A motion which repeats itself after a regular interval of time is called a *periodic motion*. A periodic motion may or may not be of harmonic nature. Displacement–time plot in Fig. 17.1 illustrates a periodic motion which is not of harmonic nature.

2. ***Time Period(τ):*** Time taken to complete one cycle is called *periodic time* or *time period*.

3. ***Frequency(f_n):*** This is defined as the number of cycles completed per unit time. Usually this is measured in terms of cycles per second (c.p.s.) which is also called Hertz (Hz).

850

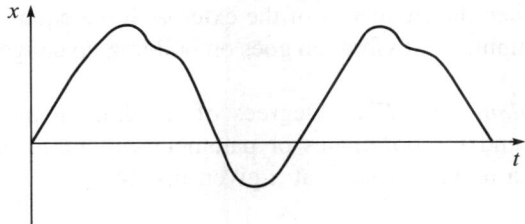

Fig. 17.1 A Plot of Periodic Motion.

4. **Simple Harmonic Motion (S.H.M.):** A 'to and fro' periodic motion of a particle whose acceleration is always directed towards the mean equilibrium position and is proportional to the displacement from the mean equilibrium position (abbreviated as m.e.p.), is called *simple harmonic motion*. Such a motion (displacement) is represented by equation (17.1)

$$x = X\sin\omega t \qquad\qquad (17.1)$$

This is represented by a sine curve as shown in Fig. 17.2. It follows that the displacement x is zero when $\omega t = 0$, π and 2π and has a maximum displacement from m.e.p. equal to X for $\omega t = \dfrac{\pi}{2}$ and $3\pi/2$.

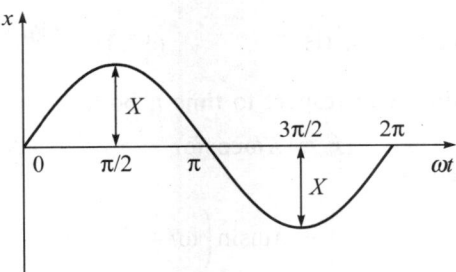

Fig. 17.2 A Simple Harmonic Motion Plot.

5. **Amplitude (X):** This is the maximum displacement of a vibrating body from the mean equilibrium position.

6. **Free Vibrations:** When external disturbance is required only to initiate vibratory motion but no external force acts thereafter and the system continues to vibrate because of its own elastic properties, the vibrations are called *free vibrations*.

7. **Natural Frequency:** Frequency at which a system executes free undamped vibrations is called its *natural frequency*. This is designated as 'ω_n' radian per second or 'f_n' cycles per second.

8. **Forced Vibrations:** The vibrations which the system executes under continuous external excitation are called *forced vibrations*. When the excitation force is periodic in nature, the system starts vibrating at the excitation frequency.

9. *Resonance:* When the frequency of the external force equals the natural frequency of the system, the amplitude of vibration goes on building up dangerously. This phenomenon is called *resonance*.

10. *Degree of Freedom (D.O.F.):* Degrees of freedom of a vibrating system implies number of independent coordinates or parameters required to specify completely the configuration of a moving system at a given instant.

17.3 SIMPLE HARMONIC MOTION AND ROTATING VECTORS

If a point P moves with uniform angular speed ω radian per second around the circumference of a circle of radius X, the foot of perpendicular M from P on a diameter AB has simple harmonic motion. Let x be the distance of point M from mean equilibrium position O along the diameter AB. As the point P rotates through 2π radian along the circle, the point M completes a cycle of oscillations about m.e.p. O (*see* Fig. 17.3). Hence, the time period of oscillations and displacement of M

$$\tau = (2\pi/\omega) \text{ s} \qquad (17.2)$$

and, $\qquad x = X \sin \omega t$

and therefore, frequency in cycles per second,

$$f = \frac{1}{\tau} = \omega/2\pi \text{ c.p.s. (Hz)} \qquad (17.3)$$

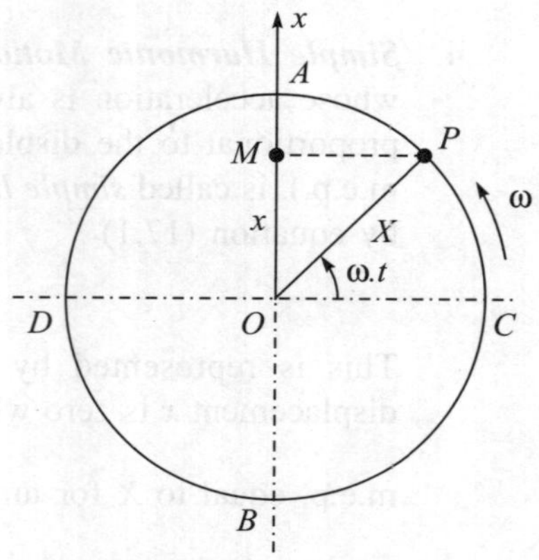

Fig. 17.3 Relation between S.H.M. and Rotating Vector.

For velocity differentiating with respect to time t, both the sides of equation (17.1),

$$\dot{x} = X\omega\cos\omega t \qquad (17.4)$$

which may also be expressed as,

$$\dot{x} = X\omega\sin\left(\omega t + \frac{\pi}{2}\right) \qquad (17.5)$$

Differentiating equation (17.4) further with respect to time for acceleration, we have

$$\ddot{x} = -X\omega^2\sin\omega t \qquad (17.6)$$

or $\qquad \ddot{x} = X\omega^2\sin(\omega t + \pi) \qquad (17.7)$

Equations (17.5) and (17.7) are respectively known as *velocity* and *acceleration equations*.

Important conclusions can be drawn by representing displacement, velocity and acceleration quantities by rotating vectors. Referring to Fig. 17.4, let $OP = X$ be the rotating displacement vector, which is inclined to OC at an angle ωt. Then component along x-axis, of the rotating vector represents displacement, given by $x = X\sin\omega t$. It follows from equation (17.5) that the amplitude $(x\omega)$ equals the length of rotating velocity vector OQ which is $\pi/2$ radian ahead of rotating displacement vector. This is shown in Fig. 17.4. Clearly, if $\omega > 1$, $X\omega$ is greater than X, on the other hand for $\omega < 1$, radius $X\omega < X$. In Fig. 17.4, ω is assumed to be greater than 1 which is generally true as N $\geq$ 10 r.p.m. Similarly, it follows from equation (17.7) that the acceleration vector $OR = X\omega^2$ is ahead of displacement vector by π radian. Also, for $\omega > 1$, the relative lengths of rotating vectors are,

$$X < X\omega < X\omega^2$$

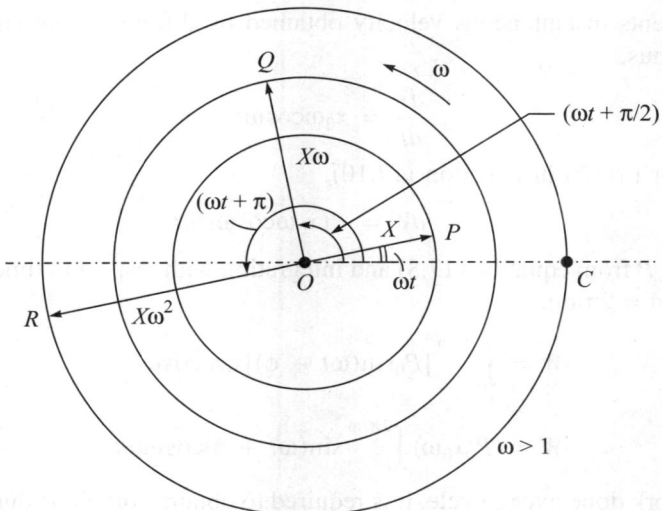

Fig. 17.4 Rotating vectors.

Thus, it may be concluded that in S.H.M.,

(a) The velocity vector leads displacement vector by $\pi/2$ radian, while acceleration vector leads displacement vector by π radian.

(b) The amplitude of velocity in S.H.M. is ω times that of displacement while amplitude of acceleration is ω^2 times that of displacement.

17.4 WORK DONE IN HARMONIC MOTION

A concept of importance in many applications is the *work done by a harmonic force in inducing a harmonic motion of the same frequency*. This is particularly useful in understanding mechanism of resonance phenomenon which is also a time-dependent phenomenon. It will be shown during discussions on forced vibrations that displacement partakes frequency of excitation but lags the forcing function by a phase angle ϕ.

Let the forcing function (i.e., excitation force) be given by,

$$P = P_0\sin(\omega t + \phi) \tag{17.8}$$

and the resulting displacement be given by

$$x = x_0\sin\omega t \tag{17.9}$$

The elemental work done by the force P, while acting over a small displacement dx, is given by

$$dW = Pdx$$

This can be rewritten as $\qquad dW = P(dx/dt)dt \tag{17.10}$

where dx/dt represents instantaneous velocity obtained by differentiating equation (17.9) with respect to time. Thus,

$$\frac{dx}{dt} = x_0\omega\cos\omega t \qquad (17.11)$$

Substituting for (dx/dt) in equation (17.10),

$$dW = P(x_0\omega\cos\omega t)dt$$

and substituting for P from equation (17.8) and integrating with respect to time over a cycle (note that the time period $= 2\pi/\omega$),

$$W = \int_0^{2\pi/\omega} [P_0\sin(\omega t + \phi)]x_0\omega\cos\omega t dt$$

or

$$W = (P_0 x_0\omega)\int_0^{2\pi/\omega} \sin(\omega t + \phi)\cos\omega t dt \qquad (17.12)$$

If instead of work done over a cycle, it is required to obtain work done during time $t_1 \le t \le t_2$, equation (17.12) modifies to,

$$W = (P_0 x_0\omega)\int_{t_1}^{t_2} \sin(\omega t + \phi)\cos\omega t dt \qquad (17.13)$$

In equations (17.12) and (17.13), the angular speed ω of rotation of the particle P has been assumed constant.

Simplifying the integrand in equation (17.14) using trigonometric identity,

$$\sin(\omega t + \phi) = \sin\omega t\cos\phi + \cos\omega t\sin\phi$$

we have,

$$W = (P_0 x_0\omega)\int_0^{2\pi/\omega} [(\sin\omega t \cos\omega t)\cos\phi + \cos^2\omega t \sin\phi]\ dt$$

Remembering that ϕ is independent of time,

$$W = (P_0 x_0\omega)\left[\frac{1}{2}\cos\phi\int_0^{2\pi/\omega}\sin 2\omega t\ dt + \frac{1}{2}\sin\phi\int_0^{2\pi/\omega}(\cos 2\omega t + 1)dt\right]$$

or

$$W = (P_0\omega)x_0\left[0 + \frac{\pi}{\omega}\sin\phi\right]$$

Therefore

$$W = (P_0 x_0\pi)\sin\phi \qquad (17.14)$$

It follows that when excitation force is either in phase (*i.e.*, $\phi = 0$) with displacement or when it is 180° out of phase with displacement, the work done by the harmonic force is zero. This condition exists in the case of undamped forced vibration when excitation frequency is either smaller than or greater than the resonating frequency. This is obvious because in the absence of damping (*i.e.*, loss of energy being absent), no additional work input is required to maintain the level of vibration amplitude. However, at resonance ($\omega \equiv \omega_n$) where $\phi = \pi/2$, maximum work input is required for building up amplitude of vibration continuously. This suggests that if the resonating frequency is crossed over rather quickly, no time is permitted for the build-up of the amplitude of vibration to a dangerous level. Note that equation (17.14) represents energy input per cycle of vibration.

17.5 ELEMENTS OF DISCRETE (LUMPED PARAMETER) VIBRATORY SYSTEM

Elements constituting discrete vibratory system are:

(i) *Mass (m):* Mass is assumed to be a rigid body and can gain or lose kinetic energy depending on whether the velocity of body increases or decreases.

(ii) *Spring:* The spring element is assumed to be massless but linearly elastic. A spring force F_s is assumed to be proportional to spring deflection. The elastic strain energy of deformation is stored in the spring and its role is similar to that of potential energy in the oscillations of pendulum. If x be the displacement between the two ends of spring, then

$$F_s = kx \tag{17.15}$$

where k is load/unit deflection and is called spring rate or stiffness in N/cm.

(iii) *Damper:* The damping element is supposed to have neither mass nor elasticity. Damping force exists because of viscous friction in a damper (*i.e.*, a dashpot) for which force of friction is assumed to be proportional to the relative velocity between the two ends of the dashpot. Thus,

$$F_d \propto \dot{x} \text{ or } F_d = c\dot{x} \tag{17.16}$$

where c = constant of proportionality called *coefficient of damping* whose unit is force per unit relative velocity.

The damper consists of a piston fitting loosely in a cylinder filled with oil or water so that the viscous fluid can flow around the piston inside the cylinder.

The work of friction in dashpot is converted into heat which is dissipated out to the atmosphere. Mere presence of a damping element renders a system non-conservative in nature. Many types of damping are encountered in practice but only the viscous damping ensures linear vibrations.

(iv) *The Excitation Force:* An excitation element is a source for supplying energy to a vibratory system. For this reason an excitation element is also non-conservative in nature. The magnitude of the excitation varies as a prescribed function of time. To simplify the analysis, excitation force is assumed to be either $F_0 \sin \omega t$ or $F_0 \cos \omega t$, where ω is called the *frequency of excitation*.

The mass, spring and damper are called *passive* (inactive) *elements* while the excitation element is called an *active element*.

In discrete parameters system, the parameters are assumed to be 'lumped together' and are symbolized by the corresponding elements. As against this, in continuous system approach these parameters are assumed to be distributed continuously along the member.

SINGLE DEGREE OF FREEDOM PROBLEMS

17.6 UNDAMPED FREE VIBRATIONS

17.6.1 Method Based on Newton's Second Law of Motion

This method is important as it is useful for conservative as well as non-conservative systems. The Newton's second law of motion may be stated as under.

"The rate of change of momentum is proportional to the impressed force and takes place in the direction in which the force acts". Mathematically, the law may be expressed as,

$$\left(\begin{array}{c} \text{Rate of change of} \\ \text{momentum in } x\text{-direction} \end{array} \right) = \sum (\text{Forces in } x\text{-direction})$$

or
$$\frac{d}{dt}(m\dot{x}) = \sum (F_x) \qquad (17.17)$$

Assuming the mass to be invariant with time,

$$m\ddot{x} = \sum (\text{Forces in } x\text{-direction}) \qquad (17.18)$$

According to the Newton's law, the direction of term $m\ddot{x}$ is commensurate with the direction of resultant force in the x-direction. Hence, the general approach is to assume tentatively a positive sign for $m\ddot{x}$ and then let the solution on right-hand side to decide actual sign for $m\ddot{x}$.

Undamped free vibration model consists of a spring and a mass (see Fig. 17.5) and represents the simplest oscillatory system. The system possess one degree of freedom if the mass is constrained to move only in the vertical direction as shown in the figure.

When disturbed in vertical direction, oscillations will take place at the natural frequency f_n, which will be shown to be a function of system parameter k and m. Thus f_n is a property of the system.

Newton's second law is the basis for examining the motion of the system in Fig. 17.5. Let line O–O represent unstretched position of spring before attaching mass at its lower end. Let line O'–O' represent the static equilibrium position achieved after attaching mass to the spring. The difference Δ in the two equilibrium positions is called static deflection and is due to the weight of the mass. Hence, for equilibrium of mass in static equilibrium position, free body diagram of mass at Fig. 17.5(b) gives,

$$k\Delta = mg \qquad (17.19)$$

Let us adopt the convention that the forces, displacement, velocity and acceleration acting downwards are positive. Measuring the displacement x from the static equilibrium position, the forces acting on mass in x-direction in Fig. 17.5(c) are $-k(\Delta + x)$ and mg. Hence applying Newton's second law of motion to the mass,

$$m\ddot{x} = \sum F_x \equiv -k(\Delta + x) + mg \qquad (17.20)$$

and since $k\Delta = mg$, cancelling the two terms we have

$$m\ddot{x} = -kx \qquad (17.21)$$

or
$$m\ddot{x} + kx = 0 \qquad (17.22)$$

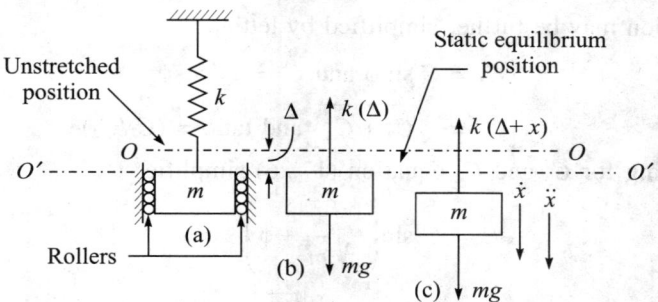

Fig. 17.5 Undamped free vibration model.

Choice of static equilibrium position as the reference for measuring displacement x therefore, eliminates the weight mg and spring force $k\Delta$ from the differential equation of motion.

Dividing equation (17.22) throughout by m, the equation reduces to

$$\ddot{x} + (k/m)x = 0 \qquad (17.23)$$

This is a homogeneous differential equation of second order with constant coefficients. Let us assume a solution,

$$x = Ce^{st} \qquad (17.24)$$

where C is an arbitrary constant

Substituting for x and $\ddot{x}$ in equation (17.23),

$$(s^2 + k/m)Ce^{st} = 0$$

As $C \neq 0$ (otherwise x is always zero), and $e^{st} \neq 0$ at all times, we have

$$s^2 + k/m = 0$$

The two roots of the above equation are,

$$s_1, s_2 = \pm j\sqrt{k/m}, \text{ where } j = \sqrt{-1}$$

Substituting the values of these two roots for s in equation (17.24), we have

$$x = C_1 e^{+j(\sqrt{k/m})t} + C_2 e^{-j(\sqrt{k/m})t}$$

Expanding, using Euler's equation, we have

$$x = C_1\left(\cos t\sqrt{\frac{k}{m}} + j\sin t\sqrt{\frac{k}{m}}\right) + C_2\left(\cos t\sqrt{\frac{k}{m}} - j\sin t\sqrt{\frac{k}{m}}\right)$$

or

$$x = (C_1 + C_2)\cos t\sqrt{\frac{k}{m}} + j(C_1 - C_2)\sin t\sqrt{\frac{k}{m}} \qquad (17.25)$$

Since the displacement is a real quantity, terms on the right-hand side of equation (17.25) must be real, which is possible only when C_1 and C_2 are complex conjugate. This requires that, if $C_1 = a + bj$ then $C_2 = a - bj$. Equation (17.25) may therefore be rewritten in terms of new arbitrary constants as,

$$x = C_3\cos t\sqrt{\frac{k}{m}} + C_4\sin t\sqrt{\frac{k}{m}} \qquad (17.26)$$

Above expression may be further simplified by letting,

$$C_3 = C'\sin\phi \text{ and } C_4 = C'\cos\phi$$

so that,
$$C' = \sqrt{C_3^2 + C_4^2} \text{ and } \tan\phi = (C_3/C_4)$$

Thus, substituting for C_3 and C_4 equation (17.26) simplifies to

$$x = C'\sin\left(t\sqrt{\frac{k}{m}} + \phi\right) \tag{17.27}$$

Equation (17.27) shows clearly that the term $\sqrt{k/m}$ represents some angular speed in radian per second. Since the solution in equation (17.27) represents a solution for free undamped model, the above term represents the natural frequency.

Thus, $\omega_n = \sqrt{k/m}$ in rad/s $\tag{17.28}$

Again, substituting for k from equation (17.19),

$$\omega_n = \sqrt{\frac{mg}{\Delta}\frac{1}{m}}$$

or
$$\omega_n = \sqrt{g/\Delta} \tag{17.29}$$

For free undamped model of a torsional system, replacing k by torsional spring stiffness k_t and m by mass moment of inertia I of the disc, the natural frequency for torsional vibration is given by –

$$\omega_n = \sqrt{k_t/I} \text{ rad/s} \tag{17.30}$$

k_t being the torsional spring stiffness of shaft $= T/\theta = \left(\dfrac{GJ}{l}\right)$ $\tag{17.31}$

and
J = polar moment of inertia of shaft
G = modulus of rigidity of shaft material
l = length of shaft

Using equation (17.3), the natural frequencies in equations (17.28), (17.29) and (17.30) respectively may be expressed in terms of cycles per second (Hz) as,

$$f_n = \frac{1}{2\pi}\sqrt{k/m} \text{ c.p.s. or Hz}$$

$$f_n = \frac{1}{2\pi}\sqrt{g/\Delta} \text{ Hz}$$

and
$$f_n = \frac{1}{2\pi}\sqrt{k_t/I} \text{ Hz} \tag{17.32}$$

Equations (17.27), (17.28) and (17.29) lead to the following important conclusions.

(1) Natural frequency ω_n is inherent in a system and is a function of system parameters k and m.

(2) Natural frequency ω_n of a system is a function of its static deflection Δ.

(3) The amplitude C' of resultant vibrations is independent of time *i.e.*, it does not decrease with time.

17.6.2 Energy Method

Differential equation of motion can also be established from energy considerations. The method, though apparently simple, is particularly suitable for a conservative system, *i.e.*, only for free undamped vibration case.

When a conservative system is set in motion, the mechanical energy associated with the system is partly kinetic and partly potential (*i.e.*, elastic strain energy). The kinetic energy is due to mass and velocity whereas the potential energy is due to spring stiffness and the relative movement between the two ends of the spring. By definition, for a conservative system,

$$T + U = \text{constant } (C) \tag{17.33}$$

where T = kinetic energy associated with the system
         U = elastic strain energy (P.E.) of the system

For a conservative system, the total energy given by equation (17.33) must not change with time. Thus,

$$\frac{d}{dt}(T + U) = 0 \tag{17.34}$$

This is the underlying principle of energy method. Choosing static equilibrium position $O'O'$ of Fig. 17.5 as the reference line, let $\dot{x}$ be the instantaneous velocity of the mass. Then assuming the spring to be massless, only the mass stores kinetic energy given by

$$T = \frac{1}{2}m(\dot{x})^2 \tag{17.35}$$

The potential energy of the system consists of two components: (a) loss/gain in P.E. of mass and (b) strain energy in the spring. To obtain expression of strain energy for a displacement x downwards of mass m, consider an elemental displacement ds at $x = s$.

Then it follows from Fig. 17.6 that spring force at displacement $x = s$,

$$F_s = k\,(s + \Delta)$$

where Δ = static deflection in spring due to weight of mass m. F_s being the average spring force, the elemental work done for elemental displacement ds is,

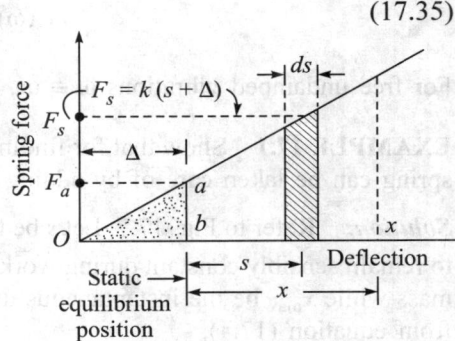

Fig. 17.6

$$dw = F_s ds = k\,(s + \Delta)ds$$

Therefore total W.D. in elongating spring by x from static equilibrium position b is

$$U = \int_0^x k(s + \Delta)ds - (\text{loss of P.E. of mass})$$

Therefore

$$U = \int_0^x (ks + mg)ds - (mg)x; \quad \text{as, } k\Delta = mg$$

Hence $\qquad U = \left(\dfrac{1}{2}kx^2 + mgx\right) - (mgx)$ or, $U = \dfrac{1}{2}\,kx^2$ $\qquad\qquad$ (17.36)

Hence, from equations (17.34), (17.35) and (17.36),

$$\frac{d}{dt}\left(\frac{1}{2}m\dot{x}^2 + \frac{1}{2}kx^2\right) = 0 \quad \text{or,} \quad (m\ddot{x} + kx)\dot{x} = 0$$

and as $\dot{x} \neq 0$ at all time, we have

$$m\ddot{x} + kx = 0 \qquad\qquad (17.37)$$

which is same as the differential equation of motion (17.22)

17.6.3 Rayleigh's Method

This is just a modified version of energy method. Note that in a conservative system, potential energy is minimum when kinetic energy is maximum and vice versa. Hence, equating maximum K.E. to maximum P.E.,

$$\frac{1}{2}m(\dot{x}_{max})^2 = \frac{1}{2}k\,(x_{max})^2 \qquad\qquad (17.38)$$

Again from equation (17.4),

$$(\dot{x})_{max} = (X\omega); \text{ when } \cos\omega t = 1$$

and from equation (17.1),

$$(x)_{max} = X; \text{ when } \sin\omega t = 1$$

Hence, from equation (17.38),

$$\frac{1}{2}m(X\omega)^2 = \frac{1}{2}k\,(X)^2 \text{ or } \omega^2 = k/m$$

For free undamped vibrations $\omega \equiv \omega_n$ and therefore, $\omega_n = \sqrt{k/m}$.

EXAMPLE 17.1 Show that for finding natural frequency of a spring mass system, mass of the spring can be taken care of by adding one-third of its mass to the main mass.

Solution: Refer to Fig. 17.7. Let s be the mass per unit axial length of spring which is assumed to remain sensibly constant during working. Also, let $\dot{x}$ be the maximum value of velocity of the mass while x_{max} be the instantaneous displacement of the mass. Then from equation (17.4),

$$\dot{x} = \omega_n(x_{max})$$

Again, with maximum displacement, maximum P.E.,

$$U_{max} = \frac{1}{2}k(x_{max})^2 \qquad\qquad \text{(a)}$$

The K.E. is contributed by mass of spring together with the mass m itself. Assuming a linear variation of velocity along the length of spring, the velocity at an element ds at a distance s from the spring support,

$$V_s = (s/l)\,\dot{x}$$

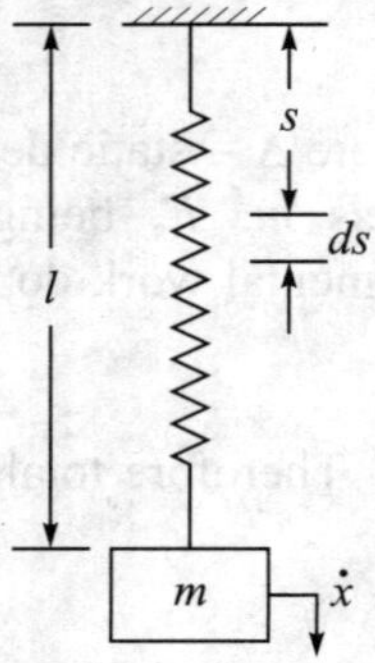

Fig. 17.7 Effect of mass of spring.

Hence maximum K.E.,

$$T_{max} = \frac{1}{2}m\,\dot{x}^2 + \int_0^l \frac{1}{2}(sds)(s\,\dot{x}/l)^2$$

or

$$T_{max} = \frac{1}{2}m\,\dot{x}^2 + \frac{1}{2}\left(\frac{s\dot{x}^2}{l^2}\right)\int_0^l s^2 ds$$

or

$$T_{max} = \frac{1}{2}m\,\dot{x}^2 + \frac{1}{2}\left(\frac{sl^3}{3l^2}\right)\dot{x}^2 = \frac{1}{2}\left(m+\frac{m_s}{3}\right)\dot{x}^2$$

Note that $sl = m_s$, the mass of spring.

Substituting $(x_{max})\omega_n$ for $\dot{x}$ and equating U_{max} with T_{max},

$$\frac{1}{2}(m+m_s/3)(x_{max}\omega_n)^2 = \frac{1}{2}k(x_{max})^2$$

or

$$\omega_n^2 = k/(m+m_s/3)$$

Therefore

$$\omega_n = \sqrt{k/(m+m_s/3)}. \quad \text{Hence shown.}$$

EXAMPLE 17.2   A shaft of 8 cm diameter and 80 cm length has one of its ends fixed and the other end carries a disc of weight 4800 N. The Young's modulus of elasticity for the material of the shaft is 196.2 kN/mm^2. Determine the frequency of longitudinal and transverse vibrations.

Solution: For the transverse arrangement like a cantilever beam, as in Fig. 17.8(a), the deflection at the free end

$$\delta = \frac{Wl^3}{3EI} = \frac{(4800)(0.8)^3}{3(1,96,200\times10^6)\times\dfrac{\pi}{64}(0.08)^4}$$

or

$$\delta = \frac{48\times(0.8)^3\times64}{3\times1962\times10^6\times\pi(0.08)^4}$$

$$= 2.0766\times10^{-3}\ \text{m} = 0.20766\ \text{cm}$$

Therefore natural frequency of transverse vibration,

$$\omega_n = \sqrt{g/\Delta} = \sqrt{\frac{981}{0.20766}} = 68.73\ \text{rad/s}$$

Therefore

$$f_n = \frac{1}{2\pi}\times\omega_n = \frac{68.73}{2\pi} = 10.94\ \text{c.p.s.} \qquad \textbf{Ans.}$$

For the longitudinal vibration, as in Fig. 17.8(b), longitudinal elongation,

$$\Delta = l \times \frac{stress}{E}$$

or

$$\Delta = 0.80 \times \frac{4800}{(\pi/4)\,(0.08)^2} \times \frac{1}{1,96,200 \times 10^6}$$

or

$$\Delta = 3.8937 \times 10^{-6} \text{ m} = 3.8937 \times 10^{-4} \text{ cm}$$

Therefore

$$f_n = \frac{1}{2\pi}\sqrt{\frac{981}{3.8937 \times 10^{-4}}} = 252.6 \text{ c.p.s.}$$ **Ans.**

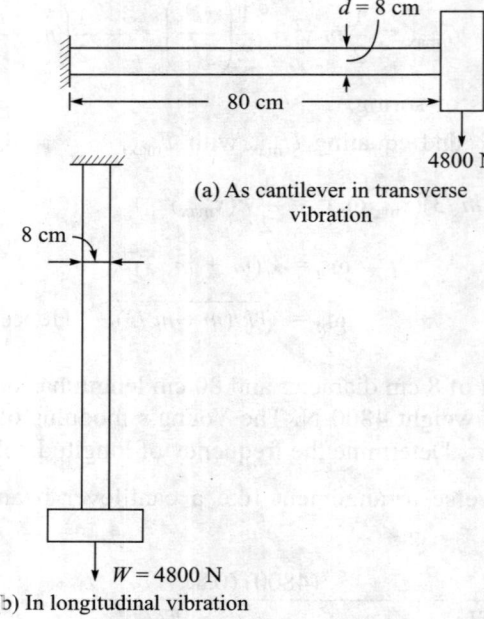

(a) As cantilever in transverse vibration

$W = 4800$ N

(b) In longitudinal vibration

Fig. 17.8

EXAMPLE 17.3 A flywheel having a mass of 35 kg was allowed to swing as pendulum about a knife-edge at the inner side of the rim, as shown in Fig. 17.9. If the measured time period of oscillation was 1.25 second, determine the moment of inertia of the flywheel about its geometric axis.

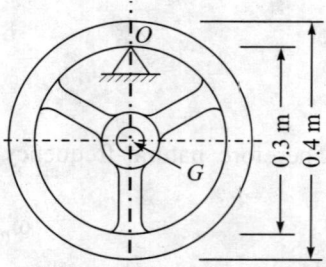

Fig. 17.9 Flywheel as pendulum.

Solution: Let I be the mass moment of inertia of the flywheel about the point of suspension O. Then, for a small angle of oscillation θ, treating the flywheel as a compound pendulum, the length of pendulum,

$$l = OG = \frac{0.3}{2} = 0.15 \text{ m}$$

and for an angle of oscillation θ from vertical line through the point of suspension, the restoring couple

$$C = W(OG)\sin\theta = (mg) \times 0.15 \sin\theta$$

and for small angles θ, $\sin\theta \approx \theta$,

Hence the differential equation of motion is

$$I\ddot{\theta} = -(mg) \times 0.15 \times \theta \quad \text{or,} \quad \ddot{\theta} + \frac{35 \times 9.81 \times 0.15}{I}\theta = 0$$

Therefore $\qquad f_n = \dfrac{\omega_n}{2\pi} = \dfrac{1}{2\pi}\sqrt{\dfrac{35 \times 9.81 \times 0.15}{I}}$. $\quad$ But $f_n = \dfrac{1}{\tau} = \dfrac{1}{1.25}$

Equating

$$\frac{1}{1.25} = \frac{1}{2\pi}\sqrt{\frac{35 \times 9.81 \times 0.15}{I}}$$

Therefore $\qquad I = \dfrac{(1.25)^2 \times 35 \times 9.81 \times 0.15}{4\pi^2} = 2.038 \text{ kg·m}^2$

Therefore, using parallel axis theorem, the mass moment of inertia about centroidal axis,

$$I_G = I - mr^2$$

or $\qquad I_G = 2.038 - (35)\left(\dfrac{0.3}{2}\right)^2 = 1.25 \text{ kg·m}^2$ $\qquad$ **Ans.**

EXAMPLE 17.4 Determine the natural frequency of the spring-mass pulley system shown in Fig. 17.10.

Solution: *Energy Method:*

$\qquad$ Total K.E. = K.E. of mass m + K.E. of pulley of mass M

Thus $\qquad$ K.E. = $\dfrac{1}{2}m\dot{x}^2 + \dfrac{1}{2}I\dot{\theta}^2$

where $\qquad \dot{x}$ = velocity of mass = $r\dot{\theta}$

Here $\dot{\theta}$ = angular velocity of pulley to produce above velocity of mass.

Therefore, $\qquad$ K.E. = $\dfrac{1}{2}m(r\dot{\theta})^2 + \dfrac{1}{2}I\dot{\theta}^2 = \dfrac{1}{2}(mr^2 + I)\dot{\theta}^2$

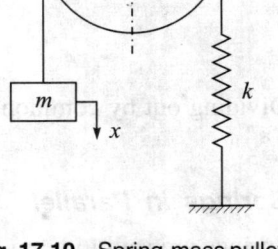

Fig. 17.10 Spring-mass pulley system.

and $\qquad$ P.E. = $\dfrac{1}{2}kx^2 = \dfrac{1}{2}k\,(r\theta)^2$

Hence from, $\qquad$ P.E. + K.E. = constant

$$\frac{1}{2}(mr^2 + I)\dot{\theta}^2 + \frac{1}{2}kr^2\theta^2 = C$$

Differentiating w.r. to time,

$$\frac{1}{2}(mr^2 + I)(2\dot{\theta})\ddot{\theta} + \frac{1}{2}(kr^2)(2\theta)\dot{\theta} = 0$$

or
$$\{(mr^2 + I)\ddot{\theta} + (kr^2)\theta\}\dot{\theta} = 0. \qquad \text{As } \dot{\theta} \neq 0 \text{ at all times,}$$
$$(mr^2 + I)\ddot{\theta} + (kr^2)\theta = 0$$

or
$$f_n = \frac{1}{2\pi}\sqrt{\frac{kr^2}{(mr^2 + I)}} \quad \text{Assuming } I = \frac{1}{2}Mr^2,$$

$$f_n = \frac{1}{2\pi}\sqrt{\frac{k}{(m + M/2)}} \quad \text{c.p.s.} \qquad\qquad \textbf{Ans.}$$

17.7 EQUIVALENT SPRINGS AND DASHPOTS

Spring mass models contain springs in various combinations. They can be placed in series or parallel to one another. It becomes necessary in all such cases to determine equivalent spring for the combination employed. Equivalent spring stiffness depends on whether the springs are arranged in series or in parallel. Similar remarks apply to combinations of dampers.

Springs in Series

When each of the two springs support the same load W and when total deflection is equal to the sum of individual deflections of the spring [see Fig. 17.11(a)], the springs are said to be in series. Thus for such a combination, as $\Delta = W/k$, writing total deflection Δ as

$$\Delta = \Delta_1 + \Delta_2$$

as
$$\frac{W}{k_{eq.}} = \frac{W}{k_1} + \frac{W}{k_2}$$

Dividing out by common load W, $\qquad \dfrac{1}{k_{eq.}} = \dfrac{1}{k_1} + \dfrac{1}{k_2}$ $\qquad\qquad$ (17.39)

Springs in Parallel

Springs are said to be in parallel when each of the two springs undergo same deflection Δ, and the combined load shared equals the external load [see Fig. 17.11(b)]. Thus,

$$W = W_1 + W_2 \qquad\qquad\qquad \text{(i)}$$

and
$$\Delta = \Delta_1 = \Delta_2 \qquad\qquad\qquad \text{(ii)}$$

Dividing equation (i) by (ii),

$$\frac{W}{\Delta} = \frac{W_1}{\Delta_1} + \frac{W_2}{\Delta_2}$$

or
$$k_{eq} = k_1 + k_2 \qquad\qquad\qquad\qquad (17.40)$$

Dampers in Series

Two dampers are said to be in series [*see* Fig. 17.11(c)] when each one transmits the same damping force *i.e.*, $F_d = F_{d1} = F_{d2}$, and also when total displacement of free end with respect to fixed end in elemental time Δt is,

$$\Delta x = \Delta x_1 + \Delta x_2$$

Then, for small interval of time, dividing out Δt and taking limits,

$$dx/dt = dx_1/dt + dx_2/dt$$

or

$$\dot{x} = \dot{x}_1 + \dot{x}_2$$

Dividing out each term of the above expression by a suitable term from amongst

$$F_d = F_{d1} = F_{d2}$$

we have

$$(\dot{x}/F_d) = (\dot{x}_1/F_{d1}) + (\dot{x}_2/F_{d2})$$

or

$$(1/c_{eq.}) = (1/c_1) + (1/c_2) \tag{17.41}$$

Dampers in Parallel

As the mass *m* moves [*see* Fig. 17.11(d)], each damper experiences same relative motion *i.e.*,

$$\dot{x} = \dot{x}_1 = \ddot{x}_2 \tag{i}$$

Further, total damping force equals the sum of damping forces provided by each dashpot. Thus,

$$F_d = F_{d1} + F_{d2}$$

Dividing by (i),

$$(F_d/\dot{x}) = (F_{d1}/\dot{x}_1) + (F_{d2}/\dot{x}_2)$$

or

$$c_{eq.} = c_1 + c_2 \tag{17.42}$$

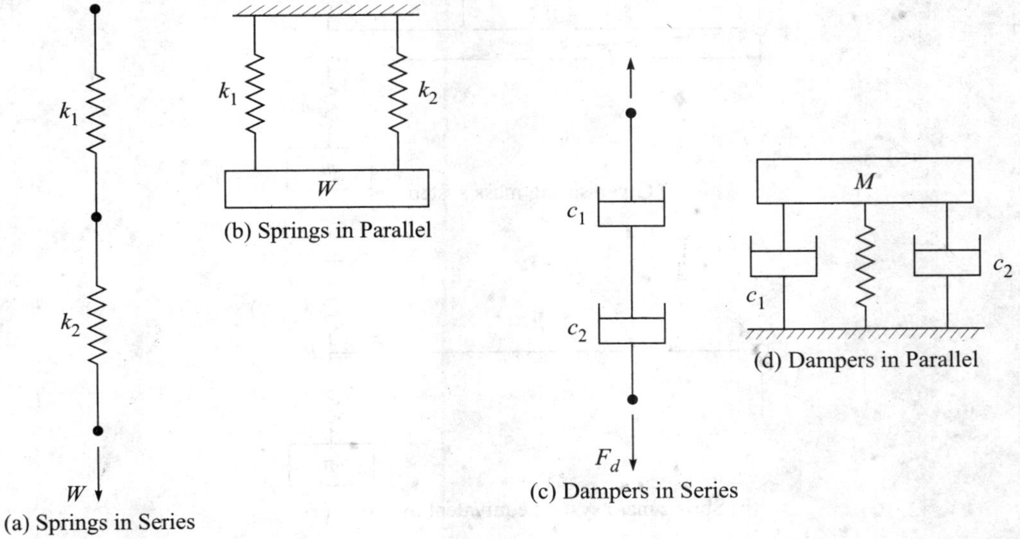

(a) Springs in Series

(b) Springs in Parallel

(c) Dampers in Series

(d) Dampers in Parallel

Fig. 17.11   Series and Parallel Arrangement.

EXAMPLE 17.5 For the system shown in Fig. 17.12, if $k_1 = 2400$ N/m; $k_2 = 1600$ N/m; $k_3 = 3600$ N/m and $k_4 = k_5 = 600$ N/m, find the mass m such that the natural frequency $f_n = 10$ Hz.

Solution: Springs k_1, k_2, k_3 are in series and have a single equivalent spring of stiffness k' given by

$$(1/k') = (1/2400) + (1/1600) + (1/3600)$$

or
$$k' = 757.89 \text{ N/m}$$

The springs of stiffness k_4 and k_5 are in parallel. Their equivalent stiffness k is

$$k = k_4 + k_5 = 600 + 600 = 1200 \text{ N/m}$$

The springs of stiffness k' and k are again in parallel and hence, a single resultant spring will have stiffness,

$$k_{eq.} = k + k' = 1200 + 757.89$$

i.e.,
$$k_{eq.} = 1957.89 \text{ N/m}$$

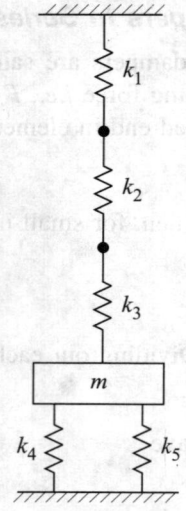

Fig. 17.12 A system with springs in series and parallel.

Therefore
$$\omega_n = \sqrt{k_{eq.}/m} \qquad \text{Hence, } m = (k_{eq.})/\omega_n^2$$

$$= (1957.89)/(2\pi \times f_n)^2$$
$$= (1957.89)/(2\pi \times 10)^2 = 0.496 \text{ kg.} \qquad \textbf{Ans.}$$

EXAMPLE 17.6 Determine the expression for natural frequency of vibration of the system shown in Fig. 17.13(a). Assume the bar AB to be rigid and weightless.

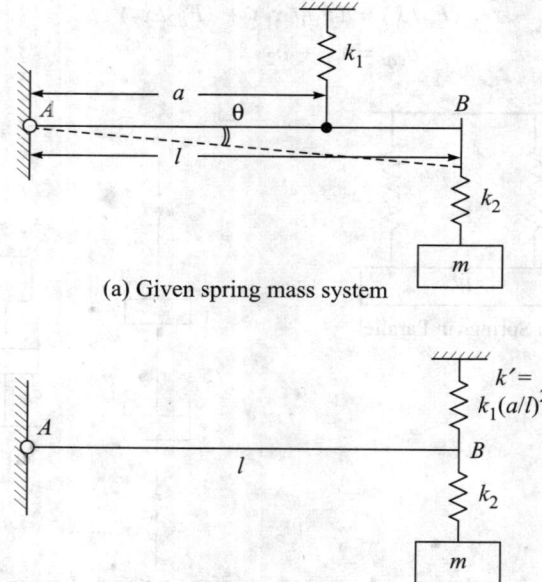

(a) Given spring mass system

(b) Spring mass system equivalent to that at (a)

Fig. 17.13

Solution: Let the bar AB rotate through a small angle θ in the plane of paper. Let the deflection at the spring k_1 in that case be $\Delta = a\theta$. The spring force resisting this deformation

$$F = k_1(a\theta)$$

This force produces a moment of $k_1(a^2\theta)$ about A. A spring k' at B which produces same moment at A as above, is given by

$$k'(l\theta)l = k_1(a\theta)a$$

or
$$k' = k_1(a/l)^2$$

This is shown in Fig. 17.13(b). The two springs of stiffness k' and k_2 are in series for sharing load mg. Hence,

$$\frac{1}{k_{eq.}} = \frac{1}{k_2} + \frac{1}{k'}$$

$$= \frac{1}{k_2} + \frac{(l/a)^2}{k_1} = \frac{k_1 + k_2(l/a)^2}{k_1 k_2}$$

Therefore
$$k_{eq.} = \frac{k_1 k_2}{k_1 + k_2(l/a)^2}$$

Therefore
$$\omega_n = \sqrt{k_{eq.}/m} = \sqrt{\frac{k_1 k_2}{m[k_1 + k_2(l/a)^2]}} \quad \text{rad/s} \qquad \textbf{Ans.}$$

17.8 EQUIVALENT LENGTH OF SHAFT

Sometimes shafts carrying rotors have different diameters at various sections. For the sake of simplicity in analysis it may be necessary to replace this shaft by a uniform shaft of known diameter. Equivalent length of such a shaft may then be evaluated. The two shafts must be torsionally equivalent.

Let the shaft shown in Fig. 17.14 have diameters d_1, d_2, d_3 and d_4 over shaft lengths l_1, l_2, l_3 and l_4 respectively. The torsional stiffness of the four segments of shaft are:

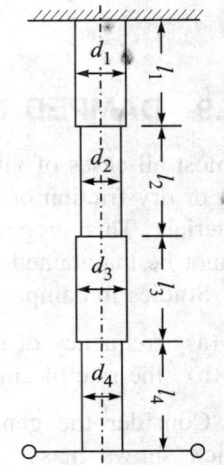

Fig. 17.14 A stepped shaft with rotor.

$$k_{t_1} = T_1/\theta_1 = GJ_1/l_1 = \frac{\pi}{32} \frac{G}{l_1} d_1^4$$

$$k_{t_2} = T/\theta_2 = GJ_2/l_2 = \frac{\pi}{32} \frac{G}{l_2} d_2^4$$

$$k_{t_3} = T/\theta_3 = GJ_3/l_3 = \frac{\pi}{32} \frac{G}{l_3} d_3^4$$

$$k_{t_4} = T/\theta_4 = GJ_4/l_4 = \frac{\pi}{32} \frac{G}{l_4} d_4^4$$

As each segment is subjected to same torque, all the corresponding springs are in series. Hence the equivalent torsional stiffness is,

$$\frac{1}{k_t} = \frac{1}{k_{t_1}} + \frac{1}{k_{t_2}} + \frac{1}{k_{t_3}} + \frac{1}{k_{t_4}} = \frac{32}{G\pi}\left[\frac{l_1}{d_1^4} + \frac{l_2}{d_2^4} + \frac{l_3}{d_3^4} + \frac{l_4}{d_4^4}\right] \qquad (17.43)$$

If the above shaft is to be replaced by a shaft of diameter d and length l of same material, then

$$k_t = \frac{\pi}{32}\frac{G}{l}d^4$$

Hence, from equation (17.43), equivalent length l is given by

$$\frac{32l}{\pi G d^4} = \frac{32}{G\pi}\left[\frac{l_1}{d_1^4} + \frac{l_2}{d_2^4} + \frac{l_3}{d_3^4} + \frac{l_4}{d_4^4}\right]$$

or $$(l/d^4) = (l_1/d_1^4 + l_2/d_2^4 + l_3/d_3^4 + l_4/d_4^4) \qquad (17.44)$$

or $$l = l_1(d/d_1)^4 + l_2(d/d_2)^4 + l_3(d/d_3)^4 + l_4(d/d_4)^4 \qquad (17.45)$$

EXAMPLE 17.7 A stepped shaft has three segments of diameters and lengths as indicated below:

$$d_1 = 50 \text{ mm}; \; l_1 = 0.4 \text{ m}; \; d_2 = 60 \text{ mm}; \; l_2 = 0.5 \text{ m}; \; d_3 = 90 \text{ mm and } l_3 = 0.6 \text{ m.}$$

Find equivalent length of shaft of uniform diameter $d = 50$ mm.

Solution: From equation 17.45,

$$l = 0.4\left(\frac{50}{50}\right)^4 + 0.5\left(\frac{50}{60}\right)^4 + 0.6\left(\frac{50}{90}\right)^4 = 0.698 \text{ m} \qquad \textbf{Ans.}$$

17.9 DAMPED FREE VIBRATIONS

Almost all cases of vibratory motion are 'non-conservative'. The energy is lost either through wet or dry friction or through air/fluid resistance or even due to mechanical hysterisis loss in materials. This suggests that in damped vibrations, steady (constant) amplitude of vibration cannot be maintained without continuous addition of energy to the system.

Studies in damped free vibrations aim at establishing,

(a) frequency of damped oscillations, and
(b) the rate of amplitude decay.

Consider the generalised model shown in Fig. 17.15 which shows mass, spring and damper elements. In the absence of excitation force, Newton's second law of motion leads to the following differential equation,

$$m\ddot{x} = -c\dot{x} - kx$$

or $$m\ddot{x} + c\dot{x} + kx = 0 \qquad (17.46)$$

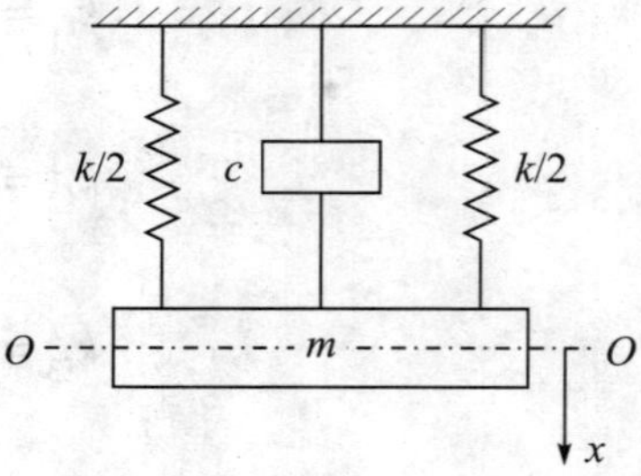

Fig. 17.15 Damped free vibrations.

This is again a homogeneous differential equation of second order with constant coefficients. Let the solution be,

$$x = De^{st} \qquad (17.47)$$

where D is an arbitrary constant.

Substituting for x, $\dot{x}$ and $\ddot{x}$ from equation (17.47) in equation (17.46), we have

$$(ms^2 + cs + k)De^{st} = 0$$

Also, as $D \neq 0$ and $e^{st} \neq 0$ at all times, the characteristic equation becomes,

$$ms^2 + cs + k = 0$$

Dividing by m, the equation becomes

$$s^2 + (c/m)s + (k/m) = 0$$

This is a quadratic in s whose roots are,

$$s_1, s_2 = -(c/2m) \pm \frac{1}{2}\sqrt{(c/m)^2 - 4(k/m)}$$

Substituting both the values of roots in equation (17.47), we have

$$x = e^{-(c/2m)t}\{D_1 e^{t\sqrt{(c/2m)^2 - (k/m)}} + D_2 e^{t\sqrt{(c/2m)^2 - k/m}}\} \qquad (17.48)$$

An examination of equation (17.48) indicates that the nature of solution depends very much on the quantities under radical sign. To understand the implications fully, we first introduce the term 'critical damping coefficient'. The critical damping coefficient c_c is defined as that value of damping coefficient c which reduces the quantity under radial sign to zero. Thus,

$$(c_c/2m)^2 - k/m = 0 \quad \text{or} \quad c_c^2 = 4m^2 (k/m)$$

or

$$c_c = 2\sqrt{km} \qquad (17.49)$$

or

$$c_c = 2\sqrt{\left(\frac{k}{m}\right)m^2} \quad \text{or, } c_c = 2m\omega_n \qquad (17.50)$$

Let us introduce at this stage one dimensionless ratio ζ (zeta), which is called *damping ratio* or *damping factor*. This is defined as the ratio of the given damping coefficient c to critical damping coefficient c_c. Thus,

$$\zeta = (c/c_c) \qquad (17.51)$$

Substituting for c_c,

$$\zeta = \frac{c}{2m\omega_n}$$

or

$$c = 2\zeta m\omega_n \qquad (17.52)$$

and

$$(c/2m) = (\zeta \times \omega_n). \quad \text{Also } k/m = \omega_n^2$$

Substituting in equation (17.48) for $(c/2m)$ and (k/m),

$$x = e^{-\zeta\omega_n t}\{D_1 e^{\sqrt{\zeta^2 - 1}\,\omega_n t} + D_2 e^{-\sqrt{\zeta^2 - 1}\,\omega_n t}\} \qquad (17.53)$$

It follows that the actual solution will depend on the value of damping ratio ζ.

Case 1: $\zeta > 1$ (Over-Damped Vibration) i.e., $c > c_c$

Rewriting equation (17.53) as

$$x = D_1 e^{[-\zeta + \sqrt{\zeta^2 - 1}]\omega_n t} + D_2 e^{[-\zeta - \sqrt{[\zeta^2 - 1]}]\omega_n t}$$

as, $\zeta > \sqrt{\zeta^2 - 1}$ the quantity in square bracket, namely, $[-\zeta \pm \sqrt{\zeta^2 - 1}]$ is negative and real in both the cases. This implies that the amplitude of vibration goes on decaying exponentially. Further as equation (17.53) cannot be expressed as a combination of sine and cosine terms, no oscillatory motion is possible. In other words the resulting motion is 'aperiodic'.

Case 2: $\zeta = 1$ (Critical Damping)

For $\zeta = 1$, the quantity under radical sign vanishes and both the roots are equal.

Thus
$$s_1 = s_2 = -\zeta\omega_n = -\omega_n$$

Solution in equation (17.53) then reduces to

$$x = (D_1 + D_2)e^{-\omega_n t}$$

or
$$x = Ce^{-\omega_n t}$$

This solution lacks in number of arbitrary constants to satisfy the two initial conditions $x(0)$ and $\dot{x}(0)$. The solution for the case of equal roots is therefore written as

$$x = (C_1 + C_2 t)e^{-\omega_n t} \tag{17.54}$$

The two arbitrary constants C_1 and C_2 may be evaluated using initial conditions,

$$x = x_0 \text{ at } t = 0$$

and
$$\dot{x} = 0 \text{ at } t = 0$$

Thus, applying the first initial condition to equation (17.54)

$$C_1 = x_0$$

and applying the second condition,

$$\dot{x} = C_2 e^{-\omega_n t} + (C_1 + C_2 t)(-\omega_n)e^{-\omega_n t}$$

Thus with $\dot{x} = 0$ at $t = 0$,

$$C_2 - \omega_n C_1 = 0; \quad \text{or} \quad C_2 = C_1 \omega_n = (x_0 \omega_n)$$

Hence, the solution is

$$x = x_0(1 + \omega_n t)e^{-\omega_n t} \tag{17.55}$$

It can be shown that as t tends to infinity, both the components in equation (17.55) tend to zero. This is again an 'aperiodic' motion and the system comes to rest at a rate faster than that in the case of over-damped vibration as shown in Fig. 17.16(a).

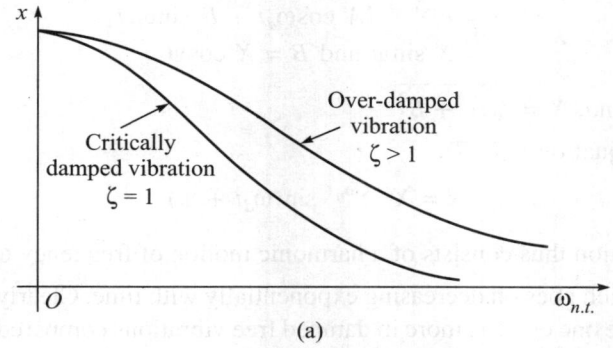

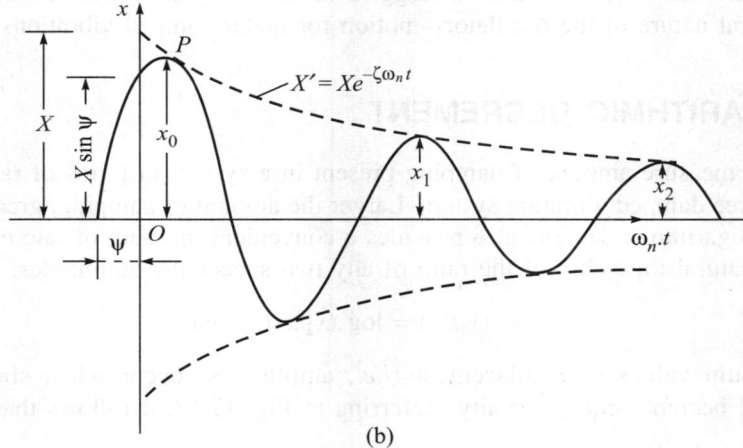

Fig. 17.16 Rate of decay for $\zeta > 1$, $\delta = 1$ and $\zeta < 1$.

Case 3 $\zeta < 1$ *(Under-Damped Vibrations)*

When $\zeta < 1$, the two roots of the characteristic equation are,

$$s_1, s_2 = \{-\zeta \pm j\sqrt{1-\zeta^2}\}\omega_n$$

Thus, the general solution is $x = e^{-\zeta\omega_n t}\{D_1 e^{j\omega_d t} + D_2 e^{-j\omega_d t}\}$ (17.56)
where ω_d = frequency of damped vibration, given by

$$\omega_d = (\sqrt{1-\zeta^2})\,\omega_n$$

Expanding equation (17.56) using Euler's formula,

$$x = e^{-\zeta\omega_n t}\{D_1 (\cos\omega_d t + j\sin\omega_d t) + D_2 (\cos\omega_d t - j\sin\omega_d t)\}$$

or $\qquad x = e^{-\zeta\omega_n t}\{(D_1 + D_2)\cos\omega_d t + j(D_1 - D_2)\sin\omega_d t\}$

For real displacements, as argued in Section (17.6 a), let

$$A \equiv (D_1 + D_2) \quad \text{and} \quad B = j(D_1 - D_2)$$

Then
$$x = e^{-\zeta\omega_n t}\{A\cos\omega_d t + B\sin\omega_d t\} \tag{17.57}$$

Letting
$$A = X\sin\psi \quad \text{and} \quad B = X\cos\psi$$

so that $\tan\psi = A/B$ and, $X = \sqrt{A^2 + B^2}$

We have from equation (17.57),

$$x = Xe^{-\zeta\omega_n t}\sin(\omega_d t + \psi) \tag{17.58}$$

The resulting motion thus consists of a harmonic motion of frequency $\omega_d = \omega_n\sqrt{1-\zeta^2}$, and amplitude $Xe^{-\zeta\omega_n t}$ which goes on decreasing exponentially with time. Clearly $\omega_d < \omega_n$ and as such time taken to complete one cycle is more in damped free vibrations compared to that in undamped free vibration. In other words, motion is sluggish in damped free vibrations. Figure 17.16(b) shows the general nature of the oscillatory motion for under-damped vibrations.

17.10 LOGARITHMIC DECREMENT

It is possible to measure amount of damping present in a system in terms of rate of decay of amplitude in a free damped vibration system. Larger the amount of damping, greater will be the rate of decay. Logarithmic decrement δ provides a convenient measure of rate of decay and is defined as the natural logarithm of the ratio of any two successive amplitudes.

Thus
$$\delta = \log(x_1/x_2) = \log(x_2/x_3) \dots \text{etc.}$$

The maximum values of displacement (i.e., amplitudes) occur when $\sin(\omega_d t + \psi)$ in equation (17.58) becomes equal to unity. Referring to Fig. 17.17, it follows that,

$$\delta = \log(x_0/x_1) \tag{17.59}$$

But
$$x_0 = Xe^{-\zeta\omega_n t_0} \quad \text{and} \quad x_1 = Xe^{-\zeta\omega_n t_1}$$

where $t_1 - t_o = $ time period τ_d of damped vibrations. Thus

$$\tau_d = (2\pi/\omega_d) = \frac{2\pi}{\omega_n\sqrt{1-\zeta^2}} \tag{17.60}$$

Substituting for x_0 and x_1 in equation (17.59),

$$\delta = \log\left(\frac{Xe^{-\zeta\omega_n t_0}}{Xe^{-\zeta\omega_n t_1}}\right) = \log[e^{-\zeta\omega_n(t_o - t_1)}]$$

or,
$$\delta = \log[e^{+\zeta\omega_n\cdot\tau_d}] = \text{the logarithmic decrement}$$

or
$$\delta = (\zeta\omega_n)\tau_d$$

Substituting for τ_d from equation (17.60),

$$\delta = \frac{2\pi\zeta}{\sqrt{1-\zeta^2}} \tag{17.61}$$

which is an exact expression for δ.

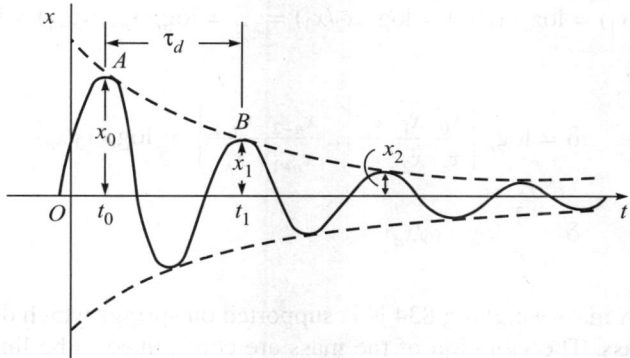

Fig. 17.17 Logarithmic decay of amplitude.

When ζ is small, $\sqrt{1-\zeta^2} \approx 1$

and the approximate expression for δ becomes,

$$\delta \approx (2\pi\zeta) \qquad (17.62)$$

Figure 17.18 shows variations between exact and approximate values of logarithmic decrement δ as a function of damping ratio ζ.

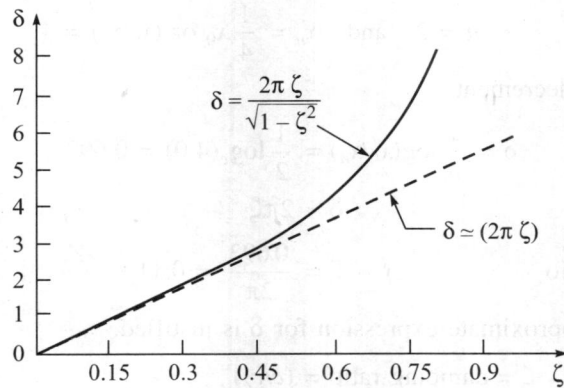

Fig. 17.18 Exact and approximate value of δ.

From the point of experimental investigations, a more convenient expression for logarithmic decrement is,

$$\delta = \frac{1}{n}\log_e(x_0/x_n) \qquad (17.63)$$

where x_0 = amplitude in any reference cycle

x_n = amplitude of vibration after completing n cycles of oscillations

The proof is as under.

It follows from the definition of logarithmic decrement that,

$$\delta = \log_e(x_0/x_1) = \log_e(x_1/x_2) = \log_e(x_2/x_3) = \ldots = \log_e(x_{n-2}/x_{n-1}) = \log(x_{n-1}/x_n)$$

Adding together,

$$n\delta = \log_e\left\{\frac{x_0}{x_1}\frac{x_1}{x_2}\frac{x_2}{x_3}\ldots\frac{x_{n-2}}{x_{n-1}}\frac{x_{n-1}}{x_n}\right\} = \log_e(x_0/x_n)$$

or

$$\delta = \frac{1}{n}\log_e(x_0/x_n)$$

EXAMPLE 17.8 A mass weighing 834 N is supported on springs which deflects 1.8 cm under the weight of the mass. The vibration of the mass are constrained to be linear and vertical and are damped by a dashpot which reduces the amplitude to one-quarter of its initial value in two complete oscillations, find

(i) magnitude of the damping force at unit speed, (ii) periodic time of damped vibration.

(AMIE, Winter 1984)

Solution: Static deflection 1.8 cm under 834 N load, then the natural frequency

$$\omega_n = \sqrt{\frac{g}{\Delta}} = \sqrt{\frac{981}{1.8}} = 23.15 \text{ rad/s}$$

Amplitude of vibration reduces to one-quarter after two cycles. Thus,

$$n = 2 \quad \text{and} \quad x_n = \frac{1}{4}x_0 \text{ or } (x_0/x_n) = 4$$

Therefore logarithmic decrement

$$\delta = \frac{1}{n}\log(x_0/x_n) = \frac{1}{2}\log_e(4.0) = 0.693$$

Again

$$\delta \approx 2\pi\zeta$$

Therefore, damping ratio

$$\zeta = \frac{0.693}{2\pi} = 0.11$$

Since ζ is small, the approximate expression for δ is justified.

But

$$\zeta = \text{damping ratio} = (c/c_c)$$

and

$$c_c = 2m\omega_n$$

or

$$c_c = 2(834/9.81) \times 23.35 = 3970 \text{ N·s/cm}$$

Therefore

$$c = \zeta \times c_c = 0.11 \times 3970 = 436.7 \text{ N·s/cm}$$

By definition coefficient damping c represents force per unit speed. Hence, damping force at unit speed = 436.7 N   **Ans.**

Again, frequency of damped vibration,

$$\omega_d = \omega_n\sqrt{1-\zeta^2} = 23.35\sqrt{1-0.11^2} = 23.208 \text{ rad/s}$$

Hence, time period of damped vibration

$$\tau_d = 2\pi/\omega_d = \frac{2\pi}{23.208} = 0.27 \text{ s} \qquad \textbf{Ans.}$$

EXAMPLE 17.9   The disc of a torsional pendulum has a moment of inertia of 0.068 kg·m² and is immersed in a viscous fluid. The brass shaft ($G = 40$ GN/m²) attached to it is of 10 mm diameter and 380 mm length. When the pendulum is vibrating the amplitudes on the same side of the rest position for successive cycles are 5°, 3° and 1.8°.

Determine (i) the logarithmic decrement, (ii) the damping torque at unit velocity, (iii) the periodic time of the vibration.

What would be the frequency of vibrations if the disc were removed from the viscous fluid?

(Univ. London)

Solution: Ratio of successive amplitudes $= \dfrac{5}{3} = \dfrac{3}{1.8}$

Therefore, logarithmic decrement $\delta = \log_e(5/3) = 0.511$

But $\qquad\qquad\qquad\qquad\qquad \delta = 2\pi\zeta$

Therefore $\qquad\qquad\qquad \zeta = \dfrac{0.511}{2\pi} = 0.0813$

A very small value of ζ justifies the use of approximate expression for ζ. The torsional stiffness of brass shaft,

$$k_t = T/\theta = GJ/l$$

or $\qquad\qquad k_t = (40 \times 10^9)\dfrac{\pi}{32}(0.01)^4/(0.38) = 103.34 \text{ N·m/rad}$

Thus, $\qquad\qquad \omega_n = \sqrt{\dfrac{k_t}{I}} = \sqrt{\dfrac{103.34}{0.068}} = 38.984 \text{ rad/s}$

Therefore $\qquad \omega_d = \omega_n\sqrt{1-\zeta^2} = 38.94\sqrt{1-0.0813^2} = 38.81$

Time period, $\qquad \tau_d = 2\pi/\omega_d = 0.1619 \text{ s}$ $\qquad\qquad\qquad$ **Ans.**

The damping torque at unit velocity $= c_t$

or $\qquad\qquad c_t = \zeta \times c_c = 0.0813 \times 2I \times \omega_n$

or $\qquad\qquad c_t = 0.0813 \times 2 \times 0.068 \times 38.984 = 0.431 \text{ N·m·s/rad}$ $\qquad$ **Ans.**

When the disc is removed from the viscous fluid, the disc will vibrate at natural frequency which was computed above as $\omega_n = 38.984$

Therefore $\qquad\qquad\qquad f_n = \dfrac{\omega_n}{2\pi} = \dfrac{38.984}{2\pi} = 6.204 \text{ Hz}$ $\qquad\qquad\qquad$ **Ans.**

EXAMPLE 17.10 A gun barrel of mass 545 kg has a recoil spring of stiffness 2,97,000 N/m. If the barrel recoils 1.2 m on firing, determine,

(a) The initial recoil velocity of the barrel

(b) The critical damping coefficient of a dashpot which is engaged at the end of the recoil stroke

(c) Time required for the barrel to return to a position 5 cm from its initial position.

Solution: (a) Let V cm/s be the initial recoil velocity. Then, equating K.E. of barrel to the elastic strain energy of the spring,

$$\frac{1}{2}(545)V^2 = \frac{1}{2}(2,97,000)(1.2)^2$$

Therefore $\qquad\qquad\qquad\qquad V = 28.013$ m/s $\qquad\qquad\qquad\qquad$ **Ans.**

(b) At the end of recoil stroke, dashpot is engaged for which,

$$c_c = 2\sqrt{km} = 2\sqrt{(2,97,000)(545)} = 25445.235 \text{ N·s/m}, \qquad\qquad \textbf{Ans.}$$

(c) For critical damping,

$$x = (C_1 + C_2 t)\, e^{-\omega_n t}$$

where $\qquad\qquad\qquad\qquad \omega_n = \sqrt{\frac{k}{m}} = \sqrt{\frac{2,97,000}{545}} = 23.344$ rad/s

The initial conditions for evaluating C_1 and C_2 are,

at $\qquad\qquad\qquad\qquad\qquad\qquad\qquad t = 0,\ x = 120$ cm

and at $\qquad\qquad\qquad\qquad\qquad\qquad t = 0,\ \dot{x} = 0$ cm/s

Thus with condition, $\qquad\qquad\qquad\qquad t = 0,\ x = 120$ cm,

$$120 = C_1$$

and $\qquad\qquad\qquad \dot{x} = C_2 e^{-\omega_n t} - (C_1 + C_2 t)\,\omega_n e^{-\omega_n t}$

Applying condition $t = 0$ at $\dot{x} = 0$

$$0 = C_2 - C_1 \omega_n$$

or $\qquad\qquad\qquad\qquad\qquad\qquad C_2 = C_1 \omega_n$

or $\qquad\qquad\qquad\qquad\qquad\qquad C_2 = 120 \times \omega_n = 120 \times 23.344$

Therefore $\qquad\qquad\qquad\qquad\qquad x = 120(1 + 23.344t)\, e^{-23.344t}$

By trial and error approach at $\qquad t = 0.1$ s; $x = 38.759$ cm

At $\qquad\qquad\qquad\qquad\qquad\qquad t = 0.2$ s; $x = 6.383$ cm

and, at $\qquad\qquad\qquad\qquad\qquad t = 0.21$ s; $x = 5.2624$ cm

at $\qquad\qquad\qquad\qquad\qquad\qquad t = 0.215$ s; $x = 4.775$ cm

At $\qquad\qquad\qquad\qquad\qquad\qquad t = 0.211$ s; $x = 5.16$ cm

also at, $\qquad\qquad\qquad\qquad\qquad t = 0.213$ s; $x = 4.9646$

or at $\qquad\qquad\qquad\qquad\qquad\qquad t = 0.2125$ s; $x = 5.013$ cm

which is sufficiently close to $x = 5$ cm.

Hence, time reqd. $t = 0.2125$ s $\qquad\qquad\qquad\qquad\qquad\qquad\qquad\qquad$ **Ans.**

17.11 FORCED VIBRATIONS WITH HARMONIC EXCITATION

Consider a spring mass system with viscous damping and subjected to a harmonic excitation $F_0\sin\omega_t$, in which F_0 is constant as shown in Fig. 17.19. Consider the mass to be displaced by x downwards with respect to static equilibrium position. Selection of static equilibrium position as the reference line, eliminates the need to consider the weight of mass (which is nullified by force due to spring deflection) in the free body diagram.

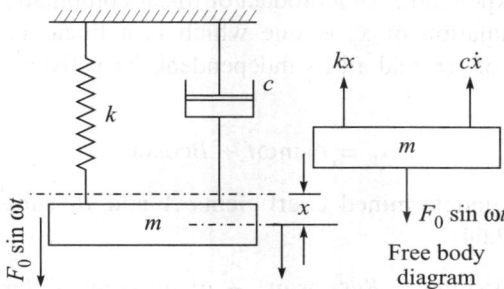

Fig. 17.19 Forced damped system.

Let us adopt the convention that forces, displacement $x(t)$, velocity $\dot{x}$ and acceleration $\ddot{x}$ are positive when measured downwards from static equilibrium position. As the spring opposes deformation and dashpot opposes relative velocity between its two ends, the spring force and force due to damping are shown acting upwards and are therefore negative. The excitation force $F_0\sin\omega t$, which causes the mass to move down, is positive. Hence, from Newton's second law of motion,

$$\left.\begin{array}{c}\text{Rate of change} \\ \text{of momentum}\end{array}\right\} = -F_s - F_d + F_0\sin\omega t$$

or
$$m\ddot{x} = -kx - c\dot{x} + F_0\sin\omega t$$

or
$$m\ddot{x} + c\dot{x} + kx = F_0\sin\omega t \tag{17.64}$$

Equation (17.64) represents a non-homogeneous, second-order linear differential equation of motion. The complete solution consists of two parts:

(i) A general solution called *complimentary function* which represents solution of corresponding homogeneous form of the above differential equation, obtained by making right hand side of equation (17.64) equal to zero. As shown in equation (17.58), the corresponding solution is,

$$x_c = Xe^{-\zeta\omega_n t}\sin(\omega_d t + \psi) \tag{17.65}$$

where
$$\omega_d = \omega_n\sqrt{1 - \zeta^2}$$

Due to exponential decay rate in equation (17.65), the motion described by complimentary function x_c die out eventually. This part of solution is therefore called *Transient Motion (solution)*.

(ii) The particular integral (solution) x_p of equation (17.64) which is also called *steady-state solution*, for reasons to be explained later.

Particular Solution x_p

A method of great importance in solving non-homogeneous, second-order differential equation motion is the method of 'undetermined coefficients'. The method is quite simple but applies only to differential equations with constant coefficients when the excitation function $f(t)$ is a sine, cosine, a polynomial, an exponential, or a product or linear combinations of these functions. The trial function for the evaluation of x_p is one which is a linear combination with constant undetermined coefficients of $f(t)$ and all its independent derivatives. Thus let us assume a trial function as,

$$x_p = A\sin\omega t + B\cos\omega t \qquad (17.66)$$

In order to evaluate undetermined coefficients A and B, substitute for $x \equiv x_p$ and its derivatives in equation (17.64),

$$m\{-A\omega^2\sin\omega t - B\omega^2\cos\omega t\} + c\{A\omega\cos\omega t - B\omega\sin\omega t\}$$
$$+ k\{A\sin\omega t + B\cos\omega t\} = F_0\sin\omega t$$

which simplifies to,

$$(kA - cB\omega - mA\omega^2)\sin\omega t + (kB + cA\omega - mB\omega^2)\cos\omega t = F_0\sin\omega t \qquad (17.67)$$

Equating coefficients of $\sin\omega t$ and $\cos\omega t$ terms separately, on the two sides of equal to sign in equation (17.67),

$$(k - m\omega^2)A - (c\omega)B = F_0$$

and
$$(c\omega)A + (k - m\omega^2)B = 0 \qquad (17.68)$$

Equation (17.68) represent homogeneous, linear, algebraic equations in A and B. Solving for A and B, using Cramer's rules,

$$A = \dfrac{\begin{vmatrix} F_0 & -(c\omega) \\ 0 & (k - m\omega^2) \end{vmatrix}}{\begin{vmatrix} (k - m\omega^2) & (-c\omega) \\ (c\omega) & (k - m\omega^2) \end{vmatrix}} = \dfrac{F_0(k - m\omega^2)}{(k - m\omega)^2 + (c\omega)^2}$$

and
$$B = \dfrac{\begin{vmatrix} (k - m\omega^2) & F_0 \\ (c\omega) & 0 \end{vmatrix}}{\begin{vmatrix} (k - m\omega^2) & (-c\omega) \\ (c\omega) & (k - m\omega^2) \end{vmatrix}} = \dfrac{-F_0(c\omega)}{(k - m\omega^2)^2 + (c\omega)^2}$$

Substituting for A and B from above in equation (17.66) and rearranging,

$$x_p = \dfrac{F_0}{(k - m\omega^2)^2 + (c\omega)^2}\{(k - m\omega^2)\sin\omega t - (c\omega)\cos\omega t\} \qquad (17.69)$$

Defining $\qquad \cos\psi = \dfrac{(k - m\omega^2)}{\sqrt{(k - m\omega^2)^2 + (c\omega)^2}}$ and $\sin\psi = \dfrac{(c\omega)}{\sqrt{(k - m\omega^2)^2 + (c\omega)^2}}$

so that $\qquad \tan\psi = c\omega/(k - m\omega^2)$ $\hfill$ (17.70)

and substituting for $\sin\psi$ and $\cos\psi$, equation (17.69) simplifies to,

$$x_p = \frac{F_0}{\sqrt{(k - m\omega^2)^2 + (c\omega)^2}} \sin(\omega t - \psi) \qquad (17.71)$$

or $\qquad x_p = X\sin(\omega t - \psi)$ $\hfill$ (17.72)

where $\qquad X = \dfrac{F_0}{\sqrt{(k - m\omega^2)^2 + (c\omega)^2}}$ $\hfill$ (17.73)

The amplitude X of the particular integral x_p does not depend on time. In other words, the amplitude of vibration represented by x_p does not change with time and this part of solution is therefore called *steady-state solution*. The body vibrates at the frequency of excitation ω and the displacement x lags behind the excitation force by ψ, the phase angle.

The amplitude of steady-state response and the phase angle ψ will now be expressed in non-dimensional form. Dividing numerator and denominator of equation (17.70) by k,

$$\tan\psi = \frac{(c/k)\omega}{1 - (m/k)\omega^2}$$

and remembering that $c/k = (2\zeta/\omega_n)$ and $m/k = 1/\omega_n^2$

$$\tan\psi = \frac{2\zeta(\omega/\omega_n)}{1 - (\omega/\omega_n)^2}$$

Defining (ω/ω_n) as the frequency ratio r, therefore,

$$\tan\psi = \frac{2\zeta r}{(1 - r^2)} \qquad (17.74)$$

Dividing numerator and denominator of equation (17.73) by k,

$$X = \frac{F_0/k}{\sqrt{\left(1 - \dfrac{m}{k}\omega^2\right)^2 + \left(\dfrac{c}{k}\omega\right)^2}}$$

or $\qquad X = \dfrac{(F_0/k)}{\sqrt{(1 - r^2)^2 + (2\zeta r)^2}}$

Further, defining $(F_0/k) = X_0$, the static deflection under the load F_0, we have

$$X = \frac{X_0}{\sqrt{(1 - r^2)^2 + (2\zeta r)^2}} \qquad (17.75)$$

or $\qquad (X/X_0) = \dfrac{1}{\sqrt{(1-r^2)^2 + (2\zeta r)^2}} = K \qquad\qquad$ (17.76)

The ratio $(X/X_0) = K$ is called *magnification factor* or *dynamic magnifier* and is a dimensionless number. It is defined as the ratio of the steady-state amplitude to the static deflection.

Dividing by r, the numerator and denominator on the right hand side of equation (17.74),

$$\tan\psi = \dfrac{2\zeta}{(1/r) - r} \qquad\qquad (17.77)$$

Conclusions:

(1) Figure 17.20(a) shows a graph plotted between magnification factor K and the frequency ratio r for different values of damping ratios ζ. Figure 17.20(b) shows a graph plotted between phase angle ψ and the frequency ratio r.

It is seen that K can be considerably greater or less than unity. For $\zeta = 0$ at $r = 1$, magnification factor (and, therefore X), theoretically shoots up to infinity. Thus at resonating condition (*i.e.*, at $r = 1.0$) the magnification factor K is limited only by damping ratio ζ.

(2) At frequency ratio $r \geq \sqrt{2}$, the magnification factor K is always less than 1, irrespective of the amount of damping present.

(3) It follows from Fig. 17.20(b) that for $\zeta = 0.0$, the phase angle is either zero or 180°. For $\zeta = 0.0$ and $r \leq 1$, the phase angle is zero and for $r > 1$ the phase angle is 180°.

(4) At $r = 1$, the phase angle $\psi = 90°$, for all values of damping ratios.

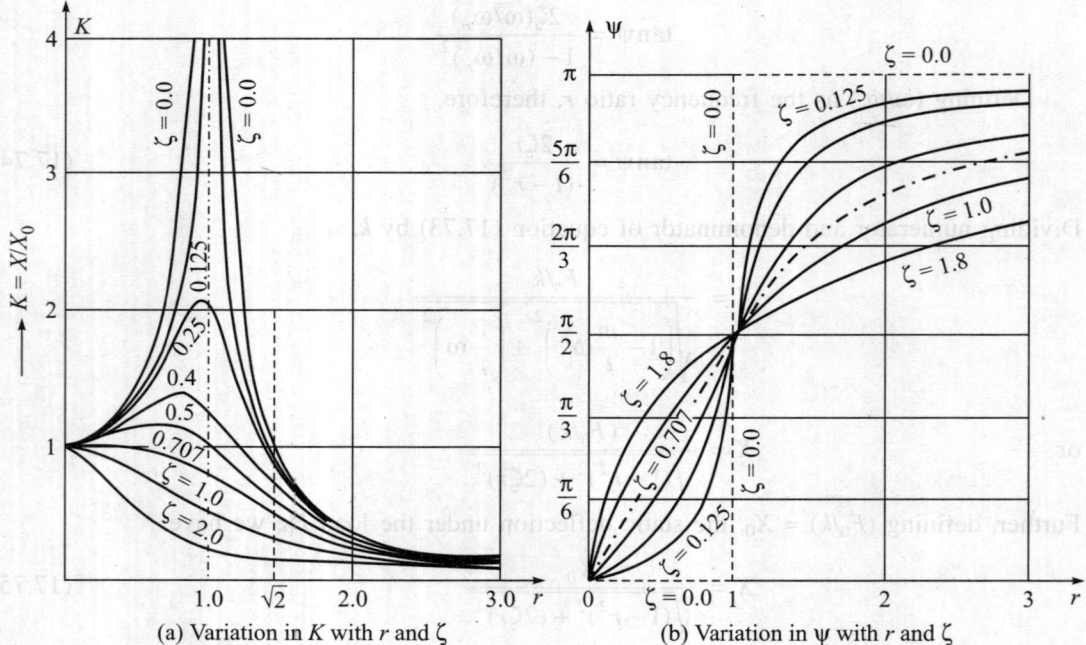

(a) Variation in K with r and ζ $\qquad\qquad$ (b) Variation in ψ with r and ζ

Fig. 17.20

Frequency Ratio r Corresponding to Peak Value of K

A close look at graph of magnification factor K v/s r reveals that when damping is present, peak value of K (and so the peak value of X) does not occur at resonating frequency but at a value $r < 1.0$. To obtain value of r at which K is maximum, differentiate equation (17.76) with respect to r and equate it to zero. Thus,

$$\frac{dK}{dr} = 0 = \frac{0 - \frac{1}{2}\{(1-r^2)^2 + (2\zeta r)^2\}^{-1/2}\{2(1-r^2)(-2r) + 8\zeta^2 r\}}{\{(1-r^2)^2 + (2\zeta r)^2\}}$$

or
$$\{(1-r^2)(-2r) + 4\zeta^2 r\} = 0$$
or
$$2r(r^2 - 1 + 2\zeta^2) = 0$$

As $r \neq 0$, we have $r = \sqrt{1 - 2\zeta^2}$ (17.78)

Thus, as damping increases, the value of r at which peak value of K occurs decreases.

EXAMPLE 17.11 A mass of 98.1 N is suspended from one end of a helical spring, the other end being fixed. The stiffness of the spring is 98.1 N/cm. Viscous damping causes the amplitude to decrease to one-tenth of the initial value in four complete oscillations. If a periodic force of 147.2 cos50t N is applied at the mass in the vertical direction, find the amplitude of the forced vibrations. What is its value at resonance? (AMIE, Summer 1980)

Solution: $k = 98.1$ N/cm; $W = 98.1$ N

$$n = 4 \text{ cycles}; x_0/x_n = 1/10$$

Therefore
$$\omega_n = \sqrt{\frac{k}{m}} = \sqrt{\frac{981 \times 98.1}{98.1}} = 31.32 \text{ rad/s}$$

Also,
$$\delta = \frac{1}{4}\log_e(10/1) = 0.5756$$

Therefore
$$\zeta \approx \delta/2\pi = 0.5756/2\pi = 0.0916$$

Since the value of damping ratio ζ works out to be quite less, we are justified in using approximate expression for δ.

The excitation force is $F = F_0\cos\omega t$, where excitation frequency $\omega = 50$ rad/s and the amplitude of excitation force $F_0 = 147.2$ N. The amplitude of steady-state response is,

$$X = X_0 K = X_0 \frac{1}{\sqrt{(1-r^2)^2 + (2\zeta r)^2}}$$

where X_0 = static deflection under load $F_0 = 147.2$ N

Therefore
$$X_0 = \frac{147.2}{k} = \frac{147.2}{98.1} = 1.5 \text{ cm}$$

As the frequency ratio $r = 50/31.32 = 1.596$

Therefore
$$X = \frac{1.5}{\sqrt{(1-1.596^2)^2 + (2 \times 0.0916 \times 1.596)^2}} = 0.952 \text{ cm}$$ **Ans.**

At resonance, $r = 1$ and the magnification factor assumes a value

$$K = \frac{1}{\sqrt{(1-r^2)^2 + (2\zeta r)^2}} = \frac{1}{0.1832} = 5.4585$$

Therefore
$$X_{max} = KX_0 = 5.4585 \times 1.5 = 8.188 \text{ cm}$$ **Ans.**

EXAMPLE 17.12   A single cylinder vertical petrol engine of total weight 2950 N is mounted upon a steel chasis frame and causes a vertical static deflection of 2 mm. The reciprocating parts of the engine weigh 200 N and move through a vertical stroke of 15 cm with simple harmonic motion. A dashpot is provided whose damping resistance is directly proportional to the velocity and amounts to 14.7 N per unit velocity in cm/s.

Considering that the steady-state vibration condition is reached, determine: (a) the speed of the driving shaft at which resonance will occur, (b) the amplitude of forced vibrations when the driving shaft of the engine rotates at (i) 480 r.p.m. and (ii) 945 r.p.m.

Solution: $W = 2950$ N; $\Delta st = 0.2$ cm; $c = 14.7$ N·s/cm

Crank radius $= \dfrac{15}{2} = 7.5$ cm; weight of reciprocating parts $= 200$ N

Here, neglecting second and higher harmonics of inertia force due to reciprocating parts,

$$F = \frac{200}{981} \times (7.5)\omega^2 \cos\theta$$

Therefore
$$F_0 = \frac{200}{981} \times 7.5 \times \left(2\pi \frac{n}{60}\right)^2 \qquad \text{or} \qquad F_0 = (0.016768)\, n^2$$

(a) The natural frequency of supporting system,

$$\omega_n = \sqrt{\frac{g}{\Delta st}} = \sqrt{\frac{981}{0.2}} = 70.036 \text{ rad/s}$$

Therefore,   resonating speed $n = \dfrac{60 \times 70.036}{2\pi} = 668.8$ r.p.m. **Ans.**

Again
$$c_c = 2m\omega_n = 2\frac{(2950)}{981}(70.036) = 421.22 \text{ N·s/cm}$$

Therefore
$$\zeta = c/c_c = 14.7/421.22 = 0.035$$

Also
$$k = \frac{2950}{0.2} = 14{,}750 \text{ N/cm}$$

(a) For $n = 480$ r.p.m., $r = 480/668.8 = 0.718$

$$X_0 = \frac{(0.016768)\,(480)^2}{14{,}750} = 0.262 \text{ cm}$$

Therefore $\qquad X_{max} = \dfrac{X_0}{\sqrt{(1-r^2)^2 + (2\zeta r)^2}}$

or, $\qquad X_{max} = \dfrac{0.262}{\sqrt{(1-0.718^2)^2 + (2 \times 0.035 \times 0.718)^2}} = 0.538 \text{ cm}$ **Ans.**

(b) For $n = 945$ r.p.m.,

$$r = \frac{945}{668.8} = 1.413; \quad X_0 = \frac{(0.016768) \times 945^2}{14{,}750} = 1.0152 \text{ cm}$$

Therefore $\qquad X_{max} = \dfrac{1.0152}{\sqrt{(1-1.413^2)^2 + (2 \times 0.035 \times 1.413)^2}} = 1.014 \text{ cm}$ **Ans.**

17.12 VIBRATION ISOLATION AND TRANSMISSIBILITY

In practice, often an unbalanced machine is to be installed in a structure where vibration is undesirable. Examples of this type are: A.C. motors installed for elevators in hospitals or hotels. Yet another example can be seen in an I.C. engine, which is a source of vibration, installed in an automobile. While installing such machines, care must be taken to ensure that no vibrations appear in the structure to which it is attached. There is yet another category of practical situations (e.g. control panels of air craft, rockets, etc.) where vibrations of structure must not be transmitted to the instruments. Naturally these panels must be isolated from the rest of the structure in respect of vibrations.

In both types of problems, effectiveness of vibration isolation is measured in terms of a ratio called *Transmissibility ratio. Thus, force transmissibility is defined as the ratio of the amplitude of the force transmitted to that of the excitation force. In a similar way, motion transmissibility is the ratio of amplitude of displacement transmitted to the displacement applied. The lesser the force/motion transmitted, greater is the isolation in a given system.*

17.13 VIBRATION ISOLATION WITHOUT DAMPERS

Problem of vibration isolation can be solved in a very simple way by mounting the machine on springs. As shown in Figs. 17.21(a) and (b), let the mass m represent a machine, and let a force $F_0 \sin \omega t$ act on it. In Fig. 17.21(a), the machine is rigidly attached to foundation and this case is equivalent to a mass mounted on a spring of infinite stiffness. In Fig. 17.21(b), the machine is mounted on springs with combined vertical stiffness of k.

Let the frequency of excitation ω be varied keeping the amplitude of excitation force F_o constant. The differential equation of motion for either case is,

$$m\ddot{x} + kx = F_o \sin \omega t$$

and the steady-state solution is obtained from equation (17.75) by letting $\zeta = 0$. Thus,

$$X = \frac{X_0}{(1-r^2)} \qquad (17.79)$$

and

$$x_p = \frac{X_0}{(1-r^2)} \sin(\omega t - \phi) \tag{17.80}$$

The magnification factor is then given by

$$K = \frac{X}{X_0} = \frac{1}{(1-r^2)} \tag{17.81}$$

The variation of magnification factor (X/X_O) with frequency ratio r is as shown in Fig. 17.21(c).

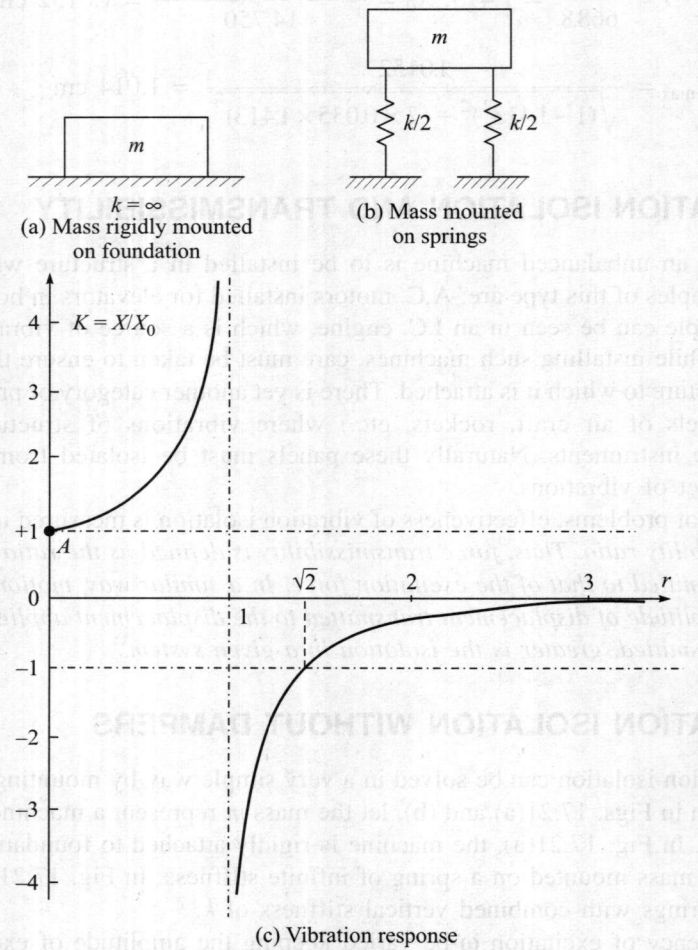

(a) Mass rigidly mounted on foundation

(b) Mass mounted on springs

(c) Vibration response

Fig. 17.21 Role of spring in vibration isolation.

Let us now consider the force transmitted to the sub-structure by the machine. The only member in physical contact with the mass is the spring element. Hence, the amplitude of force transmitted to the frame,

$$F_T = kX$$

or $\hspace{6cm} X = (F_T/k)$

Also, $\hspace{6cm} X_O = (F_O/k)$

and therefore, by dividing on corresponding sides

$$\left(\frac{X}{X_0}\right) \equiv \left(\frac{F_T/k}{F_O/k}\right) = \left(\frac{F_T}{F_O}\right)$$

Thus, $\quad \dfrac{X}{X_O} = \dfrac{\text{Force transmitted to the frame}}{\text{Amplitude of excitation force}} = \text{Transmissibility } (i.e., \text{ T.R.})$

For machines mounted on springs therefore, the magnification factor also represents transmissibility. Ideally, transmissibility should be zero but in practice an attempt is made to reduce it to as small a value as possible.

It follows from Figs. 17.2(a) and (c) that when machine is mounted directly on rigid foundation *i.e.*, when k (and hence, $\omega_n) \to \infty$, $r \to 0$ and the condition is represented in graph by a point like A. For this case, the transmissibility is 1 which means that the amplitude of force transmitted equals the amplitude of excitation force.

For the case in Fig. 17.21(b), it is seen in the graph that at $r = 0$ and $r = \sqrt{2}$, T.R. = 1 (strictly speaking, it is -1 at $r = \sqrt{2}$) and mounting the machine on springs does not make any difference. In the range $0 < r < \sqrt{2}$ it is seen in graph that the transmissibility (T.R.) is always greater than unity. In other words, amplitude of force transmitted is more than that of excitation force. Thus in the range $0 < r < \sqrt{2}$, the springs make the matter worse. For instance, at $r = 0.5$, the T.R. from equation (17.81) works out to be 1·33 and at $r = 1$ theoretically, it approaches infinity. As against this, for $r > \sqrt{2}$, the T.R. is always less than 1. Further, larger the r, smaller is the transmissibility. For instance, with $r = 4$ transmissibility is 0.067 while with $r = 8$, it is 0.0158 only.

17.14 VIBRATION ISOLATION USING DAMPERS

Adequate vibration isolation must be ensured in the design of machine mounts. This is because unbalanced forces in machines are ultimately transmitted to the foundation in the form of shaking forces via the mounts through which they are in contact with the ground. The periodic forces transmitted through the mounts to the ground can excite other machines and their elements around. The problem becomes severe, particularly when the natural frequency of any machine or machine element in the neighbourhood is close to any of the harmonic frequencies of the shaking force. In situations like this, corresponding machine or machine element will vibrate with large amplitudes.

Suitable isolators can be used to reduce or eliminate shaking forces transmitted to the ground. Such isolators are called *active isolators*. When the isolators are used to reduce or eliminate the transmission of motion from the floor to the machine, they are called *passive isolators*.

When machines are mounted on closed coil helical springs of steel alone, damping may be neglected and the resulting model is the one considered in Section 17.13. However, if the machines are mounted on leaf springs or paddings of rubber and cork, damping must be considered. In Fig. 17.22 let the mass m, representing machine, be mounted on springs and dashpot. The force transmitted to the foundation is the vector sum of spring force and the damping force. Neglecting the deflection of the foundation itself, the force transmitted F_T is given by

$$F_T = kx + c\dot{x} \qquad (17.82)$$

The steady-state response for such a system is given by,

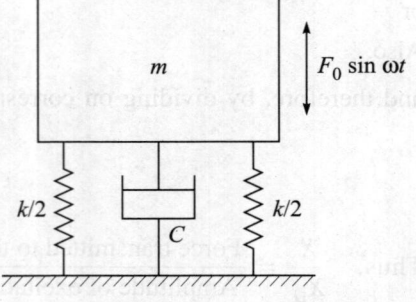

Fig. 17.22 Isolation through springs and damper.

$$x_p = X_o K\sin(\omega t - \phi) = (F_o/k)K\sin(\omega t - \phi)$$

Thus, the steady-state amplitude,

$$X = X_o K$$

and therefore,

$$(X/X_o) = K = \frac{1}{\sqrt{(1 - r^2)^2 + (2\zeta r)^2}}$$

Unlike isolation without damping, the magnification factor ($K = X/X_0$) does not represent transmissibility here. This is evident from following discussions.

Since

$$x = X\sin(\omega t - \phi)$$

and

$$\dot{x} = X\omega\cos(\omega t - \phi)$$

Thus, equation (17.82) may be rewritten as,

$$F_T = kX\sin(\omega t - \phi) + cX\omega\cos(\omega t - \phi)$$

Letting,

$$kX = P\cos\lambda \qquad \text{and,} \qquad cX\omega = P\sin\lambda$$

Substitution gives,

$$F_T = P\sin(\omega t - \phi + \lambda) \qquad (17.83)$$

where

$$\tan\lambda = (c\omega/k) = 2\zeta r \qquad (17.84)$$

and

$$P = \sqrt{(kX)^2 + (cX\omega)^2}$$

or

$$P = kX\sqrt{1 + \left(\frac{c}{k}\omega\right)^2}$$

Thus

$$P = |F_T| = kX\sqrt{1 + (2\zeta r)^2} \qquad (17.85)$$

The spring force F_s (amplitude = kX) and damping force F_d (amplitude = $cX\omega$) acts in directions opposite to those of displacement and velocity respectively. The vector sum of F_s and F_d is shown to give resultant force $P = |F_T|$ in Fig. 17.23.

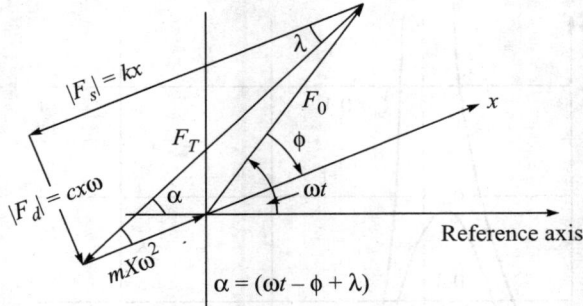

Fig. 17.23   Force vector diagram.

Thus, transmissibility

$$\text{T.R.} = \frac{|F_T|}{F_0} = \frac{kX\sqrt{1+(2\zeta r)^2}}{(kX_0)}$$

or

$$\text{T.R.} = (X/X_0)\sqrt{1+(2\zeta r)^2}$$

or

$$\text{T.R.} = K\sqrt{1+(2\zeta r)^2} \tag{17.86}$$

or

$$\text{T.R.} = \frac{\sqrt{1+(2\zeta r)^2}}{\sqrt{(1-r^2)^2+(2\zeta r)^2}} \tag{17.87}$$

It follows from equation (17.86) that graph of magnification factor vs frequency ratio does not represent graph of T.R. vs frequency ratio as in Section 17.13. Dividing numerator and denominator on right-hand side of equation (17.87) by r,

$$\text{T.R.} = \frac{\sqrt{(1/r)^2+(2\zeta)^2}}{\sqrt{(1/r-r)^2+(2\zeta)^2}} \tag{17.88}$$

Figure 17.24 shows transmissibility curves for various damping ratios and is important to us for many reasons. It may be noted that all the curves pass through points ($r = 0$, T.R. = 1) and ($r = \sqrt{2}$, T.R. = 1). In the range $0 < r < \sqrt{2}$, the T.R. value is greater than 1 for all values of damping. Further for this range, larger the damping smaller is the T.R. value. This goes to suggest that though spring makes the matter worse (*see* Section 17.13), in this frequency range, the damping helps in reducing T.R. In the frequency range $\sqrt{2} < r < \infty$, the T.R. is always less than 1 for all values of damping ratio. Further, for this frequency range larger the damping, larger is the transmissibility which is an important thing to note. This shows that though spring plays very useful role in this range, damping makes the matter worse. It follows that for operating in the frequency range $\sqrt{2} < r < \infty$, smaller the damping better it is. For effective vibration–isolation, desirably, $r \geq 5$. At $r = 5$ and $\zeta = 0.5$, T.R. $\approx 1/5$ while at $\zeta = 0.1$, T.R. $\approx 1/7$.

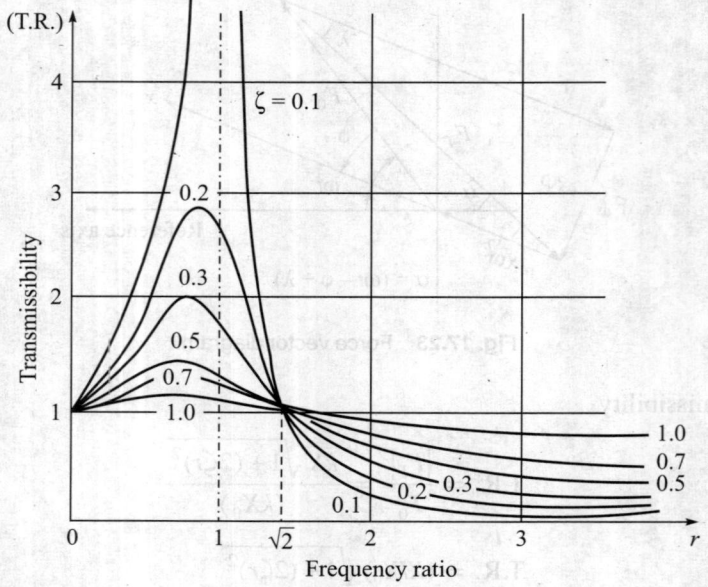

Fig. 17.24 Variation in Transmissibility with Frequency ratio.

Above discussions are true for a constant speed machine for which the amplitude of forcing function is constant. In such cases, the magnitude of the force transmitted is proportional to T.R. values. For a variable speed machine, the amplitude of forcing function ($F_{eq} = me\omega^2$) is due to unbalanced masses. This force is proportional to the square of the angular speed of rotation and the force transmitted is given by modified (*see* equation 17.86) expression.

$$P = |F_T| = F_{eq} K \sqrt{1 + (2\zeta r)^2}$$

or

$$P = (me\omega^2) K \sqrt{1 + (2\zeta r)^2} \qquad (17.89)$$

where (me) represents amount of unbalance present in a rotating system. Multiplying and dividing R.H.S. of equation (17.89) by ω_n^2 and defining $me\omega_n^2 = P_n = $ constant and rearranging, we have

$$(P/P_n) = r^2 K \sqrt{1 + (2\zeta r)^2} \qquad (17.90)$$

or

$$(P/P_n) = r^2 \times \text{T.R.} \qquad (17.91)$$

Equation (17.91) is shown plotted in Fig. 17.25. The curves drawn for different values of ζ indicate that actual force transmitted at high frequencies can be very high despite low values of T.R.

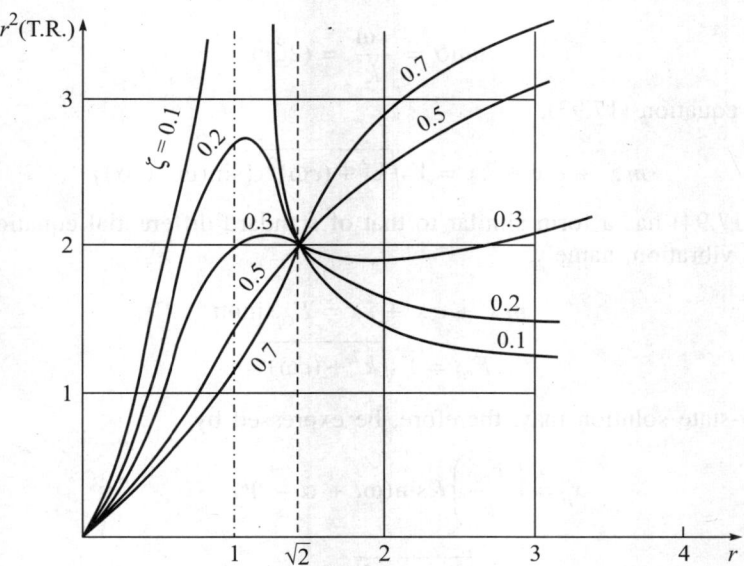

Fig. 17.25 Graphical depiction of equation (17.91).

17.15 MOTION TRANSMISSIBILITY

In many situations, machines or machine components are excited through the support. Thus, in the case of a spring-mass-dashpot system of Fig. 17.26, a sinusoidal excitation motion is shown to be applied to the support. Let the motion applied to support be represented by

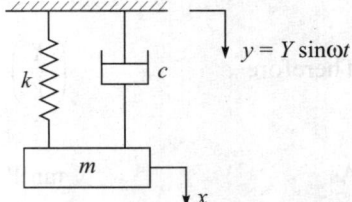

Fig. 17.26 Base excitation model.

$$y = Y\sin\omega t$$

and let x be the resultant motion so that $\dot{x}$ and $\ddot{x}$ are the velocity and acceleration of the mass. Thus, the relative displacement between the two ends of spring = $(x - y)$. Similarly, the relative velocity between the two ends of dashpot is $(\dot{x} - \dot{y})$.

Hence, the differential equation of motion may be written as,

$$m\ddot{x} = -c(\dot{x} - \dot{y}) - k(x - y)$$

or

$$m\ddot{x} + c\dot{x} + kx = (c\dot{y} + ky) \qquad (17.92)$$

Substituting for y and $\dot{y}$ above,

$$m\ddot{x} + c\dot{x} + kx = (c\omega\cos\omega t + k\sin\omega t)Y \qquad (17.93)$$

Multiplying and dividing the terms on R.H.S. by $\sqrt{k^2 + (c\omega)^2}$ and

letting $\qquad \dfrac{k}{\sqrt{k^2 + (c\omega)^2}} = \cos\alpha \quad$ and $\quad \dfrac{c\omega}{\sqrt{k^2 + (c\omega)^2}} = \sin\alpha$

so that, $$\tan\alpha = \frac{c\omega}{k} = (2\zeta r)$$

We have from equation (17.93),

$$m\ddot{x} + c\dot{x} + kx = Y\sqrt{k^2 + (c\omega)^2}\ \{\sin(\omega t + \alpha)\} \qquad (17.94)$$

Equation (17.94) has a form similar to that of standard differential equation of motion of forced damped vibration, namely,

$$m\ddot{x} + c\dot{x} + kx = F_{eq}\sin\omega t$$

with $$F_{eq} = Y\sqrt{k^2 + (c\omega)^2}$$

The steady-state solution may, therefore, be expressed by,

$$x_p = \left(\frac{F_{eq}}{k}\right) K\sin(\omega t + \alpha - \Psi)$$

or $$x_p = \frac{Y\sqrt{k^2 + (c\omega)^2}}{k} K\sin(\omega t + \alpha - \Psi)$$

or $$x_p = Y\sqrt{1 + (2\zeta r)^2}\ K\sin(\omega t + \alpha - \Psi)$$

and the amplitude of steady-state vibration is,

$$X = Y\sqrt{1 + (2\zeta r)^2}\ K$$

Therefore $$\left(\frac{X}{Y}\right) = \frac{\sqrt{1 + (2\zeta r)^2}}{\sqrt{(1 - r^2)^2 + (2\zeta r)^2}} \qquad (17.95)$$

As $$\tan\Psi = \frac{2\zeta r}{(1 - r^2)}$$

We have $$(\Psi - \alpha) = \tan^{-1}\left(\frac{2\zeta r}{1 - r^2}\right) - \tan^{-1}(2\zeta r) \qquad (17.96)$$

Equation (17.95) gives expression for motion transmissibility, which is identical to the expression for force transmissibility, as given in equation (17.87).

EXAMPLE 17.13 An air compressor of 450 kg operates at a constant speed of 1750 r.p.m. Rotating parts are well balanced. The reciprocating part is 10 kg and crank radius is 100 mm. The mounting introduces a viscous damping of damping factor 0.15. Specify the spring for the mounting such that only 20% of the unbalanced force is transmitted to the foundation. Find out the amplitude of the transmitted force. (AMIE, Summer 1993)

Solution: $M = 450$ kg; $m = 10$ kg; crank radius $= 10.0$ cm

$$\omega = \frac{2\pi \times 1750}{60} = 183.26 \text{ rad/s}$$

$$\zeta = 0.15 \quad \text{and} \quad \text{T.R.} = 20\% = 0.2$$

Let $r = \omega/\omega_n$, be the frequency ratio. The transmissibility is then given by,

$$\text{T.R.} = \frac{\sqrt{1 + (2\zeta r)^2}}{\sqrt{(1 - r^2)^2 + (2\zeta r)^2}}$$

Hence

$$0.2 = \frac{\sqrt{1 + (2 \times 0.15 \times r)^2}}{\sqrt{(1 - r^2)^2 + (2 \times 0.15 \times r)^2}}$$

Squaring on both sides and arranging as a quadratic in r^2, we have

$$r^4 - 4.16r^2 - 24 = 0$$

Solving the quadratic,

$$r^2 = 7.4 \quad \text{and hence, } r = 2.72$$

So

$$\omega_n = (\omega/r) = \frac{183.26}{2.72} = 67.375 \text{ rad/s}$$

Thus, from

$$67.375 = \sqrt{\frac{k}{450}}$$

$$k = 450 \times (67.375)^2 = 2042.7 \text{ kN/cm} \qquad \textbf{Ans.}$$

Amplitude of unbalanced (primary) force due to reciprocating parts

$$F = mR\omega^2 = (10) \times (10)(183.26)^2$$

Hence the amplitude of transmitted force,

$$F_T = (\text{T.R.}) \times mR\omega^2 = 0.2 \times 10 \times 10 \times (183.26)^2 = 671.68 \text{ kN} \qquad \textbf{Ans.}$$

EXAMPLE 17.14 A 1200 kg machine is mounted on four identical springs of total spring constant k and having negligible damping. The machine is subjected to a harmonic external force of amplitude $F_0 = 490$ N and frequency 180 r.p.m. Determine

    (a) amplitude of motion of the machine and maximum force transmitted to the foundation because of the unbalanced force when $k = 1.96 \times 10^6$ N/m

    (b) the same as in (a) above for the case when $k = 9.8 \times 10^4$ N/m

Solution: Therefore $\omega = \dfrac{2\pi \times 180}{60} = 6\pi$ rad/s

$$k = 1.96 \times 10^6 \text{ N/m}; \quad m = 1200 \text{ kg}$$

Therefore

$$\omega_n = \sqrt{\frac{1.96 \times 10^6}{1200}} = 40.41 \text{ rad/s}$$

Therefore, frequency ratio $r = \omega/\omega_n = 6\pi/(40.41) = 0.466$

For $\zeta = 0$, the magnification factor,

$$K = \frac{X}{X_0} = \frac{1}{|(1-r^2)|} = \frac{1}{|1-(0.466)^2|} = 1.277$$

and as the static deflection

$$X_0 = (F_0/k) = \frac{490}{1.96 \times 10^6} \text{ m} = 0.025 \text{ cm}$$

Thus, X, the amplitude of motion of the machine, is given as

$$X = 1.277 \ (X_0) = 1.277 \times 0.025 = 0.0319 \text{ cm} \qquad \textbf{Ans.}$$

Amplitude of force transmitted F_T and the excitation force are related as,

$$(F_T/F_O) = \frac{1}{|(1-r^2)|}$$

or

$$(F_T/490) = \frac{1}{|(1-0.466^2)|}. \quad \text{Thus} \quad F_T = 625.9 \text{ N} \qquad \textbf{Ans.}$$

(b) For $k = 9.8 \times 10^4$ N/m,

$$\omega_n = \sqrt{\frac{9.8 \times 10^4}{1200}} = 9.037 \text{ rad/s}$$

Therefore

$$r = \omega/\omega_n = 6\pi/(9.037) = 2.086$$

Also,

$$X_0 = F_0/k = \frac{490}{9.8 \times 10^4} = 0.005 \text{ m} = 0.5 \text{ cm}$$

Hence from transmissibility consideration,

$$\frac{X}{0.5} = \frac{1}{|(1-2.086^2)|}$$

Therefore

$$X = 0.2984 \times 0.5 = 0.149 \text{ cm} \qquad \textbf{Ans.}$$

and force transmitted,

$$F_T = \frac{490}{|(1-2.086^2)|} = 146.2 \text{ N} \qquad \textbf{Ans.}$$

EXAMPLE 17.15 The 4450 N vehicle, shown in Fig. 17.26a travels along a rough road at 96.0 km/h. The sine curve representing the rough road, has an amplitude of 25.4 mm and a wave length of 6.09 m. Determine (a) the spring constant k such that 96.0 km/h is the critical speed at which resonance occurs, (b) the spring constant to give a vibration amplitude of 6.36 mm of the vehicle at 96 km/h, (c) whether a rider in the vehicle would leave his seat under the conditions of (b).

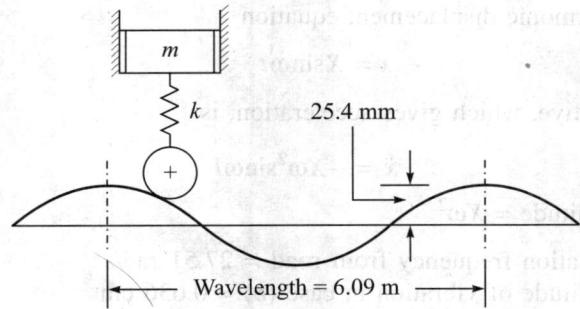

Fig. 17.26a Base excitation to vehicle from rough road.

Solution: For resonance to occur, the excitation frequency $\omega = \omega_n$. At 96 km per hour, as the speed of vehicle and wavelength of road profile as 6.09 m, the frequency of excitation from road is,

$$f = \frac{(96 \times 1000/3600)}{6.09}$$

or

$$f = 4.379 \text{ cycles/s}$$

Hence, circular frequency of excitation in radian per second is,

$$\omega = 2\pi f = 2\pi \times 4.379 = 27.51 \text{ rad/s}$$

(a)   For resonance to occur,

$$\omega_n = \omega = 27.51 \text{ rad/s}$$

Therefore $\sqrt{\dfrac{k \times 981}{4450}} = 27.51.$ Therefore $k = 3433$ N/cm   **Ans.**

(b) For a vehicle speed of 96 km per hour and $\omega = 27.51$ rad/s amplitude of vibration at this speed is given to be $X = 0.636$ cm. Neglecting damping, the transmissibility is given as (note that for X to be smaller than X_0, $r > \sqrt{2}$)

$$(X/X_O) = \frac{0.636}{2.54} = \frac{1}{(r^2 - 1)} = 0.2504$$

Therefore $r^2 = 1 + \dfrac{1}{0.2504}$ or $r = 2.23$

So from $\omega/\omega_n = 2.23$

$$\omega_n = \omega/2.23 = \frac{27.51}{2.23} = 12.34 \text{ rad/s}$$

Hence, from $12.34 = \sqrt{\dfrac{k \times 981}{4450}}$

or $k = \dfrac{4450 \times (12.34)^2}{981} = 690.75$ N/cm   **Ans.**

(c) Assuming a harmonic displacement equation

$$x = X\sin\omega t$$

The second time derivative, which gives acceleration, is

$$\ddot{x} = -X\omega^2\sin\omega t$$

Thus, acceleration amplitude = $X\omega^2$

where ω = excitation frequency from road = 27.51 rad/s
and X = amplitude of vibration in case (b) = 0.636 cm
Hence A_{max} = (0.636) × (27.51)^2 = 481.3 cm/s^2

Since, this acceleration is less than acceleration due to gravity (981 cm/s^2), the rider will not leave his seat.   **Ans.**

EXAMPLE 17.16   A trailer has 1000 kg mass when fully loaded and 250 kg when empty. The spring of the suspension is 350 kN/m. The damping factor is 0.5 when the trailer is fully loaded. The speed is 100 km/hr. The road varies sinusoidally with a wavelength of 5 m. Determine the amplitude ratio of the trailer when fully loaded and empty. (AMIE, Summer 1993)

Solution: Mass of trailer (when empty) = 250 kg
    Mass of trailer (when fully loaded) = 1000 kg
    Stiffness of suspension spring k = 350 k·N/m = (350 × 1000) N/m
Therefore natural frequency when empty,

$$\omega_{n_1} = \sqrt{\frac{350\times 1000}{250}} = 37.42 \text{ rad/s}$$

Excitation frequency due to road roughness,

$$\omega = \frac{(100\times 1000)}{3600} \times \frac{2\pi}{5} = 34.9 \text{ rad/s}$$

Therefore frequency ratio r = 34.9/37.42 = 0.933

Hence $\text{T.R.} = \dfrac{\sqrt{1+(2\zeta r)^2}}{\sqrt{(1-r^2)^2+(2\zeta r)^2}}$

or $\text{T.R.} = \dfrac{\sqrt{1+(2\times 0.5\times 0.933)^2}}{\sqrt{(1-0.933^2)^2+(2\times 0.5\times 0.933)^2}} = 1.452$

Therefore amplitude ratio, when empty = 1.452
    When fully loaded,

$$\omega_{n_2} = \sqrt{\frac{k}{m}} = \sqrt{\frac{350\times 1000}{1000}} = 18.708 \text{ rad/s}$$

Therefore frequency ratio $r_2 = \dfrac{34.9}{18.708} = 1.8655$

Hence, amplitude ratio, *i.e.*, T.R.,

$$\text{T.R.} = \frac{\sqrt{1 + (2\zeta r)^2}}{\sqrt{(1 - r^2)^2 + (2\zeta r)^2}}$$

or

$$\text{T.R.} = \frac{\sqrt{1 + (2 \times 0.5 \times 1.8655)^2}}{\sqrt{(1 - 1.8655^2)^2 + (2 \times 0.5 \times 1.8655)^2}} = 0.682$$

Thus, amplitude ratio when fully loaded = 0.682 **Ans.**

17.16 WHIRLING OF SHAFTS

Whirling motion consists of the rotation of the plane *A*, containing bent up shaft axis and the bearing centre line (*see* Fig. 17.27) about the bearing centre line. There are thus two component motions involved in the phenomenon of whirl:

(a) a spinning motion of the shaft together with disc about the bent up shaft axis

(b) the rotation of the plane *A*, containing bent up centre line of shaft and bearing centre line about the centre line of bearing.

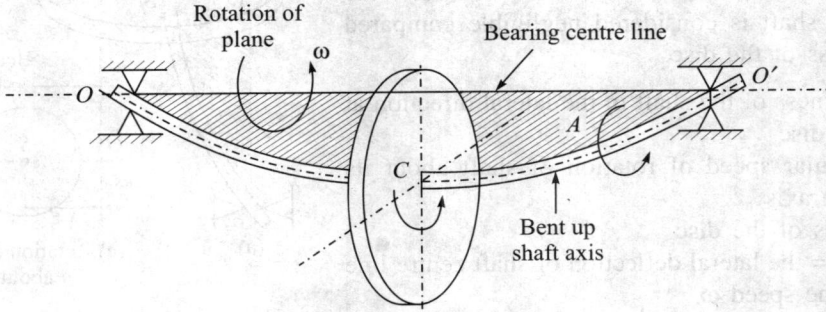

Fig. 17.27 Shaft in whirling motion.

The whirling of shaft, in particular the motion described at point (b) above, may take place in sense same as that of spinning motion as pointed out at (a) above or opposite to it. Again the speed of whirling may or may not be equal to the speed of spinning motion of the shaft. When the whirling speed equals speed of rotation of the shaft, the phenomenon is called *synchronous whirl*. In this book, only synchronous whirl will be considered.

The phenomenon of whirl results from a number of causes *e.g.*, unbalanced rotating masses, hysterisis damping, fluid-friction in bearing and gyroscopic forces. The subject of shaft whirl falls under the broad category of self-excited vibration in which the excitation forces, producing vibration, are controlled by the motion itself.

Critical Speed: At certain rotational speeds, the shafts have a tendency to bow out and whirl in a complicated manner. This speed is called *critical speed*, *whipping speed* or *whirling speed*.

At critical speed the shaft starts vibrating violently in transverse direction. The excessive vibrations associated with the critical speed may cause permanent deformation or structural damage; for instance, blades of rotor of a turbine may come in contact with the stator blades. Again, large shaft deflections associated with critical speed may produce large bearing reactions and may lead to bearing failure. The phenomenon of shaft-whirl occurs even for very accurately balanced rotors. The amplitude build up is a time-dependent phenomenon and therefore, it is advisable to cross over its critical speed as early as possible.

17.17 CRITICAL SPEED OF LIGHT VERTICAL SHAFT WITH SINGLE DISC (WITHOUT DAMPING)

Consider a light vertical shaft carrying a single disc of mass m, symmetrically located on a shaft which is supported in two bearings as shown in Fig. 17.28(a). Let G be the mass centre of the disc located at a radial distance of e from the geometric centre S of the disc at which shaft centre line intersects the disc. Further, let the bearing centre line intersect the plane of disc at O. At the synchronous whirl, the O, S and G remain fixed relative to each other. In the discussion to follow the mass of the shaft is considered negligible compared with the mass of the disc.

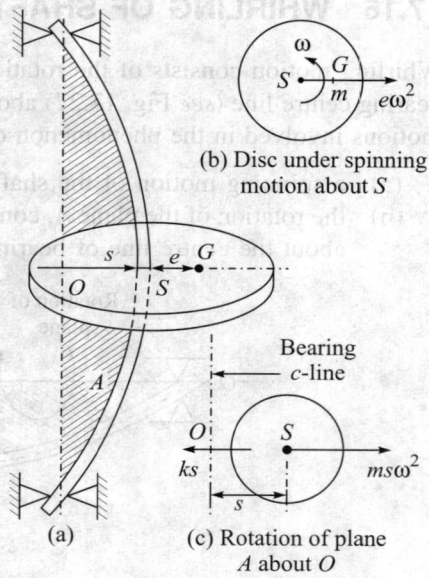

(b) Disc under spinning motion about S

(c) Rotation of plane A about O

Fig. 17.28

Let k = stiffness of the shaft in the lateral direction at the disc
   ω = angular speed of rotation of shaft about its own axis
   m = mass of the disc
   $s = OS$ = the lateral deflection of shaft centre line at the speed ω.

The damping being absent, the shaft deflection lags behind the excitation force (i.e., the centrifugal force) by a phase angle of either zero (at frequency ratio $r < 1$), $\pi/2$ (at $r = 1$) or π radian (for $r > 1$) vide Fig. 17.29(a and b). The position of the centre of gravity G with respect to S in Fig. 17.28(a) thus corresponds to frequency ratio $r < 1$.

The various forces arising out of two component motions shown in Fig. 17.28(b) and (c) during whirl and also due to flexibility of shaft, are:

(i) the centrifugal force $me\omega^2$ due to eccentricity of the mass of the disc rotating about geometric centre S (rotation of the plane A about bearing centre line considered separately)

(ii) the centrifugal force $ms\omega^2$ due to mass m of disc, assumed transferred to shaft centre line and rotating with plane A, about bearing centre line (rotation of disc about geometric centre S considered separately)

(iii) the restoring force due to stiffness of rotating shaft in lateral direction $= ks$

Thus, total inward pull at the disc centre $= s(k - m\omega^2)$

And total outward force at $S = me\omega^2$

Thus for the equilibrium of the disc,

$$s(k - m\omega^2) = me\omega^2 \tag{17.97}$$

and therefore,

$$s = \frac{me\omega^2}{(k - m\omega^2)} \tag{17.98}$$

Equation (17.98) gives deflection s produced in the centre line of the shaft at the disc due to whirling motion. Dividing numerator and denominator by k on the R.H.S., and remembering that $m/k = 1/\omega_n^2$ and $\omega/\omega_n = r$, equation (17.98) reduces to

$$s = \frac{er^2}{(1 - r^2)} \tag{17.99}$$

Dividing numerator and denominator on R.H.S. by r^2 again, equation (17.99) reduces to,

$$s = \frac{e}{(1/r^2) - 1} \tag{17.100}$$

Following conclusions can be drawn based on equation (17.100):

(1) When $\omega = \omega_n$, i.e., ($r = 1$) the deflection s at the geometric centre of disc tends to infinity. Thus the critical speed of the shaft is equal to the natural frequency of lateral vibration of the shaft.

(2) For $\omega < \omega_n$ (i.e., $r < 1$), s/e and hence the shaft deflection s is positive i.e., the deflection and the eccentricity e are in the same sense. This condition of disc is described as *heavy side on the outside* [*see* Fig. 17.29(a)]. Note that for $r < 1$, phase angle is zero degrees.

(3) For $\omega > \omega_n$ (i.e., $r > 1$), s/e is negative which means that shaft deflection s and the eccentricity e are in opposite sense. This condition corresponds to a phase angle of $180°$ and c.g. of the rotor approaches point O [*see* Fig. 17.29(b)]. When the speed ω becomes very large s approaches value $-e$ and the disc appears to revolve about the c.g. of the disc.

EXAMPLE 17.17 A shaft 12.5 mm diameter rotates in long bearings and a disc weighing 196.2 N is attached to the mid-span of the shaft. The span of the shaft between the bearings is 600 mm. The mass centre of the disc is 0.5 mm from the axis of the shaft.

Neglecting the mass of the shaft and taking the deflection as for a beam fixed at both ends, find the critical speed of the shaft. Determine the range of the speed over which the stress in the shaft due to bending will exceed 1200 kgf/cm^2. Take $E = 19.62 \times 10^6$ N/cm^2.

(AMIE, Summer 1984)

Solution: For horizontal position of the shaft centre line, the static deflection under the pull of gravity on disc,

$$\delta = \frac{Wl^3}{192EI} = \frac{196.2 \times 60^3}{192 \times 19.62 \times 10^6 \times \pi (1.25)^4/64} = 0.0939 \text{ cm}$$

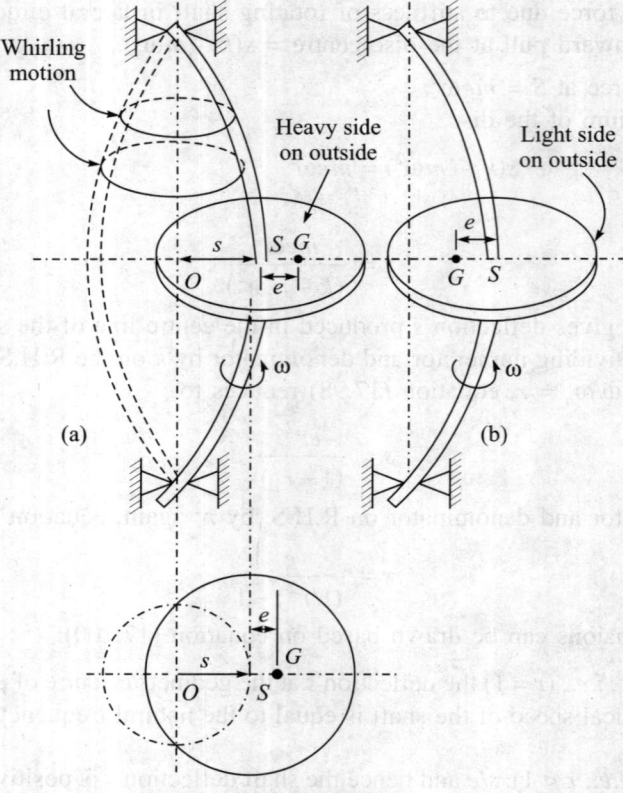

Plan view of shaft at (a)

Fig. 17.29 Whirling motion with Heavy side and Light side on outside.

Therefore

$$\omega_n = \sqrt{g/\delta} = \sqrt{\frac{981}{0.0939}} = 102.23 \text{ rad/s}$$

Therefore critical speed of shaft $N_c = \dfrac{60 \times 102.23}{2\pi} = 976.2$ r.p.m. **Ans.**

Let F be the dynamic load causing bending stress of 11772 N/cm². Then, from equation for simple bending,

$$\frac{M}{I} = \frac{\sigma_b}{y} \qquad\qquad\qquad\qquad (a)$$

But

$$M = F \times l/8 \qquad \text{and} \qquad I = \frac{\pi}{64} d^4$$

Hence from equation (a),

$$F \times l/8 = \left(\frac{\sigma_b}{d/2}\right) \times \frac{\pi}{64} d^4$$

Therefore $\qquad F = 11,772 \times \dfrac{\pi}{32}(1.25)^3 \times \dfrac{8}{60} = 300.57$ N

Maximum deflection produced under this dynamic load,

$$\delta_{max} = \frac{(300.57)\,(60)^3}{192 \times 19.62 \times 10^6 \times \dfrac{\pi}{64}\,(1.25)^4} = 0.144 \text{ cm}$$

Further for undamped case,

$$\frac{s}{e} = \frac{r^2}{1 - r^2}$$

Thus, with $s = \delta_{max} = 0.144$ cm and $e = 0.05$ cm, we have

$$\pm \frac{0.144}{0.05} = \left(\frac{r^2}{1 - r^2} \right) \qquad \text{or} \qquad 2.88(1 - r^2) = \pm r^2$$

Thus, the solutions are $\qquad\qquad 3.88 r^2 = 2.88$
and $\qquad\qquad\qquad\qquad\quad\; 1.88 r^2 = 2.88$

Hence $\qquad\qquad r_1 = \sqrt{\dfrac{2.88}{3.88}} \quad$ and $\quad r_2 = \sqrt{\dfrac{2.88}{1.88}}$

or $\qquad\qquad\qquad r_1 = 0.8615 \quad$ and $\quad r_2 = 1.238$
Therefore $\qquad\qquad N_1 = 0.8615 \times 976.2 = 841$ r.p.m. $\qquad$ **Ans.**
and $\qquad\qquad\qquad N_2 = 1.238 \times 976.2 = 1208$ r.p.m. $\qquad$ **Ans.**

EXAMPLE 17.18 A vertical steel shaft 15 mm diameter is held in long bearing 100 cm apart and carries at its middle a disc weighing 147.15 N. The eccentricity of the centre of gravity of the disc from the centre of the rotor is 0.30 mm. The modulus of elasticity for the shaft material is 19.6×10^6 N/cm^2 and the permissible stress is 6867 N/cm^2. Determine (i) the critical speed of the shaft and (ii) the range of speed over which it is unsafe to run the shaft. Neglect the weight of the shaft. (AMIE, Winter 1979; UPSC Engg. Services, 1976)

Solution: For shaft $l = 100$ cm; $d = 1.5$ cm; $E = 19.6 \times 10^6$ N/cm^2

Therefore $\qquad\qquad\qquad I = \dfrac{\pi}{64}(1.5)^4 = 0.2485$ cm^4

For computing natural frequency let the shaft be arranged horizontally and subjected to central disc weight. For long bearings, the shaft may be considered as bearing with fixed ends. Thus the static deflection under dead weight,

$$\delta = \frac{W l^3}{192 EI} = \frac{147.15 \times (100)^3}{192 \times 19.6 \times 10^6 \times \dfrac{\pi}{64}(1.5)^4} = 0.16 \text{ cm}$$

Therefore natural frequency of transverse vibration,

$$\omega_n = \sqrt{\frac{g}{\delta}} = \sqrt{\frac{981}{0.16}} = 78.3 \text{ rad/s}$$

Therefore critical speed $\qquad N_c = \dfrac{60 \times 78.3}{2\pi} = 747.7$ r.p.m. **Ans.**

Let F be the dynamic load causing bending stress of 6867 N/cm². Then from formula for simple bending,

$$M/I = f/y, \quad \text{or} \quad M = f(I/y)$$

Also, $\qquad M = Fl/8 \quad$ and $\quad y = d/2$

Therefore $\qquad F = \left(\dfrac{8}{100}\right) \times \dfrac{6867}{(1.5/2)} \times 0.2485 = 182.0$ N

Hence additional (dynamic) deflection (by proportion),

$$\delta_{max} = \dfrac{182}{147.15} \times 0.157 = 0.194 \text{ cm}$$

Therefore from $\qquad \pm \dfrac{s}{e} = \left(\dfrac{r^2}{1 - r^2}\right)$

We have $\qquad \pm \dfrac{0.194}{0.03} = \dfrac{r^2}{1 - r^2}$

or $\qquad \pm 6.467(1 - r^2) = r^2$

Thus $\qquad r_1 = 0.93 \quad$ and $\quad r_2 = 1.0876$

Therefore $\qquad N_1 = 0.93 \times 747.7 = 695.4$ r.p.m. $\qquad$ **Ans.**

$\qquad\qquad\qquad N_2 = 1.0876 \times 747.7 = 813.2$ r.p.m. $\qquad$ **Ans.**

17.18 CRITICAL SPEED OF LIGHT VERTICAL SHAFT HAVING SINGLE DISC WITH DAMPING

Usually damping is present in a whirling shaft either because of air resistance or due to structural damping. With damping present, the analysis for critical speed becomes slightly more involved. It may be recalled from Section 17.11 that, when damping is present in a forced damped vibratory system, the displacement vector lags behind the forcing function by an angle Ψ given by

$$\tan\Psi = \dfrac{2\zeta r}{(1 - r^2)} \qquad (17.101)$$

where, ζ is the damping factor and r is the frequency ratio.

The analysis for critical speed thus requires an additional force namely, the force due to damping, to be considered. Considering the equilibrium of forces acting on the disc, the different forces are:

(i) the centrifugal force $me\omega^2$ due to eccentricity of the mass centre of the disc rotating about geometric centre S of the disc with angular speed ω,

(ii) the spring force ks acting radially inward along line SO on account of flexibility of shaft,

(iii) the centrifugal force $ms\omega^2$ due to mass m of the disc transferred to the axis of shaft at S and whirling with angular speed ω, and acting in the direction OS,

(iv) the damping force due to air resistance and structural damping and assumed to be equivalent to viscous damping with equivalent coefficient of damping as c. Thus, the damping force is given by $cs\omega$, where $s\omega$ represents peripheral (tangential) velocity of shaft centre line during whirl, and acts in a direction opposite to peripheral velocity.

These forces are shown in magnitude and direction at Figs. 17.30(a) and (b). In view of finite value of damping factor ζ, the phase angle will be greater than zero for $\omega < \omega_n$ and $\dfrac{\pi}{2} < \psi < \pi$ for $\omega > \omega_n$. This is demonstrated for $\omega < \omega_n$ at Fig. 17.30(a).

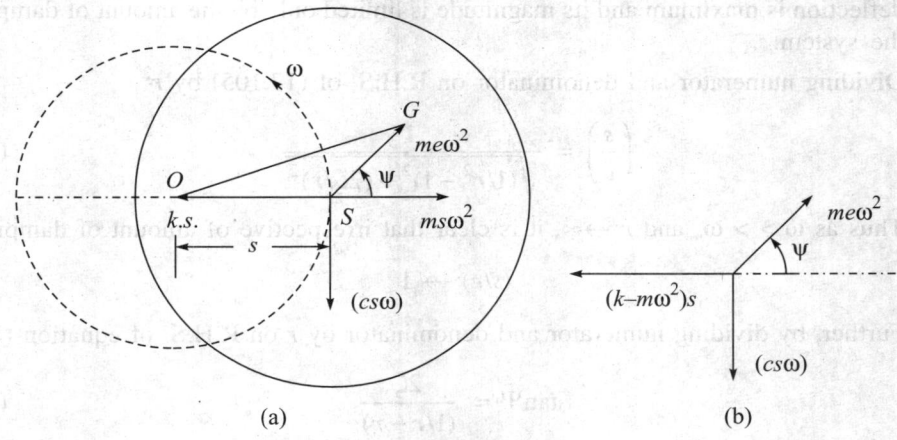

(a)   (b)

Fig. 17.30   Critical Speed: Force Analysis with Damping.

Resolving horizontally and vertically for equilibrium,

$$(k - m\omega^2)s = me\omega^2\cos\Psi \qquad (17.102)$$
$$(c\omega s) = me\omega^2\sin\Psi \qquad (17.103)$$

Squaring and adding on either sides,

$$s^2[(k - m\omega^2)^2 + (c\omega)^2] = (me\omega^2)^2$$

or

$$s = \frac{me\omega^2}{\sqrt{(k - m\omega^2)^2 + (c\omega)^2}} \qquad (17.104)$$

Dividing numerator and denominator of the R.H.S. of equation (17.104) by k and remembering that

$$\frac{m}{k} = \frac{1}{\omega_n^2} \qquad \text{and} \qquad \frac{c}{k} = \frac{2\zeta}{\omega_n}$$

we have,

$$s = \frac{er^2}{\sqrt{(1 - r^2)^2 + (2\zeta r)^2}}$$

or
$$\left(\frac{s}{e}\right) = \frac{r^2}{\sqrt{(1-r^2)^2 + (2\zeta r)^2}} \tag{17.105}$$

Again, dividing equation (17.103) by (17.102) on corresponding sides,

$$\tan\Psi = \frac{c\omega}{(k - m\omega^2)} \equiv \frac{2\zeta r}{(1-r^2)}$$

which is same as equation (17.101).

The conclusions may be summed up as under.

(a) It follows from equation (17.105) that, at $r = 1$ (*i.e.*, $\omega = \omega_n$) the amplitude of dynamic deflection is maximum and its magnitude is limited only by the amount of damping ζ in the system.

(b) Dividing numerator and denominator on R.H.S. of (17.105) by r^2,

$$\left(\frac{s}{e}\right) = \frac{1}{\sqrt{(1/r^2 - 1)^2 + (2\zeta/r)^2}} \tag{17.106}$$

Thus as $\omega >> \omega_n$ and $r \to \infty$, it is clear that irrespective of amount of damping,

$$(s/e) \to 1$$

(c) Further, by dividing numerator and denominator by r on R.H.S. of equation (17.101),

$$\tan\Psi = \frac{2\zeta}{(1/r - r)} \tag{17.107}$$

Thus for $r < 1$, $\tan\Psi$ is positive and hence,

$$0° < \Psi < 90°$$

which indicates the condition illustrated in Fig. 17.31(a) and goes under the name 'heavy side on the outside'.

(d) For $r = 1$ (*i.e.*, $\omega = \omega_n$), $\tan\Psi$ approaches a value of infinity and therefore, $\Psi = 90°$. This is illustrated in Fig. 17.31(b).

(e) For $r > 1$ (*i.e.*, $\omega > \omega_n$), $\tan\Psi$ in equation (17.107) turns out to be negative and as such

$$90° < \Psi < 180°$$

This condition goes under the name 'heavy side on the inside' and is illustrated in Fig. 17.31(c). For $\omega >> \omega_n$ *i.e.*, $r \to \infty$, it follows from equation (17.107) that,

$$\tan\Psi = -0$$

Hence, $\Psi = 180°$ (irrespective of amount of damping). In other words the centre of mass G lies on the line SO as shown in Fig. 17.31(d) and the system tends to be quite stable.

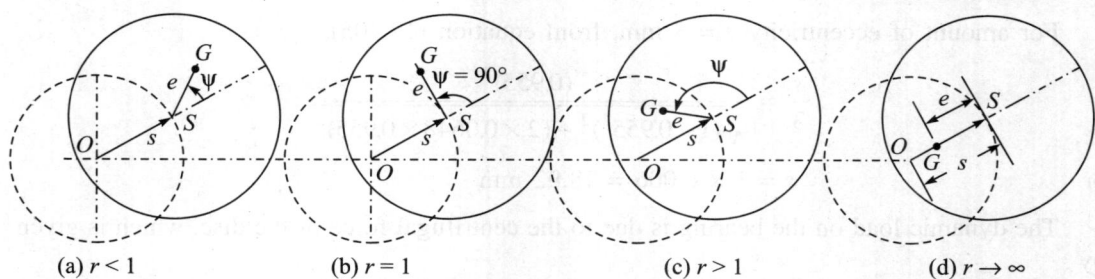

(a) $r < 1$ (b) $r = 1$ (c) $r > 1$ (d) $r \rightarrow \infty$

Fig. 17.31 Effect of frequency ratio on phase angle ψ.

EXAMPLE 17.19 A disc of mass 5 kg is mounted mid way between bearings which may be assumed to be simple supports. The bearing span is 48 cm. The steel shaft, which is horizontal, is 9 mm in diameter. The c.g. of the disc is displaced 3 mm from the geometric centre. The equivalent viscous damping at the centre of the disc shaft may be taken as 48 N·s/m. If the shaft rotates at 675 r.p.m., find the maximum stress in the shaft and compare it with dead load stress in the shaft. Also find the power required to drive the shaft at this speed.

Solution: For simply supported horizontal shaft with central disc weight (assume $E = 1.96 \times 10^7$ N/cm^2),

$$\delta = \frac{Wl^3}{48EI}$$

or
$$\delta = \frac{5 \times 9.81 \times (0.48)^3}{48 \times (1.96 \times 10^{11}) \frac{\pi}{64} (0.009)^4}$$

or
$$\delta = 1.79 \times 10^{-3} \text{ m} = 0.179 \text{ cm}$$

Therefore natural frequency of transverse vibration,

$$\omega_n = \sqrt{\frac{g}{\delta}} = \sqrt{\frac{981}{0.179}} = 74.03 \text{ rad/s}$$

Also,
$$\omega = \frac{2\pi \times 675}{60} = 70.686 \text{ rad/s}$$

Therefore
$$r = \frac{\omega}{\omega_n} = \frac{70.686}{74.03} = 0.955$$

and
$$k = \frac{W}{\delta} = \frac{5 \times 9.81}{0.00179} = 27402.2 \text{ N/m}$$

For $c = 48$ N·s/m,

$$\zeta = \text{damping factor} = \frac{c}{2m\omega_n} = \frac{48}{2 \times 5 \times 74.03} = 0.0648$$

For amount of eccentricity $e = 3$ mm, from equation (17.105),

$$\frac{s}{3} = \frac{(0.955)^2}{\sqrt{(1 - 0.955^2)^2 + (2 \times 0.0648 \times 0.955)^2}}$$

or $\qquad s = 3 \times 6.006 = 18.02$ mm

The dynamic load on the bearing is due to the centrifugal force of the disc, which is given by

$$F_d = \sqrt{(sk)^2 + (cs\omega)^2} = s\sqrt{k^2 + (c\omega)^2}$$

or $\qquad F_d = 0.01802\sqrt{(27402.2)^2 + (48 \times 70.686)^2} = 497.56$ N

The total load on bearing = static load due to pull of gravity on rotor + F_d

Thus $\qquad F_{max} = 5 \times 9.81 + 497.56 = 546.6$ N

Maximum bending stress at the mid-span in the shaft due to a central load F,

$$\sigma_b = \frac{M}{I}y = \frac{(Fl/4)}{\pi/64(d)^4} \times \frac{d}{2}$$

or $\qquad \sigma_b = \frac{8Fl}{\pi d^3} = \frac{8 \times 0.48}{\pi(0.009)^3} \times F$

or $\qquad \sigma_b = (167.67 \times 10^4)F$ N/m^2 $\qquad\qquad$ (a)

Therefore maximum stress under dynamic condition (with $F = 546.6$ N), is

$$(\sigma_b)_{max} = 167.67 \times 10^4 \times 546.6$$

or $\qquad (\sigma_b)_{max} = 91.648 \times 10^7$ N/m^2 $\qquad\qquad$ **Ans.**

Maximum bending stress under dead load (from a)

$$(\sigma'_b)_{max} = 167.7 \times 10^4 \times 49.05 = 8.224 \times 10^7 \text{ N/m}^2 \qquad\qquad \textbf{Ans.}$$

Hence, $\qquad (\sigma_b)_{max}/(\sigma'_b)_{max} = 91.648/8.224 = 11.14$ $\qquad\qquad$ **Ans.**

Also damping force = $c\omega s = 48 \times 70.686 \times 0.01802 = 61.14$ N
Therefore, torque due to damping force,

$$T = s \times (c\omega s) = 61.14 \times 0.01802 = 1.102 \text{ N·m}$$

Therefore power required to drive the shaft at this speed,

$$P = \frac{2\pi NT}{60} = \frac{2\pi \times 675 \times 1.102}{60} = 77.895 \text{ W}$$

or $\qquad P = 77.9$ W $\qquad\qquad$ **Ans.**

Longitudinal, Transverse and Torsional Vibrations

A free vibration system of this type is shown in Fig. 17.32(a). It may be made to vibrate in one of the following three ways.

(i) All the particles of the system may vibrate along paths parallel to the axis of the shaft. Such vibrations are called *Longitudinal Vibrations*. [*see* Fig. 17.32(b)]

(ii) All the particles of the system may vibrate along paths perpendicular to the shaft axis. Such vibrations are called *Transverse Vibrations*. [*see* Fig. 17.32(c)]

(iii) All the particles of the system may vibrate along circular arcs having their centres along the axis of rotation. Such vibrations are called *Torsional Vibrations* [*see* Fig. 17.32(d)]

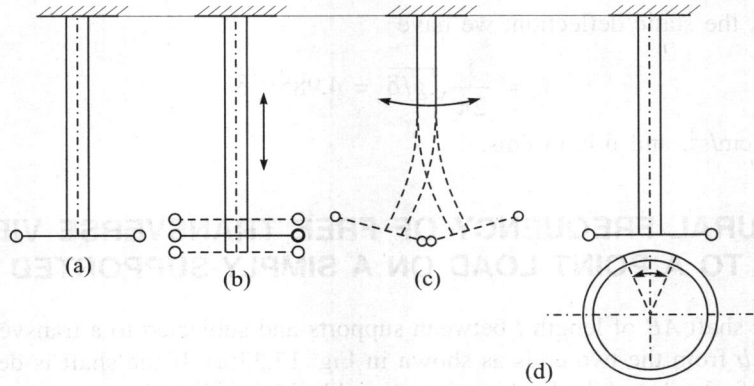

Fig. 17.32 Longitudinal, Transverse and Torsional Vibrations.

17.19 LONGITUDINAL AND TRANSVERSE VIBRATIONS

In Fig. 17.32(b) the shaft undergoes elongation and compression alternately, while in Fig. 17.32(c), the shaft assumes bent up shape on either sides. In both the cases, until limit of proportionality is not exceeded, the stress is directly proportional to strain. Hence the restoring force is proportional to the amount of displacement from mean equilibrium position. In other words, acceleration towards mean equilibrium position is directly proportional to the displacement therefrom and the vibratory motion is simple harmonic. Referring to Figs. 17.32(b) and (c), let W = weight of the flywheel, a = amplitude of vibration, k = stiffness of the shaft. In longitudinal vibration stiffness is the axial load required at the flywheel to produce unit axial deformation, while for transverse vibration it is the transverse force required at flywheel to produce unit lateral deflection at the flywheel.

Let f_n = frequency of vibrations in cycles per second.

Then, restoring force = k × deflection from mean equilibrium position.

But, as explained above, the restoring force is also equal to

$$= (W/g) \times \text{Acceleration}$$

Equating
$$(k \times \text{deflection}) = (W/g) \times \text{Acceleration}$$

or
$$\frac{\text{Acceleration}}{\text{Deflection}} = \frac{(gk)}{W}, \text{ which is constant.}$$

Again for simple harmonic motion with displacement amplitude X and frequency ω,

$$\text{Acceleration amplitude} = X\omega^2 \text{ and, displacement amplitude} = X$$

Hence

$$\frac{\text{Acceleration}}{\text{Displacement}} = \frac{X\omega^2}{X} = \frac{gk}{W} \quad \text{or} \quad \omega^2 = (gk/W)$$

or

$$\omega = \sqrt{gk/W}$$

and

$$f_n = \frac{1}{2\pi}\sqrt{gk/W} \tag{17.108}$$

and as $W/k = \delta$, the static deflection, we have

$$f_n = \frac{1}{2\pi}\sqrt{g/\delta} = 4.985/\sqrt{\delta} \tag{17.109}$$

where $g = 981$ cm/s^2, and δ is in cms.

17.20 NATURAL FREQUENCY OF FREE TRANSVERSE VIBRATIONS DUE TO A POINT LOAD ON A SIMPLY-SUPPORTED SHAFT

Consider a light shaft AB of length l between supports and subjected to a transverse load W at P distant a and b from the two ends as shown in Fig. 17.33(a). If the shaft is deflected under the application of a load, and the load is released suddenly, it will undergo transverse vibrations. The deflection of a shaft is proportional to the load acting on it. When the shaft is deflected beyond the static equilibrium position under load W, the load along with beam will vibrate with S.H.M. due to elastic properties of beam. Thus if δ be the static deflection under the load W, the natural frequency is given by

$$\omega_n = \sqrt{g/\delta}$$

or

$$f_n = \frac{1}{2\pi}\sqrt{g/\delta} = \frac{4.985}{\sqrt{\delta}} \text{ cycles/s} \tag{17.110}$$

where $g = 981$ cm/s^2 and δ = static deflection in cms.

Figures 17.33(a) through (h) show the values of static deflection for different types of beams and load conditions.

EXAMPLE 17.20 A shaft of length 0.8 m, supported freely at the ends, carries a load of mass 100 kg at 0.3 m from one end. Find the natural frequency of transverse vibration. Assume $E = 200$G N/m^2 and shaft diameter = 50 mm.

Solution: [see Fig. 17.33(a)]

Here,

$$a = 0.3 \text{ m}; \ b = 0.5 \text{ m}, \ l = 0.8 \text{ m}$$
$$W = (100 \times 9.81) \text{ N}; \ d = 0.05 \text{ m}$$

and

$$E = 200 \times 10^9 \text{ N/m}^2$$

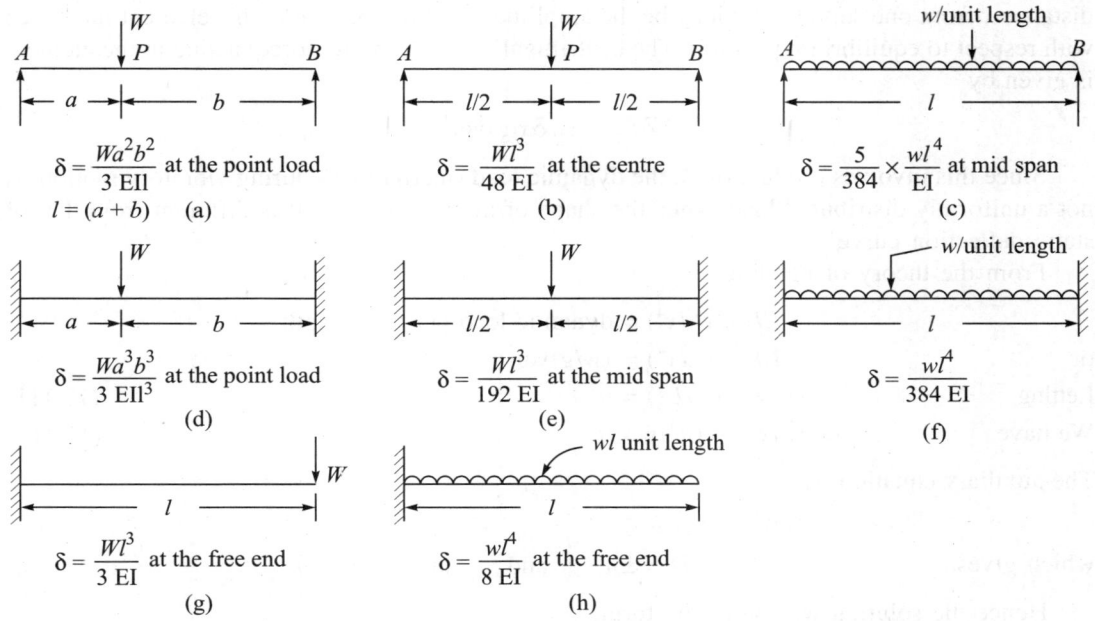

Fig. 17.33 Beam deflection under different load and End conditions.

Thus,
$$I = \frac{\pi}{64} d^4 = \frac{\pi}{64} (0.05)^4 = 3.068 \times 10^{-7} \text{ m}^4$$

and,
$$\delta = \frac{Wa^2b^2}{3EIl} = \frac{(100 \times 9.81)(0.3)^2(0.5)^2}{3(200 \times 10^9) \times (3.068 \times 10^{-7}) \times 0.8}$$

or,
$$\delta = 1.5 \times 10^{-4} \text{ m} = 1.5 \times 10^{-2} \text{ cm}$$

Hence natural frequency of transverse vibration

$$f_n = \frac{4.985}{\sqrt{\delta}} = \frac{4.985}{\sqrt{1.5 \times 10^{-2}}} = 40.7 \text{ c.p.s.} \qquad \textbf{Ans.}$$

17.21   TRANSVERSE VIBRATION OF A UNIFORMLY LOADED SHAFT

In contrast to the case of transverse vibration of a light shaft with a single concentrated load, which has a single natural frequency, a shaft with u.d.l. has theoretically infinite number of natural frequencies. *Our interest usually rests with that mode of vibration which corresponds to the lowest natural frequency. In this mode of vibration, all the particles of the shaft vibrate in phase and the corresponding frequency is called the fundamental frequency.* We discuss below what is popularly known as the *Exact Method*.

Referring to Fig. 17.34(a) let w be the load per unit length of the shaft and ω be the frequency of transverse vibration in radian per second. Consider an elemental length δx of the shaft at a distance x from one support and let y be the amplitude of displacement at this element measured with respect to equilibrium position. The centrifugal (*i.e.*, the inertia force) acting at the element is given by

$$\delta F_c = (w\delta x/g)y\omega^2$$

Since this involves product (wy), the dynamic load (inertia load) during vibratory motion, is not a uniformly distributed load. Thus the shape of the vibrating shaft is different from that of static deflection curve.

From the theory of bending,

$$EI(d^4y/dx^4) = \text{dynamic load per unit length}$$

or $$EI(d^4y/dx^4) = (w/g)y\omega^2$$

Letting $$(w/g)\,(\omega^2/EI) = n^4 \tag{17.111}$$

We have $$(d^4y/dx^4) - (n^4)y = 0 \tag{17.112}$$

The auxiliary equation is,

$$(D^4 - n^4)y = 0$$

which gives $$D = \pm n \quad \text{and} \quad \text{also, } D = \pm jn$$

Hence the solution will be of the form,

$$y = A\sin nx + B\cos nx + C\sinh nx + D\cosh nx \tag{17.113}$$

This is a general expression for the deflection of shafts with u.d.l. and constants A, B, C and D must be determined from end conditions.

Case(i) Simply Supported Ends

For simply supported beams, deflection $y = 0$ at $x = 0$ and l. Also, for simply supported end conditions $d^2y/dx^2 = 0$ at $x = 0$ and l.

Applying conditions $y = 0$ at $x = 0$ and $x = l$,

$$0 = A\sin(0) + B\cos(0) + C\sinh(0) + D\cosh(0)$$

or $$B + D = 0 \tag{17.114}$$

and $$0 = A\sin(nl) + B\cos(nl) + C\sinh(nl) + D_1\cosh(nl) \tag{17.115}$$

Differentiating equation (17.113) twice with respect to x,

$$\frac{d^2y}{dx^2} = -An^2\sin nx - Bn^2\cos nx + Cn^2\sinh nx + Dn^2\cosh nx$$

Applying the condition that $d^2y/dx^2 = 0$ at $x = 0$

$$0 = -Bn^2 + Dn^2 \text{ or } (B - D)n^2 = 0 \tag{17.116}$$

and applying the condition, $d^2y/dx^2 = 0$ at $x = l$,

$$0 = r^2(-A\sin nl - B\cos nl + C\sinh nl + D\cosh nl) \tag{17.117}$$

From equations (17.114) and (17.116),

$$B = 0 \quad \text{and} \quad D = 0$$

Equations (17.115) and (17.117) therefore reduce to,

$$A\sin(nl) + C\sinh(nl) = 0 \text{ and, } n^2(-A\sin nl + C\sinh nl) = 0$$

Since $n^2 \neq 0$, the above two equations reduce to

$$A\sin nl + C\sinh nl = 0$$

and

$$-A\sin nl + C\sinh nl = 0$$

Adding the two,

$$C\sinh(nl) = 0$$

But $\sin h(nl)$ cannot be equal to zero.

Hence $$C = 0$$

and therefore $$A\sin nl = 0$$

But A cannot be equal to zero, else y is always zero and no vibratory motion is possible.

Thus $$\sin(nl) = 0 \tag{17.118}$$

Hence $$(nl) = \pi, 2\pi, 3\pi, ..., \text{etc.}$$

Thus $$n = (\pi/l); (2\pi/l); (3\pi/l), \text{etc} \tag{17.119}$$

Again, from equation (17.111),

$$n = \left(\frac{w}{g}\frac{\omega^2}{EI}\right)^{1/4} \quad \text{or} \quad \omega^2 = n^2\sqrt{gEI/w} \tag{17.120}$$

Hence, taking lowest value of n,

$$\left(\frac{w}{g}\frac{\omega^2}{EI}\right)^{1/4} = (\pi/l)$$

Similarly, for the three lowest values of n

$$\sqrt{\omega} = \left(\frac{EIg}{w}\right)^{1/4}\frac{\pi}{l}; \left(\frac{EIg}{w}\right)^{1/4}\frac{2\pi}{l}; \left(\frac{EIg}{w}\right)^{1/4}\frac{3\pi}{l}; ..., \text{etc.}$$

or

$$\omega = \left(\frac{EIg}{w}\right)^{1/2}\frac{\pi^2}{l^2}; \left(\frac{EIg}{w}\right)^{1/2}\left(\frac{4\pi^2}{l^2}\right); \left(\frac{EIg}{w}\right)^{1/2}\left(\frac{9\pi^2}{l^2}\right); ... \tag{17.121}$$

Considering the smallest value, and rearranging

$$\omega = \left(\frac{EIg}{wl^4}\right)^{1/2}\pi^2 \tag{17.122}$$

and also,

$$\omega = \left(\frac{EIg}{w}\right)^{1/2}\left(\frac{\pi}{l}\right)^2$$

But for uniformly distributed load on simply supported ends, deflection at mid-span,

$$\delta = \frac{5}{384} \times \frac{wl^4}{EI}$$

Substituting for $\left(\dfrac{EI}{wl^4}\right) = \left(\dfrac{5}{384}\right)\Big/\delta$, in equation (17.122)

$$\omega = \left(\frac{5}{384}\frac{g}{\delta}\right)^{1/2} \pi^2 = 35.27/\sqrt{\delta}$$

Therefore

$$f_n = \frac{35.27}{2\pi}\sqrt{\frac{1}{\delta}} = 5.613/\sqrt{\delta}$$

This is the lowest natural frequency of transverse vibration and is called *fundamental frequency*. As shown in Fig. 17.34(b) it has a node at each end. Also, it follows from equation (17.120) that next higher natural frequency is four times the fundamental frequency and corresponding mode shape [*see* Fig. 17.34(c)] has two nodes at the ends and an another node at the mid-span.

Next higher natural frequency [*see* equation (17.120)] is nine times the fundamental frequency and corresponding mode shape [*see* Fig. 17.34(d)] has two nodes at the ends and a node each at a distance of $l/3$ from each end. (Readers may note that as ω is proportional to n^2, the other frequencies are 2^2, 3^2, 4^2, etc times the fundamental frequency).

The solution for differential equations with other end conditions, namely, fixed-fixed support and fixed-free support can be derived in a similar way. The results are stated below for reference.

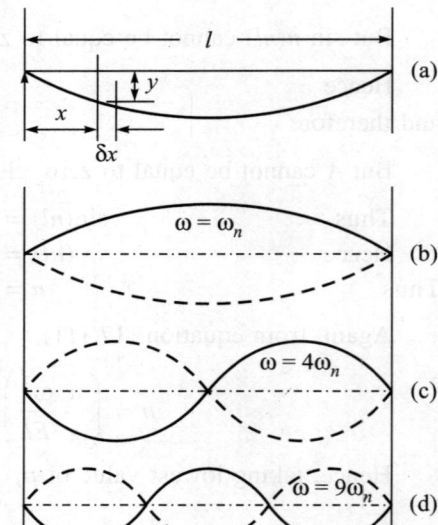

Fig. 17.34

Case (ii) Fixed-End Beam with U.D.L.

End conditions: (a) $y = 0$ at $x = 0$
(b) $y = 0$ at $x = l$
(c) $dy/dx = 0$ at $x = 0$
(d) $dy/dx = 0$ at $x = l$

$$\delta = \frac{wl^4}{384EI} \text{ at mid span}$$

The numerical values of $(nl)^2$ for first, second and third modes (*see* equation 17.118) are:

Fundamental mode: $(nl)^2 = 22.4$
Second mode: $(nl)^2 = 61.7$
Third mode: $(nl)^2 = 121.0$

Case (iii) Cantilever Beam with U.D.L.

The end conditions are:

  (i) $y = 0$ at $x = 0$
  (ii) $dy/dx = 0$ at $x = 0$
  (iii) $d^2y/dx^2 = 0$ at $x = l$
  (iv) $d^3y/dx^3 = $ shear force $= 0$ at $x = l$

$$\delta = \frac{wl^4}{8EI}, \text{ at the free end.}$$

The values of $(nl)^2$ for first, second and third modes (see equation 17.118):

  Fundamental mode: $(nl)^2 = 3.52$
  Second mode: $(nl)^2 = 22.4$
  Third mode: $(nl)^2 = 61.7$

17.22 TRANSVERSE VIBRATIONS OF SHAFT CARRYING SEVERAL LOADS

Dunkerley's Method

This semi-empirical method was suggested by Professor Dunkerley. The method is applicable only to a shaft of uniform diameter or where it can be replaced by a shaft of uniform diameter without sacrificing accuracy to any appreciable extent.

Let W_1, W_2, W_3, ..., etc. be the concentrated loads at stations 1, 2, 3, ... respectively, acting on the shaft. Let δ_1, δ_2, δ_3, ..., etc. be the static deflections of the shaft under respective load when acting alone on the shaft. Further, let w be the uniformly distributed load per unit length due to self weight and let δ_s be the maximum deflection of the shaft under its own weight.

Then, if $f_{n1}, f_{n2}, f_{n3}, ...$ be the frequency of transverse vibrations with respective point load acting alone on shaft and f_s be the frequency of transverse vibrations of the shaft under its self-weight (unaccompanied by any other point load), the frequency of transverse vibration (f_n) of the whole system is given by:

$$\left(\frac{1}{f_n}\right)^2 = \left(\frac{1}{f_{n1}}\right)^2 + \left(\frac{1}{f_{n2}}\right)^2 + \left(\frac{1}{f_{n3}}\right)^2 + ... + \left(\frac{1}{f_s}\right)^2 \qquad (17.123)$$

When point loads W_1, W_2, W_3 are acting alone, corresponding natural frequencies for simply supported ends are given using equation (17.110) as

$$f_{n1} = \frac{4.985}{\sqrt{\delta_1}}; f_{n2} = \frac{4.985}{\sqrt{\delta_{n2}}}; f_{n3} = \frac{4.985}{\sqrt{\delta_3}}, \text{ etc}$$

where, δ_1, δ_2, δ_3, ..., and δs, are in centimetre.

Also

$$f_s = \frac{5.613}{\sqrt{\delta_s}}$$

Substituting in equation (17.123),

$$\left(\frac{1}{f_n^2}\right) = \frac{1}{(4.985)^2}\{\delta_1 + \delta_2 + \delta_3 + \ldots\} + \frac{\delta_s}{(5.613)^2}$$

or

$$\frac{1}{f_n^2} = \frac{1}{(4.985)^2}\left\{\delta_1 + \delta_2 + \delta_3 + \ldots + \frac{\delta_s}{1.268}\right\}$$

or

$$f_n = \frac{(4.985)}{\sqrt{\delta_1 + \delta_2 + \delta_3 + \ldots + (\delta_s/1.268)}} \qquad (17.124)$$

where δ_1, δ_2, δ_3, ..., and δ_s, are in centimetres.

Energy Method (Rayleigh's Method)

The method is based on the assumption that calculated frequency is only slightly affected by making different assumptions for the shape of the deflected shaft during vibration, provided that the assumed shape is consistent with the end conditions. Hence the shape of deflection curve of the vibrating shaft is assumed to be similar to static deflection curve for an identical system of loading.

For a shaft with negligible mass, let m_1, m_2, m_3, ... be the masses producing point loads W_1, W_2, W_3, ... at stations 1, 2, 3, ... etc. along the shaft. Let y_1, y_2, y_3, ... be the total deflections of the shaft under respective loads when all the loads are acting simultaneously on shaft. During vibrations, which are assumed to have simple harmonic motion, maximum potential energy is gained at the extremum position. Thus,

$$\text{Maximum potential energy} = \frac{1}{2}W_1y_1 + \frac{1}{2}W_2y_2 + \frac{1}{2}W_3y_3 + \ldots$$

or

$$(U)_{\max} = (m_1y_1 + m_2y_2 + m_3y_3 + \ldots)\frac{g}{2} = \left(\frac{g}{2}\right)\Sigma my$$

Vibrating at the natural frequency ω, maximum velocity of a mass m having a maximum amplitude of displacement as y, is given by $(y\omega)$. Thus, when all the masses occupy their mean equilibrium position, the maximum kinetic energy during vibrations is given by,

$$T_{\max} = \frac{1}{2}m_1v_1^2 + \frac{1}{2}m_2v_2^2 + \frac{1}{2}m_3v_3^2 + \ldots$$

or

$$T_{\max} = \frac{1}{2}m_1(y_1\omega)^2 + \frac{1}{2}m_2(y_2\omega)^2 + \frac{1}{2}m_3(y_3\omega)^2 + \ldots$$

Thus maximum K.E. $= \frac{1}{2}(m_1y_1^2 + m_2y_2^2 + m_3y_3^2 + \ldots)\omega^2$

Hence, using Rayleigh's principle, namely,

$$(U)_{\max} = T_{\max}$$

we have

$$\left(\frac{g}{2}\right)\Sigma my = \frac{1}{2}(\Sigma my^2)\omega^2 \text{ or } \omega^2 = \frac{\Sigma my}{\Sigma my^2} \qquad (17.125)$$

EXAMPLE 17.21 For the system shown in Fig. 17.32(a), the weight of the flywheel is 2670 N, the radius of gyration is 38 cm, the shaft is 76 mm diameter and 0.92 m long to the flywheel boss. For the shaft material, young's modulus is 8.0×10^6 N/cm^2. Find the frequencies of the free longitudinal and transverse vibrations.

Solution: (a) Longitudinal vibrations:
Static elongation in the shaft under load $W = 2670$ N

$$\delta = \frac{(W/A)}{(E/l)} = \frac{2670 \times 92}{(\pi/4)\,(7.6)^2 \times 8.0 \times 10^6} = 0.000677 \text{ cm}$$

Therefore, frequency of longitudinal vibrations, from equation (17.109)

$$f_n = 4.985/\sqrt{\delta}$$

or $\qquad\qquad f_n = 4.985\sqrt{\dfrac{1}{0.000677}} = 191.59 \text{ c.p.s.} = 11{,}495 \text{ c.p.m.}$

(b) Transverse vibrations:
Static deflection at the free end of cantilever,

$$\delta = \frac{Wl^3}{3EI}$$

where $\qquad\qquad I = \dfrac{\pi}{64}\,d^4 = \dfrac{\pi}{64}\,(7.6)^4 = 163.77 \text{ cm}^4$

Therefore $\qquad\qquad \delta = \dfrac{2670 \times (92)^3}{3 \times 8.0 \times 10^6 \times 163.77} = 0.529 \text{ cm}$

Hence frequency of transverse vibration

$$f_n = 4.985\sqrt{\frac{1}{\delta}} = \frac{4.985}{\sqrt{0.529}} = 6.854 \text{ c.p.s.} = 411 \text{ c.p.m.} \qquad \textbf{Ans.}$$

EXAMPLE 17.22 A beam of I-section has a span of 3 m and is supported at the ends. The mass of the beam is 200 kg/m and the second moment of area of the section is 16×10^{-6} m^4. Two equal loads of one tonne are carried at points 1 m from each support. Find the natural frequency of transverse vibration of the system if $E = 200$ GN/m^2. (I. Mech. E.)

Solution: Due to symmetry (Fig. 17.35) of point loads of 1 tonne ($\equiv 1000$ kgf) on the beam, static deflection by each load, when acting alone, measured under the load itself is,

$$\delta_1 = \delta_2 = \frac{Wa^2b^2}{3EIl}$$

where $E = 200 \times 10^9$ N/m^2 and $W = 1000 \times 9.81$ N.

Therefore $\qquad\qquad \delta_1 = \delta_2 = \dfrac{(9.81 \times 1000) \times 1^2 \times 2^2}{3(200 \times 10^9)\,(16 \times 10^{-6})(3)}$

or $\qquad\qquad \delta_1 = \delta_2 = 0.001363 \text{ m}$

Also, maximum deflection of beam under the given u.d.l. (self weight) alone,

$$\delta_s = \frac{5}{384} \frac{wl^4}{EI} = \frac{5 \times (200 \times 9.81) \times 3^4}{384(200 \times 10^9)(16 \times 10^{-6})} = 0.0006466 \text{ m}$$

Using Dunkerley's method, equation (17.24) gives

$$f_n = \frac{4.985}{\sqrt{\delta_1 + \delta_2 + (\delta_s/1.268)}}$$

or,

$$f_n = \frac{4.985}{\sqrt{\left\{(0.001363) + (0.001363) + \left(\dfrac{0.0006466}{1.268}\right)\right\} \times 100}}$$

or

$$f_n = \frac{4.985}{0.56885} = 8.76 \text{ c.p.s. (Hz)}$$ **Ans.**

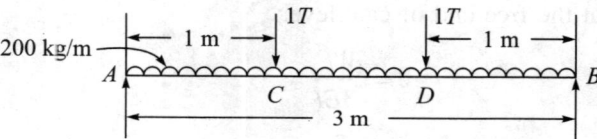

Fig. 17.35   Load diagram on steel shaft.

EXAMPLE 17.23   Determine, with the aid of Dunkerley's equation, (a) the first and (b) the second whirling speed of a steel shaft, 25 mm diameter and 0.75 m long, freely supported at both ends, carrying a concentrated mass of 20 kg at the shaft-centre. The density of the shaft material is 7.8 Mg/m³ and the value of E is 200 GN/m². (Univ. London)

Solution: The second moment of area of the shaft,

$$I = \frac{\pi}{64}(0.025)^4 = (1.917 \times 10^{-8})\text{m}^4$$

$$w = \frac{\pi}{4} \times (0.025)^2 \times 7.8 \times 10^3 \times 9.81 = 37.56 \text{ N/m}$$

Therefore static deflection under the central point load,

$$\delta = \frac{Wl^3}{48EI} = \frac{(20 \times 9.81)(0.75)^3}{48 \times (200 \times 10^9) \times (1.917 \times 10^{-8})}$$

or

$$\delta = 4.4977 \times 10^{-4} \text{ m}$$

Again, maximum deflection due to self weight

$$\delta_s = \frac{5}{384} \frac{wl^4}{EI} = \frac{5}{384} \times \frac{37.56 \times (0.75)^4}{(200 \times 10^9) \times (1.917 \times 10^{-8})}$$

or

$$\delta_s = 4.036 \times 10^{-5} \text{ m}$$

Hence, using Dunkerley's method,

$$f_n = \frac{4.985}{\sqrt{\left(4.497 \times 10^{-4} + \dfrac{4.036 \times 10^{-5}}{1.268}\right) \times 100}} = 22.72 \text{ c.p.s.}$$

The first mode of transverse vibration corresponding to the first whirling speed of 22.72 Hz is shown in Fig. 17.36(b).

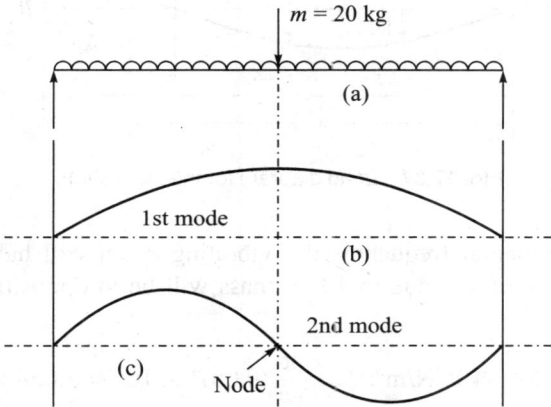

Fig. 17.36 Mode shapes of steel shaft.

For the second whirling speed, the mode shape of shaft centre line is as at Fig. 17.36(c). The central point load actually acts at the node point and therefore, does not contribute anything to the second whirling speed. Thus only self weight of the shaft is to be considered. A virtual point support may be assumed at node and u.d.l. over half the shaft length and maximum deflection due to it needs to be considered. Thus

$$\delta_{\max} = \frac{5}{384} \times \frac{37.56 \times (0.75/2)^4}{(200 \times 10^9) \times (1.917 \times 10^{-8})}$$

or,

$$\delta_{\max} = \left(\frac{1}{2}\right)^4 \times \delta_s \equiv \frac{4.036 \times 10^{-5}}{16} = (0.25225 \times 10^{-5})\text{m}$$

Therefore using Dunkerley's method,

$$f_n = \frac{4.985}{\sqrt{\dfrac{(0.25225 \times 10^{-5})}{1.268} \times 100}} = \frac{4.985}{0.0141} = 353.4 \text{ Hz}$$ **Ans.**

EXAMPLE 17.24 A steel shaft AB of 50 mm diameter and 2.5 m length, is supported in two short bearings 1.8 m apart, one being at the end A of the shaft. The shaft carries three concentrated loads with masses as indicated in Fig. 17.37.

Mass of load in kg:		80	160	40
Distance from A in m:		0.6	1.2	2.4

Calculate the deflection at each load and hence obtain a first approximation to the fundamental frequency of transverse vibration of the loaded shaft. Neglect the mass of the shaft. Take $E = 200$ GN/m^2. (Univ. London)

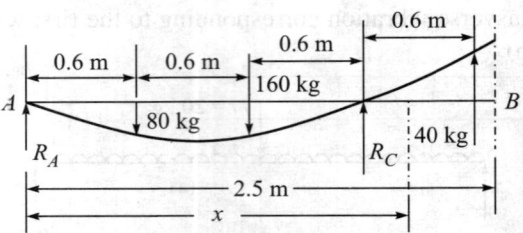

Fig. 17.37   Load diagram for the steel shaft.

Solution: At the fundamental frequency, the vibrating beam will have a form as shown in Fig. 17.37. Thus the inertia force due to 40 kg mass will be in opposition to the inertia forces of the other two masses.

$$E = 200 \times 10^9 \text{ N/m}^2; \ I = \frac{\pi}{64}(0.05)^4 \times 10^9 = 3.068 \times 10^{-7} \text{ m}^4$$

Therefore $EI = 200 \times 10^9 \times 3.068 \times 10^{-7} = 61,360$ N·m^2

Taking moments about A and using $\Sigma m_A = 0$,

$$R_c = \frac{(80 \times 0.6 + 160 \times 1.2 - 40 \times 2.4)9.81}{1.8}$$

or

$$R_c = 80 \times 9.81 \text{ N} = 784.8 \text{ N}$$

And from $\Sigma F_v = 0$,

$$R_A = (80 + 160 - 80 - 40) \times 9.81$$

or

$$R_A = 120 \times 9.81 = 1177.2 \text{ N}$$

Using Macaulay's method for deflection y at distance x,

$$EI\left(\frac{d^2y}{dx^2}\right) = 9.81\{-120x + 80(x - 0.6) + 160(x - 1.2) - 80(x - 1.8)\}$$

Integrating and applying boundary conditions, namely,

$y = 0$ at $x = 0$ and
$y = 0$ at $x = 1.8$ m,

$$y = \frac{20 \times 9.81}{61,360}\left\{-x^3 + \frac{2}{3}(x - 0.6)^3 + \frac{4}{3}(x - 1.2)^3 - \frac{2}{3}(x - 1.8)^3 + 2.44x\right\}$$

Thus, the deflection y at different mass locations are:

x m, from A:	0.6	1.2	2.4
y in m:	0.004	0.0044	-0.00616

m (kg)	y (m)	my	my^2
80	0.004	0.32	0.00128
160	0.0044	0.704	0.003098
40	-0.00616	0.2464	0.001518
Σmy	1.2704	and	$\Sigma my^2 = 0.005896$

Therefore from Rayleigh's method, equation (17.125),

$$\omega^2 = \frac{g\Sigma my}{\Sigma my^2} = \frac{9.81 \times 1.2704}{0.005896} = 2113.74$$

Therefore
$$\omega = 45.975 \text{ rad/s}$$

Therefore
$$f_n = \frac{\omega}{2\pi} = 7.32 \text{ c.p.s} \qquad \textbf{Ans.}$$

Note: No cognisance is taken of negative deflection under mass 40 kg near end B. This is because in numerator (*i.e.*, Σmy), both the load and deflection y are negative, giving positive product. Also, in denominator, as the terms correspond to K.E. and while passing through mean equilibrium position, all component K.E. must be added.

TORSIONAL VIBRATIONS

17.23 SINGLE ROTOR SYSTEM

Figure 17.38(a) represents a simple system for torsional vibrations, in which a rotor of mass moment of inertia I is attached to the free end of the shaft of polar moment of inertia J and length l. According to torsion formula, within elastic limits,

$$\frac{T}{J} = \frac{G\theta}{l} \quad \text{or,} \quad \left(\frac{T}{\theta}\right) = \frac{(GJ)}{l}$$

The torque required per unit angle of twist is called *torsional stiffness* and is given by

$$k_t \equiv \left(\frac{T}{\theta}\right) = \left(\frac{GJ}{l}\right) \tag{17.126}$$

Thus the restoring torque due to elastic property of shaft material $= k_t\theta$

Thus, restoring torque
$$= \left(\frac{GJ}{l}\right)\theta$$

Now, if the rotor is given an angular displacement of θ and then released, the angular acceleration α of the disc is given by,

$$I\alpha = \left(\frac{GJ}{l}\right)\theta, \quad \text{or} \quad \left(\frac{\alpha}{\theta}\right) = \left(\frac{GJ}{Il}\right) = \text{constant for the shaft.}$$

Thus the motion is simple harmonic and therefore angular acceleration α is given in terms of displacement θ and natural frequency ω_n as,

$$\alpha = \theta\omega_n^2 \quad \text{or} \quad \omega_n = \sqrt{\frac{\alpha}{\theta}}$$

or
$$\omega_n = \sqrt{\frac{\alpha}{\theta}} = \sqrt{\left(\frac{GJ}{Il}\right)} \tag{17.127}$$

or
$$f_n = \frac{1}{2\pi}\sqrt{\frac{(GJ)}{(Il)}} \quad \text{c.p.s.}$$

But, if the rotor is attached to the shaft between the two fixed ends, with the rotor dividing the shaft length l in two segments of length and diameters l_1, d_1 and l_2, d_2 respectively, then the torque T on the rotor is resisted by torques T_1 and T_2 in the two segments such that

$$T = T_1 + T_2 \tag{17.128}$$

At the same time, the angle of twist θ at the rotor is common to both the segments and corresponding torsional springs are in parallel. Hence, combining equations (17.126) and (17.128),

$$T = G\theta(J_1/l_1 + J_2/l_2)$$

Thus, equating accelerating torque to restoring torque, we have

$$I\alpha = T = G\theta(J_1/l_1 + J_2/l_2) \quad \text{or} \quad (\alpha/\theta) = (G/I)(J_1/l_1 + J_2/l_2)$$

Therefore
$$\omega_n = \sqrt{\{(G/I)\}(J_1/l_1 + J_2/l_2)} \quad \text{c.p.s.} \tag{17.129}$$

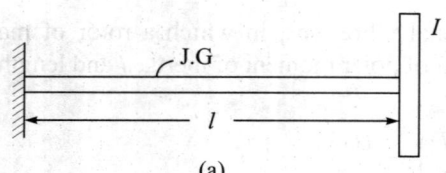 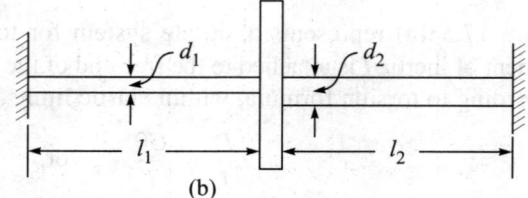

Fig. 17.38 Torsional vibration of shaft with single rotor.

17.24 FREE TORSIONAL VIBRATIONS: TWO ROTOR SYSTEM

When two heavy rotors A and B are mounted at the free ends of a uniform shaft and are allowed to have torsional vibrations, two types of distinct motions are possible.

(i) In the first case, both the rotors always move in the same direction (with same amplitude). Since the relative motion between the rotors is zero, the shaft is not subjected to twisting and no resisting torque is developed. Clearly, vibrations are not possible in this case and $\omega_n = 0$. This condition is shown in Fig. 17.39(b).

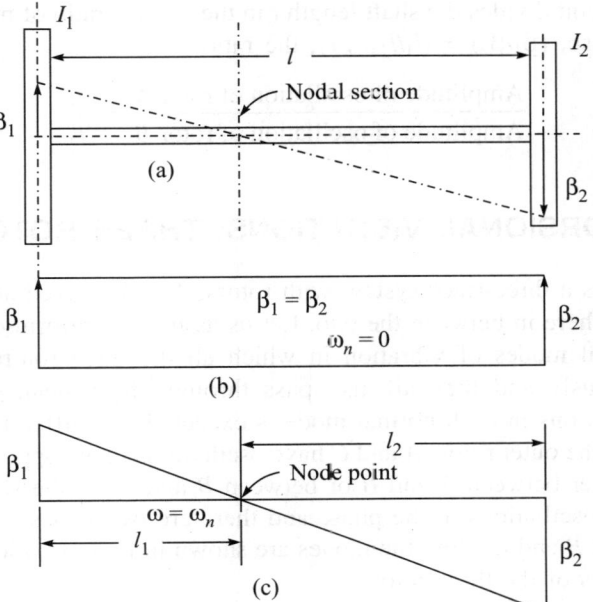

Fig. 17.39 Free torsional vibration with two end rotors.

(ii) In the second case, the two rotors always move in mutually opposite directions and pass through mean equilibrium position simultaneously. The rotors also approach extremum positions simultaneously. Thus the segmental lengths l_1 and l_2 in Fig. 17.39(c) are twisted in opposite directions. There is a section in the shaft which does not undergo any twist and is called *nodal-section*. The corresponding point on the shaft axis is called a *node*. The shafts behaves as if it is clamped at the nodal section and the two segments of the shaft vibrate with same natural frequency like two independent torsional pendulums of lengths l_1 and l_2.

Then, for equal natural frequency of torsional vibrations ω_{n1} and ω_{n2} for the two torsional pendulums, we have,

$$\omega_{n1} = \omega_{n2}$$

or

$$\sqrt{\frac{GJ_1}{I_1 l_1}} = \sqrt{\frac{GJ_2}{I_2 l_2}}$$

If shaft is of uniform diameter,

$$J_1 = J_2 = J$$

Then

$$\sqrt{\frac{GJ}{I_1 l_1}} = \sqrt{\frac{GJ}{I_2 l_2}} \qquad \text{or} \qquad I_1 l_1 = I_2 l_2 \qquad (17.130)$$

or

$$\left(\frac{I_1}{I_2}\right) = \left(\frac{l_2}{l_1}\right)$$

Thus, the node point divides the shaft length l in the inverse ratio of mass moment of inertia of the two rotors. Also, $(\beta_1/\beta_2) = (l_1/l_2)$, *i.e.*, the ratio

$$\frac{\text{Amplitude of oscillation of rotor } A}{\text{Amplitude of oscillation of rotor } B} = \frac{l_1}{l_2}$$

17.25 FREE TORSIONAL VIBRATIONS: THREE ROTORS

Figure 17.40(a) shows a three-rotor system with rotors. A and C fixed at the ends to the shaft and rotor B is somewhere in between the two. Let us assume a uniform diameter for the shaft. There are two normal modes of vibration in which all the three rotors reach their extreme positions simultaneously and they all also pass through their mean equilibrium positions simultaneously. Vibrations in each normal mode is executed at a different natural frequency.

In the first mode the outer rotors A and C have oscillations in the opposite phase and a single node point exists either between A and B or between B and C. In the second mode, the outer rotors A and C have oscillations in the phase and there are two nodes—one between A and B and the other between B and C. Both the modes are shown in Figs. 17.40(b) and (c), along with the relative amplitudes of the three rotors.

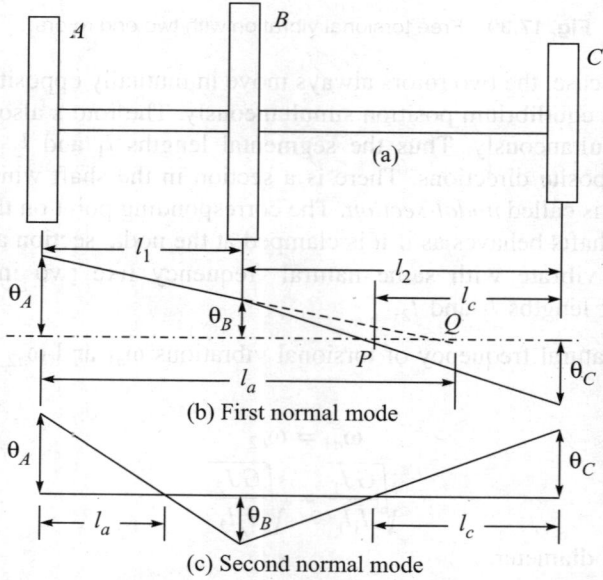

Fig. 17.40 Free Torsional Vibrations with 3 Rotors.

Assuming that both the nodes lie between B and C in Fig. 17.40(b), let l_a be the distance of one node from A and l_c be the distance of the other node from C in both Figs. 17.40(b) and (c). Further let I_a, I_b and I_c be the mass moment of inertia of the three rotors.

Then, with Q as a virtual node and P as the actual node in Fig. 17.40(b), the natural frequencies of rotors A and C using equation (17.127) for both the nodes are,

$$f_a = \frac{1}{2\pi} \sqrt{\frac{k_t}{I_a}} = \frac{1}{2\pi} \sqrt{\frac{GJ}{l_a I_a}} \qquad (17.131)$$

and

$$f_c = \frac{1}{2\pi} \sqrt{\frac{k_t'}{I_c}} = \frac{1}{2\pi} \sqrt{\frac{GJ}{l_c I_c}} \qquad (17.132)$$

For the middle rotor B,

$$f_b = \frac{1}{2\pi} \sqrt{\frac{k_t''}{I_b}} \qquad (17.133)$$

where, the torsional stiffness k_t'' represents torque required to twist rotor B through 1 radian, when the shaft is assumed to be fixed at nodes and is given by the sum of the torques required to produce unit radian of angle of twist (*i.e.*, torsional stiffnesses k_{t_1} and k_{t_2}) over lengths $(l_1 - l_a)$ and $(l_2 - l_c)$. Therefore, from equation (17.133),

$$f_b = \frac{1}{2\pi} \sqrt{\frac{GJ}{I_b} \left(\frac{1}{l_1 - l_a} + \frac{1}{l_2 - l_c} \right)} \qquad (17.134)$$

Again, when the entire shaft along with the rotors vibrate in a particular mode of vibration, each rotor has the same frequency of oscillation. And therefore, from $f_a = f_b = f_c$, we have

$$\frac{1}{2\pi} \sqrt{\frac{GJ}{I_a l_a}} = \frac{1}{2\pi} \sqrt{\frac{GJ}{I_c l_c}}$$

which gives

$$I_a l_a = I_c l_c \qquad (17.135)$$

Also, from

$$f_a = f_b,$$

$$\frac{1}{2\pi} \sqrt{\frac{GJ}{I_a l_a}} = \frac{1}{2\pi} \sqrt{\frac{GJ}{I_b} \left\{ \frac{1}{(l_1 - l_a)} + \frac{1}{(l_2 - l_c)} \right\}}$$

we have

$$\frac{1}{I_a l_a} = \frac{1}{I_b} \left\{ \frac{1}{l_1 - l_a} + \frac{1}{l_2 - l_c} \right\} \qquad (17.136)$$

Equations (17.135) and (17.136) enable a quadratic to be obtained in either l_a or l_c. The roots of this quadratic gives two values of l_a or l_c for one node [*see* Fig. 17.40(b)] and two nodes [*see* Fig. 17.40(c)]. The actual values of the frequencies are obtained by substituting the two values of l_a in the equation for f_a. It may be noted that only one of these two values of l_a may give the position of a real node, while the other gives the virtual node, obtained by extending line between A and B to cut the axis of the shaft.

It is thus seen that in a 2-rotor system, only one natural frequency could be obtained, while in 3-rotor system there are only two natural frequencies. Thus, in general, number of natural frequencies of a torsional vibration system is one less than the number of rotors.

Amplitude of Vibration

Let the amplitude of torsional oscillation of rotor A be unity. Then, using deflection lines in Figs. 17.40(b) and (c),

$$\frac{\theta_A}{\theta_B} = \frac{l_a}{(l_a - l_1)}$$

Thus,

$$\theta_B = \frac{(l_a - l_1)}{l_a}, \quad (\text{as, } \theta_A = 1) \tag{17.137}$$

Also

$$(\theta_B/\theta_C) = \frac{(l_2 - l_c)}{l_c}; \quad \text{Therefore} \quad \theta_C = \theta_B \frac{l_c}{(l_2 - l_c)}$$

and, substituting for θ_B from above,

$$\theta_C = \frac{l_c}{(l_2 - l_c)} \frac{(l_a - l_1)}{l_a} \tag{17.138}$$

Torsionally Equivalent Shaft

In contrast to the assumption made for free vibrations of two and three rotor system in Sections 17.24 and 17.25, usually the shafts do not have uniform diameter. Before proceeding for solution of two and three rotor problems, it is important that stepped shafts (*i.e.*, non-uniform shaft) be first converted into a torsionally equivalent uniform shaft of some convenient diameter. This is discussed in detail in Section 17.8. A few solved examples will make the matter more clear.

EXAMPLE 17.25 A steel shaft $ABCD$ of 1.5 m length has flywheels at its ends, A and D. The mass of the flywheel A is 5886 N and has a radius of gyration of 0.6 m. The mass of the flywheel D is 7848 N and has a radius of gyration of 0.9 m. The connecting shaft AB is 0.4 m long and has a diameter of 50 mm, a diameter of 60 mm for the portion BC which is 0.5 m long and a diameter of d mm for the portion CD which is 0.6 m long. Determine:

   (a)   the diameter d of portion CD so that the node of the torsional vibration of the system will be at the centre of the length BC
   (b)   the natural frequency of the torsional vibrations. The modulus of rigidity for the material of the shaft is 0.785×10^{11} N/m^2 (AMIE, Winter 1980)

Solution: $I_a = \dfrac{5886}{9.81} \times (0.6)^2 = 216$ kg·m^2

and $\qquad\qquad\qquad I_d = \dfrac{7848}{9.81} \times (0.9)^2 = 648$ kg·m^2

Referring to Fig. 17.41(a), let us find out length of equivalent shaft having uniform diameter of 50 mm. Thus, the equivalent length is

$$l = l_1 + l_2\left(\frac{d_1}{d_2}\right)^4 + l_3\left(\frac{d_1}{d_3}\right)^4$$

or
$$l = 0.4 + 0.5\left(\frac{50}{60}\right)^4 + 0.6\left(\frac{50}{d}\right)^4$$

or
$$l = 0.641 + \frac{375 \times 10^4}{d^4}$$

(a) Figure 17.41(b) shows rotors with torsionally equivalent shaft of diameter $d_1 = 50$ mm. The length l is assumed to be unknown to begin with. However, as is clear from expression for l in terms of d^4 above, l will be much smaller than the actual length of shaft.

Let us assume that the node N of the equivalent system lies at a distance l_a from A and at a distance l_d from D' [see Figs. 17.41(b) and (c)]. Since this position corresponds to a mode shape having a fixed natural frequency,

$$I_a l_a = I_d l_d$$

Therefore
$$l_a = (I_d/I_a)l_d$$

or
$$l_a = \left(\frac{648}{216}\right)l_d = 3l_d \qquad \text{(ii)}$$

It is required in the problem that the node of the torsional vibration of the system is to lie at the centre of the length BC of original stepped shaft. Hence, its location in the equivalent shaft from A,

$$l_a = l_1 + \frac{l_2}{2}\left(\frac{d_1}{d_2}\right)^4 = 0.4 + 0.25\left(\frac{50}{60}\right)^4 = 0.4 + 0.1206 = 0.5206 \text{ m}$$

Hence from (ii),
$$3l_d = 0.5206$$

Therefore
$$l_d = 0.1735 \text{ m}$$

and
$$l = l_a + l_d = 4l_d = 0.6940 \text{ m} \qquad \text{(iii)}$$

Hence, from (i) and (iii),

$$l = 0.694 = 0.641 + \frac{375 \times 10^4}{d^4}$$

or
$$\frac{1}{d^4} = \frac{(0.694 - 0.641)}{375 \times 10^4}$$

or
$$d = (375 \times 10^4/0.053)^{1/4} = 91.7 \text{ mm} \qquad \textbf{Ans.}$$

(b) *Natural frequency of torsional vibrations*

From Figs. 17.41(b) and (c), $l_a = 3l_d = 0.5206$ m, we have

Therefore
$$f_A = \frac{1}{2\pi}\sqrt{\frac{GJ}{I_a l_a}}$$

or
$$f_A = \frac{1}{2\pi} \sqrt{\frac{(0.785 \times 10^{11})\pi\,(0.05)^4}{(216)\,(0.52)32}} = 3.295 \text{ c.p.s.}$$

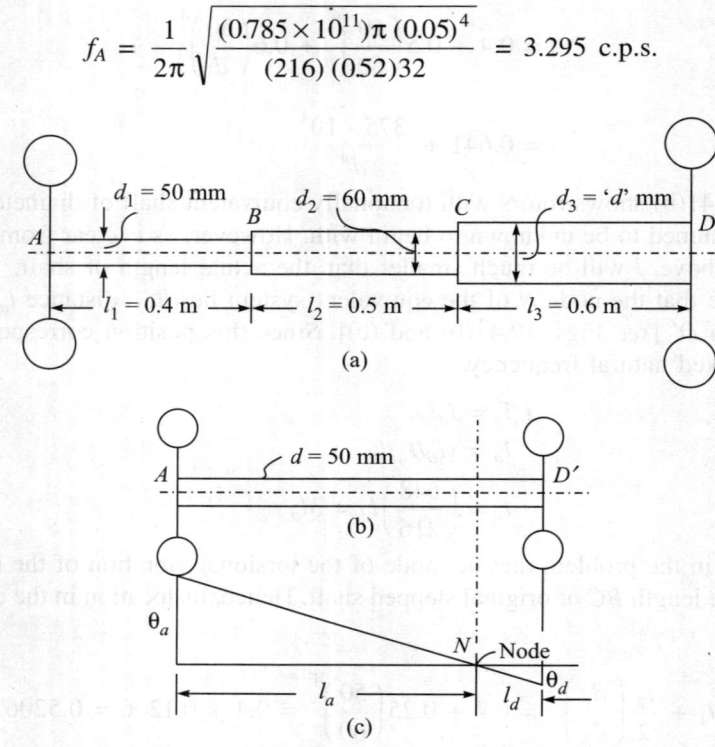

Fig. 17.41 Stepped shaft with end rotors, under Torsional Vibrations.

EXAMPLE 17.26 A motor generator set as shown in Fig. 17.42(a) consists of two armatures A and C connected with flywheel between them at B. Modulus of rigidity of the connecting shaft is 0.824×10^7 N/cm^2. The system can vibrate torsionally with one node at $x = 9.5$ cm from A, the flywheel being at anti-node. Find: (a) The position of another node (b) Natural frequency of vibrations (c) Radius of gyration of the armature C. Other data are given below:

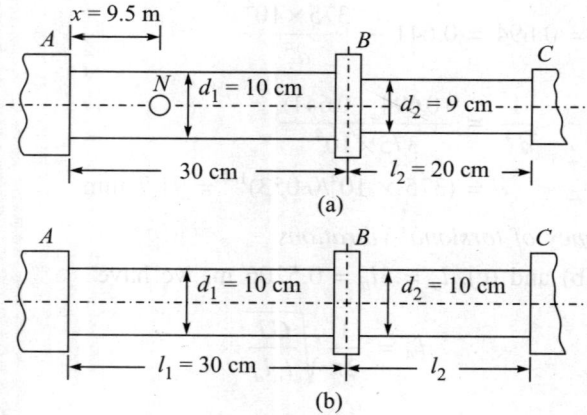

Fig. 17.42 Shaft with a rotor and connecting armatures.

Rotors: A   B   C
(i) Radius of gyration, k 30 cm   37.5 cm   –
(ii) mass, m 3924 N   4905 N   2943 N

(AMIE, Summer 1979)

Solution: $l_a = 9.5$ cm; $I_a = (3924/9.81)\,(0.3)^2 = 36$ kg·m^2
$I_b = (4905/9.81)\,(0.375)^2 = 70.3$ kg·m^2

(a) Let us first determine torsionally equivalent shaft of uniform diameter say, equal to 10 cm. The length of segment AB remains unchanged as the diameter is already equal to that of the equivalent shaft. The equivalent length,

$$l = l_1 + l_2 \left(\frac{10}{9}\right)^4$$

or
$$l = 30 + 20 \left(\frac{10}{9}\right)^4 = 30 + 30.48 \text{ cm} = 60.48 \text{ cm}$$

Therefore
$$l_2' = 30.48 \text{ cm}$$

Substituting the values for I_a, I_b, l_a, l_1 and replacing l_z in equation (17.136) by l_z' we have

$$\frac{1}{100 \times (36)\,(9.5)} = \frac{1}{(100 \times 70.3)} \cdot \left\{ \frac{1}{(30 - 9.5)} + \frac{1}{(30.48 - l_c)} \right\}$$

or
$$\frac{1}{(30.48 - l_c)} = 0.1568$$

Therefore
$$l_c = 24.1 \text{ cm from } C.$$

But distance l_c established above is on an equivalent shaft. The actual distance l_c' on given stepped shaft from C is,

$$l_c' = 24.1 \left(\frac{9}{10}\right)^4 = 15.812 \text{ cm from } C \qquad \textbf{Ans.}$$

(b) Natural frequency of torsional oscillation being same for the mode,

$$f_n = \frac{1}{2\pi} \sqrt{\frac{GJ}{I_a l_a}}$$

or
$$f_n = \frac{1}{2\pi} \sqrt{\frac{0.824 \times 10^7 \times \pi\,(0.1)^4 \times 10^4}{36 \times 32 \times (9.5/100)}} = 77.4 \text{ c.p.s.}$$

(c) From $I_a l_a = I_c l_c$
we have $I_c = I_a (l_a/l_c) = (36)\,(9.5/24.1) = 14.15$ kg·m^2

Therefore from
$$\left(\frac{2943}{9.81}\right) \times k_c^2 = 14.15$$

where k_c = radius of gyration of armature C

Therefore k_c = 0.2171 m = 21.71 cm   **Ans.**

EXAMPLE 17.27 A three-mass torsional system is shown in Fig. 17.43. Determine the torsional stiffness of the shaft BC to make the first natural frequency 6 Hz. Then calculate the corresponding second natural frequency. $G = 80$ GN/m^2. (U. Glasgow)

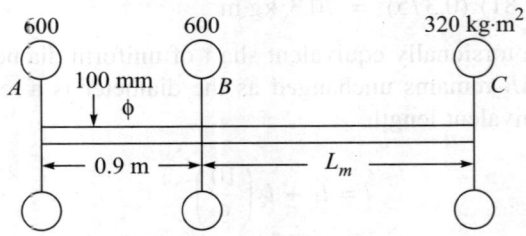

600   600   320 kg·m^2

A 100 mm B C

ϕ

← 0.9 m → | ← L_m →

Fig. 17.43 A 3-mass torsional system.

Solution: For natural frequency of torsional oscillations equal to 6 c.p.s., the node must be located at l_a from A such that,

$$f_n = \frac{1}{2\pi} \sqrt{\frac{GJ}{I_a l_a}} = 6 \text{ c.p.s.}$$

or

$$l_a = \frac{GJ}{4\pi^2 \times f_n^2 \times I_a} = \frac{80 \times 10^9 \times \pi (0.10)^4}{4\pi^2 \times (6)^2 \times 600 \times 32}$$

or $l_a = 0.921$ m

Again from $I_a l_a = I_c l_c$

$$l_c = l_a (I_a / I_c)$$

Therefore $l_c = 0.921 (600/320) = 1.727$ m

Again from equation (17.136),

$$\frac{1}{I_a l_a} = \frac{1}{I_b} \left\{ \frac{1}{l_1 - l_a} + \frac{1}{l_2 - l_c} \right\}$$

Therefore

$$\left(\frac{1}{l_2 - l_c} \right) = \left(\frac{I_b}{I_a l_a} \right) - \frac{1}{l_1 - l_a}$$

or

$$\frac{1}{l_2 - 1.727} = \left(\frac{600}{600 \times 0.921} \right) - \frac{1}{(0.9 - 0.921)}$$

or $l_2 - 1.727 = \dfrac{1}{48.7}$ Therefore $l_2 = 1.747$ m

Thus for $f_n = 6$ Hz, required length of portion $BC = 1.747$ m

Therefore,   torsional stiffness of $BC = \dfrac{G \times \pi (0.1)^4}{1.747 \times 32} = \dfrac{80 \times 10^9 \times \pi (0.1)^4}{32 \times 1.747} = 44.957 \times 10^4$

Therefore,   required torsional stiffness of $BC = 44.957 \times 10^4$ N·m/rad,   **Ans.**

The second natural frequency:

Let l_a and l_c be the locations of nodes from A and C for second mode. Then,

$$320\,l_c = 600\,l_a$$

or $l_c = (600/320)l_a = (1.875)l_a$

Again, from equation (17.136),

$$\frac{1}{I_a l_a} = \frac{1}{I_b}\left\{ \frac{1}{l_1 - l_a} + \frac{1}{l_2 - l_c} \right\}$$

And as $I_a = I_b = 600$ kg·m^2,

$$\frac{1}{l_a} = \frac{1}{(0.9 - l_a)} + \frac{1}{(1.748 - 1.875 l_a)} = \frac{(2.648 - 2.875 l_a)}{(0.9 - l_a)(1.748 - 1.875 l_a)}$$

which leads to a quadratic in l_a. One root of the quadratic is $l_a = 0.921$ m (as in first mode) and the second root is $l_a = 0.361$ m. Thus the second natural frequency is obtained with $l'_a = 0.361$ m is

$$f_n = \left(\frac{l_a}{l'_a}\right)^{1/2} f_{n1} = \left(\frac{0.921}{0.361}\right)^{1/2} \times 6 = 9.583 \text{ Hz} \qquad \textbf{Ans.}$$

17.26 TORSIONAL VIBRATION OF GEARED SYSTEM

Many torsional systems have geared pairs. Figure 17.44(a) shows diagrammatically a geared system in which a rotor A on one shaft is connected through a pinion P and the gear wheel G to the rotor B on another shaft. Let the gear ratio be N, *i.e.*, the speed of the first shaft is N times the speed of the second shaft. The first step in the analysis of such a system is to convert the original system into an equivalent two rotor system with rotors connected to a single shaft of uniform diameter.

To obtain an equivalent torsional system, the required assumptions are:

(a) There is no backlash in the gear pair,
(b) The gear teeth are rigid and are not subjected to distortion under the tooth loads,
(c) The inertia of the shafts and gears is negligible.

The basis for converting the given geared system into an equivalent system is that the kinetic energy and the potential energy for the equivalent system must be same as that for original system. If the given shafts are not loaded beyond limit of proportionality, each rotor in the geared system will execute oscillations in simple harmonic motion. A node in a situation like this will be observed either in length l_1 or in the length l_2. The equivalent system at Fig. 17.44(b) is obtained with respect to the first shaft *i.e.*, the one with shaft diameter d_1.

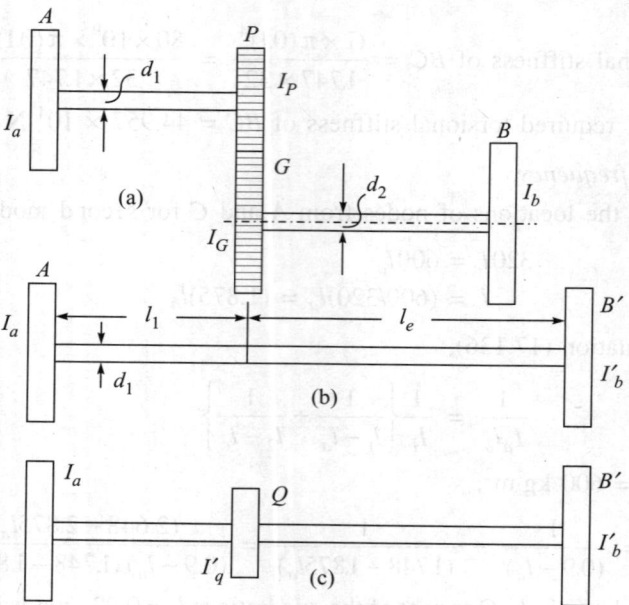

Fig. 17.44 Torsional Vibration of Geared System.

For the product $(N)(T)$ to remain same, if a torque T is applied to the rotor A, a corresponding torque $(N)(T)$ must be applied to the rotor B. Also, if θ_1 is the angle through which rotor A twists relative to pinion P and θ_2 is the angle through which the gear wheel G twists in relation to rotor B, the total angle through which rotor A twists relative to B,

$$\theta_a = (\theta_1 + N\theta_2)$$

But

$$\theta_1 = \left(\frac{Tl_1}{GJ_1}\right) \quad \text{and} \quad \theta_2 = \frac{(NT)l_2}{GJ_2}$$

Substituting for θ_1 and θ_2 above,

$$\theta_a = \left(\frac{Tl_1}{GJ_1}\right) + \frac{(N^2 T)l_2}{GJ_2} = \frac{T}{GJ_1}\{l_1 + N^2(J_1/J_2)l_2\}$$

or

$$\theta_a = \frac{T}{GJ_1}\{l_1 + N^2(d_1/d_2)^4 l_2\} = \frac{T}{GJ_1}\{l_1 + l_e\}$$

where l_e is the equivalent length of shaft of diameter d_1 which is torsionally equivalent to shaft of diameter d_2 and length l_2 when the gearing is neglected. Thus a torsionally equivalent system results when the equivalent length of shaft for position between gear and rotor B satisfies the equation

$$l_e = N^2(d_1/d_2)^4 l_2 \qquad (17.139)$$

In such situation the maximum strain energy stored in the single shaft system shown in Fig. 17.44(b) will be equal to the maximum strain energy stored in the geared system for equal amplitudes of oscillation of rotor A.

Neglecting the kinetic energy contribution of shaft and geared pair for complete equivalence in systems in Figs. 17.44(a) and (b), the kinetic energies of the two end rotors must be equal. Let ω_b be the angular speed of rotor B in the geared system at (a) and ω_b' be the angular speed in Fig. 17.44(b). Then as $\omega_b' = \omega_a$ in any mode of vibration, we have

$$\frac{1}{2} I_b \omega_b^2 = \frac{1}{2} I_b' \omega_b'^2$$

where I_b' = equivalent to M.I. of rotor B in system at Fig. 17.44(b).

Therefore
$$I_b' = I_b(\omega_b/\omega_b')^2 = (I_b/N^2) \tag{17.140}$$

Thus the single uniform shaft system in Fig. 17.44(b) is equivalent to the geared system at (a) if

(i) the additional length l_e in Fig. 17.44(b) satisfies equation (17.139) and, if
(ii) the moment of inertia I_b' of equivalent system satisfies equation (17.140).

When the inertia of the gear pair cannot be neglected, an additional rotor Q on the equivalent uniform shaft system [see Fig. 17.44(c)] must be provided at the end of length l_1 from A. The moment of inertia of this rotor is given by,

$$I_q = I_p + (I_G/N^2) \tag{17.141}$$

The resulting two-rotor or three-rotor torsional system can then be analysed as discussed earlier.

EXAMPLE 17.28   A reciprocating I.C. engine is coupled to a centrifugal pump through gearing. The shaft from the flywheel of the engine to the gear wheel is 4.5 cm diameter and 95 cm long. Shaft from the pinion to the pump is 3 cm diameter and 30 cm long. Engine speed is 1/4th the pump speed. Other particulars are as follows:

Moment of inertia of the flywheel A = 800 kg·m^2
Moment of inertia of the gear wheel G = 15 kg·m^2
Moment of inertia of the pinion P = 4 kg·m^2
Moment of inertia of the pump-impeller B = 17 kg·m^2

Determine the natural frequency of the torsional oscillation of the system. Take $G = 84 \times 10^4$ bar.

Solution: Gear ratio $N = \dfrac{n_i}{n_o} = \dfrac{1}{4}$

Therefore equivalent length of uniform shaft of 4.5 cm diameter,

$$l = l_1 + N^2(d_1/d_2)^4 l_2$$

Therefore
$$l = 95 + \left(\frac{1}{4}\right)^2 \left(\frac{4.5}{3}\right)^4 30$$

or
$$l = 95 + 9.492 = 104.49 \text{ cm}$$

and
$$l_e = 9.49 \text{ cm},$$

Also equivalent M.I. of pump impeller as reduced to uniform shaft of $d = 4.5$ cm

$$I_b' = (I_b)/G^2$$

or $\qquad I_b' = (17)/(0.25)^2 = 272$ kg·m²

Again, equivalent M.I. of pinion gear pair as reduced to uniform shaft of diameter 4.5 cm.

$$I_c = I_G + I_P/G^2$$

or $\qquad I_c = 15 + 4/(0.25^2) = 79$ kg·m²

The equivalent three-rotor system is shown in Fig. 17.45(b).

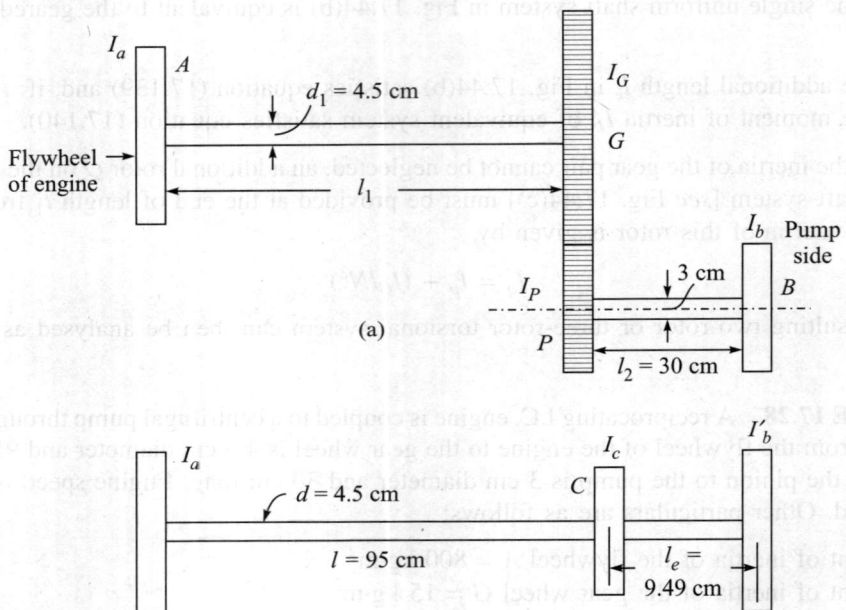

Fig. 17.45 Geared shaft connecting engine flywheel to pump.

For the equivalent system in Fig. 17.45(b),

$$\frac{1}{I_a l_a} = \frac{1}{I_b' l_b'} \qquad \text{or} \quad I_a l_a = I_b' l_b'$$

or $\qquad l_b' = (I_b'/I_a)l_a = (800/272)l_a = 2.94 l_a$

Also $\qquad \dfrac{1}{I_a l_a} = \dfrac{1}{I_c}\left[\dfrac{1}{l_1 - l_a} + \dfrac{1}{l_2 - l_b}\right]$

or $\qquad \dfrac{1}{l_a} = \dfrac{800}{79}\left[\dfrac{1}{95 - l_a} + \dfrac{1}{9.49 - 2.94 l_a}\right]$

This leads to the quadratic in l_a as under

$$\frac{1}{l_a} = 10.1 \left[\frac{104.49 - 3.94 l_a}{2.94 l_a^2 - 288.79 l_a + 901.55} \right]$$

or
$$(42.734) l_a^2 - (1344.139) l_a + 901.55 = 0$$

or
$$l_a^2 - 31.45 l_a + 21.097 = 0$$

Thus
$$l_a = \frac{1}{2} \{ 31.45 \pm \sqrt{31.45^2 - 4(21.097)} \} = 30.76 \text{ cm and } 0.686 \text{ cm}.$$

Hence, the two natural frequencies are

$$f_{n1} = \frac{1}{2\pi} \sqrt{\frac{GJ}{I_a l_a}} = \frac{1}{2\pi} \sqrt{\frac{84 \times 10^4 \times 10^5 \times \pi (4.5/100)^4}{800 \times (30.76/100) \times 32}} = 1.866 \text{ c.p.s.}$$

$$f_{n2} = f_{n1} \times \sqrt{\frac{30.76}{0.686}} = 12.495 \text{ c.p.s.} \qquad \textbf{Ans.}$$

REVIEW QUESTIONS

17.1 Assuming that the cylinder shown in Fig. 17.46 rolls on the support without slippage, determine the equation of motion of the system by energy method and also by Newton, law of motion. Also find the natural frequency.

$$\left(\textbf{Ans.} \quad f_n = \frac{1}{2\pi} \sqrt{\frac{4k(r+a)^2}{3mr^2}} \text{ c.p.s.} \right)$$

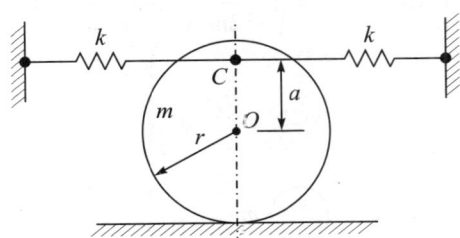

Fig. 17.46 Spring and rolling disc system.

17.2 Determine the equation of motion and the natural frequency of oscillation for the system shown in Fig. 17.47 by the energy method and also by the method based on Newton's second law of motion.

$$\left(\textbf{Ans.} \quad f_n = \frac{1}{2\pi} \cdot \sqrt{\frac{k}{(I_o/r^2) + m_1 + 4m}} \text{ c.p.s.} \right)$$

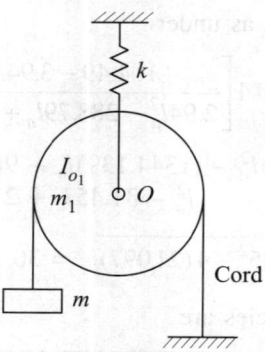

Fig. 17.47 Spring supported pulley and mass system.

17.3 Determine the differential equation of motion and natural frequency of oscillation of the system shown in Fig. 17.48, where the moment of inertia of mass m and the bar about the pivot point is I_o. Show that the system becomes unstable when $b \geq ka^2/mg$.

$$\left(Ans. \quad f_n = \frac{1}{2\pi} \sqrt{\frac{(ka^2 - mgb)}{I_o}} \text{ c.p.s.} \right)$$

17.4 A steel shaft of length L and diameter d is used as a tension spring for the wheel of a light automobile as shown in Fig. 17.49. Mass of the wheel and tyre assembly is m and its radius of gyration about its axle is r. Determine the natural frequency of the system with the wheel locked to the arm. How will the natural frequency change if the wheel is not locked to the arm and free to rotate about its axle? (AMIE, Summer 1993)

$$\left(Ans. \quad \omega_{n1} = \sqrt{\frac{G\pi d^4}{\{32mL(r^2 + a^2)\}}} \text{ rad/s}; \quad \omega_{n1} = \sqrt{\frac{G\pi d^4}{\{32ma^2 L)\}}} \text{ rad/s} \right)$$

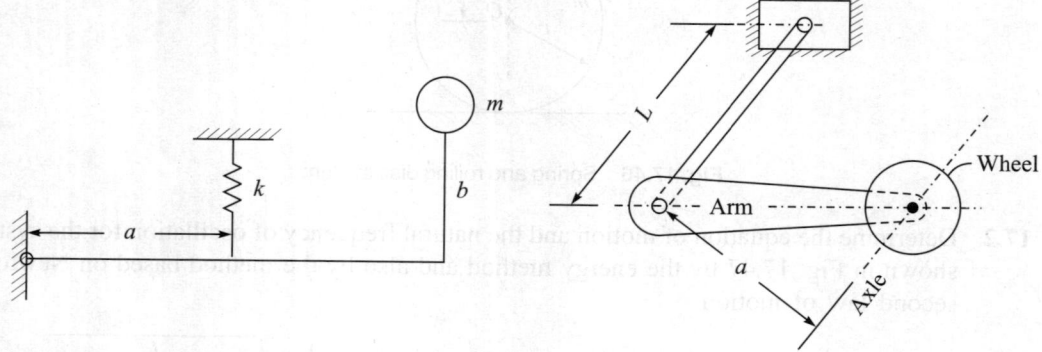

Fig. 17.48 Spring supported ever carrying mass.

Fig. 17.49

17.5 Figure 17.50 shows a system with a spring of constant k and coefficient of viscous damping c. Find out the critical damping coefficient of the system and frequency of damped oscillation. (AMIE, Summer 1993)

$$\left(\textbf{\textit{Ans.}} \quad c_c = \left(\frac{2L}{a}\right)\sqrt{mk} \; ; \; \omega_d = \left(\frac{a}{L}\right)\sqrt{\frac{k}{m} - \frac{c^2 a^2}{4m^2 L^2}} \right)$$

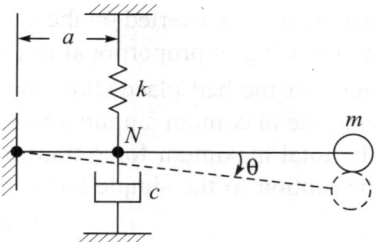

Fig. 17.50 A typical spring-mass-dashpot system.

17.6 The following data are given for a vibratory system with viscous damping: mass = 2.5 kg; spring constant = 30 N/cm, and the amplitude decreases to 0.25 of the initial value after five consecutive cycles. Determine the damping coefficient of the damper in the system. (*Ans.* $c = 0.07656$ N-s/cm)

17.7 A rod is hinged at one end and supported by a spring S at the other end (Fig. 17.51). A mass is attached at 1/3rd of the length from the hinge and a dashpot having a damping coefficient c is attached at 2/3rd of the length from the hinge. Find the equivalent mass and damping coefficient at the spring and derive an expression for the frequency of the damped free vibrations of the system. (Univ. London)

$$\left(\textbf{\textit{Ans.}} \quad f_n = \frac{1}{2\pi \cdot m} \cdot \sqrt{(9Sm - 4c^2)} \; \text{Hz} \right)$$

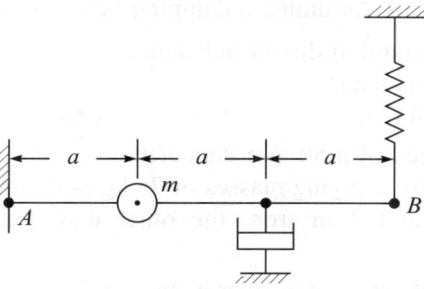

Fig. 17.51 A spring, mass and dashpot attached to a lever.

17.8 A shock absorber is to be designed such that its overshoot is 10% of the initial displacement when released. Determine the damping factor. If the damping factor is reduced to one-half this value, what will be the overshoot? (AMIE, Summer 1993)

(*Ans.* $\delta = 2.298$; $\zeta = 0.3435$; 33.45%)

17.9 A centrifugal fan weighs 445 N and has a rotating unbalance of 23 kg·cm. When dampers having damping factor $\zeta = 0.2$ are used, specify the springs for mounting such that only 10% of the unbalance force is transmitted to the floor. Also determine the magnitude of the transmitted force. The fan is running at constant speed of 1000 r.p.m.

(*Ans.* $k = 368$ N/cm; 253.6 N)

17.10 Some machinery is mounted on a bedplate which is supported on four elastic members, having a stiffness each of 3.5 mN/m. The total mass to be supported is 1 tonne. It is estimated that the total damping force exerted on the system is 20 percent of the critical, it being assumed that the damping is proportional to the velocity of motion.

Measurement of the motion of the bed place show that when the speed of rotation of the machine is 2000 r.p.m., the maximum amplitude of vertical motion of the bed plate is 0.06 mm. Calculate the total maximum force transmitted through each mounting to the ground, assuming the motion to the simple harmonic. (Univ. London).

(*Ans.* 2710 N)

17.11 A single cylinder engine has an out-of-balance force of 500 N at the engine speed of 300 r.p.m. The complete mass of the engine and base is 140 kg and this is carried on a set of springs of total stiffness 25 kN/m. Find (a) the maximum oscillating force transmitted to the ground, (b) the maximum oscillating force transmitted through the springs it a viscous damper is fitted between the base and the ground, the damping force being 840 N at 1 m/s of velocity. (*Ans.* 110.5 N; 107.6 N)

17.12 A radio set of 20 kg mass must be isolated from a machine vibrating with an amplitude of 0.05 mm at 500 c.p.m. The set is mounted on four isolators, each having a spring scale of 31,400 N/m and damping coefficient of 392 N·s/m.

(a) What is the amplitude of vibration of the radio?

(b) What is the dynamic load on each isolator due to vibration?

(*Ans.* 0.069 mm; 0.94 N)

17.13 A machine having a mass of 100 kg and supported on springs of total stiffness 7.84×10^5 N/m has an unbalanced rotating element which results in a disturbing force of 392 N at a speed of 3000 r.p.m. Assuming a damping factor of $G = 0.20$, determine,

(a) the amplitude of motion due to unbalance,

(b) the transmissibility, and

(c) the transmitted force. (*Ans.* 0.0427 mm; 0.1483; 58.2 N)

17.14 Find the whirling speed of a 50 mm diameter steel shaft simply supported at the ends in bearings 1.6 m apart, carrying masses of 75 kg at 0.4 m from one end, 100 kg at the centre and 125 kg at 0.4 m from the other end. Ignore the mass of the shaft. (U. London). (*Ans.* 9.62 r.p.s.)

17.15 A steel beam is simply supported over a span of 3.6 m and carries loads of 49,050 N at the centre and 29,430 N at 0.9 m from each end. Calculate the frequency of transverse vibrations by the energy method, assuming that the vibrating form is (a) similar to the static deflection form and (b) sinusoidal. State how it is possible to fell which of these results is the more accurate.

The static deflection at the 49,050 N load is 1.20 mm and at 29,430 N load is 0.84 mm.

(*Ans.* 15.52 Hz; 15.51 Hz)

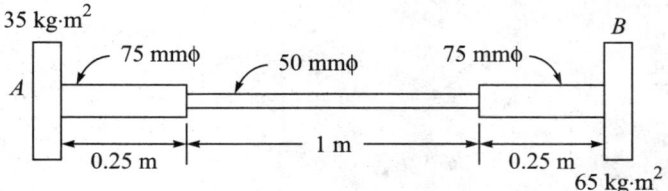

Fig. 17.52 An elastic shaft carrying two wheels.

17.16 A uniform shaft of 85 mm diameter carries three rotors, *A*, *B* and *C* having moments of inertia of 17, 40 and 24 kg·m² respectively. The distance between *A* and *B* is 0.75 m and between *B* and *C* 1.35 m. Find the frequencies of the free torsional vibration. If the rotor *A* has an amplitude of 1° in each case, find the amplitude of *B* and *C*. The modulus of rigidity of the shaft is 80 GN/m². (*Ans.* 35.6 Hz; 20.5 Hz)

17.17 Obtain an expression for the natural frequency of torsional vibration of an elastic shaft carrying two wheels. For the system shown in Fig. 17.52, calculate the natural frequency and the position of the node. Specify the dynamic shear stress in the various portions of the shaft corresponding to an amplitude of vibration of 1° at wheel *A*. (U. Glasgow). (*Ans.* 7.05 Hz; 48.9 MN/m²; 14.7 MN/m²)

17.18 The armature of an electric motor, *I* = 40 kg·m², is connected to a fly-wheel *I* = 80 kg·m², by a shaft 1.0 m long and 80 mm diameter. A shaft, 40 mm diameter and 2.0 m long connects the flywheel to a pulley, *I* = 20 kg·m². Find the frequencies of the two fundamental modes of torsional vibration, and the ratio of the angular displacements in the lower mode. (*Ans.* 3.85 Hz; 17.54 Hz; 0.1756)

17.19 An electric motor rotating at 1500 r.p.m. drives a centrifugal pump at 500 r.p.m. through a single stage reduction gearing. The moment of inertia of the pump-impeller and the electric motor are 1400 kg·m² and 400 kg·m² respectively. The pump shaft and the motor shaft are 45 cm and 18 cm long respectively and their respective diameters are 9 cm and 4.5 cm.

Determine the frequency of torsional oscillations of the system. Neglect the inertia of the gears. *G* = 8.24 × 10⁶ N/cm². (*Ans.* 41.6 c.p.s.)

17.20 An engine drives a centrifugal pump through gearing. The shaft from the engine flywheel to the gear wheel is 8 cm diameter and one metre long, while that from the pinion to the pump impeller is 6 cm diameter and 40 cm long. The pump runs at three times the engine speed and the moment of inertia in kg·m² are as under.

Engine flywheel = 90; Gear wheel = 9; Pinion = 1 and oscillations of the system if the modulus of rigidity of the material of the shaft is 0.84 million kg per sq. cm.

(*Ans.* 11 cycles/s; 63.9 cycles/s)

UNITS

Conversion from F.P.S. and M.K.S. to S.I. units

Length
1 inch = 2.54 cm
1 foot = 0.3048 metre (m)
1 yard = 0.9144 m
1 mile = 1.609344 k.m

Area
$1 \text{ in}^2 = 645.16 \text{ mm}^2$

Volume
$1 \text{ in}^2 = 16387.06 \text{ mm}^3$

Mass
1 slug = 14.5939 kg
1 pound mass = 0.45359 kg
        1 ton = 907.1847 kg

Force
1 pound force = 4.4482 N
(lb)
1 poundal = 0.138255 N
1 kgf = 9.807 N

Velocity
1 ft/s = 0.3048 m/s
1 mile/hr = 1.6096 km/hr

Acceleration
$1 \text{ ft/s}^2 = 0.3048 \text{ m/s}^2$

Pressure/stress
$1 \text{ kgf/mm}^2 = 9.8066 \text{ N/mm}^2$
$1 \text{ lb/in}^2 \text{(p.s.i.)} = 0.006896 \text{ N/mm}^2$

List of prefixes used in basic units

Prefix	Abbreviation	Standard factor by which the untis is multiplied
Exa	E	10^{18}
Peta	P	10^{15}
Tera	T	10^{12}
Giga	G	10^{9}
Mega	M	10^{6}
Kilo	K	10^{3}
Hecto	h	10^{2}
Deca	da	10
Deci	d	10^{-1}
Centi	c	10^{-2}
Milli	m	10^{-3}
Micro	m	10^{-6}
Nano	n	10^{-9}
Pico	p	10^{-12}
Femto	f	10^{-15}
Atto	a	10^{-18}

$$1 \text{ atm} = 0.10133 \text{ N/mm}^2$$
$$1 \text{ bar} = 10^5 \text{ N/m}^2$$

Work and Energy

$$1 \text{ ft pound} = 1.355818 \text{ N.m}$$
$$1 \text{ k.cal} = 4185.927 \text{ N-m}$$
$$1 \text{ kgf.m} = 9.807 \text{ J} \ (1 \text{ joule} = 1 \text{ N.m})$$
$$1 \text{ Btu} = 1055 \text{ J}$$

Power

$$1 \text{ H.P.} = 550.0 \text{ ft.lb/s}$$
$$= 75 \text{ kg}_f\text{.m/s}$$
$$= 0.7457 \text{ kW}$$
$$1 \text{ watt} = 1 \text{ J/s} = 1 \text{ N.m/s}$$
$$1 \text{ kW} = 1000 \text{ N.m/s}$$
$$1 \text{ kW} = 1.34 \text{ H.P.}$$

MATHEMATICS

(A) Trigonometrical Relations and Identities

(i) $\sin(90 \pm \theta) = +\cos\theta$
 $\sin(180 \pm \theta) = \mp \sin\theta$
 $\cos(90 \pm \theta) = \mp \sin\theta$
 $\cos(180 \pm \theta) = -\cos\theta$
 $\tan(90 \pm \theta) = \mp \cot\theta$
 $\tan(180 \pm \theta) = \pm\tan\theta$

(ii) $\sin(\alpha \pm \beta) = \sin\alpha\cos\beta \pm \cos\alpha\sin\beta$
 $\cos(\alpha \pm \beta) = \cos\alpha\cos\beta \mp \sin\alpha\sin\beta$
 $\tan(\alpha \pm \beta) = (\tan\alpha \pm \tan\beta)/(1 \mp \tan\alpha\tan\beta)$
 $\cot(\alpha \pm \beta) = (\cot\beta\cot\alpha \mp 1)/(\cot\beta \pm \cot\alpha)$
 $\sin2\alpha = 2\sin\alpha\cos\alpha$
 $\cos2\alpha = \cos^2\alpha - \sin^2\alpha = 2\cos^2\alpha - 1 = 1 - 2\sin^2\alpha$
 $\tan2\alpha = 2\tan\alpha/(1 - \tan^2\alpha)$

$$\sin\alpha + \sin\beta = 2 \sin\frac{1}{2}(\alpha + \beta)\sin\frac{1}{2}(\alpha - \beta)$$

$$\sin\alpha - \sin\beta = 2 \cos\frac{1}{2}(\alpha + \beta)\sin\frac{1}{2}(\alpha - \beta)$$

$$\cos\alpha + \cos\beta = 2\cos\frac{1}{2}(\alpha + \beta)\cos\frac{1}{2}(\alpha - \beta)$$

$$\cos\alpha - \cos\beta = 2\sin\frac{1}{2}(\alpha + \beta)\sin\frac{1}{2}(\beta - \alpha)$$

(B) Algebraic Relations and Series

(a) Progression

a = first term l = last term d = common difference
r = common ratio n = number of terms

(i) Arithmetic progression

$$l = a + (n - 1)d$$

Sum of series, $\quad S = \dfrac{n(a + l)}{2} = \dfrac{n}{2}\{2a + (n - 1)d\}$

(ii) Geometric progression

Last term $\qquad\qquad l = ar^{(n-1)}$

Sum of series, $\qquad S = \dfrac{a(r^n - 1)}{(r - 1)} = \dfrac{(lr - a)}{(r - 1)}$

when $\qquad\qquad n = \infty \quad$ and $\quad r^2 < 1,\ S = \dfrac{a}{(1 - r)}$

(b) Binomial series

$$(a \pm b)^n = a^n \pm na^{n-1}b + \frac{n(n-1)}{2!}\,a^{n-2}b^2 \pm \frac{n(n-1)(n-2)}{3!}\,a^{n-3}b^3$$

$$\text{---------} + (\pm 1)^r\, \frac{n(n-1)(n-2).....(n-r+1)}{r!}\,a^{n-r}b^r + ...$$

$$(1 \pm x)^n = 1 \pm nx + \frac{n(n-1)}{2!}\,x^2 \pm \frac{n(n-1)(n-2)}{3!}\,x^3 + \frac{n(n-1)(n-2)(n-3)}{4!}\,x^4 + ...$$

$$(1 \pm x)^{-n} = 1 \mp nx + \frac{n(n-1)}{2!}\,x^2 \mp \frac{n(n-1)(n-2)}{3!}\,x^3 + ...$$

(c) Functions expanded in series

$$e^x = 1 + x + \frac{x^2}{2!} + \frac{x^3}{3!} + \frac{x^4}{4!} + ... \qquad (-\infty < x < \infty)$$

$$\sin x = x - \frac{x^3}{3!} + \frac{x^5}{5!} - \frac{x^7}{7!} + ... \qquad (-\infty < x < \infty)$$

$$\cos x = 1 - \frac{x^2}{2!} + \frac{x^4}{4!} - \frac{x^6}{6!} + ... \qquad (-\infty < x < \infty)$$

(d) Roots of a quadratic equation, $ax^2 + bx + c = 0$

$$x = \frac{-b \pm \sqrt{b^2 - 4ac}}{2a}$$

(e) For a triangle with sides a, b and c and angles opposite to these sides as A, B and C respectively

sine rule: $\dfrac{a}{\sin A} = \dfrac{b}{\sin B} = \dfrac{c}{\sin C}$ and law of cosines, $a^2 = b^2 + c^2 - 2bc\cos A$

(f) Differential calculus

$$\frac{d}{dx}(uv) = v\frac{du}{dx} + u\frac{dv}{dx}$$

$$\frac{d}{dx}(u/v) = \left\{v\frac{du}{dx} - u\frac{dv}{dx}\right\}\Big/v^2$$

S.I. AND M.K.S. UNITS

Each kind of physical quantity has a dimension and, therefore, there can be as many dimensions as there are types of physical quantities. In practice, however, only a few basic quantities are assigned dimensions and all other quantities are expressed in terms of dimensions derived from the basic units. The choice of basic quantities is governed by convention and convenience and any set of basic quantities would be equally valid. One such set of basic units are as follows.

Quantity	Unit	Symbol
Length (L)	metre	m
Mass (M)	kilogram	kg
Time (t)	second	s
Amount of substance	mole	mol
Temperature (T)	kelvin	K
Electric current	ampere	A
Luminous intensity	candela	cd
Plane angle	radian	rad
Solid angle	steradian	sr

In addition to these, let symbol F designate unit of force. *Note that the symbols F, M, L and T are not numbers.*

When force, length and time are chosen as the basic units, mass (M) becomes a derived unit and the system is called gravitational (practical) system of units. As against this, when mass, length and time are chosen as the basic units, force becomes a derived unit and an absolute (scientific) system of units results.

Most of the fundamental equations of physics involve statements of proportionality and a constant of proportionality has to be included. For instance, Newton's three laws restated from 'Principia' are:

Law 1: Every body continues to be in its state of rest or of uniform motion in a straight line, except in so far as it is compelled to change that state by impressed forces.

Law 2: Change of motion is proportional to the moving force impressed, and takes place in the direction of the straight line in which such a force is impressed.

Law 3: Reaction is always equal and opposite to action, that is to say, the actions of two bodies upon each other are always equal and directly opposite.

The first two laws of motion may be summarised in the following equation:

$$\mathbf{F} \propto m\mathbf{A} \tag{a}$$

involving force $\mathbf{F}$ and acceleration $\mathbf{A}$ as vectors and mass m as a scalar. Equation (a) may be restated using a constant of proportionality C as,

$$\mathbf{F}C = m\mathbf{A} \tag{b}$$

The proportionality constant $(1/C)$ is a function of the units specified for the three physical quantities namely, force $\mathbf{F}$, mass m and acceleration $\mathbf{A}$. Physicists are inclined to choose units that reduce the proportionality constant to unity. System International (S.I.) unit is representative of this class in which Newton is a derived unit of force. One Newton is defined as a force producing an acceleration of one metre per second per second, in a mass of one kilogram. Thus, for the constant of proportionality

$$C = \frac{m\mathbf{A}}{\mathbf{F}} = \left(\frac{MLT^2}{\text{Newton}}\right)$$

To remain dimensionless with a value of unity, one must have,

$$1 \text{ Newton} = 1 \text{ kg·m/s}^2$$

As against this, engineers do not prefer to be so much constrained as they must use with equal facility both scientific and practical units. For instance, in M.K.S. (metre-kilogram-second) system, a kilogram force (1 kgf) is defined as the force needed to be applied to a mass of 1 kilogram for producing an acceleration equal to standard gravitational acceleration of 9.81 m/s². Thus, considering a mass of 1 kg acted upon by a gravitational force of $\mathbf{W}$, we have from equation (b),

$$\mathbf{W}C = m \times g \tag{c}$$

or

$$(1 \text{ kgf}) \times C = (1 \text{ kg}) \times (9.81 \text{ m/s}^2)$$

or

$$C = 9.81 \left(\frac{\text{kg}}{\text{kgf}}\right)(\text{m/s}^2)$$

Thus,

$$\mathbf{W} = \left(\frac{mg}{C}\right)$$

or in general,

$$\mathbf{F} = \left(\frac{m\mathbf{A}}{C}\right)$$

It follows from equation (c) that proportionality constant C is dimension and units of mass are,

$$m = \left(\frac{\mathbf{W}}{g}\right) \text{ i.e., } \text{kgf·s}^2/\text{m}$$

The unit kg·s^2/m is also called metric slug (mS1). Again, it follows from equation (c) that,

$$\left(\frac{\mathbf{W}}{g}\right) = \left(\frac{m}{\mathbf{C}}\right) \qquad\qquad\qquad\qquad \text{(d)}$$

Since the constant of proportionality C numerically equals acceleration due to gravity (g), it follows from (d) that a mass of 1 kilogram weighs 1 kilogram force. However, note that ($\mathbf{W}/g$) has dimensions of mass in metric slug (kgf·s^2/m) and not in kilogram. Hence, while solving problems in M.K.S. units, it must be remembered that the ratio $\mathbf{W}/g$ has units in kgf·s^2/m (*i.e.*, metric slug) and further that mass in metric slug must be multiplied by 9.81 to obtain mass in kilogram. Summarising above features, in M.K.S. units, numerically, 1 kgf = 1 kg mass.

and,   1 kg mass = 9.81 kgf·s^2/m

Moment of Inertia

In S.I. units, moment of inertia is given by

$$I = mk^2, \qquad\qquad\qquad\qquad \text{(e)}$$

where m is mass in kilogram and k is the radius of gyration.

Thus, unit of moment of inertia in S.I. units is kg·m^2

In M.K.S. units, as unit of mass is kgf·s^2/m, the unit of I from eqn. (e) is kgf·m·s^2.

OBJECTIVE AND SHORT ANSWER TYPE QUESTIONS

(Note: Asterisk (*) mark against options given after each objective represents correct answer.)

CHAPTER 1   MOTION, DISPLACEMENT AND VECTORIAL REPRESENTATION

(A) Short Answer Type

1.1 Briefly explain with example difference between the concept of 'motion of a point' and 'motion of a rigid body'.

1.2 Explain the term 'motion of translation'. Is it necessary that a body having motion of translation must move along a straight line?

1.3 What are the advantages of the assumption of 'motion of translation'?

1.4 Define the terms 'planar motion' and 'planar mechanisms'.

1.5 Explain briefly as to how complex polar notation, for a vector, represents magnitude, direction and sense.

1.6 What are the 'component motions' that constitute six degrees of freedom of a rigid body in space motion?

(B) Objective Type

1.7 A body in motion can be likened to the 'motion of a point' when

    (a) the body is of the size of a point

    (b) the body moves uniformly in circular path

(c)* there is no relative motion between various points in the body

(d) none of these

1.8 *Motion of translation* necessarily means

(a) reciprocating motion only

(b)* no relative motion between different points in the body

(c) pure rolling of a body along a straight path

(d) none of the above

1.9 A vector 10 ∠20 is represented by a directed line segment of 10 units length making an angle of

(a)* 20° c.c.w. with positive x-axis

(b) 20° clockwise from positive x-axis

(c) 20° c.c.w. from positive y-axis

(d) 20° clockwise from positive y-axis

1.10 Vector addition of vectors represented by $20e^{j\theta}$ and $15e^{j(\theta+\pi)}$ has a magnitude of

(a) 25   (b)* 5

(c) 35   (d) 20/15

CHAPTER 2   PLANAR MECHANISMS AND GEOMETRY OF MOTION

(A) *Short Answer Type*

2.1 Explain the terms 'Kinematics' and 'Dynamics'. How do they differ?

2.2 Define the terms Link, element and pairs. In what way a *link* differs from a *machine component*?

2.3 Distinguish clearly between terms 'element' and 'pair'.

2.4 State the conditions under which members like belt, chain and fluid can be considered a link. Explain reasons.

2.5 What is the role of a spring in a mechanism? What is the instantaneous velocity equivalent of a spring?

2.6 Distinguish with examples between a statically determinate structure, statically indeterminate structure and a closed chain.

2.7 Distinguish between closed chain and open chain, giving examples of each.

2.8 How many degrees of freedom are permitted by

(a) pair with rolling and slipping

(b) turning pair

(c) a globular pair

(d) a screw pair

(e) a cylindrical pair

2.9 What kind of relative motion is permitted between contacting points of connected links by a lower and a higher pair? Give examples.

2.10 Define the terms mechanism and machine. In what respects do the two differ.

2.11 State and explain properties of inversion.

2.12 Giving all applications, explain how principle of inversion in practice simplifies kinematic analysis.

2.13 Can higher pairs be inverted like lower pairs? Explain.

2.14 Explain advantages of using a mechanism with at least one link capable of having full rotation. State the condition which ensures the above feature.

2.15 Explain the terms 'mechanical advantage' and 'transmission angle'. How are the two related? What is the desirable range for transmission angle?

2.16 Explain the term 'constrained motion'. Does it necessarily mean a mechanism with a degree of freedom of 1?

2.17 What do you understand by the term 'multiple joint'? Why is it provided? How is it accounted for in determining degrees of freedom?

2.18 Explain briefly the inconsistencies of Grubler's equation.

2.19 Explain the term 'equivalent linkage'. For two involute gears in mesh, show an equivalent 4-bar mechanism through a sketch.

2.20 Show that in mechanisms with all turning pairs, only even number of links can produce single degree of freedom.

2.21 Show that in a simple jointed chain with all turning pairs and 'n' links, maximum possible number of elements on any link is $n/2$ when n is even.

(B) Objective Type

2.22 For a kinematic chain of n-links, total number of possible inversions are
   (a) $n + 1$
   (b) $n(n - 1)/2$
   (c)* n
   (d) n^2

2.23 A kinematic pair is formed by
   (a)   connecting rigidly two elements
   (b)   connecting not more than two elements permitting relative motion
   (c)* connecting two or more elements permitting relative motion
   (d)   two or more elements without relative motion

2.24 A quadruple joint is equivalent to _____ simple joints.
   (a)   two
   (b)* three
   (c) 4_{c_2}
   (d)   four

2.25 In a 4-bar mechanism, mechanical advantage is maximum when transmission angle is
   (a)   zero
   (b) $180°$
   (c) $45°$
   (d)* as close to $90°$ as possible

2.26 In an offset slider-crank mechanism with length of connecting rod l, crank radius r and offset e, the crank will revolve only when
   (a)* $l \geq r + e$
   (b) $l < (r - e)$
   (c) $l > (r - e)$
   (d) $l < (r + e)$

2.27 In a nine-link mechanism of d.o.f. = 2, number of joints of d.o.f. one, is given by
   (a)   *11
   (b)   10
   (c)   13
   (d)   9

2.28 Degree of freedom of a chain having n number of links and l number of lower pairs of d.o.f. = 1 is given by
(a) $F = 3n - L$ (b)* $F = 3n - 2l$
(c) $F = 3(n - 1) - 2l$ (d) $F = 3n - 3$

2.29 In a mechanism with 8 links and 9 number of lower pairs of d.o.f. = 1, the d.o.f. of the mechanism is given by
(a) 1   (b) 2
(c)* 3   (d) zero

2.30 Four links of a mechanism have lengths of 3, 7, 5, 4 units respectively. By fixing link of length 3 units resulting mechanism will be
(a) crank-crank mechanism
(b) crank-rocker mechanism
(c)* rocker-rocker mechanism
(d) a mechanism with change points

2.31 A withworth mechanism can be obtained as an inversion of slider-crank chain by fixing
(a) slotted link   (b) slider
(c) connecting rod   (d)* crank

2.32 In two inversions obtained from the same parent chain, the relative velocity between any two links
(a) changes in proportion to the lengths of the two frame links
(b) changes in proportion to the lengths of links opposite to the frame links
(c)* does not change
(d) none of the above

2.33 A globular (spherical) pair has — degrees of freedom.
(a) one   (b) two
(c)* three   (d) four

2.34 A pantograph mechanism is a
(a) double crank mechanism   (b)* double rocker mechanism
(c) crank-rocker mechanism   (d) none of these

2.35 A hand-pump mechanism is the inversion of a slider-crank mechanism by fixing
(a) crank   (b) connecting rod
(c) slotted link   (d)* slider

2.36 In a translating roller follower number of links n, number of lower pairs l and number of higher pairs h are given by
(a)* $n = 3$; $l = 2$; $h = 1$ (b) $n = 4$; $l = 3$; $h = 1$
(c) $n = 4$; $l = 4$ and   (d) $n = 4$; $l = 3$; $h = 2$

2.37 In an n-link mechanism maximum, possible number of elements on any of the n links, for odd value of n is
(a) $n/2$ (b) $(n - 1)/2$
(c)* $(n + 1)/2$ (d) none of these

2.38 In a 4-bar mechanism, the mechanical advantage is maximum when the velocity ratio is

    (a)   maximum   (b)* minimum
    (c)   1   (d)   1/2

2.39 In a 6-bar chain for constrained motion, there will be

    (a)   6 binary links   (b)* 4 binary and 2 ternary links
    (c)   5 binary and 1 ternary   (d)   3 binary and 3 ternary

2.40 A 6-bar chain can be formed to give constrained motion by using

    (a)   5 turning pairs   (b)   6 turning pairs
    (c)* 7 turning pairs   (d)   8 turning pairs

CHAPTERS 3 AND 4   VELOCITY AND ACCELERATION ANALYSIS
(Graphical Approach, Analytical Approach)

(A)   Short Answer Type

3.1 Show that relative velocity of a point A on a link with respect to another point B on the same link exists only in the transverse direction.

3.2 In a rigid body in motion, though relative velocity between any two points on the body can not exist along the line joining them, acceleration does exist along this line. Explain.

3.3 For either direction of the velocity of slider, with respect to slotted lever, for both the sense of rotations of lever show direction of coriolis component of accelerations on a neat sketch.

3.4 Explain how the instantaneous centre of rotation of the crank with respect to slider can be used to obtain angular speed of crank for a known velocity of slider in a slider-crank mechanism.

3.5 With the help of a neat sketch explain coriolis component of acceleration. What are the essential conditions for coriolis component to exist?

(B)   Objective Type

3.6 The direction of coriolis acceleration vector is

    (a)   along the sliding velocity vector
    (b)   along the tangential acceleration vector of crank
    (c)   along a line rotated 90° from the sliding velocity vector in a direction opposite to angular velocity
    (d)* along a line obtained by rotating sliding velocity vector through 90° in the sense of angular velocity

3.7 Total number of instantaneous centres of rotation for a mechanism consisting of n-links is

    (a) $n/2$ (b) $2n$
    (c)* $n(n-1)/2$ (d) $n(n-1)$

3.8 When the slider moves along a curved path, the instantaneous centre of relative motion lies at

    (a)   point of their contact

(b) at the pin connecting slider to the connected link
(c)* at the centre of curvature of curved path
(d) none of these

3.9 A coriolis component of acceleration does not exist when a point moves along a link

(a) at a uniform velocity and link oscillates
(b) and the link oscillates/rotates at a constant speed
(c)* that does not rotate
(d) (a) and (b) together

3.10 A slider sliding at 10 cm/s on a link which is rotating at 60 r.p.m. is subjected to coriolis acceleration of magnitude

(a) $40\pi^2$ cm/s^2 (b) 0.4π cm/s^2
(c)* 40π cm/s^2 (d) 20π cm/s^2

3.11 The acceleration of piston in a reciprocating steam engine is given by

(a) $\omega r^2 \left(\cos\theta + \dfrac{\cos 2\theta}{n} \right)$ (b) $r\omega^2 \left(\cos\theta + \dfrac{\cos 2\theta}{2n} \right)$

(c) $r\omega^2 \left(\sin\theta + \dfrac{\sin 2\theta}{n} \right)$ (d)* $r\omega^2 \left(\cos\theta + \dfrac{\cos 2\theta}{n} \right)$

3.12 When the crank is at the inner dead centre in a reciprocating steam engine, the acceleration of the piston will be

(a)* $r\omega^2 \left(\dfrac{n+1}{n} \right)$ (b) $r\omega^2 \left(\dfrac{n-1}{n} \right)$

(c) $r\omega^2 \left(\dfrac{n}{n+1} \right)$ (d) $r\omega^2 \left(\dfrac{n}{n-1} \right)$

CHAPTER 5 MECHANISM WITH LOWER PAIRS

(A) *Short Answer Type*

5.1 Explain the purpose of a pantograph. How does it differ from straight line motion mechanisms?

5.2 Distinguish between approximate and exact straight line motion mechanisms. Cite situations in which approximate straight line motion mechanisms suit the purpose.

5.3 Explain why Acherman steering gear is preferred over Davis steering gear in practice.

5.4 State necessary conditions to be satisfied by a double Hooke's joint.

5.5 State the condition of correct steering. What may happen if the condition of correct steering is not satisfied?

(B) Objective Type

5.6 It is a good practice for a Hooke's joint used in machinery that shaft angle of inclination be
    (a)   around $60°$ (b)   around $45°$
    (c)   around $30°$ (d)* less than or equal to $15°$

5.7 Maximum fluctuation of speed of driven shaft in a single Hooke's joint is given by

    (a) $\omega \left(\dfrac{1 - \sin^2\alpha}{\sin^2\alpha} \right)$ (b) $\omega \left(\dfrac{1 + \cos^2\alpha}{\sin^2\alpha} \right)$

    (c)* $\omega \left(\dfrac{1 - \cos^2\alpha}{\cos^2\alpha} \right)$ (d) $\dfrac{\omega^2 \cos^2\alpha}{(1 - \cos^2\alpha)}$

5.8 Maximum angular acceleration of driven shaft in a single Hooke's joint occurs at θ given by

    (a) $\cos 2\theta \approx \dfrac{2\sin\alpha}{1 - \sin^2\alpha}$ (b)* $\cos 2\theta \approx \dfrac{2\sin^2\alpha}{2 - \sin^2\alpha}$

    (c) $\cos 2\theta \approx \dfrac{2\cos^2\alpha}{2 - \cos^2\alpha}$ (d)   none of these

5.9 Condition of correct steering for a 4-wheeler is
    (a) $\cos\phi - \cos\theta = W/H$ (b) $\sin\phi - \sin\theta = W/H$
    (c)* $\cot\phi - \cot\theta = W/H$ (d) $\tan\phi + \cot\theta = W/H$
    where W = wheel base
             H = distance between pivots of front axles
             ϕ, θ angles through which axis of outer and inner wheels turn respectively.

5.10 Condition for correct steering in Davis steering gear is
    (a) $\cot\alpha = W/2H$ (b) $\cos\alpha = W/H$
    (c)* $\tan\alpha = W/2H$ (d) $\tan\alpha = W/H$
    where α = angle at which arm of bell-crank lever is inclined to longitudinal axis.

5.11 Indicate as to which one of the following is an exact straight line motion mechanism
    (a)* Hart's mechanism   (b)   Watt's mechanism
    (c)   Grasshopper mechanism   (d)   Robert's mechanism

5.12 Indicate as to which one of the following is an approximate straight line motion mechanism
    (a)   Hart's mechanism   (b)   Peaucellier mechanism
    (c)   Scott Russel mechanism   (d)* Tchebicheff's mechanism

5.13 An exact straight line motion mechanism, using sliding pair, is
    (a)   Peaucellier mechanism   (b)   Hart's mechanism
    (c)* Scott Russel mechanism   (d)   Robert's mechanism

CHAPTER 6 ELEMENTS OF KINEMATIC SYNTHESIS OF MECHANISMS

(A) Short Answer Type

6.1 Distinguish between a pole and an instantaneous centre of rotation. What does a pole triangle signify?

6.2 Explain briefly as to how a relative pole is obtained for the displacement of follower from position $O_B B_1$ to $O_B B_2$ through an angle Ψ_{12}, for the displacement of input crank through ϕ_{12}.

6.3 What are Chebyshev accuracy points? Why are they preferred in kinematic synthesis?

6.4 What do you understand by the term 'structural error'? Can this error be eliminated completely?

(B) Objective Type

6.5 For generating a function $y = \sin x$ in the interval $0 \le x \le 90°$, the three Chebyshev accuracy points are

    (a) $0°$, $45°$, $90°$ (b) $22.5°$, $45°$, $67.5°$

    (c) $15°$, $45°$, $60°$ (d)* $6°$, $45°$, $84°$

6.6 An important property of a pole triangle is that for an angle of rotation θ the interior angles of the triangle are

    (a) $\theta/4$ (b)* $\theta/2$

    (c) $\theta/3$ (d) θ

6.7 For a rigid body AB describing three positions $A_1 B_1$, $A_2 B_2$ and $A_3 B_3$, the pole P_{12} is located as the intersection point of mid-normals of lines joining

    (a) A_1 to B_2 and A_2 to B_1 (b) A_1 to B_1 and A_2 to B_2

    (c)* A_1 to A_2 and B_1 to B_2 (d)   none of the above

6.8 In a pole triangle with vertices P_{12}, P_{23} and P_{31} if M be the orthocentre then angle $P_{12} M P_{31}$ equals angle

    (a) ϕ_{12} (b)* ϕ_{23}

    (c) ϕ_{31} (d)   none of these

    where $2\phi_{ij}$ = angle of rotation of the body in going from position i to j.

6.9 The displacement equation of the four-bar linkage may be expressed as

    (a)* $\tan\Psi/2 = \dfrac{A \pm \sqrt{A^2 + B^2 - C^2}}{B + C}$ (b) $\tan\Psi/2 = \dfrac{-A \pm \sqrt{A^2 + B^2 - C^2}}{B + C}$

    (c) $\tan\Psi/2 = \dfrac{A \pm \sqrt{A + B^2 - C^2}}{B - C}$ (d) $\tan\Psi/2 = \dfrac{-A \pm \sqrt{A^2 - B^2 + C^2}}{B - C}$

where Ψ and ϕ are respectively the output and input angles and $A = \sin\phi$,

$B = (a_4/a_1) + \cos\phi$, and $C = \left((a_4/a_3) \cos\phi + \dfrac{a_1^2 - a_2^2 + a_3^2 + a_4^2}{2 a_1 a_3} \right)$. Here a_4 and a_2 are the lengths of frame and coupler while a_1 and a_3 are the input and output link lengths respectively.

6.10 The Freudenstein's displacement equation for a 4-bar mechanism is given by

(a) $k_1\cos\phi + k_2\cos\Psi - k_3 = \cos(\phi - \Psi)$

(b)* $k_1\cos\phi - k_2\cos\Psi + k_3 = \cos(\phi - \Psi)$

(c) $k_1\cos\phi + k_2\cos\phi + k_3 = \cos(\phi + \Psi)$

(d) $k\cos\phi - k_2\cos\Psi - k_3 = \cos(\phi - \Psi)$.

where ϕ and Ψ are the input and output angles respectively, while $k_1 = (a_4/a_3)$; $k_2 = (a_4/a_1)$; $k_3 = \left(\dfrac{a_1^2 - a_2^2 + a_3^2 + a_4^2}{2a_1a_3}\right)$ a_4 and a_2 are the lengths of frame and coupler links while a_1 and a_3 are the lengths of input and output links respectively.

CHAPTER 7 CAMS

(A) Short Answer Type

7.1 Define the term 'pressure angle' of a cam. How does it modify force transmission from a cam to the follower and influence the contour?

7.2 How are cams classified based on follower motion?

7.3 State the possible reasons for erroneously regarding parabolic follower motion as the best follower motion.

7.4 Explain why analysis of follower motion for velocity, acceleration and jerk is important to the designer of a cam.

7.5 Explain why constant velocity cam does not give practical solution unless it is modified. How is it modified?

7.6 Show that for a plate cam with reciprocating roller follower, the pressure angle is given by

$$\tan\phi = \frac{(dy/d\theta) - e}{\sqrt{R^2 - e^2 + y}}$$

where R is the radius of prime circle; e is the offset of follower line of action and y is the follower displacement.

7.7 Explain briefly the basis which decides minimum face width of flat faced follower. Show that face width is given by

Face width > $(dy/d\theta)_{max} - (dy/d\theta)_{min}$.

7.8 Explain briefly the roll of offset in the line of action of follower on cam action. On which side will you provide this offset?

(B) Objective Type

7.9 Maximum practical value of pressure angle for a plate cam operating roller follower is

(a)* 30–35° (b) 10–15°

(c) 45–50° (d) 15–20°

7.10 The maximum pressure angle in a cam with flat faced follower is

(a) 60° (b) 90°

(c) 45° (d)* 0°

7.11 A cam uses cycloidal motion for the rise of the follower. If the lift of the follower is d, maximum pressure angle occurs when the follower displacement is

(a) d (b) $d/4$

(c)* $d/2$ (d) $3d/4$

7.12 A plate cam operating a roller follower, the trace point is located at

(a)   the point of contact between cam and roller

(b)   cam centre of rotation

(c)* roller centre

(d)   at some point on follower axis

7.13 A plate cam operating a flat faced follower, the trace point is located at

(a)   cam centre of rotation

(b)* on the flat face of follower

(c)   somewhere on the follower axis

(d)   none of these

7.14 In parabolic follower motion, maximum jerk is experienced by follower

(a)   only at the beginning of stroke

(b)* at the beginning; end and middle of stroke

(c)   at one-fourth of the stroke length

(d)   at three-fourth of the stroke

7.15 In simple harmonic follower motion, maximum jerk is experienced by follower at

(a)   the mid-stroke

(b)   one-fourth the stroke length

(c)* the beginning and end of stroke

(d)   at three-fourth the stroke length

7.16 Circumference of pitch circle required for same pressure angle and stroke length is largest for

(a)   uniform follower motion

(b)   modified uniform follower motion

(c)   simple harmonic follower motion

(d)* cycloidal follower motion.

7.17 For a given stroke length and pitch circle diameter, a roller follower will have smallest pressure angle when its displacement is according to

(a)* parabolic motion   (b)   simple harmonic motion

(c)   modified uniform motion   (d)   uniform motion

7.18 A larger cam is undesirable as

(a)   it is difficult to be manufactured

(b)   it is difficult to mount on cam shaft

(c)   required cam is steeper

(d)* it produces more unbalance at higher speeds.

7.19 Principal problem in cam design is to compromise upon a follower motion which will have relatively mild values of velocity and acceleration as

  (a) connected inertia load on cam shaft is less
  (b) smaller force is to be transmitted to the follower
  (c) it results in lighter cam and follower
  (d)* it reduces stresses and vibration in the associated components

7.20 A flat faced follower is used for relatively steep cam curves and where space is limited because of

  (a) ease in manufacture   (b) lightness of resulting assembly
  (c)* relatively less contact stresses (d) none of these

7.21 The pressure angle Ψ and the base circle in a cam should be

  (a) both as big as possible
  (b) Ψ as low as possible and base circle as large as possible
  (c) Ψ as big as possible and base circle as small as possible
  (d)* both as small as possible

CHAPTER 8 GEARS

(A) Short Answer Type

8.1 Two involute spur gears are in perfect mesh. If the centre distance is increased slightly (say 15% of the addendum), will the fundamental law of gearing be still satisfied by the pair of gears?
  (Hint: For standard proportions of full depth tooth, addendum = 1 module, and working depth is 2 × modules. Thus, a reduction in centre distance by 0.15 m still leaves substantial working depth. As the base circles for involute tooth profiles remain unchanged, except for their centre distance, involute-involute contact again produces conjugate action.)

8.2 What is the main restriction on increasing length of addendum of a spur gear?
  (Hint: If the addendum circle of a gear cuts the common tangent to base circles beyond points of tangency, interference occurs.)

8.3 For transmitting power from a shaft to another non-intersecting shaft at right angles with large speed reduction, which type of gear pair will you use?

8.4 Length of arc of contact is greater than length of path of contact. Explain the reason with neat sketch.

8.5 Explain the term 'conjugate teeth'. Is it possible to have conjugate teeth other than involute and cycloidal?

8.6 What do you understand by the term 'Interference'? How can it be avoided?

8.7 Explain, giving reason, as to why interference is possible with involute gears but not with cycloidal gears.

8.8 Compare involute and cycloidal tooth profiles for relative advantages and disadvantages.

8.9 Does interference in involute gears depend on relative size of dedendum circle diameter and base circle diameter? Explain your answer by giving reasons.

8.10 Show that circular pitch is given in terms of module m by $p_c = m\pi$.

8.11 Explain advantages and disadvantages of helical gears over straight spur gears.

8.12 When centre distance between a pair of spur gears is slightly increased, does it amount to changing pitch point? How does it affect velocity ratio or maximum velocity of sliding?

8.13 What do you understand by the term 'contact ratio'? How does it measure merit of a gear? Explain physical significance of contact ratio of 1.3.

(B) Objective Type

8.14 The centre distance between two involute spur gears of base circle radii R_b and r_b, and pitch circle radii R and r is given by

    (a) $(R_b + r_b)$ (b) $(R + r)/\cos\Psi$

    (c)* $(R_b + r_b)/\cos\Psi$ (d) $(R_b - r_b)/\cos\Psi$

8.15 In an involute pinion meshing with involute gear, the normal to contacting involute surfaces is tangent to

    (a)   pitch circles   (b)   addendum circle

    (c)   dedendum circle   (d)* base circles

8.16 The contact ratio is given by

    (a) $\dfrac{\text{(length of path of approach)}}{\text{circular pitch}}$ (b) $\dfrac{\text{(length of path of recess)}}{\text{circular pitch}}$

    (c)* $\dfrac{\text{(length of arc of contact)}}{\text{circular pitch}}$ (d) $\dfrac{\text{(length of arc of contact)}}{\cos\Psi}$

    where Ψ is the pressure angle.

8.17 In the case of a pair of involute spur gears in mesh, maximum sliding velocity occurs at

    (a)   pitch point   (b)* point of engagement

    (c)   point of disengagement   (d)   none of these

8.18 Minimum number of teeth required on pinion to avoid interference with 20° full depth involute teeth is

    (a)   12   (b)   14

    (c)* 18   (d)   20

8.19 Usual value of contact ratio for gears is

    (a)   zero   (b)* between 1 and 2

    (c)   less than 1   (d)   greater than 2

8.20 An increase in centre distance slightly in the case of a pair of involute gears

    (a)   does not have any influence of pressure angle

    (b)* increases the pressure angle

    (c)   decreases the pressure angle

    (d)   pressure angle becomes irrelevent as the contact does not produce conjugate action

8.21 One of the disadvantages of helical gear is that

    (a)   it does not produce a gradual engagement

    (b)* it produces side thrust

    (c)   the load carrying is less than that of a spur wheel

    (d)   none of these.

8.22 The type of gears used to provide a higher reduction in angular speeds between non-intersecting shafts at right angles is

    (a) helical gears   (b) bevel gears

    (c) spiral gears   (d)* worm and worm wheel

8.23 The circular pitch p_c of a helical gear is expressed as a function of normal pitch p and spiral angle α as

    (a)* $p_c = p/\cos\alpha$ (b) $p_c = p\cos\alpha$

    (c) $p_c = p/\sin\alpha$ (d) $p_c = p\sin\alpha$

8.24 Undercutting in involute teeth is undesirable because it

    (a)* reduces strength of gear tooth

    (b) spoils the appearance of the tooth

    (c) increases minimum number of teeth required on pinion

    (d) increases interference

CHAPTER 9   GEAR TRAINS

(A) *Short Answer Type*

9.1 Show that principle of inversion is useful in solving velocity analysis problems for epicyclic gear trains.

9.2 Can higher pairs be inverted? Explain the consequent limitation on fixing any link, instead of arm, in an epicyclic gear train.

9.3 Differentiate between simple, compound and reverted gear trains.

9.4 Prove that the two rear wheels of an automobile, driven by a differential, will rotate at different speeds while rounding a curve.

9.5 Compare simple, compound and reverted gear train in respect of their applications.

9.6 State the effect of using even and odd number of intermediate gears on the relative directions of rotation of the input and output gear of a simple gear train. In what way these intermediate gears influence the gear ratio?

9.7 Explain the similarity between a reverted train, an epicyclic train and a differential train.

(B) *Objective Type*

9.8 When the axes of the input and output gears are co-axial, the gear train is called

    (a) simple gear train   (b) compound gear train

    (c)* reverted gear train and   (d) None of these

9.9 A gear train in which intermediate gears have no effect on gear ratio is

    (a)* simple gear train   (b) compound gear train

    (c) reverted gear train   (d) epicyclic gear train

9.10 A gear train in which an intermediate gear along with its shaft revolves about input and output shaft axes is called

    (a)   simple gear train   (b)   compound gear train
    (c)   reverted gear train   (d)* epicyclic gear train

9.11 Bevel gear is used in epicyclic gear trains as it makes the drive

    (a)   stronger in respect of load carrying capacity
    (b)* more compact permitting a very high speed reduction with fewer gears
    (c)   less noisy
    (d)   none of these

9.12 Advantage of compound gear train over a simple gear train is that

    (a)   it occupies less space
    (b)   it requires less maintenance
    (c)* it provides much larger speed reduction with small gears
    (d)   none of these

9.13 Gear train value is the ratio of speed of

    (a)   input to output gear
    (b)* output to input gear
    (c)   product of input and output speed to 100
    (d)   none of these

9.14 In a reverted gear train with input gear 1 of T_1 teeth meshing with gear 2 of teeth T_2 output gear 4 of teeth T_4 meshing with gear 3 of teeth T_3

    (a)* $T_1 + T_2 = T_3 + T_4$ (b) $T_2 - T_1 = T_3 - T_4$
    (c) $T_2/T_1 = T_4/T_3$ (d)   none of these

9.15 Tabular method for velocity analysis of gear trains is convenient for

    (a)   any type of planetary gear trains
    (b)   planetary gear trains with one or two input motions
    (c)* relatively simple planetary gear trains having only one input motion
    (d)   none of these

CHAPTER 10 GYROSCOPIC EFFECTS

(A) Short Answer Type

10.1 State the general expression for gyroscopic couple and mention the units of each term involved in it. In stability analysis, would you consider gyroscopic couple or reaction couple? Justify the answer in either case.

10.2 Define the terms 'Precessional motion' and 'Gyroscopic couple'. Examine critically similarities or otherwise between centripetal force/centrifugal force and gyroscopic couple/gyroscopic reaction couple.

10.3 What do you understand by the term 'Angle of heel'? What is its significance in stability analysis?

10.4 What do you understand by the term 'limiting speed' of a four-wheeler negotiating a turn?

10.5 Define the terms axis of spin, gyroscopic effects, precession, axis of precession and gyroscope.

10.6 Explain the terms with the help of a neat sketch: bow, stern, star-board side, port side, pitching, rolling and steering.

10.7 Explain 'Pitching and Rolling Motion' as applied to ships. Which one of the two is more important in ship stabilization and why?

10.8 Explain how does the gyroscopic couple tends to shear the holding down bolts when the ship pitches.

10.9 How the disturbing couple is brought into effect on sea vessels? Discuss gyroscopic effects on sea-going vessels.

10.10 Explain how gyroscopic stabilization is used in ships?

10.11 Show that as a naval ship rides over a sinusoidal wave form, the resulting rolling motion is periodic of the same time period.

10.12 Describe two possible arrangements of rotor (vis-á-vis the planes of spin) of the gyroscope for ship stabilization. Indicate precessional motion required in each case for balancing the disturbing couple due to sea.

(B) Objective Type

10.13 When a rotating disc is made to change position of the axis of spin, one must apply

    (a) centripetal force   (b)* gyroscopic couple
    (c) gyroscopic reaction couple   (d) none of these

10.14 The propeller of an aeroplane engine rotates in c.w. sense as seen from tail-end. If the aeroplane takes a turn to the left, the gyroscopic effect on the ship is to

    (a) depress the nose   (b)* raise the nose
    (c) depress the right wing   (d) raise the right wing

10.15 The rotor of a ship rotates in clockwise direction as seen from rear side and the ship takes a turn to the right. The gyroscopic effect on the ship will tend

    (a) to depress the star-board side (b) to depress port side
    (c) to depress the stern   (d)* to depress the force or bow

10.16 The pitching motion of a ship means

    (a) oscillatory motion of ship about longitudinal axis
    (b)* oscillatory motion of ship about transverse axis
    (c) turning of complete ship in a curved path to the right or left
    (d) none of these

10.17 When the pitching motion causes the bow to rise, the rotor rotating in c.w. sense (as seen from stern), the gyroscopic effect tends to

    (a) turn the ship towards port side
    (b)* turn the ship towards star-board side
    (c) depress the stern
    (d) raise the stern

10.18 Angle of heel is the

(a) angle through which precession takes place

(b) angle through which line of centres of front when is rotated with respect to horizontal

(c)* angle through which plane of two-wheeler is inclined to longitudinal vertical plane

(d) none of these

10.19 When a four-wheeler moving forward at a speed above critical takes a turn to the right the wheel (s) that tends to leave the ground is:

(a) outer front wheel   (b) outer rear wheel

(c)* both the inner wheels   (d) none of the four wheels

10.20 Pitching motion of ship, carrying a rotor that rotates in c.w. sense as seen from stern, produces couple

(a) in transverse vertical plane in c.w. sense as seen from stern

(b) in transverse vertical plane in c.c.w. sense as seen from stern

(c) in longitudinal vertical plane

(d)* in horizontal plane

CHAPTER 11   FRICTION

(A) Short Answer Type

11.1 Explain the terms friction circle and friction axis. What is their importance in force analysis?

11.2 State the laws of dry friction and fluid friction.

11.3 Distinguish between boundary friction and fluid friction.

11.4 State the conditions essential for the formation of hydrodynamic film between rubbing surfaces.

11.5 In the light of Tower's experiment draw pressure distribution diagrams in the transverse and longitudinal plane of a journal bearing.

11.6 Explain the terms: Pure rolling, rolling friction and coefficient of rolling friction.

11.7 State the assumptions of 'uniform pressure intensity' and 'uniform rate of wear'. What is the need for these two assumptions?

11.8 Which of the two assumptions for friction surfaces, namely, uniform rate of wear assumption and uniform pressure assumption, give conservative results for (a) friction drives and (b) bearings?

11.9 In a multi-plate clutch, the first and last friction plate are arranged to be outer plates. Explain the reason for the same.

11.10 Explain 'rolling friction' phenomenon and show that unlike coefficient of sliding friction, the coefficient of rolling friction is not a dimensionless quantity.

11.11 Explain the role of helical compression springs, arranged circumferentially, in the function of a multi-plate friction clutch.

11.12 "Axial load shared by each of the n collars in a horse-shoe shaped multi-collared bearing is W/n while, in the case of multi-plate clutch, axial load shared by each plate is W". Explain the reason.

11.13 Compare merits and demerits of anti-friction bearings and journal bearing.

11.14 Explain how, for a multi-plate friction clutch, number of pairs of active surfaces is $(n - 1)$ where n is the total number of friction plates.

11.15 Indicate situations indictating choice in favour of antifriction bearings rather than journal bearings.

11.16 Explain the terms 'viscosity' and 'oiliness' of a lubricant. In which cirumstances will each of these properties exert controlling influence on the friction between two lubricated surfaces?

(B) Objective Type

11.17 The friction force, which comes into play as the sliding motion of the body just begins, is called
  (a) static friction (b) dynamic friction
  (c)* limiting friction (d) none of these

11.18 The kinetic friction is the friction experienced by a body, when the body
  (a) is at rest (b) just begins to slide
  (c)* is in motion (d) none of these

11.19 A body of weight W is required to move up a rough inclined plane, whose angle of inclination with the horizontal is α. If $\mu = \tan\phi$ is the coefficient of limiting friction, the effort P required to be applied parallel to the plane is given by
  (a) $W\tan\alpha$ (b) $W\tan(\alpha + \phi)$
  (c)* $W(\sin\alpha + \mu\cos\alpha)$ (d) $W(\cos\alpha + \mu\sin\alpha)$

11.20 When α = angle of helix and ϕ = angle of friction, the effort P required in a screw jack to lift the load W is given by
  (a)* $P = W\tan(\alpha + \phi)$ (b) $W\tan(\alpha - \phi)$
  (c) $P = W\tan(\phi - \alpha)$ (d) $P = W\cos(\alpha + \phi)$

11.21 When α is the angle of helix and ϕ is the angle of friction, effort required in a screw jack to lower the load W is given by
  (a)* $P = W\tan(\alpha - \phi)$ (b) $P = W\tan(\phi - \alpha)$
  (c) $P = W\tan(\alpha + \phi)$ (d) $P = W\cot(\alpha + \phi)$

11.22 Efficiency of a screw jack is given by
  (a) $\tan(\alpha + \phi)/\tan\alpha$ (b)* $\tan\alpha/\tan(\alpha + \phi)$
  (c) $\tan(\alpha - \phi)/\tan\alpha$ (d) $\tan\alpha/\tan(\alpha - \phi)$

11.23 The radius of a friction circle for a shaft of radius r, rotating slide a bearing with coefficient of friction $\mu = \tan\phi$, is
  (a) $(r/2)\tan\phi$ (b)* $r\sin\phi$
  (c) $r\cos\phi$ (d) $\dfrac{r}{2}\sin\phi$

11.24 The efficiency of a screw jack, with helix angle α and angle of friction ϕ, is maximum when

(a) $\alpha = 45° + \phi/2$ (b)* $45° - \phi/2$

(c) $\alpha = 90° + \phi$ (d) $\alpha = 90° - \phi$

11.25 The maximum efficiency of a screw jack is

(a) $(1 - \tan\phi)/(1 + \tan\phi)$ (b) $(1 + \tan\phi/(1 - \tan\phi)$

(c)* $(1 - \sin\phi)/(1 + \sin\phi)$ (d) $(1 + \sin\phi/(1 - \sin\phi)$

11.26 For a pivot bearing with r_1 and r_2 as the outer and inner radii and W as the axial load, the friction torque, assuming uniform pressure, is given by

(a) $\dfrac{1}{2}\mu W(r_1 + r_2)$ (b) $\dfrac{1}{2}\mu W(r_1 - r_2)$

(c)* $\dfrac{2}{3}\mu W\left(\dfrac{r_1^3 - r_2^3}{r_1^2 - r_2^2}\right)$ (d) $\dfrac{2}{3}\mu W\left(\dfrac{r_1^2 - r_2^2}{r_1^2 + r_2^2}\right)$

11.27 With an assumption of uniform wear and collared bearing with r_2 and r_1 as inner and outer radii and W as the axial load, the friction torque is given by

(a)* $\dfrac{1}{2}\mu W(r_1 + r_2)$ (b) $\dfrac{1}{2}\mu W(r_1 - r_2)$

(c) $\dfrac{1}{2}\mu W\left(\dfrac{r_1^3 - r_2^3}{r_1^2 - r_2^2}\right)$ (d) $\dfrac{2}{3}\mu W\left(\dfrac{r_1^3 - r_2^3}{r_1^2 + r_2^2}\right)$

11.28 Friction torque transmitted by a disc or plate clutch is same as that of

(a) flat pivot bearing   (b)* flat collared bearing

(c) conical pivot bearing   (d) truncated conical pivot bearing

11.29 In multi-plate clutch with n_2 number of outer discs and n_1 number of inner disc, the number of pairs of active surfaces n_a is

(a) $n_1 + n_2$ (b) $n_1 + n_2 + 1$

(c)* $n_1 + n_2 - 1$ (d) $n_1 + n_2 - 2$

11.30 In a multi-collared thrust bearing consisting of n collars and subjected to axial load W, load shared by each collar is

(a) W (b)* W/n

(c) $W/(n - 1)$ (d) none of these

CHAPTER 12   BELT, ROPE AND CHAIN DRIVE

(A) *Short Answer Type*

12.1 Describe briefly various methods to increase value of limiting ratio of tensions in a flat belt.

12.2 Explain why in a horizontal belt drive, tight side is kept on the lower side.

12.3 With the help of free body diagrams of belt and pulley, show how centrifugal force comes into play and centrifugal tension is generated in the belt.

12.4 Explain what is meant by initial tension in a belt. What will happen if belt is operated without initial tension?

12.5 Angle of lap is equal on both the pulleys in crossed belt drive but unequal in the case of open belt drive. Explain how this is possible.

12.6 Explain points of similarity and dissimilarity in the case of slip and creep phenomenon.

12.7 Explain the role of idler pulley in the case of a flat-belt drive.

12.8 Compare advantages and disadvantages of chain drive over belt or rope drive.

12.9 Discuss briefly the effect of initial belt tension on maximum power transmitted by the belt.

12.10 Assuming non-linear stress vs strain diagram for belt material, show that $T_1 + T_2 > 2T_0$.

(B) Objective Type

12.11 In the expression for limiting ratio of belt tensions $T_1/T_2 = e^{\mu\theta}$

    (a) μ is the kinetic coefficient of friction and θ is the angle of lap on larger pulley.

    (b) μ is the kinetic coefficient of friction and θ is the angle of lap on smaller pulley.

    (c) μ is the limiting coefficient of static friction and θ is the angle of lap on larger pulley.

    (d)* μ is the coefficient of limiting friction (static) and θ is the angle of lap on smaller pulley.

12.12 The power transmitted by the belt is maximum when the maximum tension in the belt is

    (a)   one-half the centrifugal tension

    (b)   two-third the centrifugal tension

    (c)   two times the centrifugal tension

    (d)* three times the centrifugal tension

12.13 For a short-centre distance, one should recommend

    (a)   flat belt drive   (b)   rope drive

    (c)* V-belt drive   (d)   none of these

12.14 The included angle at V-groove in pulley, for a V–belt drive is usually

    (a)   10–20°   (b)   20–30°

    (c)* 30–40°   (d)   50–60°

12.15 The peripheral belt speed for transmitting maximum power is given by

    (a)* $\sqrt{\dfrac{T_t g}{3w}}$ (b) $\sqrt{\dfrac{T_t g}{w}}$

    (c) $\sqrt{\dfrac{(3w)}{T_t g}}$ (d) $\sqrt{\dfrac{w}{T_t g}}$

12.16 Centrifugal tension in belts
  (a) increases H.P. transmitted
  (b) decreases H.P. transmitted
  (c)* has no effect on H.P. transmitted
  (d) increases H.P. transmitted upto a certain speed and then decreases
12.17 For maximum power transmission by belt, the centrifugal tension T_c of the belt should be:
  (a) $\dfrac{2}{3}T_{max}$
  (b) $\dfrac{1}{2}T_{max}$
  (c)* $\dfrac{1}{3}T_{max}$
  (d) none of the above
12.18 The H.P. transmitted
  (a)* increases with initial tension T_o
  (b) increases with decrease in T_o
  (c) increases with T_o upto a certain value of T_o only
  (d) does not depend on T_o
12.19 The pulley on which slip will occur under the limiting condition, depends on
  (a) whether the pulley is driver or driven
  (b) only on whether the pulley is smaller of the two
  (c)* jointly on whether the pulley is smaller of the two and whether it is an open or crossed drive
  (d) none of these
12.20 The pitch circle diameter d, the pitch p and number of teeth T on sprocket of a chain are related through
  (a)* $d = p\cosec(180°/T)$
  (b) $d = p\sin(180°/T)$
  (c) $d = p\cos(180°/T)$
  (d) $d = p\sec(180°/T)$
12.21 The type of friction at the gudgeon pin and the crank pin of an I.C. engine is
  (a) dry friction
  (b)* boundary friction
  (c) viscous friction
  (d) thin film friction in unstable range

CHAPTER 13 BRAKES AND DYNAMOMETERS

(A) Short Answer Type

13.1 Distinguish between simple and differential band brakes. Can each of the two be self-energizing?
13.2 "While differential band brake can be self-energizing, the simple band brake cannot be." Explain the statement with the help of a neat sketch showing tight and slack side of band.
13.3 With the help of a neat sketch, showing tight and slack side for each of the two band brakes, explain how direction of effort on brake lever is decided.
13.4 Explain the term 'self-energization' as applied to brakes. Is this property always desirable?

13.5 What do you understand by self locking? Explain the physical significance. Is it always a desirable property? If not, where this property is made use of?

13.6 Discuss the effect of 'brake-lever-pivot' position and direction of drum rotation on self-energization in block shoe brake.

13.7 What do you understand by the terms 'long shoe' and 'short shoe' as applied to shoe brakes? In what respect does the analysis for the two types of shoes differ?

13.8 Distinguish briefly between absorption and transmission type of dynamometers.

13.9 In an external short shoe brake, the shoe is pressed externally on a drum rotating in clockwise sense. Draw free-body diagrams for (i) lever with shoe and (ii) drum, showing all the forces acting on them.

13.10 Explain why wooden blocks are used in band and block brake.

(B) Objective Type

13.11 A motor car uses
    (a)   a band brake   (b)   a band and block brake
    (c)   an external shoe brake   (d)* an internally expanding shoe brake

13.12 For a simple band brake to be effective,
    (a)   tight side is connected to the free end of lever and effort is applied at M so as to move this end away from drum
    (b)   same as (a) above but effort is so applied at M as to move this end towards the drum
    (c)   slack side is connected to the free-end of the lever and the effort is applied at M so to move it away from the drum
    (d)* same as (c) but effort is applied at M so as to move this end towards the drum

13.13 For the differential band brake shown in the figure given in Question 13.15, for c.c.w. direction of drum rotation with tension T_1 in band at A and T_2 at B, the effort required at M is

    (a)   zero   (b)* $P = \dfrac{T_1 a - T_2 b}{l}$ downwards

    (c) $P = \dfrac{(T_1 a - T_2 b)}{i}$ upwards   (d) $P = T_1 (a/l)$ downwards

13.14 For c.c.w. direction of drum rotation in the figure given in Question 13.15, brake applies if
    (a)* $a > b$ and P acts downwards   (b) $a < b$ and P acts downwards
    (c) $a = b$ and P acts downwards   (d) $a > b$ and P acts upwards

13.15 The band brake in the following figure will be self-energizing when
    (a)   drum rotates c.c.w. and P acts upwards
    (b)   drum rotates c.c.w. and P acts downwards
    (c)   drum rotates c.w. and P acts upwards
    (d)* drum rotates c.w. and P acts downwards

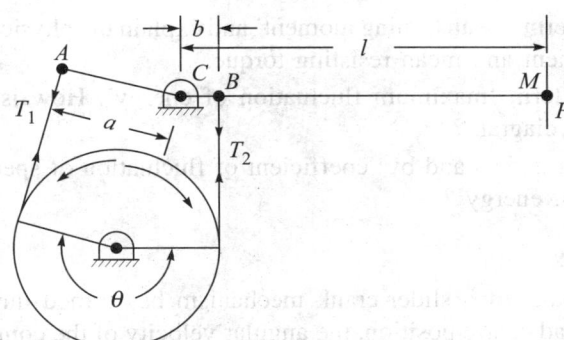

13.16 In a band and block brake, the ratio of tensions on tight and slack side of the band is

    (a) $T_n/T_o = n\mu\theta$
                            (b) $T_n/T_o = \left(\dfrac{1-\mu\tan\theta}{1+\mu\tan\theta}\right)^{1/n}$

    (c)* $T_n/T_o = \left(\dfrac{1+\mu\tan\theta}{1-\mu\tan\theta}\right)^{n}$
                       (d) $T_n/T_o = \left(\dfrac{1-\mu\tan\theta}{1+\mu\tan\theta}\right)^{n}$

13.17 An example of an absorption type dynamometer is
    (a)* prony brake dynamometer   (b)   epicyclic train dynamometer
    (c)   torsion dynamometer   (d)   belt transmission dynamometer

13.18 An example of a transmission dynamometer is

    (a)   prony brake dynamometer   (b)   rope brake dynamometer
    (c)   froude dynamometer   (d)* torsion dynamometer

CHAPTER 14 DYNAMICS OF MACHINES, T.M. DIAGRAM AND FLYWHEELS

(A) *Short Answer Type*

14.1 State the conditions of dynamically equivalent to two-mass system.

14.2 Can a connecting rod be replaced by an equivalent two-mass system, one at crank-pin and the other at gudgeon pin? Suggest a suitable remedy if this is not possible.

14.3 Explain the term 'centre of percussion' and give its physical significance.

14.4 With centre of mass G, radius of gyration k and location of one mass given, describe a geometrical method to find the other mass of dynamically equivalent two-mass system.

14.5 Explain the term 'effective force' and explain D'Alembert's principle for analysing a dynamic force/couple system.

14.6 Derive approximate analytical expressions for velocity and acceleration of the piston of a slider-crank mechanism.

14.7 Differentiate between the terms 'crank pin effort' and 'crank effort.' Derive an expression for crank pin effort in term of piston effort.

14.8 Explain the term 'mean turning moment' and explain the physical significance of mean turning moment and mean resisting torque.

14.9 Explain the term 'maximum fluctuation of energy'. How is it established from a known *T.M.* diagram?

14.10 What do you understand by 'coefficient of fluctuation of speed' and 'coefficient of fluctuation of energy'?

(B) Objective Type

14.11 When the crank of a slider-crank mechanism has turned through an angle θ with respect to dead centre position, the angular velocity of the connecting rod is given by

(a) $\dfrac{\omega\sin\theta}{(n^2 - \sin^2\theta)^{1/2}}$

(b)* $\dfrac{\omega\cos\theta}{(n^2 - \cos^2\theta)^{1/2}}$

(c) $\dfrac{\omega\sin\theta}{(n^2 - \cos^2\theta)^{1/2}}$

(d) $\dfrac{\omega\cos\theta}{\sqrt{(n^2 - \sin^2\theta)^{1/2}}}$

where ω = angular velocity of crank and

$$n = (l/r) = \frac{\text{length of connecting rod}}{\text{radius of crank}}$$

14.12 In a reciprocating steam engine, when the crank has turned from inner dead centre through an angle θ, the angular acceleration of the connecting rod is given by

(a) $\dfrac{-\omega^2(n^2 - 1)\sin\theta}{(n^2 - \cos^2\theta)^{1/2}}$

(b) $\dfrac{-\omega^2(n^2 - 1)\cos\theta}{(n^2 - \sin^2\theta)^{1/2}}$

(c)* $\dfrac{-\omega^2(n^2 - 1)\sin\theta}{(n^2 - \sin^2\theta)^{3/2}}$

(d) $\dfrac{-\omega^2(n^2 - 1)\cos\theta}{(n^2 - \sin^2\theta)^{1/2}}$

14.13 A connecting rod of length 24 cm between crank pin and gudegeon pin centres has a radius of gyration $k = \sqrt{48}$. If the C.G. of the connecting rod from the first mass at gudgeon pin is 16 cm, the second mass should be placed for dynamically equivalent system.

(a) 8 cm from C.G.
(b) 4 cm from C.G.
(c)* 3 cm from C.G.
(d) 6 cm from C.G.

14.14 If ω_1 and ω_2 are the maximum and minimum angular speeds and ω be the mean speed of the crank shaft, the coefficient of fluctuation of speed is given by

(a) ω_1/ω_2

(b) $\left(\dfrac{\omega_1 - \omega}{\omega - \omega_2}\right)$

(c)* $\left(\dfrac{\omega_1 - \omega_2}{\omega}\right)$

(d) none of these

14.15 Maximum fluctuation of energy of a flywheel is equal to
(a) $I\omega(\omega_1^2 - \omega_2^2)$ (b) $I\omega^3 c_s$
(c)* $I\omega^2 c_s$ (d)   none of these

where, ω_1, ω_2 and ω are respectively, maximum, minimum and mean speed of crank shaft and c_s is the coefficient of fluctuation of speed.

CHAPTER 15   GOVERNORS

(A) Short Answer Type

15.1 Distinguish between the functions of a flywheel and governor.
15.2 While a governor forms an essential element of a prime mover, the flywheel does not. Explain the statement giving reasons.
15.3 What are the limitations of Watt governor, which led to its obsolescence.
15.4 In what respect does a Proell governor differs from a Porter governor? How does this change gets reflected in their performance?
15.5 Compare gravity-controlled and spring-controlled governors in respect of their scope of application.
15.6 Why is it that the speed range of a Proell governor less than that of a Porter governor?
15.7 In a Wilson-Hartnell governor, show that the main spring is equivalent to two springs, each of stiffness twice that of the original stiffness.
15.8 Explain the terms stable, unstable and isochronous governor with a special reference to the controlling force curves.
15.9 Explain the terms 'Controlling force' and 'Effort' as applied to a governor mechanism.
15.10 Explain the term power of a governor. How is it determined?
15.11 Define the term 'Coefficient of Insensitiveness'. What is the effect of friction on governor performance?
15.12 Explain the term 'Hunting of Governor'.

(B) Objective Type

15.13 The height of a Watt governor is
(a)   directly proportional to speed   (b)   directly proportional to (speed)^2
(c)   inversely proportional to speed   (d)* inversely proportional to (speed)^2

15.14 A Watt governor operates satisfactorily in the speed range
(a)* 60–80 r.p.m.   (b)   80–100 r.p.m.
(c)   100–120 r.p.m.   (d)   120–180 r.p.m.

15.15 The height of Porter governor (with arms of equal length) bears a constant ratio to the height of Watt governor, given by

(a) $\dfrac{w}{w + W}$ (b) $\dfrac{W}{w + W}$

(c)* $\dfrac{w + W}{w}$ (d) $\dfrac{w + W}{W}$

where w is the weight of the ball and W is the weight of the sleeve.

15.16 A Hartnell governor is a
   (a)   dead weight type governor   (b)   pendulum governor
   (c)* spring loaded governor   (d)   inertia governor

15.17 A governor that falls in the category of pendulum governor is
   (a)* Watt governor   (b)   Porter governor
   (c)   Proell governor   (d)   Hartnell governor

15.18 When r_1 and r_2 are the minimum and maximum radii of ball rotation and x, y are the lengths of the ball arm and sleeve arm respectively, the lift of the sleeve is given by:

   (a) $(r_1 - r_2)\dfrac{x}{y}$ (b)* $(r_1 - r_2)\dfrac{y}{x}$

   (c) $(r_1 + r_2)\dfrac{x}{y}$ (d) $(r_1 + r_2)\dfrac{y}{x}$

15.19 When S_1 and S_2 are the compressive spring forces at the sleeve at maximum and minimum radii of rotation respectively, and if h be the compression of the spring, the stiffness is given by

   (a) $\dfrac{S_1 - S_2}{2h}$ (b) $\dfrac{S_1 + S_2}{h}$

   (c) $\dfrac{S_1 + S_2}{2h}$ (d)* $\dfrac{S_1 - S_2}{h}$

15.20 An example of a spring controlled governor is
   (a)   Watt governor   (b)   Porter governor
   (c)* Pickering governor   (d)   Proell governor

15.21 A governor useful in gramophone is
   (a)   Watt governor   (b)   Hartnell governor
   (c)   Hartung governor   (d)* Pickering governor

15.22 Out of the two governors A and B, if governer A has a larger lift than B for a given fractional change of speed.
   (a)* governor A is said to be more sensitive than B
   (b)   governor B is said to be more sensitive than A
   (c)   both the governors are said to be equally sensitive
   (d)   none of these

15.23 If ω_1, ω_2 and ω are respectively the maximum, minimum and mean equilibrium speeds, the sensitiveness of the governor is given by
   (a) ω/ω_1 (b) ω/ω_2

   (c) $\dfrac{\omega_1 - \omega}{\omega - \omega_2}$ (d)* $\dfrac{\omega_1 - \omega_2}{\omega}$

15.24 A governor is said to be stable if the variations in parameters (*i.e.*, radius of ball rotation r and equilibrium speed ω) is such that
    (a) r increases as ω decreases
    (b) r decreases as ω increases
    (c) r does not change as ω increases
    (d)* r increases as ω increases

15.25 In a Hartnell governor, if a spring of greater stiffness is used, the governor will
    (a)* be less sensitive
    (b)   be more sensitive
    (c)   remain unaffected in respect of sensitivity
    (d)   become isochronous

15.26 When the speed of the engine fluctuates continuously above and below the mean speed, the governor is said to be
    (a)   stable   (b)   unstable
    (c)   isochronous   (d)* hunting

15.27 Effort of a Porter governor is given by

    (a) $\dfrac{c}{w+W}$ (b) $\dfrac{c}{w-W}$

    (c) $c(w-W)$ (d)* $c(w+W)$

    where c = percentage increase in speed,
            w = weight of ball, and
            W = weight of the sleeve.

15.28 The power of a Porter governor is given by

    (a) $\dfrac{c^2}{2(1+2c)}(w+W)h$ (b) $\dfrac{c^2}{3(1+2c)}(w+W)h$

    (c) $\dfrac{3c^2}{(1+2c)}(w+W)h$ (d)* $\dfrac{4c^2}{(1+2c)}(w+W)h$

15.29 A spring controlled governor is said to be unstable when variation of controlling force F with radius r of ball rotation is such that
    (a) F remains constant at all values of r
    (b) F increases as r increases
    (c)* F decreases as r increases
    (d) F decreases as r decreases

15.30 A spring controlled governor is said to be isochronous when variation of controlling force F, with radius r of ball rotation is such that
    (a)* F remains constant at all r (b) F increases as r decreases
    (c) F decreases as r increases   (d) F decreases as r increases

15.31 A spring controlled governor is stable when the controlling for F is related to radius r of ball rotation through

(a) $F = ar$

(b) $F = ar + b$

(c)* $F = ar - b$

(d) none of these

CHAPTER 16 BALANCING

(A) Short Answer Type

16.1 Explain the meaning of the terms 'static balancing' and 'dynamic balancing'.

16.2 In the case of unbalanced masses revolving in different planes, what are the conditions to be fulfilled in order to obtain a complete balance?

16.3 Explain why two balance weights are required in two different planes to balance unbalanced masses revolving in different planes.

16.4 Explain the terms 'Primary and Secondary balancing' as used in balancing of reciprocating masses.

16.5 Why the primary disturbing forces for reciprocating parts in locomotives are only partially balanced? Explain.

16.6 What do you understand by 'Outside cylinder' and 'Inside cylinder' locomotives? What are the forces and couples that creeps into analysis because of partial balancing of reciprocating parts?

16.7 Explain the terms

(a) variation of tractive effort

(b) swaying couple

(c) hammer blow as applied to balancing of locomotives

16.8 What are coupled locomotives and why are they invariably used?

16.9 Explain the term hammer blow. How can it be reduced in locomotives?

16.10 Explain the principle of direct and reverse cranks as used in balancing.

16.11 What are the conditions of balance in V-engines?

16.12 What are In-line engines? How are they balanced?

(B) Objective Type

16.13 A disturbing mass of weight W_1 rotating at a radius r_1 can be balanced by another mass of weight W_2 placed at radius r_2 in the same diametral plane such that

(a)* $W_1 r_1 = W_2 r_2$

(b) $W_1 r_2 = W_2 r_1$

(c) $W_1 \sqrt{r_1} = W_2 \sqrt{r_2}$

(d) none of these

16.14 The primary unbalanced force due to reciprocating parts of weight W and with crank radius r and connecting rod as l is

(a) $(W/g) r w^2 \dfrac{\sin 2\theta}{n}$

(b) $(W/g) r w_2 \dfrac{\cos 2\theta}{n}$

(c) $\dfrac{W}{g}rw^2\sin\theta$ (d)* $(W/g)rw^2\cos\theta$

where $n = (l/r)$.

16.15 The secondary unbalanced force due to reciprocating parts of weight w and with crank radius of r and length of connecting rod as l, is given by

(a) $(W/g)r\omega^2\sin\theta$ (b) $(W/g)r\omega^2\cos\theta$

(c)* $(W/g)r\omega^2\,\dfrac{\cos2\theta}{n}$ (d) $(W/g)r\omega^2\,\dfrac{\sin2\theta}{n}$

where $n = (l/r)$.

16.16 The secondary unbalanced force has a magnitude that is k times the amplitude of primary unbalanced force, where k equals

(a) half   (b) one-third

(c) n (d)* $1/n$ times (where $n = l/r$)

16.17 The secondary unbalanced force assumes maximum value N times in one revolution where

(a) N = 2   (b) N = 3

(c)* N = 4   (d) none of these

16.18 The primary and secondary unbalanced forces are

(a) constant in magnitude and direction

(b) constant in magnitude but variable in direction

(c)* constant in direction but variable in magnitude

(d) varies in magnitude and direction both

16.19 In order to facilitate starting of locomotive in any position the cranks of a two cylinder locomotive, are arranged at

(a) 45 to each other   (b) 120 to each other

(c)* 90 to each other   (d) 180 to each other

16.20 Ratio of connecting rod length to crank radius is kept large in locomotives to

(a) minimise primary forces

(b)* minimise secondary forces

(c) to enable starting of the locomotive in any position

(d) achieve complete balancing

16.21 The tractive force in a locomotive with two cylinders is

(a) $c\dfrac{W}{g}r\omega^2\cos\theta$ (b) $\dfrac{cW}{g}r\omega^2\sin\theta$

(c)* $(1-c)\dfrac{W}{g}r\omega^2(\cos\theta-\sin\theta)$ (d) $\dfrac{W}{g}r\omega^2(\cos\theta-\sin\theta)$

where c = fraction of reciprocating points per cylinder,

       W = weight of reciprocating parts,

       ω = angular speed of crank

       r = radius of the crank, and

       θ = angle of inclination of crank to the line of stroke.

16.22 The maximum and minimum value of the tractive effort is given by

(a) $\pm \dfrac{cW}{g} \times r\omega^2$

(b) $\pm\sqrt{2} \times \dfrac{cW}{g} r\omega^2$

(c)* $\pm\sqrt{2}\,(1 - c)\dfrac{W}{g}\,r\omega^2$

(d) $\pm a(1 - c)\dfrac{W}{g}\,r\omega^2$

16.23 Maximum and minimum value of the swaying couple is

(a) $\pm \dfrac{cW}{g}\,r\omega^2$

(b)* $\pm\sqrt{2}\,a(1 - c)\dfrac{W}{g}\,r\omega^2$

(c) $\pm \dfrac{a}{\sqrt{2}}\,(1 - c)\dfrac{W}{g}\,r\omega^2$

(d) $\pm a(1 - c)\dfrac{W}{g}\,r\omega^2$

16.24 In a locomotive where primary forces are partially balanced, the maximum magnitude of the unbalanced force along the perpendicular to the line of stroke is known as

(a) tractive effort
(b) swaying couple
(c)* hammer blow
(d) none of the above

16.25 In a locomotive, resultant unbalanced force on account of two cylinders is known to cause

(a)* variation in tractive effort
(b) hammer blow
(c) swaying couple
(d) none of these

CHAPTER 17 VIBRATIONS

(A) Short Answer Type

17.1 A spring-mass system has a mass m and spring stiffness k and the length of the spring is L. If the spring is cut down to half its length, how will the natural frequency change?

17.2 What do you understand by natural frequency of vibration? What is its importance?

17.3 Define the terms 'under-damped vibrations', 'over-damped vibrations' and 'critically damped vibrations'. Explain the terms with the help of amplitude vs time plot.

17.4 Explain the difference between transient and steady-state response.

17.5 Why is it customary to make an assumption of small amplitude of oscillations? Give justification.

17.6 A spring mass system (k, m) has a natural frequency of ω_1. If a second spring k' is added in series to the first spring, the natural frequency is lowered to $\dfrac{1}{2}\omega_1$. Determine k' in terms of k.

$$\left[\text{Hint: } \frac{kk'}{m(k + k')} = \frac{1}{4}\frac{k}{m}; \text{ Hence } \left(\frac{k}{k'} + 1\right) = 4 \right]$$

17.7 A spring-mass-dashpot system with viscous damping is known to have a natural frequency of 3 c.p.s. and a damped frequency of vibration of 2 c.p.s., find out the damping factor. (**Ans.** = 0.745)

17.8 A spring-mass-dashpot system with natural frequency 5 c.p.s., when subjected to forced vibration test, develops maximum amplitude at 4.12 c.p.s. What is the damping factor. (**Ans.** 0.4)

17.9 For a damping factor $\zeta = 0.6$, and natural frequency of 10 rad/s, what is the difference between the damped and natural frequency? (**Ans.** 2.00 rad/s)

17.10 What is meant by logarithmic decrement. Why is it determined experimentally?

17.11 What do you understand by whirling motion? Does it involve transverse or torsional vibrations?

17.12 Compare a closed coiled helical spring with leaf spring in respect of transmissibility. Is their choice affected by the range of frequency ratio? Explain how.

17.13 Define the term critical speed of a rotating shaft. Explain the behaviour of a whirling shaft below and above the critical speed.

17.14 A stepped shaft has a diameter d over half the shaft length l, and '$2d$' over remaining half. Show that a torsionally equivalent shaft of uniform diameter d will have a length of $\left(\dfrac{17}{32}\right)l$.

(B) Objective Type

17.15 The natural frequency of vibration of a spring mass system is

(a)* $\dfrac{1}{2\pi}\sqrt{g/\delta}$ c.p.s.

(b) $2\pi\sqrt{g/\delta}$ c.p.s.

(c) $\dfrac{1}{2\pi}\sqrt{\delta/g}$

(d) none of these

where δ = static deflection in spring.

17.16 In a spring-mass system with k = stiffness of spring; M = main mass and m as the mass of spring, the natural frequency is given by

(a) $\sqrt{\dfrac{k}{M+m}}$ rad/s

(b) $\sqrt{\dfrac{k}{M+m/2}}$

(c)* $\sqrt{\dfrac{k}{M+m/3}}$

(d) $\sqrt{\dfrac{3k}{M+m}}$

17.17 A periodic motion in which body comes to static equilibrium position in smallest possible time, when released from a displaced postion, is called

(a) under-damped vibration

(b)* critically damped vibration

(c) over-damped vibration; and

(d) none of these

17.18 Logarithmic decrement is expressed by

(a) $\log_{10}\left(\dfrac{x_2}{x_1}\right)$

(b) $\log_{10}\left(\dfrac{x_1}{x_2}\right)$

(c) $\log_e\left(\dfrac{x_2}{x_1}\right)$

(d)* $\log_e\left(\dfrac{x_1}{x_2}\right)$

where x_1 and amplitudes of vibration in the first and second cycle respectively.

17.19 Amplitudes of oscillations in successive cycles is observed to be 13.5, 11.0, 8.5, 6.0 mm, etc. the type of damping is
- (a)   viscous damping
- (b)* coulomb damping
- (c)   structural damping
- (d)   slip damping

17.20 For a simply-supported beam carrying a u.d.1. of w per unit length over a span l the frequency of transverse vibration is
- (a) $0.4985/\sqrt{\delta}$
- (b) $0.571/\sqrt{\delta}$
- (c) $0.563/\delta$
- (d)* $0.621/\delta$

where δ is in metres and represents static deflection of simply supported beam due to u.d.l.

17.21 For a simply supported beam of length 1m and carrying a central point load frequency of vibration is
- (a) $0.5623/\sqrt{\delta}$
- (b)* $0.4985/\sqrt{\delta}$
- (c) $0.4317/\sqrt{\delta}$
- (d) $0.6253/\sqrt{\delta}$

where δ = static deflection in metres for a simply supported beam due to central point load.

17.22 The transmissibility of vibration-isolation system $TR = 1$, for all values of damping factor at
- (a) $\omega/\omega_n \rightarrow 2$
- (b) $\omega/\omega_n = 1$
- (c)* $\omega/\omega_n = \sqrt{2}$
- (d) $\omega/\omega_n \rightarrow \infty$

17.23 In the range $0 < (\omega/\omega_n) < \sqrt{2}$, the transmissibility of a vibration isolation for various values of damping is
- (a)   less than one only for a few values of ζ
- (b)   less than one for all values of ζ
- (c)* greater than one for all values of ζ
- (d)   greater than one for only a few values of ζ

BIBLIOGRAPHY

Beggs, J.S., *Mechanism*, McGraw-Hill, New York, 1955.

Bevan, Thomas, *Theory of Machines*, Indian ed., Longman, London, 1984.

Dudley, D.W. (Ed.), *Gears Hand Book*, McGraw-Hill, New York, 1962.

Erdman, A. and Sandor, G., *Mechanism Design: Analysis and Synthesis*, Vols. I and II, Prentice-Hall of India, New Delhi, 1988.

Green, W.G., *Theory of Machines*, Blackie, London, 1962.

Hall, Allen S., Jr., *Kinematics and Linkage Design*, Prentice-Hall of India, New Delhi, 1964.

Ham, C.W., Crane, E.J. and Rogers, W.L., *Mechanics of Machinery*, McGraw-Hill, 1958.

Hartenber, R. and Denavit, J., *Kinematic Synthesis of Linkages*, McGraw-Hill, New York, 1964.

Hinkle, R.T., *Kinematics of Machines*, Prentice Hall, Englewood Cliffs, New Jersey, 1960.

Hirschhorn, Jeremy, *Kinematics And Dynamics of Plane Mechanisms*, McGraw-Hill, 1962.

Hunt, K.H., *Kinematic Geometry of Mechanisms*, Clarendon Press, Oxford, 1978.

Kimbrell, Jack, T., *Kinematics Analysis And Synthesis*, McGraw-Hill, 1991.

Kurt Hain, *Applied Kinematics*, McGraw-Hill, (English Translation) 1967.

Louis, Toft and Kersey, A.T., *Theory of Machines*, Sir Issac Pitman, 1949.

Mabie, H.H. and Quirk, F.W., *Kinematics and Dynamics of Machinery*, John Wiley, 1978.

Martin, G.H., *Kinematics and Dynamics of Machines*, McGraw-Hill, New York, 1969.

Paul, B., *Kinematics and Dynamics of Planar Machinery*, Prentice Hall, Englewood Cliffs, New Jersey, 1979.

Rosenauer, N. and Willis, A.H., *Kinematics of Mechanisms*, Dover Publication, Inc, New York, 1967.

Rothbart, H.A., *Cams*, John Wiley, New York, 1956.

Shigley, I.E. and Uicker Jr., J.J., *Theory of Machines and Mechanisms*, McGraw-Hill.

Shigley, J.E., *Theory of Machines*, McGraw-Hill, 1958.

Soni, A.H., *Mechanism Synthesis and Analysis*, McGraw-Hill, 1974.

Thomson, W., *Theory of Vibration with Applications*, Prentice-Hall of India, New Delhi, 1975.

Tse, F., Morse, I. and Hinkle, R., *Mechanical Vibrations*, Prentice-Hall of India, New Delhi, 1974.

Tutle, S.B., *Mechanisms for Engineering Design*, John Wiley, New York, 1967.

Index

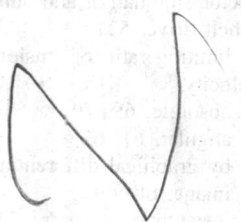